AUTOMATIC TRANSMISSIONS and TRANSAXLES

THEORY, OPERATION, DIAGNOSIS AND SERVICE

FOURTH EDITION

MATHIAS F. BREJCHA
RONALD A. TUURI

Online Services

Tuuri/TSA Online

To relay messages to Ronald A Tuuri/Technical Services and Assessments, inc. (TSA), concerning this text, NATEF evaluation/certification, or ASE certification, use any of the following sources:

e–mail: tuurir@cot01.ferris.edu
http://www.netonecom.net/~silco/TSA/
Ferris State University (FSU), AC-101,
708 Campus Drive, Big Rapids, MI 49307-2281
FSU Phone: (616) 592-2354, FAX: (616) 592-5982
TSA/Technical Services and Assessments, inc.
Post Office Box 1132
Big Rapids, MI 49307-1132
TSA Phone/FAX: (616) 796-8772

Online Services

Delmar Online

To access a wide variety of Delmar products and services on the World Wide Web, point your browser to:

http://www.delmar.com/delmar.html
or email: info@delmar.com

thomson.com

To access International Thomson Publishing's home site for information on more than 34 publishers and 20,000 products, point your browser to:

http://www.thomson.com
or email: findit@kiosk.thomson.com

A service of I(T)P®

AUTOMATIC TRANSMISSIONS and TRANSAXLES

THEORY, OPERATION, DIAGNOSIS AND SERVICE

FOURTH EDITION

MATHIAS F. BREJCHA
Professor Emeritus
Transportation and Electronics Department
Ferris State University
Big Rapids, Michigan

RONALD A. TUURI
Associate Professor
Transportation and Electronics Department
Ferris State University
Big Rapids, Michigan

Delmar Publishers

An International Thomson Publishing Company 

Albany • Bonn • Boston • Cincinnati • Detroit • London • Madrid • Melbourne
Mexico City • New York • Pacific Grove • Paris • San Francisco • Singapore • Tokyo
Toronto • Washington

NOTICE TO THE READER

Publisher does not warrant or guarantee any of the products described herein or perform any independent analysis in connection with any of the product information contained herein. Publisher does not assume, and expressly disclaims, any obligation to obtain and include information other than that provided to it by the manufacturer.

The reader is expressly warned to consider and adopt all safety precautions that might be indicated by the activities herein and to avoid all potential hazards. By following the instructions contained herein, the reader willingly assumes all risks in connection with such instructions.

The publisher makes no representation or warranties of any kind, including but not limited to, the warranties of fitness for particular purpose or merchantability, nor are any such representations implied with respect to the material set forth herein, and the publisher takes no responsibility with respect to such material. The publisher shall not be liable for any special, consequential, or exemplary damages resulting, in whole or part, from the readers' use of, or reliance upon, this material.

Delmar Staff
Senior Administrative Editor: Vernon R. Anthony
Production Manager: Mary Ellen Black
Art and Design Coordinator: Michael Prinzo

The ITP logo is a trademark under license.

Printed in the United States of America

For more information, contact:

Delmar Publishers
3 Columbia Circle, Box 15015
Albany, New York 12212-5015

International Thomson Publishing Europe
Berkshire House 168-173
High Holborn
London, WC1V 7AA
England

Thomas Nelson Australia
102 Dodds Street
South Melbourne, 3205
Victoria, Australia

Nelson Canada
1120 Birchmont Road
Scarborough, Ontario
Canada, M1K 5G4

International Thomson Editores
Campos Eliseos 385, Piso 7
Col Polanco
11560 Mexico D F Mexico

International Thomson Publishing GmbH
Königswinterer Strasse 418
53227 Bonn
Germany

International Thomson Publishing Asia
221 Henderson Road
#05-10 Henderson Building
Singapore 0315

International Thomson Publishing—Japan
Hirakawacho Kyowa Building, 3F
2-2-1 Hirakawacho
Chiyoda-ku, Tokyo 102
Japan

2 3 4 5 6 7 8 9 10 XXX 02 01 00 99 98

Library of Congress Cataloging-in-Publication Data

Brejcha, Mathias F.
Automatic Transmissions and Transaxles : theory, operation, diagnosis and service / [Mathias F. Brejcha and Ronald A. Tuuri]—4th ed.
p. cm.
Includes index.
ISBN 0-8273-8038-0 (pbk.)
1. Automobiles—Transmission devices. I. Tuuri, Ronald A.
II. Title.
TL262.T88 1997
629.2'44–DC21

97-4014
CIP

Contents

CHAPTER 13 Torque Converter Clutch Diagnosis 318

CHAPTER 14 Electronic Shift Diagnosis 334

CHAPTER 15 Transmission/Transaxle Removal and Installation 374

Preface

Automatic Transmissions and Transaxles, Fourth Edition, is a comprehensive, fully revised text that presents up-to-date material in an easy-to-understand format. Fundamentals of operation, diagnosis procedures, and service practices are presented in a blend ranging from a basic/introductory level to an in-depth/advanced status. The text is designed to meet all levels of instruction.

The text is configured to meet the requirements of today's high-caliber automatic transmission/transaxle technician training. It provides the essential technical information for ASE technician and NATEF training program certifications. SAE J1930 terminology, prevailing transmission designations, and OBD II (on-board diagnostics, generation two) procedures are used throughout the text. The detailed presentations of diagnostic and service procedures promote manufacturer recommendations and traditional field service practices.

Current three- and four-speed automatic transmissions and transaxles designed with planetary gearsets are the focus of this text. Considering the wide assortment of vehicle models requiring maintenance and service, exposure to other designs is essential. Therefore, attention is given to parallel-shaft/helical gear automatic transaxles, CVT (continuously variable transmission) powertrains, and five-speed electronically controlled automatic transmissions.

The most current technical information available is provided. Coverage of common electronic controls and powertrain designs of transmissions/transaxles used in the 1990s is emphasized. The domestic transmission/transaxle summary chapters (Chapters 22, 23, and 24) include reviews of the recently introduced Chrysler 42RE, 46RE, and 42LE Autostick versions, Ford 4R70W, 4R44E, 4R55E, 5R55E, AX4S, AX4N, and CD4E models, and GM Hydra-Matic 4L60-E, 4T60-EHD, 4T65-E, 4T40-E, and 4T80-E models.

Chapter 25 offers abundant coverage of the significant import and foreign automatic transmission/ transaxle market. Over twenty-five European and Asian vehicle manufacturers are reviewed. The unit illustrates the interrelationships of vehicle manufacturers and cross-usage of transmissions and transaxles. Also, many popular import automatic transmission/transaxle designs and operating features are exhibited.

The authors, applying a combined total of over forty years of teaching automatic transmissions at the collegiate level, many years of field service experience, and participation in numerous manufacturer and trade association training programs, have assembled the Fourth Edition to be "learner friendly." Twenty-five logically sequenced chapters segment the topic matter into practical units of study. The simplified text write-up, supplemented with over one thousand illustrations and photographs, facilitates the learning process. In addition, many helpful features are incorporated throughout the book. These are identified and described in the following Format section.

Automatic Transmissions and Transaxles, Fourth Edition, has been published to provide you with the most up-to-date and thorough coverage of the topic matter available. As you progress in the text and in this fascinating subject, we are confident that you will discover the book's content to be complete and presented in an uncomplicated style.

Ronald A. Tuuri and Mathias F. Brejcha

FORMAT

To assist you in developing additional skills and knowledge of automatic transmissions and transaxles, this text was developed with the following features.

Contents

The text is divided into twenty-five chapters, beginning with an introduction to automatic transmissions, the powertrain, and safety. This is followed by units on torque converters, planetary gear systems, hydraulics, electronic controls, diagnostic procedures, and service practices. The closing four chapters deliver domestic and import/foreign transmission and transaxle summaries and can be utilized as reference sources. Use the Contents in the front of the book to locate general topic matter.

Challenge Your Knowledge

Each of the first twenty-one chapters begins with a Challenge Your Knowledge section. This feature will allow you to determine how much you know about the subject matter and identify what you should pay special attention to as you study the chapter.

Chapter Introduction

The introduction provides a general overview of the purpose of the chapter and highlights some of the important concepts that will be discussed.

Cautions, Warnings, and Notes

Cautions and Warnings are located throughout the book to identify specific safety hazards. Pay special attention to these to ensure your well-being and the safety of those working around you. Notes provide information to help you perform repairs/service tasks accurately.

Key Terms/Acronyms

Important terms and abbreviations are highlighted in bold print and described in their first usage within the text. It is essential to become familiar with these Key Terms and Acronyms to progress in your understanding of automatic transmissions and transaxles operation, diagnosis, and service.

Figures, Charts, Tables, and Guides

While reading, refer to Figures, Charts, Tables, and Guides when so directed by the text. Figures include illustrations and photographs to clarify concepts, display transmissions and related components, and demonstrate procedures. Charts and Tables provide information in the form of component specifications and functions. Guides may be used to record data while performing diagnosis procedures.

Chapter Summaries

Each chapter discussion concludes with a Summary highlighting important topics within the unit. In addition, the summaries in Chapters 22, 23, 24, and 25 provide insight to the anticipated trends in automatic transmission/transaxle design and application.

Chapter Reviews—Key Terms/Acronyms and Questions

Use the Chapter Reviews, located in Chapters 1 through 21, to determine how well you comprehend the material contained within those units. A listing of the Key Terms and Acronyms is furnished for your review. Questions are provided in a variety of formats including ASE style, true/false, multiple choice, matching, and fill-in-the-blank. To determine the level of your progress, it is suggested that you revisit the Challenge Your Knowledge section at the beginning of each chapter.

Domestic Automatic Transmission/Transaxle Summaries

Full coverage of Chrysler, Ford, and General Motors automatic transmission/transaxle summaries are provided in Chapters 22, 23, and 24, respectively. In addition to the Hydra-Matic units built by GM, Saturn parallel-shaft/helical gear automatic transaxles are represented in Chapter 24. These chapters can be used as a reference source to study specific domestic manufacturer transmission/transaxle models available in the 1990s. Model designs, applications, operating characteristics, and special features are explained and illustrated.

Import and Foreign Automatic Transmission/Transaxle Summaries

Over twenty-five European and Asian vehicle manufacturers are reviewed in Chapter 25. Discussion reveals the interrelationships of vehicle manufacturers and cross-usage of transmissions and transaxles. Popular import transmission powertrains and operation designs are emphasized. In addition, Honda parallel-shaft/helical gear automatic transaxles and several CVTs (continuously variable transmissions) are featured.

Appendix

The Appendix contains supplemental references. Included are English-metric conversion charts, a "NATEF (National Automotive Technicians Education Foundation) Task List and Correlation to Chapters," and ASE automatic transmission/transaxle test preparation material.

Glossary

The Glossary begins by identifying common automatic transmission/transaxle related abbreviations and acronyms. This is followed by a complete listing of key terms and their definitions.

Index

Refer to the Index to efficiently determine the location of specific subject matter used within the text.

Instructor's Guide

An Instructor's Guide is available and is designed to supplement the text. It contains over 150 transparency masters of figures and charts used within the book. Also provided are the answers to Challenge Your Knowledge and Chapter Review questions and problems, copies of diagnostic guides, and numerous automatic transmission/transaxle lab worksheets.

ACKNOWLEDGMENTS

Automatic transmissions and transaxles have evolved rapidly during the past few years. Assembling an up-to-date, comprehensive text representing the wide range of product designs and operations requires a great deal of effort from a variety of sources.

The authors are grateful for the enthusiastic support and encouragement of our families. We wish to thank and acknowledge the following individuals, groups, and manufacturers/industries for their assistance in creating this edition:

Charlotte Tuuri—manuscript typing and editing

Ferris State University
- Jerry Sholl—Media Production Head/Special Photography
- Jeff Eck—Media Production/Special Art Illustration
- William Papo—Professor, Printing Management
- Patrick Klarecki—Assistant Professor, Printing Management
- Fawn Hockett (deceased)—Special Art Illustrations
- David Wininger—Special Art Illustrations

American Honda Motor Company, Inc.
- Mike Clark—District Service Representative

Chrysler Corporation
- Mike Collins—Training Program Design Supervisor
- Ron Watkins—Training Program Design Manager

Ford Motor Company
- Joe Barney—Training Developer, Ford Parts and Service Division
- Earl Peterson—Training Developer, Ford Parts and Service Division
- Matthew Lee—Training Developer, Powertrain Operations

General Motors Corporation, Service Technology Group
- Gregory Stanfel—Training Program Developer

Saturn Corporation

Subaru of America, Inc.
Toyota Motor Corporation
Automotive Trades Division, 3M Corporation
Fluke Corporation
OTC Division, SPX Corporation
TRU-COOL, Long Mfg. Ltd., Canada
Certified Transmissions, Inc., Mishawaka, Indiana
Don Kulwicki—President
Bruce Norris—Technician and Shop Foreman
Automatic Transmission Rebuilders Association
2472 Eastman Avenue, #23
Ventura, California 93003
Automotive Service Association
P.O. Box 929
Bedford, Texas 76095-0929
Society of Automotive Engineers

Technical Reviewers

Albert L. Dent, South Puget Sound Community College
Thomas Jenkins, MoTech Education Center
Lawrence F. Legree, Montcalm Community College
Dana K. Brewer, Wyoming Technical Institute
Norris W. Martin, Texas State Technical College, Waco
Jack Erjavec, Columbus State Community College
John Bradford, Renton Technical College

Introduction to Automatic Transmissions/Transaxles, the Powertrain, and Safety

OBJECTIVE:

The objective of this unit is for you to become proficient with the terms, concepts, and tasks contained in this chapter. Completion of this objective is essential to becoming a successful automatic transmission technician.

CHALLENGE YOUR KNOWLEDGE

Define the following key terms:

Powertrain, transmission, transaxle, torque, torque converter, differential, cyclic vibrations, tractive effort, and coefficient of friction.

List the factors needed to calculate the following:

Axle torque and tractive effort.

Describe the purpose and operation of the following:

Rear-wheel drive powertrain; front-wheel drive powertrain; torque converter; final drive and differential; and all-wheel drive/four-wheel drive configurations.

INTRODUCTION

Among the major car manufacturers, there are over thirty different automatic transmissions in production in the mid-1990s. These transmissions represent a variety of four-speed (overdrive) families designed to meet the powertrain requirements of both front- and rear-wheel drive (FWD/RWD) systems. A significant portion of these units include electronic control for converter lockup, pressure modulation, and shift scheduling. Typical new-generation design models are the General Motors Hydra-Matic 4L60-E, 4L80-E, 4T40-E, 4T45-E, 4T60-E/4T60-EHD, 4T65-E, and 4T80-E; Chrysler 41TE (Ultradrive A-604), 42LE, 42LE Autostick, 40RH/42RH (A-500), 46RH/47RH (A-518), 42RE, and 46RE; Ford E4OD, AOD-E/4R70W, AXOD-E/AX4S, AX4N, CD4E, 4R44E/4R55E, and 5R55E. European and Asian transmission developments continue to meet the needs of the import market.

When the import automatic transmissions are added to the domestic units, the numbers are overwhelming. There is no reason to be afraid of imports—Asia and Europe are not building them differently. Shift systems and planetary geartrains typically work the same worldwide. Many of the import transmission geartrains are identical to those used in domestic units. The three-speed Simpson and Ravigneaux planetary trains are classic examples: the hydraulics and even the electronic controls are comparable. To limit the scope of this book, however, domestic automatic transmissions are given the main thrust.

The text objective is to present a simple, comprehensive study of the fundamentals of automatic transmissions, covering operation, diagnosis, and servicing. Furthermore, the book illustrates that automatic transmissions are more alike in design and operation than they are different. The differences that do exist are primarily in component sizes, design configuration, and minor variations in product hydraulic/electronic controls. Front-wheel drive units work the same as rear-wheel drive units. When getting involved with import transmissions, one can think about them as being like domestic units in operation, diagnosis, and servicing. The same thought process applies to light- and heavy-duty truck applications. Application of fundamentals applies worldwide to all transmissions.

The study of this text is one step in a series of learning experiences needed for a complete understanding of automatic transmissions. With a solid background in fundamentals, the technician needs to learn the product specifics of each transmission by studying the manufacturer's or equivalent operation, service, and diagnosis manuals. As part of the education process, driving vehicles equipped with the various transmissions will enable the technician to feel how a normal-operating engine and transmission team performs. And finally, the technician must get involved in the actual diagnosis and servicing of transmissions.

There is a lot of ongoing excitement in the area of transmissions, both with manual and automatic units. Car owners are rediscovering the manual transmission after it almost disappeared from the scene. Shifting on the floor is "sporty" and makes a manual transmission easy and fun to operate. Four- and five-speed transmissions are the trend, with a few six-speed units now on the road. In some applications, an electronic dash display tells the driver when to shift, or electronic controls shift the transmission.

Even more exciting is what is happening in the area of domestic automatic transmissions: four-speed overdrives, lockup clutch converters, and automotive transmission electronics have arrived. Electronics have been incorporated to optimize engine performance, fuel economy, and enhance shift feel and scheduling. Some transmissions have as many as sixteen wires plugged into a case connector for transmission "electronification." An array of solenoids and hydraulic electronic switches are now mounted on the valve body. In place of the traditional hydraulic/mechanical governor, magnetic speed sensors are used to measure both input and output rpm.

The new technology requires that technicians be knowledgeable regarding both computer-controlled engine management systems and computer-controlled transmission management systems, as these systems interface with each other. The technician must be capable of testing these systems without destroying a perfectly functioning computer or circuit diodes. The automatic transmission is probably the most complicated unit on the vehicle, and the technicians must have a high level of mechanical, electronic, and diagnostic skills to maintain and service these units. The new generation of transmissions offers a challenging adventure in electronics. Technicians need to know more and work smarter than ever before.

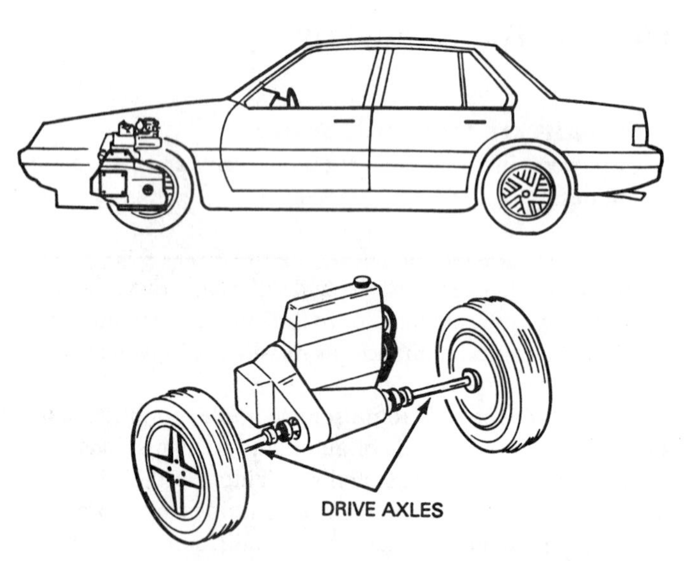

FIGURE 1–1 Transaxle bolts to the rear of the transverse-mounted engine. Two drive axles connect the transaxle to the front wheels and the vehicle is pulled.

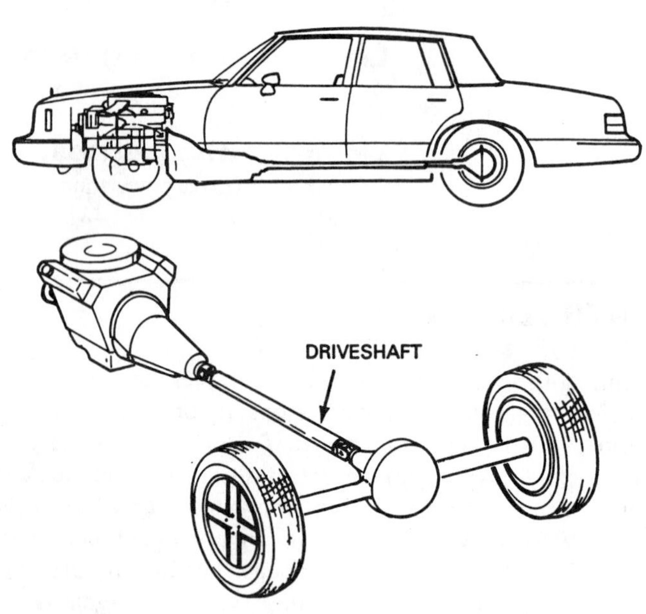

FIGURE 1–2 Drive shaft connects transmission to rear drive axle. The rear wheels push the vehicle.

FIGURE 1–3 Left: typical rear-wheel drive transmission. Right: typical front-wheel drive transaxle.

THE POWERTRAIN

To appreciate the role of the engine and the transmission in the scheme of powering the vehicle, the technician needs to review some basic powertrain fundamentals of operation. Although the engine is the source of power, the term **powertrain** includes all the drive components between the engine flywheel and the drive wheels. As the term implies, its main function is power transmission. It is the link between the engine and drive wheels that permits the engine to start the vehicle moving and keep it moving.

Typically, for an automatic **transmission** application, the powertrain system includes a flex or drive plate, fluid torque converter, transmission, universal joints and drive shaft, final drive/differential, and the drive axles and wheels. Two types of popular powertrains are currently utilized: the **transaxle** front-wheel drive and the transmission rear-wheel drive (Figures 1–1 to 1–3). The front-wheel drive systems use either the favored transverse (east/west) or longitudinal (north/south) engine installation (Figures 1–1 and 1–4). Some longitudinal installations use a bolt-on traditional ring and pinion final drive/differential (Figure 1–4). In another format, used by Chrysler in their 42LE transaxles, the hypoid gearset is an integral component of the transaxle (Figure 1–5).

❖ **Powertrain:** All drive components from the flywheel to the drive wheels; defined, at times, to include the engine.

❖ **Transmission:** A powertrain component designed to modify torque and speed developed by the engine; also provides direct drive, reverse, and neutral.

❖ **Transaxle:** A transmission and final drive unit combined into a single unit; may be designed in a longitudinal or transverse configuration for either a front- or rear-wheel drive vehicle.

The engine and powertrain are engineered together. They work as a balanced team for best utilization of available engine power to meet changing road load and driver demands. Available axle **torque** to drive the vehicle, therefore, is a function of engine torque multiplied by the converter torque and the gear reduction through the transmission and final drive:

$$T_a = T \times R_c \times R_t \times R_a$$

where

T_a = axle torque
T = engine torque
R_c = converter ratio
R_t = gear ratio through transmission for the particular gear engaged; in direct drive, R_t equals 1
R_a = drive axle ratio

EXAMPLE: What is the drive axle torque in first gear on a vehicle having 195 lb-ft engine torque teamed to a 2.0:1 converter at stall, a transmission ratio of 2.92:1, and a final drive ratio of 2.84:1?

$$T_a = 195(264 \text{ N-m}) \times 2.0 \times 2.92 \times 2.84$$
$$= 3234 \text{ lb-ft } (4382 \text{ N-m})$$

❖ **Torque:** A twisting effort or turning force applied to a radius.

The final torque always passes through the differential, where it is split equally between the drive axles. In the example calculation, the drive torque at the axle ends = 3234/2 = 1617 lb-ft (2191 N-m).

NOTE: Calculations do not compensate for driveline mechanical efficiency, which is generally 0.85.

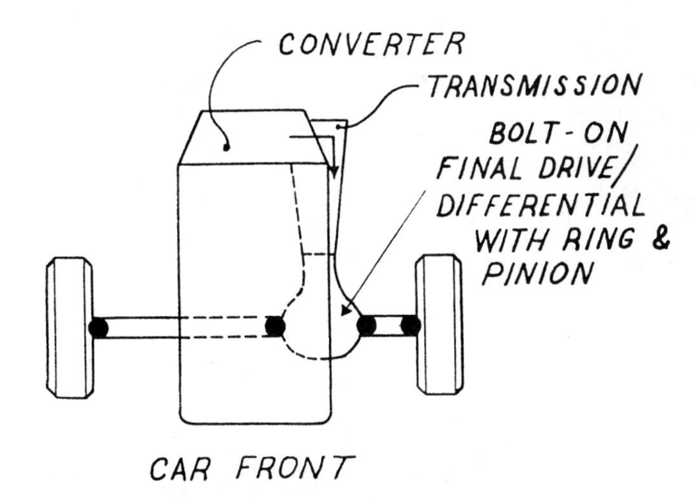

FIGURE 1–4 Front-wheel drive with longitudinal north/south powertrain.

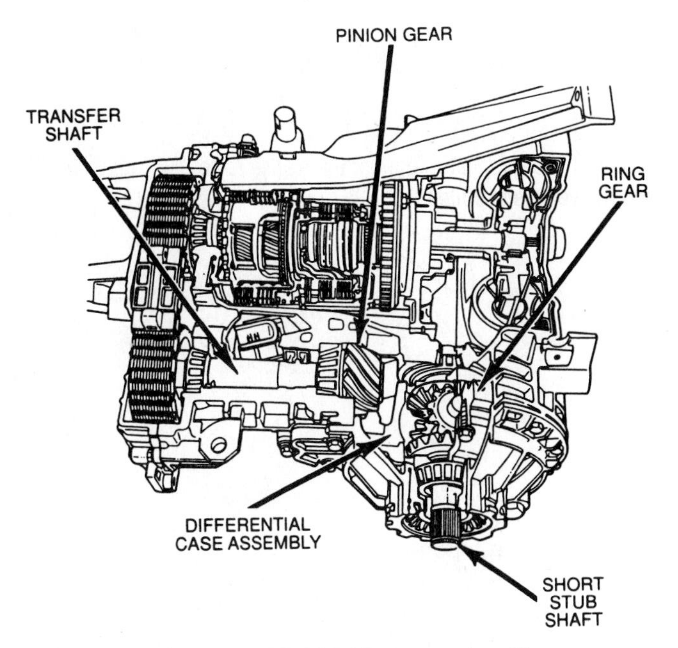

FIGURE 1–5 Chrysler 42LE final drive geartrain utilizes a hypoid ring and pinion gearset. (Courtesy of Chrysler Corp.)

Torque Converter

In the powertrain, the **torque converter** is the fluid link between the engine and the transmission (Figure 1–6). It acts as both a torque multiplier and fluid coupling. It is bolted to the engine flywheel, commonly referred to as the flex plate, and turns at engine rpm (Figure 1–7). Within the converter assembly are three basic bladed elements: the impeller or pump, turbine, and stator (Figure 1–8). The impeller is part of the converter housing and puts the energy or fluid thrust into the fluid that drives the turbine output member. The stator mounted on a one-way roller clutch is locked to a reaction (stationary) shaft during the torque phase of operation and freewheels during the coupling phase of operation. Depending on road load and engine throttle opening, the converter torque output is variable between its maximum rated or engineered torque and coupling torque of 1:1. Converter operation is discussed in more detail in Chapter 2.

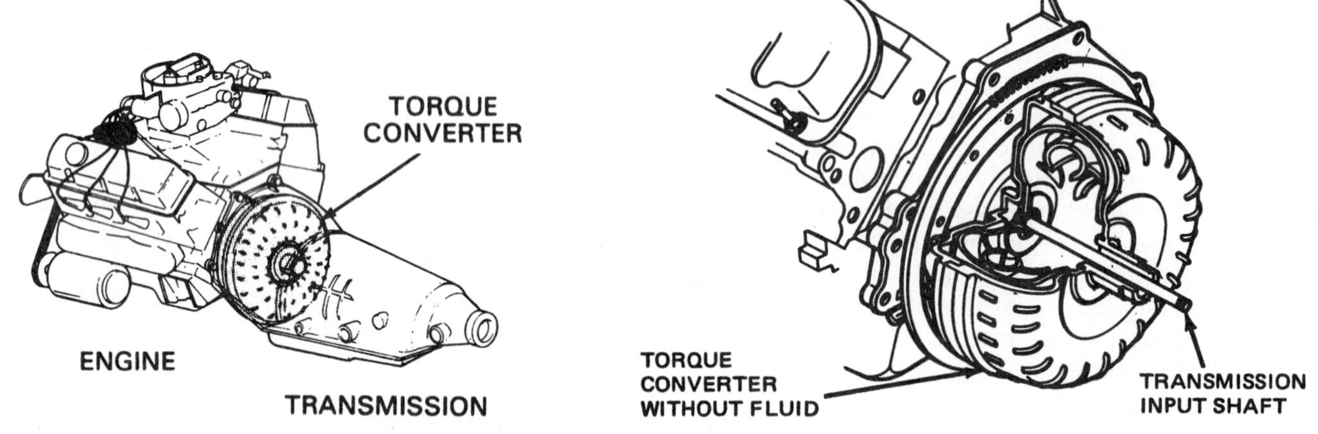

FIGURE 1–6 Converter acts as a fluid link between the engine and the transmission. (Courtesy of General Motors Corporation, Service Technology Group; reprinted with the permission of Ford Motor Company)

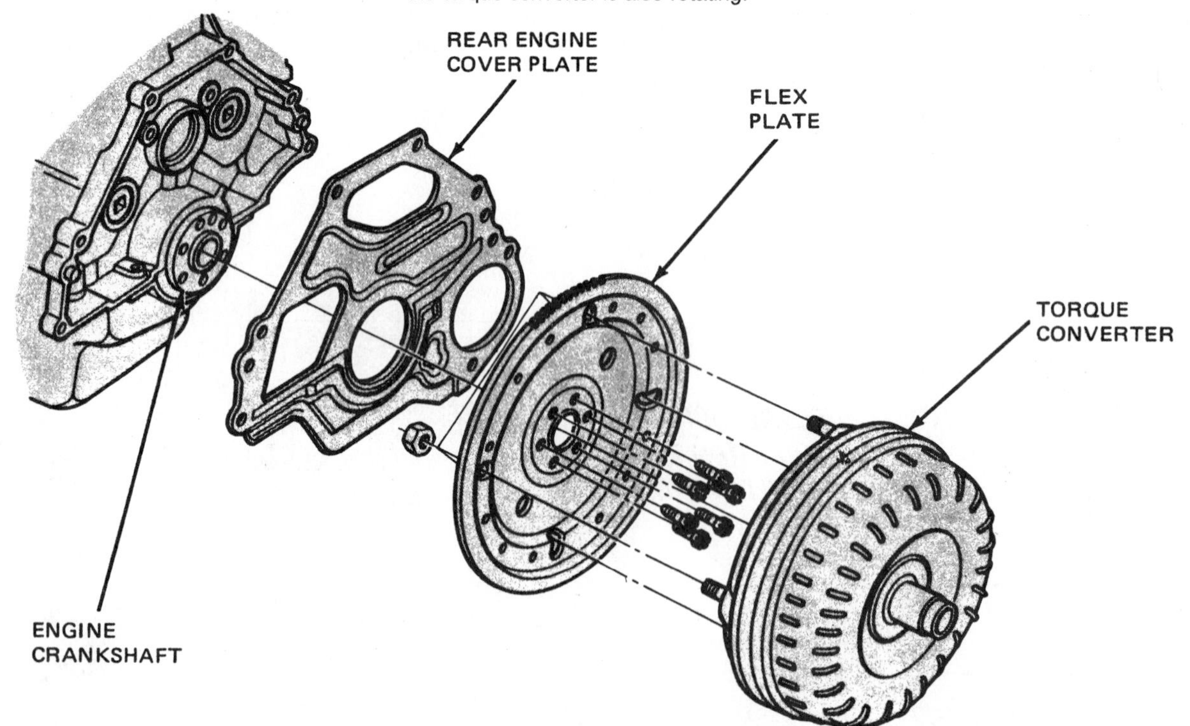

FIGURE 1–7 Attached to the end of the crankshaft is a flex plate and torque converter. Whenever the engine is running, the torque converter is also rotating. (Reprinted with the permission of Ford Motor Company)

❖ **Torque converter:** A fluid-coupling device designed to transmit and multiply engine torque into the transmission.

Important to automatic transmission operation is the ability of the converter to disconnect the engine output from the transmission at idle. At engine idle, there is no significant energy thrust in the fluid to move the vehicle, and the transmission can remain in gear with the vehicle stopped. The nuisance of a mechanical clutch is eliminated.

Transmission and Engine

The transmission is a machine featuring either simple or planetary gear arrangements for adapting available engine power to road and load conditions (weight, speed, and road terrain). It acts as a torque and speed changer by providing suitable gear ratio changes—three-, four-, and five-speed systems—that permit the engine to move the load efficiently. For vehicle operating purposes, it must also provide for reverse, neutral, and engine braking during deceleration.

The role of the transmission can be better appreciated by examining a few fundamental characteristics of an internal combustion engine. Torque output or crankshaft twisting ability is the prime performance factor of the engine. The engine, however, develops very little torque at low rpm and, therefore, must run fairly fast before it produces a significant torque output. As can be observed in Figure 1–9, however, an engine does not continue to turn out more and more torque the faster

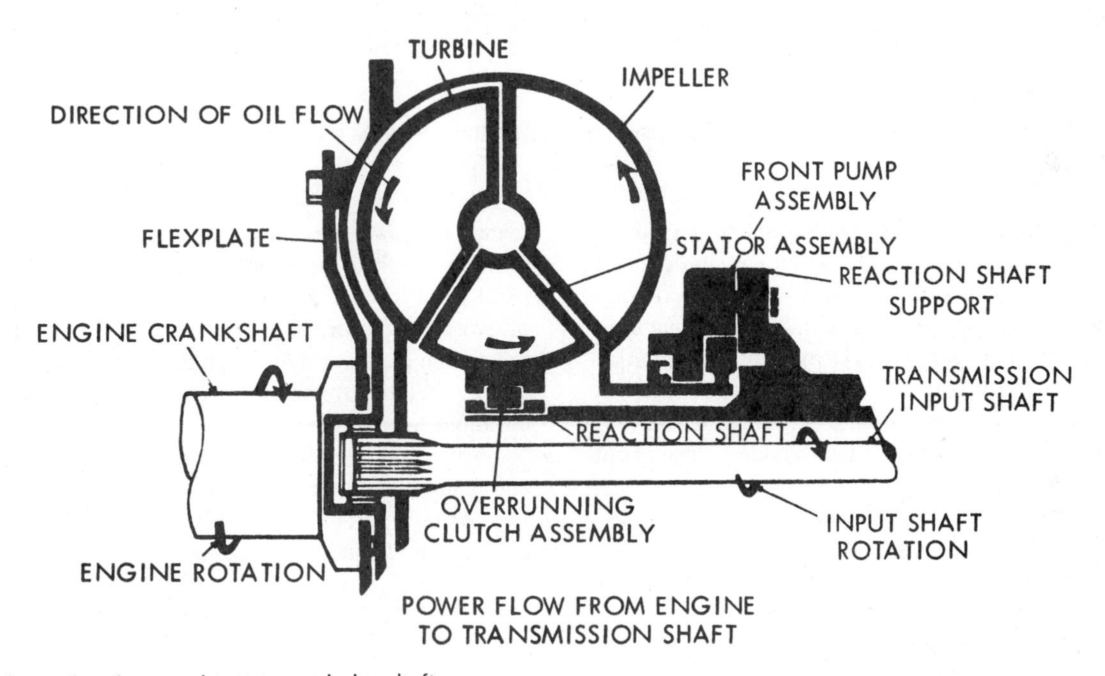

FIGURE 1–8 Powerflow from engine to transmission shaft.

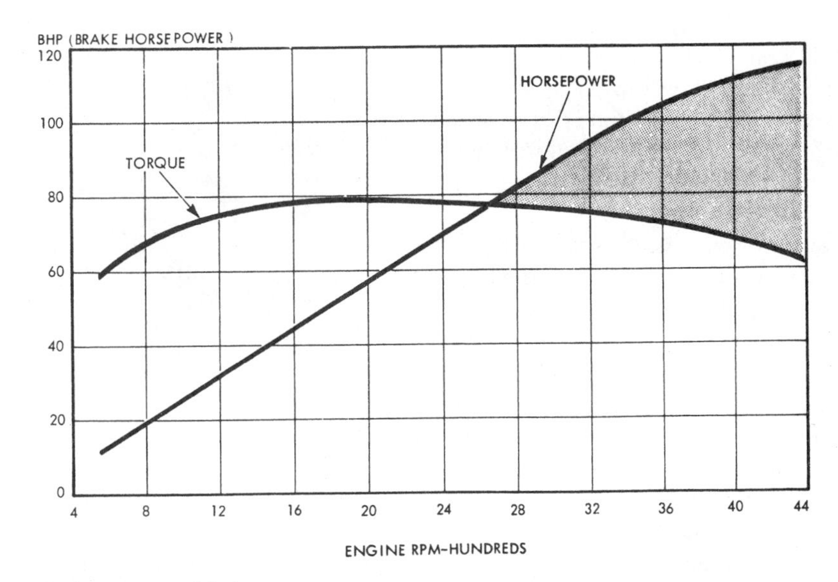

FIGURE 1–9 Relationship of brake horsepower (bhp) to torque.

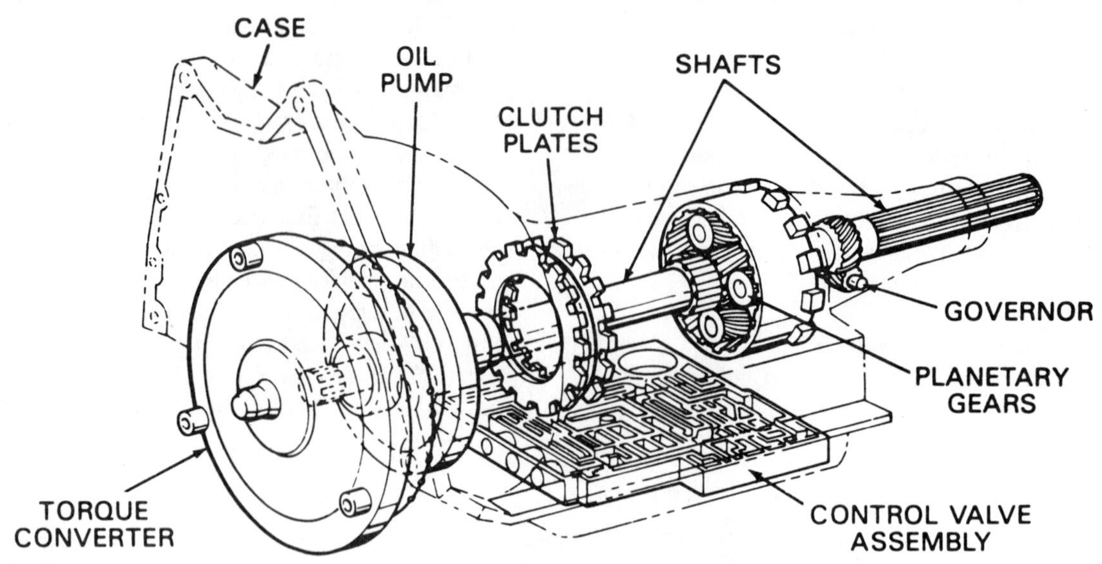

FIGURE 1–10 Typical automatic transaxle components. (Courtesy of General Motors Corporation, Service Technology Group)

it runs. As the engine rpm increases, so does the torque output until it peaks at approximately the midrange of operation. It is at the midrange that engine output efficiency is at its maximum. Any further increase in rpm results in a falloff of torque until at maximum rpm it is again very low. The range of engine torque output from its slowest rpm (idle) to its maximum possible rpm is referred to as the torque curve (Figure 1–9). Engine torque is the average torque produced by all the cylinders throughout their cycle.

The engine torque characteristics can be largely explained by looking at the breathing ability or volumetric efficiency (the ability of the atmosphere to charge the cylinders with a quantity of air/fuel mixture). At slow rpm's, cylinder intake valves are held open for relatively long intervals and permit the cylinders to take on a heavy gulp of air/fuel; however, the air intake is restricted at the throttle valve, resulting in low torque and high engine vacuum. As the throttle valve opens, more air/fuel is available for the cylinders, increasing the torque to midrange rpm. After midrange, more air/fuel is available to the cylinders but gets stacked in the manifold because of the very short interval opening time of the intake valves. Volumetric efficiency and torque decrease with a resulting high manifold pressure or low vacuum.

In an automobile, the torque or crankshaft twist developed by the engine is what turns the drive wheels. Remembering that an internal combustion engine develops very little torque at low rpm, it should be easy to understand that the engine needs assistance to start the vehicle moving. It cannot do it by itself. The transmission has the job of assisting the engine by permitting it to run at fairly high rpm and by multiplying its available torque to overcome the motionless deadweight of the vehicle. Once the vehicle starts moving, it takes less torque to keep it moving, so there is a chance for the engine to increase its torque output. As the engine becomes less dependent, the transmission has the ability to change gear ratios or torque assist for efficiency of power transmission. Finally, the engine reaches a point at which it can handle the vehicle and road load all by itself. That is when the transmission preferred ratio is direct 1:1 or overdrive. The transmission is available at all times to assist the engine in powering the vehicle, not only to get the vehicle moving but also for conditions requiring extra power, such as climbing steep grades.

Typical automatic transmission rear- and front-wheel drive automatic transaxle component arrangements are illustrated in Figures 1–10 to 1–13. In the transaxle units, note that the final drive assembly and transmission are combined.

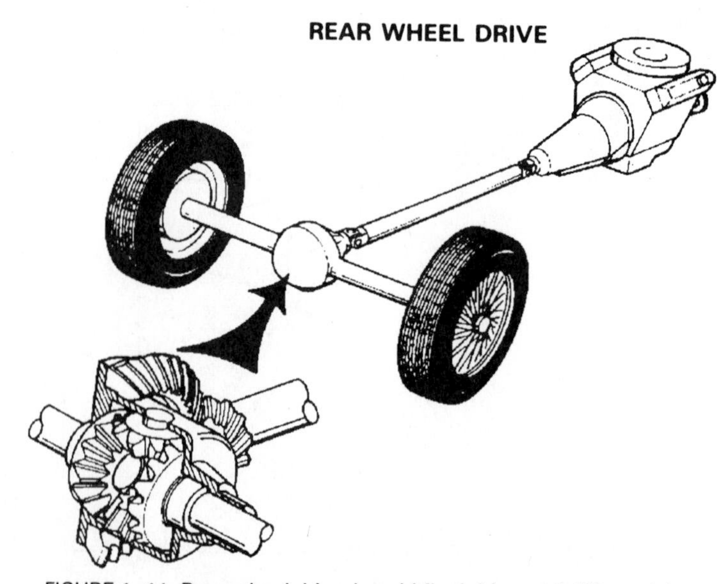

FIGURE 1–11 Rear-wheel drive: hypoid final drive and differential location. (Courtesy of General Motors Corporation, Service Technology Group)

Final Drive and Differential

The final drive and differential are normally packaged into one assembly, and although their functions are different, they work together. The purpose of the final drive is to receive the torque and speed output of the transmission and give one last gear reduction to the powertrain. In other words, the drive torque again undergoes multiplication and further speed reduction. The final drive has only one ratio, and the same reduction is applied for each gear ratio used in the transmission. In rear-wheel drive (Figure 1–11) and longitudinal front-wheel drive applications (Figure 1–5, page 3) where a hypoid ring and pinion gear arrangement is used, the final drive has the additional job of changing the powerflow at a right angle (ninety degrees) to match the axle shaft and wheel rotation. Shown in Figure 1–14 is a hypoid ring and pinion final drive. The gears are a special spiral bevel arrangement. In transaxles, the drive pinion gear is located above the ring gear center (Figure 1–5, page 3). This allows for a lower and more desirable front axle (half-shaft) installation. In other transaxle applications, a planetary or helical gearset is used (Figures 1–13 and 1–15).

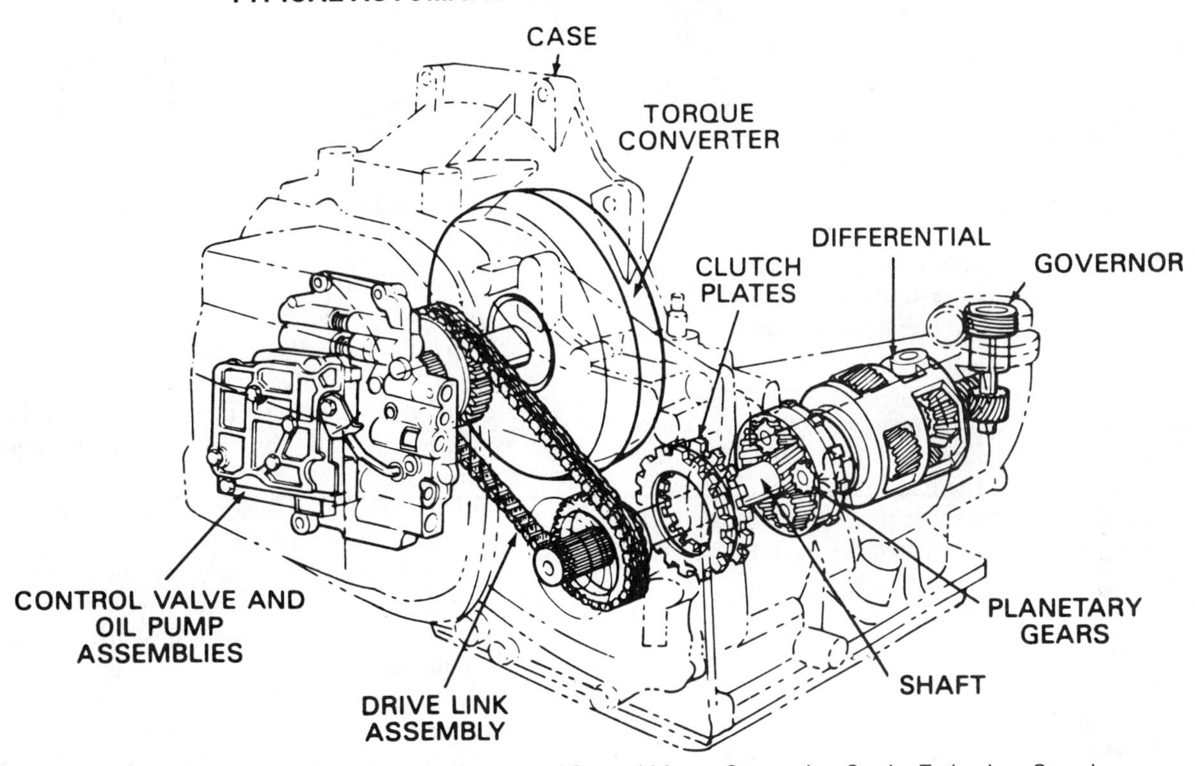

FIGURE 1–12 Typical automatic transaxle components. (Courtesy of General Motors Corporation, Service Technology Group)

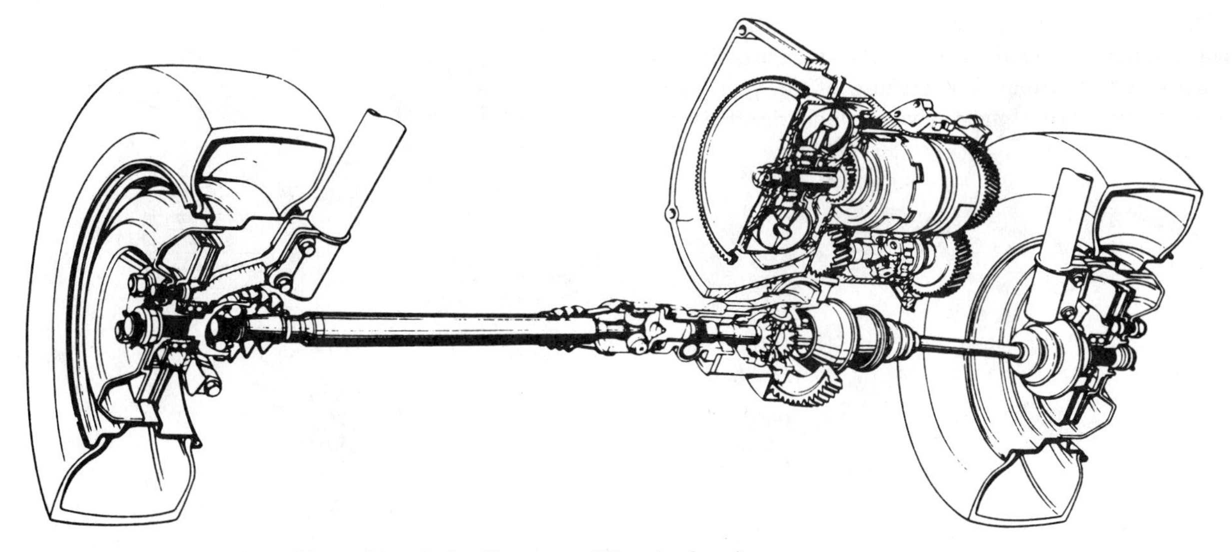

FIGURE 1–13 Automatic transaxle and front-drive shafts. (Courtesy of Chrysler Corp.)

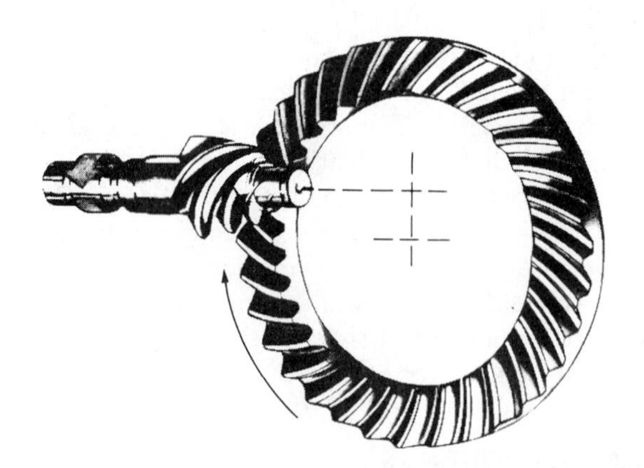
FIGURE 1–14 Hypoid ring and pinion final drive. (Courtesy of General Motors Corporation, Service Technology Group)

Note that the final drive output member is attached or fastened to the differential case and does not drive the axle directly.

❖ **Differential:** Located within the final drive unit, the differential provides power to the drive wheels for straight-ahead driving and turning conditions.

The final drive torque must pass through the differential unit where it always (1) divides the torque equally to the drive wheels and (2) divides the final drive speed equally or unequally to the drive wheels. The function of the differential is to allow the two drive wheels to rotate at different speeds, which is required when turning a corner in order to still have both wheels driving.

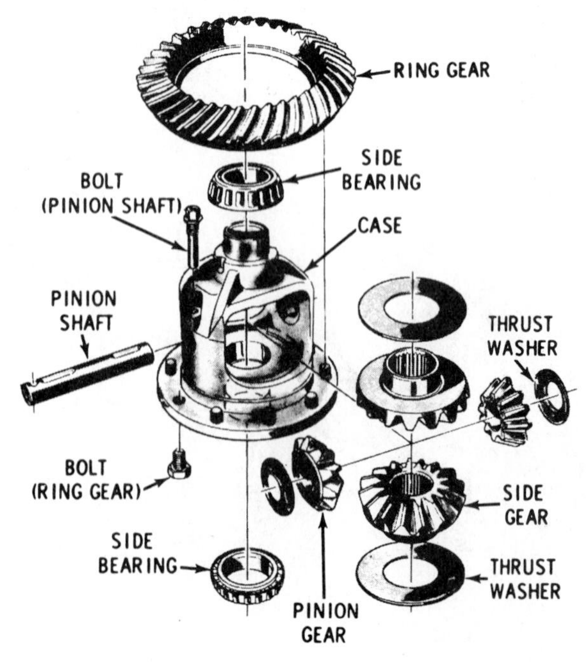

FIGURE 1–16 Differential case components. (Courtesy of General Motors Corporation, Service Technology Group)

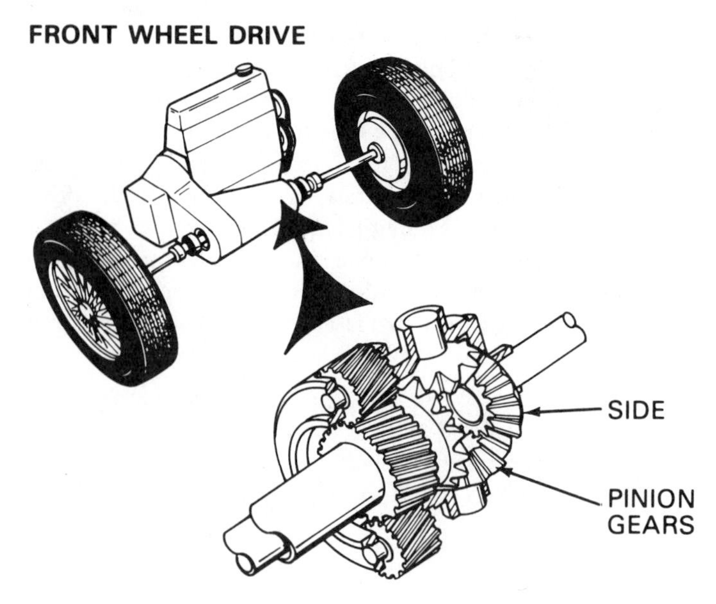

FIGURE 1–15 Planetary final drive and differential location. (Courtesy of General Motors Corporation, Service Technology Group)

Regardless of the final drive gear arrangement, the differential makeup is the same. It consists of the case, two side gears that spline to the axle shafts, and usually two pinion gears that mount on a differential case pinion shaft (Figure 1–16). The shaft is pinned and turns with the case, with the pinion gears free to spin on the shaft.

The differential is a mechanism whose action is difficult to describe but easy to understand when it is observed working. The following text highlights the operational modes and tries to keep it simple, using a rear-wheel drive unit for illustration. Shown in Figure 1–17 is the flow of final drive torque through the differential, where it splits the torque equally to each drive wheel. Even when one wheel is spinning on ice, the wheel with good traction does not get any more torque than the spinning wheel. Therefore, the wheel with good traction cannot move

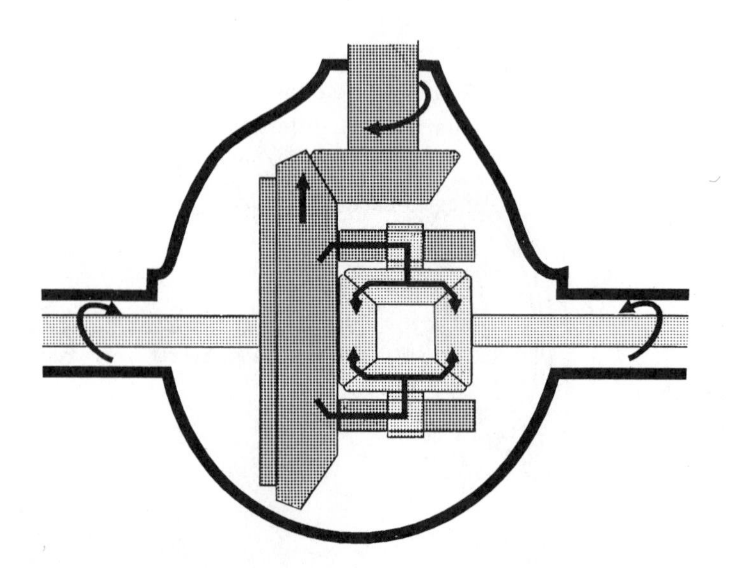
FIGURE 1–17 Differential splits the final drive torque.

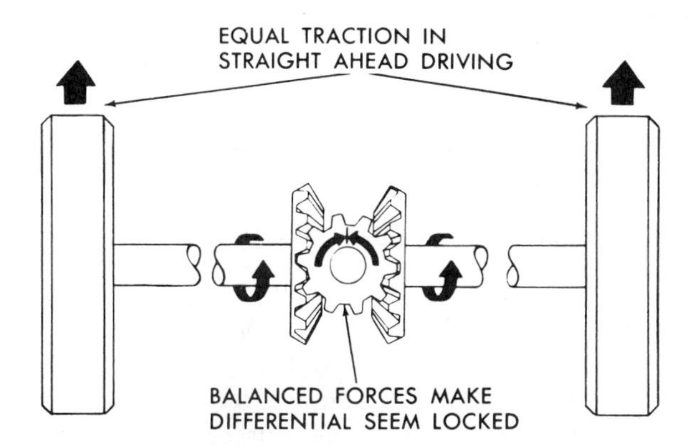

FIGURE 1–18 Differential not active for straight driving. (Courtesy of General Motors Corporation, Service Technology Group)

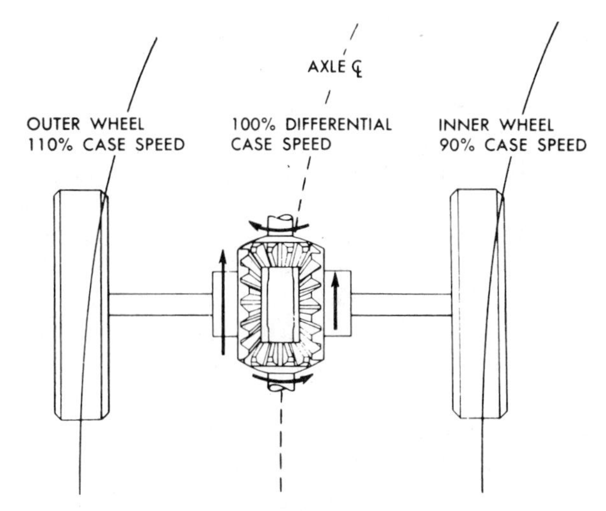

FIGURE 1–19 Differential action on turns. (Courtesy of General Motors Corporation, Service Technology Group)

the vehicle. The powerflow sequence starts with the drive pinion gear and then goes to the ring gear, differential case, differential pinion shaft, pinion gears, side gears, and to the axle shafts.

During straight-ahead driving, when traction is equal, both side gears offer equal resistance to the pinion gear rotation. Observing the thrust forces in Figure 1–18, the pinion gears and side gears are locked together, and therefore, no differential gear action occurs.

To prevent tire scuffing and vehicle skid when turning corners, the differential becomes effective and the axle shafts can then turn at different speeds (Figure 1–19). The side gear connected to the inside wheel slows down and causes the pinions to activate and spin on their centers or differential shaft. While maintaining equal tooth loads on the side gears, the pinion gear action permits the side gears to turn the axle shafts at unequal speeds. Should the inner wheel slow down to ninety percent of ring gear speed, the outer wheel will turn at one hundred and ten percent. One stopped wheel will cause the other wheel to turn at two hundred percent. Differential action is summarized in Figure 1–20.

The powertrain discussion has focused mainly on the engine, transmission, and final drive systems. In concluding the discussion, some comment should be made about the propeller shaft of an RWD versus the half-shafts of an FWD. The FWD powertrain is less susceptible to **cyclic vibrations** because (1) it uses true constant-velocity universal joints, (2) it has short half-shaft lengths, and (3) half-shaft speeds run at approximately sixty-six percent less rpm. The latter factor is due to running the transmission output directly through the final drive gear ratio, where the powertrain rpm is reduced before driving the half-shafts and wheels. In the RWD system, the propeller shaft is the link between the transmission and rear drive and, therefore, always turns at transmission output speed.

❖ **Cyclic vibrations:** The off-center movement of a rotating object that is affected by its initial balance, speed of rotation, and working angles.

- Gear 1 = ring gear
- Gear 2 = L.H. differential side gear
- Gear 3 = R.H. differential side gear

• Same speed both wheels straight driving	• Gears 1, 2, and 3 same speed
• L.H. wheel faster right turn	• Gear 2 faster than gear 1 Gear 3 slower than gear 1
• R.H. wheel faster left turn	• Gear 3 faster than gear 1 Gear 2 slower than gear 1
• L.H. wheel spinning	• Gear 3 fixed gears 1 and 2 rotating
• R.H. wheel spinning	• Gear 2 fixed gears 1 and 3 rotating
• Gear 1 fixed Engine stopped	• Rotate R.H. wheel forward Gear 3 rotates forward-gear 2 and L.H. wheel rotate in reverse

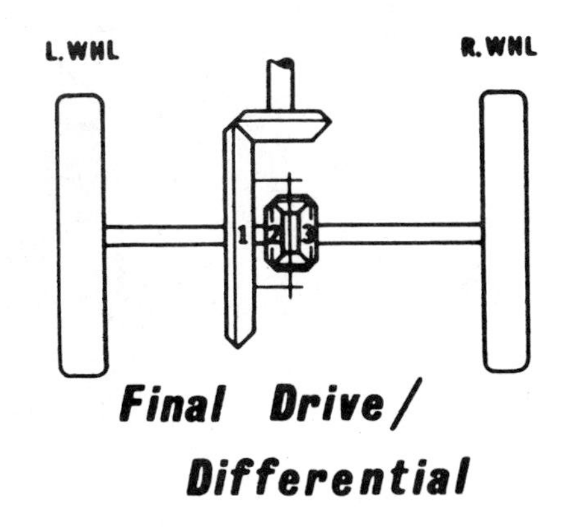

FIGURE 1–20 Differential action.

All-Wheel Drive/Four-Wheel Drive

All-wheel drive (AWD) or four-wheel drive (4WD) means that all four wheels, front and rear, are driving. For most driving conditions, two-wheel drive does an excellent job. However, when encountering roads that are rough or muddy, have loose sand, or are icy, two driving wheels have limitations. When the **tractive effort** (force available at the drive wheels to move the vehicle) exceeds the tire-to-road traction, or the ability of a tire to transmit the drive force, the tire slips. The factors determining traction are the **coefficient of friction** on the road and the effective vehicle weight on the tire. Using a formula to express the relationship:

$$\text{Traction} = VW \times u$$

where *VW* is the vehicle weight on the tire and *u* is coefficient of friction of the tire on the average road surface, generally 0.6. The factors determining tractive effort are the engine torque, gear ratios, and rolling radius of the driving tires.

$$TE = \frac{T \times C \times R \times e \times 12}{r}$$

where

T = gross engine torque in lb-ft
C = correction factor to determine net engine torque at flywheel; generally 0.85
R = overall torque effort of converter, transmission, and final drive
e = mechanical efficiency of powertrain; generally 0.85
r = rolling radius of loaded driving tire in inches
12 = a constant converting, lb-ft to lb-in

EXAMPLE: What is the tractive effort when the gross engine torque is 110 lb-ft, the converter torque 2:1, the trans ratio 3:1, the final drive ratio 2.5:1, and the rolling radius of driving tires is 10 in?

$$TE = \frac{110 \times 15 \times 0.85 \times .85 \times 12}{10} = 1430 \text{ lb } (6361 \text{ N})$$

$$= 715 \text{ lb } (3180 \text{ N}) \text{ at each of two drive wheels}$$

$$\text{Required traction} = \frac{715 \text{ lb}}{0.6u} = 1191 \ (540 \text{ kg})$$

- ❖ **Tractive effort:** The amount of force available to the drive wheels, to move the vehicle.
- ❖ **Coefficient of friction:** The amount of surface tension between two contacting surfaces; identified by a scientifically calculated number.

In examining the results of the preceding example, the drive wheels would each need to be weighted with 1191 lb (540 kg) or more to avoid slipping under ideal conditions. When all four wheels are made the driving wheels, the drive torque exiting the transmission is split between the front and rear drive axles (axle ratios are the same or within one tooth of being equal). In the example, 357.5 lb (1590 N) of *TE* force would be distributed to each wheel. This doubles the traction and considerably cuts back on potential tire slip.

There are a number of AWD/4WD powertrain configurations used in passenger cars and trucks. The objective at this time is to bypass the powerflow and operation details of these systems and deal with the concept only. These units are either an all-gear or a combination gear and chain design. In the latter system, the driving chain supplies power to the front output shaft.

Whatever the configuration, to get all four wheels into the act, a transfer case is coupled to the transmission. The job of the transfer unit is to split and transfer the transmission output shaft torque to the rear and front drive axles (Figure 1–21). In a conventional truck system (Figure 1–22), it allows the driver to select two- or four-wheel drive operation. An additional reduction in the transfer case, for deep pulling power, is avail-

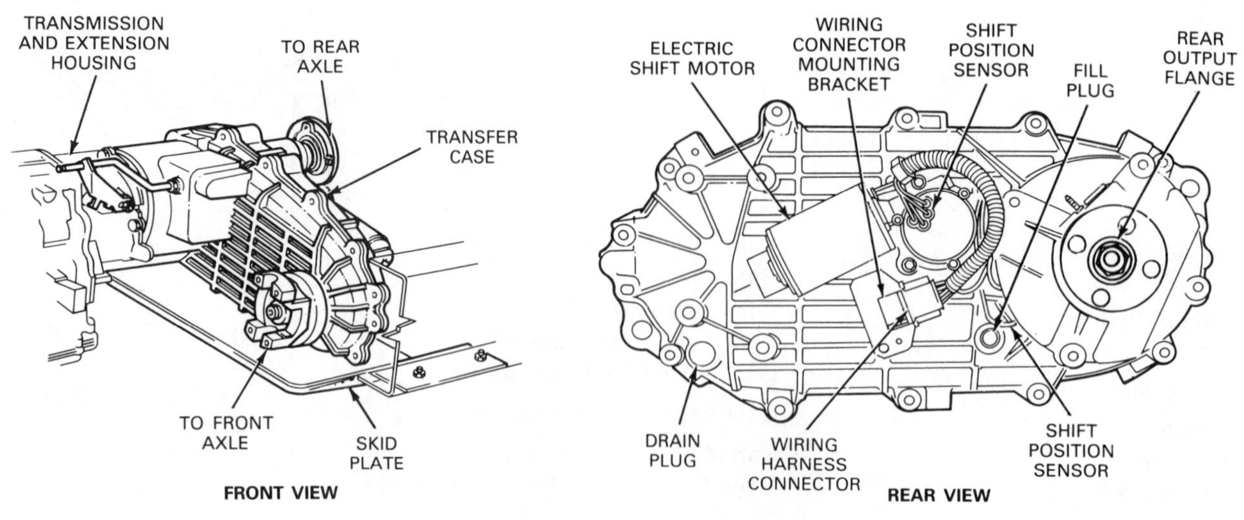

FIGURE 1–21 Electronic shift transfer case. (Reprinted with the permission of Ford Motor Company)

able in four-wheel drive only. A two-speed transfer case doubles the number of available gear ratios from the transmission.

To get power to the front driving axle, a short drive shaft connects the axle to the transfer case. For the front wheels to turn and steer the vehicle at the same time they are being driven, universal joints are coupled to the steering spindles. Since the two drive shafts in this system are connected to the transfer case, the front and rear axle ratios plus the tire sizes must be closely matched to prevent drivetrain tie-up.

Another style of all-wheel drive features full-time operation. Four wheels are always driving. This system has been offered for some time in many imports and is gradually being introduced to American passenger cars in limited production. The 1980 American Motors Eagle pioneered the first United States four-wheel drive passenger car. Other full-time system versions are now in production for Chrysler, Ford, and General Motors.

The following is a brief look at one of the newer system developments. For 1988, General Motors introduced its passenger car full-time, all-wheel drive system (Figure 1–23) on the Pontiac STE AWD. The system was adapted with some modification to the 3T40 (125C) transaxle and given the product designation 3T40-A. The final drive section of the transmission was removed and incorporated into the transfer case unit. The transfer unit bolts to the 3T40 and contains three planetary gearsets for the following functions:

- *Center differential:* receives its input directly from the 3T40 and splits the driving torque 60:40 between the front and rear wheels. In addition, it allows the front and rear drives to turn at different speeds. In a turn, the rear wheels run a tighter circle than the front wheels, resulting in slower rear-wheel speed and faster front-wheel speed. This enhances the vehicle maneuverability.
- *Final drive:* serves the same function as the final drive in the regular 3T40. It gives a final gear reduction of 2.84:1 to power the front drive wheels.
- *Front differential:* splits the drive torque between the two front wheels and allows them to turn at different speeds.

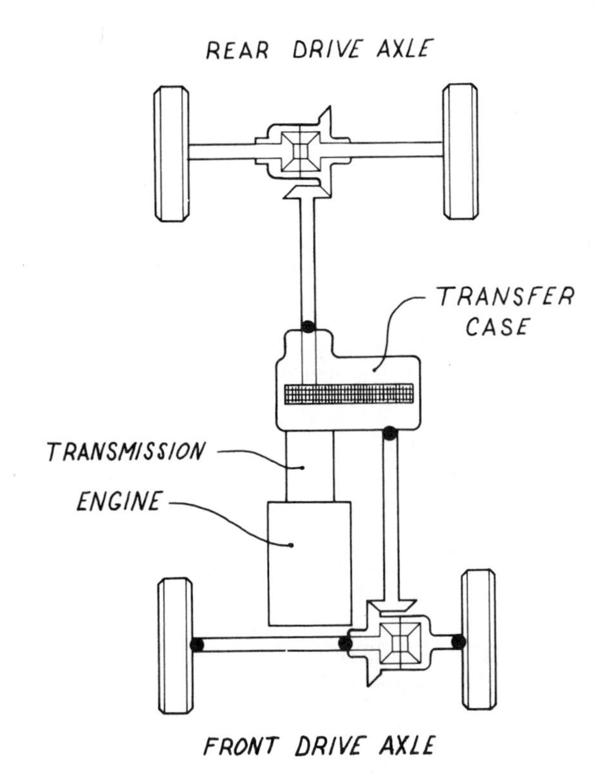

FIGURE 1–22 4WD concept.

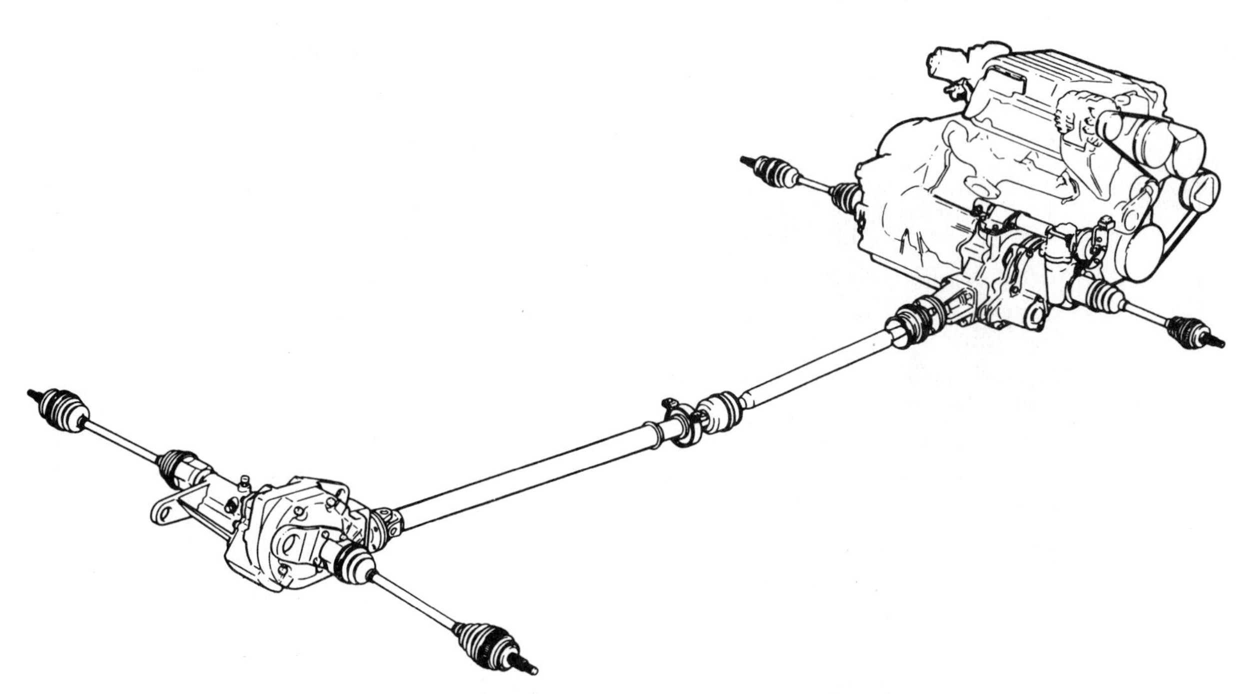

FIGURE 1–23 AWD drivetrain. (Courtesy of General Motors Corporation, Service Technology Group)

The transfer unit also contains a spiral bevel gearset providing for a minor gear reduction and a ninety-degree change in powerflow direction. The bevel set then drives the propeller shaft to the rear drive axle. The rear drive uses a conventional ring and pinion final drive/differential unit.

The rotational speeds of the front and rear wheels must be matched to avoid a powertrain tie-up. The balance is achieved within the transfer case by selecting appropriate ratios for the helical transfer and spiral bevel gearsets. The rear drive/differential unit is located in the rear cradle assembly and secured with a three-point mounting arrangement.

To ensure that power is provided to the front and rear axle drives for low-traction situations when the vehicle is hung up, a differential "lock" activates an electrovacuum circuit that initiates a vacuum motor action. This moves a shaft and fork assembly to lock the center differential in the transfer unit. At least one front and one rear wheel then have power.

Important Safety Notice

Appropriate service methods and proper repair procedures are essential for the safe, reliable operation of all motor vehicles as well as the personal safety of the individual doing the work. This Service Manual provides general directions for accomplishing service and repair work with tested, effective techniques. Following them will help assure reliability.

There are numerous variations in procedures, techniques, tools, and parts for servicing vehicles, as well as in the skill of the individual doing the work. This Manual cannot possibly anticipate all such variations and provide advice or cautions as to each. Accordingly, anyone who departs from the instructions provided in this Manual must first establish that he compromises neither his personal safety nor the vehicle integrity by his choice of methods, tools or parts.

Notes, Cautions, and Warnings

As you read through the procedures, you will come across NOTES, CAUTIONS, and WARNINGS. Each one is there for a specific purpose. NOTES give you added information that will help you to complete a particular procedure. CAUTIONS are given to prevent you from making an error that could damage the vehicle. WARNINGS remind you to be especially careful in those areas where carelessness can cause you personal injury. The following list contains some general WARNINGS that you should follow when you work on a vehicle.

- Always wear safety glasses for eye protection.
- Use safety stands whenever a procedure requires you to be under the vehicle.
- Be sure that the ignition switch is always in the OFF position, unless otherwise required by the procedure.
- Set the parking brake when working on the vehicle. If you have an automatic transmission or automatic transaxle, set it in PARK unless instructed otherwise for a specific operation. If you have a manual transmission or manual transaxle, it should be in REVERSE (engine OFF) or NEUTRAL (engine ON) unless instructed otherwise for a specific operation. Place wood blocks (4" x 4" or larger) against the front and rear surfaces of the tires to provide further restraint from inadvertent vehicle movement.
- Operate the engine only in a well-ventilated area to avoid the danger of carbon monoxide.
- Keep yourself and your clothing away from moving parts when the engine is running, especially the drive belts.
- To prevent serious burns, avoid contact with hot metal parts such as the radiator, exhaust manifold, tail pipe, three-way catalytic converter and muffler.
- Do not smoke while working on a vehicle.
- To avoid injury, always remove rings, watches, loose hanging jewelry, and loose clothing before beginning to work on a vehicle.
- If it is necessary to work under the hood, keep hands and other objects clear of the radiator fan blades! Your vehicle may be equipped with a cooling fan that may turn on, even though the ignition switch is in the OFF position. For this reason care should be taken to ensure that the radiator electric motor is completely disconnected when working under the hood when engine is not running.

FIGURE 1–24 Ford Service Manual safety notice. (Reprinted with the permission of Ford Motor Company)

INTRODUCTION TO SAFETY

Diagnosing and repairing transmissions and transaxles can be a rewarding experience. Product knowledge, helpful resources and references, and the proper tools contribute to reliable servicing. While working with vehicles, it is important to work efficiently and make wise decisions.

Of all the decisions that a person makes while diagnosing and repairing vehicles, the most important ones involve safety. It must be remembered that the vehicle is a powerful machine. Hand tools can break. Testers and service equipment fail. The technician must always be aware of the hazards within the work environment. The vehicle must be operated and the tools and equipment used as designed. The condition of these items must be maintained to make sure they are in good working order.

Efficient service practices never involve jeopardizing the safety of the technician or others. Shown in Figure 1–24 is the safety notice Ford Motor Company uses in the introductory section of the 1994 service manuals. The contents of the "Important Safety Notice" should be read, including the "Notes, Cautions, and Warnings." The safety points listed should always be followed.

This textbook also provides notes, cautions, and warnings to highlight important concerns. The format is similar to that used by Ford in the safety notice (Figure 1–24), and special attention should be paid to these concerns. A successful transmission technician works intelligently and consistently employs safe diagnosis and service practices.

SUMMARY

Automatic transmissions and transaxles are an integral part of a vehicle's powertrain and serve many functions. Most importantly, they transmit and modify engine torque and speed. By altering the engine's output, the transmission generates adequate torque to move the vehicle from a stopped position. By changing gear ratios, highway speeds are obtainable and fuel economy is enhanced.

The transmission's torque converter and the final drive unit are important factors in the overall torque and speed output at the drive wheels. In a transaxle, the final drive unit is internally contained in the transmission case. Typically, a transmission is accompanied with a rear axle housing to produce the necessary gear reductions needed. In most conventional four-wheel drive vehicles, both front and rear axle assemblies provide the final drive gear ratios. Whichever type of final drive unit is used, it must provide differential action for cornering.

Transmissions and transaxles share similar operating characteristics. Thus, the term *transmission* is used to describe both design types in this text when applicable.

The following chapters expose the student to common transmission mechanical powerflows, hydraulic systems, electronic shift controls, diagnostic procedures, and overhaul techniques. The fundamental principles of transmission operation and servicing techniques provide the foundation for being a successful automatic transmission technician.

❑ REVIEW

Key Terms

Powertrain, transmission, transaxle, torque, torque converter, differential, cyclic vibrations, tractive effort, and coefficient of friction.

Questions

1. Arrange in sequence the following powertrain components from the crankshaft to drive wheels (FWD).

1. A	A. Crankshaft
2. ____	B. Propeller shafts
3. ____	C. Differential
4. ____	D. Converter
5. ____	E. Flex plate
6. ____	F. Final drive
7. ____	G. Transmission
8. H	H. Drive wheels

2. Arrange in sequence the powerflow through the parts makeup of a final drive/differential assembly unit (RWD).

1. A	A. Drive pinion gear
2. ____	B. Differential case
3. ____	C. Side gears
4. ____	D. Ring gear
5. ____	E. Pinion gears
6. ____	F. Pinion case shaft
7. F	G. Drive axles and wheels

Multiple Choice

____ 1. In a planetary final drive used in FWD automatic transmissions:

I. the internal gear drives the differential carrier.
II. the transmission output shaft drives the sun gear.

A. I only C. both I and II
B. II only D. neither I nor II

____ 2. Helical final drives for passenger car applications are found in:

I. rear-wheel drives.
II. longitudinal front-wheel drives.

A. I only C. both I and II
B. II only D. neither I nor II

____ 3. (Fluid torque converter) The engine crankshaft drives:

A. the turbine. C. the stator.
B. the impeller. D. the input shaft.

_____ 4. In a fluid torque converter operation:
I. the converter acts as a torque multiplier.
II. the converter acts as a fluid coupling.

A. I only C. both I and II
B. II only D. neither I nor II

_____ 5. The transmission:

A. changes engine torque.
B. changes engine speed.
C. provides a neutral.
D. provides a reverse.
E. all of the above.

_____ 6. A four-wheel drive powertrain:
I. doubles the wheel traction.
II. doubles the wheel tractive effort.

A. I only C. both I and II
B. II only D. neither I nor II

_____ 7. When compared to the drive shaft in RWD applications, the FWD half-shafts:
I. turn at fifty percent less rpm.
II. must use true constant velocity U-joints.

A. I only C. both I and II
B. II only D. neither I nor II

Completion

1. Name the three factors that determine the road load on the engine.

A. ____________________

B. ____________________

C. ____________________

2. The ability of the atmosphere to charge the engine cylinders with a quantity of air/fuel mixture is called

____________________.

Problems

1. Calculate the axle torque from the following data. Show your work. Compensate for driveline efficiency.

Net engine torque = 180 lb-ft (244 N-m)
Converter torque at stall = 2.0:1
Transmission first-gear ratio = 2.84:1
Final drive ratio = 2.75:1

Ta = ____________________

How much axle torque is delivered to each drive wheel?

T = ____________________

2. Calculate the force available at the drive wheels to move the vehicle from the following data. Show your work. Compensate for driveline efficiency.

Net engine torque = 110 lb-ft (149 N-m)
Converter torque at stall = 2.0:1
Transmission first-gear ratio = 2.4:1
Final drive ratio = 2.84:1
Rolling radius of driving tires = 10.5 in

TE = ____________________

TE at each drive wheel = ____________________

3. Calculate the final drive ring gear speed from the following data. Show your work.

Left-wheel speed = 800 rpm

Right-wheel speed = 650 rpm

Ring gear speed = ____________________

2 Torque Converters

OBJECTIVE:

The objective of this unit is for you to become proficient with the terms and concepts contained in this chapter. Completion of this objective is essential to becoming a successful automatic transmission technician.

CHALLENGE YOUR KNOWLEDGE

Define the following key terms/acronyms:

Torque converter, fluid coupling, hydrodynamic drive units, member, element, torsional, impeller, turbine, stator (reactor), rotary flow, vortex flow, stage, torque phase, one-way roller clutch, torque rating, stall torque, coupling phase, lockup converter, torque converter clutch, viscous clutch, PCM (powertrain control module), and ECM (engine control module).

List the following:

Three components of a simple torque converter; four desirable characteristics of torque converter operation; three conditions needed for converter clutch apply to occur; and four conditions that release a converter clutch.

Describe the following:

The design, construction, and operation of a torque converter; purpose and operation of the reactor/stator; dual stator operation; and lockup (converter clutch) operation.

INTRODUCTION

Torque converters and **fluid couplings** are closely related and are termed **hydrodynamic drive units** by engineers. The fluid coupling and torque converter are European, not American, developments. The first experimental fluid drive unit, a torque converter, was built by Germans in 1908. In the years between World Wars I and II, much of the work on fluid drives took place in England and Germany. In the early stages, the coupling and converter were designed as two distinct and separate units and were not combined as a single unit as is common in current practice.

- **Torque converter:** Positioned between the engine and the transmission input shaft, the torque converter is a hydrodynamic drive unit that acts as a fluid coupling and has the ability to multiply torque.
- **Fluid coupling:** The simplest form of hydrodynamic drive, the fluid coupling consists of two look-alike members with straight radial vanes referred to as the impeller (pump) and the turbine. Input torque is always equal to the output torque.
- **Hydrodynamic drive units:** Devices that transmit power solely by the action of a kinetic fluid flow in a closed recirculating path. An impeller energizes the fluid and discharges the high-speed jet stream into the turbine for power output.

As a key to developing semi- and fully automatic transmissions for passenger cars, the two-**member** fluid coupling was eventually adopted in the United States in 1938 by Chrysler (Figure 2–1). In 1948, the Buick Motor Division of General Motors Corporation developed the first converter-coupling for use in an American passenger car. It was a five-**element** configuration designed for its new Dynaflow transmission (Figure 2–2).

- **Member:** An independent component of a hydrodynamic unit such as an impeller, a stator, or a turbine. It may have one or more elements. (Figure 2–3)
- **Element:** A device within a hydrodynamic drive unit designed with a set of blades to direct fluid flow. (Figure 2–3)

Following the Dynaflow development, most success with converter work has occurred in the United States. Converter applications are now used extensively in domestic and import passenger cars, trucks, and buses. A wide variety of heavy-duty construction machinery also uses converters, such as wheeled and track-laying tractors, loaders, cranes, lift trucks, graders, earth movers, and large farm tractors.

In modern passenger cars, the torque converter has gradually taken over as the prime fluid drive unit. The fluid coupling has not been used in American automatic transmissions since 1965. When the term *torque converter* is used in connection with automatic transmissions, it is understood that the torque converter has a dual function. It must act as a torque multiplier

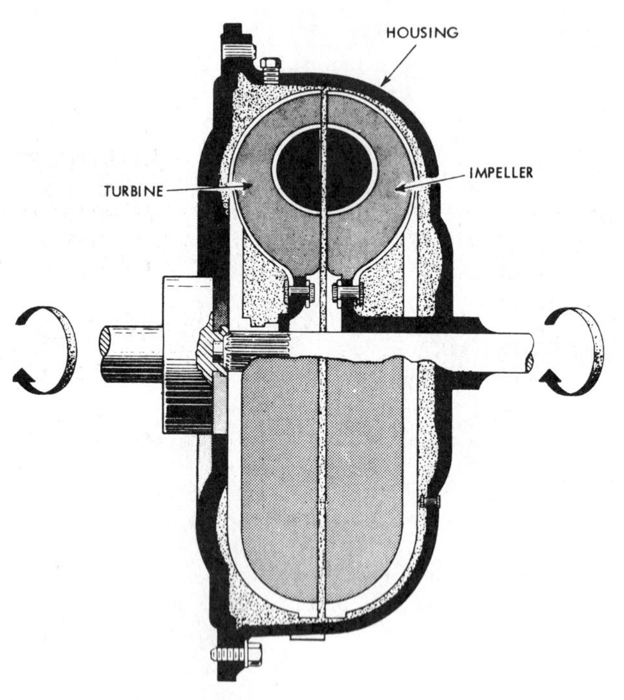

FIGURE 2–1 Simplified fluid coupling.

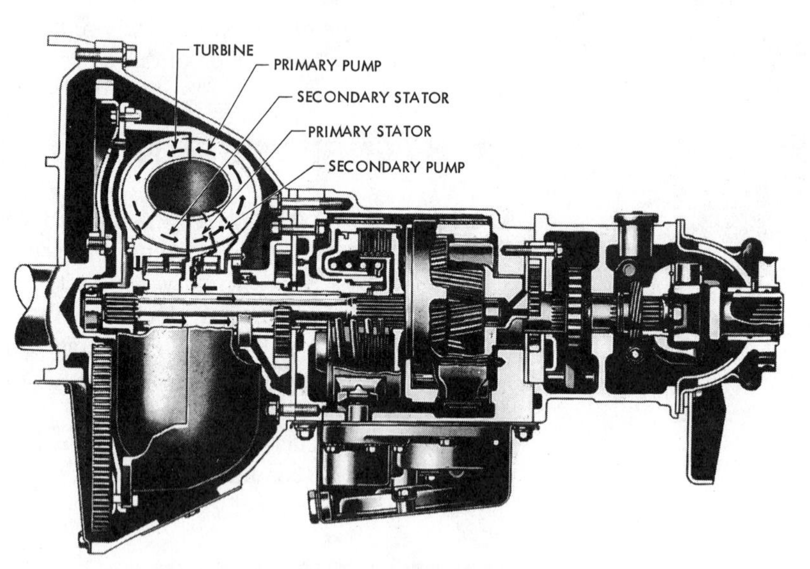

FIGURE 2–2 Dynaflow 1948 model, the first passenger car application of a converter-coupling.

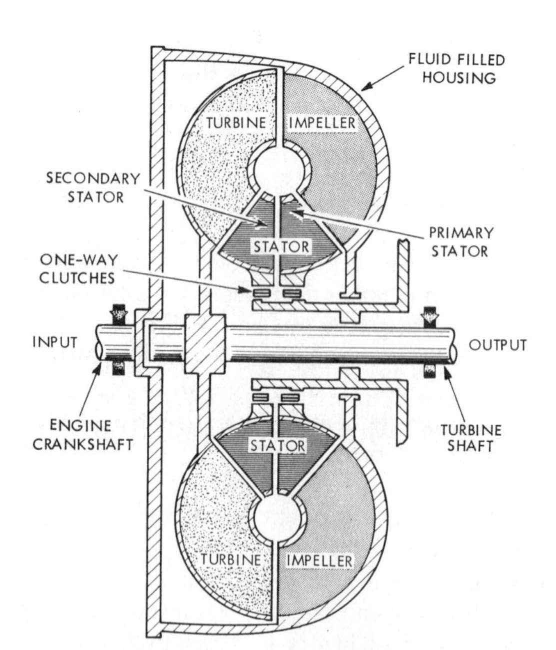

FIGURE 2–3 Three-member, four-element converter. The stator member is made of two elements mounted on separate overrunning clutches that can freewheel independently. Used in early converters, this converter design was reintroduced in the 1990s.

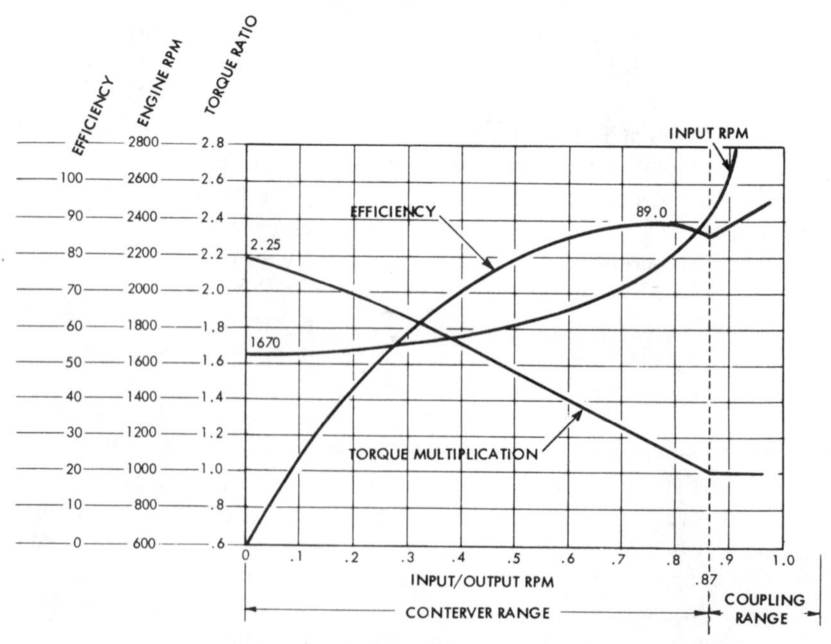

FIGURE 2–4 A torque converter performance curve. (Courtesy of Chrysler Corp.)

with infinitely variable ratios from its maximum engineered torque output to unity (1:1), where it acts as a fluid coupling (Figure 2–4).

The converter offers several desirable operating features. It is a simple, rugged unit that operates as a clutch in a constant oil bath that gives unlimited life and requires no maintenance, in contrast to the foot-operated friction clutch that it eliminated. Because it is a fluid unit, it provides a silky-smooth attachment of the engine power to the vehicle that eliminates any sudden engagement shock to the powertrain components and results in their longer life and reduced repair costs. Another dividend is the excellent dampening of engine **torsional** vibration that is taken up by the fluid before it extends into the transmission and driveline. The converter can be likened to a cushion that protects the powertrain from shocks and vibrations and prevents lugging or stalling of the engine.

❖ **Torsional:** Twisting or turning effort; often associated with engine crankshaft rotation.

Although a variety of fluid torque converter designs have been used in automatic transmissions, the simple three-element unit, consisting of **impeller**, **turbine**, and **stator (reactor)**, provides the basic ingredients needed for torque converter action to occur.

❖ **Impeller:** Often called a pump, the impeller is the power input (drive) member of a hydrodynamic drive. As part of the torque converter cover, it acts as a centrifugal pump and puts the fluid in motion.

❖ **Turbine:** The output (driven) member of a fluid coupling or fluid torque converter. It is splined to the input (turbine) shaft of the transmission.

❖ **Stator (reactor):** The reaction member of a fluid torque converter that changes the direction of the fluid as it leaves the turbine to enter the impeller vanes. During the torque multiplication phase, this action assists the impeller's rotary force and results in an increase in torque.

FLUID COUPLING

Function and Construction

Because the fluid coupling is the simplest form of fluid drive unit, the principles of fluid coupling operation serve as an ideal introduction to how a fluid force can put a solid object into motion. Some of these working principles will apply later to our discussion of the fluid converter-coupling operation. The construction of a fluid coupling is very simple. It consists of impeller and turbine members of identical structure contained in a housing filled with oil (Figures 2–5 and 2–6). The coupling members face one another closely with the impeller driven by the engine and the turbine driving the wheels through the transmission and axle.

Operation

The following describes what happens inside a fluid coupling when the car is started or when acceleration under heavy load takes place. As the engine drives the impeller, it sets the fluid mass into motion, creating a fluid force. The path of the fluid force strikes on a solid object, the turbine. The impact of the fluid jet stream against the turbine blades sets the turbine in motion. An energy cycle has been completed: mechanical to fluid and back to mechanical.

The fluid action that takes place between the impeller and turbine is an interesting science. When the impeller spins up, two separate forces are generated in the fluid. One is **rotary flow**, which is the rotational effort or inertia of the impeller rotation. The other is **vortex flow**, which circulates the fluid between the coupling members and is caused by the centrifugal pumping action of the rotating impeller (Figure 2–7). The vortex flow is the fluid exit velocity from the impeller.

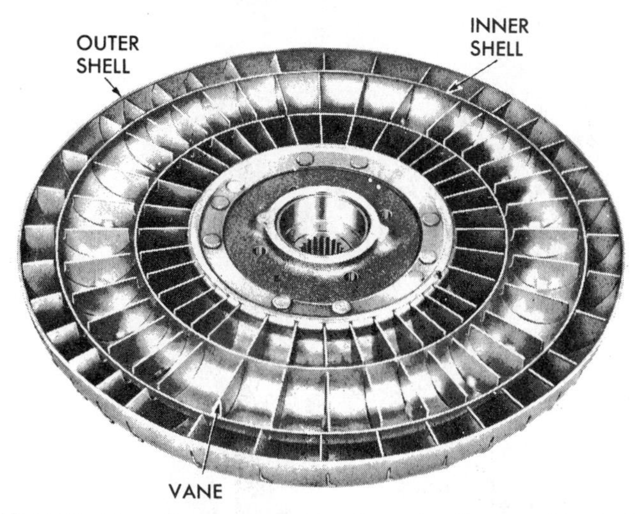

FIGURE 2–5 Coupling members (impeller and turbine) use identical straight radial vanes. (Courtesy of General Motors Corporation, Service Technology Group)

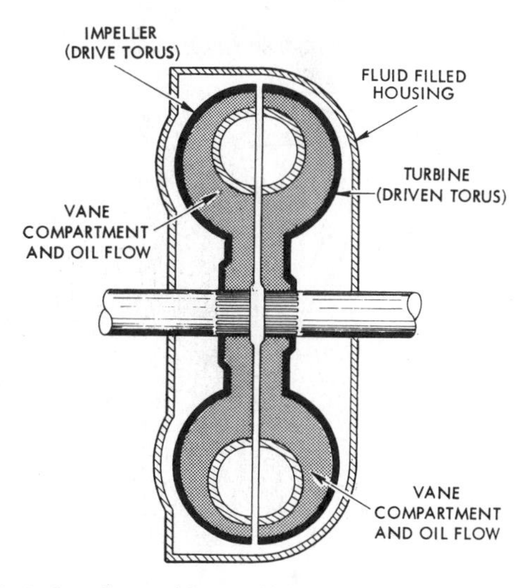

FIGURE 2–6 Coupling turbine and impeller member in a housing. (Courtesy of General Motors Corporation, Service Technology Group)

- **Rotary flow:** The path of the fluid trapped between the blades of the members as they revolve with the rotation of the torque converter cover (rotational inertia).
- **Vortex flow:** The crosswise or circulatory flow of oil between the blades of the members caused by the centrifugal pumping action of the impeller.

The rotary and vortex flows can also be explained by using a bucket of water as an example. Any time there is a spinning mass, the mass (in the case of a transmission, the oil) follows the rotational movement and creates a centrifugal force. This dual effect is illustrated by swinging a bucket of water in a circle (Figure 2–8). As the bucket is swinging, the water is following the circular path of the bucket. At the same time, it is developing a centrifugal force that keeps the water in the bucket as it passes through the overhead position. The water is confined by the solid side and bottom of the bucket and cannot discharge (fly) outward, so it is forced to follow the bucket rotation only.

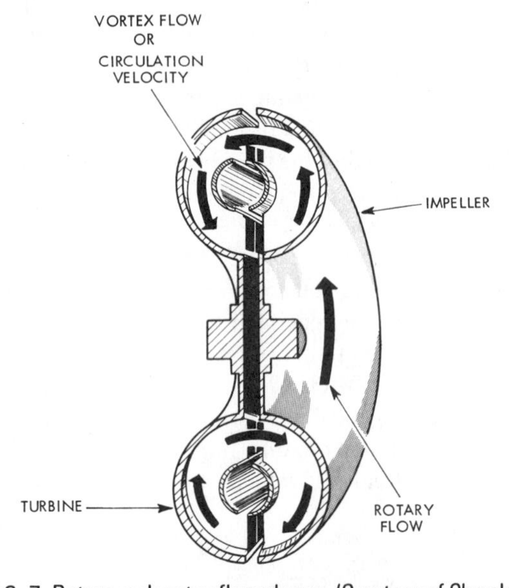

FIGURE 2–7 Rotary and vortex flow shown. (Courtesy of Chrysler Corp.)

FIGURE 2–8 Swinging bucket shows both rotary and vortex forces.

In a fluid coupling, however, the centrifugal force of the oil is not confined. The oil at the center of the spinning impeller follows the curved shell and is discharged along the outer diameter and into the turbine to establish a circular path between the impeller and the turbine (Figure 2–9). The combination of rotary flow and vortex flow causes an oil motion that follows the course of a rotating corkscrew. It is like watching the blade tips of a rotating pinwheel at the end of a stick, with the stick itself turning about a center (Figure 2–10). With the corkscrew oil action created by the rotation of the engine-driven impeller, the turbine is pushed around ahead of the oil, striking on the turbine blades. A fluid clutch is thus established, with the turning torque on the turbine never exceeding impeller input torque.

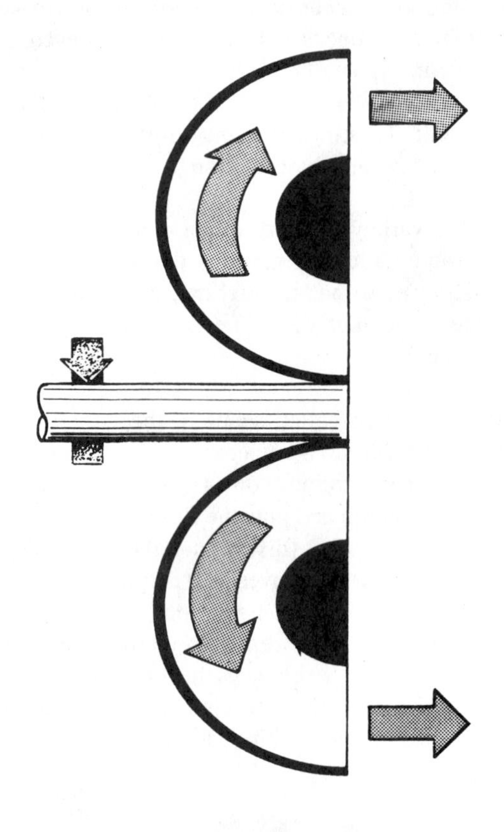

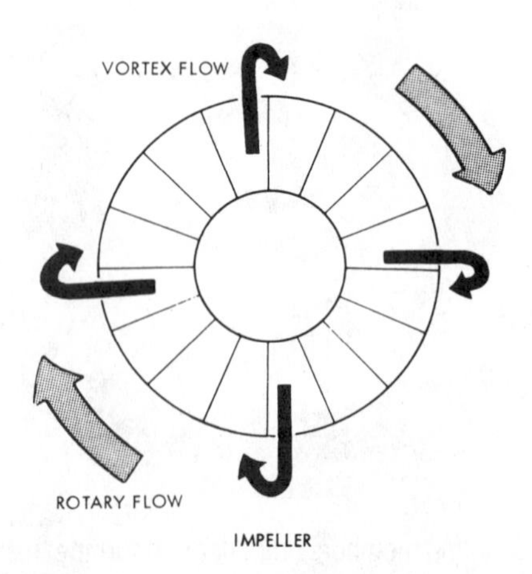

FIGURE 2–9 View of oil discharged from impeller-vortex flow; side and front views of impeller showing pumping action on fluid.

Speed Ratio

The striking force of the fluid on the turbine can be explained from a slightly more technical viewpoint. In doing this, it becomes necessary to define another term that reflects the efficiency of a fluid coupling or torque converter. The number of revolutions that the turbine makes relative to one rotation of the impeller is its speed ratio, which is expressed in percentages. For example, if the impeller rotation is 1000 rpm and the turbine rotation is 900 rpm, the speed ratio is 90%:

$$\text{speed ratio} = \frac{\text{turbine rpm}}{\text{impeller rpm}} \text{ or } \frac{900}{1000} = 90\%$$

Just when the car starts to move, there is an instant when the impeller is rotating and the turbine has not begun to move, a condition of zero speed ratio. During this situation, the following rotary and vortex flow conditions are in effect:

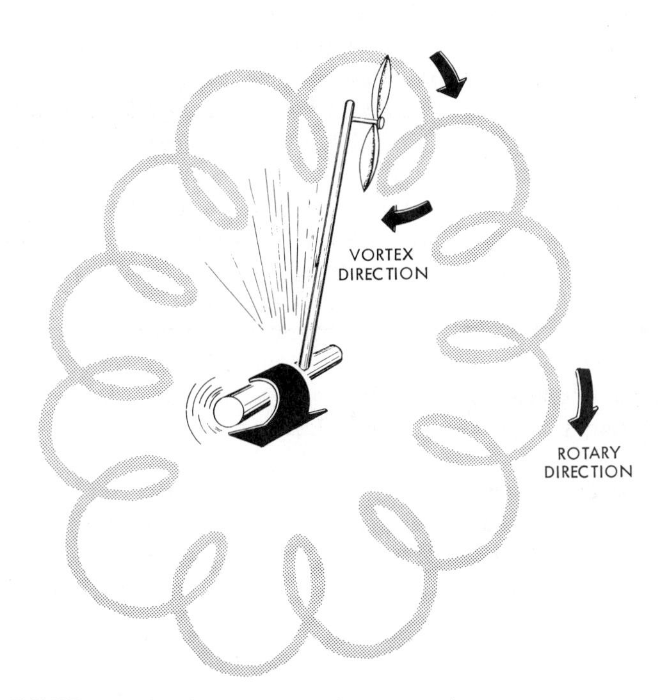

FIGURE 2–10 Combined rotary and vortex motion.

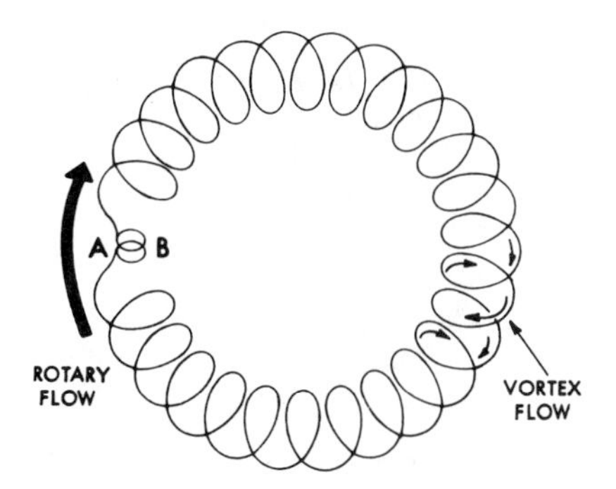

FIGURE 2–11 Rotary and vortex flow with turbine stationary.

1. Because the turbine is stationary, the vortex flow cycles through the turbine unopposed, giving a massive cross circulation between the coupling members (Figure 2–11).
2. The stationary turbine also opposes the rotary flow and the moving oil does not favor rotary flow (Figure 2–11).

The direction of the oil striking on the turbine is determined by the strength of the respective oil flows. This is illustrated by the vector diagram in Figure 2–12. The diagram shows the movement of both the vortex flow and the rotary flow and the obvious fact that the direction of impact cannot be in two directions at the same time. The direction of fluid thrust that results from the two flows will be a resultant angle to the rotary and the vortex action. This is determined by the speed ratio, or drive condition.

Going back to the coupling start condition, it is evident that the high rate of vortex flow does not have a favorable oil impact on the turbine blades. It is sufficient, however, to get the turbine moving, as shown in Figure 2–13. Considerable slipping takes place because the fluid impact is striking only a glancing blow on the straight blades of the turbine. As the turbine begins to rotate and catch up to impeller speed, the vortex flow gradually slows down because of the counter pumping or centrifugal action of the turbine (Figure 2–14). This permits the rotary action to become the greater influence on the fluid, and the resultant thrust becomes more effective in propelling the turbine. Finally, at 90% plus speed ratio, the rotary inertia or momentum of the fluid and coupling members forms a hydraulic lock or bond, and the coupling members turn at unity. In other words, the impeller and turbine get buried into the rotational inertia mass of the fluid, and the fluid coupling now operates at its coupling point and maximum efficiency. Figures 2–15 and 2–16 illustrate the vortex and rotary flow conditions and their effect on the resultant fluid thrust at the coupling point.

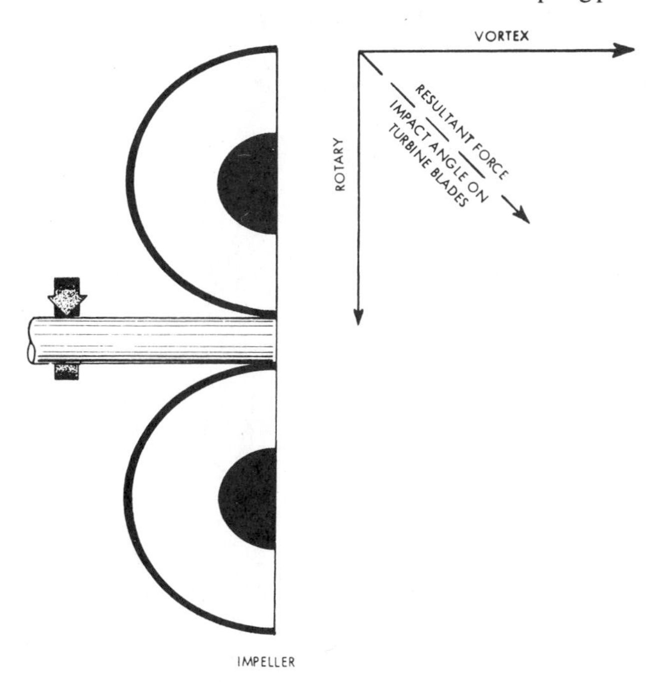

FIGURE 2–12 Vector diagram of impact force of rotary and vortex flow.

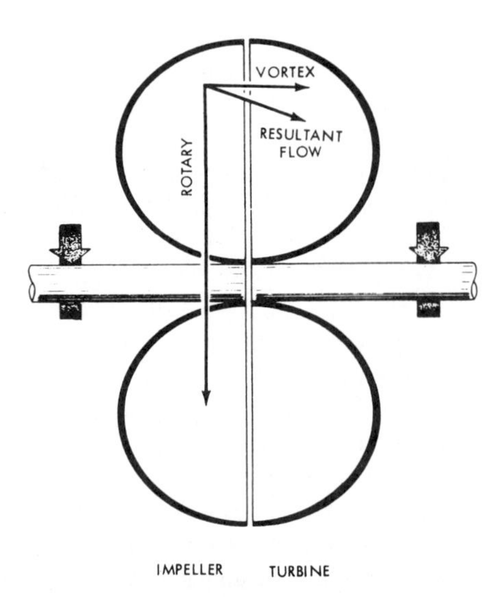

FIGURE 2–13 Vortex and rotary impact angle with turbine stationary.

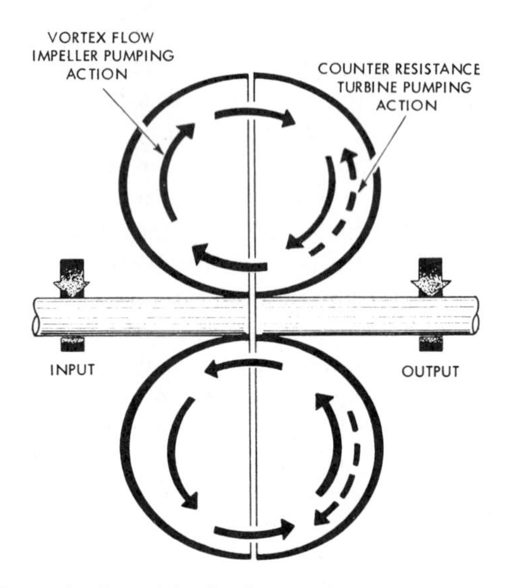

FIGURE 2–14 As the turbine begins to turn, a counter pumping action builds up.

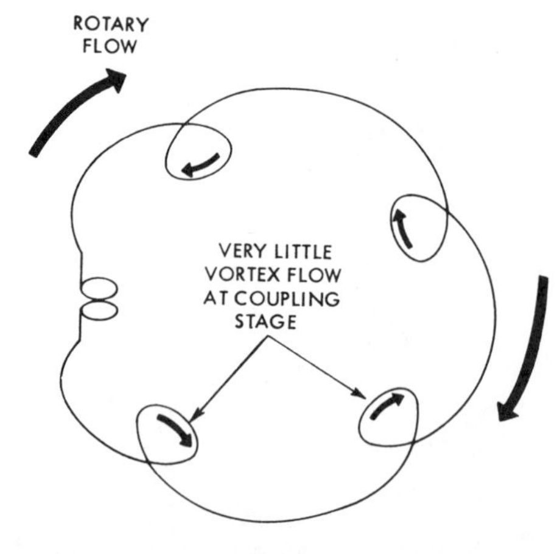

FIGURE 2–15 At the coupling point.

The change from a state of coupling slip to a state of coupling full torque occurs quite rapidly and is effective in controlling excessive engine run-up in the process. The nature of any fluid drive unit is that it will always have slip. Even at the coupling point, the maximum efficiency is only in the 90% range. For practical purposes, however, it is considered to transmit torque at 1:1.

Another feature of a coupling or converter is the ability to keep the vehicle drive wheels coupled to the engine for braking action. On deceleration, the turbine becomes an impeller and drives the oil against the normal impeller rotation and engine compression and intake strokes. The impeller fluid thrust is ineffective at idling speed, which results in a desirable automatic disengagement of engine power to the drive wheels.

FLUID TORQUE CONVERTER

Function and Construction

The behavior of the fluid action in a coupling can be applied to the operation of a fluid torque converter. The same vortex and rotary forces are generated by a moving impeller that puts the fluid in motion. To multiply torque, however, a different makeup of the fluid unit is necessary. Although more complicated designs have been used, the current practice in automatic transmissions is to use a basic three-element/single-**stage** converter-coupling unit: an impeller, a turbine, and a stator (Figure 2–17). A split guide ring (Figure 2–18) is built into the impeller and turbine for greater operational efficiency. Because the center of the vortex flow sets up a turbulence that results in a loss of efficiency, the guide ring is used to provide a smooth, uniform flow between the impeller and turbine.

❖ **Stage:** The number of turbine sets separated by a stator. A turbine set may be made up of one or more turbine members. A three-element converter is classified as a single stage.

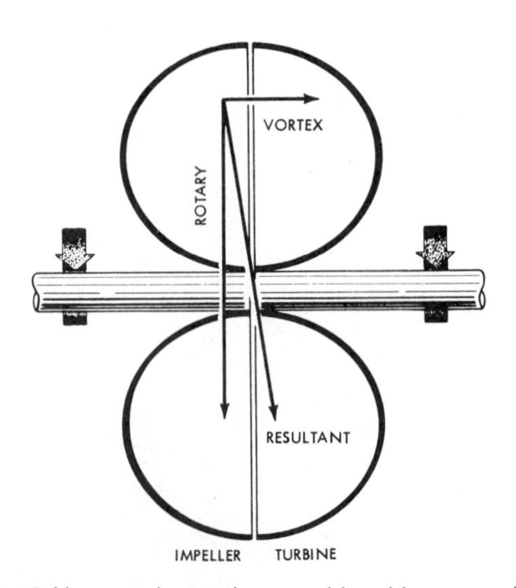

FIGURE 2–16 Vortex and rotary impact with turbine at coupling point.

Figure 2–19 shows the impeller as an integral part of the converter housing, which in manufacturing production is welded to the cover half. Enclosed within the welded housing is the turbine and the stator. The stator incorporates a one-way clutch and mounts on a stationary support shaft that is grounded to the transmission case directly or indirectly through the transmission pump assembly.

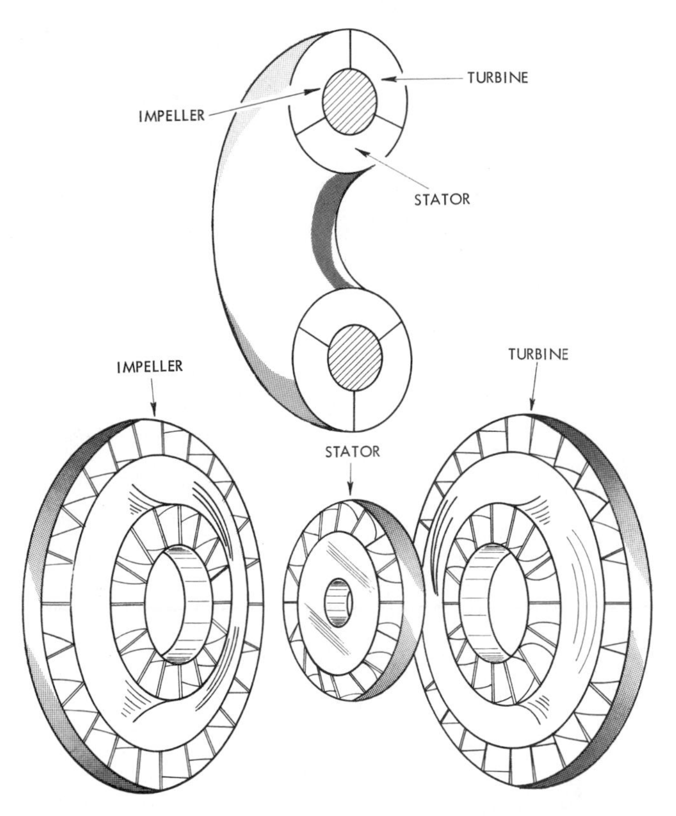

FIGURE 2–17 Three-element design torque converter.

Another construction feature is the design of the impeller and turbine blades, as illustrated in Figure 2–20. The curved shape of the impeller blades in a backward direction gives added acceleration and energy to the oil as it leaves the impeller, while the curved shape of the turbine vanes is designed to absorb as much energy as possible from the moving oil as it passes through the turbine.

Turbine vane curvature has two functions that give the turbine its excellent torque-absorbing capacity. It reduces shock losses due to sudden change in oil direction between the impeller and turbine (Figure 2–21). It also takes advantage of the hydraulic principle that the more the direction of a moving fluid is diverted, the greater the force that the fluid exerts on the diverting surface. The two functions are illustrated in Figure 2–22. A fluid jet stream directed against a flat surface exerts force on the plate, but not without a shock loss caused by the breakdown of the smooth fluid flow (Figure 2–22, left). By curving the inlet side (Figure 2–22, center), the shock loss is reduced considerably, but the force on the flat surface remains the same. The plate surface (Figure 2–22, right) is curved at both inlet and outlet, keeping the fluid flow smooth and greatly increasing the force of the fluid jet stream on the plate. Figure 2–23 shows this effect on a curved turbine vane. Note that the fluid impact is absorbed along the full length of the vane surface as the fluid reverses itself.

The third bladed member of the converter is the stator. During **torque phase**, its function is to redirect the fluid flow as it leaves the turbine and reenters the impeller. This assists the impeller rotation and gives a thrust boost to the fluid discharge (Figure 2–24).

❖ **Torque phase:** Sometimes referred to as *slip phase* or *stall phase,* torque multiplication occurs when the turbine is turning at a slower speed than the impeller, and the stator is reactionary (stationary). This sequence generates a boost in output torque.

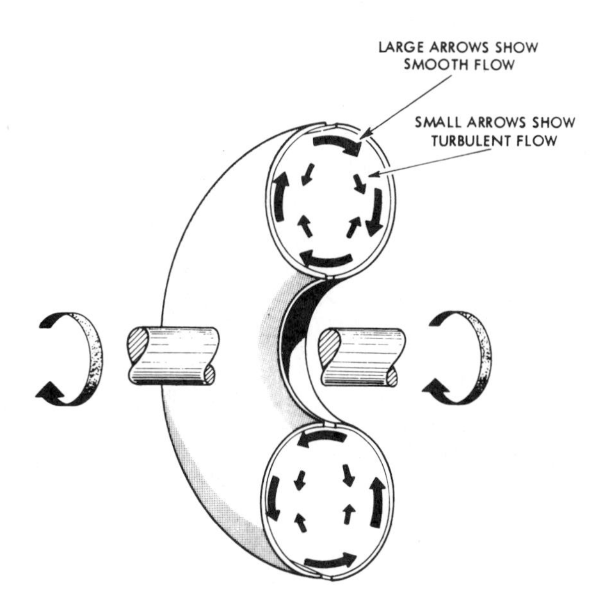

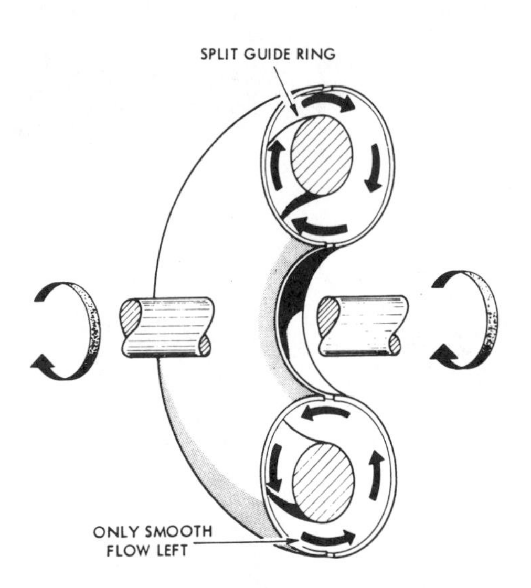

FIGURE 2–18 Converter guide ring function.

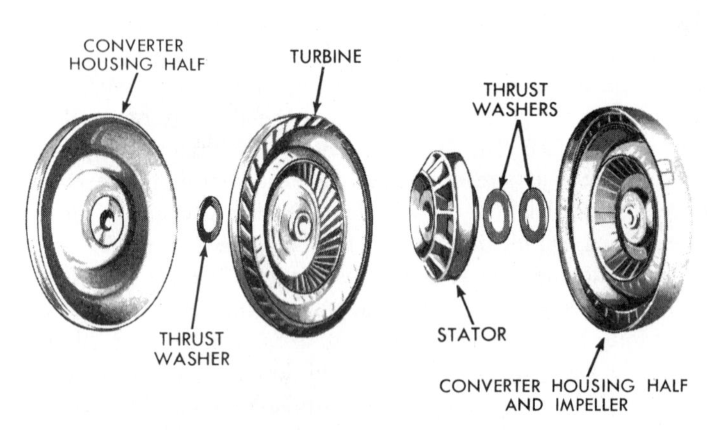

FIGURE 2–19 Impeller as part of housing. (Reprinted with the permission of Ford Motor Company)

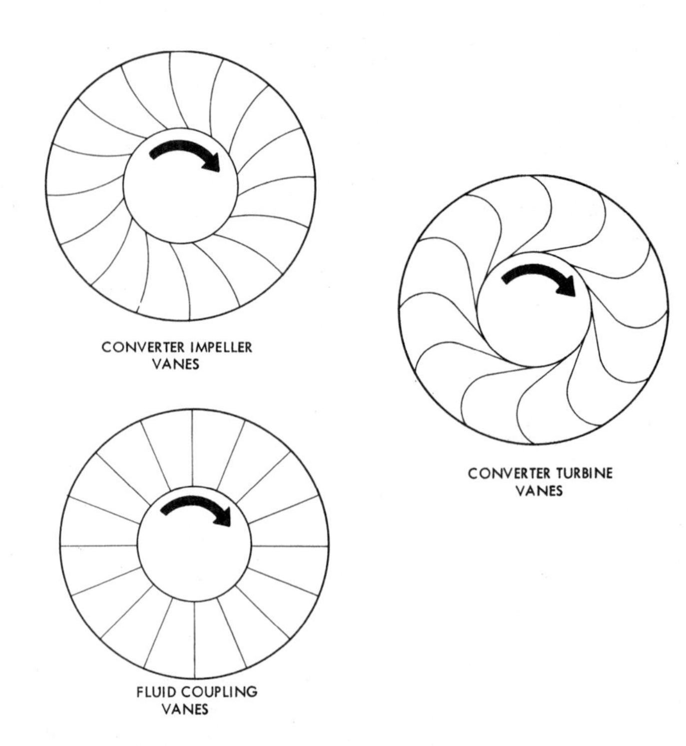

FIGURE 2–20 Comparison of converter impeller and turbine blades to coupling blades. (Courtesy of General Motors Corporation, Service Technology Group)

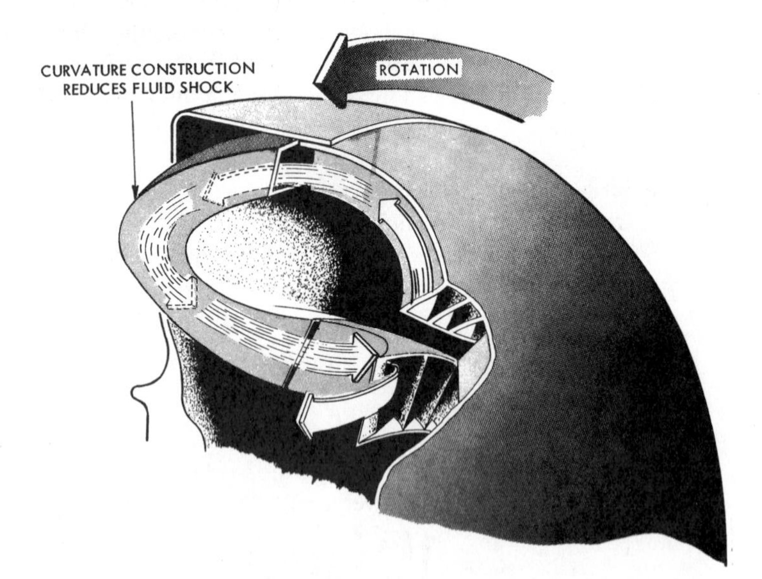

FIGURE 2–21 Oil flow between impeller and turbine modified by vessel curvature. (Courtesy of Chrysler Corp.)

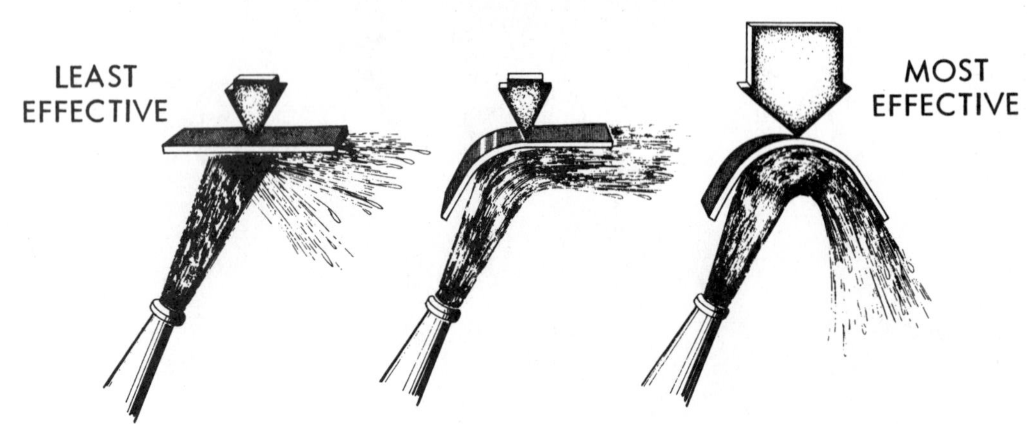

FIGURE 2–22 Effect of various blade configurations on diverting and absorbing the fluid jet thrust.

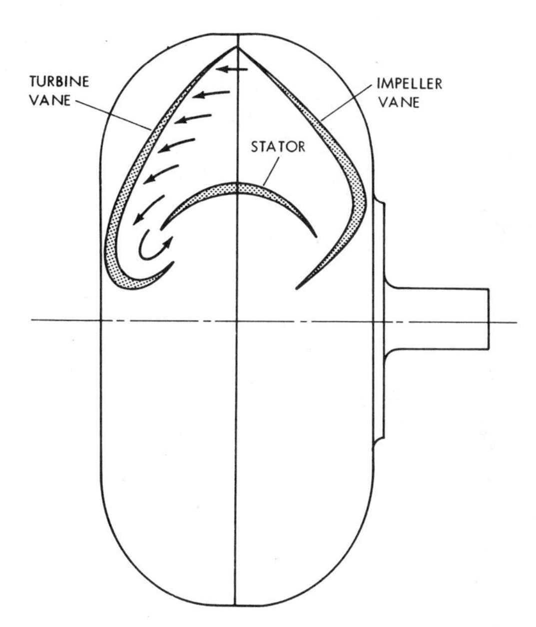

FIGURE 2–23 Curvature of turbine blade designed for maximum thrust absorption; fluid thrust uses the entire blade surface for turbine propulsion—entry to exit.

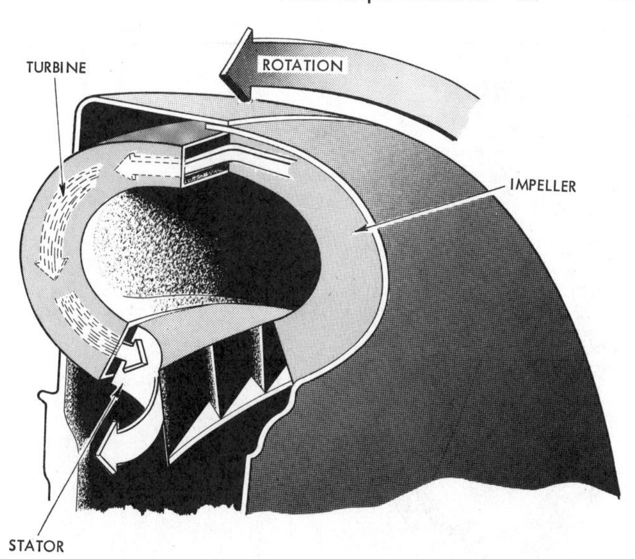

FIGURE 2–24 Stator design redirects fluid flow to assist impeller rotation. (Courtesy of Chrysler Corp.)

Operation

To assist in the discussion, the power flow relationship between the engine crankshaft, fluid torque converter, and transmission input shaft is illustrated in Figure 2–25.

Impeller and Turbine. Converter operation starts with the impeller putting the fluid in motion, with the engine furnishing the energy input. The rotating impeller creates a centrifugal pumping head or vortex flow. At the same time, the fluid must follow the rotational inertia or effort of the impeller. These two fluid forces combine to produce a resultant force in the form of an accelerated jet stream against the turbine vanes (Figure 2–26).

Figure 2–27 show that at this point of operation, the impeller and turbine alone are attempting to act as a more effective fluid coupling by featuring curved impeller and turbine vanes rather than a straight radial design. Note that the turbine vanes have reversed the fluid direction. Although the curved turbine vanes provide for efficient energy transfer, reentry of the remaining fluid thrust back to the impeller works against the impeller and crankshaft rotation and lugs the engine. It is necessary, therefore, to introduce the stator element to make the converter work.

Stator (Reactor). The stator (reactor) is employed between the turbine outflow and impeller inflow to reverse the direction of the fluid and make it flow in the same direction as impeller rotation (Figure 2–28). Instead of the fluid opposing the impeller, the unexpended fluid energy now assists the crankshaft and impeller rotation. In effect, an rpm boost results. This allows the impeller to accelerate more and recycle the fluid with a greater thrust against the turbine vanes. The process of using the remaining fluid energy to drive the impeller is referred to as regeneration gain. The stator mounts on a **one-way roller clutch** (Figures 2–29 and 2–30). During the torque phase, the stator remains locked. At coupling speed, it overruns.

❖ **One-way roller clutch:** A mechanical device that transmits or holds torque in one direction only.

Torque and Coupling Phases

Recycling of the fluid permits more of the impeller input from the engine crankshaft to be used in increasing the jet stream velocity and the turning effort on the turbine. It should be noted that by helping the impeller to accelerate the fluid thrust against the turbine, the stator provides the basis for torque multiplication. The curved turbine blades absorb energy from the impeller discharge until the force of fluid is great enough to overcome the turbine resistance to motion (4000-lb vehicle). The force on the turbine is actually working on the end of a series of rotating lever arms attached to the transmission input shaft. The lever arms represent the mean or working radius of the turbine (Figure 2–31). The converter torque is therefore equal to the effective fluid force times the working radius of the turbine (torque = force × lever arm). The converter torque output is a product of the working turbine lever arms, which is similar to torque multiplication by gear reduction (Figure 2–32).

The maximum torque multiplication occurs with the engine at wide open throttle (WOT) and zero turbine speed. This is commonly referred to as the **torque rating**, or **stall torque**, of the converter. The turbine lever arms are a stationary target at this point and absorb the maximum fluid force or thrust. For best efficiency, engineering design of the three-element converter keeps the maximum torque ratings within the general range of 2:1 to 2.5:1. Although higher ratios can be achieved, it would be difficult to control overheating, especially when long steep grades are encountered. The higher torque ranges also decrease the efficiency of the converter **coupling phase** that is delayed into the higher vehicle speeds, making fuel economy suffer.

TURBINE

STATOR

IMPELLER (pump)

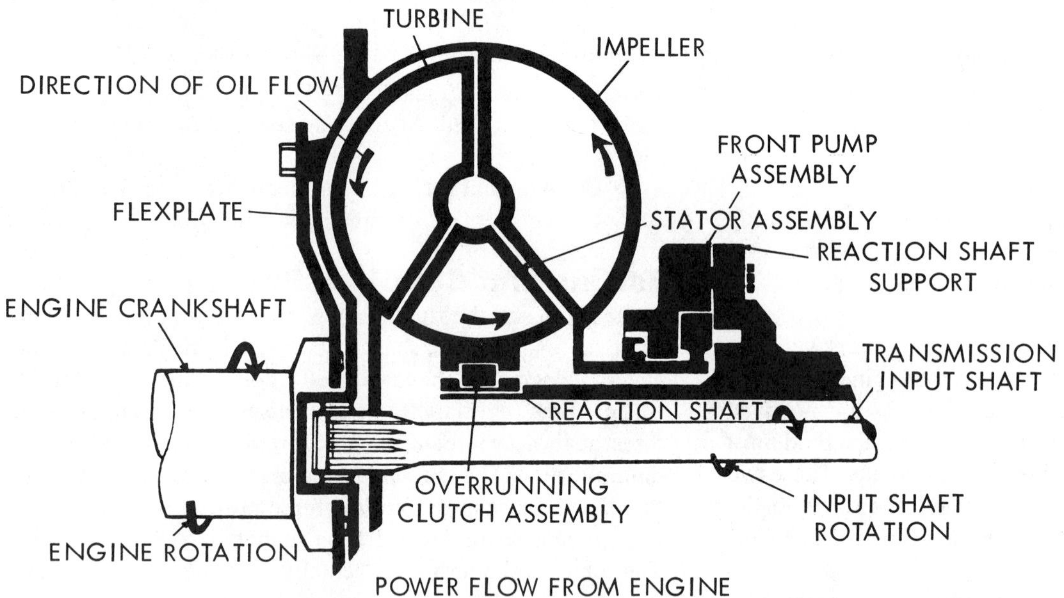

FIGURE 2–25 Power flow from engine to transmission shaft.

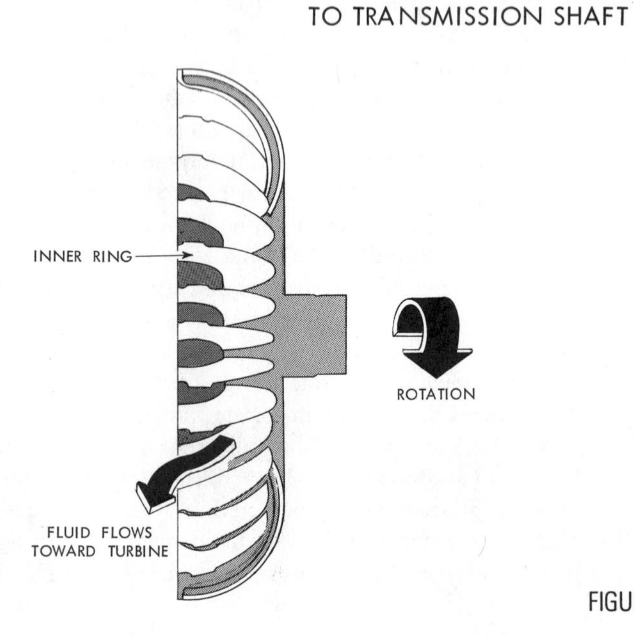

FIGURE 2–26 Impeller operation—fluid flow and thrust.

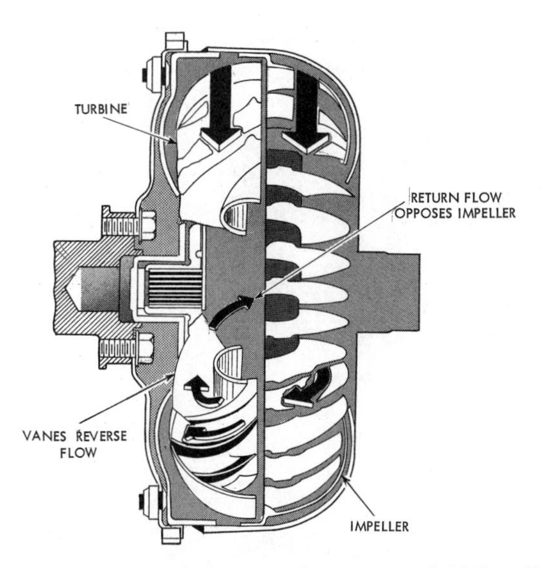

FIGURE 2–27 Turbine vane configuration reverses fluid flow. The fluid thrust opposes impeller rotation.

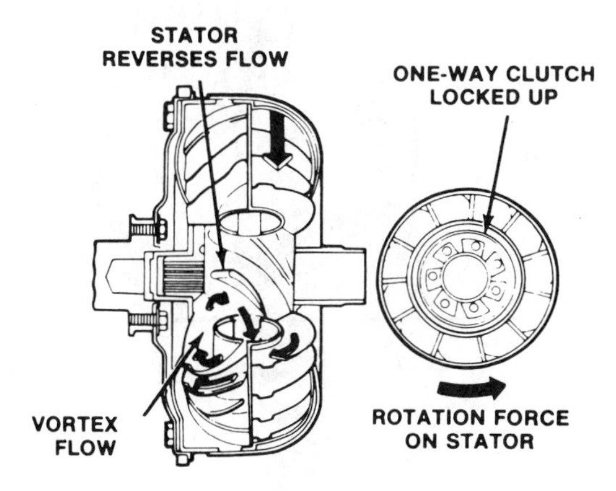

FIGURE 2–28 Stator operation with converter in torque multiplication stage.

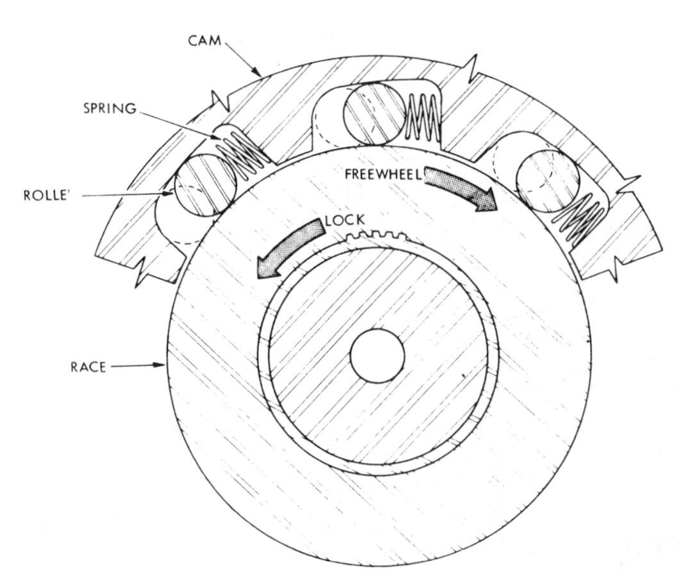

FIGURE 2–29 Viewed from the front, typical cross section of one-way roller clutch. (Courtesy of General Motors Corporation, Service Technology Group)

- **Torque rating/stall torque:** The maximum torque multiplication that occurs during stall conditions, with the engine at wide open throttle (WOT) and zero turbine speed.
- **Coupline phase:** Occurs when the torque converter is operating at its greatest hydraulic efficiency. The speed differential between the impeller and the turbine is at its minimum. At this point, the stator freewheels, and there is no torque multiplication.

The stall torque stabilizes at its design point when the impeller reaches its capacity for pumping—it takes horsepower to pump the fluid against the vehicle load. The engine crankshaft finally reaches a point where it no longer can boost the

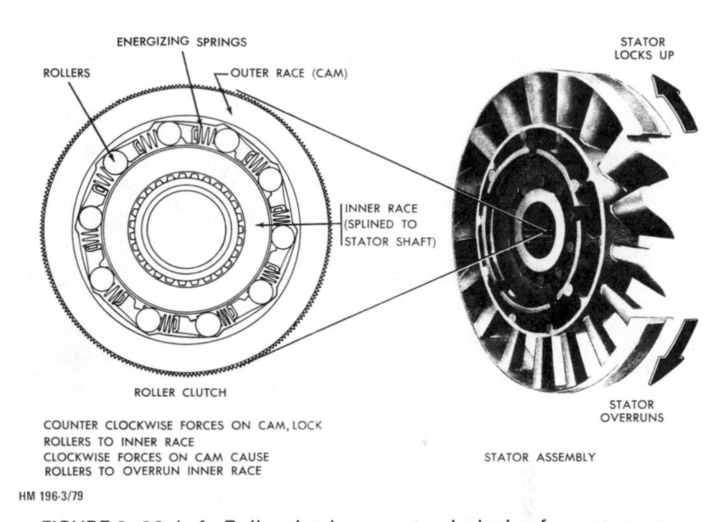

FIGURE 2–30 Left: Roller clutch—counterclockwise forces on cam lock rollers to inner race, clockwise forces on cam cause rollers to overrun inner race. Right: Stator assembly. (Courtesy of General Motors Corporation, Service Technology Group)

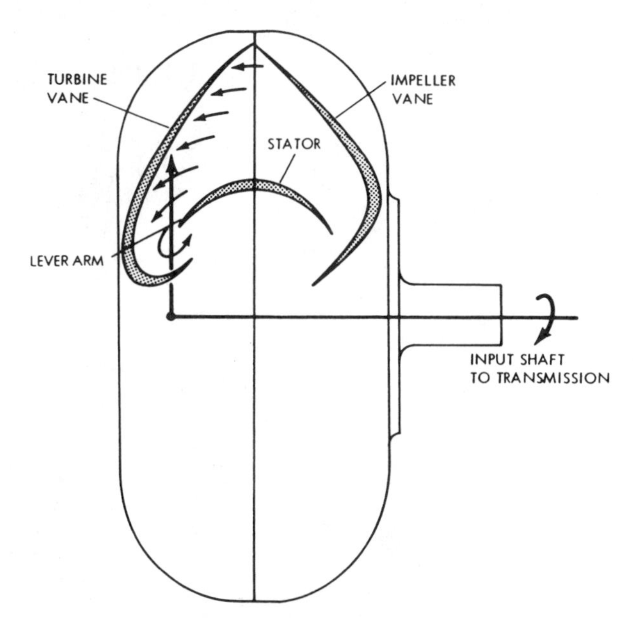

FIGURE 2–31 Turbine lever arm represented by mean radius of turbine.

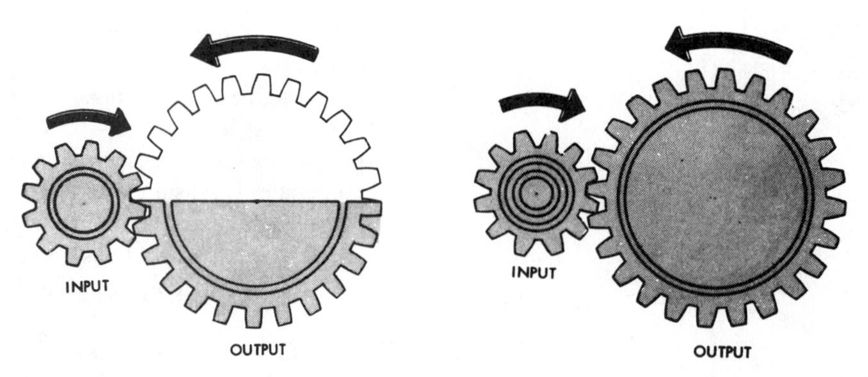

FIGURE 2–32 Fluid torque multiplication gives the same effect as gear reduction.

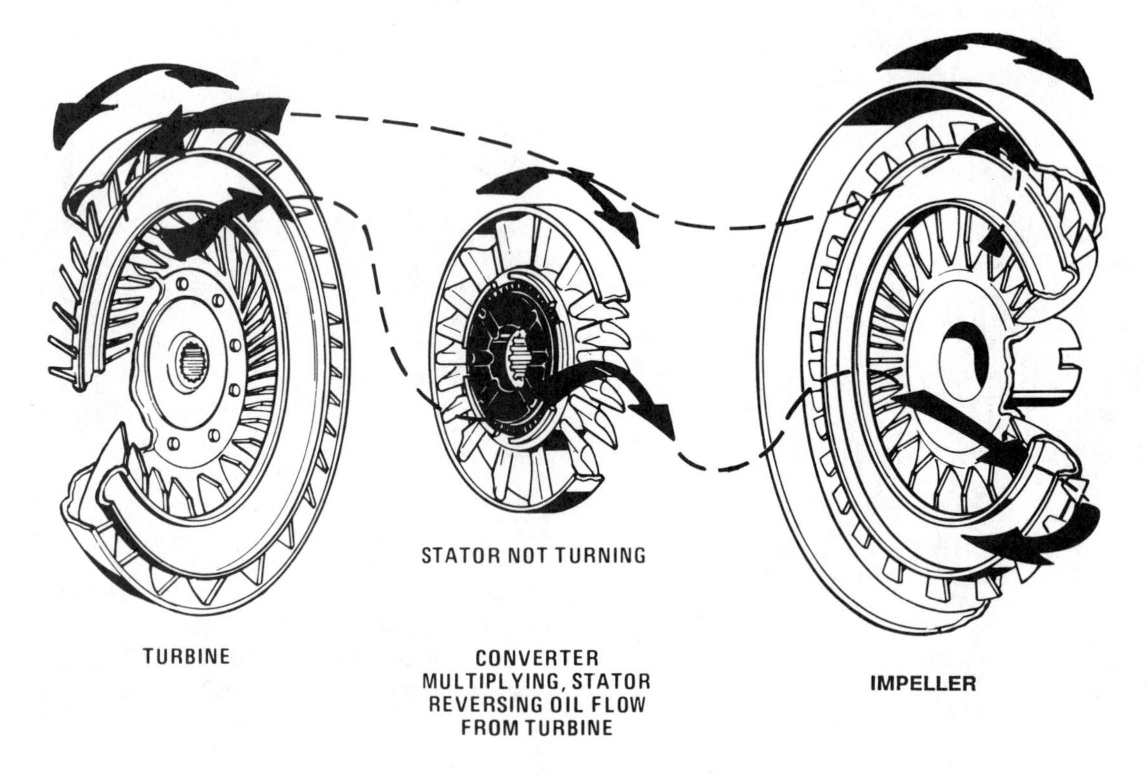

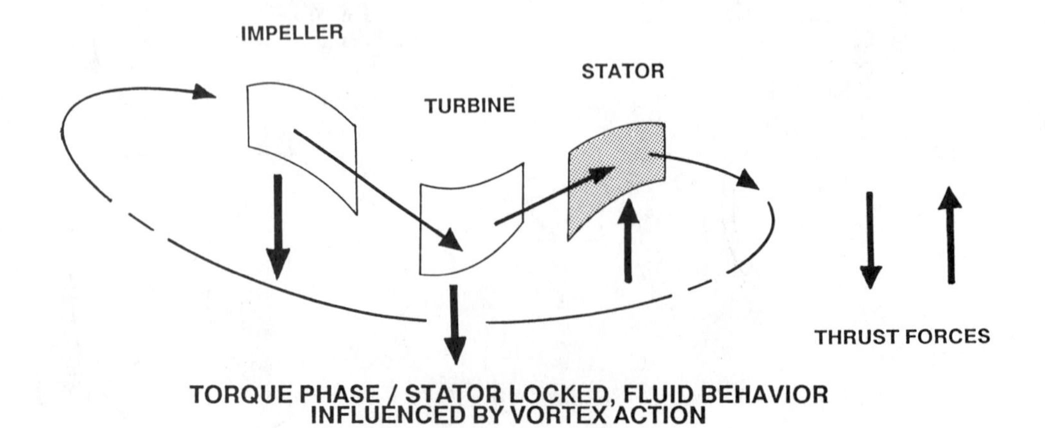

FIGURE 2–33 Top: Converter multiplying, stator reversing oil flow from turbine. Bottom: Torque phase; stator locked, fluid behavior influenced by vortex action. (Top print courtesy of General Motors Corporation, Service Technology Group)

impeller rpm or fluid thrust against a stationary turbine load. Ideally, this occurs with the crankshaft rpm stabilized at the peak of the engine torque curve.

During torque phase, vortex flow is the predominant force in the fluid, and therefore the fluid cycles like a continuous chain from the impeller, to the turbine, through the locked stator, and back to the impeller (Figure 2–33). This action is continuous until the turbine speed is at nine-tenths of the impeller speed, at which point the converter has achieved a 90% plus speed ratio. After a moment at stall, the turbine and vehicle load begin to move. The inertia of the stationary load is overcome, and the turbine becomes a moving target to the fluid thrust. Once the turbine levers and vehicle motion start, it becomes easier and easier for the fluid force to drive the turbine and the vehicle. The turbine rpm actually starts to gain and approach impeller speed. As the turbine gains in speed, the lever arms are absorbing less and less of the fluid force, and converter torque output gradually drops. The fluid thrust under vortex influence is trying to hit a moving target that is moving away from it faster and faster.

Finally, when the turbine speed is at nine-tenths of impeller speed, which means the converter has reached a 90% plus speed ratio, the converter enters its coupling phase of operation. The stator is no longer needed and must freewheel with the rotary flow (Figure 2–34). The vortex effect on the fluid has dropped significantly, and the rotary flow is now the main force. The rotating inertia of the fluid mass, impeller, and turbine form a hydraulic lock or bond (Figure 2–35). The converter is now in coupling phase, and the torque ratio is at 1:1.

Figure 2–36 summarizes the converter torque/coupling operation. The input angles to the stator are shown at various stages as the converter torque progressively drops from maximum to 1:1. Study the torque converter performance curve in Figure 2–4. The converter drops out of coupling phase and reenters torque phase anytime the vehicle load condition changes from cruising to engine torque demand. It is not usually practical to use the torque converter as the sole means of power transmission in automotive vehicles. From an efficiency standpoint, it cannot satisfy the need for high starting torques and all-around wide range of torque multiplication. It lends itself well, however, in combination with a gear transmission.

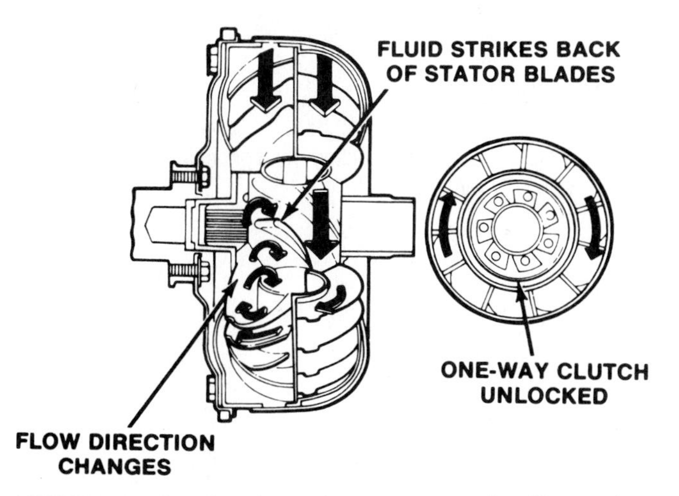

FIGURE 2–34 Coupling phase of converter operation. Three units turn together. (Reprinted with the permission of Ford Motor Company)

Most automatic or manual transmissions are sequence, or fixed-gear-ratio, transmissions. The fixed-ratio torque curve in Figure 2–37 shows the sudden torque changes that occur between shifts. The ideal engine crankshaft torque would produce smoother transitions for optimum performance and efficiency. A combination of the converter and gear transmission is the best way to provide the needed high torque and smooth progressive torque changes (Figure 2–38).

Newton's Law. The ability of the torque converter to multiply torque can be approached by applying Newton's law of physics: "For every action there is an equal and opposite reaction." In the converter, the impeller, the turbine, and the reactor are points of action and reaction with respect to oil flow. During the period of torque multiplication, the reaction of the stationary stator (reactor) blades to the oil is in the same direction as the impeller rotation (Figure 2–39). In accordance with Newton's law, the reaction of the turbine blades on the oil must be equal to the combined reactor and impeller torque ($A + B = C$). Turbine torque is therefore greater than impeller torque by the amount of the reactor (stator) reaction torque. Therefore, $C - B = A$. Using Newton's law again, it is obvious why a coupling cannot increase torque. Without a stator or reactor, the fluid coupling has only two points of action and reaction: the impeller and the turbine. The impeller action on the fluid is opposed only by the reaction of the turbine, so the action-reaction between the two must always be equal.

Dual Stator Converter

Unique in modern times, but used in some early automatic transmissions, is the dual stator converter. General Motors uses the dual stator concept in certain 4L80-E units (Figure 2–40). Depending on engine application, it can permit a stall torque ratio of 3.5:1. The converter is described as a three-member, four-element converter. The stator member is made up of two elements mounted on separate overrunning clutches that can freewheel independently. The primary stator faces the impeller, and the secondary stator faces the turbine. During heavy-load conditions, when high converter torque is demanded, both stators are locked to produce a high-angle fluid thrust as the fluid is redirected back to the pump. As the torque demand from the converter drops, it is important to tighten the slip action with more efficiency. When the resultant converter forces swing toward rotary motion, the secondary stator freewheels, cutting back on converter torque and increasing converter efficiency. The primary stator with a lower-pitched angle stays locked until the converter speed ratio reaches 90% or more. In effect, the stators provide a two-phase converter operation: high torque for extra power takeoff, and low torque for efficiency.

Cooling the Hydraulic Fluid

When the converter is multiplying torque, a shearing action of the fluid occurs between the impeller and turbine because of

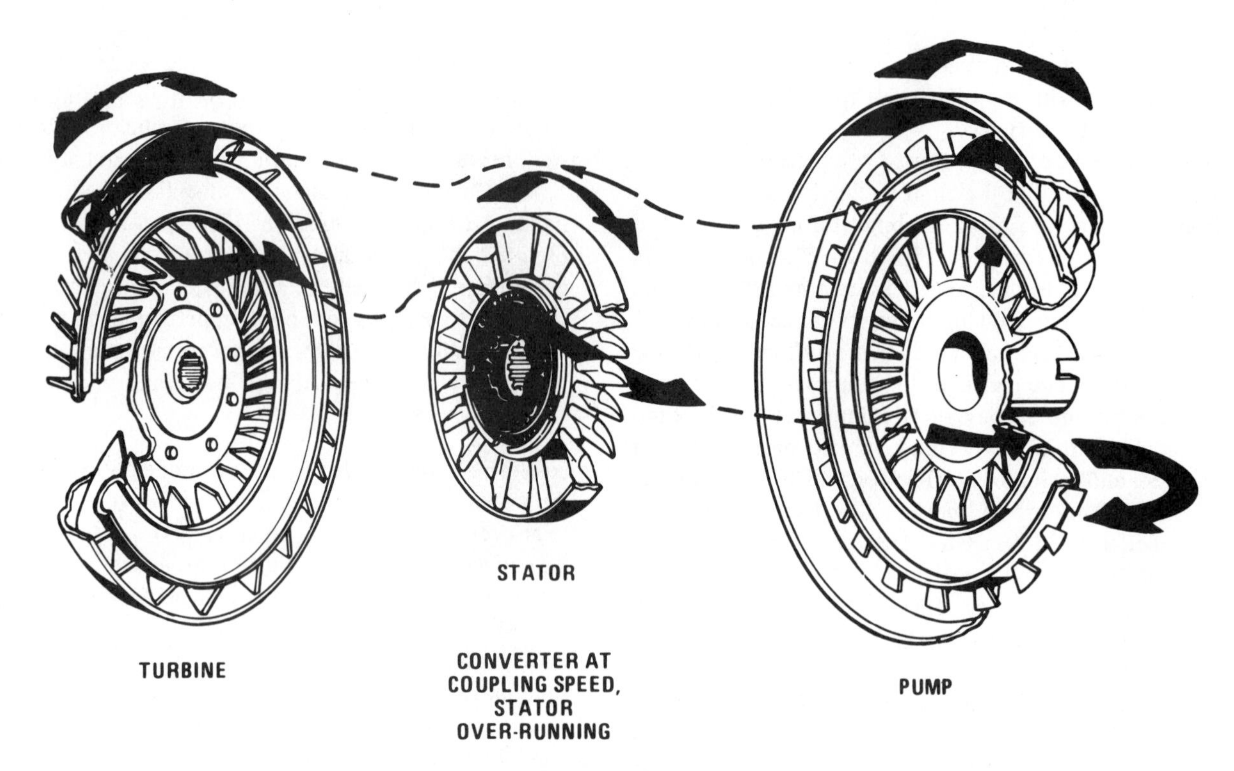

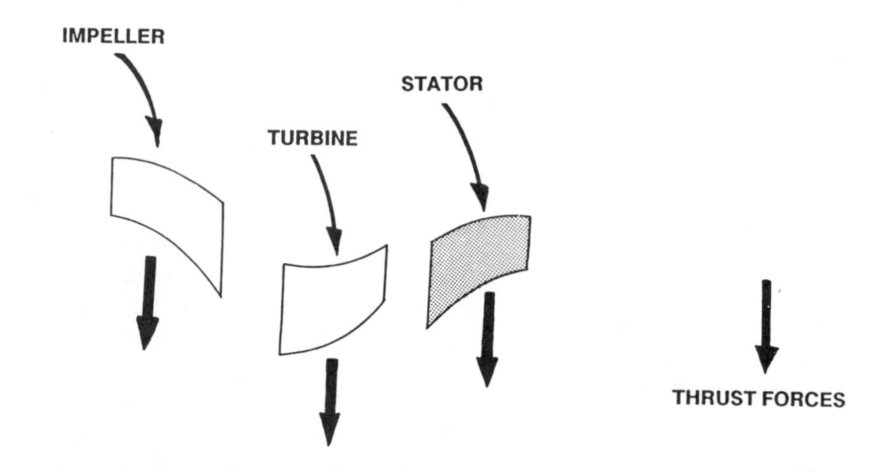

FIGURE 2–35 Top: Converter at coupling speed, stator overrunning. Bottom: Coupling phase; stator freewheeling, fluid behavior influenced by rotary inertia action. (Top print courtesy of General Motors Corporation, Service Technology Group)

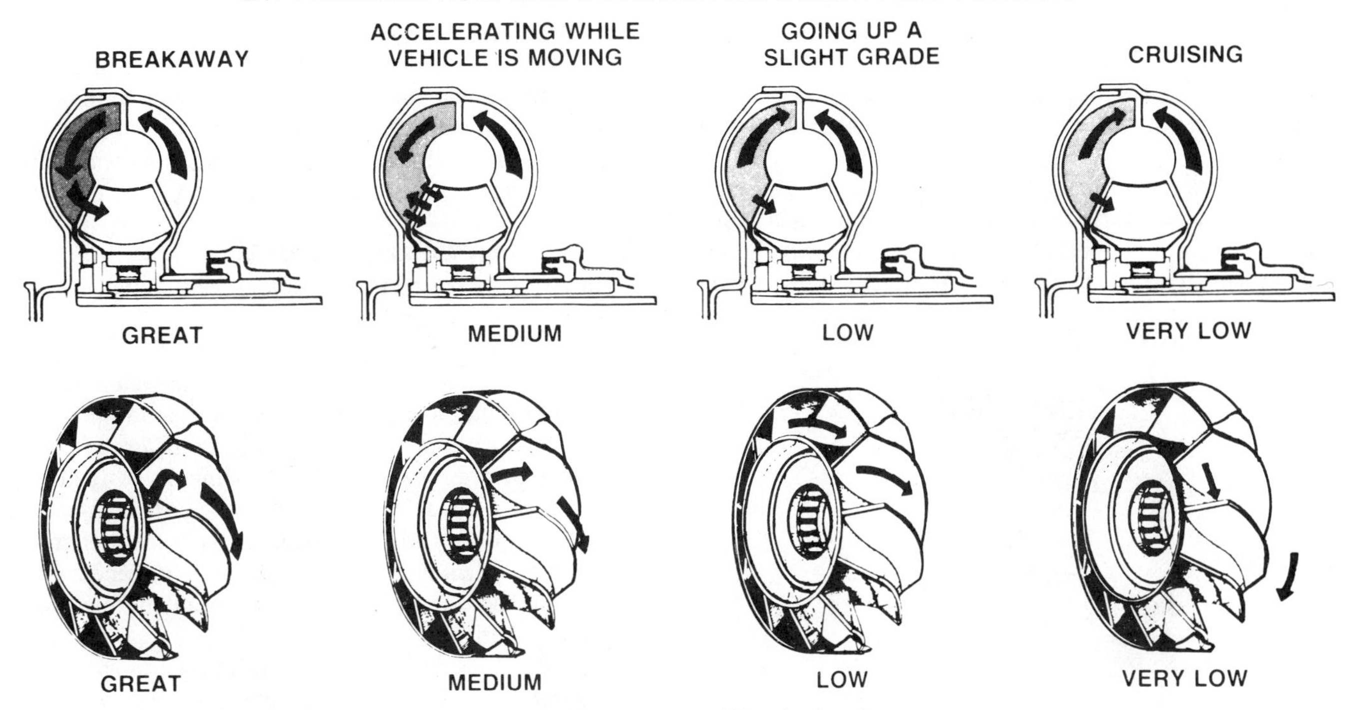

FIGURE 2–36 Difference in speed between impeller and turbine. (Courtesy of Chrysler Corp.)

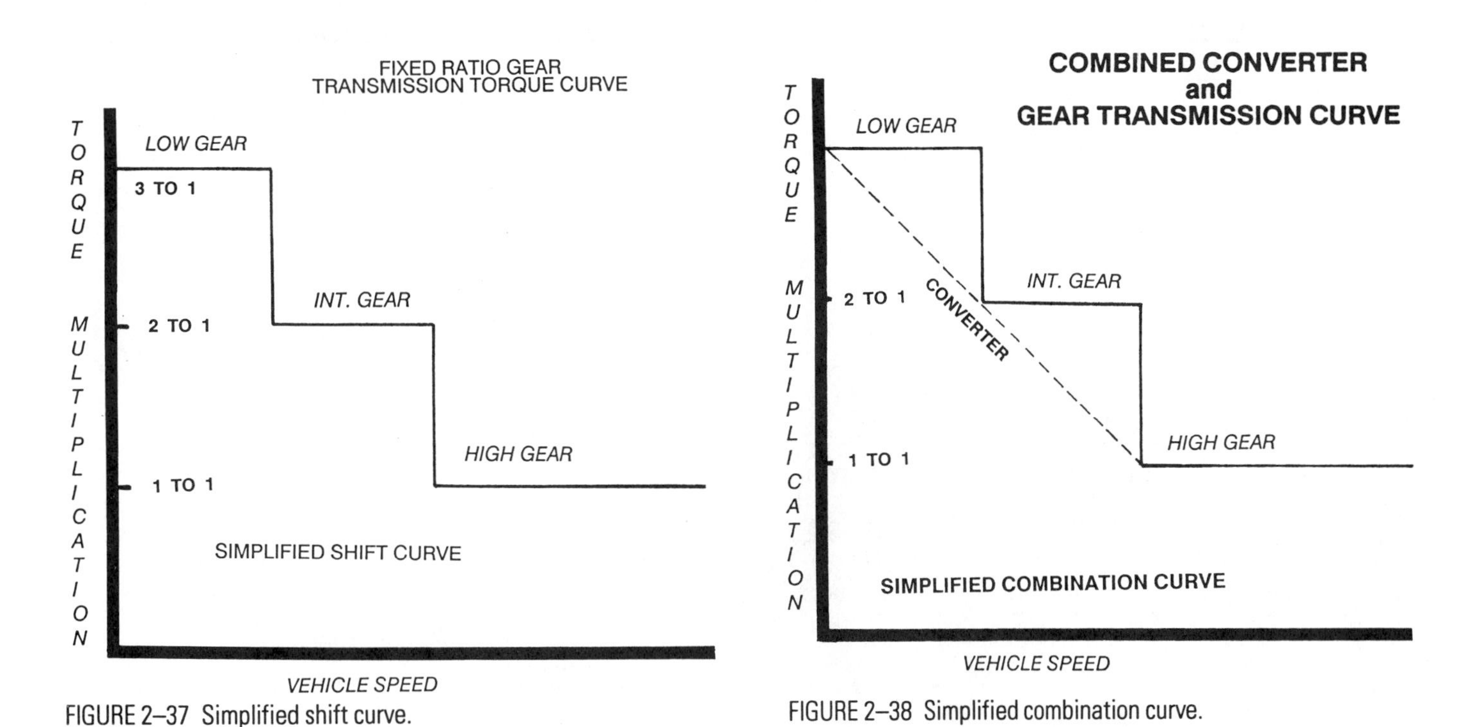

FIGURE 2–37 Simplified shift curve.

FIGURE 2–38 Simplified combination curve.

the strong recirculating vortex flow and converter slip. Considerable heat is generated, and it becomes necessary to provide some type of cooling to keep the fluid from overheating. This differs from the coupling phase, in which heat is not a major factor. Two types of cooling systems are used: oil-to-water and oil-to-air.

In both types of cooling systems, feed oil from the transmission pump regulator valve constantly cycles oil into and out of the converter, with the outflow returning for transmission lubrication and sump return. In the oil-to-water system, which has always been the prevailing system, the oil is simply routed through the water cooler lower tank or side tank (Figures 2–41 and 2–42). The oil-to-air system requires an auxiliary cooler unit to supplement the heat exchanger located in the radiator. This miniature transmission fluid radiator is installed in series with the existing oil-to-water system. Add-on coolers are described in greater detail in Chapter 8.

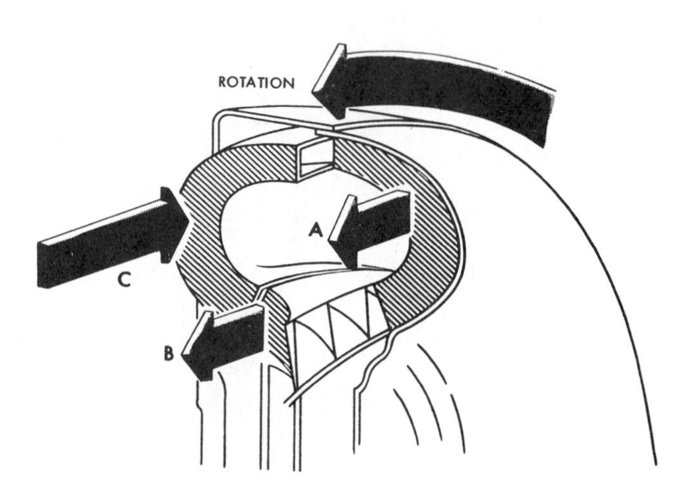

FIGURE 2–39 Newton's law as applied to converter torque multiplication. (Courtesy of Chrysler Corp.)

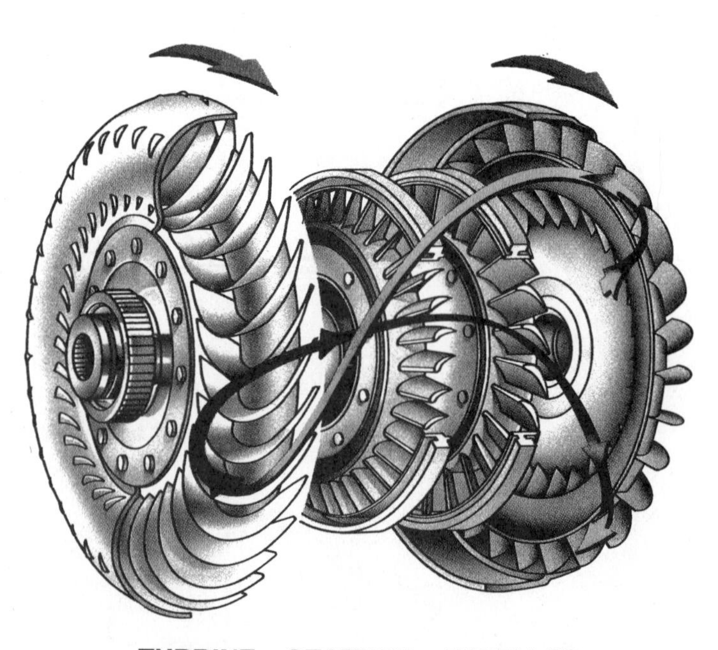

FIGURE 2–40 Dual element stator. (Courtesy of General Motors Corporation, Service Technology Group)

Operating Characteristics

In operation on the road, the converter provides effortless driving characteristics.

1. At engine idle, the converter acts as an automatic clutch and permits the engine to run and the car to stand still.
2. The converter automatically adjusts its torque output to drive shaft torque requirements within its design limits. It acts as a fluid coupling for level-road, constant-speed conditions, but when performance is needed for acceleration or hill climbing, it responds with the necessary extra torque dictated by the slowdown of the turbine from increased drive shaft torque.
3. As the converter is a fluid unit, it acts as a natural shock absorber during gear ratio changes and adds to shift smoothness. The fluid also absorbs the torsional vibrations from the engine crankshaft.
4. The converter permits continuous shock-free acceleration, extending the life of drivetrain components—which is especially important for stop-and-go driving.

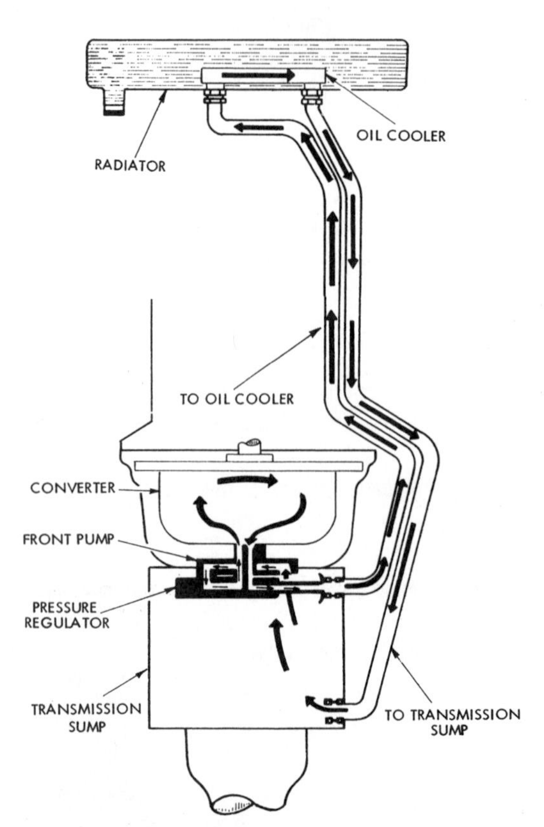

FIGURE 2–41 Oil cooling circuit for the converter, bottom tank of radiator. (Reprinted with the permission of Ford Motor Company)

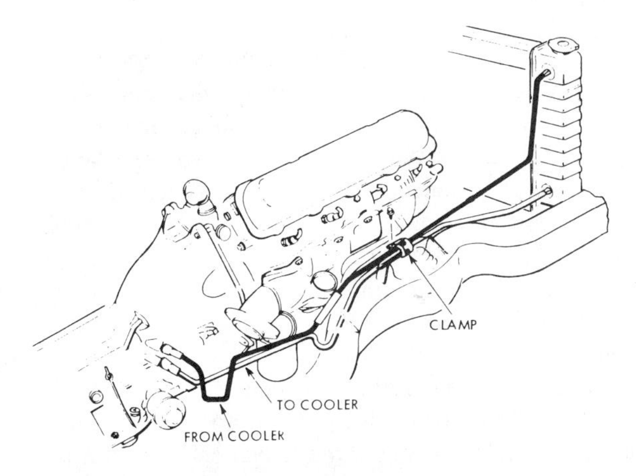

FIGURE 2–42 Converter oil cooler circuit, side tank design. (Courtesy of General Motors Corporation, Service Technology Group)

Converter/Engine Match

The torque converter must be engineered to balance its performance with the engine rpm and vehicle load. For a particular converter torque output and vehicle load, the engine throttle opening and rpm should be producing an optimum crankshaft torque for the drive load. The converter acts like a variable dynamometer load on the end of the engine. The converter and engine are matched when the WOT stall rpm meets engineered specifications (Figure 2–43), which is achieved basically by choosing converter elements with the correct diameter sizing and internal blade angularity.

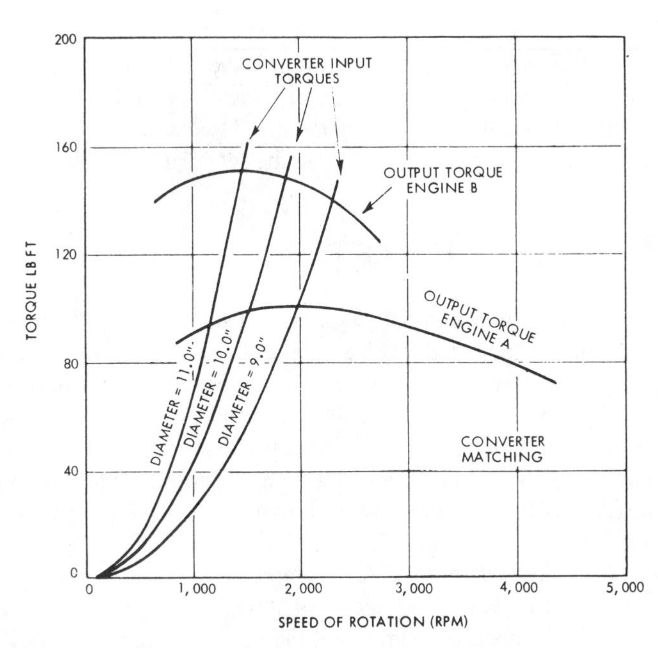

FIGURE 2–43 Matching of converter design to the engine.

If converter size is too small or the converter has too high a stall speed for a particular application, the engine operates at higher rpm's and lower torque efficiency for the drive load. Where a converter is too large or has a low stall speed, the engine lugs and starves the torque output to meet the vehicle loads. Both converter extremes result in undesirable overspeed or underspeed conditions that downgrade engine performance

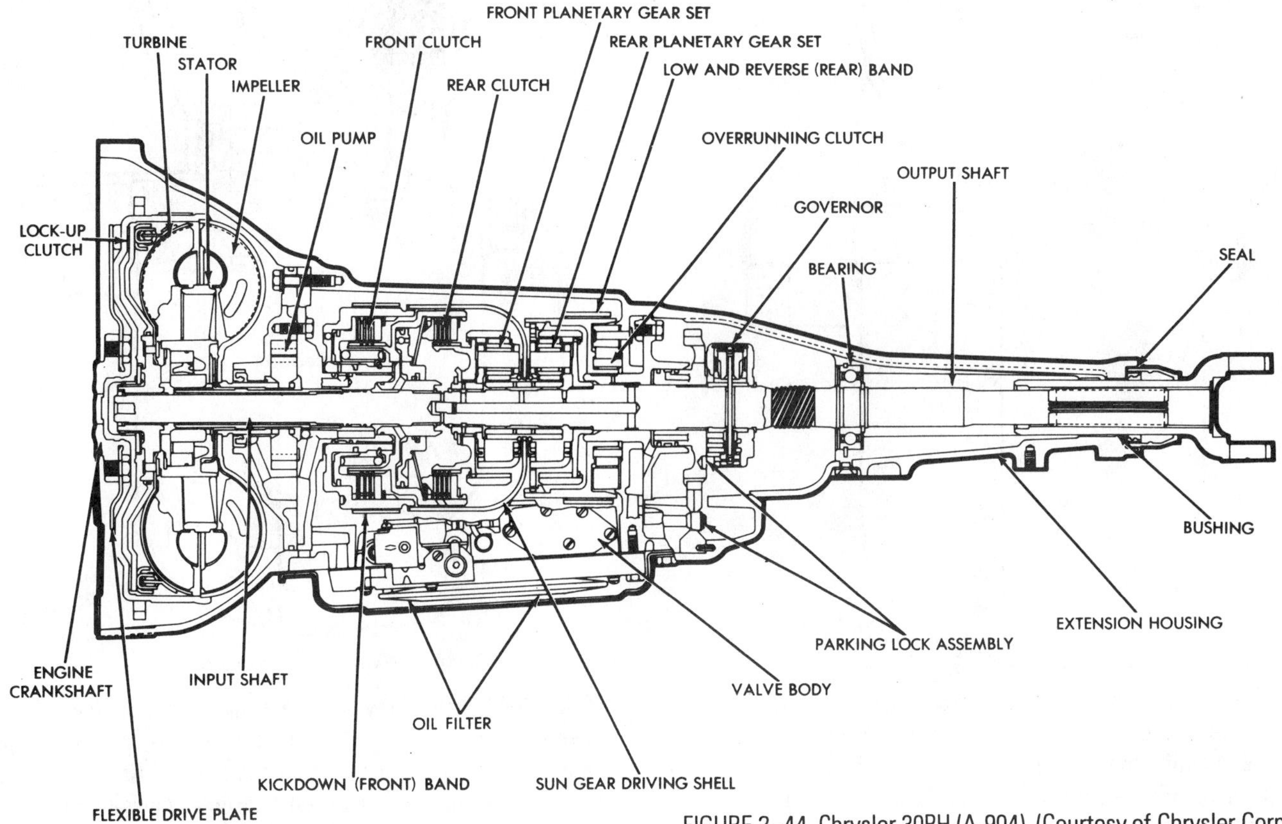

FIGURE 2–44 Chrysler 30RH (A-904). (Courtesy of Chrysler Corp.)

and efficiency. It should be apparent to the field technician that the engineered converter-to-engine match should not be altered. A mismatch is almost guaranteed to cause a driveability complaint and, worse yet, turn on the "check engine" light.

LOCKUP CONVERTER

With the current cost of gasoline and the scramble to meet government fuel economy standards, the **lockup converter** has become a standard feature built into most converter applications for passenger cars and light trucks. Although not a new concept, Chrysler made the first comeback move in 1978 and introduced the lockup converter with an internal clutch for the A-727 (36RH) and A-904 (30RH) transmissions (Figure 2–44).

❖ **Lockup converter:** A torque converter that operates hydraulically and mechanically. When an internal apply plate (lockup plate) clamps to the torque converter cover, hydraulic slippage is eliminated.

The lockup converter enables the converter to transmit torque in two ways: (1) hydraulically when normal converter action is demanded (Figure 2–45), or (2) mechanically for cruising speeds. The mechanical lock bypasses the converter fluid drive and connects the engine crankshaft directly to the transmission input shaft and planetary gearset (Figure 2–46). Normal slippage that occurs between the converter impeller and turbine during coupling phase is eliminated. The improved efficiency in crankshaft torque transmission boosts fuel economy.

There are a number of lockup converter designs using their own unique control circuitry that automatically dictates the unlock and lockup modes of the converter. The control circuitry may be purely hydraulic or mechanical, or a combination electro/hydraulic using either a computer- or noncomputer-controlled electrical circuit. The objective at this time is to limit the lockup theory to an overview of the various converter internal design configurations. Lockup (converter clutch) control circuitry is featured in Chapter 10.

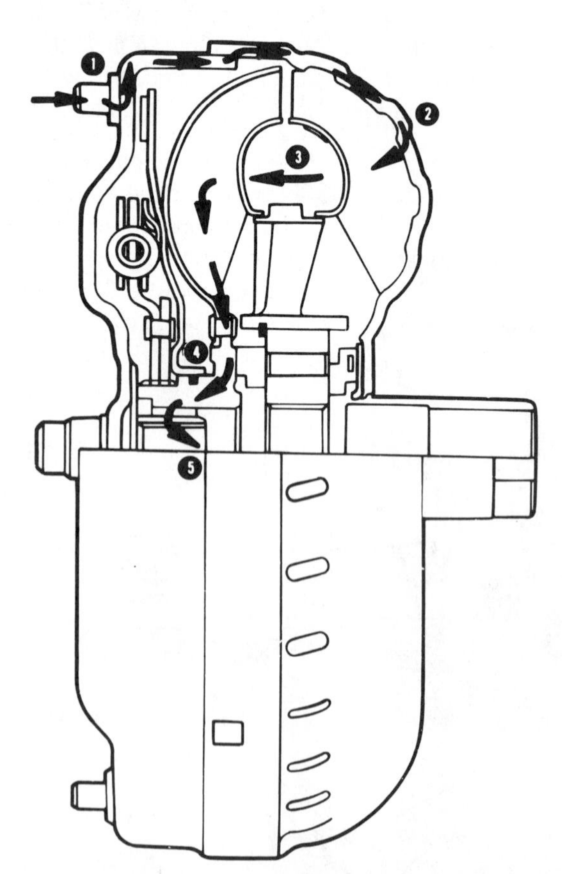

FIGURE 2–45 Power flow with hydraulic input. (1) The engine drives the converter cover. (2) The converter cover drives the impeller. (3) The impeller drives the turbine hydraulically. (4) The turbine drives the input shaft. (5) The input shaft transmits power into the geartrain. (Reprinted with the permission of Ford Motor Company)

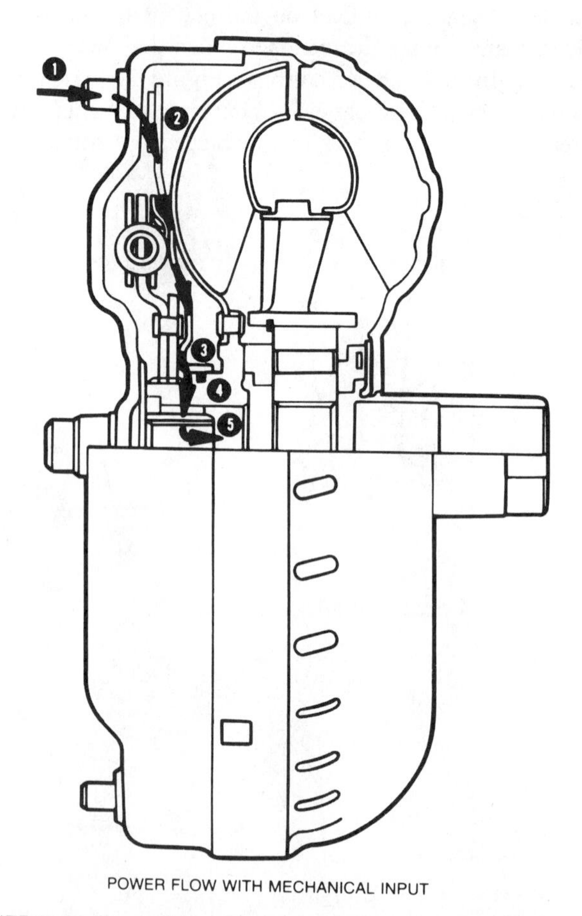

FIGURE 2–46 Power flow with mechanical input. (1) The engine drives the converter cover. (2) The converter cover drives the piston plate clutch. (3) The clutch drives the turbine. (4) The turbine drives the input shaft. (5) The input shaft transmits power into the geartrain. (Reprinted with the permission of Ford Motor Company)

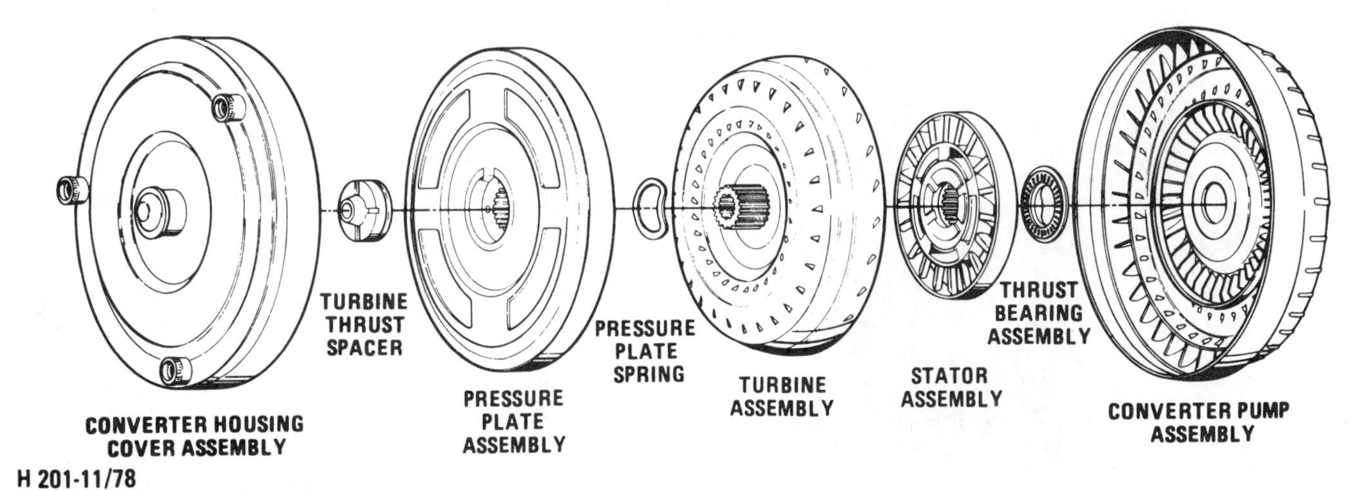

FIGURE 2–47 Torque converter clutch assembly.

FIGURE 2–48 TCC pressure plate damper assembly. Inner splined hub fits on front side of turbine. Damper assembly attached to back side of pressure plate.

Hydraulically Applied Clutch Piston

A hydraulically applied clutch piston is a torque converter design used in both domestic and import transmissions. The converter clutch assembly is referred to as either the lockup clutch or **torque converter clutch**. A typical converter clutch assembly is shown in Figure 2–47. The clutch (piston) or pressure plate assembly is splined to the turbine with a slip fit. On the back side (turbine side), it has a torsional spring assembly or damper unit attached at the hub (Figure 2–48). During lockup this allows the clutch plate to pivot independently of the hub section and work against the springs. Compression of the springs cushions the engagement of the clutch and absorbs the engine torsional vibrations from the crankshaft. This is the same action that takes place in the clutch plate torsional spring damper during engagement of the clutch in standard transmission applications. The mechanical power flow must pass through the damper springs before entering the input shaft to the transmission. On the front side (Figure 2–49) of the clutch

FIGURE 2–49 Left: Converter cover contacting face. Right: Pressure plate friction facing.

FIGURE 2–50 Clutch disc lining bonded to the converter cover.

plate is a band of friction facing that engages the turbine to the metal contacting surface of the converter housing cover.

- **Torque converter clutch:** The apply plate (lockup plate) assembly used for mechanical power flow through the converter.

The converter clutch assembly shown in Figure 2–49 is typical of the units used in all the General Motors family of automatic transmissions and some of the Ford family. Chrysler lockup clutch converters have used a slightly different physical arrangement but have achieved the same results. Note that in Figure 2–50, the clutch friction lining is bonded to the converter cover, and in Figure 2–51, the lockup piston provides the metal contacting surface. Although it was the early practice to bond

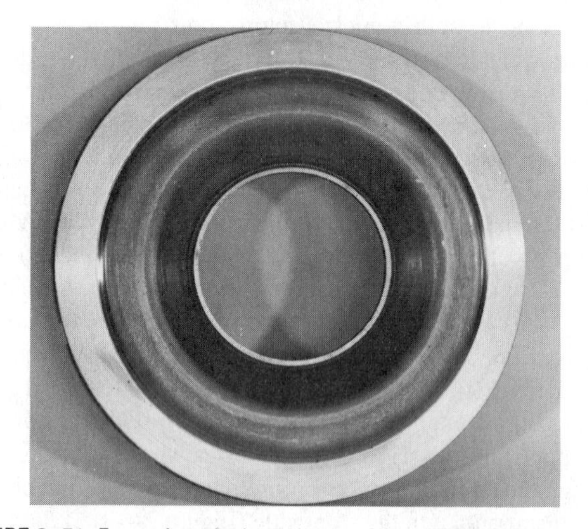

FIGURE 2–51 Front view; lockup piston with machined clutch contacting surface.

FIGURE 2–53 Front side of turbine with torsional damper spring inserts. The input shaft lip seal in the turbine hub prevents internal leakage past the input shaft.

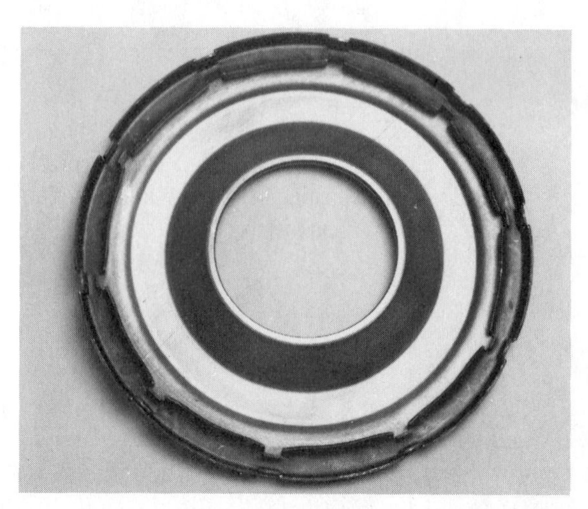

FIGURE 2–52 Rear view; lockup piston with torsional spring cage.

FIGURE 2–54 Rear view; lockup piston fit to turbine damper springs.

FIGURE 2–55 Close-up of turbine damper spring fit to lockup piston damper cage. The linkup is like a floating brake caliper.

the friction lining to the converter cover, Chrysler changed from the bonded lining to a floating friction disc between the cover and piston contacting surfaces. The free-floating ring improves the torque carrying capacity of the clutch due to better conformability to the mating parts. The free-float arrangement, however, does not increase the number of contacting surfaces, which relates to torque holding power.

The torsional dampening is located at the rim area between the piston and the turbine. The piston side facing the turbine (Figure 2–52) has a series of fixed spring cages that fit over the torsional damper spring inserts on the front side of the turbine (Figures 2–53 and 2–54). Mechanical linkup of the piston to the turbine through the damper springs is like a floating brake caliper (Figure 2–55). The lockup piston mounts on the smooth turbine hub (Figure 2–56) with an O-ring inserted in the inset diameter to provide a pressure seal. The hub mount and seal allow for both lateral and pivotal piston movement.

All piston-style lockup clutch converters typically use the same hydraulic circuitry into and out of the converter for clutch release and apply. The lockup and unlock modes are hydraulically finalized by a converter clutch or switch valve. Simplified early-style Chrysler illustrations show the converter circuit and clutch action (Figures 2–57 and 2–58).

In the unlock mode, normal converter feed oil flows through the input or turbine shaft and enters the converter between the cover and piston. This keeps the clutch piston separated from the cover and permits the converter feed oil to flow around the piston and charge the converter. The converter outflow then continues to the cooler and lubrication circuits.

In the lockup mode, the converter clutch valve switches the circuit and relays mainline or modified mainline pressure through what was the converter out-circuit to the apply side of the piston. Venting of the fluid between the clutch piston and converter cover is provided back through the turbine shaft to allow clutch engagement. The turbine is now locked to the engine crankshaft, and the converter fluid drive is bypassed. To ensure firm holding power, the clutch contacting surfaces plus the piston center seal and turbine shaft seal must keep the apply oil from bleeding into the exhaust side of the circuit.

FIGURE 2–56 Front view; lockup piston mounts on turbine hub with O-ring at the inside diameter, providing for a pressure seal and piston movement.

In electro/hydraulic lockup control circuits, the converter clutch application is based on a combination of transmission hydraulics and either an on-board computerized or noncomputerized electrical control circuit. Both electrical circuit types use a network of sensors and switches that control an internal transmission solenoid for converter apply and release. (See Chapter 10 for typical circuit references.) The sensors and switches work together and determine the best time to lock up the clutch. When the engine rpm, vehicle speed, throttle position, manifold vacuum, and engine temperature data are right, the electrical circuit is completed and the transmission hydraulics takes over.

The converter clutch typically applies when:

- The engine temperature is at least 130°F (54°C) to achieve an operating level to handle the extra load imposed by the direct connection to the transmission.
- The vehicle speed and load will not lug the engine or cause it to pulsate.
- The transmission hydraulics is in the proper gear.

The converter clutch typically releases when:

- Converter torque multiplication is needed for part throttle or forced downshifts.
- During closed throttle or coast conditions when emissions are affected.
- During downshifting at closed throttle to avoid unwanted engine torsional bumping.
- When brakes are applied to reduce the feel of engine braking.

Where the lockup circuitry is controlled completely by the transmission hydraulics, the system is less complicated, but it must live with the side effects of cold engine performance. It also lacks other necessary input data for complete precision sensitivity to what the engine is doing. Deceleration and application of the brakes do not disengage the clutch for the avoidance of engine braking and torsional shock feel.

Minimum clutch engagement speed depends upon the engine size, final drive ratio, and vehicle application. Typically it occurs in the range of 30 to 45 mph (48 to 72 km/h). In three-speed applications, clutch engagement is usually limited to third gear. Most four-speed overdrive automatics feature clutch lockup engagement in third and fourth gears, with some transmissions having a second-gear apply capability.

Viscous Converter Clutch. Another innovative design of the converter piston clutch incorporates a **viscous clutch** (Figure 2–59). The viscous clutch, which operates in a stiff silicone fluid sealed between the cover and the body of the clutch assembly, takes the place of torsional damper assembly and provides an extra-smooth apply feel during clutch assembly lockup to the converter cover (Figure 2–60). Some minimal slippage occurs within the viscous unit, but good fuel economy is maintained under cruising conditions with the clutch assembly in the apply mode.

❖ **Viscous clutch:** A specially designed torque converter clutch apply plate that, through the use of a silicone fluid, clamps smoothly and absorbs torsional vibrations.

Like the conventional converter clutch plate system, the viscous converter clutch system activates the clutch apply with a solenoid controlled by a powertrain control module (**PCM**) or an engine control module (**ECM**). The PCM/ECM evaluates one additional piece of data: transmission fluid temperature through an internal thermistor threaded into the pump body. Fluid temperature influences the PCM/ECM decision when to engage the viscous converter clutch starting with second gear. Should fluid temperature exceed 315°F (157°C), an external temperature switch threaded into the channel plate opens a

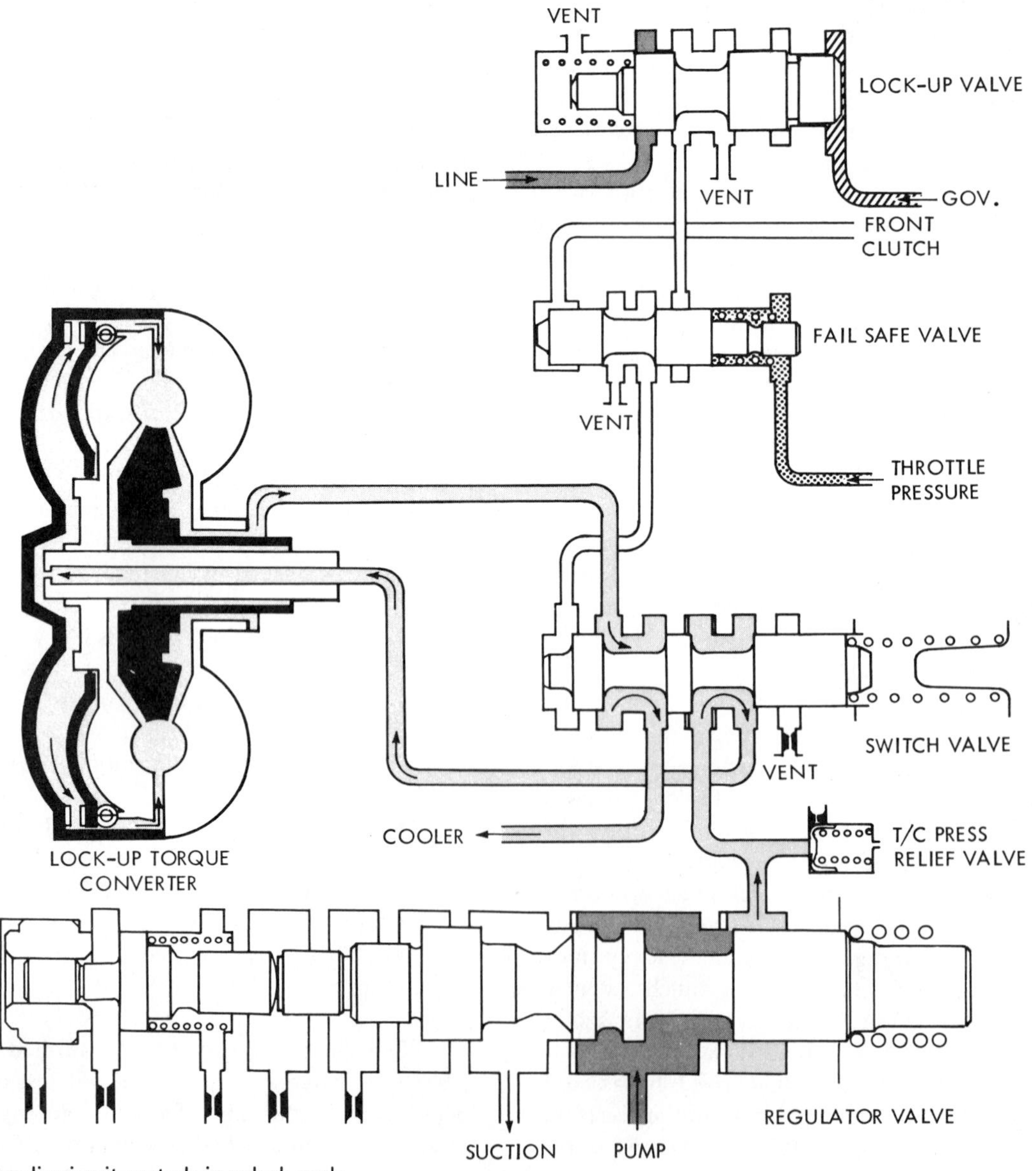

FIGURE 2–57 Hydraulic circuit controls in unlock mode.

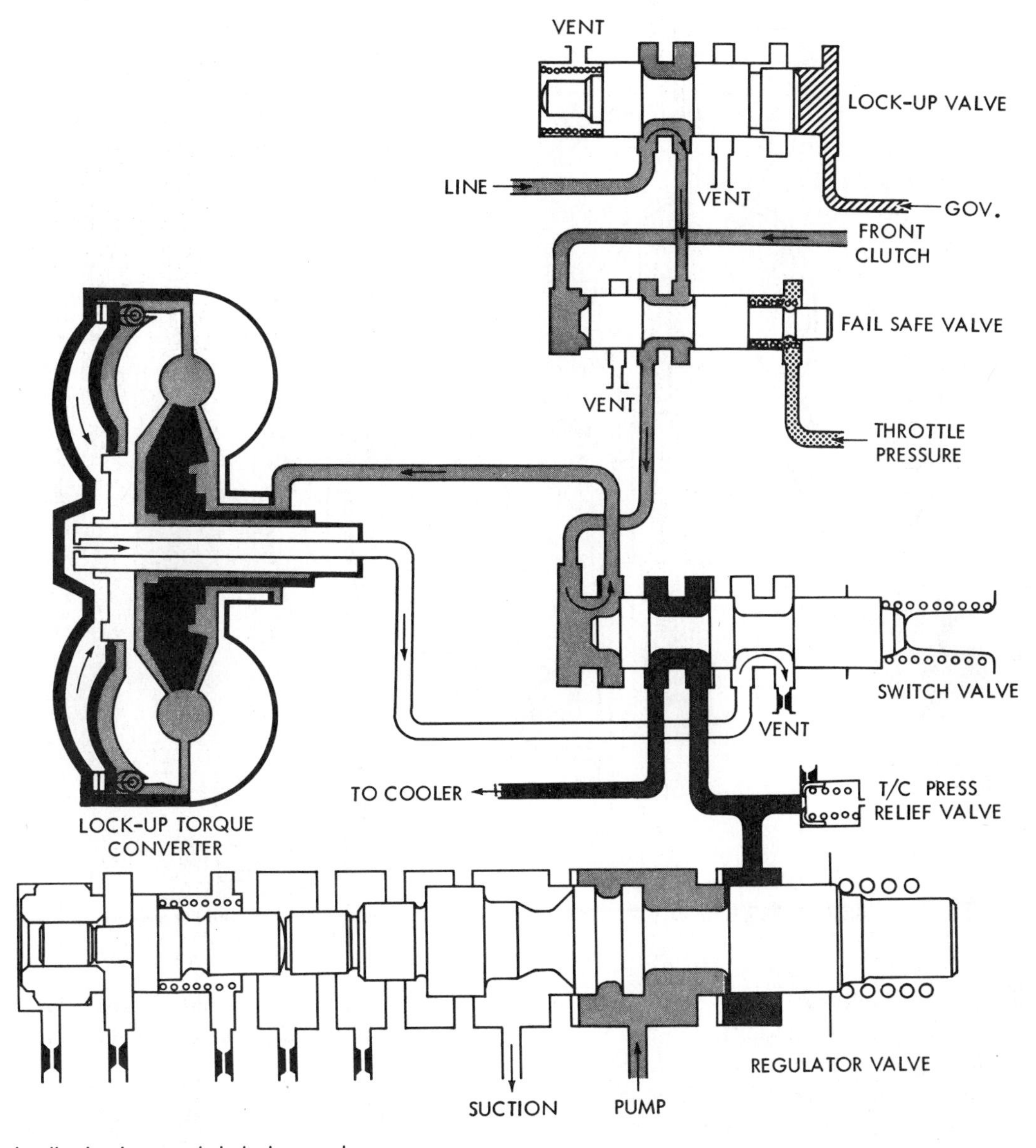

FIGURE 2–58 Hydraulic circuit controls in lockup mode.

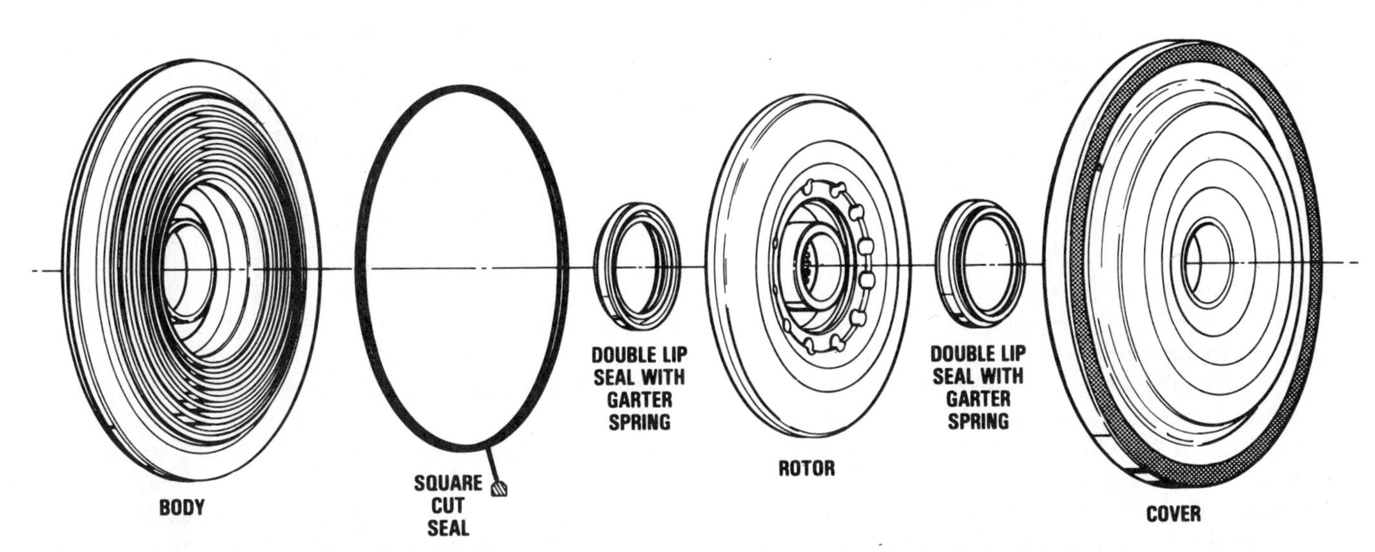

FIGURE 2–59 Viscous converter clutch assembly. (Courtesy of General Motors Corporation, Service Technology Group)

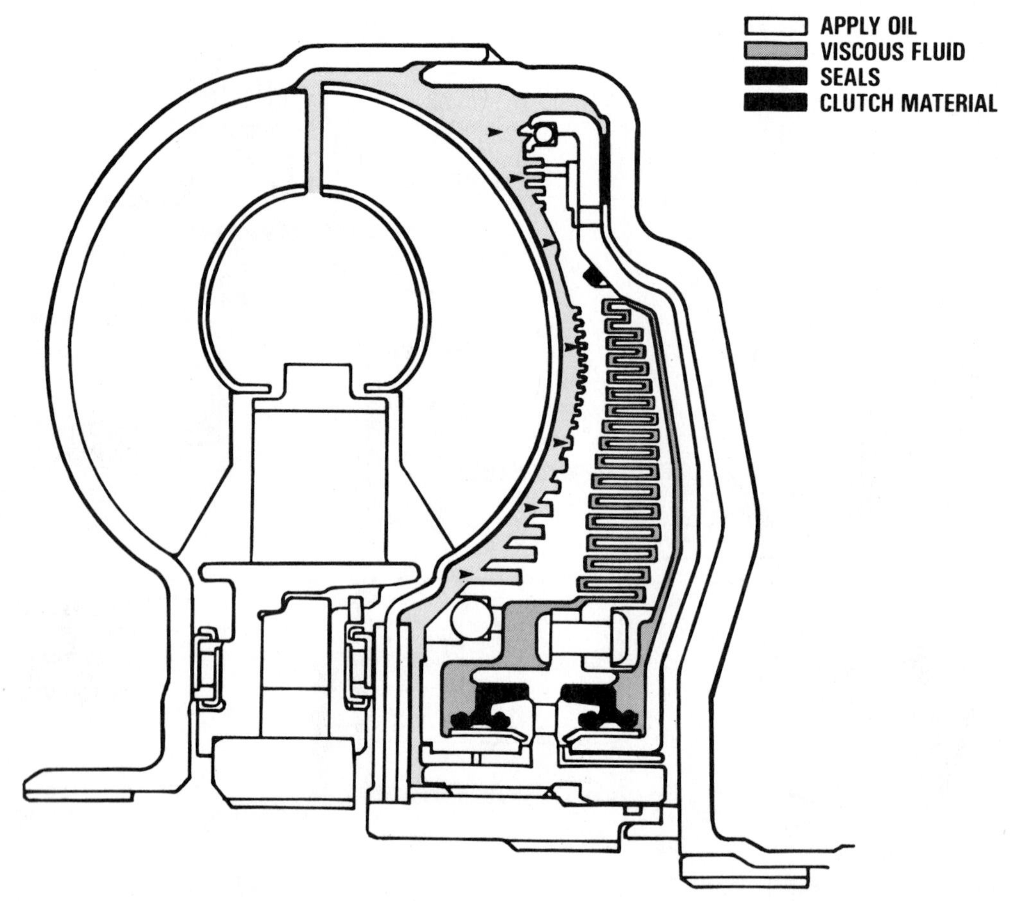

FIGURE 2–60 Viscous converter clutch applied. (Courtesy of General Motors Corporation, Service Technology Group)

hydraulic circuit exhaust and releases the viscous clutch. In the unlock mode, converter oil is switched back to cooling, and the transmission is protected from overheating. Model versions of the General Motors Hydra-Matic 4T60 (THM 440-T4), 4T60-E, and 4T80-E transaxles use either a spring- or viscous-dampened converter clutch.

- **PCM (powertrain control module):** Manages the engine and transmission, including the output control over the torque converter clutch solenoid.
- **ECM (engine control module):** Manages the engine and incorporates output control over the torque converter clutch solenoid. (Note: Current designation for the ECM in late model vehicles is PCM).

Direct Mechanical Converter

This technique provides a simple method for bypassing the converter fluid drive and mechanically driving the transmission. Unique to the Ford AOD four-speed automatic transmission, a solid second input shaft, referred to as the direct-drive shaft, splines into the converter cover and spring-style damper assembly (Figures 2–61 and 2–62). The mechanical input is keyed to the transmission direct clutch application and planetary gear powerflow in third and fourth gears. First, second, and reverse gears are all driven only by the converter fluid drive. In third gear, the torque split through the converter is 60% mechanical and 40% hydraulic. Fourth-gear drive through the converter is 100% mechanical via the direct-drive shaft.

SUMMARY

Torque converters provide the means of transmitting the power of the engine into the transmission. In this hydraulic coupling, the impeller blades thrust fluid against the turbine blades to drive the transmission input shaft. The stator is held stationary and directs the fluid flow in the proper direction as it leaves the turbine and reenters the impeller chambers.

The simple three-element torque converter can be in two different phases. During stall phase, often referred to as slip phase, torque multiplication takes place. This condition benefits overall powertrain torque output. When coupling phase occurs, the speed of the turbine is nearly that of the impeller—approximately 90%. To be able to accomplish this, the stator turns freely with the turbine and impeller.

A clutch plate is used in most currently produced torque converters. When applied, it provides a mechanical connection between the engine and the transmission input shaft. With a reduction in torque converter slippage, fuel economy is improved. Though some transmissions use multiple input shafts to provide a mechanical connection, applications of torque converters using clutch plates are considered typical in modern transmission designs.

FIGURE 2–61 Ford AOD input shafts. Pointer on solid direct-drive shaft followed by a hollow turbine shaft and stator shaft.

FIGURE 2–62 Ford AOD converter damper assembly.

❑ REVIEW

Key Terms/Acronyms

Torque converter, fluid coupling, hydrodynamic drive units, member, element, torsional, impeller, turbine, stator (reactor), rotary flow, vortex flow, stage, torque phase, one-way roller clutch, torque rating, stall torque, coupling phase, lockup converter, torque converter clutch, viscous clutch, PCM (powertrain control module), and ECM (engine control module).

Completion

1. The name given to the input or drive member of a fluid coupling. ______________________

2. The name given to the output or driven member of a fluid coupling. ______________________

3. The name given to the input or drive member of a fluid torque converter. ______________________

4. The name given to the output or driven member of a fluid torque converter. ______________________

5. The name given to the torque multiplier of a fluid torque converter. ______________________

6. The reference made to the fluid flow from the centrifugal pumping action of the converter or coupling input member.

   ____

7. The reference made to the fluid flow from the inertia effect of the rotating converter or coupling input member.

   ____

8. The reference made to the effective directional thrust of the fluid between the drive and driven members of a coupling or converter. ______________________

9. List in sequence the converter members through which the fluid cycles in a torque converter during the torque phase.

   (a) ______________________
   (b) ______________________
   (c) ______________________ and back to the
   (d) ______________________

10. The nonrotating converter shaft is referred to as the ______________ shaft.

11. The torque converter member or element that is driven by the engine. ______________________

12. The member of the torque converter that puts the fluid into motion. ______________________

13. The type of cooling system used with current converter applications is oil-to-______________________.

14. Fluid torque converters in their simplest form are usually described by using the following terminology: ______________ element, ______________ stage.

15. The input shaft of an automatic transmission is attached to the member of the torque converter called the ________________________.

16. Name the converter element that makes up part of the converter housing. ________________

17. A fluid coupling may also be correctly called a fluid ____________.

18. A fluid coupling cannot transmit a torque ratio of more than ______________.

19. A fluid torque converter acts as both a ______________ and ______________.

20. A term used to reflect coupling or converter efficiency. ________________

21. The converter front is supported and aligned to the engine crankshaft by a ____________________ in the crankshaft end.

22. The converter drive hub is typically supported and aligned by a ____________________ located in the transmission pump body. (RWD)

23. The transmission pump is driven by the converter ______________ or a pump ______________ splined to the converter cover.

24. The crankshaft is attached to the converter cover with a plate referred to as the ______________ plate.

25. The stator one-way roller clutch locks when the fluid effort is attempting to rotate the stator in a ______________ ____________________ direction.

Multiple Choice

____ 1. The function of the flex plate is to:
I. support and align the converter to the crankshaft.
II. mechanically connect the converter to the crankshaft.

A. I only
B. II only
C. both I and II
D. neither I nor II

____ 2. At engine idle, transmission in drive, and vehicle stationary:
I. The stator is locked.
II. The turbine is stationary.

A. I only
B. II only
C. both I and II
D. neither I nor II

____ 3. At engine idle, transmission in neutral, and vehicle stationary:
I. The impeller turbine and stator turn together.
II. The speed ratio is zero.

A. I only
B. II only
C. both I and II
D. neither I nor II

____ 4. At wide open throttle (WOT), transmission engaged, and vehicle stationary:
I. Converter torque is at maximum.
II. Rotary fluid force is at its maximum.

A. I only
B. II only
C. both I and II
D. neither I nor II

____ 5. During the unlock mode of a lockup converter:
I. Converter feed oil releases the clutch.
II. The fluid drive of the converter is unaffected.

A. I only
B. II only
C. both I and II
D. neither I nor II

____ 6. During the lockup mode of a lockup converter:
I. The clutch makes a mechanical connection between the crankshaft and the turbine.
II. The mechanical lockup torque must pass through a torsional damper assembly.

A. I only
B. II only
C. both I and II
D. neither I nor II

____ 7. In place of a spring-type damper, some converter clutches may use:
I. a silicone damper.
II. a solid converter clutch.

A. I only
B. II only
C. both I and II
D. neither I nor II

____ 8. Which of the following transmissions use a pure mechanical converter bypass and utilize a solid shaft connection, crankshaft-to-planetary set?

A. GM 4T60 (THM 440-T4)
B. Ford AOD
C. Ford AXOD
D. Chrysler A-904 (30RH)

____ 9. In an electro/hydraulic control system, the torque converter clutch (three-speed transmission):

A. normally locks up in third gear.
B. disengages when the brakes are applied.
C. disengages during engine torque demand.
D. cannot engage when the engine is cold.
E. all of the above.

_____ 10. Converter clutch circuit (Figure 2–58):

I. Clutch apply oil enters the converter through the input/turbine shaft.

II. In the lockup mode, converter oil cooling is bypassed.

A. I only
B. II only
C. both I and II
D. neither I nor II

_____ 11. Torque multiplication is at its peak in a torque converter when:

I. rotary flow has maximum velocity.

II. the turbine has reached its maximum rpm.

A. I only
B. II only
C. both I and II
D. neither I nor II

3 Planetary Gears in Transmissions/Transaxles

OBJECTIVE:

The objective of this unit is for you to become proficient with the terms and concepts contained in this chapter. Completion of this objective is essential to becoming a successful automatic transmission technician.

CHALLENGE YOUR KNOWLEDGE

Define the following key terms:

Gear ratio, planetary gearset, reaction member, gear reduction, band, multiple-disc clutch, one-way clutch, sprag clutch, roller clutch, freewheeling, overdrive, direct drive, and compound planetary.

List the following:

The three members of a planetary gearset; four advantages of planetary gearsets; characteristics of one-way clutches; four methods to hold reactionary members stationary; and three ways to drive an input member.

Describe the following:

The law of neutral; law of reduction; law of overdrive; law of direct drive; law of reverse; and the layout of both Simpson and Ravigneaux planetary geartrains.

Calculate:

Gear ratios.

INTRODUCTION

Planetary gears have been associated with the American automobile since its very beginning at the turn of the century. They were among the first type of gearing used in passenger car and light-duty truck transmissions and offered the advantage of minimum driver skill when changing gears (gear ratios). The sliding gearboxes in the early years did not feature synchronized shifts, and it was a skillful art to change gears on the move. Design restrictions confined the planetary gearboxes to two speeds and a reverse, although it is on record that the 1906 Cadillac used a three-speed planetary transmission. These early versions were not without their problems. They were noisy, had short bearing life (the practice at that time was to use bushing-mounted pinions), and at times chattered or grabbed during shift changes because of uneven brake band application.

Advances in the sliding gear transmission design eventually led to its popularity over the planetary designs and the almost universal use of sliding gear transmissions in passenger cars and trucks. The Ford Model T, however, used a planetary transmission until 1928. Planetary gears staged a comeback in the 1930s with the introduction of the Borg-Warner automatic overdrive and the General Motors Hydra-Matic transmission. Research and development in helical gears, alloy steels, heat treatment of metal, and needle bearings eliminated many of the deficiencies of the early types of planetary gears. Planetary gears today have a wide range of applications, varying from passenger car and truck automatic transmissions, to steering and final drive mechanisms of track-and-wheel-driven construction machinery, to reduction gears for aircraft propeller drives. These are just a few modern examples of planetary gear applications in power transmission.

GEARS AND GEAR RATIOS

Gears are spinning lever arms that control the delivery of engine crankshaft speed and torque to the drive wheels. The transmission gearbox, whether manual or automatic, is an example of how a gear assembly acts as a mechanical machine (wheel and axle) to transfer power with ability to change crankshaft input torque, speed, and direction. A gear works on the lever arm principle to change torque and speed.

- *Torque:* A twisting or turning effort or force acting through a lever arm

Torque = force × length of lever arm

Correctly expressed as pound-feet or pound-inches. The usual vernacular, however, is foot-pounds or inch-pounds.

- *Ratio:* A fixed numerical relationship between two similar items.

$$\text{Ratio} = \frac{\text{length of output lever}}{\text{length of input lever}}$$

Expressed as 1:1, 2:1, 3.43:1, and so on.

An ordinary 20-in lever can be used to illustrate transfer of force and torque about a fulcrum or pivot point. With the fulcrum centered, the 100 lb of force acting on *A* exerts an equal and opposite force at *B*:

$$100 \text{ lb} \times 10 \text{ in} = 100 \text{ lb} \times 10 \text{ in}$$

100 LB
(A)
100 LB
(B)

The lever arm ratio is 1:1. A 1-ft movement at *A* moves *B* 1 ft.

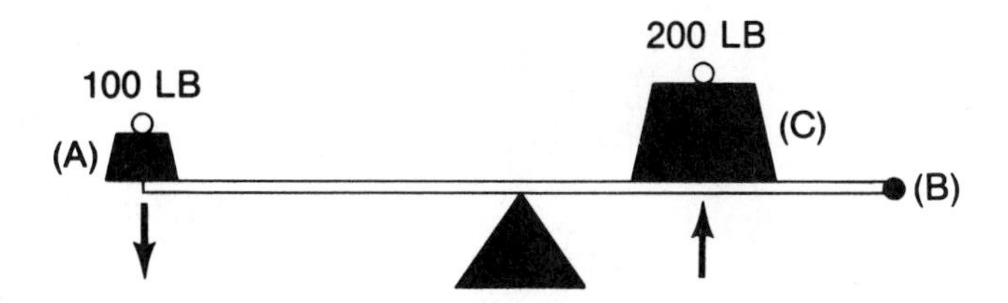

With point *B* located 5 in from the fulcrum, the 100 lb of force at *A* exerts a 200-lb and opposite force at *B*.

$$100 \text{ lb} \times 10 \text{ in} = 200 \text{ lb} \times 5 \text{ in}$$

The lever arm ratio is 2:1. A 1-ft movement at *A* moves *C* 1/2 ft.

In place of the fulcrum, the lever can be fixed to a shaft. When a force is applied on one end of the lever, a twisting or turning effort is applied to the shaft.

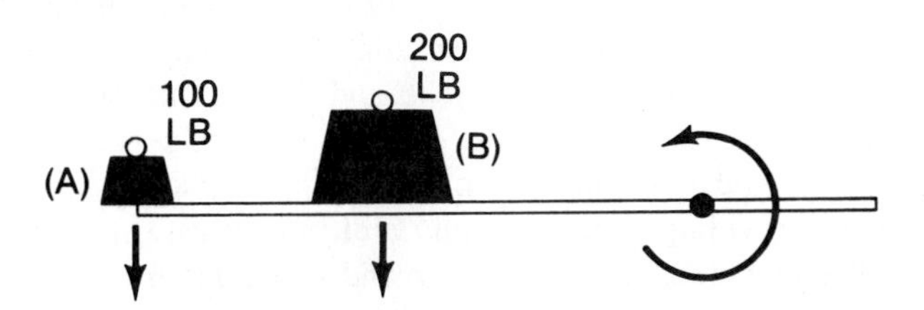

A 100-lb force applied at *A* exerts 1000 lb-in of torque on the shaft:

$$\text{torque} = 100 \text{ lb} \times 10 \text{ in} = 1000 \text{ lb-in}$$

200 lb of force applied at *B* and located 5 in from the shaft also exerts 1000 lb-in of torque on the shaft.

$$\text{torque} = 200 \text{ lb} \times 5 \text{ in} = 1000 \text{ lb-in}$$

If the shaft was exerting the torque, it follows that 1000 lb-in at the shaft would result in 100 lb of force at *A* and 200 lb at *B*:

$$\text{torque} = \frac{\text{torque}}{\text{lever arm}}$$

Although the simple lever arm demonstrates various capabilities of transmitting force and torque, it lacks the ability of continuous motion. Gears are used, therefore, to provide the continuous lever arm action necessary for the drivetrain power transmission.

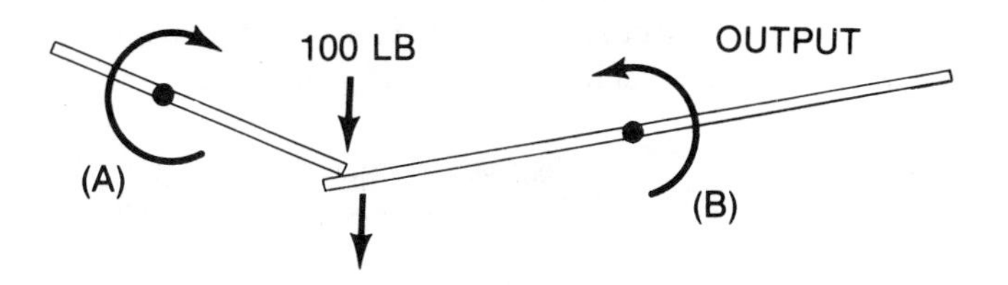

Lever *A* is 10 in long with an input shaft torque of 500 lb-in. This results in a force of 100 lb applied to the tip of lever *B*:

$$\text{force} = \frac{\text{torque}}{\substack{\text{lever arm} \\ \text{pivot length}}} = \frac{500 \text{ lb-in}}{5 \text{ in}} = 100 \text{ lb}$$

Lever *B* is 20 in long, and therefore, the 100 lb of apply force develops a torque of 1000 lb-in on the second shaft:

$$\begin{aligned} \text{torque} &= \text{force} \times \text{lever arm pivot length (10 in)} \\ &= 100 \text{ lb} \times 10 \text{ in} = 1000 \text{ lb-in} \end{aligned}$$

Because lever *B* is longer, the output torque is greater than the input torque of lever *A*.

In observing the two levers, it is obvious that their travel is very limited before they disengage. To fill the void, a series of lever arms is used to provide continuous motion.

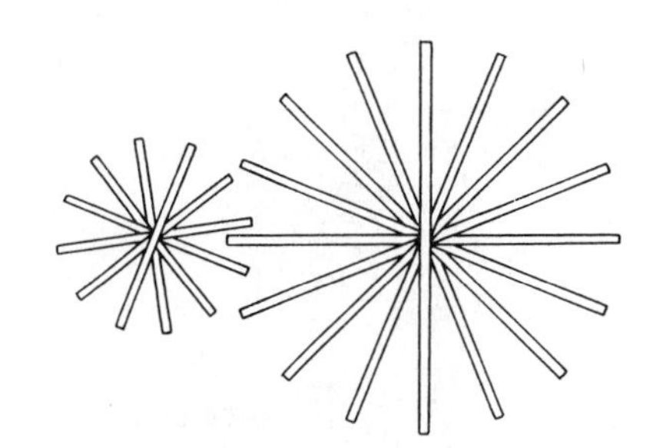

For practical application, the lever arms are designed into a matching gearset configuration.

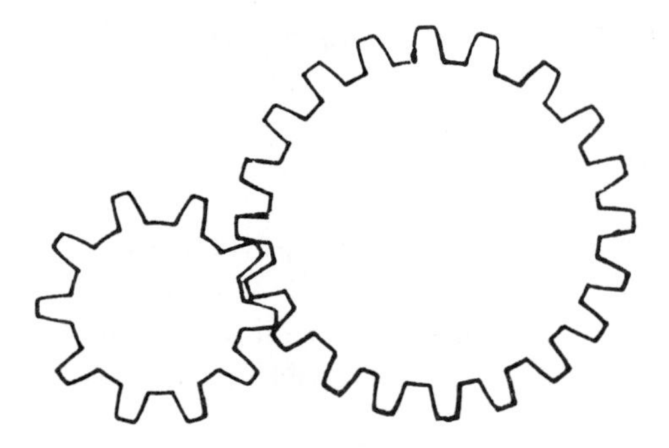

When dealing with levers, whatever is gained in force results in a proportional loss in distance. Conversely, a loss of force results in a proportional gain in distance. The amount that torque is increased depends on the relative size of the gears. If the diameter of the driven gear is twice the diameter of the drive gear, the input torque will be doubled and the speed cut in half. Should the driving gear be twice the diameter of the driven gear, the input torque is cut in half and the speed doubled. This torque-speed relationship is reflected in the design gear ratio of the gearset and can be determined by the gear sizes:

$$\text{gear ratio} = \frac{\text{driven diameter}}{\text{drive diameter}}$$

Comparing gear diameter sizes is a theoretical approach that works fine for academic problem calculation, but for the technician, a **gear ratio** is calculated by comparing the number of teeth on the gears. The teeth of two mating gears must be of the same size if they are to fit together properly. If the diameter of the large gear is twice the size of the small gear, the large gear will have twice as many teeth:

$$\text{gear ratio} = \frac{\text{driven teeth}}{\text{drive teeth}} = \frac{20}{10} = 2{:}1$$

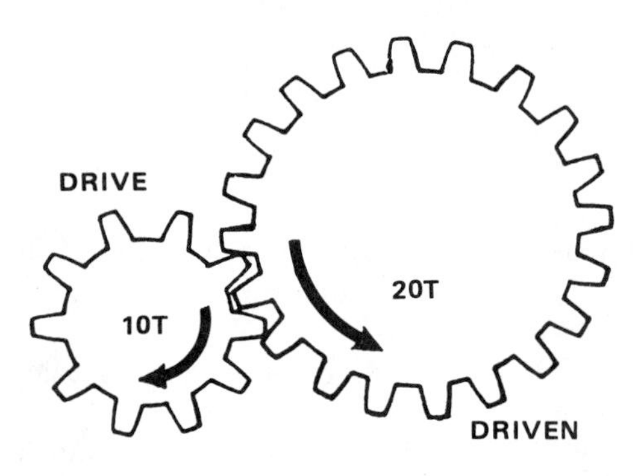

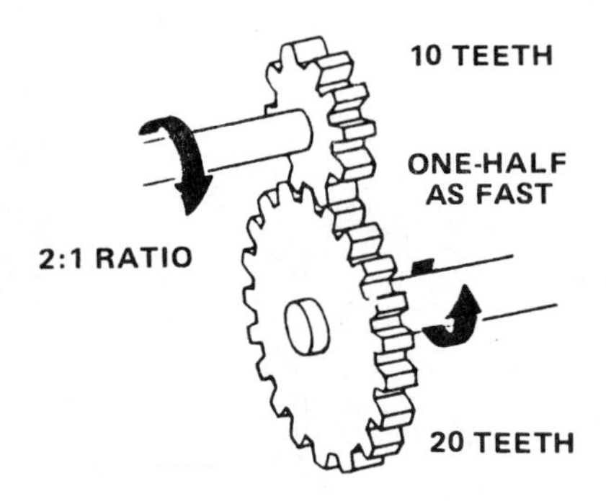

A gear ratio is defined as the number of turns the drive gear makes to one revolution of the driven gear. In following our 2:1 ratio example, for every two revolutions of the small drive gear, the large driven gear makes one turn. If the crankshaft drives the small gear at 1000 rpm, the large driven gear speed is 500 rpm. Whatever is gained in torque is sacrificed in speed. It is not a sacrifice, however, if it is exactly what is desired.

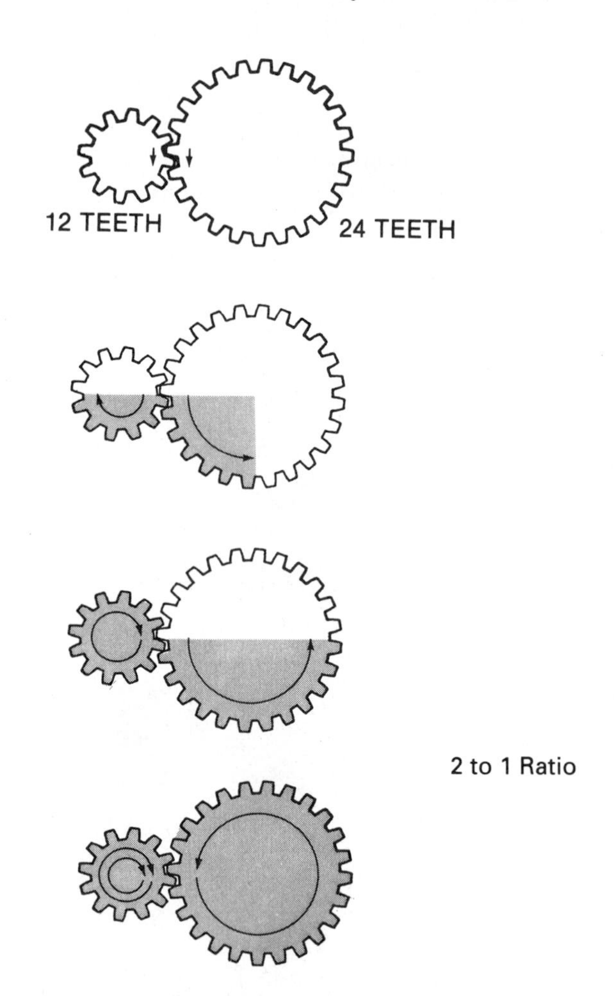

❖ **Gear ratio:** The number of revolutions the input gear makes to one revolution of the output gear. In a simple gear combination, two revolutions of the input gear to one of the output gear gives a ratio of 2:1.

Gears are used to control engine torque output and speed. In the transmission, the built-in gear ratios are available to match the engine output to the road load and desired performance demand by the driver. In Figure 3–1, a simplified power summary shows how the transmission adapts the engine to the road load by providing suitable torque multiplying ratios. Notice how the gear sizes influence torque and speed output.

Gears can change torque and speed but never power. Power is a function of time and motion or the rate at which work is done. One horsepower is the ability to move 33,000 lb one foot in one minute or 550 lb one foot in one second. When working with power transmission through gearing, pulleys, or sprocket/chain drives, the power constant can be shown by using the applicable horsepower formula:

$$hp = \frac{torque \times rpm}{5252}$$

Engine torque at 2000 rpm is 180 lb-ft, so

$$\text{drive gear hp} = \frac{180 \times 2000}{5252} = 68.5$$

Using a gear ratio of 2:1, we have

$$\text{driven gear hp} = \frac{360 \times 1000}{5252} = 68.5$$

Gears come in many different forms. Whether the geartrain in the transmission is a conventional gear-designed system or a planetary system, the power transmission fundamentals remain the same. Gears are all a form of lever action.

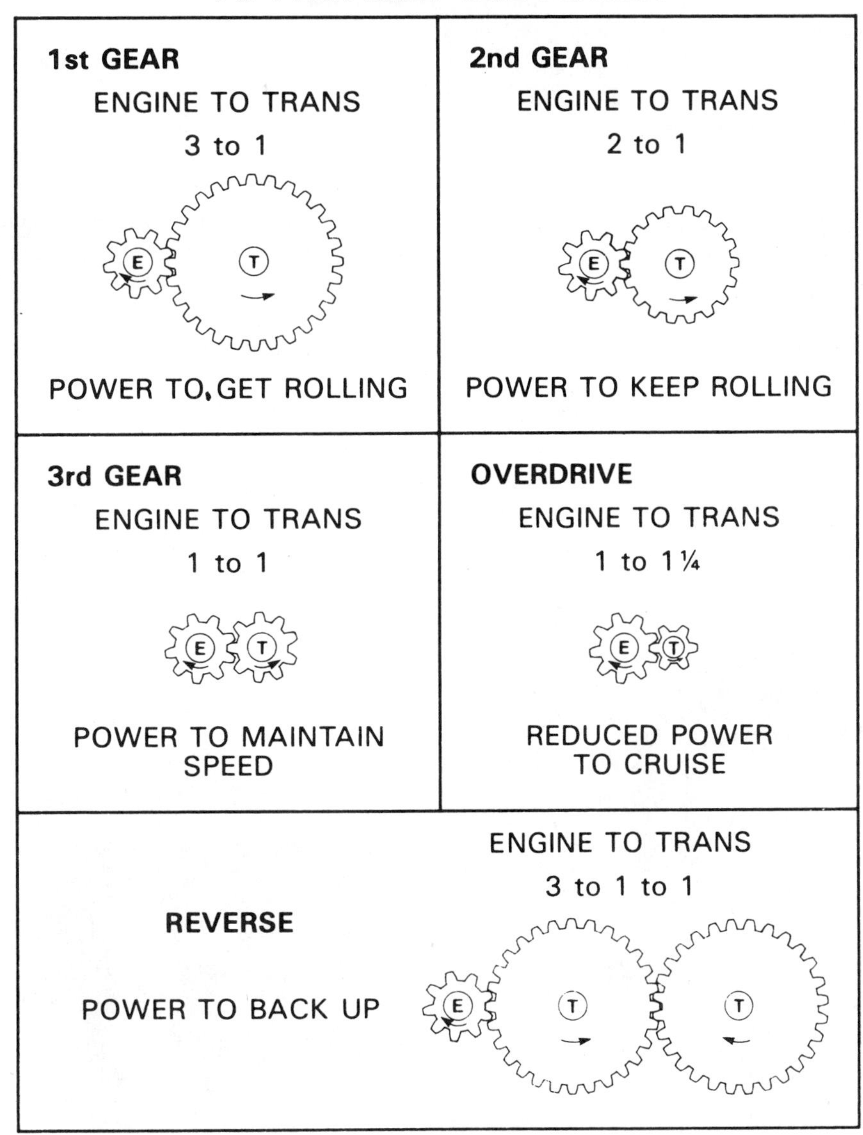

FIGURE 3–1 Forward and reverse gears. (Courtesy of General Motors Corporation, Service Technology Group)

PLANETARY GEAR ASSEMBLY AND OPERATION

The heart of the automatic transmission is the planetary gear system. It is essential, therefore, to review the basic construction of a simple **planetary gearset** as an introduction to how planetary gears operate. A simple planetary gearset (Figures 3–2 and 3–3) consists of a sun gear or center gear that is surrounded and in constant mesh with the planet pinion gears. The pinion gears are mounted and free to rotate on their support shafts, which are pinned to a planetary pinion carrier. The internal gear, also referred to as the ring or annulus gear, is in constant mesh with the planetary pinions and surrounds the entire assembly. It should be noted that the sun gear, planet carrier, and internal gear rotate on a common center, while the planet pinion gears rotate on their own independent centers. For clarification, the planet pinion gears are considered to be part of the planet carrier.

❖ **Planetary gearset (simple):** An assembly of gears in constant mesh consisting of a sun gear, several pinion gears mounted in a carrier, and a ring gear. It provides gear ratio and direction changes, in addition to a direct drive and a neutral.

The planetary gearset gets its name from the action of the planet pinion gears. As we will see later, these gears have the ability to turn on their own centers and, at the same time, revolve around the sun gear. This is similar to the earth turning on its axis and rotating around the sun.

Advantages. Figures 3–2 and 3–3 illustrate several major advantages of planetary gears:

1. All members of the planetary gearset share a common axis, which results in a structure of compact size.
2. Planetary gears are always in full and constant mesh, eliminating the possibility of gear tooth damage from gear clash or partial engagement. The full and constant mesh feature also permits automatic and quick gear ratio changes without power flow interruption.
3. Planetary gears are strong and sturdy and can handle larger torque loads, for their compact size, in comparison to other gear combinations in manual transmissions. This is because the torque load as it passes through the planetary set is distributed over the several planet pinion gears, which in effect allows more tooth contact area to handle the power transmission.
4. The location of the planetary members makes it relatively easy to hold the members or lock them together for ratio changes.

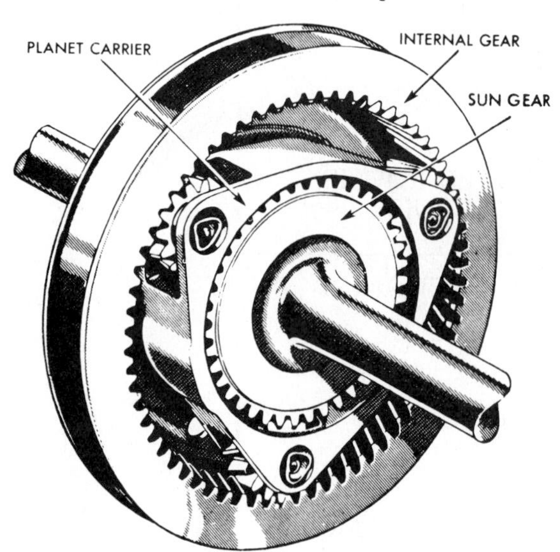

FIGURE 3–2 Simple planetary gear assembly. (Courtesy of General Motors Corporation, Service Technology Group)

LAWS OF PLANETARY GEAR OPERATION

The operation of a planetary geartrain is governed by five basic laws that provide the key to understanding the various gearing power flows in all automatic transmissions, regardless of differences in planetary systems: the laws of neutral, reduction, overdrive, direct drive, and reverse. Study them carefully, one at a time. Several rules govern planetary member rotation that will assist you in understanding the power flow in each of the planetary operating modes:

- When the internal gear and carrier pinions are free to rotate at the same time, the pinions always follow the same direction as the internal gear.
- The sun gear always rotates opposite carrier pinion gear rotation.
- When the planet carrier is the output, it always follows the direction of the input gear.
- When the planet carrier is the input, the output gear member always follows carrier direction.

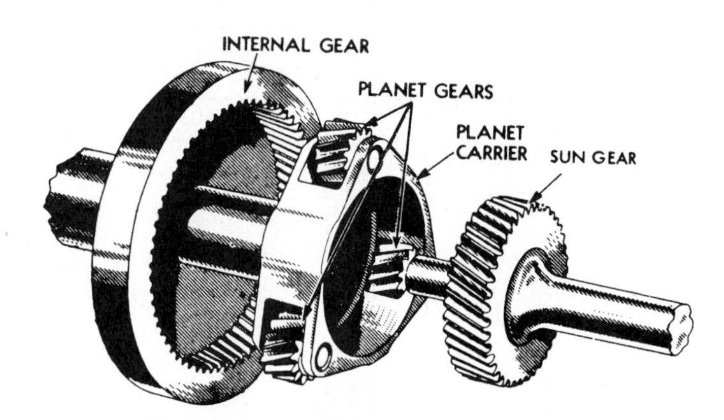

FIGURE 3–3 Planetary gear assembly—exploded view. (Courtesy of General Motors Corporation, Service Technology Group)

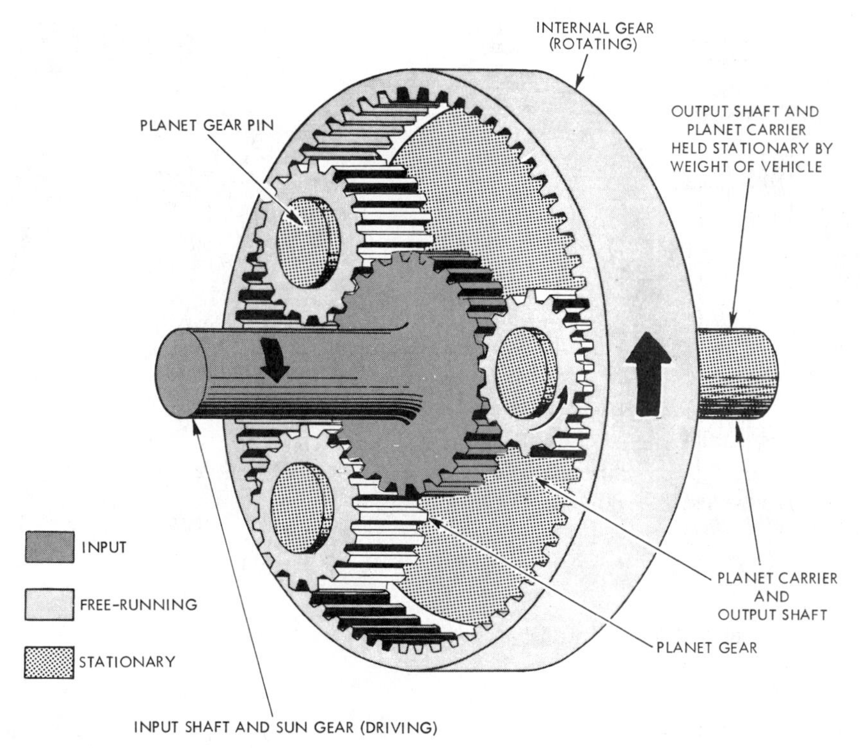

FIGURE 3–4 Planetary gearset during neutral operation.

Law of Neutral

When there is an input but no **reaction member**, the condition is neutral. In Figure 3–4, the sun gear serves as the driving input member, and the internal gear is free to rotate because it is not grounded to any part of the transmission. The planet carrier is held stationary by the weight of the car on the rear wheels. This causes the planet pinion gears to rotate on their pins and drive the internal gear opposite the sun gear or input direction.

- ❖ **Reaction member:** The stationary planetary member, in a planetary gearset, that is grounded to the transmission case through the use of friction and wedging devices known as bands, disc clutches, and one-way clutches.

In many automatic transmissions, neutral may be achieved by another method. In these units, the input shaft is declutched from the gearset. Whether or not another gearset member is held stationary, the result is neutral.

Law of Reduction

When there is a reaction member and the planet carrier is the output, the condition is **gear reduction**. There are two reduction possibilities that meet the requirements for the law of reduction. These are illustrated by using a brake or friction **band** to clamp around a rotating drum and ground it to the transmission case (Figure 3–5). Either the internal gear or sun gear attached to the drum becomes the stationary or reactionary member. The band is applied by a servo unit that is nothing more than a bore in the case containing an apply piston-and-pin assembly. Hydraulic force is changed to a mechanical force.

- ❖ **Gear reduction:** Torque is multiplied and speed decreased by the factor of the gear ratio. For example, a 3:1 gear ratio changes an input torque of 180 lb-ft and an input speed of 2700 rpm to 540 lb-ft and 900 rpm, respectively. (No account is taken of frictional losses, which are always present.)
- ❖ **Band:** A flexible ring of steel with an inner lining of friction material. When tightened around the outside of a drum, a planetary member is held stationary to the transmission case.

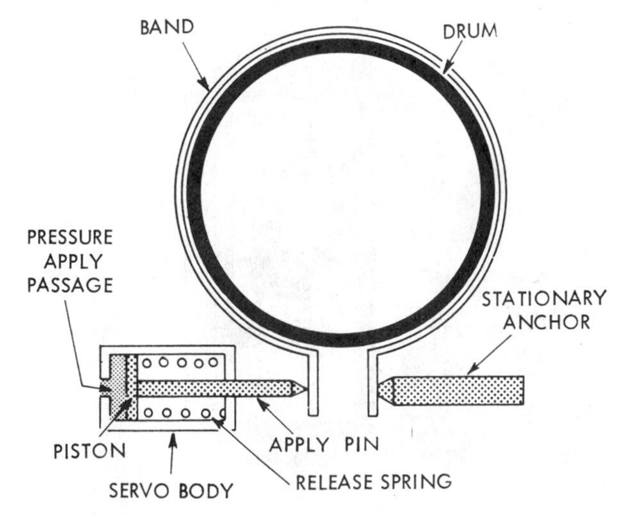

FIGURE 3–5 Hydraulic servo clamps the band around the drum to hold a planetary member.

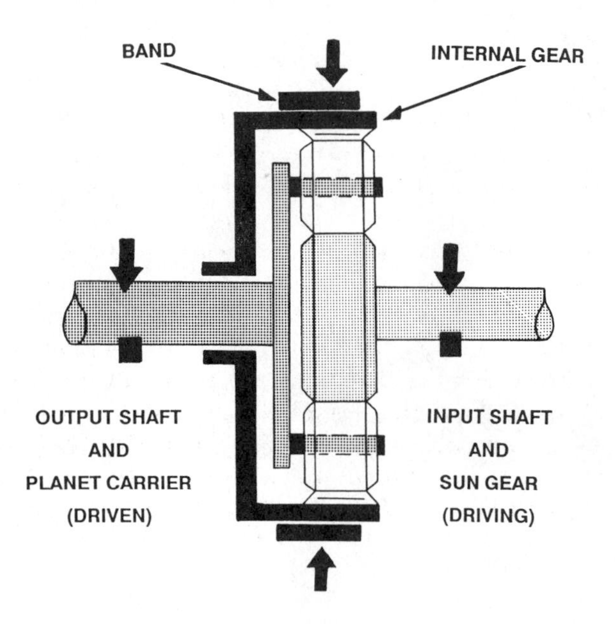

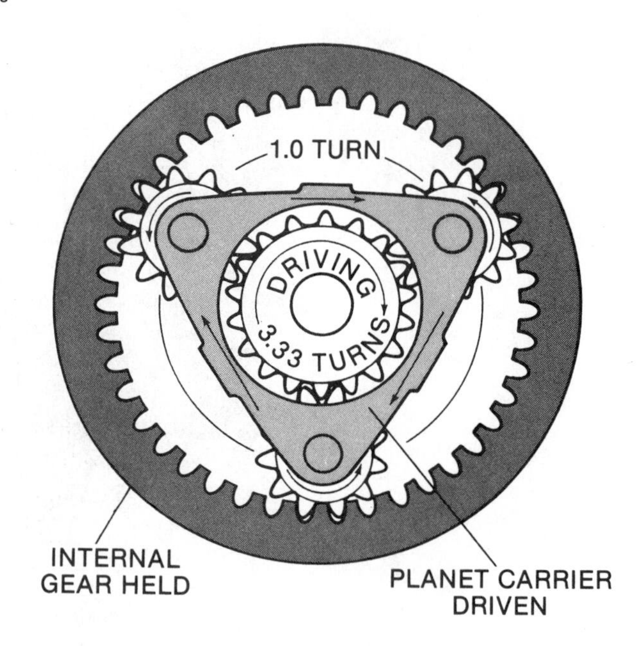

FIGURE 3–6 Planetary gear unit in reduction with internal gear held.

The first method of reduction shows the input sun gear driving the pinion gears on their pins opposite the input direction (Figure 3–6). Because the pinion gears cannot move the stationary internal gear, a reaction force is created between the two gears that causes the pinions to push off the internal gear teeth and walk around the internal gear as they rotate on their centers. This moves the carrier in a forward direction at a reduced speed. If the input torque is 100 lb-ft (135.5 N-m) and the gear reduction is 3.33:1, the output torque is increased to 333 lb-ft (451.2 N-m).

The second reduction method is set up with the sun gear stationary and power input applied to the internal gear (Figure 3–7). The planet pinions now rotate on their centers, push off the stationary sun gear, and walk around the sun gear to produce another forward reduction effect on the carrier. This planetary gear reduction is widely used for second-gear operation in many automatic transmissions. When comparing the two reduction possibilities, note that driving the sun gear and holding the internal gear give the deepest reduction.

Calculating the gear ratios of single planetary sets still involves counting the gear teeth—sun gear and internal gear. Because of concept differences in planetary structure and operation versus the conventional matching of drive and driven gear members, modified ratio formulas are used. When the planet carrier is the driven (output) member during forward reduction:

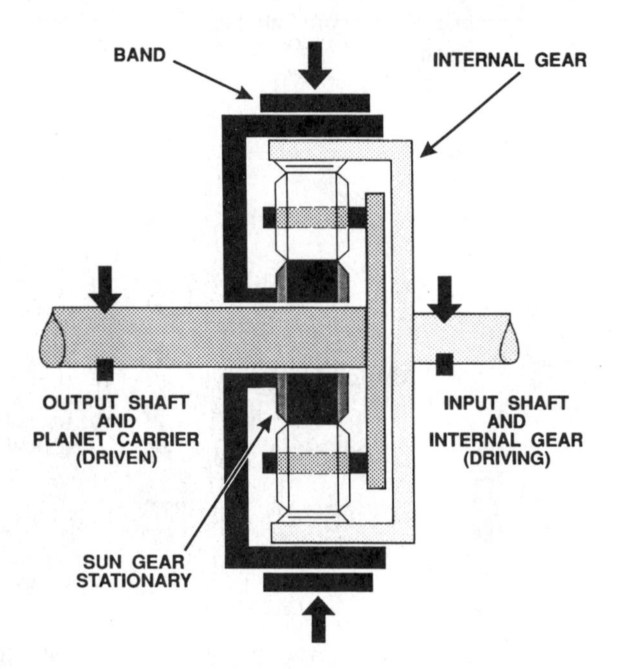

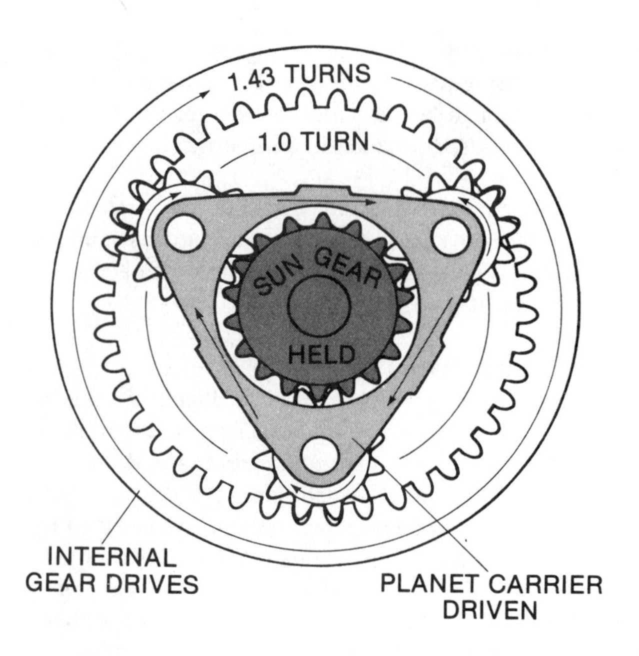

FIGURE 3–7 Planetary gear unit in reduction with sun gear held.

$$\frac{\text{reduction}}{\text{(carrier driven)}} = \frac{Ns + Ni}{N \text{ driving}}$$

Using the planetary illustrations in Figures 3–6 and 3–7, we have

Ns = number of teeth on the sun gear (18)

Ni = number of teeth on the internal gear (42)

Figure 3–6, sun gear driving/internal gear held:

$$\text{reduction} = \frac{18 + 42}{18} = 3.33{:}1 \text{ ratio}$$

Figure 3–7, internal gear driving/sun gear held:

$$\text{reduction} = \frac{18 + 42}{42} = 1.43{:}1 \text{ ratio}$$

Instead of using a band to connect a selected planetary reaction member solidly to the transmission case, **multiple-disc clutches** and **one-way clutches** can be used as grounding components for reduction. One-way clutches are classified as **sprag clutches** (Figure 3–8) or **roller clutches** (Figure 3–9) and are rather interesting devices. They are self-initiating in the exact timing of their holding and **freewheeling** actions and require no hydraulic controls. They operate solely by mechanical means, and both use a wedging action when they self-apply. This self-operating feature permits simplified and space-saving transmission design with no band adjustment requirement.

- **Multiple-disc clutch:** A grouping of steel and friction lined plates that, when compressed together by hydraulic pressure acting upon a piston, lock or unlock a planetary member.
- **One-way clutch:** A mechanical clutch of roller or sprag design that resists torque or transmits power in one direction only. It is used to either hold or drive a planetary member.
- **Sprag clutch:** A type of one-way clutch design using cams or contoured-shaped sprags between inner and outer races.
- **Roller clutch:** A type of one-way clutch design using rollers and springs mounted within an inner and outer cammed race assembly.
- **Freewheeling:** There is a power input with no transmission of power output. The power flow is ineffective.

The freewheeling action of the one-way clutch is used to an advantage to improve shift quality. Changes in ratio eliminate a timing problem during band-to-band or band-to-clutch shifts. A study of Figures 3–8 and 3–9 shows the wedging (ON position) and freewheeling (OFF position) action of these clutches. Both illustrations could be reversed to show the inner race as the stationary member and the outer race as the rotating or drive member. In actual practice, the stationary race may be bolted, riveted, or clutched to the transmission case.

Another behavior pattern is peculiar to both sprag and roller clutch operation. These clutches are effective as long as the engine powers the vehicle. If coasting occurs when gear reduction is in effect, the sprags and rollers will unwedge and allow freewheeling to take place. The planetary gearset is then in neutral, making it impossible to relay the power flow to the torque converter and engine for braking.

The illustrated lock and freewheel action is based on a planetary system in line with the torque converter and crankshaft—lockup with counterclockwise effort, and freewheel in

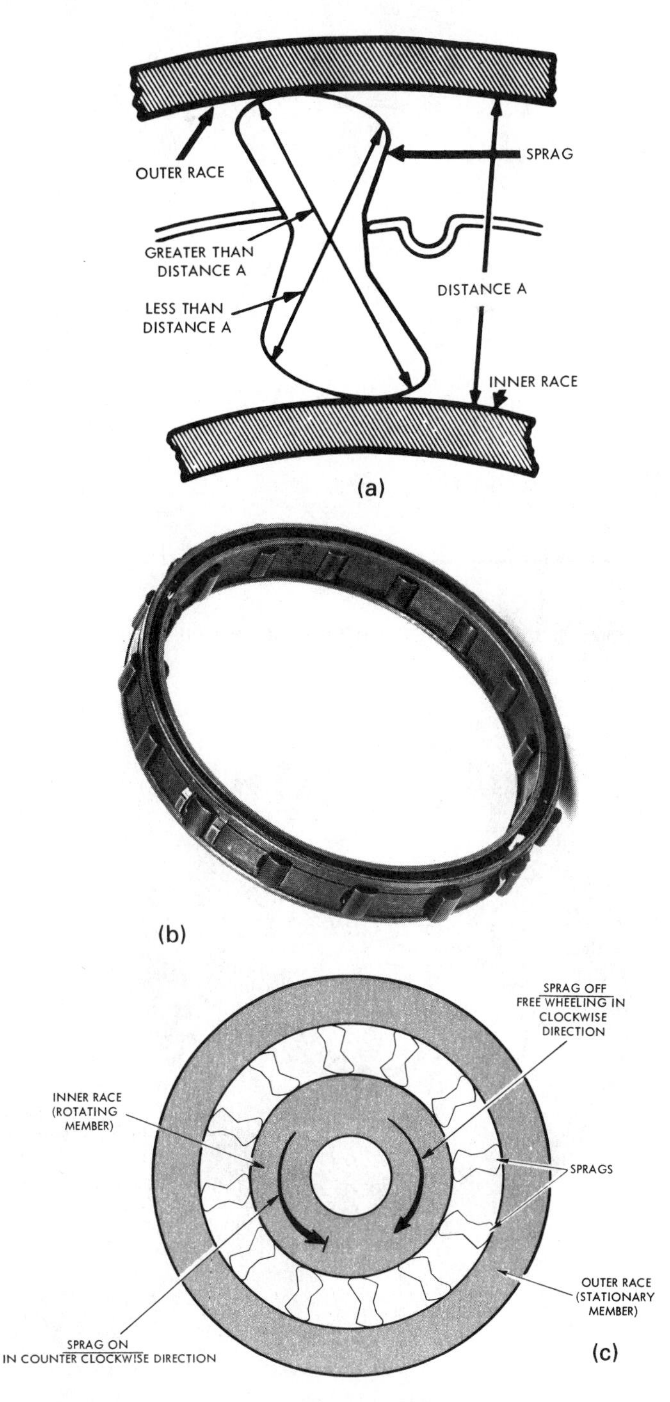

FIGURE 3–8 (a) Sprag configuration; (b) sprag and cage assembly; (c) sprag clutch action with stationary outer race—front view.

clockwise or input direction. This holds true for all rear drive transmissions and some transaxle drives. In transaxle applications where the converter output is transferred with a chain drive to the planetary system, the input to the system is counterclockwise. The one-way clutch action is reversed. Lockup is clockwise, and freewheel is counterclockwise. All rotation of transmission gears and related parts is based on entry direction front to rear.

A one-way clutch is set up in a single planetary to show a simplified illustration of how the drive and coast action can be controlled by the planetary input and rotation (Figures 3–10 and 3–11). Actual setups in automatic transmissions vary, but the

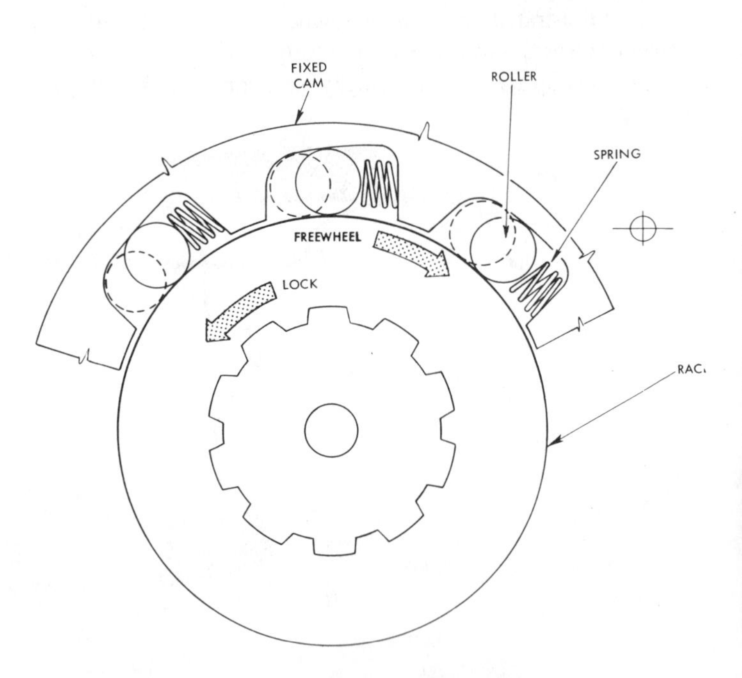

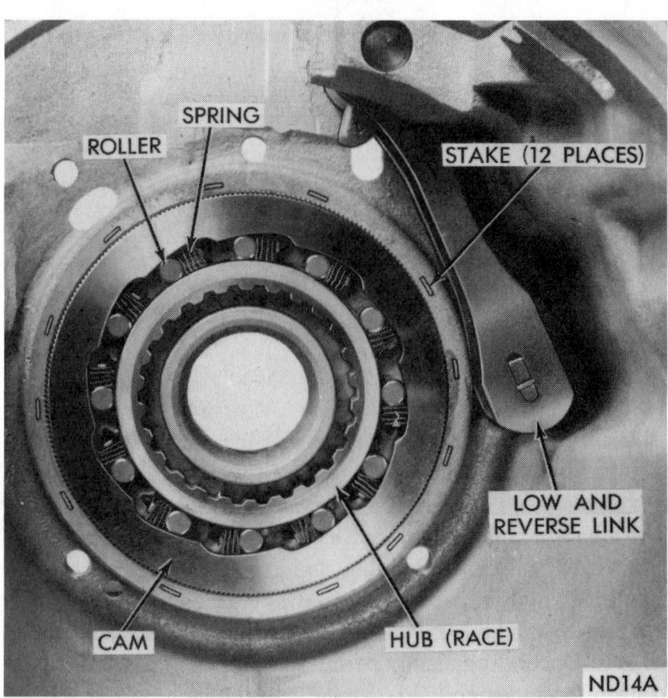

FIGURE 3–9 Roller clutch action with stationary outer cam race—front view.

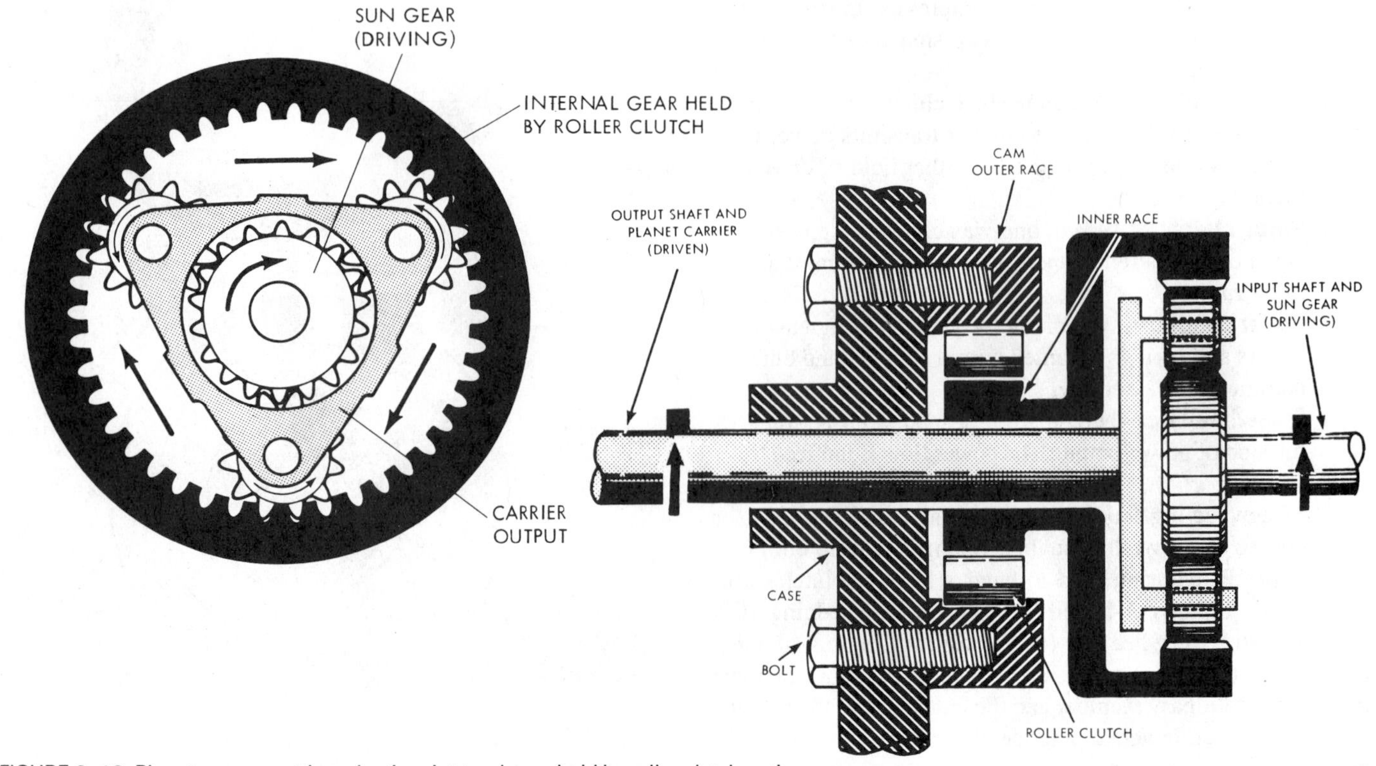

FIGURE 3–10 Planetary gearset in reduction, internal gear held by roller clutch action.

principle of operation remains the same. During a drive condition, the sun gear is the power input to the planetary (Figure 3–10). The weight or load of the vehicle on the output shaft and planet carrier allows the sun gear to drive the internal gear counterclockwise (CCW) through the planet carrier pinions. The roller clutch is keyed to the internal gear, however, and locks it to the transmission case. The internal gear becomes reactionary and puts the planetary unit in reduction to move the vehicle. The planetary remains in reduction as long as the sun gear is driving and the internal gear is attempting to rotate CCW.

During a deceleration condition such as closed-throttle, the output shaft and planet carrier are now the power input to the planetary unit. The planet carrier input causes the pinion gears to rotate the internal gear clockwise and effect a release of the roller clutch. The gear unit is now in neutral, and the connection between the drive wheels and engine is broken. Although the freewheeling action of a one-way clutch is an asset to closed-throttle downshift quality, it is definitely unsafe when the transmission gear reduction needs to be used for controlling vehicle speed on descending steep hills or mountain grades. To overcome this deficiency, the driver can select an operating range, such as manual low, that eliminates the freewheeling and returns positive gearing action to the engine for braking and a safe descent.

Figure 3–11 shows the same gear unit as Figure 3–10, with the addition of a coast band that is applied in the case of manual low operation. While the vehicle is under power from the engine, such as in climbing a steep grade or while using manual low for maximum performance, the applied band is not really holding against the drive torque because the roller clutch is effective. During grade descent or deceleration, the band is already applied and holds against coasting torque to keep the internal gear reactionary. With the planet carrier serving as the input to the gearset during coasting, it is now possible to transfer an overdrive speed to the converter turbine. With the turbine now acting as a pump, the turbine fluid discharge attempts to overspeed the impeller against engine compression, plus the vacuum draw on the intake stroke.

Not all one-way clutches have a permanently grounded race. Engineering design strategy may call for clutching/declutching the race to the transmission ground (Figure 3–12). The one-way clutch in this setup cannot be effective unless the multiple-disc clutch is applied and grounds the outer race. A follow-through on the planetary rotations shows a CCW effect on the sun gear, resulting in roller clutch lockup. In addition to using the band and one-way clutch for the reduction function, a multiple-disc clutch can be incorporated with the planetary set for the purpose of holding the internal gear or sun gear. Figure 3–13 has the internal gear clutched to the transmission case and the sun gear acting as the input drive.

Law of Overdrive

When there is a reaction member and the planetary carrier is the input, the condition is **overdrive**. Because overdrive gives the opposite effect of gear reduction, the planet carrier serves as the input rather than the output member, with either the sun gear or the internal gear held stationary (Figures 3–14 and 3–15). It should be noted that even with the planet carrier as the power input, the pinions are still free to rotate on their centers and walk as they push off and react to the fixed planetary

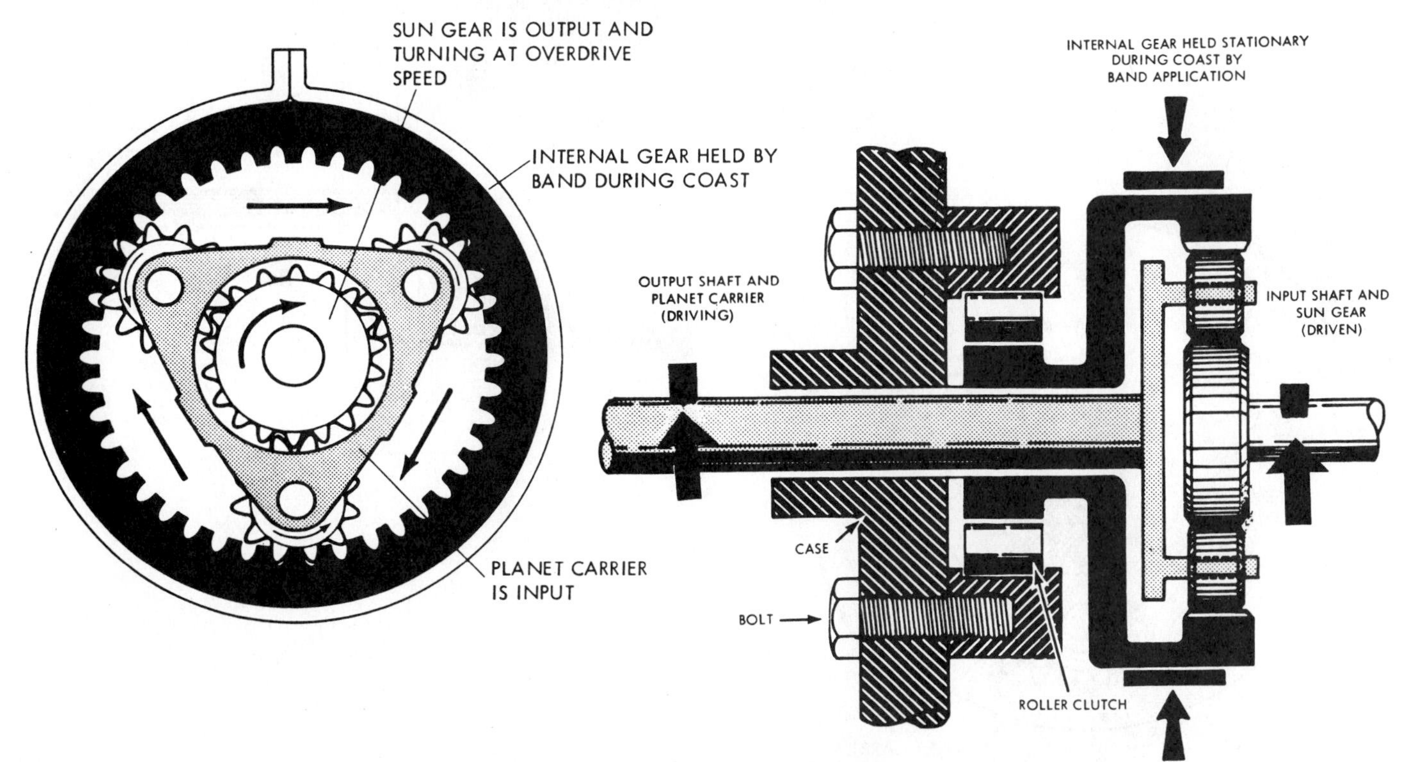

FIGURE 3–11 Planetary gearset in reduction showing band apply for braking during coast. Band is normally applied at all times during drive or coast conditions where coast braking is needed in intermediate and manual low operation.

member. This time, the turning and walking action moves the output member at an increased speed and reduced torque.

- ❖ **Overdrive:** Produces the opposite effect of a gear reduction. Torque is reduced, and speed is increased by the factor of the gear ratio. A 1:3 gear ratio would change an input torque of 180 lb-ft and an input speed of 2700 rpm to 60 lb-ft and 8100 rpm, respectively.

Regardless of differences in planetary systems, always look for the planet carrier as the input to achieve overdrive. The usual setup for automatic overdrive uses the stationary sun gear with the internal gear as the output.

The gear ratio formula for overdrive:

$$\frac{\text{Overdrive}}{\text{(carrier driving)}} = \frac{N \text{ driven}}{Ns + Ni}$$

Using the planetary illustrations in Figures 3–14 and 3–15, we have

Ns = number of teeth on the sun gear (18)
Ni = number of teeth on the internal gear (42)

Figure 3–14, sun gear held/internal gear driven:

$$\text{Overdrive} = \frac{42}{130} = 0.7{:}1 \text{ ratio}$$

18 + 42

Figure 3–15, internal gear held/sun gear driven:

$$\text{Overdrive} = \frac{18}{18 + 42} = 0.33{:}1 \text{ ratio}$$

Overdrive planetary outputs for automatic transmissions were avoided and deemed not cost effective or necessary at one time, but unpredictable fuel costs and government mileage and emission standards influenced the introduction of a wide variety of four-speed overdrives in the late 1970s and early 1980s, for both rear-wheel and front-wheel drive systems, by domestic

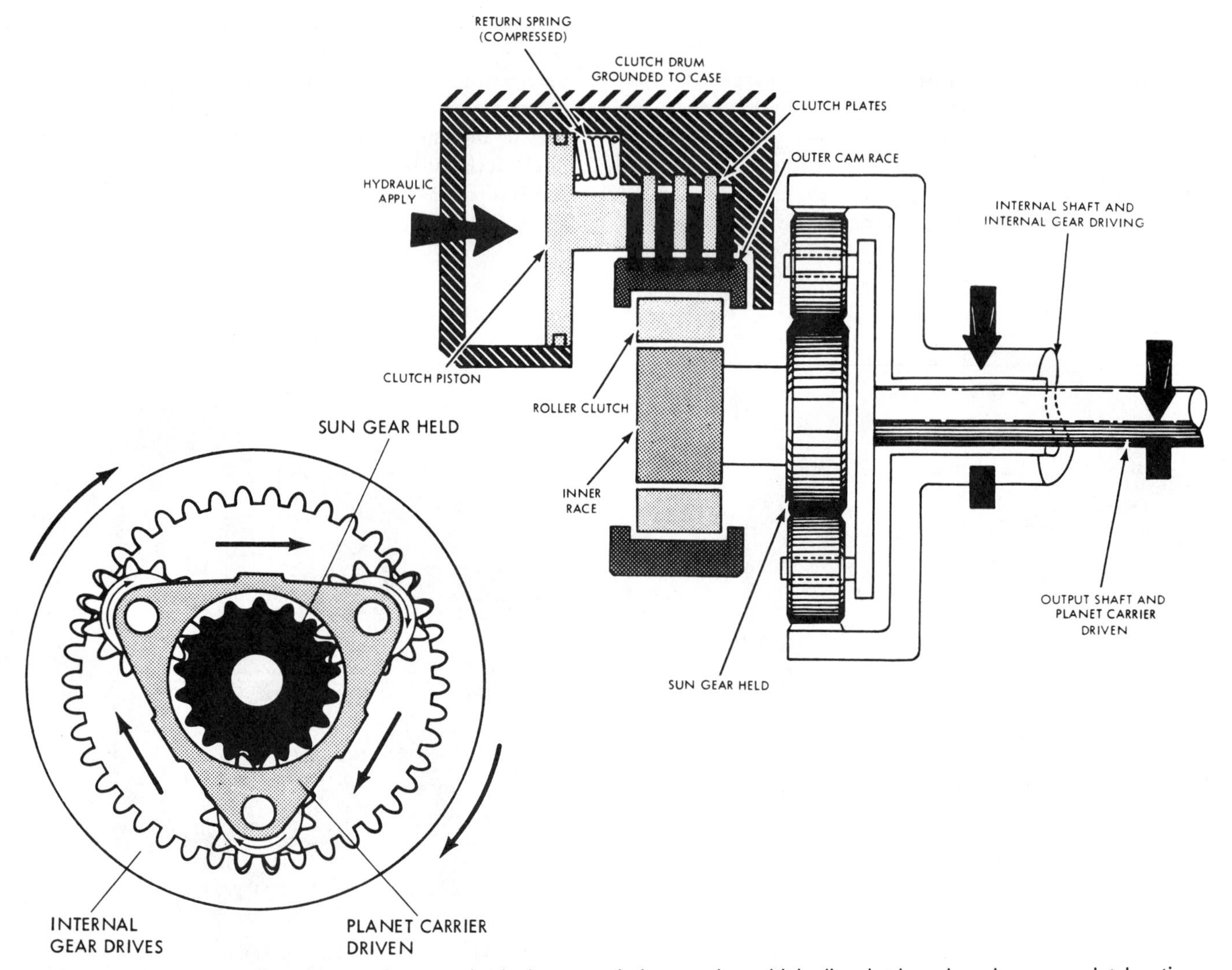

FIGURE 3–12 Planetary gearset in reduction. Sun gear clutched to transmission case by multiple-disc clutch apply and one-way clutch action.

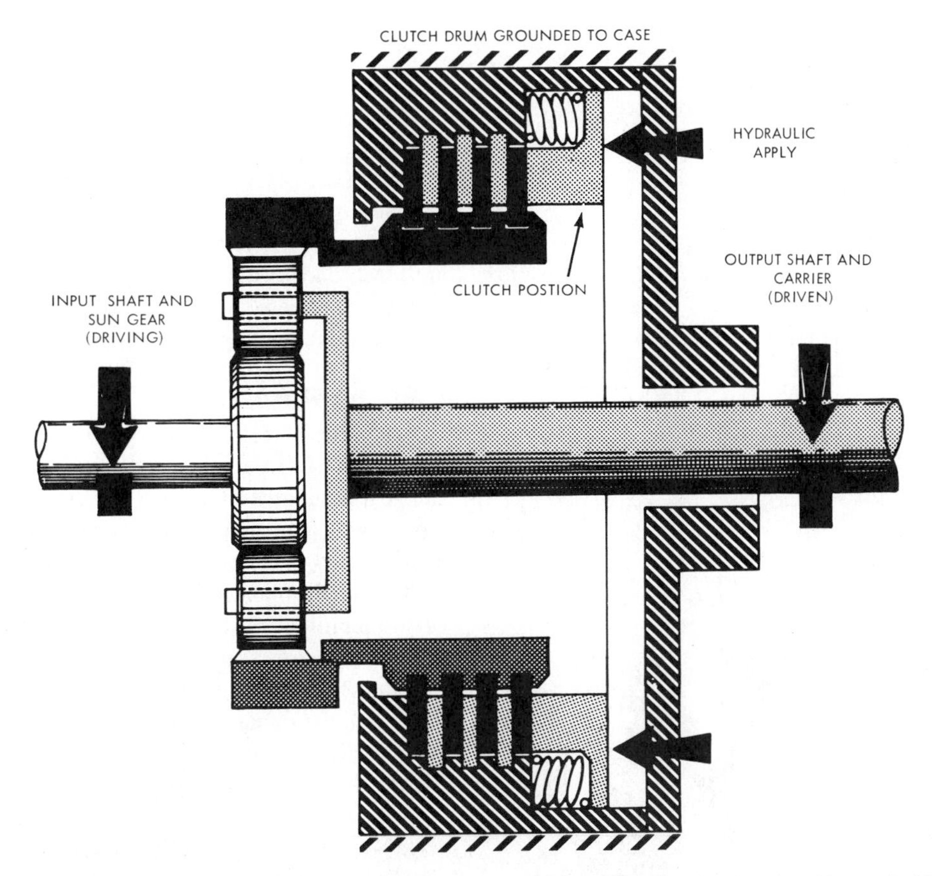

FIGURE 3–13 Internal gear clutched to the transmission case, with sun gear driving. No roller clutch or band is needed for drive or coast.

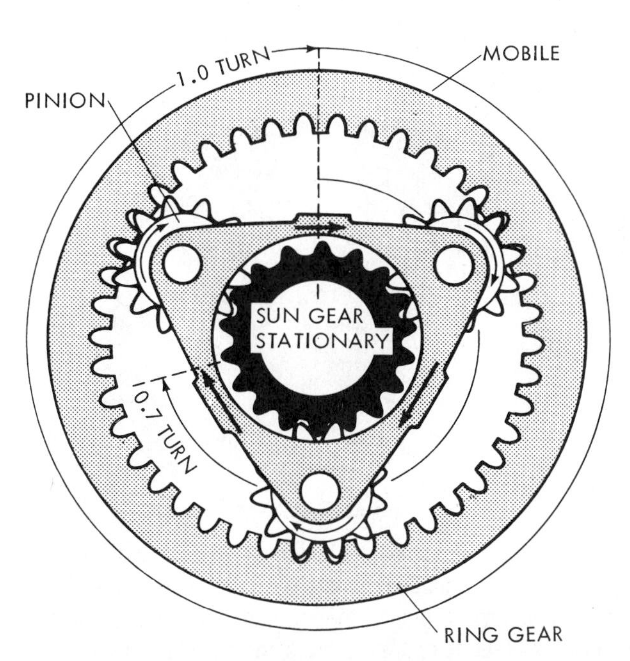

FIGURE 3–14 Planetary gearset in overdrive, with the sun gear stationary, showing pinion carrier versus ring gear travel.

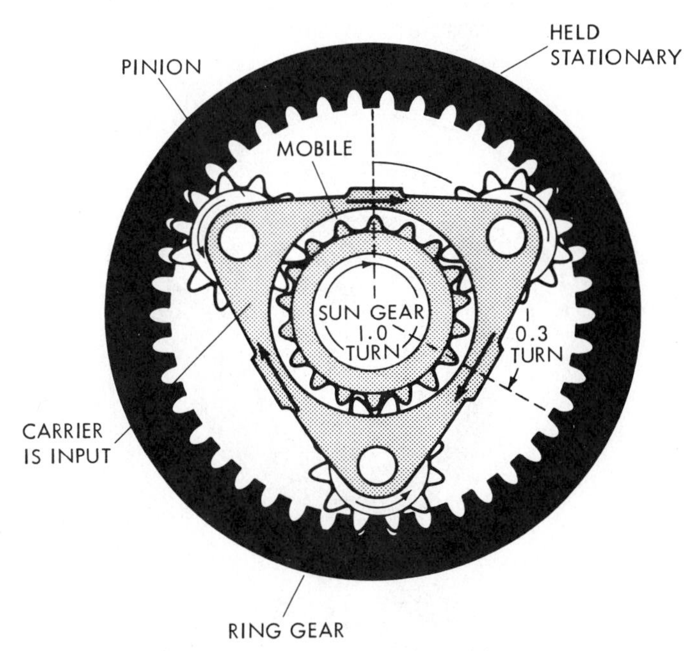

FIGURE 3–15 Planetary gearset in overdrive, with sun gear stationary, showing pinion carrier versus sun gear travel.

and import car manufacturers. It is common to find a three-speed automatic transmission integrated with an overdrive planetary to achieve fourth gear. The overdrive planetary is either at the input or output end of the three-speed system (see Chapter 4).

Law of Direct Drive

Direct drive is obtained by clutching or locking any two members of the gearset together. Driving any two members at the same relative speed and in the same direction gives the same effect. The principle of direct drive is shown and explained in Figures 3–16 to 3–18.

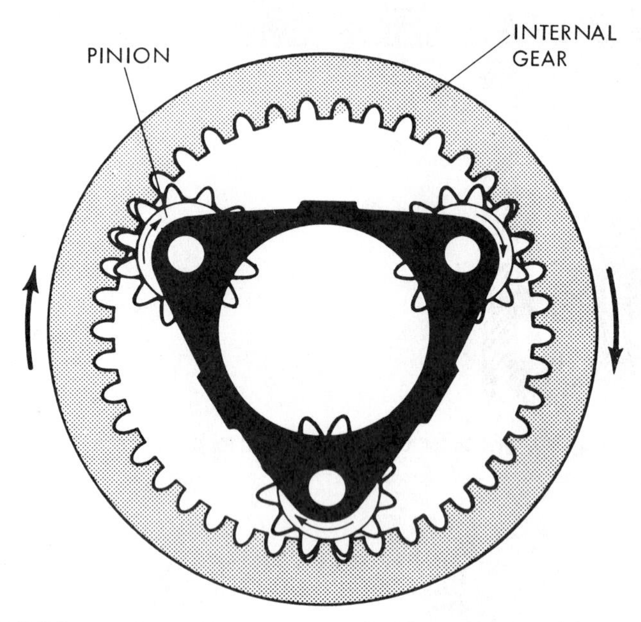

FIGURE 3–16 Planetary gearset in direct-drive sequence, with internal gear driving. As clockwise power is applied to the internal gear, the planet pinions are rotated clockwise.

FIGURE 3–17 Planetary gearset in direct-drive sequence, with sun gear driving. Clockwise power applied to planet pinions.

- **Direct drive:** The gear ratio is 1:1, with no change occurring in the torque and speed input/output relationship.

The usual practice in direct drive is to use multiple-disc clutches to lock two members of a planetary gearset to the input shaft. Figure 3–19 shows a typical multiple-disc clutch, which in this case locks the internal gear and planet carrier together. Apply and release of the clutches are controlled by the transmission valve body.

Law of Reverse

When the planet carrier is held against rotation with either the sun gear or internal gear driving, the result is reverse. The sun gear input gives a reverse reduction, while an internal gear input gives a reverse overdrive. Reverse reduction, the one usually used in automatic transmissions, is shown in Figure 3–20. With the planet carrier held by a band application, the sun gear input rotation is reversed by the carrier pinion gears acting as reverse idlers. The counterclockwise pinion rotation drives the internal gear and output shaft at a reverse reduction. In place of a band, the transmission design can incorporate a multiple-disc clutch to ground the carrier to the transmission case. If Figure 3–20 were rearranged with the internal gear driving and the sun as the output, the result would be reverse overdrive—hardly of any practical value in a transmission.

The gear ratio formula for reverse:

$$\begin{array}{c}\text{Reverse}\\ \text{(carrier held)}\end{array} = \frac{N \text{ driven teeth}}{N \text{ driving teeth}}$$

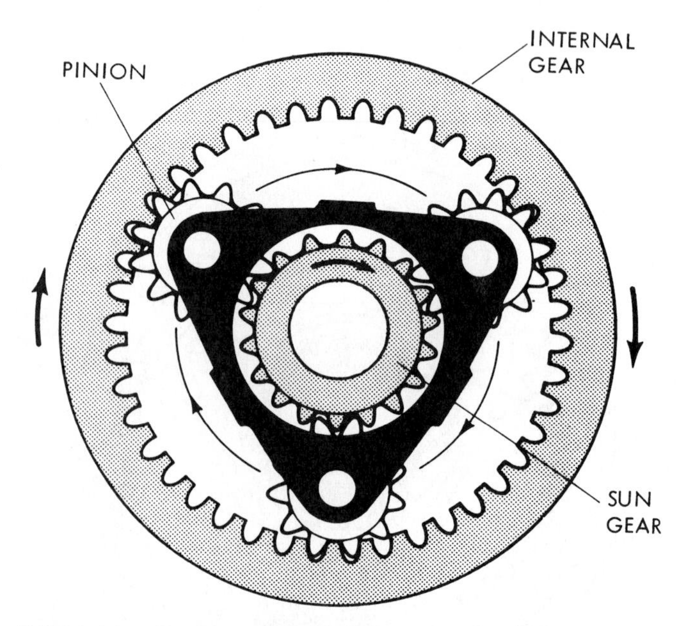

FIGURE 3–18 Planetary gearset in direct drive. Clockwise power applied to the sun gear and the internal gear meets at the planet pinion, preventing the pinions from rotation on their centers. If the pinions cannot rotate on their centers, the planet pinions are trapped solidly between the sun gear and the ring gear and must turn with them at the same relative speed.

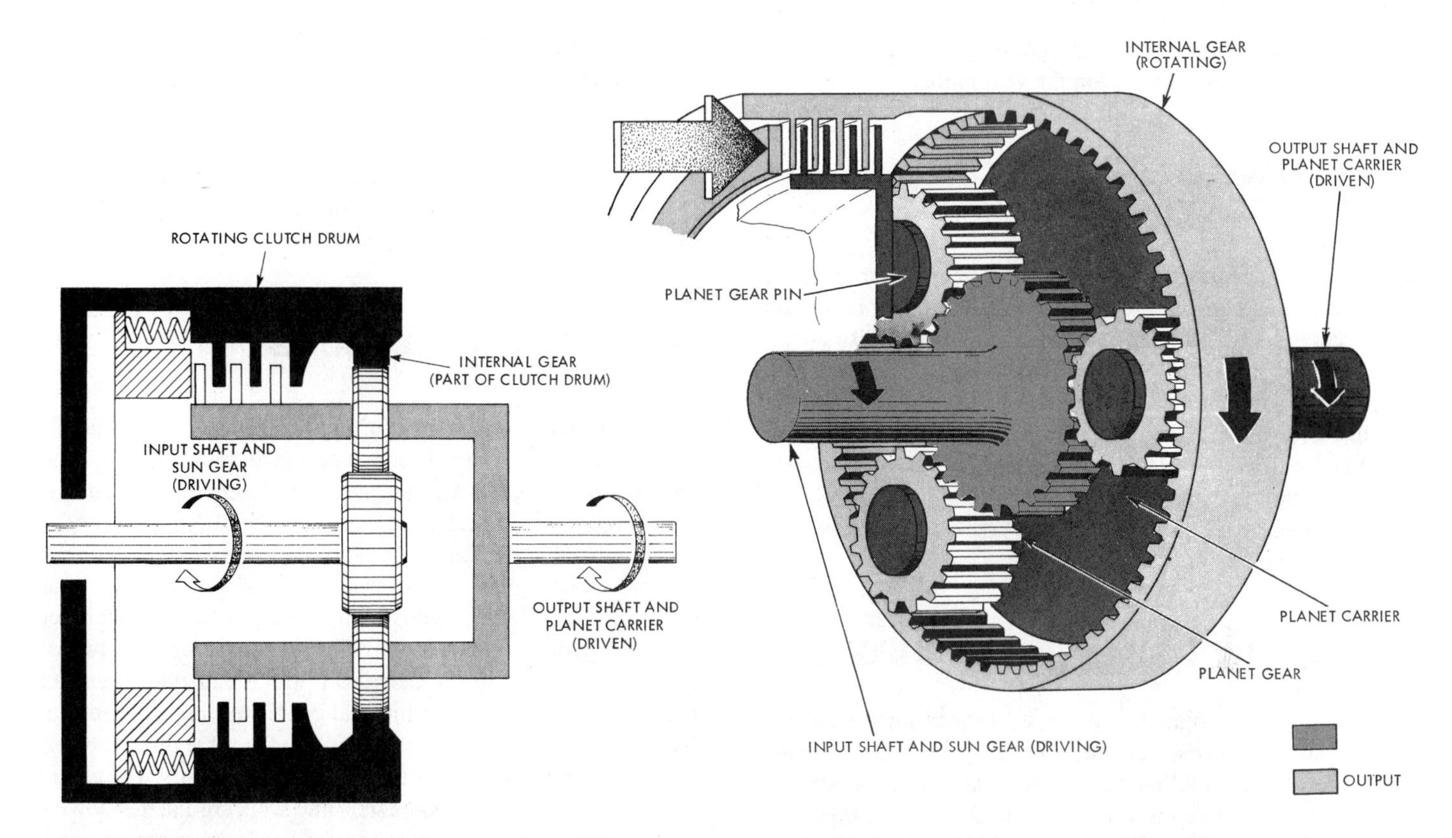

FIGURE 3–19 Multiple-disc clutch is used to lock the internal gear and the planet carrier for a direct-drive effect. The clutch is oil-applied and spring-released.

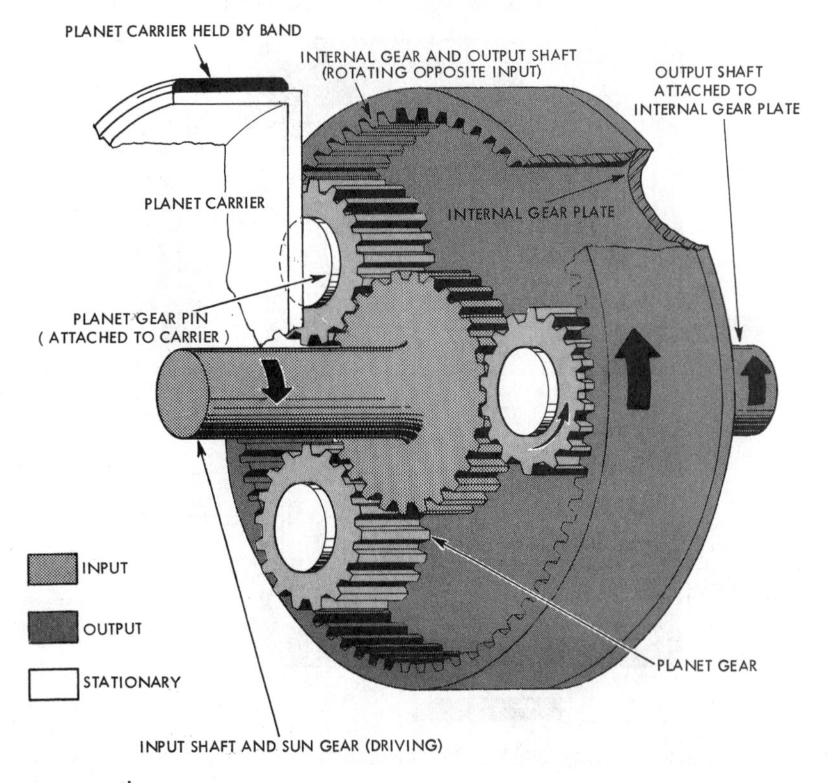

FIGURE 3–20 Typical reverse gear operation.

Since the carrier is stationary, the pinion gears act as reverse idlers only and have no effect on the gear ratio.

COMBINING GEAR REDUCTION AND DIRECT DRIVE

It is possible to use a single planetary gear assembly as a two-speed transmission if the basic requirements for reduction and direct drive are satisfied. A band and multiple-disc clutch are combined in Figure 3–21 to illustrate the possibility. The band is used for holding the internal gear for reduction, and the multiple-disc clutch ties the internal gear and planet carrier together for direct drive. The shift controls in the valve body coordinate the apply and release of these friction units so that they do not resist one another or cause a power flow interruption during a gear change.

PLANETARY DRIVING CLUTCHES

In the discussion of planetary gears, it has been shown how (1) multiple-disc clutches, (2) one-way clutches, and (3) brake bands are used as holding devices for planetary gear control, with no attention given to how the planetary gets its input. To cause an output motion, an input needs to be provided to drive the planetary gearset. This is achieved by using a driving clutch unit that connects the planetary to the fluid converter output. The driving clutch unit used for this function is either (1) a hydraulically controlled multiple-disc clutch, (2) a mechanical one-way clutch, or (3) a multiple-disc clutch and one-way clutch combination. Simplified concepts of these techniques are illustrated in Figures 3–22 to 3–24. One-way drive clutches are designed to lock up with input rotation working against the road load. This differs from holding one-way clutches, which normally lock up opposite input rotation.

PLANETARY VARIATIONS

Planetaries with two sources of input widen their ratio possibilities in power transmission. For discussion purposes, the internal and sun gears are assigned as independent input members (Figures 3–25 and 3–26). Should they be turning at the same rpm and in the same direction, the known outcome on the carrier output is direct drive. If the internal gear and sun gear are turning at different speeds and in the same direction, another behavior pattern is developed. The member with the fastest speed is the direction which the pinions and output carrier follow. In Figure 3–25, the internal gear is turning at a constant fixed speed with the sun gear turning in the same direction, but slower, and having a variable-speed input capability. The planet pinion gears rotate on their centers and walk around the slower-moving sun gear, always rotating the carrier at an rpm exactly halfway between the two input speeds. The planetary torque ratio also changes correspondingly with the speed differences between the input members.

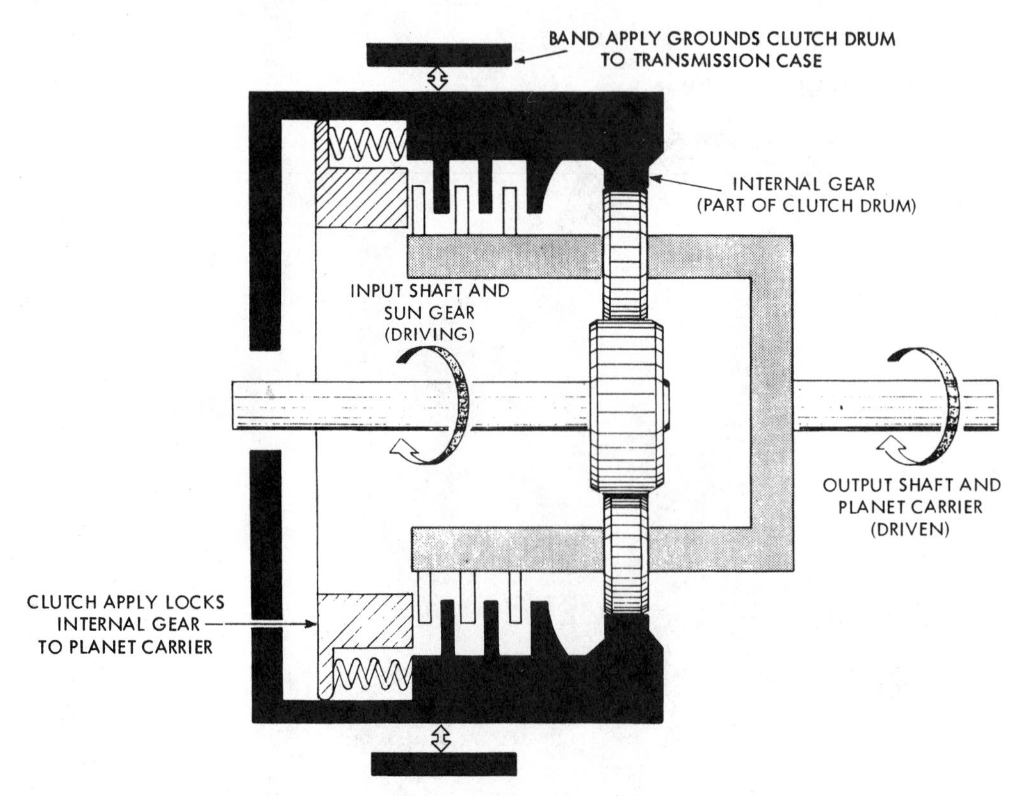

FIGURE 3–21 A multiple-disc clutch is used in a drum assembly to lock the internal gear to the planet carrier for direct drive. The clutch is oil-applied and spring-released. For gear reduction, the band is APPLIED and the clutch is OFF.

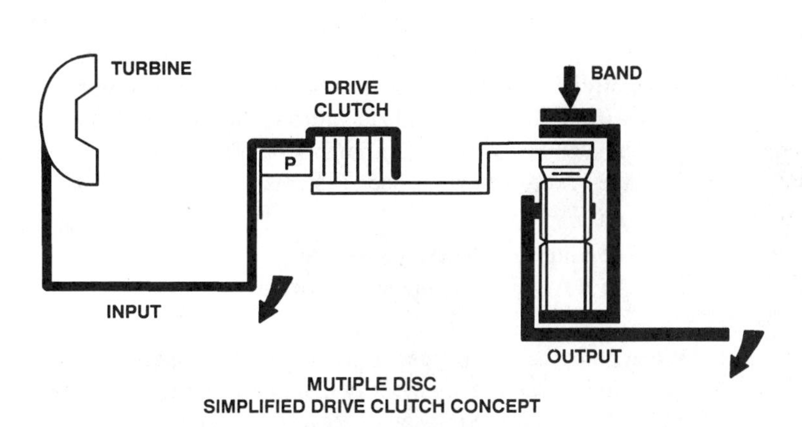

FIGURE 3–22 Multiple-disc simplified drive clutch concept.

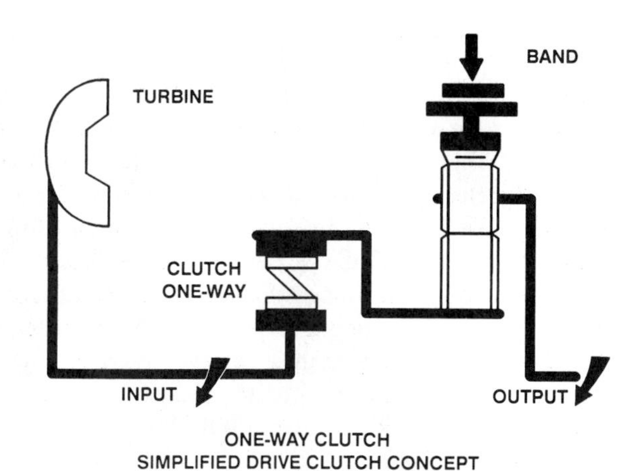

FIGURE 3–23 One-way clutch simplified drive clutch concept.

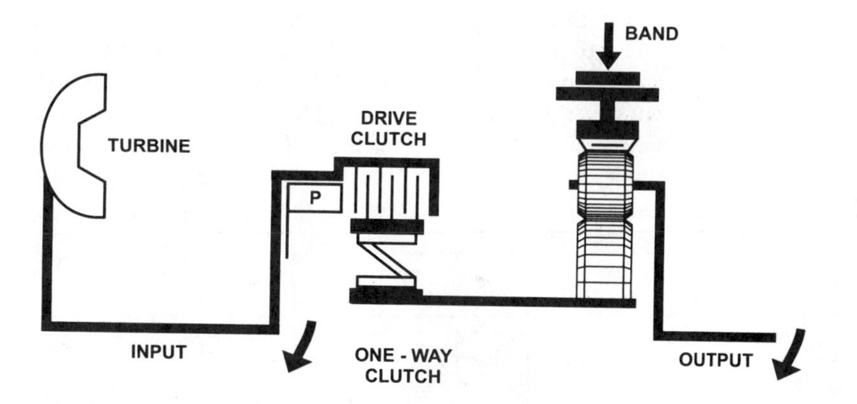

FIGURE 3–24 Multiple-disc and one-way clutch combination simplified concept.

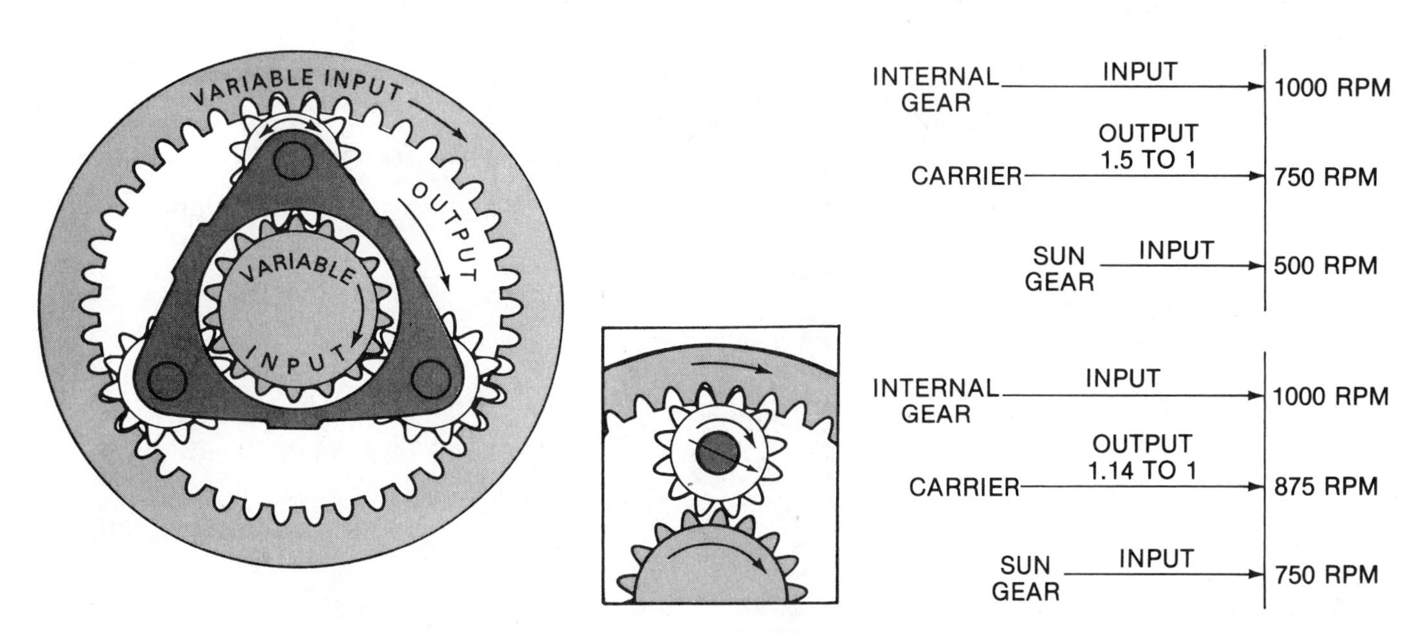

FIGURE 3–25 To achieve a wide range of ratios, a planetary unit can utilize two inputs.

Another variation in using the internal and sun gears as independent inputs features the variable-speed sun gear and constant-speed internal gear turning in opposite directions (Figure 3–26). Tooth speed now becomes the factor that controls planetary behavior. With a slower-moving sun gear and constant, but faster-moving internal gear, the pinions follow the internal gear and walk the carrier in the same direction. As the sun gear tooth speed gains on the internal gear tooth speed, the carrier output slows down. When the tooth speeds become equal, the pinions turn on their centers with no carrier output movement. If the sun gear tooth speed becomes faster, the carrier output direction reverses itself.

PLANETARY SYSTEMS

In the real world, automatic transmission geartrains need a more functional or flexible planetary system than that which can be provided by a single-set planetary. To satisfy engineering design requirements to achieve a three-, four-, or five-speed geartrain, planetary gears and gearsets can be arranged mechanically in a variety of ways. Regardless of how the systems are designed, they all operate according to the basic planetary laws and use the same devices (bands, multiple-disc clutches, and one-way clutches) to control the planetary output.

A wide variety of planetary arrangements are currently in use, and many of these are examined throughout the text—especially in Chapters 4, 22, 23, 24, and 25. At this time, the discussion is limited to a brief introduction to two types of planetary geartrain configurations popularly used worldwide for three-, four-, and five-speed systems:

- The Simpson planetary geartrain
- The Ravigneaux planetary geartrain

When studying other types of three- and four-speed systems, the technician can easily identify that they are simply modified concepts of a basic Simpson or Ravigneaux arrangement.

Simpson Gear System

The Simpson gear system is an arrangement that integrates two single planetary gearsets placed in series (Figure 3–27). For planetary control, it features forward and direct-driving clutches complemented by two holding bands and a holding one-way clutch. The two planetaries share a sun gear and output shaft. The sharing or integration of common planetary elements gives an additional classification of the arrangement, referred to as a **compound planetary**. Another identifying mark of the unit is the drive shell or bell-shaped housing that couples the

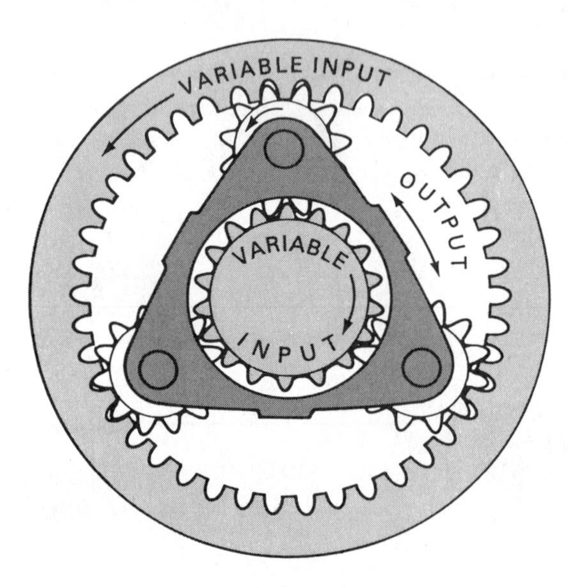

FIGURE 3–26 Two inputs turning in opposite directions can be used to achieve ratio changes and reverse the power output.

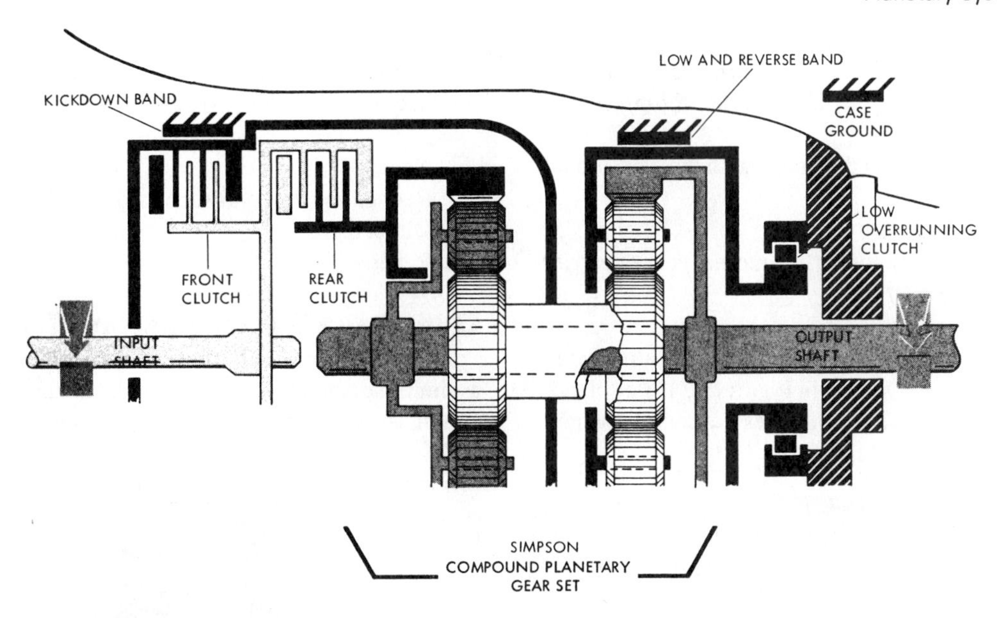

FIGURE 3–27 Typical Simpson planetary system.

sun gear to the direct clutch drum. The Simpson design in its basic configuration produces the following power flows:

- Neutral
- Reduction (first and second gears)
- Direct drive
- Reverse

❖ **Compound planetary:** A gearset that has more than the three elements found in a simple gearset and is constructed by combining members of two planetary gearsets to create additional gear ratio possibilities.

The Simpson system is used for most three-speed automatic transmissions. For a fourth gear, many automatic transmissions use a single overdrive planetary incorporated into the input or output end of the three-speed planetary geartrain.

Ravigneaux Gear System

The Ravigneaux gear system is another example of planetary compounding that has major differences in design compared with the Simpson gear system. The Ravigneaux design includes (Figure 3-28[a]):

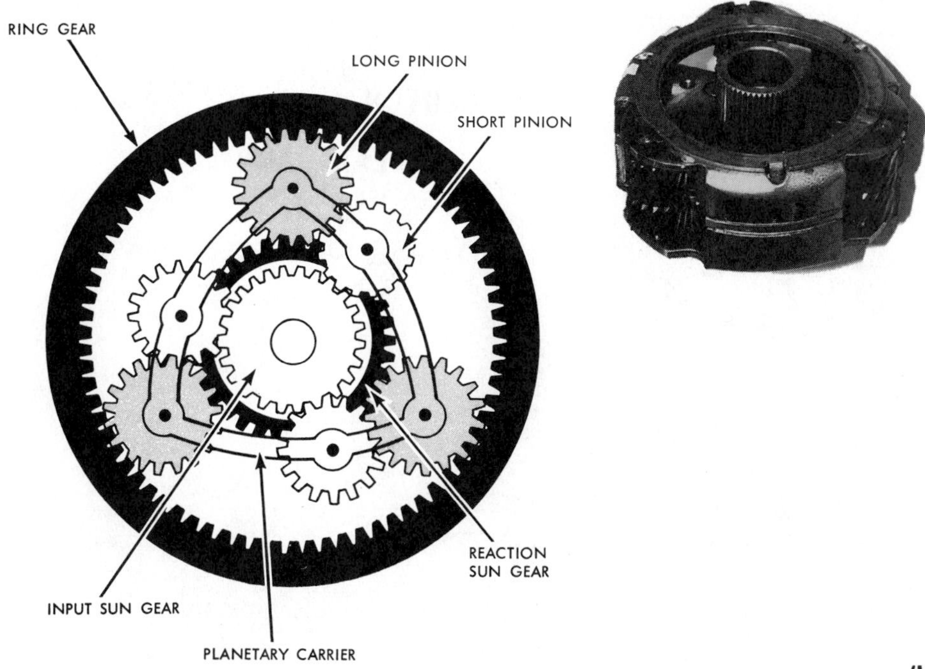

FIGURE 3–28 (a) Ravigneaux planetary unit—rear view; (b) Ravigneaux planet carrier.

- One carrier with three sets of dual pinions. Each dual pinion set is made up of one short and one long pinion (Figure 3-28[b]).
- Two independent sun gears that mesh with the carrier dual pinions.
- One internal gear.
- One output member only, either the internal gear or carrier (Figures 3–29 and 3–30).

A sectional view of a basic Ravigneaux system featuring the internal gear as the output is shown in Figure 3–31. For planetary control, front and rear driving clutches are complemented by two holding bands and a holding one-way clutch. The system produces the following power flows:

- Neutral
- Reduction (first and second gears)
- Direct drive
- Reverse

Overdrive Transmissions/Transaxles

Fourth gear in many overdrive transmissions/transaxles uses a similar planetary gearset layout. Essentially, an overdrive clutch is employed to drive the planetary carrier. Planetary

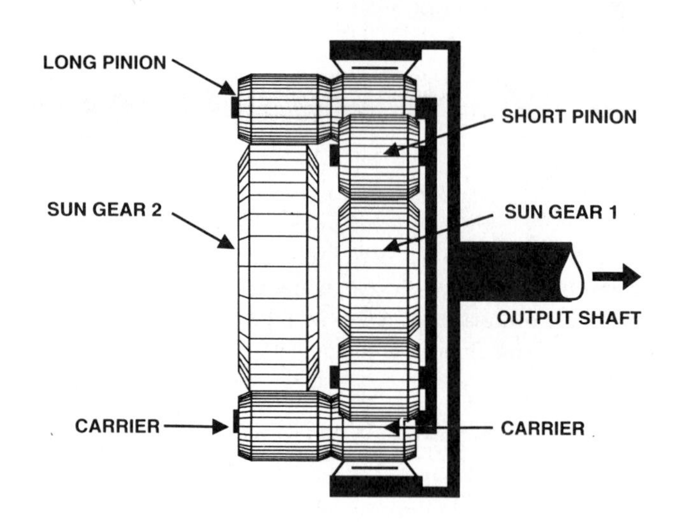

FIGURE 3–29 Simplified Ravigneaux planetary, ring gear→output.

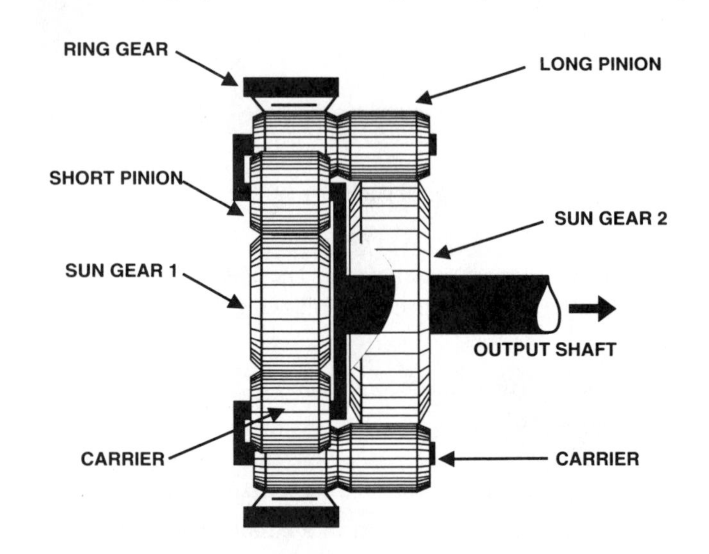

FIGURE 3–30 Simplified Ravigneaux planetary, dual pinion carrier→output.

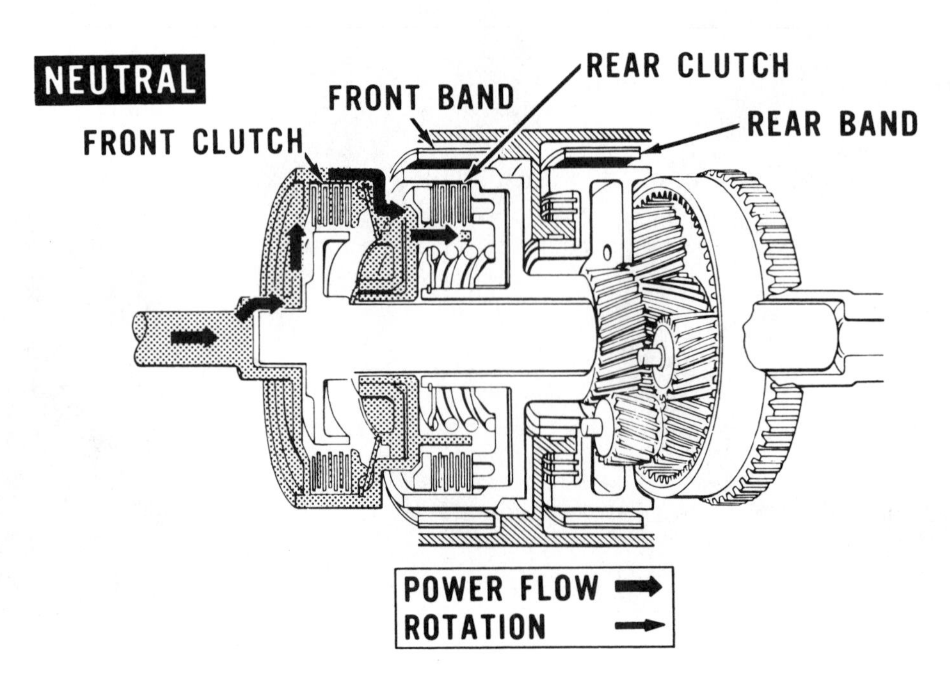

FIGURE 3–31 Basic Ravigneaux planetary system.

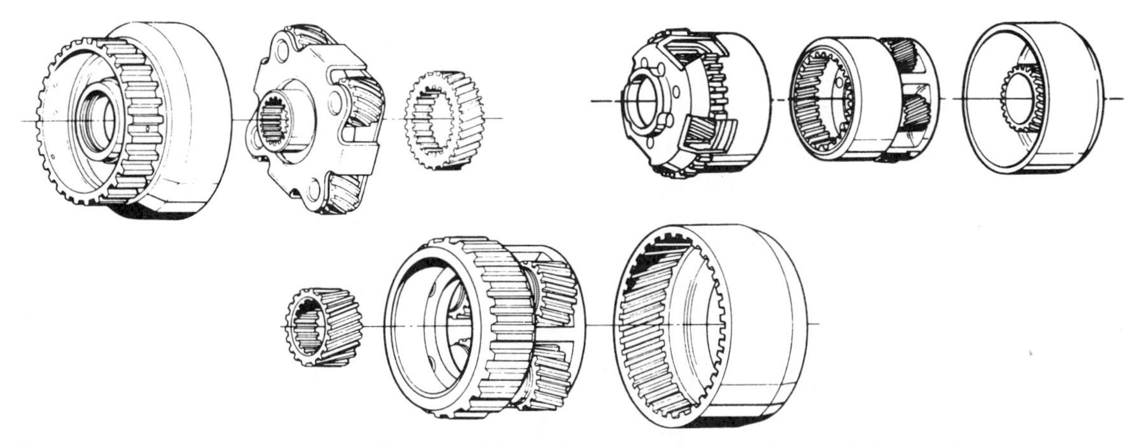

FIGURE 3–32 Representative planetary geartrains. (Courtesy of General Motors Corporation, Service Technology Group)

geartrains come in a variety of designs, but they all work on the same principle of operation and perform the same functions (Figure 3–32).

SUMMARY

A simple planetary gearset is the basic mechanism used in most automatic transmissions to provide a mechanical means to obtain various gear ratios. Its three members—the sun gear, planet pinion carrier, and ring gear—transmit and modify the engine torque and speed. By changing which member is the input device, which is held reactionary, and which is the output, numerous functions are obtainable.

Different mechanisms are used to control the action of a planetary gearset. Clutch packs and one-way clutches drive components. Clutch packs, bands, and one-way clutches can hold planetary members stationary. Usually, output devices are splined directly to output shafts.

One-way clutches can be either of the sprag or roller design. They are torque-sensitive, holding force in one direction and releasing it in the other. This feature contributes to smooth upshifts and is used to provide freewheel action during deceleration.

A variety of different combinations of driving and holding devices are used within transmissions. When applied to either a Simpson or Ravigneaux gearset, the two common planetary configurations, a reverse and three forward gears are present. Regardless of the power flow layout, the laws of planetary gearset operation are always present.

❑ REVIEW

Key Terms

Gear ratio, planetary gearset, reaction member, gear reduction, band, multiple-disc clutch, one-way clutch, sprag clutch, roller clutch, freewheeling, overdrive, direct drive, and compound planetary.

Completion

1. Identify the basic parts makeup of a single-set planetary gear assembly.

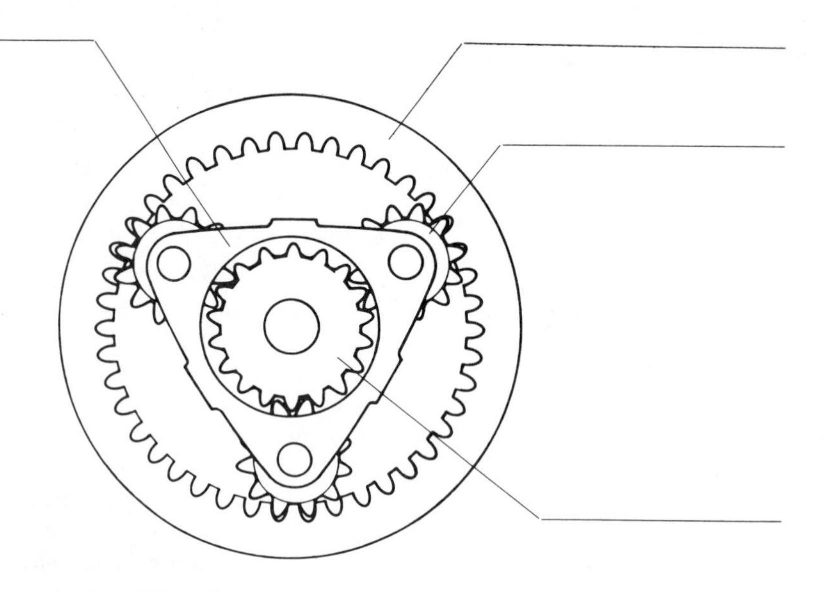

NOMENCLATURE
PLANETARY GEAR ASSEMBLY

2. In the reduction, overdrive, direct, and reverse reduction figures following, indicate with arrows the rotational direction of the planetary members. Color code all planetary members turning at input speed with solid red. Stationary members are shown in solid black. Other planetary members remain plain.

REDUCTION: RING GEAR DRIVING

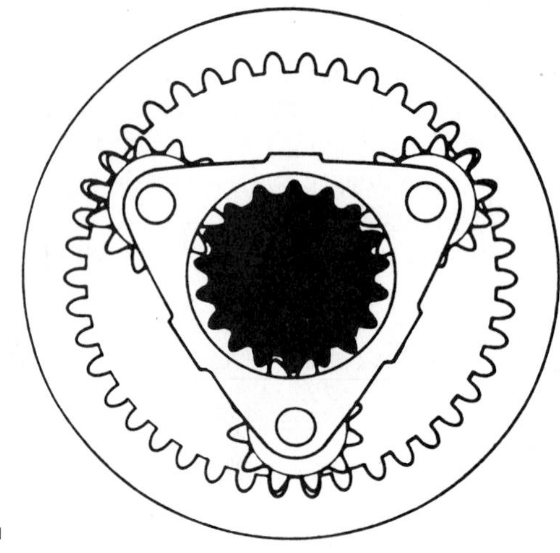

FIGURE 1

OVERDRIVE: SUN GEAR REACTIONARY

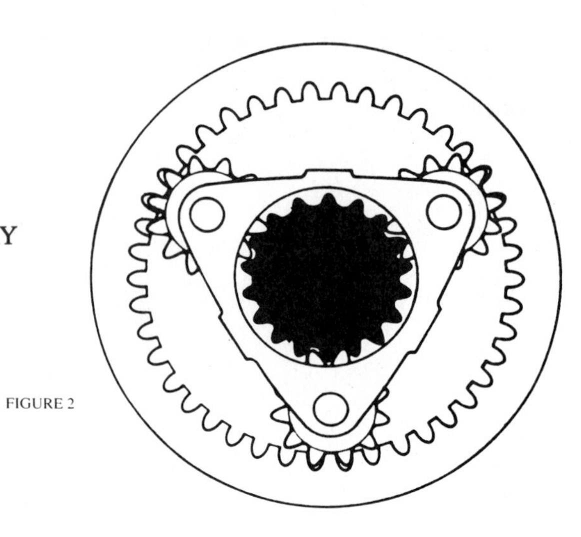

FIGURE 2

DIRECT: SUN GEAR AND RING GEAR LOCKED TOGETHER

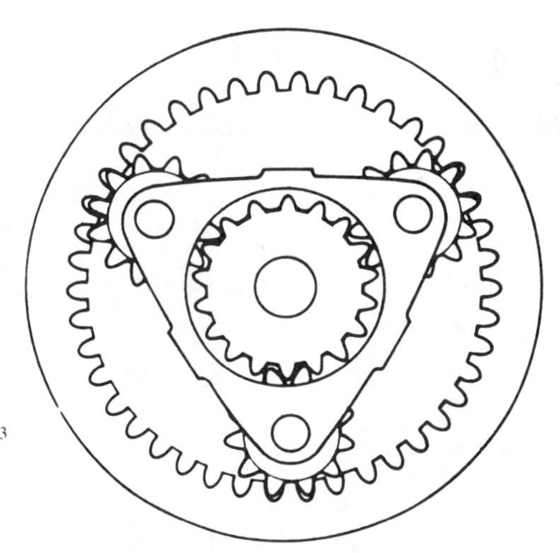

FIGURE 3

REVERSE REDUCTION: SUN GEAR

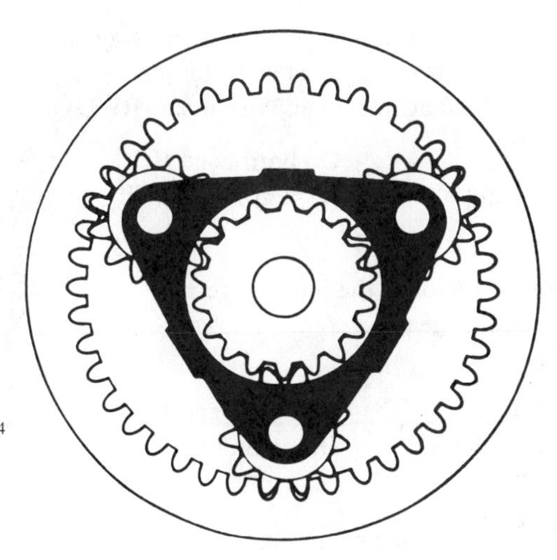

FIGURE 4

3. When there is a reaction member and the planet carrier is the output, the planetary unit is operating in __________ ______________________________.

4. When there is a reaction member and the planet carrier is the input, the planetary unit is operating in __________ ______________________________.

5. Two types of one-way clutch designs used in automatic transmissions:

   (a)______________________________

   (b)______________________________

6. When any two planetary members are clutched together, the resulting effect is ______________________________ ______________________________.

7. Reduction ratios ______________________________ speed and ______________________________ torque.

8. Overdrive ratios ______________________________ speed and ______________________________ torque.

9. The reaction of the rotating planetary carrier pinions to a fixed sun or internal gear causes a pinion gear __________ ______________________________ motion.

10. A servo changes a ______________________________ force into a mechanical force for the purpose of a band apply.

Multiple Choice

____ 1. Used as a device for holding a planetary member:
I. one-way clutch
II. multiple-disc clutch

A. I only C. both I and II
B. II only D. neither I nor II

____ 2. Used as a driving clutch to drive a planetary member:
I. one-way clutch
II. multiple-disc clutch

A. I only C. both I and II
B. II only D. neither I nor II

____ 3. When used as a holding clutch, a one-way clutch:
I. will freewheel against coast torque.
II. will hold against clockwise input (RWD).

A. I only C. both I and II
B. II only D. neither I nor II

____ 4. Another name for the internal gear is:
I. annulus gear.
II. ring gear.

A. I only C. both I and II
B. II only D. neither I nor II

____ 5. The function of the planet carrier pinions in reverse:
I. act as reverse idlers
II. determine the reverse ratio

A. I only C. both I and II
B. II only D. neither I nor II

____ 6. A Ravigneaux planetary system:
I. has one common sun gear.
II. uses a conventional three-pinion carrier.

A. I only C. both I and II
B. II only D. neither I nor II

____ 7. A Ravigneaux planetary system:
I. cannot be adapted for four-speed overdrive.
II. has one output planetary member only.

A. I only C. both I and II
B. II only D. neither I nor II

____ 8. A Simpson planetary system:
I. uses a one-way holding clutch.
II. uses two drive clutches.

A. I only C. both I and II
B. II only D. neither I nor II

____ 9. In a Simpson planetary system:
I. front and rear planetaries have integrated sun gears.
II. front and rear planetaries share a common output gear.

A. I only C. both I and II
B. II only D. neither I nor II

____ 10. Gear ratios change speed plus:
I. power.
II. torque.

A. I only C. both I and II
B. II only D. neither I nor II

____ 11. A Simpson planetary system:
I. The front planet carrier is attached to the output shaft.
II. The rear planet carrier is attached to the output shaft.

A. I only C. both I and II
B. II only D. neither I nor II

____ 12. A Ravigneaux planetary system:
I. can be arranged with the planet carrier as the output.
II. can be arranged with the internal gear as the output.

A. I only C. both I and II
B. II only D. neither I nor II

Planetary Gear Systems in Automatic Transmissions/Transaxles

OBJECTIVE:

The objective of this unit is for you to become proficient with the terms and concepts contained in this chapter. Completion of this objective is essential to becoming a successful automatic transmission technician.

CHALLENGE YOUR KNOWLEDGE

Define the following key terms/acronym:

Range reference and clutch/band apply chart, breakaway, annulus gear, overrun clutch, overdrive planetary gearset, solenoid, and SMEC (single-module engine controller).

List the following:

Three purposes/characteristics of manual range positions; and two locations in which overdrive planetary gearsets are positioned.

Describe the following:

Basic layout of a three-speed Simpson planetary gearset; arrangement and usage of a range reference and clutch/band apply chart; basic layout of a three-speed Ravigneaux planetary gearset; and the operating characteristics of an overdrive planetary gearset.

INTRODUCTION

Chapter 3 served as the basis for understanding the fundamentals of planetary gears: their function, construction, and operation. The objective of this unit is to expand upon those fundamentals and make direct application to modern three-speed transmissions and transaxles and to four-speed automatic transmissions that utilize an overdrive gearset.

Most domestic and import transmissions use either a Simpson or a Ravigneaux planetary gearset configuration. Knowing the basic operation and layout of these two designs, a technician can easily identify the planetary system type and understand how it works. While analyzing the contents of this chapter, study the product variations, use the accompanying charts, and visualize the powerflow within the various transmissions presented.

When troubleshooting an automatic transmission problem that is related to a power flow problem, it is an advantage to know how the geartrain operates in each of the ratios and clutch/band combinations that control these ratios. With this knowledge, noise and friction element failures can logically be isolated to the problem area.

Three-speed automatic transmissions and transaxles were quite popular in earlier decades but have steadily been replaced with four-speed overdrive units. Three-speed RWD transmissions are still being produced, but typically they are used in special applications. A few domestic three-speed transaxles, such as GM's 3T40 and Chrysler's 31TH, are found in some 1996 model year passenger vehicles. Foreign manufacturers also continue to use three-speed transaxles. Toyota's A-131L/A-132L, used in 1996 Corollas and Tercels, is an example. Chapter 4 begins with the study of modern three-speed transmissions and transaxles. The power flow concepts of most three-speed units are similar. Once the fundamentals of operation of these basic units are understood, the operating concepts of four-speed transmissions and transaxles will be easier to grasp.

Four- and five-speed transmissions and transaxles are designed using a variety of different gearset configurations. Chapter 4 examines units that are directly evolved from three-speed layouts. For additional coverage of planetary systems, refer to Chapters 22, 23, 24, and 25 where a wide variety of domestic, European, and Asian transmissions and transaxles are featured. Some interesting variations in the basic Simpson and Ravigneaux systems are illustrated, including mechanical operation and electrical/electronic shift fundamentals.

Problems

1. Calculate the geartrain torque output when:

   Engine torque = 180 lb-ft
   Converter stall torque = 2:1
   Trans gear ratio = 2.5:1
   Final drive ratio = 2.93:1

   Geartrain torque output ________________

2. Calculate the gear ratio, output rpm, and torque output from the following data:

   Gear ratio ________________

   Output torque ________________

   Output rpm ________________

INPUT
2,000 RPM
180 LB-FT
13
OUTPUT
42

3. Calculate output rpm and output torque from the f
   data:

   Output rpm ________________

   Output torque ________________

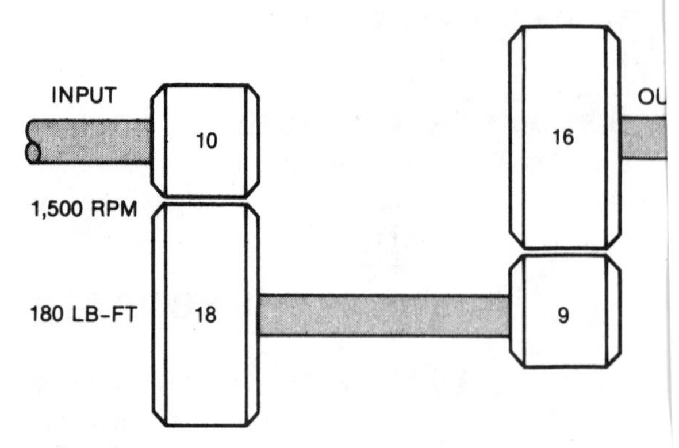

4. Calculate the planetary gear ratio from the followi

   Internal gear = 50 teeth
   Sun gear = 20 teeth

   (a) Gear Reduction/Internal Gear Driving

   Gear ratio ________________

   (b) Reverse Reduction

   Gear ratio ________________

SIMPSON PLANETARY SYSTEM: THREE-SPEED CHRYSLER TORQUEFLITE

Powertrain Assembly

The TorqueFlite transmission combines a three-element torque converter with a fully automatic controlled Simpson planetary gearset system. The basic Simpson design is used worldwide in many three-speed transmissions.

Chrysler first used this geartrain in the TorqueFlite family beginning in 1960, and it has been applied to both RWD and FWD applications (Figures 4–1 and 4–2). Popular three-speed RWD units include the 36RH (A-727), 30RH (A-904), and 32RH (A-999).

In FWD applications, the A-400 series [30TH (A-404), 31TH (A-413/A-670), and A-470] were prevalent in the 1980s and common into the 1990s, due to Chrysler's commitment to FWD passenger car development. The 31TH, a three-speed, light-duty, transverse, hydraulically shifted transaxle, is used in some 1996 Chrysler FWD vehicles.

The three-speed TorqueFlite transmission powertrain consists of the following:

- A three-element fluid torque converter.
- Two multiple-disc driving clutches, referred to as the front and rear clutches.
- Two holding bands with hydraulic servos, referred to as the kickdown (front) band and the low and reverse (rear) band.
- A holding one-way roller clutch for a first gear.
- A compound planetary system made up of two single planetary sets, front and rear in series, integrated by a common sun gear shaft and output shaft.

The planetary design provides neutral, three forward gear ratios, and reverse.

Operating Characteristics

The transmission operating range is programmed by the manual control at the steering column or on-the-floor console selector. The selector quadrant has six positions in the following order: P-R-N-D-2-1.

P Park position locks the transmission output shaft mechanically, preventing the vehicle from rolling forward or backward. An engine start is provided in the park position, and the geartrain is in neutral.

R Reverse has a separate ratio and enables the vehicle to be moved backward.

N Neutral provides for engine start and operation without transferring power to the output shaft. The output shaft is not locked.

D Drive range is used for all normal driving conditions and maximum economy. This range provides three ratios: first gear, second gear, and third gear (direct drive). The trans-

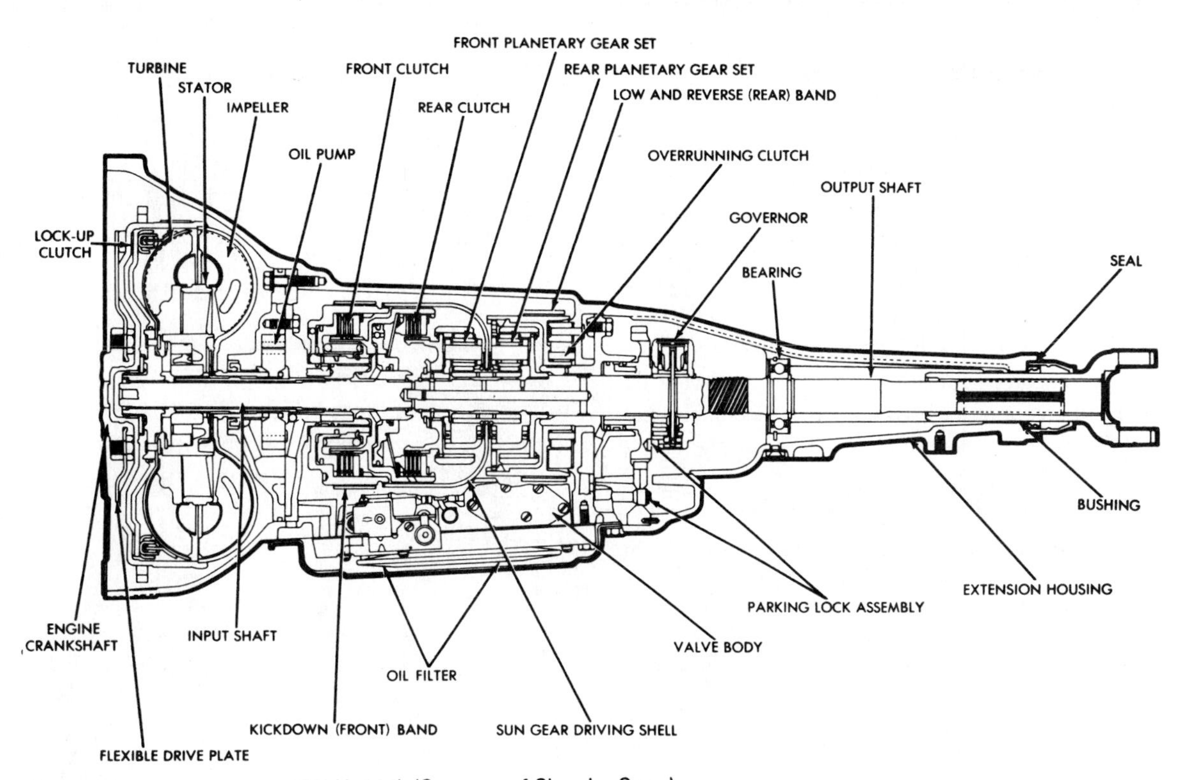

FIGURE 4–1 TorqueFlite transmission: 30RH (A-904). (Courtesy of Chrysler Corp.)

mission starts in first gear and automatically shifts to second and then to direct. Shift points are determined by vehicle speed and throttle position. If added performance is desired, the transmission can be forced to downshift 3-2, 3-1, or 2-1 by depressing the accelerator to wide-open throttle (WOT). The WOT downshift depends on the vehicle speed range and available crankshaft torque for acceleration. For controlled acceleration at lower vehicle speeds, a part throttle 3-2 downshift is available. On closed throttle downshifts, the transmission shifts 3-1, skipping the band apply and taking advantage of the smooth freewheeling action from the overrunning roller clutch during coast (deceleration).

2 This position is used to operate the transmission in the first two gears only. Third gear is inhibited. The 1-2 automatic shift occurs in the same manner as in D. This range is suitable for heavy city traffic, when the driver may desire to use part-throttle second-range operation for more precise control. It may also be used on long downgrades, when additional engine braking is needed, or for extra pulling power.

1 Manual low range keeps the transmission in first gear only. It provides engine braking for handling ease in mountainous terrains and on steep grades. If low range is selected at an excessive vehicle speed, the transmission shifts to second gear and remains there until the vehicle is slowed to a safe speed range. Once low gear is engaged, the transmission does not upshift.

Planetary Power Flow

Before proceeding with the details of operation, some basic facts need to be mentioned as a background for understanding the planetary system power flow. The power input to the transmission comes from the engine crankshaft torque through the

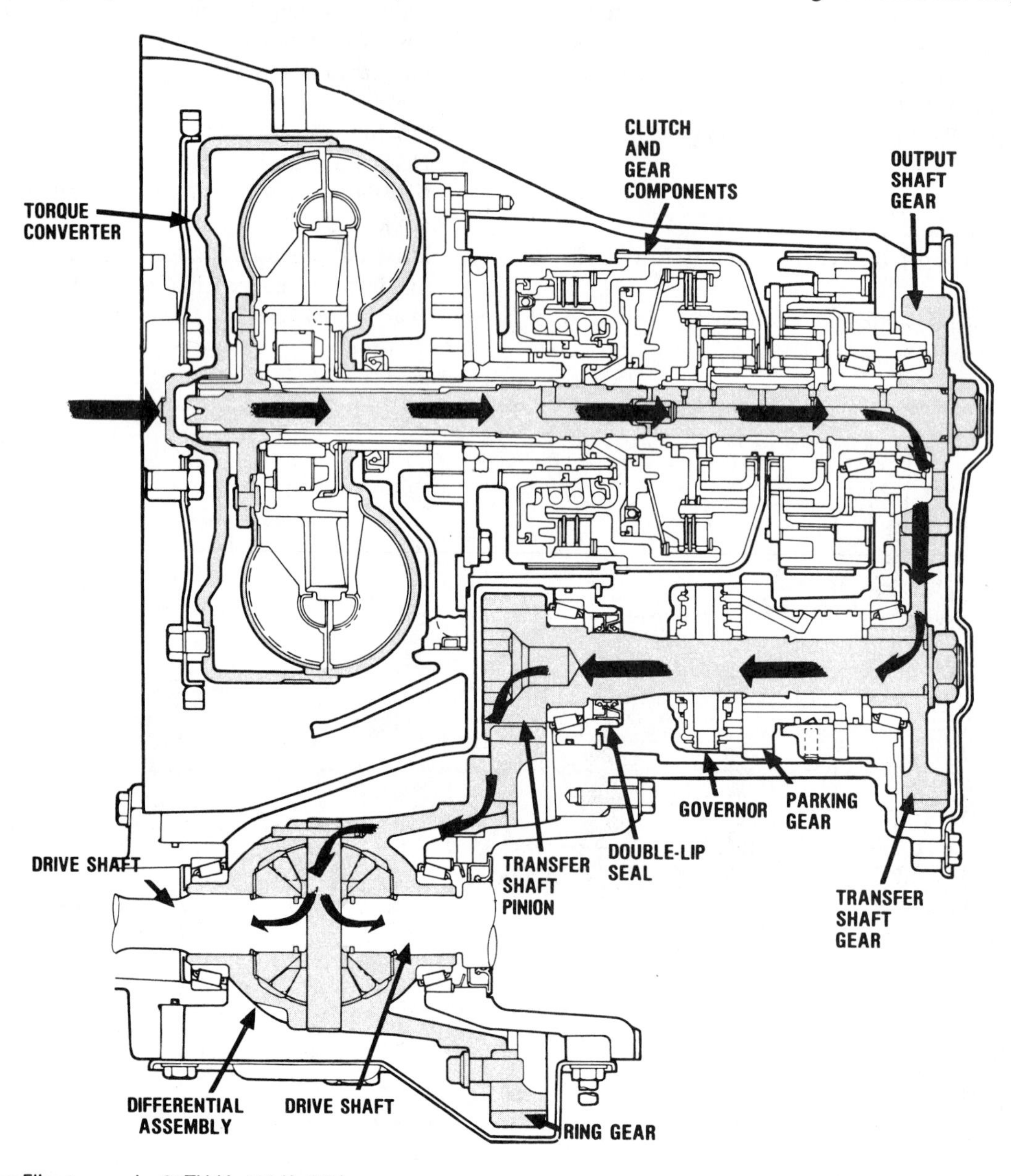

FIGURE 4–2 TorqueFlite transaxle: 31TH (A-413/A-670).

converter to the transmission input shaft and the input drive clutch units in the transmission. The power flow through the planetary geartrain depends on the clutch/band combination for each of the gear ratios. Refer to the **range reference and clutch/band apply chart** (Chart 4–1).

❖ **Range reference and clutch/band apply chart:** A guide that shows the application of a transmission's/transaxle's clutches and bands for each gear, within the selector range positions. These charts are extremely useful for understanding how the unit operates and for diagnosing malfunctions.

NOTE: Of special concern to the reader is that all reference to direction of rotation in describing the gearing power flows is based on facing the planetary input, front to rear.

Neutral. In neutral (N), none of the clutches are engaged or bands applied. There is a power feed to the input shaft from the converter turbine. Since the front and rear input drive clutches are disengaged, no power is transmitted to the planetary gearset and output shaft. The power flow dead-ends at the clutches.

First (Breakaway). With the selector in either D or 2 range, the rear clutch is engaged and the overrunning clutch holds (Figures 4–3 and 4–4). The rear clutch can also be referred to as the forward clutch, since it is an input drive to the planetary system in all forward gears. The torque flows from the converter turbine through the input shaft to the rear clutch. With the rear clutch engaged, the torque flow is able to drive the **annulus gear** of the front planetary in a clockwise direction. Because the front planet carrier is splined to the output shaft and the weight of the car, the annulus gear drives the front planetary pinions, rotating them in the same direction. The pinions, in mesh with the sun gear, rotate the sun gear counterclockwise (Figures 4–5 and 4–6).

❖ **Breakaway:** Often used by Chrysler to identify first-gear operation in D and 2 ranges. In these ranges, first-gear operation depends on a one-way roller clutch that holds on acceleration and releases (breaks away) on deceleration, resulting in a freewheeling coastdown condition.

❖ **Annulus gear:** Preferred by Chrysler to describe the internal (ring) gear.

Lever Position	Start Safety	Parking Sprag	Clutches			Bands	
			Front	Rear	Over-Running	(Kickdown) Front	(Low-Rev.) Rear
P—Park	X	X					
R—Reverse			X				X
N—Neutral	X						
D—Drive							
First				X	X		
Second				X		X	
Direct			X	X			
2—Second							
First				X	X		
Second				X		X	
1—Low (First)				X	X		X

Transmission	Model Year	Gear Ratio			
		1 or Drive Breakaway	2 or Drive Second	Direct	Reverse
Auto Transaxle (FWD)	78–80 81–Current	2.48:1 2.69:1	1.48:1 1.55:1	1:1 1:1	2.10:1 2.10:1
904 Family (RWD)	60–79 *80–Current	2.45:1 2.74:1	1.45:1 1.54:1	1:1 1:1	2.22:1 2.22:1
A727 Family (RWD)	62–Current No Change	2.45:1	1.45:1	1:1	2.22:1

*Some 80 & 81 model year A904 family transmissions have the previous style gear ratios

CHART 4–1 TorqueFlite RWD/FWD Series Range Reference and Clutch/Band Apply Summary.

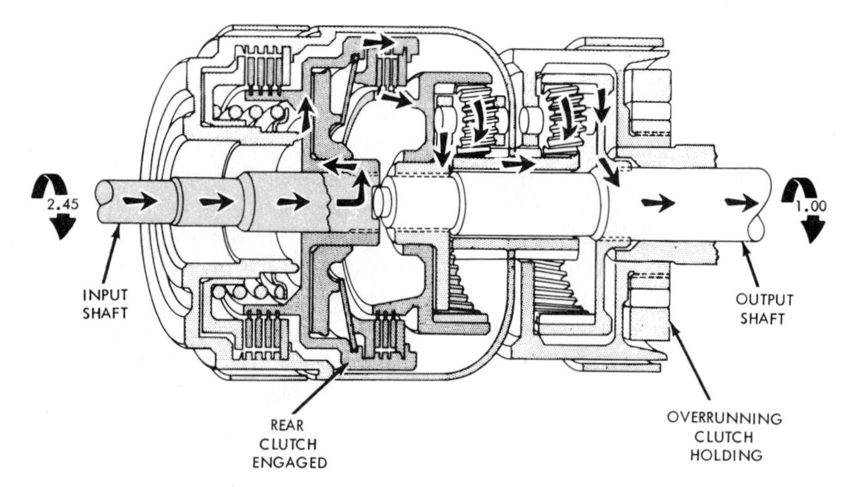

FIGURE 4–3 Power flow in breakaway. (Courtesy of Chrysler Corp.)

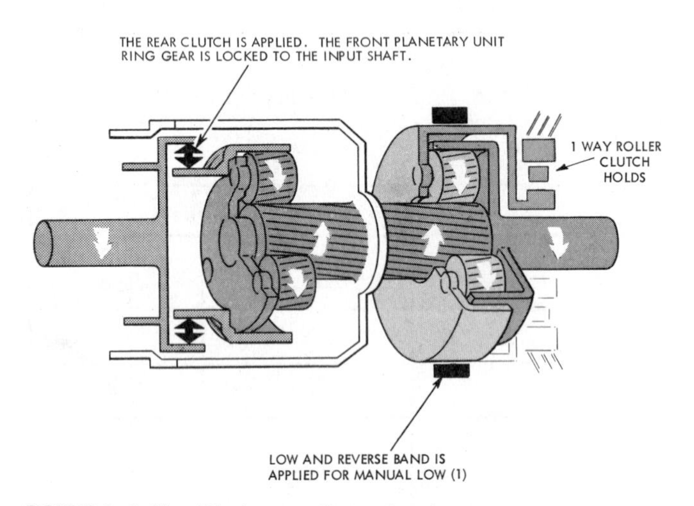

FIGURE 4–4 Simplified power flow—breakaway.

The front and rear planetary sets share the same sun gear shaft. Therefore, the sun gear turns the rear planet pinions in clockwise rotation. The planet pinions, in mesh with the rear annulus gear, turn the annulus gear and output shaft in a clockwise direction. The rear planet carrier is held by the overrunning clutch because of the action of the pinions as they rotate the rear annulus gear. The pinions must rotate the annulus gear against the weight of the car, which makes the annulus gear reactionary. This results in the pinions attempting to walk the rear carrier counterclockwise, causing a lockup of the roller clutch (Figures 4–4 and 4–6).

Lockup of the **overrun clutch** (one-way clutch) to the rear planetary carrier is effective as long as the engine is driving the planetary system and the vehicle load. On coast, with the rear wheels and the weight of the car working against the engine compression, the output shaft drives the planetary system. The resulting planetary action causes the rear carrier to turn clockwise and release the roller clutch. This results in a neutral gear condition. Engine braking is not possible during coast breakaway in the D and 2 ranges.

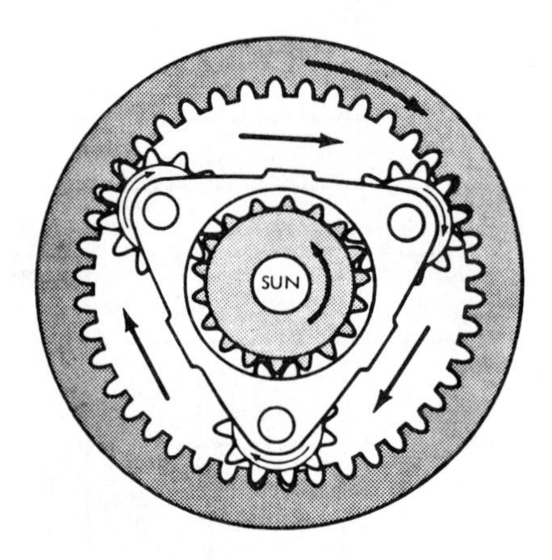

FIGURE 4–5 Front planetary breakaway. The carrier rotates with the output shaft once the gearing action begins, thus the annulus gear and the planet carrier are inputs to the front planetary. This results in a reverse overdrive effect on the sun gear, which is the input to the rear planetary.

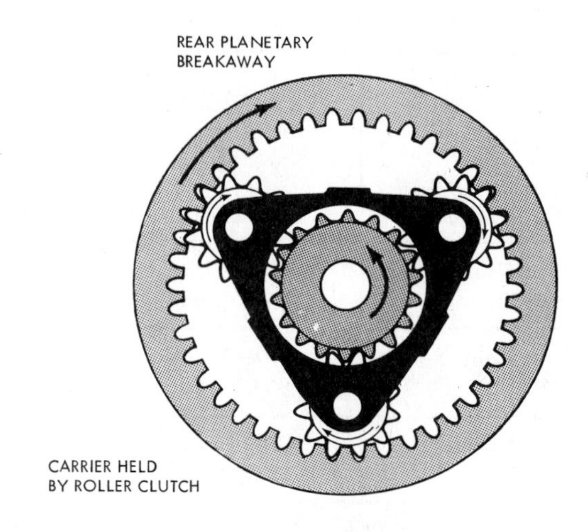

FIGURE 4–6 Rear planetary breakaway, with carrier held by the roller clutch.

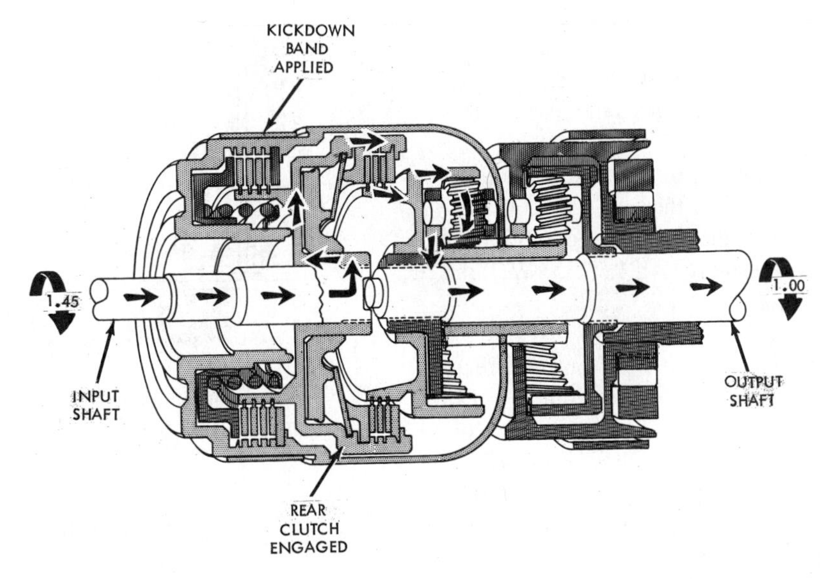

FIGURE 4–7 Power flow in second gear. (Courtesy of Chrysler Corp.)

- **Overrun clutch:** Another name for a one-way mechanical clutch. Applies to both roller and sprag designs.

In summary, first gear is a result of the compound gearing action of both the front and rear planetary gearsets working together. The effective compound torque ratio is 2.48/2.96:1 (FWD) or 2.45/2.74:1 (RWD).

Second. With the selector in D or 2, the rear clutch remains engaged while the kickdown band is applied to hold the driving shell and sun gear stationary (Figures 4–7 to 4–9). With the rear clutch engaged and driving, the front annulus gear and the front planetary pinions rotate in a clockwise direction. The rotating planet pinions react on the stationary sun gear and walk the front planetary carrier and output shaft at a reduced speed and increased torque (Figure 4–10).

The rear planetary gearset in second gear goes along for the ride. The rear annulus gear attached to the output shaft drives

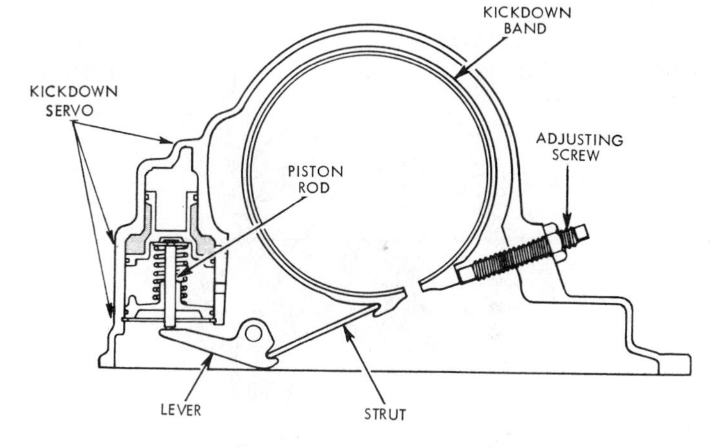

FIGURE 4–9 TorqueFlite servo and linkage mechanism is used for applying the kickdown around the clutch drum.

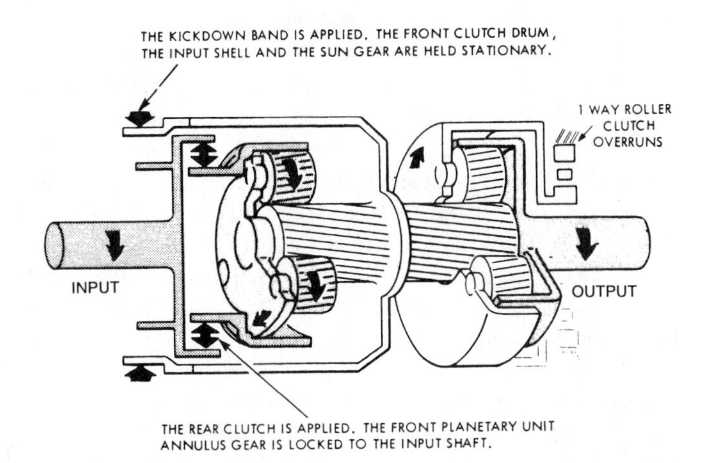

FIGURE 4–8 Simplified power flow—second.

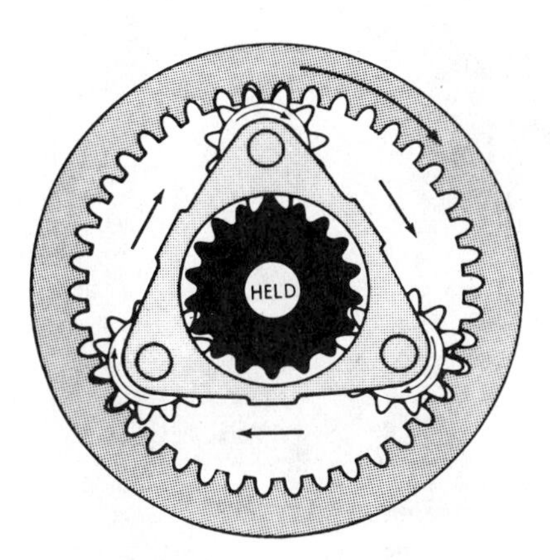

FIGURE 4–10 The front planetary carrier is splined to the output shaft and turns at a reduced speed. The sun gear is held by the kickdown band.

the rear planet pinions in a clockwise direction. These pinions react on the stationary sun gear and walk the carrier in a clockwise direction. The overrun clutch allows the rear carrier to freewheel (Figure 4–11).

In summary, second gear is a result of a simple reduction action from the front planetary. When there is a reactionary member and the planetary carrier is the output, the condition is gear reduction. The torque ratio result is 1.48/1.55:1 (FWD) or 1.45/1.54:1 (RWD).

Direct (Third). Direct (D) occurs when the front and rear clutches are engaged and both bands are released (Figures 4–12 and 4–13). The front clutch is referred to as the direct clutch in many Simpson power flows. With both drive clutches engaged, the torque from the input shaft takes a dual path—one through the front clutch to the drive shell and sun gear, and the other through the rear clutch to the front planetary annulus gear. The sun gear and annulus are essentially locked to each other and to the input shaft. This prevents the planetary pinions from rotating and locks the front planetary to the output shaft at a 1:1 ratio.

In direct drive, the rear sun gear (driven by the front clutch) and rear annulus gear (connected to the output shaft) each turn at input shaft speed. The rear planetary carrier, therefore, is forced to turn at the same speed and in the same rotation as the other planetary members. The overrun clutch continues to allow the rear planetary carrier to freewheel.

In summary, the planetary geartrain behaves in accordance with the law of direct drive. When two members of a planetary gearset are clutched together or driving at the same speed, the pinion gears are trapped and kept from rotating on their centers. This action results in a direct drive and is the method used in most automatic transmissions for third gear.

Direct drive locks up the drive clutches and planetary gears so that all motion within the powertrain is essentially at the same speed. When troubleshooting noise problems that are audible in direct drive or high gear, it is not likely that the planetary gears or clutch assemblies are at fault. It is interesting to note that neither the front nor rear clutch handles the full input torque to the planetary set. Because of the dual-path input, the torque to the planetary set is divided between the clutch units. The holding power of the front and rear clutch units is more than adequate for high-gear operation.

Manual low. In manual low (1), the rear clutch is engaged and the low and reverse band applied (Figures 4–14 and 4–15). A comparison of Figures 4–3 (page 70) and 4–14 shows that the power flow in first gear and manual low is the same. Although the

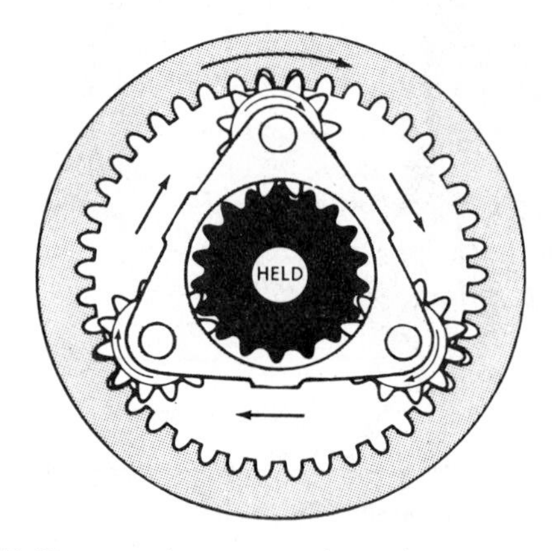

FIGURE 4–11 The rear planetary carrier and overrun clutch freewheel, and the planetary is ineffective. The sun gear, also common to the front planetary, is held by the kickdown band.

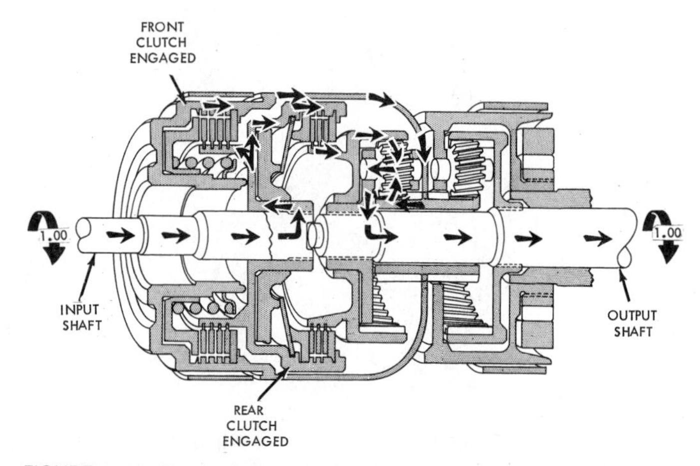

FIGURE 4–12 Power flow in D (drive) range—third gear, direct drive. (Courtesy of Chrysler Corp.)

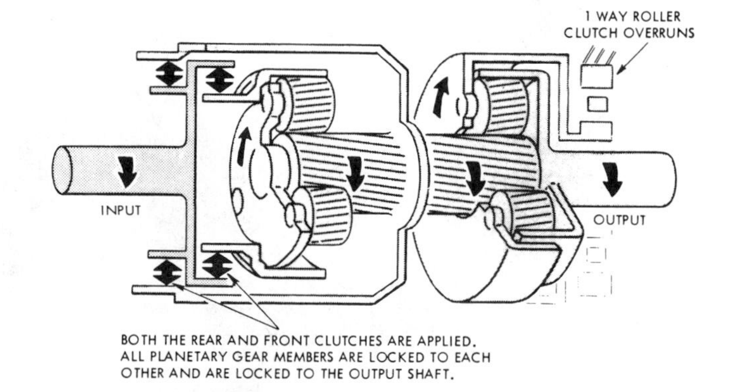

FIGURE 4–13 Simplified power flow in D (drive) range—third gear, direct drive. (Courtesy of Chrysler Corp.)

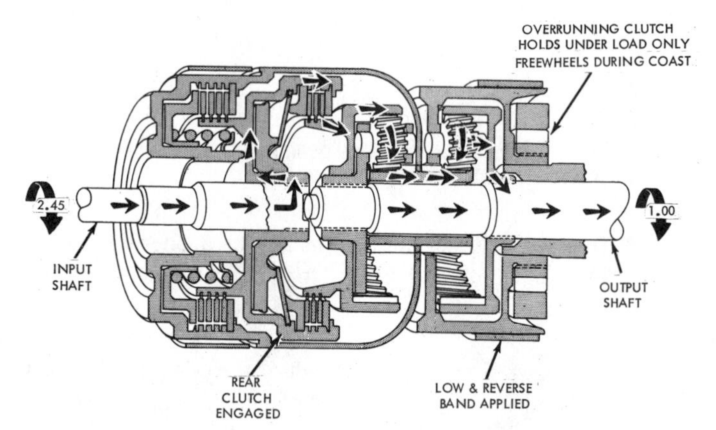

FIGURE 4–14 Power flow in (1) manual low—first gear. (Courtesy of Chrysler Corp.)

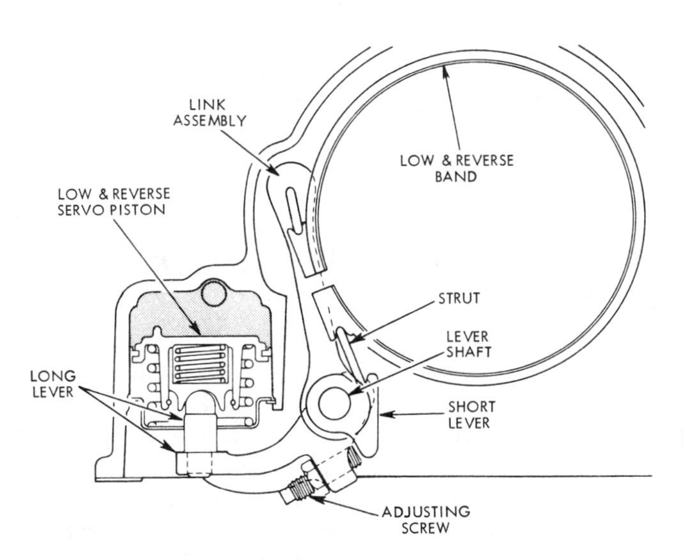

FIGURE 4–15 TorqueFlite servo and linkage used for applying the low/reverse band around the low/reverse drum, which is part of the rear planet carrier.

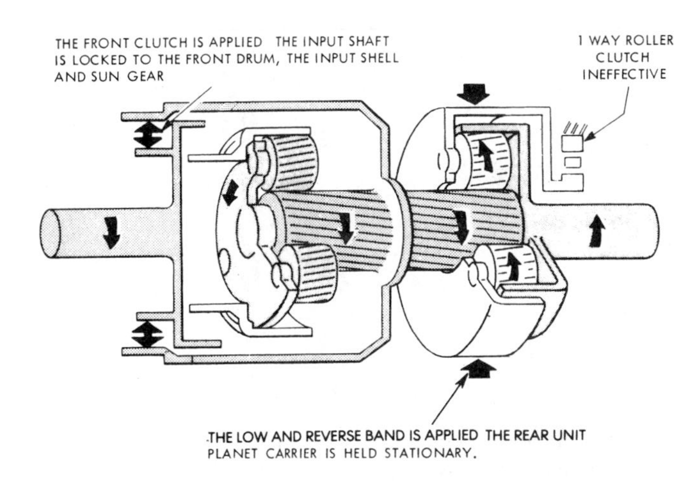

FIGURE 4–17 Simplified power flow—reverse position.

low-reverse band is applied, it is not effective when manual low is used for sustained driving power. The roller clutch is torque-sensitive and responds prior to the application of the rear band. The band is applied and holds the rear carrier during a coast/deceleration condition. The overrun clutch is unable to hold the rear planetary carrier and is considered to be ineffective during coast/deceleration. Due to the rear band being applied, the planetary gear unit remains effective for engine braking during deceleration or when descending steep grades.

As previously mentioned, on a high-speed drop to manual low range, second gear engages. Its operation is the same as the second gears in D and 2 ranges.

Reverse. In reverse (R), the front clutch is engaged to the drive shell and the sun gear. The low and reverse drum and rear planetary carrier are held stationary by the low-reverse band (Figures 4–16 and 4–17). Engagement of the front clutch drives the shell and sun gear in a clockwise direction. With the rear carrier held stationary, the sun gear drives the rear planet pinions counterclockwise. This reversing motion of the planet pinions also drives the rear annulus gear (splined to the output shaft) in the reverse direction at a reduced speed (Figure 4–18).

During reverse operation, the front planetary gears are active but do not enter into the power flow picture. The rear clutch is OFF, and the powerflow dead-ends (Figure 4–19).

In summary, reverse is a result of a simple reduction action from the rear planetary gearset with the pinion gears acting as reverse idlers. When the planetary carrier is held and the sun gear is the input member, a reverse reduction results—2.10:1 (FWD) and 2.22:1 (RWD).

Park. In park (P), the transmission geartrain remains in neutral. A manually activated linkage engages a spring-loaded pawl in the parking lock gear assembly. It is splined to the output shaft to lock the drive wheels to the transmission case as shown in Figure 4–20.

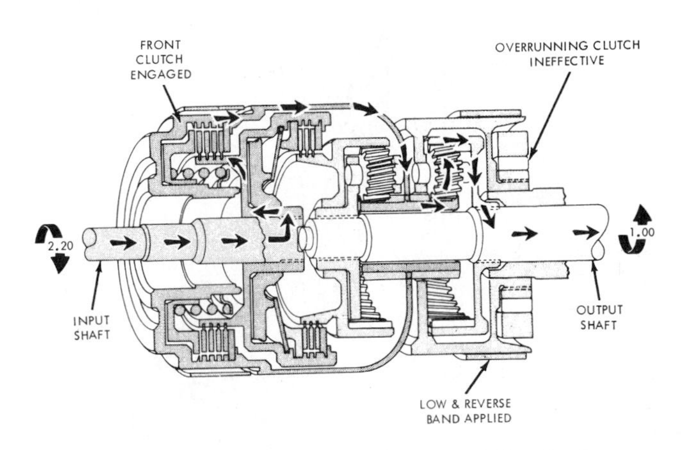

FIGURE 4–16 Power flow in reverse. (Courtesy of Chrysler Corp.)

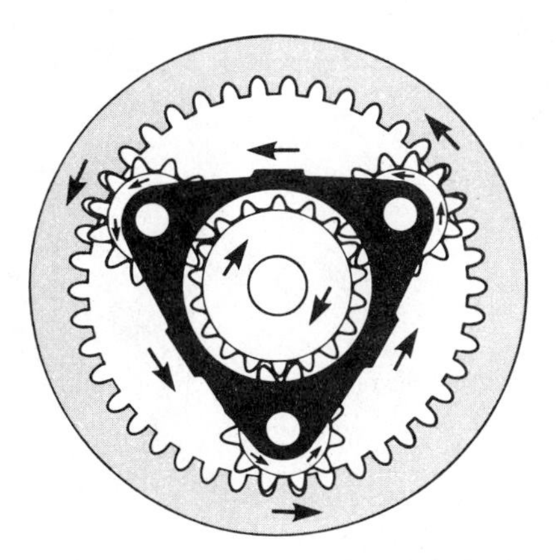

FIGURE 4–18 Reverse—rear planet carrier held by low-reverse band. Planet pinions reverse the motion of the sun gear and drive the rear annulus gear and output shaft in reverse.

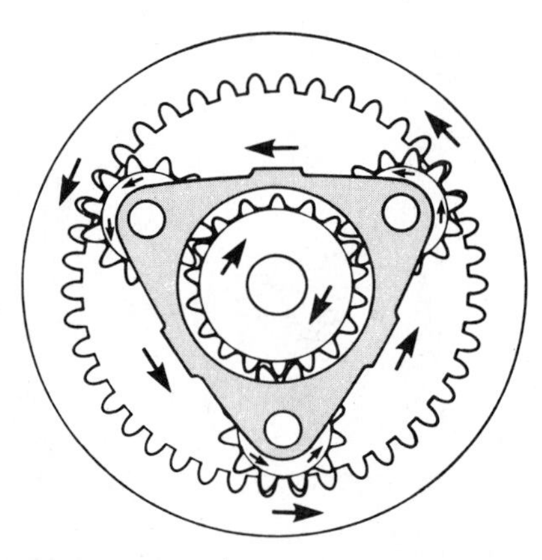

FIGURE 4–19 Reverse—front planetary action. Power flow dead-ends at the annulus gear, which is rotating at a reverse overdrive speed. The rear clutch is off.

Simpson Low-Gear Formula (Compound Reduction)

$$\text{First Gear Compound Ratio} = \left[\frac{I_2}{S_2}\right]\left[\frac{\left(\frac{S_2}{I_2}\right)(I_2 + S_2) + S_1}{I_1}\right]$$

where, for Planetary Set 1:

I_1 = number of teeth on internal gear
S_1 = number of teeth on sun gear

and for Planetary Set 2:

I_2 = number of teeth on internal gear
S_2 = number of teeth on sun gear

This formula is accurate if the carrier of Set 1 is connected to the internal gear of Set 2, with the sun gears coupled, and the Set 2 carrier is held reactionary. This formula works even if the planetary gearsets are not identical.

SIMPSON PLANETARY SYSTEM: THREE-SPEED FORD

The Ford family of automatic transmissions utilizing the traditional Simpson three-speed planetary dates back to 1964 with the introduction of the C-4 transmission for cars and trucks (1964–1982). In the following years, the Simpson gearset was incorporated in several Ford transmission developments. It was introduced as the C-6 for cars (1966–1980), C-6 for Ford trucks (1966–current), JATCO for cars (1977–1979), JATCO for Courier (1974–current), C-3 (1974–1986), and the C-5 for cars (1982–1985). These are all identified with the Ford Cruise-O-Matic group of automatic transmissions and used in RWD applications.

Powertrain Assembly

The Cruise-O-Matic geartrains are essentially identical in operation and layout, with one major difference in the clutch/band control. For manual low and reverse, either a low-reverse band or a multiple-disc clutch is used.

The C-6 transmission is highlighted in this section and shown in Figure 4–21. The transmission powertrain, comparable with the Chrysler TorqueFlite, consists of the following:

- A three-element fluid torque converter.
- Two multiple-disc driving clutches, referred to as the reverse-high and forward clutches.

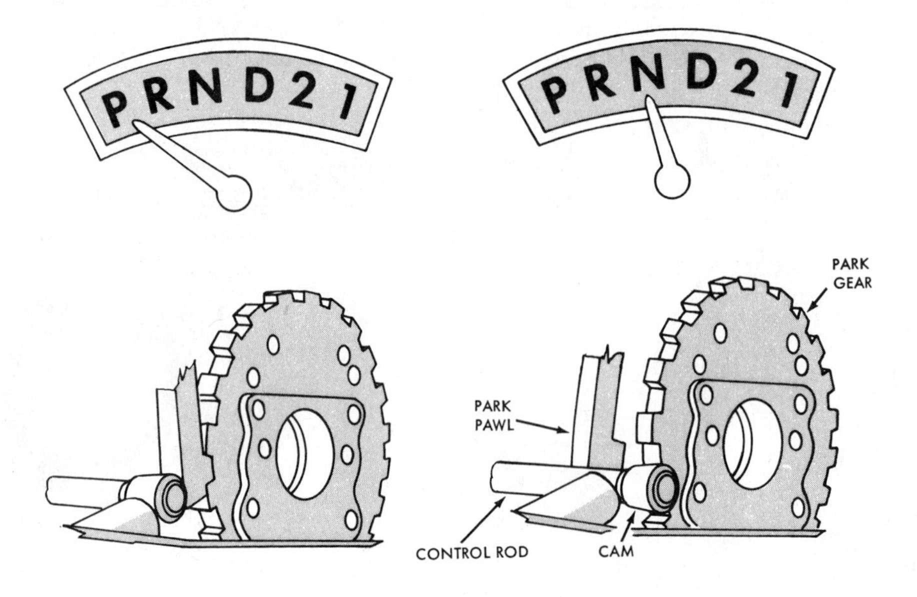

FIGURE 4–20 Engine running neutral and park. (Courtesy of Chrysler Corp.)

- One holding band with a hydraulic servo, used for second gear and referred to as the intermediate band.
- A multi-disc clutch for holding the rear planetary carrier in reverse and manual low ranges.
- A holding one-way roller clutch for first gear.
- A compound planetary system made up of two single planetary sets, front and rear in series, integrated by a common sun gear shaft and output shaft.

The planetary design is a typical Simpson setup providing neutral, three forward gear ratios, and reverse.

Operating Characteristics

A standard Cruise-O-Matic shift selector with six positions is used as follows:

P Park. The geartrain is in neutral and the output shaft is mechanically locked to the case by an internal locking pawl and parking gear. An engine start is provided.

R Reverse. A single reduction ratio provides the transmission with a reverse mode.

N Neutral. The geartrain is in neutral and the output shaft is not locked. An engine start is provided.

D Drive. This is the normal driving range for economy. A first-gear start is provided, with automatic shifts to intermediate (second) and high (direct) determined by road speed and throttle opening. Forced downshifts 3-2, 3-1, and 2-1 are available at WOT kickdown, depending on the vehicle speed and engine crankshaft torque reverse. A 3-2 part-throttle downshift provides for controlled acceleration at lower speeds. At closed throttle, the transmission downshifts 3-1 at about 10 mph (16 km/h), except the C-3

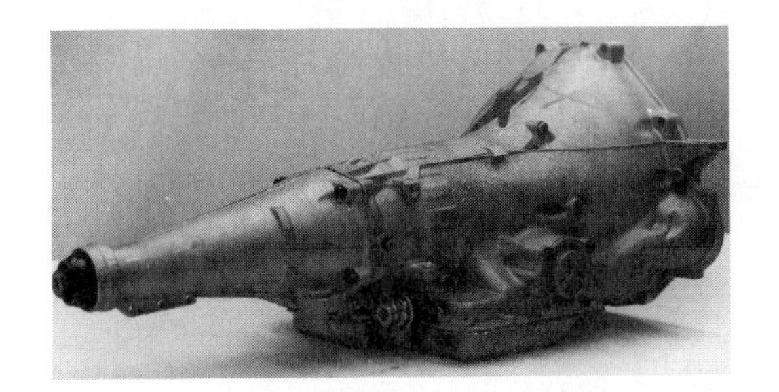

A

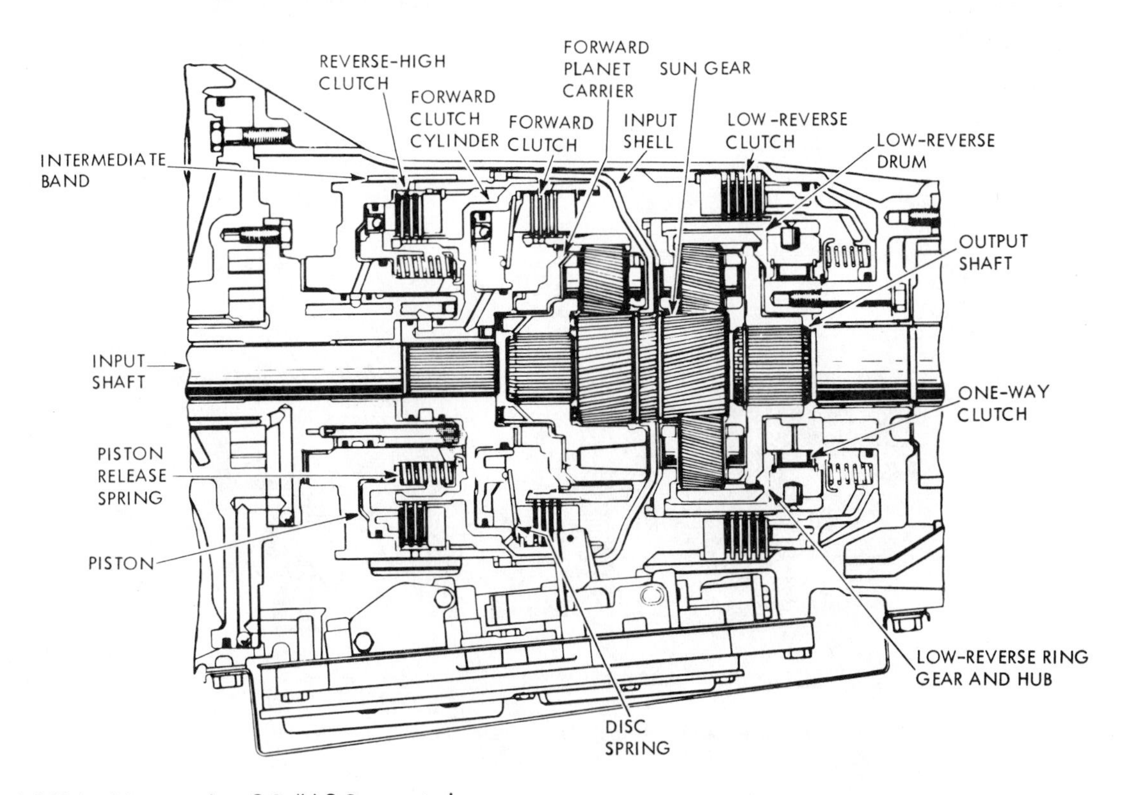

B

FIGURE 4–21 (a) Right side case view C-6; (b) C-6 powertrain.

sequence, which is 3-2 at about 15 mph (24 km/h) and 2-1 at about 7 mph (11 km/h).

2 Manual second gear. This operating range has second gear only, no first-gear start. This range can be selected for use on slippery road surfaces, when pulling away from a stopped position. Second gear has less torque than first gear, and the possibility of breaking traction is reduced.

1 Manual low gear. The transmission is locked in low gear, with no upshifts. If the range selector is moved from D to 1 at excessive speed, the transmission shifts to intermediate until the vehicle slows to a safe speed. Once low gear is engaged, the transmission cannot upshift.

Planetary Power Flow

The technician should quickly recognize that the three-speed Simpson planetary system used in this Ford family is identical to the Chrysler TorqueFlite. The geartrain operation is also the same (see the discussion and supporting illustrations in the preceding text). The Ford planetary power flow sequence is illustrated in Figures 4–22 to 4–26 with a brief commentary.

In making a comparison to the TorqueFlite, note some of the differences in nomenclature. The reverse-high clutch and forward clutch are the driving clutches and are referred to as the front and rear clutches in the TorqueFlite. The intermediate band is comparable to the TorqueFlite kickdown band. The clutch/band combinations for each of the ratios are summarized in the range reference and clutch/band apply chart (Chart 4–2).

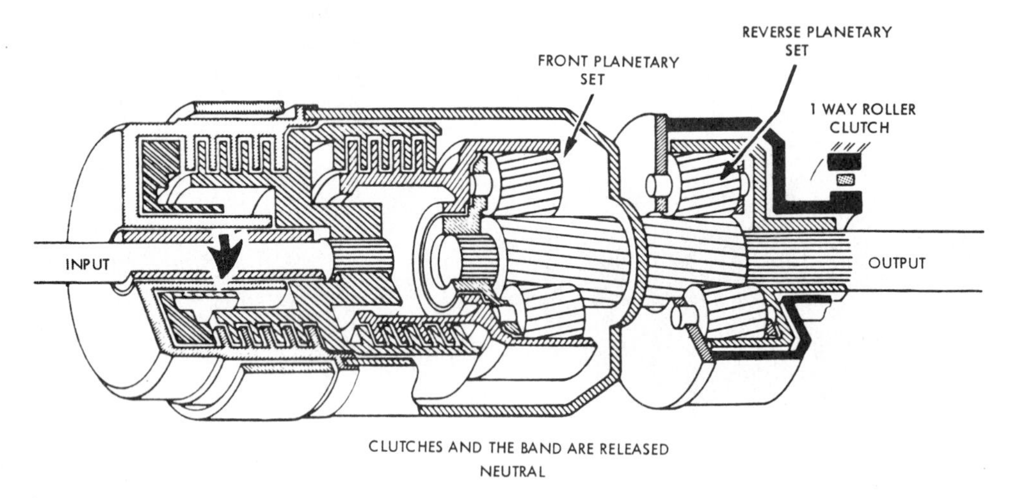

FIGURE 4–22 Neutral—with bands and clutches in the OFF position, there is no input to the gearsets. (Reprinted with the permission of Ford Motor Company)

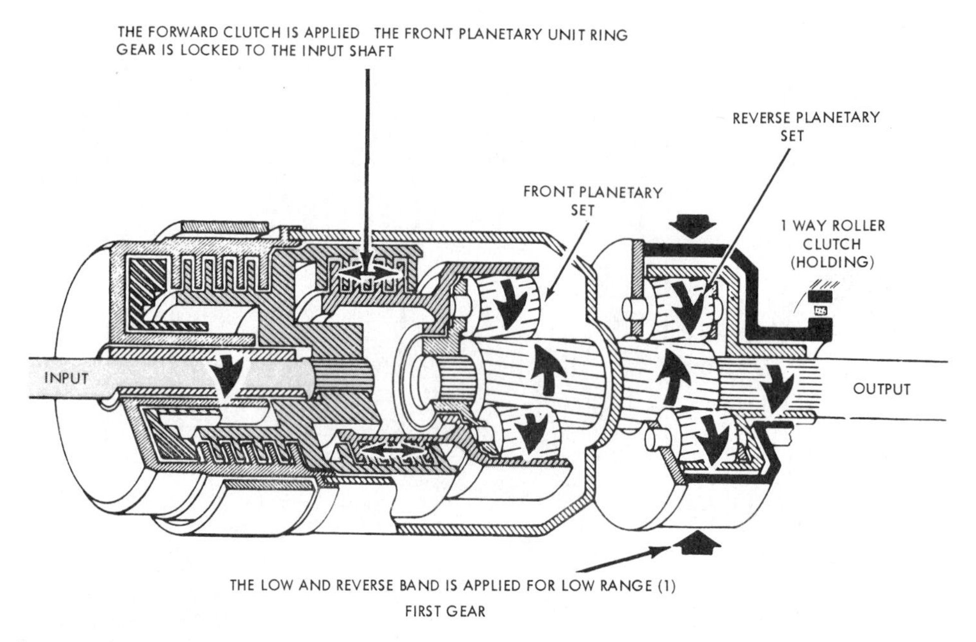

FIGURE 4–23 Drive—first gear is a function of the front and the reverse planetary sets. (Reprinted with the permission of Ford Motor Company)

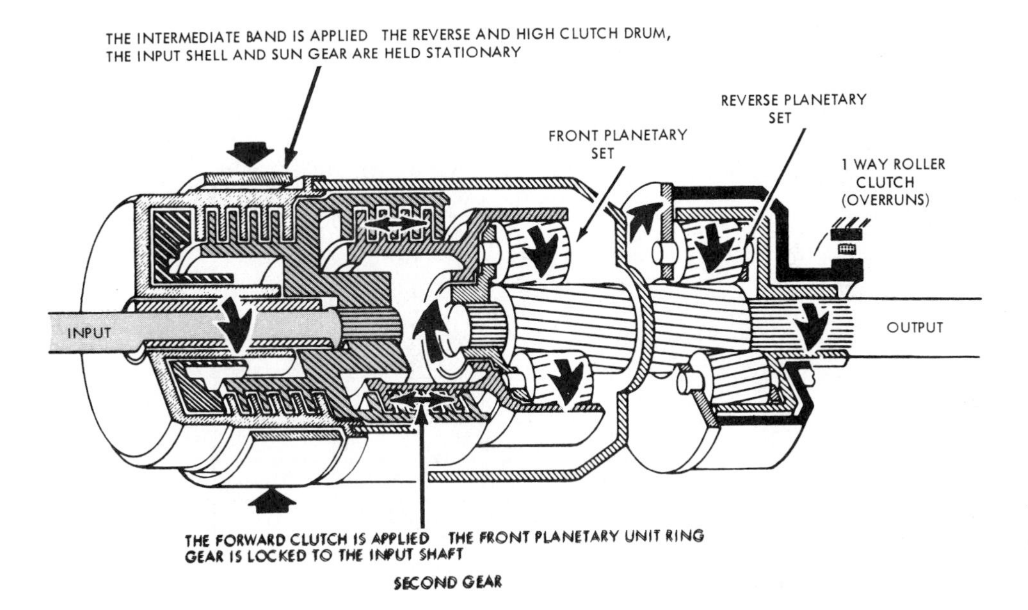

FIGURE 4–24 Drive—second gear is a function of the front planetary gearset. (Reprinted with the permission of Ford Motor Company)

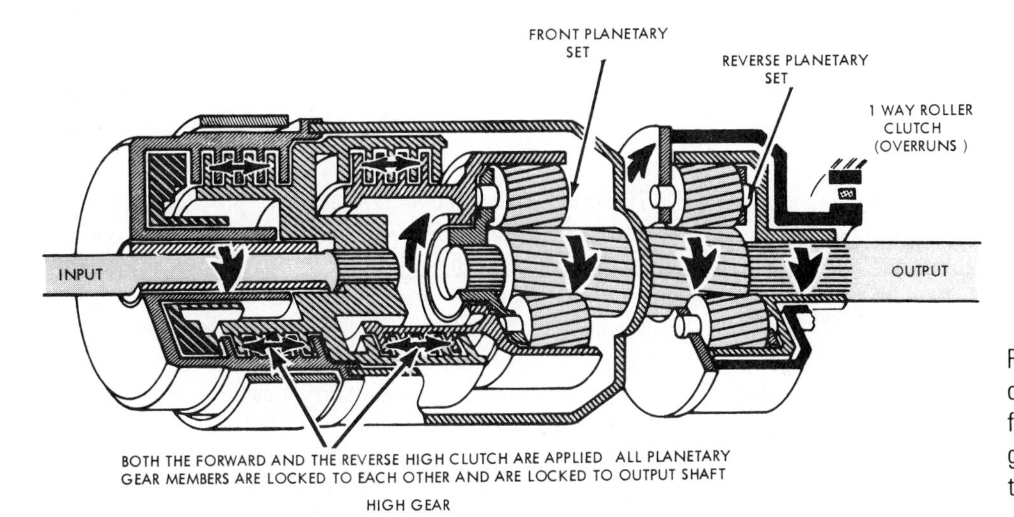

FIGURE 4–25 Drive—in high the locking of the planetary gears is a result of the forward clutch driving the front internal gear and the reverse-high clutch driving the sun gear.

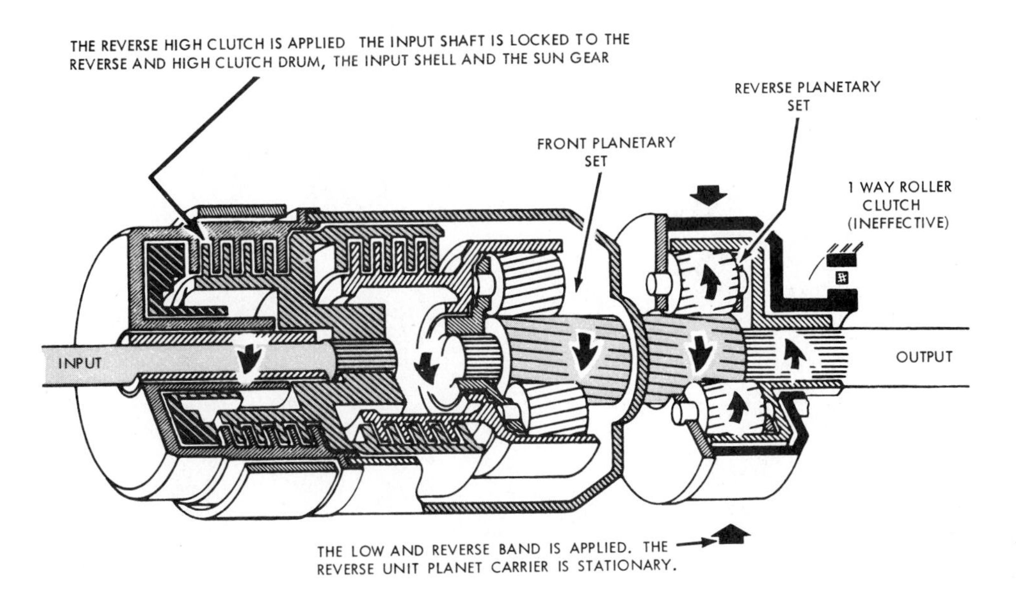

FIGURE 4–26 Reverse is the function of the reverse planetary unit. (Reprinted with the permission of Ford Motor Company)

Ford C-6 Range Reference and Clutch/Band Apply Summary
Selector Pattern P R N D 2 1

Range	Gear	Forward Clutch	Reverse High Clutch	Intermediate Band	Low Reverse Clutch	One-Way Clutch	Park Pawl
Park							In
Reverse			On		On		
Neutral							
D-Drive	First	On				Holds	
	Second	On		On			
	Third	On	On				
2-Drive	Second	On		On			
1-Low	First	On			On	Holds	

CHART 4–2 Ford C-6 Range Reference and Clutch/Band Apply Summary—Selector Pattern: P R N D 2 1.

SIMPSON PLANETARY SYSTEM: THREE-SPEED GENERAL MOTORS PRODUCTS

Hydra-Matic 3T40 (THM 125/125-C)

For the purpose of comparisons with the Chrysler and Ford products, this text features GM's 3T40 (THM 125/125C). The 3T40, introduced for model year 1980, is a popular three-speed transaxle used in GM FWD passenger cars and APV minivans. A former application of the 3T40 was in the mid-engine, RWD, Pontiac Fiero (1986–1988). A sectional view is shown in Figure 4–27.

The transverse arrangement utilizes sprockets and a chain assembly to provide power input from the torque converter and turbine shaft to the geartrain. Referring to the power flow schematic shown in Figure 4–28, observe that the planetary geartrain is the Simpson design with two gearsets in series, sharing a common sun gear and output shaft. The output from the compound planetary gearset provides the power input to the final drive planetary set.

The sprocket and chain assembly causes a change in power flow direction. This change results in a counterclockwise rotation of the power input to the Simpson compound planetary

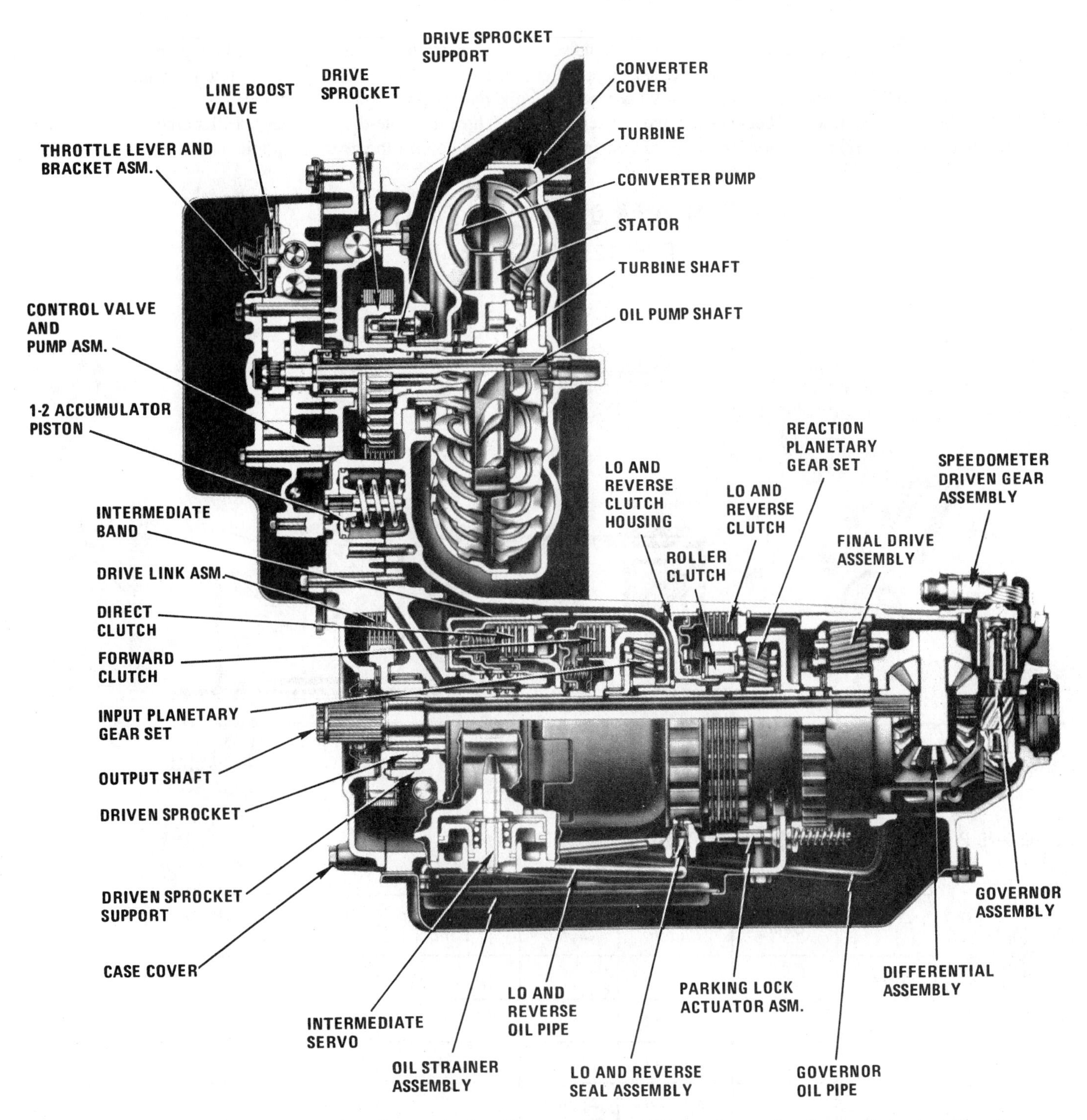

FIGURE 4–27 Hydra-Matic 3T40 (THM 125/125-C) cutaway view. (Courtesy of General Motors Corporation, Service Technology Group)

gearset. The roller clutch, responding to opposite rotational forces as those used within transmissions and transaxles previously examined, holds and release the reactionary carrier as needed. The roller clutch is straddled between the two gearsets. This is another variation from the Chrysler and Ford systems, which place the roller clutch behind the planetary assembly at the rear of the case. The one-way clutch remains positioned to the rear carrier for first-gear operation.

Three multiple-disc clutches, a roller clutch, and a band are used to control the gear operation. The clutch/band combina-

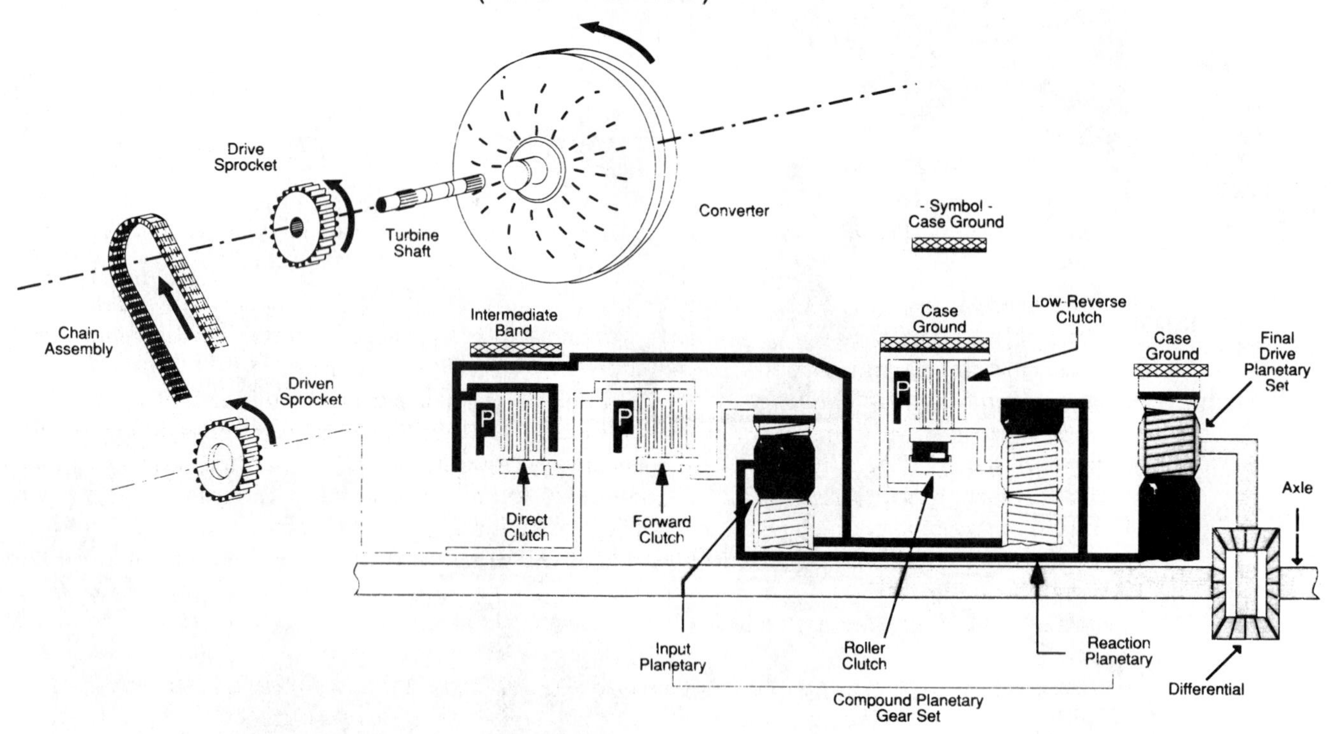

FIGURE 4–28 Power flow schematic: Hydra-Matic 3T40 (THM 125/125-C).

Range	Gear	Forward Clutch	Direct Clutch	Intermediate Band	Roller Clutch	Lo-Reverse Clutch
P-N						
D	1st	Applied			Holding	
	2nd	Applied		Applied		
	3rd	Applied	Applied			
2	1st	Applied			Holding	
	2nd	Applied		Applied		
1	1st	Applied			Holding	Applied
	2nd	Applied		Applied		
R	Reverse		Applied			Applied

Note: THM 125C product designation has been changed to 3T40.

CHART 4–3 GM Hydra-Matic 3T40 (THM 125/125-C), Range Reference and Clutch/Band Apply Summary—Typical Selection Pattern P R N D 2 1.

tions for each of the ratios are summarized in the range reference and clutch/band apply chart (Chart 4–3).

Operating Characteristics. The manual range selector has the six quadrant positions arranged in the typical order for three-speed transmissions: P-R-N-D-2-1. The operating characteristics in each range are almost identical to the Chrysler TorqueFlite family covered in detail at the beginning of this chapter. The only deviation is in the closed throttle downshift sequence. General Motors units typically follow a 3-2, 2-1 pattern instead of 3-1.

Other General Motors three-speed RWD automatic transmissions that incorporated a similar Simpson design are the THM-250/250C RWD (1974–1984), THM-350/350C RWD (1969–1983), and THM-325 FWD (1979–1981). GM's 3L80/3L80-HD (THM-400/475) (1964–1990) used a Simpson planetary gearset, but with a modified layout. The 3L80/3L80-HD incorporates some important operating characteristics—concepts used by many manufacturers in designing late model transmissions and transaxles.

General Motors Hydra-Matic 3L80/3L80-HD (THM-400/475)

General Description. The Hydra-Matic 3L80/3L80-HD (THM-400/475) is a fully automatic, three-speed, high-torque-capacity transmission with a long history of usage in RWD car, truck, and special application vehicles (Figure 4–29). The transmission was introduced in Buicks and Cadillacs in 1964. For 1965, it was added to Chevrolet, Oldsmobile, Pontiac, and truck product lines in two- and four-wheel drive versions.

In the following years, adaptations of the basic design were engineered for varied vehicle and engine applications known as the 375 and 425. The 375, 400, and 475 models essentially are alike in appearance, with some internal changes to adjust the transmission torque-handling ability. The 425 model was a longitudinal FWD design that was used in Cadillac Eldorados and Oldsmobile Toronados through production year 1978.

Although eliminated from the domestic passenger car scene by the late 1970s, the 3L80/3L80-HD continued to be used in 1990 GM domestic trucks, vans, motor homes, and school

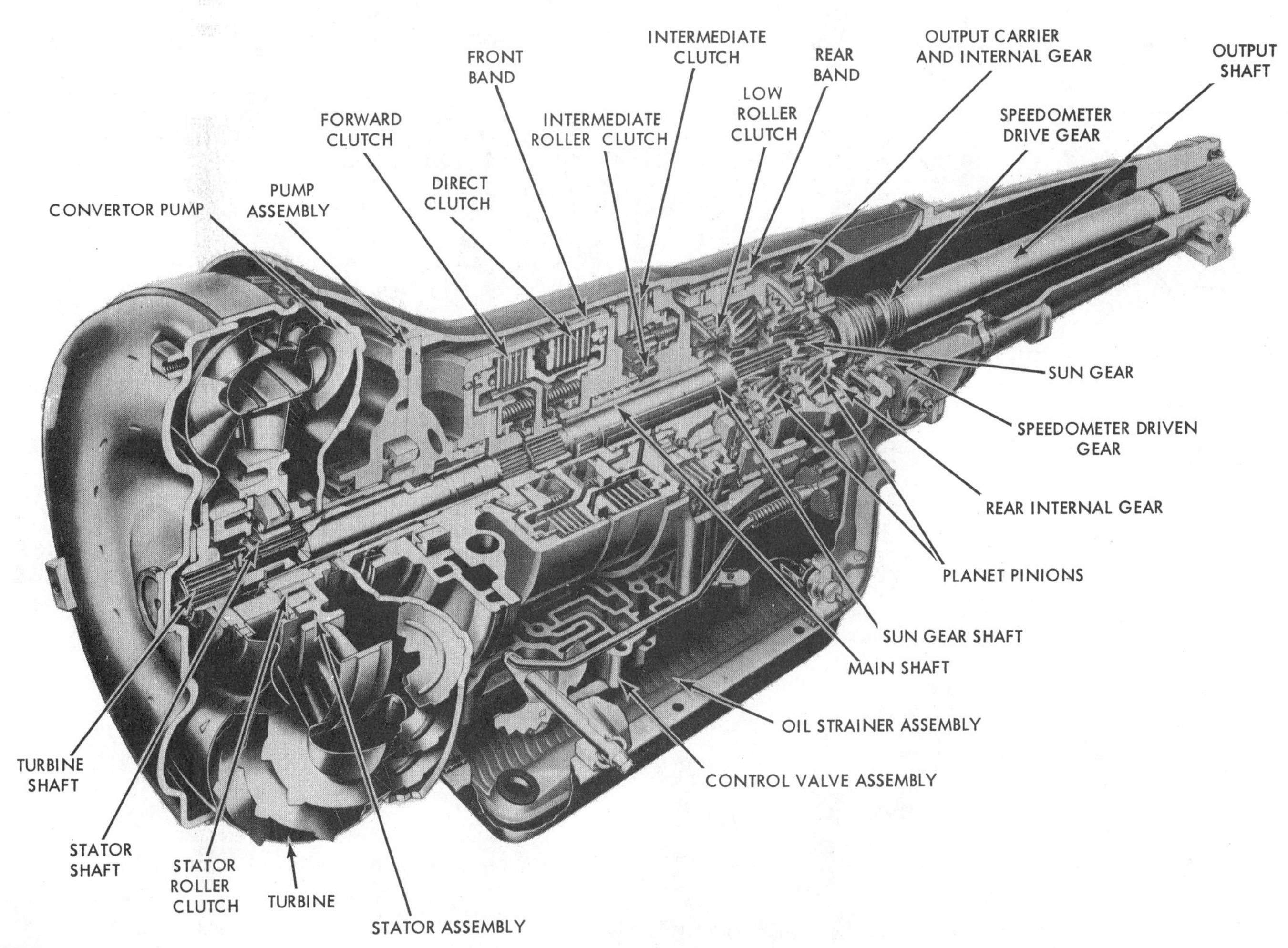

FIGURE 4–29 GM 3L80 (THM-400) sectional view. (Courtesy of General Motors Corporation, Service Technology Group)

buses. During this same period, some foreign car manufacturers, such as Ferrari, Jaguar, and Rolls Royce, used 3L80/3L80-HD transmissions. AM General Hummer continued to use the 3L80/3L80-HD transmission into the mid-1990s. The heavy-duty (HD) version is identified internally by straight-cut planetary gear teeth. Though noisier, this design characteristic eliminates destructive end thrust forces generated by helical-cut gear teeth during heavy load applications.

The 3L80/3L80-HD series basic powertrain consists of a three-element nonlockup torque converter and a three-speed Simpson planetary (Figure 4–30). The physical layout of the input and reaction gearsets is different than conventional three-speed units previously discussed, but the function is the same.

Three multiple-disc clutches, two one-way clutches (early versions used two roller clutches, while late model versions use a sprag and a roller clutch), and two bands provide the means to control the geartrain operation. The use of an intermediate clutch and intermediate sprag (or roller) clutch in drive range, instead of an intermediate band, provides a means to hold the sun gears reactionary for second gear. This feature allows for a freewheel coast in second gear and a quick upshift to third. This concept continues to be used in new generation transmissions and transaxles.

Operating Characteristics. The operating range selector quadrant has six positions, typically P-R-N-D-2-1 or P-R-N-D-I-L.

P/N Park/Neutral. These two quadrant positions provide for engine start. The input drive clutches are off, and there is no input to the geartrain. In the P position, the output shaft is mechanically locked by a parking pawl anchored in the case.

R Reverse. The vehicle is operated in a reverse direction.

D Drive. The drive range is used for normal driving and contributes to maximum fuel economy. Fully automatic shifting and three gear ratios are provided. Third gear is a direct drive. The driver can overrule the shift system by depressing the accelerator pedal. Within the proper vehicle speed and engine torque range, three WOT detent downshift combinations are available: 3-2, 3-1, and 2-1. For controlled part-throttle acceleration, the transmission will downshift 3-2. The closed-throttle downshift sequence is 3-2 and 2-1.

2 Intermediate. Sometimes marked I, S, or L_2, this range provides for added performance in congested traffic or hilly terrain. It starts in first gear and has a normal 1-2 shift pattern. The transmission remains in second gear, however, for the desired acceleration and engine braking. Intermediate can be selected at any high-gear vehicle speed and the transmission shifts to second. A WOT detent downshift (2-1) is available. The closed-throttle downshift pattern is 2-1.

1 Manual low. Sometimes marked L, or L_1, low range locks the transmission in first gear for continuous heavy-duty

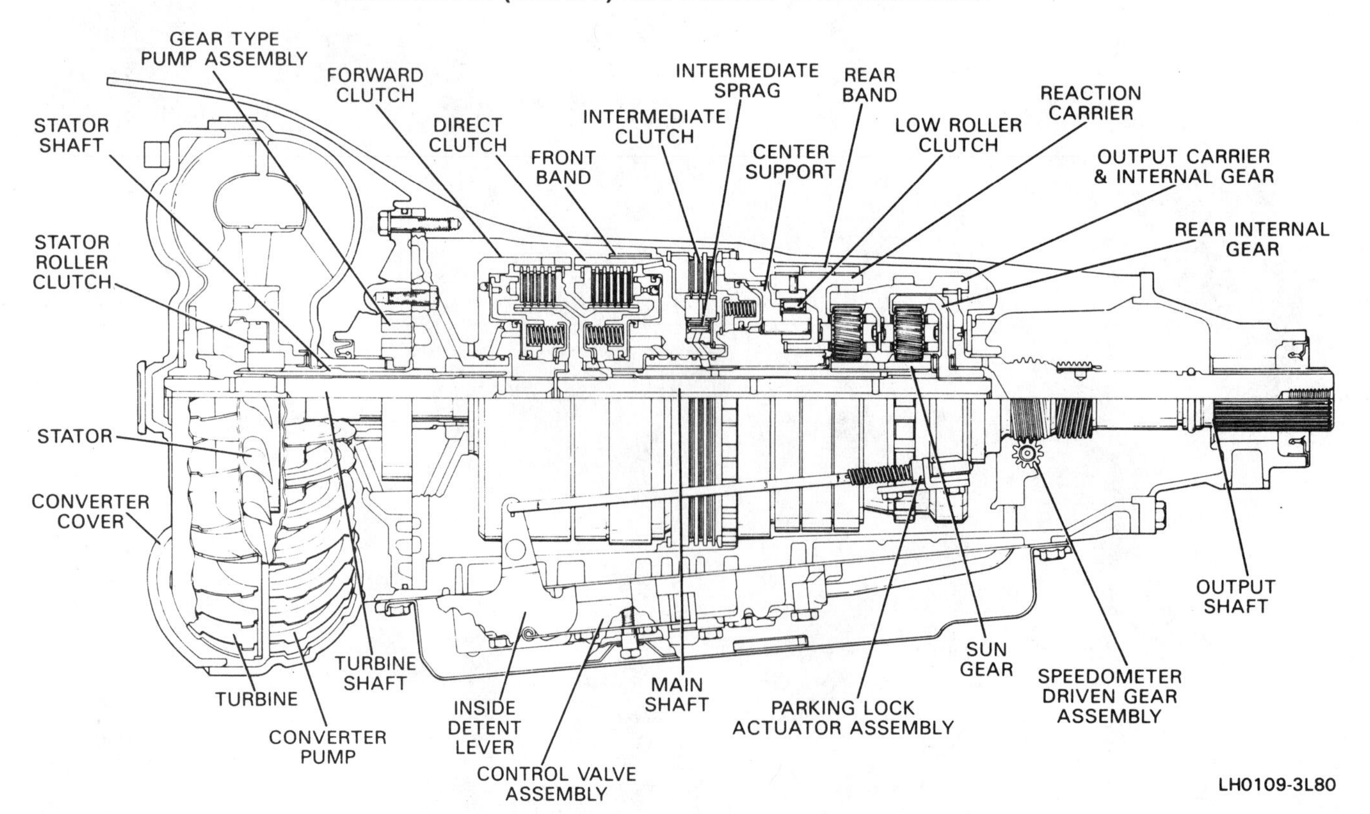

FIGURE 4–30 GM 3L80/3L80-HD (THM-400/475) sectional view. (Courtesy of General Motors Corporation, Service Technology Group)

pulling power and engine braking that is demanded for steep grades, especially in mountainous terrain. Manual low can be selected at any vehicle speed range but shifts to second gear if the vehicle speed is too high. When vehicle speed is below 30 mph (48.5 km/h), first gear engages, and the transmission does not upshift regardless of throttle opening and vehicle speed.

Powerflow Summary. A powerflow schematic drawing of the geartrain and clutch/band elements is shown in Figure 4–31. Unlike other Simpson compound planetary geartrains found in automatic transmissions, the 3L80/3L80-HD uses a third shaft, referred to as the mainshaft. When the forward clutch is applied, the power input to the geartrain is through the rear planetary ring gear rather than through the front planetary ring gear. The planetary unit is driven through the back instead of the front of the Simpson gearset. Notice that the forward, direct, and intermediate clutch units are positioned in reverse order.

Refer to Chart 4–4, the range reference and clutch/band apply chart. The various clutches and bands identified, and their functions, are as follows:

- *Forward clutch:* input drive clutch that locks the mainshaft and rear ring gear to the turbine shaft in all forward gears.
- *Direct clutch:* input drive clutch that locks the front and rear sun gears to the turbine shaft in third and reverse gears.
- *Intermediate clutch:* reaction clutch that grounds the outer race of the intermediate one-way sprag (or roller), which allows the one-way clutch to mechanically lock the sun gears to the transmission case whenever the sun gears want

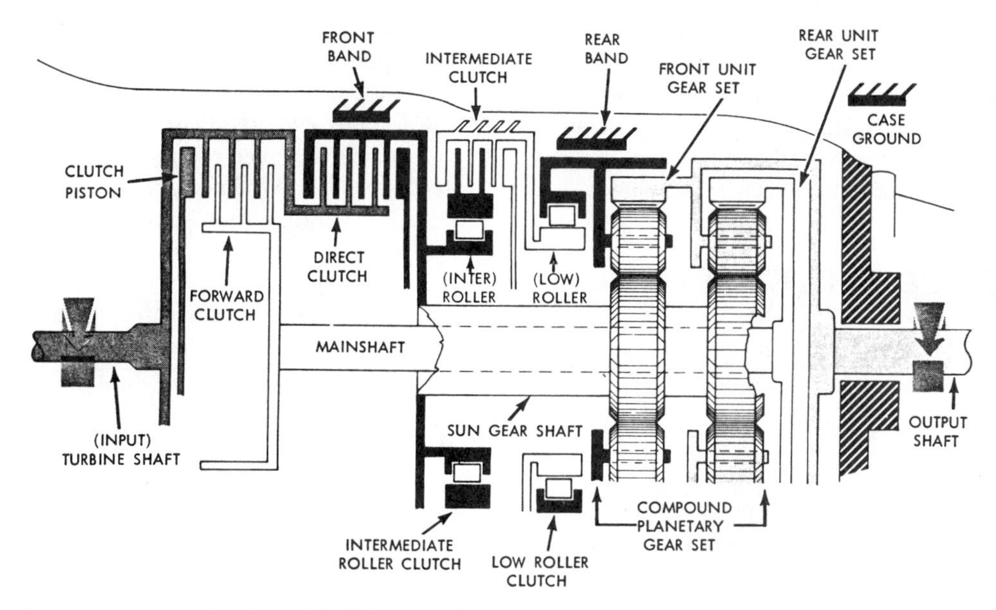

FIGURE 4–31 Power flow schematic: 3L80/3L80-HD (THM-400/475).

3L80(400) Operating Range and Clutch/Band Apply Summary

RANGE	GEAR	FORWARD CLUTCH	DIRECT CLUTCH	FRONT BAND	INTERMEDIATE CLUTCH	INTERMEDIATE SPRAG	LO-ROLLER CLUTCH	REAR BAND
P-N								
D	1st	APPLIED					HOLDING	
	2nd	APPLIED			APPLIED	HOLDING		
	3rd	APPLIED	APPLIED		APPLIED			
2	1st	APPLIED					HOLDING	
	2nd	APPLIED		APPLIED	APPLIED	HOLDING		
1	1st	APPLIED					HOLDING	APPLIED
	2nd	APPLIED		APPLIED	APPLIED	HOLDING		
R	REVERSE		APPLIED					APPLIED

CHART 4–4 3L80/3L80-HD (THM-400/475) Range Reference and Clutch/Band Apply Summary.

to rotate counterclockwise. The one-way clutch allows for clockwise rotation of the sun gears with the intermediate clutch engaged, as in third-gear operation.

- *Front band:* provides engine braking for second gear in the manual 2 and 1 ranges. The front band wraps around the direct clutch drum and holds the sun gears during second-gear deceleration in the intermediate and manual low ranges.
- *Rear band:* wraps around the front carrier drum and holds the carrier reactionary in reverse. The rear band is effective during coast/deceleration in manual low.
- *Intermediate one-way clutch (sprag or roller):* mechanically locks the sun gears to the transmission when the intermediate clutch is engaged and sun gear rotational effort is counterclockwise (second-gear acceleration). Clockwise sun gear rotation releases the one-way clutch.
- *Low roller one-way clutch:* has a permanently grounded inner race to the transmission case through the center support and mechanically locks the front carrier to the case during first-gear acceleration when the front carrier is attempting to turn counterclockwise. During deceleration in first gear, clockwise rotation of the front carrier releases the one-way clutch and provides freewheeling action.

PLANETARY OVERDRIVE SYSTEMS

To meet federal standards for emissions and fuel economy, four-speed overdrive automatic transmissions began to appear in the production model year 1980. They have gradually displaced many established three-speed units for light and heavy-duty service. Third gear is at 1:1, with fourth gear providing the overdrive. The overdrive is achieved by a variety of approaches in geartrain design.

- An additional single **overdrive planetary gearset** is placed at the front or the rear of a three-speed Simpson or Ravigneaux planetary geartrain.
- The conventional three-speed Simpson planetary is redesigned with a new configuration and clutch/band support system. Utilizing independent sun gears, two planetary gearsets are combined by joining internal gears and planetary carriers. Completely new planetary geartrain systems have been introduced.
- The conventional Ravigneaux three-speed configuration is adapted to a clutch/band support system that has the capabilities to drive the pinion carrier (internal gear the constant output). No additional gearing is needed.

❖ **Overdrive planetary gearset:** A single planetary gearset designed to provide a direct drive and overdrive ratio. When coupled to a three-speed transmission configuration, a four-speed/overdrive unit is present.

There is no attempt at this time to present all these systems, but additional coverage is found in Chapters 22, 23, 24, and 25. The overdrive operation in all the systems is based on the law

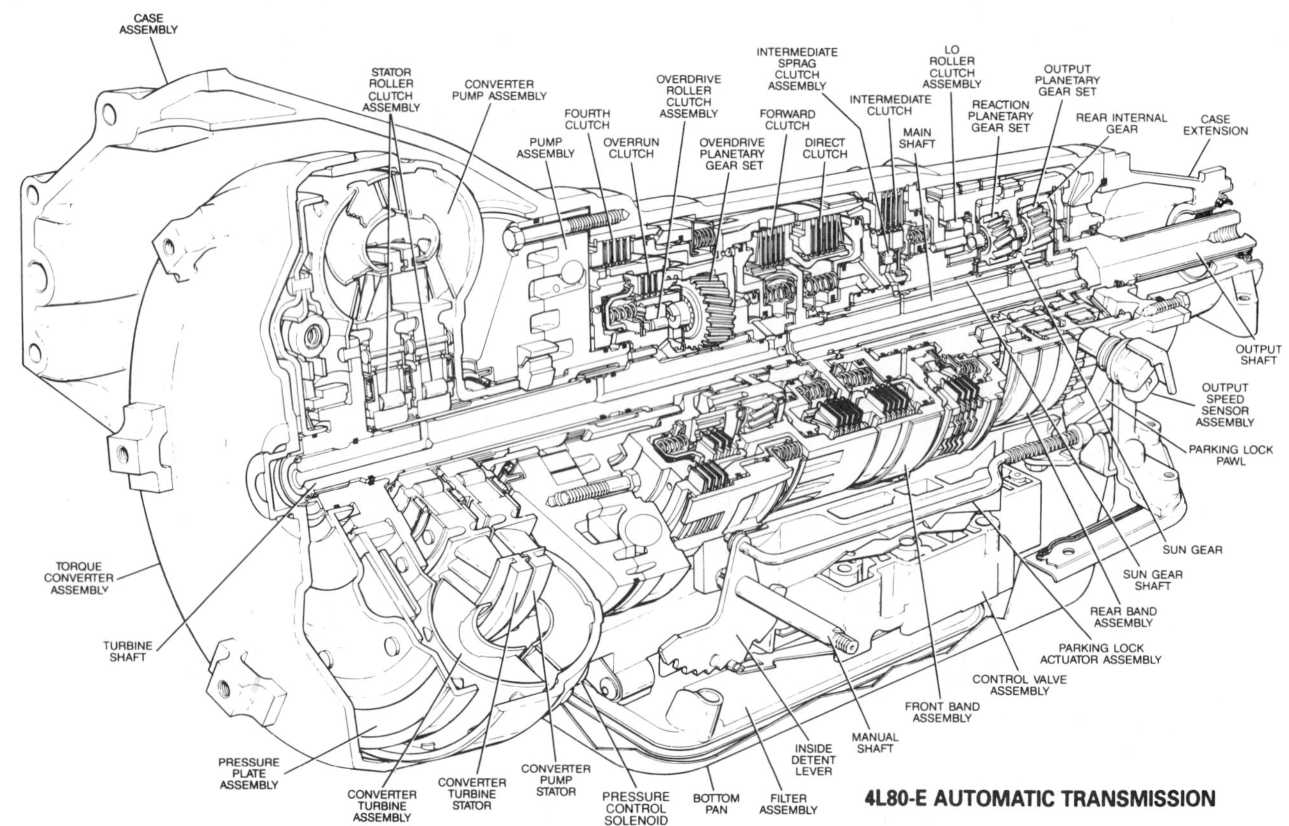

FIGURE 4–32 Sectional view: 4L80-E transmission. (Courtesy of General Motors Corporation, Service Technology Group)

of overdrive. The carrier is the input, and the sun gear is the reactionary member. As an introduction to the overdrive concept, we will explore domestic transmissions using an overdrive planetary gearset added to the front or rear of a three-speed Simpson compound planetary gearset.

General Motors 4L80-E/4L80-EHD

The GM 4L80-E/4L80-EHD transmission (1991–current) is illustrated in a cutaway view in Figure 4–32. It essentially uses a 3L80/3L80-HD design with an overdrive planetary upfront and is available in two- and four-wheel drive versions. Converter output is relayed through the overdrive planetary before driving the three-speed Simpson planetary system.

The THM 200-4R (1981–1989) and the 4L30-E use an overdrive gearset similar in design and operation to the one used in a 4L80-E/4L80-EHD. The THM 200-4R, a four-speed RWD overdrive transmission used in GM passenger cars, was based on the THM 200/200C three-speed Simpson planetary gearset transmission. It was GM's first automatic four-speed overdrive transmission.

Built in Strasbourg, France, by General Motors, the 4L30-E transmission (1989–current) is based on the three-speed, Ravigneaux planetary gearset, 3L30 (THM 180/180-C) (1967–current). Applications of the 4L30-E include usage in these foreign vehicles: Opel Omega, Opel Senator, Isuzu Trooper/Rodeo, and BMW. The 4L30-E and the 3L30 (THM-180/180-C) are discussed with other import/foreign automatic transmissions in Chapter 25.

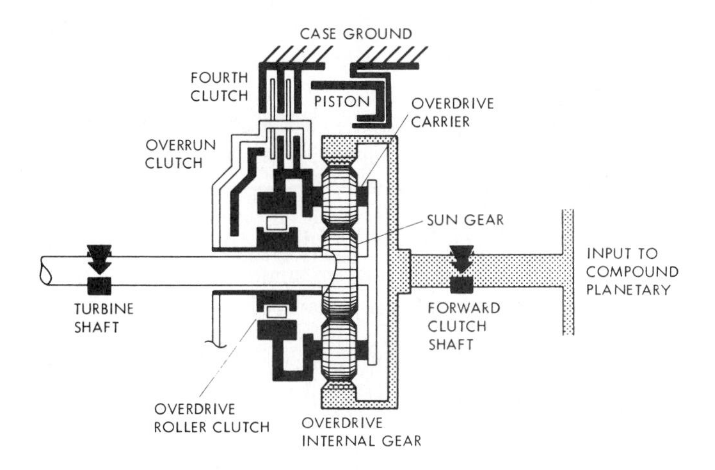

FIGURE 4–33 Schematic view—overdrive assembly.

Overdrive Planetary Assembly. The overdrive assembly is shown in a simplified illustration in Figure 4–33. The assembly relationship of the gearset components is shown in Figure 4–34.

- *Overdrive planet carrier:* four-pinion carrier that is splined to the turbine shaft. The carrier is the input to the overdrive planetary and is always turning at converter turbine speed. It is integral with the overrun clutch hub and roller clutch outer race.
- *Overdrive sun gear:* in constant mesh with the planet carrier pinions and splines to the overdrive roller cam, the sun gear can be driven at turbine speed through the one-way roller clutch.
- *Overdrive internal gear:* in constant mesh with the planet carrier pinions, the internal gear is the output member of the planetary unit and splines to the forward clutch shaft and housing of the compound planetary gearset.

For control of the overdrive planetary operation, three clutches are used for the following functions:

- *One-way roller clutch:* drive clutch that locks the sun gear to the overdrive planet carrier for direct-drive operation.
- *Overrun clutch:* locks the sun gear to the overdrive planet carrier to prevent roller clutch freewheeling in ranges 3, 2, or 1.
- *Fourth clutch:* holds the sun gear stationary for overdrive operation in O/D range.

Overdrive Planetary Operation: Direct Mode. The direct mode is functional in first, second, and third gears, plus reverse and neutral/park.

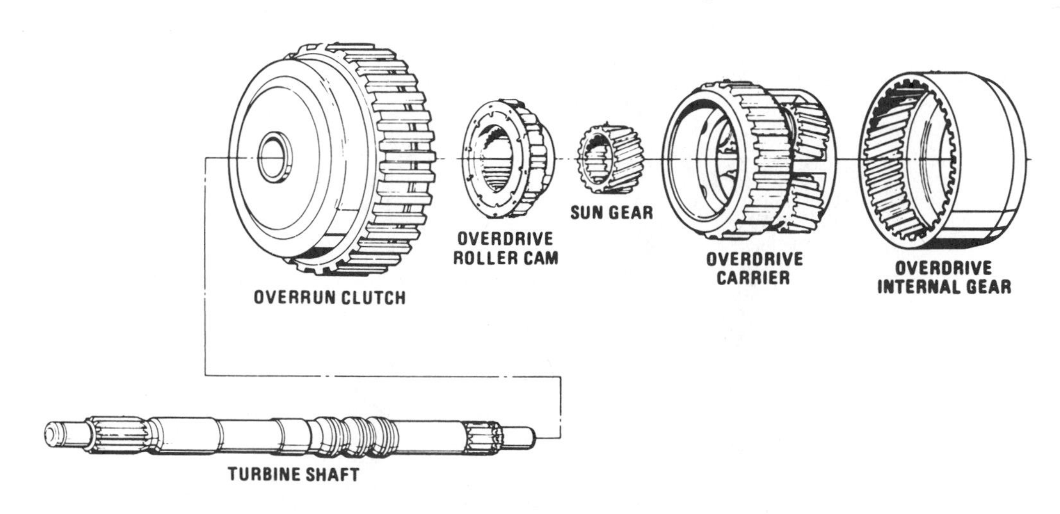

FIGURE 4–34 Overdrive planetary components. (Courtesy of General Motors Corporation, Service Technology Group)

- *One-way roller clutch:* locks the sun gear to the overdrive planet carrier during acceleration.
- *Overrun clutch:* applied in the 3, 2, and 1 operating ranges, the clutch provides engine braking on deceleration for second and third gears, and for first gear in manual 1.

Operational Sequence: Direct Drive.

- The turbine shaft drives the planet carrier clockwise.
- The turning effort of the planet carrier results in a counterclockwise motion of the carrier pinions.
- The carrier pinions want to turn the internal gear and sun gear clockwise.
- The internal gear is connected to the compound planetary gearset and resists movement due to the vehicle weight on the driveline.
- The resistance of the vehicle weight on the internal gear causes the pinions to push counterclockwise. The pinions, therefore, attempt to turn the sun gear faster than the planet carrier. The rotational effort of the input forces is shown in Figure 4–35.
- Because the sun gear wants to turn faster than the planet carrier, the roller clutch locks the sun gear to the carrier.
- The pinions are trapped and cannot rotate and force the internal gear to turn with the carrier and sun gear for direct drive (Figure 4–36).

A one-way clutch is generally used to hold a planetary member stationary. In the GM overdrive planetary assemblies, however, it is used to lock two planetary members together and transmit drive torque. Also note that a clockwise turning effort by the sun gear, not the usual counterclockwise effort, locks the roller clutch.

The transmission must provide engine braking in the top gear in the manual ranges D(3), 2, or 1. This is essential to maintain control over the vehicle for closed throttle stopping or when descending hills. Under these driving conditions, the rear-wheel coast torque from the output shaft drives through the transmission and overspeeds the turbine shaft. With the overdrive internal gear acting as the input to the overdrive assembly, the turning forces in the planetary spin the sun gear counterclockwise. This releases the roller clutch and disengages the engine from the driveline (Figure 4–37). To prevent this action, application of the overrun clutch keeps the sun gear locked to the carrier and maintains the direct-drive connection.

Overdrive Planetary Operation: Overdrive. The overdrive mode is functional in fourth-gear O/D operating range only. The three-speed Simpson planetary operates in direct drive.

- *Fourth clutch:* applied and holds the sun gear.

Operational Sequence: Overdrive.

- The turbine shaft and carrier continue to drive the gearset.
- The pinions, turning clockwise, are walking around the stationary sun gear and driving the internal gear at a faster speed than that of the carrier and turbine shaft (Figure 4–38).
- The input speed to the transmission compound gearset is now faster than engine speed. The overdrive ratio in a 4L80-E/4L80-EHD is 0.75:1.

The clutch/band combinations for each of the 4L80-E/4L80-EHD gear ratios are summarized in the range reference and clutch/band apply chart (Chart 4–5, page 88).

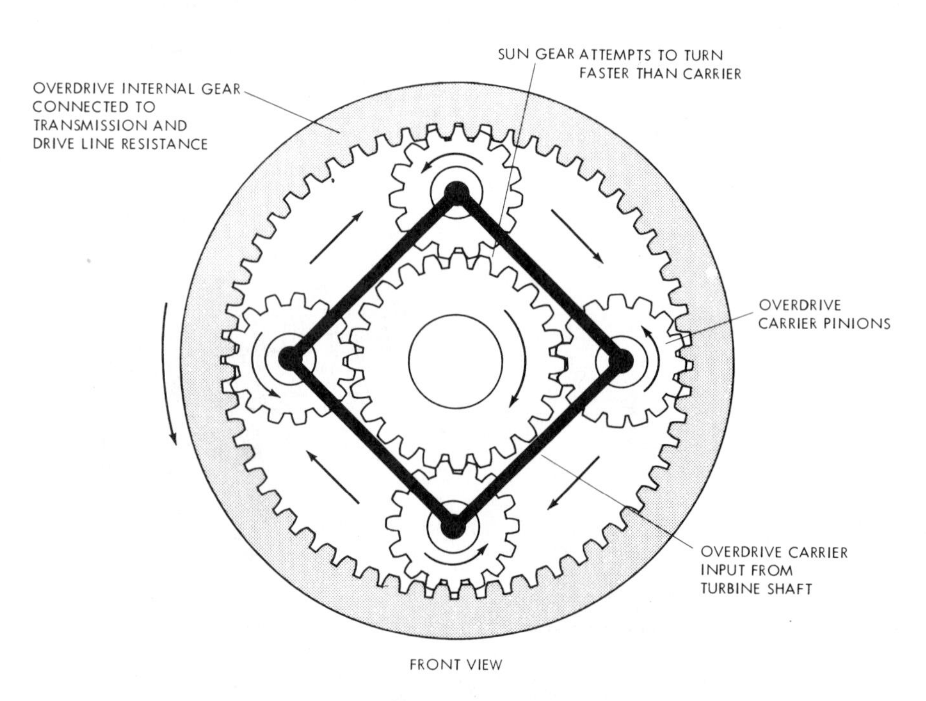

FIGURE 4–35 Direct drive—directional effort of input forces.

Ford A4LD (RWD)

The A4LD (1985–1995) featured in Chart 4–6 (page 89) is an automatic four-speed, longitudinal drive transmission. It is used in light-duty truck/utility vehicles and passenger cars, and is produced by Ford in Bordeaux, France. The A4LD basically employs an overdrive planetary in front of the C-3, three-speed Simpson compound planetary geartrain and is available in two- and four-wheel drive versions.

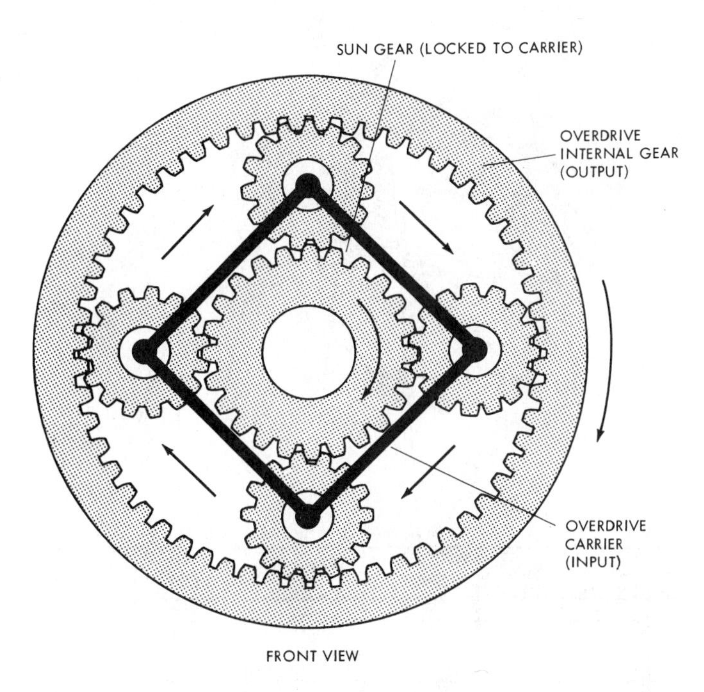

FIGURE 4–36 Direct drive—roller clutch holding.

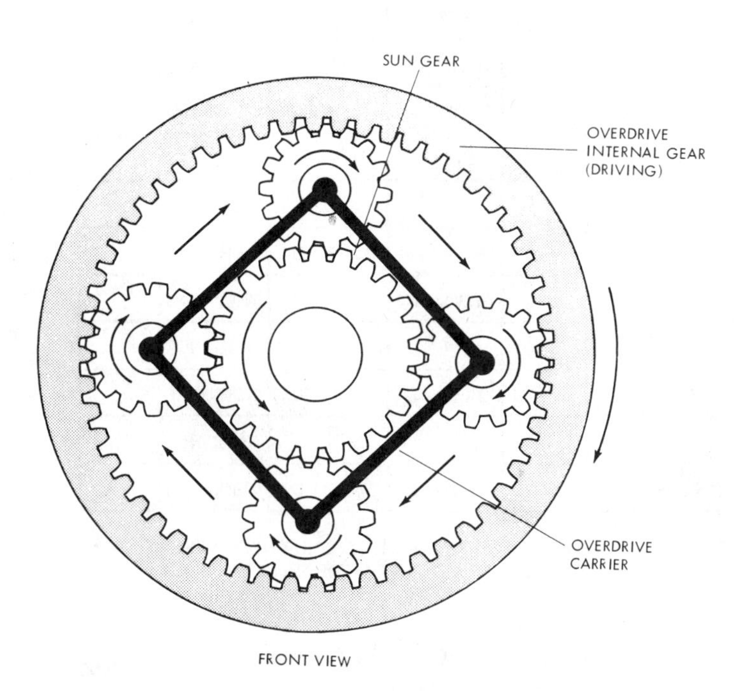

FIGURE 4–37 Direct drive—directional effort during coast/deceleration.

Overdrive Planetary System. Featured in Figure 4–39 is a simplified schematic of the overdrive assembly. For control of the overdrive planetary operation a one-way roller clutch, a multiple-disc clutch, and a band are used for the following functions:

- *One-way roller clutch:* drive roller clutch that locks the turbine shaft directly to the three-speed input shaft. At the same time, this locks the planet carrier to the ring gear for direct-drive lockup of the planetary.
- *Overdrive clutch:* locks the sun gear to the overdrive planet carrier in reverse, D, 2, and 1 ranges. This produces an engine braking condition during deceleration in reverse, second, and third gears, and first gear in manual 1 range.
- *Overdrive band:* holds the sun gear stationary for fourth-gear overdrive in the O/D range.

Overdrive Planetary Operation. The coverage of the GM 4L80-E/4L80-EHD overdrive planetary should not make it necessary at this time to repeat another detailed overdrive operation. There are only slight design differences in the style of clutch/band control. The planetary operation theme is essentially the same. To simplify understanding of how the planetary works, the technician should review the following power flow highlights.

- The converter turbine shaft drives the planet carrier and one-way roller clutch inner race.

Direct Mode. The direct mode is functional in first, second, and third gears, plus reverse and neutral/park.

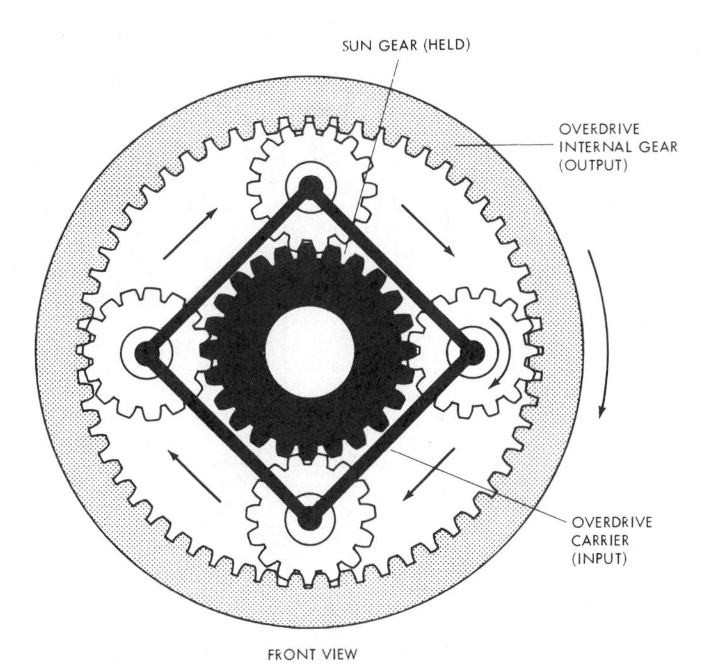

FIGURE 4–38 Overdrive operation.

HYDRAMATIC 4L80-E - GEAR RATIOS

FIRST	2.48	FOURTH	.75
SECOND	1.48	REVERSE	2.08
THIRD	1.00		

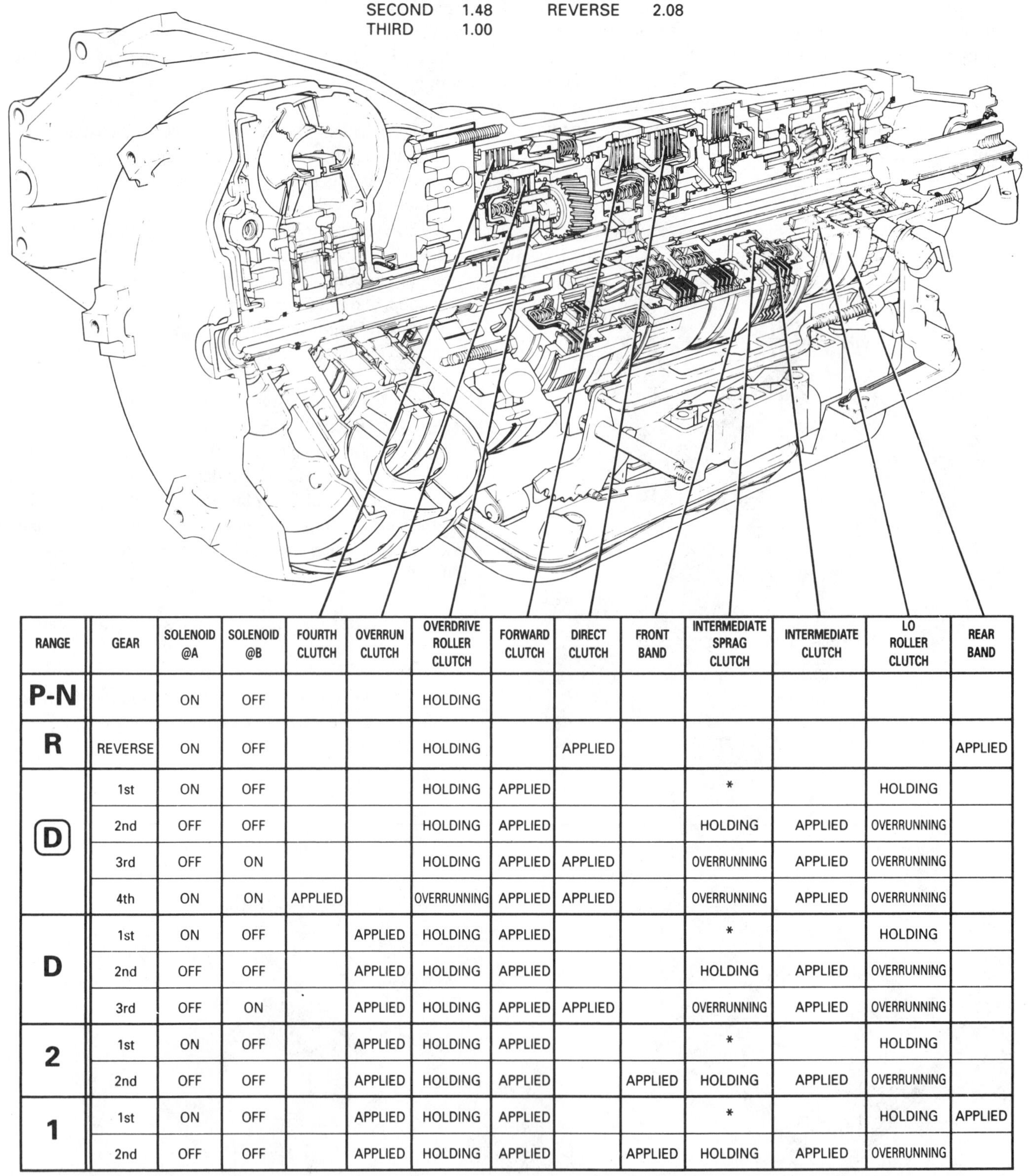

RANGE	GEAR	SOLENOID @A	SOLENOID @B	FOURTH CLUTCH	OVERRUN CLUTCH	OVERDRIVE ROLLER CLUTCH	FORWARD CLUTCH	DIRECT CLUTCH	FRONT BAND	INTERMEDIATE SPRAG CLUTCH	INTERMEDIATE CLUTCH	LO ROLLER CLUTCH	REAR BAND
P-N		ON	OFF			HOLDING							
R	REVERSE	ON	OFF			HOLDING		APPLIED					APPLIED
(D)	1st	ON	OFF			HOLDING	APPLIED			*		HOLDING	
	2nd	OFF	OFF			HOLDING	APPLIED			HOLDING	APPLIED	OVERRUNNING	
	3rd	OFF	ON			HOLDING	APPLIED	APPLIED		OVERRUNNING	APPLIED	OVERRUNNING	
	4th	ON	ON	APPLIED		OVERRUNNING	APPLIED	APPLIED		OVERRUNNING	APPLIED	OVERRUNNING	
D	1st	ON	OFF		APPLIED	HOLDING	APPLIED			*		HOLDING	
	2nd	OFF	OFF		APPLIED	HOLDING	APPLIED			HOLDING	APPLIED	OVERRUNNING	
	3rd	OFF	ON		APPLIED	HOLDING	APPLIED	APPLIED		OVERRUNNING	APPLIED	OVERRUNNING	
2	1st	ON	OFF		APPLIED	HOLDING	APPLIED			*		HOLDING	
	2nd	OFF	OFF		APPLIED	HOLDING	APPLIED		APPLIED	HOLDING	APPLIED	OVERRUNNING	
1	1st	ON	OFF		APPLIED	HOLDING	APPLIED			*		HOLDING	APPLIED
	2nd	OFF	OFF		APPLIED	HOLDING	APPLIED		APPLIED	HOLDING	APPLIED	OVERRUNNING	

*HOLDING BUT NOT EFFECTIVE

@ THE SOLENOID'S STATE FOLLOWS A SHIFT PATTERN WHICH DEPENDS UPON VEHICLE SPEED AND THROTTLE POSTION. IT DOES NOT DEPEND UPON THE SELECTED GEAR.

ON = SOLENOID ENERGIZED
OFF = SOLENOID DE-ENERGIZED

NOTE: DESCRIPTIONS ABOVE EXPLAIN COMPONENT FUNCTION DURING ACCELERATION.

RH0006-4L80-E

CHART 4–5 Hydra-Matic 4L80-E/4L80-EHD Range Reference and Clutch/Band Apply Summary. (Courtesy of General Motors Corporation, Service Technology Group)

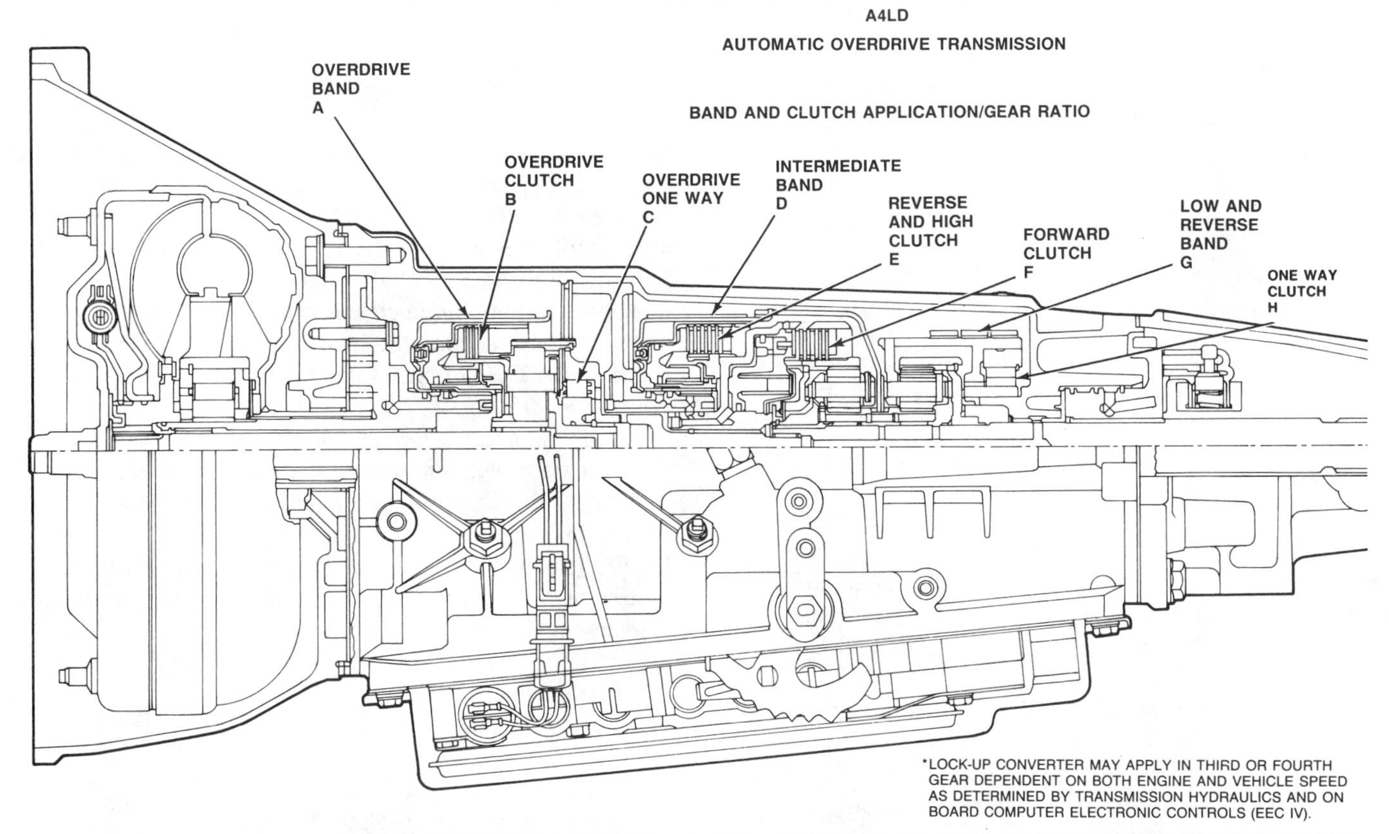

GEAR	OVERDRIVE BAND A	OVERDRIVE CLUTCH B	OVERDRIVE ONE WAY CLUTCH C	INTERMEDIATE BAND D	REVERSE AND HIGH CLUTCH E	FORWARD CLUTCH F	LOW AND REVERSE BAND G	ONE WAY CLUTCH H	GEAR RATIO
1 — MANUAL FIRST GEAR (LOW)		APPLIED	HOLDING			APPLIED	APPLIED	HOLDING	2.47:1
2 — MANUAL SECOND GEAR		APPLIED	HOLDING	APPLIED		APPLIED			1.47:1
D — DRIVE AUTO — 1ST GEAR		APPLIED	HOLDING			APPLIED		HOLDING	2.47:1
D — O/D AUTO — 1ST GEAR			HOLDING			APPLIED		HOLDING	2.47:1
D — DRIVE AUTO — 2ND GEAR		APPLIED	HOLDING	APPLIED		APPLIED			1.47:1
D — O/D AUTO — 2ND GEAR			HOLDING	APPLIED		APPLIED			1.47:1
D — DRIVE AUTO — 3RD GEAR		APPLIED	HOLDING		APPLIED	APPLIED			1.0:1
D — O/D AUTO — 3RD GEAR			HOLDING		APPLIED	APPLIED			1.0:1
D — OVERDRIVE AUTOMATIC FOURTH GEAR	APPLIED				APPLIED	APPLIED			0.75:1
REVERSE		APPLIED	HOLDING		APPLIED		APPLIED		2.1:1

D6417-B

CHART 4–6 Ford A4LD Range Reference and Clutch/Band Apply Summary. (Reprinted with the permission of Ford Motor Company)

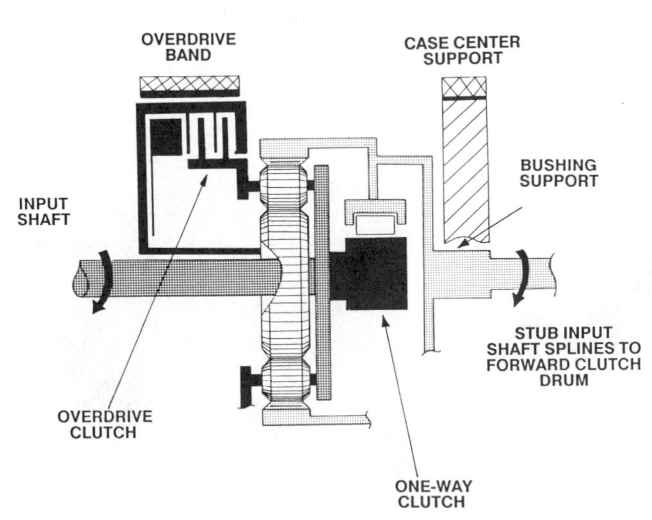

FIGURE 4–39 Power flow schematic: A4LD overdrive planetary unit.

- The turbine shaft is driving against the vehicle load during acceleration, and the one-way roller clutch locks the turbine shaft and stub input shaft together for direct drive. The one-way clutch also couples the planet carrier and ring gear together and locks the planetary in direct.
- With the exception of fourth-gear overdrive, all power flow is relayed to the transmission through the one-way roller clutch.
- Overdrive clutch is applied in D, 2, 1, and R. This causes an engine braking condition during deceleration in reverse, second, and third gears, and first gear in manual 1 range.

Overdrive Mode. The overdrive mode is functional in fourth-gear O/D operating range only. The three-speed Simpson planetary is operating in direct drive.

- Overdrive band applies and holds the sun gear, and the overdrive clutch is off.

- Planet carrier is the input and drives the ring gear and stub shaft at an overdrive ratio 0.75:1.
- The ring gear and stub shaft turn faster than the turbine shaft. The one-way roller clutch, therefore, is ineffective.

The clutch/band combinations for each of the A4LD ratios are summarized in the range reference and clutch/band apply chart (Chart 4–6).

Ford E4OD

Ford's heavy-duty, electronic shift, four-speed overdrive transmission incorporates an overdrive planetary gearset coupled to a three-speed Simpson planetary gearset. The overdrive unit is similar in design and operation to the one used in GM's 4L80-E/4L80-EHD.

The three-speed unit is designed with a conventional Simpson planetary gearset layout. For a free wheeling coast in second gear and quick upshifts into third gear in O/D range, the E4OD uses an intermediate clutch and intermediate sprag. This combination is similar in concept to GM's 3L80/3L80-E (THM 400/475) transmission.

The E4OD, first introduced during the 1989 production, began replacing numerous applications of the C-6. It is available in two- and four-wheel drive versions. Refer to Chapter 23 for an in-depth study of this overdrive transmission.

Chrysler 42RH/46RH (A-500/A-518)

The 42RH (A-500) RWD overdrive transmission, pictured in Figure 4–40, is a four-speed version of the 32RH (A-999) three-speed LoadFlite. The 32RH (A-999) is a heavy-duty design of the 30RH (A-904). The transmission was introduced in production year 1988 for use in the truck line with 3.9-L, V-6 and 5.2-L, V-8 engine applications—namely, the Dakota and D-Ram pickups and the B-Ram vans and wagons.

The 46RH (A-518) RWD overdrive transmission is based on the three-speed 36RH (A-727) TorqueFlite transmission. It is similar in design and operation to the 42RH (A-500).

The rear-mounted overdrive unit is placed into an extra-large, long extension housing (Figure 4–40). To adapt the three-speed system to the overdrive operation, the traditional output shaft is now referred to as the intermediate shaft and is the drive input for the overdrive unit. An output shaft extends from the rear of the overdrive unit.

Operating Characteristics. The range selector quadrant has the usual six positions (P-R-N-D-2-1). The driver can choose to drive the transmission as a three-speed or a four-speed, with an ON/OFF overdrive switch on the instrument panel. A switch indicator light turns on when the system is in the OFF or lockout position. An electronic controller is bypassed, and the transmission operates in a regular TorqueFlite three-speed mode. The switch automatically resets to OD when the ignition is turned off.

To achieve a fourth gear, a 3-4 shift valve has been added to the valve body, with an overdrive **solenoid** for electronic control of the shift. The shift electronics are keyed to the on-board engine management computer, referred to by Chrysler as **SMEC** (single-module engine controller). The SMEC determines the 3-4 shift by monitoring the following information from the input sensors:

- Coolant temperature
- Engine rpm
- Vehicle speed
- Throttle position
- Manifold vacuum/absolute pressure

❖ **Solenoid:** An electromagnetic device containing a coil winding and movable core. As current is sent through the winding, the core is moved. When applied to hydraulic oil circuits, pressure in that circuit can be controlled electrically.

❖ **SMEC (single module engine controller):** In certain Chrysler applications, it interfaces with the transmission to control selected shifting strategies.

When the vehicle is in third gear, above 25 mph (40 km/h), and the input sensor signals meet the computer's program criteria, the 3-4 solenoid is energized. The SMEC controls all 3-4 upshift and 4-3 downshift events. The SMEC also controls the converter lockup. Once in fourth gear, the lockup solenoid

A

B

C

FIGURE 4–40 Chrysler 42RH (A-500) transmission—compound planetary intermediate shaft and rear-mounted overdrive unit.

is energized and triggers the lockup hydraulics. In the regular three-speed mode, the transmission shift control is all hydraulic without any electronic assist. It operates similarly to the regular TorqueFlite transmissions.

Overdrive Planetary Assembly. A simplified schematic in Figure 4–41 shows the basic makeup of the overdrive assembly. For control of the planetary operation, two multiple-disc clutches and a double-row, one-way roller clutch are used for the following functions:

- *Overrunning roller clutch:* drive clutch that locks the intermediate shaft to the output shaft in forward gears.
- *Direct clutch:* drive clutch that locks the sun and annulus gears together, producing a direct drive within the overdrive unit. Applied in reverse and all forward gears except fourth. The direct clutch produces an engine braking condition in reverse, second, and third gears, and first gear in manual 1 range.

CAUTION: The direct clutch spring is a special large-diameter, heavy-duty, single-coil spring rated at 800 lb (3,558 N) that applies the clamping force to the direct clutch.

- *Overdrive clutch:* reactionary clutch that holds the sun gear for fourth-gear overdrive.

Overdrive Planetary Operation. The direct mode is functional in first, second, and third gears, plus reverse.

- The intermediate shaft drives the planet carrier and overrunning clutch.
- The intermediate shaft is driving against the vehicle load during acceleration. In forward ranges, the overrunning clutch locks the intermediate shaft to the output shaft in first, second, and third gears. This produces a direct drive through the overdrive unit.
- The power flow also takes a dual path through the planetary unit (Figure 4–42). The heavy-duty direct clutch spring acting through the sliding hub applies the direct clutch and clamps the annulus and sun gears together. This locks the planetary unit and output shaft at 1:1, with the carrier driving.
- The overrunning clutch, located between the intermediate shaft and output shaft, is ineffective to produce engine braking through the overdrive unit during deceleration in forward gears. However, since the direct clutch is applied in reverse, first, second, and third gears, it maintains the connection between the two shafts through the overdrive planetary unit. This results in engine braking for reverse, second, and third gears, and first gear in manual 1 range.

On the shift to fourth-gear overdrive, the overdrive piston is applied and the following sequence occurs:

- The piston is matched to both the overdrive clutches and the sliding hub (splined to sun gear shaft).

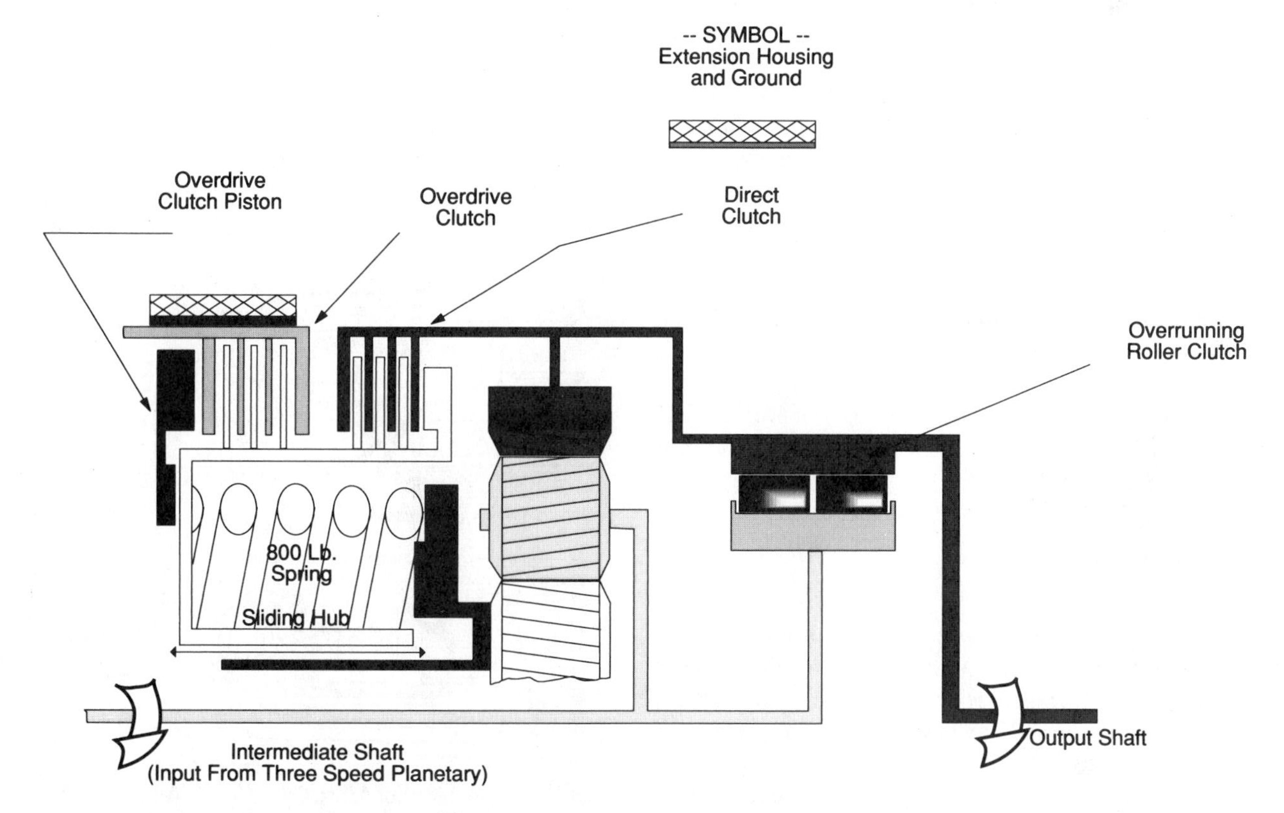

FIGURE 4–41 Power flow schematic: Chrysler 42RH/46RH (A-500/A-518) overdrive unit.

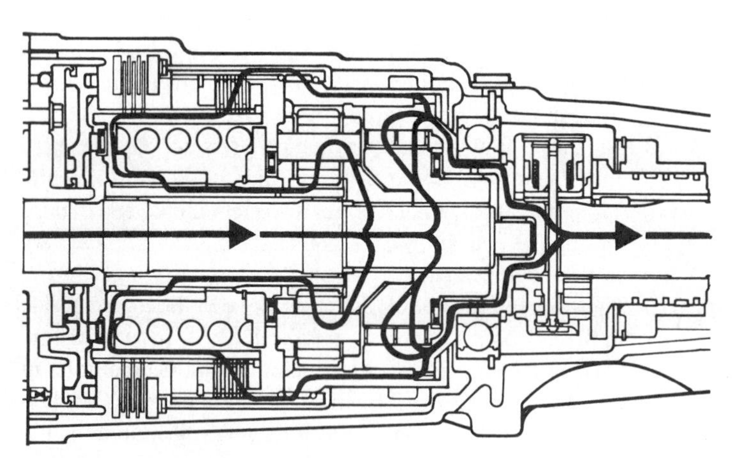

FIGURE 4–42 Chrysler 42RH/46RH (A-500/A-518) overdrive unit power flow in direct drive for first, second, and third gears. (Courtesy of Chrysler Corp.)

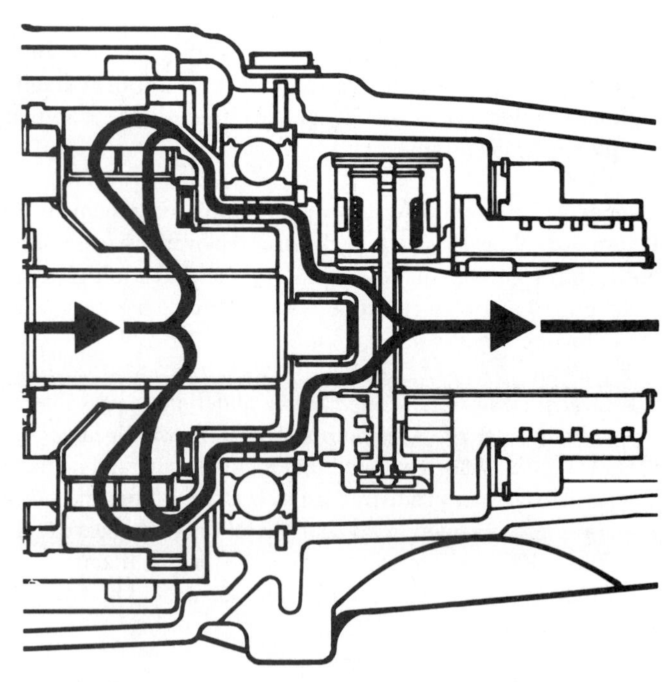

FIGURE 4–43 Chrysler 42RH/46RH (A-500/A-518) power flow during 3-4 shift transition. (Courtesy of Chrysler Corp.)

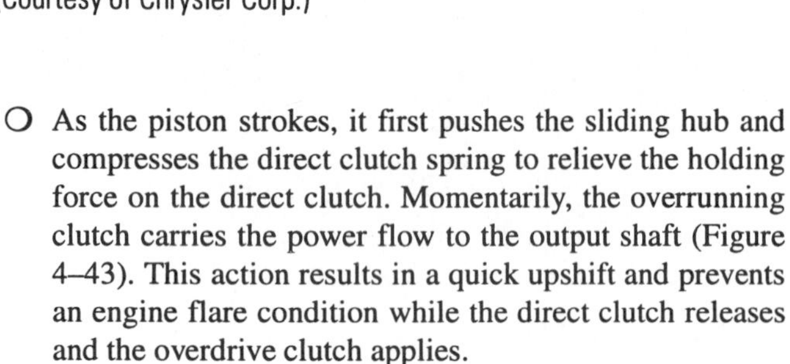

- As the piston strokes, it first pushes the sliding hub and compresses the direct clutch spring to relieve the holding force on the direct clutch. Momentarily, the overrunning clutch carries the power flow to the output shaft (Figure 4–43). This action results in a quick upshift and prevents an engine flare condition while the direct clutch releases and the overdrive clutch applies.
- As the piston continues to stroke and slide the hub, the overdrive clutch engages. This locks the sliding hub and sun gear to the overdrive clutch and extension housing.
- With the planet carrier driving around the sun gear, the annulus gear and output shaft are driven at an overdrive ratio of 0.69:1. The power flow is entirely through the overdrive planetary (Figure 4–44).
- With the output shaft turning faster than the intermediate shaft, the overrunning clutch freewheels.

The clutch/band combinations for each of the 42RH/46RH (A-500/A-518) ratios is summarized in the range reference and clutch/band apply chart (Chart 4–7).

SUMMARY

When two planetary gearsets are combined, forming a compound planetary gearset, reverse and three or more forward gears are possible. These include gear reductions, direct drives, and overdrives. Simpson and Ravigneaux planetary gearsets are the two common designs used worldwide.

Simpson compound planetary gearsets are commonly used in three-speed transmissions and transaxles. The operation of the gearsets, to achieve the ratio changes, is accomplished by clutch and band applications. The names of these driving and holding devices vary from one manufacturing group to another, but their functions are quite standard.

Clutch and band application charts, also referred to as range reference charts, identify which clutches and bands are on for a specific gear in a range position. They are beneficial when diagnosing malfunctions and illustrate engine braking concepts in the manual drive gear settings.

A variety of planetary drive layouts are used in four-speed transmissions. Some transmissions use an additional planetary gearset, referred to as an overdrive planetary, to achieve a fourth

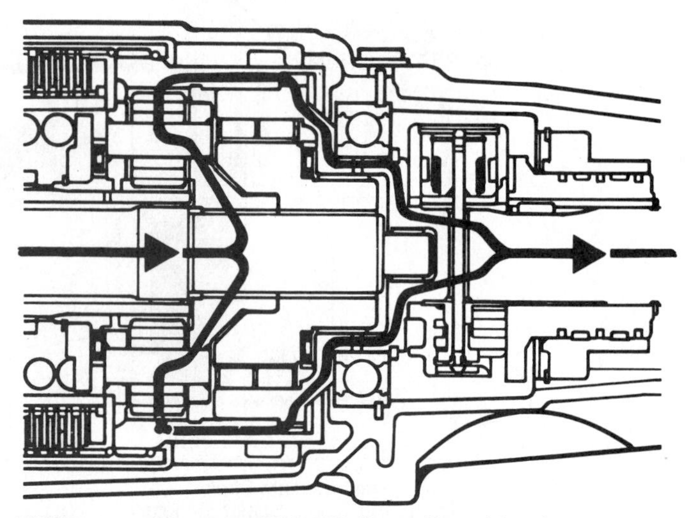

FIGURE 4–44 Chrysler 42RH/46RH (A-500/A-518) fourth-gear power flow. (Courtesy of Chrysler Corp.)

SHIFT LEVER POSITION	TRANSMISSION CLUTCHES AND BANDS					OVERDRIVE CLUTCHES		
	FRONT CLUTCH	FRONT BAND	REAR CLUTCH	REAR BAND	OVERRUN. CLUTCH	OVERDRIVE CLUTCH	DIRECT CLUTCH	OVERRUN. CLUTCH
Reverse	X			X			X	
Drive Range								
First			X		X		X	X
Second		X	X				X	X
Third	X		X				X	X
Fourth	X		X			X		
2-Range (Manual) Second)		X	X		X		X	X
1-Range (Manual Low)			X	X	X		X	X

J9421-218

CHART 4–7 Chrysler 42RH/46RH (A-500/A-518) Range Reference and Clutch/Band Apply Summary—Selector Pattern P R N D 2 1.

gear. Others acquire four forward speeds by utilizing independent sun gears. Ravigneaux gearsets, though not as common as Simpson gearsets, are used in some three- and four-speed transmissions.

The many advantages associated with planetary gearsets will ensure the application of these gearsets in automatic transmissions into the foreseeable future.

❑ REVIEW

Key Terms/Acronym

Range reference and clutch/band apply chart, breakaway, annulus gear, overrun clutch, overdrive planetary gearset, solenoid, and SMEC (single-module engine controller).

Completion

1. The name given to a compound single planetary configuration integrating a dual pinion planet carrier and two sun gears in a single assembly. ________________

2. The name given to a compound planetary configuration using two single planetary sets in series sharing the same sun gear shaft and output shaft. ________________ ________________

Matching

Mechanics of Operation for Simpson Three-Speed Planetary with TorqueFlite Terminology

Use the given choices and select the one best answer for the following questions. Indicate your choice by placing the letter code in the question blank. A choice may be used more than once.

GROUP I

A. Front annulus gear
B. Rear annulus gear
C. Front planet carrier
D. Rear planet carrier
E. Sun gear
F. Front planetary unit
G. Rear planetary unit
H. Compound planetary unit

_____ 1. Provides the second-gear ratio.

_____ 2. Provides the reverse-gear ratio.

_____ 3. The stationary planet member in second gear.

_____ 4. The stationary planet member in first gear.

_____ 5. The stationary planet member in reverse.

_____ 6. The output planetary member in first gear.

_____ 7. The output planetary member in second gear.

_____ 8. The output planetary member in reverse gear.

_____ 9. The planet member motion related to the overrun clutch.

GROUP II

A. Front clutch
B. Rear clutch
C. Kickdown band
D. Low-reverse band
E. Overrunning clutch

_____ 1. Holds the sun gear.

_____ 2. Drives the sun gear.

_____ 3. Drives the front annulus gear.

_____ 4. Holds the stationary planet member first-gear acceleration.

Four-Speed Planetary Systems with Overdrive General Motors GM 4L80-E/4L80-EHD Planetary Assembly

1. Label the planetary members in the illustrated GM 4L80-E/4L80-EHD overdrive planetary unit.

2. Shade the gears using the following color code to indicate the input, output, and stationary members as they apply to the overdrive and direct modes of the 4L80-E/4L80-EHD overdrive planetary.

Red: member or members turning at input rpm.
Blue: member turning at overdrive output rpm.
Black: stationary member.

NOTE: Leave carrier pinions plain.

3. Indicate with arrows the direction of rotation of the planetary members. Be sure to show carrier rotation.

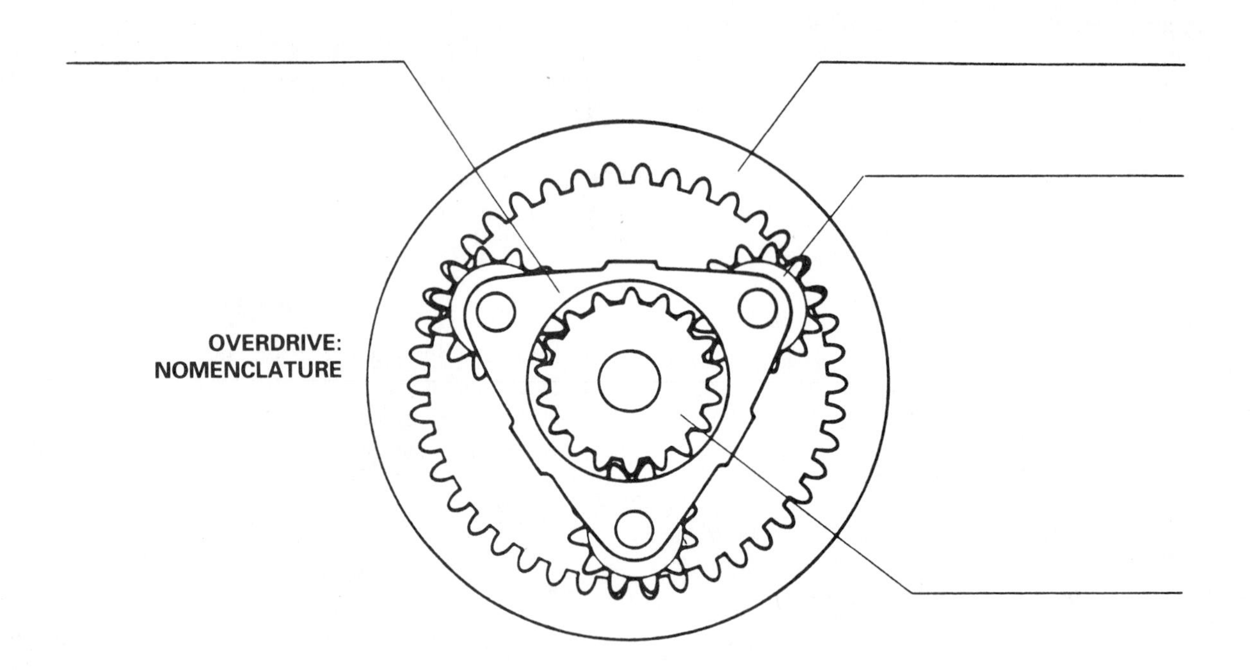

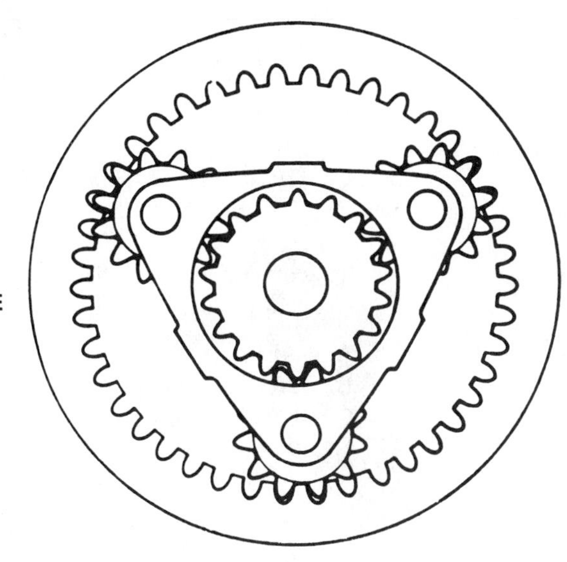

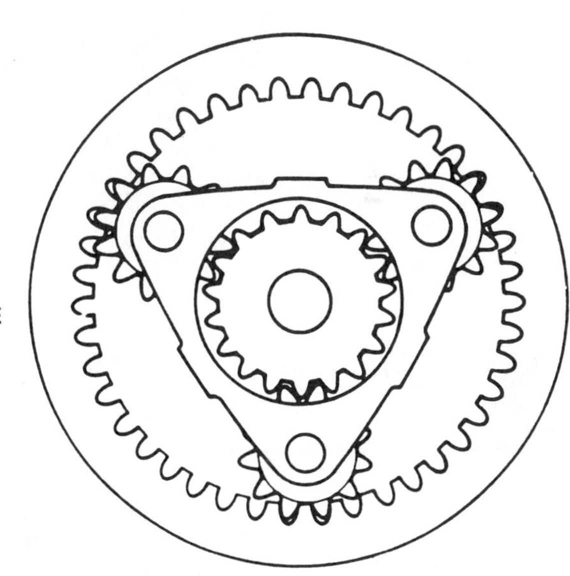

Overdrive Planetary Components

From the following list of overdrive planetary components, select the one best answer to match the question.

A. Overdrive sun gear
B. Overdrive internal gear
C. Overdrive planet carrier
D. Roller clutch
E. Overrun clutch
F. Fourth clutch

____ 1. Splined to the turbine shaft and always turns at converter turbine speed.

____ 2. Allows the drive torque to lock the planet carrier and sun gear together.

____ 3. Locks the sun gear to the transmission case.

____ 4. The output member of the overdrive planetary gearset.

____ 5. Splines to the forward clutch shaft and housing.

____ 6. Locks the sun gear and carrier together.

Multiple Choice

____ 1. To achieve a forward planetary overdrive:
I. The sun gear is fixed.
II. The internal gear is driving.

A. I only
B. II only
C. both I and II
D. neither I nor II

____ 2. The overdrive roller clutch will not hold:
I. in third gear.
II. in fourth gear.

A. I only
B. II only
C. both I and II
D. neither I nor II

____ 3. During overdrive operation:
I. The internal (annulus) gear is turning faster than the input/turbine shaft.
II. The roller clutch is not holding.

A. I only
B. II only
C. both I and II
D. neither I nor II

_____ 4. Uses a band application instead of a holding clutch to hold the sun gear fixed in the overdrive planetary:

I. 42RH/46RH (A-500/A-518)
II. A4LD

A. I only
B. II only
C. both I and II
D. neither I nor II

_____ 5. Incorporates an overdrive planetary at the output end of the three-speed power flow:

A. A4LD
B. 4L80-E/4L80-EHD
C. 42RH/46RH (A-500/A-518)
D. E4OD

_____ 6. Uses a heavy-duty, 800-pound (3558-N) spring to apply the overdrive direct clutch:

A. A4LD
B. E4OD
C. 42RH/46RH (A-500/A-518)
D. 4L80-E/4L80-EHD

_____ 7. To achieve a lockup in the overdrive planetary, the sun gear is clutched to the internal (annulus) gear:

A. A4LD
B. 42RH/46RH (A-500/A-518)
C. 4L80-E/4L80-EHD
D. all of the above

5 Hydraulic Fundamentals

OBJECTIVE:

The objective of this unit is for you to become proficient with the terms and concepts contained in this chapter. Completion of this objective is essential to becoming a successful automatic transmission technician.

CHALLENGE YOUR KNOWLEDGE

Define the following key terms:

Hydraulics, force, pressure, reservoir, pump, valves, actuating mechanism, atmospheric pressure, and vacuum.

Describe the following hydraulic principles:

Pascal's law; and the relationships between force, area, and pressure.

INTRODUCTION

The principal objective in the development of automatic transmission units was to relieve the driver of the physical effort and coordination required to operate a clutch pedal and a shift lever for gear changes. For fully automatic shifting, the transmission must start the car in motion smoothly, swiftly, and silently. It must select the proper gear ratio for any given engine torque/vehicle speed combination. It must also respond immediately to the will of the driver.

To provide the automatic gear ratio changes, fluid energy from a hydraulic pump is routed through a valve control system to apply the band and clutch combinations that work the planetary gearset. The hydraulic system used for this automatic control consists of the following (Figure 5–1):

- Fluid
- Fluid reservoir (sump)
- Pressure source from a hydraulic pump
- Hydraulic operating units (actuators) to apply the bands and clutches
- Control or valve system for regulating pressures and directing fluid flows for automatic and manual gear engagements
- Mechanical control to allow for driver selection of the operating ranges

Operational details of the hydraulic system involve the use of gear and vane pumps, clutch and servo pistons, accumulator pistons, regulating and relay valves, metered orifices, and a host of supporting parts and controls. Before the manner in which all these system devices work is discussed, the technician needs to know the science of pressure hydraulics.

FLUID AND THE HYDRAULIC LEVER

The word *hydraulic* comes from the Greek word for water. For many years, the science of **hydraulics** was nothing more than storing water, moving it from place to place, and operating water wheels. The machine age changed all of this simplicity. Modern hydraulics involves a lot more science and machine applications than ever dreamed by the ancient Greeks, although

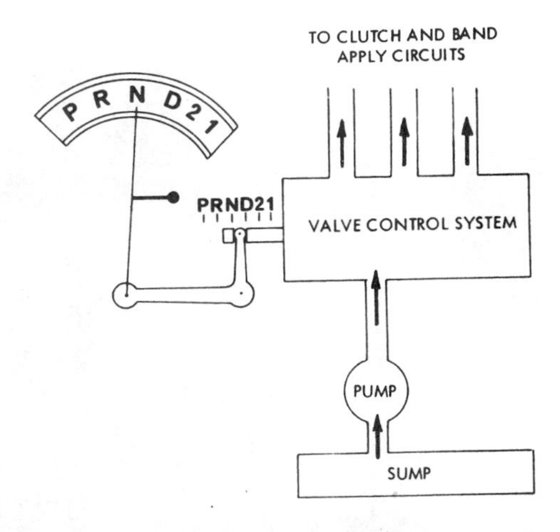

FIGURE 5–1 To clutch and band apply circuits.

hydraulics does involve the use of a fluid. We can begin to "de-Greek" hydraulics by looking at the definition of a fluid.

❖ **Hydraulics:** The use of liquid under pressure to transfer force of motion.

Fluid

A fluid can be either a liquid or a gas. Because our subject matter is hydraulics as it applies to automatic transmissions, the transmitting of force and motion takes place through the use of a liquid. In automatic transmissions, a mineral oil fortified with additives is used.

An essential part of pressure hydraulics is understanding the basic behavior of a fluid and how it acts as a lever arm in transmitting force and motion. This is defined in Pascal's law.

Pascal's Law

In the seventeenth century, Pascal, a French scientist, discovered the hydraulic lever. In laboratory experiments, he proved that force and motion could be transferred by means of a confined liquid. Experimenting with weights and pistons of varying size, Pascal found that mechanical advantage or force multiplication could be obtained in a pressure system. He noted that the relationships between force and distance were exactly the same as with a mechanical lever (Figure 5–2). From the data Pascal collected, he formulated a law, which states:

> Pressure on a confined fluid is transmitted equally in all directions and acts with equal force on equal areas.

To the novice learning hydraulics, Pascal's law may sound like a mass of complicated words. Therefore, we will break it down into easy-to-understand parts, demonstrating it with the kind of equipment Pascal used in his experiments. To simplify the discussion of Pascal's law, it is important to review two terms that are commonly used when talking about hydraulics: *force* and *pressure*. Force and pressure are actually units of measurement used in hydraulics.

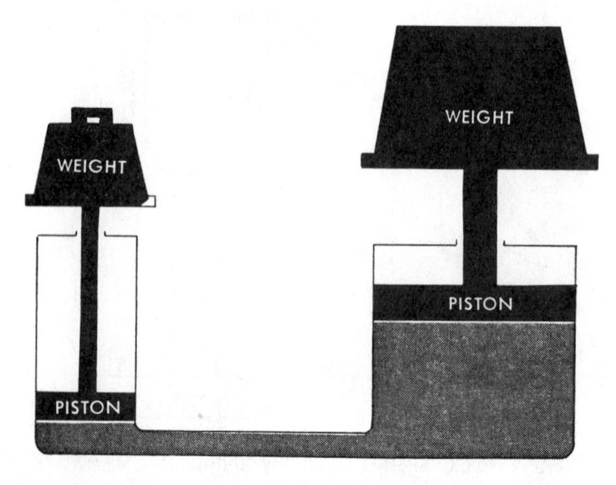

FIGURE 5–2 Pressure is transmitted by fluids. (Courtesy of Chrysler Corp.)

Force. For our purposes, **force** can be defined as a push or pull "effort" measured in pounds (lb) or Newtons (N). It has the ability to cause motion, although motion does not necessarily take place. A classic example of a kind of force is gravity. The force of gravity is nothing more than the weight of an object. If you weigh 175 lb (79 kg), you exert a downward force of 175 lb (778 N) on the floor on which you are standing. (In metrics, weight or mass is measured by kilograms [kg] and force by Newtons [N]. Weight is the result of the earth's gravity versus force, as a push or pull effort, in any direction.)

❖ **Force:** A push or pull effort measured in pounds or Newtons.

In hydraulics, a common type of force encountered is spring force (Figure 5–3). Spring force is the tension in the spring when it is compressed or stretched. The engineering and household unit for force is the pound (lb). Force can be measured on any scale designed to measure weight. In the metric-English conversion, force is measured in Newtons. Pounds of force can be converted to equivalent Newtons by multiplying pounds by a factor of 4.448: 100 pounds of force equals 444.8 Newtons. When weight or mass is measured, the metric-English conversion from pounds to kilograms is calculated by multiplying pounds by a factor of 0.453: 100 pounds equals 45.3 kilograms.

Pressure. Pressure is force divided by area, or force applied per one unit area:

$$\text{pressure} = \frac{\text{force}}{\text{area}}$$

To illustrate pressure (Figure 5–4), a uniform weight of 1000 lb (453 kg) rests on a surface 100 in^2 (645 cm^2). The total force is 1000 lb (4448 N), the weight of the object. The force on each square inch of area is 1000 lb (4448 N) divided by 100 in^2 (645 cm^2), or 10 lb (44.5 N) of pressure on the surface per square inch: pressure = 10 psi (69 kPa).

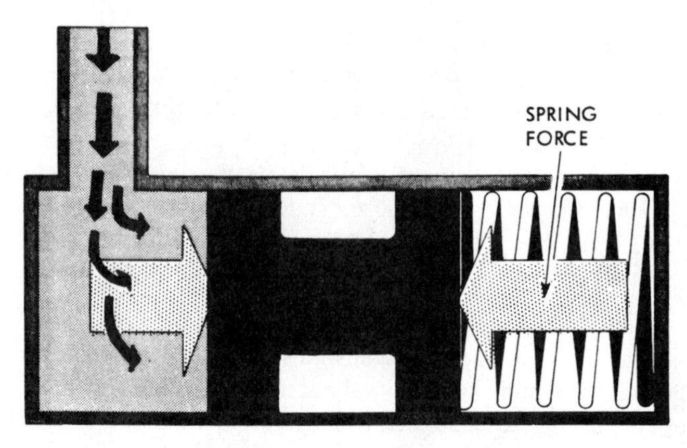

FIGURE 5–3 Hydraulic force versus spring force. (Courtesy of Chrysler Corp.)

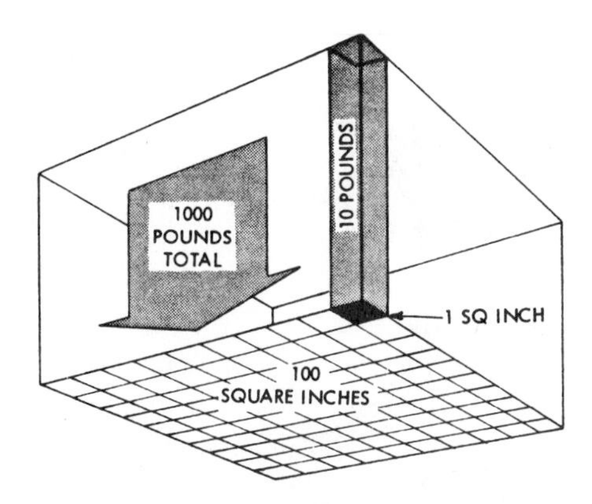

FIGURE 5–4 Pressure is force divided by area. (Courtesy of Chrysler Corp.)

❖ **Pressure:** The amount of force exerted upon a surface area.

The unit for measuring pressure is pounds per square inch (psi). In the metric-English conversion, pounds per square inch is measured by kilopascals (kPa) and can be converted to kilopascals by multiplying psi by a factor of 6.895: 100 psi equals 689.5 kilopascals.

Force/Pressure Relationship

When discussing hydraulics, it is helpful to think of force as the "working component" of pressure. It produces the movement in the system when pressure is applied. In place of a mechanical system, pressure acting through a noncompressible liquid is the means of transmitting force from one location to another:

$$\text{force} = \text{pressure} \times \text{area}$$

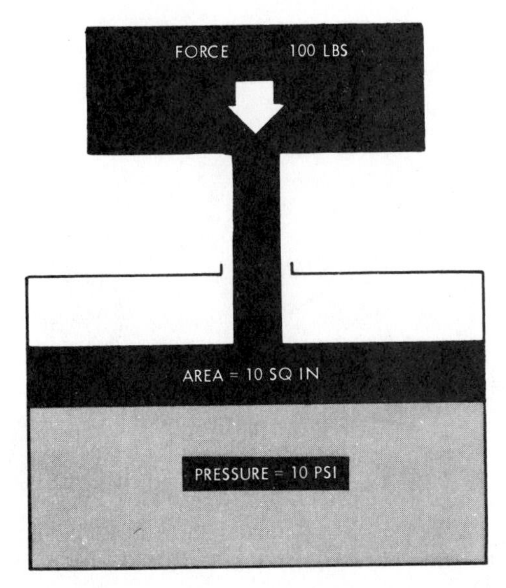

FIGURE 5–6 Pressure is equal to the force applied. (Courtesy of Chrysler Corp.)

FIGURE 5–5 Pressure is force on a confined liquid. (Courtesy of Chrysler Corp.)

Pressure on a Confined Fluid. Pressure is exerted on a confined fluid by applying a force to some area in contact with the fluid. For example, if a cylinder is filled with fluid, and a piston closely fitted to the cylinder has a force applied to it, pressure will be created in the fluid (Figures 5–5 and 5–6). If the fluid is not confined, no pressure will be created, but a fluid flow will result. There must be a resistance to flow to create pressure. In the illustrations, the applied force used is that of gravity, or weight applied downward. The principle is the same, no matter what direction the force is applied.

Figures 5–5 and 5–6 show that the pressure created in the fluid is equal to the applied force divided by the piston area. If the force is 100 lb (444.8 N) and the piston area is 10 in^2 (64.5 cm^2), pressure equals 10 psi (69 kPa). According to Pascal's law, the pressure of 10 psi (69 kPa) is equal everywhere in the trapped fluid (Figure 5–7). Pressure on a confined fluid is transmitted and undiminished in all directions.

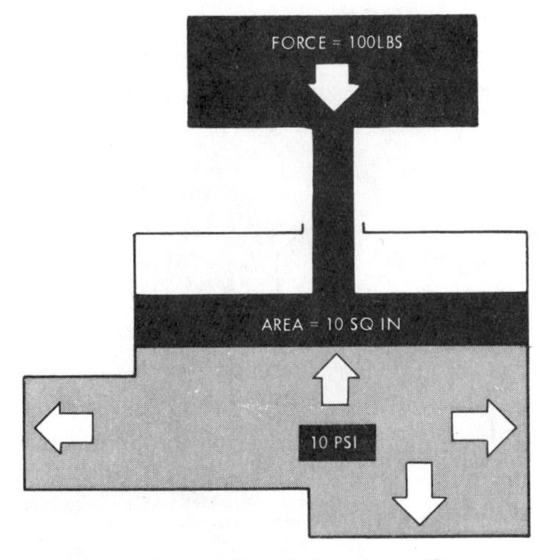

FIGURE 5–7 Pressure is equal in all directions. (Courtesy of Chrysler Corp.)

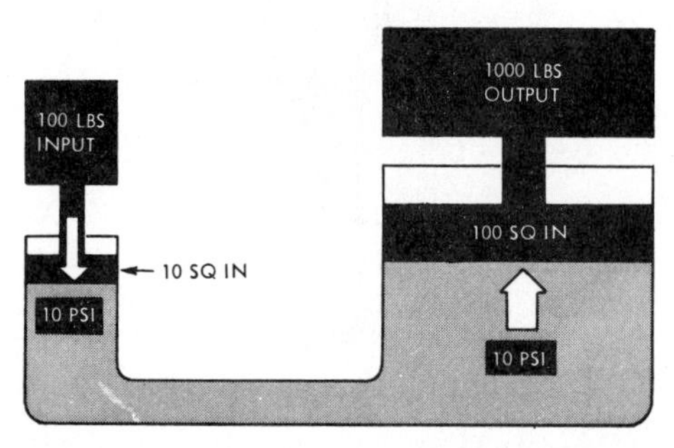

FIGURE 5–8 Force is pressure multiplied by area. (Courtesy of Chrysler Corp.)

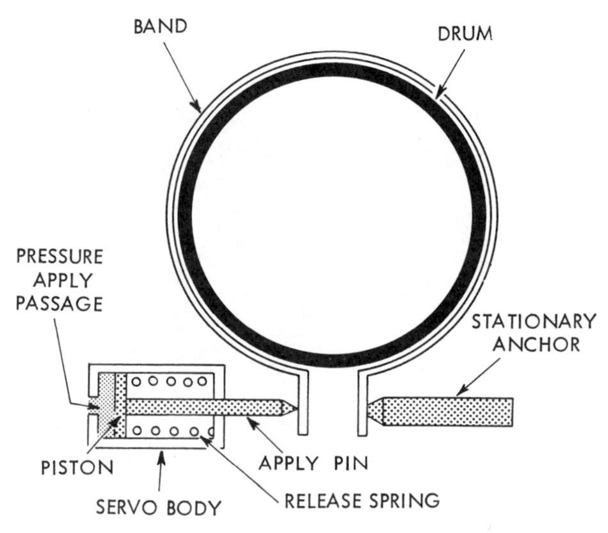

FIGURE 5–9 A servo is a hydraulic piston used for band application.

No matter what the shape of the container, and no matter how large the container is, this pressure will be maintained throughout, as long as the fluid is confined.

Force on an Area. Another part of Pascal's law states:

Pressure acts with equal force on equal areas.

The greater the area, the greater the force. In fact, the total force on any area equals the pressure multiplied by the area (Figure 5–8). In the illustration, 10 psi (69 kPa) is created and applied to a piston with an area of 100 in^2 (645 cm^2). Thus, a total of 10 times 100, or 1000 lb (4448 N), of force is exerted. Input force may be multiplied 100:1, or even 1000:1, by increasing the size of the output piston.

The servo and clutch pistons in an automatic transmission are working examples of how hydraulic pressure is transformed into a mechanical force to apply the bands and clutches (Figures 5–9 and 5–10). If the transmission apply pressure is 100 psi (689.5 kPa), a 5-in^2 (32.25-cm^2) servo piston and a 10-in^2 (64.5-cm^2) clutch piston would develop 500 lb (2224 N) and 1000 lb (4448 N) of apply force, respectively.

Conservation of Energy and the Hydraulic Lever

The law of conservation of energy says that "energy can neither be created nor destroyed." The only way to get a large output force with a small input force is to make the input force travel farther. In a mechanical setup using the fulcrum and lever principle (Figure 5–11), a 100-lb (45-kg) weight is used to move a 1000-lb (453-kg) weight with a lever. It should be obvious that to get an output ten times the input lever, the input lever arm has to be ten times as long. The long input arm produces 1000 ft-lb of work. (Work involves force and motion and is expressed in foot-pounds. The term *foot-pounds* is commonly used for reference to "torque," which technically should be expressed in pound-feet. The incorrect use of this terminology, however, has not caused any harm.)

$$\text{Work} = \text{effective force} \times \text{distance moved}$$
$$= 100 \text{ lb} \times 10 \text{ ft} = 1000 \text{ ft-lb}$$

$$\text{Torque} = \text{effective force} \times \text{lever arm length}$$
$$= 100 \text{ lb} \times 10 \text{ ft} = 1000 \text{ lb-ft } (1356 \text{ N-m})$$

For every foot the 1000-lb (453-kg) weight moves, the 100-lb (45-kg) weight has to move 10 ft. The energy transfer in this

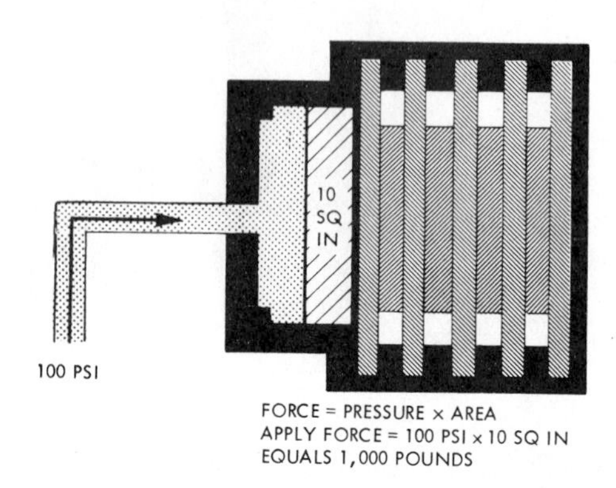

FIGURE 5–10 A hydraulic piston used for clutch application.

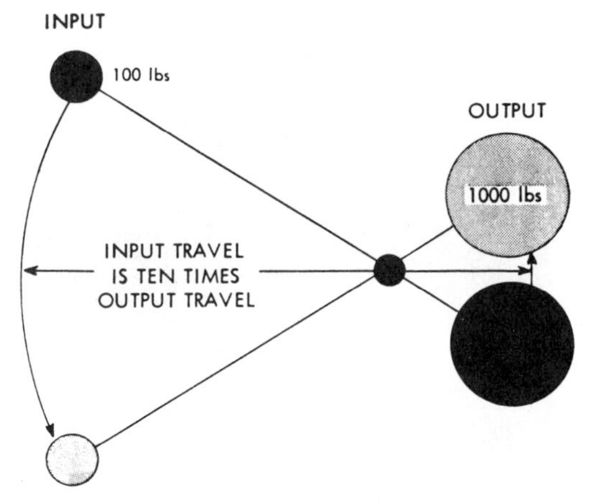

FIGURE 5–11 Distance and force comparisons. (Courtesy of Chrysler Corp.)

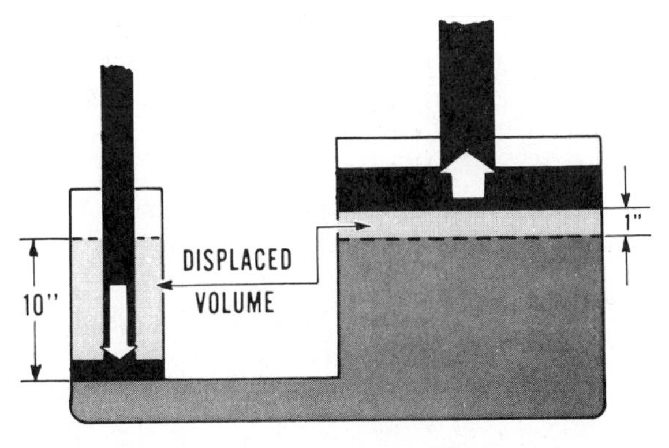

FIGURE 5–12 Hydraulic piston movement. (Courtesy of Chrysler Corp.)

system can be easily measured in foot-pounds (ft-lb). The 1000-lb (453-kg) weight moves 1 ft, and therefore it also has 1000 ft-lb of energy. In summation, if friction losses are ignored, 1000 lb (453 kg) of energy is transferred in the system with no loss or gain, thus proving the law of conservation of energy.

The law of conservation of energy can be related to hydraulics. Returning to a small and large piston illustration (Figure 5–12), the same weight-to-distance relationship is established as with the lever. In this case, if the values are used from Figure 5–8, the input piston has to travel 10 in to displace enough fluid to move the output piston 1 in, a lever ratio of 10:1. In this example, the energy transfer is calculated in inch-pounds (in-lb). It works out to 1000 in-lb at each piston if you ignore friction loss in the system.

As a final example of the hydraulic force/pressure relationship, a three-servo-piston hydraulic system is illustrated in Figure 5–13. A 1.0-in^2 (6.45-cm^2) piston with an apply force of 10 lb (44.5 N) produces a system pressure of 10 psi (69 kPa). The use of three sizes of servo piston areas illustrates how the applied output force and piston travel is controlled.

SUMMARY OF FUNDAMENTAL LAWS OF HYDRAULICS

- A fluid always seeks a common level.
- For all practical purposes, a liquid fluid is noncompressible, which gives it the ability to transmit force. It would take 32 tons (64,000 lb or 332,800 N) to compress the volume of 1 in^3 (16.387 cm^3) of water by 10%.
- The pressure applied to a fluid in a closed system is transmitted equally in all directions and to all areas of the system.
- A fluid under pressure may change its natural state or character (liquid versus vapor).

BASIC HYDRAULIC SYSTEM

With an understanding of Pascal's law and the principles of the hydraulic lever, let us look at how a simple hydraulic system works. Every pressure hydraulic system has certain basic components (Figure 5–14). Even the seemingly complex control system in an automatic transmission has the same basic units.

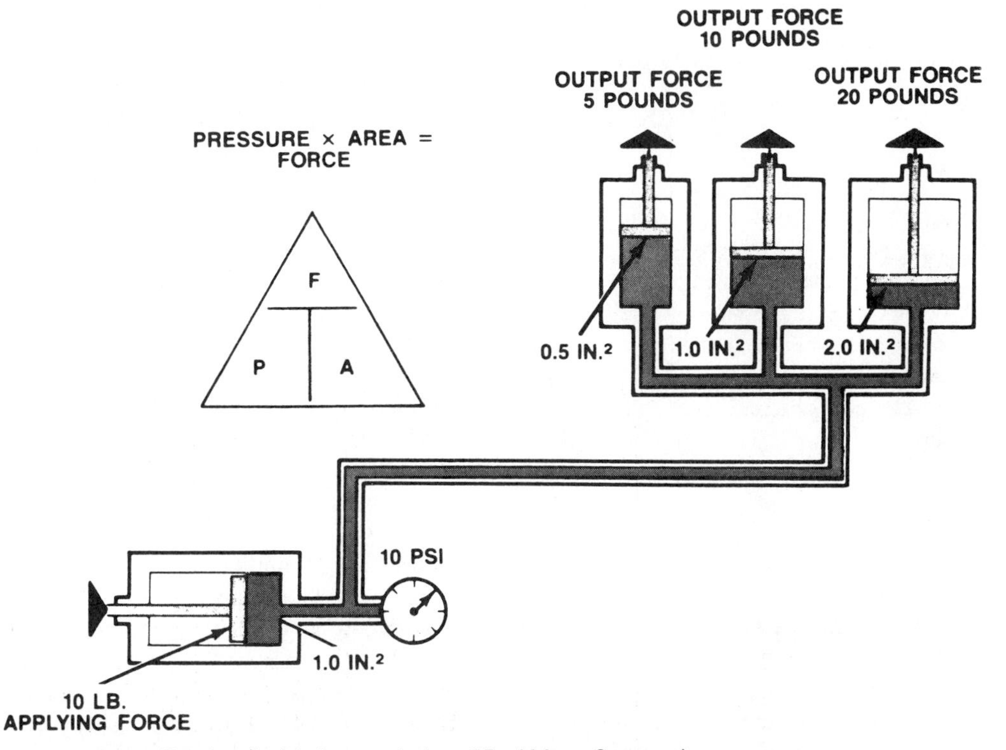

FIGURE 5–13 Pressure versus force. (Reprinted with the permission of Ford Motor Company)

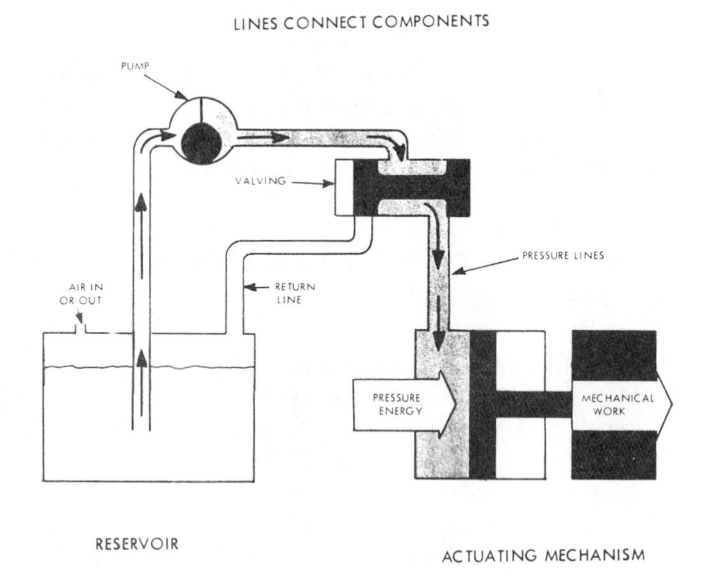

FIGURE 5–14 Basic hydraulic system. (Courtesy of Chrysler Corp.)

Reservoir

The **reservoir** (sump) is a storage area for fluid until it is needed in the system. In some systems in which there is constant circulation of the fluid, the reservoir aids cooling by transferring heat from the fluid to the container and ultimately to the atmosphere. In a brake system, the reservoir is part of the master cylinder (Figure 5–15). In a power steering system, the reservoir surrounds the pump housing. In automatic transmissions, the control system reservoir is the oil pan.

❖ **Reservoir:** The storage area for fluid in a hydraulic system; often called a sump.

Pump

The **pump** creates flow and applies force to the fluid. It pushes the fluid into the system, and pressure is built up when the fluid encounters resistance. Here is an important point to remember. The pump cannot create pressure by itself—it can only create flow. If the flow does not meet any resistance, it is referred to as a free flow, and there is no pressure buildup. There must be a blind alley, a dead end, or a resistance to flow in the system to create pressure.

❖ **Pump:** A mechanical device designed to create fluid flow and pressure buildup in a hydraulic system.

Pumps can be the reciprocating piston type (as in a brake master cylinder), or they can be rotary (like the pump in an automatic transmission). In hydraulics, piston pumps are usually operated manually and rotary pumps are driven by an engine or an electric motor. In automatic transmissions, the pump is driven by the engine through the converter external hub or an internal pump drive shaft splined to the converter cover.

Valving

Valving regulates and directs the fluid. Some **valves** simply interconnect passages, telling the fluid where to go and when. Other valves control or regulate pressure and flow. In a brake system, for instance, valving is made up of the piston and ports in the master cylinder. In automatic transmissions, the valving for controlling the shifts and shift quality is housed in a control valve body.

❖ **Valves:** Devices that can open or close fluid passages in a hydraulic system and are used for directing fluid flow and controlling pressure.

Actuating Mechanism

The working or **actuating mechanism** changes pressure energy to mechanical force. This is where the flow from the pump runs into a dead end and causes pressure to build up. The pressure works against a surface and causes a force to be applied. In a brake system, there are eight actuating mechanisms: the wheel cylinder pistons (Figure 5–15). In automatic transmissions, there are servo pistons and clutch pistons to apply the bands and clutches.

❖ **Actuating mechanism:** The mechanical output devices of a hydraulic system, for example, clutch pistons and band servos.

Lines

The individual components of a hydraulic system need to be tied together for the system to operate. These components are interconnected by tubing, hoses, or passages that are machined or cast in the system housing and attachments. Pressure lines carry fluid from the pump to the actuating mechanisms, and return lines release fluid to the sump when pressure is released. In many systems, such as brakes and automatic transmissions, the same lines perform both functions.

Complete System

The preceding basic components make up any hydraulic system, whether it is simple or complex. An understanding of how these components work, along with some basic facts about hydraulic systems, will let you approach any system with confidence. No matter how many lines and how many valves are in the system, each has a basic function that can be studied apart from the rest.

Flow in a Circuit

Flow has been mentioned several times. Flow is what comes out of the pump when it is pumping and is commonly measured in gallons per minute (gpm). If the system output is force and motion, there is continuous flow to the actuating mechanism. If the output is force only, there is very little flow in the system, only enough to maintain pressure and make up for normal

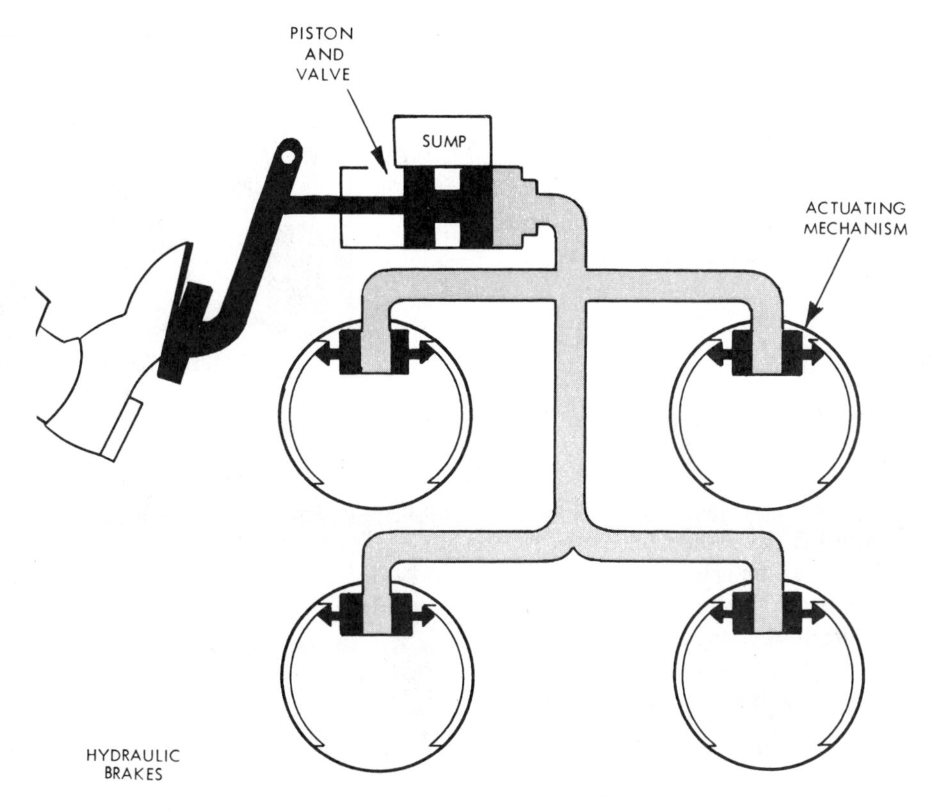

FIGURE 5–15 Brake system as an application of hydraulics. (Courtesy of Chrysler Corp.)

leakage. The pump is still delivering fluid, but it is bypassed to the sump by a regulator valve. In any pressure hydraulic system, the components are kept full of fluid at all times. Thus, response to flow, or pressure buildup, is instantaneous. This is one reason that air, which is compressible, cannot be tolerated in the pressure system and must be bled out for proper operation of the system.

HYDRAULIC JACK

Now let us look at the basic component makeup of a hydraulic jack circuit and examine how it works (Figure 5–16). The small piston is the pump, and the large piston is the actuating mechanism. The large piston is used to raise a load. Its output, therefore, is force and motion. The system makes use of pressure and flow: pressure to supply force and flow to supply motion. A reservoir and valving are needed to permit repeated stroking of the pump, which results in raising the output piston another notch with each stroke. Two check valves are needed, one to keep the load from lowering on the intake stroke, and the other to prevent pressure loss on the power stroke.

The component makeup of the hydraulic jack circuit meets the basic hydraulic system requirements. Before we examine the manner in which it operates, however, we will look at two terms that need to be defined.

Atmospheric Pressure. Atmospheric pressure is the force exerted on everything around us by the weight of the air. If a 1-in^2 column of air as high as the atmosphere goes could be isolated, it would weigh 14.7 lb at sea level. Because air has mass and is confined to the earth's atmosphere, 14.7 psi is exerted equally over everything on the earth's surface at sea level. A pressure of 14.7 psi, therefore, is often referred to as 1 atmosphere. Atmospheric pressure varies with altitude and weather conditions. Denver, the mile-high city, has an atmospheric pressure of less than 1 atmosphere.

- ❖ **Atmospheric pressure:** The force exerted on everything around us by the weight of the air. At sea level, atmospheric pressure is 14.7 psi (pounds per square inch), or in the metric system, 1.0355 kg/cm^2 (kilograms per square centimeter).

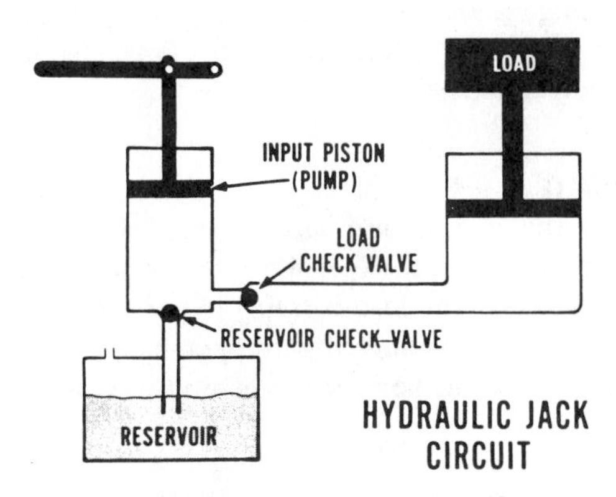

FIGURE 5–16 Diagram of hydraulic jack circuit. (Courtesy of Chrysler Corp.)

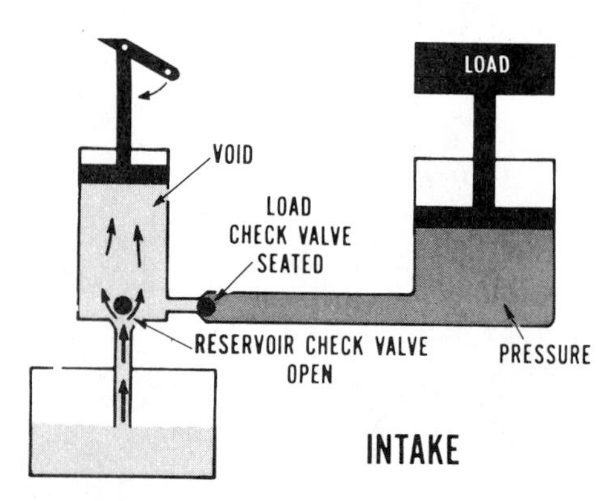

FIGURE 5–17 Intake stroke in a jack. (Courtesy of Chrysler Corp.)

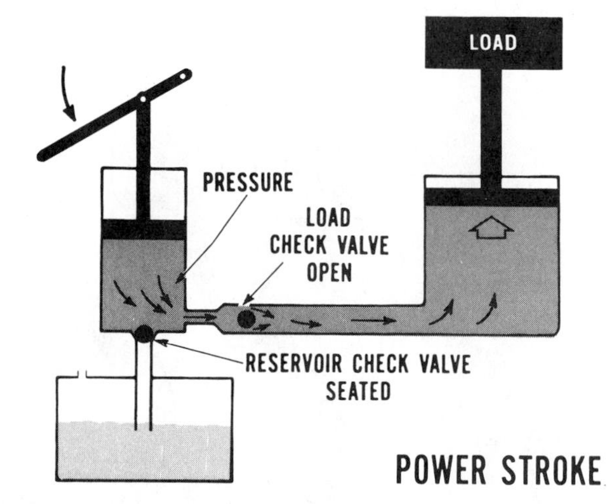

FIGURE 5–18 Pressure stroke in a jack. (Courtesy of Chrysler Corp.)

Vacuum. Technically, a **vacuum** is the absence of pressure. Actually, any condition in which pressure is less than 1 atmosphere is referred to as a vacuum condition. For example, when you sip on a straw, the pressure in the straw is less than 1 atmosphere. The liquid in your glass is still at atmospheric pressure—it is the pressure difference that forces the liquid up through the straw.

❖ **Vacuum:** A negative pressure; any pressure less than atmospheric pressure.

Atmospheric pressure and vacuum play an important part in the later discussion of hydraulic pump operation.

Intake Stroke

As we study the pump system in Figure 5–17, we see that as the pump piston is stroked upward, a partial vacuum is created below it. Atmospheric pressure in the reservoir forces fluid past the reservoir check valve, which is unseated by the flow. The load is prevented from coming down by high pressure seating the load check valve and preventing any back flow.

Power Stroke

When the pump piston is stroked downward (Figure 5–18), the pressure builds up below it, seating the reservoir check valve and preventing return of the fluid to the sump. The load check valve opens, and fluid is forced under the large piston, raising the load another notch.

Lowering

To lower the load, a third valve is connected, a manually controlled needle valve between the large piston and the reservoir (Figure 5–19). The load is trying to push fluid back past the needle valve to the sump. Slightly opening the needle valve meters the fluid back to the reservoir, permitting gravity to bring the load down.

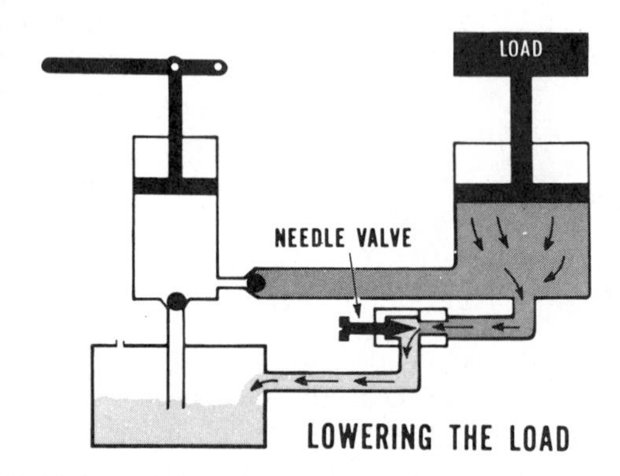

FIGURE 5–19 Lowering a jack. (Courtesy of Chrysler Corp.)

SUMMARY

Within transmissions, hydraulic systems provide the force necessary to provide automatic gear changes and many other essential operations. The systems function on common laws of fluid hydraulics.

Pascal, a seventeenth century French scientist, studied and developed many of the general principles associated with hydraulic fluid power. He determined that pressure on a confined fluid is transmitted equally in all directions and acts with equal force on equal areas. The amount of pressure generated in a confined area is related to the amount of force and the size of the piston or valve. Pressure is force divided by area. Conversely, the amount of hydraulic force available on a movable piston or valve is directly related to the fluid pressure and the surface area of the piston or valve. Force is pressure multiplied by area.

All basic hydraulic systems contain similar components. These include a reservoir, a pump, valving, an actuating mechanism, and fluid lines. Through the use of fluid hydraulic systems, force and motion can be transmitted and controlled.

REVIEW

Key Terms

Hydraulics, force, pressure, reservoir, pump, valves, actuating mechanism, atmospheric pressure, and vacuum.

Completion

1. A fluid can either be a liquid or a ________________. In automatic transmissions, a ________________ is used as the fluid medium for transmitting force.

2. Force is defined as a ____________ or ____________ effort with the ability to cause motion. The English unit of measure for force is the ____________ and is abbreviated ____________. In the metric-English conversion, force is measured in ______________ and is abbreviated ________.

3. Weight or mass is determined by the earth's __________. The English measurement for weight is pounds, which is abbreviated ____________ and the metric measurement is ___________ and is abbreviated ____________.

4. Pressure is ____________ per unit area. The English unit for measuring pressure is ____________ per square inch and is abbreviated _________. In the metric-English conversion, pressure is measured in ____________ and is abbreviated __________.

5. Force is the working component of ______________.

6. Work involves force and ____________. The correct English unit for measuring work is ______________.

7. Pressure on a confined fluid is transmitted ___________, everywhere, in all directions.

8. In a hydraulic system, the ______________ creates the fluid flow. If the flow does not meet any resistance, it is referred to as a __________________, and there is no pressure buildup. To create a hydraulic pressure head, the fluid flow must encounter a ____________________, such as a dead end against a hydraulic piston.

9. To convert hydraulic force to mechanical force, _______ ________ pistons and ______________ pistons are used in automatic transmissions to apply the bands and clutches.

10. The pressure exerted on everything around us because of the weight of the air is referred to as ____________ pressure.

11. An absence of atmospheric pressure—a pressure that is less than atmospheric pressure—is referred to as a ______ ____________________.

12. Every hydraulic system needs a reservoir or sump for fluid storage. In an automatic transmission, the fluid supply for the system pump is stored in the ______________.

13. In addition to hydraulic force, a common type of force often encountered in hydraulics is ____________ force.

Problems

1. Convert a 750-lb weight to its metric equivalent. Show your work.

2. Convert a 500-lb force to its metric equivalent. Show your work.

3. Convert 60 psi to its metric equivalent. Show your work.

4. Calculate the pressure on the confined fluid in the following illustration. Show your work.

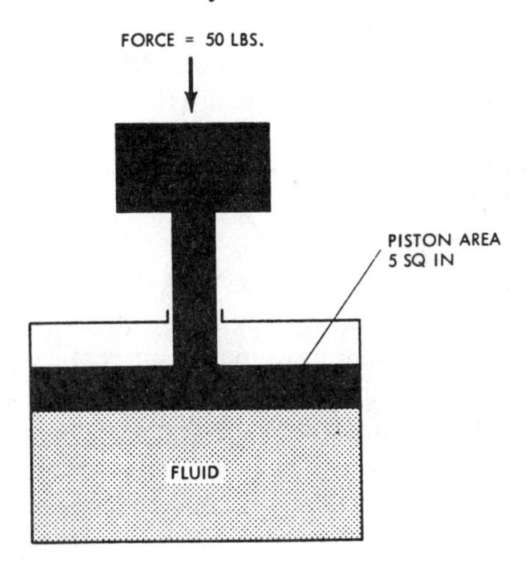

5. Using the data in the illustration below, calculate the following. Show your work.
   (a) Pressure on the confined fluid
   (b) Pressure on piston (B)
   (c) Force on piston (B)
   (d) Piston (A) travels 1 in. How far does piston (B) travel?

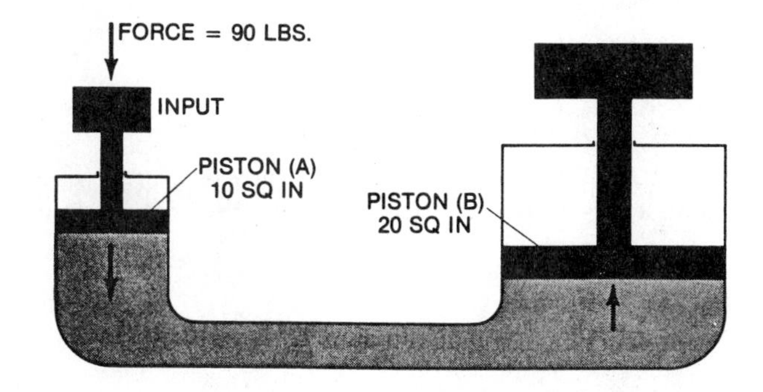

6 Hydraulic System Fundamentals: Pumps and Valves

OBJECTIVE:

The objective of this unit is for you to become proficient with the terms and concepts contained in this chapter. Completion of this objective is essential to becoming a successful automatic transmission technician.

CHALLENGE YOUR KNOWLEDGE

Define the following key terms:

IX rotary lobe pump, IX rotary gear pump, variable displacement (variable capacity) vane pump, orifice, fluid viscosity, relay valve, regulator valve, check valve, relief valve, and spool valve.

List the following:

Two methods to drive a pump; an undesirable characteristic of fixed displacement pumps; an advantage of variable displacement vane pumps; two basic purposes of valves; three materials used for "check balls"; and three forces that act upon and position spool valves.

Describe the operation of the following:

IX pumps (lobe and gear type); variable displacement vane pump; regulator valve; and "check balls."

INTRODUCTION

The hydraulic oil system is the "lifeblood" circuitry of the automatic transmission. A steady supply of oil, routed to internal components at the proper time and appropriate pressure, is essential to the smooth operation of the transmission.

To provide full function of the gear shift range positions, including automatic upshift and downshift capabilities, numerous hydraulic circuits are needed. Within automatic transmissions, many of these circuits are designed to pressurize and move actuating mechanisms for engaging clutch packs and bands to cause the needed shifting and gear ratio changes.

Integral components of hydraulic systems include a pump and assorted valves. They work together to create pressurized fluid flow and to direct the fluid to the appropriate internal devices. This chapter discusses the three common pump configurations and basic valve types and designs that are used in modern transmissions.

HYDRAULIC PUMPS

Basic Operation

The hydraulic pump is the heart of any pressure hydraulic system. It must provide a flow of fluid and develop an operating pressure head from which force and motion can be transmitted. When it fails to operate to specifications, the system encounters partial or total failure. In theory, the pump only creates a flow. There must be a resistance to the flow for pressure to develop. Without resistance, the pump flow is referred to as a free flow, and there is no pressure.

For automatic transmission applications, a variety of rotary pumps are used, and they all work on similar operating principles. Fluid is trapped in chambers that are constantly expanding and contracting—expanding at the pump inlet to allow fluid into the pump, and contracting at the outlet to force fluid into the system under pressure (Figure 6–1).

Many pumps have two round members, with the inner drive member turning inside the outer (Figure 6–2). The members are on different centers; therefore, at one point of rotation, they are in close mesh and provide no clearance. As the members continue to turn, they will separate to a point of maximum clearance and then come back together. The pumping mechanisms (lobes, gears, or vanes) form sealed chambers between the members. These pumping chambers are carried around by rotation of one or both members.

Figure 6–1 shows how the pumping action takes place. At the point where clearance begins to increase, the pumping chamber expands in size, creating a void or low-pressure zone.

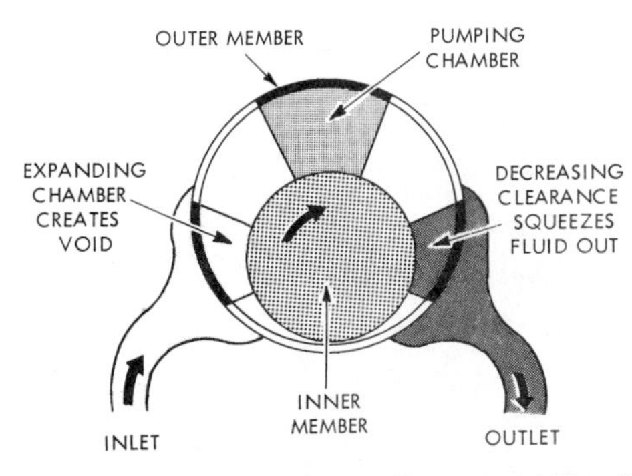

FIGURE 6–1 Rotary pumping chambers. (Courtesy of Chrysler Corp.)

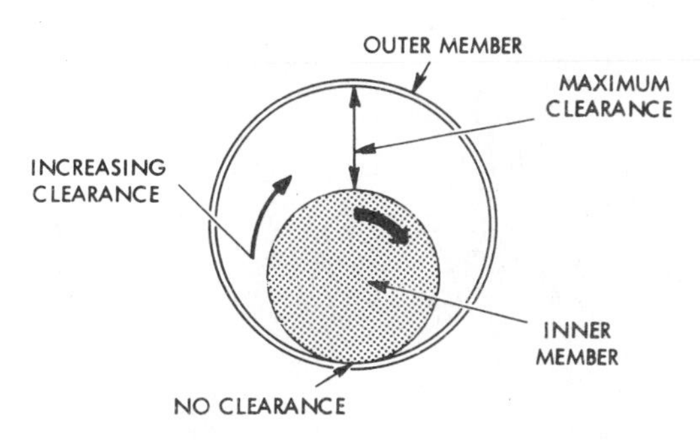

FIGURE 6–2 Rotary pump members are arranged to rotate on different centers. (Courtesy of Chrysler Corp.)

The inlet is located at this void, and atmospheric pressure in the sump forces fluid into the void. The chamber continues to increase in size and take on fluid until rotation carries it past the inlet to the point of maximum clearance. For a few degrees of travel at maximum clearance, the chamber neither increases nor decreases but keeps the fluid trapped and isolated from the inlet side. The outlet is located where the clearance is decreasing. Here the pumping chamber decreases in size, and fluid is squeezed out of the chamber and into the system under pressure. Succeeding chambers follow each other closely so a continuous output is maintained.

In summary, the pumping action is a constant process of expanding and contracting chambers that create low- and high-pressure zones. Although it would appear that the pump operation produces a smooth delivery, the chamber action has a pulsating effect on the fluid.

Pump Drives

A hydraulic pump requires an external source of power to drive the pump. In automatic transmissions and transaxles, hydraulic pumps are driven mechanically by the engine crankshaft through the torque converter assembly by two techniques:

- externally, off the converter drive hub which is designed to engage two lug slots or flats in the pump drive member (Figures 6–3 and 6–4). This pump drive system is currently used on all rear-wheel drive and some front-wheel drive applications.
- internally, by a hex or splined end shaft that fits into the center of the converter cover and to the pump drive member (Figure 6–5). This pump drive style is widely used in FWD transaxles.

The pump operates whenever the engine is operating and satisfies the hydraulic demands of the transmission.

IX Rotary Pumps

The abbreviation IX refers to the internal/external design of the members used in the two styles of IX pumps, rotary lobe and rotary gear (Figures 6–6 and 6–7). The outer pump (driven) member has the internal cut lobe or gear, while the companion inner (drive) member has the external cut, hence the expression IX.

In the **IX rotary lobe pump** (sometimes referred to as a gerotor-type pump) both rotor members turn together. The inner rotor is the drive member and carries the outer rotor by meshing of the lobes (Figure 6–8). A pumping chamber is formed between the lobes. As the lobes separate from the mesh, the chamber expands at the inlet to create the low-pressure zone. At the outlet, the lobes are meshing again, decreasing the chamber size and squeezing the fluid into a high pressure zone and out into the system (Figure 6–9). The inlet is sealed from the outlet by the close clearance between the lobe tips at the point of maximum displacement.

❖ **IX rotary lobe pump:** Sometimes referred to as a gerotor-type pump. Two rotating members, one shaped with internal lobes and the other with external lobes, separate and then mesh to cause fluid to flow.

The pumping action of the **IX rotary gear pump** (Figure 6–10) is similar to that of the IX rotary lobe pump. The inner gear drives the outer gear, and the space between the gears expands as the gear teeth separate and pass the inlet port. The space then decreases as the gear teeth come together and pass the outlet port. There is a crescent-shaped divider between the two gears. Oil is trapped between the divider and the gear teeth of both gears and is prevented from bleeding back into the inlet side.

❖ **IX rotary gear pump:** Contains two rotating members, one shaped with internal gear teeth and the other with external gear teeth. As the gears separate, the fluid fills the gaps between gear teeth, is pulled across a crescent-shaped divider, and then is forced to flow through the outlet as the gears mesh.

The IX lobe and IX gear pumps are classified as fixed displacement pumps, which means that the pumps have a continuous delivery characteristic: the volume is in proportion to the drive speed. The pump output normally exceeds the hydraulic demands of the transmission, especially at high rpm's. The excessive volume is bled off at the system pressure regulator valve to protect against pressure overloads that would

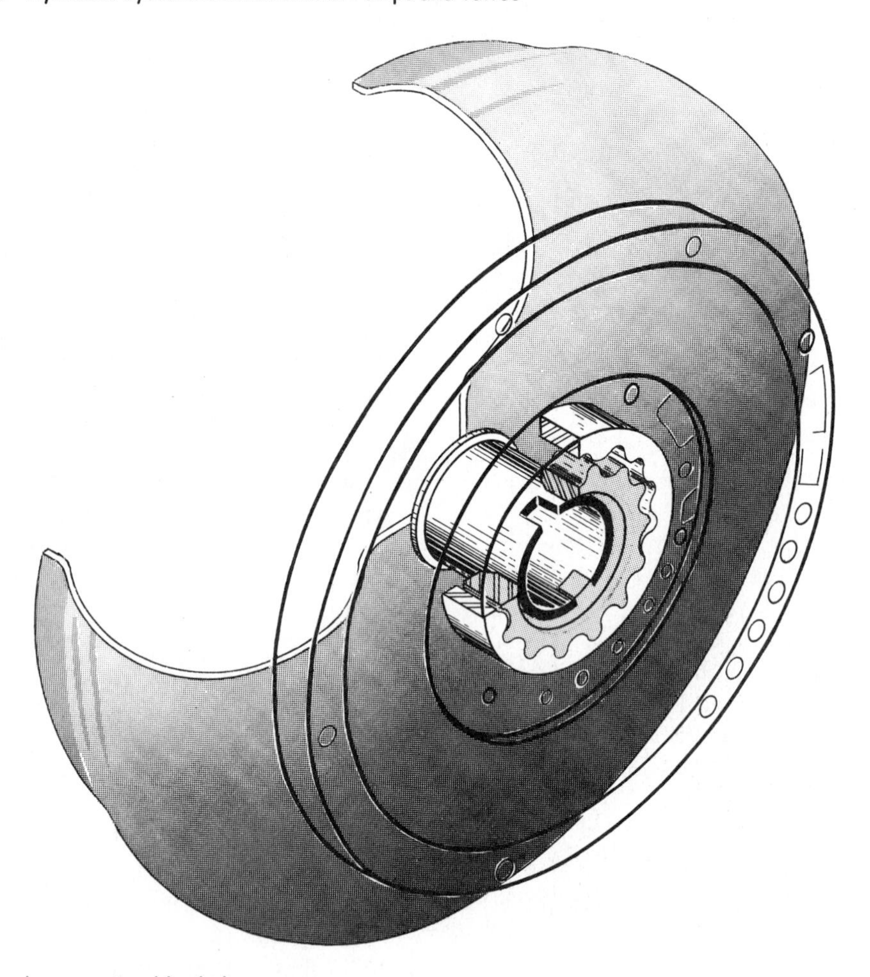

FIGURE 6–3 Oil pump driven by converter drive hub.

A

B

FIGURE 6–4 (a) Typical converter drive hub with drive slots; (b) typical converter drive hub with drive flats.

FIGURE 6–5 (a) Pump drive shaft, converter end; (b) converter cover drives pump shaft (cutaway view); (c) pump drive shaft and pump.

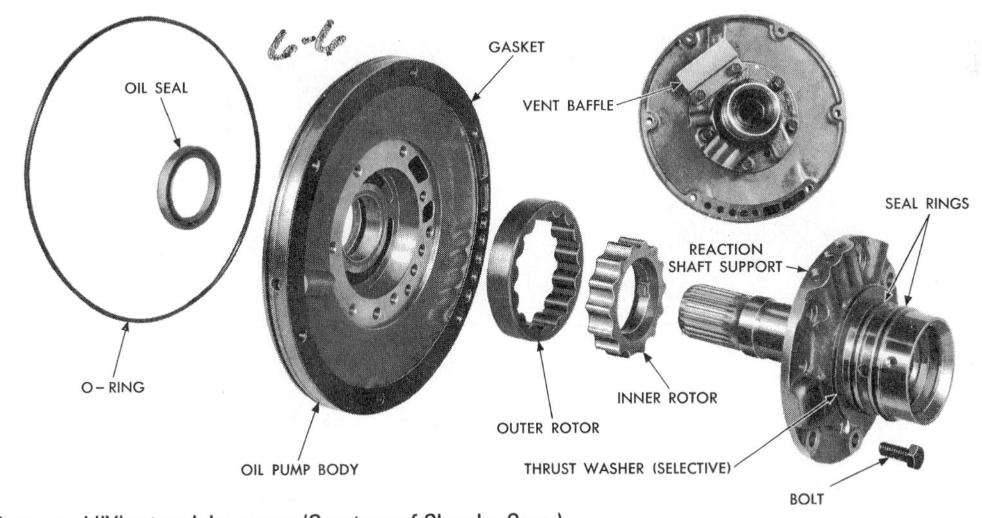

FIGURE 6–6 Internal-external (IX) rotary lobe pump. (Courtesy of Chrysler Corp.)

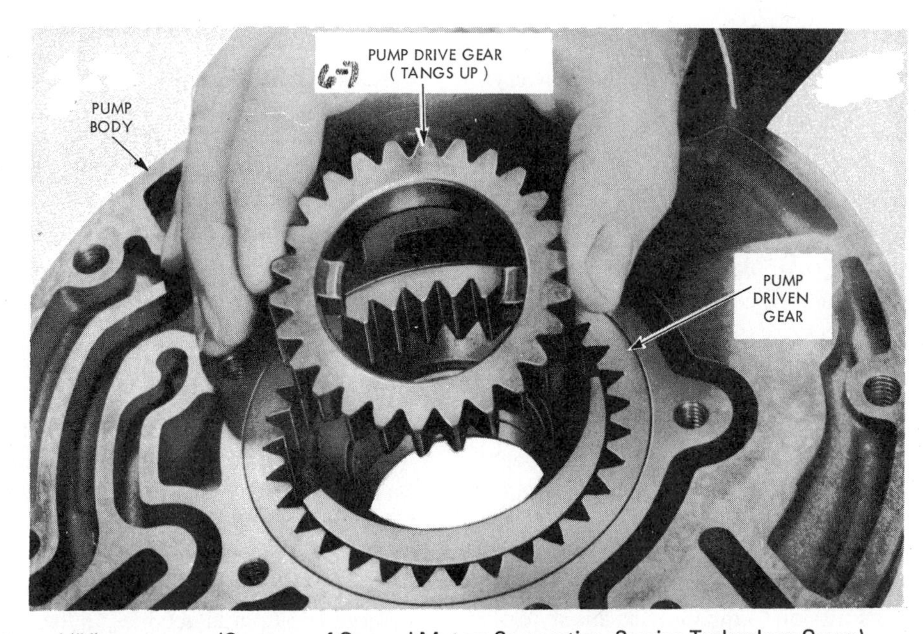

FIGURE 6–7 Internal-external (IX) gear pump. (Courtesy of General Motors Corporation, Service Technology Group)

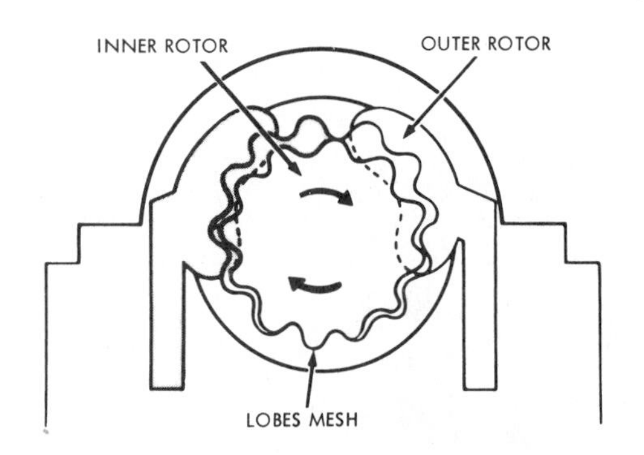

FIGURE 6–8 IX rotary lobe design. (Courtesy of Chrysler Corp.)

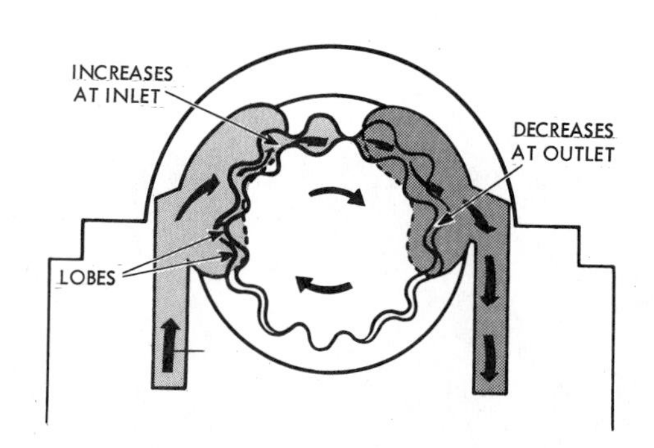

FIGURE 6–9 IX rotary lobe pump flow pattern. (Courtesy of Chrysler Corp.)

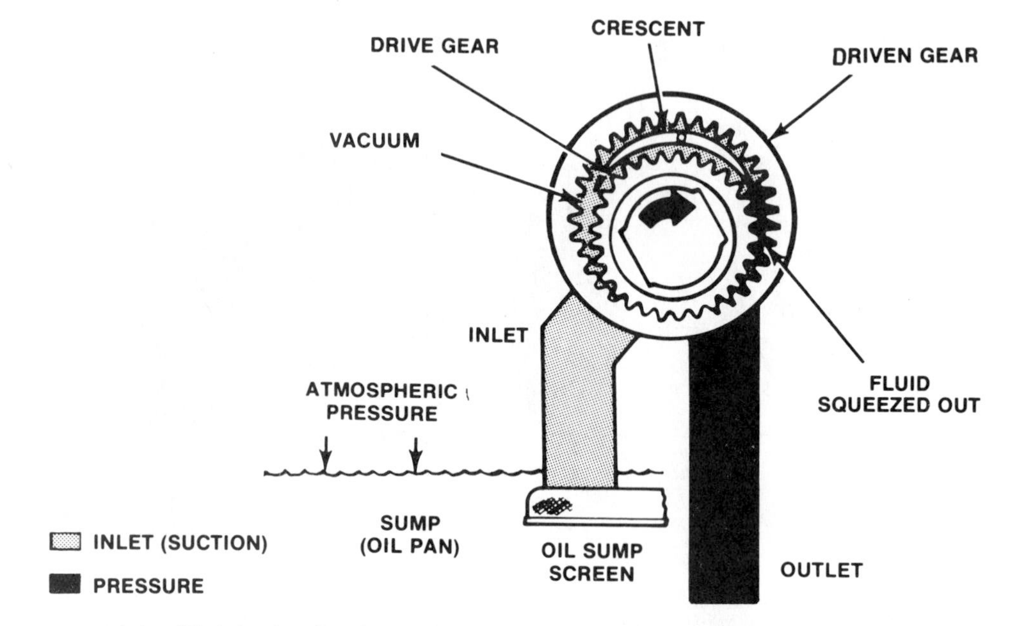

FIGURE 6–10 IX gear pump and simplified circuitry. (Reprinted with the permission of Ford Motor Company)

be damaging to the transmission. Because of the fixed delivery characteristic of the IX pumps, energy or horsepower is wasted in moving fluid that is not needed.

Variable Displacement (Variable Capacity) Vane Pump

A **variable displacement (variable capacity) vane pump** varies its output according to the requirements of the transmission, thereby conserving energy. It offers the advantage of delivering a large volume of fluid when the demand is great, especially at low pump speeds. At high speeds, the pump load is usually not high, and the transmission volume demand is low. Compared to a fixed displacement pump, the variable displacement pump requires very little effort to drive at high speeds. Once the hydraulic demands of the transmission are met, the pump idles along, delivering only makeup oil to maintain the regulated pressure head. General Motors, Ford, and JATCO (Japanese Automatic Transmission Company) use this style of pump in many of their automatic transmission designs.

❖ **Variable displacement (variable capacity) vane pump:** Slipper-type vanes, mounted in a revolving rotor and contained within the bore of a movable slide, capture and then force fluid to flow. Movement of the slide to various positions changes the size of the vane chambers and the amount of fluid flow. Note: GM refers to this pump design as variable displacement, and Ford terms it variable capacity.

A typical variable displacement (variable capacity) vane pump design is shown in Figure 6–11. The pump rotor is engine-driven by a converter pump shaft and carries slipper-type vanes. The rotor and vanes are contained within the bore of a moveable slide that can pivot on a pin. The pivot position of the slide determines the pump output. When the slide pivot position changes, the rotor vanes have the ability to slip in their slots and adjust to the new bore location.

Variable output is obtained in the following manner. When the priming spring pivots the slide in the fully extended position, the slide and rotor are not on the same centers (Figure 6–12[a])—slide and rotor vanes are at maximum output. As the rotor and vanes rotate within the slide bore, the expanding and contracting areas form inlet and pressure chambers. Fluid trapped between the vanes at the inlet side is moved to the pressure side. When the slide moves toward the center, a greater quantity of fluid is allowed to recycle from the pressure side back to the inlet side (Figure 6–12[b]). When the slide and rotor are both centered, a neutral or no-output condition is attained.

The priming spring keeps the slide in the fully extended position so that when the engine is started, full output is immediate. Movement of the slide against the priming spring occurs when the pump pressure regulator valve reaches its predetermined value. At the regulating point, the pressure regulator valve opens a port feed and sends a hydraulic signal (signal oil) to the pump slide outer cavity. This causes the slide to move against the priming spring to cut back on the volume delivery and maintain regulated pressure. The slide can assume an infinite number of positions.

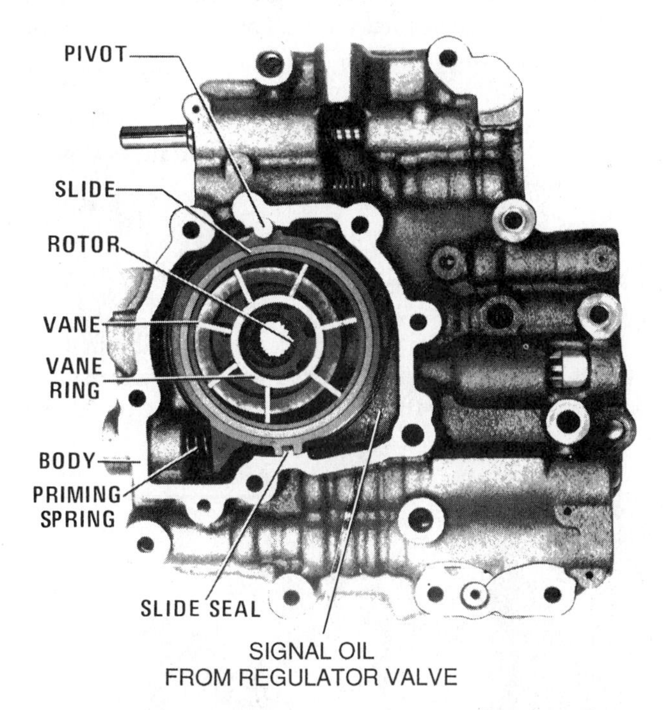

FIGURE 6–11 Variable displacement vane pump. (Courtesy of General Motors Corporation, Service Technology Group)

HYDRAULIC ORIFICE

The simplest means of controlling flow and pressure is by an **orifice.** An orifice is a restriction. It slows down fluid flow to either create back pressure or delay pressure buildup downstream. When fluid is pumped to an orifice, there is not enough room for it to go through all at once, and a back pressure is created on the pump side. If there is a flow path on the downstream side, a pressure difference is maintained across the orifice. Pressure is lower on the downstream of the orifice (Figure 6–13) as long as fluid continues to flow in the circuit. In hydraulic circuits, the orifice is a simple means of lowering pressure. When flow is blocked on the downstream side, Pascal's law applies and pressure equalizes on both sides of the orifice (Figure 6–14). The pressure does not equalize until flow across the orifice stops. The orifice then is used to delay pressure buildup or control fluid flow volume.

❖ **Orifice:** Located in hydraulic oil circuits, it acts as a restriction. It slows down fluid flow to either create back pressure or delay pressure buildup downstream.

Figure 6–15(a) shows a clutch piston with two apply chambers, inner and outer. Note how the inner chamber gets a rapid feed to apply the clutch for initial engagement while the outer chamber apply is delayed by an orifice. By not engaging the clutch with full force, a harsh clutch apply is not experienced by the driver. Eventually, the flow across the orifice ceases, and

A

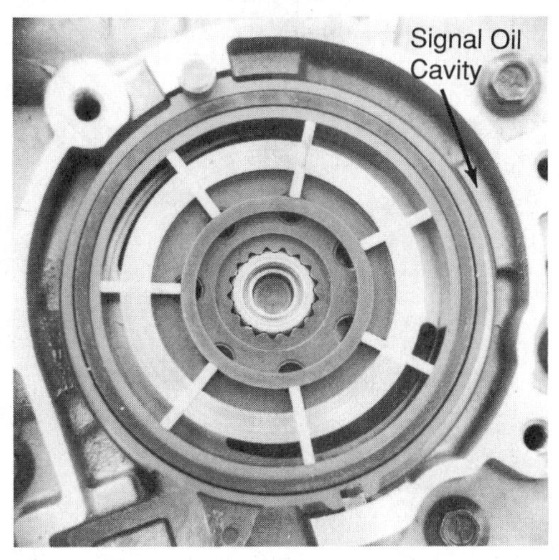

B

FIGURE 6–12 (a) Maximum pump output; (b) minimum pump output.

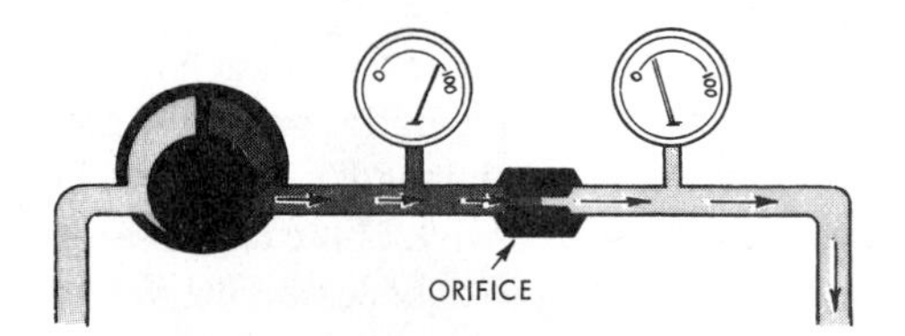

FIGURE 6–13 An orifice acts as a fixed or metered opening. (Courtesy of Chrysler Corp.)

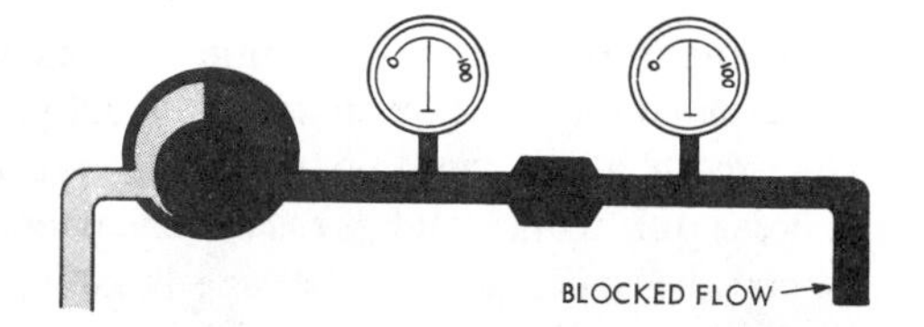

FIGURE 6–14 Pressure equalizes when flow through orifice stops on the blocked side. (Courtesy of Chrysler Corp.)

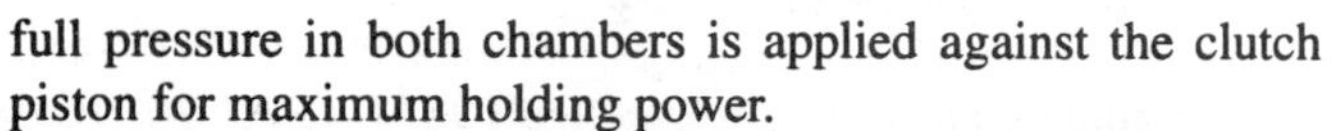

full pressure in both chambers is applied against the clutch piston for maximum holding power.

In automatic transmissions, the change from one gear ratio to another requires delicate control. Orifices are generally used in the hydraulic circuitry to obtain a desired timed release and apply of clutches and bands for quality shifts. Orifices are also used in the fine-tuning of regulating valve circuitry.

Many new generation transmissions incorporate a variable orifice thermal valve along with a fixed orifice. It compensates for cold and hot **fluid viscosity** differences that would affect the flow rate through a fixed orifice, and consequently the timing and engagement feel of a clutch or band. The objective is to have constant quality timing and engagement feel, cold or hot, by using a variable orifice thermal valve. A simplified circuit setup is shown in Figure 6–15(b). With the fluid temperature at 0°F (-18°C), the thermo orifice is wide open, and two flow paths charge the circuit to compensate for cold fluid viscosity. As the fluid gets warm, the thermal valve gradually closes, and circuit apply oil flows through the fixed orifice only. The variable orifice thermal valve is favored for use in forward and reverse clutch circuits.

❖ **Fluid viscosity:** The resistance of a liquid to flow. A cold fluid (oil) has greater viscosity and flows slower than a hot fluid (oil).

HYDRAULIC VALVES

Valves are used for the control of hydraulic circuits. In automatic transmissions, valves are used for pressure regulation and directing fluid traffic to the appropriate servo and clutch

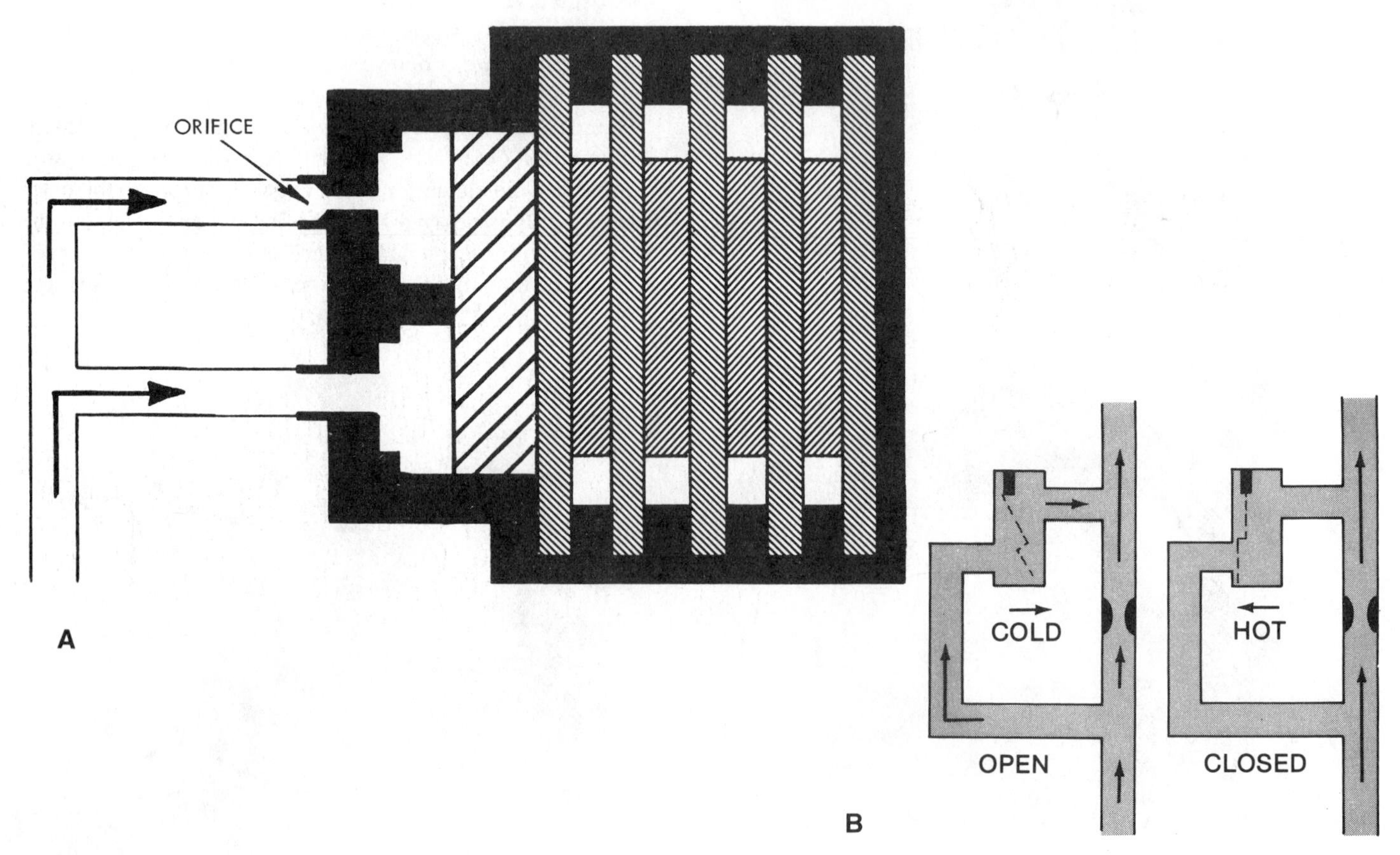

FIGURE 6–15 (a) An orifice gives delaying action. (Courtesy of General Motors Corporation, Service Technology Group); (b) combined variable orifice and fixed orifice control circuit.

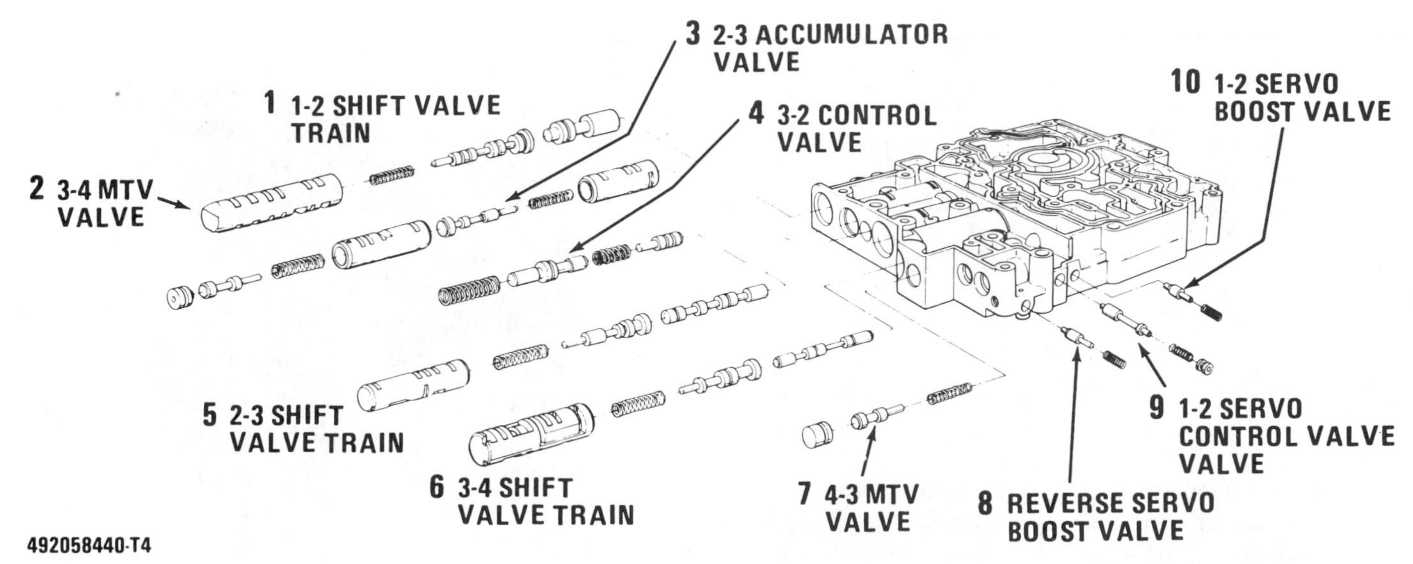

FIGURE 6–16 Partial makeup of typical valve body assembly. (Courtesy of General Motors Corporation, Service Technology Group)

circuits required for each gear ratio change. With only a few exceptions of manual control, the valves are automatically operated by programmed hydraulics. The valves are of a spool valve design and are grouped together with their springs in an assembly called the valve body (Figure 6–16). Some transmission designs incorporate the pressure regulator and converter clutch valve trains in the pump cover (Figure 6–17).

Hydraulic valves can be divided generally into two classes: those that direct flow and pressure, and those that regulate or control flow and pressure. Valves that direct flow and pressure are like ON/OFF switches. They simply connect or disconnect interrelated passages without restricting the fluid flow or changing the pressure. These valves are usually called switch or **relay valves** and perform the simple function of turning the circuit on or off. **Regulator valves** are valves that change the pressure of the oil as it passes through the valve by bleeding off (or exhausting) some of the volume of oil supplied to it.

- **Relay valve:** A valve that directs flow and pressure. Relay valves simply connect or disconnect interrelated passages without restricting the fluid flow or changing the pressure.
- **Regulator valves:** A valve that changes the pressure of the oil in a hydraulic circuit as the oil passes through the valve by bleeding off (or exhausting) some of the volume of oil supplied to the valve.

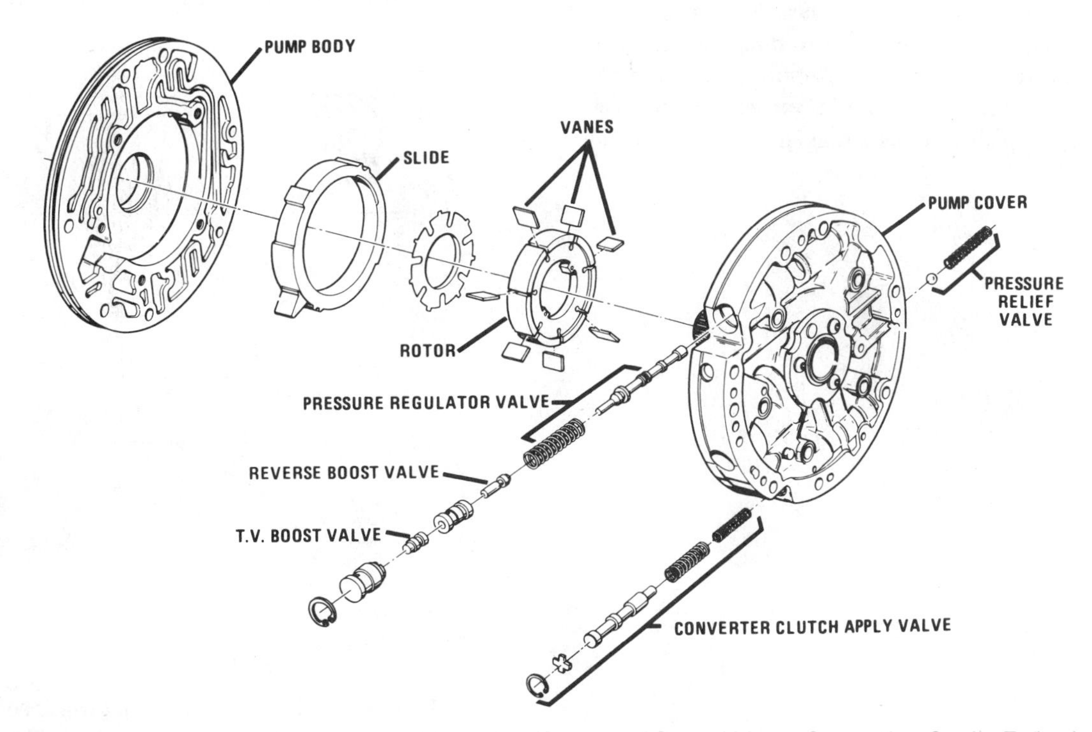

FIGURE 6–17 Control valves incorporated in pump assembly casting. (Courtesy of General Motors Corporation, Service Technology Group)

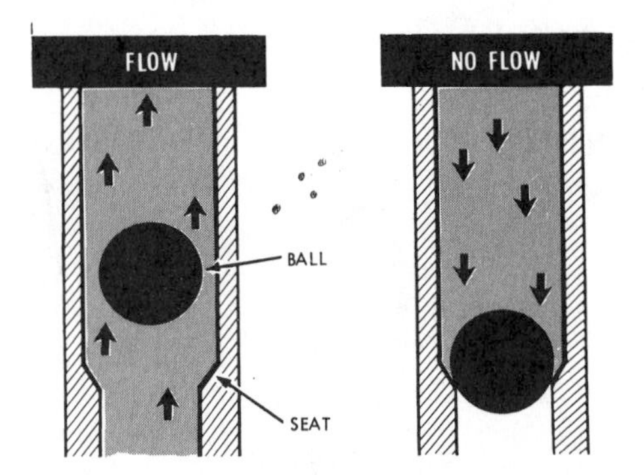

FIGURE 6–18 One-way ball check valve. (Courtesy of Chrysler Corp.)

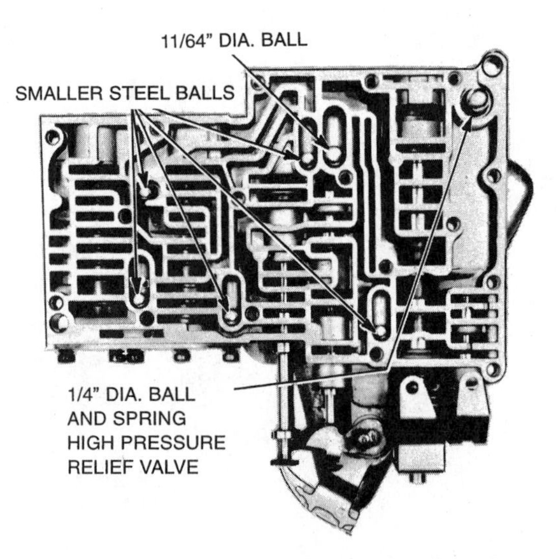

FIGURE 6–19 Ball-type check valves are used to direct flow patterns in transmissions.

Let us look at some of the relay control and regulating valves typically used in automatic transmissions. Understanding hydraulic circuits will be easier if we determine which valves function as relay or regulating components.

Check Valves

A **check valve** is a simple one-way directional valve. It routes the hydraulic circuit oil flow in the correct direction. The ball check–style valve, often referred to as "check ball" and seat, is commonly used in transmissions (Figure 6–18). It consists of a steel, viton, or nylon ball and seat arrangement. When pressure is applied on the seat side, the ball is forced off the seat and permits flow. Pressure on the opposite side holds the ball against its seat and blocks flow in the reverse direction.

❖ **Check valve:** A one-way directional valve in a hydraulic circuit, commonly using a "check ball" and seat arrangement.

In automatic transmission valve control body assemblies, the ball check plays an important part in directing the many flow patterns of the fluid (Figure 6–19). Some of the ball checks have two seats to control the fluid traffic between interconnecting hydraulic circuit paths. In this arrangement, the check ball is seated against the dead-pressure line by the high-pressure circuit (Figure 6–20).

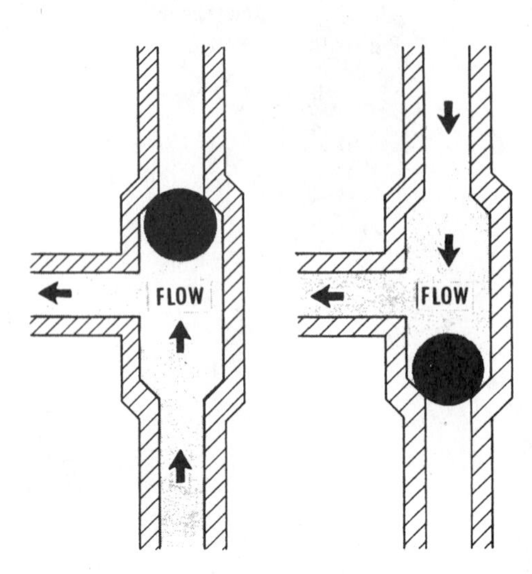

FIGURE 6–20 Two-seat ball check valve.

Relief Valves

A **relief valve** is a spring-loaded, pressure-operated valve that limits circuit pressure to a predetermined maximum value. A check ball–style relief valve with a spring load is featured in Figure 6–21. The spring tension can either be fixed or adapted to an adjusting screw to vary the spring tension. Once the spring tension is set, it cannot be varied in operation. System pressure is determined by the spring tension. In operation, the spring holds the valve seated against system pressure. Should the system pressure exceed its engineered safe limits, the valve

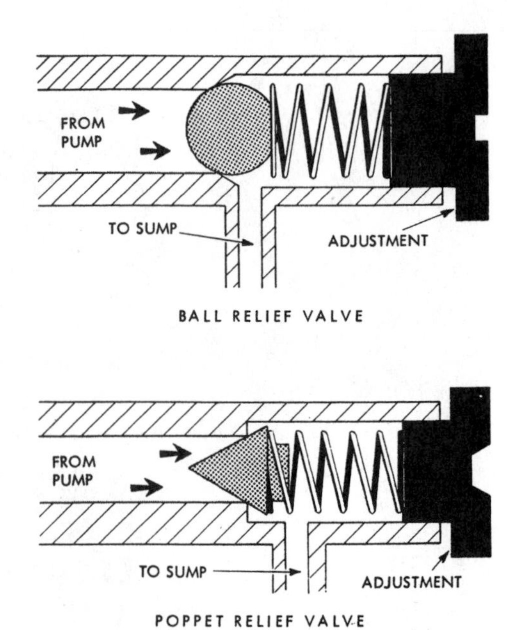

FIGURE 6–21 A relief valve is a check valve with a heavy spring exerting a fixed pressure to one side. (Courtesy of Chrysler Corp.)

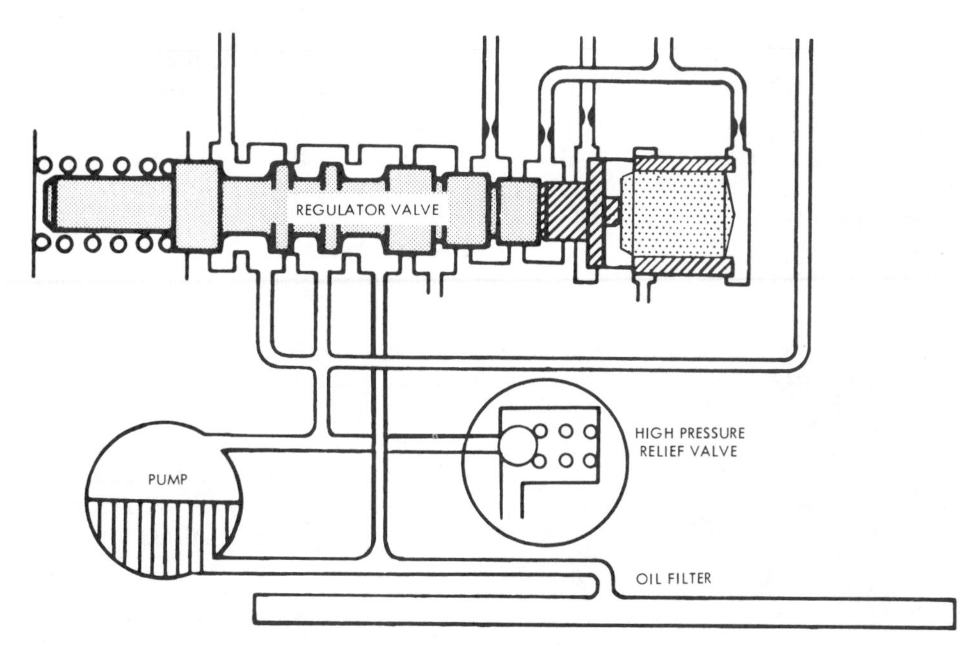

FIGURE 6–22 High-pressure relief valve used in TorqueFlite to protect against excessive pressure.

unseats and bleeds fluid back to the sump to relieve the excess pressure. It is not unusual to find relief valves protecting the transmission against high-pressure damage in the mainline and the converter feed circuits. Relief valves are sometimes used in low- reverse, modulator/throttle valve (TV), and other circuits to limit oil pressure in those circuits.

❖ **Relief valve:** A spring-loaded, pressure-operated valve that limits oil pressure buildup in a hydraulic circuit to a predetermined maximum value.

The pressure regulation in an automatic transmission requires a highly responsive balanced regulation to avoid deep fluctuations in control pressure. It must also have a built-in capacity to regulate at variable values. The relief valve does not meet these requirements and, therefore, cannot be used for pressure regulation. In Figure 6–22, a typical relief valve setup is shown in the regulated mainline circuit.

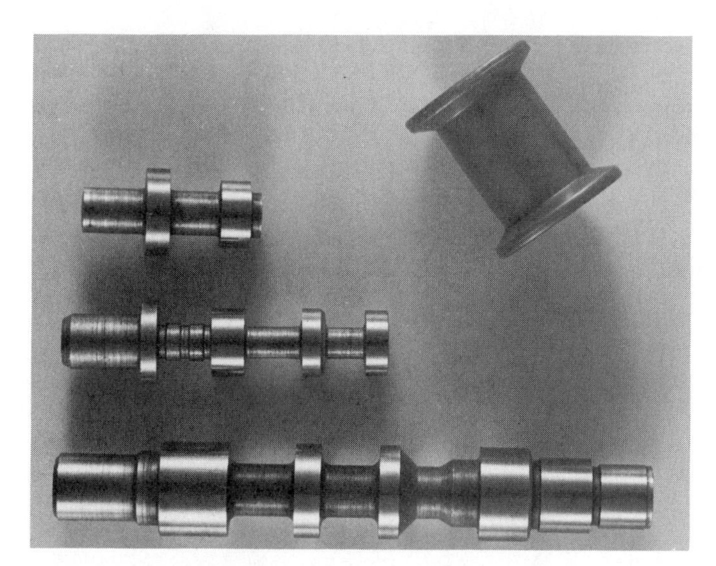

FIGURE 6–23 Typical spool valves.

Spool Valves

The check valve and relief valve are the simplest kinds of hydraulic valves, both in construction and operation. They are limited, however, in the means of control and the number of flow paths that can be handled. When a valve must interconnect several passages or react to more than one pressure, a **spool valve** is usually used.

❖ **Spool valve:** A precision-machined, cylindrically shaped valve made up of lands and grooves. Depending on its position in the valve bore, various interconnecting hydraulic circuit passages are either opened or closed.

A spool valve (Figure 6–23) is a precision-machined, cylindrically-shaped valve made up of lands and grooves (Figure 6–24). The valve is closely fitted to a round bore, with the valve lands sliding on a thin film of fluid. Fluid passages are interconnected to each other, depending on the position of the valve lands. The fluid is passed through the valve at the grooves (Figure 6–25). In the transmission, spool valves are positioned manually, by springs, or by oil pressure (Figure 6–26).

When a valve is acted on by a spring and by hydraulic pressure, the spring exerts a force on one end of the valve and holds it in position (Figure 6–27). On the opposite end, the pressure opposes the spring by acting on a reaction area. A reaction area refers to any spool surface or facing area that provides an advantage for hydraulic force to oppose an outside force on the valve, in this case the spring. The valve always moves in the direction of the greater force:

$$\text{force} = \text{pressure} \times \text{reaction area}$$

For example, a valve has a reaction area of 1/2 in^2 (3.23 cm^2) opposed by a spring with a 50-lb (222-N) force. If the hydraulic reaction pressure is 50 psi (345 kPa), the force from this pressure is 25 lb (111 N). This is less than spring force, and the spring holds the valve position. To move the valve hydraulically, the pressure must rise above 100 psi (690 kPa), and the reaction force must be greater than 50 lb (222 N).

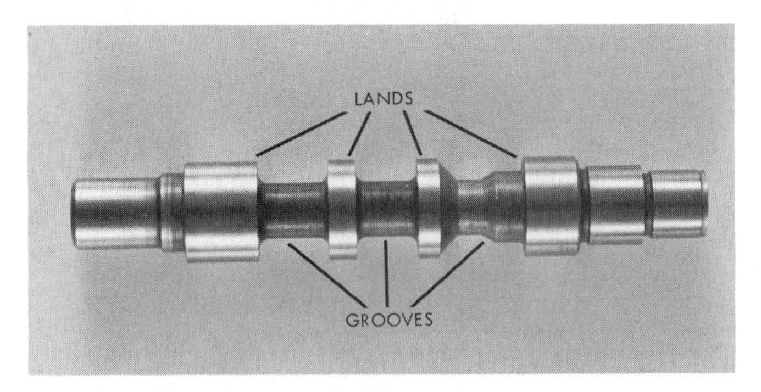

FIGURE 6–24 Spool valve lands and grooves.

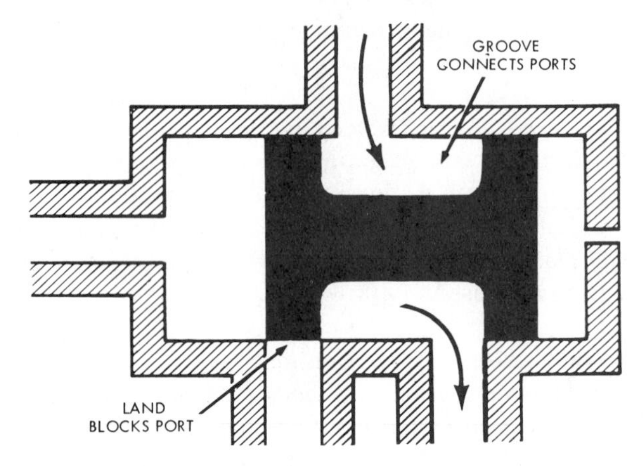

FIGURE 6–25 Spool valves are widely used to control flow direction. (Courtesy of Chrysler Corp.)

When two adjacent spools have different diameters and pressure is applied between the spools, a differential force results (Figure 6–28). The force on the larger spool is greater than on the smaller spool, which results in a valve movement in the direction of the large spool.

Regulator Valves

Regulator valves are used extensively in a transmission to control a variety of pressure schedules. The classic regulator valve that is the key to transmission hydraulic control is referred to as the pressure regulator valve. It regulates the output of the pump, supplying the transmission with its basic operating pressure. All the other regulating valves in the system depend on the regulated mainline fluid as their hydraulic feed source.

Regulator valves work on the balanced valve principle, which means that the regulated output of the valve is balanced

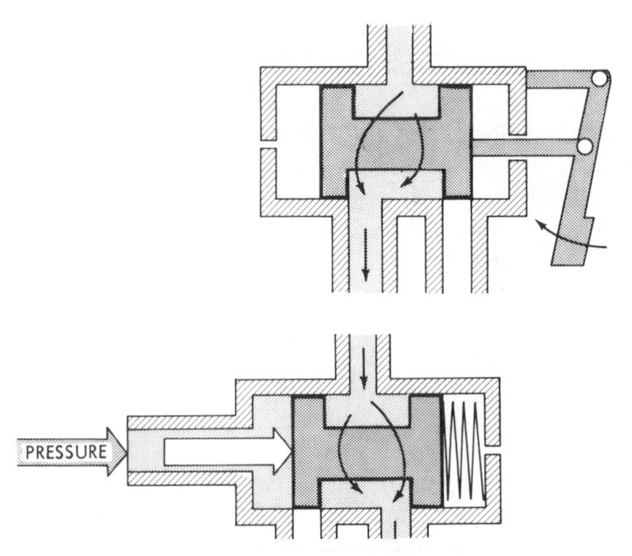

FIGURE 6–26 Spool valves may be operated manually or positioned by hydraulic pressure and/or spring force. (Courtesy of Chrysler Corp.)

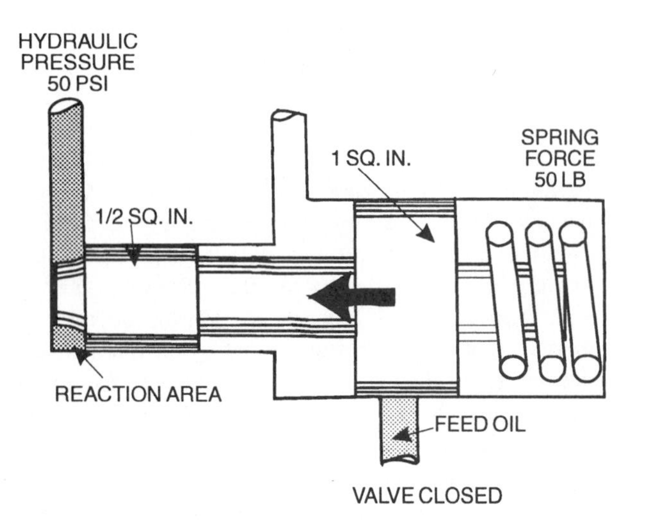

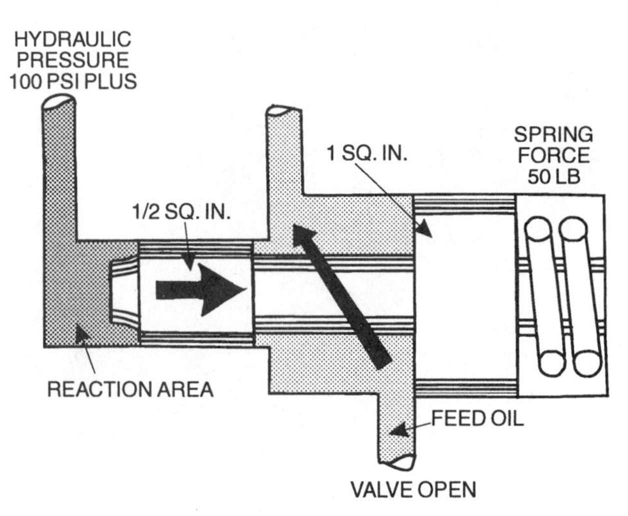

FIGURE 6–27 Left: Valve closed. Right: Valve opens as hydraulic pressure in the reaction area is increased to overcome the spring force.

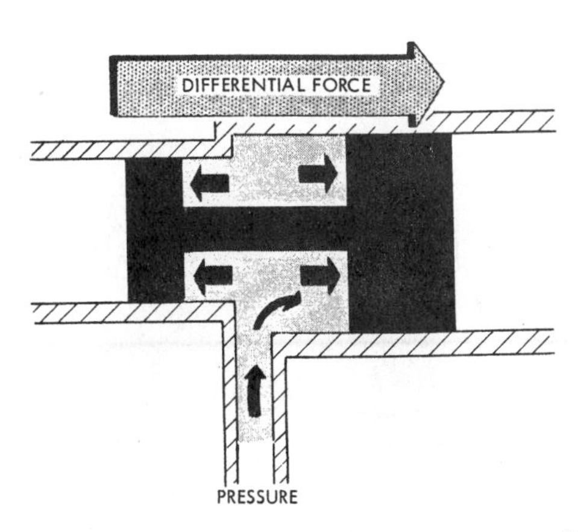

FIGURE 6–28 Spool valves can be made to operate on a differential force. (Courtesy of Chrysler Corp.)

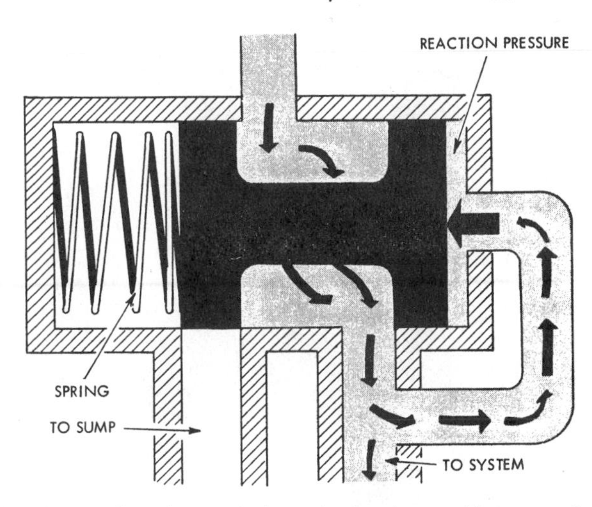

FIGURE 6–29 Regulator or balanced valve system. (Courtesy of Chrysler Corp.)

against a mechanical outside force or combination mechanical/hydraulic force. The simplified regulating valve in Figure 6–29 shows the typical valve circuitry for a balanced operation and demonstrates how it can control pump pressure. System pressure opposes spring force.

Before fluid begins to flow (Figure 6–30), there is no reaction pressure. The spring is the only positioning force, and it holds the valve in the extreme position of a wide-open feed to the system and a completely blocked sump port. All the pump delivery goes to the system until it dead-ends and pressure begins to build up. As pressure builds up in the system, it reacts against spring force. When it starts to exceed the spring force, the valve moves and bleeds off part of the pump flow back to the sump (Figure 6–31). The bleed-off causes the pressure buildup to drop. Because the spring force is sensitive to the pressure drop, it immediately moves the valve back to close down the bleed-off. The system pressure recovers very rapidly, and a regulated pressure output equal to the spring force is maintained.

The valve is poised between the spring force and reaction pressure. It is therefore cycling back and forth continuously, acting as a variable bleed orifice to maintain an equilibrium or balance between the two forces (Figure 6–32). If the reaction area of the valve is 1 in^2 (6.46 cm^2) and a spring force of 90 lb (400 N) is used, the valve balances or regulates at 90 psi (621 kPa).

Depending on pump speed and transmission demands, the valve cycling may not always call for closed porting of the exhaust port or sump return. When high-volume relief is demanded for pressure control, the valve cycles with open port bleeding—the exhaust port never closes completely. Some regulating valves operate at a constant setting, the spring alone acting as the determining factor. In most applications, the regulating valve has the capability of providing variable pressure schedules. When the transmission needs an operating pressure boost, an auxiliary pressure is added to the spring (Figure 6–33).

Valve bore circuitry is commonly provided with a combination of port openings in the valve body casting and separator plate. The port openings allow for the valve circuit oil to work between the valve spools and valve ends. Occasionally, this design system does not directly provide for fluid to reach the

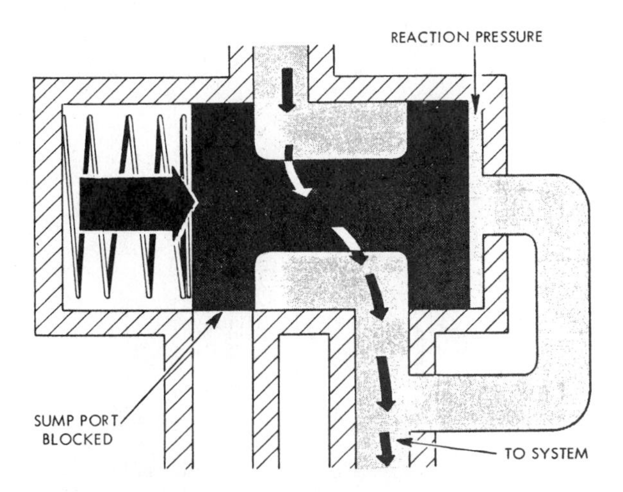

FIGURE 6–30 Regulator valve is closed by spring force before fluid flow. (Courtesy of Chrysler Corp.)

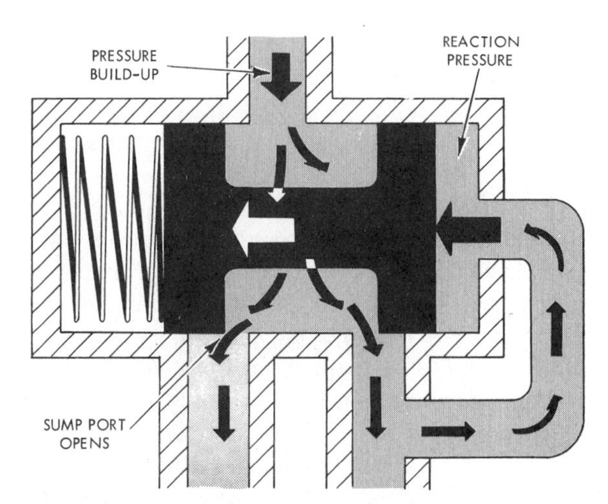

FIGURE 6–31 Regulator valve opens when pressure in system builds up and overcomes spring tension. (Courtesy of Chrysler Corp.)

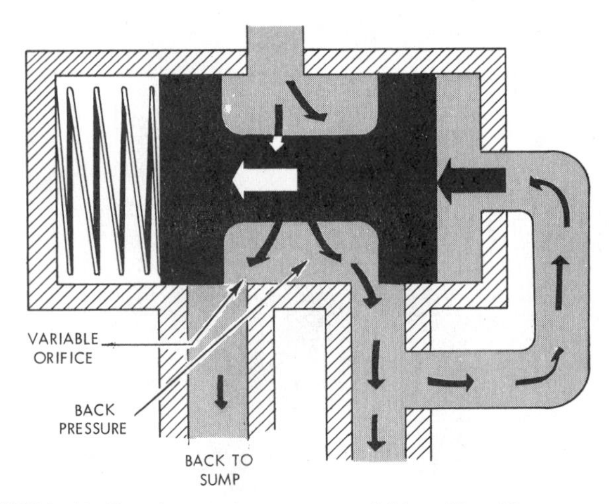

FIGURE 6–32 Regulator valve acts as variable orifice. (Courtesy of Chrysler Corp.)

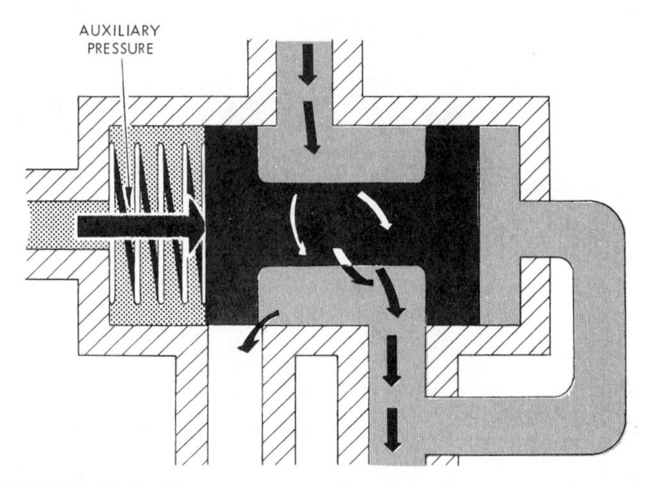

FIGURE 6–33 Auxiliary pressure assists the spring to control regulator valve. (Courtesy of Chrysler Corp.)

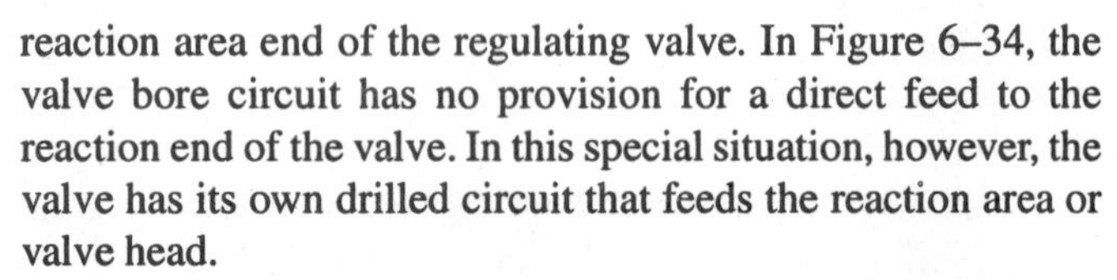

reaction area end of the regulating valve. In Figure 6–34, the valve bore circuit has no provision for a direct feed to the reaction end of the valve. In this special situation, however, the valve has its own drilled circuit that feeds the reaction area or valve head.

It is a common practice to use mainline operating pressure as the auxiliary source to the regulator valve. Applications of this hydraulic principle include reverse, manual low, and intermediate boost circuits. It may appear that this strategy would result in an uncontrolled increase of mainline pressure that would damage the transmission, but this dilemma is avoided by designing the effective boost valve spool area at a smaller dimension than the regulator valve. Figure 6–35 explains how the system develops a balance point.

A typical pressure regulator valve assembly used in automatic transmissions is illustrated in Figure 6–36. Note that the reaction area is on the top end of the valve and that line pressure is regulated according to a fixed spring force and auxiliary fluid pressures on the boost valve. In automatic transmission hydraulic control systems, the pressure regulator valve, modulator and throttle valves, and the governor valve are classic examples of regulating valves. Regulating valves are also used for fluid control in accumulator, clutch, and servo circuits. The balanced valve principle is critical to an understanding of how these regulating valves work. They are discussed further in Chapter 7.

Relay Valves

The relay valve is a circuit control valve having two positions, ON and OFF. Relay valves are used in hydraulic control systems to give direction to fluid traffic without changing pressure. The valve is held in one position by spring force or by spring force plus an auxiliary pressure. When the reaction pressure opposing the spring rises high enough, the valve shifts and interconnects the porting for proper circuit flow. The relay valve is not designed for metering—it either opens or closes like a circuit switch. Figure 6–37 shows a relay valve, triggered by hydraulic pressure, overcoming spring force and auxiliary pressure.

The manual valve and shift valves are examples of relay valve applications. A manual valve establishes the operating range of the transmission (Figure 6–38). Movement of the manual valve interconnects the line pressure with the various range circuits.

SUMMARY

Basic hydraulic componentry includes pumps and valves to generate fluid flow, create oil pressure, and control the numerous oil circuits. These essential items are needed to create the correct amount of fluid force to activate clutches and bands.

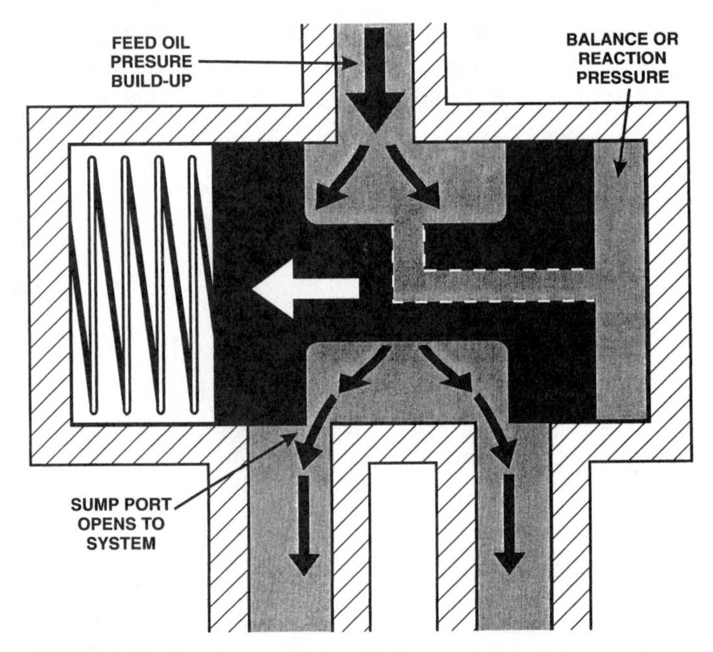

FIGURE 6–34 Regulator valve is drilled to provide fluid feed to the reaction end of the valve.

SIMPLIFIED PRESSURE REGULATOR VALVE DESIGN WITH REVERSE AUXILIARY BOOST SYSTEM

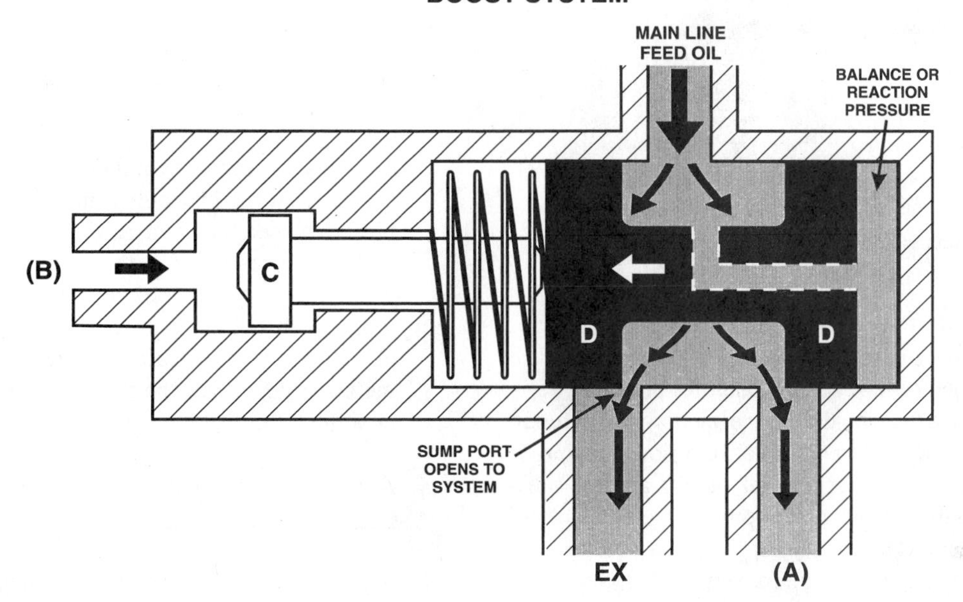

(A) REGULATED MAINLINE (OPERATING) PRESSURE

(B) MAINLINE PRESSURE INTRODUCED AS AUXILIARY PRESSURE TO BOOST OPERATING PRESSURE FOR REVERSE. NORMALLY RELAYED FROM MANUAL VALVE

(C) BOOST VALVE SPOOL AREA 1/2 SQUARE INCH

(D) PRESSURE REGULATOR VALVE SPOOL AREA 1 SQUARE INCH

* SPRING FORCE 100 LBS.

* SYSTEM BOOST DESIGNED FOR 200 psi

BOOST VALVE FORCE (LBS)		SPRING FORCE (LBS)	MAINLINE PRESSURE (psi)
		100	100
50	plus	100	150
75		100	175
87.5		100	187.5
93.75		100	193.75
96.875		100	196.875
98.437		100	198.437
99.218		100	199.218
ETC		BALANCE POINT →	200psi

FIGURE 6–35 Mainline pressure used as an auxiliary boost oil.

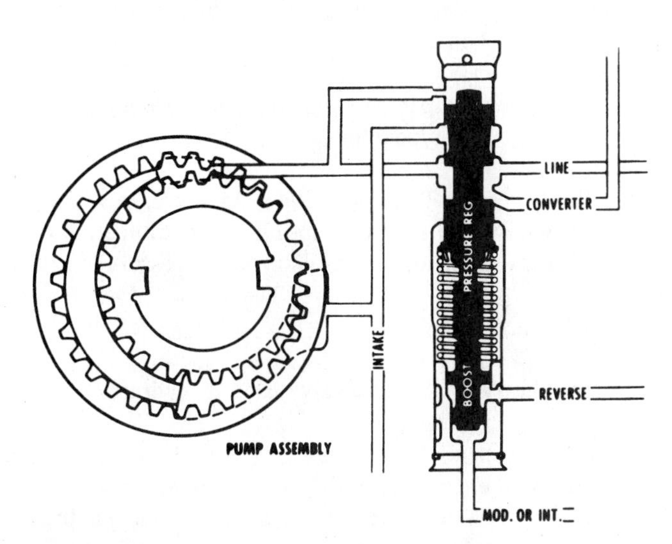

FIGURE 6–36 Typical pressure regulator valve. (Courtesy of General Motors Corporation, Service Technology Group)

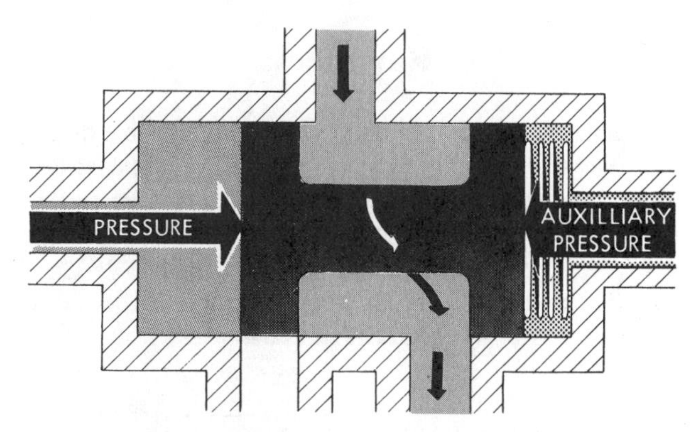

FIGURE 6–37 Relay valve ON/OFF position is controlled by opposing forces. (Courtesy of Chrysler Corp.)

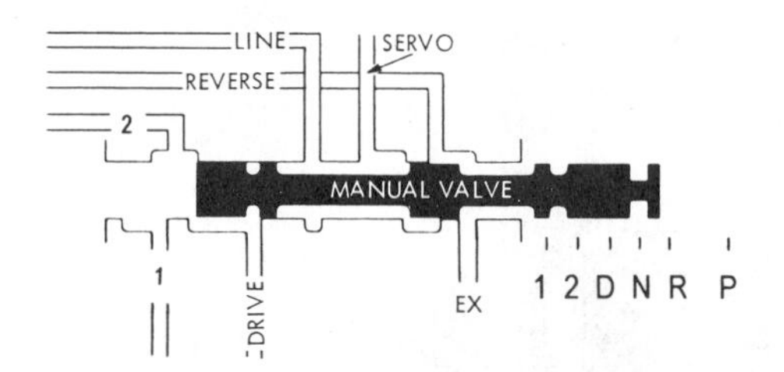

FIGURE 6–38 Manual relay valve is used to establish operating range in automatic transmissions. (Courtesy of General Motors Corporation, Service Technology Group)

Also, pumps and valves are responsible for the shift timing and shift quality of the gear ratio changes that take place.

Pumps, which are driven by the engine and torque converter, create the fluid flow. By containing the fluid flow, pressure is generated. The two main categories of pumps are fixed displacement and variable displacement pumps. Fixed displacement pumps, IX gear or IX lobe styles, provide a fixed amount of fluid flow for each engine/torque converter revolution. Variable displacement pumps, sometimes referred to as variable capacity vane pumps, have the ability to change the volume of fluid being pumped. The volume is adjusted automatically by the needs of the transmission, and not determined by the speed of the engine.

Hydraulic valving in transmissions can range from simple devices, such as orifices, to complex pressure regulator valves. Relay valves open and close hydraulic circuits to direct flow and pressure. Check ball–style valves and spool valves are examples of relay valves. Spool valve movement can be caused by manual linkage operation, spring force, and hydraulic oil pressure. Regulator valves control and meter circuit pressures. Relief valves are a simple type of regulator valve.

Various types of transmission valves provide the means to apply the proper clutch and band combinations at the right time and at the correct intensity, which is necessary for the transmission to operate smoothly through the different gear ranges.

❑ REVIEW

Key Terms

IX rotary lobe pump, IX rotary gear pump, variable displacement (variable capacity) vane pump, orifice, fluid viscosity, relay valve, regulator valve, check valve, relief valve, and spool valve.

Completion

1. In a hydraulic system, the ____________ creates flow and applies force to the fluid when it dead-ends.
2. Name three types of rotary pump designs common to automatic transmissions.

   (a) ____________________________
   (b) ____________________________
   (c) ____________________________

3. The rotation of the hydraulic pump produces a pumping action from the expanding and contracting action taking place between the inner and outer pump members. When the clearance between these members is expanding, a ________________ is created, and ________________ pressure in the sump forces the fluid into the pumping chambers. As the clearance between the inner and outer members decreases, the fluid is squeezed into the system, creating a fluid ____________.
4. A pump design that uses a crescent-shaped divider is ____________________________.
5. A pump design that uses a close tolerance tip clearance to isolate the inlet and outlet sides of the pump is ______________________.
6. A pump that has a continuous delivery characteristic is called a positive ______________________ or positive ______________ pump.
7. A variable displacement (variable capacity) vane pump has the ability to adjust its output volume to the transmission needs. For maximum output, a ______________ spring moves the pump slide in the fully extended position. To cut back on pump output, the ______________ valve relays a hydraulic signal to the pump slide, which results in a slide movement against the ______________ spring. To cut back on the pump output, the slide will pivot toward the center of the ______________. The degree of the centering movement determines the amount of pump output that is cycled back to the ______________ side of the pump.
8. A rotary lobe pump is sometimes referred to as a ________ ______________-type pump.
9. A fluid pressure that can move a spool valve against a mechanical force such as a spring or a combination of a mechanical force and an auxiliary fluid pressure is called a ______________ pressure. The spool valve area on which it acts is called a ______________ area.
10. A popular type of check valve used in hydraulics is the ____________ check.
11. To boost the pressure point on a regulating valve, an auxiliary hydraulic pressure is added to the calibrated ______________ force.

Matching

Use the choices below to select the one best answer for the following questions. Indicate your choice by placing the letter in the question blank. A choice may be used more than once.

A. Spool valve
B. Relay valve
C. Regulator valve
D. Relief valve
E. Check valve
F. Orifice

____ 1. A one-way directional valve used for directing fluid traffic between interconnecting hydraulic circuits.

____ 2. Engineered to control the rate of fluid flow and pressure buildup time.

____ 3. A valve design that is cylindrically shaped with two or more lands separated by annular grooves.

____ 4. A simple spring-loaded ball check used as a safety device to protect the hydraulic system when system pressure exceeds engineered values.

____ 5. Controls the porting that connects and disconnects a hydraulic circuit from its fluid supply source. It could easily be called a "switch" valve.

____ 6. Works on the balanced valve principle.

____ 7. Provides a variable pressure control.

Indicate which of the following valves used in an automatic transmission hydraulic control system are classified as regulating valves.

______ Pressure regulator valve
______ Manual valve
______ Governor valve
______ Throttle or modulator valve
______ Shift valve

Multiple Choice

____ 1. The engine is at all times coupled to and drives:
I. the converter impeller.
II. the transmission pump.

A. I only
B. II only
C. both I and II
D. neither I nor II

____ 2. The transmission pump can be set up to be driven by:
I. the converter turbine shaft.
II. the converter drive hub.

A. I only
B. II only
C. both I and II
D. neither I nor II

____ 3. The type of pump that is identified with a variable displacement (variable capacity) characteristic is the:
I. rotary gear design.
II. rotary lobe design.

A. I only
B. II only
C. both I and II
D. neither I nor II

____ 4. An IX pump configuration is identified with:
I. the rotary gear design.
II. the rotary lobe design.

A. I only
B. II only
C. both I and II
D. neither I nor II

____ 5. Which of the following applies to a variable displacement (variable capacity) vane pump?
I. The pump uses one rotary member only.
II. The pump uses fixed vanes.

A. I only
B. II only
C. both I and II
D. neither I nor II

____ 6. A relief valve is typically used to protect against excessive hydraulic pressures:
I. in the pump mainline circuit.
II. in the converter feed circuit.

A. I only
B. II only
C. both I and II
D. neither I nor II

____ 7. Spool valves are positioned in their bores by:

A. manual control.
B. hydraulic pressure.
C. spring force.
D. all of the above.

____ 8. Fluid always passes through the spool valve area referred to as the:

A. port groove.
B. annular groove.
C. reaction area.
D. none of the above.

Hydraulic Control System: Fundamentals of Operation

OBJECTIVE:

The objective of this unit is for you to become proficient with the terms and concepts contained in this chapter. Completion of this objective is essential to becoming a successful automatic transmission technician.

CHALLENGE YOUR KNOWLEDGE

Define the following key terms:

Mainline pressure, throttle pressure/modulator pressure, boost valve, vacuum modulator, manual valve, governor, centrifugal force, valve body assembly, and accumulator.

List the following:

Three methods to relay throttle and modulator pressures; three governor designs; five factors that affect clutch and band shift quality; and three return spring designs to release clutch pistons.

Describe the operating principles of the following:

Pressure regulator valve; converter-cooling-lubrication circuit; compensated vacuum diaphragm control unit; accumulator; shift valve; part-throttle 3-2 downshift; and full-throttle forced downshift.

INTRODUCTION

The hydraulic control system is responsible for a variety of major activities that involve pressure and performance circuits that determine the clutch/band combinations for the select action of the planetary geartrain. Although the design of the system appears to be complex, it is made up of basic interrelated components and subsystems that are easy to understand. The hydraulic control system typically consists of the following:

- A pump to provide a steady supply of regulated mainline oil to meet the hydraulic demands of the transmission.
- A regulator valve to control the variable mainline pressure requirements and the feed oil to the converter-cooling-lubrication circuits.
- A control valve assembly that allows for manual or automatic control of the gear ratio clutch/band combinations.
- A governor to produce a regulated vehicle speed sensing hydraulic pressure. It is used primarily as one of two pressures to control the shift schedule. In some systems, it may also be used to modify the mainline pressure, to cause a pressure drop.
- A throttle or modulator valve to produce an engine torque sensing hydraulic pressure. It is used primarily as one of two pressures to control the shift schedule. It has the added job of always modifying the mainline pressure, to cause a pressure increase.

There is always a support cast of auxiliary valves and check balls to augment the primary components of the hydraulic control system, and these vary between transmission products. To simplify discussion, these product variances are avoided, and the main concentration is directed at the basic working of the primary controls in the hydraulic system, which includes the fundamentals of automatic shifting. Complete hydraulic system features of product specifics can be found in the manufacturers' service and training manuals. We reemphasize that domestic and import systems typically have components similar in function, construction, and operation.

Based on the hydraulic fundamentals from Chapters 5 and 6, we gradually build an automatic transmission hydraulic control system. To make it easy to understand, we introduce the technician to the several basic systems that make up a typical hydraulic control package. The technician learns about their job function and examines how they work. Then we illustrate how they all work together in a total system. Learning the concepts of the traditional hydraulic controls first simplifies the progression to transmission electronics.

PUMP AND PRESSURE REGULATION SYSTEM

The transmission hydraulic operation is keyed to a hydraulic pump that is engine-driven through the converter. The pump must essentially deliver a steady supply of oil under controlled pressure to meet the demands of the transmission circuits. It provides the basic fluid supply and operating pressure. The pump output is sensitized by a pressure regulator valve that controls the pressure head. The controlled pump pressure is referred to as **mainline pressure**. In addition to pressure control, the regulator valve controls the feed to the converter-cooling-lubrication circuit.

> ❖ **Mainline pressure:** Often called control pressure or line pressure, it refers to the pressure of the oil leaving the pump and is controlled by the pressure regulator valve.

In a typical pump circuit (Figure 7–1), the pump output is sensed by a pressure regulator valve. The regulator valve works on the balanced valve principle and regulates the mainline pressure according to a fixed spring force and auxiliary oil pressures acting through a boost valve. Note in Figure 7–1 how the reaction force of the fluid on top of the regulator valve moves the valve against the spring and the boost valve forces. A bleed-off opening (port) back to the inlet side of the circuit is provided for valve balancing.

The metered orifice at the reaction end of the regulator valve acts as a dampener and isolates the reaction fluid from the pump pulsations. This keeps the valve from making deep cycles and cuts back the pulsation peaks.

Returning the bleed oil to the intake allows for quick makeup oil to the pump and prevents overburdening of the intake filter.

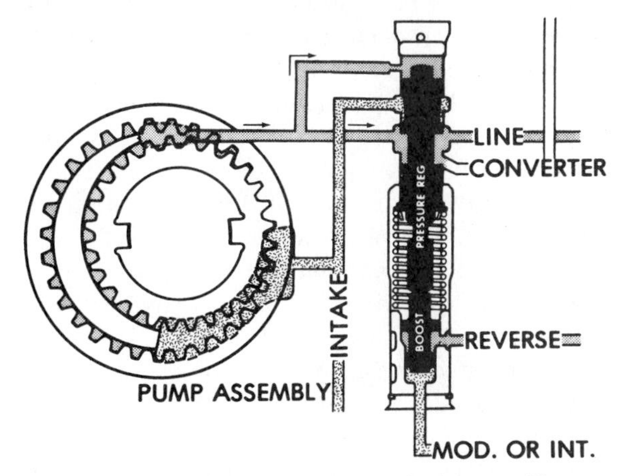

FIGURE 7–1 Pressure supply system with single pump. The differential fluid force on top of the valve balances against the spring tension and boost valve auxiliary pressures. (Courtesy of General Motors Corporation, Service Technology Group)

The intensity of the bleed oil, if directed back to the sump, would aerate the sump oil.

Depending on product requirements, the mainline oil pressure can vary between 60 psi (412 kPa) and 280 psi (1930 kPa). The regulator valve usually balances against the following force or combination of forces:

- Spring tension only
- Spring tension plus throttle boost oil
- Spring tension plus throttle and mainline boost oil for reverse operation

Where the pump design is a variable displacement (variable capacity) vane pump, movement of the pressure regulator valve opens a port to mainline oil (Figure 7–2). The mainline oil, acting as a signal oil, enters a cavity zone on the side of the slide opposite the priming spring. Mainline oil acts against the slide and priming spring to change the slide position and control the pump output. The regulator valve remains sensitive to the usual spring and boost oil forces.

By using **throttle pressure**, or **modulator pressure**, in some transmissions, as a boost oil, the transmission line pressure can be adjusted to the torque performance of the engine and converter. For power performance at wide open throttle (WOT), the line pressure is boosted to maximum, and a delayed shift schedule is accompanied by aggressive clutch/band engagements, whereas light throttle conditions require lower operating pressures. Operating at economy throttle openings means a lower horsepower loss in driving the pump and earlier shift schedules rewarded by quality clutch/band engagements.

> ❖ **Throttle pressure/modulator pressure:** A hydraulic signal oil pressure relating to the amount of engine load, based on either the amount of throttle plate opening or engine vacuum.

A **boost valve** is also used in reverse for increasing the line pressure to ensure adequate fluid pressure for additional torque-holding requirements. Shown in Figures 7–1 and 7–2, reverse pressure (line pressure) is directed to the boost valve in addition to the throttle or modulator boost signal to provide the necessary increase in line pressure. Increased line pressure for reverse operation is required for all automatic transmissions to keep the reverse band or clutch from slipping under full engine torque.

> ❖ **Boost valve:** Used at the base of the regulator valve to increase mainline pressure.

In the forward gear ratios, the crankshaft assists the clockwise rotation of the powertrain output. Therefore, the torque reaction handled by the holding friction element is equal to $(r-1)$ times the engine/converter torque (r is the gear ratio). Using a forward gear ratio of 2.69 with the engine at full torque, the holding friction element is only required to handle a torque reaction of 1.69 times engine/converter torque.

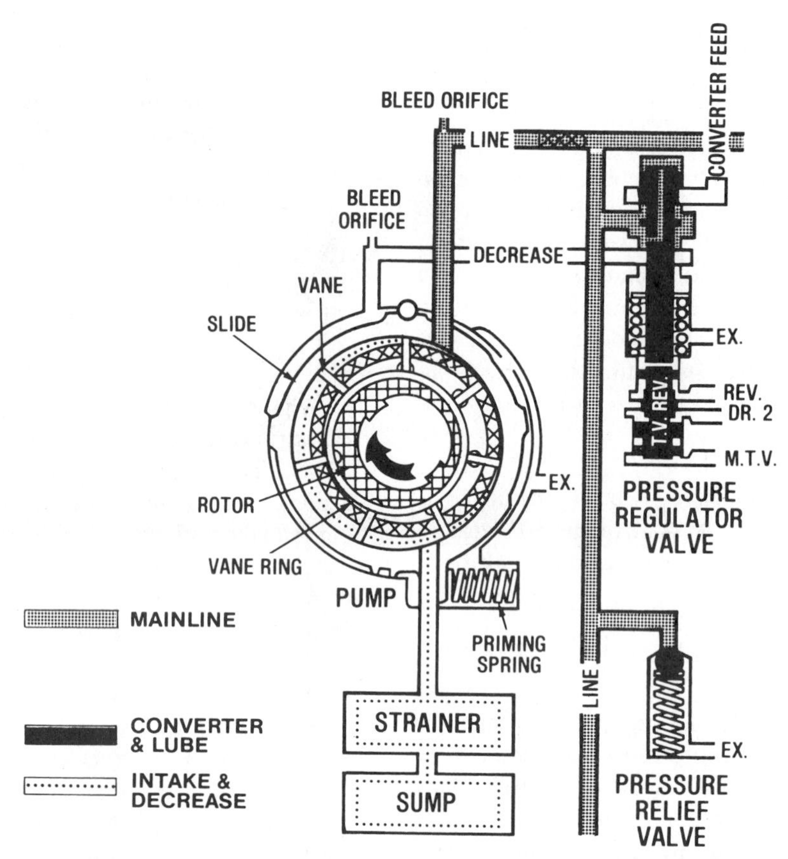

FIGURE 7–2 Pressure regulator valve operation; variable capacity pump system. (Courtesy of General Motors Corporation, Service Technology Group)

In reverse, the crankshaft rotation and powertrain output rotation are working in opposite directions. The reverse torque reaction on the reverse carrier is much greater than the reaction torques in first and second gears. Torque reaction in reverse is ($r + 1$) times the engine/converter torque. Using the same gear ratio, 2.69, the holding friction element on the reverse carrier would need to handle a reaction torque of 3.69 times engine/converter torque. The maximum line pressure for reverse is boosted to two or three times the value required in the forward operating ranges.

Since the regulator valve feeds the converter-cooling-lubrication circuit, it must prevent oil from draining back out of the converter when the engine is stopped (Figure 7–3). In the at-rest position, the regulator valve is positioned by the spring, and the port feed is sealed. Should a converter drain-back occur, there would be a time delay in positive converter action and temporary starving of the lubrication circuit.

CONVERTER-COOLING-LUBRICATION SYSTEM

A typical converter-cooling-lube system is illustrated in Figure 7–3. It shows three separate circuits, linked in series. The system can get charged by mainline pressure acting through an open port at the regulator valve, or by a separate converter oil pressure from an additional regulating valve (Figures 7–3 and 7–4). The converter regulating valve still receives a mainline pressure feed from the main regulator valve. The addition of a converter regulator valve gives closer control of converter oil pressure.

As one might conclude from this discussion, the standard practice is to tap the regulator valve as the source of oil feed to the converter-cooling-lube circuits. These circuits are unaffected by the manual valve position and are continuously functional in each operating range of the transmission. When the regulator valve moves from its rest position, and before regulation begins, it opens the converter feed port and charges the system. The transmission priority is to stabilize the converter-cooling-lube circuit before other hydraulic requirements can be met.

The converter-cooling-lube circuit is not only continuous but also open-ended. The fluid flow does not stop moving, and therefore, the system pressure is less than mainline. A minimum oil pressure must be maintained to prevent converter cavitation and provide adequate fluid movement for cooling and lube. Pressure drop control is dependent on a simple arrangement of engineered restrictions in the flow circuitry. Lower and steady drop pressures are maintained.

The cooler and lube circuits on some systems have working pressures as low as 10 to 15 psi (69 to 103 kPa). The mainline oil pressure to the converter drops off at the inlet but remains

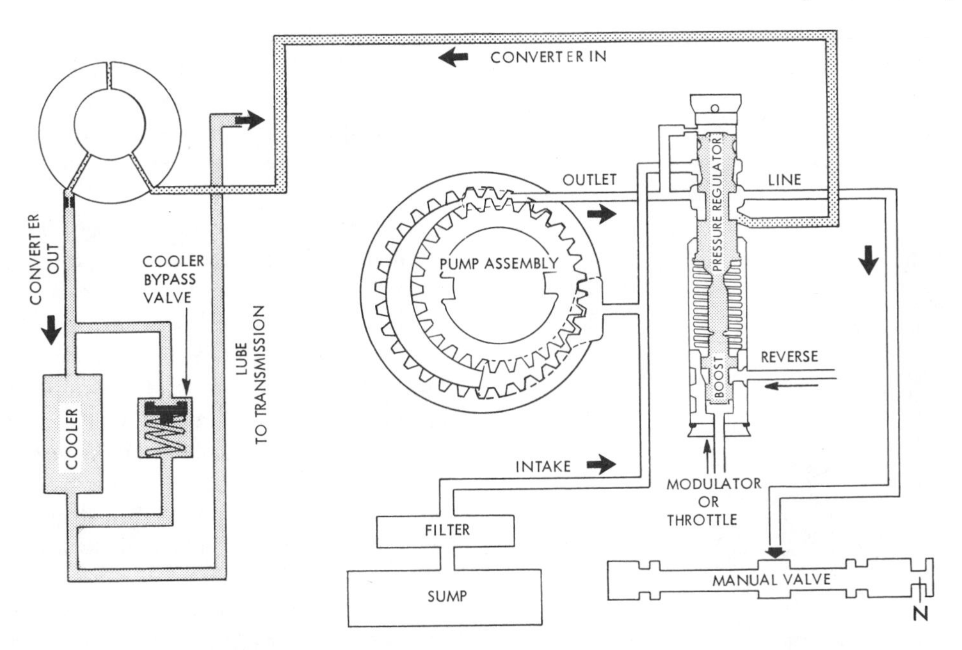

FIGURE 7–3 Typical pressure regulator valve and converter-cooling-lube circuit. Cooler bypass valve is not always included.

on the high side. Some converter circuits include a pressure relief valve to prevent excessive pressure buildup in the torque converter (Figure 7–5). This prevents converter ballooning.

When the engine is off, the regulator valve assumes its rest position and seals the converter port feed to prevent converter drain-back. It may appear that the converter fluid could drain out through the open-ended lube circuit, but this should not happen. The regulator valve has the other end of the circuit sealed off, and a circuit air bleed is unavailable. Consequently, fluid drain is not possible from a gravitation or siphoning

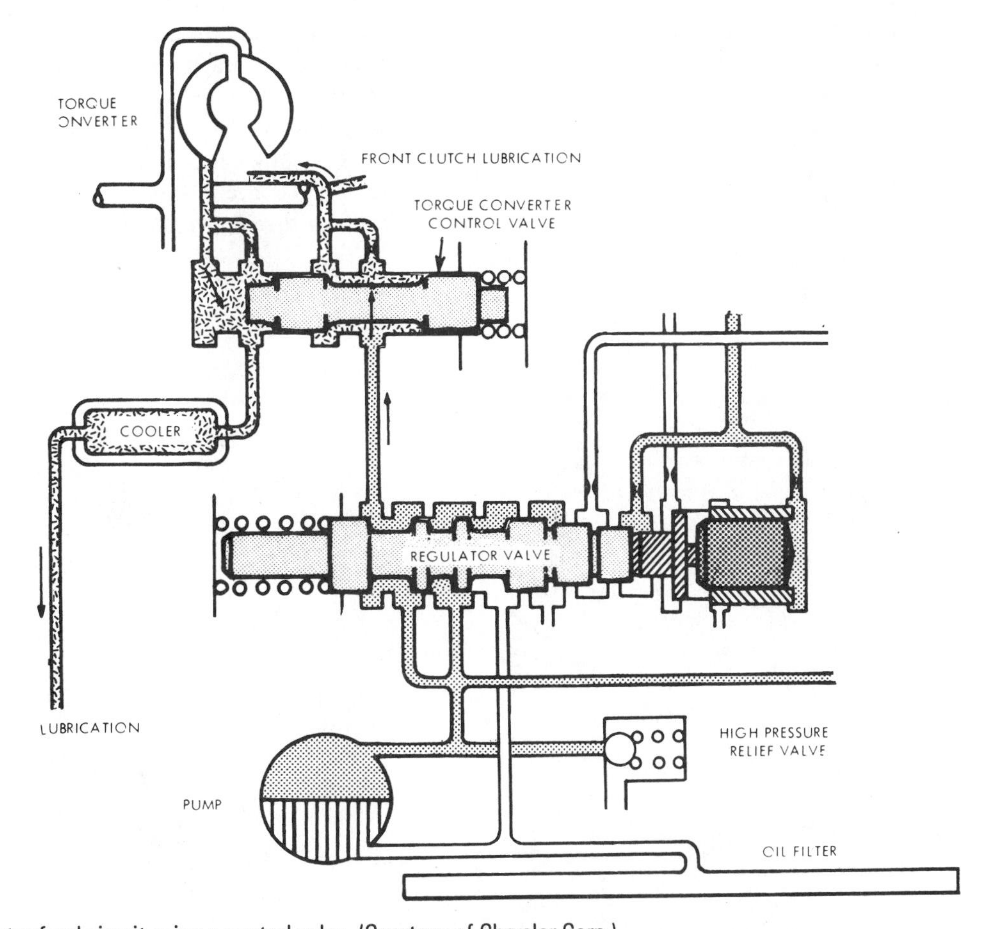

FIGURE 7–4 Converter feed circuit using a control valve. (Courtesy of Chrysler Corp.)

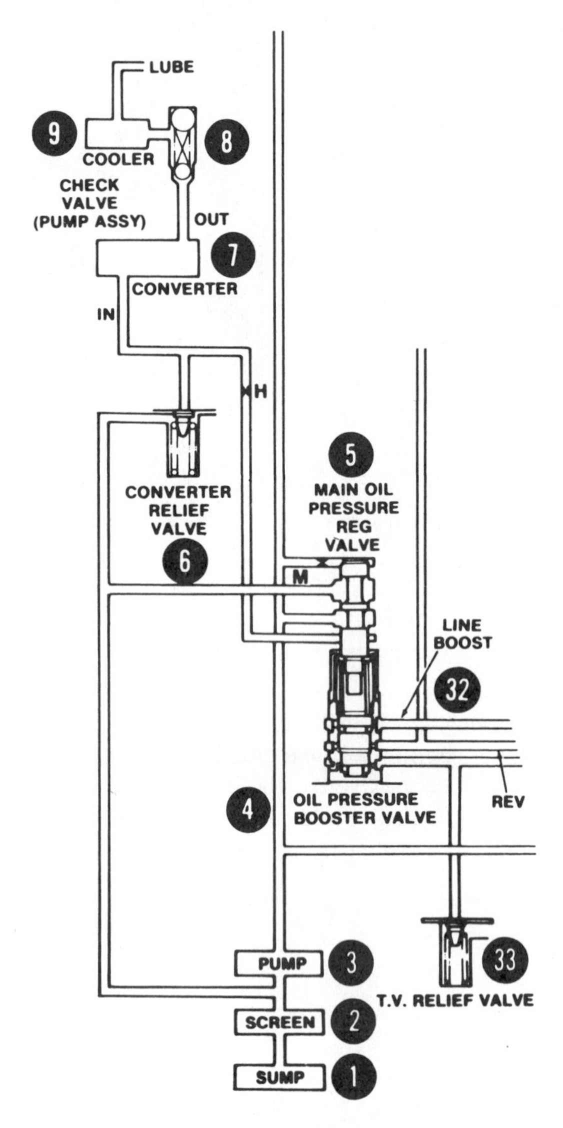

FIGURE 7–5 Converter circuit with pressure relief valve. (Reprinted with the permission of Ford Motor Company)

action. Should an unwanted air bleed be induced by a worn stator support or pump body bushing, converter drain-back does happen. Some converter circuits use a check valve, however, to prevent converter draining through the cooler-lube circuits.

Getting the oil flow in and out of the converter is quite an engineering feat. In a nonlockup torque converter, the converter fluid enters through the clearance between the stator support and converter pump drive hub. For converter outflow, the fluid goes past the outside of the front bushing of the stator support, through the pump housing and transmission case. From the transmission case, it is piped to and from the cooler through steel tubing (Figure 7–6). The return line from the cooler is coupled to the transmission case, where the fluid is often used for lubrication of the planetary geartrain, clutch plates, bushings, thrust washers, sprag, and roller clutches. When the job of lubrication is completed, the fluid gravitates to the sump. Notice in Figure 7–6 that the line to the cooler enters at the bottom to prevent cavitation. Auxiliary coolers are always added in series with the return line to the transmission.

In Figure 7–4, the front lubrication of the transmission is supplied from a metered tap in the converter charge line. To satisfy rear lubrication, the cooler return line is connected at the rear of the case (Figure 7–7). The lube oil is then distributed through the output shaft. Another circuit variation is to connect the cooler return line to the front of the transmission, where lube oil enters the stator support or pump cover and lubricates the entire transmission front-to-rear (Figure 7–8). The lube oil is distributed through the stator support and the input and output shafts.

Figure 7–9 shows the typical circuit porting used for the converter-in and converter-out circuit routing for a FWD application with a torque converter clutch.

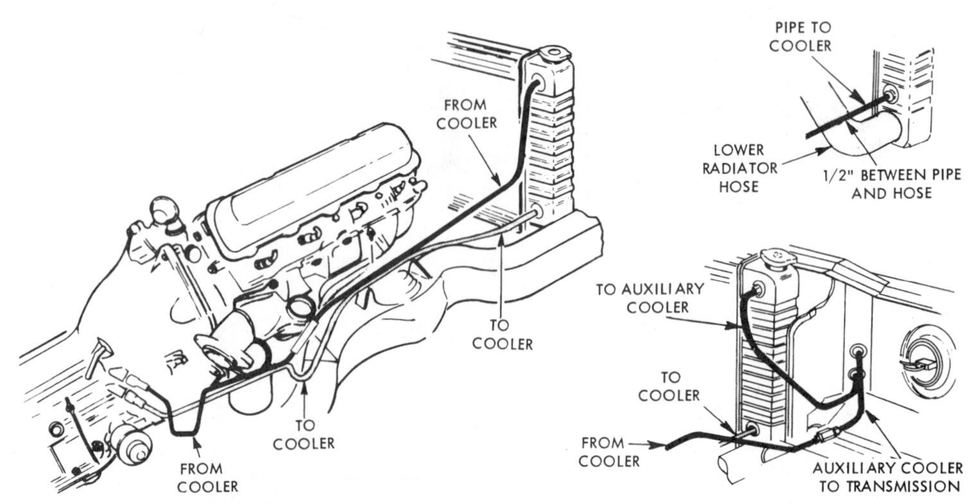

FIGURE 7–6 Converter cooler lines. (Courtesy of General Motors Corporation, Service Technology Group)

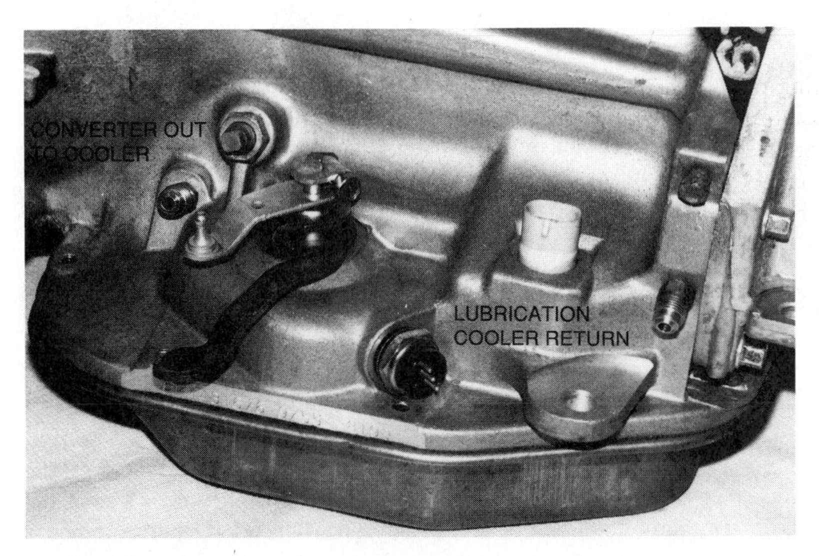

FIGURE 7–7 Lubrication cooler return connector at rear of case.

FIGURE 7–8 Lubrication cooler return connector at front of case.

THROTTLE/MODULATOR SYSTEM

The terms **throttle** and **modulator** have the same meaning, and the systems perform similar functions. The use of these terms depends on the language used by the car manufacturer. Either system generates a regulated hydraulic engine torque signal for the following functions:

- To supply a boost pressure to the primary or main regulator valve train. When engine breathing or throttle opening changes, the torque signal changes along with the mainline operating pressure. This is an instant chain reaction.
- In the shift system, to provide for a wide range of automatic shift points by delaying the shifts in relation to engine performance.
- In an accumulator system, to control the shift aggressiveness on a clutch or band engagement. For accumulator control, a modified throttle oil is usually used.

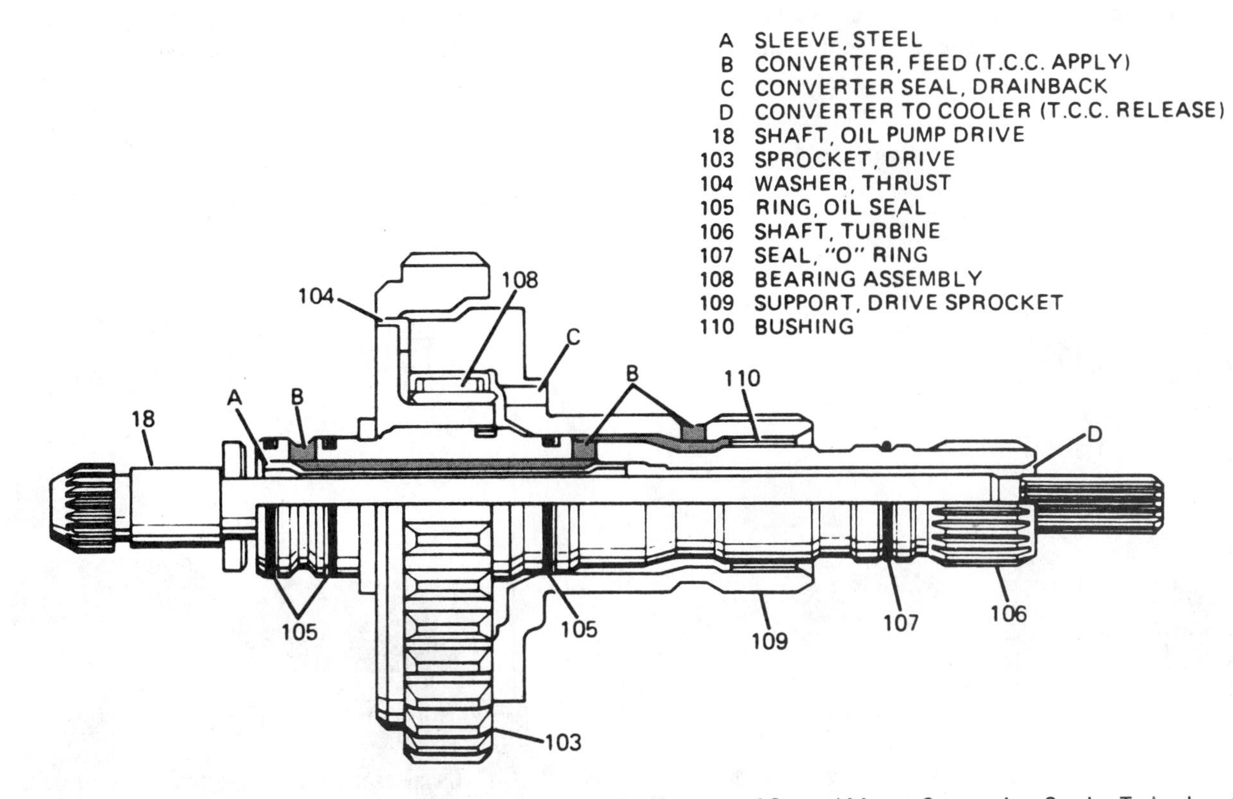

FIGURE 7–9 Converter oil passages—parts cutaway view of a transaxle. (Courtesy of General Motors Corporation, Service Technology Group)

Throttle Valve Control

The throttle valve is another example of a regulating or balanced valve application. Regardless of the system control, the regulated throttle oil always balances against a variable spring force. The engine torque message to the transmission throttle valve assembly control is relayed by several methods:

- Manifold vacuum control
- Mechanical linkage control
- Mechanical cable control
- Combined mechanical linkage and cable control

A typical vacuum-controlled throttle system arrangement is shown in Figure 7–10. The function of the vacuum unit is to respond to instant changes in manifold vacuum and convert the manifold vacuum signal to a proper spring force on the throttle valve. To balance the valve action against the spring force, an exhaust port is provided at the valve to bleed off excess pressure. The exact spring force working against the throttle valve is determined by the atmospheric pressure and manifold vacuum working on opposite sides of a flexible diaphragm. With the spring located in the airtight vacuum chamber, manifold vacuum acts to decrease the effective spring force and reduce throttle pressure. The atmospheric pressure flexes the diaphragm against the spring force. At engine idle, the throttle pressure is at zero or close to zero.

When the engine is under load, the vacuum drops, offsetting the atmospheric pressure. This permits a higher effective diaphragm force to the left, causing an increase in throttle pressure. Throttle pressure usually peaks below 3-in of vacuum. In this range, it is equal to full mainline pressure unless a limit valve is used to either cap the mainline feed to the throttle valve or cap the maximum TV output. (Throttle pressure is usually abbreviated as TV pressure.) It is important for the technician to understand that regulated throttle oil output is inversely proportional to the manifold vacuum. High engine vacuum produces a low throttle oil pressure, and low vacuum causes a high throttle oil pressure.

The vacuum throttle unit just described is generally referred to as a **vacuum modulator** control unit. It is noncompensated for the effects that high altitudes have on its operation. With an increase in altitude, atmospheric pressure, as well as the engine manifold pressure (vacuum), is lower. Referring again to Figure 7–10, it is apparent that when identical engine throttle conditions are compared in Chicago and Denver, the throttle oil pressure output increases with the altitude. This effect raises the minimum shift points and oil pressure schedule, causing the shift quality to be more aggressive. The engine performance is no longer matched to the shift system. At high altitudes, the engine torque output drops, and shifts occur after the engine torque has peaked. For ideal performance, gear ratio changes should occur just before the peaks of the engine torque output.

❖ **Vacuum modulator:** Generates a hydraulic oil pressure in response to the amount of engine vacuum.

To compensate for the high-altitude effect, an altitude-compensated vacuum diaphragm is used for some transmissions (Figure 7–11). This unit incorporates an evacuated spring bellows that is sensitive to barometric pressure. The left side is anchored to a rigid backing plate and the other end to the flexible diaphragm. A box frame that works through two openings in the backing plate relays the spring force to the pushrod.

The throttle oil pressure is the same at sea level for both noncompensated and altitude-compensated vacuum units. At

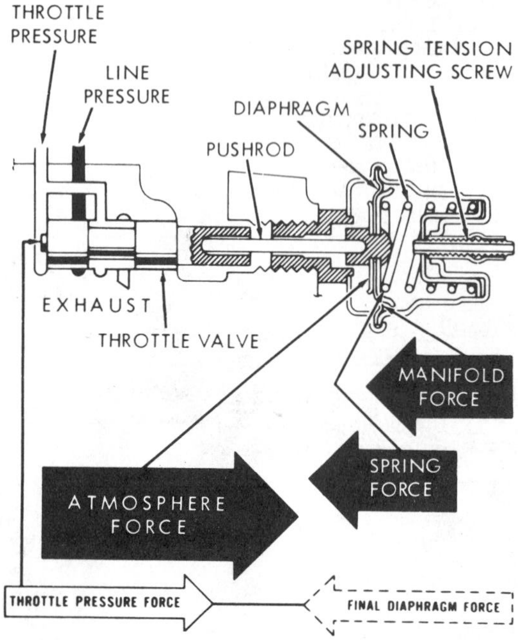

FIGURE 7–10 Vacuum diaphragm control unit. (Reprinted with the permission of Ford Motor Company)

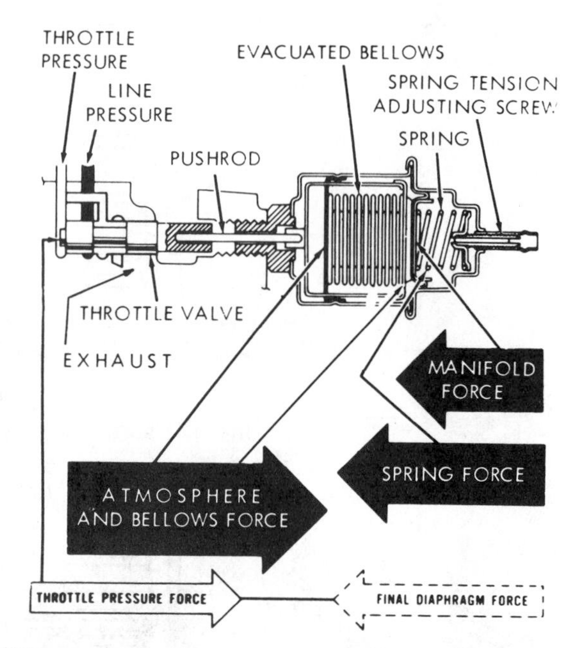

FIGURE 7–11 Compensated vacuum diaphragm control unit. (Reprinted with the permission of Ford Motor Company)

higher altitudes, however, the barometric pressure decreases, and the crush effect on the bellows is less. This permits the bellows to expand and assist the atmospheric force on the diaphragm to decrease the effective spring force on the throttle valve. In effect, the bellows reprograms the throttle system so that shifts occur slightly sooner to match the lower peak engine torques related to high altitude.

In the mechanical version of throttle pressure control, a linkage or cable system actuated by the accelerator pedal movement is used to vary the spring load on the end of the throttle valve. A throttle valve lever and shaft arrangement within the transmission is rotated in proportion to the amount of throttle opening of the manifold air intake. This means that a mechanical linkage hookup between the transmission and air intake throttle movement must be accurately coordinated (Figure 7–12).

As the air intake throttle opening increases, the throttle valve spring exerts a greater force on the throttle valve and a corresponding increase in throttle pressure. In the wide-open position, the throttle valve (TV) is mechanically bottomed out in the bore, and line pressure passes through the valve unregulated. Throttle pressure is equal to line pressure unless a limit valve is used to either cap the mainline pressure feed to the throttle valve or cap the TV output. This gives a better control of line pressure boost and shift quality.

Illustrated in Figure 7–13 is a TV limit valve inserted in the TV feed line. Note that a line bias valve is included in the system to modify the TV boost to the pressure regulator valve. At closed throttle, TV pressure is at or close to zero.

The cable-controlled TV system works similarly to the mechanical linkage TV system (Figure 7–14). The cable movement coordinates the manifold air intake throttle opening with the transmission TV plunger movement. In this particular TV system, a spring-loaded lifter and exhaust check ball are featured (Figure 7–15). If the cable becomes disconnected, broken, or extremely misadjusted, the TV-exhausted lifter pin drops and permits the check ball to seal the TV exhaust passage.

Without the exhaust check ball system, a disconnected or broken cable would result in 0 psi throttle pressure and low operating line pressure regardless of engine throttle opening. Medium and heavy throttle performance would allow bands and clutches to slip and burn, damaging the transmission. Sealing the TV exhaust forces the TV pressure to maximum, regardless of engine throttle opening. This causes delayed shift system problems, but the transmission is saved until the problem is resolved.

An incorrect linkage or cable adjustment can cause low line pressure and permit clutch or band slippage. If not corrected immediately, the excessive friction element slippage leads to transmission failure.

Modulator System

Some vacuum-operated TV systems are referred to as modulator systems, generally identified with General Motors transmissions. The modulator works similarly to other vacuum-controlled TV systems. It produces a torque-sensing oil in response to manifold vacuum and a spring force (Figure 7–16). As a special feature, governor pressure is used to assist the modulator to balance against the spring force. The governor pressure assist acts to decrease the modulator pressure relative to vehicle speed (modulator output is modified to retard the line pressure boost). The transmission oil pressure requirements to hold the bands and clutches decrease as the car speed increases. The system uses both noncompensated and altitude-compensated modulator control units. General Motors used this system on the 3L80 (THM-400) and 3L80-HD (THM-475) transmissions.

Modulator System Variation

On the General Motors 4T60 (THM 440-T4), the technician will notice a vacuum modulator unit plugged into the transmission case, as well as a throttle cable control. This transmission produces two separate throttle-sensing oils:

- A cable-controlled TV system used exclusively for the shift system and auxiliary functions
- A vacuum-controlled modulator oil used exclusively for mainline regulator valve boost

BYPASS
REGULATED THROTTLE VALVE PRESSURE
THROTTLE VALVE
EX
THROTTLE VALVE PLUNGER
MAIN LINE PRESSURE
THROTTLE VALVE LINKAGE
ACCELERATOR
PART THROTTLE

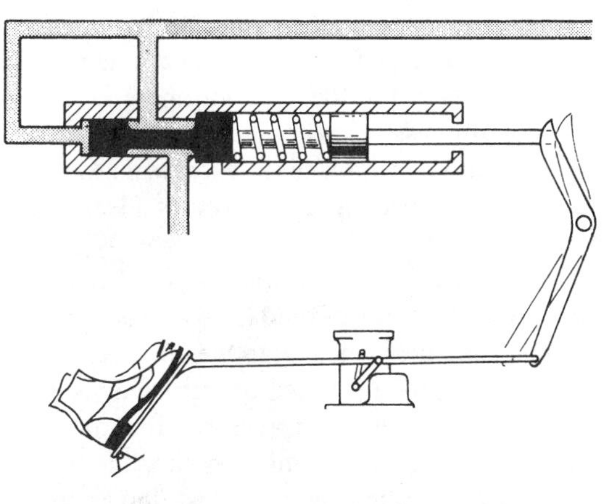

FIGURE 7–12 TV pressure in part-throttle and wide-open positions.

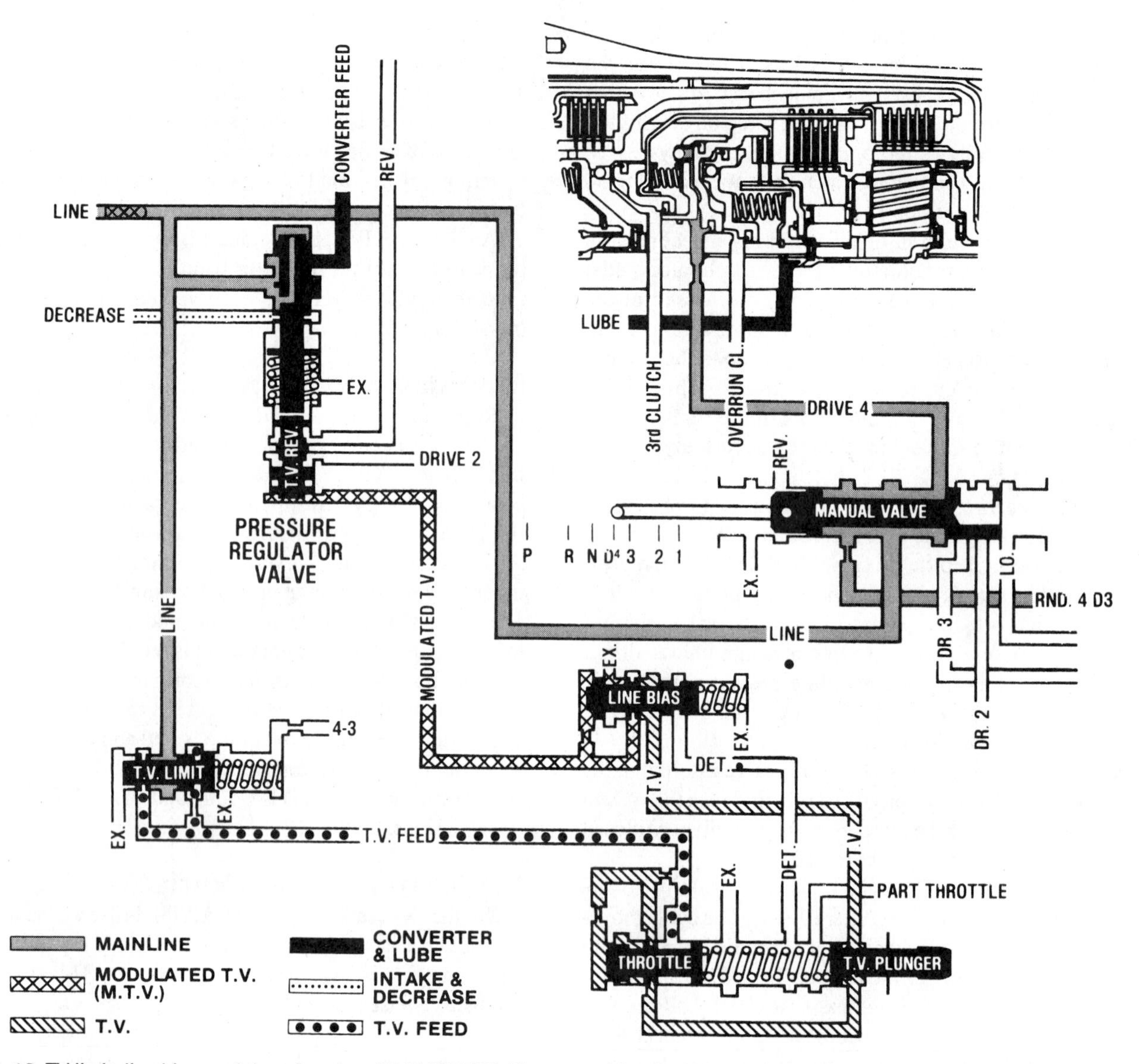

FIGURE 7–13 TV limit, line bias, and throttle valves (4L60/700-R4). (Courtesy of General Motors Corporation, Service Technology Group)

The cable adjustment does not affect the mainline pressure. Should a cable disconnection occur or the cable be adjusted too long, line pressure remains normal, and clutch/band slipping is avoided. Only the shift schedule is affected—a minimum shift schedule results regardless of throttle aggressiveness.

In the vacuum control system, periodic adjustments of the system are not required after the transmission leaves the factory. If the diaphragm fails or the vacuum line develops a leak, high line pressure will result. A leaky diaphragm will actually permit the manifold vacuum to pull the transmission fluid into the engine for consumption. This loss of fluid is usually not picked up until transmission slippage is evident to the driver, and that might be too late. In addition, a pinched or restricted vacuum line will cause a temporary delay in line pressure buildup during quick acceleration, resulting in transmission slippage, especially during heavy acceleration. The restricted line prevents the vacuum unit from sensing an immediate change in manifold pressure.

MANUAL CONTROL SYSTEM

The manual control system programs the operating ranges of the transmission. It consists of a driver select lever with gate stops and external linkage to the transmission manual valve lever (Figure 7–17). This indexes the manual valve position with the selector position. A spring-loaded detent arrangement holds the valve in each operating position (Figure 7–18).

For the control of the clutch/band combinations required in each of the operating circuits, the **manual valve** distributes regulated line pressure to the shift valves or directly to the friction element (Figure 7–19). Other basic distribution includes line pressure feed to the throttle and governor systems. The valve is designed to pressurize only the necessary circuits required of the operating range—inactive lines are kept exhausted. Park (P) and neutral (N) cut off all operating circuits and inhibit gear engagement.

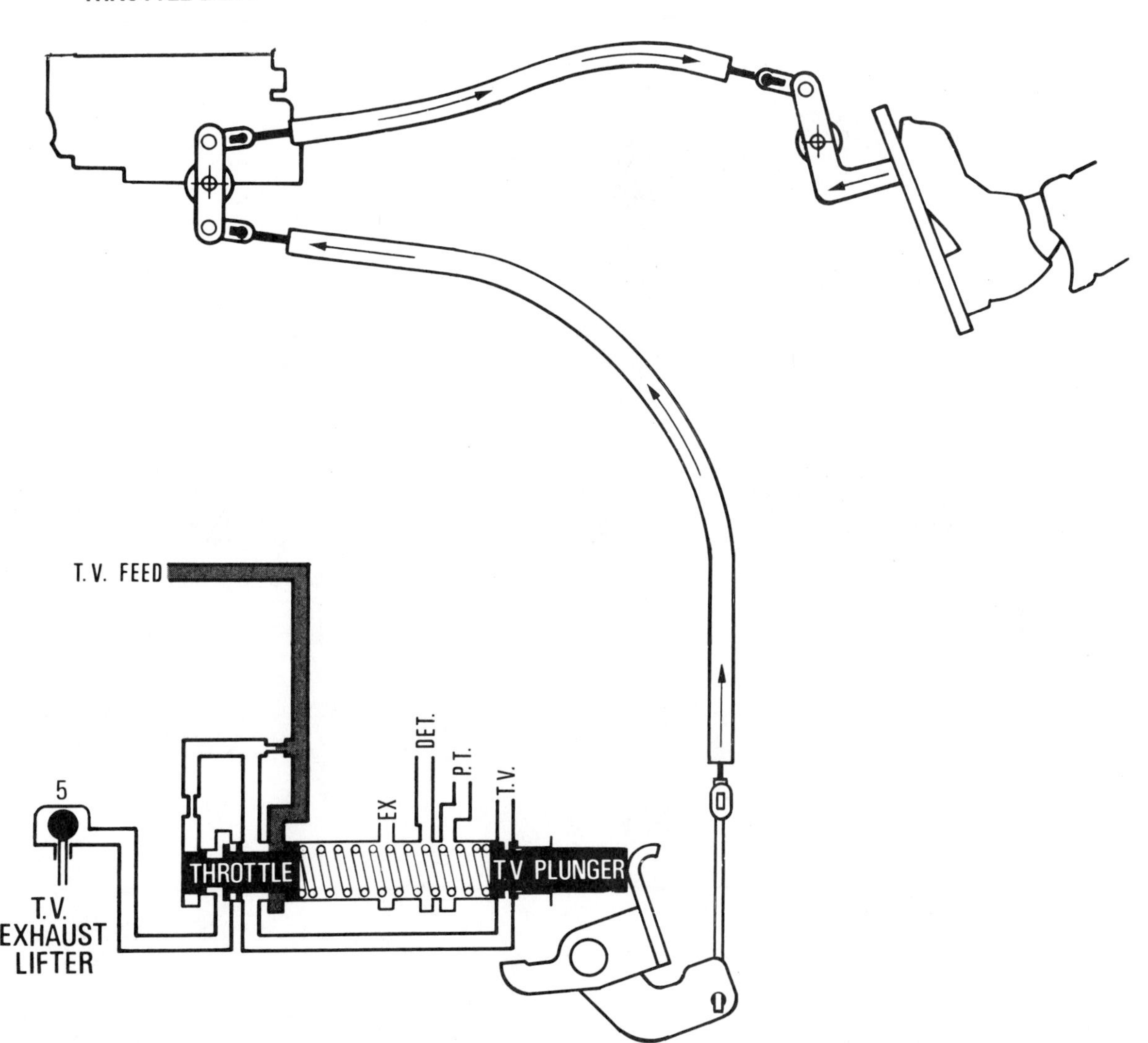

FIGURE 7–14 Accelerator, throttle, and cable control. (Courtesy of General Motors Corporation, Service Technology Group)

In park, an internally operated park mechanism engages a park pawl in the output shaft parking gear and locks it to the case. The park pawl is spring-loaded and does not lock until fully extended into one of the parking gear slots. In situations where the pawl is not aligned with a park gear slot, the selector is permitted to move against the spring load and stop gate until a slight vehicle roll arranges the correct alignment. The spring load on the pawl drops it into place for a mechanical lock. The spring-load feature also prevents the park pawl from engaging when the vehicle is in motion. It allows the pawl to rachet and produce a "clicking" noise.

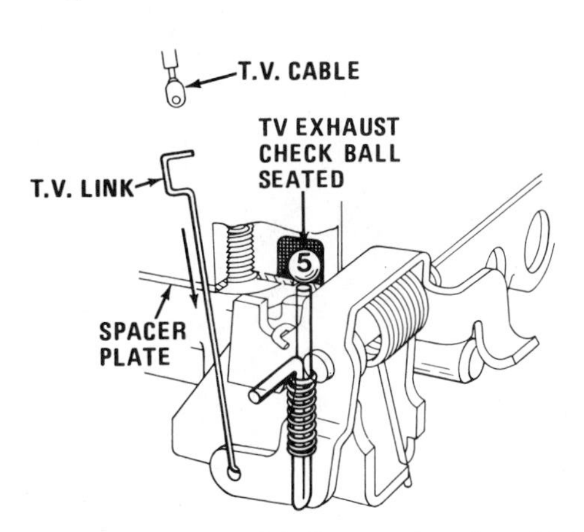

FIGURE 7–15 TV exhaust check ball seats when the cable is disconnected or adjusted too long. (Courtesy of General Motors Corporation, Service Technology Group)

- **Manual valve:** Located inside the transmission/transaxle, it is directly connected to the driver's shift lever. The position of the manual valve determines which hydraulic circuits will be charged with oil pressure and the operating mode of the transmission.

GOVERNOR SYSTEM

The **governor** is a hydraulic speedometer that is driven by the output shaft of the transmission (Figure 7–20). It receives fluid from the mainline pressure and produces a regulated governor

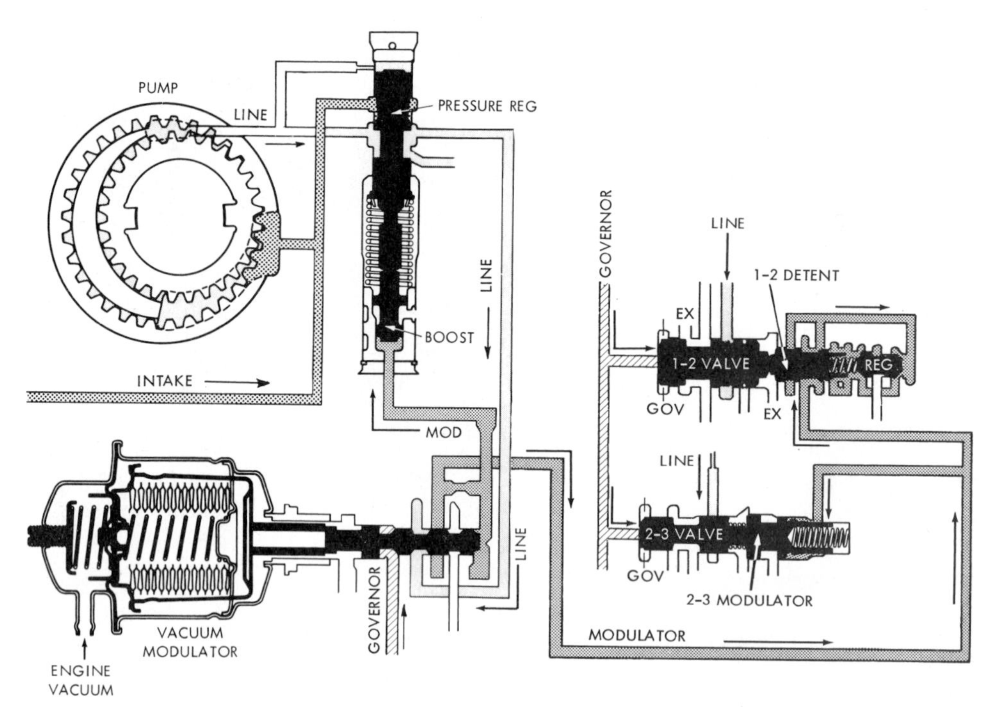

FIGURE 7–16 Modulator system used for pressure control and shift scheduling. (Courtesy of General Motors Corporation, Service Technology Group)

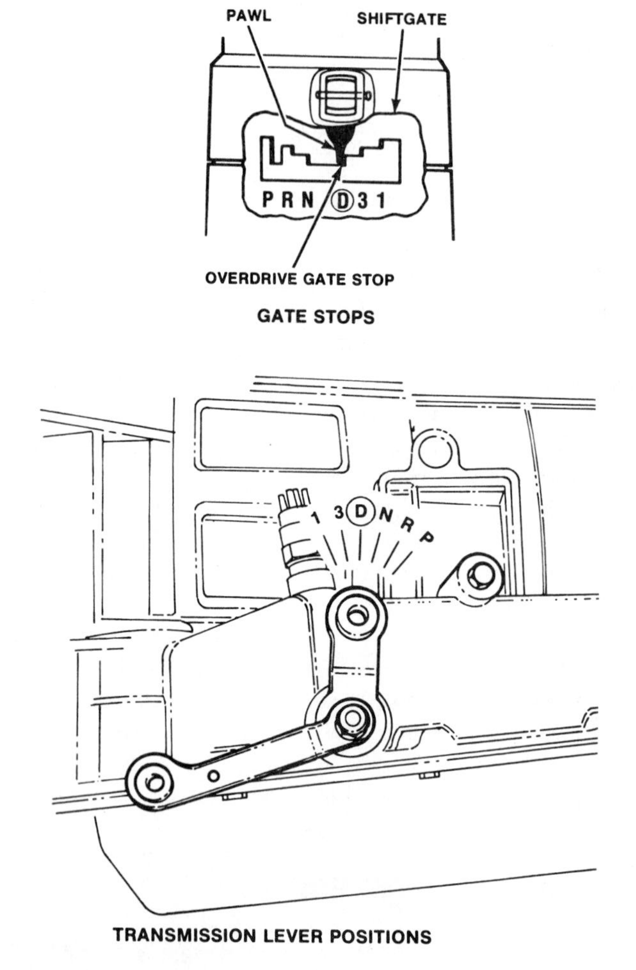

FIGURE 7–17 Manual linkage gate stops and lever positions. (Reprinted with the permission of Ford Motor Company)

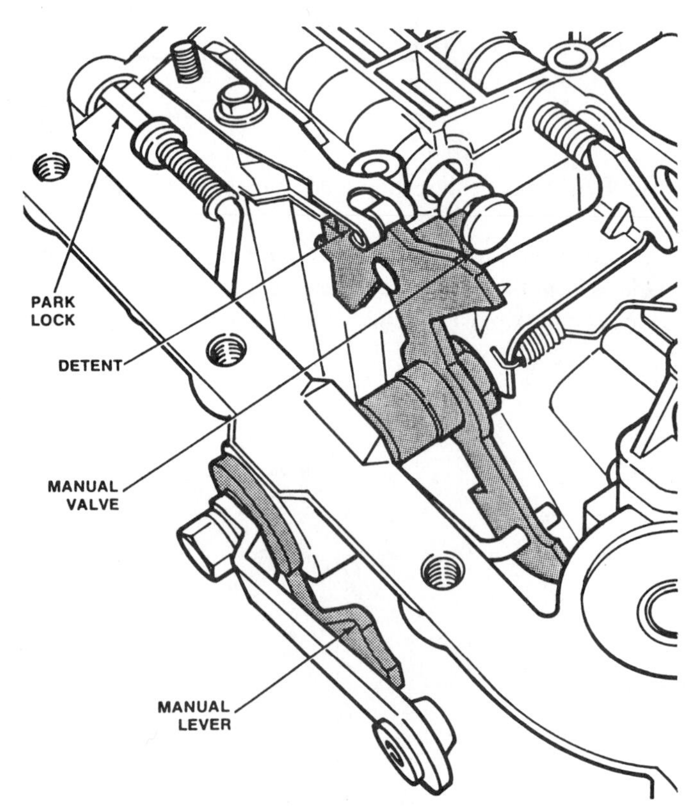

FIGURE 7–18 Manual valve location detented in valve body for each operating position. (Reprinted with the permission of Ford Motor Company)

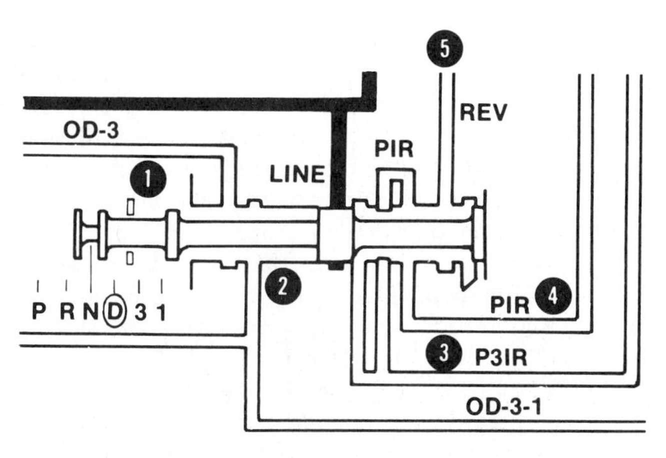

FIGURE 7–19 Manual valve control passages. (Reprinted with the permission of Ford Motor Company)

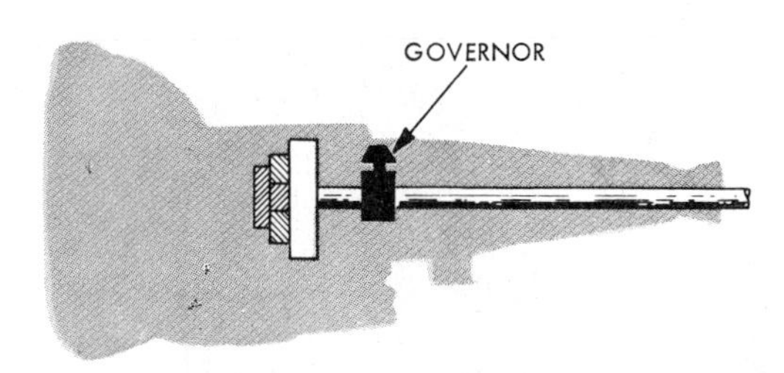

FIGURE 7–20 The governor, which generates a hydraulic road speed signal, is often driven by or mounted on the transmission output shaft.

pressure signal that is proportional to vehicle speed. It is used primarily for scheduling the transmission shifts along with throttle or modulator pressures. It is also used as a pressure signal for auxiliary or supporting control valves in the valve body.

In some transmissions, the governor oil acts to reduce the line pressure with increasing vehicle speed. We concentrate on its use for shift scheduling, in which capacity it opens the shift valve and causes the shift to happen.

There are several types of governor valve assembly designs in current use, but they all rely on the centrifugal effort of a rotating mass (weights).

❖ **Governor:** A device that senses vehicle speed and generates a hydraulic oil pressure. As vehicle speed increases, governor oil pressure rises.

Illustrated in Figure 7–21 is a case-mounted governor assembly. When the transmission output shaft drives the governor assembly, the governor weights fly outward and exert a **centrifugal force** on the governor valve. Drive oil, which is actually regulated mainline oil from the manual valve, feeds the governor valve until sufficient reaction pressure buildup on top of the valve balances the centrifugal force of the weights.

The greater the vehicle speed, the greater the centrifugal force of the weights, and hence the greater the governor pressure necessary to balance the centrifugal force. Eventually, vehicle speed reaches a point at which the governor valve cannot balance itself against the centrifugal force of the weights. When this happens, the governor valve is permanently in the open position, and governor pressure equals the mainline supply pressure.

❖ **Centrifugal force:** The outward pull of a rotating object, away from the center of revolution. Centrifugal force increases with the speed of rotation.

The governor weight assembly is made up of two sets of weights, primary and secondary, plus two springs. These parts are combined to produce a two-stage output curve (Figure 7–22). The primary (heavy) and secondary (light) weights have their own independent action. The primary weights, however, are arranged to work together initially with the secondary weights. The springs hold the secondary weights against the primary weights. The weights are arranged so that the lighter secondary weights act directly on the regulating valve.

At low speeds, the heavy mass is needed to generate a regulated governor oil for the 1-2 shift. The weights move in and out together as the governor valve regulates. Since the centrifugal force of a rotating mass increases by the square of the speed, it becomes difficult to schedule a 2-3 shift with good spacing.

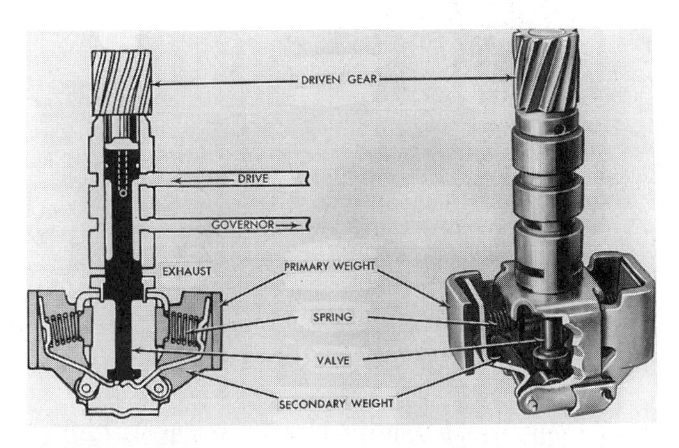

FIGURE 7–21 Case-mounted governor assembly. (Courtesy of General Motors Corporation, Service Technology Group)

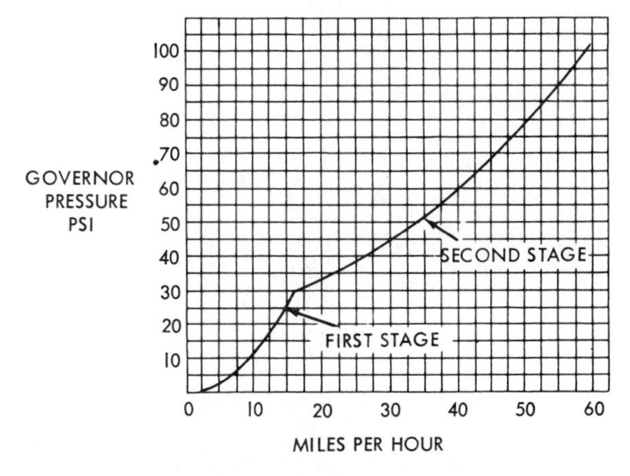

FIGURE 7–22 Typical two-stage governor pressure curve.

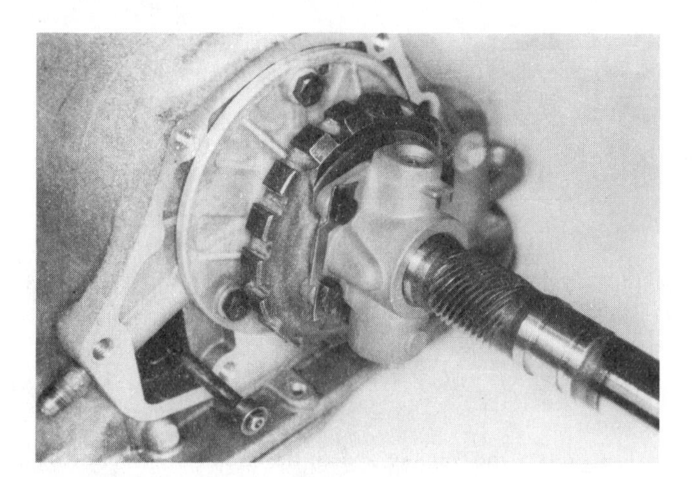

FIGURE 7–23 Output shaft–mounted governor.

Doubling the rpm increases the centrifugal force by four times. Therefore, it becomes necessary to arrange for a cutback in the rate of centrifugal advance. At approximately 20 mph (16 km/h), the centrifugal force of the primary heavy weight exceeds the spring force and permanently stays moved out against a stop. The primary weights are now separated from the secondary (light) weights and are ineffective. The governor valve now balances against the centrifugal force of the secondary weights plus the spring force. The two-stage governor action results in a more even governor output distribution with increases in vehicle speed.

Governor assemblies are also designed to mount over the output shaft (Figures 7–23 and 7–24). The governor valve is connected to the weights by a rod passing through the output shaft. The governor is two-stage, with the light and heavy weights working together for low speeds below 20 mph (16 km/h), and with the light weight plus the spring force providing the cutback in the rate of regulated governor increase in the second stage. The heavy weight at this time stays bottomed against the outside stop.

Another style of governor design features two check balls that seat in two pockets opposite each other in the governor shaft. The complete assembly and circuit concept is illustrated in Figure 7–25. Notice the absence of a governor valve. Regulated governor pressure is determined by the amount of drive oil (regulated line) that bleeds past the check balls. The centrifugal weight mass determines the value of the regulated governor oil. The primary (heavy) weight and secondary (light) weight each have a J-hook that goes around the shaft and captivates the check ball on the opposite side.

During low-speed operation, the centrifugal force of the secondary (light) weight plus the spring force outguns the primary (heavy) weight. The secondary check ball is seated, and bleed-off of excess governor oil to balance the system is through the primary check ball. As vehicle speed increases, the centrifugal force on the primary weight greatly increases over the centrifugal and spring force of the secondary weight. The primary check ball stays seated, and the governor bleed is only at the secondary check ball.

SERVO AND CLUTCH ASSEMBLIES

The use of hydraulic power to apply the bands and clutches is confined to a single device, the hydraulic piston. The pistons are housed in cylinder units known as servo and clutch assemblies. The function of the piston is to convert the force of fluid into a mechanical force capable of handling large loads. Hydraulic pressure applied to the piston strokes the piston in the cylinder and applies its load. During the power stroke, a mechanical spring or springs are compressed to provide a means of returning the piston to its original position. The springs also determine when the apply pressure buildup will stroke the piston. This is critical to clutch/band life and shift quality.

Servo Assembly

A servo unit provides the method of application and disengagement of the bands. The band is energized hydraulically by

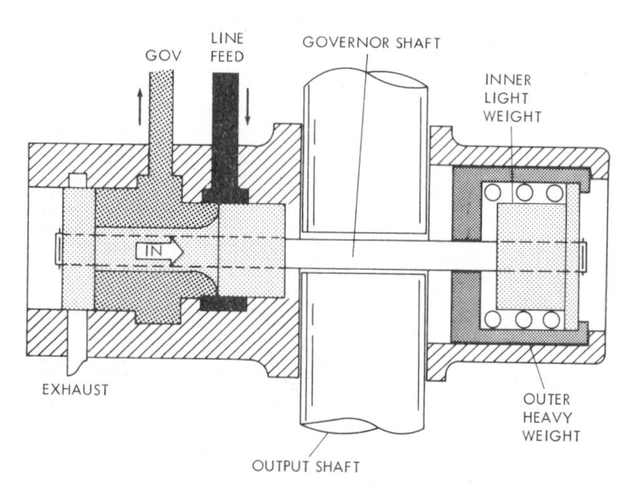

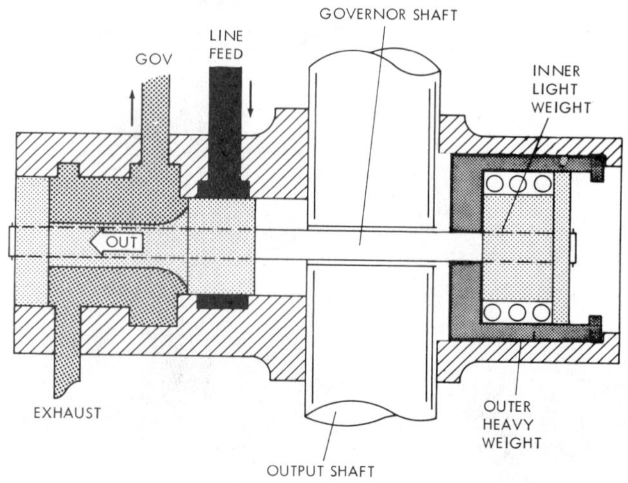

FIGURE 7–24 Cross-sectional views of governor operation in secondary phase. Left: Governor valve open to line feed by centrifugal force. Right: Governor valve moved out over exhaust port to balance pressure. Note that heavy weight remains bottomed out against the bore stop. Light weight and heavy weight are no longer moving together.

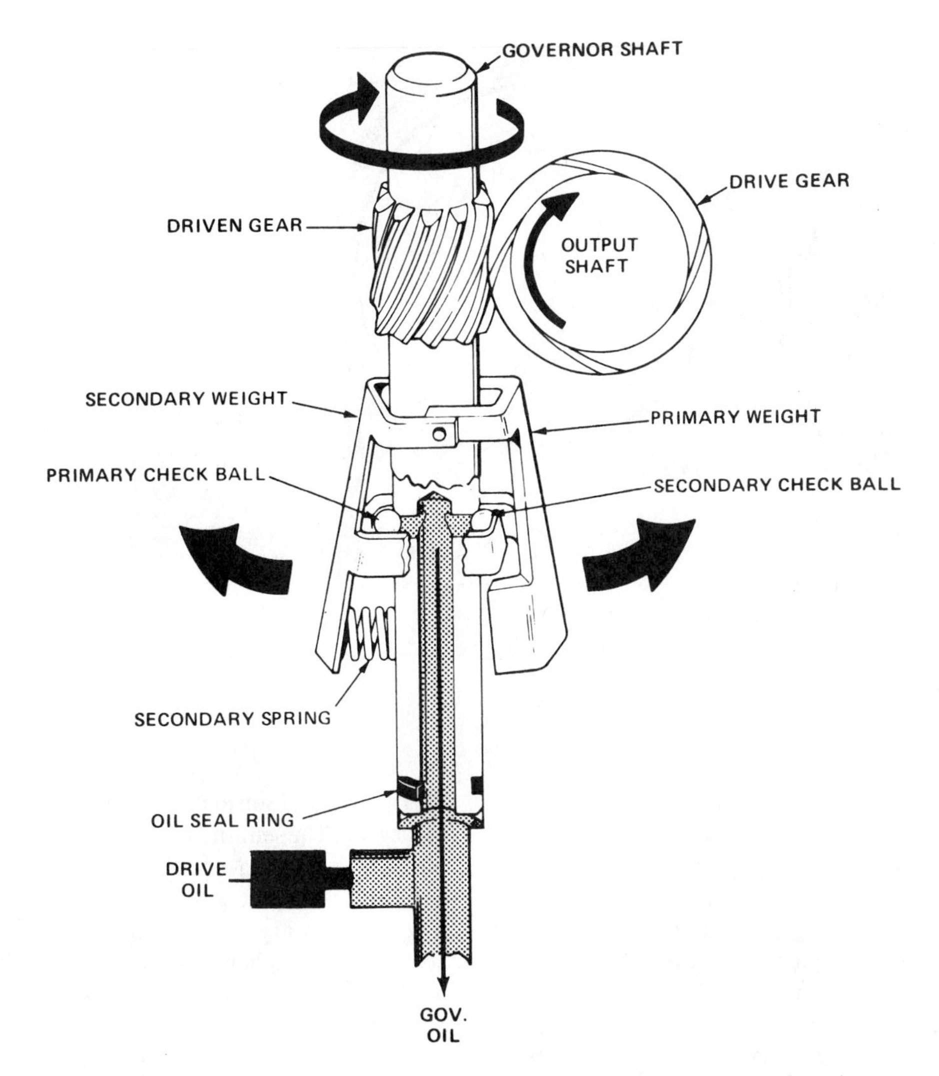

FIGURE 7–25 Check ball style of governor. (Courtesy of General Motors Corporation, Service Technology Group)

the piston force acting on one end of the band while the other end is anchored to the transmission case and absorbs the reaction force of the gear ratio (Figure 7–26). The servo unit consists of a piston in a cylinder and a piston return spring. It may be a separate cylinder assembly bolted to the transmission case, or the cylinder can be designed as an integral part of the case (Figure 7–27). Through suitable linkage and lever action, the servo is connected to the band it operates. The servo force acts directly on the end of the band with an apply pin or through a lever arrangement that provides a multiplying force. These arrangements are illustrated in Figures 7–27 and 7–28.

The servo unit band application must rigidly hold and ground a planetary gear member to the transmission case for forward or reverse reduction. To assist the hydraulic and mechanical apply forces, the servo and band anchor are positioned in the transmission to take advantage of the drum rotation. In Figure 7–28, the drum rotational effort is in the counterclockwise direction. When the band is applied, it becomes self-energized and wraps itself around the drum in the same direction as drum rotation. This self-energizing effect reduces the force

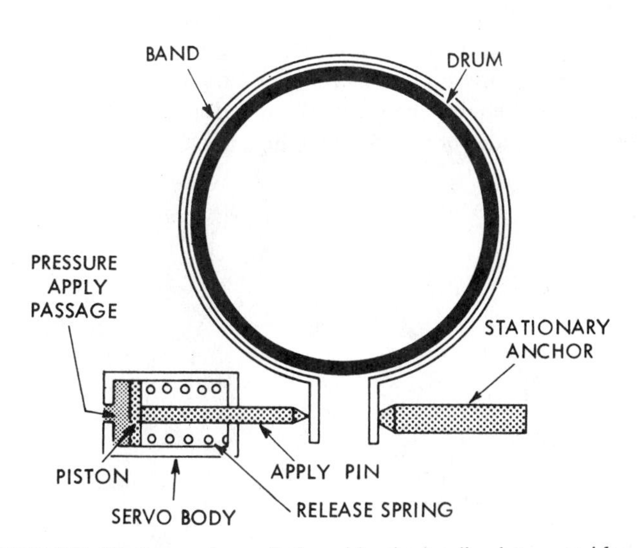

FIGURE 7–26 A servo is a cylinder with a hydraulic piston used for band application. (Courtesy of General Motors Corporation, Service Technology Group)

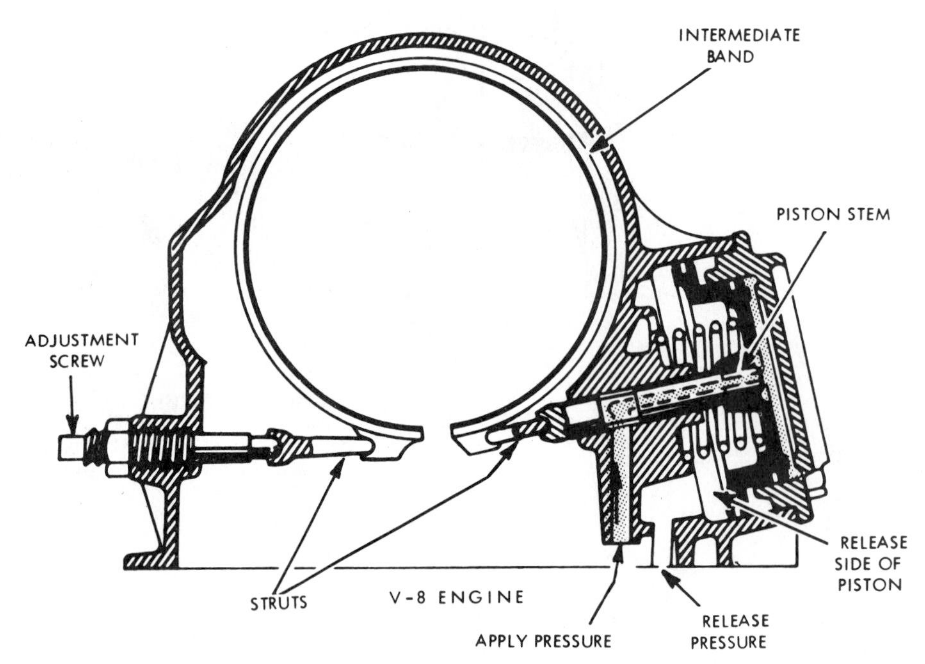

FIGURE 7–27 Direct-acting servo unit as part of the case. (Reprinted with the permission of Ford Motor Company)

that the servo must produce to hold the band. The principle is the same one used to describe the action of self-energized drum brakes.

To release the servo apply action on the band, the servo apply oil is exhausted from the circuit or a servo release oil is introduced on the servo piston that opposes the apply oil. The servo piston in Figure 7–29 is shown in the apply position. When servo release oil is introduced on a 2-3 shift, the hydraulic mainline pressure acting on the top side of the piston plus the servo return spring will overcome the servo apply oil and move the piston downward.

Band Design

Band designs in automatic transmissions use a flexible contracting band and are classified as single wrap (Figure 7–30) or double wrap (Figure 7–31). The double wrap offers the advantage of having the greater holding force should the drum rotation provide for a self-energizing effect.

The composition of the friction lining is either semi-metallic or organic. The semi-metallic materials can withstand high unit pressures but have a tendency to scrape away the drum surface. Semi-metallic applications, therefore, are limited to conditions that require a high static torque and a minimum of dynamic service. This explains why the semi-metallic band is usually confined to use as a reverse band. A favored organic material is paper pulp-based or cellulose-based compound, which is a soft material that has very little wear effect on a drum and shows better uniform contact to the drum surface. Grooving of the friction surface provides for a controlled escape of fluid and vapor during engagement.

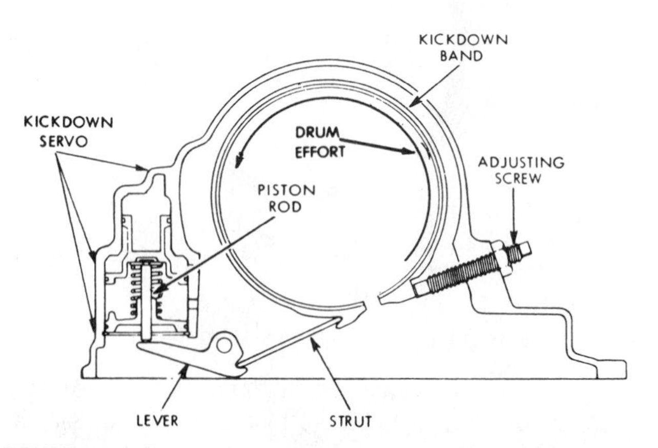

FIGURE 7–28 Servo unit lever arrangement: self-energizing.

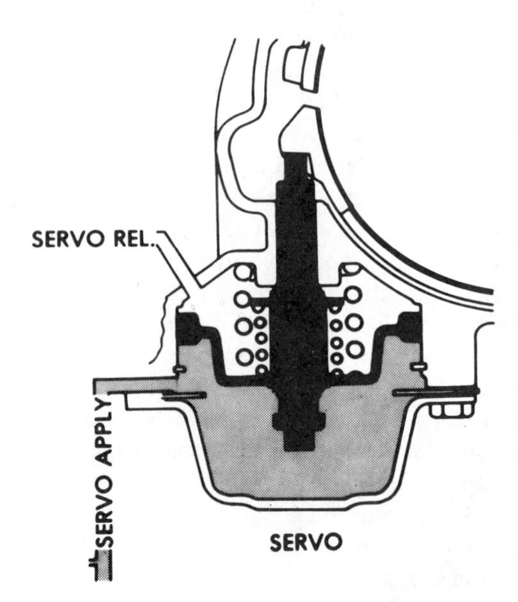

FIGURE 7–29 Servo piston in apply position.

FIGURE 7–30 Single wrap bands.

Soft organic cellulose materials are preferred where the band is operating under dynamic conditions. Drum surface speeds over 6000 ft/min can be encountered prior to band application, with initial engagement surface temperatures at times reaching 600–800°F (316–427°C). These temperatures quickly vaporize the oil from the surface pores of the paper material, and it would appear the band is doomed to failure. The material compresses under force, however, and oil stored in the inner micro cells is squeezed to the surface with a sponge-like action and replenishes the vaporized surface oil. A cooling effect is maintained to keep the band alive for the next cycle. Paper friction material has negligible wear and offers considerable durability.

Critical to the proper band action is the drum material, hardness, and surface finish. Where drum material is a soft iron, a high surface finish cannot be used. This combination would have a tendency to cause glazing of the band and drum. In service practice, it is recommended that any polished surface on a soft iron drum be deglazed with 120- to 180-grit emery cloth. In contrast, a smoother finish on a hard-surface drum is optimal for the right band action.

Consistent shift quality depends on the following factors:

FIGURE 7–31 Double wrap band.

- Friction material
- Grooving in the friction material
- Drum material and hardness
- Drum surface finish
- Proper adjustment of band clearance
- Transmission fluid type and condition
- Band apply force
- Adequate fluid circulation to dissipate heat

Clutch Assembly

The multiple-disc clutch is the favored clutch unit used in automatic transmissions. It offers the following features:

- Multiple discs give the clutch a sufficient area of frictional or torque-holding capacity in an overall small diameter drum. The number of contacting surfaces is a factor in determining holding torque.
- Unlike bands, disc clutches can easily be used as rotating engagement members.
- Once the proper running clearance has been established during factory or field service assembly, there is no adjustment requirement for wear.
- The disc clutch can be used as a reaction or holding member by connecting a planetary component to the case ground. It performs the same function as a band. The low-reverse clutch used in a variety of transmission designs is a classic example. The clutch connects the rear planet carrier to the case.

A typical rotating clutch drive unit is made up of the following components and is illustrated in Figure 7–32.

- Friction plates or drive discs with internal splines that fit on a torque-transmitting hub.
- Steel discs or plates that mate with the friction plates. The external drive lugs of the steel plates fit into a torque-transmitting drum or cylinder.
- A reaction or pressure plate at the end of the clutch pack.
- A hydraulic apply piston to engage the clutch pack.

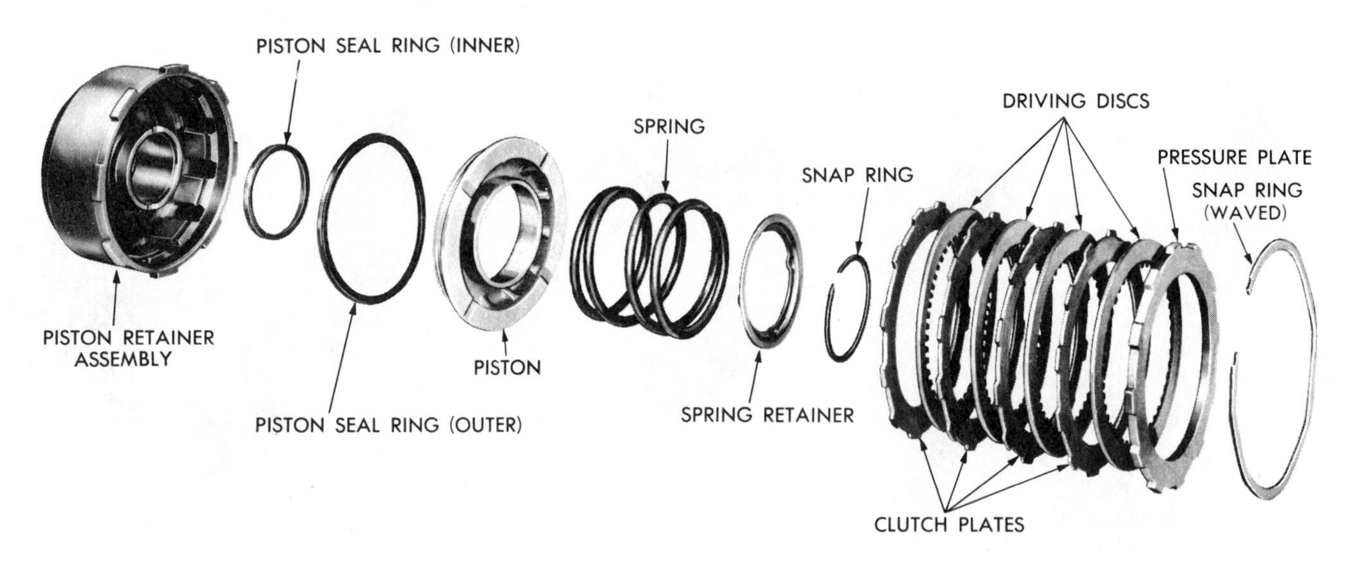

FIGURE 7–32 Disc clutch assembly. (Courtesy of Chrysler Corp.)

- A spring-loaded piston release.
- A retainer or drum that houses the complete assembly.

When the piston is applied, it squeezes the clutch pack together against the pressure plate and snap ring. The snap ring fits in front of the pressure plate and into a snap-ring groove in the clutch drum.

In Figure 7–33, a pair of multiple-disc clutch assemblies shows a more exact relationship to a transmission power flow. Both clutches are oil-applied and spring-released. With both clutches OFF, there is no power to the gearset. When the front clutch is applied, the primary sun gear is driven. When the rear clutch is applied, the secondary sun gear is driven. This is how power enters the gearset for forward speeds and reverse. In high gear, the front and rear clutches are applied and drive both sun gears together.

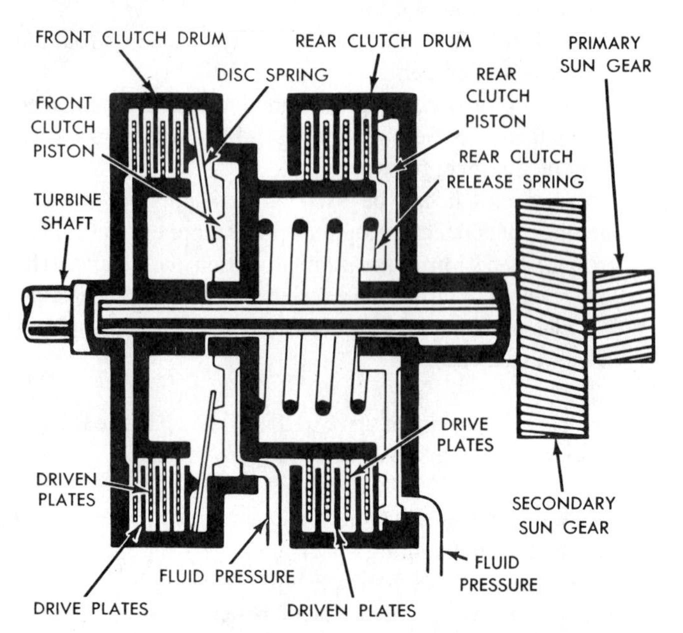

FIGURE 7–33 Front and rear disc clutch assemblies controlling power input to planetary geartrain. (Reprinted with the permission of Ford Motor Company)

To return or release a clutch piston, several spring designs are used:

- A large single coil spring (Figure 7–34)
- A series of small coil springs (Figure 7–35)
- A belleville or disc spring (Figure 7–36)

Notice the cross-sectional view of the disc spring in the front clutch shown in Figure 7–33. The spring acts as a lever arm to gain additional apply force as it moves about the pivot ring, or fulcrum point, of the inner pressure plate. This arrangement requires the piston to take a longer apply stroke, which results in cushioning the clutch apply. The disc spring effect is detailed in Figure 7–37. Some clutch units may use a steel-waved plate or waved snap ring to cushion the clutch apply (Figure 7–36).

In rotating clutch units, a problem usually arises when the clutch is not engaged. With the clutch OFF, the clutch drum or housing still spins. As a matter of fact, the spin-up may be at an overdrive speed. The high-speed rotation could create sufficient centrifugal force in the residual or remaining oil in the clutch apply cylinder to partially engage the clutch. This creates an unwanted drag between the clutch plates. To prevent this problem, a centrifugal check ball relief (Figure 7–38) is built into the clutch drum or clutch piston.

The steel check ball operates in a cavity with a check ball seat. A small hole or orifice is tapped from the seat for pressure relief. In the clutch drum example, when the clutch unit is applied, oil pressure holds the ball on its seat and blocks the orifice. In the released position, the centrifugal force created by the rotating clutch drum moves the ball off its seat, allowing the residual oil behind the clutch piston to be discharged.

Many modern clutch housings use another technique to allow the residual oil to escape when not under pressure. A small bleed hole is drilled into the housing or the piston. The

FIGURE 7–34 Top: Clutch piston return spring—large single coil. Bottom: Return spring positioned into housing.

FIGURE 7–35 Clutch piston return springs—small coil springs shown without spring seat and retainer in place.

size of the hole is engineered so that the amount of oil leaking through it during clutch apply does not drop below system requirements. When a shift valve turns the clutch housing off and the oil pressure drops to release the clutch plates, oil remaining in the housing can escape through the bleed hole. This design concept is referred to as a "feed and bleed" system.

Clutch Plate Design

The clutch composition or friction discs are made up of a steel plate core, faced with a friction material that is fully metallic, semi-metallic, or paper cellulose. The sintered, fully

FIGURE 7–36 Disc spring installation. (Courtesy of Chrysler Corp.)

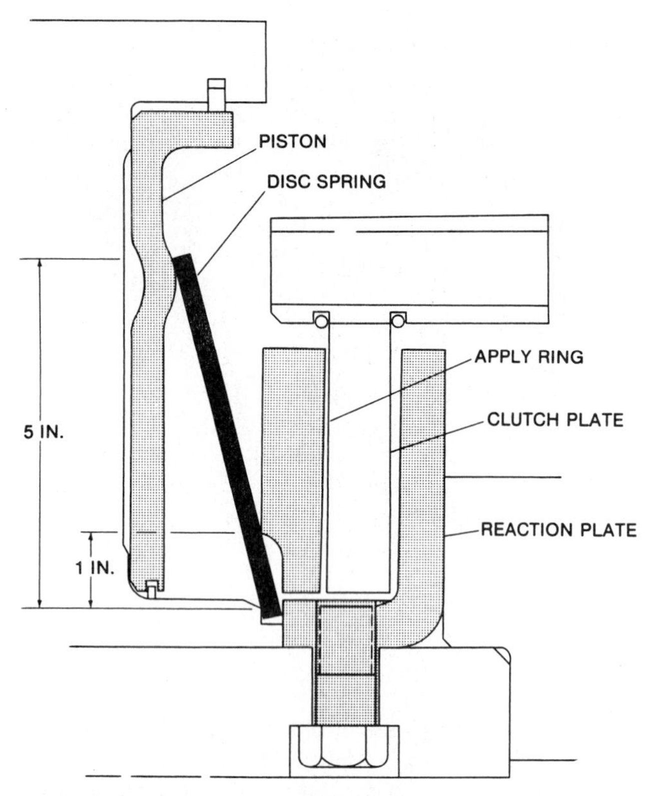

FIGURE 7–37 Disc spring effect.

metallic friction material is usually made from a dry powder mix of copper and friction-modifying agents containing graphite. Semi-metallic materials usually consist of copper powder compounded with lead powders, asbestos, and resin binders. Paper friction material is processed from cellulose fibers saturated with a liquid binder consisting of formulated organic resins.

Depending on the desired friction and durability requirements, ceramic or graphite powder materials are sometimes added to the paper materials. Advanced technology in formulating paper cellulose has made this friction material the favored choice for automatic transmission clutch plates. Its ability to absorb oil and give it up during engagement keeps the friction material cool and ensures durability—wear is negligible. This material also provides good surface conformity and excellent coefficient-of-friction stability over a wide range of sliding velocities. These are important factors contributing to quality engagement feel and the cost of production.

To improve the cooling during the engagement cycle, the friction plates are often grooved to provide a controlled fluid wiping action and vapor escape (Figure 7–39). The fluid flows and cools the mating steel plates that actually absorb the heat generated from the engagement. The friction plates stay cool by giving up heat to the fluid that is squeezed out of the fibrous material upon clutch application. The steel plates and the fluid provide the clutch cooling. A clutch engagement should take no longer than 0.5 to 0.7 of a second. Longer engagement time means prolonged slipping of the clutch surfaces and excessive heat generation that cannot be adequately handled by the available cooling: the reason for clutch failure.

The steel plates are made from carbon steel rolled stock. The rolled stock is run through straightening rolls and then blanked into clutch discs. As a final process, the blanked discs are tumbled to produce a dull or mat surface finish. The surface finish breaks in the mating composition friction-faced discs. After the break-in period, the steel discs become polished. These discs should never be used with replacement friction discs, however, as the new friction plates are likely to develop surface glazing and cause lazy or slipping shifts, resulting in short clutch life.

Consistent clutch engagement quality depends on the following factors:

- Friction material
- Friction plate grooving
- Proper clutch plate clearance
- Clutch apply force
- Transmission fluid type and condition
- Adequate fluid circulation for heat dissipation
- Steel plate surface finish

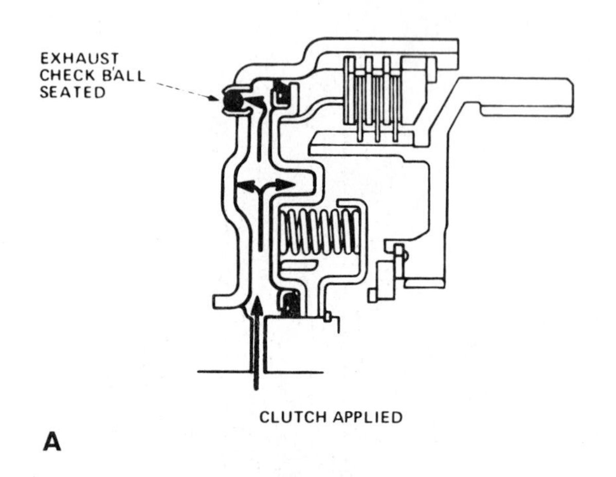

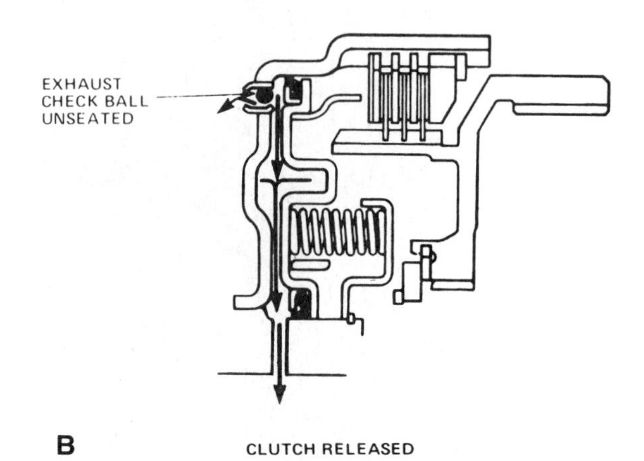

FIGURE 7–38 Centrifugal check ball relief. Left: Seated when clutch is applied. Right: Unseated when clutch is released. (Courtesy of General Motors Corporation, Service Technology Group)

FIGURE 7–39 Typical friction disc surface patterns.

VALVE BODY ASSEMBLY

The **valve body assembly** is the heart of the hydraulic control system. It is an intricate network of interrelated passages, precision valves, springs, check balls, and orifices (Figures 7–40 to 7–43). The assembly normally contains the manual valve, the throttle and forced-downshift valves, shift valves, and sometimes the pressure regulator valve. In response to external messages and driver demands, the valve body controls the hydraulic circuits to apply the clutch/band combinations for the planetary gear ratios.

❖ **Valve body assembly:** The main hydraulic control assembly of the transmission that contains numerous valves, check balls, and other components to control the distribution of pressurized oil throughout the transmission.

The valve body works like a hydraulic computer. It is programmed for the operating range by the manual valve and monitors two hydraulic signals, one from the throttle system and another from the governor system. These signals are evaluated by the automatic shift system to determine the shift points and by the pressure regulation system to determine line pressure modulation (Figure 7–44). A third signal is provided by the forced-downshift or detent mechanism. The system is driver-controlled and provides a hydraulic command to overrule the shift valves for either a 3-2 or 3-1 forced downshift in three-speed units. Four-speed overdrive units have an additional 4-3 and 4-2 capability.

ACCUMULATOR

Accumulators are used in servo and clutch apply circuits for the purpose of controlling the shift feel or quality. This is done by controlling the slip time or aggressiveness at which the band or clutch fully applies.

❖ **Accumulator:** A device that controls shift quality by cushioning the shock of hydraulic oil pressure being applied to a clutch or band.

Two types of accumulators are used in transmissions. The piston-style accumulator is a popular choice of transmission design engineers. The other type, the valve-style accumulator, is used in some familiar applications.

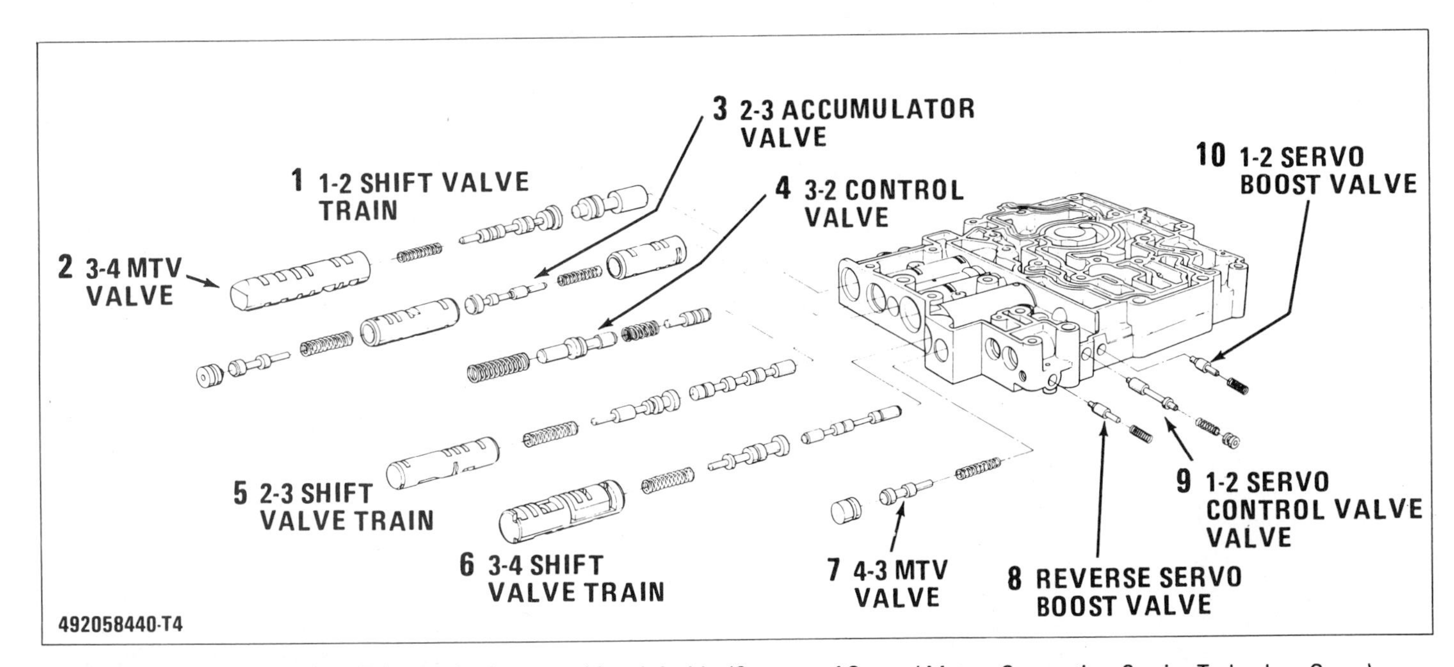

FIGURE 7–40 Typical 4T60 (440-T4) control valve assembly—left side. (Courtesy of General Motors Corporation, Service Technology Group)

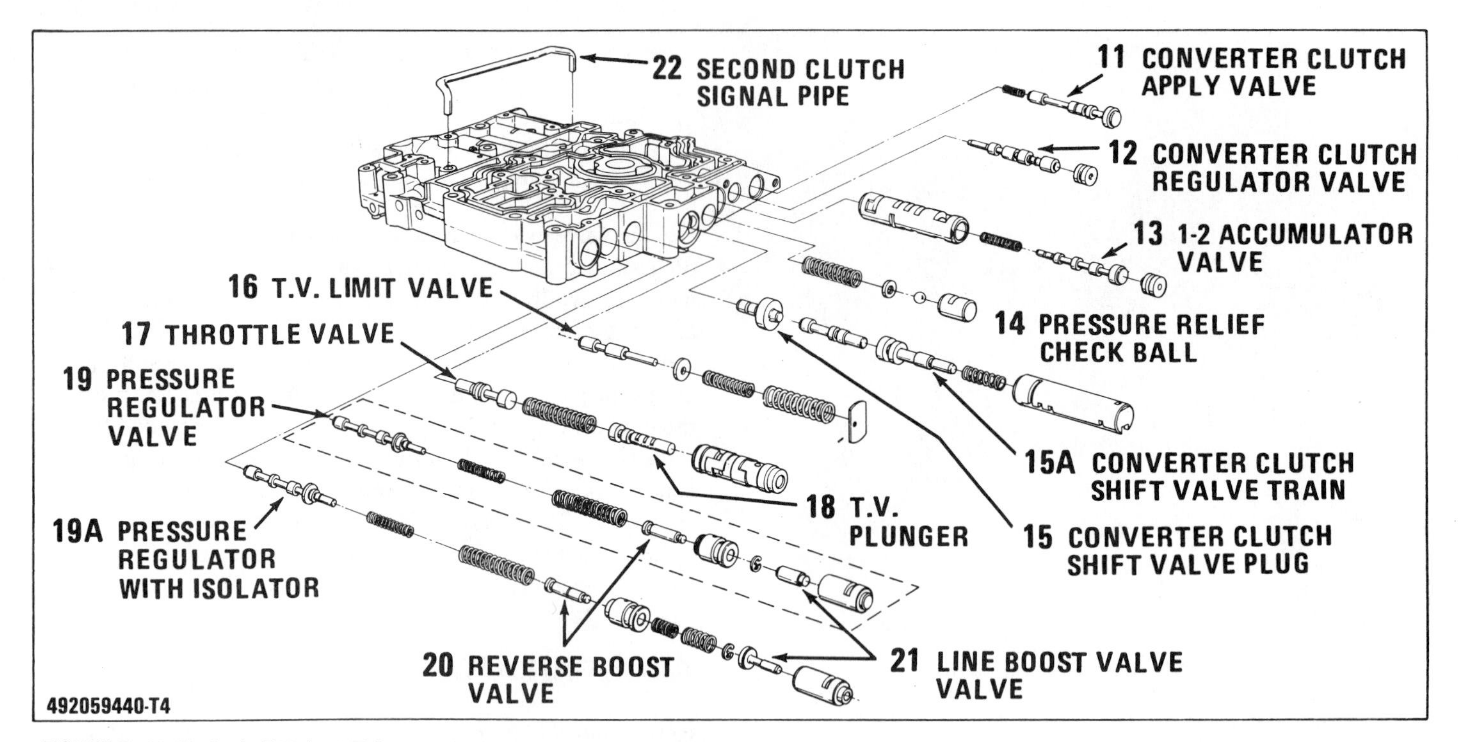

FIGURE 7–41 Typical 4T60 (440-T4) control valve assembly—right side. (Courtesy of General Motors Corporation, Service Technology Group)

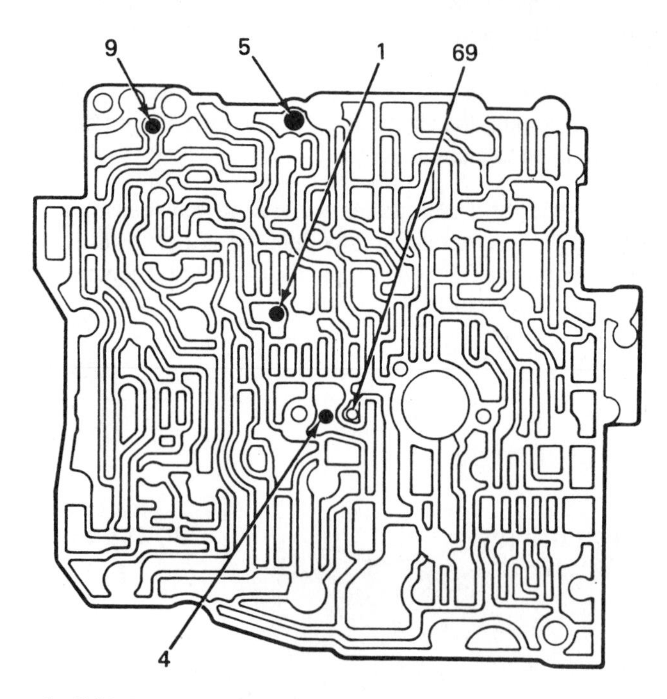

1 FOURTH CLUTCH CHECK BALL
4 THIRD CLUTCH CHECK BALL
5 2-3 ACCUMULATOR FEED CHECK BALL
9 REVERSE SERVO FEED CHECK BALL
69 SCREEN, 3RD CLUTCH EXHAUST

NO. 7 CHECK BALL LOCATED IN A CAPSULE (137) IN CASE

JH0091-440T4

FIGURE 7–42 Typical check ball locations—control valve assembly 4T60 (440-T4). Note: Check ball locations and number can vary from year to year. (Courtesy of General Motors Corporation, Service Technology Group)

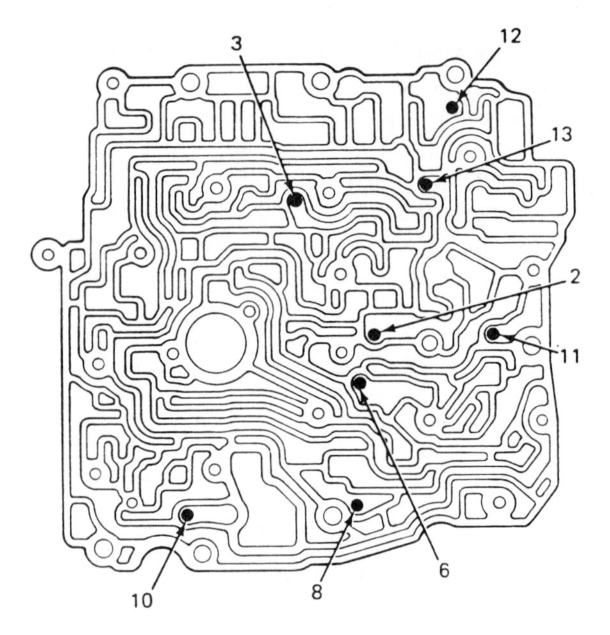

2 3-2 CONTROL CHECK BALL
3 PART THROTTLE AND DRIVE 3 CHECK BALL
6 2-3 ACCUMULATOR EXHAUST CHECK BALL
8 SECOND CLUTCH CHECK BALL
10 CONV. CLUTCH RELEASE/APPLY CHECK BALL
11 THIRD/LO-1ST CHECK BALL
12 1-2 SERVO FEED CHECK BALL
13 INPUT CLUTCH/REVERSE CHECK BALL

NO. 7 CHECK BALL LOCATED IN A CAPSULE (137) IN CASE.

JH0044-440T4

FIGURE 7–43 Typical check ball locations—channel plate 4T60 (440-T4). Note: Check ball locations and number can vary from year to year. (Courtesy of General Motors Corporation, Service Technology Group)

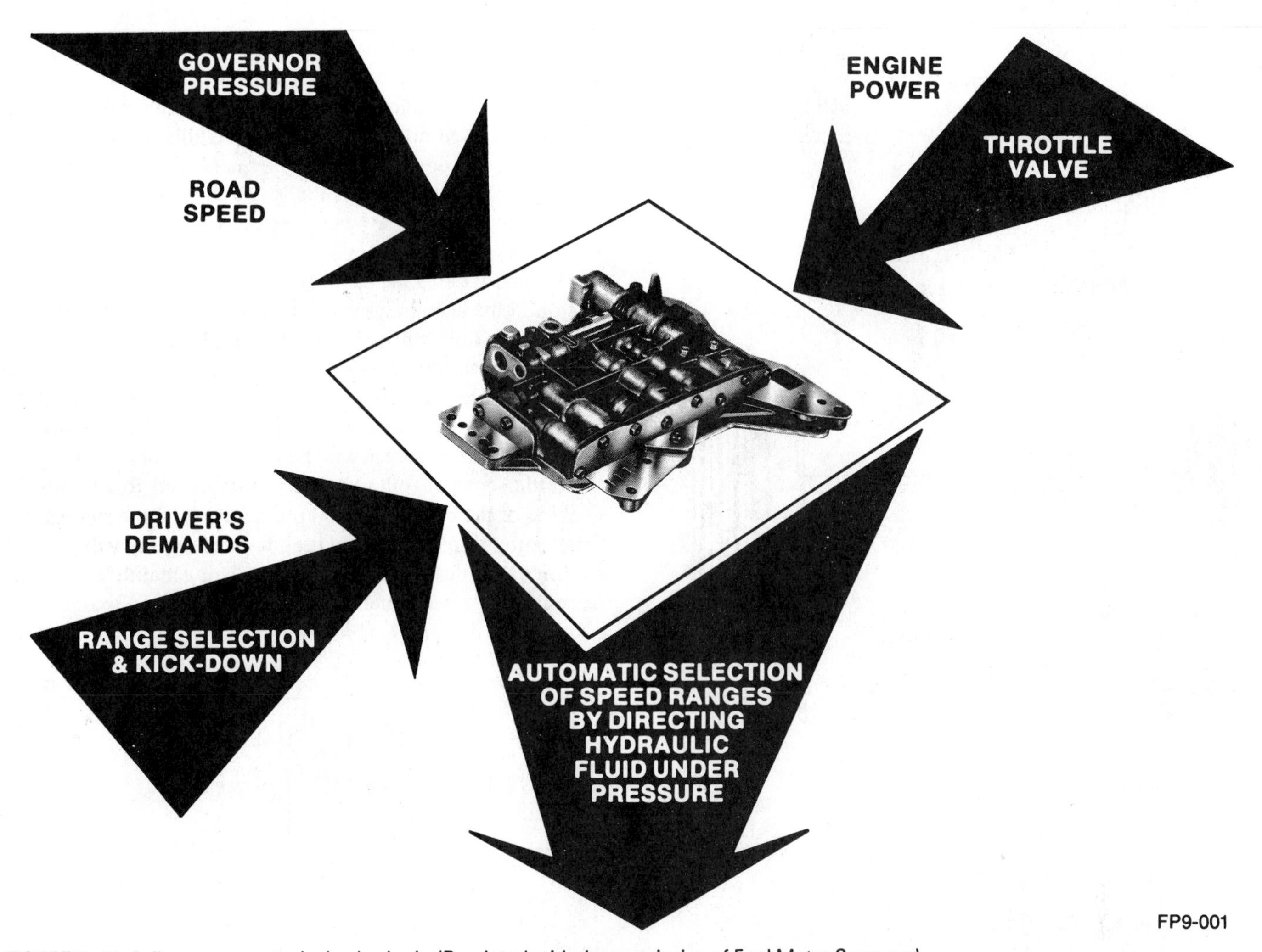

FIGURE 7–44 Influences on a typical valve body. (Reprinted with the permission of Ford Motor Company)

Piston-style Accumulator System

The piston-style accumulator uses a spring-loaded piston device located in a cylinder bore that cushions the shift engagement according to engine torque output (Figure 7–45). It does this by absorbing a certain amount of fluid flow in the circuit during a band or clutch application.

A series of illustrations shows how a typical piston-style accumulator system works. Figure 7–46 shows a simple clutch circuit without provisions to cushion the automatic shift. With this arrangement, the clutch engagement is sudden, and the car jerks forward. The shift sensation is similar in effect to "popping" the clutch pedal on a manual transmission. The driver feels a kicking or bucking motion.

To smooth the shift, an accumulator is designed into the clutch apply circuit (Figure 7–47). In the first stage, initial rapid flow of apply oil and pressure buildup is permitted in the circuit. The clutch piston return springs are overcome, and the slack or clearance is taken up between the discs. As soon as the clutch slack is removed and the discs begin to tighten, the pressure buildup in the circuit is rapid. At a point just before maximum pressure is reached, the accumulator piston is stroked against the spring force (Figure 7–48). Because the accumulator absorbs fluid as the piston strokes, the final pressure buildup in the clutch circuit is more gradual, and a smooth engagement is attained. The complete sequence of events is shown in Figure 7–49.

FIGURE 7–45 Chrysler 41TE (A-604) accumulators.

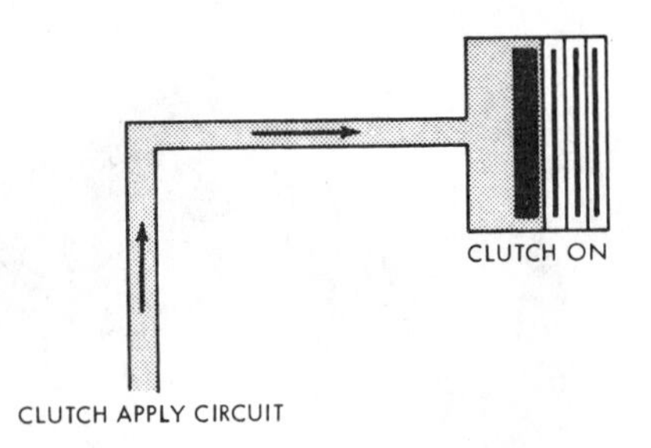

FIGURE 7–46 Clutch circuit without accumulator.

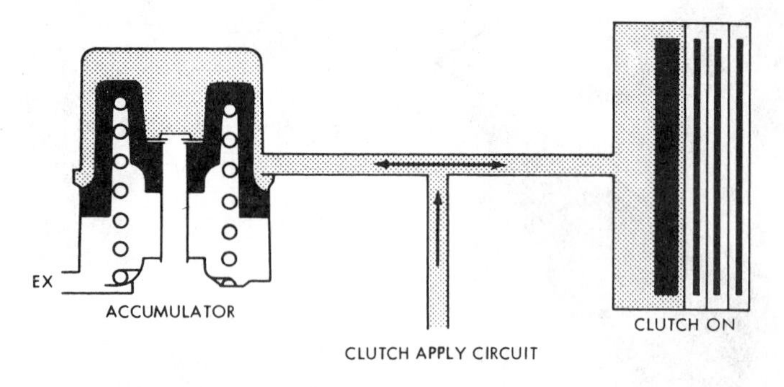

FIGURE 7–47 Simple accumulator with spring.

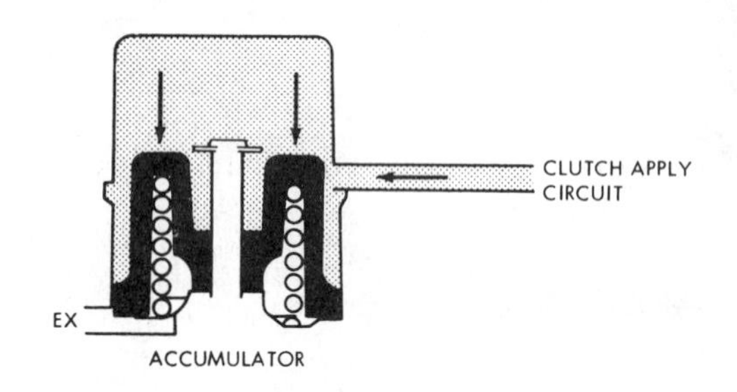

FIGURE 7–48 Simple accumulator with piston fully stroked.

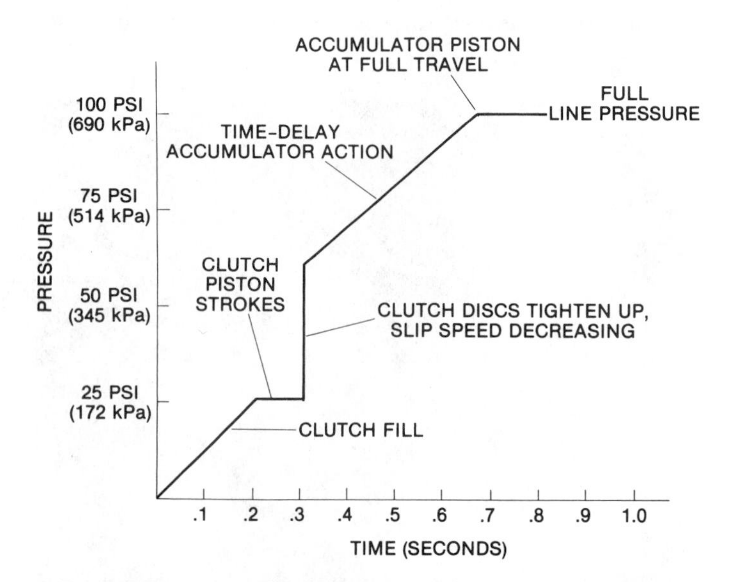

FIGURE 7–49 Typical pressure rise and accumulator action in the clutch circuit.

The accumulator action can be compared to engaging the clutch with a manually operated transmission. Through driver foot control, the clutch pedal is let out rapidly until the clutch slack is removed and starts to apply. At this point, the pedal travel is slowed down to control the final lockup in accordance with crankshaft torque. Essentially, clutch lockup slip needs to be controlled during the tightening phase.

The accumulator in Figure 7–47 would be adequate if all shifts were made at the same engine torque, so that the spring tension in the accumulator would provide a fixed rate of pressure buildup time for final clutch lockup. When shifts are made under heavy throttle conditions, it is necessary to shorten the apply pressure buildup time on the clutch discs before the accumulator piston strokes—otherwise, clutch spin-up occurs.

To vary the pressure at which the accumulator action begins, a throttle-sensitive oil pressure is introduced to oppose the stroking of the piston. This oil pressure can be the same as (1) the existing throttle pressure used for shifting the valves, (2) a modified throttle pressure, or (3) the existing mainline pressure. Accumulator arrangements using throttle oil and mainline oil pressures are illustrated in Figures 7–50 and 7–51. When mainline

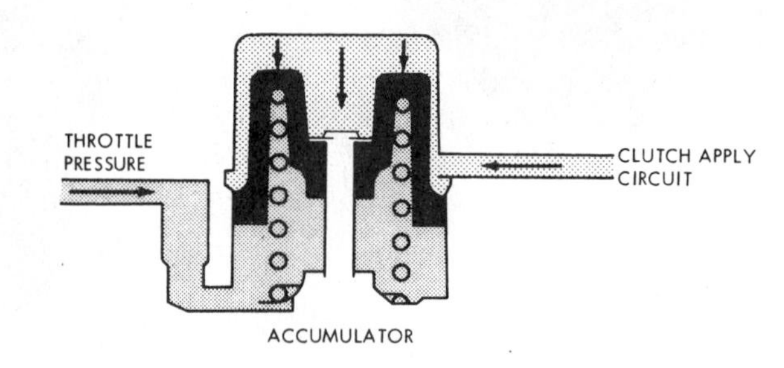

FIGURE 7–50 Throttle-sensitive accumulator.

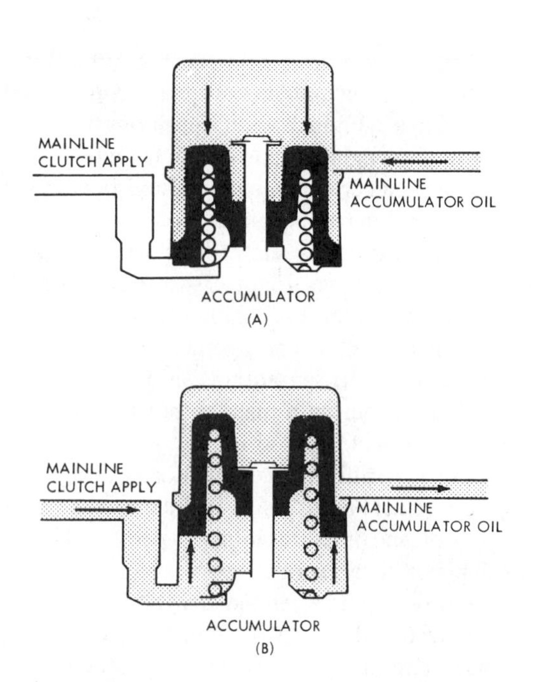

FIGURE 7–51 Accumulator action with mainline used as the accumulator oil.

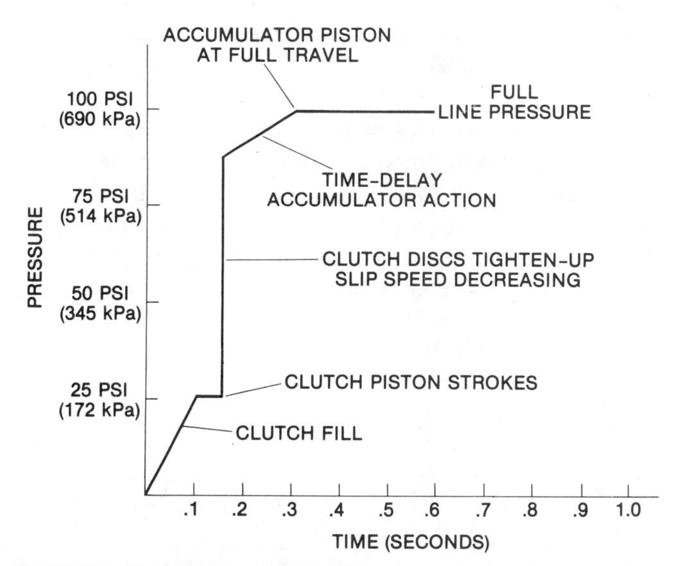

FIGURE 7–52 Typical heavy throttle pressure rise and accumulator action in clutch circuit. (Compare to Figure 7–49.)

oil is used as the variable, the accumulator works in a different manner from a throttle-sensitive arrangement. When the accumulator piston is stroked, the spring force decreases. This permits a controlled rise in oil pressure during the piston stroke or clutch apply time.

Using a throttle-sensitive oil in the accumulator makes it behave like a variable shock absorber. With light throttle shifts, the accumulator allows for a soft engagement of the clutch or band. Under heavy throttle, the shift feel is quick, firm, and aggressive. (See Figure 7–52.)

It is common practice to build in an accumulator action with the servo unit. A servo piston in the apply position for intermediate gear operation is shown in Figure 7–53. On the 2-3 shift, direct clutch apply oil is routed into the servo release side of the piston. The release oil plus the release springs overcome the servo apply force, and the piston strokes downward. This action releases the band while the clutch is being applied. Note that the servo release, or downward piston movement, provides an accumulator action for the direct clutch.

The intermediate band adjustment is critical to the accumulator action because it controls the servo release spring load. A loose band adjustment, for example, results in a longer piston apply stroke. The band apply is not affected on a 1-2 shift because of the transition from a holding one-way clutch, but the release spring load is increased. During direct clutch apply, therefore, the servo is released too early, and the accumulator action occurs too early. The end result is an engine flare-up on the 2-3 shift.

Former practice was to use accumulators for the dynamic shift circuits only. The new generation of transmissions often incorporates accumulators in forward and reverse clutch circuits. Harsh "garage shifts"—N to D, or N to R—are subdued.

Valve-style Accumulator System

Valve-style accumulator systems are sometimes used instead of the more conventional piston-style system. Figure 7–54 (A–D) illustrates the four phases of operation of a valve style application in the intermediate system of the Ford C-6.

- **Phase A.** Mainline pressure at the small end of the accumulator valve holds the valve against the spring.
- **Phase B.** When the shift valve opens, mainline pressure is routed through the capacity modulator valve and charges the spring side of the 1-2 accumulator valve and the servo apply line.
- **Phase C.** The orificed connection to the spring ends of both valves momentarily prevents a pressure buildup until the band contacts the drum. As pressure begins to build up past the orifice, the accumulator valve begins to stroke. The orifice control and stroking of the piston lowers the pressure above the valves. During Phase C, the capacity

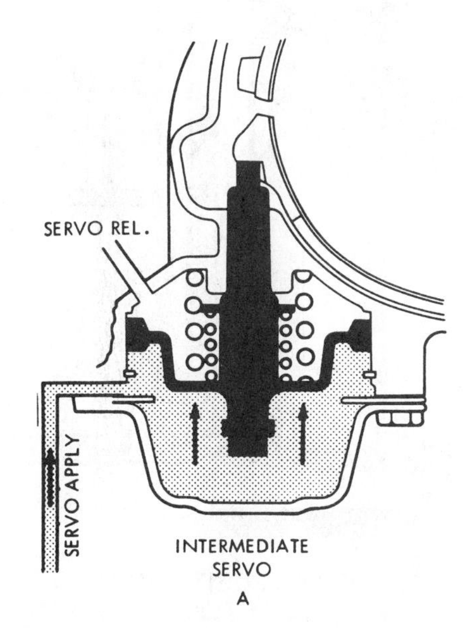

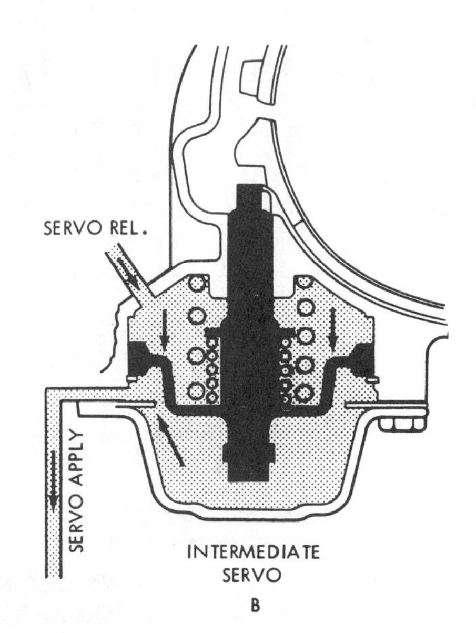

FIGURE 7–53 (A) Servo piston applied for intermediate gear operation; (B) servo releasing for upshift, providing accumulator action.

modulator valve becomes a balanced valve. Pressure at the reaction end of the valve balances against spring tension plus the lower pressure above the valve. The result is a controlled reduction of the intermediate servo apply pressure while the accumulator valve is stroking.

- **Phase D.** When the accumulator valve bottoms out, fluid flow through the orifice stops, and mainline servo apply pressure equalizes throughout the circuit.

SHIFT SYSTEM

The shift system provides both automatic and nonautomatic gear selection. The nonautomatic selections are provided by the manual valve and include a neutral, a reverse, and forward gears, including provisions for engine braking. Automatic gear selection provides for gear ratio changes that are compatible to the vehicle speed and available performance desired by the

ACCUMULATOR VALVE AND CAPACITY MODULATOR VALVE

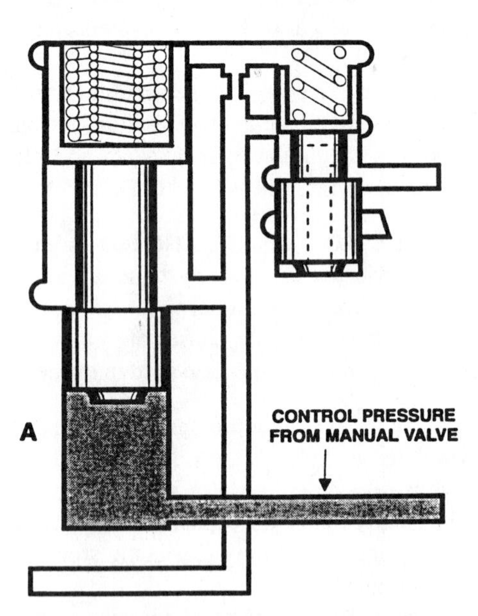

VALVE POSITIONS BEFORE SERVO APPLY SYSTEM IS CHARGED

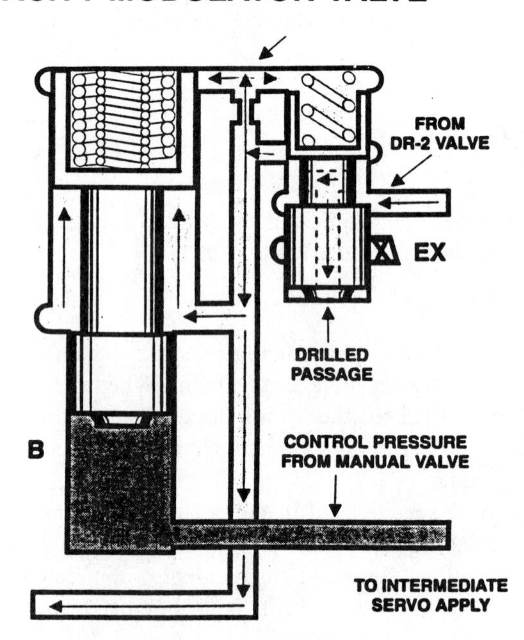

VALVE POSITIONS DURING FILL PHASE AND INITIAL ENGAGEMENT

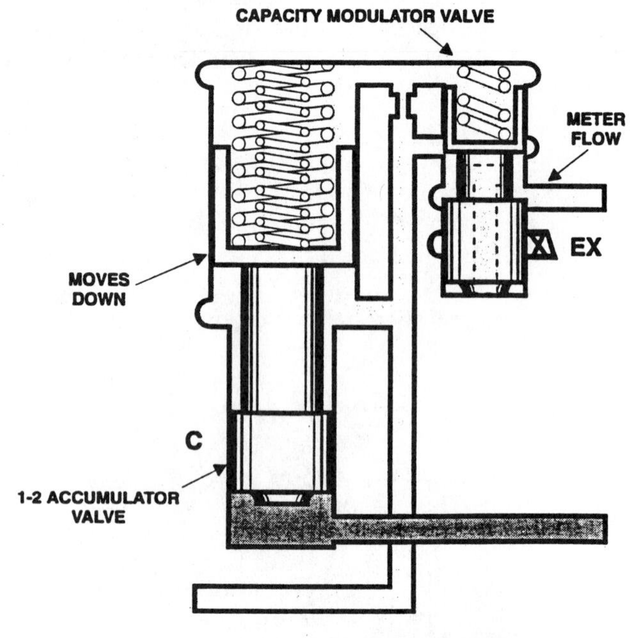

VALVE POSITIONS DURING THE APPLICATION PHASE

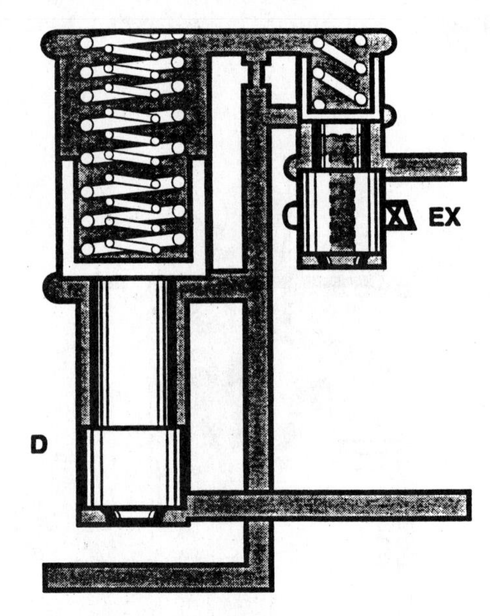

VALVE POSITIONS AT FULL-APPLY PHASE

FIGURE 7–54 Accumulator valve and capacity modulator valve operation.

driver. In a three-speed system, the shift schedule is programmed for a 1-2 and a 2-3 shift in the fully automatic drive range. For ease of understanding the shift system, a three-speed transmission is examined in the following discussion.

The shift system can be viewed as a hydraulic computer. Through driver selection, the manual valve is properly indexed in the valve body and programs the transmission for the desired operating range. In fully automatic drive range, the two shift valve arrangements respond to a hydraulic road speed signal (governor pressure) and a hydraulic engine torque signal (throttle/modulator pressure). The shift valves evaluate the information and automatically trigger clutch/band circuits ON or OFF to provide the best gear ratio for the driving load.

The essential hydraulic support systems for transmission operation have already been discussed: the pressure supply system, converter-cooling-lube system, throttle and modulator systems, and the governor system. Add to these the manual valve, servos and bands, clutches, and accumulators, and you are ready to tie it all together with the shift system operation. A gradual buildup of a three-speed hydraulic control system shows how a seemingly complex subject is made up of a number of simple circuits and assorted devices that are easy to understand. Our control system uses the standardized range selection P-R-N-D-2-1 as it relates to a typical Simpson planetary geartrain.

Neutral/Park

The manual valve blocks the line pressure to the clutch and band apply circuits (Figure 7–55). With the forward and direct clutches both disengaged, the transmission is isolated from the engine and torque converter. The line pressure control continues to function. A steady supply of regulated operating oil is maintained, and the constant requirement to feed the converter and lubrication circuits and maintain cooling is satisfied (Figure 7–56).

Automatic Drive Range (D)

When the selector lever is positioned in D, the manual valve connects the line pressure to the forward clutch circuit. The oil is usually metered through an orifice to help control overaggressiveness of the clutch apply (Figures 7–57 and 7–58). Should the piston seals leak—from wear or heat brittleness, for example—the orifice prevents a mainline pressure drop in the transmission.

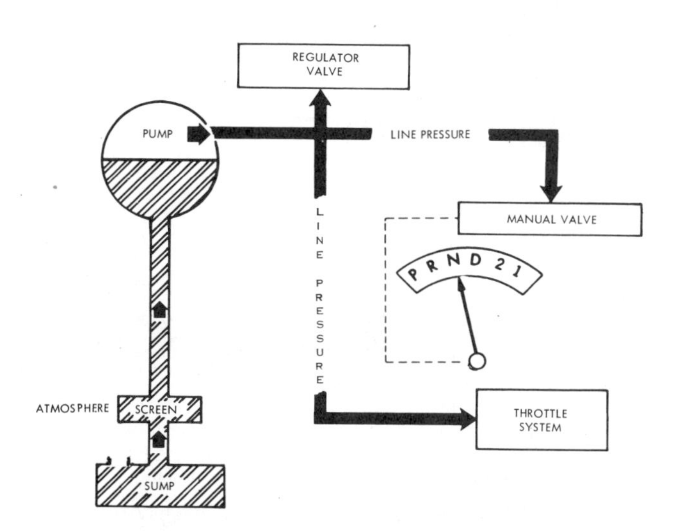

FIGURE 7–55 Simplified line pressure routing in neutral.

Application of the forward clutch mechanically connects the geartrain to the engine and torque converter. The compound planetary gear action causes the low roller clutch assembly to lock up and initiate first-gear operation.

As the vehicle accelerates in first gear, progressive shifts to second and third gears are determined by automatic shift valve movement. A shift valve is a relay valve and acts as a simple type of hydraulic circuit ON/OFF switch.

First Gear—Simplified Hydraulics. Figure 7–59 shows hydraulic oil routed through the manual valve and to the forward clutch to engage first gear. Line pressure oil is also directed to the shift valves but is blocked at the valves. With the vehicle stopped and in first gear, the 1-2 and 2-3 shift valves are in the closed position due to the spring force exerted on them. At this point, there is no governor or TV pressure acting on the valves.

As the vehicle moves forward, TV oil pressure increases and decreases in relationship to accelerator pedal position and engine load. The TV oil assists the springs in keeping the shift valves closed.

Governor pressure begins to rise with an increase in vehicle road speed and exerts a force upon the base of the shift valves. For the duration of first gear, both shift valves remain closed. The combined force of the spring and TV oil pressure on each shift valve is greater than the force of the governor oil.

Second Gear—Simplified Hydraulics. As the vehicle speed increased during first-gear operation, the governor oil pressure also increased. When the force from the governor oil pressure was greater than the combined force of the 1-2 shift valve spring and TV oil pressure, the 1-2 shift valve was pushed into the open position (Figure 7–60). When this happened, line pressure oil was allowed to pass through the 1-2 shift valve to the intermediate band servo, and second gear was engaged.

The 2-3 shift valve is in the closed position during second-gear operation. Observe in Figure 7–60 that a stronger spring is used on the 2-3 shift valve. Also notice in the figure that the TV/spring end of the 2-3 shift valve is larger than the end of the 1-2 shift valve. Because of these design features, the governor pressure was great enough to move the 1-2 shift valve, but it is not strong enough to reposition the 2-3 shift valve.

Third Gear—Simplified Hydraulics. As the vehicle continued to increase in speed, governor pressure continued to exert greater force upon the 2-3 shift valve. When the governor force was greater than the combined force of the 2-3 shift valve spring and the TV oil pressure, the 2-3 shift valve was moved to the open position (Figure 7–61). When this happened, line pressure oil was allowed to pass through the 2-3 shift valve into the direct clutch circuit, and third gear was engaged.

Since the 1-2 shift valve remains open, the oil passing through the 2-3 shift valve has two tasks to accomplish. The direct clutch circuit oil from the 2-3 shift valve must first release

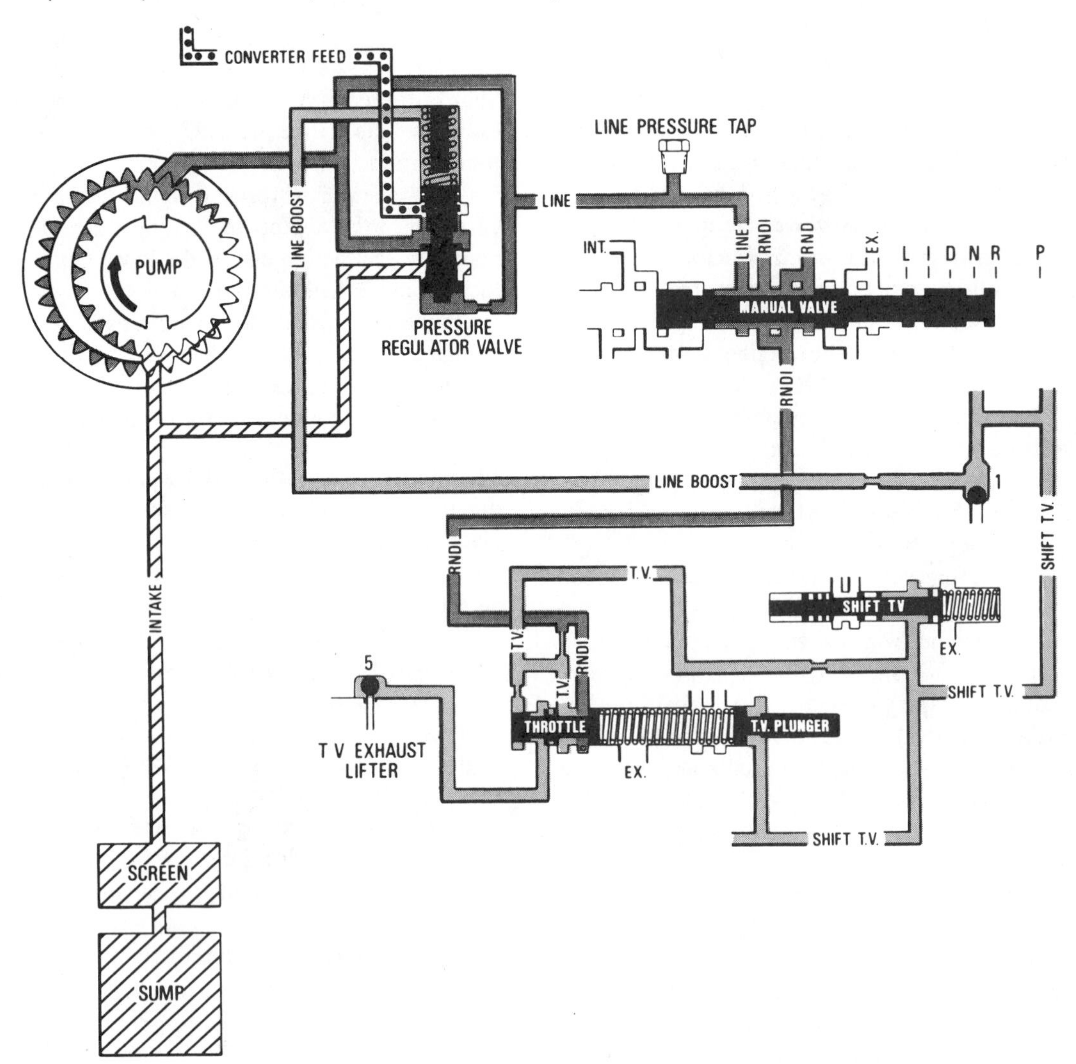

FIGURE 7–56 Line pressure control and manual valve in neutral. (Courtesy of General Motors Corporation, Service Technology Group)

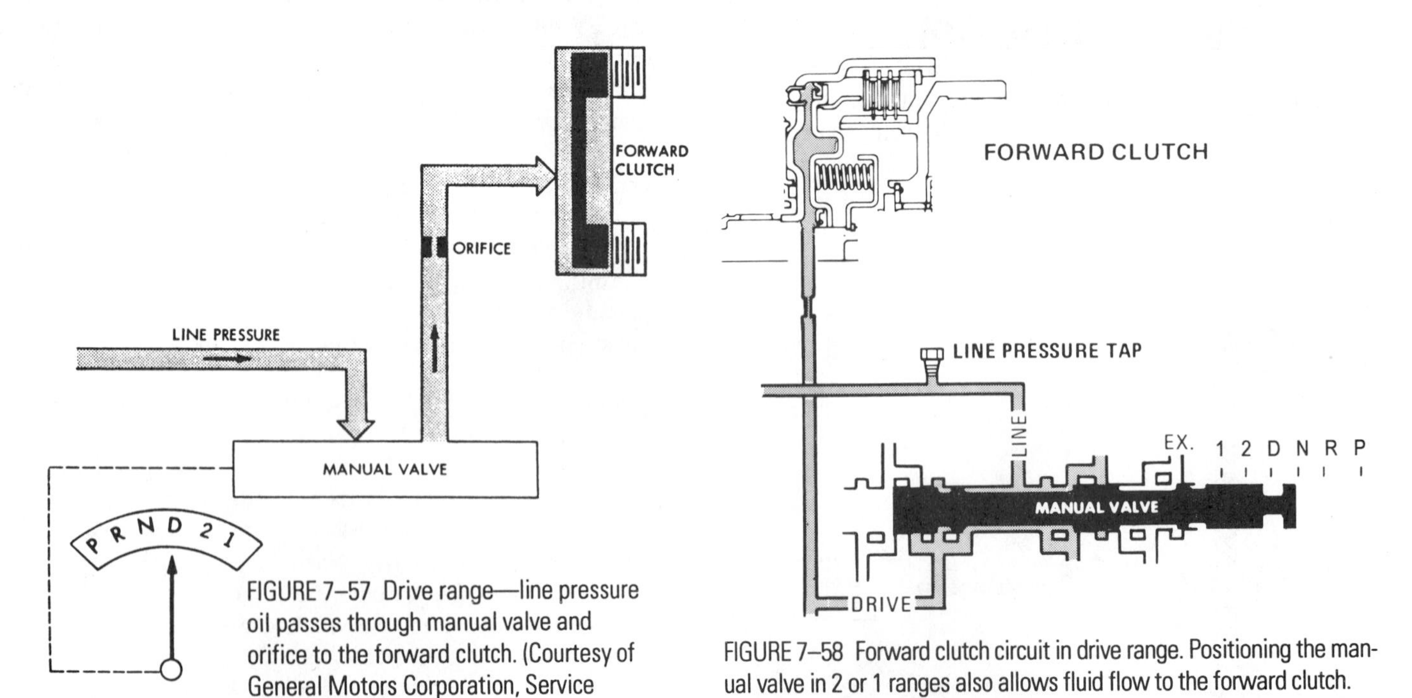

FIGURE 7–57 Drive range—line pressure oil passes through manual valve and orifice to the forward clutch. (Courtesy of General Motors Corporation, Service Technology Group)

FIGURE 7–58 Forward clutch circuit in drive range. Positioning the manual valve in 2 or 1 ranges also allows fluid flow to the forward clutch. (Courtesy of General Motors Corporation, Service Technology Group) the intermediate servo and band, and then apply the direct clutch to engage third gear. (The operation of the intermediate servo release circuit is explained later in this section.)

Fourth Gear—Simplified Hydraulics. The 1-2 and 2-3 shift valves of a hydraulically controlled four-speed transmission operate similarly to those found in the three-speed unit just described. An additional valve, however, the 3-4 shift valve, is needed. It opens and closes the hydraulic circuit that engages fourth gear. To ensure that the 3-4 shift valve will open after the 2-3 shift valve, a stronger shift valve spring is required, and the TV/spring end of the valve may be larger. These two design characteristics delay the opening of the 3-4 shift valve until governor pressure is great enough to reposition the valve and hydraulically engage fourth gear. (During fourth gear operation, both the 1-2 and 2-3 shift valves remain in their open position.)

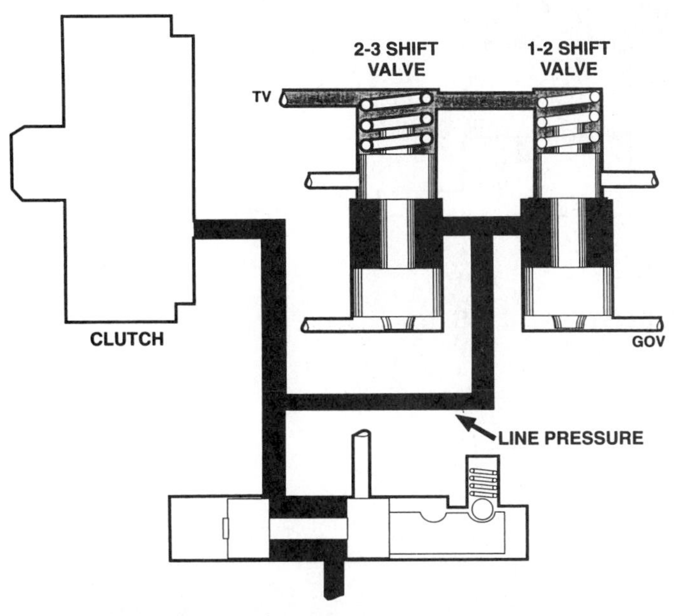

FIGURE 7–59 First-gear hydraulics—simplified.

Summary: Simplified Upshift Sequence. In the simplified upshift sequence just described, the shift valves were forced open, one at a time. When vehicle speed increased and governor pressure was great enough to overcome the TV pressure and spring force, the shift valves were moved, and an upshift occurred. In high gear, all the shift valves were in the open position.

Simplified Downshift Sequence. An opened shift valve was moved to that position because the effective force of the governor oil was greater than the combined forces of the TV oil and spring. A downshift, closing of the valve, can be caused by two different methods: a reduction in governor pressure, or an increase in TV pressure.

When decelerating the vehicle (coasting), TV pressure drops to zero, and the governor pressure is reduced. The lowering of governor pressure is in relationship to the decreasing vehicle speed. Referring to Figure 7–61 and then to Figures 7–60 and 7–59, a downshift sequence can be depicted. Spring force causes the valves to close, with the strongest spring responding first.

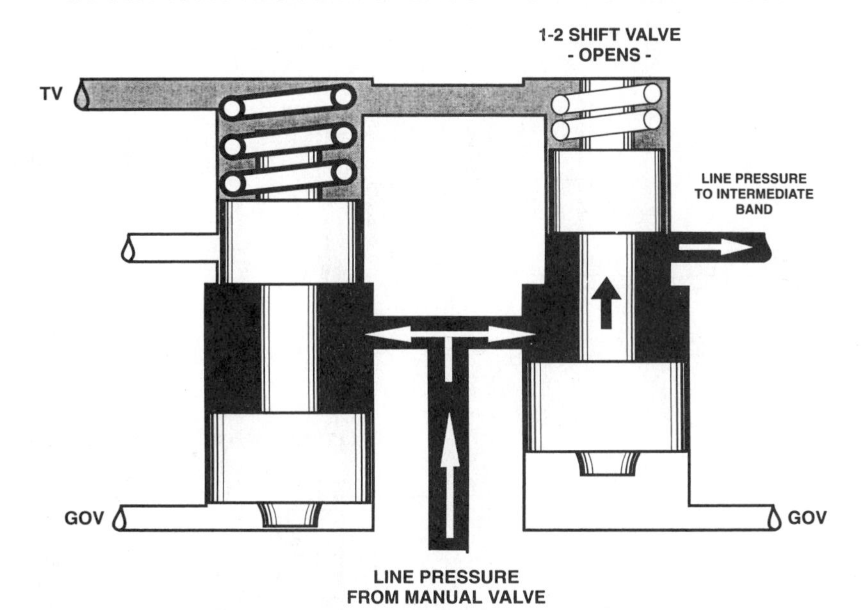

FIGURE 7–60 Second-gear hydraulics—simplified.

GOVERNOR PRESSURE OVERCOMES THROTTLE PRESSURE
TRANSMISSION UPSHIFTS TO THIRD

2-3 SHIFT VALVE - OPENS -
1-2 SHIFT VALVE - OPENS -
TV
LINE PRESSURE TO DIRECT CLUTCH
LINE PRESSURE TO INTERMEDIATE BAND
GOV
GOV
LINE PRESSURE FROM MANUAL VALVE

(SIMPLIFIED) THIRD GEAR HYDRAULICS

FIGURE 7–61 Third-gear hydraulics—simplified.

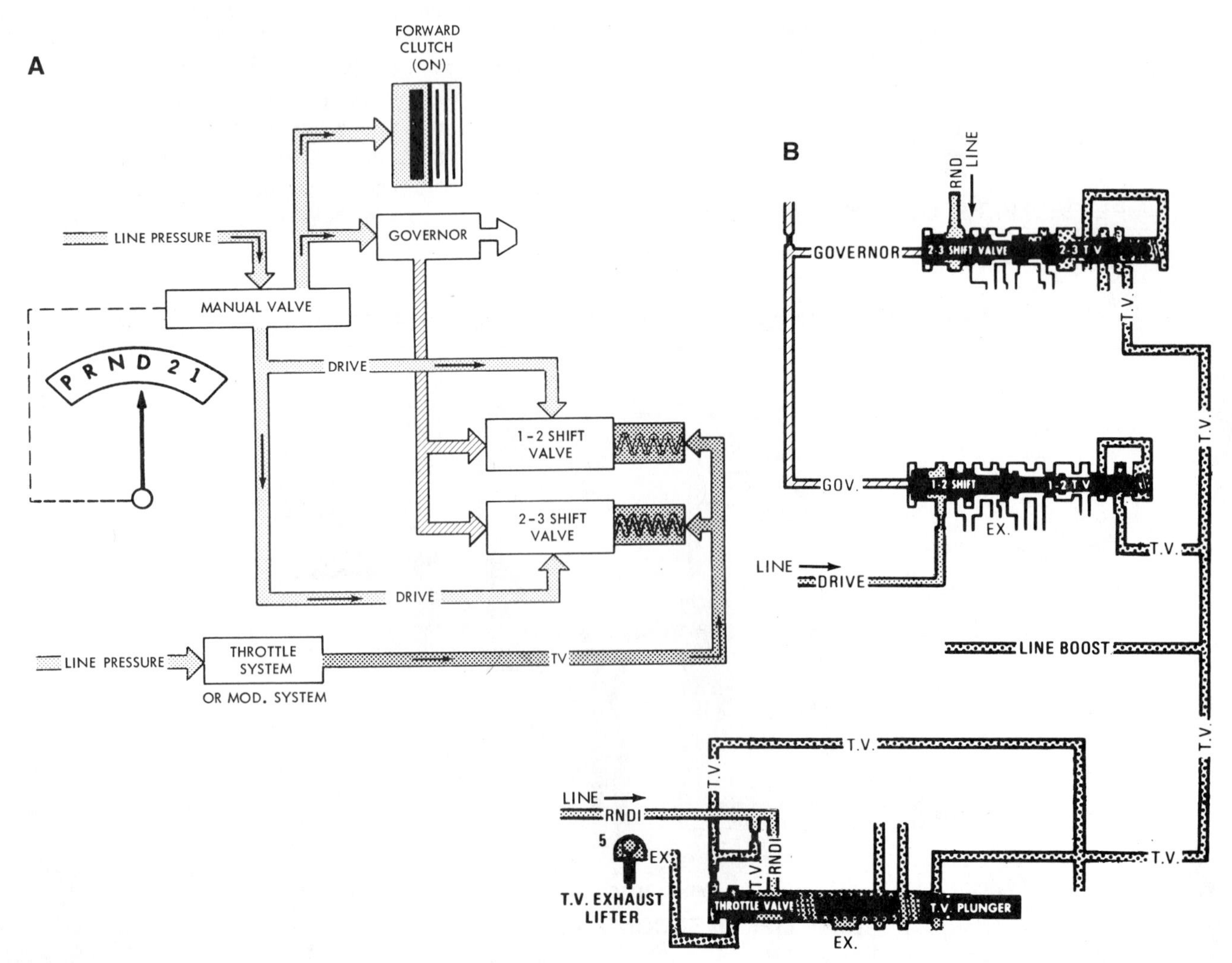

FIGURE 7–62 (a) Simplified shift system; (b) detailed hydraulic schematic. (Courtesy of General Motors Corporation, Service Technology Group)

Forced downshifts can occur when the driver pushes down on the accelerator pedal to increase vehicle speed or to go up a steep grade. The TV oil pressure increases. If the combined TV oil pressure and valve spring force is greater than the governor pressure, an opened shift valve is closed.

Review of Shift Valve Movement. Shift valve movement or position is determined by pressures from its supporting controls: the governor and throttle systems (Figure 7–62). Governor oil pressure acts on one end of the shift valve and works to open (upshift) the valve. Opposing governor pressure is throttle pressure, a regulated engine torque sensitive signal. The throttle pressure acts to delay the upshift movement of the valve. In most cases, throttle pressure is modified by a regulating valve in the shift valve train before acting on the shift valve. In addition, a fixed spring load is used on the throttle side of the shift valve to determine the closed throttle downshift point. The line pressure at the shift valve is distributed from the manual valve and serves as the prime mover of the entire shift when it occurs.

Shift Timing and Feel

Shift Timing. A gear change is determined by the acting forces of governor pressure versus throttle pressure. At light throttle openings, a small throttle signal is received by the shift valve. To overcome the throttle pressure resistance, only a low vehicle speed is required to generate a sufficient governor force to shift the valve. Therefore, an early shift takes place.

If the driver wants medium and heavy throttle performance, the shift is delayed. The throttle opening produces a correspondingly larger throttle signal, which in turn dictates higher vehicle speeds to generate the necessary governor signal for the shift. The maximum shift point, therefore, occurs at wide open throttle and within the engineered safe speed limit. Typically, for high-torque engine applications, the 1-2 shift spread is 10 to 50 mph (16 to 80 km/h), and the 2-3 shift spread is 18 to 70 mph (29 to 112 km/h).

Shift Feel. The 1-2 shift circuit is illustrated in Figures 7–63 and 7–64. In Figure 7–64, note how the accumulator regulates the buildup time of line pressure on the intermediate servo piston for proper shift feel. Stroking of the intermediate servo piston compresses the servo spring and applies the intermediate band. The band apply holds the planetary sun gear and causes the low roller clutch to release. The second-gear ratio is provided by the front planetary only.

The 2-3 shift circuit is illustrated in Figures 7–65 and 7–66. When the shift valve opens (Figure 7–66), line pressures flow into the circuit. Line pressure applies the direct clutch and releases the intermediate servo piston. The direct clutch apply oil, acting on larger servo piston surface area, and the compressed servo spring force overcome the intermediate apply oil. The servo piston absorbs direct clutch oil and provides an accumulator action against the torque-sensitive intermediate apply oil that is mainline.

Coast or Closed-Throttle Downshift

In reviewing the 1-2 and 2-3 shift valve movement, notice that the TV pressure is cut off from the end of the shift valves when they open. This sudden exhaust of TV pressure gives the shift valves a snap action on the upshift and eliminates any hunting or indecision of the valves to shift and stay shifted, especially if the vehicle speed is maintained at or near the shift point. Once the shift is made, spring tension alone opposes governor pressure on the shift valves. As the vehicle slows down on closed throttle, governor pressure is reduced at the shift valves. When governor pressure is less than the opposing fixed spring force, the shift valves close with a snap action. The direct and intermediate circuits are then exhausted at the shift valves. At closed throttle, TV pressure is 0 psi and has no effect on the shift valves when they close.

In a shift system in which the closed-throttle sequence is 3-2 and 2-1, the 3-2 shift generally occurs between 8 and 12 mph (13 to 19 km/h), and the 2-1 shift occurs between 3 and 5 mph (5 to 8 km/h). On Ford shift systems using a 3-1 closed-throttle pattern, the governor cuts out at 10 mph (16 km/h), and both

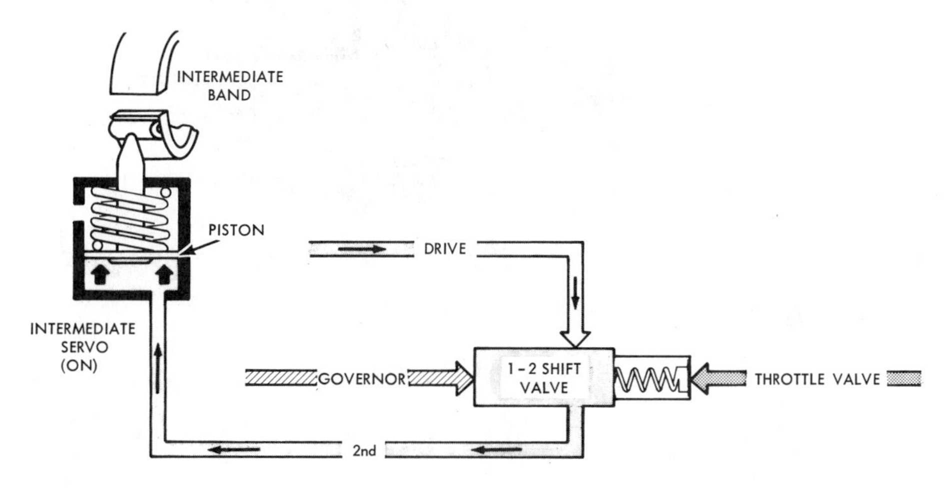

FIGURE 7–63 Simplified 1-2 shift circuit.

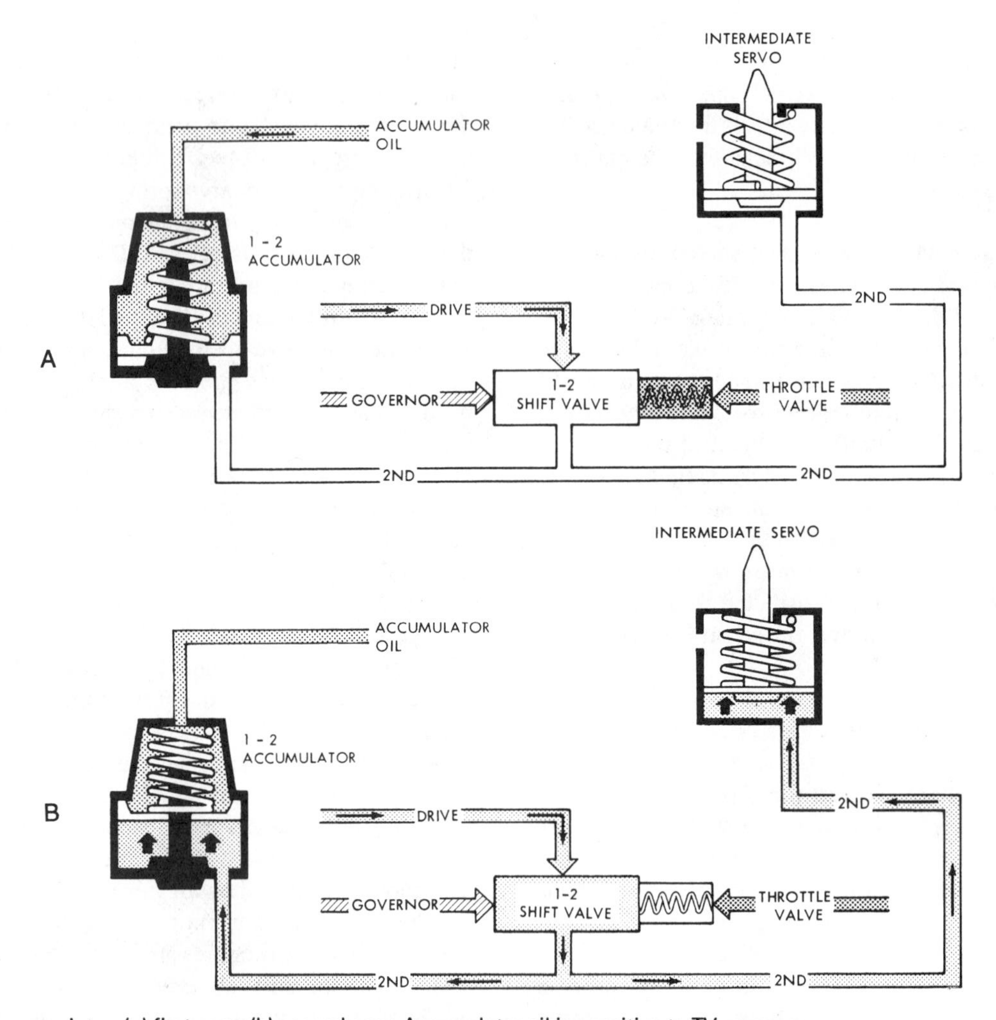

FIGURE 7–64 1-2 accumulator: (a) first gear; (b) second gear. Accumulator oil is sensitive to TV pressure.

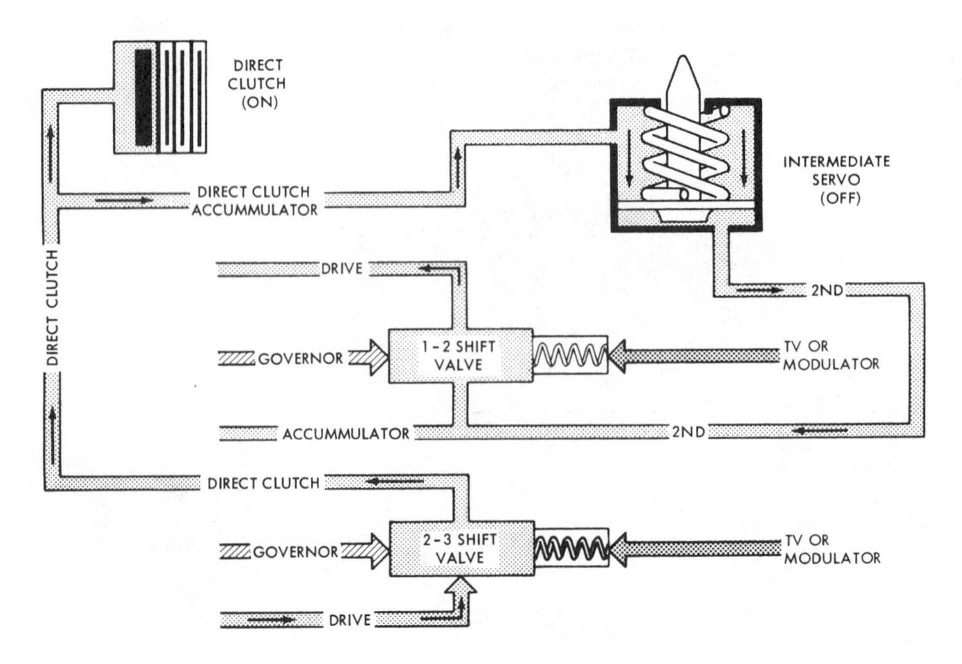

FIGURE 7–65 Third-gear oil applies clutch and strokes the intermediate servo piston to release the band.

shift valves close simultaneously. Chrysler closed-throttle shift patterns are also 3-1. A hydraulic interlock between the shift valves does not permit the 2-3 shift valve to close until the 1-2 shift valve closes. At approximately 5 mph (8 km/h), both shift valves close together.

Part-Throttle 3-2 Downshift

Most shift systems feature a part-throttle 3-2 downshift for extra performance during moderate acceleration at low speeds. The accelerator pedal needs to be only partially depressed to cause the shift. The transmission automatically returns to third gear as car speed increases or the accelerator is relaxed. Depending on the vehicle make and engine application, a part-throttle 3-2 downshift can be accomplished at speeds as high as 50 mph (81 km/h), while others may require the road speed to drop below 30 mph (48.5 km/h).

Although each automatic transmission has its own part-throttle design system, they all provide a throttle pressure bypass circuit to the 2-3 shift valve. Illustrated in Figures 7–67 and 7–68 are two part-throttle systems with different circuit approaches. Featured in Figure 7–67 is a mechanically operated throttle system tied in with the part-throttle 3-2 downshift circuit. By depressing the accelerator pedal far enough, the TV plunger opens the PT (part-throttle) circuit to TV pressure.

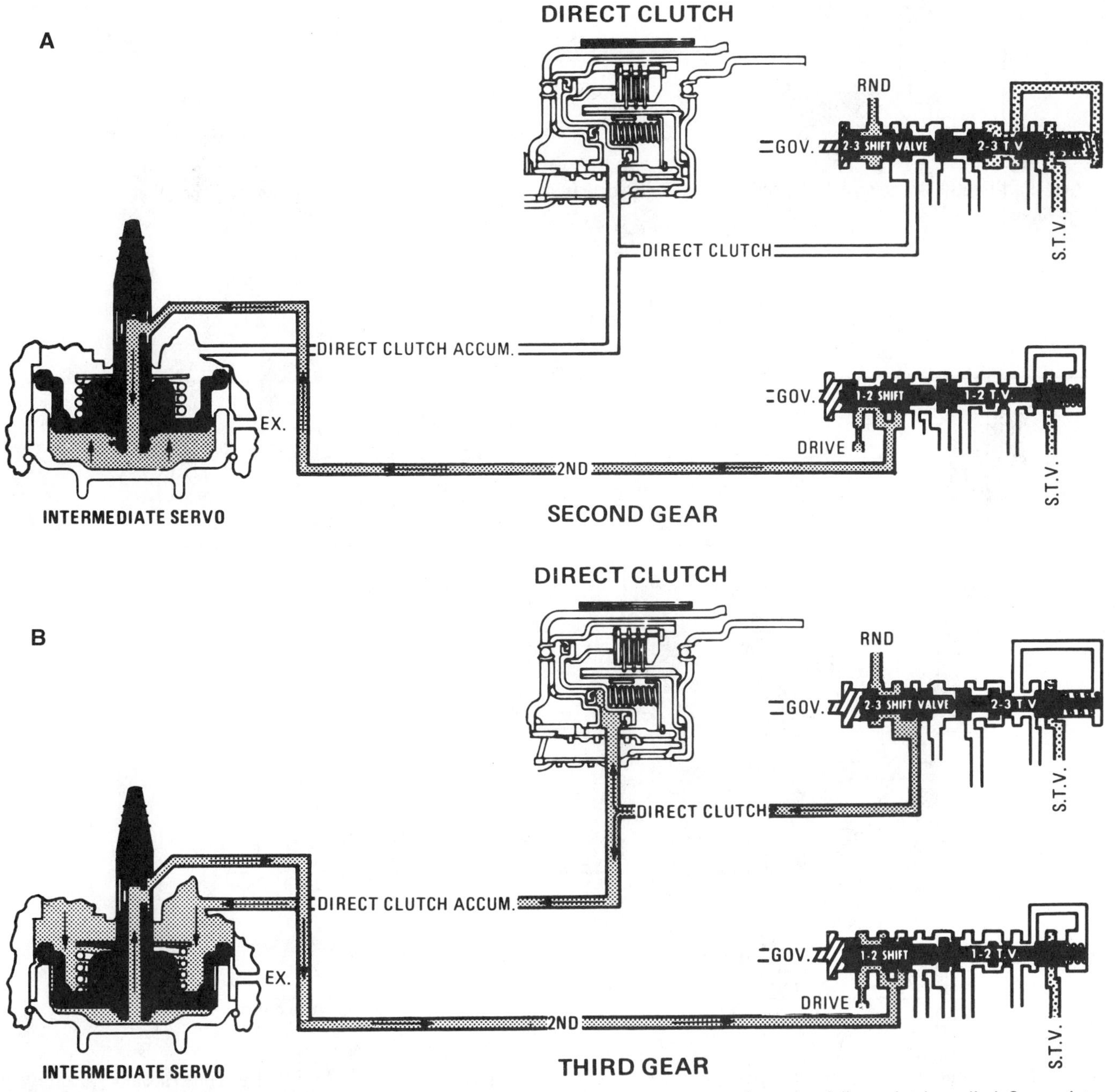

FIGURE 7–66 (a) Second gear, intermediate servo applied; (b) third gear, intermediate servo released and direct clutch applied. Servo piston release stroke provides 2-3 accumulator action.

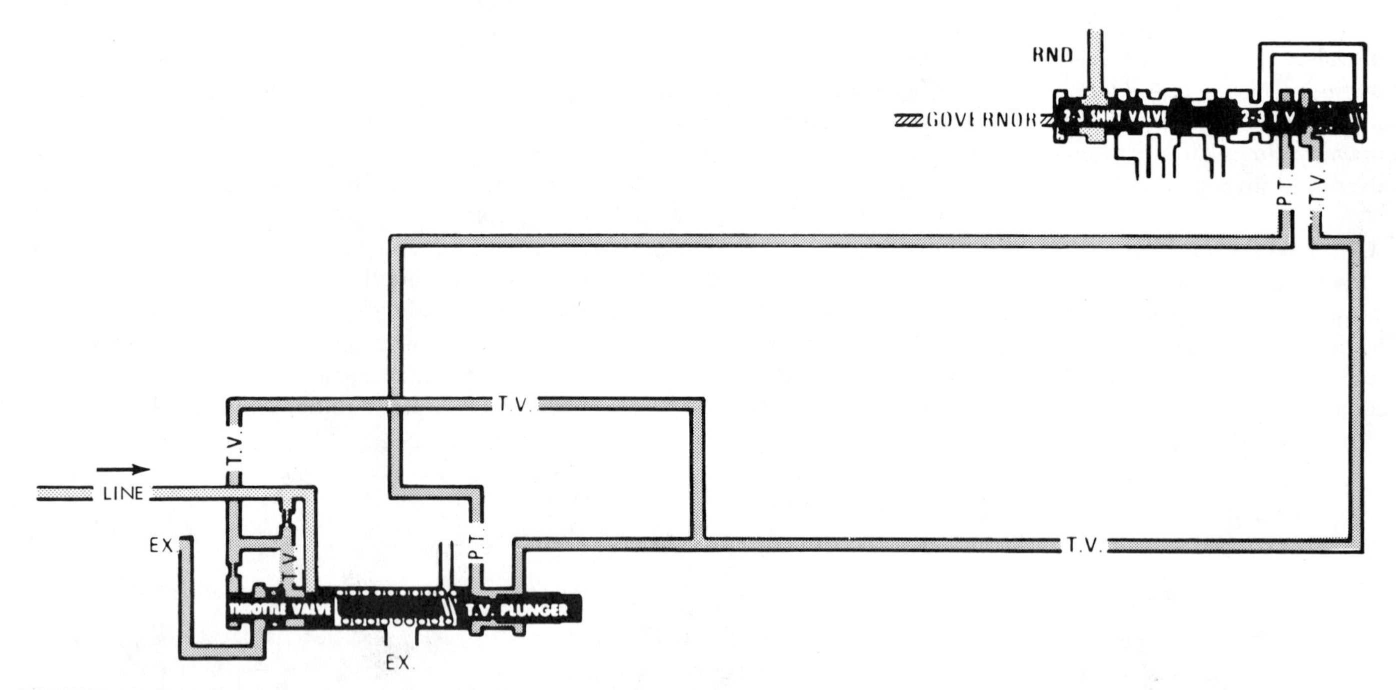

FIGURE 7–67 Part-throttle 3-2 downshift. 2-3 shift valve shown in downshift position. (Courtesy of General Motors Corporation, Service Technology Group)

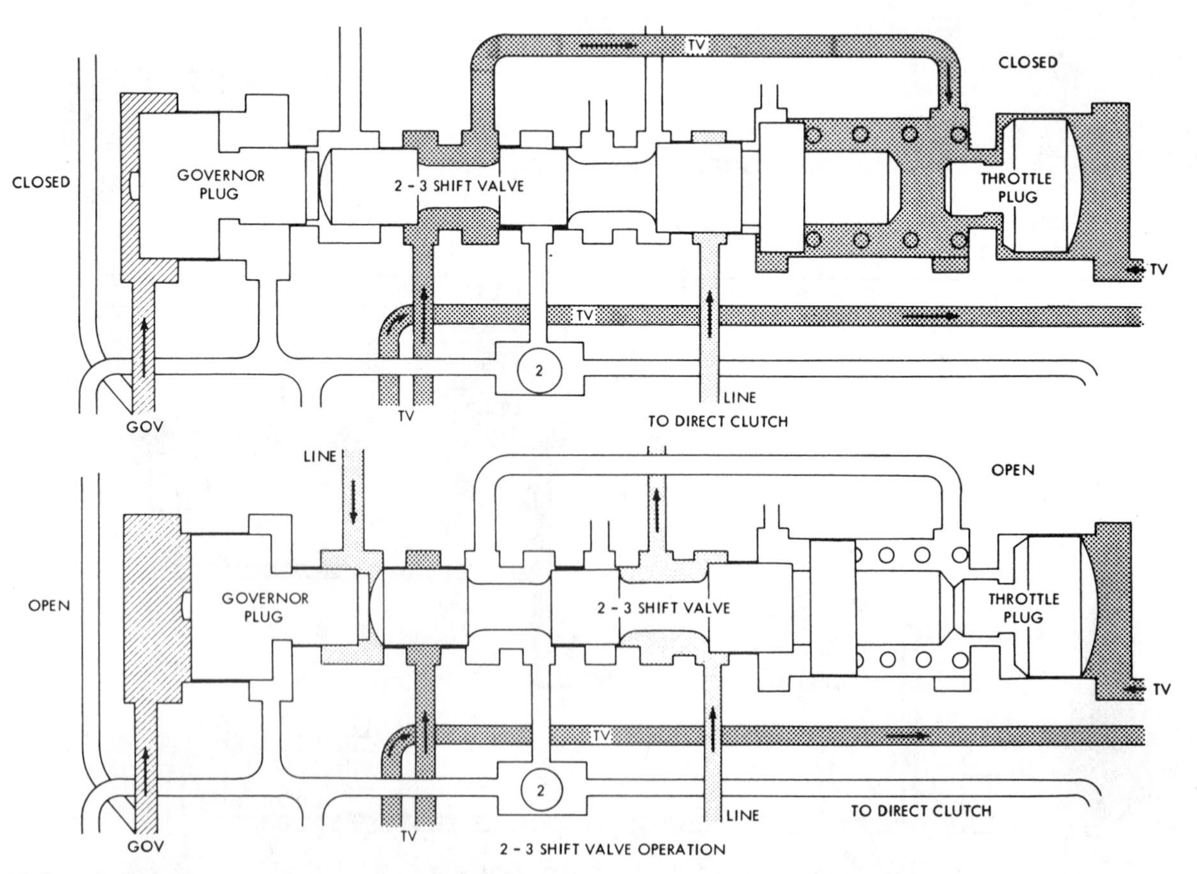

FIGURE 7–68 Part-throttle system with throttle plug. Top: 2-3 shift valve in closed position. Bottom: 2-3 shift valve in open position. (Courtesy of Chrysler Corp.)

When the TV plunger moves, it also raises the TV pressure, which can now act on the 2-3 shift valve train. If the vehicle speed is right, the TV pressure moves the 2-3 valve train against governor pressure, and the transmission shifts to second gear.

The part-throttle system shown in Figure 7–68 also ties into the throttle system. A throttle plug is designed into the 2-3 shift valve train and can keep the 2-3 shift valve throttle pressure sensitive. After the valve has made the 2-3 upshift, it is positioned against the end of the 3-2 throttle plug.

In Figure 7–68, the 3-2 throttle plug is shown with a throttle pressure load. The throttle pressure load at the 3-2 throttle plug is switched ON and OFF by a limit valve (Figures 7–69 and 7–70). The limit valve is a relay valve that is controlled by spring force and governor pressure. At low road speeds, the spring force on the limit valve is greater than the governor force and the valve is moved to its ON position (Figure 7–69). This permits the throttle pressure supply to load the 3-2 throttle plug.

Any moderate accelerator pedal depression raises the throttle pressure and pushes the throttle plug against the 2-3 shift valve and opposing governor pressure. The 2-3 shift valve closes, and the direct clutch circuit is exhausted for a 3-2 part-throttle downshift.

The limit valve prevents annoying part-throttle 3-2 downshifts at higher speeds. As vehicle speed increases, the governor pressure builds to a point where its force on the limit valve is greater than the spring force. This moves the limit valve to the OFF position, and the throttle pressure supply to the 3-2 throttle plug is cut off (Figure 7–70). The 3-2 part-throttle plug is now inactive. At high road speeds, the only 3-2 downshift that can occur is a wide open throttle forced downshift.

Full-Throttle Forced Downshift

Full-throttle forced downshift is also called detent, kickdown, or, simply, forced downshift. The function of the forced downshift is to overrule the governor system control of the shift valves by driver demand. At full throttle, the downshift system provides a bypass circuit to the shift valves loaded with either throttle or line pressure. By adding temporarily to the normal effective throttle force at the shift valves, the WOT upshifts can be delayed longer, or a forced downshift can be promoted. Maximum upshifts are

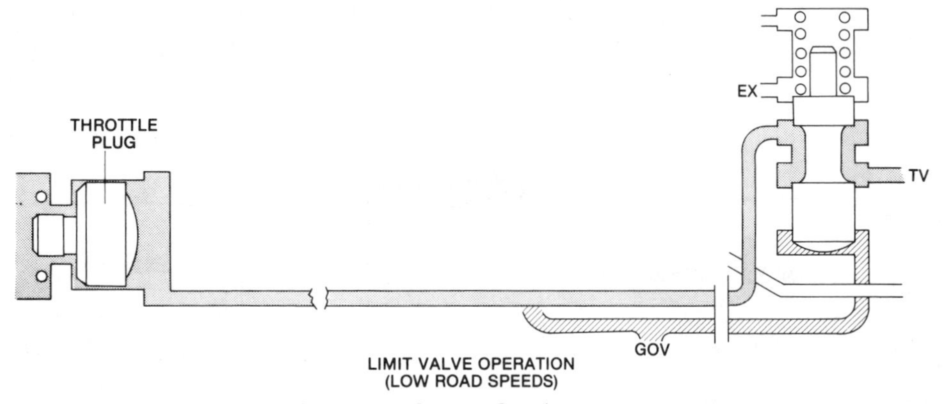

FIGURE 7–69 2-3 shift valve in part-throttle 3-2 mode. (Courtesy of Chrysler Corp.)

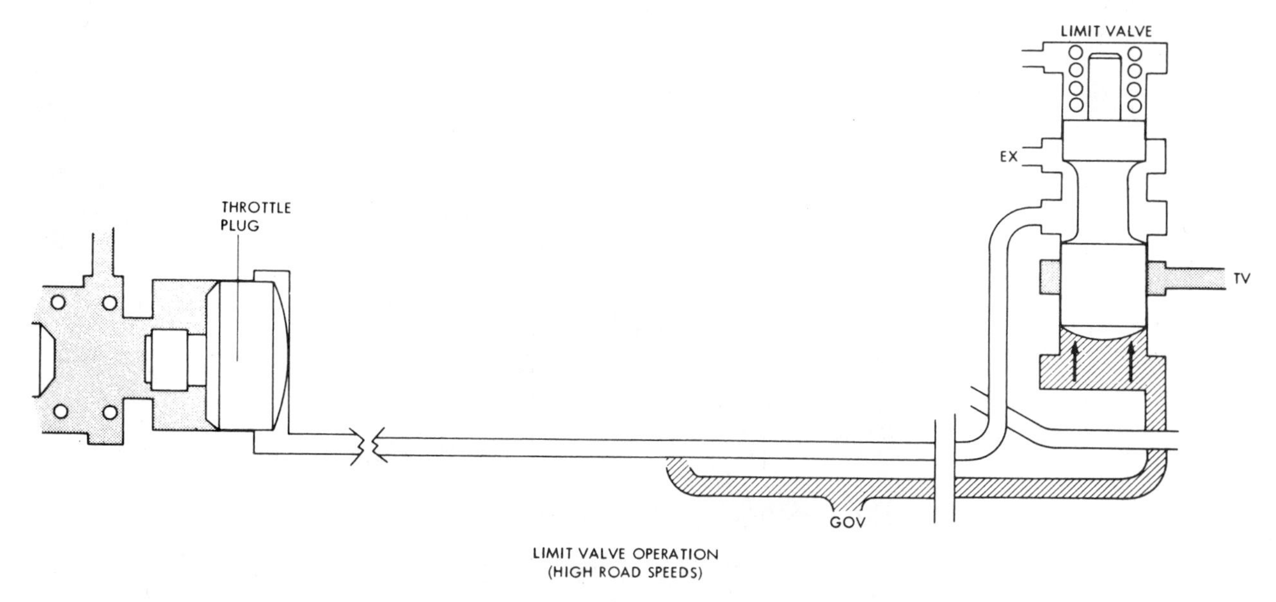

FIGURE 7–70 2-3 shift valve in open position. (Courtesy of Chrysler Corp.)

usually delayed by 5 to 10 mph (8 to 16 km/h), and depending on the vehicle speed range, forced 3-2 and 3-1 downshifts occur.

In a mechanically operated throttle system that is linkage- or cable-controlled, a downshift valve or plunger works in tandem with the throttle valve (Figure 7–71). With the accelerator pedal fully depressed, the TV plunger movement bottoms the throttle valve to provide maximum TV pressure. The detent and part throttle circuits are also opened to full TV pressure, which is directed to the shift valves to signal a downshift. Full TV pressure can now act on both shift valves, even though they are in the upshifted position.

Figure 7–71 shows the 2-3 shift valve closed against the governor pressure and the direct clutch circuit exhausted. The 1-2 shift valve remains open simply because the vehicle speed is too high and the opposing governor pressure is in command, inhibiting the downshift. When the vehicle speed is slowed to the safe range, the 1-2 shift valve is forced to close for a 3-1 or 2-1 downshift. Should the vehicle speed be too high, the downshift or detent system fails to close the 2-3 shift valve against governor pressure, and no forced downshift occurs.

Although the speed ranges at which the forced downshifts occur vary with the engine application, axle ratio, and tire size, the following guideline can be used for high-torque engines: high-speed 3-2 shifts are usually available up to 65 to 70 mph (105 to 113 km/h), and lower-speed 3-1 shifts occur up to 40 to 45 mph (65 to 73 km/h). In summary, the forced detent shifts do not occur if the engine is going to overspeed or there is no available crankshaft torque reserve for acceleration.

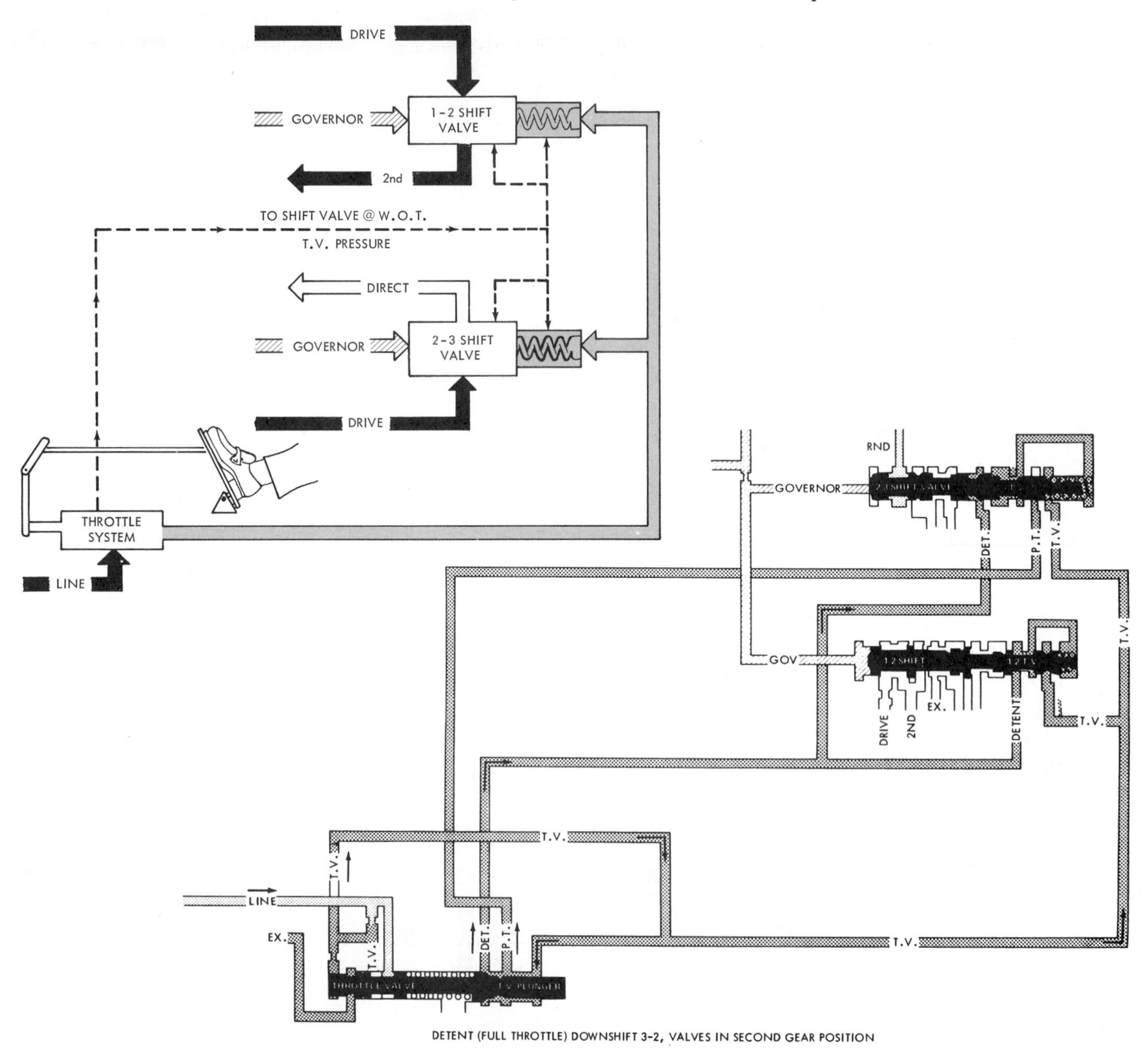

FIGURE 7–71 Detent (full-throttle) downshift 3-2, valves in second-gear position. Top: Simplified view. Bottom: Detailed schematic. (Courtesy of General Motors Corporation, Service Technology Group)

If the throttle pressure control uses engine vacuum, an independent downshift valve must be used. This valve is usually controlled by an external mechanical connection to the accelerator linkage. Some transmissions, however, including the GM 3L80 (THM-400) and JATCO (Japanese Automatic Transmission Company) units, have used an electric solenoid control on the valve.

Illustrated in Figures 7–72 and 7–73 are two simplified block diagrams showing variations of forced downshift systems used when the throttle system is vacuum-controlled. Figure 7–72 shows that the TV oil is the supply oil to the downshift valve. At wide open throttle, the mechanical linkage triggers the valve, and the TV pressure charges the downshift circuit and signals the shift valves. The same action takes place in Figure 7–73, but line pressure is substituted for TV pressure.

Intermediate and Manual Low

Intermediate and manual low operations are functions of the manual valve. The manual valve controls the hydraulic loading of the shift valve trains when the selector lever is placed in the 1 or 2 position. This arrangement can prevent the 1-2 shift valve from initiating a change into second gear, or the 2-3 shift valve can be prevented from initiating a change into high gear.

The simplified block diagrams in Figures 7–74 to 7–78 provide a summary of how the shift valves can be inhibited. Figure 7–74 is representative of an intermediate range system that allows a first-gear start and an automatic shift into second gear. In this instance, the manual valve opens the 2 circuit to line pressure that charges the 2-3 shift valve train and hydraulically locks the valve closed. If the car is in high gear when the 2 range is selected, the transmission downshifts to second regardless of vehicle speed. The 1-2 shift valve performs the same as it did in automatic drive range. A variation of the same intermediate system finds the manual valve cutting off the line pressure feed to the 2-3 shift valve and producing the same results (Figure 7–75).

The intermediate range circuit can also be designed to provide a second-gear start (Figure 7–76). The manual valve cuts off the line pressure feed to the 2-3 shift valve and allows line pressure to charge the 2 circuit leading to the 1-2 shift valve train. The line pressure rules and the 1-2 shift valve is immediately blocked open.

The manual low (1) range system, shown in Figure 7–77, has the 2 and 1 circuit lines charged with line pressure and leading to the 1-2 and 2-3 shift valve trains. With the TV line at the shift valves also carrying line pressure, both shift valves are locked closed hydraulically, and upshifts cannot occur. Should the vehicle speed be too high when manual low is engaged, the 1 circuit in the shift valve train is ineffective until a safe vehicle speed is reached. Once first gear is locked in, the transmission cannot upshift unless it is part of a "shift-kit" package or specially designed to avoid abusive situations. Under these circumstances, some systems allow a 1-2 shift between 60 and 90 mph (96 to 145 km/h).

The foregoing manual low system can be modified to produce the same results, as illustrated in Figure 7–78. The TV pressure in this setup remains effective at the shift valves. The 1-2 shift valve is hydraulically locked closed by the combination of line pressure and throttle pressure. The 2-3 shift valve is permitted to open, but the manual valve eliminates the line pressure feed to the direct clutch circuit. Although not illustrated in the diagrams, some intermediate and manual low systems send a line pressure boost to the pressure regulator valve. During coast conditions, TV pressure is at O psi, and very little line pressure is available to hold the friction elements against coast torque.

Reverse

A simplified illustration (Figure 7–79) shows a typical reverse circuit. When the manual valve is moved into R position, the reverse and direct clutch circuits are charged with line pressure, and the clutches are applied. Note that a line boost is

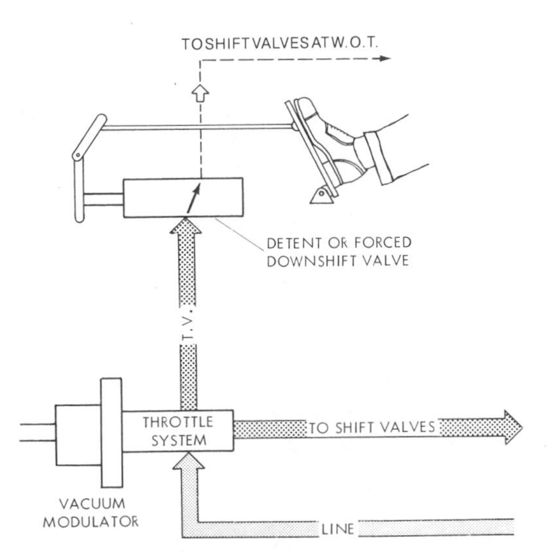

FIGURE 7–72 Forced downshift system, TV-modulator controlled.

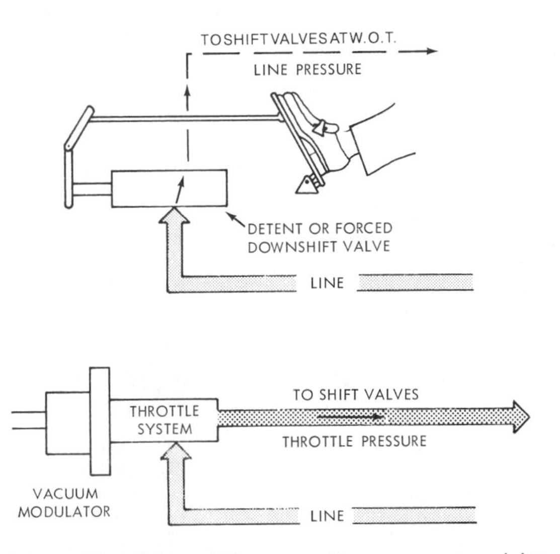

FIGURE 7–73 Forced downshift system, line pressure–modulator controlled.

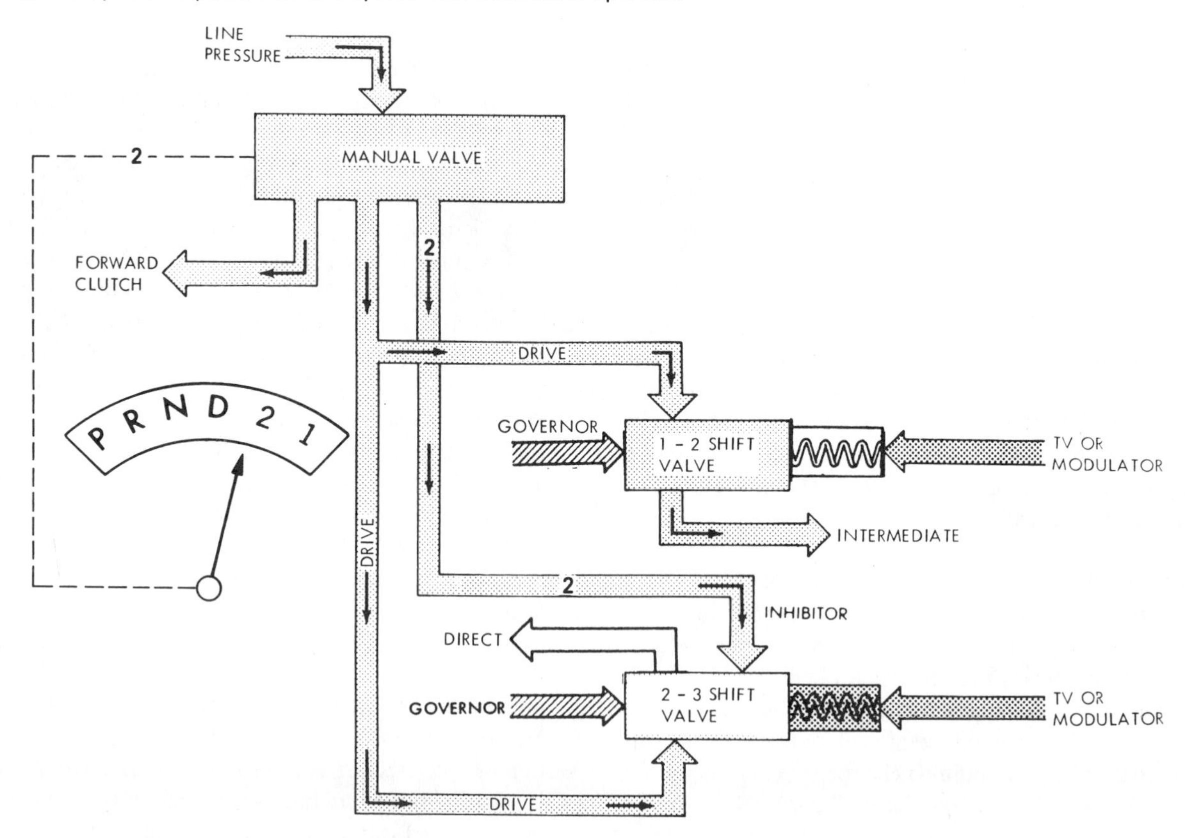

FIGURE 7–74 Intermediate range allows a first-gear start and an automatic upshift into second gear.

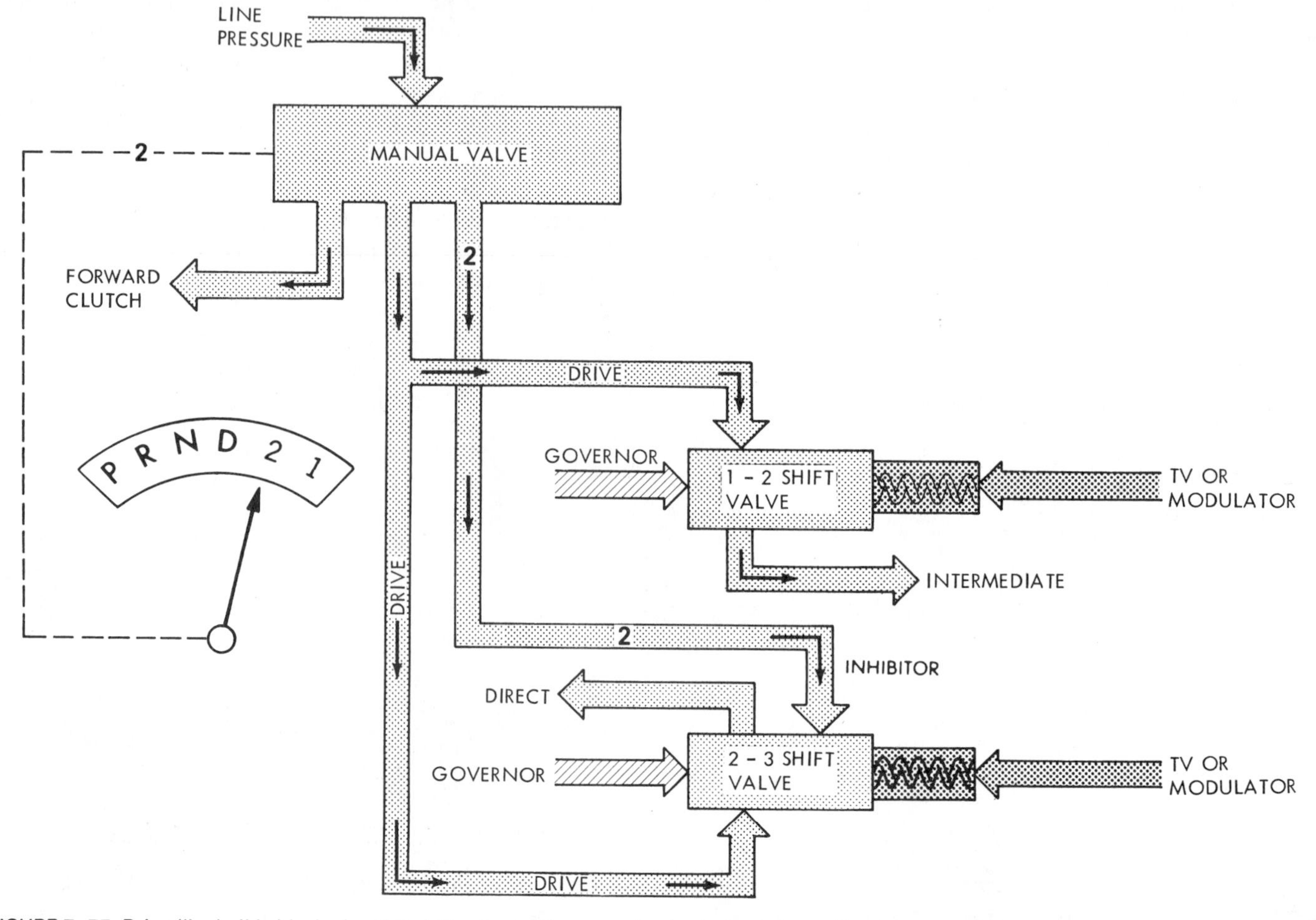

FIGURE 7–75 Drive (line) oil is blocked to 2-3 shift valve by the manual valve in the intermediate range position.

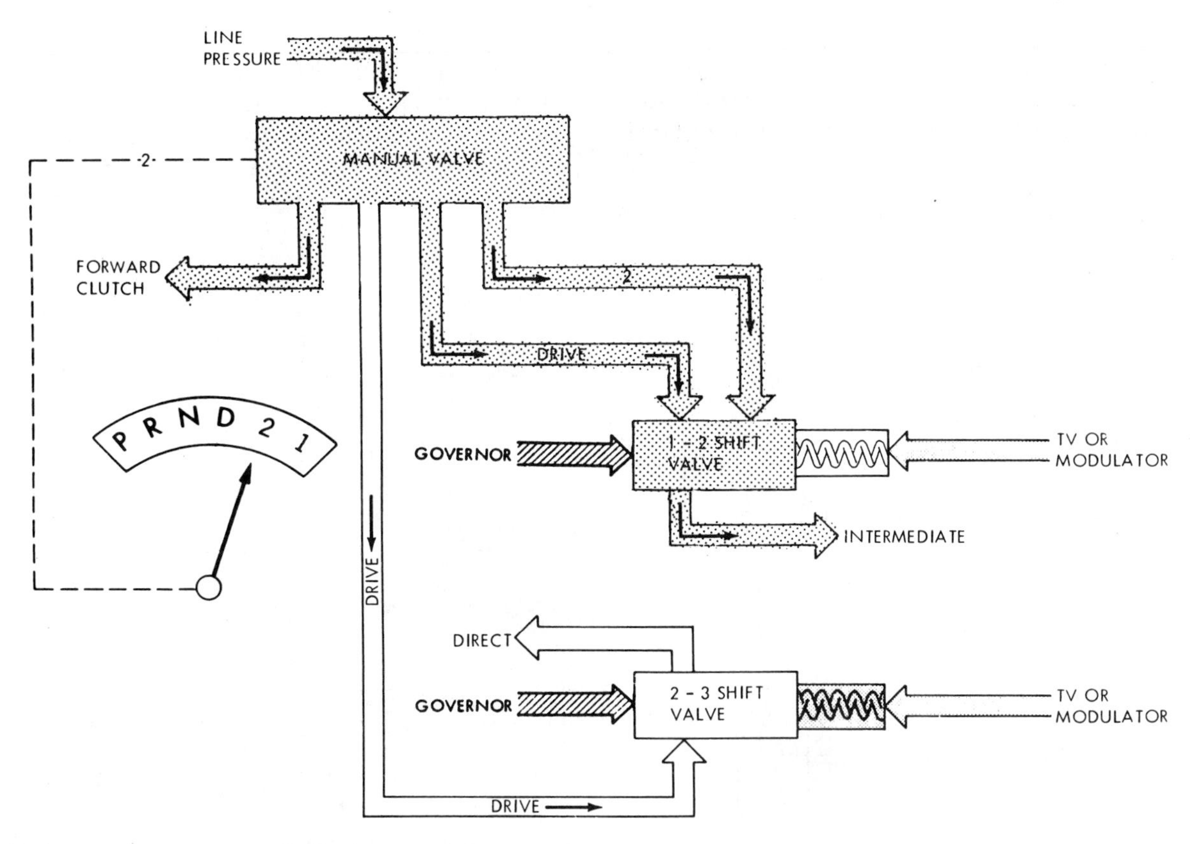

FIGURE 7–76 Intermediate range circuit allowing a second-gear start.

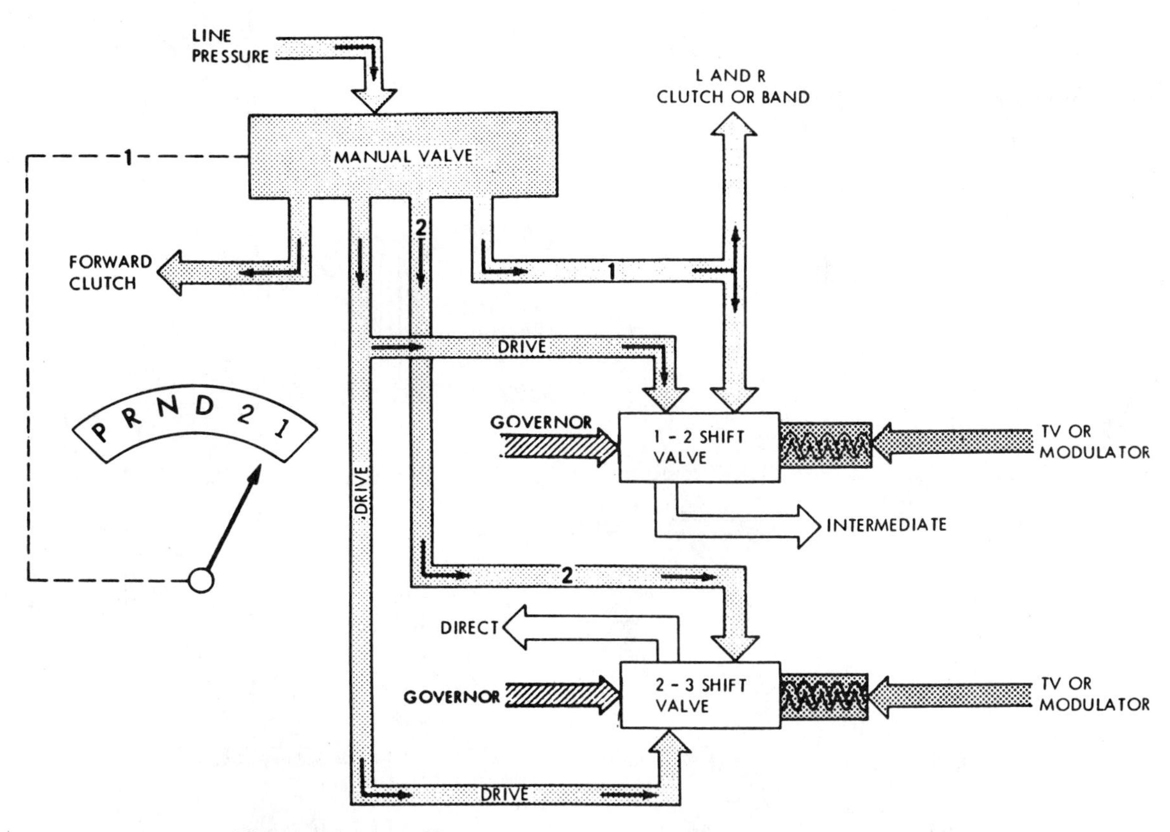

FIGURE 7–77 TV pressure closes shift valves in manual low.

routed to the pressure regulator valve to double the regulated line pressure. Reverse boost pressures can run as high as 250 to 300 psi (1,725 to 2,070 kPa). The extra pressure is required to prevent slipping of the reverse clutch or band. In reverse, the torque reaction on the planet carrier exceeds three times the engine torque. Although first gear has the largest gear ratio, the torque reaction on the very same planet carrier is much less. Forward-gear reductions produce less torque reaction, so the stress on a band or clutch to hold it is lower.

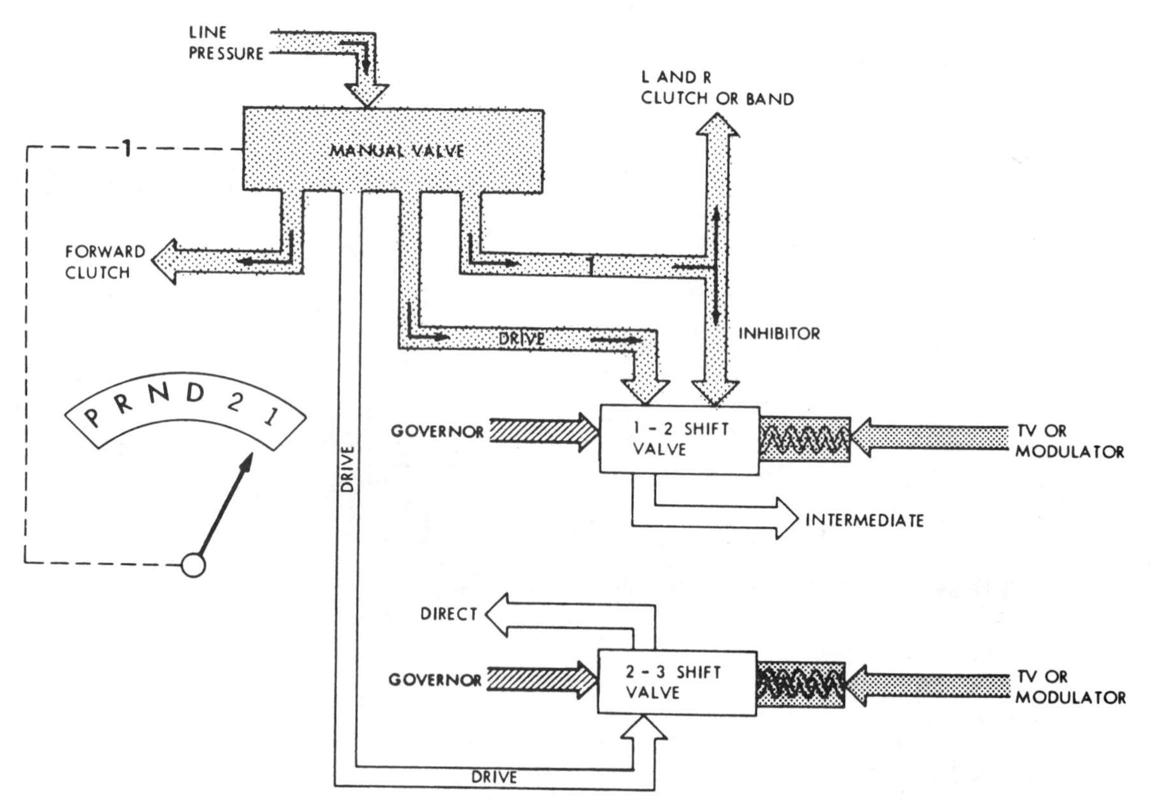

FIGURE 7–78 Manual low, upshifts locked out.

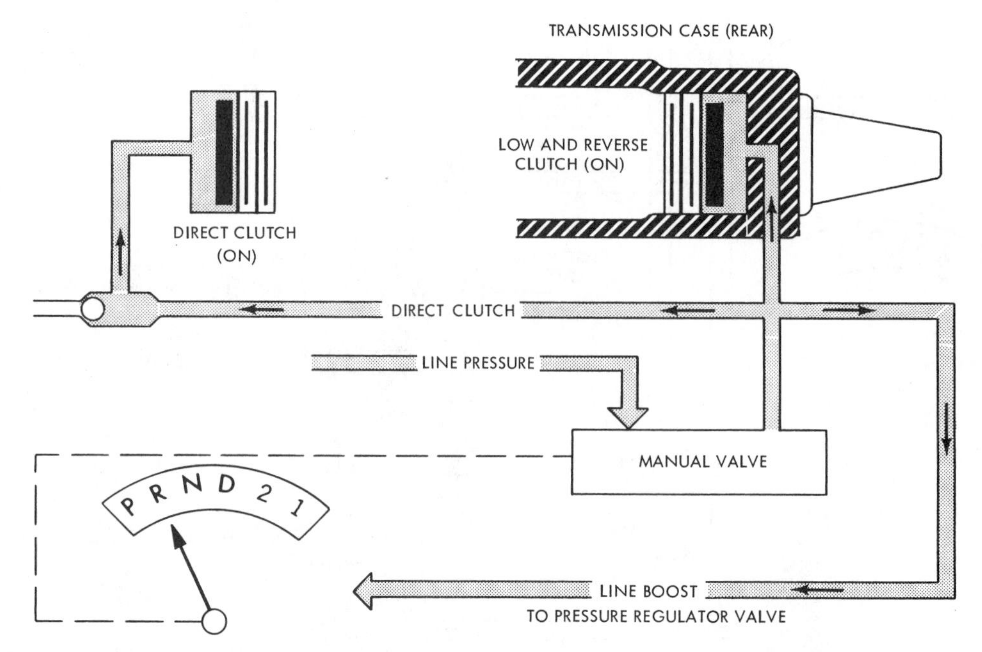

FIGURE 7–79 Typical reverse circuit.

Four-speed Overdrive Transmissions/Transaxles

The shift systems used in four-speed overdrive transmissions operate on the same hydraulic concepts as those just covered. Though there are additional clutches and bands, hydraulic valves and circuits, and each model has its own unique methods to provide automatic shifting, the fundamentals of operation are based upon those found in three-speed transmissions and transaxles.

SUMMARY

Transmission shift timing and quality is determined by a variety of different valving mechanisms. Although the system appears to be complex, each valve operates on basic regulating and/or relay valve principles.

Two important hydraulic pressures utilized in hydraulic-controlled transmissions are TV/modulator pressure and governor pressure. The TV/modulator pressure is an engine load signal that is determined through a throttle linkage, TV cable, or vacuum modulator. This pressure is responsible to boost mainline pressure, influence shift quality, and adjust shift timing. Governor pressure is a hydraulic signal of road speed. Its primary purpose is to act on shift valves to move them into an open position. To move the shift valves, governor pressure must overcome the shift valve springs and TV/modulator pressure.

As shift valves are moved, oil circuits to the appropriate clutches and bands are hydraulically energized. To control the aggressiveness of gear changes, accumulators are often used to cushion the oil flow. The cushioning effect is often directly related to the amount of engine load.

Forced downshifts occur due to additional oil pressures acting on open shift valves. When the force of the shift valve spring and additional oil pressure is greater than the governor pressure, downshifts occur.

The position of the manual valve determines which hydraulic circuits can be energized. Understanding hydraulic circuit operation fundamentals is important for accurately diagnosing and repairing automatic transmissions.

❑ REVIEW

Key Terms

Mainline pressure, throttle pressure/modulator pressure, boost valve, vacuum modulator, manual valve, governor, centrifugal force, valve body assembly, and accumulator.

Completion

Pressure Supply System

1. The mainline operating pressure is controlled by the ________ ________________ valve.
2. The regulator valve works on the ____________ valve principle. The regulated pressure is determined by a fixed ____________ force and hydraulic auxiliary boost forces usually from the ______________________________ and ________________ pressures.
3. When the regulating valve bleeds off excess fluid to control line pressure, the fluid is dumped back into the __________ ________________ side of the pump circuit.
4. For increased demand on the clutch band holding power, extra mainline pressure boost is provided for ____________ ____________________ gear. This is achieved by using ________________ oil as a boost force along with the throttle oil.

Throttle and Modulator Systems

1. Throttle pressure may be referred to as ________________ pressure in other transmissions.
2. The throttle system produces a regulated hydraulic pressure that reflects engine ________________.
3. In every automatic transmission, throttle pressure is distributed to the shift valves to ____________ the shifts and to the pressure regulator valve train to ____________ the line pressure.
4. With closed throttle or high manifold vacuum conditions, throttle pressure output should be close to or at __________ psi (____kPa).
5. As altitude increases, atmospheric pressure decreases with a resulting (increase) (decrease) in manifold vacuum.
6. In some vacuum modulator systems, governor pressure is used to (increase) (cutback) the modulator output relative to car speed.
7. A leak in the vacuum diaphragm causes the throttle valve to regulate at a (higher) (lower) than normal output.
8. A punctured aneroid bellows puts the throttle output at its extreme (high) (low) end.

Governor System

1. The governor system produces a regulated hydraulic pressure that is proportional to ____________ speed.
2. In every automatic transmission, governor pressure is distributed to the shift valves, where its function is to oppose throttle pressure and (open) (close) the shift valves.
3. A governor usually has two sets of weights. The heavy weights are called the ________________ weights, and

the light weights are called the __________________ weights.

Servo and Clutch Assemblies

1. The purpose of a servo unit is to change a hydraulic force into a __________________ force.

2. Band designs are classified as single wrap or __________ wrap.

3. A planetary gear member can be locked to the transmission case by either a band, ________________________, or ________________.

4. Multiple-disc clutches are used for two functions in controlling the planetary geartrain—either as a holding clutch or a ___________ clutch.

5. The clutch is hydraulically applied and released by ______ __________________ force.

6. The thick reaction plate that absorbs the end force of the clutch apply piston is usually referred to as the ________ _____________ plate.

7. To assist in cushioning the clutch apply, clutch units sometimes use a ___________ steel plate as part of the clutch disc pack.

Accumulator

1. The purpose of an accumulator is to control the shift feel by controlling the rate of pressure buildup in the circuit during a ____________ or ____________. The pressure buildup rate is keyed to engine _________ output.

2. When the accumulator piston strokes, it ___________ fluid from the clutch or band circuit and delays the final pressure buildup in the circuit.

3. The accumulator piston starts to stroke (before) (after) final clutch lockup.

Shift System

1. The shift valve spring is located on the _____________ pressure side of the shift valve.

2. On very light throttle upshifts, throttle pressure is usually at or near _____ psi (____kPa). Therefore, the minimum shift points are determined by the resistance of the shift valve _____________ to the governor oil.

Intermediate, Manual Low, and Reverse Ranges

1. The intermediate and manual low operating ranges are established by the __________ valve.

2. Should the vehicle speed be too high when moving the selector from D to 1, the transmission will function in _________ gear until a safe vehicle speed is reached for manual low engagement.

3. In reverse, the torque reaction on the stationary reaction carrier exceeds ______ times the engine torque. For extra clutch/band holding power, maximum reverse boost pressures can run the mainline pressure as high as __________ psi (____kPa).

Matching

Use the list of lettered terms in completing the following questions. Place the letter of the term you use in the blank space next to the question. Use a single answer only.

A. Governor
B. Throttle
C. Manual
D. Spring
E. Mainline
F. Vacuum
G. Mechanical

____ 1. Provides for the mainline pressure boost in D.

____ 2. The force that works to open the shift valve.

____ 3. The control system that enables the driver to program the transmission operating range.

____ 4. The force that closes the shift valve on closed-throttle downshifts.

____ 5. The control system that senses vehicle speed.

____ 6. Generally determines the mainline pressure value in P and N.

____ 7. Determines the minimum shift points.

____ 8. The force that works to delay the opening of the shift valve.

____ 9. The force that determines the regulated output of the throttle valve.

____ 10. The control system that senses engine torque.

8 Fluids and Seals

OBJECTIVE:

The objective of this unit is for you to become proficient with the terms and concepts contained in this chapter. Completion of this objective is essential to becoming a successful automatic transmission technician.

CHALLENGE YOUR KNOWLEDGE

Define the following key terms:

Oxidation stabilizers, viscosity index improvers, dispersants, antiwear agents, corrosion inhibitors, friction modifiers, seal swell controllers, antifoam agents, compatibility, Type F, Dexron, Mercon, Type 7176, auxiliary add-on coolers, static, dynamic, positive sealing, nonpositive sealing, lip seal, square-cut seal, O-ring seal, metal sealing rings, Teflon sealing rings, and Vespel sealing rings.

List the following:

Four main functions of ATF (automatic transmission fluid); eight additives blended into ATF; three types of synthetic rubber seals; two metal seal ring end joint designs; two Teflon seal designs; and one application for each of the following types of seals: lip, square-cut, O-ring, metal, Teflon, and Vespel.

Describe the following:

The main difference between Dexron and Type F fluid; fluid type recommendations for domestic transmissions; effectiveness of aftermarket fluid conditioners; the relationship between fluid life and fluid temperature; fluid change recommendations for normal and severe driving conditions; purpose of auxiliary add-on coolers; and characteristics of positive and nonpositive sealing techniques.

INTRODUCTION

Today's vehicle owners desire transmissions that provide responsive and smooth shifting, contribute to good fuel economy, and give many miles of trouble-free travel. As automatic transmissions have become more sophisticated, improved hydraulic fluids and methods of sealing the fluid passages and chambers have been developed. Features such as lockup torque converters, overdrive clutch and band units, and electronic shift controls have contributed to the need for high-quality fluid blends and modern sealing techniques.

The first portion of Chapter 8 discusses the functions and characteristics of automatic transmission fluid (ATF). Highlighted topics include manufacturer recommendations, additives, fluid change recommendations, and add-on coolers.

To provide quality shifts at the right time and with the desired degree of aggressiveness, fluid must be routed to many different internal components. A variety of seal designs are needed to ensure the stability of the pressurized hydraulic circuits to both stationary and moving parts. The common methods used to seal the fluid to these devices are explored in the second half of this chapter.

FUNCTIONS OF AUTOMATIC TRANSMISSION FLUID

Automatic transmission fluid (ATF) is the lifeblood of the transmission and plays an extremely important part in determining transmission life and performance. Automatic transmission fluids are among the most complex lubricants in the petroleum industry: select mineral oil fortified with a precise blend of additives. These additives comprise approximately ten percent of the ATF volume and provide the special properties needed for transmission operation. Some of these additives are the same types used in engine oils. Automatic transmission fluid, however, must be provided with special friction properties and added oxidation stability. In terms of viscosity, ATF closely falls in the category of SAE 20 grade oil with exceptionally good low temperature properties.

To avoid confusion with ordinary engine and lubricating oils, the term *fluid* is applied to the liquids used in hydraulics. The most important job of any hydraulic fluid is to transmit force and do it immediately. Therefore, a liquid is used because it is virtually noncompressible. Automatic transmission fluid performs four main functions. It:

1. transfers power from the engine to the drive line through the torque converter.
2. absorbs heat from the torque converter, from friction elements such as clutches and bands, and from bushings, bearings, and other moving components.
3. transmits hydraulic pressure through a hydraulic control system, which utilizes the fluid in the complex array of pumps, valves, lines, servos, and clutch cylinders.
4. is a multipurpose lubricant for the gears, thrust bearings, bushing supports, clutches, bands, and so on. Like an engine oil, it must lubricate, cool, seal, and clean, but it is subject to more severe service.

ATF BLEND ADDITIVES

To meet total service requirements, ATF is blended with additives to give additional properties needed for transmission operation. These additives include **oxidation stabilizers, viscosity index improvers, dispersants, antiwear agents, corrosion inhibitors, friction modifiers, seal swell controllers,** and **antifoam agents. Compatibility** with the entire transmission is also an important factor of the fluid blend.

- **Oxidation stabilizers:** Absorb and dissipate heat. Automatic transmission fluid has high resistance to varnish and sludge buildup that occurs from excessive heat that is generated primarily in the torque converter. Local temperatures as high as 600°F (315°C) can occur at the clutch plates during engagement, and this heat must be absorbed and dissipated. If the fluid cannot withstand the heat, it burns or oxidizes, resulting in an almost immediate destruction of friction materials, clogged filter screen and hydraulic passages, and sticky valves.
- **Viscosity index improvers:** Keep the viscosity nearly constant with changes in temperature. This is important at low temperatures, when the oil needs to be thin to aid in shifting and for cold-weather starting. Yet it must not be so thin that at high temperatures it will cause excessive hydraulic leakage so that pumps are unable to maintain the proper pressures.
- **Dispersants:** Suspend dirt and prevent sludge buildup.
- **Antiwear agents:** Zinc agents that control wear on the gears, bushings, and thrust washers.
- **Corrosion inhibitors:** Prevent corrosion of bushings, thrust washers, and oil cooler brazed joints.
- **Friction modifiers:** Change the coefficient of friction of the fluid between the mating steel and composition clutch/band surfaces during the engagement process and allow for a certain amount of intentional slipping for a good "shift-feel."
- **Seal swell controllers:** Control swelling, hardness, and tensile strength of synthetics and keep the seals pliable.
- **Antifoam agents:** Minimize fluid foaming from the whipping action encountered in the converter and planetary action.
- **Compatibility:** The ATF blend must be compatible with the materials used in the transmission. It cannot react chemically with the metals, seal materials, or friction materials.

DEVELOPMENT OF TRANSMISSION FLUIDS

A great quantity of fluid of various kinds has been poured into automatic transmissions since they were introduced to American passenger cars in 1939. Over the years, much research has gone into the development and improvement of hydraulic fluids to meet the changing requirements of the various transmission designs and the types of operational service encountered.

In the beginning, automatic transmissions were generally fed straight mineral oil, exactly the type used in the engines. In some cases, an oxidation inhibitor was added to the motor oils. Development of new friction and seal materials and the demand for the fluid to handle higher temperatures resulted in the adoption of the first automatic transmission fluid specification for the industry. The fluid was developed by General Motors in 1949 and labeled Type A. Suppliers of Type A fluid had to be approved and licensed by General Motors, which is still the practice for fluids developed by General Motors. In 1957, an improved Type A fluid was introduced, called Type A–Suffix A. The Type A fluids were universally used by all the car manufacturers in their automatic transmissions.

The next phase of ATF formula change and upgrading produced the **Type F** and **Dexron** fluids. Type F was developed in 1965 by Ford Motor Company for use in its transmission products, while General Motors replaced Type A–Suffix A with Dexron in 1968. The Dexron formula is still widely used in the industry for domestic and import applications.

Type F and Dexron basically use the same fluid formulation—they differ only in their friction characteristics. Type F is nonfriction-modified, and Dexron is friction-modified. Due to a combination of friction material composition and use of a short count in the clutch packs, Type F provided the necessary high coefficient of friction, or "grab," to avoid excessive engagement slip. Dexron is used where the paper-based composition friction materials have a high coefficient of friction, producing an aggressive shift grab. The friction modifier, under pressure, lowers the coefficient of friction and provides controlled slip during final lockup of the clutch discs to ensure smooth engagement.

The characteristics of nonfriction- and friction-modified fluids are compared in Figure 8–1. While Type F was gradually

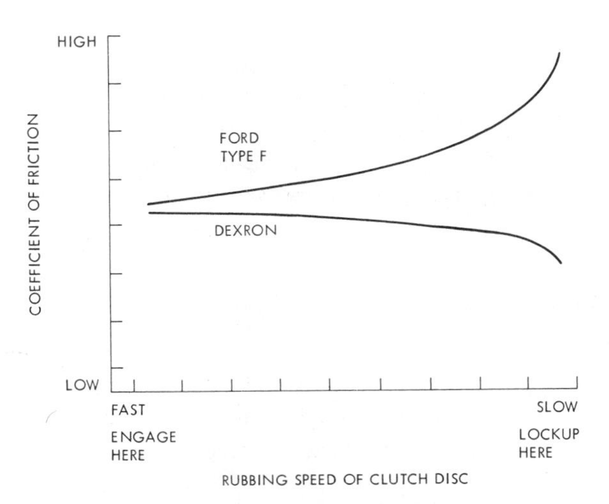

FIGURE 8–1 Dexron and Type F friction curves. Type F produces a higher friction rise, especially during final lockup.

phased out completely by Ford from its automatic transmission product line, Dexron remains a favored fluid. Technicians usually recognize the updated Dexron, referred to as Dexron II-D. A new advanced formula, Dexron III, originally introduced as Dexron II-E, is currently required for all General Motors electronically controlled automatic transmissions.

- **Type F:** Transmission fluid developed and used by Ford Motor Company up to 1982. This fluid type provides a high coefficient of friction.
- **Dexron:** Transmission fluid developed by General Motors and used by numerous other vehicle manufacturers. This fluid contains a friction modifier.

Starting in 1977, Ford began to use a variety of fluid formulas new to their automatic transmissions, including Dexron II. Technicians might remember the Type CJ and Type H fluids that followed and were recommended for specific Ford transmission applications. The CJ fluid was very closely formulated to Dexron II specifications. Since it was not available in bulk quantity, Dexron II was acceptable as a service fill. Beginning with 1988 production, a new Ford ATF called **Mercon** was introduced. Mercon fluid exceeds Dexron and Type H requirements. For Ford automatics, Mercon can be used in all previous Dexron, Dexron II, CJ, and Type H ATF applications. It cannot, however, be substituted for Type F applications.

- **Mercon:** A fluid developed by Ford Motor Company in 1988. It contains a friction modifier and closely resembles operating characteristics of Dexron.

Poor quality, low-cost ATF is available in the aftermarket and should not be substituted for Dexron or Mercon. Oil companies packaging Dexron and Mercon must have their ATF formulas tested by General Motors and Ford. If the sample fluids meet specifications, the companies are given a certification number and the fluids can be packaged under the Dexron and Mercon trademark (Figure 8–2). Inferior ATF cannot use

FIGURE 8–2 Examples of industry-certified ATFs.

the Dexron or Mercon name. Even though the different brands of ATFs may meet the same specifications (Dexron III, Mercon, and so on), they can have significant differences in frictional properties and produce totally different shift-feel characteristics. The formulation may even decrease the static holding capacity of a band or clutch. The technician needs to be selective in the choice of ATF brands.

Power Steering

At one time, ATF was used as the fill for power steering. This is no longer the requirement. Power steering systems now use their own special fluid formula. Automatic transmission fluid is not compatible with rack and pinion seals and will cause severe bleeding.

ATF RECOMMENDATIONS

Chrysler Recommendations

Dexron was the long-standing ATF recommendation for Chrysler automatic transmissions until **Type 7176** was released in 1986 for service fill. In terms of their friction characteristics, some of the aftermarket fluids affected lockup clutch performance, resulting in shudder and/or slippage. Although Type 7176 was used for a long time as a factory fill, it is now the Chrysler preferred choice for replacement fluid in all front- and rear-wheel drive applications with lockup torque converter dating back to 1978.

- **Type 7176:** The preferred choice of ATFs for Chrysler automatic transmissions and transaxles. Developed in 1986, it closely resembles Dexron and Mercon. Type 7176 is the recommended service fill fluid for all Chrysler products utilizing a lockup torque converter dating back to 1978.

The Type 7176 ATF is available only through Chrysler dealerships. Dexron III may be used, however, when Type 7176 is not available. Dexron III is also the service fill recommendation on early Chrysler transmissions using Type A–Suffix A or Dexron fluids. If a lockup shudder condition is encountered, Chrysler recommends checking the manual valve and TV linkage adjustments. If the condition still exists, the transmission should be drained and refilled with Type 7176 fluid prior to other repair considerations. If the fluid has been changed recently, this is a probable cure.

Ford Recommendations

Mercon is the primary service fill from 1985 model year onward. It replaces Type H and Dexron II fluids. Product automatic transmissions affected are AOD, AXOD, ATX, A4LD, 4EAT, and E4OD. Mercon also may be used in all previous Dexron, Dexron II, and Type H applications prior to 1985. Do not use a Dexron fluid in place of Type H, as it will not satisfy transmission requirements. Type F is completely phased out from the Ford automatic transmission product line. It was used primarily in selected passenger car/light truck applications from 1964 to 1982.

Type F usage included the following Ford transmissions for the years listed:

JATCO for Courier Only	1974–82
FMX	1968–81
C3	1974–80 only
C4	1964–79 only
C6	1966–76 only

Use Type F in Ford automatic transmissions built prior to 1964 and not listed above.

General Motors Recommendations

The Dexron formula has had several updates to meet the rapid transmission/transaxle design changes in recent years, especially those involving electronic controls and four-speed geartrains. For the production years 1974 to 1991, Dexron II-D was the primary fill for all Hydra-Matic automatic transmission/transaxle applications for passenger cars and light trucks. It was also used as the service fill on earlier units using Type A–Suffix A or earlier Dexron fluids.

In further response to the special requirements for the new electronic transmissions/transaxles, an advanced formula Dexron II-E was introduced for all 1992–1994 Hydra-Matic passenger car and light truck applications. The most notable changes occurred in the areas of (1) friction modulation, where more attention is given to the requirements of the lockup torque converters; (2) antiwear, to provide more acceptable wear in the final drive, chain drive, and sprag clutches; (3) low-temperature fluidity of −30°F (−2°C), to improve the quality of the electronic control responses under extreme cold conditions; (4) oxidation; and (5) compatibility with new seal material.

Dexron II-E was upgraded and replaced by Dexron III for all Hydra-Matic applications from 1995 to 1997 with probable extension usage. Dexron III basically retains the qualities of Dexron II-E with more stringent requirements in the area of oxidation stability and frictional performance. Dexron III replaces all previous Dexron service fills.

Import ATF Recommendations

European and Japanese original equipment manufacturers (OEM) use an ATF with higher shear stability and low-temperature properties. The differences, however, remain minor with European and Japanese OEM North American imports allowed to use Dexron III/Mercon ATF for service fills. There is still a limited use for Type F fluid. The technician should make sure the right ATF service fill is being used by checking the lubrication section of the service manual or equivalent.

DEXRON VERSUS MERCON

There are some minor differences between Dexron III and Mercon test requirements and formula makeup. There is enough compatibility, however, between the two ATF fluids so that suppliers usually market Dexron III and Mercon in a single package, meeting OEM standards or better.

Future ATF product improvements will feature Dexron IV and Mercon V with improved low-temperature fluidity and oxidation stability. Both fluids will probably feature a fill for life, regardless of vehicle service usage. When Dexron and Mercon ATFs are upgraded, they replace the service fill on previous transmissions/transaxles where Dexron or Mercon was used.

ATF AFTERMARKET CONDITIONERS

The use of transmission conditioners to rejuvenate tired ATF or cure transmission leaks is, at best, questionable. A conditioner may temporarily improve or eliminate a shift-feel problem or shift-squawk noise only to cause premature transmission failure. Conditioners that are specifically aimed at fluid leaks and contain seal swell compounds can sometimes cure minor external seepage problems in older transmissions. Conditioner additives cannot compensate for friction plate deterioration or seals that have become worn or brittle. The technician should be concerned that the addition of conditioner additives can produce unpredictable results, causing component damage—mainly to friction elements and seals.

The use of limited-slip differential additive, a friction modifier, has been suggested as a solution for torque converter clutch shudder problems. Although it may give instant results, it has a negative effect on the frictional properties of the fluid. For example, lockup or slip time to apply a clutch or band is increased by fifteen to twenty-five percent. Even static or load-holding capacity of the friction elements is cut by more than one-half. In addition, the additive attacks the silicone seals. Although an immediate solution is found to eliminate lockup clutch shudder, the aftereffects cause clutch plates to glaze or burn.

There really is no way to recondition deteriorated ATF. The additives in the fluid progressively wear out and lose their function, especially the friction modifier. The best solution for tired ATF is replacement, along with a filter change. New ATF can do wonders in bringing back a good shift feel and eliminating clutch squawks. The latest method of dealing with lockup converter clutch shudder is to first try a fluid change, especially if the problem is related to a recent service fill.

FLUID CHANGE PRACTICES

Unfortunately, too many vehicle owners think the transmission fluid is sealed for life and needs no attention until the transmission self-destructs. As the fluid ages, the additive formula breaks down. The service life of an ATF fill is determined largely by the running temperatures to which it is exposed, created by a combination of driving conditions and atmospheric temperatures.

Heat is the major factor in deteriorating the fluid. Once the transmission starts operating beyond its design temperature of 170 to 175°F (77 to 79°C), the fluid life rapidly goes down. The design temperature gives a fluid life of 100,000 miles (160,000 km). Most passenger cars, however, run at 190 to 195°F (88 to 90°C). The 20°F (7°C) temperature increase cuts the life of the fluid in half. At 390°F (195°C), the fluid is good for forty miles (64 km).

High fluid temperatures cause the following:

Varnishes form	240°F (116°C)
Seals harden	260°F (127°C)
Plates slip	295°F (146°C)
Seals and clutches burn out, and oil forms carbon	315°F (157°C)

The following general guidelines are the usual recommendations. Every ATF and filter change should include band adjustments where applicable:

- Normal usage: Every 100,000 miles (160,000 km). This means that the fluid has always been operating at the optimum temperature. Since this is not usually the case, give the transmission a treat at 50,000 miles (80,000 km).
- Severe usage: Every 15,000 miles (24,000 km). Severe usage is defined as:
  (1) More than 50% operation in heavy city traffic during hot weather above 90°F (32°C).
  (2) Police, taxi, limousine, commercial operation, trailer towing, or mountain driving.
  (3) Rocking the vehicle out of snow, during which temperatures can reach 300°F (149°C) or higher.

Many service facilities suggest to their average customers that the fluid and filter be changed every 30,000 miles (48,000 km) or every two years. For owners of vehicles involved in trailer towing, four-wheeling, and other severe service usage, they recommend changing the fluid and filter every 15,000 miles or every year, whichever comes first.

ADD-ON COOLERS

There has already been considerable discussion of the effect of heavy-stress driving conditions on fluid temperature and the effect of the temperature on fluid and transmission life. Next to the engine, an automatic transmission generates more heat than any other vehicle component. To cope with the transmission heat, the fluid is routed through a cooler located in the radiator. This cooling arrangement is expected to control the temperature at a normal range of 170 to 175°F (77 to 79°C) under light-duty normal driving conditions. Since nine out of ten transmissions fail because of overheating, it is obvious that light-duty driving conditions are not the norm.

The transmission lives in a hostile heat environment from converter torque and under-the-hood temperatures that can reach close to 300°F (150°C), especially in front-wheel drive vehicles. Adding to the heat problem is the aerodynamic styling that cuts wind resistance but reduces the airflow to an already downsized radiator. Higher engine temperature requirements leave limited cooling capability left over for the ATF. Many radiators operate in the 230°F (110°C) range, and the transmission fluid can actually end up getting heated rather than cooled. Because of the interrelationship of engine and transmission cooling, do not overlook the fact that an overheated engine can roast a good transmission, and an overheated transmission can boil the radiator coolant.

It is obvious that the transmission fluid cooling needs help, and this is where **auxiliary add-on coolers** come into the picture. The technician does not need to go crazy installing an auxiliary cooler on every transmission, but needs to evaluate the circumstances under which the transmission is asked to perform. Any vehicle driving that involves significant heavy-duty service should be a candidate. Front-wheel drive vehicles especially can benefit.

❖ **Auxiliary add-on coolers:** A supplemental transmission fluid cooling device that is installed in series with the heat exchanger (cooler), located inside the radiator, to provide additional support to cool the hot fluid leaving the torque converter.

Auxiliary coolers typically look and act like small radiators. One popular design style consists of U-shaped tubes that are finned to dissipate heat into the airflow (Figure 8–3[a]). The tubes are designed to give a swirl effect to stimulate flow and to make sure that all the hot fluid turns over and gets exposed to the tube walls. Without proper turbulence, fluid becomes layered. A stream of hot fluid never cools if layered by fluid clinging to the walls of the tube. Illustrated in Figure 8–3(b) is a stack plate design cooler that also creates a corkscrew-type oil action for maximum turbulence. There are a number of styles of good auxiliary coolers on the market, and choice is a matter left to the technician choice. Before you install one, do your homework.

An aftermarket transmission cooler can usually be installed in thirty minutes. To meet factory warranty requirements on the vehicle, the cooler is generally installed in series between the heat exchanger located in the radiator and the return line to the transmission. Be sure to use the recommended mounting location for maximum cooling on the front side of the radiator. Also make sure that the auxiliary cooler hoses are not kinked or the cooler lines misrouted. Do not allow cooler lines to lean against radiator hoses, exhaust manifold, or the engine block. This can increase fluid temperature by thirty percent. Installing auxiliary

A

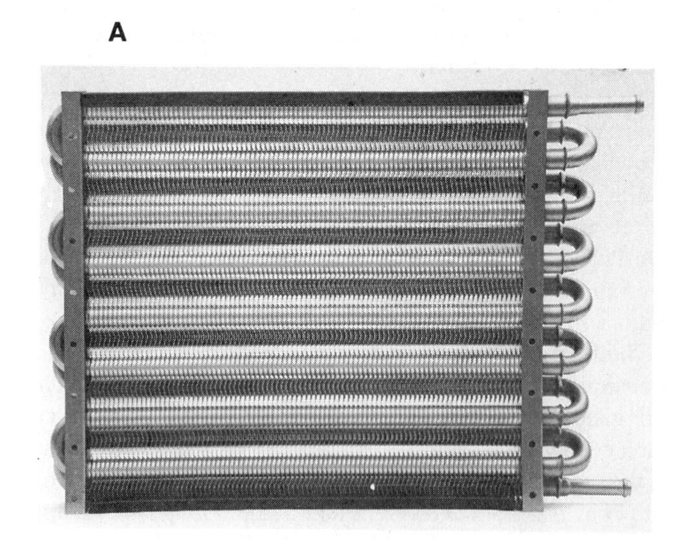

B

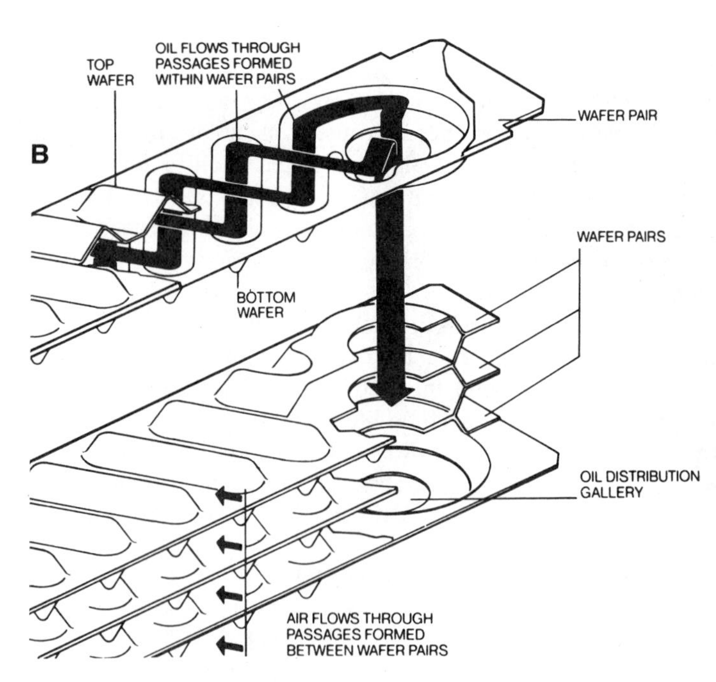

FIGURE 8–3 (a) Tube and fin cooler design; (b) oil path–stack plate cooler design. (Courtesy of TRU-COOL)

coolers is not difficult, but a careless installation will contribute to premature transmission torque converter failure.

For transmissions operating in extreme hot/cold seasonal conditions +90/+100°F (+32/38°C) summer versus –30/–40°F (–34/–40°C) winter temperatures, auxiliary coolers are available with a thermal switch valve to bypass the cooler in extreme cold temperatures. Installing a cooler that is too large or not mounted against the radiator may cause the fluid to become too cold when the torque converter is locked up and little heat is being generated. Running an ATF that never warms up can be dangerous to transmission life and performance. The lube circuit to the transmission especially is affected.

PURPOSE OF AUTOMATIC TRANSMISSION/TRANSAXLE SEALS

Transmission overhaul/rebuild procedures involve the replacement of the wide variety of seals that contain the fluid. Any fluid spill on the garage floor or driveway is classified as a transmission failure. Internal leakage is less obvious and usually tolerated by the owner as long as the vehicle moves forward reasonably well and provides a reverse.

Seals are a source of concern in transmission operation and fluid control. Because of their simplicity, seals often are not taken very seriously. If seals are not handled properly during installation, transmission failure is imminent. Knowing the function of the various styles of seals and how they work is helpful when reconditioning subassemblies and reassembling the transmission. It is not unusual for a transmission to require over thirty seal replacements during overhauling/rebuilding.

Sealing applications are classified as static or dynamic. **Static** and **dynamic** conditions are illustrated in Figures 8–4 to 8–6. Used in these two settings, seals provide **positive** or **nonpositive sealing**.

- **Static:** A sealing application in which the parts being sealed do not move in relation to each other.
- **Dynamic:** A sealing application in which there is rotating or reciprocating motion between the parts.
- **Positive sealing:** A sealing method that completely prevents leakage.
- **Nonpositive sealing:** A sealing method that allows some minor leakage, which normally assists in lubrication.

SYNTHETIC RUBBER SEALS

Synthetic rubber seals are used extensively throughout the transmission. The synthetic material in most cases is either a

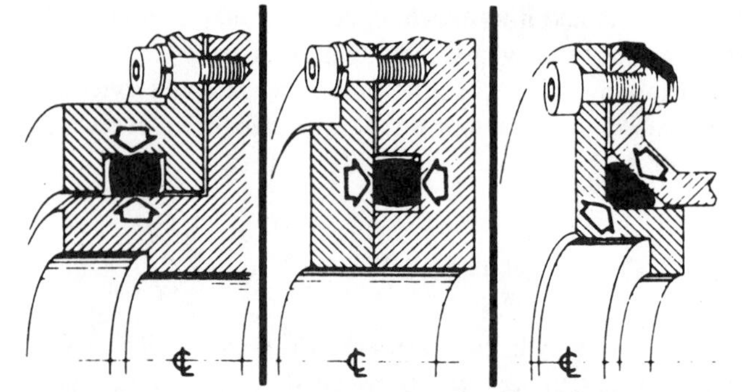

FIGURE 8–4 Static seal application for sealing case to cover. (Courtesy of SAE)

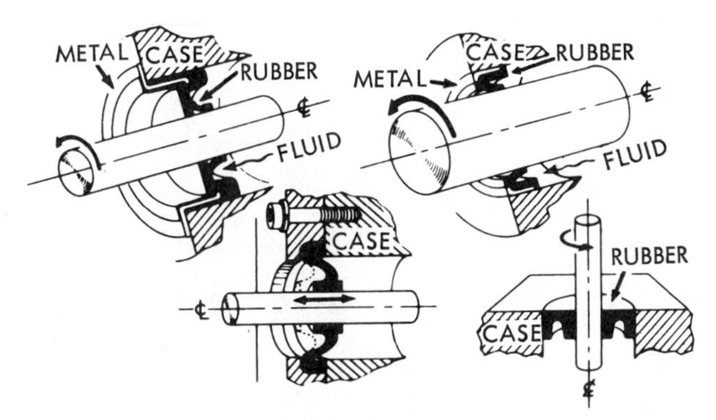

FIGURE 8–5 Dynamic seal application to rods/shafts. (Courtesy of SAE)

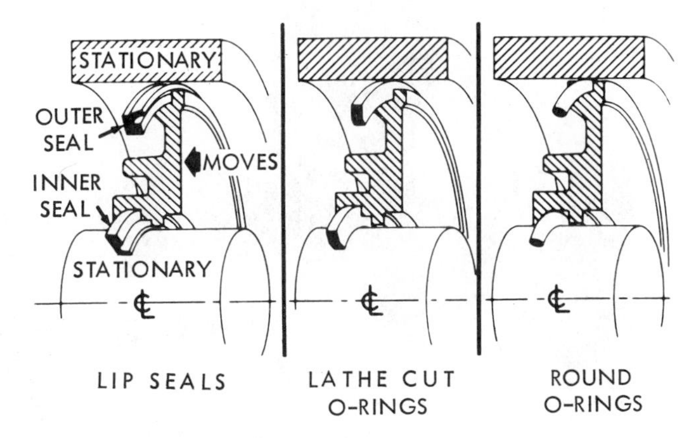

FIGURE 8–6 Dynamic piston seal applications.

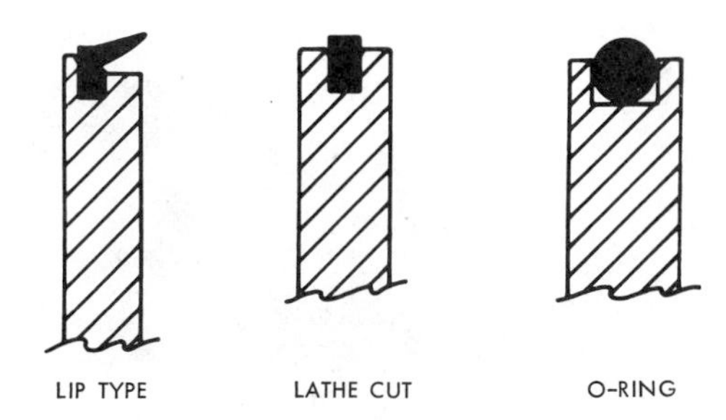

FIGURE 8–7 Basic types of synthetic rubber seals.

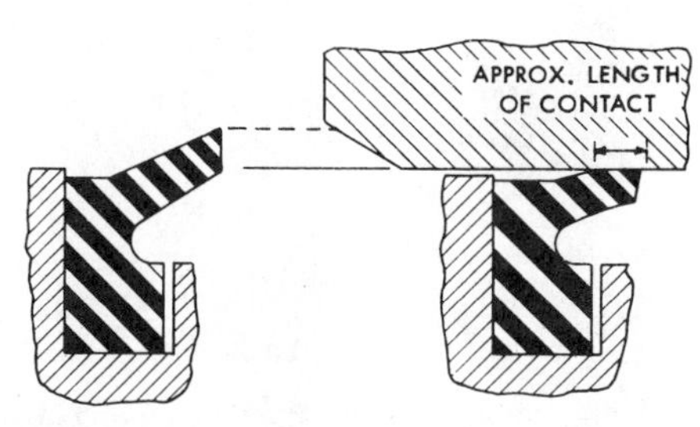

FIGURE 8–8 Lip seal. (Courtesy of SAE)

neoprene or silicone rubber. Compared with pure rubber, these materials better maintain their elasticity over a wide range of temperatures and withstand ATF chemical contact. The three basic types of synthetic rubber seals used in automatic transmissions are (Figure 8–7):

- Lip
- Square-cut
- O-ring

Lip Seal

The **lip seal** is a molded seal used where rotational and axial forces are present. It can seal against high pressure and, therefore, is very effective in piston sealing applications involving the clutch and servo units. The lip seal works on the deflection principle (Figure 8–8). The installed lip tip diameter, which is larger than the cylinder bore, is deflected during installation into the cylinder, resulting in lip tension against the cylinder wall. The compression of the free lip position (maximum diameter) thus gives an oil-tight seal. Furthermore, in piston applications, the seal lip is installed toward the pressure source, and the piston apply fluid exerts a pressure against the lip. This flares the lip out and aids in sealing. In Figure 8–9, note the flexibility of the lip seal in adapting itself to changes in bore and piston clearances. The seal easily slides in the cylinder bore and permits effortless piston movement in either direction.

❖ **Lip seal:** Molded synthetic rubber seal designed with an outer sealing edge (lip) that points into the fluid containing area to be sealed. This type of seal is used where rotational and axial forces are present.

Lip seals are also used as rotating shaft seals (Figure 8–10). In automatic transmission applications, they are used for sealing

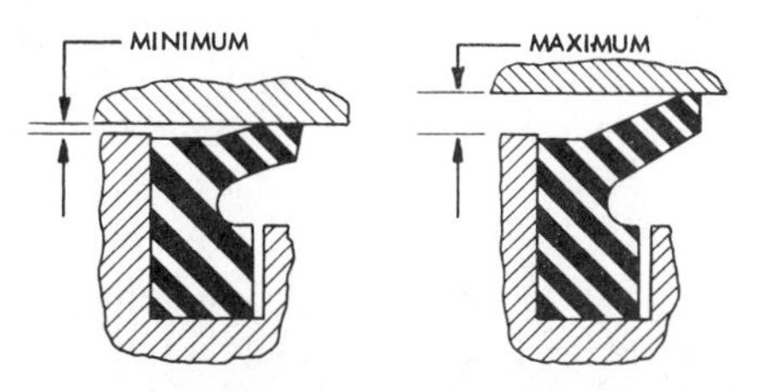

FIGURE 8–9 Lip seal provides flexibility. (Courtesy of SAE)

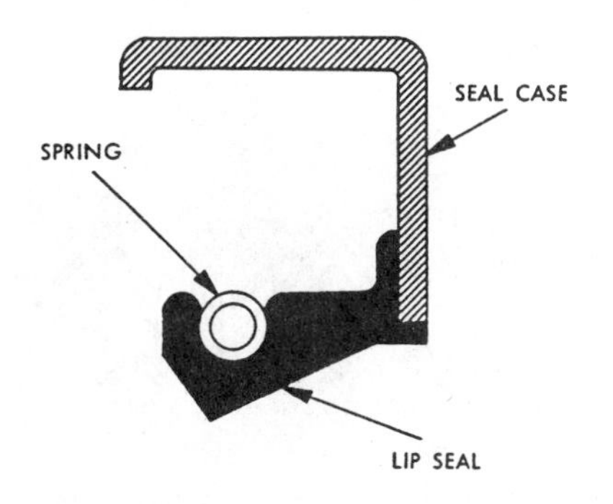

FIGURE 8–10 Lip seal design.

the drive shaft yoke to the output shaft, CV joints, manual valve shaft, and converter hub. Sealing is accomplished by the spring-loaded flex section maintaining a lip contact pressure on the rotating hub or shaft. The spring load is the only pressure on the lip, and therefore, the seal design cannot handle high pressure. The seal is designed primarily to retain the fluid in the transmission and keep external contaminants from entering.

Square-cut Seal

The **square-cut seal** is an O-ring design with a square cross-section used for both dynamic and static applications. The seal is made from a tube-shaped mold that is cut into individual rings on a lathe. In dynamic applications, the square-cut seal is used on servo, accumulator, and clutch pistons. To seal, the lathe-cut seal requires a squeeze when installed in its cylinder bore. Note in Figure 8–11 how the seal is squeezed into the piston groove because its outside diameter is larger than the cylinder bore into which it is installed. Complete installation in the cylinder bore shows that the seal is squeezed and under compression. The reaction force created by the compression tries to return the seal to its original size and shape, thereby creating a sealing effect on the contact area of the cylinder bore and bottom of the piston groove.

❖ **Square-cut seal:** Molded synthetic rubber seal designed with a square- or rectangular-shaped cross-section. This type of seal is used for both dynamic and static applications.

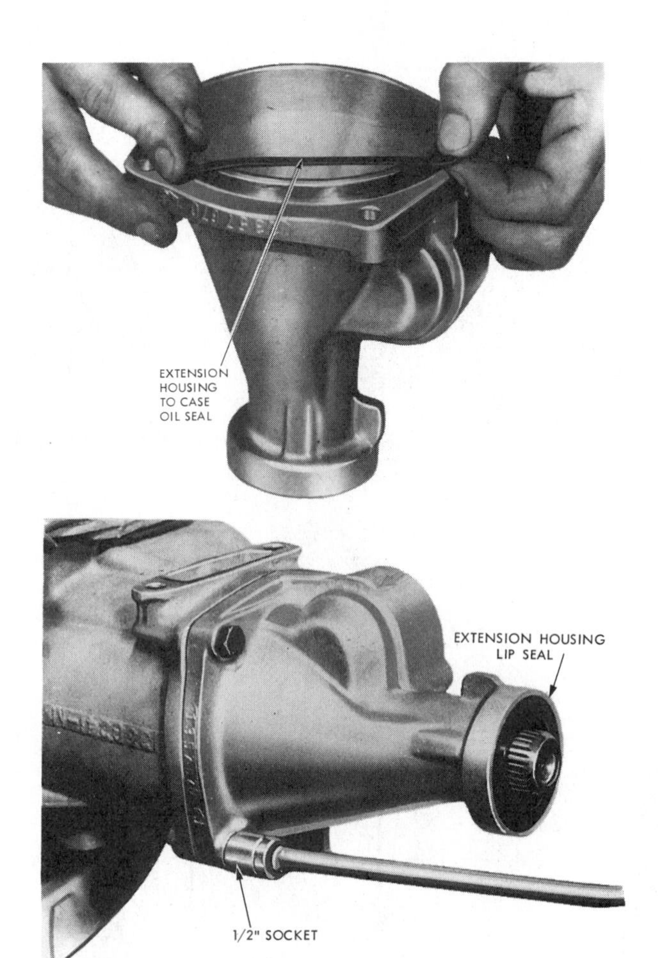

FIGURE 8–12 Lathe/square-cut seal used on extension housing. (Courtesy of General Motors Corporation, Service Technology Group)

Static applications of square-cut seals on an extension housing and pump housing assembly are shown in Figures 8–12 and 8–13. Although the square-cut rubber seal is not normally used for dynamic rotational applications, some new developments have occurred in this area. As an example, Ford uses square-cut rubber rings for rotational applications. Examples include the A4LD high clutch seals on the center support hub and the sealing rings on the governor body (Figures 8–14 and 8–15).

O-Ring Seal

The **O-ring seal** is a round molded seal that has a circular cross-sectional area. Like a lathe-cut seal, it is squeezed during installation, forming a seal against the bore contacting surfaces and holding groove (Figure 8–16). The O-ring seal is

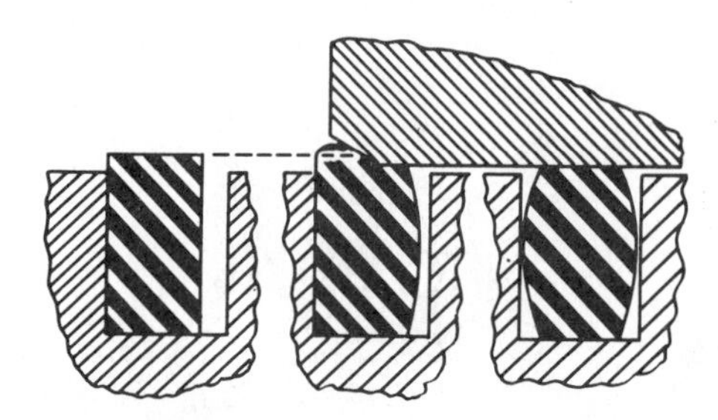

FIGURE 8–11 Square-cut seal squeezed into piston groove. (Courtesy of SAE)

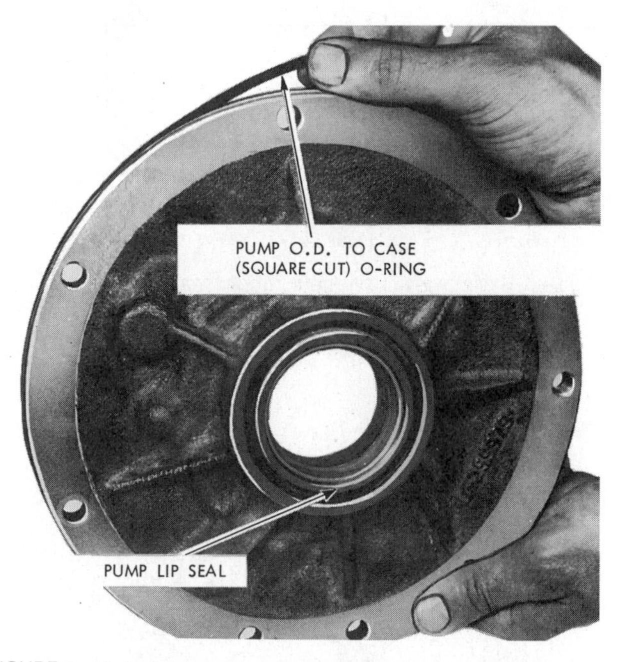

FIGURE 8–13 Lathe/square-cut seal used on pump housing. (Courtesy of General Motors Corporation, Service Technology Group)

FIGURE 8–14 Square-cut rubber rings—rotational application.

used as a static seal for pipes, tubes, shafts, and hydraulic cylinder covers (Figures 8–17 and 8–18).

In dynamic applications, the O-ring seal cannot hold up to rotational forces, but it can be used where axial movement is limited to short, slow strokes. The O-ring seal has a tendency to roll or twist in its groove during stroking. If a piston moves too far inside the cylinder, the seal will actually twist, get damaged, and fail to seal the hydraulics.

❖ **O-ring seal:** Molded synthetic rubber seal designed with a circular cross-section. This type of seal is used primarily in static applications.

Figure 8–19 illustrates how a typical O-ring seal behaves in a cylinder bore. Note that the seal is compressed, forming a seal against the contacting surface of the bore and bottom of the piston groove. When pressure is applied, it forces the seal against the side of its groove and, in effect, packs it into a corner. The end result is a positive seal on three sides capable of withstanding very high pressure. The O-ring also has the ability to seal effectively from both directions (Figure 8–20).

METAL SEALING RINGS

Metal sealing rings are usually made from cast iron, or sometimes from aluminum. They are used to seal rotating shafts or clutch drums that carry fluid under pressure and where some fluid leakage is needed to lubricate the shaft journals and bushings (Figure 8–21). Although the largest usage is found in rotary applications, metallic seals sometimes are used in reciprocating applications on accumulator pistons and servo pistons (Figures 8–22 and 8–23). The metal ring also may be used as a static seal, although this application is very limited.

The basic metallic ring used in automatic transmissions is designed or shaped like the common engine piston ring. Depending on the application, these piston-type rings are made from cast iron or aluminum. Cast iron rings are made from the same kind of piston ring iron used for engine piston rings or small castings. When required, they are coated with nickel, chrome, or tin. Updated aluminum governor rings on the Ford AXOD call for Teflon-coated aluminum. The sealing effect of the metallic ring is achieved by the fluid pressure head working the ring against the side of the groove and bore wall (Figure 8–24).

FIGURE 8–15 (a) A4LD—governor rubber rings; (b) A4LD center support hub rubber rings.

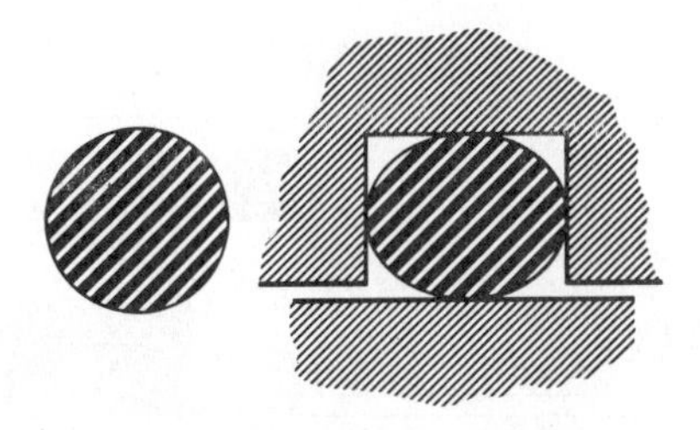

FIGURE 8–16 O-ring seal compressed when installed.

- **Metal sealing rings:** Made from cast iron or aluminum, their primary application is with dynamic components involving pressure sealing circuits of rotating members. These rings are designed with either butt or hook lock end joints.

Transmission metal oil rings use two types of joint ends: (1) butt joint, and (2) hook lock joint (Figures 8–23 and 8–25). The butt joint is used where small leakage past the joint is not important or where the leakage is utilized for lubrication. A lock joint is used where improved control of fluid leakage is required

FIGURE 8–17 Modular O-ring seal.

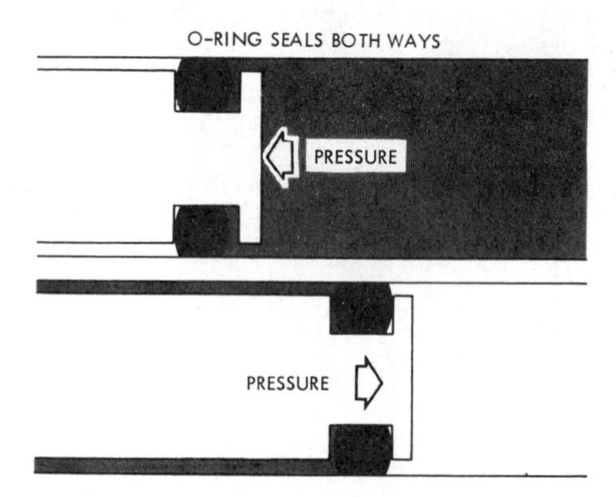

FIGURE 8–20 O-ring is forced against walls and seals from either direction.

FIGURE 8–18 Servo cover O-ring.

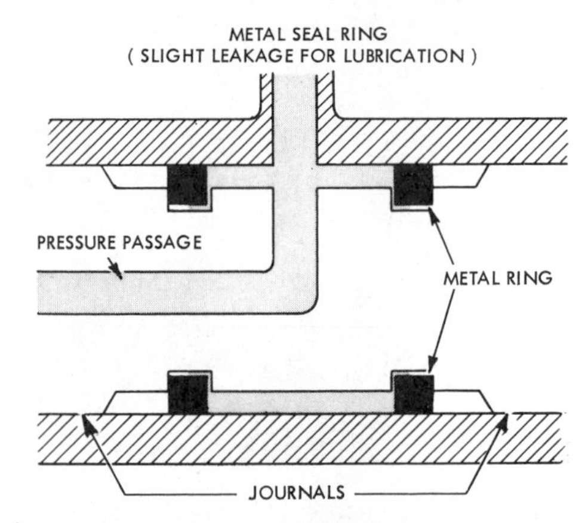

FIGURE 8–21 Metal seals allow leakage for lubrication.

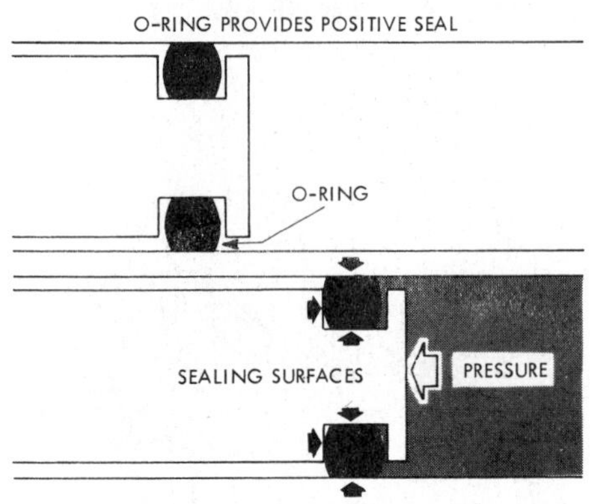

FIGURE 8–19 O-ring is compressed on three sides.

FIGURE 8–22 Butt-type metal seal rings used on servo and accumulator pistons (Chrysler TorqueFlite).

and when compression is needed initially to fit the ring into its bore during installation. The lock joint is designed with small tangs that hold the ring in compression. At one time, metal sealing rings were popular in automatic transmissions, but many modern transmission designs use Teflon, Vespel, and other rubber sealing rings in areas once dominated by the metal ring.

TEFLON SEALING RINGS

Teflon sealing rings are the popular choice in many current transmission applications. They have almost entirely replaced metal rings. Teflon is a soft, durable, plastic-like material that is highly heat-resistant, provides excellent sealing, reduces wear in rotating components, and offers a cost reduction over metal rings. Because Teflon is pliable, it can adapt itself to conform to the out-of-round tolerance given to a machine-bore versus a metal ring. The material, however, is very sensitive to nicks and scratches and penetration from metallic particles.

FIGURE 8–23 Nonreciprocating butt-type metal rings—governor application.

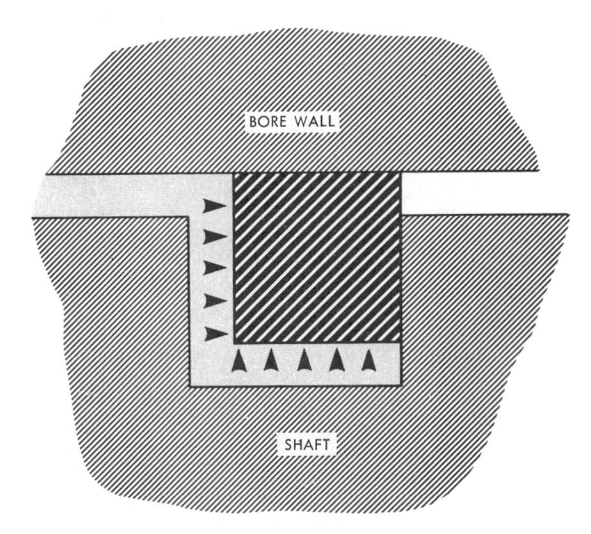

FIGURE 8–24 Sealing of metallic ring against the side of the groove and bore wall.

- ❖ **Teflon sealing rings:** Teflon is a soft, durable, plastic-like material that is resistant to heat and provides excellent sealing. These rings are designed with either scarf-cut joints or as one-piece rings. Teflon sealing rings have replaced many metal ring applications.

Teflon seals either have a scarf-cut joint or come as a solid one-piece ring (Figures 8–26 and 8–27). The one-piece ring requires special installation tooling. The Teflon rings are designed to fit loosely into their ring grooves with considerably more clearance than metal rings. Once installed in the bore, the oil pressure creates almost a no-leak seal. The solid Teflon seal provides better sealing than the scarf-cut Teflon seal and is better than the metal ring.

VESPEL SEALING RINGS

Vespel sealing rings were developed for special situations. To improve the durability and shift feel in the GM 4T60 (440-T4) second clutch, Vespel rings are used on the sprocket support. The revised sprocket support uses a deeper groove to accept the ring (Figure 8–28). Note that a four-lobe synthetic rubber O-ring expander fits the groove under the Vespel ring. The ring itself is made from a new extra-durable synthetic material that takes on the characteristics of a hard plastic but maintains flexibility. The ring uses a step joint and has two tabs to keep it locked in the groove (Figure 8–29).

- ❖ **Vespel sealing rings:** Hard plastic material that produces excellent sealing in dynamic settings. These rings are found in late versions of the 4T60 and in all 4T60-E and 4T80-E transaxles.

FIGURE 8–25 Metal hook–type oil rings (bottom three). (Courtesy of General Motors Corporation, Service Technology Group)

The Vespel ring has excellent wear and sealing qualities. Because of these desirable characteristics, GM uses Vespel rings in certain automatic transaxles. In the 4T60-E transaxle, Vespel rings are used in the same location as in the 4T60 (440-T4). In the 4T80-E, they are used in a similar type of setting—between the driven sprocket support and the reverse clutch housing.

SUMMARY

Modern automatic transmission fluid (ATF) and seals work together to provide numerous hydraulic functions. The confined fluid is used to transfer engine power into the transmission, absorb and transmit heat, operate clutch and band mechanisms, and lubricate internal components.

Quality transmission fluids contain many important additives, such as dispersants, antiwear agents, corrosion inhibitors,

FIGURE 8–26 Scarf-cut Teflon rings; center support seals—3L80 (THM-400).

FIGURE 8–27 Solid Teflon rings; input housing shaft seals—4T60 (440-T4).

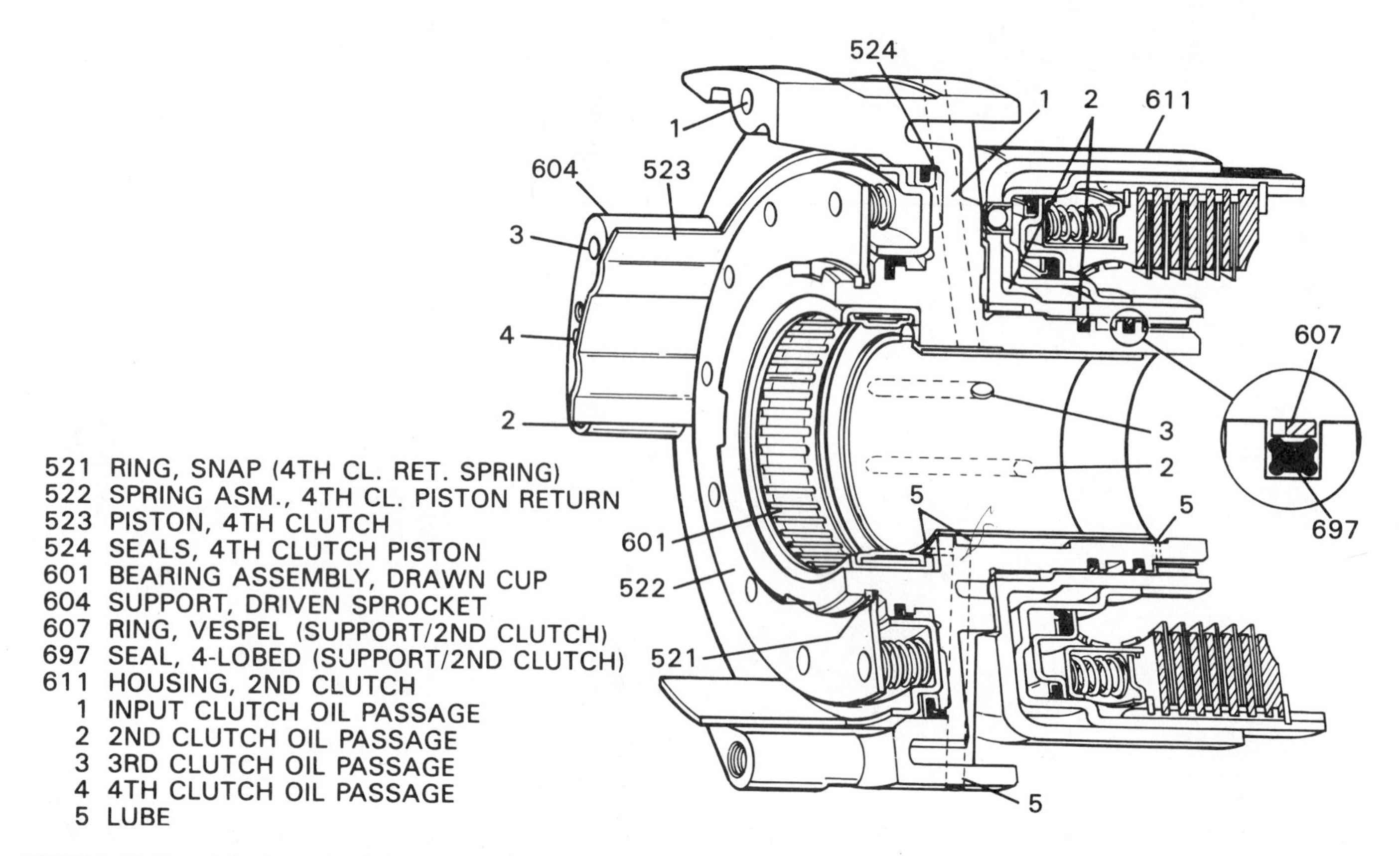

FIGURE 8–28 Vespel ring (#607/697) shown in sprocket support—4T60 (440-T4). (Courtesy of General Motors Corporation, Service Technology Group)

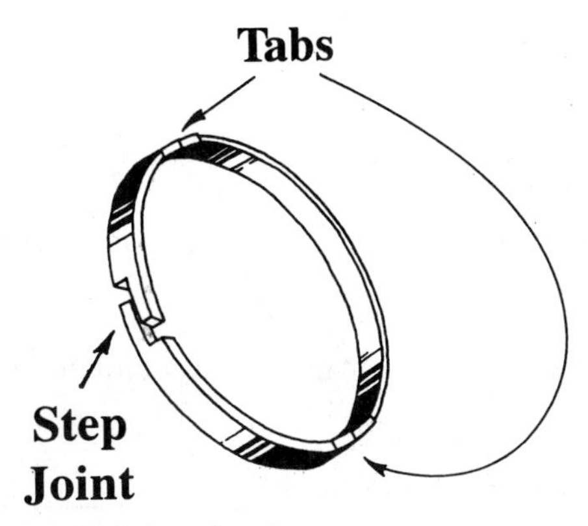

FIGURE 8–29 Vespel sealing ring.

friction modifiers, seal swell controllers, and antifoam agents. In addition, the fluid must be compatible to the materials used inside the transmission.

Because of fluid contamination, the breakdown of its additives, and overheating, ATF occasionally must be drained and replaced. Use the manufacturers' suggested fluid change time and mileage interval as a minimum guideline. Always follow their recommendations when choosing which type of ATF to use.

Seals used to contain the transmission fluid can be classified as either static or dynamic. Various seal designs are used, depending on their applications. Care must be exercised when replacing seals to position them correctly and avoid damaging them. The lip of a seal must point toward the pressure chamber it is sealing.

When servicing or repairing transmissions, always use fresh fluid and renew the appropriate seals. This will help ensure transmission operation and longevity.

❑ REVIEW

Key Terms

Oxidation stabilizers, viscosity index improvers, dispersants, antiwear agents, corrosion inhibitors, friction modifiers, seal swell controllers, antifoam agents, compatibility, Type F, Dexron, Mercon, Type 7176, auxiliary add-on coolers, static, dynamic, positive sealing, nonpositive sealing, lip seal, square-cut seal, O-ring seal, metal sealing rings, Teflon sealing rings, and Vespel sealing rings.

Questions

ATF Fluids

1. (Ford) Mercon replaces Dexron II and Type H fluids for automatic transmission applications. ___T/___F
2. (Ford) Mercon is the replacement fluid for Type F. ___T/___F
3. (Ford) Mercon is recommended as a power steering fluid. ___T/___F
4. ATF has a controlled swelling effect on the rubber seals. ___T/___F
5. Type F formula basically lacks a friction modifier. ___T/___F
6. Type F formula used in place of Dexron will cause drawn-out, soft shifts. ___T/___F
7. Dexron II-E formula is the recommended ATF for electronically controlled GM passenger car and light truck automatic transmission applications. ___T/___F
8. If a lockup shudder condition is encountered with a Chrysler TorqueFlite, drain and refill the unit with Dexron II. ___T/___F
9. (Ford) The FMX transmission ATF recommendation is Mercon. ___T/___F
10. Many imports use the same type of automatic transmission fluids available for domestic use. ___T/___F
11. Adding limited-slip differential additive to ATF is an acceptable practice. ___T/___F
12. When dealing with lockup converter shudder, a fluid change may be the remedy. ___T/___F

Fluid Change Practices/Add-on Coolers

____ 1. The optimum operating temperature for ATF is:

A. 150°F (66°C)
B. 175°F (80°C)
C. 220°F (104°C)
D. 240°F (116°C)

____ 2. Which one of the following conditions is not considered heavy-duty service?

A. stop-and-go city traffic
B. police/taxi operation
C. highway cruising
D. rocking the vehicle

____ 3. For severe usage conditions, the ATF should be replaced based on the following time or mileage interval, whichever occurs first:

A. 90 days/10,000 miles (16,000 km)
B. 120 days/10,000 miles
C. 6 months/15,000 miles (24,000 km)
D. 1 year/15,000 miles

_____ 4. Which is true?
I. Overheated ATF can cause an overheated engine.
II. An overheated engine can cause overheated ATF.

A. I only C. both I and II
B. II only D. neither I nor II

_____ 5. When installing an auxiliary cooler:
I. It normally is in series with the inlet line to the transmission cooler.
II. It normally is installed on the back side of the radiator.

A. I only C. both I and II
B. II only D. neither I nor II

Sealing the Fluid

1. Name the three types of synthetic rubber seals.

(a) ____________________
(b) ____________________
(c) ____________________

2. Where lip seals are used as rotating shaft seals, lip contact pressure is provided by a ____________ load.

3. Transmission metal oil rings use two types of joint ends.

(a) ____________________
(b) ____________________

4. Metal sealing rings used in automatic transmissions are made primarily from ____________ metal.

5. Teflon is a soft synthetic material that bruises very easily. ___T/___F

6. Jointed Teflon seal rings use a ______________ joint.

7. Teflon seal rings typically have considerable ring groove clearance. ___T/___F

8. The solid Teflon seal ring provides better sealing than a metal ring. ___T/___F

9. A synthetic hard plastic material used for a new sealing ring development is called ________________.

10. Name three modern sealing ring materials that are replacing metal rings in many transmission designs.

(a) ____________________
(b) ____________________
(c) ____________________

9 Principles of Electronification

OBJECTIVE:

The objective of this unit is for you to become proficient with the terms and concepts contained in this chapter. Completion of this objective is essential to becoming a successful automatic transmission technician.

CHALLENGE YOUR KNOWLEDGE

Define the following key terms/acronym:

Electronification, voltage, current, electronics, EMF (electromotive force), resistance, Ohm's law, electromagnetism, residual magnetism, electromagnetic induction, self-induction, mutual induction, and isolation/clamping diodes.

List the following:

Three factors that determine the magnetic pull of an electromagnet; six components of a basic electrical circuit; two types of computer input signals; and two computer output actuator circuit designs.

Describe the following:

The relationship between voltage, current, and resistance; the relationship between electricity and magnetism; the adverse effects of self-induction and mutual induction to computer circuitry; conventional current flow theory; electron current flow theory; and the three common circuit paths: series, parallel, and series-parallel.

INTRODUCTION

Electronics is playing an ever-expanding role in automotive design. It is predicted that by the year 2000, electronics will account for twenty percent of the price of an automobile. As shown in Figure 9–1, a multitude of in-vehicle electronic control systems are currently available in many models as standard features. These systems reflect the demand for improved performance, safety, environmental controls, and service diagnostics.

With electronic engine management getting the most attention, the transmission technician was temporarily spared from any significant tie-in with electronics. That has now changed. Most of today's late model automatic transmissions feature computer control of the converter lockup clutch and shift control system. Transmission electronics lends itself to optimum engine performance, better fuel economy, reduced engine emissions, greater shift system reliability, and improved shift feel.

Understanding automatic transmission electronics is an important factor in the servicing and repairing of vehicles. Electronic transmission problems are not solved by snipping wires, unplugging connectors, or switching transmissions. Rather than being intimidated by new technology, the technician needs to learn how automatic transmission electronic systems operate in order to diagnose and service related concerns efficiently.

- Transmission electronics operate on basic laws of electricity and electronics.
- Electronic circuits that appear complicated may be broken down into simple, easy-to-understand principles.
- Though the movement of electricity through an electronic circuit cannot be directly observed, its action can be tested and measured. Malfunctions can be isolated and then corrected.

This chapter provides a short introductory course in electrical and electronic concepts that apply to automatic transmissions. At times, examples using common automotive circuits are presented to clarify these topics. Chapters 10 and 11 directly apply these concepts in an examination of torque converter clutch and electronic shift control circuits. Diagnosis of these systems is addressed in Chapters 13 and 14, respectively. Possessing a clear understanding of **electronification** in late

1. Automatic transmission control
2. Ignition control
3. Carburetion control
4. Fuel injection control
5. Exhaust emission control
6. Power-steering control
7. Fuel pump driver
8. Fuel pump control
9. Overall engine and fuel consumption control
10. Voltage regulation
11. Voice warning and recognition
12. Time control
13. Trip control
14. Lighting control
15. Electronic navigation/trip computer
16. Turn signals
17. Windshield wiper/washer
18. Fuel meter
19. Clock
20. Speedometer
21. Tachometer
22. Digitally tuned radio
23. Car radio and car stereo
24. Coolant temperature meter
25. Safety and malfunction monitoring system
26. Air-bag activating system
27. Vehicle collision prevention system
28. Anti-skid control system
29. Auto door-lock control system
30. Seat-belt control system
31. Automatic suspension control system
32. Air conditioning control system
33. Optical fiber wiring system

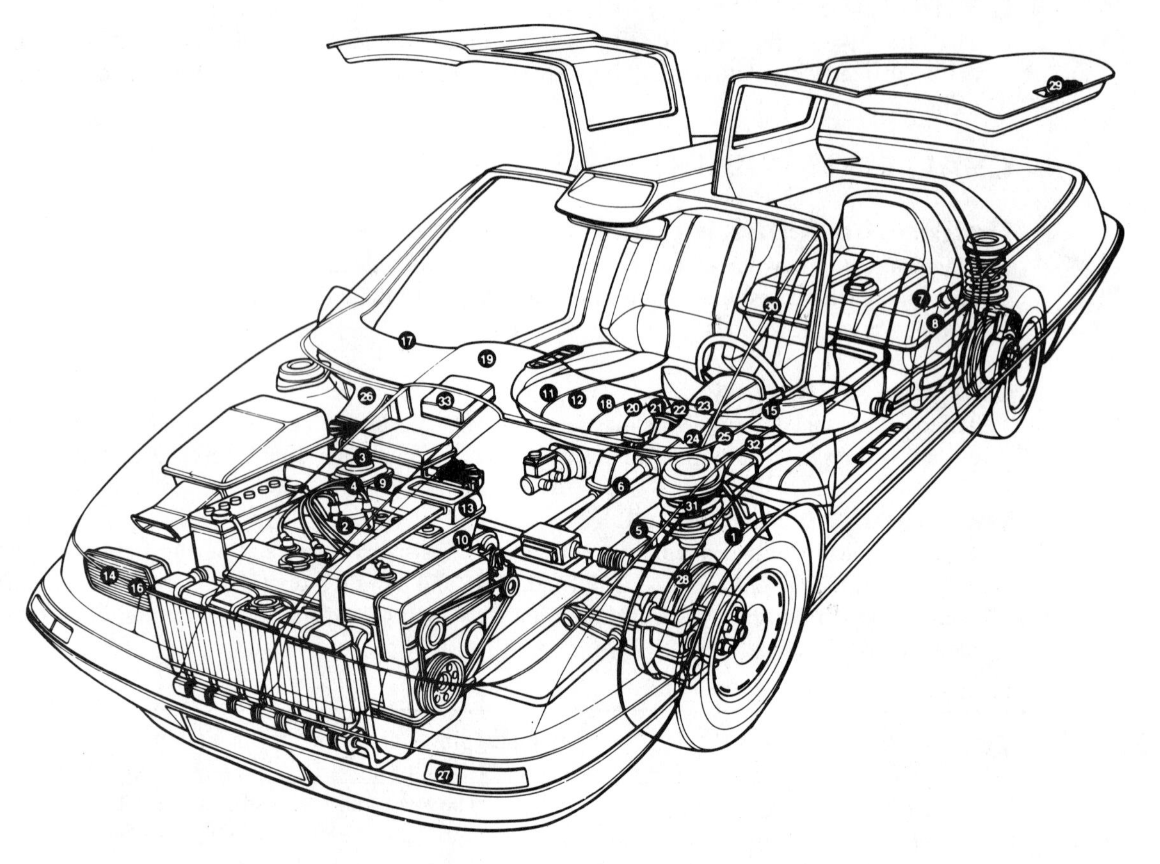

FIGURE 9–1 Significant electronic control systems. (Courtesy of SAE)

model automatic transmissions is necessary for accurate diagnosis and quality repairs.

- ❖ **Electronification:** The application of electronic circuitry to a mechanical device. Regarding automatic transmissions, electronification is incorporated into converter clutch lockup, shift scheduling, and line pressure control systems.

ELECTRICITY IS ELECTRONICS

Electricity normally deals with circuit loads using relatively high **voltage** and **current** flow. In the automobile, a full twelve volts (12 V) is used to push the current flow through circuits that include interior and exterior lighting, accessory drive motors, starter motor, and numerous other circuits. Excluding the starter motor, the circuits typically draw five to twenty amperes (5–20 amp).

- ❖ **Voltage:** The electrical pressure that causes current to flow. Voltage is measured in volts (V).
- ❖ **Current:** The flow (or rate) of electrons moving through a circuit. Current is measured in amperes (amp).

Electronics refers to miniaturized electrical circuits utilizing semiconductors, solid-state devices, and printed circuits. Electronic circuits utilize small amounts of power. They operate with low voltage (0 to 5-V or 0 to 8-V range) and with low current flow, measured in milliamperes (0.001 A).

- **Electronics:** Miniaturized electrical circuits utilizing semiconductors, solid-state devices, and printed circuits. Electronic circuits utilize small amounts of power.

The automobile uses both electrical and electronic circuitry in various combinations. The electronics of a circuit handle information and decision-making processes. The 12-V circuitry typically is involved with high-power output device operation. The distinctions are usually ignored and grouped under the single term *electronics*.

ELECTRICAL UNITS

Voltage

Voltage is the electrical pressure that exerts an **electromotive force (EMF)** to cause free electrons or current to move through a conductor. It is the potential difference between the ends of a conductor and an electrical/electronic circuit. The current always flows from the high potential to the low potential.

- **EMF (Electromotive force):** The force or pressure (voltage) that causes current movement in an electrical circuit.

In automotive electricity, current flow is from the battery positive post (+) to the battery negative post (-). The battery and charging circuits provide the voltage power for all the electrical circuits. This relationship is illustrated in Figure 9–2. The volt is the measurement unit of electrical pressure. Voltage is measured in an electrical circuit with an instrument called a voltmeter.

Current

The flow of electrons through a conductor is called current. It is the direct result of an EMF (voltage). The rate of electron flow is measured in amperes or electrons per second, which is similar to measuring the flow of water through a pipe in gallons per minute. It takes approximately 6.3 billion billion electrons passing a given point per second to measure one ampere. Amperage is measured in an electrical circuit with an instrument called an ammeter.

Resistance

Resistance is the opposition to the current flow and therefore limits the current (amperes). The resistance to current flow can be compared to a large-diameter water pipe versus a small-diameter pipe (Figure 9–3). The internal resistance of the small diameter pipe reduces the water flow. All conductors offer some degree of resistance to current flow. Metals such as copper, aluminum, and silver are used as conductors because they offer low resistance. When replacing an originally installed wire, it is important to remember that the diameter and length of the wire affect the circuit resistance. A short, large-diameter wire offers less resistance than a long, small-diameter wire.

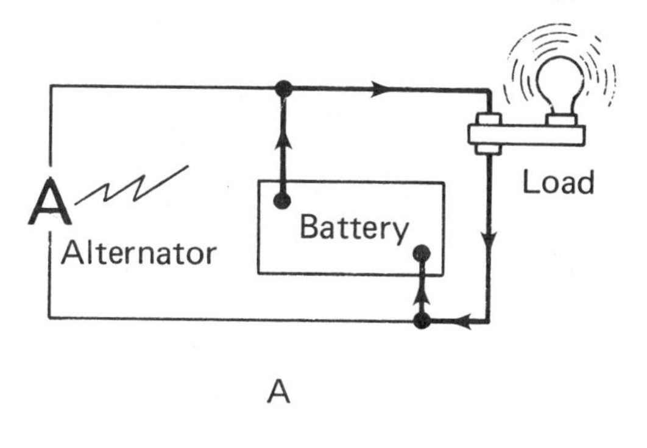

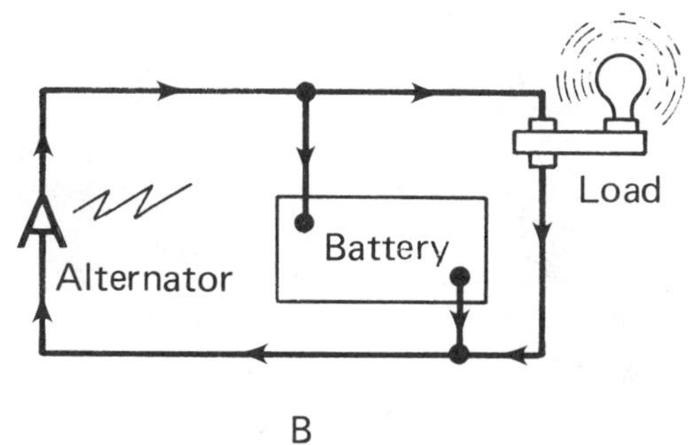

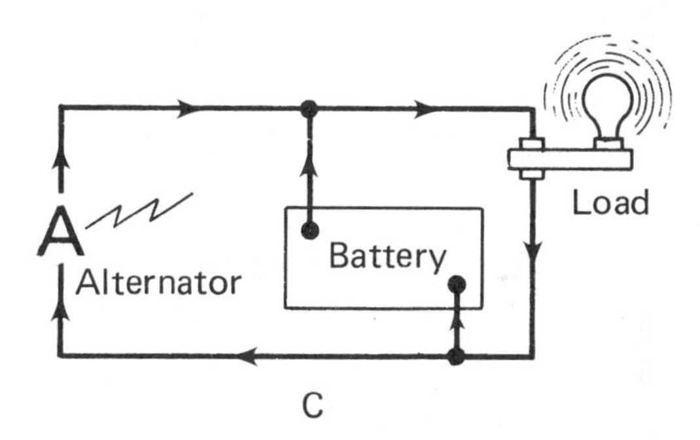

FIGURE 9–2 (a) Battery supplying load current; (b) Alternator supplying load current; (c) alternator and battery supplying load current.

- **Resistance:** The property of an electrical circuit that tends to prevent or reduce the flow of current. Resistance is measured in ohms.

A by-product of resistance is heat. The electron movement through the conductor is slowed with countless collisions occurring between the electrons. These collisions result in the formation of heat. Resistance always consumes voltage and causes a drop in or loss of voltage in a circuit. The unit of measurement for resistance is the ohm, measured with an instrument called an ohmmeter.

OHM'S LAW

The flow of current through an electrical circuit is determined by the applied voltage and effective resistance. The relationship between current, voltage, and resistance conforms to a definite behavior pattern known as **Ohm's law**. The law states that the

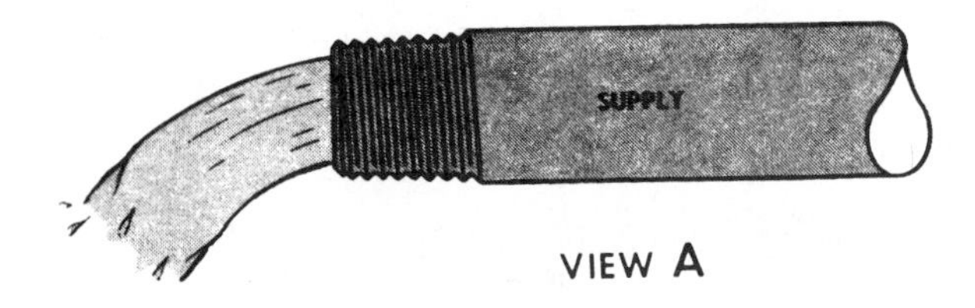

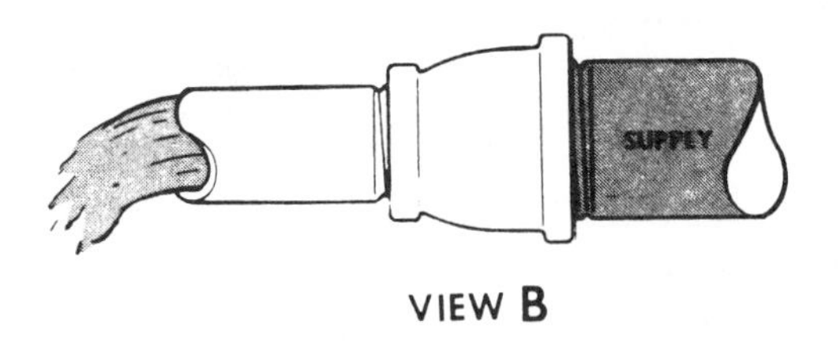

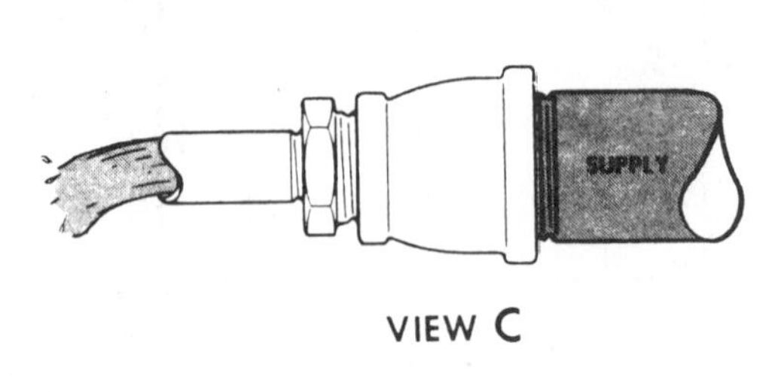

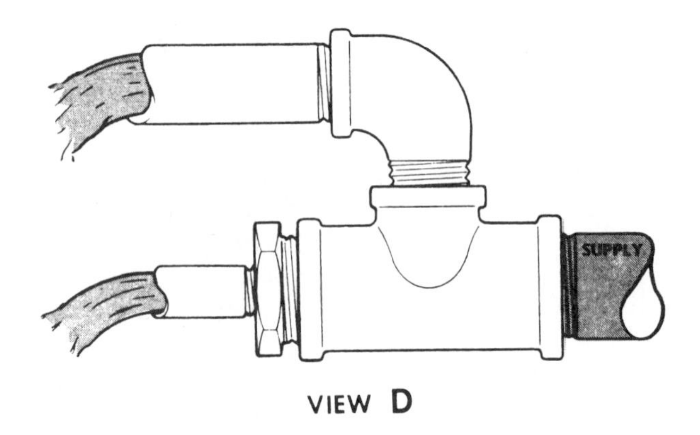

FIGURE 9–3 Pipe size affects water flow.

amount of current flow through an electrical circuit is determined by the amount of EMF working against a resistance value.

❖ **Ohm's law:** A law of electricity that states the relationship between voltage, current, and resistance.

$$\text{current} = \frac{\text{EMF}}{\text{resistance}}$$

In terms of electrical units, Ohm's law can be stated as:

$$\text{amperes} = \frac{\text{volts}}{\text{resistance}}$$

The usual formula symbols are:

$$I = \frac{E}{R}$$

It takes one volt applied to one ohm to produce a current flow of one ampere. Three forms of Ohm's law can be used to calculate an unknown value if the other two values are known. Formula variations are as follows:

$$I = \frac{E}{R} \qquad \text{amperes} = \frac{\text{volts}}{\text{ohms}}$$

$$E = I \times R \qquad \text{volts} = \text{amperes} \times \text{ohms}$$

$$R = \frac{E}{I} \qquad \text{ohms} = \frac{\text{volts}}{\text{amperes}}$$

There is no attempt here to solve any Ohm's law problems. The technician with some basic math background can find additional sources that are readily available which deal with solving Ohm's law problems. These problem exercises are a definite asset in understanding how electricity behaves in various circuit flow situations, but technicians rarely need to work out Ohm's law problems. Rather, they utilize the relationships between the various factors to help determine the cause of an electrical problem. The use of meters to measure current, voltage, and resistance is an effective means to observe the action taking place in a circuit. Comparing these readings to normal conditions helps in isolating a circuit problem or verifying that a circuit is normal.

Current flow in a circuit is directly proportional to the voltage and inversely proportional to the resistance. Chart 9–1 (page 181) displays the relationships between changes in voltage, resistance, and amperage.

CREATING ELECTRICITY

Electricity does not just happen. Something has to excite the electrons to get them moving and keep them moving. An EMF (electromotive force) is required to create an electric current. Electromotive force can be thought of as the "electron moving force." It is the pressure used in a circuit to get an electric

Ohm's Law Relationships.

Voltage	Resistance	Amperage
Up	Down	Up
Up	Same	Up
Same	Down	Up
Same	Same	Same
Same	Up	Down
Down	Same	Down
Down	Up	Down

- As resistance decreases, current flow increases.
- As resistance increases, current flow decreases.
- As voltage decreases, current flow decreases.
- As voltage increases, current flow increases.

CHART 9–1 Ohm's law relationships.

current flowing. It does this by creating a difference of potential between the front and back ends of a circuit. The front end is stacked with an excess of electrons that move to the back-end area, which has a void in electrons. In other words, the EMF moves the electrons or current from high potential to low potential.

Electromotive force can be generated through a variety of phenomena:

- Chemical
- Magnetism
- Friction
- Heat
- Light
- Pressure (piezoelectric)

MAGNETISM AND ELECTRICITY

Electricity and magnetism share a close relationship. Electricity can produce magnetism, and magnetism can produce electricity. Like electricity, magnetism is an invisible force. If it were not for magnetism, electrical energy would be hard to come by and would have very limited use.

In automobile applications, magnetism is essential to the operation of the ignition, charging, and starting systems, and the various electrical circuits that use solenoids, relays, buzzers, clutches, speed sensors, and small motors. Magnetic forces, invisible lines of force that can be useful in generating electricity, also produce a hostile environment that can penetrate and influence the operation of electronic circuits and components. A background in **electromagnetism** is essential to the technician working with electronics.

◆ **Electromagnetism:** The effects surrounding the relationship between electricity and magnetism.

Permanent Magnets

Although magnetism is an invisible force, we can understand it by examining the effects it produces. The simplest type of magnet is the bar magnet. The space affected by a magnet, called a field of force, has a designated north pole (N) and south pole (S) (Figure 9–4). A strong magnet produces many lines of force, while a weak one produces fewer lines of force.

The concentration of the magnetic lines of force is called the magnetic flux density (Figure 9–5). The stronger magnet has the greater flux density. Invisible lines of force leave the magnet at the north pole and enter again at the south pole. Inside the magnet, the lines of force travel from the south pole to the

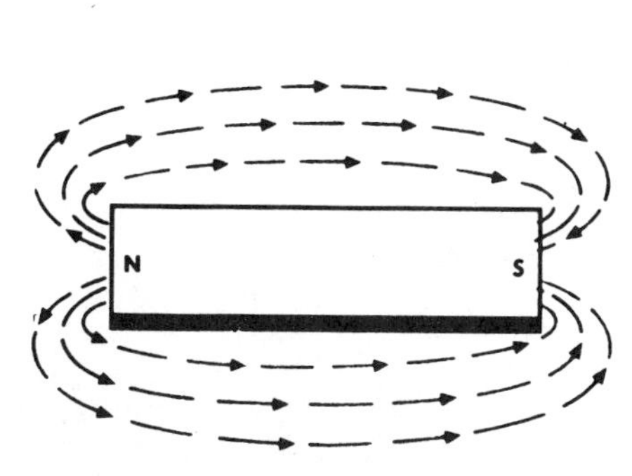

FIGURE 9–4 Lines of force = magnetic field.

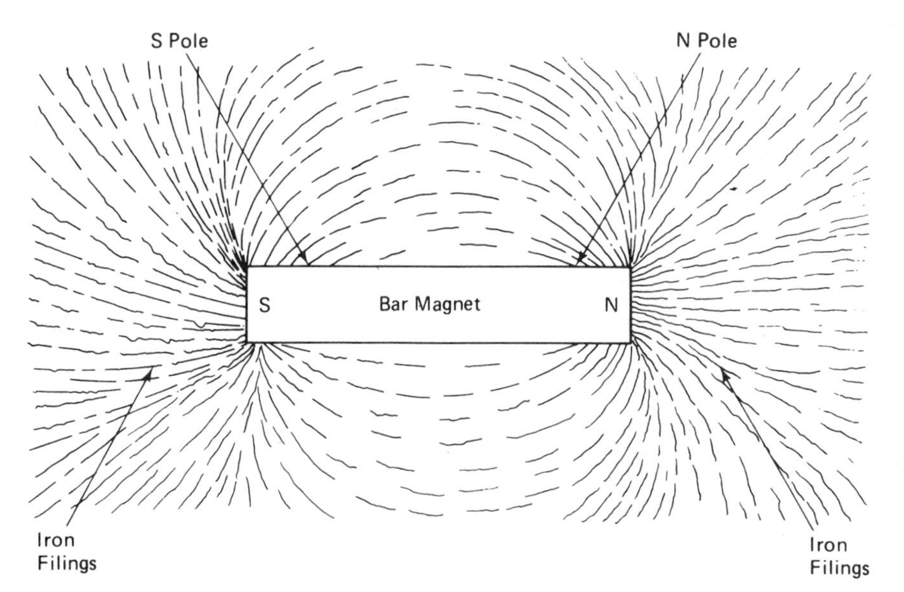

FIGURE 9–5 Iron filings illustrate the concentration of the invisible magnetic lines of force that surround a magnet. (Courtesy of General Motors Corporation, Service Technology Group)

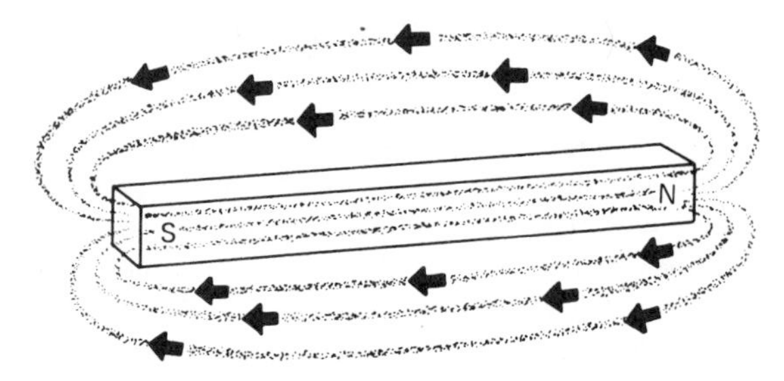

FIGURE 9–6 Magnetic lines of force.

north pole (Figure 9–6). Each line of magnetic force, therefore, is a continuous line. The magnetic field, or the field of force, is all the space outside the magnet that contains lines of magnetic force.

Permanent magnets are made of hardened steel alloys that can retain their magnetic properties for a long time. The original magnetism is induced by exposing the magnet to a strong electromagnetic field. The magnetic strength stored in the magnetic material after the magnetizing field has been removed is called **residual magnetism.**

❖ **Residual magnetism:** The magnetic strength stored in a material after a magnetizing field has been removed.

Soft iron and soft low-alloy steels retain only a small amount of magnetic strength. For this reason, they are commonly used in electromagnets where residual magnetism is undesirable.

Most materials are nonmagnetic and have no effect on a magnetic field. Materials such as iron, nickel, cobalt, and their alloy mixtures magnify or concentrate the magnetic field. These substances are called magnetic materials. Magnetic lines of force penetrate all substances. They are deflected or distorted only by other magnetic materials or by another magnetic field. There is no known insulator against magnetic lines of force.

Electromagnetism

One form of magnetism is electromagnetism. When an electrical current flows through a conductor, a magnetic field is set up around the conductor (Figure 9–7). These magnetic lines of force are concentric circles formed around the length of a straight conductor. They have no north or south poles (that is, no polarity).

The number of lines of force and the strength of the magnetic field produced increases in direct proportion to increased current flow. The lines of force (as expanding circles) are more dense at the surface of the conductor and increasingly less dense as the distance from the conductor increases (Figure 9–8).

Electromagnetic Fields

If a current-carrying conductor is formed into a loop, the lines of force around the conductor all pass through the center of the loop. This creates a weak electromagnet with a north and south pole (Figure 9–9). The magnetic lines leave the inside of the loop at the north pole, then flow around the outside of the loop and enter at the south pole. This produces the same field pattern as that of a bar magnet. If more loops are added to form a coil, the magnetic effect of the conductor is greatly increased because the magnetic lines of force become more concentrated (Figure 9–10). This field force, however, is still not strong enough to use in most electrical equipment. It is therefore necessary to add an iron core to an electromagnet, which greatly increases its magnetic strength (Figure 9–11). The iron core offers very little magnetic resistance (reluctance) compared to an air core.

The magnetic pull at the core of an electromagnet depends on the following:

- amount of current flowing in the coil
- number of turns in the coil
- size, length, and type of core material

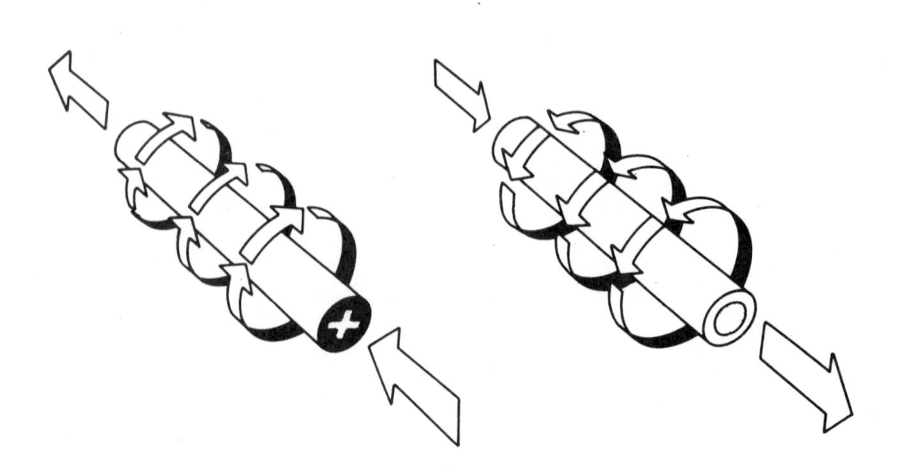

FIGURE 9–7 A magnetic field surrounds a current-carrying conductor.

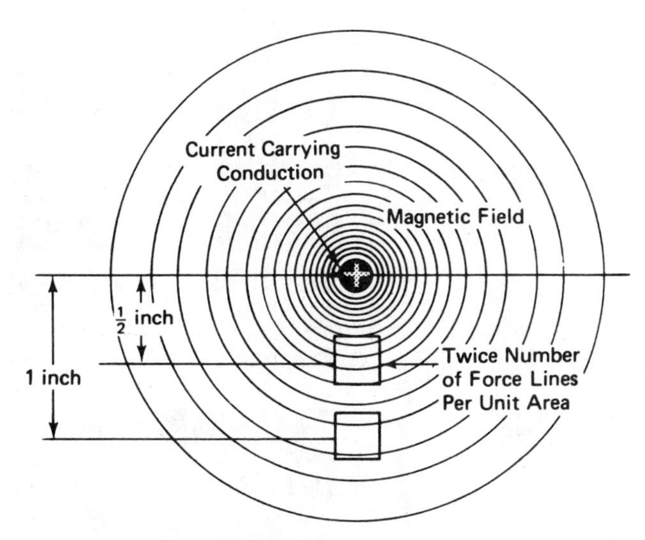

FIGURE 9–8 Cross-sectional view of magnetic field density encircling a current-carrying conductor. (Courtesy of General Motors Corporation, Service Technology Group)

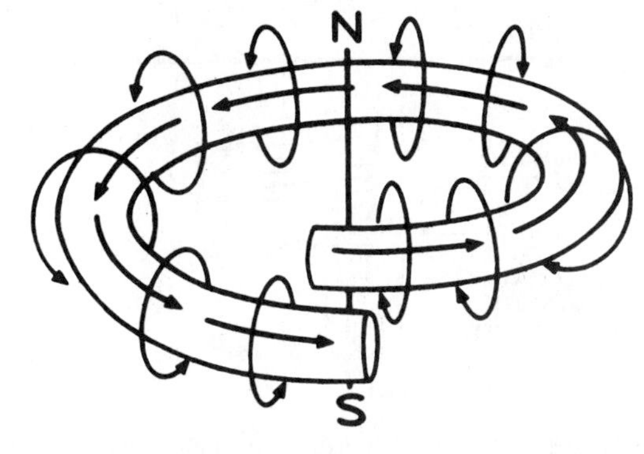

FIGURE 9–9 Single loop produces a weak electromagnetic field. (Courtesy of Chrysler Corp.)

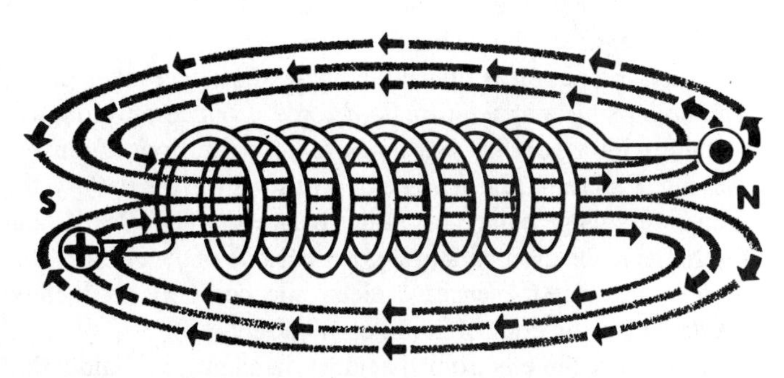

FIGURE 9–10 Additional loops create a stronger electromagnetic field. (Courtesy of General Motors Corporation, Service Technology Group)

Often the strength of an electromagnet is talked about in terms of ampere-turns:

$$\text{ampere-turns} = \text{amperes} \times \text{number of turns}$$

By using this formula, we can compare electromagnets. Shown in Figure 9–12, an electromagnet having 1000 turns of fine wire carrying 1 amp should have the same magnetizing force as an electromagnet having 100 turns of heavy wire carrying 10 amp.

Electromagnetic Induction

A magnetic field can be used to produce an EMF and make electrons move through a conductor. The generating of electricity by magnetism is called **electromagnetic induction**. To induce a voltage potential in a conductor, magnetic lines of force must cut across the conductor. Current then flows when an external circuit is completed.

◆ **Electromagnetic induction:** A method to create (generate) current flow through the use of magnetism.

There are three basic methods used to induce an EMF or voltage potential electromagnetically:

1. The conductor can be moved through the magnetic field (Figure 9–13). Direct-current (DC) generators use this principle by rotating a series of multiple conductors inside a stationary magnetic field.

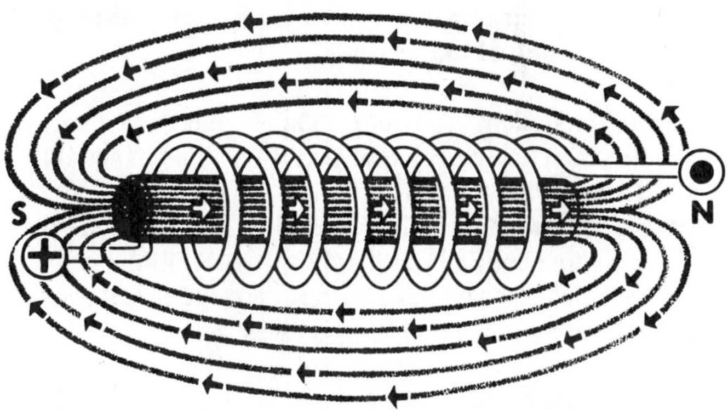

FIGURE 9–11 An iron core strengthens a magnetic field.

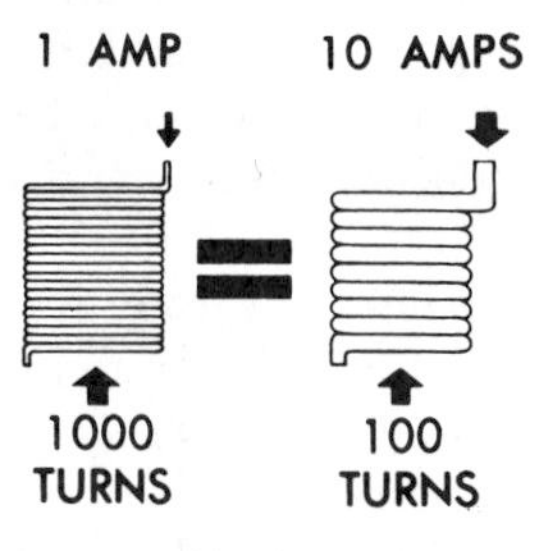

FIGURE 9–12 Comparison of the magnetic field strengths of two coils. (Courtesy of General Motors Corporation, Service Technology Group)

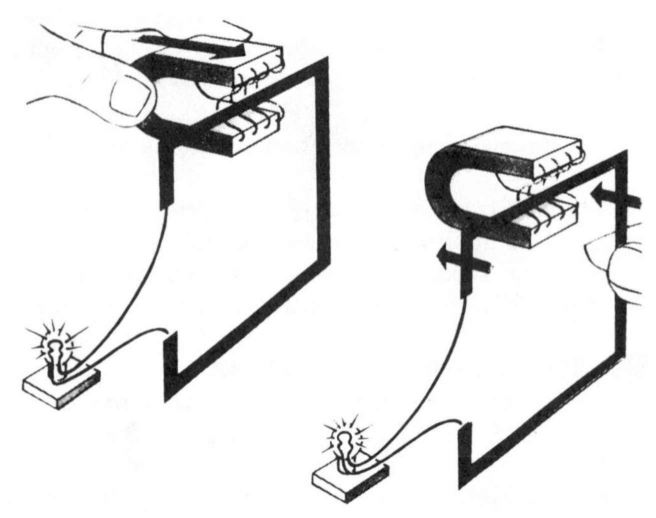

FIGURE 9–13 Electromagnetic induction. (Courtesy of Chrysler Corp.)

2. The magnetic field can be moving to cut lines across a stationary conductor (Figure 9–13). Alternators use this principle by rotating a magnetic field inside a stationary stator frame holding a series of multiple conductors. The direction in which electromagnetic induction moves the current flow in a closed circuit is determined by the movement of the conductor or magnetic field (Figure 9–14).
3. Induction can occur without a mechanical movement of either the magnetic field or the conductor. The magnetic field can simply be collapsed across a straight wire or a wire coil. This principle is used in explaining **self-induction** and **mutual induction**.

- **Self-induction:** The generation of voltage in a current-carrying wire by changing the amount of current flowing within that wire.
- **Mutual induction:** The generation of current from one wire circuit to another by movement of the magnetic field surrounding a current-carrying circuit as its ampere flow increases or decreases.

MOVING CONDUCTOR; STATIONARY FIELD

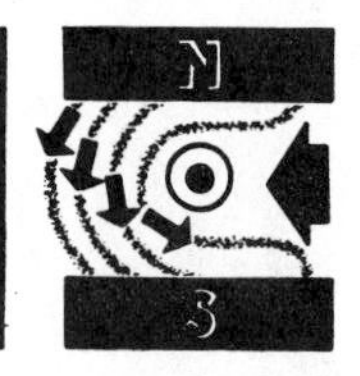

CURRENT IN CURRENT OUT

MOVING FIELD WITH STATIONARY CONDUCTOR

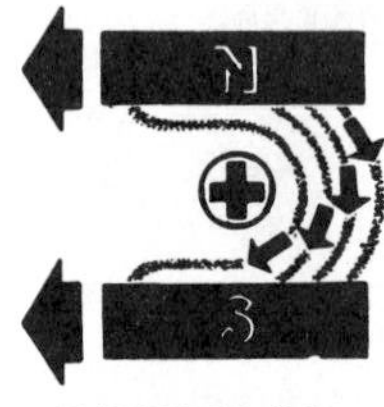

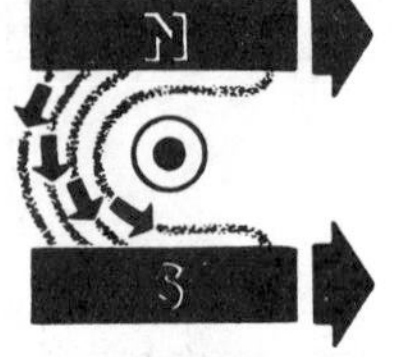

FIELD MOVING TO RIGHT REVERSES CURRENT

CURRENT IN CURRENT OUT

FIGURE 9–14 Direction of current flow affected by movement of a conductor and movement of a magnetic field. (Courtesy of General Motors Corporation, Service Technology Group)

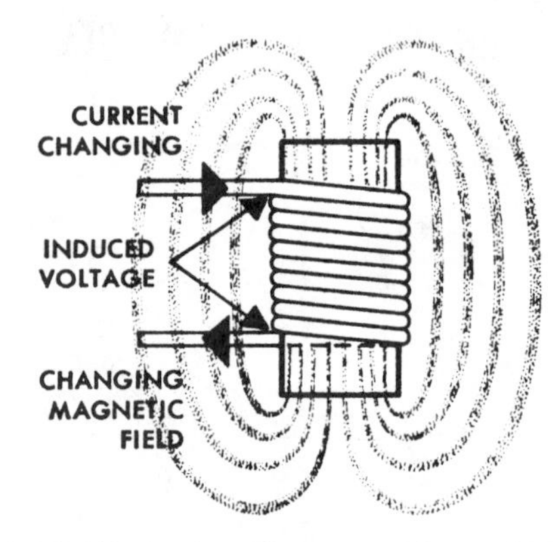

FIGURE 9–15 Self-induction. (Courtesy of General Motors Corporation, Service Technology Group)

Self-induction

Self-induction is a phenomenon that occurs in a carrying conductor. This generates a magnet field of concentric circles around the wire. When the current within a conductor increases or decreases, the magnetic field cuts across the conductor and induces a voltage within the conductor itself. As will be explained later, self-induction can be very disruptive to electronic systems.

The self-induction effect is greatly amplified in a wire coil, especially when wound around an iron core (Figure 9–15). The self-induction that occurs from the coil windings produces damaging voltage spikes.

When an electrical circuit designed with a solenoid (wire coil) is switched ON, the current increases from zero to the maximum design limits. The expanding magnetic lines of force produce an inductive voltage with a polarity that opposes the battery voltage and current (Figure 9–16). The inductive effect of the coil lasts about one-third of a second and then the circuit stabilizes. When the circuit is closed, the induced voltage remains near zero and is harmless.

When the circuit is opened, an opposite effect takes place. The collapsing magnetic lines of force produce an inductive voltage with a polarity that tends to keep the flow of current moving (Figure 9–17). With no opposition from the battery, the inductive voltage can reach values between 50 and 150 V and discharge to the ground or low-potential side of the circuit. This is very harmful to sensitive electronic components located inside electronic control modules (Figure 9–18).

To protect the electronics against damaging self-induction from solenoids, **isolation,** or **clamping, diodes** are popularly used. A diode is a semiconductor that permits flow in one direction only. It acts like a one-way check valve. The direction

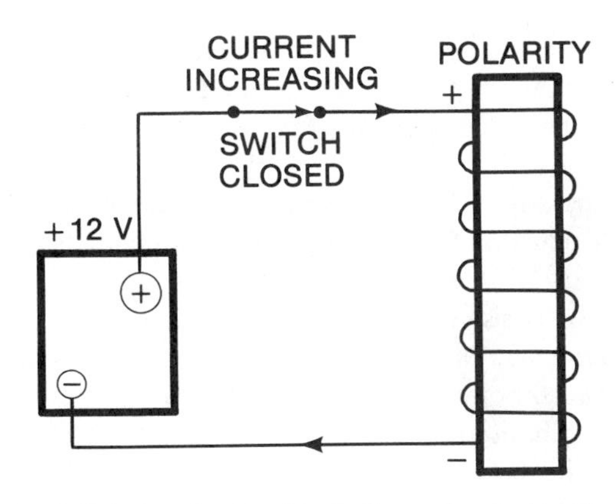

FIGURE 9–16 Polarity of solenoid winding—circuit closed.

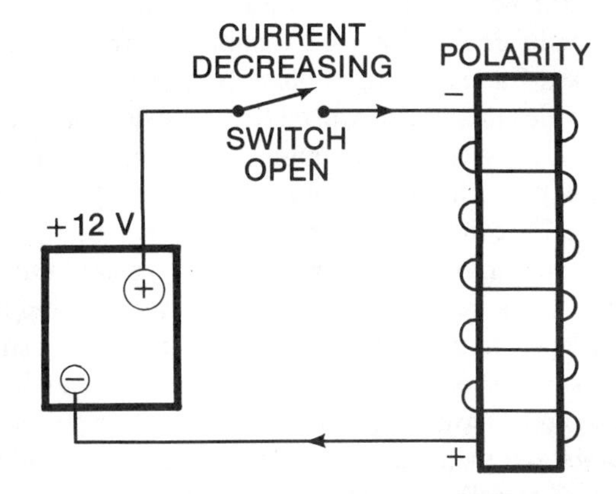

FIGURE 9–17 Polarity of solenoid winding—circuit opened.

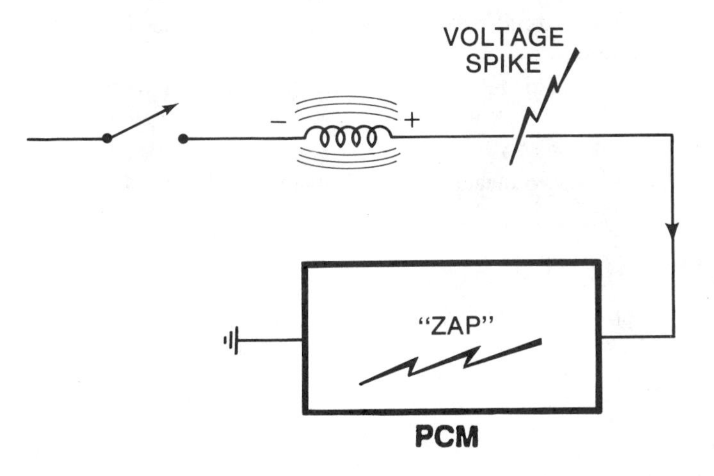

FIGURE 9–18 Voltage spike induced by circuit opening.

the diode allows the current to flow is referred to as forward bias, while the direction in which the diode blocks the current is reverse bias. A diode can be either positive or negative, depending on the forward bias (Figure 9–19).

❖ **Isolation/clamping diodes:** Diodes positioned in a circuit to prevent self-induction from damaging electronic components.

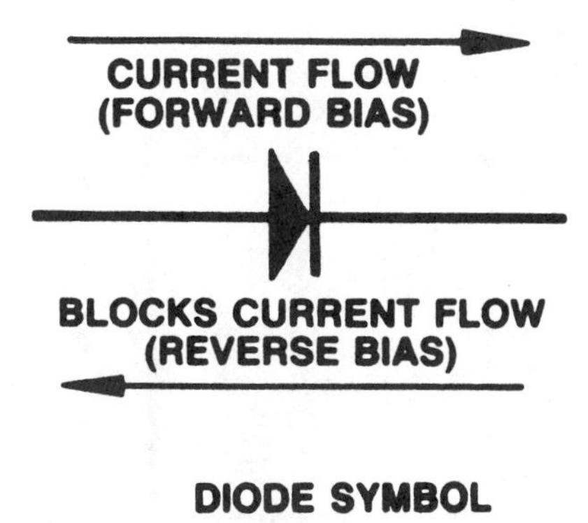

FIGURE 9–19 Forward bias/reverse bias.

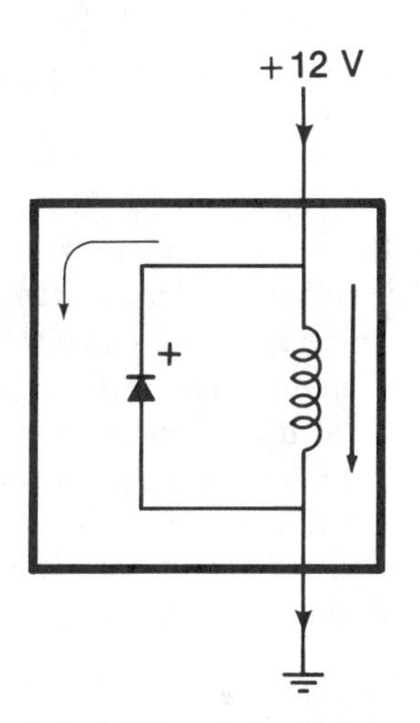

FIGURE 9–20 Reverse bias—effects to normal circuit flow.

The clamping or isolation diode is connected in parallel to the wire coil (Figure 9–20). The diode is usually clamped at the solenoid (wire coil) or sometimes located in the circuit harness connector. During normal circuit flow, the diode is reverse-biased and is not really needed. When the circuit is switched OFF, the diode becomes forward-biased and allows the voltage spike to dissipate itself at the source in preference to damaging sensitive circuit electronics (Figure 9–21).

Mutual Induction

Like self-induction, mutual induction produces an EMF without any moving parts. Mutual induction can be intentionally caused by placing two stationary windings, referred to as the primary and secondary windings, around a common iron core. The primary winding is connected to a source voltage (12 V) and carries the flow that excites a magnetic field. This links or couples both windings together magnetically. An EMF (voltage potential) is induced in the secondary winding anytime the primary current flow changes (Figure 9–22). This is the basis of ignition coil and power transformer operation.

Mutual induction can take place between two conductors running in parallel. When a second conductor runs along side a current-carrying conductor, it is under the influence of a

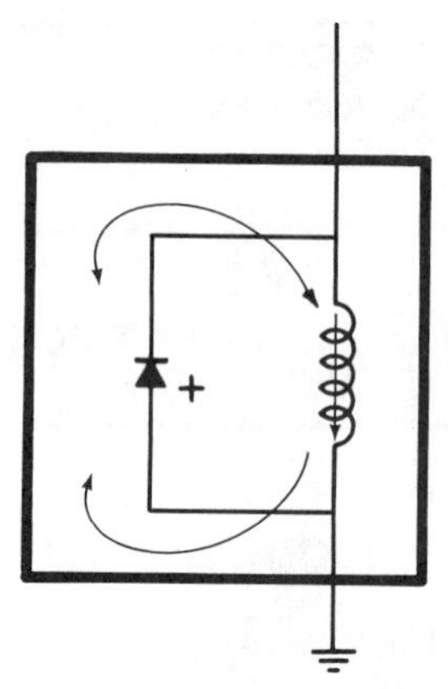

FIGURE 9–21 Forward bias—effects to self-induced current flow.

magnetic field from the primary conductor (Figure 9–23). Induction takes place in the second wire when the magnetic field is building or collapsing during changes in current flow.

Technicians need to be aware that electrical/electronic circuit wire routing is an important factor in proper circuit operation. Electromagnetic interferences can penetrate electronic circuits and components, causing numerous signal disturbances and system failures. Sensor wiring should never run in parallel next to a high-current-carrying wire, such as those near alternators and air-conditioner compressor clutches.

ELECTRICAL CIRCUITS

A circuit refers to a circle. An electrical circuit is a path that provides for the routing of current flow, starting at the voltage source and returning to the starting point. Current flow in a circuit is always from the high-potential (voltage) side to the low-potential (voltage) side.

Basic Circuit Components

A basic automotive electrical circuit is made up of the following circuit components (Figure 9–24):

- *Power source.* The battery and alternator must maintain a 12-V potential for all the electrical circuit loads to operate properly.
- *Switch.* The switch opens and closes the circuit.
- *Conductors.* These are the wires or metal components that connect the circuit parts together and carry the current flow to the circuit load and back to the battery. The wire going to the load is called the "feed" or "hot side" of the circuit. The return side of the circuit carrying the current flow back to the battery is called the ground circuit. Rather than using a wire for the circuit's ground side, automotive circuits use the metal part of the car for a common return to the battery: car body, frame, or engine block (Figure 9–25).
- *Load.* The circuit load is an electrical device that offers the major resistance in the circuit. It consumes most of the current flow and applied circuit voltage. A load device essentially converts electrical energy into light, heat, or mechanical movement. Examples of typical automotive load devices are light bulbs, solenoids, motors, and radios.
- *Circuit protection.* When an electrical circuit transmits an excessive current flow, the circuit overheats and can be damaged. Overheating problems can include melting conductors and their plastic insulation, possibly resulting in damage to other circuits/components or even starting a fire. To prevent circuit damage should the current flow exceed engineered design limits, protection devices such as fuses, circuit breakers, and fusible links are used to open the circuit.
- *Ground.* This returns the electricity back to the battery.

Current Flow

There are two means of describing current flow:

MUTUAL INDUCTANCE BETWEEN COILS

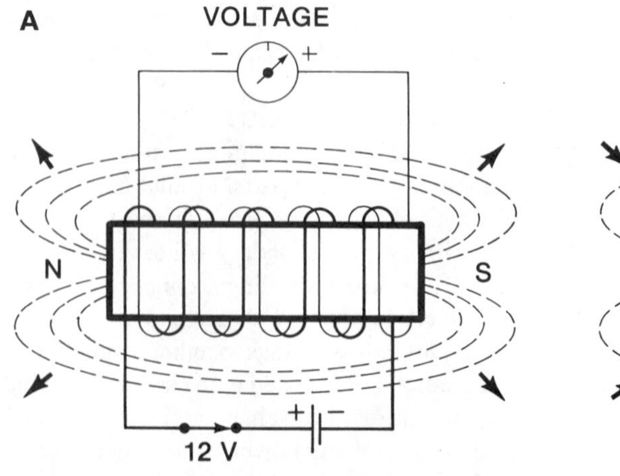

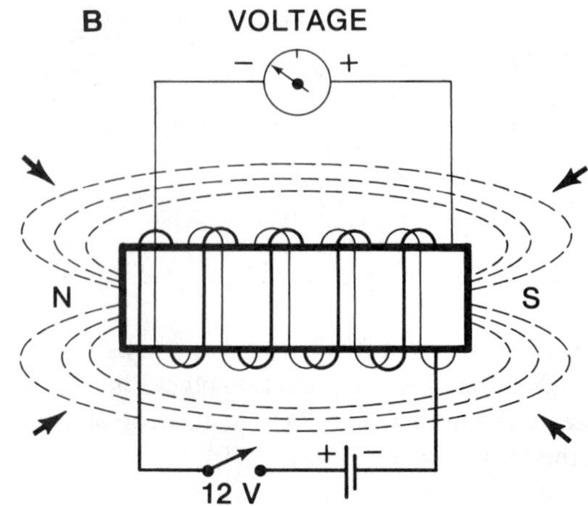

FIGURE 9–22 (a) Switch closed—field buildup; (b) switch open—field collapses.

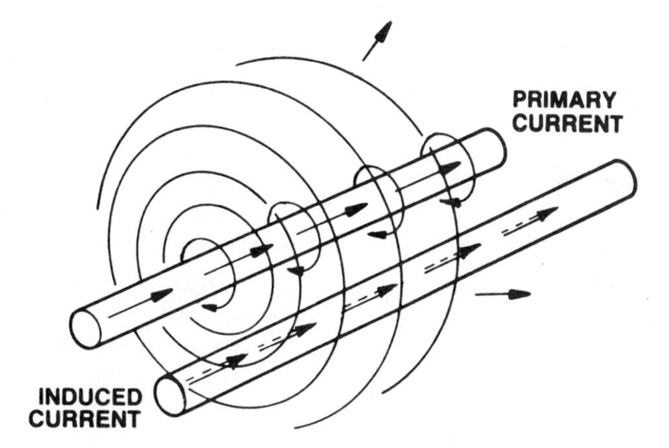

FIGURE 9–23 Mutual induction between parallel wires.

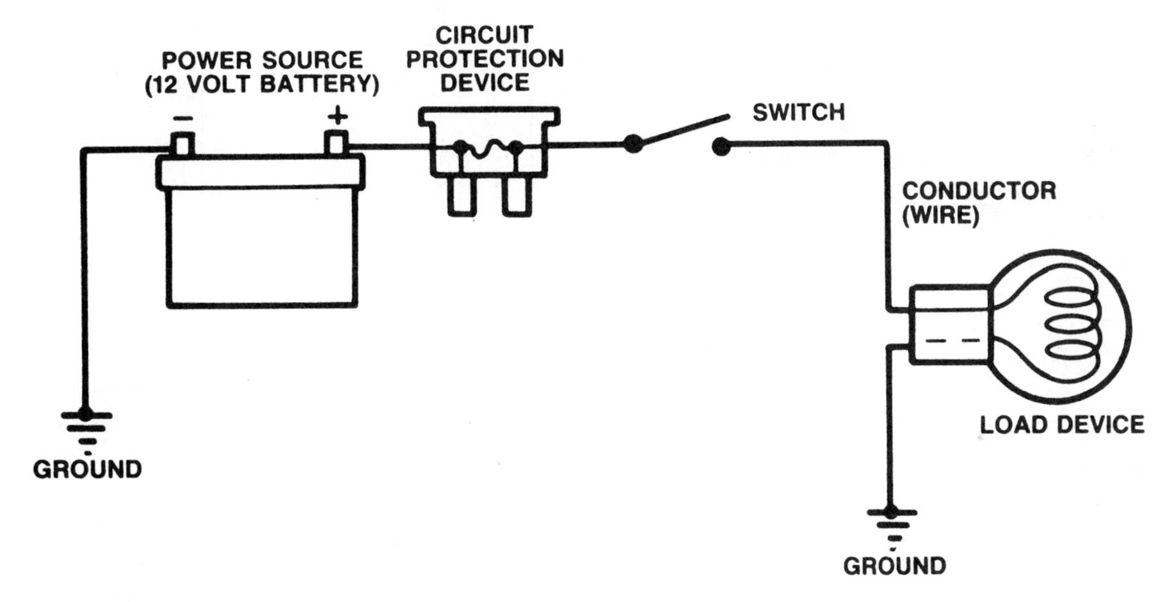

FIGURE 9–24 Typical circuit components. (Courtesy of Chrysler Corp.)

1. *Conventional current flow theory.* In this theory, the direction of flow is arbitrarily chosen to be from the positive terminal of the voltage source, through the external circuit, and then back to the negative terminal of the source.
2. *Electron flow theory.* This theory states that current flows from the negative terminal, through the external circuit, then back to the positive terminal of the source. This is based on the principle that electrons have a negative charge, and the surplus of electrons are at the negative terminal. Since the positive terminal lacks electrons, the current flow is from negative to positive.

Either theory can be used. The conventional theory is widely used in the domestic industry, while the import industry typically uses the electron flow theory.

Circuit Designs

Regardless of the theory used, automotive electrical circuits can be identified by the path the current follows. The three common circuit designs are series, parallel, and series-parallel. A technician needs to recognize the different designs and the effects the three designs have on voltage, amperage, and resistance.

Series Circuits

A series circuit provides only one continuous path for current flow (Figure 9–26). An opening in any part of the circuit

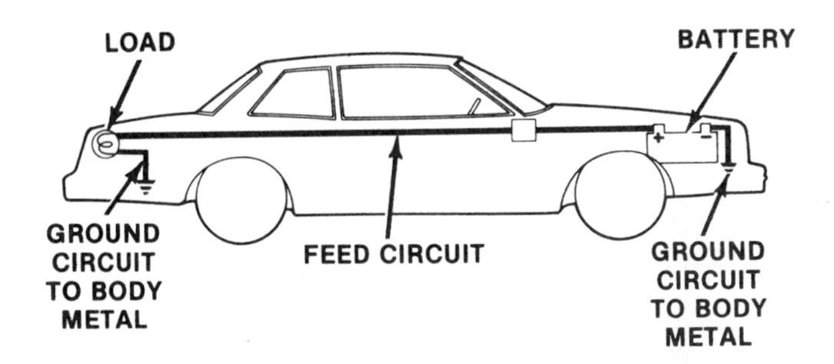

FIGURE 9–25 Basic automotive circuit. (Courtesy of General Motors Corporation, Service Technology Group)

stops the flow throughout the entire circuit. The circuit is dead, and the load devices stop working. A series circuit has three characteristics:

1. The current flow is the same everywhere in the circuit (Figure 9–27).
2. The source voltage divides itself among the individual circuit resistances. The sum of the individual voltage drops is equal to the source voltage (Figure 9–28). Voltage is always reduced by the circuit resistance. This reduction is called voltage drop. It is the difference in voltage between two points in the same circuit. Voltage drop meter readings are used in diagnosis to locate abnormally high resistance in the circuit.
3. The total effective circuit resistance is the sum of all the individual circuit resistances. Figure 9–29 illustrates a typical heater blower motor in series with a three-pole switch and resistor block to control motor speed.

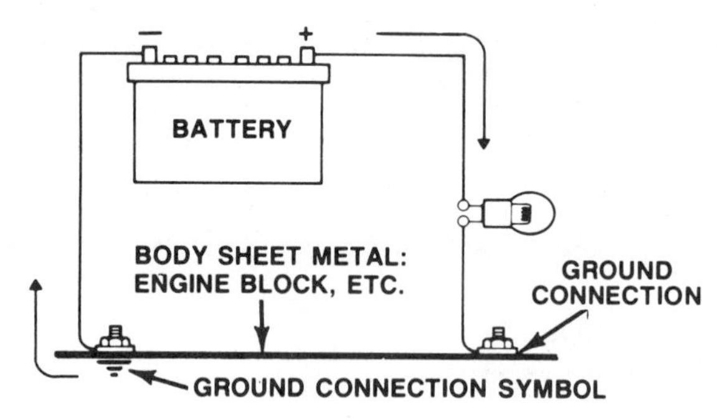

FIGURE 9–26 A series circuit provides one continuous path for current.

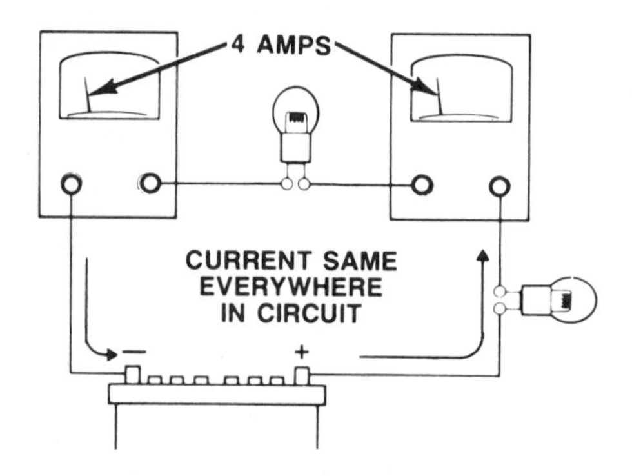

FIGURE 9–27 Current flow is the same throughout a series circuit. (Reprinted with the permission of Ford Motor Company)

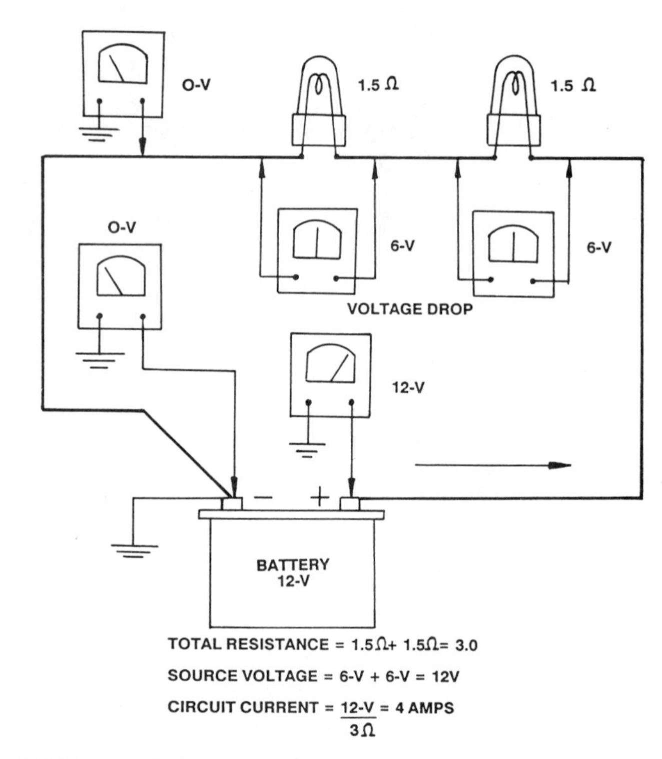

FIGURE 9–28 Voltage drop readings in a series circuit.

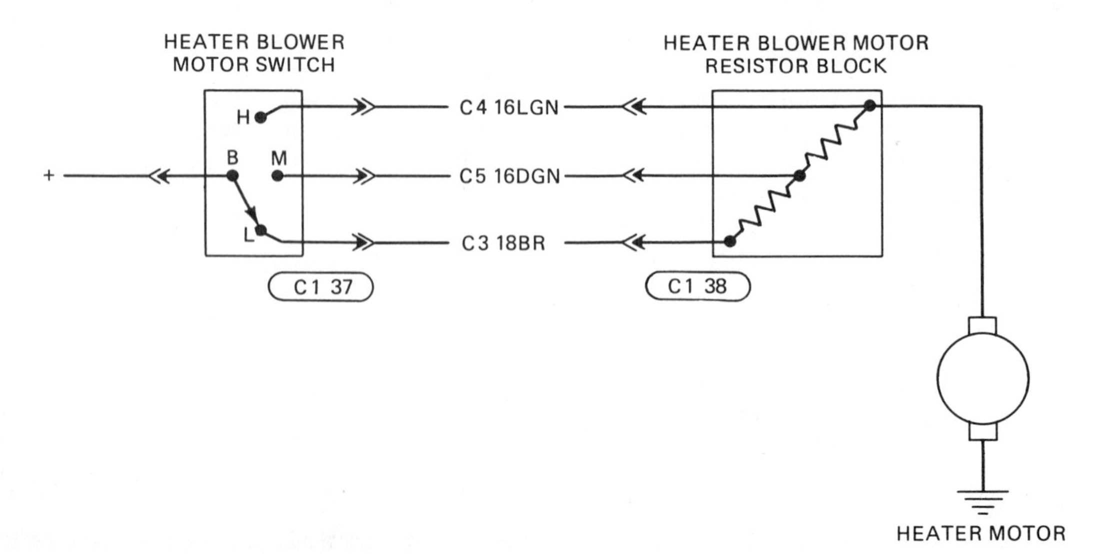

FIGURE 9–29 Resistors in series—blower motor circuit. (Courtesy of Chrysler Corp.)

Parallel Circuits

Normally, 12-V lamps are not connected in series because they would glow dimly and the failure of one would cause the other to lose current. Two 12-V lamps with the same candlepower would have equal voltage drops of 6 V. Without a full 12 V applied to each lamp, they would not glow with full brilliance. Therefore, a parallel circuit is needed.

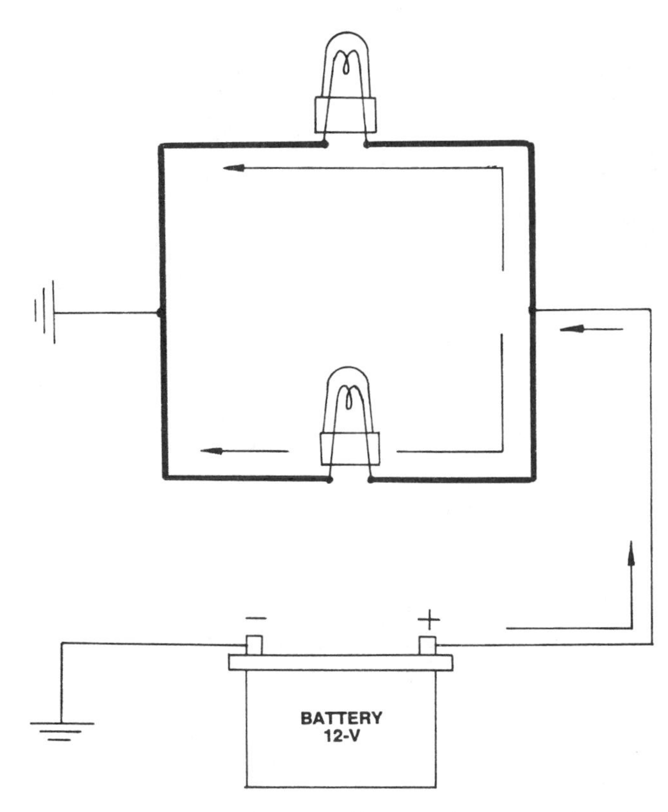

FIGURE 9–30 A parallel circuit provides more than one path for the current.

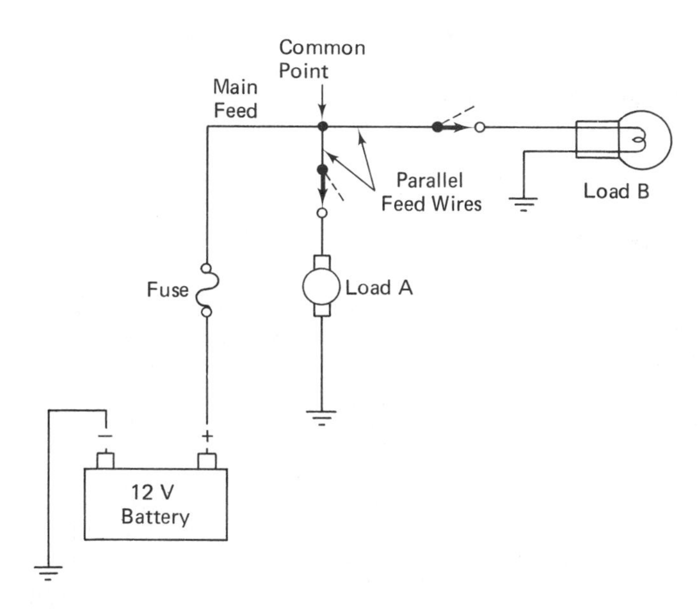

FIGURE 9–31 Parallel circuit with switches.

In a parallel circuit, current flows in two or more paths or branch circuits (Figure 9–30). If one circuit branch opens and interrupts its flow, all the other branches still continue to draw a current and operate their load devices. A parallel circuit always starts with a common point, where the hot wire splits to supply power to more than one circuit.

In Figure 9–31, the common point provides current to load A and load B. If load A switch is open and load B switch is closed, load B will operate, and vice versa. If the fuse blows open, neither A nor B will operate.

A parallel circuit has three characteristics:

1. The applied voltage is the same for all the parallel branches (Figure 9–32).
2. The main circuit flow divides itself among the several parallel branches according to the resistance value of each branch load. The sum of the separate currents equals the total circuit flow (Figure 9–33).
3. The total effective circuit resistance is always less than the parallel branch offering the smallest resistance (Figure 9–33).

Series-Parallel Circuits

Series-parallel circuits combine the characteristics of pure series circuits and pure parallel circuits. In most cases, series-parallel arrangements use a circuit switch in series with parallel branch loads (Figure 9–34).

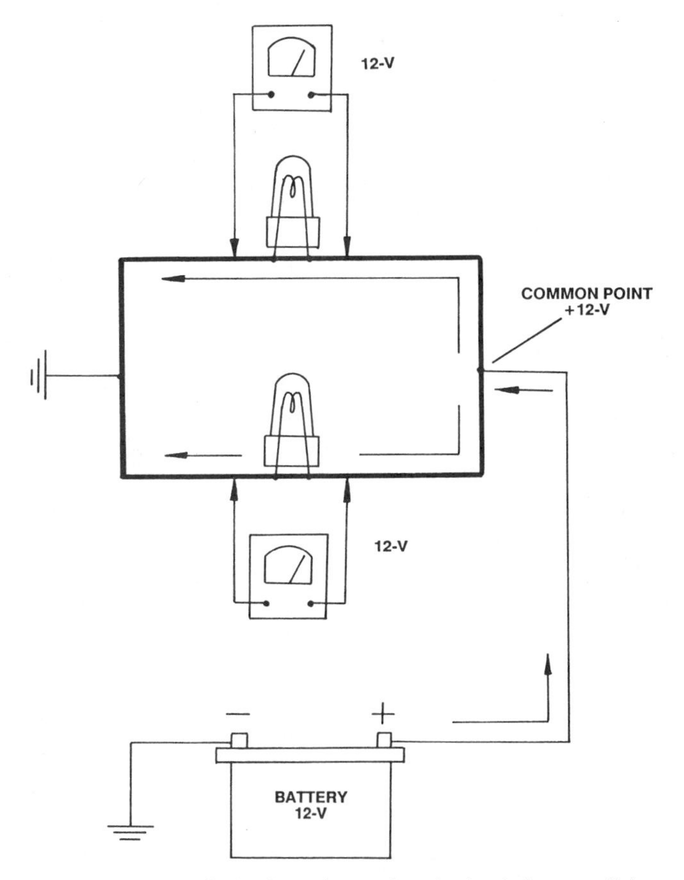

FIGURE 9–32 Applied voltage is equal at the loads in a parallel circuit.

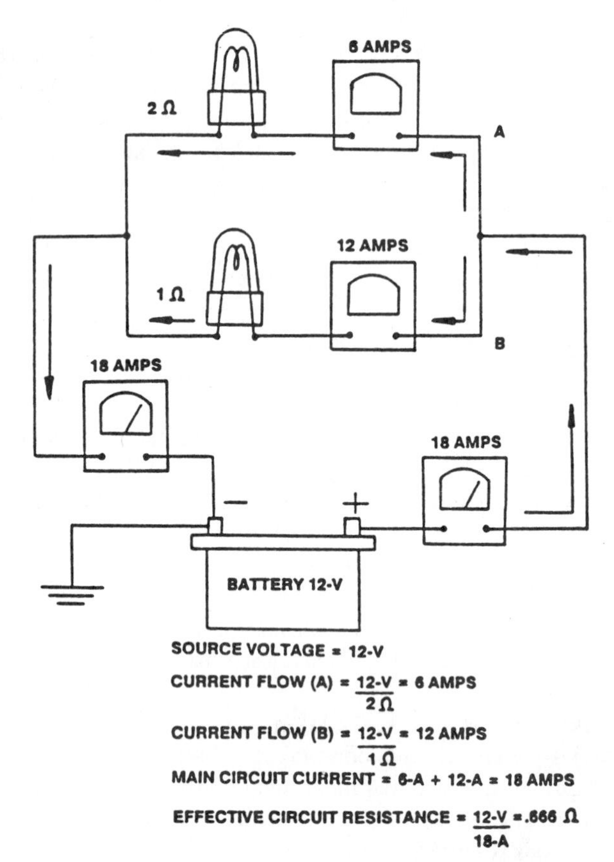

FIGURE 9–33 Current flow in a parallel circuit.

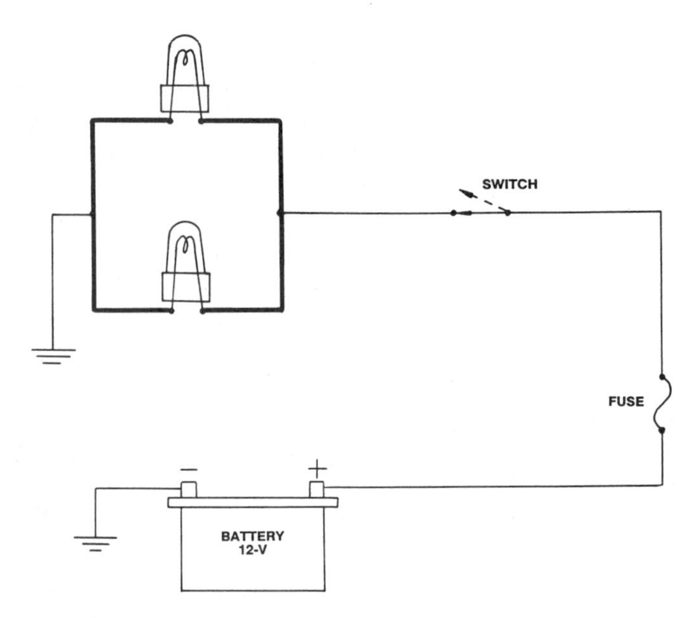

FIGURE 9–34 Series-parallel circuit; switch in series to parallel load.

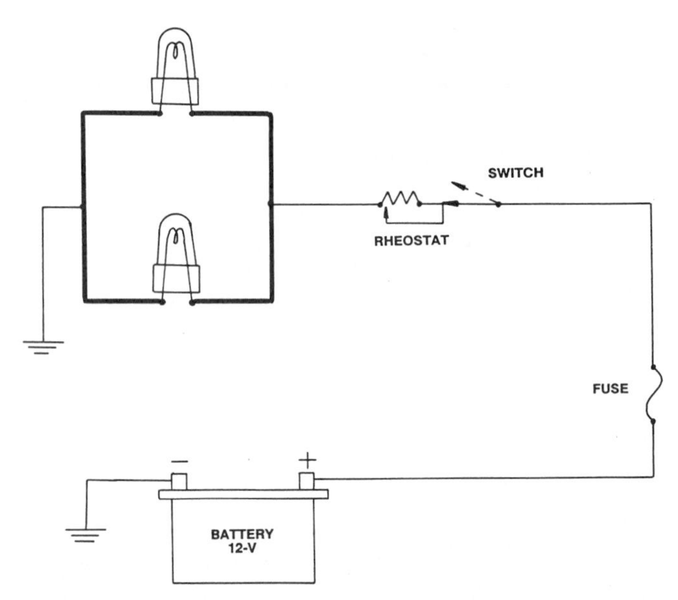

FIGURE 9–35 Series-parallel circuit with a rheostat in series with parallel lamp circuits.

Another version of a series-parallel circuit is when a rheostat (variable resistor) is put in series with the parallel lighting (Figure 9–35). The rheostat can vary its resistance and adjust the voltage drop from 0 to 12 V. In essence, the brilliance of the lights can be controlled. If the rheostat is turned for a 4-V drop, 8 V is left for the lights, and so on. This is similar to dimming the dash lights.

Wiring Diagrams

The modern technician has to deal with complex electrical circuitry. Wiring diagrams have become a necessary and valuable diagnostic tool, as well as an important tool in understanding the operation of an electrical circuit.

A wiring diagram is a symbolic map of all the electrical circuits on the vehicle. Circuit wires, switches, fuses, breakers, flashers, motors, solenoids, connectors, splices, and all other vehicle electrical components are represented on paper or a computer monitor. Circuit components, connectors, grounds, and splices are even given location on the vehicle. The gauge number and color codes of the circuit wires are shown. This makes it possible to trace circuit wiring through the wiring harness, multiple connectors, switches, and splices. Circuits can be traced from the front to the rear of the vehicle with no problem.

Working with wiring diagrams allows you to do your troubleshooting on paper first, and to think out a diagnostic strategy before going to the vehicle.

COMPUTER ELECTRONICS

Computer electronics control numerous circuits and devices on modern vehicles. The transmission technician deals with on-board computers and computer circuits on a regular basis to diagnose and repair powertrain problems relating to torque converter control circuits, shift timing, and shift quality. A basic understanding of how the computer system operates is advantageous for accurate, efficient powertrain servicing.

On-board automotive computers, often referred to as processors or control modules, are essentially decision makers. They receive and evaluate input signals from numerous switches and sensors and compare these signals to prepro-

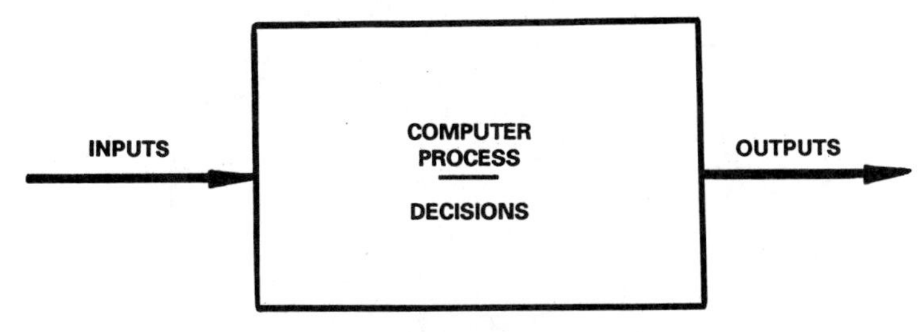

FIGURE 9–36 Simplified computer circuit concept.

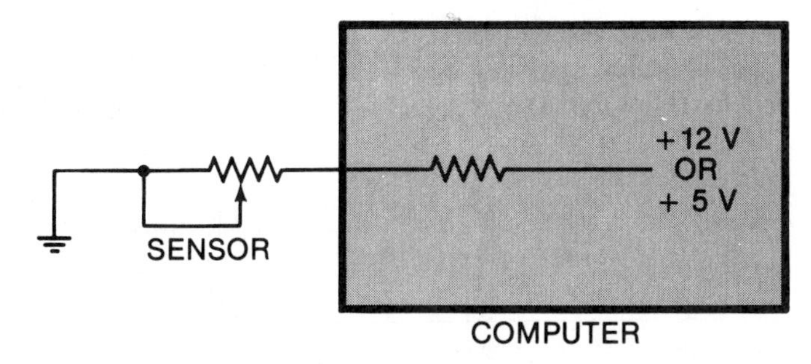

FIGURE 9–37 Ground-side sensor—variable resistor; operates as an input signal.

grammed instructions. Once the inputs are evaluated, the information is processed, and the decisions are made, the computer provides control over the output devices (Figure 9–36).

- The computer gathers data from various input sensors that monitor system concerns.
- As the data is scanned and digested, the computer makes an output decision based on instructions and information stored in its PROM (programmable read-only memory). The PROM is a miniaturized solid-state component, designed for the internal computer circuitry, that contains the instructions for the computer to make decisions.
- Computer decisions convert into electrical outputs that energize output devices or actuators such as solenoids, relays, and valves.

Computer Inputs

The inputs are the actual data that is relayed from sensors and switches to the computer. The computer uses voltage signals from the sensors and switches for its data. Most sensors are in the form of variable resistors that alter voltage signals from the computer (Figure 9–37). In some instances, the sensor actually produces a voltage signal. The O_2 sensor and vehicle speed sensor are examples in this category.

The computer has been programmed to know what voltage readings to expect from its sensors within a given parameter. An example can be made using the coolant temperature sensor. The sensor works similarly to those used in a conventional temperature gauge circuit. The computer in this case expects to see a voltage parameter of 0.5 V (hot engine) to 4.5 V (cold engine). Any voltage input reading within this parameter is accepted by the computer as normal. Voltage values or data above or below the parameter range are rejected by the computer and stored as a trouble code, and the "check engine" light comes on. Sensors of special concern to automatic transmission electronics include manifold absolute pressure (MAP), vehicle speed (VS), throttle position (TP), and engine coolant temperature (ECT).

Switch Sensors

A switch sensor is a simple device that tells the computer a certain mode of activity is taking place. An example illustrating mechanical activity includes the physical positioning of a park/neutral switch (Figure 9–38). The switch identifies to the computer when the transmission is or is not in a park or neutral range.

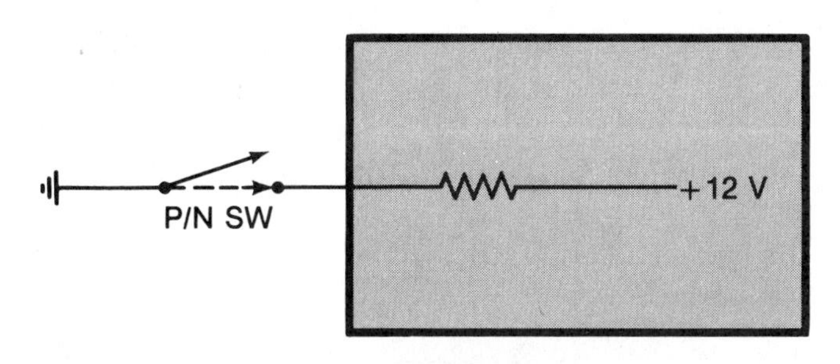

FIGURE 9–38 Input signal—park/neutral; ground-side switch.

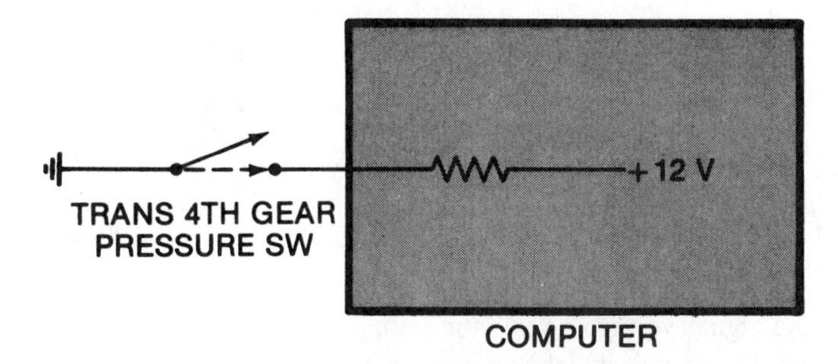

FIGURE 9–39 Input signal—fourth-gear pressure switch; ground-side switch.

A switch sensor can also be in a form to identify hydraulic activity. A fourth gear clutch or band pressure switch tells the computer when the transmission is or is not in fourth gear (Figure 9–39). When the switch is closed, the ground side of the circuit is completed.

Grounding pressure switches typically found in automatic transmissions are supplied a reference voltage of 5 or 12 V through a current-limiting resistor. When the pressure is below the switch setting, the switch is open and no current flows (Figure 9–40). The signal voltage equals the reference voltage.

When the oil pressure closes the switch, current flows and the resistor drops most of the reference voltage (Figure 9–41).

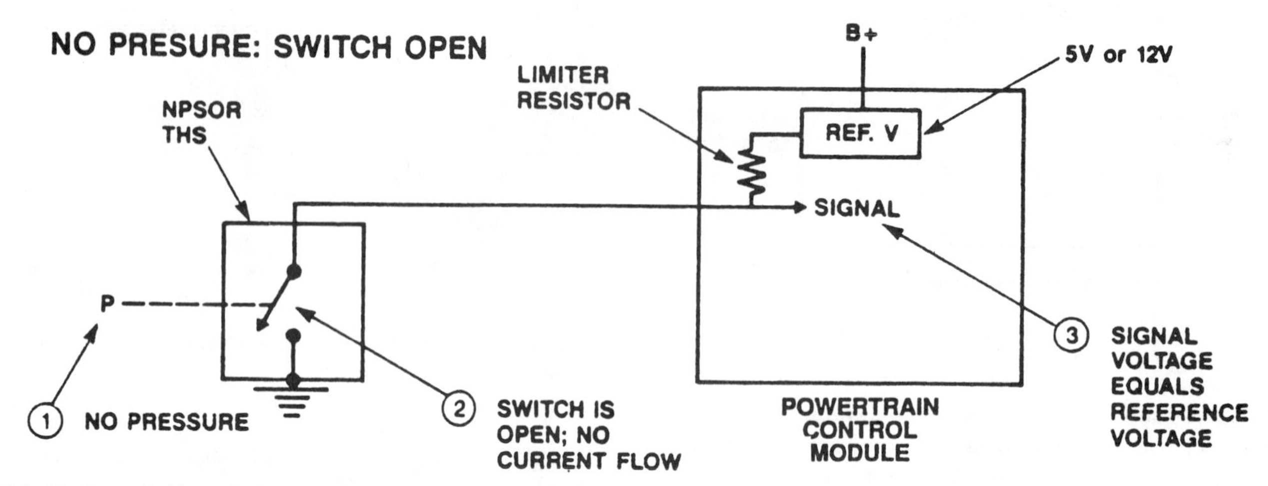

FIGURE 9–40 Ground-side switch operation—switch open. (Reprinted with the permission of Ford Motor Company)

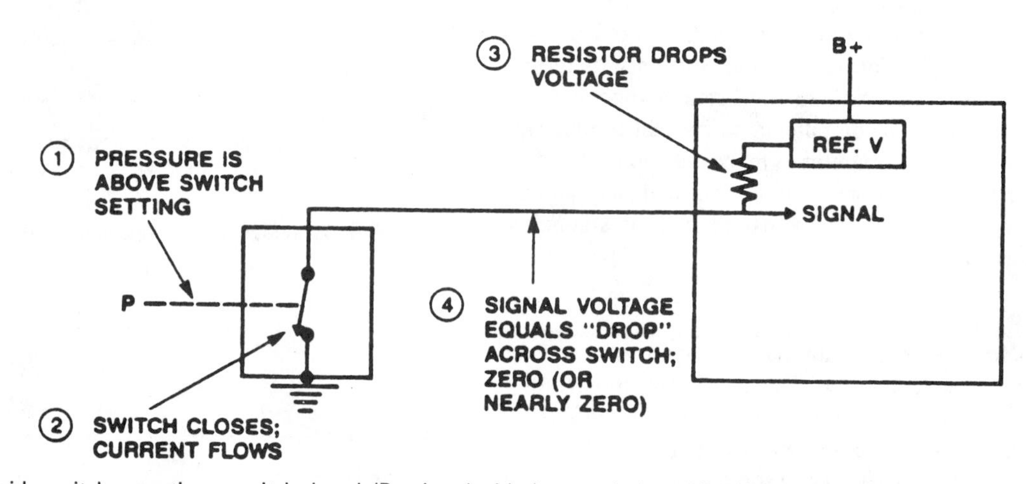

FIGURE 9–41 Ground-side switch operation—switch closed. (Reprinted with the permission of Ford Motor Company)

The signal voltage equals the voltage drop across the switch and the ground, which is normally zero.

Failed switch sensors and wiring leads can send incorrect signals to the computer. If the ground switch fails and remains open or the lead is open, the control module receives a constant high-voltage signal. If the switch fails and remains closed or the lead is grounded, the control module senses a constant ground signal. In either case, the control module follows its own internal instructions and, based upon the input data received, responds accordingly.

Computer Outputs

The output side of the computer controls the energizing of actuators, such as shift solenoids, torque converter clutch (TCC) solenoids, and electronic (oil) pressure controllers (EPC). The computer scans the input data and reads the various signals. When the input signals that provide data for specific output device operation are in the proper ranges, the computer activates the output device. The computer, through internal electronic circuitry, acts like a switch and turns the appropriate actuator circuit (output device) ON or OFF as needed.

There are two types of actuator circuits referred to as hot-side and ground-side switching. In hot-side switching, the actuator device is already grounded, with the computer supplying the 12 V (Figure 9–42). This type of actuator circuit is commonly used in import/foreign vehicles. In ground-side switching, the actuator is directly fed 12 V, and the grounding of the circuit occurs in the computer (Figure 9–43). This type is commonly used in domestic vehicles.

SUMMARY

Electronification in transmissions enhances vehicle performance, improves fuel economy, and reduces vehicle exhaust emissions. Electrical and electronic circuits appear quite complicated when not understood. To be able to more easily diagnose and repair these systems, the technician must realize that individual circuits operate according to the basic principles of electricity.

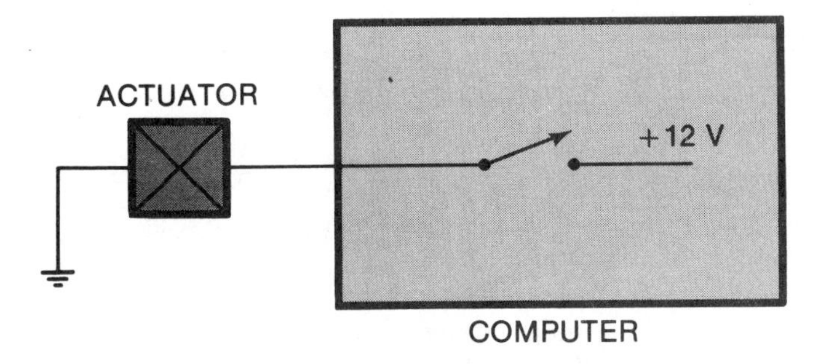

FIGURE 9–42 Hot-side circuit switching.

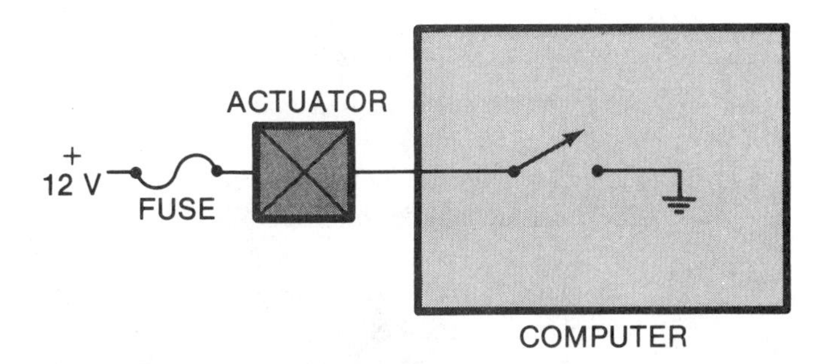

FIGURE 9–43 Ground-side circuit switching.

The basic operating principles of electricity are the building blocks of transmission electronification. The relationship between voltage, current, and resistance are shown in Ohm's law (volts = amperes × ohms). Two other important points to remember are that electricity can be created by magnetism and that magnetism is generated by electricity.

These concepts are used in various transmission electrical/electronic systems. Electromagnetism is a fundamental feature used to operate transmission solenoids. Because of self-induction of high-voltage spikes from a winding such as a solenoid, isolation (clamping) diodes are needed to prevent component damage. Mutual induction is another wiring concern that must be avoided.

Wiring circuits can be either series, parallel, or series-parallel in design. Schematics and wiring diagrams are extremely useful in better understanding and diagnosing electrical/electronic circuits.

Transmission electronics are controlled by the PCM/ECM or TCM, which makes decisions based on a number of computer inputs. The input signals can be resistance values, voltage readings, and those signals caused by open, closed, or grounded switches. The PCM/TCM processes this information and controls the output devices in the transmission. Most often these output devices are solenoids that act upon hydraulic circuits to control the torque converter, shifting of gears, and in some cases mainline pressure.

Electronification in transmissions continues to be an increasing factor in modern vehicles. Understanding the basic principles of electricity is fundamental to diagnosing and repairing today's transmissions.

❑ REVIEW

Key Terms/Acronym

Electronification, voltage, current, electronics, EMF (electromotive force), resistance, Ohm's law, electromagnetism, residual magnetism, electromagnetic induction, self-induction, mutual induction, and isolation/clamping diodes.

Questions

Electricity/Electronics

____ 1. What is true about the conventional current flow theory?
I. Flow is from positive to negative.
II. It is the usual flow theory used for both domestic and import vehicle circuitry.

A. I only C. both I and II
B. II only D. neither I nor II

____ 2. Which is true?
I. Electricity can produce magnetism.
II. Magnetism can produce electricity.

A. I only C. both I and II
B. II only D. neither I nor II

____ 3. Which of the following influences the magnetic pull of an electromagnet?

A. the amount of flowing in the coil.
B. the number of turns in the coil.
C. use of an iron core.
D. all of the above.

____ 4. Which statement is false?

A. Self-induction and mutual induction phenomena are the same.
B. Mutual induction requires no moving parts.
C. Self-induction requires no moving parts.
D. Mutual induction magnetically links together two parallel coils or two parallel conductors.

____ 5. An electromagnetic phenomenon that is disruptive to electronic systems is:
I. self-induction.
II. mutual induction.

A. I only C. both I and II
B. II only D. neither I nor II

____ 6. A purpose of a clamping or isolation diode is to:

A. maintain magnetism in a coil.
B. dampen voltage spikes.
C. delay magnetic buildup in a coil.
D. boost the magnetism in a coil.

____ 7. Technician I says, "The wire going to the circuit load is called the hot wire." Technician II says, "The return side of a circuit is called the ground side." Who is right?

A. I only
B. II only
C. both I and II
D. neither I nor II

____ 8. Electrical circuits are protected from excessive circuit flow by:

A. fuses.
B. fusible links.
C. circuit breakers.
D. all of the above.

PROBLEMS

SERIES CRCUIT PROBLEM

R_T = ____________ A_4 = ____________

A_1 = ____________ V_1 = ____________

A_2 = ____________ V_2 = ____________

A_3 = ____________ V_3 = ____________

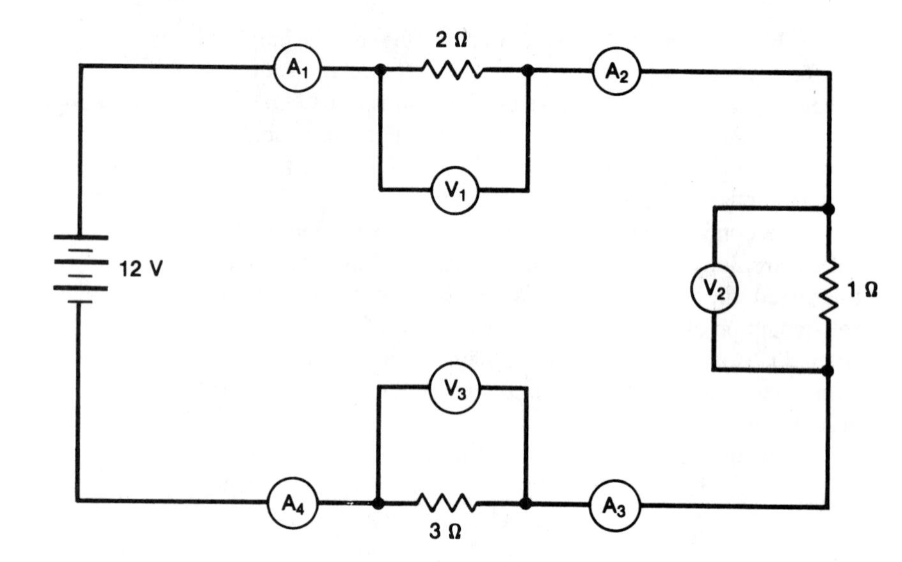

PARALLEL CIRCUIT PROBLEM

R_T = ____________ A_2 = ____________

A_1 = ____________ A_3 = ____________

A_4 = ____________ V_2 = ____________

V_1 = ____________

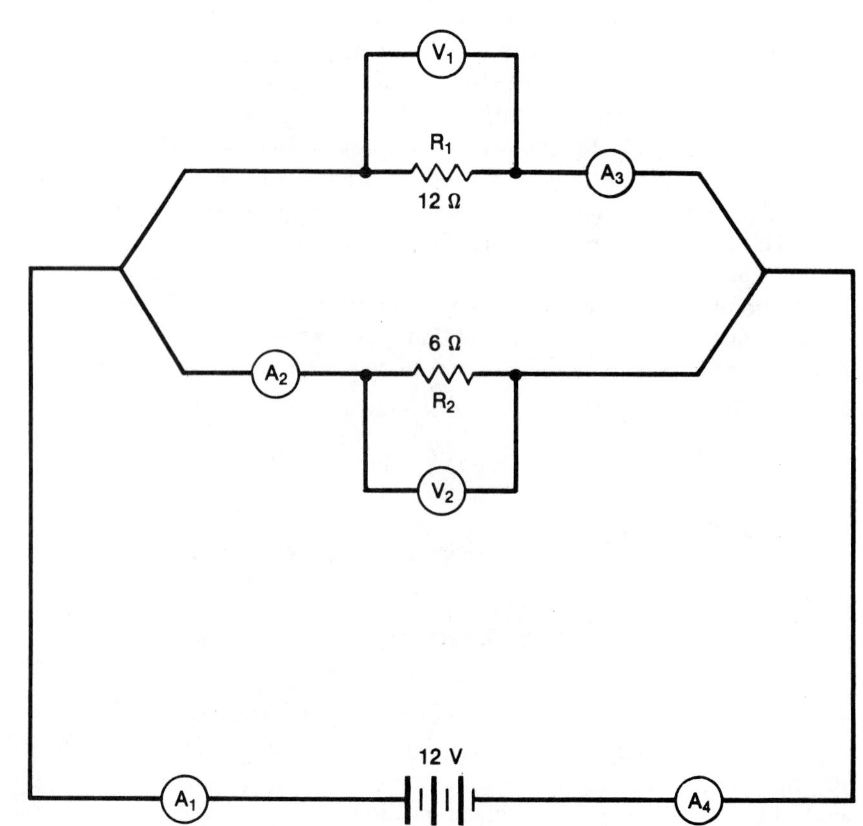

Definitions of Prefixes

Kilo (k) is a prefix meaning ______________________.

Mega (M) is a prefix meaning ______________________.

Milli (m) is a prefix meaning ______________________.

Micro (μ) is a prefix meaning ______________________.

8 MΩ (megohms) = ______________________ ohms.

20 kV (kilovolts) = ______________________ volts.

50 mV (millivolts) = ______________________ volts.

18 μA (micro amperes) = ______________________ amperes.

SERIES-PARALLEL CIRCUIT PROBLEM

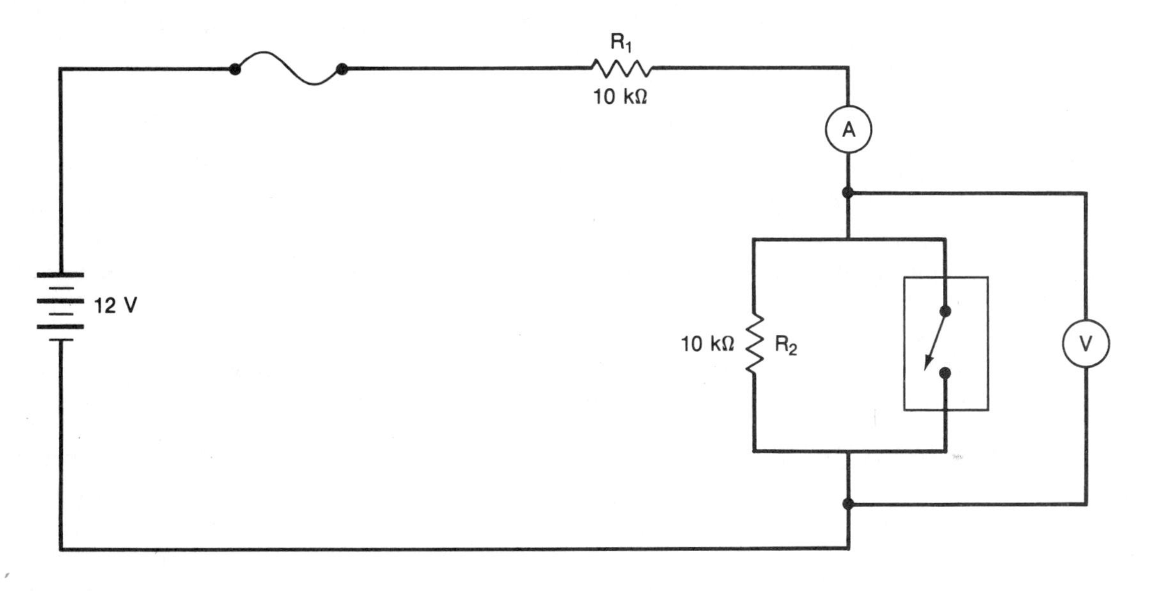

R_T (switch open) ____

A reading (switch open) ____

V reading (switch closed) ____

V reading (switch open) ____

R_T (swich closed) ____

A reading (switch closed) ____

Computer Electronics

1. Sensors furnish input data to the computer. ___T/___F
2. Name four sensors of special concern to automatic transmission electronics.

   (a) ______________________
   (b) ______________________
   (c) ______________________
   (d) ______________________

3. In ground-side switching, the actuator completes the circuit ground. ___T/___F

10 Converter Clutch Circuit Controls

OBJECTIVE:

The objective of this unit is for you to become proficient with the terms, acronyms, and concepts listed in this chapter. Completion of this objective is essential to becoming a successful automatic transmission technician.

CHALLENGE YOUR KNOWLEDGE

Define the following key terms/acronyms:

transmission control module (TCM), powertrain control module (PCM), modulated, pulse width duty cycle (pulse width modulated) solenoid, EEC-IV module, ECM (electronic control module), ECT (engine coolant temperature) sensor, TP (throttle position) sensor, MAP (manifold absolute pressure) sensor, VSS (vehicle speed sensor), TRS (transmission range selector) switch, PSA (pressure switch assembly), TFT (transmission fluid temperature) sensor, TIS (transmission input speed) sensor, BARO (barometric pressure) sensor, ALDL (assembly line diagnostic link), and DLC (data link connector).

List the following:

Nine inputs (switches and sensors) that control application and release of a converter clutch; and two purposes for modulated solenoid action.

Describe the following:

The solenoid action causing a relay valve to control/direct converter clutch apply oil; the function of a converter clutch regulator valve; and the operation of a pulse width modulation solenoid.

INTRODUCTION

In Chapter 2, various styles of converter lockup clutch designs were discussed. In principle, they all provide a direct mechanical drive—engine crankshaft to transmission input shaft. The normal fluid drive of the converter is bypassed. To accomplish this change in power flow, a hydraulic circuit containing a relay valve directs fluid to either the apply or release side of the converter clutch. Early versions of this oil circuit were controlled by hydraulic and vacuum-electric systems. Most late model converter clutch hydraulic circuits are managed by an electronic control module to produce a lockup condition that contributes to improved fuel economy and reduced emissions. In addition, electronic controls are able to operate the clutching and declutching of the apply plate smoothly, often unnoticeably to the driver.

There are many variations of the hydraulic circuitry and electronic control systems from year to year, car to car, transmission to transmission, and engine to engine. This chapter features typical circuits and controls used in the 1990s to illustrate the common operating principles of converter clutch controls.

CHRYSLER CORPORATION

Chrysler lockup circuit controls come under the two categories of hydraulic- or computer-operated. Pure hydraulic circuit control was used with the 1978 lockup introduction in the 36RH (A-727) and 30RH (A-904) transmissions. This style of lockup control continued through 1985 production year for RWD trucks and vans, and through 1986 production year for RWD passenger cars. For 1986 production, the 900 series lockup was changed to electronic control in the RWD trucks and vans, followed by the RWD car line in 1987. The 36RH (A-727) incorporates lockup for passenger car applications only. Chrysler FWD automatic transaxles initially used nonlockup converters until electronic lockup was introduced in 1987 for most FWD applications. All Chrysler automatic transmissions and transaxles now use electronic lockup.

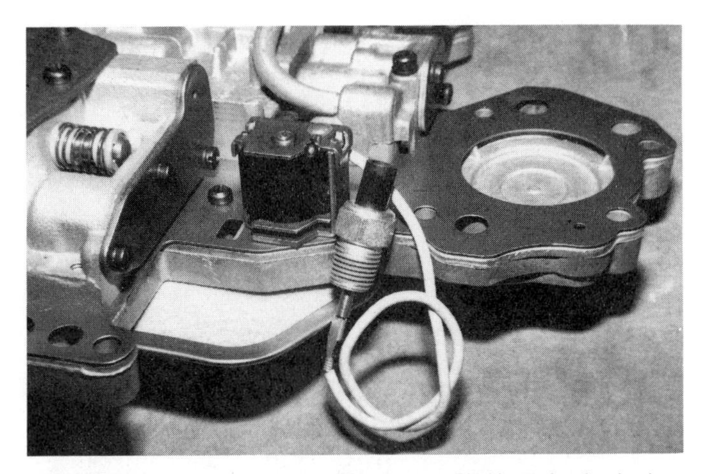

FIGURE 10–1 Lockup solenoid—Chrysler 30RH (A-904) valve body.

Electronic Lockup

The electronic lockup control uses an electric solenoid mounted on the valve body for all RWD and FWD applications (Figure 10–1), except for the 41TE (A-604) and the 42LE in which the electric solenoid is located in the solenoid assembly. The operation of the solenoid is controlled by the engine computer—either by the ignition spark control, SMEC, or **transmission control module (TCM)**. The computer controls the lock and unlock states of the converter based on the following sensor inputs. Lockup occurs when:

- Coolant temperature is above 150°F (66°C).
- Engine load: manifold vacuum is above 4 in and below 19 in.
- Vehicle speed is above 40 mph (64 km/h).
- Throttle is off idle.

❖ **Transmission control module (TCM):** Manages transmission functions. These vary according to the manufacturer's product design but may include converter clutch operation, electronic shift scheduling, and mainline pressure.

Basically, the electronic transmission lockup control is interfaced with the existing engine computer program and uses inputs from already existing sensors. This allows for more precise engagement cut-in speed of the lockup. The computer control also provides for diagnostic fault code memory.

Electronic RWD Lockup

The RWD electronically controlled lockup torque converter is similar in design and operation to the early hydraulically controlled units. The lockup valve body is retained with only a minor change in the hydraulic circuitry (Figure 10–2). The governor pressure is eliminated from the end of the lockup valve and replaced with mainline oil from the 1-2 shift valve circuit. Note that the 1-2 oil to the lockup valve is orificed and keyed to the solenoid. The solenoid is normally deenergized and vents the 1-2 shift oil to the lockup valve. When the solenoid is energized, it closes the vent, and mainline pressure shifts the valve for lockup. The unlock and lockup solenoid hydraulic modes are illustrated in Figures 10–3 and 10–4.

In Figure 10–5, the electronic circuitry is shown with the hydraulic circuitry. The distance (speed) sensor is a magnetic pulse style placed in series with the speedometer cable at the transmission couple or in the cable line (Figure 10–6). In the 42RH (A-500), the SMEC controls the ground circuit to activate or deactivate the lockup solenoid. When the criteria for lockup has been met, the solenoid is energized for converter lockup. Lockup occurs in fourth gear only. The lockup hydraulics use a three-valve control group similar to the A-900 series transmissions. A lockup timing valve replaces the fail-safe valve and ensures that the torque converter is unlocked before the 4-3 downshift takes place.

Electronic FWD Lockup

In three-speed FWD applications, the lockup solenoid is controlled by engine electronics. Except for 1987 models, the SMEC controls the ground circuit to activate or deactivate the solenoid (Figure 10–7). When the criteria for lockup have been met, the solenoid is energized for converter lockup. Lockup occurs in third gear only. The lockup hydraulics is achieved with the use of a single valve. Converter clutch circuits designed with switch valves have eliminated the need for lockup and fail-safe valves (Figure 10–8).

The 41TE (A-604) and 42LE four-speed electronic shift transaxles use similar lockup systems. Unlike three-speed transaxles, their control is handled by a transmission control module (TCM) interfaced with a **powertrain control module (PCM).**

❖ **Powertrain control module (PCM):** Current designation for the engine control module (ECM). In many cases, late model vehicle PCMs manage the engine as well as the transmission. In other settings, the PCM controls the engine and is interfaced with a TCM to control transmission functions.

The lockup solenoid is grounded in the computer circuitry, and the hydraulics has only one valve, a switch valve. The lockup operation, however, is more sophisticated. A partial lockup is possible in second and third gears, and full lockup in both third and fourth gears. The decision for partial or full lockup depends on these input conditions: shift lever position, current gear range, engine coolant temperature, input speed, and throttle angle.

For full lockup, the electronically **modulated** converter clutch (EMCC) apply is delayed until the turbine speed is within 60 to 100 rpm of the crankshaft speed. Then full lockup occurs. Normally, lockup is not felt by the driver. Release, dependent upon driving conditions, can be accomplished

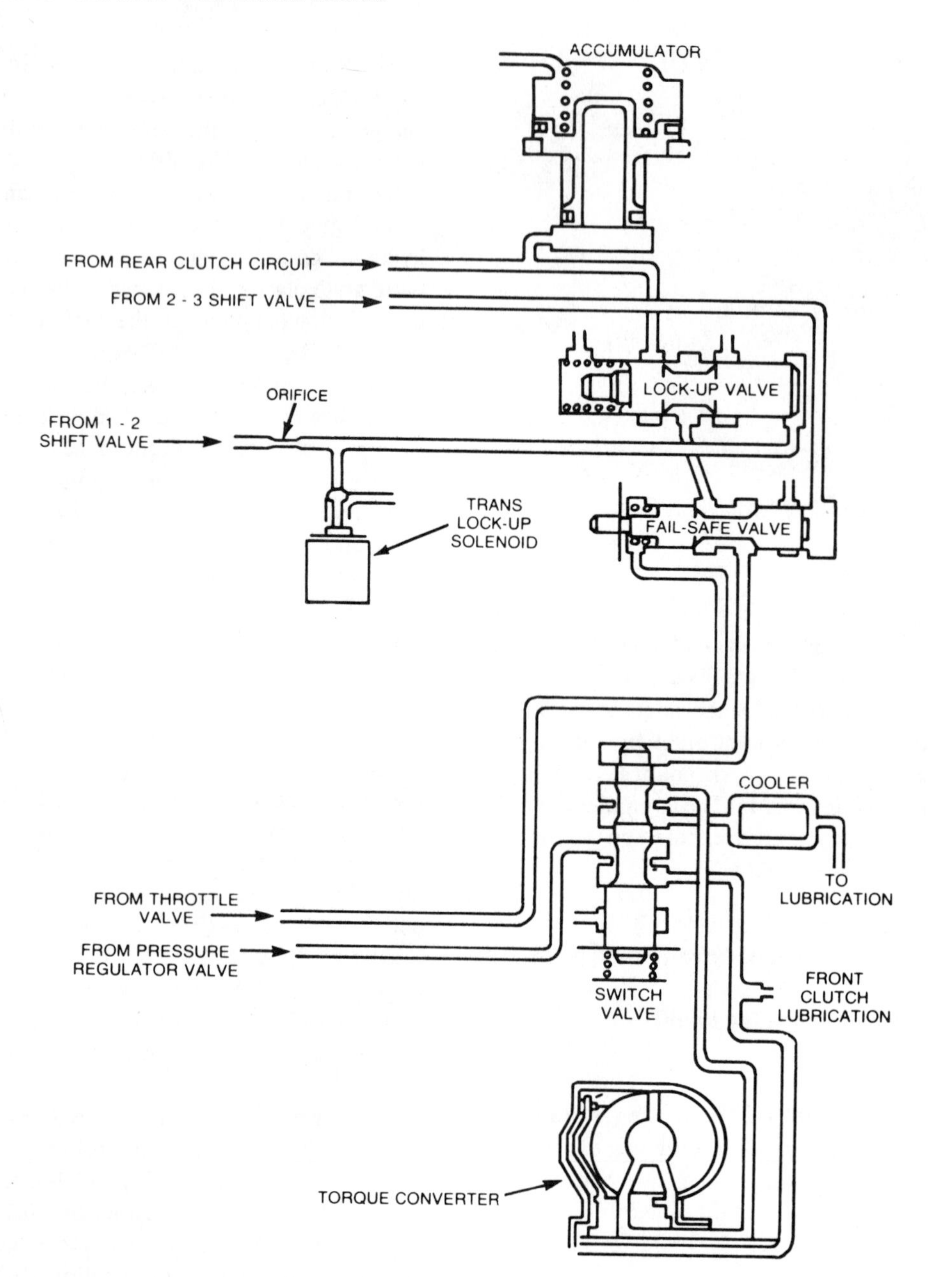

FIGURE 10–2 Hydraulic lockup system. (Courtesy of Chrysler Corp.)

quickly or gradually. This is accomplished with the use of a **pulse width duty cycle (pulse width modulated) solenoid.** Figure 10–9 illustrates the relationships between the different torque converter clutch operation phases.

- **Modulated:** In an electronic-hydraulic converter clutch system (or shift valve system), the term *modulated* refers to the pulsing of a solenoid, at a variable rate. This action controls the buildup of oil pressure in the hydraulic circuit to allow a controlled amount of clutch slippage.
- **Pulse width duty cycle (pulse width modulated) solenoid:** A computer-controlled solenoid that turns on and off at a variable rate producing a modulated oil pressure; often referred to as a pulse width modulated (PWM) solenoid. Employed in many electronic automatic transmissions and transaxles, these solenoids are used to manage shift control and converter clutch hydraulic circuits.

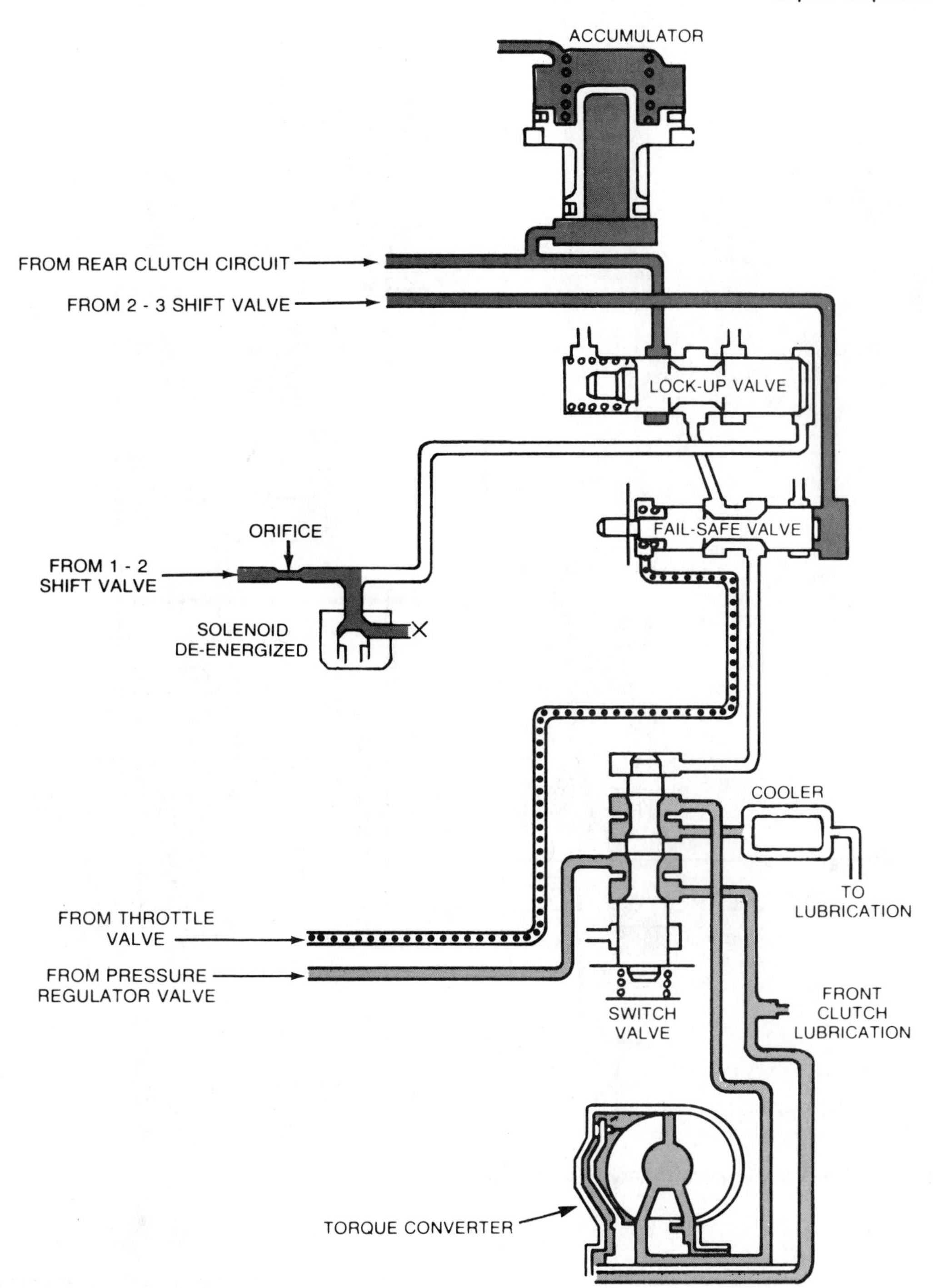

FIGURE 10–3 Converter clutch—unlock mode. (Courtesy of Chrysler Corp.)

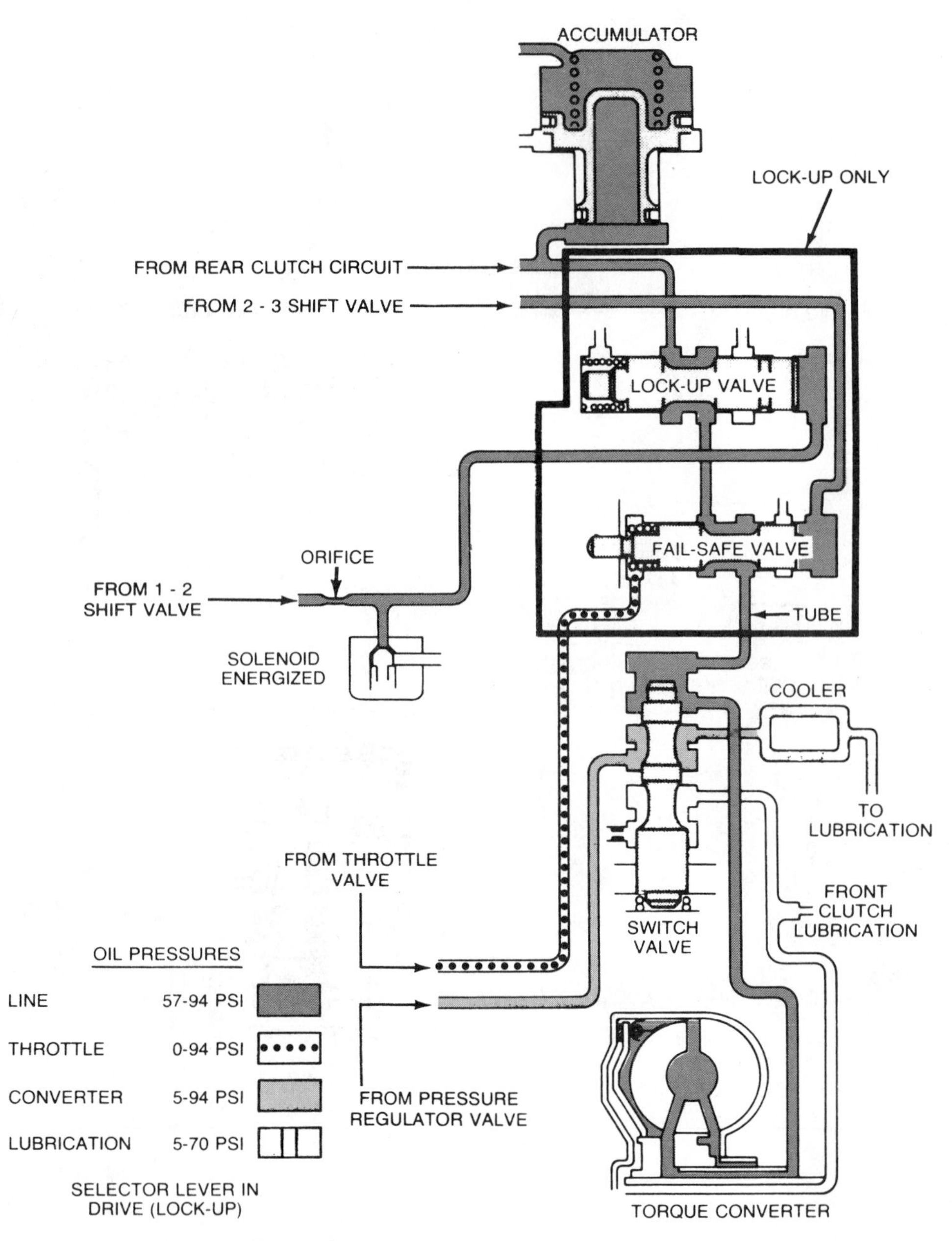

FIGURE 10–4 Converter clutch—lockup mode. (Courtesy of Chrysler Corp.)

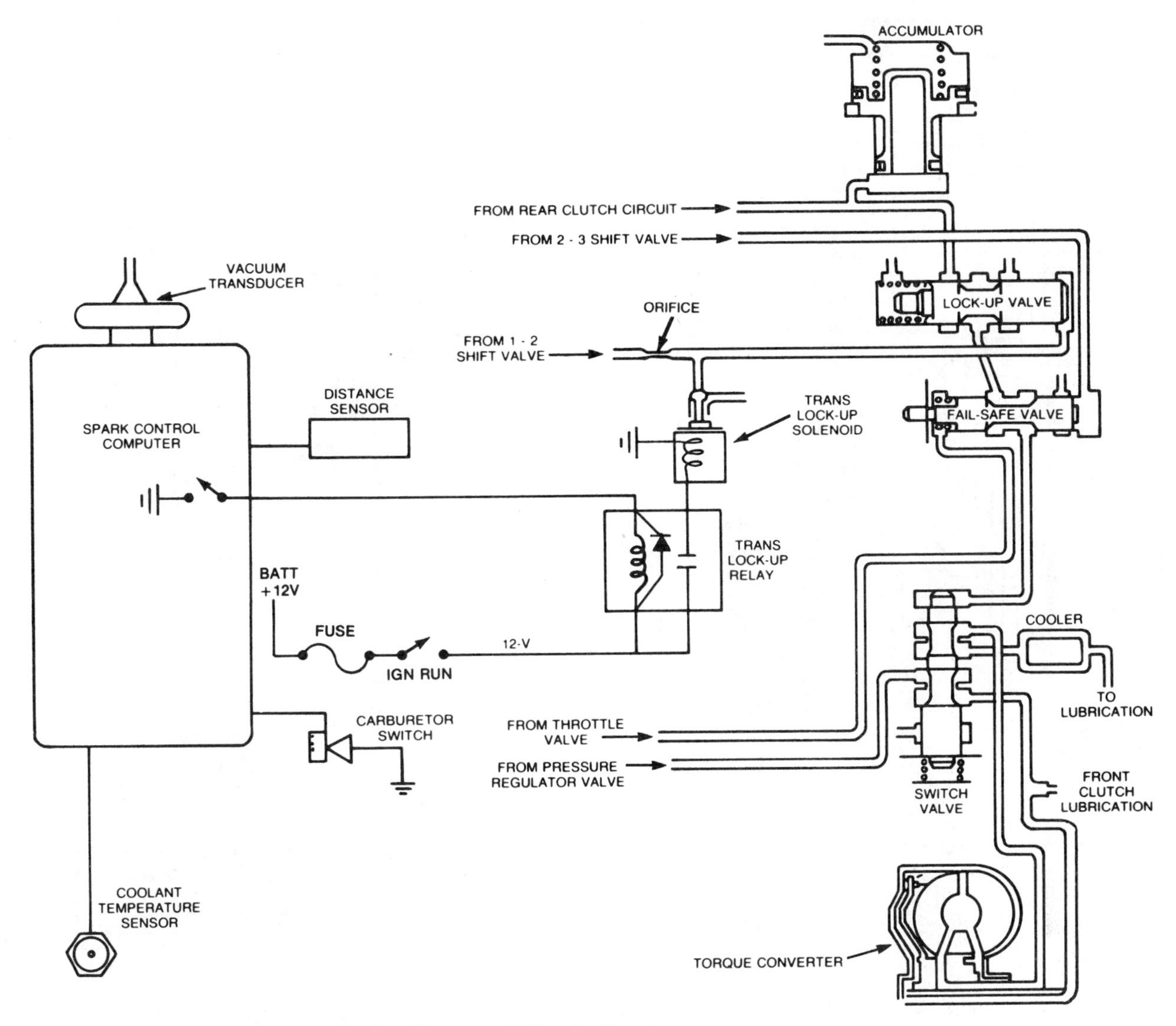

FIGURE 10–5 Typical hydraulic-electronic circuitry. (Courtesy of Chrysler Corp.)

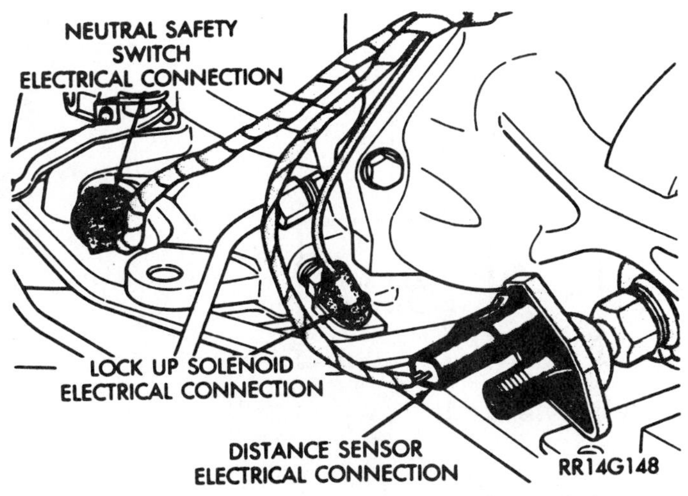

FIGURE 10–6 Distance sensor and other external transmission wiring connections. (Courtesy of Chrysler Corp.)

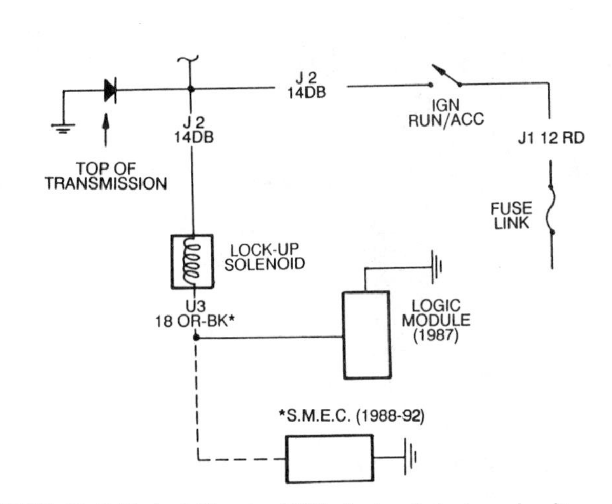

FIGURE 10–7 Typical Chrysler FWD electronic lockup circuit.

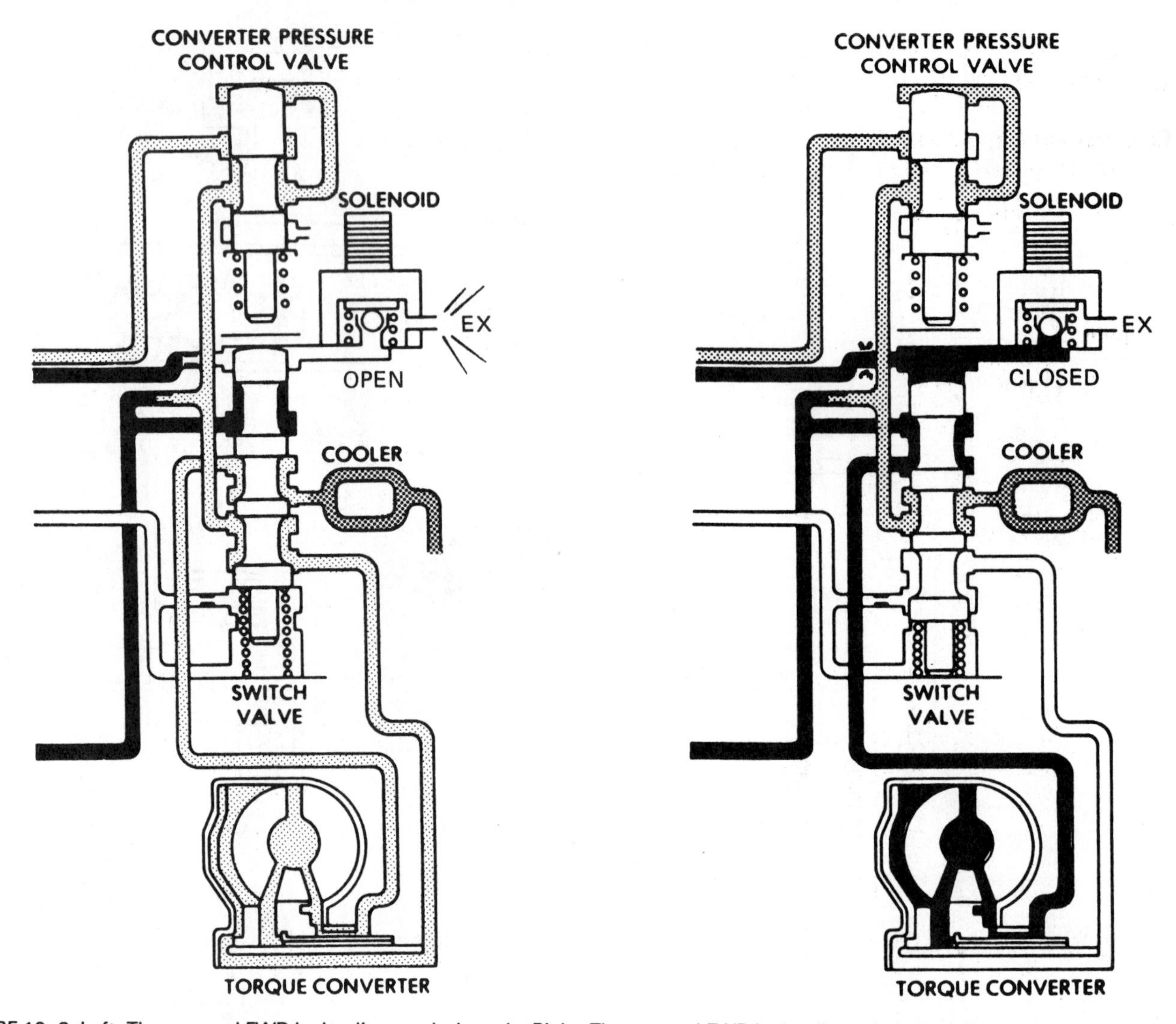

FIGURE 10–8 Left: Three-speed FWD hydraulics—unlock mode. Right: Three-speed FWD hydraulics—lockup mode.

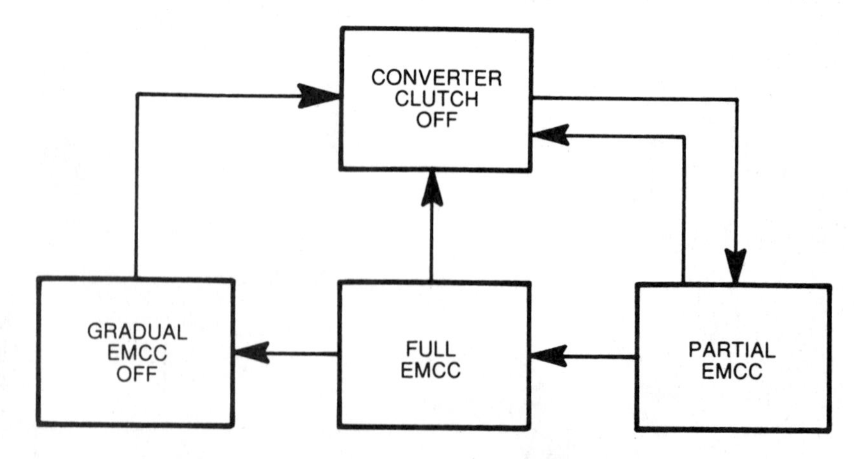

FIGURE 10–9 Chrysler 42LE torque converter clutch operation phases. (Courtesy of Chrysler Corp.)

FORD MOTOR COMPANY

Ford Motor Company uses the lockup converter clutch in most of its late model automatic transmissions and transaxles—mainly the A4LD, AXOD, AXOD-E/AX4S, AODE, E4OD, 4EAT, and CD4E. In each of these, except for the A4LD, converter clutch engagement and release is scheduled electronically. The A4LD uses hydraulic scheduling with an electronic override. All current Ford electronic shift transmissions and transaxles utilize computer control modules to provide diagnostic trouble code retrieval.

A4LD Converter Clutch Control

Figures 10–10 to 10–12 illustrate the operation of the A4LD converter clutch system. Observe how the hydraulics schedule the converter clutch unlock and lockup modes and how an electronically controlled solenoid overrides or inhibits its engagement.

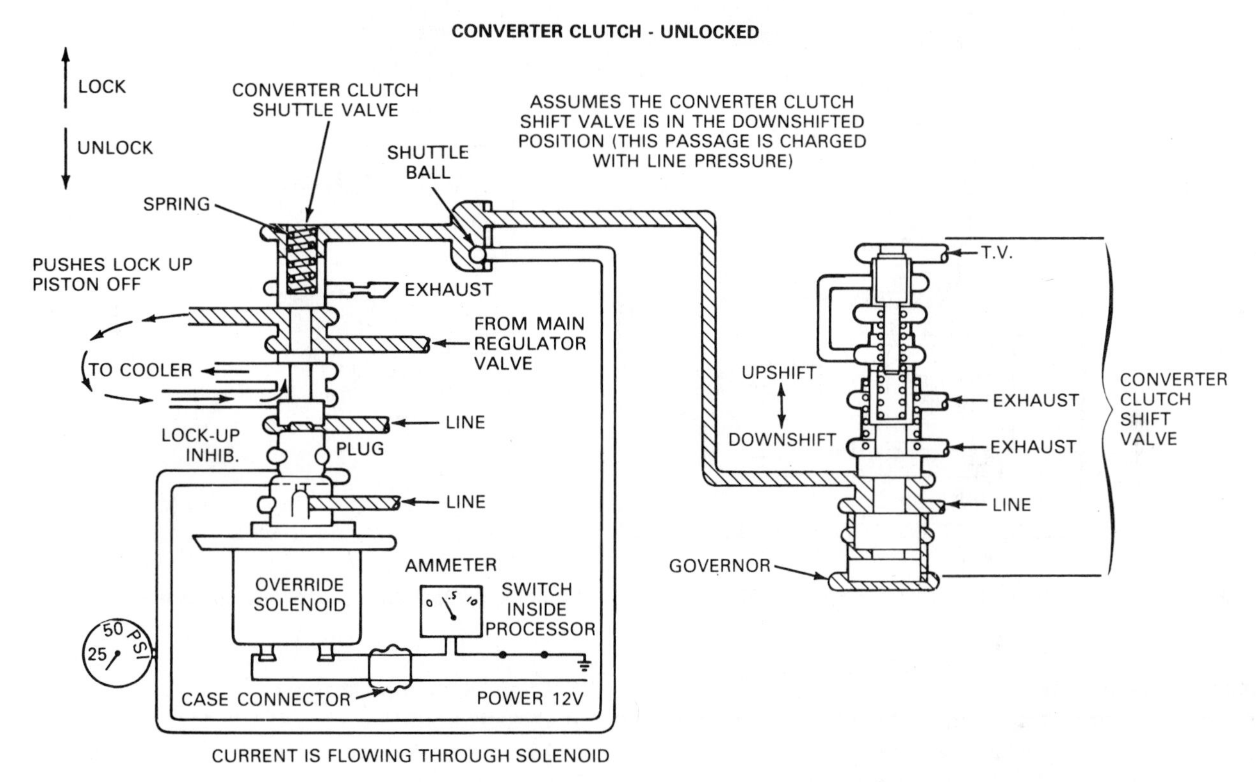

FIGURE 10–10 A4LD—Current is flowing through solenoid. (Reprinted with the permission of Ford Motor Company)

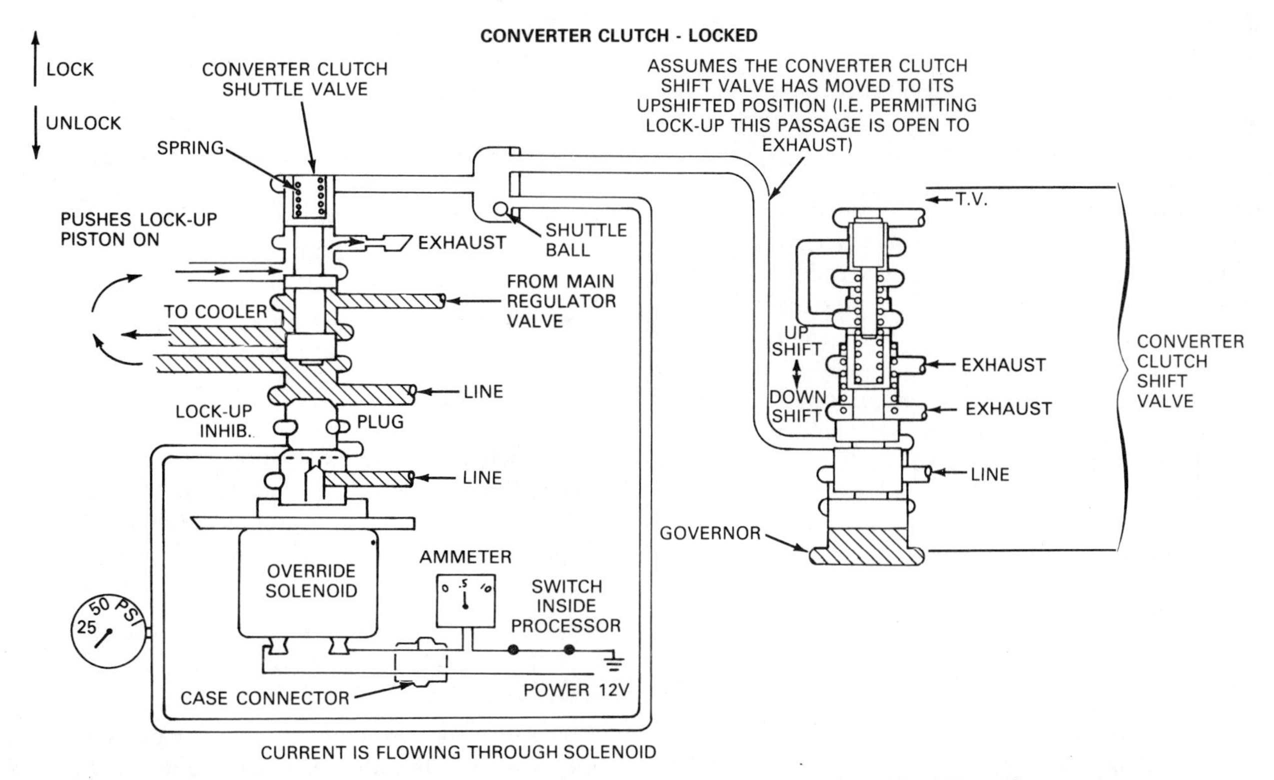

FIGURE 10–11 A4LD—Current is flowing through the solenoid. (Reprinted with the permission of Ford Motor Company)

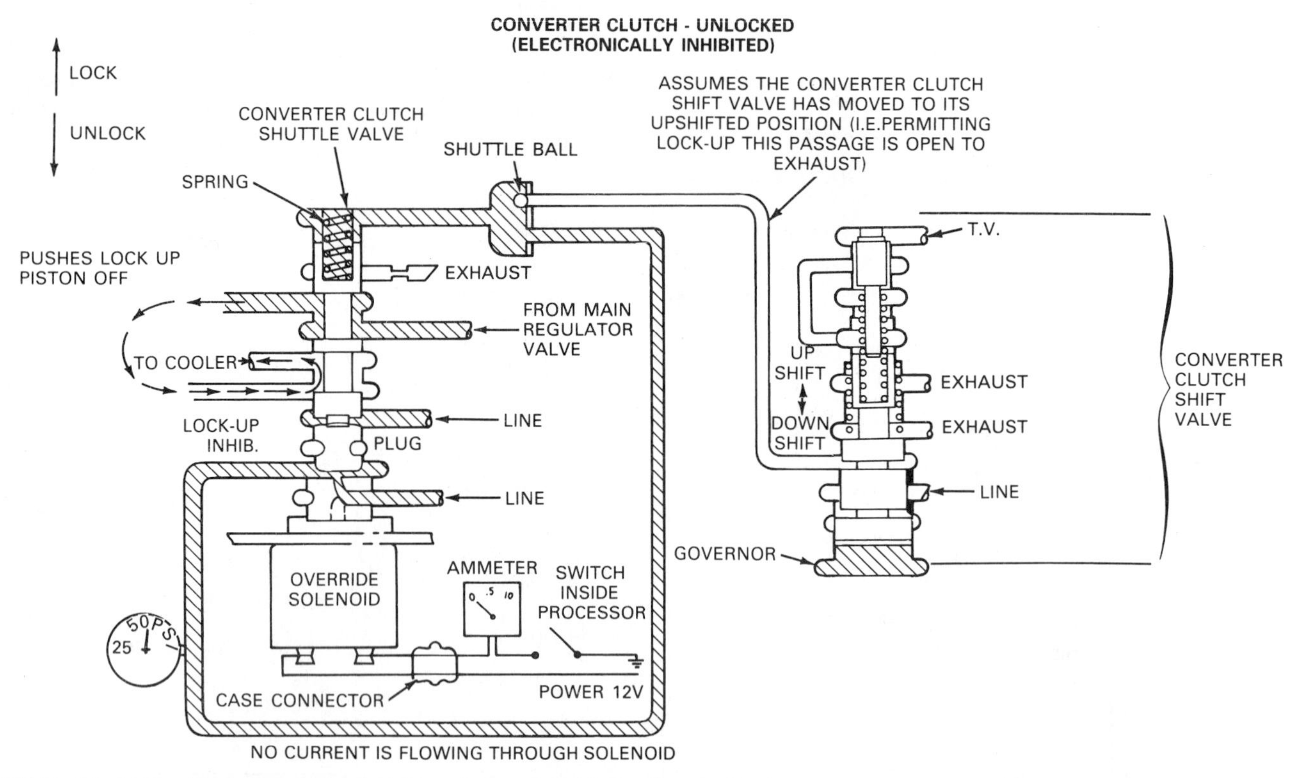

FIGURE 10–12 A4LD—No current is flowing through the solenoid. (Reprinted with the permission of Ford Motor Company)

Converter Clutch Unlocked.

- Governor pressure acting on the end of the converter shift valve has not yet moved the valve to the upshifted position.
- The converter shift valve relays line pressure from the 2-3 circuit to the spring side of the shuttle valve and holds the shuttle valve against the plug. This is the unlocked position.
- The shuttle valve indexes the mainline pressure from the regulator valve to the converter feed circuit in a flow path that keeps the lockup piston released.

Converter Clutch Lockup.

- When the vehicle speed generates enough governor pressure to overcome combined spring force and TV pressure, the converter clutch valve shifts. Line pressure is cut off, and the feed line to the spring side of the shuttle valve is open to exhaust.
- The shuttle valve moves against the spring and assumes a new position. The converter release circuit is open to exhaust, and a line pressure feed charges the converter clutch apply circuit.

Converter Clutch Override. Under certain operating conditions hydraulic lockup can be overruled and inhibited by the electronically controlled override solenoid. The solenoid is keyed to the PCM (**EEC-IV module**) and is normally energized. When the following operating conditions are sensed by the EEC-IV module, power to the solenoid is cut off.

- Engine coolant below 128°F (54°C) or above 240°F (116°C).
- Application of brakes.
- Closed throttle.
- 23.5 in hg or below absolute barometric pressure.
- Quick tip-ins.
- Quick tip-outs.
- When actual engine speed is below a certain value at lower vacuums (this ensures that all 4-3 torque demands will be made on an unlocked converter).
- With the solenoid deenergized, a line pressure bypass flows through the solenoid to the spring side of the shuttle valve.
- The shuttle valve moves against the plug and is held in the converter unlock position.

◆ **EEC-IV module:** The computer processing unit used in Ford's electronic engine control system, fourth generation.

Depending on engine operating parameters and transmission hydraulics, converter lockup can be applied in both third and fourth gears. The solenoid mount to the valve body and simplified wiring diagram are shown in Figure 10–13.

Ford Electronic Converter Clutch Control—AXOD, AXOD-E/AX4S, AODE, E4OD, and 4EAT

This transmission group uses the more traditional electronically controlled lockup system: a computer-operated solenoid keyed to an electronic module assembly controls the converter lockup hydraulics. The typical lockup circuitry is illustrated in Figures 10–14 and 10–15. In an AXOD circuit version, the lockup solenoid controls the exhaust porting in reverse order of the E4OD and 4EAT. However, the same results are achieved.

- AXOD unlock mode: Solenoid is OFF and exhaust port open.
- AXOD lockup mode: Solenoid is ON and exhaust port closed.

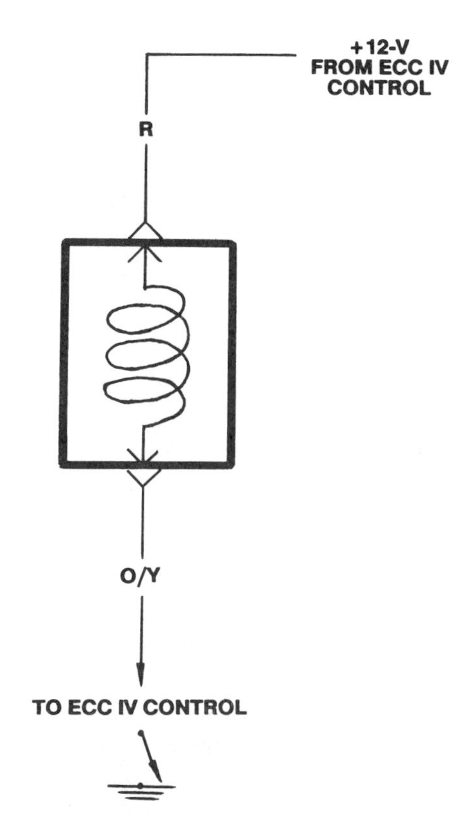

FIGURE 10–13 Lockup solenoid mounted to the valve body and simplified wiring diagram—Ford A4LD.

Converter clutch engagement depends on engine operating parameters and transmission hydraulics. In the AXOD and 4EAT, lockup can be achieved in third and fourth gears.

In the AXOD, a group of three pressure switches located on the valve body interfaces the transaxle converter clutch operation with the PCM (EEC-IV module) (Figure 10–16). These switches act as input sensors to the module, identifying the current status of the transaxle.

- Neutral pressure switch (NPS): signals the EEC-IV module when the transaxle is shifted from N or P into drive or reverse.
- 3-2 pressure switch: signals the EEC-IV module when the transaxle shifts from second to third or from third to second gear.
- 4-3 pressure switch: signals the EEC-IV module when the transaxle shifts from third to fourth or from fourth to third gear.

The transmission high-pressure switch (THS) and neutral/park switch (NPS) sensors are grounding switches and are normally open until acted on by pressure. The EEC-IV module engages the converter clutch when the pressure sensors and transaxle are in the following modes:

- Transaxle in third gear: (NPS ON, THS 3-2 ON, THS 4-3 ON) and engine operating within required performance parameter.
- Transaxle in fourth gear: (NPS OFF, THS 3-2 ON, THS 4-3 OFF) and engine operating within required performance parameter.

AXOD-E/AX4S. There are two variations of converter clutch operation used on AXOD-Es. On Taurus and Sable vehicles, the operation of the lockup solenoid in the AXOD-E is the same as in the AXOD. On Continental vehicles, a modulated lockup solenoid is used to provide a desirable combination of fuel economy and performance.

Continental's lockup solenoid is pulse width modulated, for clutch apply and release, by the PCM (Figure 10–17). By varying the pulse width (on/off time) at a steady frequency, the apply and release oil pressure buildup is controlled. This produces a desirable amount of clutch slippage.

AODE. A modulated converter clutch control (MCCC) solenoid is used in the AODE transmission. Clutch apply typically occurs in third and fourth gears but may occur in second gear under certain conditions. To control the pressure buildup to the converter bypass clutch control valve and the amount of clutch slippage, the duty cycle of the solenoid ranges from zero (clutch off, zero pressure) to one hundred percent (clutch on, full pressure) (Figure 10–18).

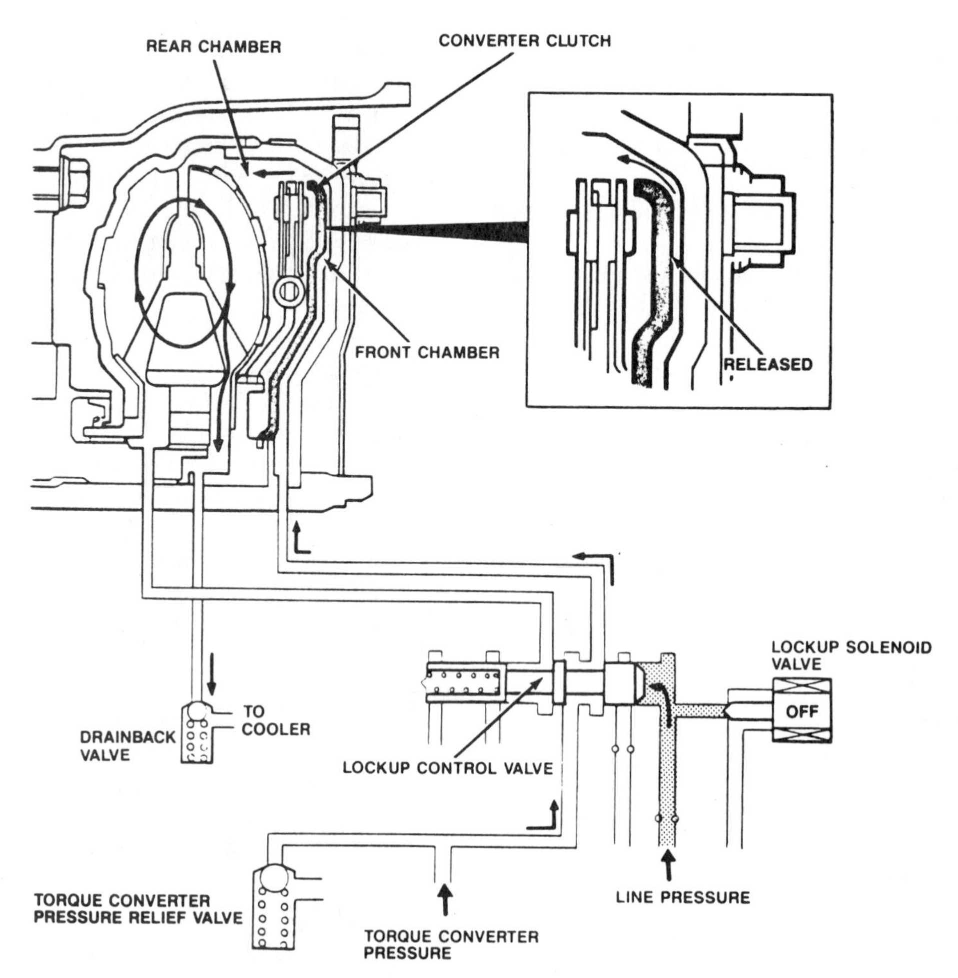

FIGURE 10–14 Converter clutch released—4EAT hydraulic circuit. (Reprinted with the permission of Ford Motor Company).

E4OD. The E4OD has a sophisticated converter clutch operation, with lockup possible in first, second, third, and fourth gears. The converter clutch engagement is inhibited under the usual conditions of initial cold operation, closed throttle, brake apply, high tip-in or tip-out rates, and low barometric pressure. Release is also demanded for forced downshifts and certain upshift conditions.

GENERAL MOTORS CORPORATION

The torque converter clutch (TCC) was introduced in the General Motors family of automatic transmissions for the 1980 production year and used in the 350C, 250C, and 200C transmissions. In the following production years, GM expanded the TCC coverage to include all passenger cars, and by 1991 all light-duty truck applications.

Like the Chrysler and Ford systems, the release and apply of the TCC is basically controlled by a hydraulic converter clutch valve and solenoid-operated check valve. When the solenoid is energized, the converter clutch valve switches the TCC hydraulics to lockup mode. The solenoid itself is controlled by a computer electronic circuit interfaced with the electronic control module (**ECM**), PCM, or a TCM. The computer controller also provides for diagnostic trouble code retrieval.

- **ECM (electronic control module):** GM's electronic control module incorporates output control over the TCC solenoid. (Note: Current designation for the ECM in late model vehicles is PCM).

There are numerous variations in GM TCC control circuitry because of the wide variety of three- and four-speed automatic transmission models used with different engine and car/truck applications. For ease of illustrating GM converter clutch circuit

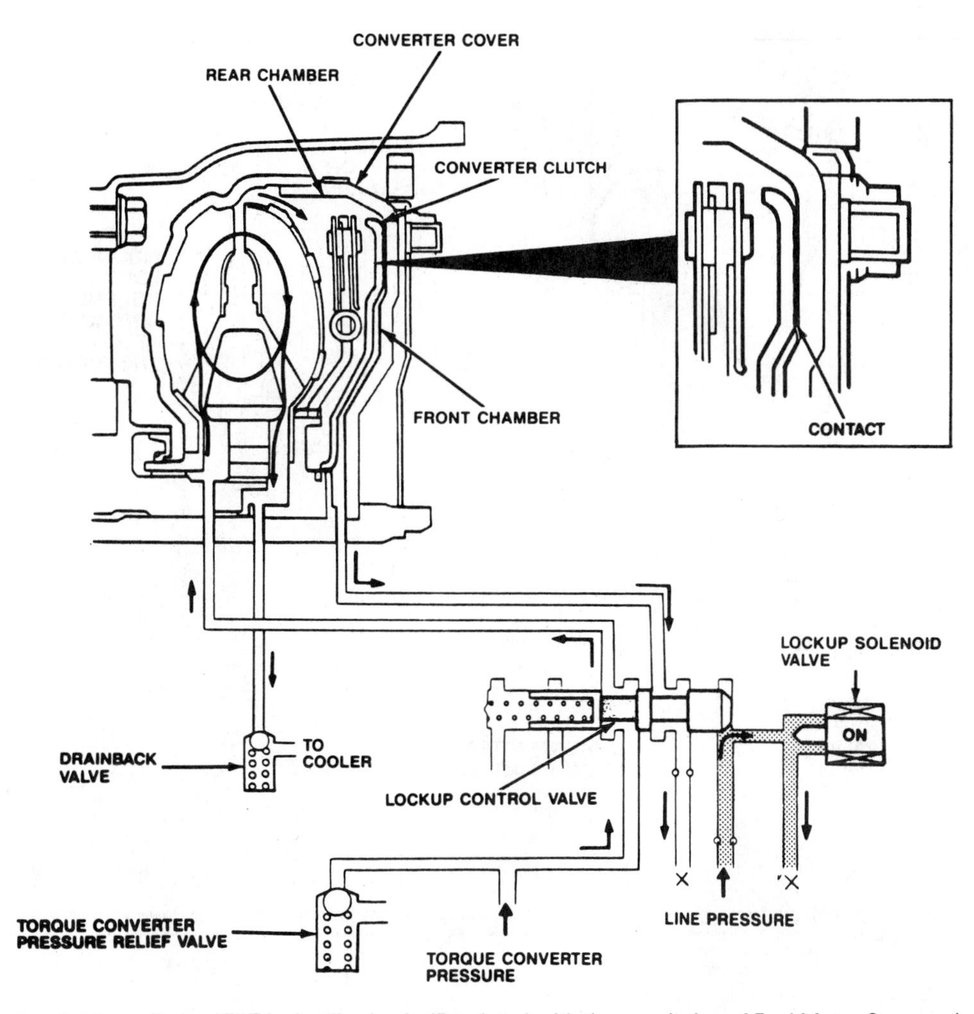

FIGURE 10–15 Converter clutch applied—4EAT hydraulic circuit. (Reprinted with the permission of Ford Motor Company)

FIGURE 10–16 Pressure switch group—Ford AXOD.

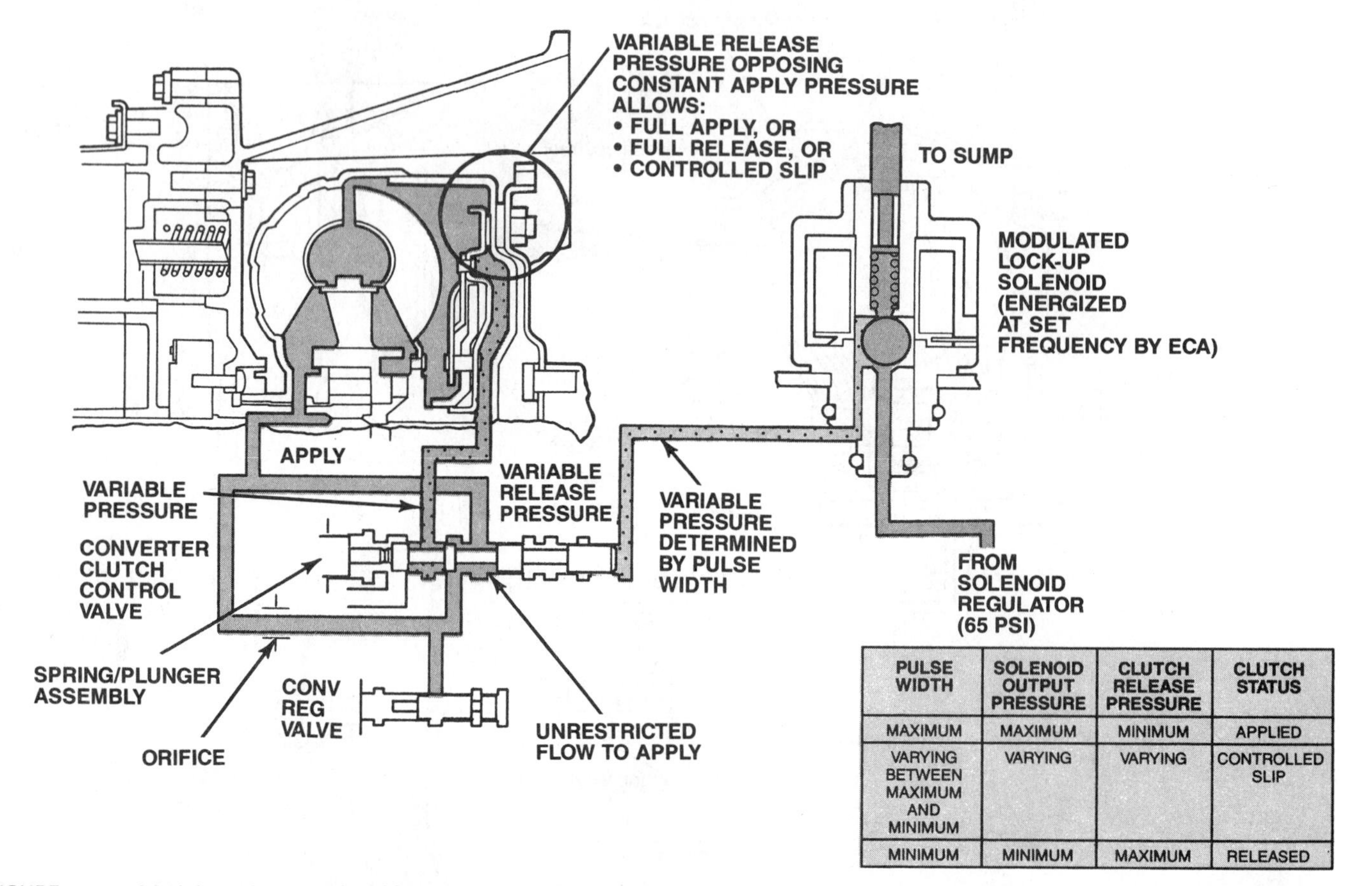

PULSE WIDTH	SOLENOID OUTPUT PRESSURE	CLUTCH RELEASE PRESSURE	CLUTCH STATUS
MAXIMUM	MAXIMUM	MINIMUM	APPLIED
VARYING BETWEEN MAXIMUM AND MINIMUM	VARYING	VARYING	CONTROLLED SLIP
MINIMUM	MINIMUM	MAXIMUM	RELEASED

FIGURE 10–17 Modulated lockup solenoid operation—AXOD-E/ AX4S (AXODE), Continental. (Reprinted with the permission of Ford Motor Company)

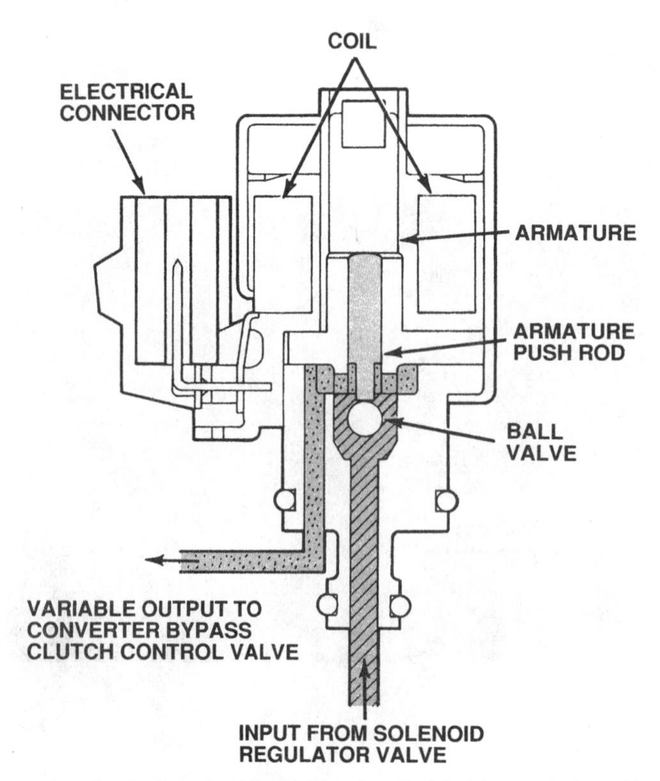

FIGURE 10–18 AODE—modulated converter clutch control solenoid. (Reprinted with the permission of Ford Motor Company)

controls, some basic examples are highlighted in this section. In brief, most three-speed, TCC-equipped units lock up in third gear only.

There are variables in the four-speed, TCC-equipped units. They all have third and fourth gear lockup capability. Some 4L60 (700-R4) models have a second-gear lockup, while all 4T60 (440-T4) include second gear.

The TCC should engage when:

- Brakes are released.
- Transmission is in the proper gear.
- Engine coolant meets minimum specified temperature.
- Manifold vacuum is right.
- Vehicle speed is high enough to avoid engine pulse sensations in the powertrain.

The TCC should disengage when:

- Additional torque demand is needed from the converter.
- Emissions would be affected during closed-throttle coast.
- Brakes are applied.

Basically, the release and engagement modes are largely dependent on engine performance parameters, transmission hydraulics, and vehicle speed.

The hydraulic apply and release of the TCC is achieved by using a converter clutch valve acting as the relay or switch valve. The position of the valve controls whether the converter feed oil charges the release side or apply side to the converter.

3T40 (125C) TCC Hydraulics

Typical release and apply hydraulic circuits are illustrated in Figures 10–19 and 10–20.

TCC Release Circuit.

- Solenoid is deenergized and check valve is open to exhaust.
- Line pressure holds the converter clutch control valve in the release position.
- Converter feed oil tapped from the pressure regulator valve is indexed to charge the release circuit in a flow path that keeps the TCC released.
- The feed oil cycles in and out of the converter for normal cooling and lubrication.

TCC Apply Circuit.

- Solenoid is energized, and the check valve seals the exhaust port bleed.
- Line pressure through the orifice cup plug builds up in the solenoid circuit and shifts the converter clutch control (CCC) valve in the apply position.
- Converter feed oil is cut off from the apply circuit and is short-circuited directly to cooling and lubrication.
- Regulated line or converter clutch oil is indexed to charge the apply circuit in a path flow that engages the TCC to the converter cover.

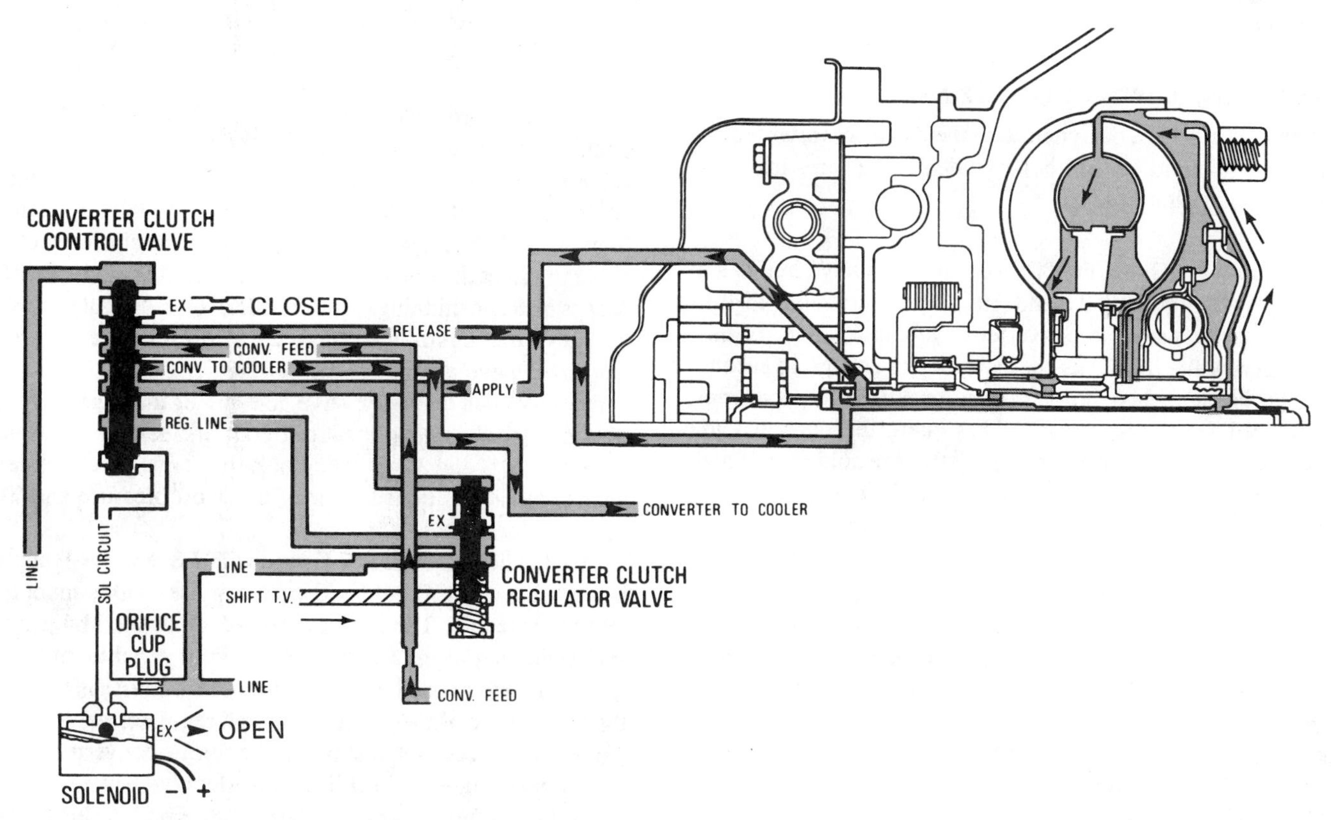

FIGURE 10–19 Oil flow—TCC-released 3T40 (125C). (Courtesy of General Motors Corporation, Service Technology Group)

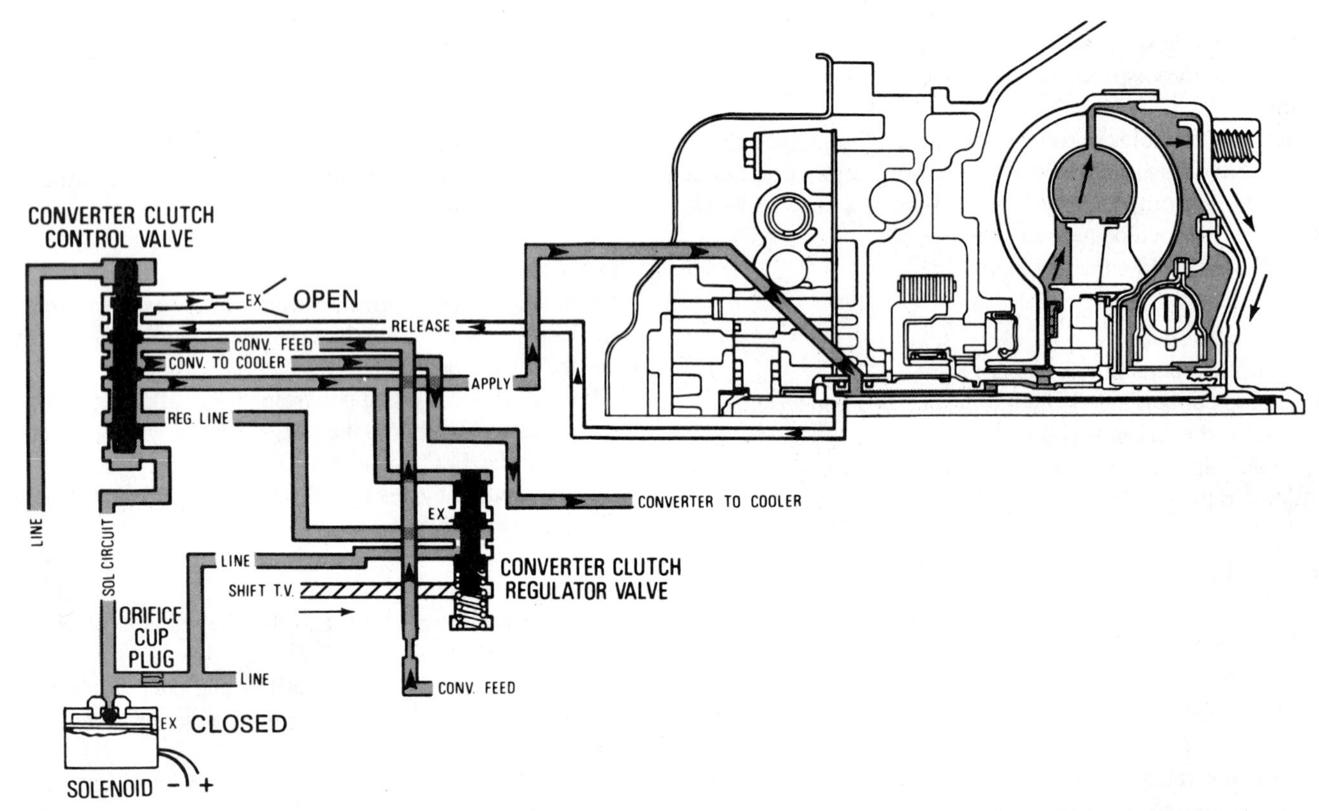

FIGURE 10–20 Oil flow—TCC-applied 3T40 (125C). (Courtesy of General Motors Corporation, Service Technology Group)

- Release circuit oil is exhausted from the converter at the converter clutch valve through an orifice to help prevent sudden engagement.

Hydraulic Circuit Controls

These circuits are designed with the basic converter clutch valve and solenoid control but include some interesting variations in the hydraulics.

4L60 (700-R4). The 4L60 TCC release and apply circuits are featured in Figures 10–21 and 10–22. The highlight in this circuitry is the use of a converter clutch shift valve. In this design setup, the TCC has the potential to engage in second gear, provided the governor oil has upshifted the converter clutch shift valve and the PCM has energized the solenoid. Second-gear oil is used for charging the solenoid circuit and switching the converter clutch valve via the converter clutch shift valve.

4L60-E. The TCC solenoid and hydraulic circuitry in the 4L60-E function slightly differently than in a 4L60 (700-R4). Normally the converter clutch applies in fourth gear only while in overdrive range, and in third gear only while in drive/manual third range. Note: Under heavy throttle conditions, at high speeds, the PCM energizes the solenoid in third-gear–overdrive range. Also, if the transmission fluid temperature exceeds 275°F (135°C), the converter clutch is applied all the time in fourth gear. This is done to lower the fluid temperature.

4T60 (440-T4). The 4T60 TCC circuit incorporates a converter clutch accumulator in addition to the converter clutch regulator valve on the apply side of the circuitry (Figure 10–23). This gives an improved cushion to the converter clutch apply.

4T60-E. Two versions of the 4T60-E TCC hydraulic/electronic circuit controls are used. Many 4T60-Es simply use a TCC solenoid system that functions like a 4T60. The other design has an additional solenoid, a pulse width modulator solenoid. In both, the TCC solenoid opens and closes the converter clutch valve to direct the release and the apply circuit oil. The difference is in the positioning of the converter clutch regulator valve, the valve that controls the amount of oil pressure fed to the converter valve and clutch apply circuit. The position of the converter clutch regulator valve determines the aggressiveness of converter clutch application. In the first setting mentioned, used with 3.1- and 3.4-liter engines, the converter clutch regulator valve is positioned by modulator oil pressure and TCC accumulator oil pressure.

Applications of the 4T60-E used with 3.8- and 4.9-L engines incorporate a TCC solenoid and a pulse width modulator (PWM) solenoid. The PWM solenoid, controlled by ground-side switching in the PCM, operates on a variable duty cycle rate from zero to one hundred percent. Its purpose is to control the positioning of the converter clutch regulator valve and the aggressiveness of clutch apply in the torque converter.

In both designs, the 4T60-E converter clutch can be applied in third and fourth gear in overdrive range and in third gear in drive/manual third range.

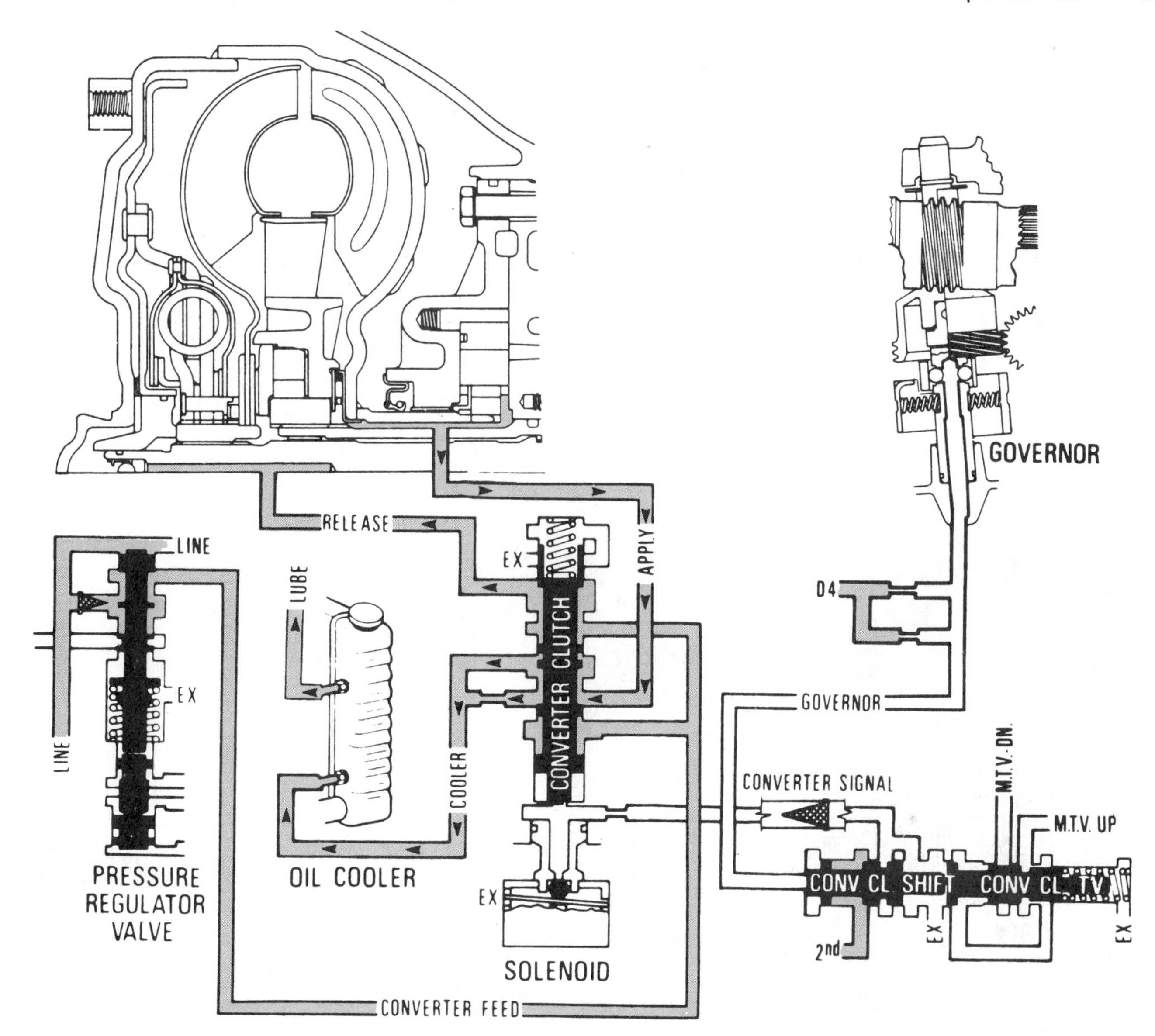

FIGURE 10–21 Oil flow—TCC-released 4L60 (700-R4). (Courtesy of General Motors Corporation, Service Technology Group)

4L80-E and 4T40-E. The TCC solenoid used in a 4L80-E and 4T40-E is a pulse width modulation unit that controls the rate of TCC apply and release. By varying the duty cycle, the movement of the TCC apply valve is controlled for a smooth application and release (Figure 10–24). Under certain conditions, such as depressing the brake pedal or releasing or fully applying the accelerator pedal, the TCC is released immediately. Like the 3.8- and 4.9-L engine versions of the 4T60-E, a TCC accumulator is not needed.

Computer-Controlled Solenoid

Starting in 1982, the TCC solenoid control was interfaced with all ECM/CCC-equipped vehicles. The brake switch controls the circuit applied voltage (12 V), while the PCM controls the circuit ground. The basic hydraulics and criteria for TCC engagement remain the same. The ECM/PCM determines when to complete the ground-side circuit by scanning the parameter inputs from numerous devices. Depending on the transmission and its application to a variety of engines, vehicle models, and years of production, the ECM/PCM is programmed to scan readings from certain input devices. The selection of input sensors and switches can include: the **ECT, TP, MAP, VSS, TRS, PSA, TFT, TIS, BARO,** the TCC brake switch, and engine speed.

- **ECT (engine coolant temperature) sensor:** Prevents converter engagement with a cold engine.
- **TP (throttle position) sensor:** Reads the degree of throttle opening; its signal is used to analyze engine load conditions. The PCM decides to apply the TCC, or to disengage it for coast or load conditions that need a converter torque boost.
- **MAP (manifold absolute pressure) sensor:** Reads the amount of air pressure (vacuum) in the engine's

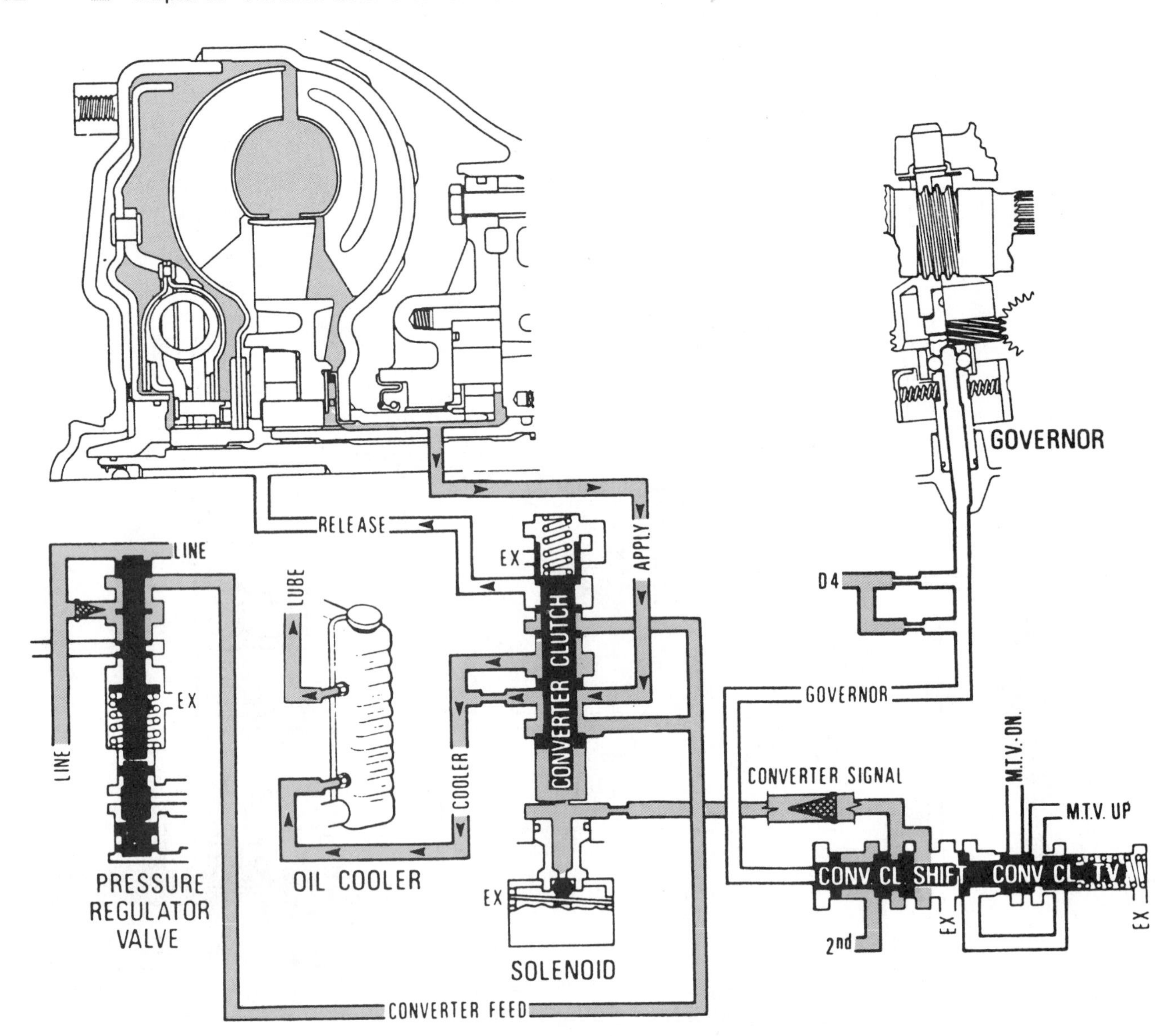

FIGURE 10–22 Oil flow—TCC-applied 4L60 (700-R4). (Courtesy of General Motors Corporation, Service Technology Group)

intake manifold system; its signal is used to analyze engine load conditions.

- **VSS (vehicle speed sensor):** Provides an electrical signal to the computer module, measuring vehicle speed, and affects the torque converter clutch engagement and release.
- **TRS (transmission range selector) switch:** Tells the module which gear shift position the driver has chosen.
- **PSA (pressure switch assembly):** Mounted inside the transmission, it is a grouping of oil pressure switches that inputs to the PCM when certain hydraulic passages are charged with oil pressure.
- **TFT (transmission fluid temperature) sensor:** Used to determine "hot mode" TCC operation and modify the transmission line pressure.
- **TIS (transmission input speed) sensor:** Measures turbine shaft (input shaft) rpm's and compares to engine rpm's to determine torque converter slip. When compared to the transmission output speed sensor or VSS, gear ratio and clutch engagement timing can be determined.
- **BARO (barometric pressure) sensor:** Measures the change in the intake manifold pressure caused by changes in altitude.

When the PCM reads desirable inputs while scanning the sensors, the TCC solenoid gets energized. To illustrate the concept, observe the 3T40 (125C) TCC solenoid control circuit shown in Figure 10–25. The third-gear pressure switch has no signal function to the PCM. It does, however, permit TCC engagement in third gear. The circuit test lead is on the ground side of the circuit and connected to terminal F of the

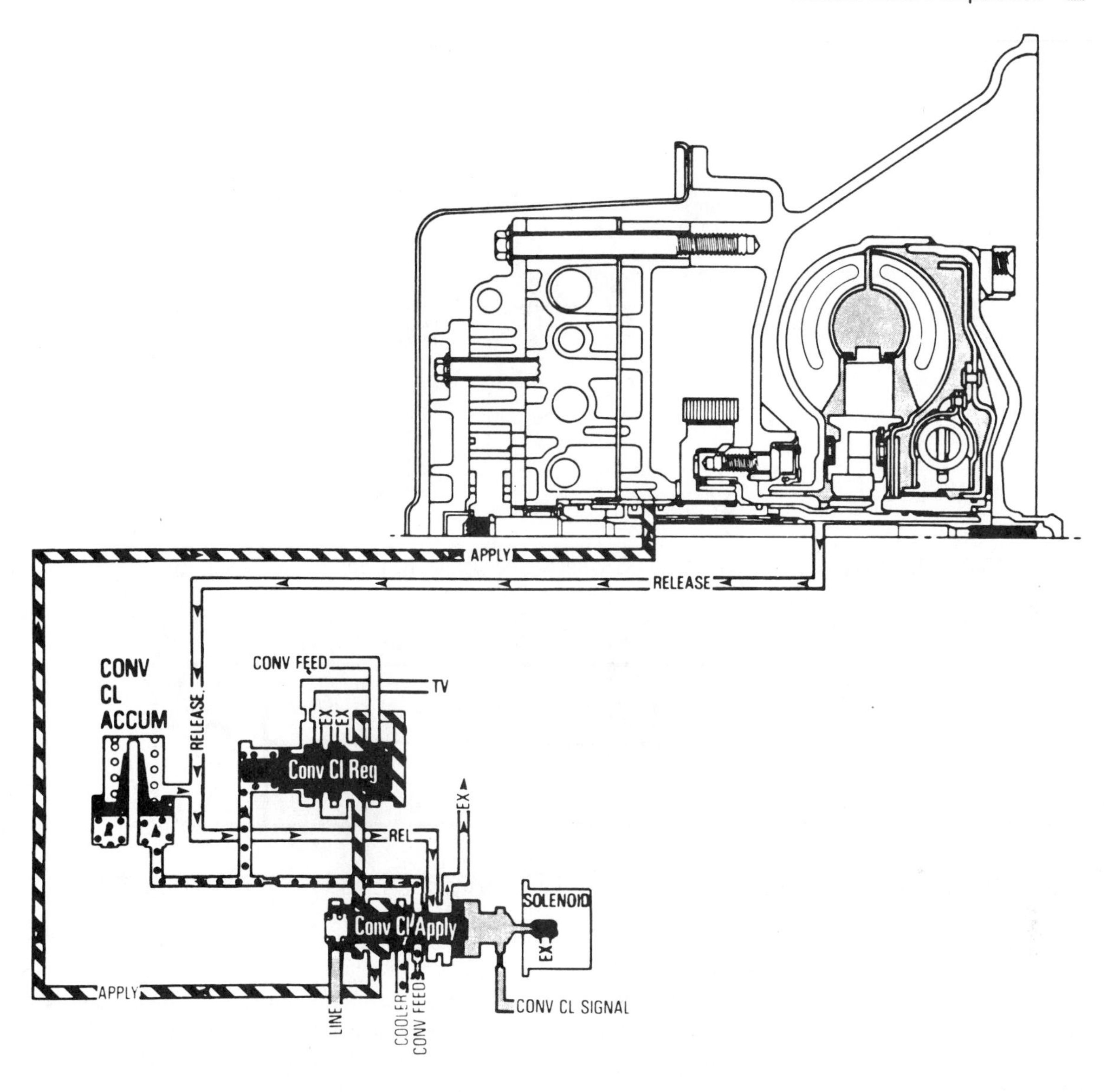

FIGURE 10–23 Converter clutch apply feel, 4T60 (440-T4). (Courtesy of General Motors Corporation, Service Technology Group)

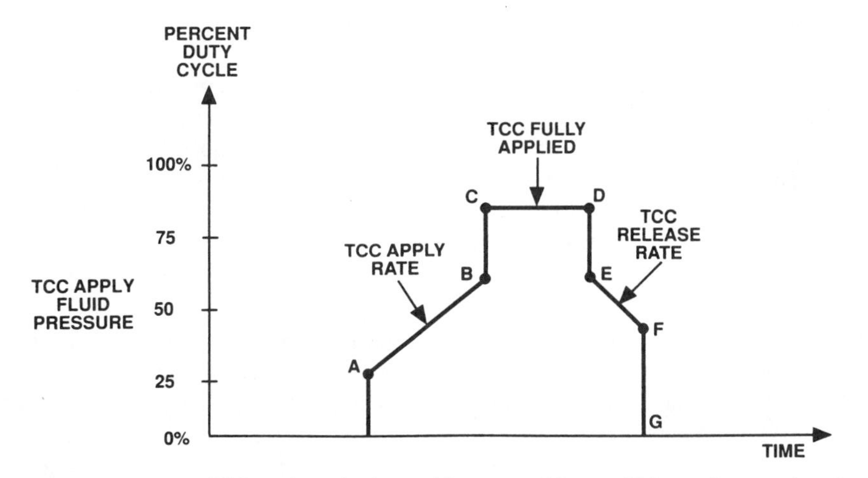

FIGURE 10–24 4L80-E duty cycle percentage—TCC apply and release. (Courtesy of General Motors Corporation, Service Technology Group)

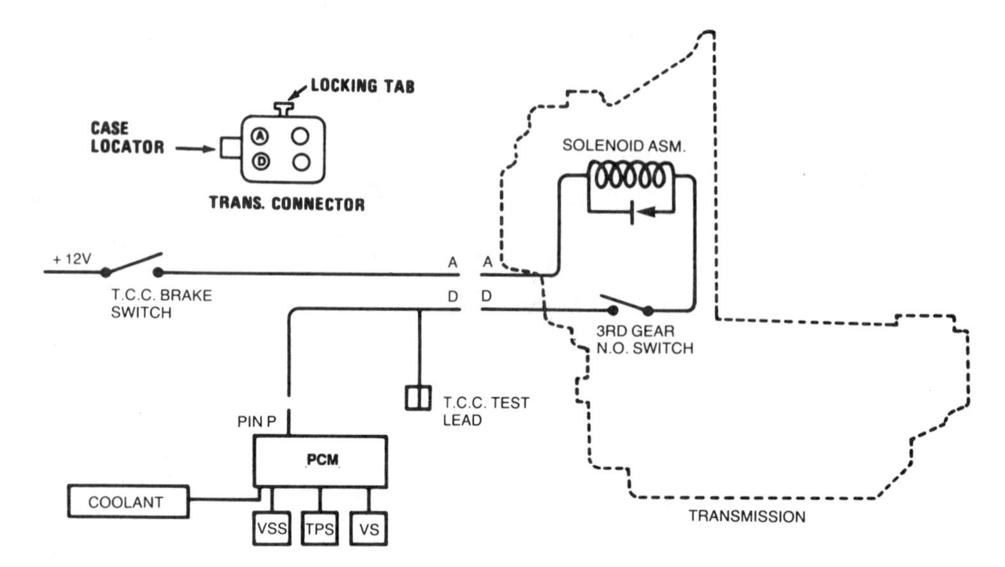

FIGURE 10–25 3T40 (125C)—TCC solenoid circuit. (Courtesy of General Motors Corporation, Service Technology Group)

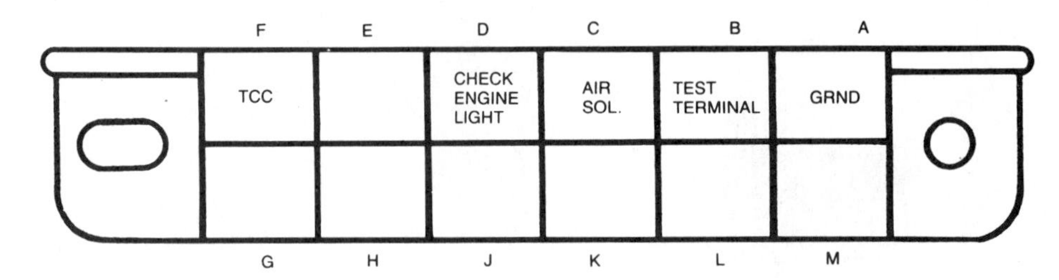

FIGURE 10–26 ALDL/DLC terminal F used for diagnostics. (Courtesy of General Motors Corporation, Service Technology Group)

ALDL/DLC (assembly line diagnostic link/data link connector) (Figure 10–26). The F terminal can be used for diagnostics using a high-impedance test light. This same circuit is basic to other third-gear TCC circuits.

- **ALDL (assembly line diagnostic link):** Electrical connector for scanning ECM/PCM/TCM input and output devices.
- **DLC (data link connector):** Current acronym/term applied to the federally mandated, diagnostic junction connector that is used to monitor PCM/TCM inputs, processing strategies, and outputs including diagnostic trouble codes (DTCs).

Figure 10–27 shows one of various TCC solenoid circuits used on 4L60 (700-R4) applications with PCM control.

One of the more common TCC solenoid circuits with PCM control used on the 4T60 (440-T4) is diagramed in Figure 10–28. The vehicle speed sensor (VSS) can be a magnetic pulse style with the rotor driven by the transmission governor shaft (Figure 10–29), or it can be an optical sensor with a light-emitting diode and phototransistor assembly mounted at the speedometer head (Figure 10–30). The speedometer is still cable-driven.

The optical speed sensor works as follows:

- The light-emitting diode is lit when the ignition is turned ON. It produces an invisible infrared light.
- A reflective blade is part of the speedometer cable head assembly. As the vehicle starts moving, the cable starts rotating the reflective blade.
- When the blade interrupts the light beam provided by the LED, the light beam is reflected back to the photo transistor where it is converted into a low-power electrical signal.
- The number of electrical signals sent to the PCM per unit of time represents the vehicle speed.

With this style of vehicle speed sensing, a broken or disconnected speedometer cable prevents engagement of the TCC. If the sensor assembly or reflector blade gets dirty, the TCC applies erratically.

SUMMARY

Torque converter clutch mechanisms have become a standard feature in most transmission designs. Through the application of the converter clutch, fuel economy is improved without sacrificing powertrain performance. In addition, when locked on, converter slippage is eliminated, and transmission fluid temperature is reduced.

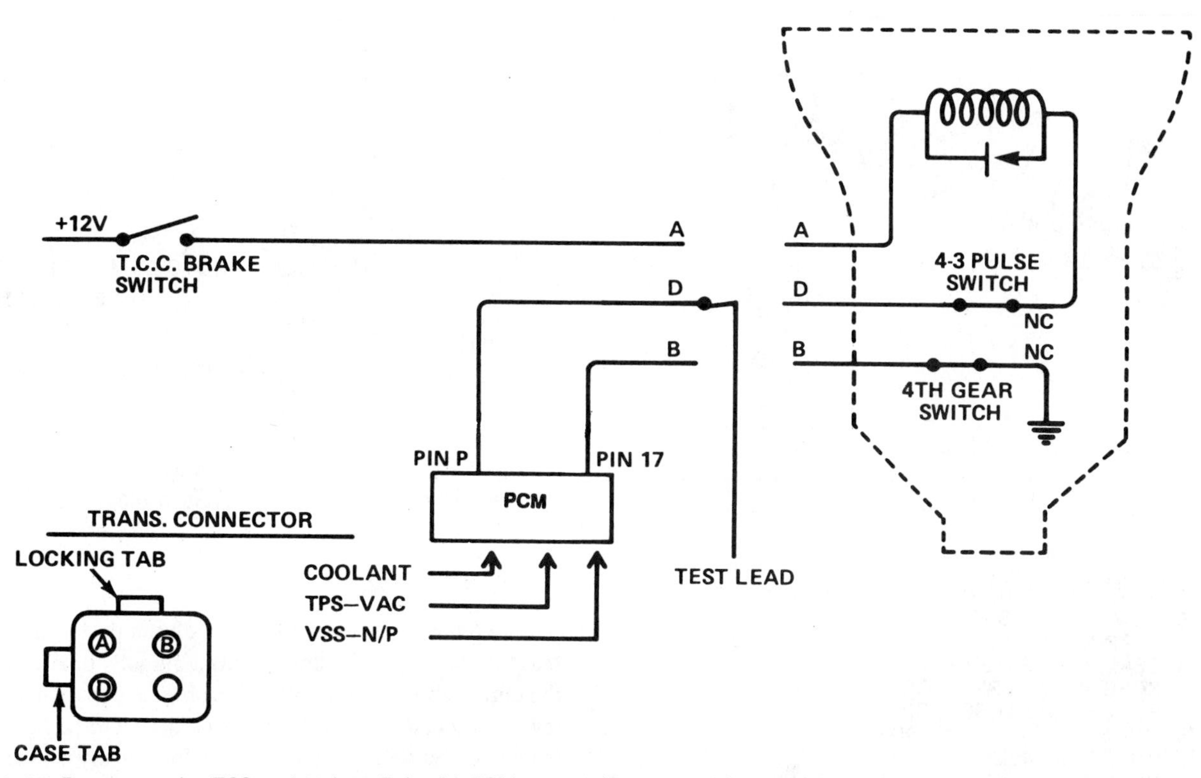

FIGURE 10–27 Representative TCC—4L60 (700-R4), with PCM control. (Courtesy of General Motors Corporation, Service Technology Group)

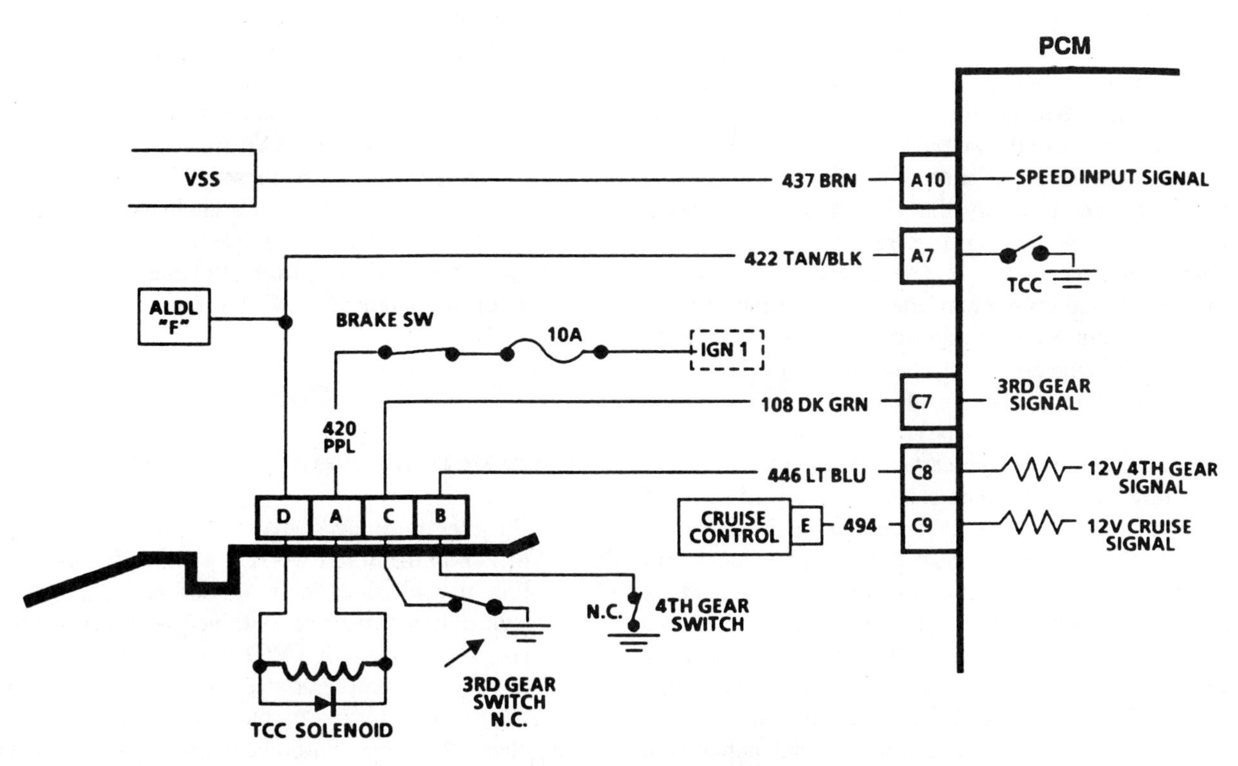

FIGURE 10–28 Representative TCC—4T60 (440-T4), with PCM control. (Courtesy of General Motors Corporation, Service Technology Group)

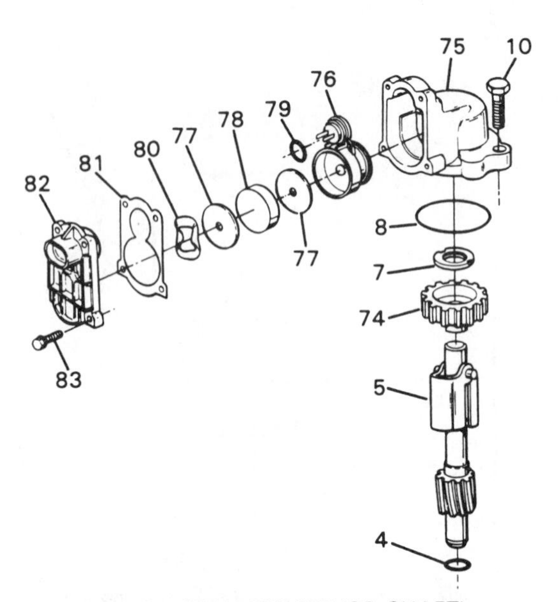

4 RING, OIL SEAL (GOVERNOR SHAFT)
5 GOVERNOR ASSEMBLY
7 BEARING ASM., THRUST (SPEEDO GEAR)
8 SEAL, "O" RING (GOVERNOR COVER)
10 SCREW, GOVERNOR COVER/CASE
74 ROTOR, SPEED SENSOR
75 HOUSING, SPEED SENSOR
76 COIL ASSEMBLY
77 WASHER
78 MAGNET
79 SEAL, "O" RING
80 WASHER, WAVE SPRING
81 GASKET, COVER
82 COVER, HOUSING
83 BOLT

FIGURE 10–29 Vehicle speed sensor (VSS). (Courtesy of General Motors Corporation, Service Technology Group)

The timing of clutch apply is determined by numerous electrical input signals to the PCM/TCM. The control module evaluates the signals and determines when the clutch should be applied and released. Typically, the clutch should be released when the engine is cold, during acceleration or closed throttle, and when braking.

Application of the torque converter clutch in the forward gears varies in different transmission designs. Most three-speed transmissions are designed to apply the clutch only in third gear. In most four-speed units, clutch application is present in third and fourth gear. In some four-speed units, clutch apply may also occur in second gear under certain conditions.

Manufacturers usually use an exhaust port styled solenoid to operate the clutch system. When commanded by the PCM/TCM to be electrically energized, the solenoid closes off the exhaust port. This allows oil pressure to build up and act upon a relay valve to reroute the torque converter oil. When this happens, the clutch apply plate is held tightly to the converter cover, and mechanical power through the converter drives the turbine shaft of the transmission. Additional valving is often present to control the aggressiveness of the clutch application process.

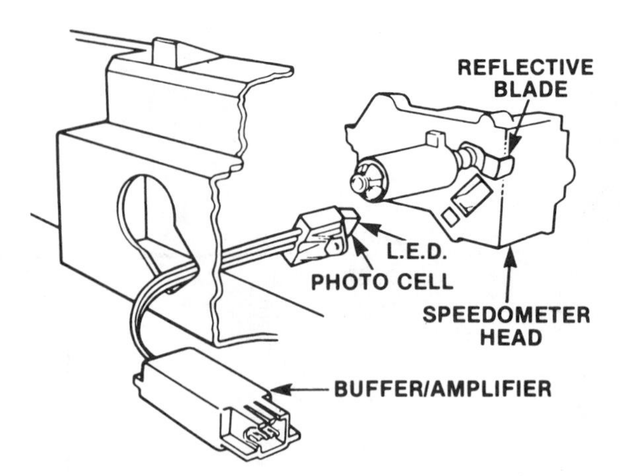

FIGURE 10–30 Optical sensor in speedometer head. (Courtesy of General Motors Corporation, Service Technology Group)

In diagnosing torque converter clutch problems, remember that the PCM/TCM relies on input signals to determine when the clutch should apply. When problems exist, check the fluid level and condition. If the fluid is okay, check for diagnostic trouble codes with a scan tool and correct any that are present.

❑ REVIEW

Key Terms/Acronyms

Transmission control module (TCM), powertrain control module (PCM), modulated, pulse width duty cycle, (pulse width modulated) solenoid, EEC-IV module, ECM (electronic control module), ECT (engine coolant temperature) sensor, TP (throttle position) sensor, MAP (manifold absolute pressure) sensor, VSS (vehicle speed sensor), TRS (transmission range selector) switch, PSA (pressure switch assembly), TFT (transmission fluid temperature) sensor, TIS (transmission input speed) sensor, BARO (barometric pressure) sensor, ALDL (assembly line diagnostic link), and DLC (data link connector).

Questions

Chrysler Corporation

1. Chrysler introduced hydraulically controlled converter lockup in the 36RH (A-727) and 30RH (A-904) TorqueFlite transmissions in production year __________. This style of lockup control continued through the __________ production year for RWD trucks and vans and through __________ production year for RWD passenger cars.

2. Name the valve that prevents lockup at speeds below 40 mph (64 km/h). ____________________

3. Name the valve that switches the circuitry for the unlock and lockup modes. ___________________

4. During the unlock mode, converter feed oil enters the converter through the ___________________ and exits through the converter ___________________.

5. The lockup control system is unaffected by engine coolant temperature. ___T/___F

6. The lockup control system is unaffected by closed throttle or brake apply. ___T/___F

7. Converter lockup is restricted to third gear. ___T/___F

Chrysler Electronic Lockup

1. Chrysler introduced electronic converter lockup control in the 900 series for RWD trucks and vans in production year __________. RWD passenger car applications of the 900 series and 36RH (A-727) followed in production year __________. Initially, Chrysler FWD automatic transaxles used nonlockup converters until electronic lockup was introduced in production year __________.

2. Chrysler 41TE (A-604) and 42LE transaxles use an independent transaxle control module (TCM) for converter lockup control. ___T/___F

3. The lockup control system is unaffected by engine coolant temperature. ___T/___F

4. In the four-speed 42RH (A-500), the lockup solenoid is hot-side switched. ___T/___F

5. The lockup solenoids are normally vented solenoids. ___T/___F

Ford A4LD Hydraulic-Electronic Lockup

_____ 1. In this system:
I. The converter clutch unlock and lockup modes are scheduled hydraulically.
II. The function of the solenoid is to inhibit converter lockup.

A. I only C. both I and II
B. II only D. neither I nor II

_____ 2. The position of the converter clutch shuttle valve:
I. is controlled by the solenoid.
II. is controlled by the converter clutch shift valve.

A. I only C. both I and II
B. II only D. neither I nor II

_____ 3. What can be said about the override solenoid?
I. It is controlled by the EEC-IV processor.
II. It is hot-side switched.

A. I only C. both I and II
B. II only D. neither I nor II

_____ 4. Converter lockup is possible in second, third, and fourth gears. ___T/___F

Mark the following conditions that inhibit or unlock the converter clutch:

_____ Engine coolant is below 128°F (53°C).
_____ Engine coolant is above 240°F (116°C).
_____ Heavy or WOT acceleration.
_____ Quick tip-ins and tip-outs.
_____ Application of brakes.
_____ Closed throttle.
_____ Low speed with a low vacuum condition.

Ford AXOD, E4OD, & 4EAT Electronic Lockup

Mark the transmission or transaxles that apply to the following statements about the converter lockup control.

1. The solenoid is interfaced with the EEC-IV module.
_____ AXOD _____ E4OD-Gas _____ 4EAT

2. The solenoid valve is a normally vented valve.
_____ AXOD _____ E4OD _____ 4EAT

3. The solenoid is hot-side switched.
_____ AXOD _____ E4OD _____ 4EAT

4. The lockup electronics uses a group of pressure switches mounted to the valve body to monitor the current status of transmission/transaxle operation.
_____ AXOD _____ E4OD _____ 4EAT

5. Lockup is possible in third and fourth gears only.
_____ AXOD _____ E4OD _____ 4EAT

6. When the solenoid is energized and the transmission/transaxle is in proper gear, the converter clutch engages.
_____ AXOD _____ E4OD _____ 4EAT

General Motors TCC Control

_____ 1. The first GM TCC-equipped automatic transmissions were introduced in production year:

A. 1978 C. 1981
B. 1980 D. 1983

_____ 2. The TCC release and apply is controlled by:
I. a hydraulic converter clutch valve.
II. a solenoid-operated check valve.

A. I only C. both I and II
B. II only D. neither I nor II

____ 3. The TCC solenoid:
I. is normally closed.
II. controls the oil to the converter clutch valve.

A. I only C. both I and II
B. II only D. neither I nor II

____ 4. The TCC applies:
I. in third gear only in most three-speed applications.
II. in all gears in most four-speed applications.

A. I only C. both I and II
B. II only D. neither I nor II

____ 5. In the electronically controlled solenoid system:
I. terminal D at the transmission connector is the ground-side return to the PCM.
II. the PCM keeps the solenoid at a constant ground.

A. I only C. both I and II
B. II only D. neither I nor II

____ 6. Which is true about the TCC solenoid?
I. It is ground-side switched.
II. When energized, it vents hydraulic oil.

A. I only C. both I and II
B. II only D. neither I nor II

The torque converter clutch (TCC) engages when:

1. ____________________
____

2. ____________________
____

3. ____________________
____

4. ____________________
____

5. ____________________
____

The TCC disengages when:

1. ____________________
____

2. ____________________
____

3. ____________________
____

11 Electronic Shift Control Circuits: Principles of Operation

OBJECTIVE:

The objective of this unit is for you to become proficient with the terms and concepts contained in this chapter. Completion of this objective is essential to becoming a successful automatic transmission technician.

CHALLENGE YOUR KNOWLEDGE

Define the following key terms/acronyms:

CCD/C^2D Bus, barometric manifold absolute pressure (BMAP) sensor, adaptive memory/adaptive strategy, EATX relay, MVLPS (manual valve lever position sensor), pressure control solenoid (PCS), MLPS (manual lever position switch), output speed sensor (OSS), and pulse generator.

List the following:

Two basic electrical circuit designs used to operate shift, converter clutch, and pressure control solenoids; two basic hydraulic circuit designs using solenoids to control oil pressure; two applications of pulse width duty cycle solenoids; three applications of pulse generators; ten electrical input signals to a transmission control module; and four output devices controlled by a transmission's control module.

Describe the operation of hydraulic circuits that use the following solenoids:

Pressure/exhaust solenoid; circuit flow control solenoid; pulse width duty cycle solenoid; and pressure control solenoid.

INTRODUCTION

Electronic shift control systems operate using basic hydraulic and electrical fundamentals. Chapter 10 highlighted the use of electronically controlled solenoids to operate torque converter clutch control systems. The principles employed to cause a converter clutch solenoid to move a valve and direct oil pressure into the proper hydraulic circuits are essentially the same ones used in electronic shift control systems.

Electronic shift transmissions come in many levels of sophistication. Typically, shift electronification is computer-controlled and eliminates the use of the hydraulic governor, with the throttle valve system often retained for line pressure control. Advanced state-of-the-art designs completely eliminate the governor and throttle valve. Shifting takes place by energizing and de-energizing solenoid valves (Figure 11–1). The block diagrams shown in Figures 11–2 and 11–3 compare the concept of hydraulic versus electronic shift control.

This chapter emphasizes the fundamental principles of electronic shift control and its influence on the hydraulic system to engage clutches and bands at the right time and intensity.

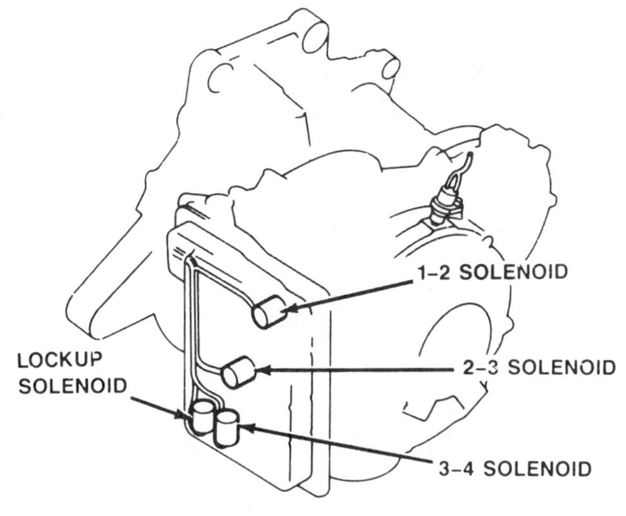

FIGURE 11–1 4EAT solenoid locations. (Reprinted with the permission of Ford Motor Company)

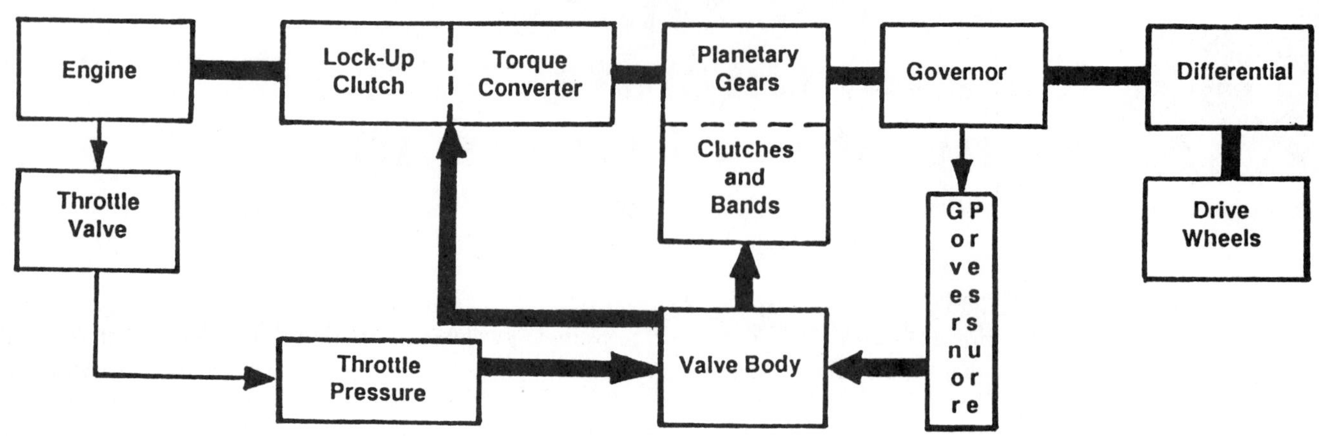

FIGURE 11–2 Shift system—hydraulic control.

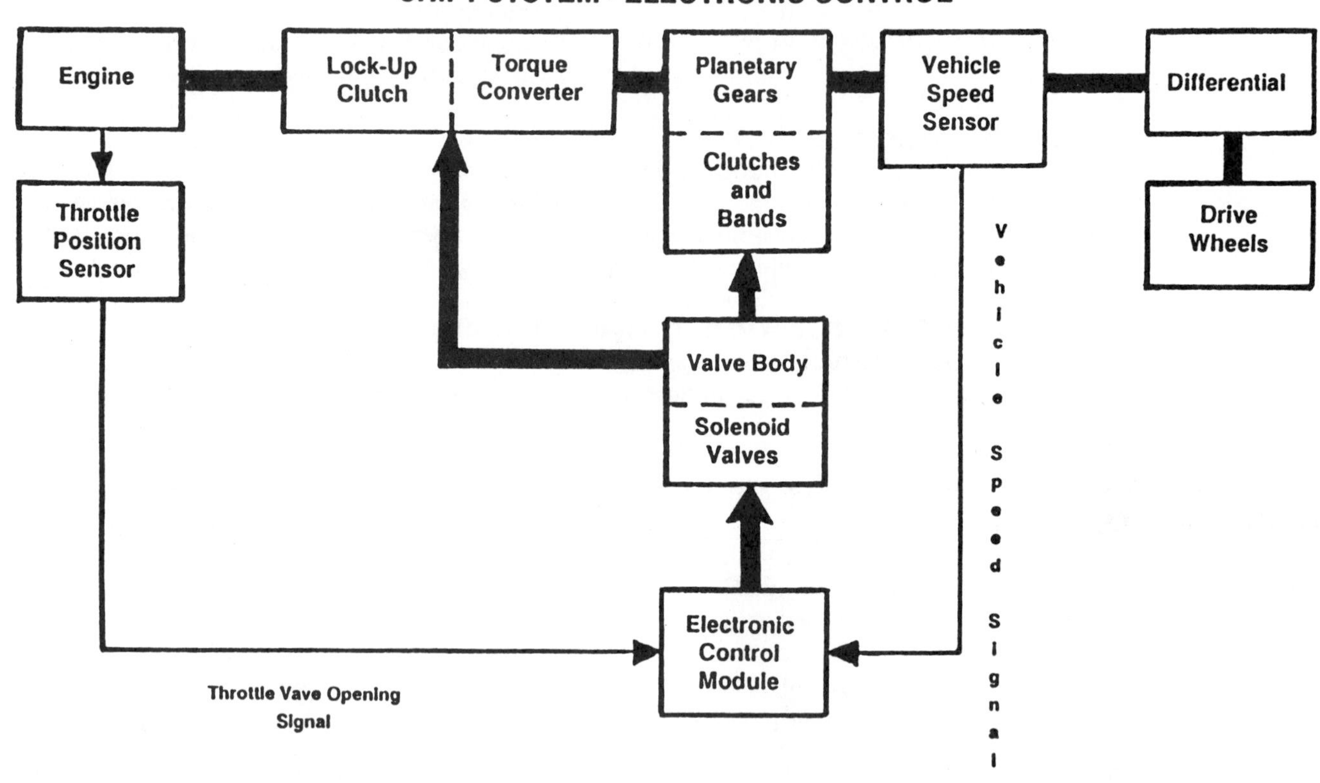

FIGURE 11–3 Shift system—electronic control.

Through the use of electronics, improved shift timing and feel is possible.

ELECTRONIC SHIFT BASICS

On gasoline engines, the operation of the shift solenoids, including the TCC solenoid, is usually interfaced directly with the powertrain control module (PCM). The PCM makes the shift decisions. Many of the engine management concerns are also concerns of the transmission. A few examples are throttle position, manifold absolute pressure, vehicle speed, coolant temperature, and engine rpm.

Where the electronic control is more complex, such as with the Chrysler 41TE (A-604) and 42LE, the transaxle control module (TCM) is linked with the ECM (Figure 11–4). This allows the two controllers to share information through a two-wire system referred to as the **CCD/C²D Bus**. With this system, the transaxle controller makes the shift decisions.

- **CCD/C²D Bus:** Chrysler's Computer Communication Data link, which is an electronic communication link

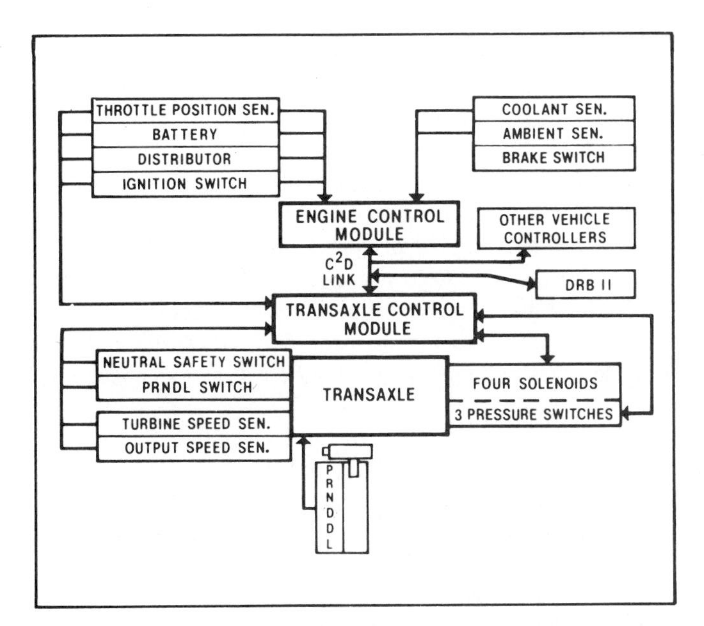

FIGURE 11–4 41TE (A-604) multiplexing CCD circuit. (Courtesy of Chrysler Corp.)

used to transmit information signals to electronic components with only two wires. It operates on multiplexing principles of electronics.

On most diesel engine applications, the transmission is controlled by a single TCM. Most of the input information is comparable to a gasoline engine PCM with some variations. In place of the throttle position sensor, a potentiometer similar to the TP sensor mounts to the fuel injection pump lever to indicate the amount of fuel delivery to the engine. A **barometric manifold absolute pressure (BMAP) sensor** that is similar in operation to a MAP sensor takes the place of that sensor. The shift schedule is adjusted for altitude. The output components are the same solenoids as those used with gasoline engines.

- **Barometric manifold absolute pressure (BMAP) sensor:** Operates similarly to a conventional MAP sensor; reads intake manifold pressure and is also responsible for determining altitude and barometric pressure prior to engine operation.

Shift Solenoid Operation

The shift solenoids are usually mounted to the valve body, or in the special design of the Chrysler 41TE (A-604), they are externally case-mounted within a solenoid assembly housing. The solenoid circuits are either ground-side or hot-side switched inside the computer (Figures 11–5 and 11–6). The solenoid actuates either a needle- or ball-style one-way check valve that controls fluid flow. Two common solenoid designs are used: pressure/exhaust solenoids and circuit flow control solenoids.

Pressure/Exhaust Port Solenoid Operation. The check valve and exhaust porting is designed to be either normally closed or normally open when a pressure/exhaust port solenoid is de-energized. The simplified illustration in Figure 11–7 shows how a normally closed solenoid controls the shift valve. Note that governor pressure and throttle pressure are absent from the shift valve, and mainline pressure keeps the shift valve closed against spring tension. When the computer scans the shift-related inputs and determines that a shift should take place, it energizes the solenoid, and the check valve opens the exhaust port (Figure 11–8). Mainline pressure, beyond an orifice, drains out at the exhaust port. With a loss of hydraulic oil pressure opposing the spring force, the spring opens the shift valve to a clutch or band apply circuit. Normally closed solenoids are spring-loaded (Figure 11–9).

A series of illustrations from the Ford Probe 4EAT show how normally closed solenoids control the shifting (Figures 11–10 to 11–12). Note that the 1-2 and 3-4 solenoids are

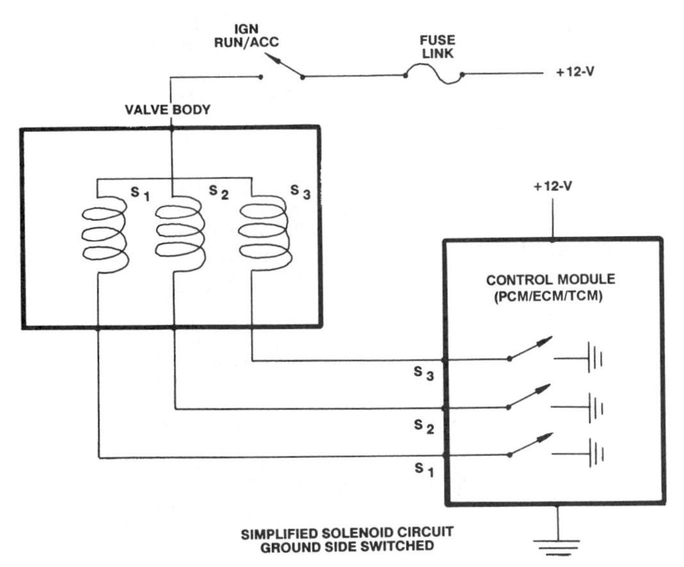

FIGURE 11–5 Simplified solenoid circuit ground-side switched; used in most domestic vehicles.

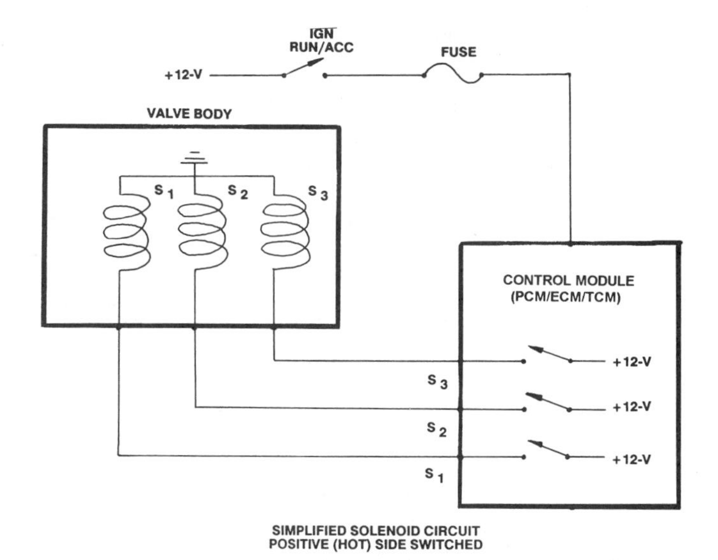

FIGURE 11–6 Simplified solenoid circuit hot-side (positive-side) switched; common in import/foreign vehicles.

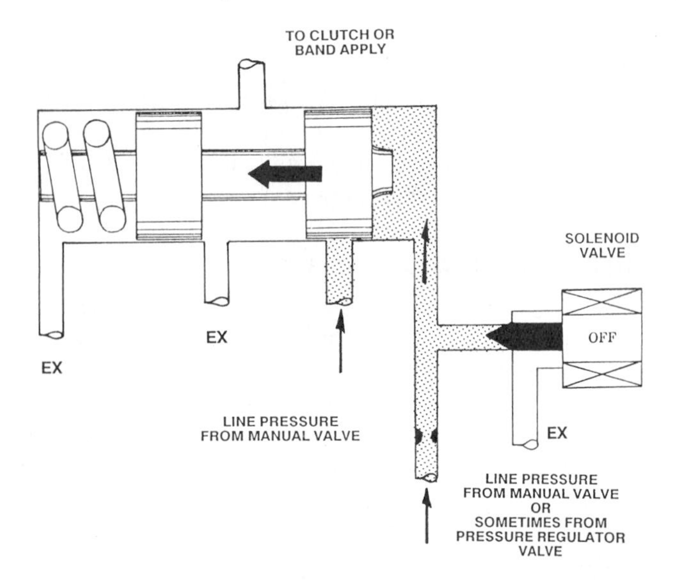

FIGURE 11–7 Simplified solenoid-controlled shift valve—closed.

normally closed, and the 2-3 solenoid is normally vented. Normally closed or normally vented solenoids are used in switching the clutch valve in TCC circuitry. Switching of the converter clutch valve ON/OFF is no different than switching a shift valve.

Circuit Flow Control Solenoid Operation. Figure 11–13 illustrates a simplified version of a circuit flow control solenoid. It can also be classified as a pulse width duty cycle solenoid and is used in Chrysler 41TE (A-604) and 42LE transaxles. In these units, the solenoids actually replace shift valves.

The computer can read actual transaxle input and output speeds from the turbine speed and output speed sensors (Figure 11–14). It determines the length of time needed to complete each shift. Through the use of pulse width duty cycle solenoids, it can alter the fill rate and pressure buildup time for each clutch to optimize vehicle shift quality.

The TCM automatically adapts for engine performance and friction element torque variations. This concept is referred to as **adaptive memory** or **adaptive strategy**. It does this by determining the exact fluid volume and time it will take to apply the clutch for given engine torque. Once it learns the parameters of engine torque and transmission hydraulics, the computer stores the data in memory. Should the parameters change, the computer relearns.

> ❖ **Adaptive memory/adaptive strategy:** The learning ability of the TCM or PCM to redefine its decision-making process to provide optimum shift quality.

The 41TE (A-604) and 42LE use four pulse width duty cycle solenoids. In the de-energized position, two solenoids block oil circuits, and the other two open oil circuits.

Figure 11–15 shows the 41TE (A-604) hydraulic schematic for direct (third) gear operation (lockup converter clutch off), obtainable because S2 and S4 solenoids are energized. Note that in the upper right corner, the two clutch packs needed for direct gear, the OD (overdrive) and UD (underdrive), are

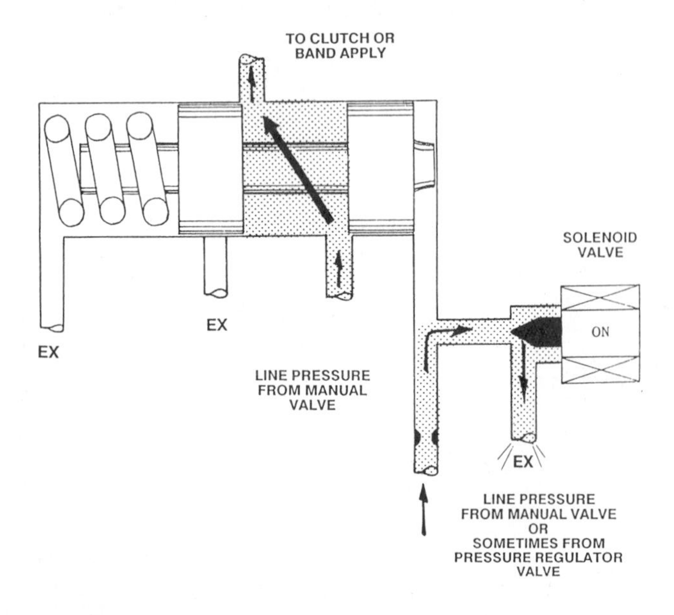

FIGURE 11–8 Simplified solenoid-controlled shift valve—open.

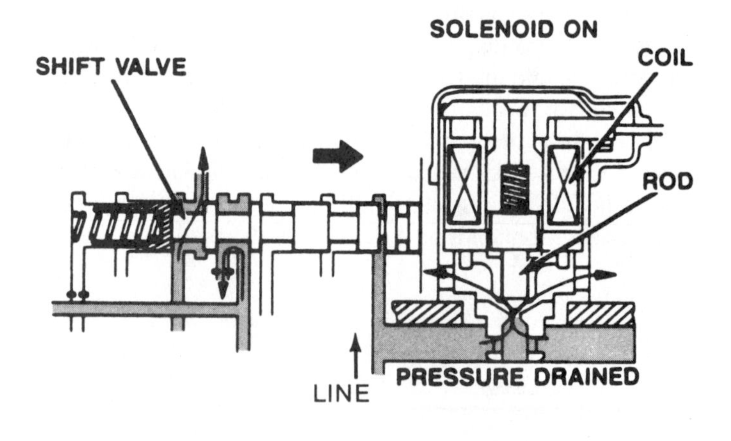

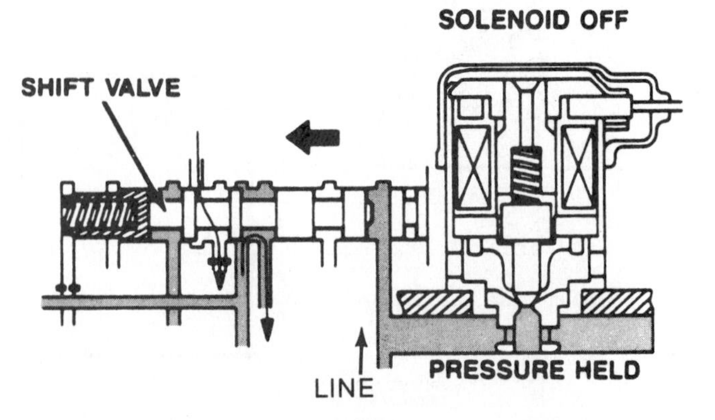

FIGURE 11–9 Solenoid valve operation. (Reprinted with the permission of Ford Motor Company)

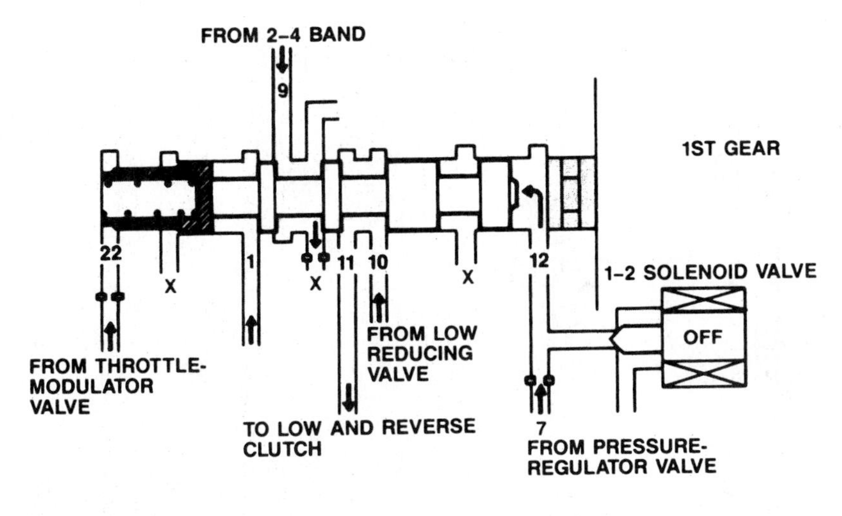

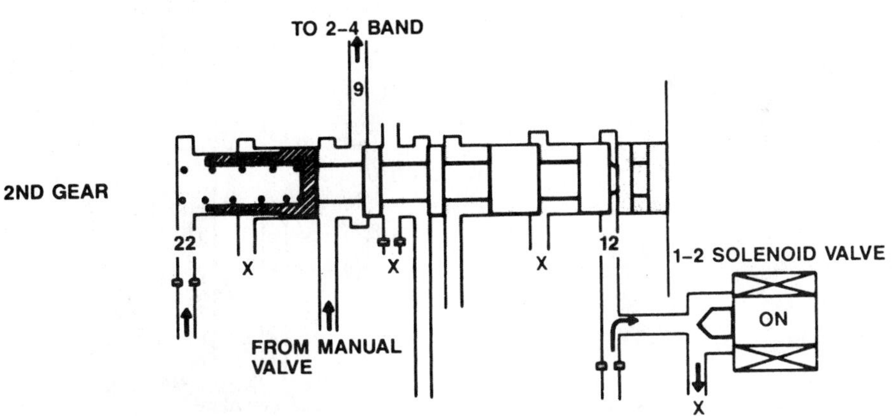

FIGURE 11–10 1-2 shift. (Reprinted with the permission of Ford Motor Company)

charged with oil pressure because solenoids S1 (underdrive) and S2 (overdrive) allow oil to flow. Also, observe how the other two solenoids, S3 (low-reverse/lockup converter clutch) and S4 (2-4/low-reverse), block oil passages.

When a solenoid is energized, the plunger pushes the check ball valve down, which either opens a blocked passage or closes an opened passage, depending upon the design.

Venting oil pressure in the clutch apply circuit occurs at the solenoid when blocking flow. The top of solenoids S2 and S3 have a tapered seat to allow fluid to exhaust when de-energized. These solenoids are referred to as "normally vented" (Figure 11–16). When solenoids S1 and S4 are de-energized, the clutch apply circuit is charged with oil pressure. These solenoids are referred to as "normally applied" (Figure 11–17).

In direct gear, S3 and S4 are blocking flow and venting clutch circuits. At the same time, S1 and S2 are pressurizing the clutch apply circuits (Figures 11–15 to 11–17).

What makes this system a hybrid is the ability of the computer to cycle the solenoids ON and OFF at variable rates referred to as pulse width duty cycles (percent ON/OFF time). A solenoid's duty cycling, to modulate apply oil, is adjusted by the TCM to any value between zero (solenoid de-energized closed—zero duty cycle) and full line pressure (energized—one hundred percent duty cycle). The computer adjusts the pulse width duty cycles in response to input data.

The 41TE (A-604) electronic circuit is diagramed in Figure 11–18. The basics of operation are as follows:

- There is always uninterrupted battery voltage to the controller (TCM) to keep the memory alive. Loss of battery voltage results in loss of TCM memory, which means the TCM must then relearn the characteristics of the transaxle for optimum shifting. It takes approximately ten shift cycles for the TCM to relearn shift characteristics on early production units, but only two shift cycles on later production units. Refer to a service manual for the exact steps of the procedure used to reset the adaptive memory.
- The ignition switch powers up the controller. When the switch is turned ON, the controller immediately makes a self-check before activating the system.
- The controller reads the input voltage. Voltages above 24 V or below 8 V could damage the computer, so at those points, the computer puts itself into default.
- While confirming the voltage, the controller simultaneously makes an internal self-test to determine whether the controller, **EATX relay** (solenoid shut down relay), and

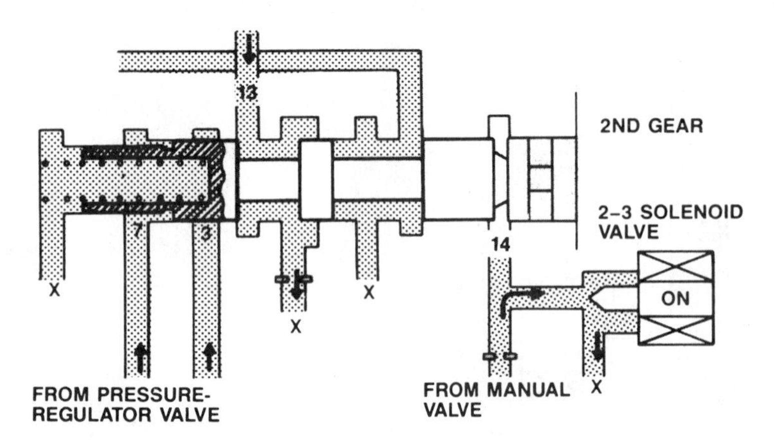

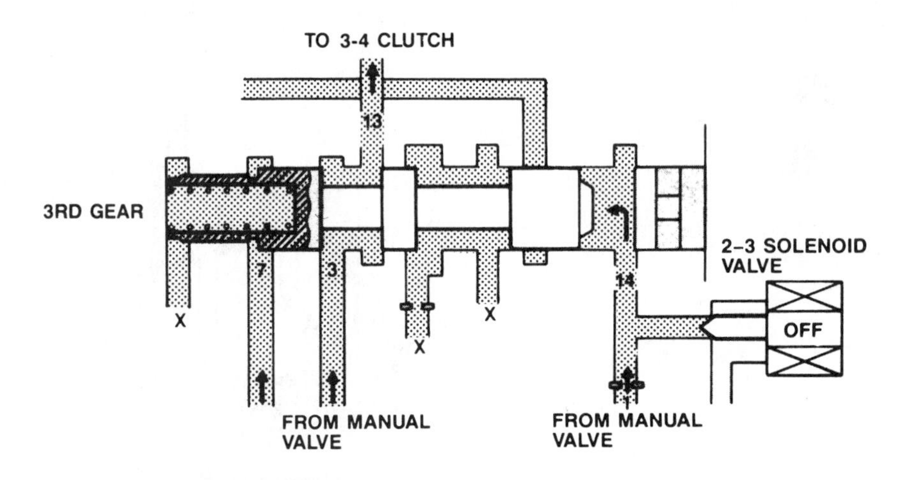

FIGURE 11–11 2-3 shift. (Reprinted with the permission of Ford Motor Company)

solenoid assembly are ready to go. The backup lamp relay is also keyed to the controller. It is basically operated by the neutral safety switch and the PRNDL switch.

- **EATX relay:** Chrysler's Electronic Automatic Transaxle Relay that, when energized, provides an electrical power input to the transaxle's TCM (controller). When certain malfunctions are noticed, the relay shuts down and de-energizes solenoid assembly, preventing a hazardous driving situation and/or transaxle damage.

The relays are located next to one another. As shown in Figure 11–19, they are often mounted on a bracket that either bolts to the 41TE (A-604) controller or on the right inner fender shield next to the strut tower.

- If the self-test confirms the system to be ready, the controller energizes the EATX relay. Operating voltage is now available to the solenoids and switches in the solenoid assembly.
- Whenever the controller goes into default, it de-energizes the EATX relay and cancels the operating voltage to the solenoids. The solenoids in their de-energized position put the transaxle permanently into second gear for all forward drive ranges. Neutral and reverse remain functional.

The C²D communication link between the 41TE (A-604) and 42LE transaxle controllers and engine controllers permits rapid exchange of information through a network called multiplexing. This concept is similar to numerous telephone conversations occurring over two wires—receiving and sending messages to and from different locations. For the pulse duty cycle solenoids to provide optimum shift quality, the transaxle and engine controllers need to continuously update one another.

The 41TE (A-604) and 42LE controllers put into memory numeric diagnostic trouble codes (DTCs) when malfunctions are detected in the major electronic circuits. Some hydraulic and mechanical malfunctions also produce a DTC (diagnostic trouble code) that is retrievable with a scanning device.

Refer to Chapter 22 for additional coverage of the 41TE (A-604) and 42LE electronic control systems, including adaptive memory.

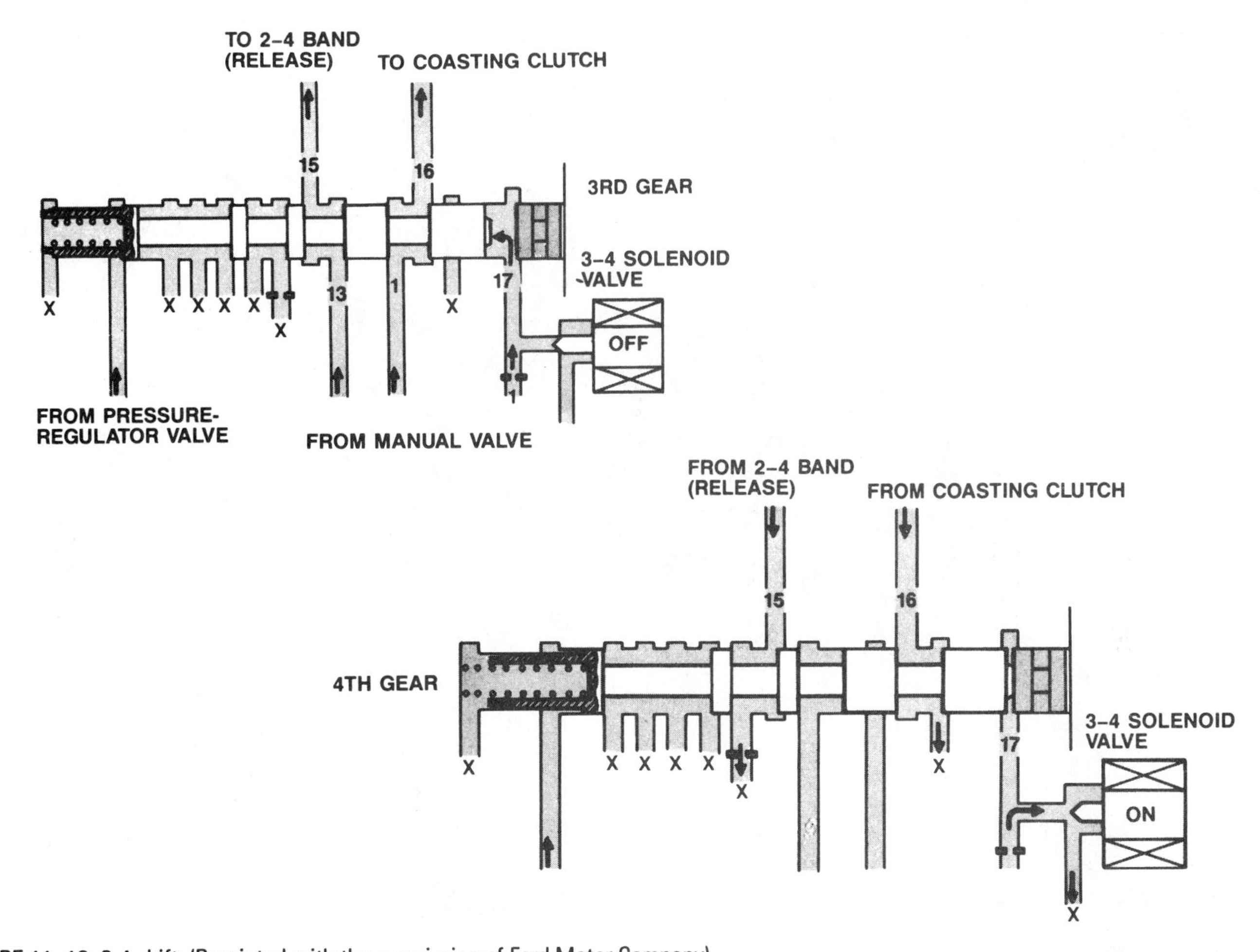

FIGURE 11–12 3-4 shift. (Reprinted with the permission of Ford Motor Company)

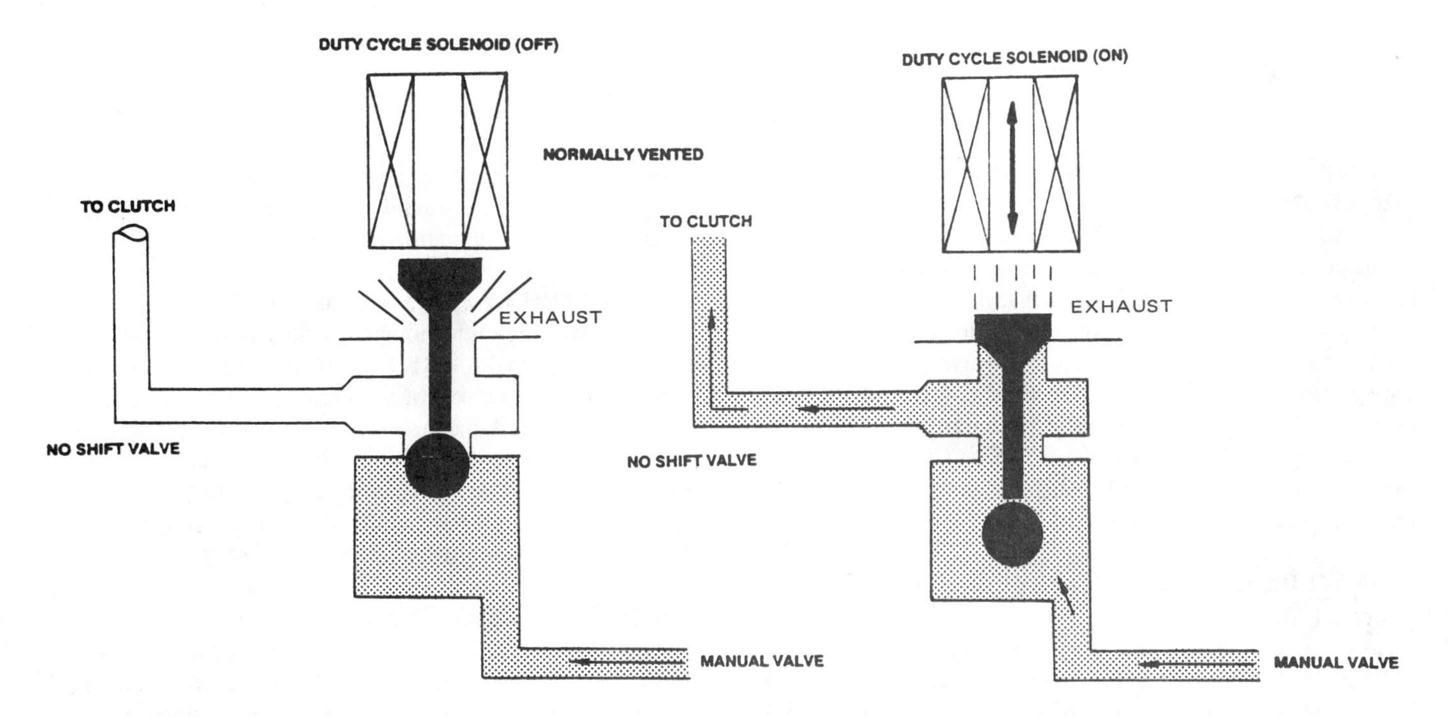

FIGURE 11–13 Simplified circuit flow control solenoid operation; pulse width duty cycle design.

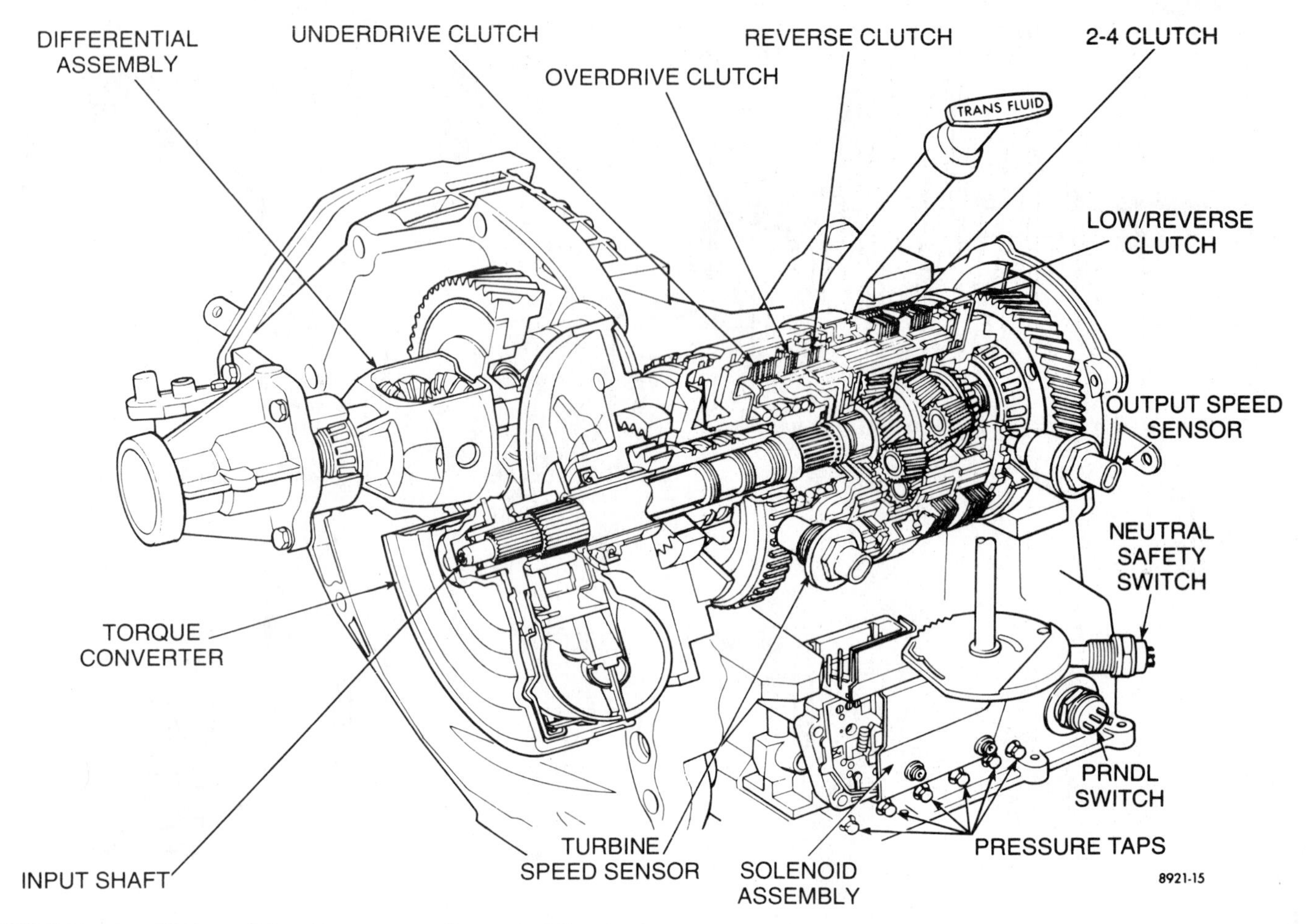

FIGURE 11–14 41TE (A-604) Ultradrive cutaway. (Courtesy of Chrysler Corp.)

Combined Hydraulic/Electronic Shift Systems

These systems are usually found where the manufacturer has incorporated an overdrive planetary with one of their traditional three-speed transmissions. The normal three-speed hydraulic shift system control remains the same. However, for the 3-4 shift control, a solenoid-operated shift valve is designed into the system. Examples include the Ford A4LD and Chrysler 40RH (A-500) and 42RH(A-518) transmissions. These units are basically C-3, 30RH (A-904), and 36RH (A-727) converted to overdrive transmissions. (The mechanical operation of these units was highlighted in Chapter 4.)

Ford A4LD Shift Control. In A4LD production units built prior to 1987, the 3-4 shift circuit was all hydraulic control. Newer versions incorporate electronics. The A4LD has two solenoids mounted in tandem on the valve body (Figure 11–20). They are referred to as the converter clutch override and 3-4 shift solenoids in the A4LD. The solenoids are wired to a three-terminal case connector with the center terminal connected to the 12-V power source. The outside terminals are independent solenoid ground-side returns to the computer. A simplified circuit illustration is shown in Figure 11–21.

Chrysler 40RH (A-500) Shift Control. The 40RH (A-500) also uses two solenoids mounted in tandem on the valve body (Figure 11–22). In the 40RH (A-500), the solenoids are referred to as the lockup and overdrive solenoids. Like the A4LD, the solenoids are wired to a three-terminal case connector with the center terminal connected to the 12-V power source. They are ground-side switched by the computer (Figure 11–23).

Refer to Chapter 22 for a 40RH (A-500) transmission circuit diagram, with the overdrive module and SMEC identified.

Shift Solenoid Sequencing

Common electronic shift control systems use two, three, or four solenoids. The sequencing of these solenoids in their ON and OFF status with one another provide the correct combinations to hydraulically activate the clutches and bands for planetary gear ratio engagement. Chart 11–1 (page 230) shows the

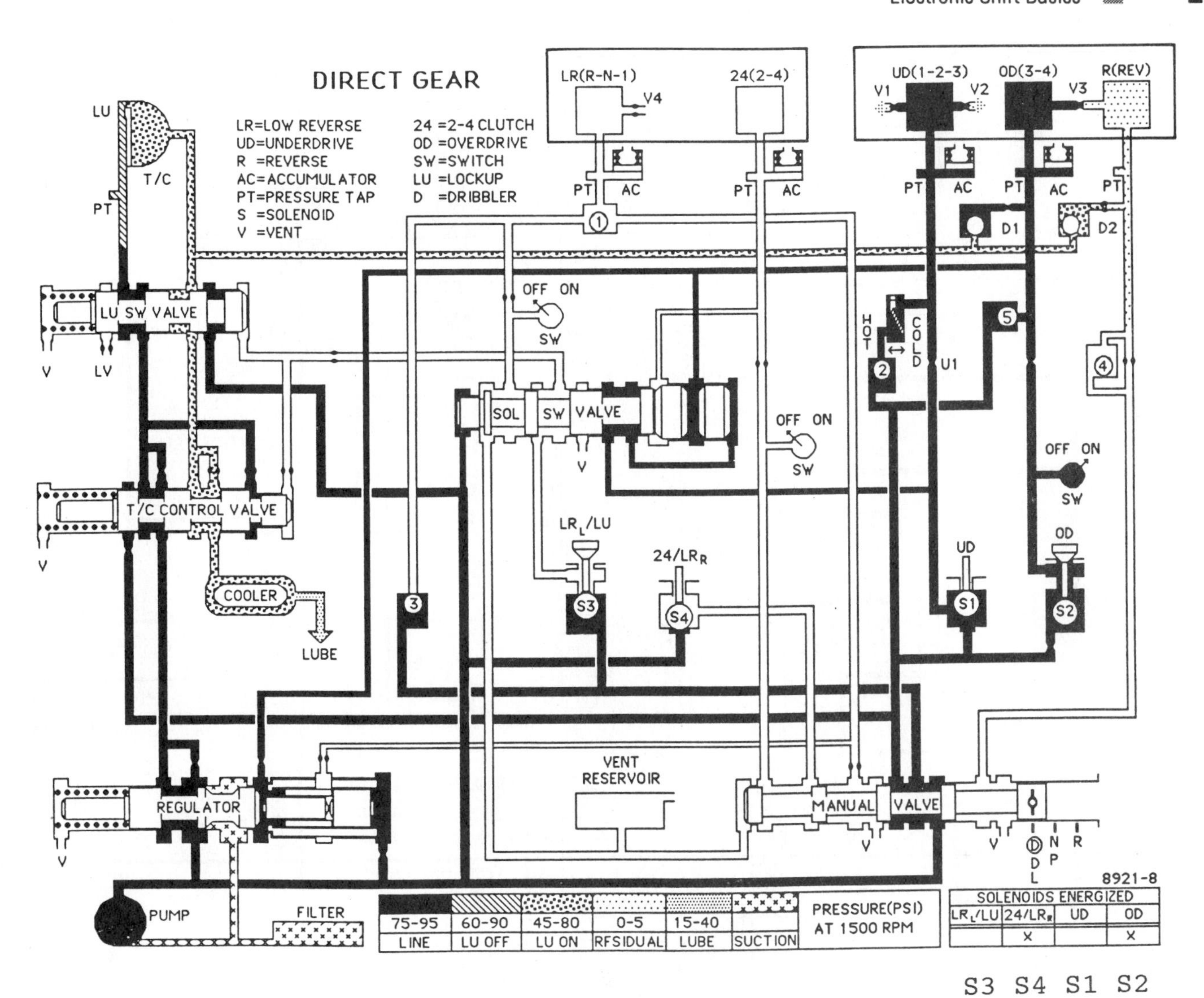

FIGURE 11–15 41TE (A-604)—direct gear. (Courtesy of Chrysler Corp.)

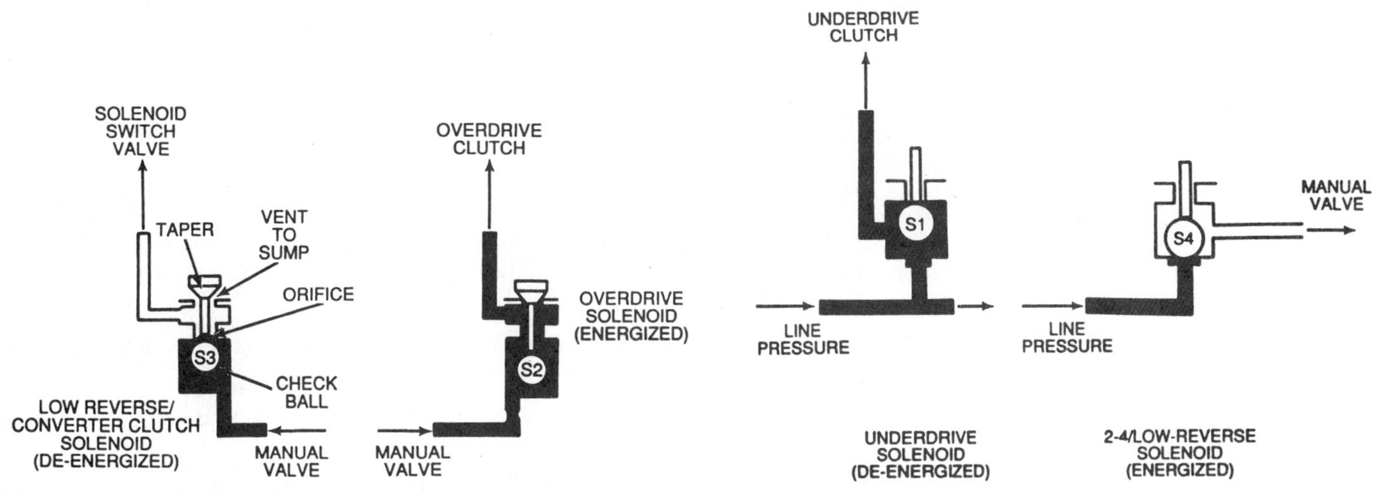

FIGURE 11–16 Low reverse/lockup converter clutch and overdrive solenoids. (Courtesy of Chrysler Corp.)

FIGURE 11–17 2-4/low reverse and underdrive solenoids. (Courtesy of Chrysler Corp.)

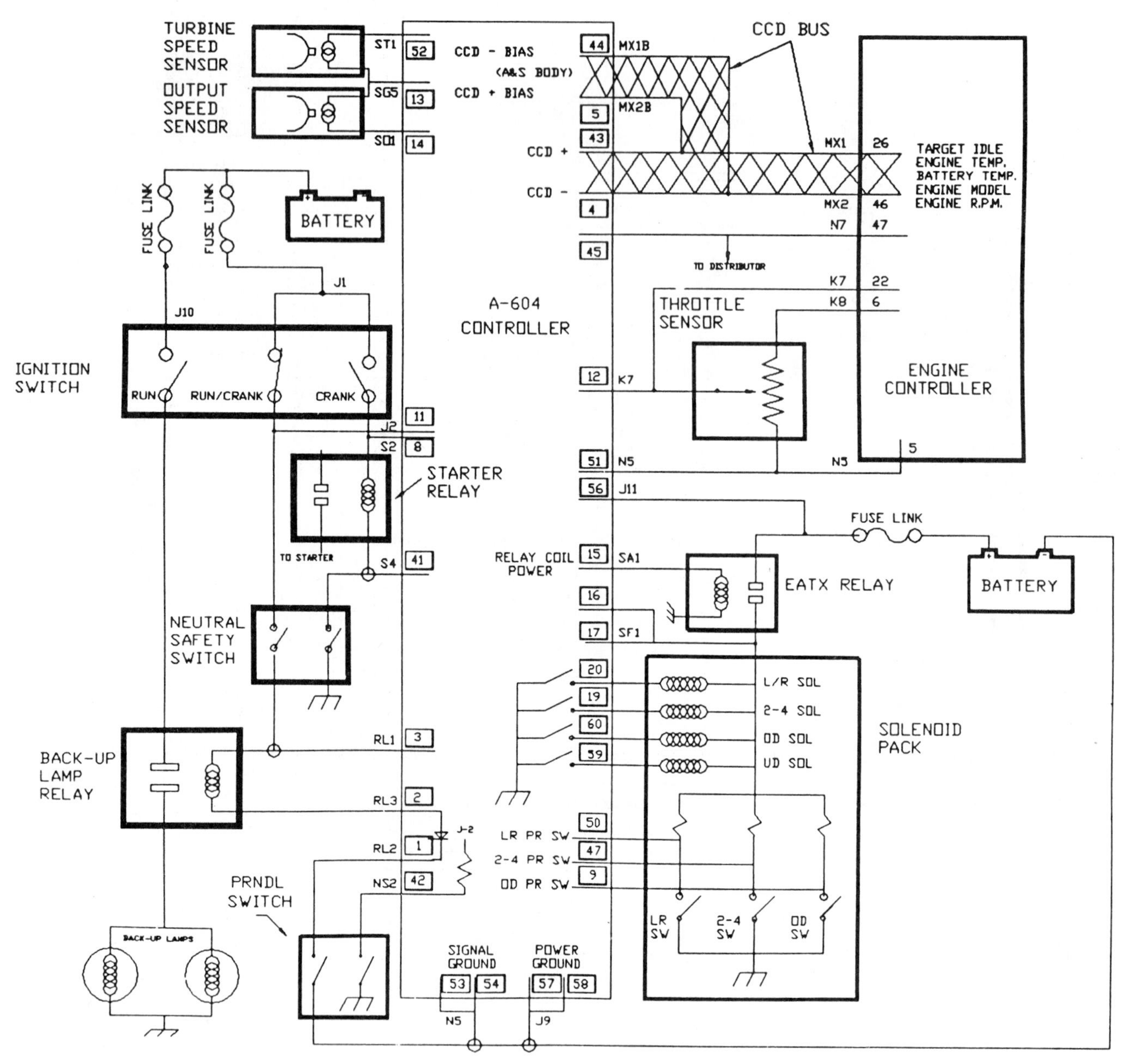

FIGURE 11–18 41TE (A-604) electronic circuit. (Courtesy of Chrysler Corp.)

combinations of the three shift solenoids (SS1, SS2, and SS3) used in a Ford AX4S (AXODE).

A wide variety of solenoid combinations exists throughout the automobile industry. This situation is compounded by the fact that virtually all transmission designs use hydraulic systems that are different from one another. The following discussion highlights a few common examples of shift solenoid sequencing.

Ford AXODE 1-2 Shift. The AXODE requires the same solenoid sequencing for first-gear operation in overdrive range and drive range as the AX4S: SS1 OFF, SS2 ON, and SS3 OFF (Chart 11–1, page 230). Figure 11–24 shows the basic solenoid and hydraulic action that occurs for first gear.

- Solenoid 1 (OFF) allows oil pressure to move the 1-2 shift valve to the right.
- Solenoid 2 (ON) blocks oil flow and needs to stay closed until third gear.
- Solenoid 3 (OFF) allows oil pressure to flow to the forward clutch.

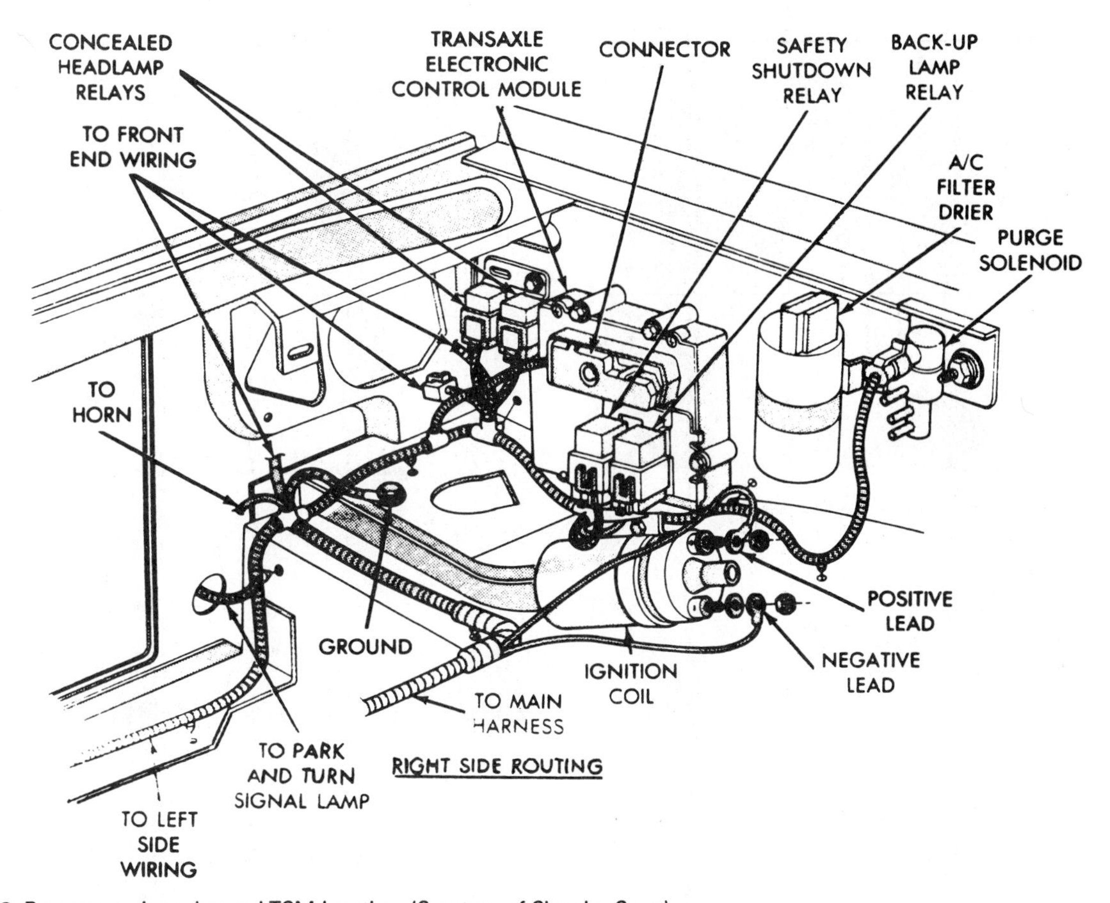

FIGURE 11–19 Representative relay and TCM location. (Courtesy of Chrysler Corp.)

FIGURE 11–20 Ford A4LD converter clutch override and 3-4 shift solenoids.

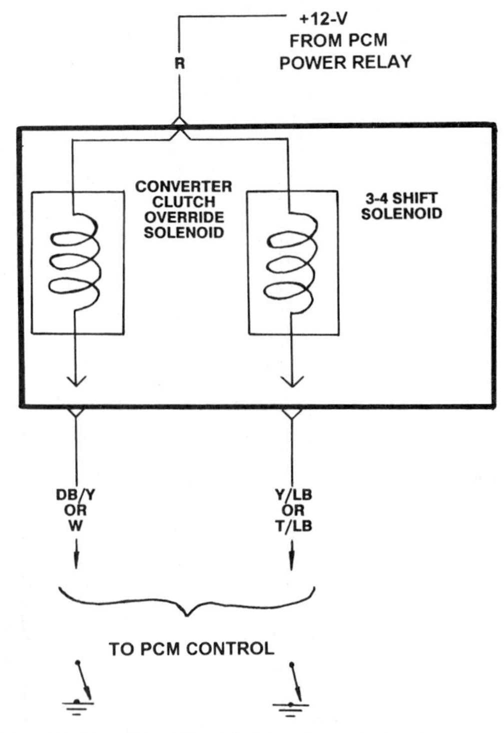

FIGURE 11–21 Simplified Ford A4LD solenoid circuitry.

FIGURE 11–22 Chrysler 40RH (A-500) lockup and overdrive solenoids.

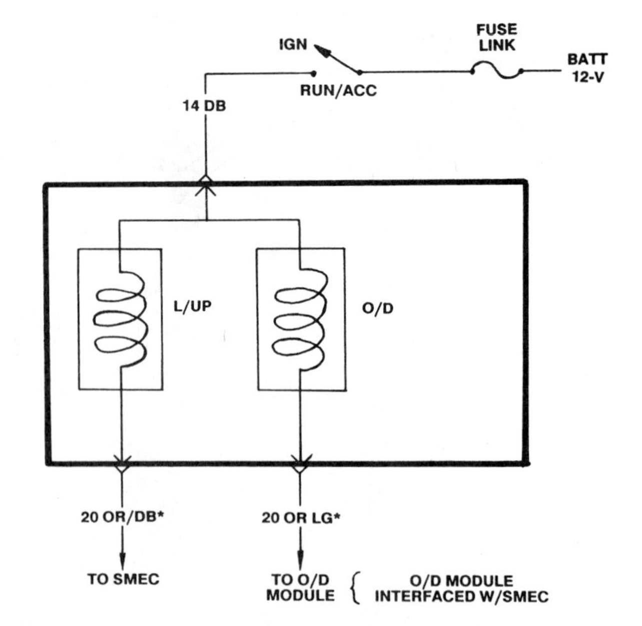

FIGURE 11–23 Simplified Chrysler 40RH (A-500) solenoid circuitry.

SOLENOID OPERATIONS CHART—AX4S

GEAR SELECTOR POSITION	POWERTRAIN CONTROL MODULE (PCM) COMMANDED GEAR	AX4S SOLENOIDS			
		ENG BRAKE	SS1	SS2	SS3
P/R/N	P/R/N	NO	OFF[a]	ON[a]	OFF
Ⓓ	1	NO	OFF	ON	OFF
Ⓓ	2	YES	ON	ON	OFF
Ⓓ	3	NO	OFF	OFF	ON
Ⓓ	4	YES	ON	OFF	ON
D or 3rd w/ Ⓓ OFF (SHO)					
1	1	NO	OFF	ON	OFF
2	2	YES	ON	ON	OFF
3	3	YES	OFF	OFF	OFF
SHO ONLY MANUAL 2	2	YES	ON	ON	OFF
2[b]	3[b]	YES	OFF	OFF	OFF
MANUAL 1	1	YES	OFF	ON	OFF
1[B]	2	YES	OFF	OFF	OFF
1	3	[c]	[c]	[c]	[c]
1	4	[c]	[c]	[c]	[c]

a Not contributing to powerflow.
b When a manual pull-in occurs above a calibrated speed the transaxle will downshift from the higher gear until the vehicle speed drops below this calibrated speed.
c Not allowed by hydraulics.

CHART 11–1 Solenoid Operations Chart—AX4S (AXODE). (Reprinted with the permission of Ford Motor Company)

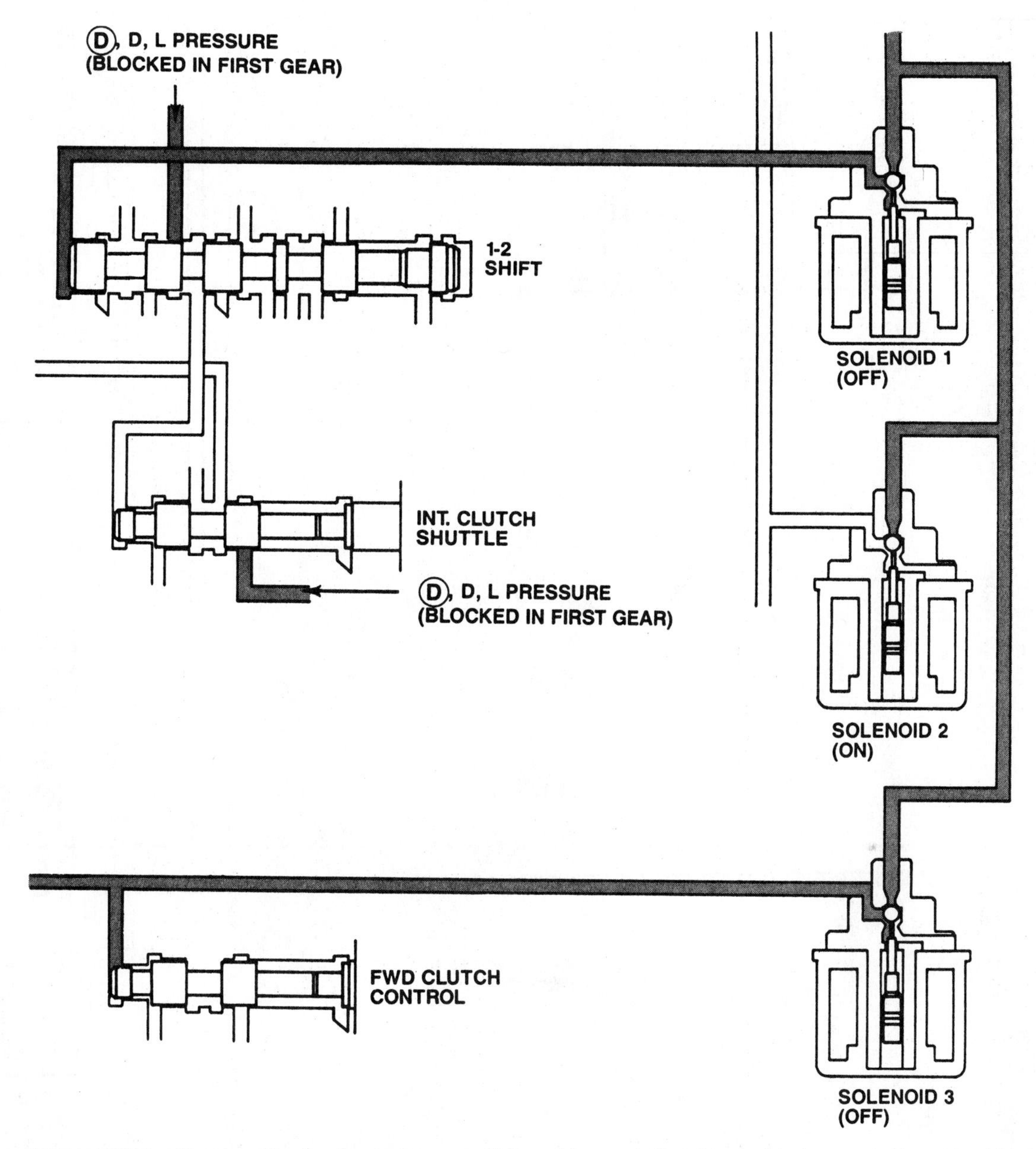

FIGURE 11–24 AX4S (AXODE) shift solenoid operation in first gear (O/D and D range). (Reprinted with the permission of Ford Motor Company)

An upshift from first to second gear is caused, as shown in Figure 11–25, by the following:

- Solenoid 1 turns ON, while solenoid 2 remains ON and solenoid 3 remains OFF.
- Solenoid 1 (ON) now blocks oil pressure to the left end of the 1-2 shift valve. The 1-2 shift valve moves left, and oil flows through the valve and applies pressure to the intermediate clutch shuttle valve, causing it to move right. Oil is now able to flow through the intermediate clutch shuttle valve and into the circuit to apply the intermediate clutch for second gear.

GM 4T60-E Solenoid Sequencing—Forward Gears. The 4T60-E uses two solenoids (A and B) to provide shift valve movement to obtain the four forward gears (Chart 11–2). The solenoid action and hydraulic flow for each gear are as follows:

- First gear (Figure 11–26). Both solenoids are ON and the manual valve directs fluid to the necessary components.

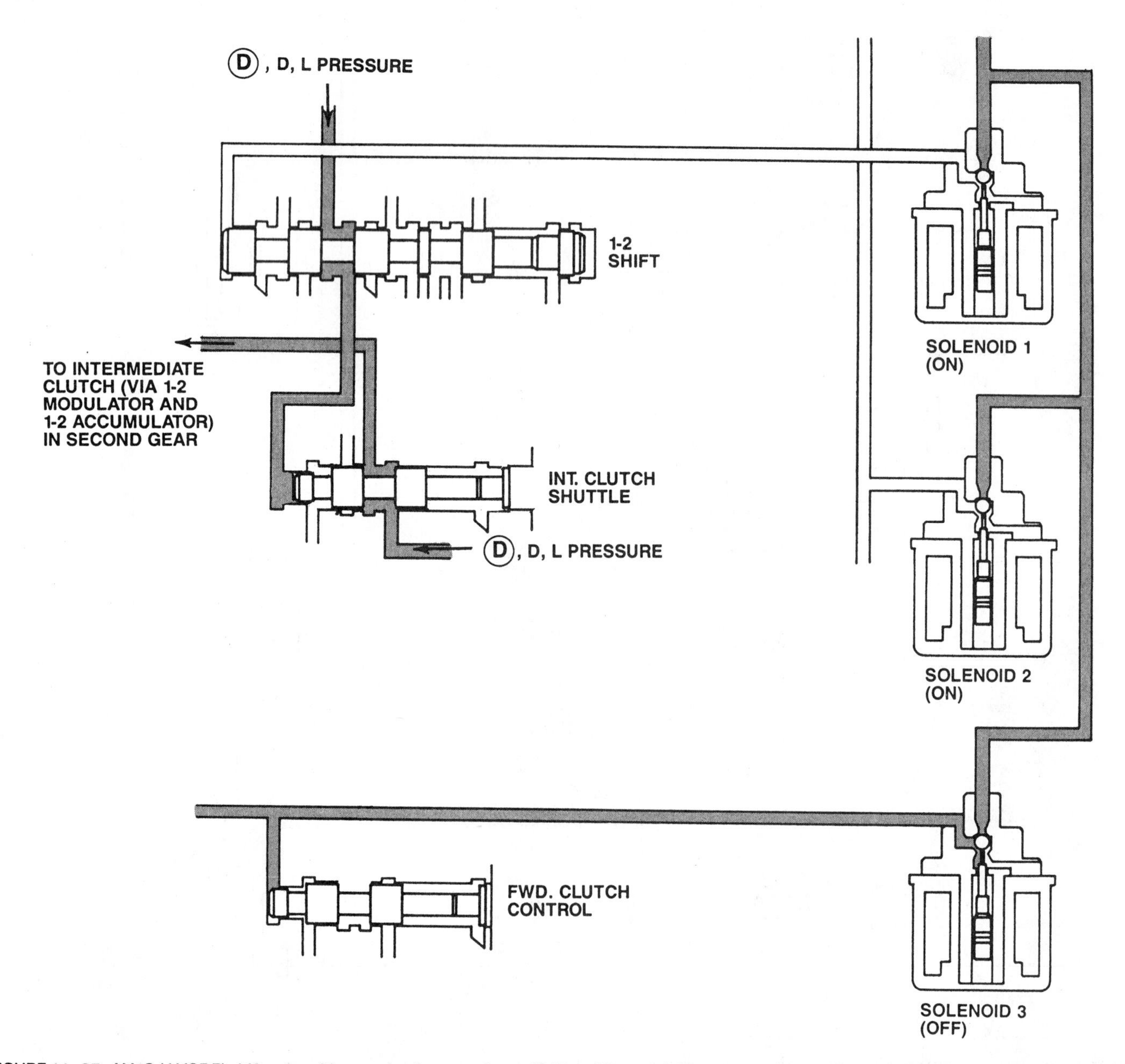

FIGURE 11–25 AX4S (AXODE) shift solenoid operation in second gear (O/D and D range). (Reprinted with the permission of Ford Motor Company)

Gear	Solenoid	
	A	B
1st	On	On
2nd	Off	On
3rd	Off	Off
4th	On	Off

CHART 11–2 4T60-E shift solenoid operation. (Courtesy of General Motors Corporation, Service Technology Group)

- Second gear (Figure 11–27). Solenoid A is OFF and solenoid B is ON. Solenoid A exhausts fluid pressure on the left end of the 1-2 shift valve. Spring force causes the valve to move left, opening the hydraulic circuit to second-gear components.
- Third gear (Figure 11–28). Both solenoids are OFF. Solenoid B exhausts fluid pressure, causing the 2-3 shift valve and the 3-2 and 4-3 downshift valves to move right. Oil flows through the 2-3 shift valve and activates third gear.
- Fourth gear (Figure 11–29). Solenoid A is ON and solenoid B is OFF. Solenoid A causes the 3-4 shift valve to move to the right. Oil flows through the 3-4 shift valve, and fourth-gear components are hydraulically pressurized.

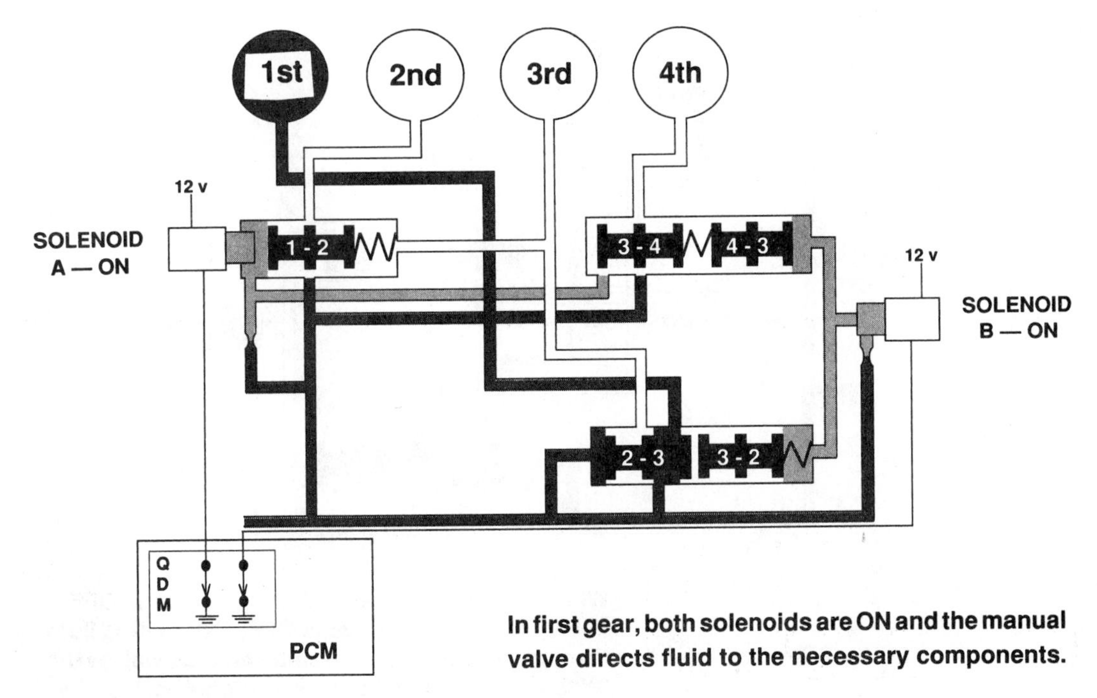

FIGURE 11–26 4T60-E first-gear hydraulic flow. (Courtesy of General Motors Corporation, Service Technology Group)

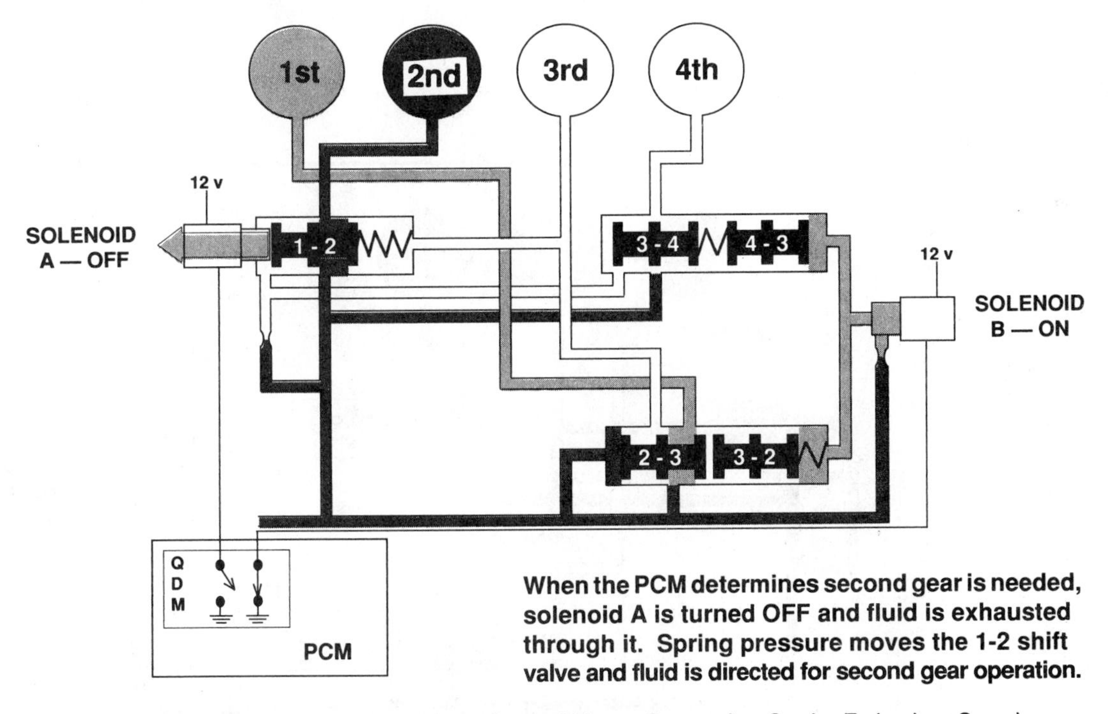

FIGURE 11–27 4T60-E second-gear hydraulic flow. (Courtesy of General Motors Corporation, Service Technology Group)

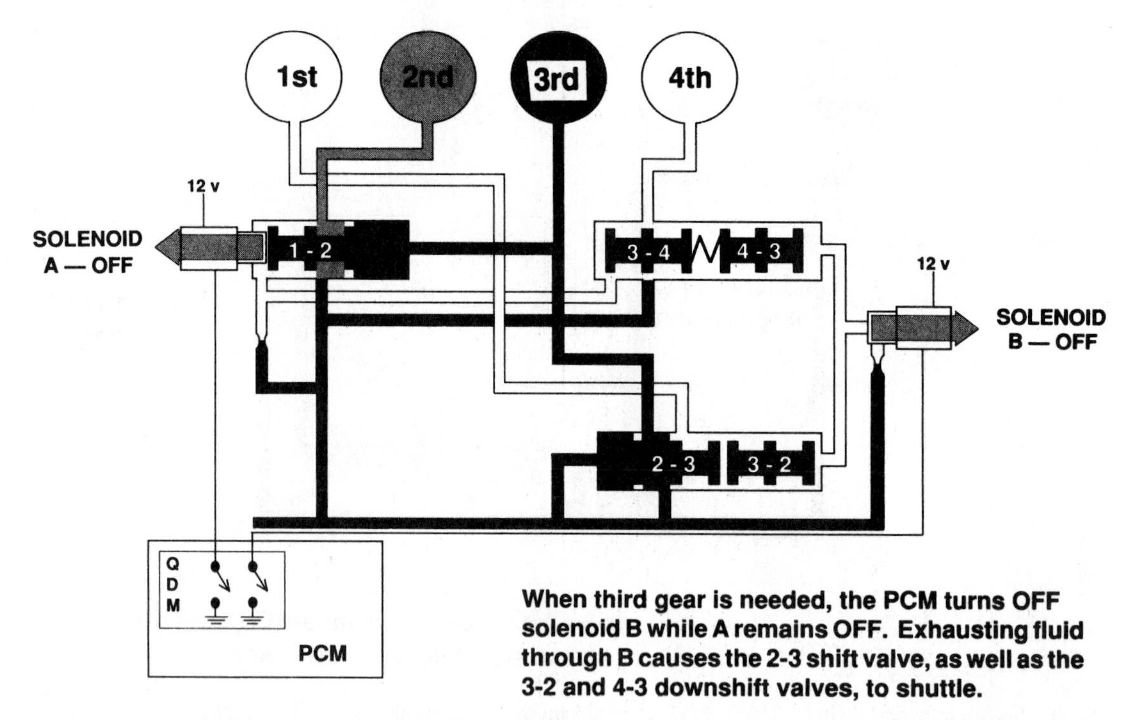

FIGURE 11–28 4T60-E third-gear hydraulic flow. (Courtesy of General Motors Corporation, Service Technology Group)

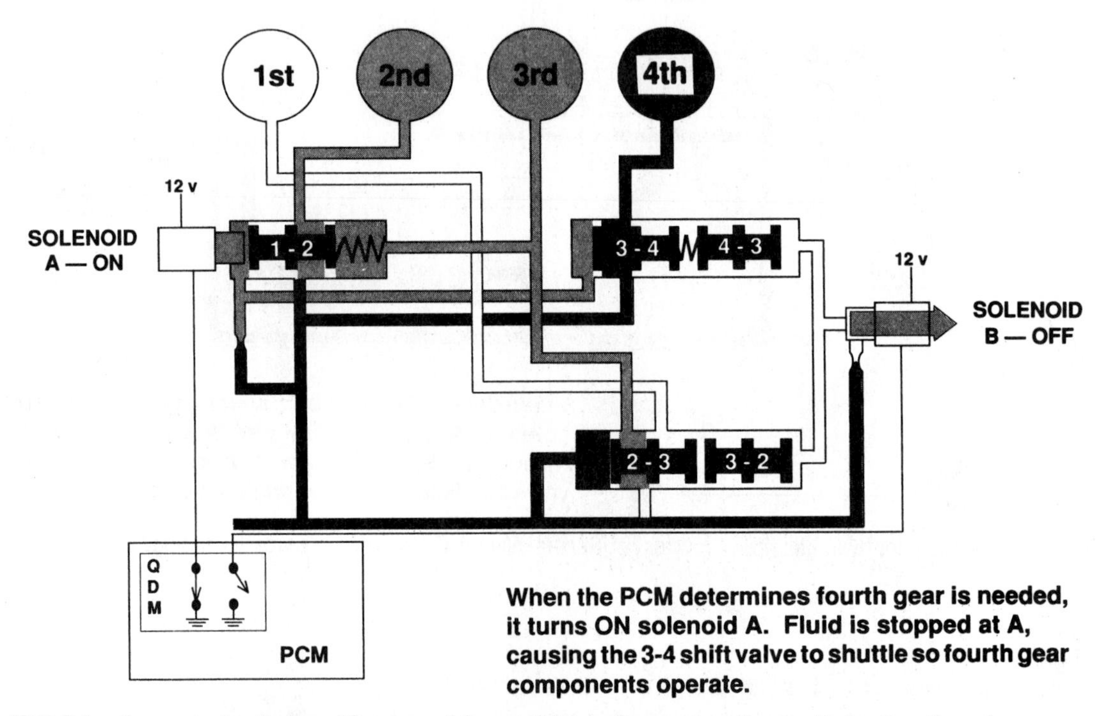

FIGURE 11–29 4T60-E fourth-gear hydraulic flow. (Courtesy of General Motors Corporation, Service Technology Group)

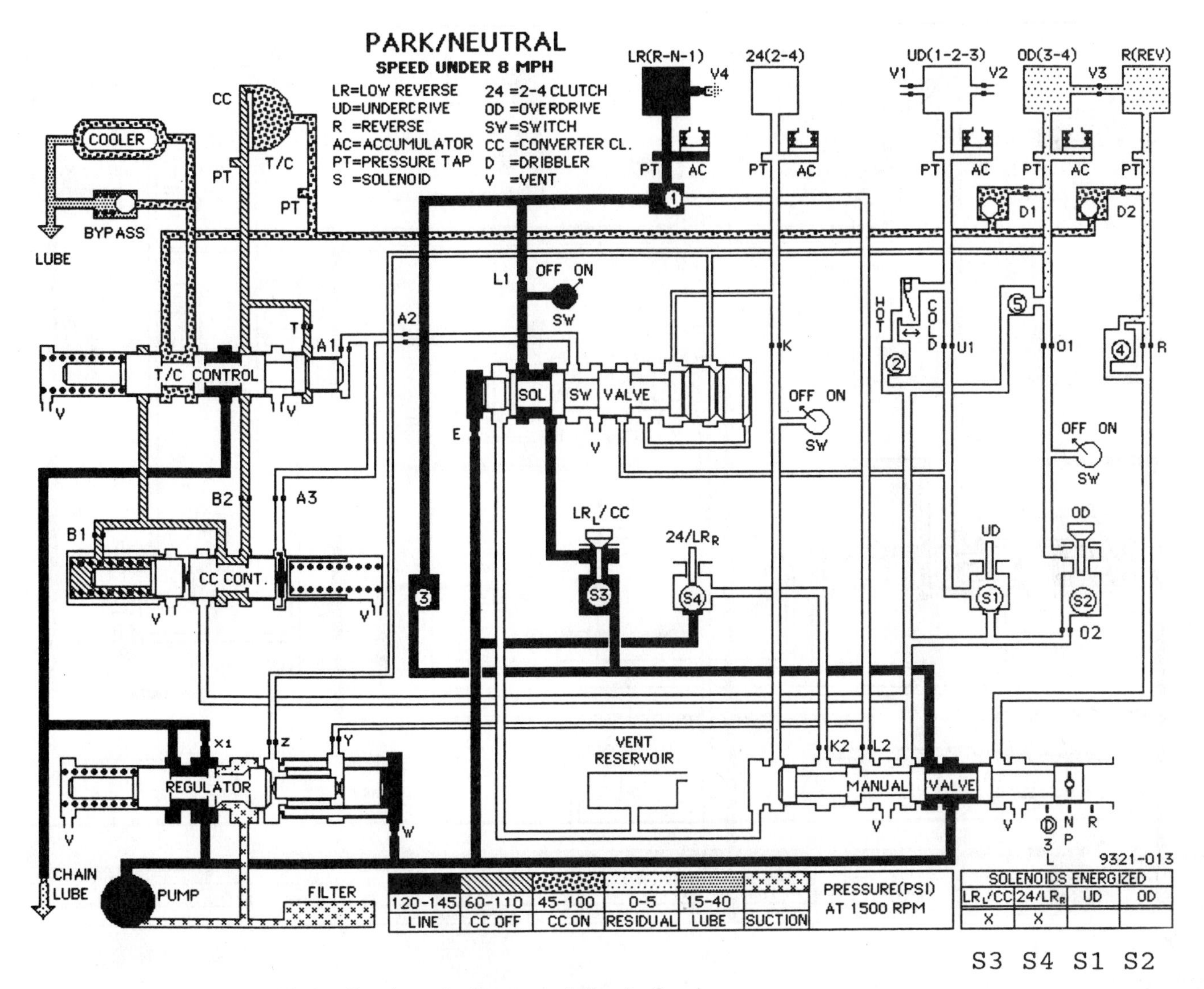

FIGURE 11–30 42LE park/neutral hydraulic schematic. (Courtesy of Chrysler Corp.)

Chrysler 42LE Solenoid Sequencing. Four solenoids operate in various combinations to provide oil flow to the necessary clutch housings to obtain park, neutral, reverse, and the forward gears. The 42LE hydraulic operation and schematics are similar to those of the 41TE and are described below.

Park/Neutral (Figure 11–30). Solenoids 3 and 4 are ON. Hydraulically, park and neutral are identical. The manual shaft mechanism moves the manual valve into the same position for park and neutral. In park position, the parking pawl locks into lugs attached to the rear carrier/output shaft unit.

The five clutch units are shown along the top right of Figure 11–30. Both the LR (low reverse) and 2-4 clutch units hold planetary members to the case. There are three input clutch units to drive planetary members. They are identified as the UD (underdrive), OD (overdrive), and R (reverse).

In the lower right corner of Figure 11–30 is a chart titled "Solenoids Energized." The two solenoids energized are marked with an "X"—the S3 (low reverse/converter clutch) and S4 (2-4/low reverse). Observe that the S3 and S4 check balls are pushed down. This combination allows the LR (low reverse) clutch unit to be pressurized. Since none of the three input clutch units (UD, OD, or R) are applied, a neutral condition exists.

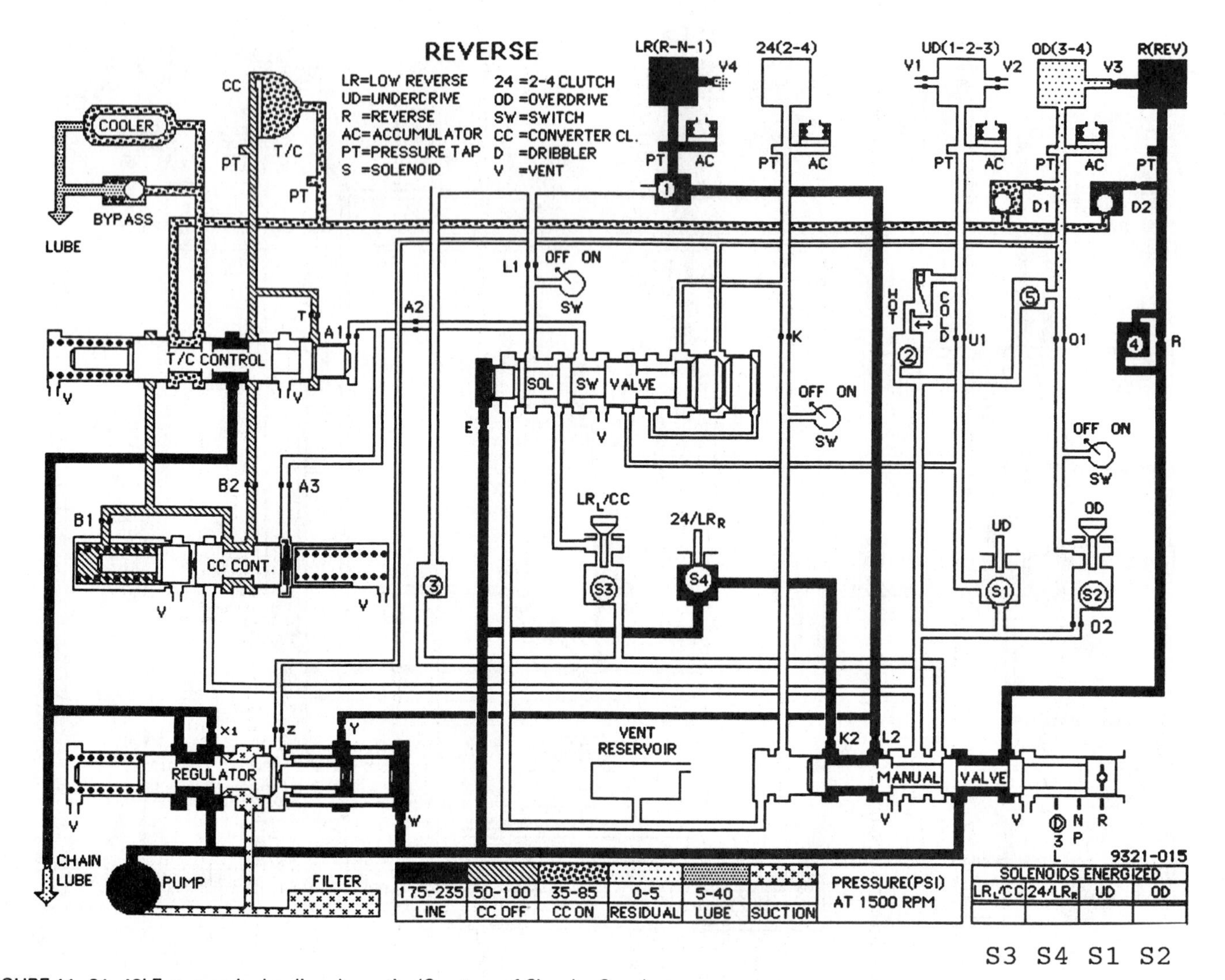

FIGURE 11–31 42LE reverse hydraulic schematic. (Courtesy of Chrysler Corp.)

Reverse (Figure 11–31). All solenoids are OFF. The manual valve routes oil to the LR (low reverse) and R (reverse) clutch housings. Note: If the vehicle is moving forward above 8 mph (13 km/h) and reverse range is selected, solenoid 4 is energized by the TCM. The solenoid blocks flow to the LR (low reverse) clutch housing, and a neutral condition occurs, rather than reverse. This action prevents damage to the transaxle.

First Gear (Figure 11–32). Solenoid 3 and 4 are ON. Oil flows through solenoids 3 and 1 to apply the LR (low reverse) and UD (underdrive) clutch housings.

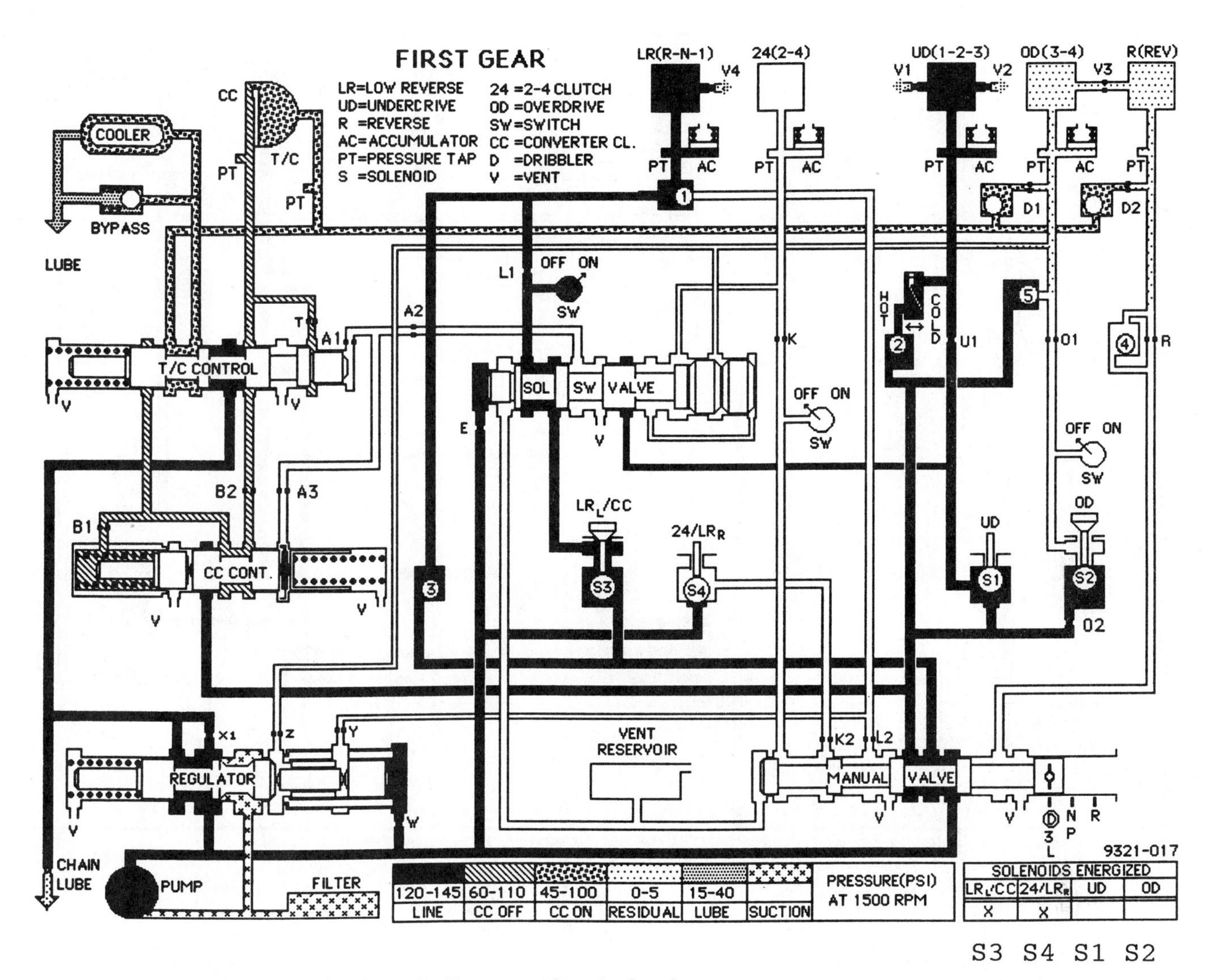

FIGURE 11–32 42LE first-gear hydraulic schematic. (Courtesy of Chrysler Corp.)

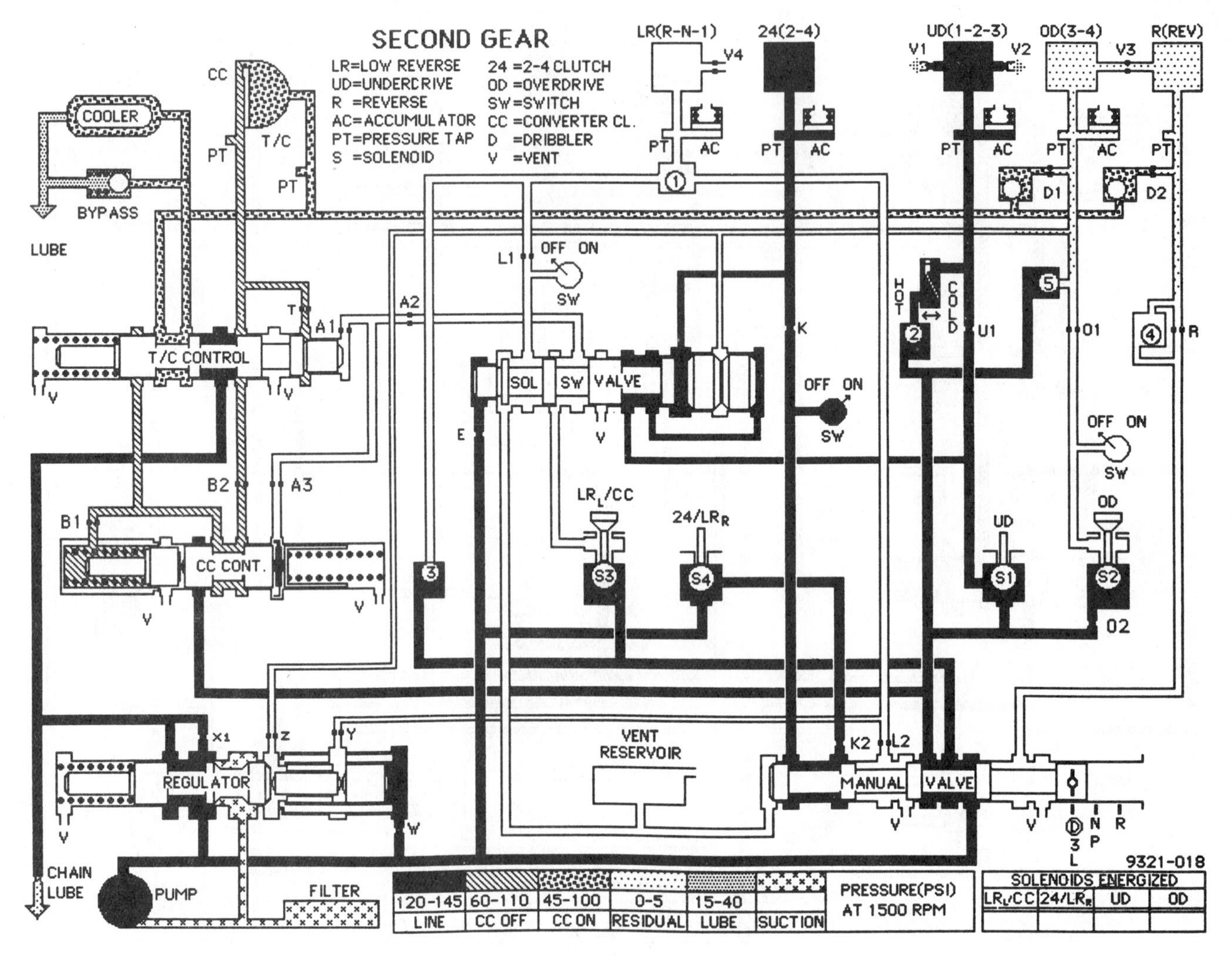

S3 S4 S1 S2

FIGURE 11–33 42LE second-gear hydraulic schematic. (Courtesy of Chrysler Corp.)

Second Gear (Figure 11–33). All solenoids are OFF. Oil flows through solenoids 4 and 1 to activate the 2-4 and UD (underdrive) clutch units.

The torque converter clutch is released in Figure 11–33. Under certain conditions in second gear, the TCM modulates apply oil for partial lockup. Modulation is accomplished by pulsing solenoid 3 (low reverse/converter clutch).

Direct (Third) Gear (Figure 11–34). Solenoids 4 and 2 are energized. Oil flows through S1 and S2 and into the UD (underdrive) and OD (overdrive) clutch housings. Both of these are input devices and together produce direct gear.

In direct gear, the converter clutch can be released, partially applied, or fully engaged. In Figure 11–34, the converter clutch is partially applied as S3 modulates the oil flow to the torque converter control and converter clutch control valves. A partial lockup condition is referred to as electronically modulated converter clutch (EMCC).

Overdrive (Fourth) Gear (Figure 11–35). Solenoids 1 and 2 are ON. Oil flows through S4 and S2 to activate the 2-4 and OD (overdrive) clutch housings.

In overdrive gear, the converter clutch can be released, partially applied, or fully engaged. Figure 11–35 shows the converter clutch fully engaged. Solenoid 3 is energized and allows full line pressure to act upon the torque converter control and converter clutch control valves.

Manual Range—Drive/3. Unlike overdrive range, where all four forward gears are available, in drive/3 range, fourth gear is eliminated. Refer to Figure 11–35 and observe that the manual valve has three positions. Ranges O/D, 3, and L are achieved with the manual valve in the same position. The TCM recognizes the selector lever position, based on inputs from the manual valve lever position sensor **(MVLPS)** (Figure 11–36), which determines which gears and shift pattern to use.

❖ **MVLPS (Manual valve lever position sensor):** The input from this device tells the TCM what gear range was selected.

In drive/3 range, the 1-2 upshift occurs at the same speed as it does in OD range. The 2-3 shift occurs above 45 mph (72 km/h).

Manual Range—Low/L. The only difference between the low/L and drive/3 ranges is when the shifts occur. The upshifts are significantly delayed and occur to protect the engine from being damaged from overspeed. The 1-2 shift occurs at 38 mph (61 km/h), and the 2-3 shift at 70 mph (113 km/h). Downshifts also occur at higher speeds.

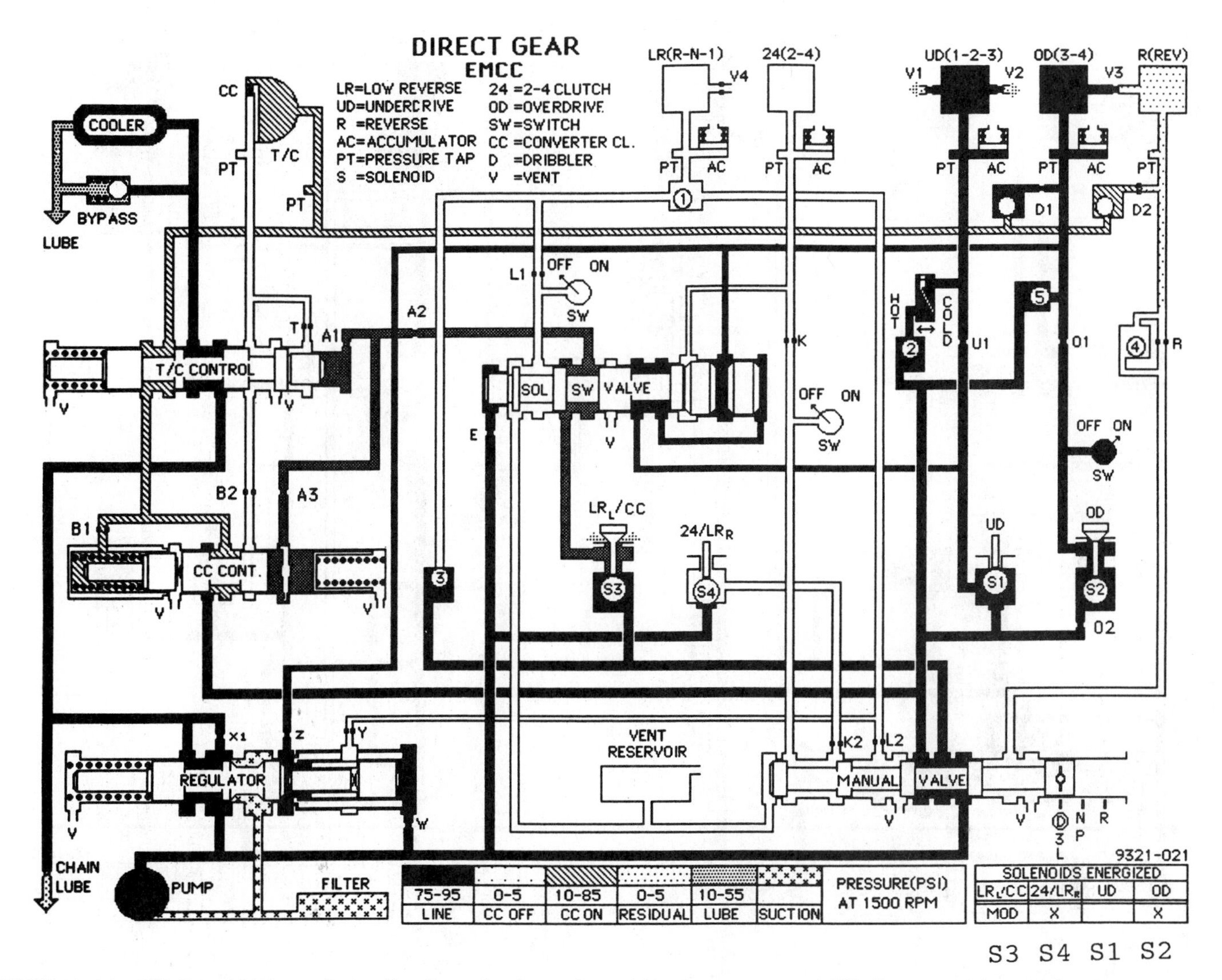

FIGURE 11–34 42LE direct (third) gear hydraulic schematic; electronic modulated converter clutch ON. (Courtesy of Chrysler Corp.)

Electronic Shift Feel and Quality

Many different methods are used to control the intensity of clutch and band application in electronically controlled transmissions and transaxles. These include accumulators, pulse width duty cycle shift solenoids, vacuum modulators, throttle valve cables, and electronic **pressure control solenoids (PCSs)**. Line pressure is directly related to shift feel and quality.

❖ **Pressure control solenoid (PCS):** An output device that provides a boost oil pressure to the mainline regulator valve to control line pressure. Its operation is determined by the amount of current sent from the PCM.

The regulation of line pressure can be controlled electronically with a PCS (pressure control solenoid). Within the PCS is a moveable spool valve (Figure 11–37). Its position is determined by the magnetic strength of the solenoid. With no current applied to the solenoid, spring force moves the spool valve out, and maximum line pressure flows through the valve. As more current is sent to the solenoid from the PCM, the spool valve is pulled in, and output pressure is reduced.

The output oil leaving the PCS is routed to the base of the main regulator valve. As a result, the output oil acts as a boost pressure on the main regulator valve and provides changes in mainline oil pressure (Figure 11–38).

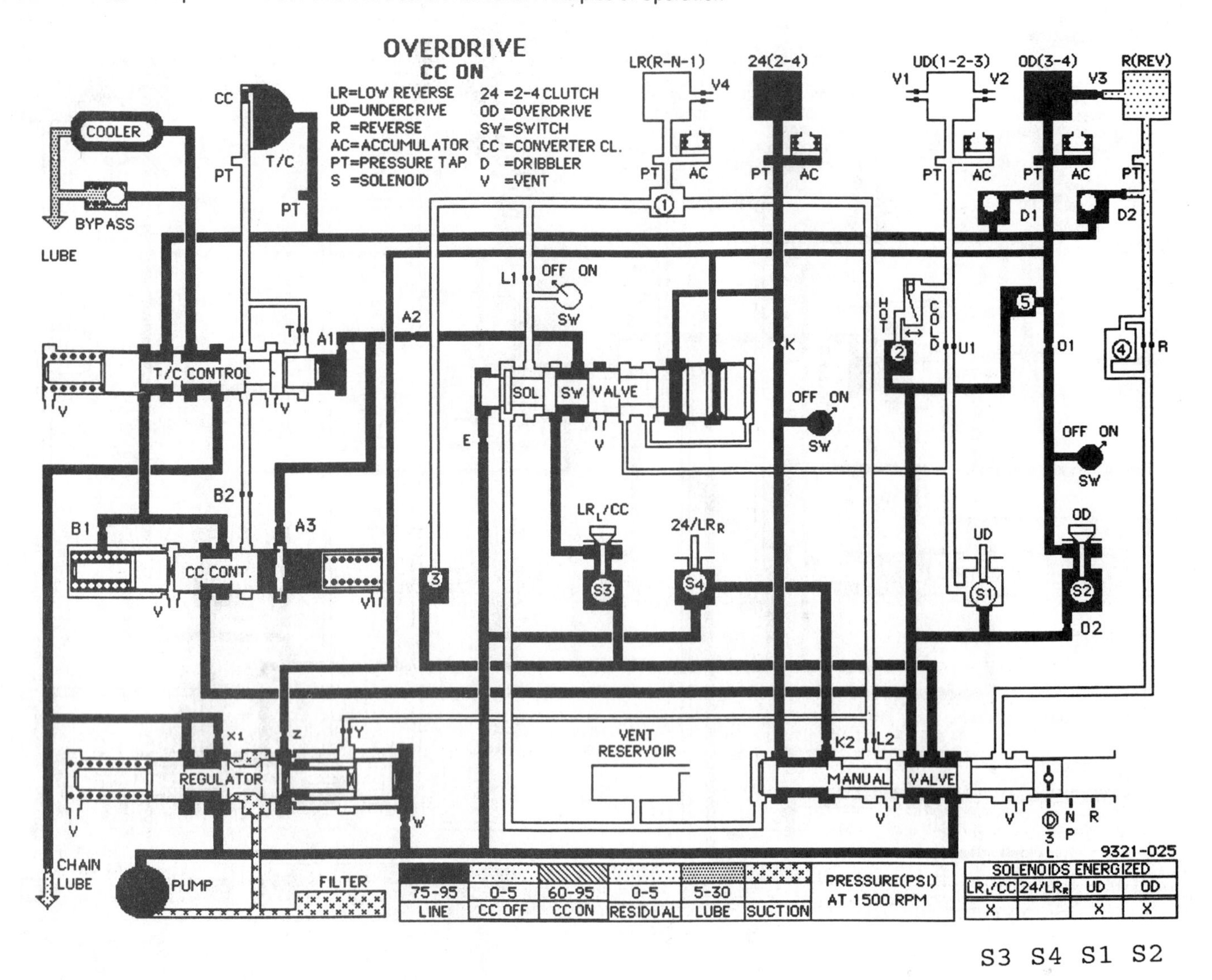

FIGURE 11–35 42LE overdrive (fourth) gear schematic; converter clutch ON. (Courtesy of Chrysler Corp.)

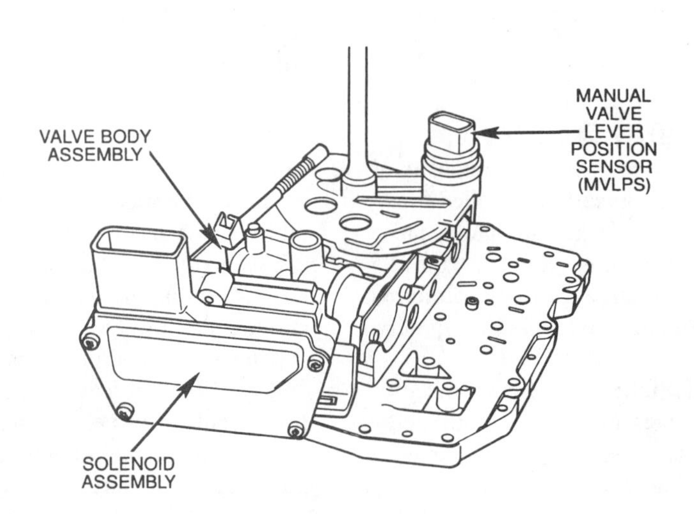

FIGURE 11–36 42LE manual valve lever position sensor (MVLPS) and solenoid assembly mounted on valve body. (Courtesy of Chrysler Corp.)

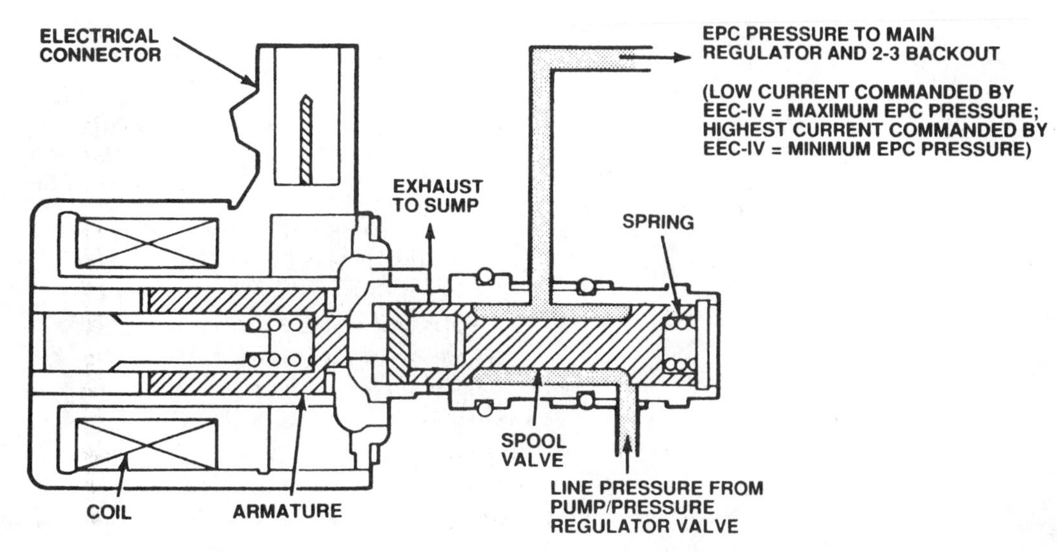

FIGURE 11–37 Pressure control solenoid (PCS), AODE. (Reprinted with the permission of Ford Motor Company)

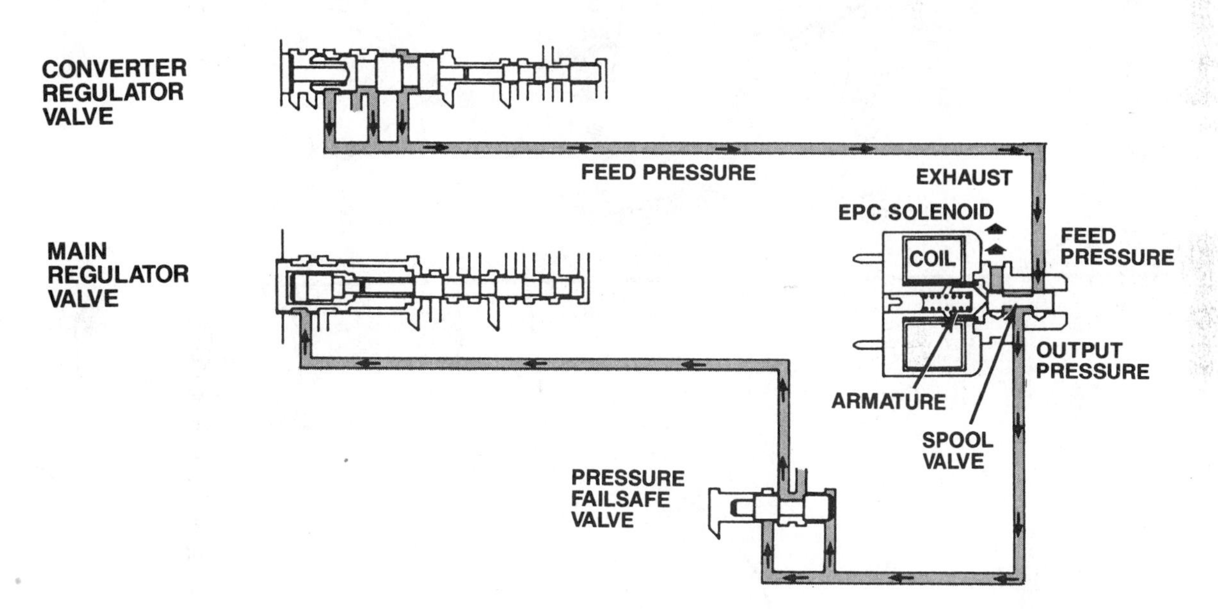

FIGURE 11–38 Pressure control solenoid (PCS) operation—AX4S (AXODE). (Reprinted with the permission of Ford Motor Company)

ELECTRONIC TRANSMISSION/TRANSAXLE CONTROL SYSTEMS

Energizing the solenoids to activate the gear ratio changes is the responsibility of the PCM/TCM, which determines when an upshift or downshift is desirable. The PCM/TCM then turns the related solenoids ON and OFF in various combinations to achieve the needed results.

The decision-making process of the PCM/TCM is based on the messages it receives from various input devices. Signals from switches and sensors located under the hood and from the transmission are constantly being read and evaluated.

The controlling of shift solenoids occurs in a manner similar to converter clutch solenoid operation. Input data is sent to the PCM/TCM. The information is processed and evaluated. Comparisons are made to its internal programming. The PCM/TCM continually makes decisions and controls the output devices—the solenoids.

Electronic shift control has rapidly evolved over the past decade. General Motor's 4L60-E transmission is typical of the high level of sophistication used in modern powertrains (Figure 11–39). Its many inputs, and their interfacing with the PCM/TCM to control the output devices, are described later in this chapter.

The operation and control of the electronic shift transmission are dependent upon the decision-making capabilities of its

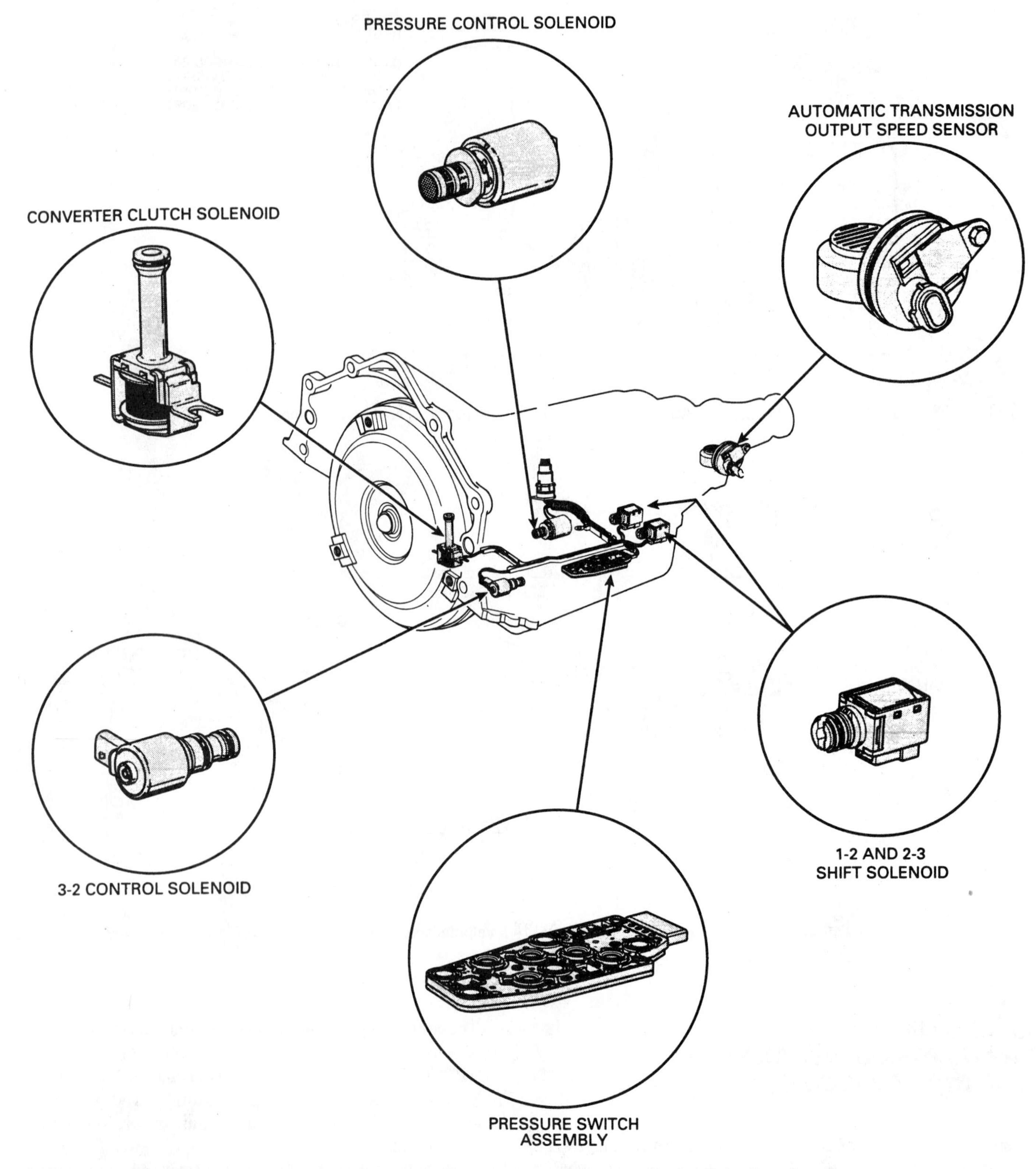

FIGURE 11–39 4L60-E electronic components. (Courtesy of General Motors Corporation, Service Technology Group)

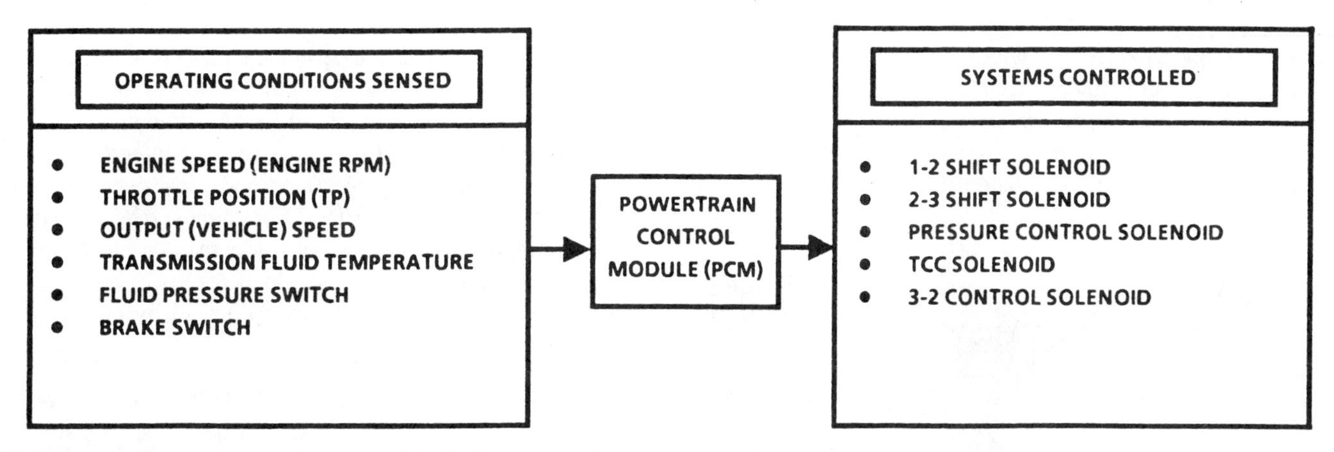

FIGURE 11–40 Electronic control system, 4L60-E. (Courtesy of General Motors Corporation, Service Technology Group)

PCM/TCM. Figure 11–40 shows some of the main inputs that are sensed by a 4L60-E PCM and the output devices the PCM controls. The inputs listed are common to those found in many electronically controlled transmissions and transaxles.

Common Inputs

Engine Speed Input. Engine RPM is used by the PCM for transmission control. In the 4L60-E, the speed is determined by the pickup coil of the ignition module.

Throttle Position (TP) Sensor. This measures the amount of throttle plate opening and indicates engine load.

Engine Coolant Temperature (ECT) Sensor. Through the use of a thermistor, a resistor that changes value based on temperature, the engine coolant temperature is measured. The PCM uses this information to delay TCC engagement until the engine is warmed up. In many transmissions, the data is also used to adjust line pressure.

Transmission Fluid Temperature (TFT) Sensor. The PCM uses this input for TCC regulation and shift quality. Like an ECT, a thermistor is used to measure fluid temperature.

Fluid Pressure Switch. This switch informs the PCM whether or not a hydraulic circuit is pressurized with oil. It may be grouped with other switches to form a fluid pressure switch assembly (PSA). A PSA tells the PCM which gear range position has been selected.

Many transmissions use mechanical switch units, operated by the manual lever and shaft linkage, to inform the PCM of the chosen gear range. A **manual lever position switch (MLPS)** is often externally mounted to a transmission. Chrysler uses a manual valve lever position switch (MVLPS) internally in the 42LE transaxle.

❖ **Manual lever position switch (MLPS):** A mechanical switching unit that is typically mounted externally to the transmission to inform the PCM/ECM which gear range the driver has selected.

Brake Switch. The brake switch is normally closed when the brake pedal is released. When the pedal is depressed, the switch opens and interrupts the electrical signal to the PCM for TCC control.

BARO Input/Manifold Absolute Pressure (MAP) Sensor. The measurement of altitude or barometric pressure is done by the MAP sensor prior to engine operation. This signal is used by the PCM for engine and transmission control.

Air Conditioning Input. The A/C compressor clutch signal is identified for transmission control. When the A/C clutch is ON, less engine torque is available to the transmission input shaft.

Cruise Control Input. A signal from the cruise control module tells the PCM that the driver has selected cruise control. The PCM modifies the shift pattern strategy to reduce shift busyness.

Output Speed Sensor (OSS)/Vehicle Speed Sensor (VSS). Attached to the transmission and/or transfer case, an **output speed sensor (OSS)** determines the speed of output shaft and vehicle speed. The OSS and VSS, as well as an input speed sensor (used in many transmissions and transaxles), operate on the principle of pulse generation.

❖ **Output speed sensor (OSS):** Identifies transmission output shaft speed for shift timing and may be used to calculate TCC slip; often functions as the VSS (vehicle speed sensor).

A **pulse generator** is a two-wire magnetic pickup device used to sense rotational speed. Depending on the electronic control system used for engine and transmission management, a transmission may have one or two speed sensors. They are used to sense input and output. For example, the Chrysler 41TE (A-604) and 42LE transaxles and the GM 4L80-E transmission each use a turbine input speed and an output speed sensor. General Motor's 3T40 (125C) uses only an output sensor (Figure 11–41).

- **Pulse generator:** A two-wire pickup sensor used to produce a fluctuating electrical signal. This changing signal is read by the controller to determine the speed of the object and can be used to measure transmission input speed, output speed, and vehicle speed.

FIGURE 11–41 Vehicle speed sensor (VSS)—GM 3T40 (125C).

The pickup sensor uses a permanent magnet with a coil wire wrap. It detects speed by responding to a rotating rotor. The rotor may be an independent rotor or integrated with a rotating clutch drum. In Figure 11–42, the rotor is integrated with a rotating drum. As the projections and dips sweep past the sensor, the magnetic field from the permanent magnet expands and contracts. This produces an induction or electromotive force with changing polarities, which results in an AC wave effect. The computer determines the speed by counting the waves per time cycle.

4WD Low Switch. This switch tells the PCM that the driver has shifted to low gear with the transfer case lever. The signal is used to modify the output speed sensor signal/VSS to respond to the gear reduction taking place.

A wide variety and combination of various electronic shift control input and output devices are used by transmission manufacturers. A discussion of GM's 4L60-E follows to help the reader better understand the interrelationship of the inputs and outputs of a single transmission design.

GM 4L60-E Transmission Electronic Control System

The 4L60-E inputs and outputs are identified in Figure 11–43. Many of the inputs are mounted on or within the

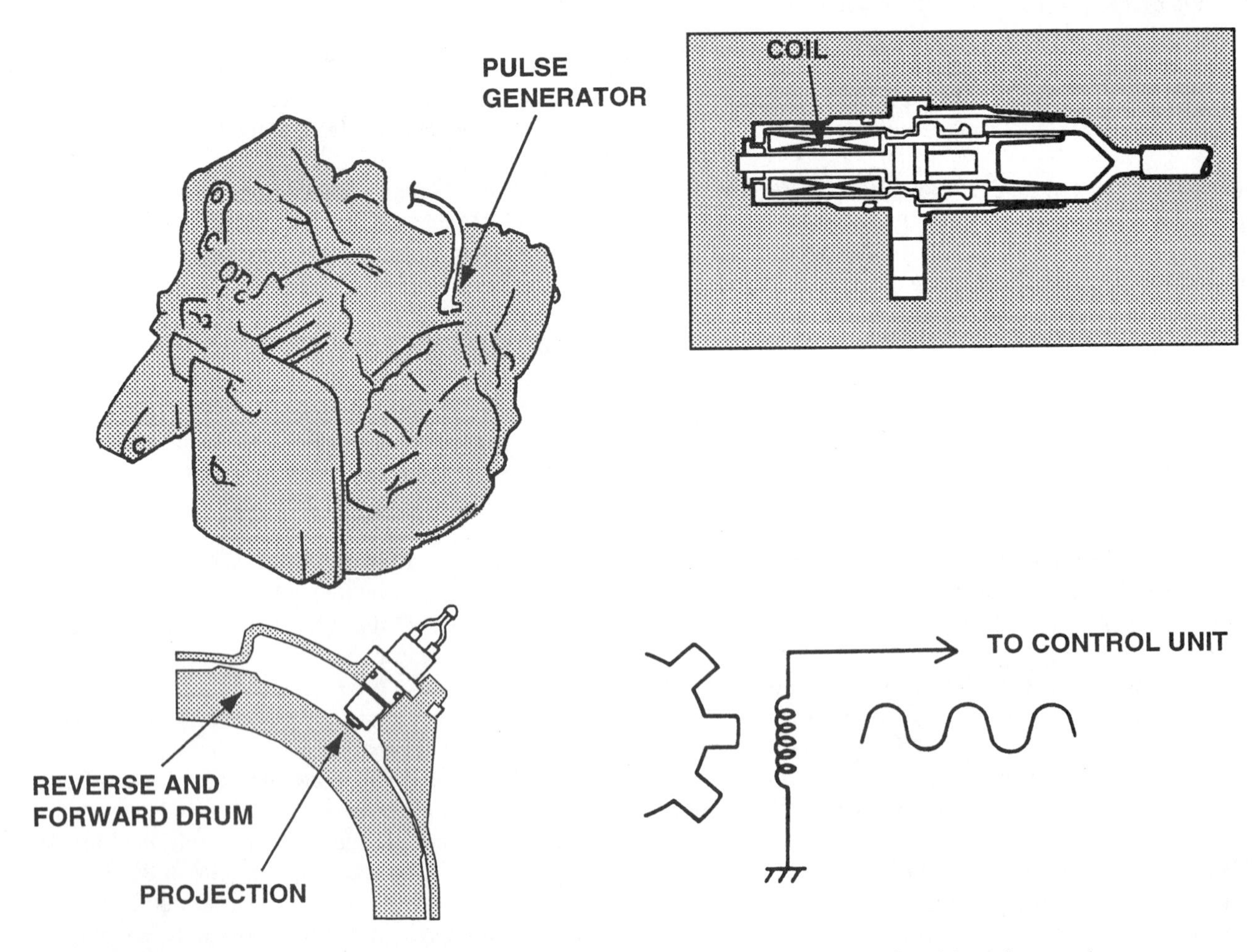

FIGURE 11–42 Pulse generator (VSS) operation concept—4EAT. (Reprinted with the permission of Ford Motor Company)

4L60-E ELECTRONIC CONTROL – OVERALL

Inputs		Outputs
OUTPUT SPEED SENSOR	POWERTRAIN CONTROL MODULE (PCM)	1-2 SHIFT SOLENOID
4WD LOW SWITCH		2-3 SHIFT SOLENOID
TP SENSOR		PRES CONT SOLENOID
PRESSURE SWITCH ASSEMBLY		3-2 CONT SOLENOID
TRANS TEMP SENSOR		TCC SOLENOID
ENGINE SPEED INPUT		DATA LINK CONNECTOR
BARO INPUT		
ECT SENSOR		
A/C INPUT		
C/C INPUT		
BRAKE SWITCH		
DATA LINK CONNECTOR		

FIGURE 11–43 4L60-E electronic control—input and output devices. (Courtesy of General Motors Corporation, Service Technology Group)

transmission, and their operation is explained in this section. All of the output devices, except for the data link connector (DLC), are located within the transmission, as shown in Figure 11–39.

4L60-E Inputs.

Output Speed Sensor (OSS)/VSS. Through pulse generation, the OSS sensor/VSS informs the PCM of the transmission's output shaft speed and vehicle speed. The sensor is used to control shift points and calculate TCC slip (Figure 11–44).

Transmission Fluid Pressure Switch Assembly (PSA). The PSA uses five pressure switches to inform the PCM which gear range has been selected and hydraulically activated (Figures 11–39, 11–45, and 11–46). Each gear range uses a different combination of closed switches (Figure 11–45). Three signal lines are monitored by the PCM. Voltage at each signal line is either 12 V or zero.

Transmission Fluid Temperature (TFT) Sensor. The TFT sensor is used to help control TCC apply and shift quality. Figure 11–46 shows that the TFT sensor/thermistor is located

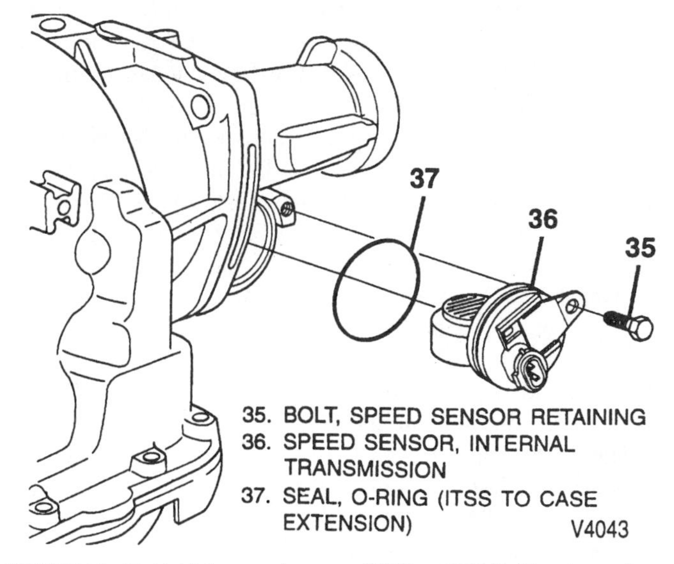

FIGURE 11–44 Vehicle speed sensor (VSS)—4L60-E. (Courtesy of General Motors Corporation, Service Technology Group)

RANGE INDICATOR	OIL PRESSURE				
	REV	D4	D3	D2	LO
PARK					
REVERSE	■				
NEUTRAL					
D4		■			
D3		■	■		
D2		■	■	■	
D1		■	■	■	■

■ OIL PRESSURE PRESENT

VALID TPS COMBINATION CHART

	"A"	"B"	"C"
PARK	12	0	12
REVERSE	0	0	12
NEUTRAL	12	0	12
D4	12	0	0
D3	12	12	0
D2	12	12	12
D1	0	12	12
ILLEGAL	0	12	0
ILLEGAL	0	0	0

EXPECTED VOLTAGE READINGS

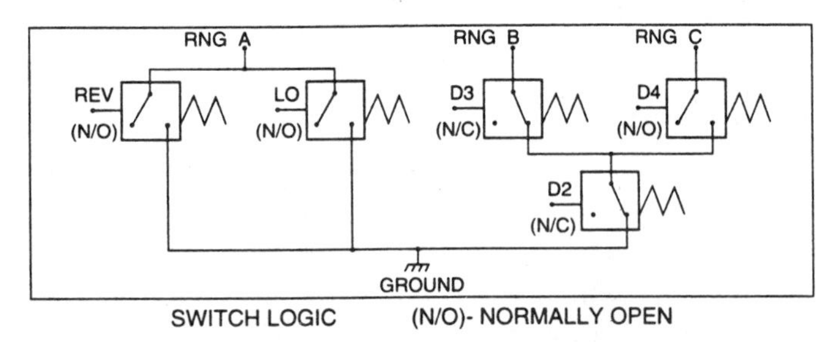

FIGURE 11–45 Pressure switch assembly (PSA) operation—4L60-E. (Courtesy of General Motors Corporation, Service Technology Group)

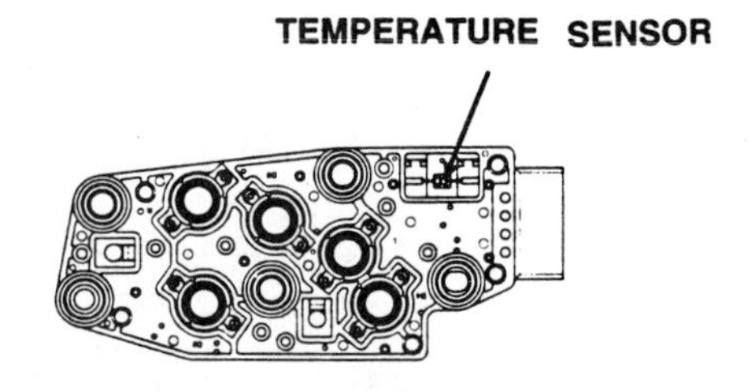

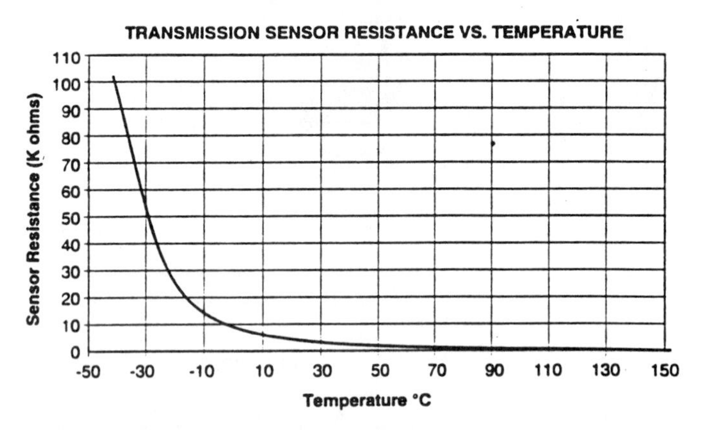

FIGURE 11–46 Transmission fluid temperature (TFT) sensor located on the pressure switch assembly (PSA) and TFT sensor resistance/temperature graph—4L60-E. (Courtesy of General Motors Corporation, Service Technology Group)

on the PSA. Also shown is the drop in resistance as the fluid temperature rises.

4L60-E Outputs.

1-2 and 2-3 Shift Solenoids (Figure 11–47). The 1-2 and 2-3 shift solenoids control shift valve movement. Observe their locations in Figure 11–48. The 1-2 solenoid (A) is at the end of the 1-2 shift valve (#366) and the 2-3 solenoid (B) is at the end of the 2-3 shift valvetrain (## 368/369). When the solenoids are OFF, oil pressure is able to exhaust, and the valves move to the right.

The shift solenoids operate ON and OFF in four combinations to produce the necessary valve movement to engage the four forward gears (Chart 11–3). The timing of the upshifts and downshifts are controlled by the PCM. The following inputs influence the PCM decision-making process to control the shift solenoids: output speed sensor, 4WD low switch, TP sensor, PSA, TFT sensor, baro reading, and cruise control signal.

Pressure Control Solenoid (PCS) (Figure 11–49). Attached to the valve body (Figure 11–48), the PCS controls line pressure by moving its internal regulator valve against spring pressure (Figure 11–50). The oil that is allowed to pass through the PCS acts on a boost valve at the base of the main pressure regulator valve.

The PCM changes line pressure in response to engine load. By changing the amount of amperage to the PCS, line pressure is altered. Low amperage, .1 ampere, produces high pressure, while an increase to 1.1 ampere causes the PCS to provide a low pressure. The solenoid operates on a duty cycle and in response to the amount of current flow.

The PCM determines the amount of amperage to send the PCS to control shift quality. The TP sensor is an important

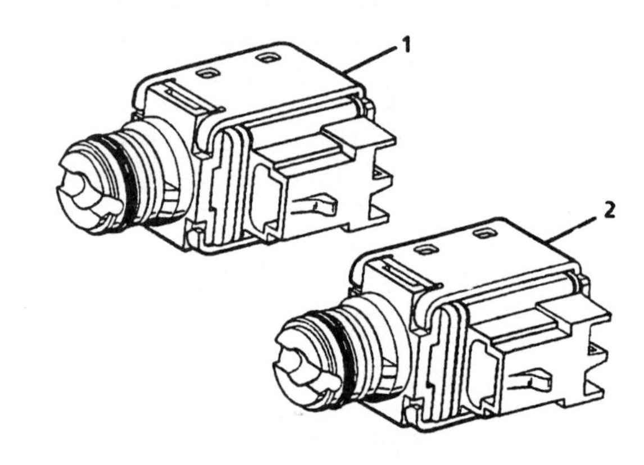

FIGURE 11–47 4L60-E shift solenoids. (Courtesy of General Motors Corporation, Service Technology Group)

factor in the logic of this circuit. The inputs that influence the PCM decision-making process for shift quality are the output speed sensor, 4WD low switch, TP sensor, PSA, TFT sensor, engine speed, baro reading, and A/C signal.

3-2 Control Solenoid (Figure 11–51). This duty cycle solenoid is responsible for controlling the 3-2 downshift. Mounted on the valve body (see Figure 11–48, #394), it regulates the timing and release of the 3-4 clutch and application of the 2-4 band.

Like other 4L60-E solenoids, the PCM controls the solenoid circuit through ground-side switching (Figure 11–52). The

CHART		
Gear	1-2 Shift Solenoid	2-3 Shift Solenoid
1	ON	ON
2	OFF	ON
3	OFF	OFF
4	ON	OFF

CHART 11–3 4L60-E shift solenoid operation. (Courtesy of General Motors Corporation, Service Technology Group)

three inputs that are used by the PCM to regulate the duty cycle of the 3-2 control solenoid are the output speed sensor, TFT sensor, and A/C signal.

Torque Converter Clutch (TCC) Solenoid (Figure 11–53). The TCC solenoid operates as a pressure exhaust valve when de-energized. When commanded ON by the PCM, the TCC solenoid stops exhausting converter signal oil and allows pressure to build against the converter clutch apply valve. The apply valve is repositioned, and the converter clutch engages.

Many inputs are evaluated by the PCM for TCC solenoid application. These include the output speed sensor, 4WD low switch, TP sensor, PSA, TFT sensor, engine speed signal, baro reading, ECT sensor, A/C signal, and brake switch.

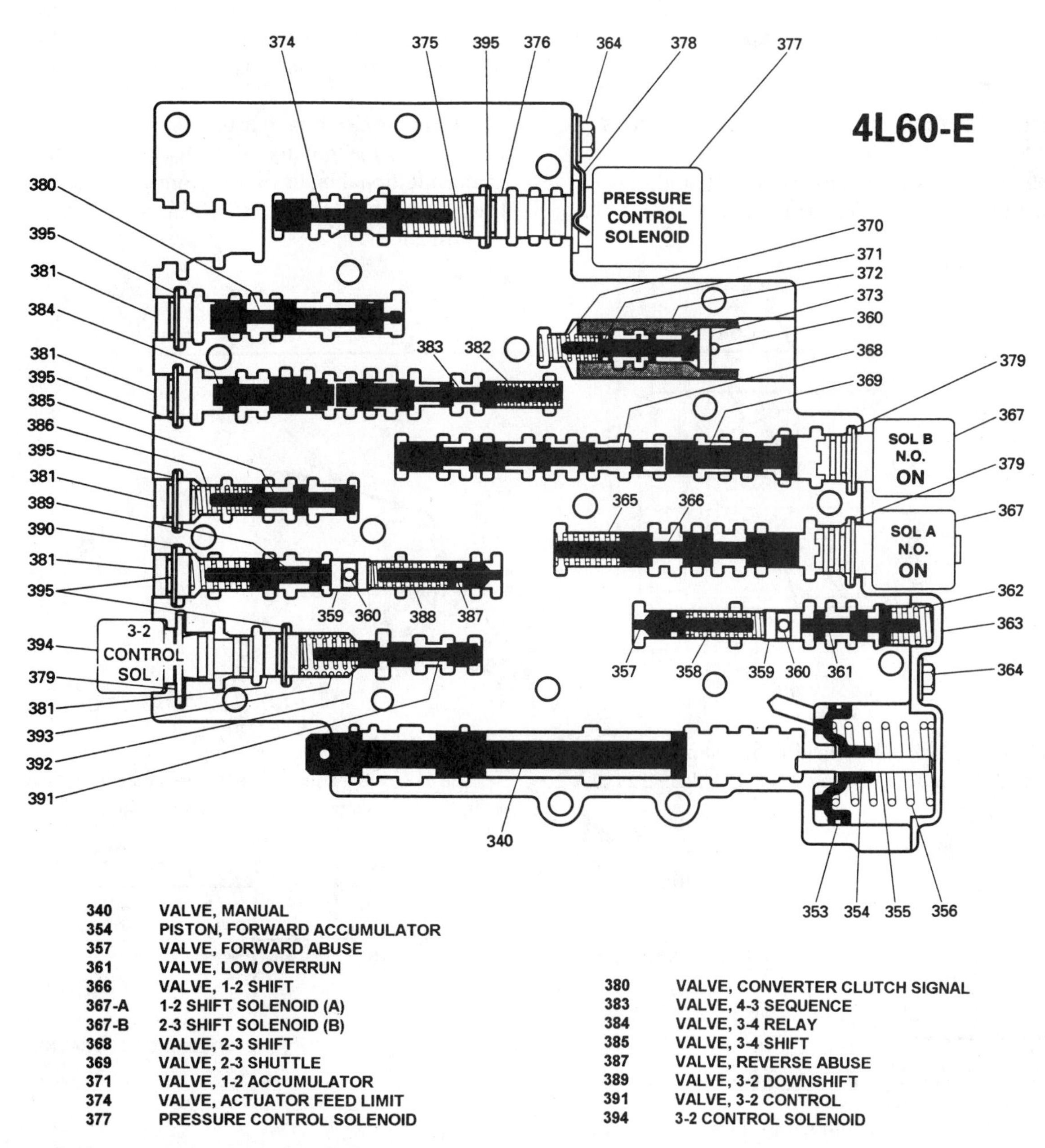

FIGURE 11–48 4L60-E valve body components. (Courtesy of General Motors Corporation, Service Technology Group)

Data Link Connector (DLC). The DLC, shown in Figure 11–54, is located in the driver's compartment under the left side of the instrument panel. Through this junction point, a scan tool can be connected to read PCM serial data and diagnostic trouble codes (DTCs).

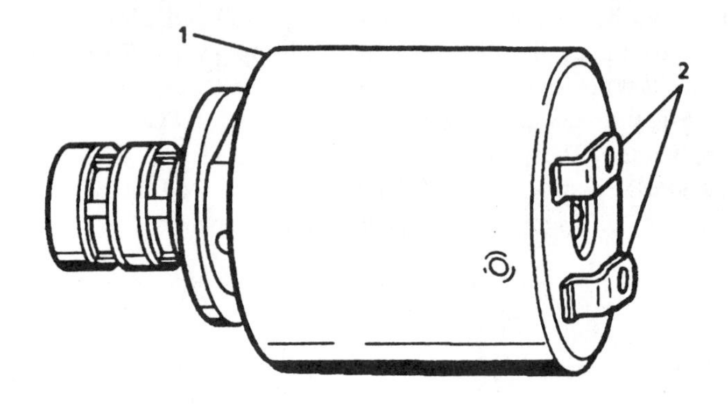

FIGURE 11–49 4L60-E pressure control solenoid (PCS). (Courtesy of General Motors Corporation, Service Technology Group)

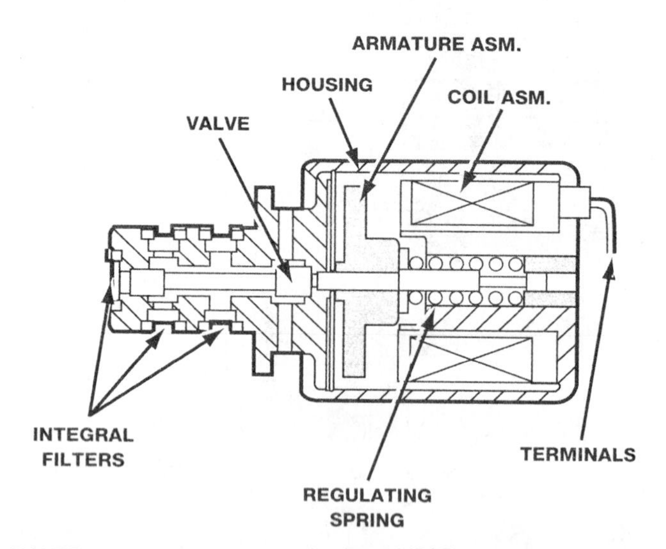

FIGURE 11–50 pressure control solenoid (PCS), cutaway view—4L60-E. (Courtesy of General Motors Corporation, Service Technology Group)

SUMMARY

Electronic shift control systems provide precise operation of the transmission. Through the use of electrical input signals, the electronic control module activates the output solenoids to shift the transmission at the proper time.

Shift solenoids can be of two common designs: pressure/exhaust solenoids and circuit flow control solenoids. Pressure/exhaust solenoids allow an oil circuit to vent fluid or contain it to build up pressure. When pressurized, the circuit is able to act upon a shift valve. Circuit flow control solenoids essentially open and close a fluid passage leading to a clutch or band. In some applications, solenoids are pulse width modulated to control the buildup rate of circuit pressure. This feature controls shift quality.

Many transmissions now use electronic pressure control solenoids. The mainline pressure is managed by the PCM/TCM by sending varying amounts of current to the pressure control solenoids.

As electronic shift transmissions have become more sophisticated, the quality of their operation has improved. Using additional input signals from under the hood and from the transmission, the PCM/TCM is able to provide superior performance.

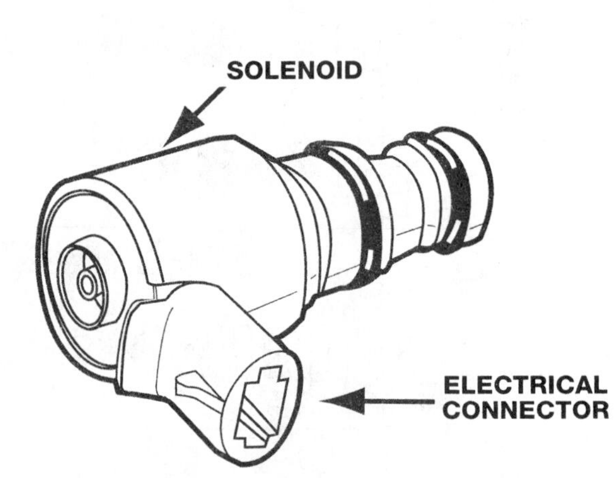

FIGURE 11–51 3-2 control solenoid—4L60-E. (Courtesy of General Motors Corporation, Service Technology Group)

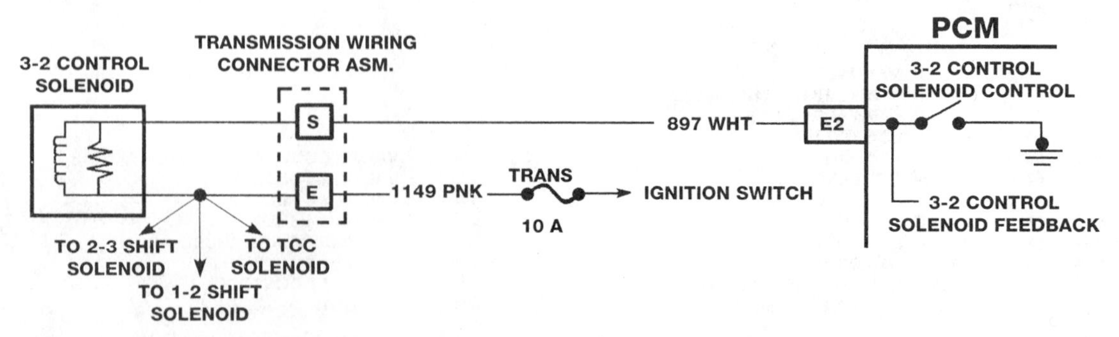

FIGURE 11–52 3-2 control solenoid and PCM circuit—4L60-E. (Courtesy of General Motors Corporation, Service Technology Group)

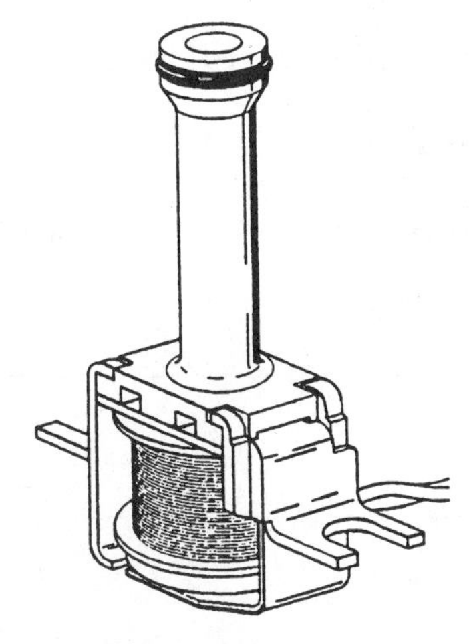

FIGURE 11–53 TCC solenoid—4L60-E. (Courtesy of General Motors Corporation, Service Technology Group)

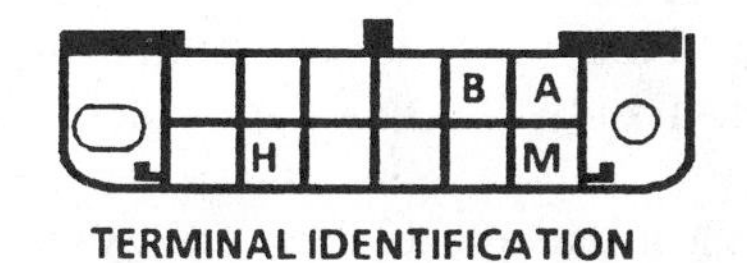

A GROUND

B DIAGNOSTIC TERMINAL

H BRAKE

M SERIAL DATA

FIGURE 11–54 Data link connector (DLC/ALDL)—OBD I. (Courtesy of General Motors Corporation, Service Technology Group)

❑ REVIEW

Key Terms/Acronyms

CCD/C^{2}D Bus, barometric manifold absolute pressure (BMAP) sensor, adaptive memory/adaptive strategy, EATX relay, MVLPS (manual valve lever position sensor), pressure control solenoid (PCS), MLPS (manual lever position switch), output speed sensor (OSS), and pulse generator.

Multiple Choice

____ 1. Electronic shift control systems feature:

A. interfacing with the on-board PCM.
B. a single transmission control module.
C. a transmission control module linked with the PCM.
D. two of the above.
E. all of the above.

____ 2. In all fully electronic shift systems:
I. governor oil is eliminated.
II. shift valves are eliminated.

A. I only C. both I and II
B. II only D. neither I nor II

____ 3. Which is true about pressure/exhaust shift solenoids?
I. When energized, they always allow pressure to build up.
II. When energized, they always exhaust hydraulic oil.

A. I only C. both I and II
B. II only D. neither I nor II

____ 4. In Chrysler 41TE (A-604)/42LE transaxles:
I. the shift valves are eliminated.
II. the throttle valve is eliminated.

A. I only C. both I and II
B. II only D. neither I nor II

____ 5. In Chrysler 41TE (A-604) transaxles:
I. the solenoids are valve body mounted.
II. the solenoids are classified as pulse width duty cycle.

A. I only C. both I and II
B. II only D. neither I nor II

____ 6. When using pulse width duty cycle solenoids:
I. the computer can customize the pressure buildup or apply time for each friction element with engine performance.
II. the computer needs to compare transmission input rpm to output rpm.

A. I only C. both I and II
B. II only D. neither I nor II

____ 7. (Chrysler 41TE) Should the backup light relay solenoid fail:
I. the backup lights will not work.
II. the transaxle will default into second gear.

A. I only C. both I and II
B. II only D. neither I nor II

____ 8. (Chrysler 41TE) If the EATX relay fails to close:
I. it could mean low battery voltage.
II. the transaxle will fail to move forward.

A. I only C. both I and II
B. II only D. neither I nor II

____ 9. (Chrysler 41TE) Which is true?
I. The solenoids in the solenoid assembly are ground-side switched.
II. The solenoid assembly receives its operating voltage from the EATX relay.

A. I only C. both I and II
B. II only D. neither I nor II

____ 10. When the solenoids in a Chrysler 41TE are de-energized:
I. Two solenoids block fluid flow and vent the clutch circuit.
II. Two solenoids open the clutch apply circuit.

A. I only C. both I and II
B. II only D. neither I nor II

____ 11. Pressure control solenoids:
I. are used in all electronic shift transmissions.
II. increase line pressure when sent more current from the PCM.

A. I only C. both I and II
B. II only D. neither I nor II

____ 12. Which transmission uses both hydraulic and electronic shift controls?

A. Ford A4LD.
B. Chrysler 40RH (A-500).
C. Chrysler 42RH (A-518).
D. Two of the above.
E. All of the above.

True-False

1. Adaptive strategy is unaffected by removing and replacing the vehicle's battery. ___T/___F
2. The manual valve in a Chrysler 41TE or 42LE has three positions. ___T/___F
3. The four solenoids in a Chrysler 41TE or 42LE are each referred to as "normally vented" solenoids. ___T/___F
4. The GM 4T60-E transaxle uses four pressure/exhaust port shift solenoids. ___T/___F
5. Only first gear is available in low/L range in a Chrysler 42LE transaxle. ___T/___F
6. PSA refers to a pulse solenoid assembly and is controlled by the PCM. ___T/___F
7. As the fluid warms up, the resistance of a TFT sensor increases. ___T/___F
8. In a Chrysler 42LE transaxle, partial lockup of the converter clutch can occur in second, direct (third), and overdrive (fourth) gears. ___T/___F

Completion

PCM Inputs

List ten PCM inputs that relate to electronic shift transmissions/transaxles.

1. ______________________
2. ______________________
3. ______________________
4. ______________________
5. ______________________
6. ______________________
7. ______________________
8. ______________________
9. ______________________
10. ______________________

PCM Outputs

List four output devices located in transmissions/transaxles that are controlled by the PCM.

1. ______________________
2. ______________________
3. ______________________
4. ______________________

12 Diagnostic Procedures

Section 1 Introduction to Diagnostic Procedures

OBJECTIVE:

The objective of this unit is for you to become proficient with the terms, concepts, and procedures contained in this chapter. Completion of this objective is essential to becoming a successful automatic transmission technician.

CHALLENGE YOUR KNOWLEDGE

Define the following key terms/acronym:

Throttle positions: minimum throttle, light throttle, medium throttle, heavy throttle, wide open throttle (WOT), full throttle detent downshift, zero-throttle coast down, and engine braking; shift conditions: bump, chuggle, delayed (late or extended), double bump (double feel), early, end bump (end feel or slip bump), firm, flare (slipping), garage shift, harsh (rough), hunting (busyness), initial feel, late, mushy, shudder, slipping, soft, surge, and tie-up.

List the following:

Seven failure conditions; and the seven-step diagnostic procedure.

INTRODUCTION

Diagnosis involves a scientific or logical step-by-step plan. It is an orderly procedure that produces accurate information that the technician can use as clues to pinpoint the most probable cause of trouble. A diagnostic procedure follows a definite sequence of steps that accomplishes the most results with the least amount of unnecessary time and labor.

From the technician's viewpoint, diagnosis cannot use a scattered approach. The procedure must cover the likeliest causes of trouble first. It must eliminate as many causes as possible before the transmission is opened. Once the transmission is removed and taken apart, many disgnostic clues are no longer available. Careful and complete diagnosis leads to quality repairs.

Technicians thinking "diagnosis" must avoid the domestic/import mental block. Automatic transmissions work the same worldwide. In addition to mechanical power flows and hydraulics, every technician needs to know both engine and transmission electronics. There is a definite relationship between engine performance and proper transmission operation.

Diagnosis is a mental process of tying together information or data collected about the transmission and engine performance and applying theory to a logical problem solution. Problems need to be isolated to engine driveability, internal mechanical or friction element failure, hydraulic logic controls, or electronic failure—the computer, sensors, or shift solenoids.

The new technology requires the use of scanners, analyzers, digital and analog meters, and transmission testers to help diagnose problems. The technician needs to maintain an updated technical library, attend training seminars and schools, and join a professional technical group.

Automatic Transmission Rebuilders Association
2472 Eastman Avenue #23
Ventura, CA 93003

Automotive Service Association
PO Box 929
Bedford, TX 76095-0929

The industry is moving swiftly and there are too many transmissions for a single person to know it all. Today's transmission technician cannot go it alone or survive on past knowledge. Although the transmission has an intricate maze of hydraulic circuits, clutch/band units, and electronic wiring and switches, diagnosis can be simplified by:

- Emphasizing what is right about the transmission instead of taking the usual approach of emphasizing what is wrong.
- Confining or isolating the problem to the individual hydraulic clutch/band unit or electronic circuit with the fault.

If the problem area is not isolated, diagnosis becomes a hit-and-miss proposition, time-consuming and frustrating. Diagnosis represents a key phase in the service cycle (Figure 12–1). It not only isolates the problem but also serves as a basis for the type of service work to be performed. Note: It is most important that the service performed always be followed up and tested to validate the correction of the problem. Did the diagnosis and service work do what it was intended to do? If not, why not? *Fixing* a problem implies a permanent cure, which means more than any warranty.

Modern transmission diagnostics essentially involves the following process:

1. Read and separate engine and transmission fault codes with a scan diagnostic tester.
2. Make a notation of both engine and transmission fault codes.
3. Proceed to correct engine-related faults.
4. If corrected engine faults do not solve the customer's complaint, proceed with transmission testing. Always eliminate engine fault codes first (Figure 12–2).

In the final analysis, it is the technician who must make the decision about what is the problem, what caused it, and what must be done to repair it. The technician must also depend on his or her built-in diagnostic tools that nature provides: eyes, ears, nose, and some common sense. There is no substitute for knowledge, experience, and imagination. Treat diagnosis as a challenge and as being fun. Knowledge of theory will lead to solutions every time.

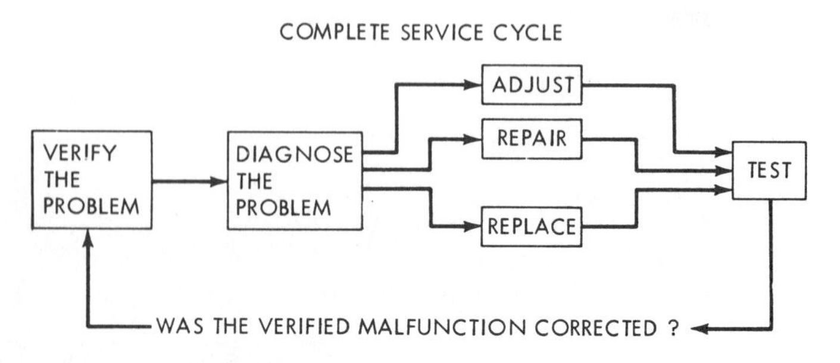

FIGURE 12–1 Service cycle.

DIAGNOSTIC FUNDAMENTALS

Automatic transmission problems usually are caused by one or more of the following conditions:

- Hydraulic failure
- Mechanical failure
- Friction element failure
- Converter failure
- Electrical or electronic fault
- Poor engine performance or improper engine signal to transmission
- Improper external adjustments

Diagnosis of these conditions and related problems follows an exact procedure that gives excellent results. The success of the procedure, however, depends on the accuracy of the data or information collected. The technician's task is to stick with the procedure until enough information is gathered for a diagnostic decision (see Guide 12–1 A and B, pages 253 and 254).

TERMINOLOGY

When discussing automatic transmission diagnosis, it is important to use a common language to describe related operating conditions. The following are terms used to describe **throttle positions** and **shift conditions**.

❖ **THROTTLE POSITIONS:**

Minimum throttle: The least amount of throttle opening required for upshift; normally close to zero throttle.

Light throttle: Approximately one-fourth of accelerator pedal travel.

Medium throttle: Approximately one-hawlf of accelerator pedal travel.

Heavy throttle: Approximately three-fourths of accelerator pedal travel.

Wide open throttle (WOT): Full travel of accelerator pedal.

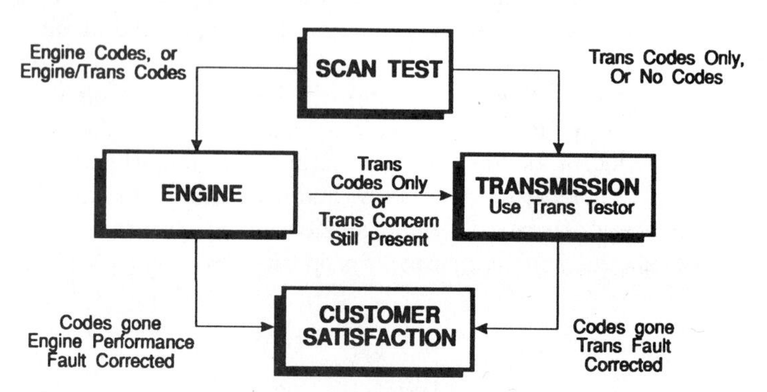

FIGURE 12–2 Powertrain diagnostic step diagram.

Full throttle detent downshift: A quick apply of accelerator pedal to its full travel, forcing a downshift.
Zero-throttle coast down: A full release of accelerator pedal while vehicle is in motion and in drive range.
Engine braking: Use of engine to slow vehicle by manually downshifting during zero-throttle coast down.

❖ **SHIFT CONDITIONS:**
Bump: Sudden and forceful apply of a clutch or band.
Chuggle: Bucking or jerking condition that may be engine-related and may be most noticeable when converter clutch is engaged; similar to the feel of towing a trailer.
Delayed (late or extended): Condition where shift is ex-

VIN ____________ MILEAGE ____________ R.O.# ____________

MODEL YEAR ____________ VEHICLE MODEL ____________ ENGINE ____________

TRANS MODEL ____________ TRANS DATE CODE/OR SERIAL # ____________

CUSTOMER'S CONCERN

____

____

____

____

WHEN:	OCCURS:	RECENT WORK:		
___ VEHICLE WARM	___ ALWAYS	COLLISION	Yes __	No __
___ VEHICLE COLD	___ INTERMITTENT	TRANSMISSION	Yes __	No __
___ ALWAYS	___ SELDOM	ENGINE	Yes __	No __
___ NOT SURE	___ FIRST TIME	ACCESSORY INSTALLATION	Yes __	No __

Notes: ____

____

PRELIMINARY QUICK CHECKS

. Fluid Level	_____ Pass	_____ High	_____ Low
. Fluid Condition	_____ Pass	_____ Fail	
. Cooler Lines Visual	_____ Pass	_____ Fail	

Notes: ____

____

GUIDE 12–1A Transmission/transaxle diagnosis guide.

pected but does not occur for a period of time, for example, where clutch or band engagement does not occur as quickly as expected during part throttle or wide open throttle apply of accelerator or when manually downshifting to a lower range.
Double bump (double feel): Two sudden and forceful applies of a clutch or band.
Early: Condition where shift occurs before vehicle has reached proper speed, which tends to labor engine after upshift.
End bump (end feel or slip bump): Firmer feel at end of shift when compared with feel at start of shift.
Firm: A noticeable quick apply of a clutch or band that is considered normal with medium to heavy throttle shift; should not be confused with *harsh* or *rough.*
Flare (slipping): A quick increase in engine rpm accom-

. Engine Idle ____ Pass ____ Fail

. Accelerator W.O.T. ____ Pass ____ Fail

. Engine Codes ____ Pass ____ Fail

Notes: ______________________________

____

. Electrical Visual ____ Pass ____ Fail

. Vacuum Lines Visual ____ Pass ____ Fail

. Vacuum Flow to Modulator ____ Pass ____ Fail

Notes: ______________________________

____

. T.V. Linkage/Cable Visual ____ Pass ____ Fail

. T.V. Linkage/Cable Free Movement ____ Pass ____ Fail

. T.V. Adjustment ____ Pass ____ Fail

. Manual Linkage/Cable Visual ____ Pass ____ Fail

. Manual Linkage/Cable Operation ____ Pass ____ Fail

. Manual Linkage Adjustment ____ Pass ____ Fail

Notes: ______________________________

____

. Road Test to Verify Customer Concern

Notes: ______________________________

____

GUIDE 12–1B Transmission/transaxle diagnosis guide *(continued).*

panied by momentary loss of torque; generally occurs during shift.
Garage shift: Initial engagement feel of transmission, neutral to reverse or neutral to a forward drive.
Harsh (rough): An apply of a clutch or band that is more noticeable than a *firm* one; considered undesirable at any throttle position.
Hunting (busyness): Repeating quick series of upshifts and downshifts that causes noticeable change in engine rpm, for example, as in a 4-3-4 shift pattern.
Initial feel: A distinct firmer feel at start of shift when compared with feel at finish of shift.
Late: Shift that occurs when engine is at higher than normal rpm for given amount of throttle.
Mushy: Same as *soft*; slow and drawn out clutch apply with very little shift feel.
Shudder: Repeated jerking or stick-slip sensation, similar to *chuggle* but more severe and rapid in nature, that may be most noticeable during certain ranges of vehicle speed; also used to define condition after converter clutch engagement.
Slipping: Noticeable increase in engine rpm without vehicle speed increase; usually occurs during or after initial clutch or band engagement.
Soft: Slow, almost unnoticeable clutch apply with very little shift feel.
Surge: Repeating engine-related feeling of acceleration and deceleration that is less intense than *chuggle*.
Tie-up: Condition where two opposing clutches are attempting to apply at same time, causing engine to labor with noticeable loss of engine rpm.

STEPS OF DIAGNOSIS

1. Listen to the customer's concern.
2. Check the fluid level and condition.
3. Quick check engine idle, vacuum line, wiring connectors, mechanical linkage/cable control, and throttle body accelerator for WOT.
4. Check PCM/ECM or TCM for codes.
5. Stall test.
6. Road test.
7. Pressure test.

❑ SECTION 1 REVIEW

Key Terms/Acronym

Throttle positions: minimum throttle, light throttle, medium throttle, heavy throttle, wide open throttle (WOT), full throttle detent downshift, zero-throttle coast down, and engine braking; shift conditions: bump, chuggle, delayed (late or extended), double bump (double feel), early, end bump (end feel or slip bump), firm, flare (slipping), garage shift, harsh (rough), hunting (busyness), initial feel, late, mushy, shudder, slipping, soft, surge, and tie-up.

Questions

Diagnosis Procedure

Outline the basic seven-step approach to the diagnosis of automatic transmission problems.

1. ______________________________

2. ______________________________

3. ______________________________

4. ______________________________

5. ______________________________

6. ______________________________

7. ______________________________

Matching

Terminology

A. WOT
B. Garage shift
C. Hunting
D. Flare
E. Chuggle
F. Soft

_____ 1. Busy shift cycling pattern such as 4-3-4.

_____ 2. Wide open throttle.

_____ 3. A "bucking" condition that may be related to engine or converter clutch.

_____ 4. Engagement feel in N to D or N to R.

_____ 5. A sudden rise in engine rpm during a shift.

_____ 6. A drawn-out and almost unnoticeable shift.

Section 2 Diagnostic Steps 1–4: Preliminary Checks

OBJECTIVE:

The objective of this unit is for you to become proficient with the terms, concepts, and procedures contained in this chapter. Completion of this objective is essential to becoming a successful automatic transmission technician.

CHALLENGE YOUR KNOWLEDGE

Define the following key terms/acronyms:

Thermostatic element, malfunction indicator lamp (MIL), and diagnostic trouble codes (DTCs).

Summarize the following procedures:

Steps 1 through 4 of the diagnostic procedure; manual linkage adjustment; TV cable/linkage adjustment; and ECM/PCM diagnostic trouble code identification.

Describe the procedures needed to examine/test the following:

Fluid level and fluid condition.

DIAGNOSTIC STEP 1: LISTEN TO THE CUSTOMER'S CONCERN

An often overlooked but important aspect of diagnosis is to find out the exact nature of the customer's concern. It is the technician's job to listen kindly to the customer (Figure 12–3). The customer gives the technician an impression or clues about the transmission's performance. Some customers need assistance. The technician should converse with customers at their level and ask questions to bring out the whole story.

If the verbal description is vague or the concern appears to be a normal condition, a short preliminary test drive with the customer may be needed. (Check fluid level before taking the test drive.) Questioning the customer and taking a short road test both help the technician find the cause. The test drive may simply show that the customer has misunderstood how the transmission should operate and that the condition is normal. When the vehicle will not move, questioning the customer about the failure is extremely important, along with a pressure test follow-up.

While the customer is a captive audience, do some detective work. A trailer hitch or odometer mileage that is higher than average for the age of the car can alert the technician to the fact that the transmission has been under heavy-duty service. An inspection of the drive wheel tires or spare tire may indicate a hot rodder.

Get the customer to talk about the problem. You may need to establish a transmission preventative maintenance program with the customer to fit the type of service that the particular transmission needs. Some common sense driver education pertaining to transmission life may be included. This not only adds up to trouble-free miles for the customer but also protects the reliability of the repair, especially when band and clutch failures have occurred.

The objective of this first step—listening to the customer's concern—is to avoid the same plight as the dentist who just pulled the wrong tooth. It ensures that the technician and the customer are on the same track.

DIAGNOSTIC STEP 2: CHECK THE FLUID LEVEL AND CONDITION

Fluid level and fluid condition are very basic to proper transmission operation. Checking the dipstick can give the techni-

FIGURE 12–3 Discuss the customer's concern.

cian some immediate and important clues to the general condition of the transmission. In many cases, the diagnostic procedure stops here, and a service recommendation is made to cure the problem. Improper fluid level alone can be responsible for over twenty malfunctions. The transmission will act up when only one pint low on fluid.

Fluid Level

The fluid level check is technically not a difficult task, yet a significant number of car owners, and even technicians, do not accurately follow the recommended factory fluid check procedure. This results in false dipstick readings and contributes to misdiagnosis. The first rule is: never assume that someone else's fluid level check is right. Do it yourself, and you will be surprised at some of the differences between what you are told and what you actually find.

Ideally, the fluid level should be checked with the transmission at normal operating temperature. Normal operating range is approximately 180°F (85°C). To reach this temperature range takes at least twenty minutes of expressway driving or equivalent heavy city driving. This operating temperature cannot be reached with the transmission in neutral or park and the engine set at a fast idle. The fluid is at operating temperature when the sample fluid on the dipstick is too hot for the fingers to handle. Because it is not always convenient to check the fluid at operating temperature, most dipsticks give a hot and cold measurement (Figure 12–4).

The fluid level can change as much as 1 in from cold to hot. When the fluid is overheated from severe service operation, expect a reading over the full mark. Lukewarm readings are usually at the add mark.

There is variation in checking procedure among different transmissions. For accurate results, follow these guidelines. For safety measures, be sure to set the park brake.

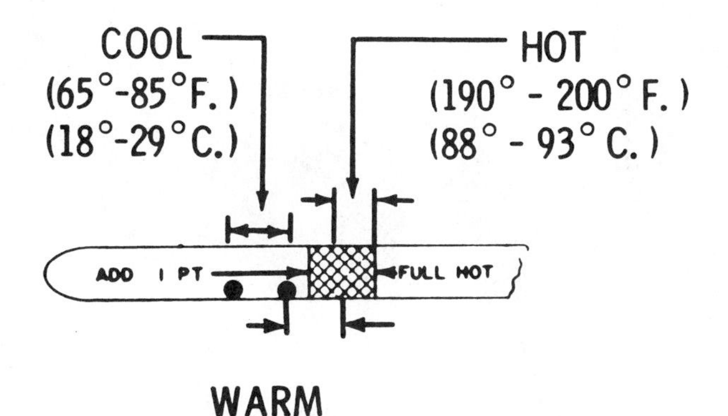

NOTE: <u>DO NOT OVERFILL</u> IT TAKES ONLY ONE PINT TO RAISE LEVEL FROM ADD TO FULL WITH A HOT TRANSMISSION.

FIGURE 12–4 Fluid level range—cool and hot.

Chrysler 900 Series, 36RH (A-727) AND 500 Series RWD Transmissions

- ❍ Vehicle level and engine at curb idle. Allow three minutes to stabilize the fluid level in N.
- ❍ Move the selector level through all the ranges to fill all clutch and servo cavities.
- ❍ Place the selector in neutral only and check.

NOTE: The manual valve opens the pump circuit to exhaust in the park position, which results in flooding the pan area with extra fluid. If the correct level is based on the park position, the transmission will be operating over one quart low on fluid. If the transmission was running in P, allow three minutes for the fluid level to stabilize in N before checking.

WARNING: **When operating the vehicle in neutral and checking the transmission fluid level, set the parking brake. Also, place blocks against the front and rear surfaces of the tires to provide additional resistance to vehicle movement.**

Chrysler 400 Series, 41TE (A-604) AND 31TH (A-670) FWD Transaxles

The transmission and differential areas share a common oil sump with an opening between the two areas. To ensure an accurate dipstick reading:

- ❍ Run the engine at idle for three minutes to stabilize the fluid level between the transmission and differential areas.
- ❍ When the transaxle has a fluid and filter change or a fluid fill after an overhaul, allow eight minutes for the fluid to stabilize with engine at idle.
- ❍ Vehicle level and engine at curb idle.
- ❍ Move the selector lever through all the operating ranges.
- ❍ Place the selector, preferably, in the park position. Both P and N in the transaxles both provide converter fill. The pump circuit is not open to exhaust in P.

NOTE: The transaxle units used through production year 1982 had separate sumps for the transmission and differential areas. The differential uses transmission fluid but takes a separate fill and level check. The differential cover is designed with a filler plug.

Ford—RWD Units Plus FWD, ATX, AND 4EAT Transaxles

- ❍ Vehicle level and engine at curb idle. Allow three minutes to stabilize the fluid in P.
- ❍ Move selector lever through all the operating ranges.
- ❍ Place the selector in park and check.

Ford—AXOD Series FWD Transaxles

Due to the dual sump design characteristics of the AXOD, the fluid level must be checked at operating temperature only. Cold or lukewarm readings can give a false assurance that the fluid level is okay. Other procedures remain the same.

General Motors—RWD Units

- Vehicle level and engine at curb idle. Allow three minutes to stabilize the fluid level in P.
- Move the selector lever through all the ranges to fill all clutch and servo cavities with residual fluid.
- Place the selector in park and check.

General Motors—FWD Units

These transaxle models use a dual sump system (Figure 12–5). Due to the design characteristics, the fluid level must be checked at operating temperature. Cold or lukewarm readings can give a false assurance that the fluid level is okay. If a cold check must prevail, follow the directions in Figure 12–6. Other procedures remain the same.

Dual Sump Systems

In the dual sump system used in Ford's AXOD, AX4S (AXODE), and AX4N and in GM's 4T60, 4T60-E, and 3T40 transaxles, a small-capacity shallow lower sump or pan is used for ground clearance, which means that the full transaxle fluid capacity cannot be stored permanently in the lower pan. Storage is shared with the valve body side cover (Figure 12–5).

Fluid can transfer between the two sump areas through a **thermostatic element** and valve plate installed on the case or case cover (Figure 12–7). In the 4T60 (440-T4) and 4T60-E, it is installed on the transaxle case. The element and valve plate control the fluid volume exchange between the lower and upper sumps, depending on fluid temperature. At the same time, the fluid levels in the sumps adjust to the volume exchange.

❖ **Thermostatic element:** A heat-sensitive, spring-type device that controls a drain port from the upper sump area to the lower sump. When the transaxle fluid reaches operating temperature, the port is closed and the upper sump fills, thus reducing the fluid level in the lower sump.

1. *Cold operation.* During cold operation, all the fluid is retained in the lower sump to prevent aeration at the filter inlet. The high viscosity and density of the fluid cause the thermostatic element to relax and the valve plate to unseat. Any fluid accumulating in the upper sump returns to the lower sump unrestricted.
2. *Hot operation.* As the fluid temperature rises, the thermostatic element gradually applies pressure to the valve plate and the fluid begins to back up and remain in the upper sump. The amount of fluid retained in the upper sump is determined by the rate of return to the lower sump, which is directly related to the following factors:

- Viscosity of the fluid.

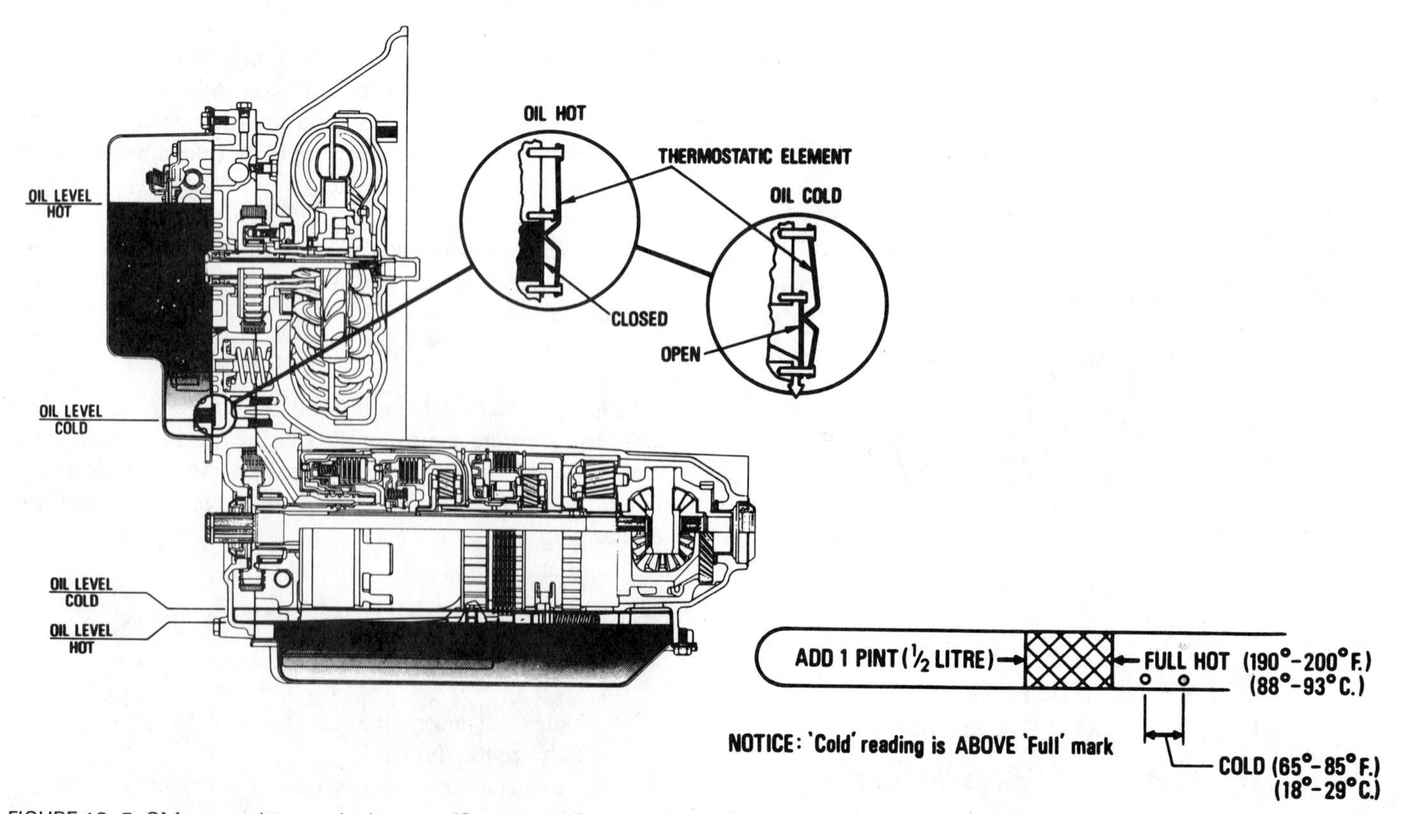

FIGURE 12–5 GM transaxles use dual sumps. (Courtesy of General Motors Corporation, Service Technology Group)

1 FLUID LEVEL INDICATOR (125C)
2 LEVEL TO BE IN CROSS-HATCHED AREA ON FLUID LEVEL INDICATOR BLADE. CHECK AT OPERATING TEMPERATURE.
3 COLD LEVEL ENGINE OFF

1 FLUID LEVEL INDICATOR (440-T4)
2 LEVEL TO BE IN CROSS-HATCHED AREA ON FLUID LEVEL INDICATOR BLADE. CHECK AT OPERATING TEMPERATURE.
3 COLD LEVEL ENGINE OFF

FIGURE 12–6 3T40 (125) and 4T60 (440-T4) fluid level indicators. (Courtesy of General Motors Corporation, Service Technology Group)

- Temperature of the thermostatic element.
- Height adjustment of the thermostatic element.

If the fluid level becomes excessive in the upper sump, its weight overcomes the thermal tension on the valve plate and increases the drain back to the lower sump. If the lower sump is not relieved of its excess fluid when hot, it rises into the powertrain area and aerates. The fluid, therefore, will read above the full mark when cold.

Some problems arise in reading the fluid level. When the fluid is new and very clear, it is sometimes difficult to read where it is on the dipstick. A trick of the trade is to rub the end of the dipstick with typewriter carbon paper, which allows the technician to easily pick up the level reading. The effect of the carbon treatment is illustrated in Figure 12–8, using the same dipstick on the same transmission fluid.

On occasion the wrong dipstick is used. This may occur in both new and used cars. A slipup at the factory or on a replacement can play tricks on the diagnosis. The wrong dipsticks are usually too long, and therefore, a full reading can mean the transmission is one or more quarts low on fluid. If this is suspected, add a quart of fluid and perform a road test. A short dipstick results in an overfill and causes fluid to spew out the filler tube.

As part of a fluid level check, include a visual examination of the cooler lines for kinks, especially in the area where an add-on cooler is installed. Cooler lines should not be touching any part of the chassis.

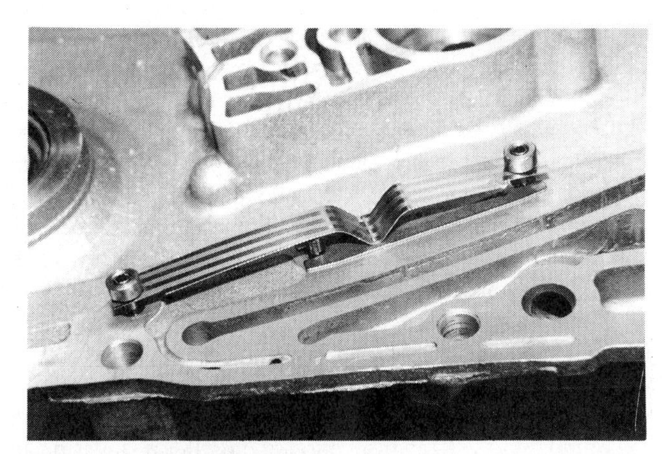

FIGURE 12–7 Case cover thermostatic element.

Low Level. Low fluid level can result in the oil pump taking in air along with the fluid. In any hydraulic system, air bubbles will make the fluid spongy and compressible. This can result in a slow buildup of pressure in the transmission hydraulic system. Low fluid level causes delayed engagement in drive and reverse and slipping on upshifts. It may also cause disturbing noises, such as pump whine and regulator valve buzz. The slipping action adds to the troubles by causing overheating and rapid wear of clutches and bands. With air in the fluid, the pump cannot provide an adequate supply of fluid for converter feed and lubrication, so more overheating is generated, and other transmission parts wear to destruction.

The add mark on the dipstick means add one pint of fluid only, but this does not mean that being one pint short is insignificant as the cause of problems. When the transmission is one pint short and the selector is dropped into drive, the forward clutch circuit fills and moves the car with no problem. On the shift to second gear, the intermediate circuit takes on more volume of fluid, and still there is no apparent problem. On the shift to third, however, the transmission that is one pint short is struggling to satisfy the direct clutch circuit, and it is common to get a drawn-out shift or slight engine flare during direct clutch apply.

If the fluid level is down, the low level may be the result of an external leak, a leaky vacuum modulator diaphragm, or improper filling. If there is no evidence of an external leak, or inspection and testing of the vacuum modulator show no sign of a leaky diaphragm, then in all probability someone has failed to refill and check the fluid level properly. If the fluid condition looks normal, simply add enough fluid to reach the full cold or hot mark, and it is a good bet that the transmission will perform normally again. Do not forget to perform a road test before giving the final okay.

High Level. When the fluid level is too high, the planetary geartrain is riding in the sump fluid, which is not necessary. The geartrain has its own lubrication feed system. The geartrain action actually whips the fluid into a foam and typically causes the same conditions as low fluid. The combination of foam (aeration) and overheating causes rapid oxidation and fluid varnishing, which interferes with normal valve, clutch, and servo operation. Therefore, overfilling must be avoided. High fluid level causes the fluid to push out through the filler tube and transmission case breather.

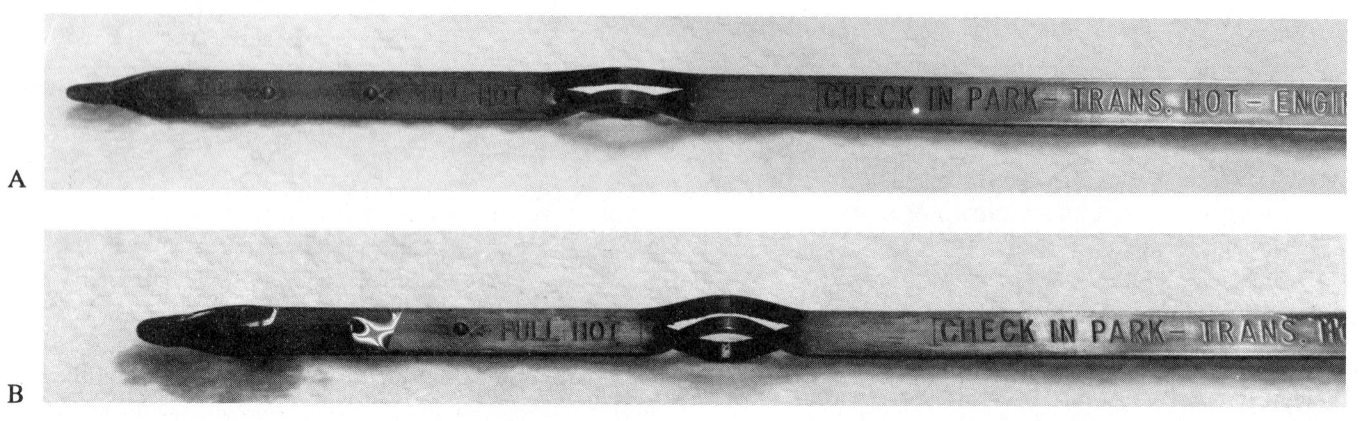

FIGURE 12–8 (a) Fluid level check without carbon-treated dipstick; (b) fluid level check with carbon-treated dipstick.

To avoid overfills, check with the manufacturer's refill recommendations when replacing the fluid during an overhaul or fluid change. Overhaul fluid requirements are based on the total transmission, which includes the converter and the oil pan. In most cases, the pan fluid is all that can be changed for preventative maintenance. Therefore, oil pan fill requirements need to be followed.

In the past, drain plugs on the torque converter were common. General Motors has not provided converter drain plugs for many years, and since midyear 1977, Chrysler has eliminated converter drain plugs. Ford is still using converter drain plugs in most applications. The fluid from the converter should be drained as part of its preventative maintenance package.

When refilling for an overhaul or preventative maintenance schedule, consider the manufacturer's recommendation a guideline only. It is impossible to determine how much fluid has been added for lubrication during an overhaul or how much has not drained out for a fluid change. Therefore, always stop at least one quart short of the recommended refill and check the dipstick. Because this is a cold check, bring the fluid into the cold zone or no further than the add mark. Recheck and adjust the fluid level when the transmission oil is at operating temperature.

A frequent cause of overfill is simply improper judgment when the transmission is cold or warm. The dipstick normally shows that fluid should be added. If fluid is added at this time, especially one quart instead of the usual one pint, the hot reading ends up greatly above the full mark. As a final tip when adding fluid: Always allow the fluid level to stabilize over a three-minute period before making the dipstick check.

GM 4T40-E Fluid Level Check Procedure. Unlike other transmissions and transaxles, the 4T40-E does not use a dipstick. With the engine running and warmed up, the fluid level is checked by removing the oil level screw located below the stub shaft. The fluid should be level with the hole. When needed, add fluid through the vent tube/fill cap (Figure 24–39, page 601).

Fluid Condition

With the current transmission fluids, color and smell are no longer absolutes when examining fluid condition. Fresh fluid retains its red or green dye color at first but soon changes and takes on an appearance of a dark clear varnish. In the recent past, this change in color indicated fluid failure. The smell of transmission fluid is also a false indicator with the current fluids. The fluid now smells burned when it is fresh from the container, so smell should not be used as a basis for transmission overhaul.

The smell and color changes in current fluids are caused by the chemical formulation of the new fluids and the added resin content in the friction material for bonding. These fluids were developed after the use of asbestos fibers was discontinued. A dark clear brown or dark clear red fluid without other failure indicators is now okay, as is a burned smell.

Some failure clues can still be detected from the fluid. To help analyze fluid condition, place a sample of the fluid on a white paper shop cloth (Figure 12–9). Should any of the following conditions exist, a transmission disassembly is required.

- Fluid discoloration takes on a black appearance, possibly accompanied by residue particles of friction material.
- Fluid is contaminated with residue particles of metal and/or friction material.
- Fluid is obviously heavily varnished, losing its clear characteristic. It becomes tacky and may even stick onto the dipstick. For a simple test of fluidity, compare the soak rate of the actual transmission fluid with new fluid using a paper towel. A good fluid easily soaks into the paper towel, while a questionable fluid pools up and absorbs slowly.
- Fluid has a milky appearance, which means that engine coolant has leaked into the fluid. The water and glycol content swells the transmission seals and softens the friction material. It is not uncommon to find friction facing material unglued from its backing. The only cure is to replace the transmission cooler element, flush cooler lines and nonlockup converter, and completely overhaul the transmission. Nonlockup converters can be flushed, while lockup converters need to be replaced.
- Fluid looks foamy or bubbly. This can be caused by either low or high fluid level conditions. If the level appears to be correct, suspect an air leak on the suction side of the pump. Start with an inspection of the filter installation and

pay particular attention to whether a gasket or O-ring is missing, damaged, or off-location.

If in doubt about the fluid condition and residue, the transmission pan should be removed for inspection and evaluation during the diagnosis process (Figure 12–10). There is always some normal residue in the pan, but watch for the following evidence:

FIGURE 12–9 Analyzing fluid condition.

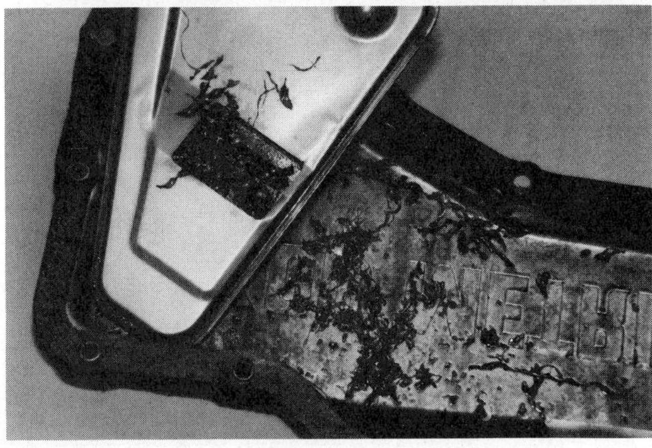

FIGURE 12–10 Inspection: bottom pan and filter.

- Significant particles of steel, bronze, or plastic, indicating damage to bushing, thrust washer, or internal hard parts.
- Composition friction material, indicating clutch or band failure.
- Steel particles or friction material that cannot be associated with internal worn or damaged parts, which indicates possible converter damage.

Excessive deposits may not always be obvious in the pan area, and the fluid may look healthy. The design of the new filters tends to conceal these deposits. Remove the filter and open it up for inspection (Figure 12–11). Do not get too excited over a plastic "lollipop" that is found in some Ford pans (Figure 12–12). This is a shipping plug used for blocking the filler/dipstick tube cavity during shipment, and you may discard it.

If a customer's concern about drawn-out shifts is attended to at an early stage, the problem can often be corrected by a fluid and filter change. TorqueFlite transmissions use an extra-fine filtration screen that is noted for suddenly shutting off the pump intake and causing a "no forward or reverse" condition. Even though the fluid looks healthy, a fluid and filter change works wonders.

Anytime the fluid looks healthy and residue deposits in the pan or filter are not significant, hold tight. In terms of probability, the solution to the problem will not be resolved by removal and disassembly of the transmission. In many cases

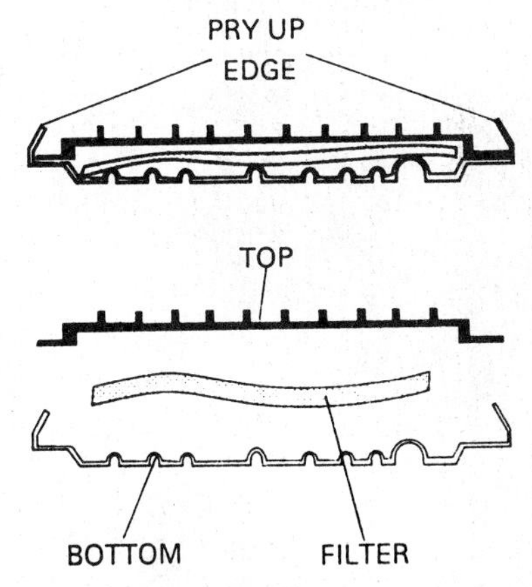

FIGURE 12–11 Filter unit disassembly. (Courtesy of General Motors Corporation, Service Technology Group)

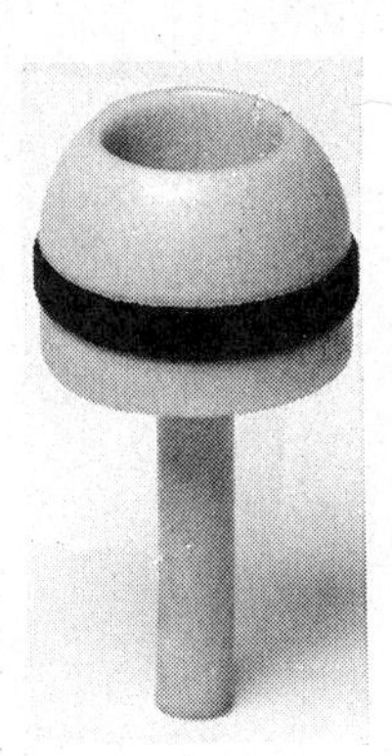

FIGURE 12–12 Shipping plug used in Ford dipstick tube cavity.

the fluid condition identifies a significant internal problem and the diagnostic process stops here. Generally, no further information is needed.

DIAGNOSTIC STEP 3: QUICK CHECK ENGINE IDLE, VACUUM LINE, LINKAGE/CABLE CONTROL, THROTTLE BODY ACCELERATOR FOR WOT, AND WIRING

In keeping with the diagnostic objective of reaching the end result with the least amount of effort, use your eyes, ears, and touch for some quick checks in preparation for a complete diagnostic road test. Quick checking simply means inspecting the communication between the chassis/engine and transmission. Are the communication links in place, and do they appear to be working and in correct adjustment? Make all necessary corrections exposed by the quick check before performing a complete road test.

NOTE: A technician may wish to make a judgment call and decide whether to make minor repairs and adjustments prior to a road test. If the repairs and adjustments are made first, it may be impossible to verify the original concern and validate that those efforts have fixed the problem.

Engine Idle

Listen to the engine hot idle and include the cold idle if necessary. Is it reasonable? If not, set the idle to the manufacturer's specifications using a reliable tachometer (Figure 12–13). When the engine idle is too high, the forward or reverse clutch engagement is harsh and usually associated with a driveline "clunk." Settling down the idle rpm can also eliminate that closed-throttle downshift clunk. Dropping the idle 25 or 50 rpm can sometimes make a noticeable difference.

Manual Linkage

This check takes place from the driver's seat and is to be done with the engine off. It is a simple matter of running the selector through all the ranges and feeling for the spring-loaded transmission detent dropping into the notches of the manual valve. Are they coordinated with the markings on the shift selector? This is accurately illustrated in Figure 12–14.

Each operating range has a gate-stop position in the selector mechanism that indexes with the markings on the selector. If the markings are slightly off with the detents in the transmission, it is still considered acceptable. If the selector indexing is noticeably off target, an adjustment is necessary.

Drivers by nature usually move the selector to the gate-stop position, where the transmission detents should be properly positioned. If the detents are not synchronized (Figure 12–15), the manual valve will be out of position and cross-feed hydraulic oil into another circuit. Observe in Figure 12–16 how the

FIGURE 12–13 Checking engine idle setting.

drive circuit line pressure is partially exhausting out the neutral circuit. This results in a loss of pressure apply and subsequent clutch and band failure. When the manual linkage is not properly adjusted, it can even cause the vehicle to creep forward or backward with the selector in neutral.

Manual linkage adjustment is simple but too often neglected. The linkage adjustment is usually coordinated with the selector against the drive gate stop and transmission manual valve in its drive detent. Refer to the car manufacturer's service manual, or equivalent, for the specific procedures needed. Always counter check the adjustment for park engagement and neutral safety switch operation. Do all the shift gate stops give a positive detent feel? Because of the neutral safety switch used in Chrysler TorqueFlite units, the manual valve adjustment is considered correct anytime both the P and N positions allow "engine crank" (Figure 12–17).

Throttle Linkage, Vacuum Line, and Wire Connections

This is the area in which most of the problems that cause erratic shift conditions and even complete transmission failure are found. Engine performance and the torque signal to the transmission must be right if the transmission is to work properly. Too many transmissions have been removed and disassembled unnecessarily when the cause of the problem was actually in the engine torque signal.

This subject is not discussed in detail here but is reserved for a separate discussion involving erratic shift situations and pressure testing later in the chapter. The emphasis here is to treat throttle linkage and vacuum line connections as a quick check item, unless complete road testing and pressure testing point to a detailed involvement. A quick check involves mostly a close visual inspection to verify that the connections between the engine and transmission are properly attached, not worn or loose, and show no evidence of tampering.

On TV linkage or cable control, check for freedom of movement and positive response to the throttle opening—especially on closed-throttle movement. Any delays by the linkage or cable in following the action cannot be tolerated. Where an external throttle arm at the transmission is involved, make sure that the TV arm is freely moving full stroke between closed and WOT. Where the TV cable connects internally to the transmission TV assembly, give close attention to the cable follow-up at the throttle body (Figure 12–18). Although a cable might work freely with the engine stopped and cold, always recheck after the engine is hot.

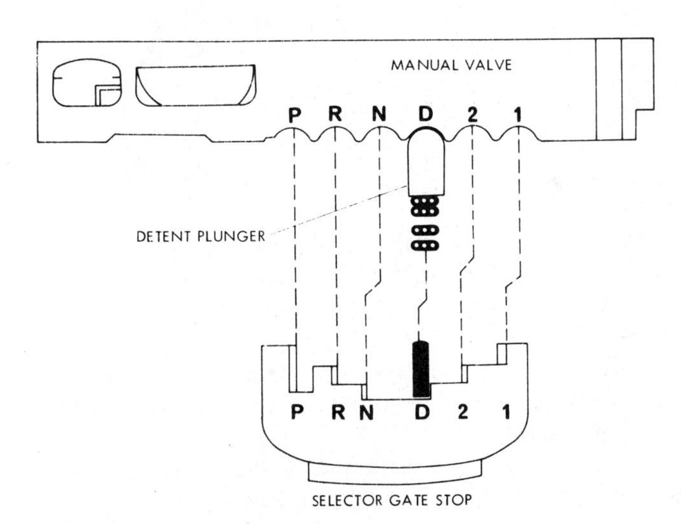

FIGURE 12–14 Correct shift gate to manual valve position.

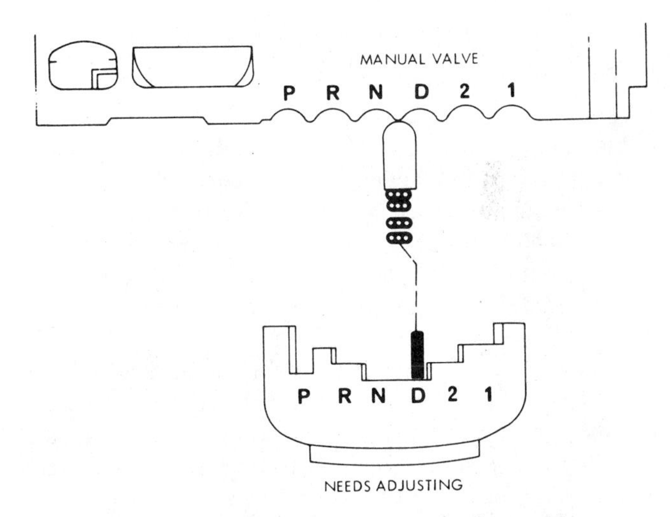

FIGURE 12–15 Incorrect shift gate to manual valve position.

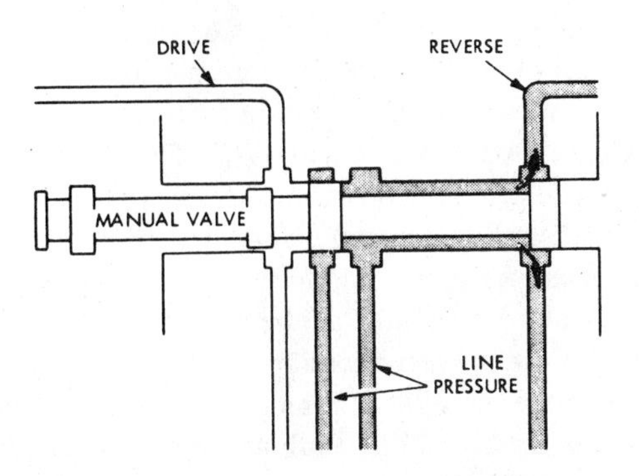

FIGURE 12–16 Manual valve not in adjustment; line pressure bleeding into reverse circuit (selector lever in neutral position).

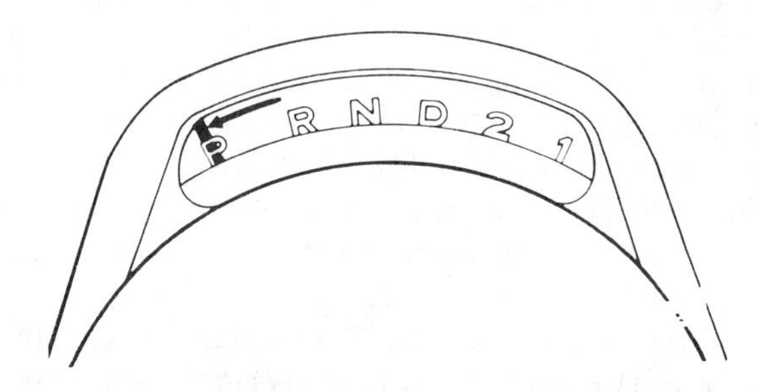

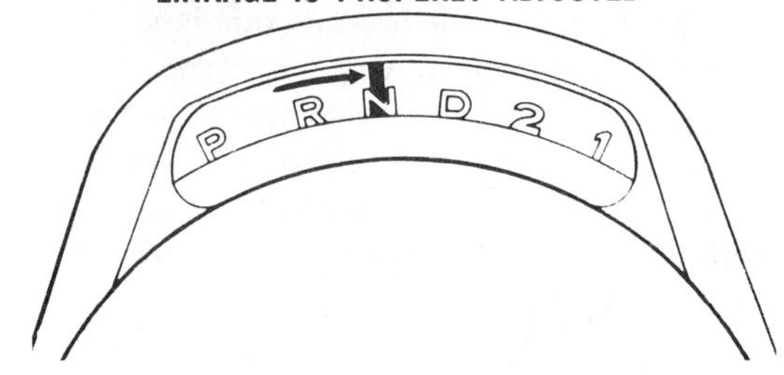

FIGURE 12–17 Detent "feel" locates stopping point. (Courtesy of Chrysler Corp.)

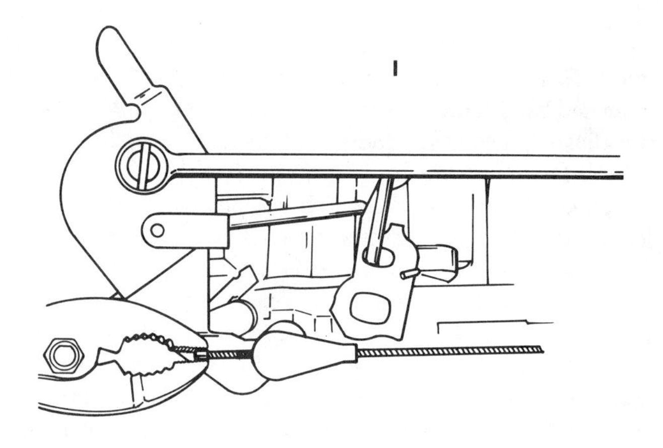

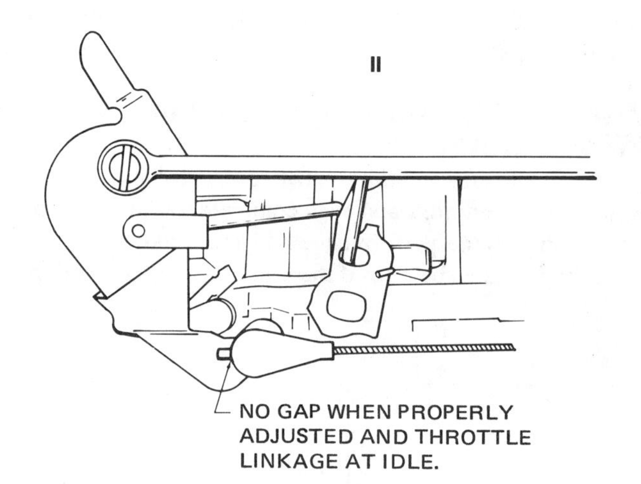

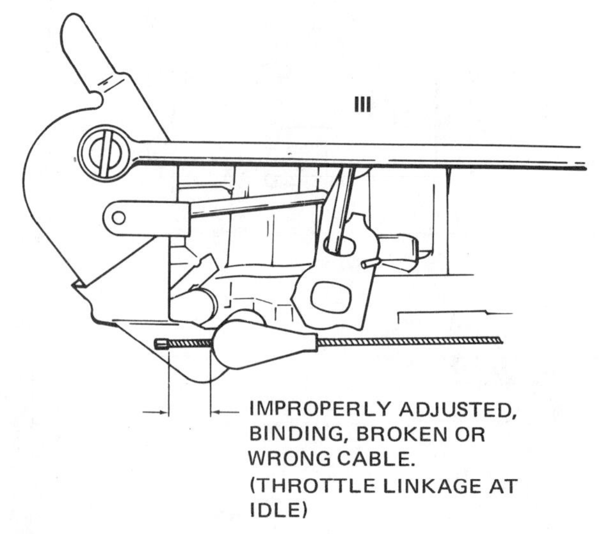

FIGURE 12–18 (I) Pulling TV cable; (II) TV cable released; (III) sticking TV cable. (Courtesy of General Motors Corporation, Service Technology Group)

Prior to any adjustment to the TV cable, the accelerator movement at the throttle body must be checked for WOT by fully depressing the accelerator pedal. Use an assistant for this operation while making the observation (Figure 12–19). If full throttle is not achieved, the accelerator system must be corrected. Throttle body travel and the transmission TV lever travel must be synchronized.

Because of the varied procedures for TV linkage/cable, the technician needs to refer to the manufacturer's service manual or equivalent for details. Be sure to follow directions in removing linkage/cable slack. Slight wear occurs at linkage points and cables stretch, so periodic adjustment is necessary.

Some of the new generation transmissions require the TV circuit to be adjusted with a slightly advanced travel or output at closed throttle. Required beginning pressure can be as high as 25 psi (172 kPa). This differs from the past practice of setting the TV output at zero or close to zero psi for closed throttle. Some of the transmissions requiring an advanced TV setting are the AXOD, AOD, 4L60, and 4T60. The TV pressure test ports are provided should it become necessary to countercheck against actual starting adjustment and full throttle pressures.

On the vacuum line leading to the modulator unit, look for kinks and leaks, especially at the manifold and modulator connections. Give special attention to the vacuum hose nipple connection at the modulator. It tends to become disconnected or leak from deterioration. With the maze of vacuum line plumbing under the hood, it is essential to make sure the transmission is attached to the proper vacuum source fitting.

On manifold vacuum trees, be positive that the vacuum line is not coupled to a metered fitting. Direct manifold vacuum is a normal source.

To quickly verify a vacuum flow to the modulator, disconnect the line at the modulator. The engine idle rpm should increase.

FIGURE 12–19 Checking for full throttle travel with digital volt ohm meter (DVOM).

Quick check the wiring under the hood and at the transmission. The wire connectors need to be in place and the wires not cut or totally missing.

Forced Downshift System

This system is sometimes referred to as the detent or kickdown. When the system is triggered by external transmission controls, three types of controls are used:

1. When the throttle and forced downshift valving are located in tandem in the same valve body bore, the downshift system is mechanically triggered by the TV cable or linkage.
2. When the throttle valve is vacuum-operated through a modulator, the downshift valve is located in its own independent valve body bore. The downshift system is mechanically triggered by an independent cable or linkage.
3. In some independent downshift systems, an electrical switch is mechanically closed and a downshift solenoid is energized. Essentially, the solenoid closes an exhaust port and allows downshift oil to switch the valve. Because there is wide variety in checkout procedures, it is best to consult the car manufacturer's service manual when in doubt about the procedure.

An electrically controlled system can easily be checked out as follows:

- Turn on the radio and tune in to the local AM radio station with the final setting de-tuned on the high-frequency side of the broadcast, engine off.
- Depress the accelerator pedal to energize the solenoid.
- Release the accelerator pedal and listen for a speaker sound caused by a magnetic induction from the solenoid.

NOTE: This check only verifies that the electrical circuit is working.

Step 3 is very vital in diagnosis and can lead to a quick correction of a transmission problem without getting involved with major service. Conditions found to be improper must be corrected at this time.

DIAGNOSTIC STEP 4: CHECK POWERTRAIN/ENGINE CONTROL MODULE FOR CODES

The engine and transmission are a matched team and their performance must be synchronized. An important part of transmission diagnosis, therefore, is to verify the proper operation of the engine. The transmission and engine are linked together and synchronized by an engine torque signal through a vacuum line, throttle cable, or electrical signals from devices such as throttle position and manifold absolute pressure sensors.

If the engine is underpowered, it sends a false signal through the communication link to the transmission. The transmission has no way to determine if the information is right or wrong, but only reacts to inputs. Depending on the transmission, this results in harsh shifting or a combination of harsh/late shifting that adds to the already underperforming engine.

Inconsistent shift schedules also can be experienced, especially with a clogged exhaust or misindexed timing gear. Adding to complications is the ability of the engine control module to cover up most problems and keep the engine running smoothly while the transmission operation is erratic.

Before proceeding, it is essential to investigate the **malfunction indicator lamp (MIL)** and correct the driveability fault indicated by **diagnostic trouble codes (DTCs)**. Any suspicion of engine performance as a problem needs to be corrected at this point. The technician can be fooled and disassemble the transmission only to find that the problem goes back to the engine or engine exhaust.

- **Malfunction indicator lamp (MIL):** Previously known as a *check engine light,* the dash-mounted MIL illuminates and signals the driver that an emission or driveability problem with the powertrain has been detected by the ECM/PCM. When this occurs, at least one diagnostic trouble code (DTC) has been stored into the ECM's/PCM's memory.
- **Diagnostic trouble codes (DTCs):** A digital display from the ECM's/PCM's memory that identifies the input, processor, or output device circuit that is related to the powertrain emission/driveability malfunction detected. Diagnostic trouble codes can be read by commanding the MIL to flash any codes or by using a handheld scanner.

❑ SECTION 2 REVIEW

Key Terms/Acronyms

Thermostatic element, malfunction indicator lamp (MIL), and diagnostic trouble codes (DTCs).

MULTIPLE CHOICE

____ 1. Which of the following fluid levels cannot be checked with the selector in P?

A. Chrysler RWD series
B. Chrysler FWD series
C. Ford RWD series
D. General Motors RWD series

____ 2. As a standard practice, when checking fluid levels:

I. allow the fluid level to stabilize for one minute at engine idle.
II. move the selector through all the operating ranges.

A. I only   C. both I and II
B. II only   D. neither I nor II

____ 3. Which of the following FWD transaxles require that the fluid level check be made at operating temperature only?

A. AXOD   C. 4T60
B. 3T40   D. All of the above

____ 4. When adding automatic transmission fluid (ATF), and before rechecking the level, allow the fluid level to stabilize for a time period of:

A. three minutes   B. two minutes
C. one minute   D. none of the above

____ 5. When the fluid looks like a milk shake:

I. Engine coolant has contaminated the fluid.
II. The transmission can be kept in service by flushing with new ATF.

A. I only   C. both I and II
B. II only   D. neither I nor II

____ 6. Technician I says that contaminants in the pan provide important clues for diagnosis. Technician II says that filters should be opened for contaminant inspection. Who is right?

A. I only   C. both I and II
B. II only   D. neither I nor II

Questions

1. TV cable movement should be checked with both a cold and a hot engine. ___T/___F

2. The modulator vacuum line should be connected to a metered orifice manifold vacuum supply. ___T/___F

3. The preferred shift selector position when adjusting manual linkage is park. ___T/___F

4. Final manual linkage adjustment should always be checked against the neutral safety switch operation. ___T/___F

5. The vacuum modulator on the Hydra-Matic 4T60 (440-T4) controls the TV system. ___T/___F

6. The AXOD TV and kickdown systems share a common cable control. ___T/___F

7. Before proceeding with transmission diagnosis, it is necessary to correct any engine driveability fault indicated by computer diagnostic codes. ___T/___F

Section 3 Diagnostic Step 5: Stall Test

OBJECTIVE:

The objective of this unit is for you to become proficient with the procedures relating to stall tests. Completion of this objective is essential to becoming a successful automatic transmission technician.

CHALLENGE YOUR KNOWLEDGE

Define the following key term:

Stall test.

Summarize the following:

Stall test procedures.

Identify the warning associated with:

Performing a stall test.

Interpret the following:

Stall test results.

DIAGNOSTIC STEP 5: STALL TEST

The **stall test** works full engine power against the stationary weight of the vehicle. The purpose of the test is to determine band and clutch holding ability, torque converter operation, and engine performance. This procedure must be used with caution, however, because the test produces severe heat and twisting stress. Vehicles with neglected cooling systems could burst a hose. Broken engine mounts are also an area of concern.

NOTE: Due to the cautions and hazards associated with stall testing, some manufacturers do not recommend this procedure for diagnosing transmission problems.

❖ **Stall test:** A procedure recommended by many manufacturers to help determine the integrity of an engine, the torque converter stator, and certain clutch and band combinations. With the shift lever in each of the forward and reverse positions and with the brakes firmly applied, the accelerator pedal is momentarily pressed to the wide open throttle (WOT) position. The engine rpm reading at full throttle can provide clues for diagnosing the condition of the items listed above.

Perform the stall test when you feel that it is needed as an additional diagnostic tool, especially when the complaint is power performance. When slippage is not a factor, the test can be very useful in isolating power loss to either the engine or converter. The stall test can be compared to a full power dynamometer test.

Stall Test Procedure

The engine coolant, engine oil, and transmission fluid levels must be correct and at normal operating temperatures.

WARNING: The full weight of the vehicle must be on the wheels, and the vehicle must be positively prevented from moving. It will be necessary to apply the park and service brakes and block the vehicle. No person should stand in front of or behind the vehicle during the test. Refer to the manufacturer's service manual for specific details concerning stall test procedures.

The stall test procedure is as follows:

1. Connect the tachometer and position it so you can read it from the driver's seat.
2. Shift the selector lever to the D range.
3. Accelerate the engine to full throttle.
4. Note the maximum rpm attained by the engine and record this information (Figure 12–20).

CAUTION: A stall test in one specific gear must not take over five seconds or severe overheating and damage to the transmission may result.

5. Return the selector lever to neutral and operate the engine at fast idle for a minimum of thirty seconds for cooling.

FIGURE 12–20 Stall test tachometer reading.

6. Repeat the above procedure for each test run.

If engine speed exceeds the maximum limits of the stall specification, release the accelerator immediately, because slippage is indicated and further damage to the transmission must be avoided. The test also should be stopped when unusual noises occur. A churning noise or even a whining noise from the converter fluid action is normal during a stall test, but any metallic-type noise in the assembly indicates a defective torque converter.

Engine Liter	Transaxle Type	Stall Speed Engine rpm
2.2 EFI	A-413	2050-2250
2.2 EFI (turbocharged)	A-413	3150-3350
2.5 EFI	A-413	2250-2450
3.0 EFI	A-413	2500-2700

CHART 12–1 TorqueFlite 31TH (A-413) transaxle stall speed chart.

RANGE	ACTUAL	CONDITION AS TESTED		
		LOW	NORMAL	HIGH
DRIVE				
(2)				
(1)				
REVERSE				

OPERATING RANGE	HOLDING MEMBERS APPLIED
D	REAR CLUTCH AND ONE-WAY CLUTCH
2	REAR CLUTCH AND ONE-WAY CLUTCH
1	REAR CLUTCH AND LOW-REVERSE BAND ONE-WAY CLUTCH EFFECTIVE
R	FRONT CLUTCH AND LOW-REVERSE BAND
KICKDOWN BAND IS NOT USED DURING STALL TESTS.	

GUIDE 12–2 TorqueFlite stall test summary.

Stall Test Interpretation

For interpretation of test results, a comparison is made against the manufacturer's test specifications. In Chart 12–1, typical stall speed specifications are listed for the TorqueFlite transaxle engine group. Allowances should be made for altitude changes, as engine power decreases at higher altitude levels. Be sure to record test results and compare them to the manufacturer's specifications (Guide 12–2).

Variations in test speed results are interpreted as follows:

- Stall speed is 200 to 300 rpm higher than normal. The transmission is not holding and the probable cause is a slipping friction element. Aerated transmission fluid and low operating pressure can also be suspect areas.
- Stall speed is extremely high and runs away, and the test run has to be stopped. A likely suspect is still a slipping clutch or low operating pressure, but the possibility of sheared turbine splines on the input shaft should not be overlooked.
- Stall speed is 100 to 200 rpm lower than normal. This is an indication of engine power loss. Typically, the customer has already noticed a drop in engine performance, and driveability service is in order. If a stall torque engine miss is experienced, the problem is usually in the spark plugs or secondary wiring.
- Stall speed is 33% lower than normal. A freewheeling stator is the suspect. The converter during a stall test is producing full torque, at which point the stator is normally in a locked condition. If the stator roller clutch fails to take hold, the fluid action works against the impeller and engine rotation. The engine power is drastically lost in working against the converter fluid. Car performance in the low-speed range, 0 to 30 mph (0 to 48 km/h), is dead and has almost no pulling power or acceleration.
- Stall speed is normal. This means that the engine power is normal, the transmission is holding, and the converter stator is holding. It does not determine, however, if the stator can release and permit the converter to act as a fluid coupling.

A frozen stator can be suspected when the top vehicle speed at WOT (wide open throttle) is greatly reduced (usually by 33%). Low-speed acceleration and performance is normal, but when the converter wants to operate in coupling phase, the frozen stator acts as a dam or brake against the rotary fluid motion. The side effects are overheating of transmission fluid and possible loss of transmission pump pressure. The car usually travels 15 to 20 miles (24 to 32 km) down the road, and then the transmission loses pressure and holding power.

As another test for a frozen stator, place the manual selector in N and push the accelerator to WOT. If the engine indicates a normal no-load condition and easily exceeds 3000 rpm, stop the test—the stator is freewheeling and turning together with the impeller and turbine in the fluid flow. If the engine cannot exceed 3000 rpm at WOT, the stator roller clutch is frozen. The stator acts as a hydraulic brake against the rotary fluid force and lugs the engine.

❑ SECTION 3 REVIEW

Key Term

Stall test.

Questions

_____ 1. The purpose of the stall test is to determine:

A. transmission band/clutch holding ability.
B. engine power output.
C. torque converter operation.
D. all of the above.

_____ 2. Each stall test run must be limited to:

A. five seconds C. twenty seconds
B. ten seconds D. thirty seconds

_____ 3. Between each stall test run, the transmission must be run in N with the engine at fast idle for at least:

A. 5 seconds C. twenty seconds
B. ten seconds D. thirty seconds

_____ 4. A frozen stator causes:
I. very poor low speed acceleration or performance.
II. extreme ATF temperature.

A. I only C. both I and II
B. II only D. neither I nor II

_____ 5. When testing for a frozen stator, engine rpm should not exceed:

A. 1500 rpm C. 2500 rpm
B. 2000 rpm D. 3000 rpm

_____ 6. The frozen stator test is made:
I. with the range selector in N.
II. with the engine at WOT.

A. I only C. both I and II
B. II only D. neither I nor II

_____ 7. A freewheeling stator:
I. will reduce the WOT top vehicle speed.
II. will cause extreme ATF temperatures.

A. I only C. both I and II
B. II only D. neither I nor II

_____ 8. A customer complains that her vehicle will only go 60 mph (96 km/h) at WOT. Technician I says the torque converter could be defective. Technician II says the exhaust system could be restricted. Who is right?

A. I only C. both I and II
B. II only D. neither I nor II

9. Briefly explain why a frozen stator cannot be determined with a stall test.

____

____

____

10. The stall test specification for an engine is given as 1800 to 2000 rpm. The actual stall test reading is 950 rpm. What is the likely problem?

____

____

____

Section 4 Diagnostic Step 6: Road Test

OBJECTIVE:

The objective of this unit is for you to become proficient with the terms, concepts, and procedures contained in this chapter. Completion of this objective is essential to becoming a successful automatic transmission technician.

CHALLENGE YOUR KNOWLEDGE

Define the following key terms/acronym:

Garage shift feel, shift busyness, deceleration bump, O_2 sensor, PROM (programmable read-only memory), brinneling, and spalling.

Summarize the following procedure:

Road test.

Interpret and demonstrate application of the following:

Shift-speed chart; road test summary report; and clutch/band apply chart.

Describe the techniques needed to analyze the following:

Torque converter clutch operation; mechanical failures; noise concerns; and erratic shift problems.

Describe the procedures needed to examine/test the following:

Garage shift feel; clutches and bands; and torque converter clutch.

DIAGNOSTIC STEP 6: ROAD TEST

Road Test Procedure

The engine coolant, engine oil, and transmission fluid levels must be correct. Test drive the vehicle under exact conditions of concern such as:

- ❍ Cold only
- ❍ Hot only
- ❍ Always
- ❍ Startup only
- ❍ Morning only

To road test the transmission, it should be operated in each selector range to check for clutch/band slipping and engagement quality and any abnormal variations in the shift schedule. Performance testing should be made at minimum-, medium-, and heavy-throttle modes. During the performance testing, observe whether the shifts are harsh or long and drawn out. Check the speeds at which the upshifts and downshifts occur. Look for an engine flare-up during a shift, especially on medium- and heavy-throttle testing. How does the converter clutch behave?

To attain your minimum shift points, the throttle should be teased just enough to keep the vehicle moving through 20 mph (32 km/h). Engine manifold vacuum must be kept high. For best results, find a flat stretch of road or one that has a slight downgrade. What is usually considered mild acceleration from a stop is too aggressive for a checkout on the minimum shifts. The minimum shift pattern checkout requires a careful and very easy throttle manipulation by the technician for accurate test results—almost at closed throttle. It takes a little practice to master the technique.

The diagnostic road test requires collecting information and data. Therefore, reference must be made to the manufacturer's applicable shift speed specifications (Chart 12–2). Because of the many items being tested, it is best to record the test results rather than to depend on memory. A comprehensive road test program for three-speed transmissions is outlined in the test summary guide, Guide 12–3, pages 272 and 273. The test summary guide can be modified and adapted as necessary for four-speed transmissions.

Sometimes the transmission shift feel is difficult to sense, especially at the lower shift schedules. To assist the technician, a special tester called a Shiftalizer can be plugged into the cigar lighter. It acts as an audio tachometer by picking up the alternator rpm. The shift points plus the shift quality are easily detected. As an alternative technique, tune the local AM radio station to the off-beat high-frequency side of the broadcast.

Garage Shift Feel

Garage shift feel is assessed before moving the car onto the road. With the brake firmly applied, move the shift lever into each of the range positions. Wait for the gear to engage and then analyze the results. You are checking for power transmission as well as engagement quality in the forward and reverse gears. No delay, no slip, and no chatter means that the pump is turning, line pressure is there, and the converter is full.

❖ **Garage shift feel:** A quick check of the engagement quality and responsiveness of reverse and forward gears. This test is done with the vehicle stationary.

Clutch and Band Slip Analysis

Any slipping or engine flare condition during the road test may indicate clutch, band, or overrunning clutch problems. By the process of elimination, any friction units that slip can be verified, as well as those that are in good working order. It must be emphasized that the clutch/band slip analysis during the road test is meaningless unless steps 2 and 3 have been completed in the diagnosis process. Conditions not found or corrected in steps 2 and 3 can cause slipping and flare-up. The idea is to find and correct external conditions that may cause the slip before it is too far advanced and an overhaul is needed to restore normal operation.

The key to a band and clutch analysis is in using a summary clutch/band apply chart. Experienced technicians need only make mental notes. To illustrate how the technique works, a TorqueFlite clutch/band apply summary is given in Chart

	Overall Top Gear Ratio							
	2.78		3.22		3.02 (except turbocharged)		3.02 (turbocharged)	
	mph	km/h	mph	km/h	mph	km/h	mph	km/h
Throttle minimum								
1-2 Upshift	13–16	21–27	13–16	21–26	13–17	21–27	15–19	24–31
2-3 Upshift	17–21	27–34	17–21	27–34	18–22	29–35	20–25	32–40
3-1 Downshift	12–15	19–24	12–15	19–24	13–16	21–26	15–19	24–31
Throttle wide open								
1-2 Upshift	35–42	56–68	34–42	55–68	36–44	58–71	38–42	61–68
2-3 Upshift	61–68	98–103	59–66	95–106	63–71	101–114	70–80	113–129
Kickdown limit								
3-2 WOT downshift	56–64	90–103	55–62	89–100	58–66	93–106	64–74	103–119
3-2 Part throttle downshift	44–52	71–84	44–51	71–82	46–54	74–87	48–59	77–95
3-1 WOT downshift	31–38	50–61	30–37	48–60	32–39	51–63	37–40	60–64
Governor pressure [a]								
15 psi	23–25	37–40	24–27	39–43	24–27	39–43	28–31	45–50
50 psi	59–65	95–105	56–63	92–101	61–68	98–109	69–76	111–122

[a]Governor pressure should be from zero to 3 psi at standstill or downshift may not occur.

Note: Changes in tire size or tire pressure will cause shift points to occur at corresponding higher or lower vehicle speeds.

CHART 12–2 Typical TorqueFlite three-speed transaxle automatic shift speeds and governor pressure chart (approximate miles and kilometers per hour at road load).

Many three-speed automatic transmissions use the same basic combinations.

Rear (Forward) Clutch Slipping. The rear clutch is in constant apply for all forward gears. With a slipping condition:

- There will be a slip or no drive in all forward gears: D, 2, and 1.
- If the rear clutch is the only unit affected, the transmission will have a working R.
- Should the clutch greatly delay engagement when cold and work normally when hot, the suspect area is hard or worn clutch piston seals.

Front (Direct/Reverse) Clutch Slipping. With a slipping front clutch, high gear and reverse are affected.

- The slip is first noticeable in high gear, with an engine flare-up on the 2-3 shift. When the condition is advanced, reverse slip also becomes noticeable. Reverse works at almost three times the line pressure required in D and, therefore, may not slip in early stages of clutch failure.
- Worn or hard piston seals first act up under cold conditions. The shift into high is drawn out or may not occur until the fluid is warm. Advanced conditions prevent total engagement of high gear, cold or hot.
- Clutch failure may not always give a definite slip action—it may result in a reverse shudder complaint. Be alert and

USE THE MANUFACTURER'S PUBLISHED SHIFT SPEED SPECIFICATIONS

Drive The Vehicle In Each Range, And Through All Shifts, Including Forced Downshifts, Observing Any Irregularities In Transmission Performance.

INITIAL GARAGE SHIFT FEEL AT ENGINE IDLE								
RANGE	POWER ENGAGES			QUALITY				
		YES	NO	HARSH	DELAYED	ENGINE STOPS	NORMAL
O/D							
DRIVE							
(2)							
(1)							
REVERSE							
NEUTRAL							

CAR CREEPS ____ O.K. ____

NOTES: ____________________

Vehicle Acceleration Pass __________ Fail __________

MINIMUM THROTTLE UPSHIFTS AND CLOSED THROTTLE DOWNSHIFTS								
RANGE	SHIFT POINTS				SHIFT QUALITY			
	Early	Late	No Shift	O.K.	Harsh	Drawn Out	Slips	O.K.
DRIVE 1-2								
2-3								
3-2 or 3-1								
2-1								
(2) 1-2								
2-1								

NOTES: ____________________

TRAFFIC THROTTLE UPSHIFTS WITH MODERATE THROTTLE				
RANGE	SHIFT QUALITY			
	Too Harsh	Drawn Out	Slips	O.K.
DRIVE 1-2				
2-3 At 30-40 MPH				
1-2 At 15-25 MPH				

NOTES: ____________________

GUIDE 12–3 Three-speed transmission/transaxle road test summary.

WIDE OPEN THROTTLE UPSHIFTS								
NOTE: This test is limited to prevailing road and traffic conditions								
RANGE	SHIFT POINTS				SHIFT QUALITY			
	Early	Late	No Shift	O.K.	Too Harsh	Drawn Out	Slips	O.K.
DRIVE 1-2								
2-3								

NOTES: ____________________

(KICKDOWN) WIDE OPEN THROTTLE FORCED DOWNSHIFTS								
NOTE: This test is limited to prevailing road and traffic conditions								
RANGE	SHIFT POINTS				SHIFT QUALITY			
	Early	Late	No Shift	O.K.	Too Harsh	Drawn Out	Slips	O.K.
DRIVE 3-2								
3-1 or 2-1								

NOTES: ____________________

	PART THROTTLE 3-2 DOWNSHIFT							
RANGE	SHIFT POINTS				SHIFT QUALITY			
	Early	Busy	No Shift	O.K.	Harsh	Drawn Out	Slips	O.K.
DRIVE 3-2								

NOTES: ____________________

MANUAL (2)
MANUAL CLOSED THROTTLE DOWNSHIFT FROM DRIVE AT 40 MPH (64 KM/H) TO TEST BRAKING ACTION O.K. ____________ NO BRAKING ____________
MANUAL LOW
MANUAL CLOSED THROTTLE DOWNSHIFT FROM (DRIVE) OR (2) AT 25 MPH (40 KM/H) TO TEST BRAKING ACTION O.K. ____________ NO BRAKING ____________
REVERSE
HOLDS ____________ SLIPS ____________ CHATTERS ____________

NOTES: ____________________

GUIDE 12–3 Three-speed transmission/transaxle road test summary *(continued).*

LEVER POSITION	START SAFETY	PARKING SPRAG	CLUTCHES FRONT	REAR	OVER-RUNNING	BANDS (KICKDOWN) FRONT	(LOW-REV.) REAR
P—PARK	X	X					
R—REVERSE			X				X
N—NEUTRAL	X						
D—DRIVE							
First				X	X		
Second				X		X	
Direct			X	X			
2—SECOND							
First				X	X		
Second				X		X	
1—LOW (First)				X			X

CHART 12–3 Chrystler TorqueFlite RWD and FWD operating range and clutch/band apply summary.

check for loose or worn transmission and engine mounts. Also, a problem with the reverse boost operating pressure may exist.

- First and second gears are not affected.

Kickdown (Front/Intermediate) Band Slipping. Kickdown band trouble is easy to spot because the band applies only in second gear.

- When the band completely fails to hold, second gear is skipped and a 1-3 upshift occurs. With the kickdown band not applying, the overrunning clutch holds in first gear (breakaway) until the road speed is high enough for the shift into direct (Figure 12–21).
- If the kickdown band slip is not in an advanced condition, there is a short upshift delay accompanied by a bump as the band takes over from the overrunning clutch.
- When the selector is moved to 2 from the D range or a 3-2 forced downshift is attempted, there is a slipping or no downshift condition. On the 3-2 forced downshift attempt, the engine races, trying to catch up with the first-gear ratio.

Overrunning Clutch Slipping. This affects the D and 2 ranges.

- The transmission fails to engage first gear, and therefore, a vehicle startup is not possible in D or 2.
- In manual 1, the vehicle can move forward in first gear. The rear (low/reverse) band holds the rear carrier reactionary for first gear operation. However, the design of the rear band is not engineered to hold against pulling power in this operating range and its life may be shortened.

Low-Reverse (Rear) Band Slipping. Manual 1 and R use the low-reverse band. Indications of a slipping band are:

- A slipping, chattering, or complete loss of reverse drive.
- No engine braking in manual 1 during coast. The holding ability of the band cannot be checked under acceleration or power. Even though the band is applied, the low roller clutch does the holding in manual 1 under drive conditions. For a band checkout in 1, accelerate to 30 mph (48 km/h) and quickly decelerate and feel for engine braking (Figure 12–22).

Rear (Forward) Clutch Fails to Release. This does not happen often, but it causes the following conditions to prevail:

- The transmission operates normally in all forward drive ranges.
- The transmission moves forward in neutral and locks up in reverse.

The clutch/band road test analysis just described can be adapted to any automatic transmission provided that the technician is aware of operating characteristics and construction differences. For example, a clutch may be substituted for the low-reverse band or an intermediate roller clutch system substituted for the intermediate band. In Ford's three-speed Cruise-O-Matic series, the 2 range starts out in second gear.

Torque Converter Clutch Analysis

The technician needs to know what is considered normal converter clutch operation. The converter clutch engagement can produce some driveability effects that are normal and can be explained.

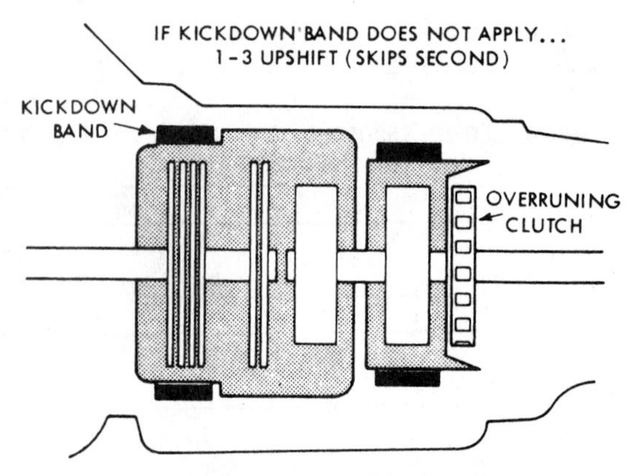

FIGURE 12–21 1-3 upshift skips second gear if kickdown band does not apply; overrunning clutch remains effective on 1-2 shift.

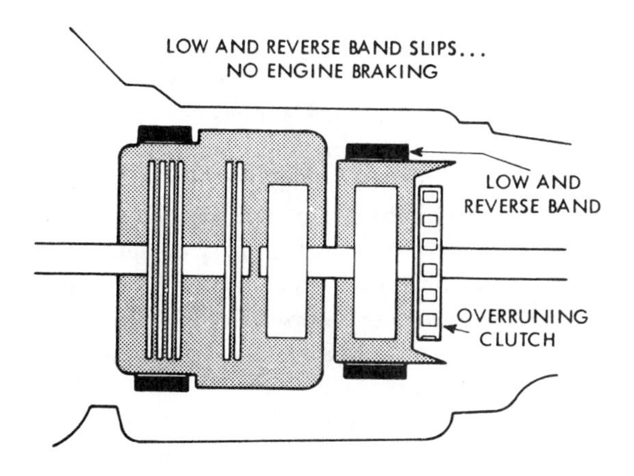

FIGURE 12–22 When low and reverse band slips, there is no engine braking or reverse.

- **Shift busyness** is recognized when the transmission produces an extra shift feel at inappropriate times caused by the actual apply and release of the torque converter. The driver feels the torque switching between hydraulic and mechanical drive. The condition occurs during various phases of acceleration and other TCC system component controls related to different engine conditions. Shift busyness is more active in hilly terrain or when trailer hauling. Some busyness situations may require attention, such as engine power performance or transmission throttle setting.
- **Deceleration bump** occurs with a sudden release of the accelerator pedal by the driver. With the engine at closed throttle/low torque, the drivetrain abruptly sends power to the engine before the clutch releases. The shock produced by the torque reversal causes a deceleration bump.

- **Shift busyness:** When referring to a torque converter clutch, it is the frequent apply and release of the clutch plate due to uncommon driving conditions.
- **Deceleration bump:** When referring to a torque converter clutch in the applied position, a sudden release of the accelerator pedal causes a forceful reversal of power through the drivetrain (engine braking), just prior to the apply plate actually being released.

The TCC apply and release activity should be sequenced into the road test for overall transmission performance (Guide 12–4). The objective at this time is to identify conditions of concern that may be normal or that may need diagnostic attention. Some of the customer concerns might be expressed as follows:

- "The car stalls when I put it in gear."
- "My car is getting noticeably lower gas mileage."
- "The car jerks when I'm on the highway."
- "My car lacks power at low speeds, right after a shift."
- "I get a dash vibration right after a low-speed shift."
- "After warm-up, my converter clutch stays on at all times in high gear. The car almost stalls and jerks just before coming to a stop."

In some transmissions, it is sometimes very difficult to feel that extra shift when the TCC engages and determine if it is holding. If the TCC is holding, there should be an engine rpm drop in the range of 200 to 300 rpm. Using a scanner is the most accurate method of determining whether the TCC does actually engage and hold.

Although it does not give a readout, an audio tachometer lets you hear the shift and engine rpm drop. An advantage of the

	________ No release	________ Harsh apply
	________ No apply	________ Over busyness
	________ Early apply	________ Shudders/chatters
	________ Late apply	________ Chuggle
• Shift point	________ Pass	________ Fail
• Part-throttle downshift release	________ Pass	________ Fail
• WOT forced downshift release	________ Pass	________ Fail
• Brake release	________ Pass	________ Fail
• Closed throttle release	________ Pass	________ Fail

Notes: ____

Note: Deceleration bump and a certain amount of shift busyness is a normal characteristic.

GUIDE 12–4 Torque converter clutch road test summary.

audio tachometer is its ability to pick up an engine miss that might be causing a chuggle or clutch shudder. As soon as the clutch engages, a sputtering noise is picked up, and the drive-ability problem is separated from the converter. An alternative, though sometimes less effective, method is the use of the car radio as an audio tachometer.

There can be a fine line between "chuggle" and "shudder." Both are related to a repetitive jerking sensation. In shudder, however, the action is more severe and rapid and can even cause dash vibration. Shudder and chuggle are conditions that can occur after converter clutch engagement. To determine if the problem is actually converter-related, hold the vehicle power steady and lightly hit the brake pedal. If chuggle or shudder continues, the problem is not in the converter—it is chassis-related, with defective U-joints or motor mounts as high-suspect items.

Another method that can be used to achieve the same results is to drop the selector into manual 2 from third gear D. This only works if the transmission does not have a built-in second-gear lockup.

Converter clutch shudder diagnosis also can be done by monitoring the oxygen sensor action with a scan tool while driving the vehicle. Sensor voltage switches between high and low voltage readings. When the vehicle speed reaches the critical range of the shudder complaint, the switching action of the **O_2 sensor** should be observed. If voltages stay high or low for longer than normal periods of time, the problem is engine-related. The suspect area is usually the secondary ignition system.

❖ **O_2 sensor:** Located in the engine's exhaust system, it is an input device to the ECM/PCM for managing the fuel delivery and ignition system. A scanner can be used to observe the fluctuating voltage readings produced by an O_2 sensor as the oxygen content of the exhaust is analyzed.

Torque converter lockup at slow speeds under some driving conditions can produce a shudder or chuggle that is usually cured by changing the cut-in speed. This usually requires a new **PROM (programmable read-only memory)** package or governor switch recommended by the manufacturer.

❖ **PROM (programmable read-only memory):** The heart of the computer that compares input data and makes the engineered program or strategy decisions about when to trigger the appropriate output based on stored computer instructions.

Do not forget to clear any diagnostic trouble codes. If the coolant temperature sensor voltage, for example, remains high, the computer can see only an extremely cold condition even though it is a hot summer day. The computer acting on the false input signal keeps the engine running rich and the TCC disengaged. Refer to the Torque Converter Clutch Road Test Summary (Guide 12–4, page 275).

Other strange TCC operating circumstances sometimes occur. For example, the TCC operation may prove to be normal on a road test in heavy traffic but does not function properly during cruising. Before getting trapped into changing the converter, the converter solenoid, and possibly the valve body or computer, think about the cause more carefully. Could it be a coolant temperature problem or a faulty coolant temperature sensor? Keep in mind that heavy traffic raises the coolant temperature while cruising lowers it. Coolant temperature must reach approximately 180°F (83°C) for the computer to ground the TCC solenoid circuit.

Occasionally someone thinks that cooler is better for the engine and changes the thermostat. The engine never gets to its engineered operating temperature and experiences difficulty reaching or exceeding the critical 180°F (83°C) level during road cruise. A malfunctioning thermostat itself can cause the same condition. For a quick check during a diagnostic road test, use a piece of cardboard to cover part of the radiator. If the problem persists, concentrate on the engine coolant temperature sensor and circuitry. To avoid false information to the computer, be sure that the engine coolant level is correct. To work properly, it is essential that the temperature sensor rides in the coolant.

Mechanical Failure Analysis

This area mainly involves internal conditions of sheared splines or broken welds. Diagnosis depends on knowing a combination of both theory and transmission construction. Some common failures and resulting transmission behavior are examined in the following.

Sun Gear Drive Shell Splines Sheared. The sheared splines are illustrated in Figure 12–23. The planetary sun gears can no longer be held for second gear or driven for third and reverse in a Simpson power flow. This includes both three- and four-speed automatics. The result is first gear only, and no reverse.

Sheared Front Carrier Splines. The photograph in Figure 12–24 shows the stripped splines of a front planet carrier from a three-speed Simpson planetary train. Since the planet carrier splines to the output shaft, this produces a condition of reverse only, and no forward speeds.

FIGURE 12–23 Sheared sun shell splines.

FIGURE 12–24 Sheared front carrier splines.

Sheared Converter Turbine Splines. This gives a condition of no forward speeds and no reverse. Transmission operating pressures are normal. If there is no input, there can be no output.

Sheared or Broken Welds. With the use of lightweight metal stamping construction, it is common practice to use electronic beam weld joints to complete the unit construction of the part. Typically, the weld bonds a hub, shaft, or critical machined part to the stamping. Although the welds are highly reliable, they can get "stressed out" and shear or break. Several examples come to mind.

- In the Ford AOD, the sun gear to drive shell weld shears (Figure 12–25). The transmission reacts with a 1-3 shift only, with no fourth gear or reverse.
- A sheared weld section is exhibited in a photograph series of an A4LD overdrive planet carrier assembly (Figure 12–26). The input shaft splines and overrun clutch inner race hub have been separated from the overdrive planetary. This results in no fourth gear.
- A broken weld on the overdrive housing in the GM 4L80-E results in no fourth gear (Figure 12–27).
- A broken weld on a forward clutch housing results in separating the input shaft from the drum (Figure 12–28).

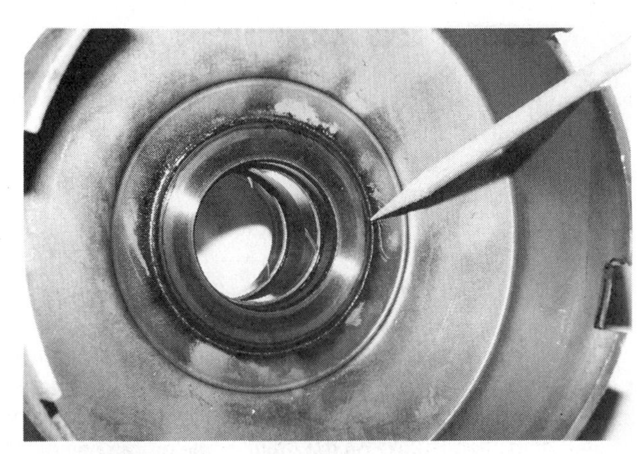

FIGURE 12–25 Weld broken inside drive shell.

- Input to the planetary geartrain is cut off, and there is no forward or reverse gear. Transmission operating pressures are normal.
- In Figure 12–29, the Ford ATX planet carrier (output) splines are sheared and cannot drive the input gear to the final drive. If the input to the final drive is cut off, the vehicle cannot respond to forward or reverse drive. The transmission operating pressures and clutch/band functions are normal.

A

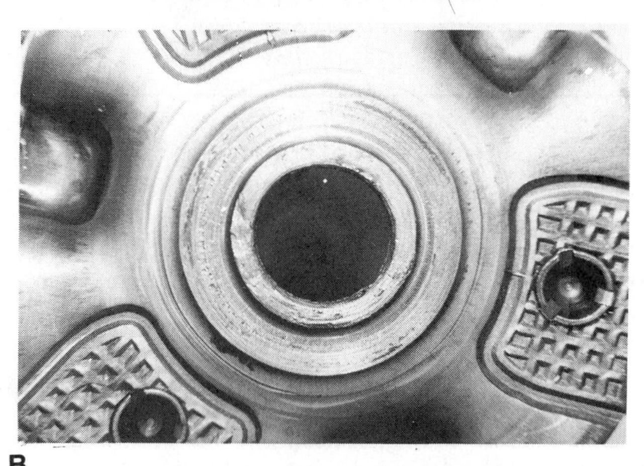

B

C

FIGURE 12–26 (a) A4LD overdrive carrier assembly; (b) weld break—overdrive carrier side; (c) weld break—roller clutch side.

FIGURE 12–27 Broken weld on an overdrive housing.

FIGURE 12–28 Broken weld on a forward clutch housing.

Noise Analysis

Analyzing noises is a talent that involves the total vehicle and a special knowledge of the mechanical and hydraulic moving parts in the torque converter and transmission. What makes this a tricky area on a transmission noise concern is the ability of the transmission to absorb and telegraph a real noise entirely unrelated to the transmission. Many a converter and front pump assembly have been replaced without solving the noise. Even entire transmissions have been wrongfully disassembled in pursuing noise concerns.

To be good at noise analysis, the technician especially needs to know the under-car operation. In FWD transaxle units, for example, worn or damaged CV joints can be responsible for a "clunk" when accelerating, decelerating, or when putting the transaxle into drive. Is the clunk concern an actual CV joint problem or transaxle problem? Even worn, loose, or damaged suspension bushings in front-drive systems can cause mysterious clunks on acceleration/deceleration. Be careful when trying to interpret transmission noise. What might appear to be a transmission noise may be the transmission responding to a problem elsewhere.

Noise analysis usually includes a combination of in-the-stall and on-the-road testing, listening for the noise, and visual inspection. The noises can cover a wide variety of items that include rare cases of clutch plate rattle in P and N, hydraulic valve buzz or moan, cooler line vibration, engine accessory drives, and even the exhaust system. Valving noises from emission control devices sometimes are the culprit. Removing the accessory drive belts, checking the exhaust system for isolation from the chassis, and cooler line routing may be a necessary part of the investigation. Use the stall test as a diagnostic noise procedure. Most of the noise problems take place in the converter and transmission powertrain. These will be the areas of concentration.

Converter Noise. There is a need to know that a whining or siren-like noise from the fluid action is normal during stall operation. The technician should become familiar with this normal noise by stall-testing vehicles with known good converters. To avoid having the normal noise interfere with problem noises, work the converter stall under light throttle in D. This procedure picks up most metallic noises from loose parts or interference between rotating elements. During stall, the needle-type thrust bearing between the stator and impeller is under pressure and can be noise-detected if the needles and race surfaces are the problem.

FIGURE 12–29 Carrier output splines sheared.

Transmission Powertrain Noise. Transmission powertrain noises come from the planetary gears, needle-type thrust bearings, overrunning clutch rollers, and clutch assemblies. In transaxles, there is the added area of the final drive to consider. The technician needs to have a good mental picture of parts relationships and what is happening inside the transmission in each gear. It is important to isolate the noise to a suspect area. When the transmission is disassembled, it gives the technician a specific area for location, inspection, and correction of the problem.

Guide 12–5 summarizes items to identify and be concerned about during the noise analysis process. Notice that external factors in the drivetrain and steering become important, especially "clunks" and "clicks" in FWD vehicles. It is also important to listen to the rhythm and noise level as the transmission changes gears. Is the noise on the input or the output end of the transmission powertrain?

Type:

_____	Resonating	_____	Squawk	_____	Vibration
_____	Whine	_____	Buzz	_____	Moan/Drone
_____	Rattle	_____	Clunk	_____	Click
_____	Whirring/rasping				

When noticed:

_____	Steering sensitive	_____	Acceleration
_____	Engine rpm sensitive	_____	Coast
_____	Trans ratio rpm sensitive	_____	Float

_____ Always _____ Intermittent

Vehicle speed range _____________

Operating range P___ R ___ N ___ D ___ 2 ___ 1 ___

In ____________________________ gear/gears

Pitch		Intensity level	
_____	Low	_____	Light
_____	Medium	_____	Medium
_____	High	_____	Heavy

Notes: ____

____

GUIDE 12–5 Noise analysis.

Planetary geartrain noise is related to speed changes that take place with changes in the gear ratios. Planetaries run slowly in first and second, and much faster in overdrive. The ratio changes also affect the relative motion between the planetary input and output. First gear, for example, provides the greatest speed difference. Noises also can appear and change their pitch and intensity level during acceleration or deceleration. During acceleration or road load, the transmission end thrust forces shift toward the rear of the case, while deceleration reaction transfers the end thrust toward the front. Clutch noises are usually confined to their apply activity, with the clutch plates being the usual problem.

The following examples illustrate some basic analysis of internal transmission noise.

- If the noise factor is most noticeable in first gear, cuts back in second gear, and then eliminates itself in direct drive, the problem is confined to the geartrain unit. In direct drive, the clutch and planetary units are locked at 1:1, and any internal planetary assembly noise should disappear. It is not uncommon to find the cause related to worn or brinneled races used with needle-type thrust bearings employed in the gear assembly. Rough needles and thrust races typically respond to ratio changes with a change in the "rasping" or "whizzing" noise rhythm and intensity.
- If the unusual noise is not a factor in first gear or reverse but starts in second gear and continues with more intensity and faster rhythm in direct drive, the low roller clutch assembly is the high-suspect area.

The low roller clutch has no overrunning action in first and reverse. In second gear, its freewheel speed is less than the output shaft speed, and with the shift into direct drive, the freewheel speed gets a sudden jump to equal output shaft speed.

Let us take a look at what NOT to do from an example taken from an actual field problem involving a Ford AOD.

- Customer concern: noise in all gears including P and N.
- Preliminary checks: all normal, with the fluid looking good.
- Noise was audible in P and N during an in-the-stall check; engine rpm related.
- A road test indicated reasonable shift scheduling and quality with the throttle opening. When testing for the noise in all the forward and reverse ranges, the noise was present in all gears, with third gear exhibiting the least intensity. The noise was a constant factor during acceleration and deceleration and did respond to ratio changes. During stall testing, the same type of noise predominated. The noise was a "whizzing" type, appearing to come from the converter.

Based on the diagnostic information, the technician installed a new converter, but the noise remained. What went wrong? In the Ford AOD, the technician needs to know that a splined third shaft or torsional shaft connects the converter cover directly to the direct clutch drum, and a needle thrust bearing assembly is positioned between the direct drum and output shaft flange. Review the photographs in Figure 12–30.

During the no-load acceleration test in P and N, the converter is normally in coupling phase, and noise from fluid action or the rear stator needle thrust washer should not be a big factor. On the road, there is always a speed difference between the direct drum and output shaft flange related to ratio changes, even in third gear. Because of the 60/40 split-torque feature between the converter fluid coupling input and direct mechanical input from the torsional shaft, some slip shows up in the planetary unit output. The two inputs do not turn at exactly the same speed. A slight speed difference always remains between the direct drum and output shaft flange in third gear.

A second opinion diagnosed the direct clutch drum as the problem area, with the number 8 needle thrust washer as the prime target. Upon disassembly, the needles were found to be pitted, and the matching microfinish on the output shaft was severely damaged by a combination of **brinneling** and **spalling** (observe the photos in Figure 12–31). The brinneling can be an indication of excessive transmission endplay, so be sure that the transmission endplay is adjusted on the low end before installing the pump.

- **Brinneling:** A wear pattern identified by a series of dents/grooves/indentations at regular intervals. This condition is caused by a lack of lube, overload situations, and/or vibrations.
- **Spalling:** A wear pattern identified by metal chips flaking off the hardened surface. This condition is caused by foreign particles, overloading situations, and/or normal wear.

Approach noise concerns cautiously, and avoid jumping to conclusions. The place where the noise appears to be coming from may not be the cause. In the particular problem just discussed, the torsional shaft probably transmitted the thrust bearing noise into the converter. Finally, know your theory and transmission construction, including every thrust bearing and surface location.

Erratic Shift Analysis

Erratic shifts simply mean that the shifting does not happen or is happening at incorrect shift points, either too high or too low. The shifts may be accompanied by incorrect shift quality: slipping or harsh. Again, the complete diagnostic road test can be used to verify the complaint and analyze the probable shift system cause.

Road testing for erratic shift analysis requires that the technician be knowledgeable about theory of shift system operation. Two block diagrams in Figures 12–32 and 12–33 illustrate the makeup of two typical shift systems. The only variance shown is in the style of throttle system control: either vacuum or mechanical.

The essentials of shift system operation are as follows:

1. A throttle or modulator system produces an engine torque-sensitive pressure:

FIGURE 12–30 Top: Direct clutch drum and needle-type thrust bearing. Bottom: Output shaft flange.

- Used at the shift valve to delay shifting.
- Used at the pressure regulator valve for line pressure boost.

2. A governor system produces a vehicle speed-sensitive pressure used at the shift valve to open the valve or cause the shift to happen.
3. In most cases, once the shift is made, only spring tension opposes the governor force. Throttle pressure is cut off from the shift valve assembly. Some shift valve assemblies may not entirely eliminate the TV oil but do greatly reduce its effectiveness on the shift valve.
4. The shift valve spring determines the closed-throttle downshifts against governor force.

Most erratic shifts occur from an out-of-balance governor or throttle system, which causes the following problems:

- The upshifts are delayed or, in extreme cases, do not occur regardless of vehicle speed.
- The upshifts occur too soon, or in extreme cases, the transmission starts out already in second or high gear.

The throttle and governor systems can be analyzed on a diagnostic road test to determine which system is at fault, and then the technician must locate the problem within the system.

FIGURE 12–31 Top: Thrust bearing needles flaked. Bottom: Brinneled and flaked output shaft flange thrust surface.

What will be found is entirely unknown, but at least the problem is confined to a system. Some of the throttle and governor system fault analysis is obvious. (A three-speed system is used for illustration purposes.)

- When the transmission starts out in high gear, the 2-3 shift valve is open. This means that the governor system has not been able to shut down the line pressure feed to the shift system. Governor pressure equals line pressure at all times. The throttle system is ruled out because its function is to delay shift.
- When the transmission will not upshift at any speed, the throttle system is ruled out and a governor system problem is indicated. The simple reasoning is that even maximum throttle pressure permits an upshift if the car is driven fast enough.

But what about those in-between shift problems when the transmission quick-shifts or the shifts are moderately but noticeably delayed? Here is how the road test can be used. The use of shift data provided by the manufacturer is preferable, but in the absence of that data, use the following guidelines:

Minimum Upshifts at Almost-Closed Throttle
1-2 upshift at 8 to 12 mph (13 to 19 km/h)
2-3 upshift at 15 to 20 mph (24 to 32 km/h)

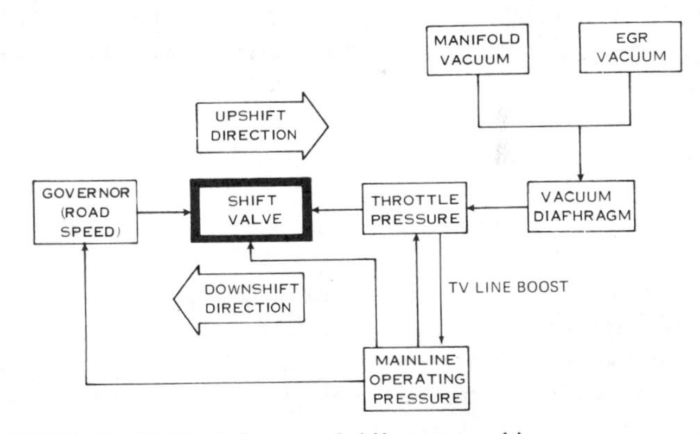

FIGURE 12–32 Block diagram of shift system with vacuum-controlled throttle system.

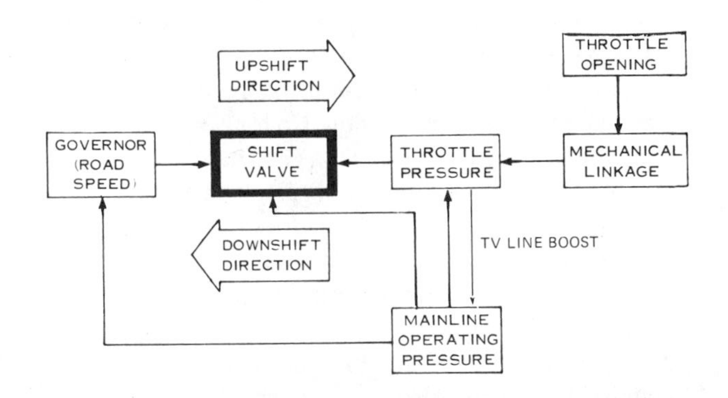

FIGURE 12–33 Block diagram of shift system with mechanically controlled throttle system.

Closed-Throttle Downshift
3-2 at 10 to 12 mph (16 to 19 km/h)
2-1 at 5 to 8 mph (8 to 13 km/h)
3-1 at 5 to 8 mph (8 to 13 km/h)

Case Problem. Minimum upshift pattern is delayed. The 1-2 shift occurs at 20 mph (32 km/h) and the 2-3 shift at 35 mph (56 km/h). On the closed-throttle test run, the downshift 3-1 happens at 10 mph (16 km/h), which can be considered close enough to normal. Consider the following:

- Could it be low governor system output?
- Could it be high throttle system output?

Because throttle pressure is not a factor at the shift valve once the shift happens, the closed-throttle downshift 3-1 is determined by spring tension working against governor pressure. If the closed-throttle shift is correct, it verifies that the governor system is proper. The problem is isolated to the throttle system. If, in the very same problem, the closed-throttle downshift 3-1 happens at 18 mph (29 km/h), the throttle system is ruled out. The early downshift is caused by the shift valve springs overcoming governor system output, which is too low for the road speed.

Case Problem. Minimum upshift pattern is too early. The 1-2 shift occurs at 5 mph (8 km/h) and the 2-3 shift at 10 mph (16 km/h). A medium- and heavy-throttle shift sequence shows that the shifts are delayed with no transmission slipping, but that they still occur early. On the closed-throttle test run, the downshift 3-1 happens at almost a dead stop, which is very late. It may even fail to downshift to first gear, resulting in a high gear start.

Cause: Governor system output too high for road speed.

On closed throttle, the shift valve springs must close the valves against governor pressure. If the governor unit is regulating too high, the vehicle speed must slow down more than normal before the springs can close the valves. Low throttle system pressure is ruled out because of two factors:

- TV pressure at minimum shift conditions is close to or at zero psi. The governor pressure only needs to overcome shift valve spring tension. In this case, it overcomes the shift valve springs too soon.
- Even though the shifts occur early, there is a delay pattern and no transmission slipping. No transmission slipping at medium and heavy throttle is a good indication that the TV boost signal to the pressure regulator valve is right.

Case Problem. Minimum upshift pattern is correct. The 1-2 shift occurs at 10 mph (16 km/h) and the 2-3 shift at 18 mph (29 km/h). Under medium- and heavy- throttle testing, the shift system does not respond and the shift points remain the same. Transmission slipping also becomes part of the problem with medium and heavy throttle. A check on the closed-throttle shift 3-1 finds that it occurs correctly at 5 mph (8 km/h).

Cause: The throttle system output is remaining close to or at zero psi. It does not respond to changes in engine performance.

Because the minimum shift points and closed-throttle 3-1 shift are right, the governor system has to be right. Without a shift valve throttle signal:

- The governor pressure needs only to overcome the shift valve springs, which offer the same fixed resistance from closed to wide open throttle.
- The line pressure regulator valve fails to receive a torque signal for pressure boost. The pressure regulating point stays at a minimum fixed value as determined by the regulator valve spring.

When testing the Hydra-Matic 4T60 (440-T4) for erratic shifts, it is important to know that it uses independent systems to control shift timing and shift firmness. The vacuum modulator and modulator valve modify the mainline pressure and control the shift firmness (Figures 12–34 and 12–35). Shift point control is separately regulated by the traditional cable-operated throttle system. In this setup, the following erratic shift characteristics can prevail:

- Shift points normal with continuous soft or slipping condition; modulator condition.
- Shift points normal with continuous harsh feel; modulator condition.
- Shift points not correct but usually have a normal shift feel; TV condition.

As another road test trick to determine governor circuit malfunction, drive the vehicle at 45 mph (73 km/h) and move the range selector from D to 1. If manual low engages, the governor circuit is at fault. The governor pressure at the 1-2 shift valve determines the safe vehicle speed at which the transmission engages manual low. Once downshifted, the shift valve is hydraulically locked in place by line pressure.

On four-speed systems, give special attention to shift system hydraulics. The 3-4 shift circuit usually has one or two auxiliary valves that must switch into position in addition to the 3-4 shift valve if fourth gear is to happen.

FIGURE 12–34 4T60 vacuum modulator, nonaltitude-compensated.

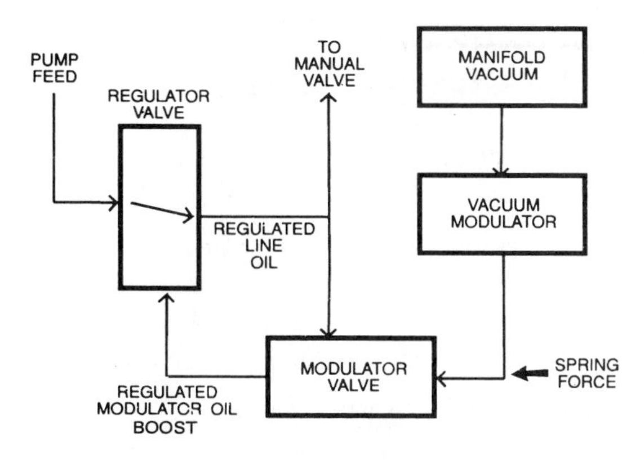

FIGURE 12–35 Block diagram of modulator system—4T60.

The Ford AOD uses an orifice control valve. If pressure data shows that the direct clutch circuit oil is 15 psi or more below line pressure, the orifice control valve does not switch on the 2-3 shift. If the switch does not happen, line pressure from the manual valve keeps the 3-4 shift valve hydraulically locked against governor oil. The shift valve does not see normal TV reactionary oil. Third gear typically does not slip because the direct clutch handles only sixty percent of the crankshaft torque.

❑ SECTION 4 REVIEW

Key Terms/Acronym

Garage shift feel, shift busyness, deceleration bump, O_2 sensor, PROM (programmable read-only memory), brinneling, and spalling.

Questions

Diagnostic Road Test: Clutch/band Failure Analysis

Problem 1

- ❍ Transmission engagement is soft in all forward ranges (D, 2, and 1).

TYPICAL THREE-SPEED CLUTCH/BAND APPLICATION CHART

Operating Range	*Forward Clutch*	*Direct Clutch*	*Intermediate Band*	*L & R Band*	*Low Roller Clutch*
Park/Neutral					
Reverse		ON		ON	
Drive—1st	ON				Holding
2nd	ON		ON		
3rd	ON	ON			
(2) —1st	ON				Holding
—2nd	ON		ON		
(1)	ON			ON	Holding

CHART 12–4

- ❍ Transmission slips in all forward ranges (D, 2, and 1), and in advanced stages, there is no drive in the forward ranges.
- ❍ The transmission performs okay in R.

Likely failure? ______________________________

Problem 2

- ❍ Transmission slips or fails to move the vehicle in D or 2 ranges only.
- ❍ Manual 1 provides a good start.
- ❍ The transmission performs okay in R.

Likely failure? ______________________________

Problem 3

- ❍ Transmission forward ranges (D, 2, and 1) test okay.
- ❍ Vehicle moves forward in N.
- ❍ Transmission locks up in R.

Likely failure? ______________________________

Problem 4

- ❍ The transmission has a positive first-gear start in D, 2, and 1.
- ❍ The transmission skips second gear (D range) and shifts 1-3. Should the 1-2 shift occur, it may be drawn out, slip, or bump.
- ❍ When the selector is moved from D range to 2 range, there is no downshift to second gear. A freewheeling coast sensation is felt in "the seat of the pants."
- ❍ 3-2 forced downshifts slip entirely or result in a slip or bump affair while engaging second gear.
- ❍ The transmission performs okay in R.

Likely failure? ______________________________

Problem 5

- ❍ The transmission first and second gears give good positive action.
- ❍ On the 2-3 shift in D, an engine flare-up might occur before engagement, the shift may be long and drawn out, or it may not occur.
- ❍ Reverse may slip, chatter, have a no-drive condition, or even appear to work normally.

Likely failure? ______________________________

Problem 6

- ❍ Transmission starts out in second gear (D range) and locks up on the 2-3 shift.

- The transmission also locks up in R range and 1 range.
- 2 range starts out in second gear.
- N range is okay.

Likely failure? ______________________________

Problem 7

- Slip, chatter, or no drive in R.
- Manual low (1 range) has no braking during coast.

Likely failure? ______________________________

Diagnostic Road Test: Mechanical Noise Analysis

Problem 1

- Transmission's unusual noise factor is a very audible rasping or whirring sound in first gear, all forward ranges.
- In second gear the noise factor cuts back, and in direct drive it disappears.
- Reverse operation may or may not produce the noise.

Likely problem area? ______________________________

Problem 2

- Transmission produces no unusual noise in first gear, all forward ranges, and in R range.
- The noise factor is a rasping or whirring sound that starts in second gear and continues with increased intensity and speed rhythm in direct.

Likely problem area? ______________________________

Diagnostic Road Test: Torque Converter Clutch Analysis

____ 1. Sudden or brief intermittent jerks after the TCC is applied is referred to as:

A. chuggle
B. bump
C. shutter
D. none of the above

____ 2. Chuggle can be caused by:

A. an engine spark plug miss
B. loose motor mounts
C. early TCC engagement
D. all of the above

____ 3. If the converter clutch fails to engage, the problem could be in:

I. the engine coolant temperature sensor.
II. an internal converter.

A. I only
B. II only
C. both I and II
D. neither I nor II

Diagnostic Road Test: Mechanical Failure Analysis

____ 1. (Three-speed Simpson) The front planet carrier output shaft splines are sheared.

I. The transmission will not move in any of the forward ranges.
II. The transmission will not have reverse.

A. I only
B. II only
C. both I and II
D. neither I nor II

____ 2. (Three-speed Simpson) The sun gear drive shell splines are sheared.
I. The transmission will "skip" second gear; first and third response only.
II. The transmission will not have reverse.

A. I only
B. II only
C. both I and II
D. neither I nor II

____ 3. The transmission does not move forward or in reverse. Technician I says that the input shaft is separated from the forward clutch drum. Technician II says that the problem could be in the converter. Who is right?

A. I only
B. II only
C. both I and II
D. neither I nor II

Diagnostic Road Test: Erratic Shift Analysis

____ 1. A customer is concerned about an unusually busy 4-3-4 transmission activity. Technician I says that it could be an engine performance problem. Technician II says the TV cable may need adjustment.

A. I only
B. II only
C. both I and II
D. neither I nor II

____ 2. The governor filter screen is clogged with debris. Technician I says that in a valve-style governor circuit this will cause the transmission to start in high gear. Technician II says that in a check ball-style governor circuit this will also cause the transmission to start in high gear. Who is right?

A. I only
B. II only
C. both I and II
D. neither I nor II

Problem 1

❍ The minimum 1-2 shift speeds are higher than normal.
❍ The closed throttle downshift 3-1 occurs at 5 mph (8 km/h).

Likely shift system problem area? ____________________

Problem 2

❍ Transmission starts out in high gear.

Likely shift system problem area? ____________________

Problem 3

❍ Transmission stays in first gear and fails to upshift regardless of vehicle speed.

Likely shift system problem area? ____________________

Problem 4

❍ (Hydra-Matic 4T60/440-4) Transmission shift timing is normal, but unusually aggressive.

Likely shift system problem area? ____________________

Section 5 Diagnostic Step 7: Pressure Test

OBJECTIVE:

The objective of this unit is for you to become proficient with the concepts and procedures contained in this chapter. Completion of this objective is essential to becoming a successful automatic transmission technician.

CHALLENGE YOUR KNOWLEDGE

Define the following keywords:

Pressure control solenoid (PCS).

Summarize the following procedures:

Mainline pressure test for a transmission controlled by TV cable/linkage, vacuum modulator, and pressure control solenoid (PCS).

Interpret and demonstrate application of the following:

Fluid pressure test chart.

Describe the techniques needed to analyze the following:

Governor operation; and vacuum modulator concerns.

DIAGNOSTIC STEP 7: PRESSURE TEST

Pressure testing is usually avoided by most technicians. In most cases, this can be attributed to a lack of expertise in hydraulic system operation or a misunderstanding of what pressure testing can and cannot tell the technician. Pressure testing is most valuable for use in special problem situations, especially diagnosis of shift complaints and slip conditions that remain unresolved. When a malfunction still exists after ordinary diagnostic checks, the final step is to make oil pressure checks. The pressure tests will pinpoint problem areas such as the pump, throttle, and governor systems.

Avoid the error of pulling the transmission from the vehicle for disassembly and inspection on a chance that you might find the problem. What are you going to find if you do not know where to look? It is too late once the transmission is apart. Let the pressure tests tell you where to look. In most cases, you will find that the solution to the problem can be handled by an in-the-car repair.

Pressure tests do not usually identify if servos or clutches work properly. They do eliminate key hydraulic systems as a possible cause of trouble when the readings are right. Specifically, pressure testing checks the hydraulic integrity of the following systems:

- Pump
- Throttle system
- Governor in some cases
- Stuck shift valves in some cases
- Clutch circuit apply oil in some cases

Pressure testing consists of reading the transmission pressures in each of the operating ranges under specific conditions spelled out by the manufacturer. The validity of the test results depends on the following factors:

- Transmission fluid level and condition are good.
- Transmission fluid is at operating temperature.
- Manual linkage is set properly.
- Throttle adjustment to transmission is right.
- Modulator vacuum flow is right.
- Exact test procedures are followed.
- Test instruments are known to give accurate readings.
- Test results are recorded for comparison with specifications and not left to memory.

Test Equipment

A variety of professional diagnostic tools are shown in Figure 12–36. An automatic transmission tester can offer the advantage of combining several instruments in a portable metal case for quick and accurate pressure testing (Figure 12–37). It provides easy readings of engine rpm, vacuum, and hydraulic pressures and can be used for on-the-road testing. Readouts can determine idle speed settings, TV cable settings, vacuum modulator system condition, hydraulic circuit pressures, transmission stall speeds, and transmission shift points. The TCC shift circuit and engagement also can be verified as working.

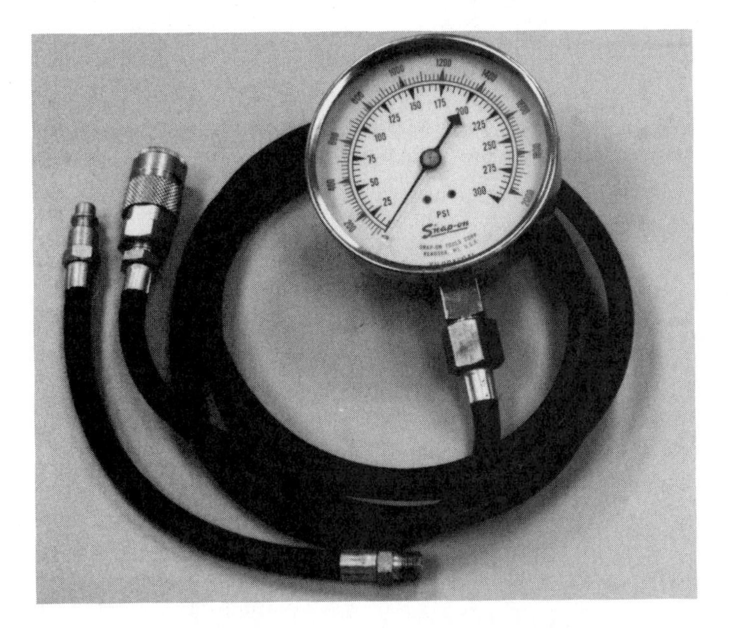

A

B

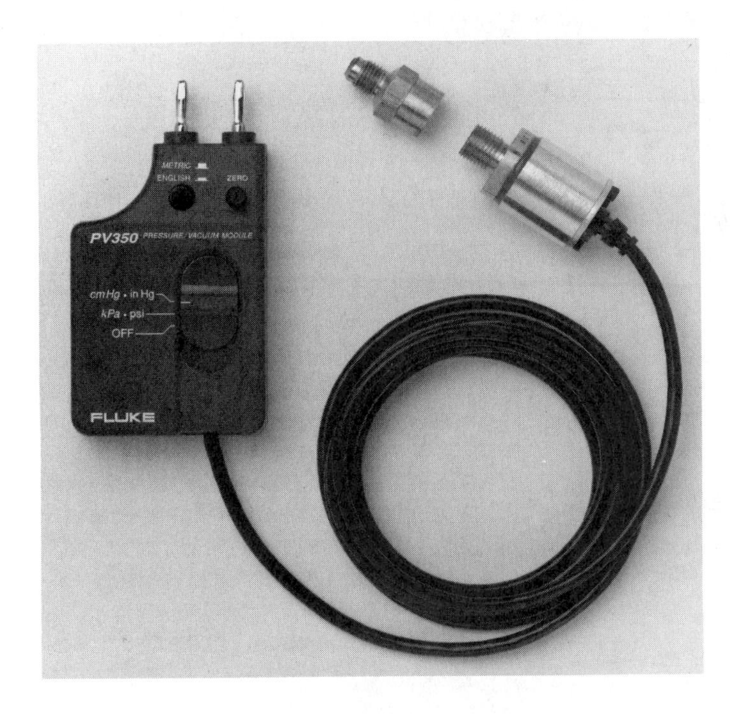

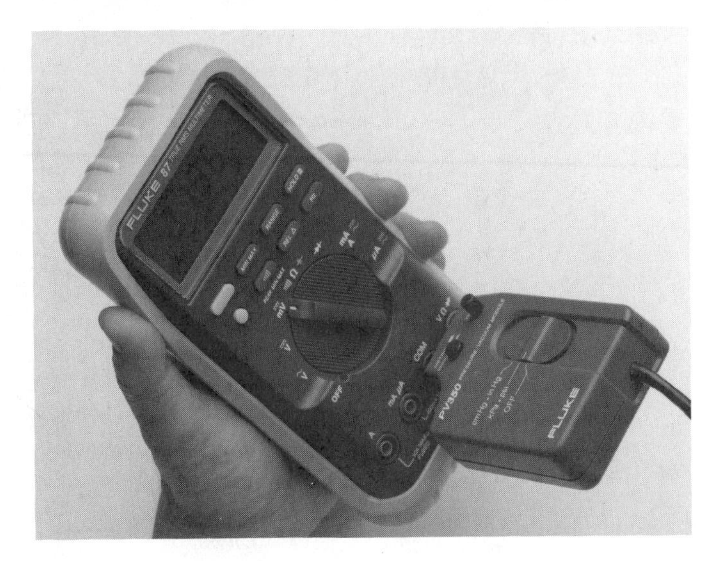

C

FIGURE 12–36 Diagnostic tools: (a) oil pressure gauge with quick coupler; (b) vacuum pump/gauge; (c)Fluke DVOM with pressure-vacuum module (transducer) (Courtesy of Fluke Corporation).

A pressure gauge hose and adapter are connected to the main pressure tap of an automatic transmission. In most cases, this is the only pressure takeoff available. If you recall the basic theory of pressure control, you know that operating line pressure is always sensitive to engine torque and sometimes to vehicle speed. Therefore, from the one mainline pressure tap, the pump, modulator, and governor systems can be monitored, as they will be reflected in the mainline pressure test readings.

Although the single pressure tap is used in most transmissions, you will find that some models, for extra convenience, provide several pressure gauge outlets. Examples of these transmissions are the Chrysler TorqueFlite family (Figure 12–38); the General Motors 4L60 (700-R4), 3T40 (125-C), and 4T60 (440-T4) (Figures 12–39 and 12–40); and the Ford AOD (Figure 12–41).

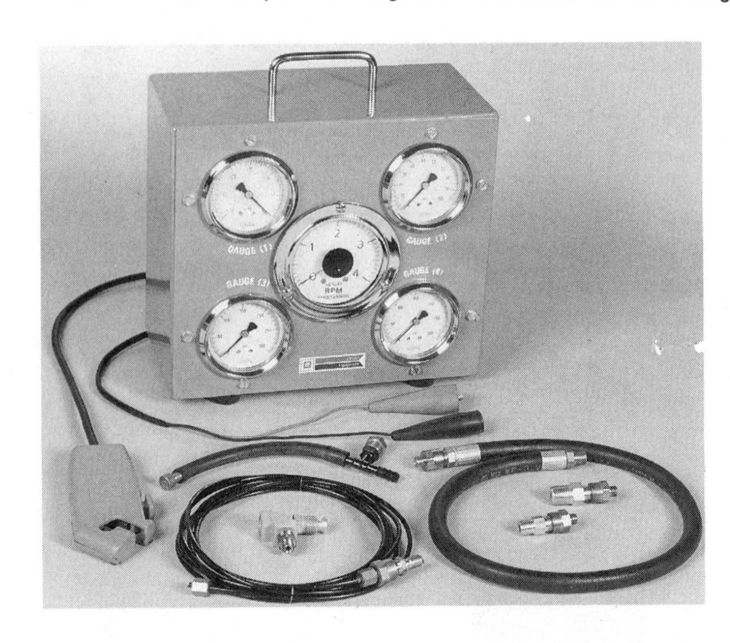

FIGURE 12–37 Nuday automatic transmission tester.

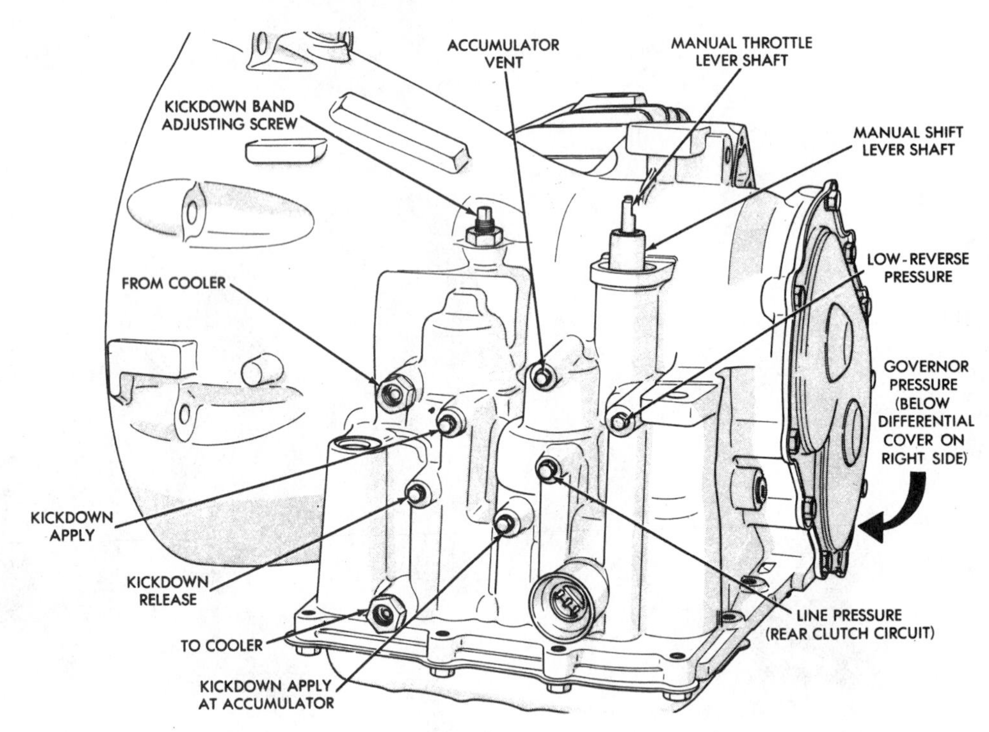

FIGURE 12–38 Pressure test plugs—31TH (TorqueFlite 400) series. (Courtesy of Chrysler Corp.)

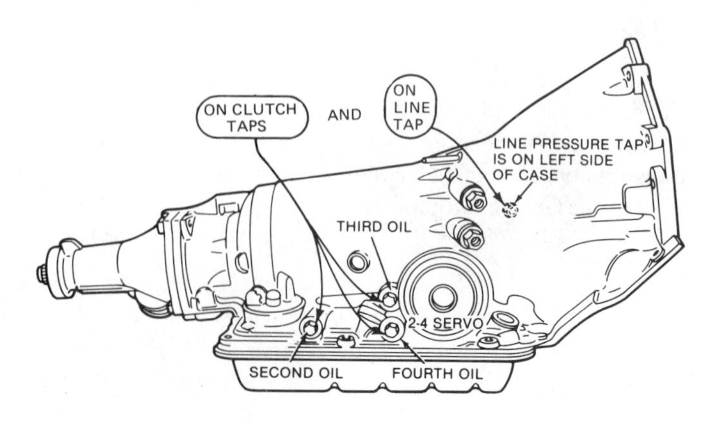

FIGURE 12–39 4L60 (700-R4) pressure test plug locations. (Courtesy of General Motors Corporation, Service Technology Group)

FIGURE 12–40 GM 4T60 (440-T4) governor test plug location. The 3T40 (125-C) governor test plug is in similar location.

Never connect a 0 to 100 psi (0 to 690 kPa) gauge into a high pressure port. Reverse line pressure, for example, can reach 300 psi (2070 kPa). The mainline pressure parameter in D range is usually between 65 and 175 psi (448 to 1208 kPa).

On transmissions equipped with vacuum modulators, the vacuum gauge hose is tied into the manifold vacuum line at the modulator with a tee fitting. This is done to ensure that we are working with the exact vacuum at the modulator for accurate test results. If the vacuum gauge is hooked directly to the intake manifold, you will have no way to determine if the vacuum signal is interrupted on its way to the modulator.

Manifold vacuum represents the engine torque or load signal to the transmission. If the signal is not right, the modulator regulated output is not right. This means that the transmission will definitely have the wrong line pressure, and the shift schedule and shift quality also will be upset. Many slipping transmissions have been overhauled when the actual cause was in the vacuum supply. It makes sense to do a first-class investigation and correction of the vacuum supply if the pressure tests are to reveal true internal transmission problems.

Modulator and Vacuum Supply Checkout Procedures

Vacuum Source at the Manifold. Evaluate the vacuum source at the manifold. Acceptable vacuum depends on altitude, engine design, and emission control equipment. In D range (hot idle), most engines should pull a minimum of 14 in hg (47.5 kPa). Perform engine tune-up if necessary to correct engine breathing problem.

Modulator Vacuum Line and Hoses. On the engine top side, examine the modulator vacuum line and hoses for the following defects:

- Loose connections
- Hard, brittle, and cracked hoses, especially on high- mileage cars
- Collapsed, spongy, stretched, or pinched hoses
- Restricted manifold porting

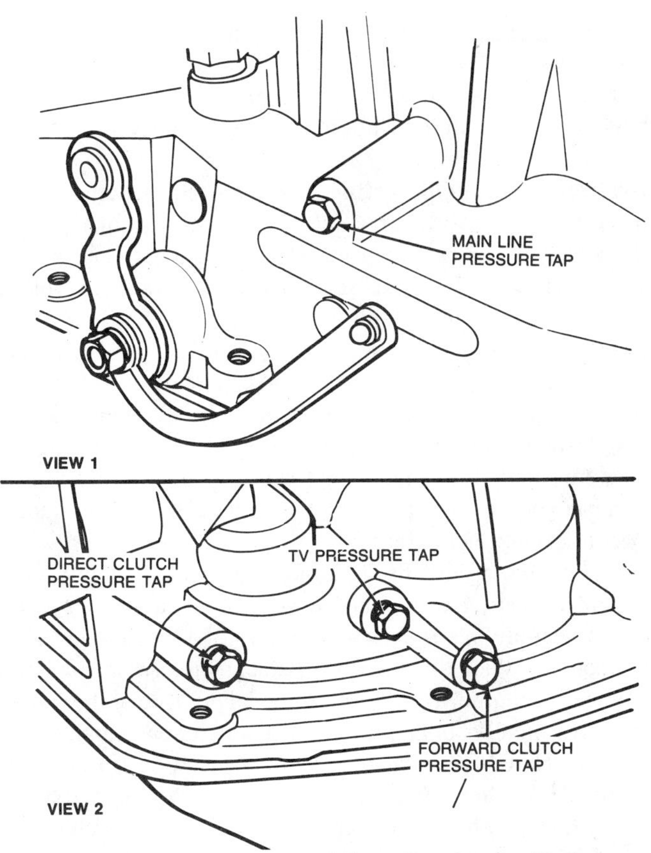

FIGURE 12–41 AOD test plug locations. (Reprinted with the permission of Ford Motor Company)

When hoses are connected to a fire-wall–mounted vacuum tree, they need enough slack to prevent stretching and closing when the engine rolls on its mounts during heavy throttle shifting. Be aware of the fact that the hose line to the intake manifold vacuum tree may not collapse and close until the engine gets hot. Depending on when it occurs, a low- or high-vacuum signal can be trapped on the modulator side of the fault. A low-vacuum trap produces late and harsh shifts, while a high-vacuum trap causes sudden quick shifts and transmission slipping.

Modulator vacuum hoses must be a good grade of reinforced hose to prevent collapse problems. Absolutely avoid the use of windshield washer hose.

Check the vacuum hoses for correct routing. Are they attached to the proper vacuum fittings? For example, on a manifold tree, is the modulator hose connected to a large open fitting, and not an orificed or metered fitting?

The vacuum supply to the modulator can be either easy to find or very elusive. The nature of the vacuum plumbing on top of the engine makes all vacuum circuits interrelated. A leak in a vacuum circuit entirely unrelated to the transmission modulator can affect how the modulator controls the transmission. For example, a leak in an engine control vacuum circuit or an accessory device may cause a transmission vacuum modulator not to operate properly.

Restricted Manifold Porting. The manifold vacuum port can be carboned and form a restriction. Use a correct drill bit size to check the opening dimension and resize the port if necessary. The engine load signal can always be bypassed with a "cheater." A combined hand-operated vacuum pump and gauge applied to the modulator simulates any desired vacuum condition (Figure 12–42). The instrument can be extended into the front seat of the car and manipulated by the technician for diagnostic road testing and pressure checks. When the normal vacuum supply to the modulator is in doubt, vacuum substitution is a good backup procedure to verify if the normal supply system is malfunctioning.

In concluding the discussion on the vacuum supply, let us look at the restricted-line and iced-line effect on the load signal. A clear supply line is shown in Figure 12–43, delivering 15 in of vacuum. When the line is restricted, there is a time delay to changes in engine vacuum on the modulator side of the restriction. Eventually, both sides of the restriction equal engine vacuum. Here is what can happen. While waiting at a stoplight, both sides of the restriction are stabilized. With a medium or heavy throttle on the green light, engine vacuum is close to 0 in, and the modulator is temporarily stuck with 15 in hg. (50.9 kPa) of vacuum. This delayed vacuum action is illustrated in Figure 12–44. The same theory can be applied with forced downshifts.

Upon icing, the condensation or water accumulation in the vacuum line and modulator obviously freezes after the engine is shut down. This holds the modulator in the high mode. Modulator diaphragm movement is restricted, and the vacuum flow is at O in until the engine and transmission warm up and thaw the ice (Figure 12–45). In the meantime, line pressure is high, and the shift schedule is working at its maximum limits.

Modulator icing can cause the line pressure to exceed its engineered maximum limits and result in a cracked transmission case or servo cover. Icing starts to happen on that first frosty morning when the outside temperature drops to +15°F (–10°C).

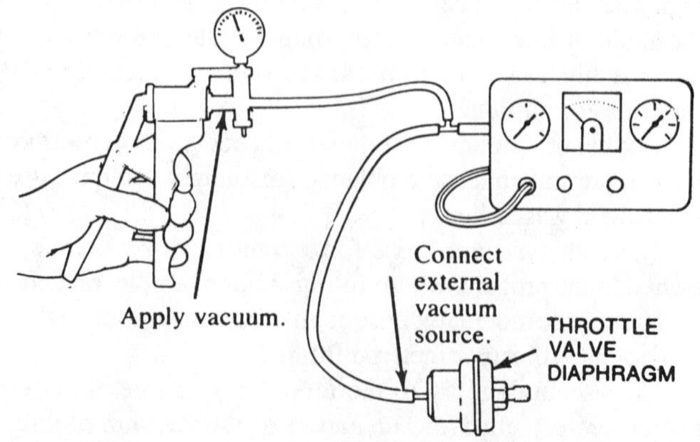

FIGURE 12–42 Vacuum pump and pressure gauge unit. (Reprinted with the permission of Ford Motor Company)

Vacuum Tube and those Connections. At the modulator, check the physical condition of the vacuum tube and hose connections. Examine the nipple connector between the metal vacuum tube and the modulator (Figure 12–46). Connector ends should be tight and the hose firm but not brittle. If the hose is brittle and cracked or soft and spongy, a replacement is necessary. Do not use straight hose where molded elbows are required for proper routing (Figure 12–47). A straight hose may become pinched where it bends at the corner.

Modulator Vacuum Supply. At the transmission, check the modulator vacuum supply. With the engine running at hot idle, pull the nipple connection off the modulator and listen to the engine response—the engine rpm should speed up. If the engine fails to respond, a restriction exists in the modulator line back to the source of vacuum. If the metal tube line is pinched, it must be replaced. Note: Some tubes are designed with a factory- built restriction to dampen manifold air induction wave oscillations. Should the tube appear satisfactory, send a shot of air

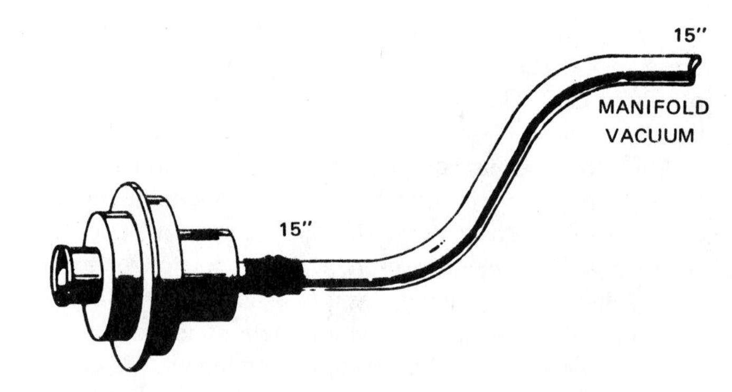

FIGURE 12–43 Vacuum modulator hose in good condition.

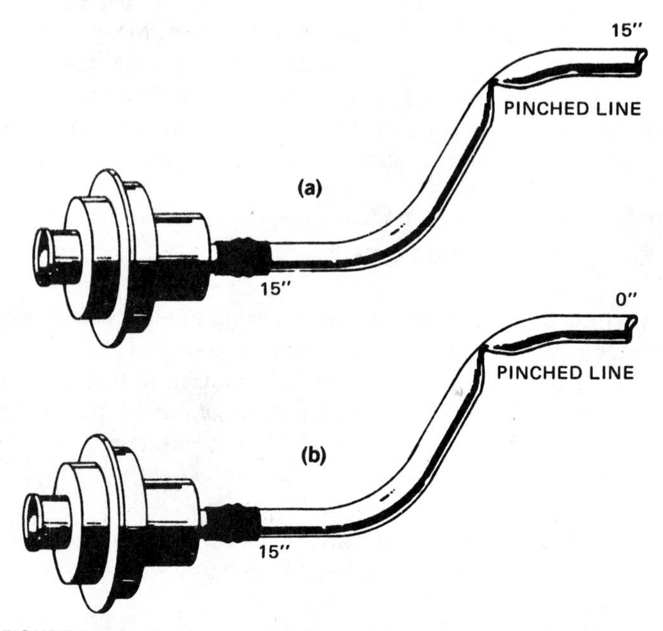

FIGURE 12–44 (a) Vacuum stabilized; (b) acceleration or load change—delayed vacuum signal.

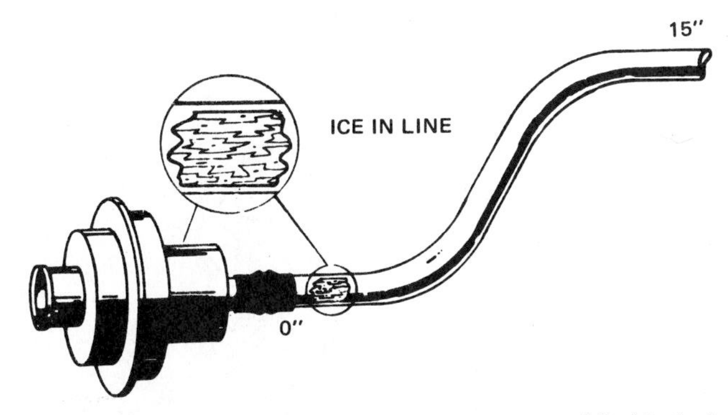

FIGURE 12–45 Modulator at constant zero vacuum while blocked by ice.

through the tube line and recheck the results. The unwanted restriction must be located and eliminated.

The most effective method for testing the vacuum response at the modulator is the use of the vacuum gauge. It represents a more scientific approach and leaves no margin of error. With the gauge connected in the modulator line, the idle vacuum at the modulator should be the same as manifold vacuum (Figures 12–48 and 12–49). This reading should be taken with the transmission in D, drive wheels locked with the brake pedal, and engine at hot curb idle. If the gauge does not read manifold vacuum, a misrouted line or vacuum leak is the probable cause. Snap the throttle and observe the gauge needle. The gauge needle must drop and recover with instant response to engine breathing. If the needle action lags behind throttle action, a vacuum restriction is indicated.

Another concern that can cause some strange shift behavior is a restricted exhaust system. The vacuum gauge can tell the story. With the transmission selector in D, increase the engine rpm to 1300 while holding the brake. Observe the vacuum gauge. If the vacuum level is initially normal but begins a gradual drop to as low as 4 in (13.56 kPa) of vacuum, there may be a restriction in the exhaust system. This is a mini stall test, so be careful about the time: five-second limit for each test run, with cool-down time between runs. The vacuum gauge checks are illustrated in Figure 12–50.

Modulator Unit. At the transmission, check the modulator unit. At the vacuum port, make an analysis of contaminants that may have settled in the diaphragm area. A pipe cleaner inserted into the vacuum port (Figure 12–51) is effective for gathering contaminant samples.

- Transmission fluid indicates that the diaphragm leaks and must be replaced.
- Water may have accumulated from condensation or been pulled through a loose or cracked hose during a rainstorm. Once water enters the diaphragm area, the spring begins to rust and lose its calibrated tension. A modulator replacement is in order.

FIGURE 12–46 Inspect vacuum hose between tube and modulator.

FIGURE 12-47 Molded vacuum hose attached to modulator.

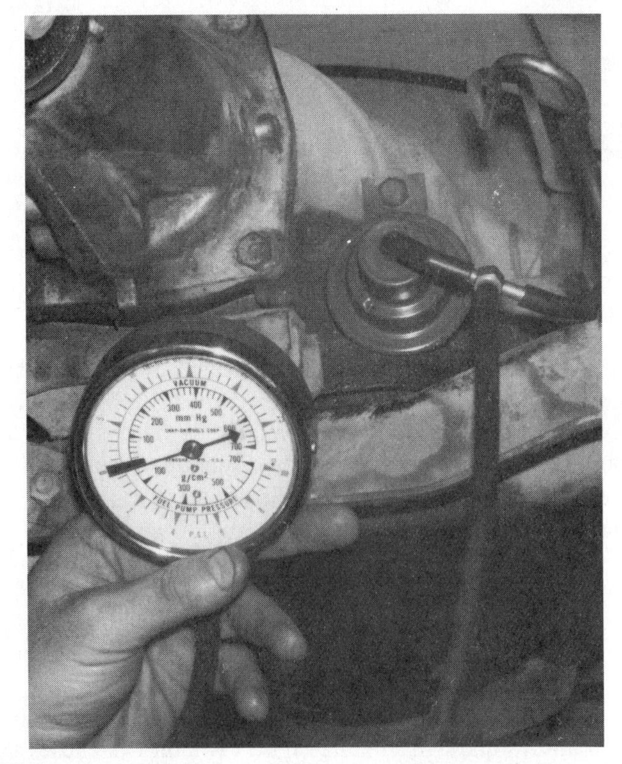

FIGURE 12–48 Checking manifold vacuum with tee and gauge. (Reprinted with the permission of Ford Motor Company)

MANIFOLD VACUUM TO DIAPHRAGM

RESULTS REQUIRED

- AT IDLE — Steady engine vacuum as specified for test altitude.
- MOMENTARY ACCELERATION — Drop quickly; return to idle vacuum.

CHECK VACUUM LINES FOR BREAKS, RESTRICTION, ROUTING IF RESULTS NOT OBTAINED.

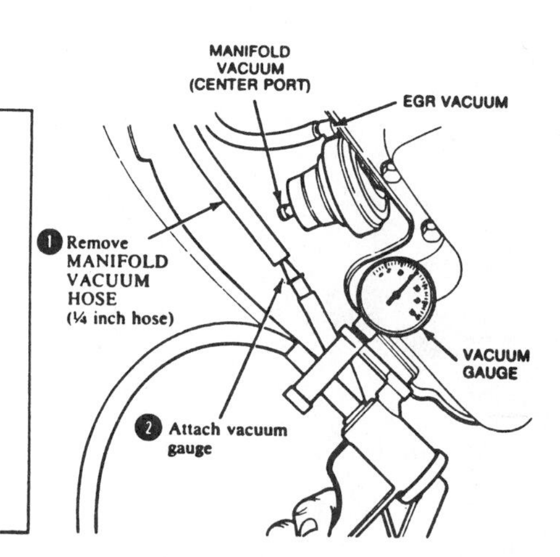

FIGURE 12–49 Checking manifold vacuum to modulator with a gauge.

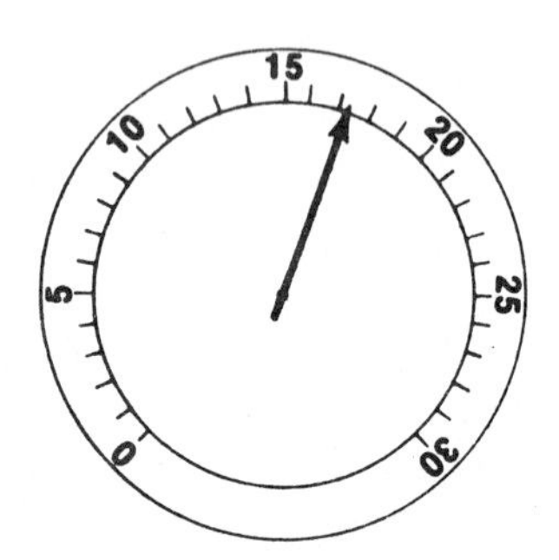

NORMAL READING
14-20" at idle.

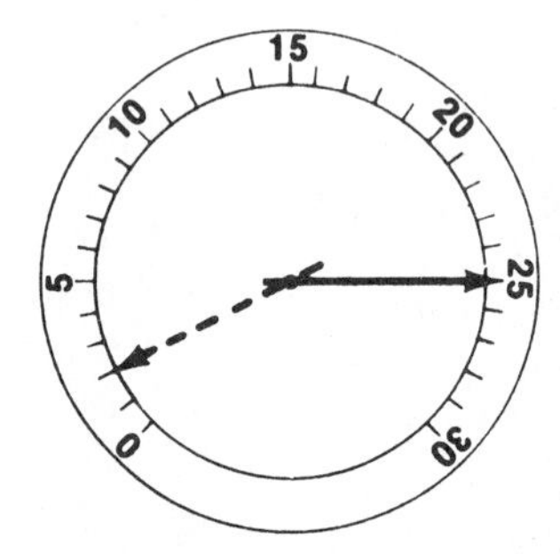

NORMAL READING
Drops to 2", then rises to 25" when accelerator is rapidly depressed and released.

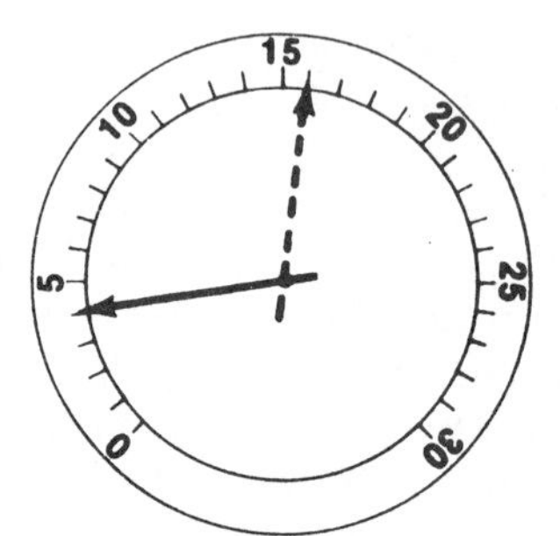

RESTRICTED EXHAUST SYSTEM
Normal when first started. Drops to as low as 4" and stabilizes.

FIGURE 12–50 Modulator vacuum checks for restricted exhaust.

- If gasoline is found without the presence of transmission fluid, the modulator should not be changed. The gasoline gravitated into the area from either an oversoaked manifold or failure of the modulator tube to be routed uphill from the manifold (Figure 12–52).

If transmission fluid contamination is not found, do not assume the diaphragm is not leaking. Attach a vacuum tester to the modulator and draw 20 in of vacuum. The diaphragm should hold 20 in for one minute (Figure 12–53). Then test the diaphragm at 10 in for one minute, even though it held at 20 in. Should the vacuum diaphragm checkout prove positive, you may then proceed to pressure testing.

Even though a modulator passes a diaphragm condition analysis, this is not conclusive evidence that the modulator is right. Other considerations are the modulator-calibrated spring tension and diaphragm diameter. They also must be correct for the transmission application.

A tight modulator spring results in a higher-than-normal hydraulic torque signal for the engine vacuum. This causes high line pressure, late/harsh shifts, and downshift clunks. A soft modulator spring produces an opposite effect and causes low line pressure and early shifts. Also, medium- and heavy-throttle shifts result in transmission slipping or sometimes a chatter and, consequently, burned clutches.

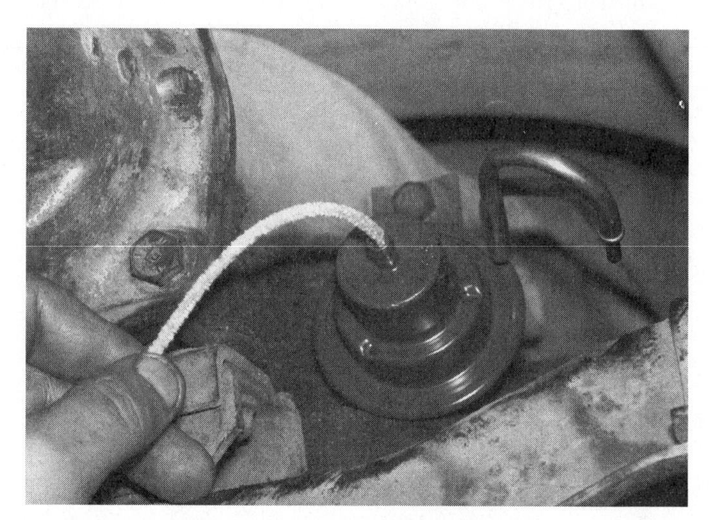

FIGURE 12–51 Check for contamination in the vacuum modulator with a pipe cleaner.

FIGURE 12–52 Uphill routing of vacuum supply hose prevents gasoline and moisture from seeping to the modulator.

FIGURE 12–53 Vacuum check the modulator.

Diaphragm diameters make a difference. When the diameter is too large for the engineered modulator system, the applied vacuum has an extra draw on the diaphragm. This makes the modulator spring work too light for the engine vacuum signal. End results are low line pressures and early/slipping shifts.

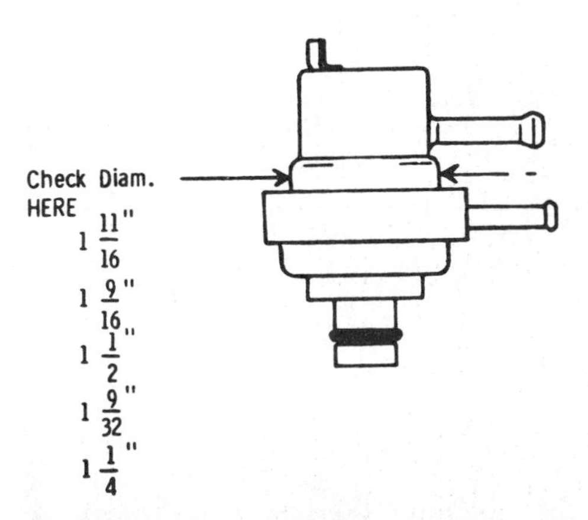

FIGURE 12–54 Measuring Ford vacuum modulator diameter.

Diaphragm diameters on the small side have the opposite effect—high line pressures accompanied by harsh/late shifts.

How can you absolutely be sure that the modulator spring tension and diameter are right for the transmission? Simply measure the diaphragm diameter on the vacuum can and weigh the spring tension. Ford modulators use four diameter sizes: 1-11/16 in, 1-9/16 in, 1-9/32 in, and 1-1/2 in. General Motors uses two diameter sizes: 1-11/16 in and 2 in. The measurement checkpoints for Ford and GM modulators are illustrated in Figures 12–54 and 12–55.

Using GM modulators as an example, all 1-11/16–in and 2-in vacuum cans do not necessarily have the same calibrated spring tension. Adding to the confusion, the 1-11/16–in and 2-in vacuum cans are interchangeable. It may not be right for the transmission, but they do interchange. Ford has similar problem conditions.

The spring tension is weighed on a common kitchen scale. The technique consists of preconditioning the spring by flexing the diaphragm on the bench with your thumb on the vacuum port (Figure 12–56). Then push the modulator down on the

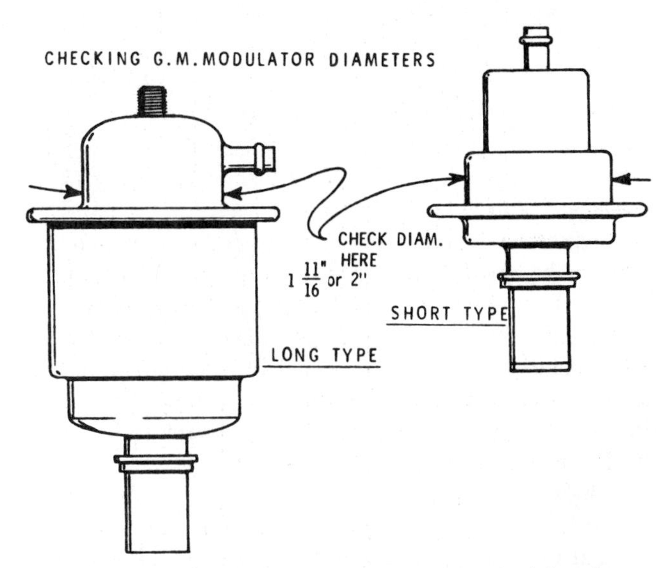

FIGURE 12–55 GM vacuum modulator diameter check locations.

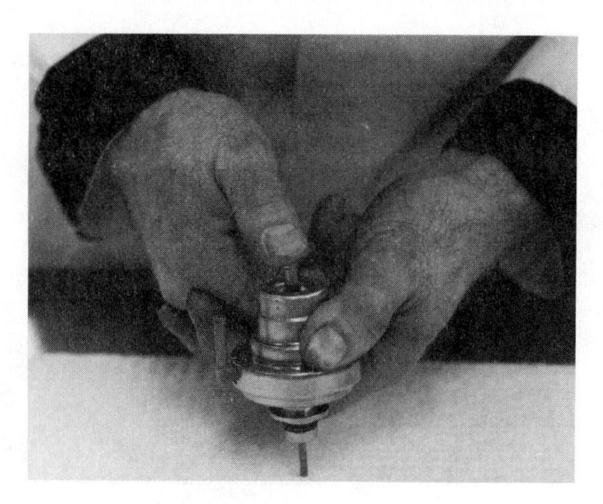

FIGURE 12–56 Preconditioning Ford vacuum modulator spring.

scale until the push rod just begins to move (Figure 12–57). The illustrated modulator tips the scales at 12 lb.

Measuring and weighing vacuum modulators is a development of Trans Go/Research (P.O. Box 3337, El Monte, CA 91733-0337). This company provides the modulator specifications for all Ford and GM transmission applications. For GM modulators, a factory-recommended comparison test can be used to check the load of the modulator in question. A Kent-Moore comparison gauge J-24466 or an equivalent self-made gauge, as shown in Figure 12–58, can be used on both the long and short modulators.

To check the modulator load using a self-made gauge, follow these steps:

1. Use a known good modulator with the same part number as the modulator in question. The part numbers are located on the back of the vacuum can.
2. Install the gauge between the ends of the known good modulator and the questionable modulator.
3. Hold the modulators in a horizontal position and gradually squeeze them together on the gauge pin until either modulator sleeve end just touches the center line of the gauge. The gap between the opposite modulator sleeve and the gauge line should be 1/16 in or less (Figure 12–59). A distance greater than 1/16 in means that the modulator in question must be replaced (Figure 12–60).

The use of comparison gauge J-24466, shown in Figure 12–61, produces the same results.

Just like the vacuum supply systems, modulator vacuum unit faults can be elusive. Even new replacement modulators have proven to be faulty. You might diagnose a faulty modulator, but the replacement might create the same problem or a new problem in the system. You will even find that some adjustable modulators cannot be adjusted to specifications. For reliable results, make it a practice to check each replacement modulator for (1) vacuum leaks, (2) correct diameter and transmission application, and (3) correct spring load.

Absolutely avoid setting the diaphragm spring load (adjustment type) out of the specification range for a "softer" shift feel.

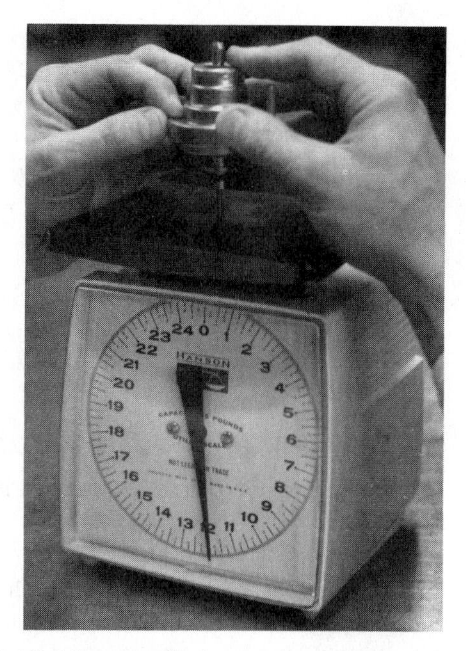

FIGURE 12–57 Measuring Ford vacuum modulator spring tension.

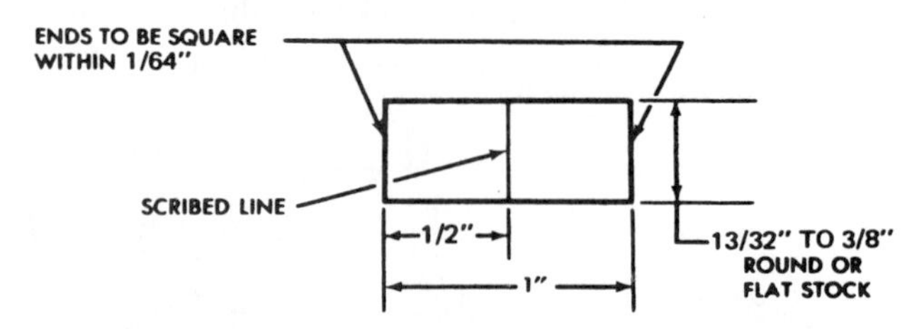

FIGURE 12–58 Self-made gauge dimensions. (Courtesy of General Motors Corporation, Service Technology Group)

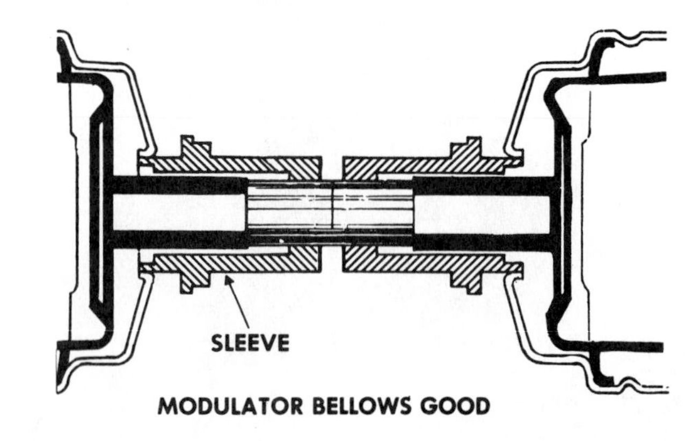

FIGURE 12–59 Vacuum modulator gauge centered: good.

Any change in the modulator setting affects not only shift feel but also transmission operating pressure. A light spring load results in low operating pressure and major clutch/band failure. The transmission is likely to slip under medium- and heavy-throttle operation. Be careful about backing off the spring load to correct a delayed shift problem. If low governor pressure is the real problem, again, you end up building in a lower-than-normal line pressure boost schedule, and the transmission is subject to failure.

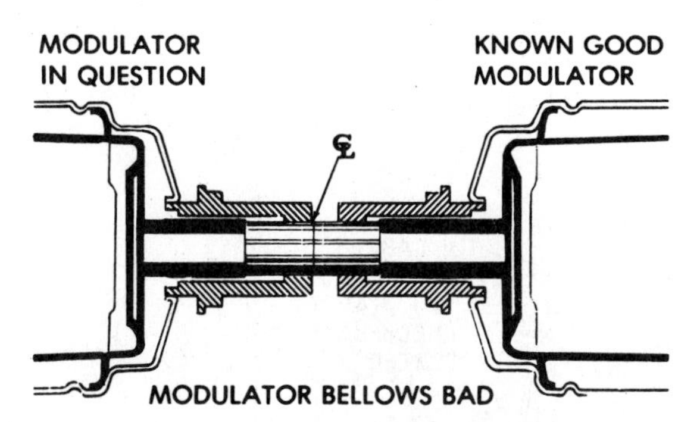

FIGURE 12–60 Vacuum modulator gauge off-center: bad.

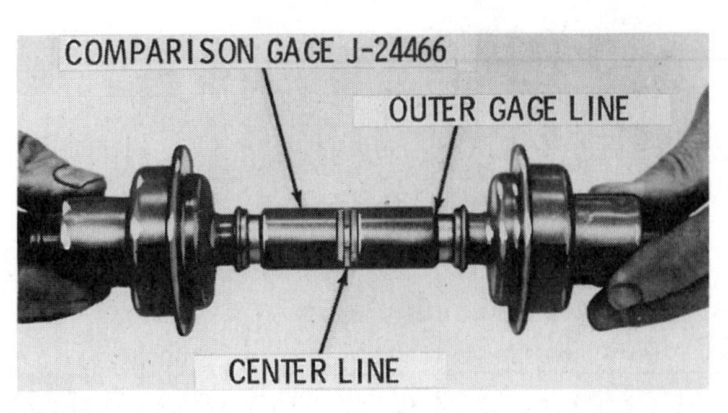

FIGURE 12–61 Kent-Moore vacuum modulator gauge. (Courtesy of General Motors Corporation, Service Technology Group)

When working with the Ford A4LD and C-3, the modulators look identical and install interchangeably. Should the C-3 modulator get installed in the A4LD, there will be no automatic upshifts—only a manual 1-2 shift will be available. The C-3 has a deeper valve travel bore. An A4LD modulator, however, can be used on a C-3. Finally, do not swap turbocharged modulators with regular modulators. On either the C-3 or A4LD, a regular modulator used with the turbo engine gives the following result: light throttle shifts normal, but no shifts at moderate and heavy throttle.

Pressure Test Procedure

It has already been emphasized that pressure tests must follow the manufacturer's procedures. Because instrument hookups and test procedures vary between transmission makes, there is no attempt here to discuss individual transmission test requirements except for selected examples. Always consult the applicable manufacturer's service manual or equivalent for exact procedures and specifications. Some transmission models have as many as three or four pressure schedules, and these can vary from year to year.

In the following pressure test discussions, the displayed pressure schedules do not apply to all transmissions. They are merely examples used to illustrate common situations.

Pressure Test Interpretation/Diagnosis

To interpret pressure test readings, the technician needs to be familiar with the specifics of how the line pressure is controlled for the transmission being tested—there are variations between transmissions. Line pressure is basically controlled by the pressure regulator valve spring and boost oil that acts upon the regulator valve and adds to the spring tension. The boost oil is supplied by the TV output and mainline oil. Depending on the throttle opening and operating range, line pressure values are determined by:

- Spring tension only; boost pressures isolated from the regulator valve
- Spring tension plus TV oil boost
- Spring tension plus mainline oil boost
- Spring tension, plus TV oil boost, plus mainline oil boost.

Any time the regulator valve is sensitive to TV oil boost, the line pressure rises and falls with engine breathing. This makes it possible to evaluate what the TV system is doing from a mainline pressure reading. Some mainline pressure regulator systems are designed to be governor-oil sensitive. The governor oil is designed to cut back on the line pressure at discrete vehicle speeds. What the governor is doing shows up in the mainline pressure check. Examples of this type of system are the Ford Cruise-O-Matic series and the GM 3L80 (400) and 3L30 (180).

Special Pressure Test Considerations. Pressure regulator valves are usually sensitive to the same throttle oil as the shift valves, but the technician should be alert to exceptions. In the GM Hydra-Matic 4T60 (440-T4), the vacuum modulator control provides a torque-sensitive boost oil for shift feel. The vacuum modulator system works independently of the cable-controlled throttle system that controls shift timing. The TV cable movement has no effect on transmission line pressure.

Where the TV cable internally connects to the throttle valve mechanical control, the TV system is usually designed with a linkage and check valve arrangement that controls the TV bleed port (Figure 12–62). When the TV cable is properly connected and adjusted, the bleed port is open, and the TV valve regulates normally.

If the TV cable is disconnected or misadjusted too long, the spring-loaded linkage lets the check valve seat, and the TV regulated output is put into high mode. This protects the transmission from low line pressure and damage to the friction elements during heavy throttle, which is the usual result of a TV disconnect or an extremely retarded adjustment. Examples of transmissions with fail-safe TV check valve exhaust porting are the GM Hydra-Matic 3T40 (125) and 4L60 (700-R4).

Ford AOD, ATX, and AXOD units also incorporate a fail-safe TV feature by using a heavy-duty spring load to move the TV plunger to WOT in case of a TV disconnect or extremely retarded adjustment.

In the AXOD, activating the TV fail-safe feature results in a hydraulic lock on the TV plunger. With the engine running, the driver cannot depress the gas pedal for acceleration, but with the engine OFF, pedal movement is normal. Excessive force on the gas pedal results in a bent throttle cable bracket and makes normal cable adjustment difficult. Use a pressure gauge on the

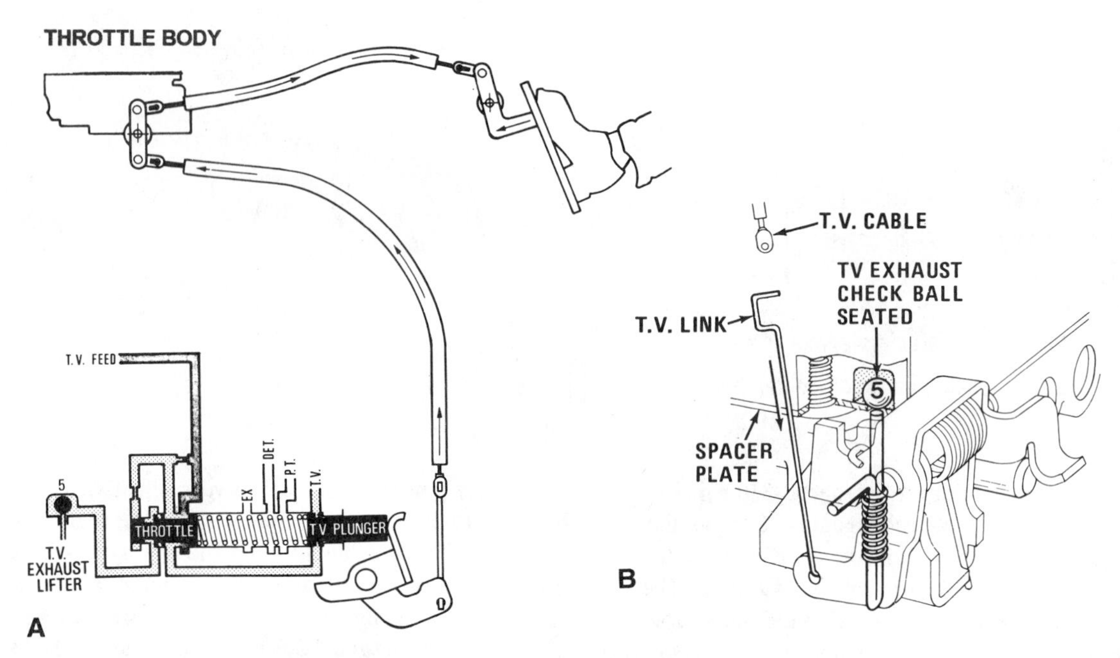

FIGURE 12–62 (a) TV system with check valve bleed port; (b) TV cable disconnected, causing exhaust check ball to block bleed port. (Courtesy of General Motors Corporation, Service Technology Group)

TV tap to make sure the bracket and cable adjustment get back to normal.

The TV fail-safe feature boosts line pressure to maximum at closed throttle. If the engine is running, it does not respond to external corrective action. The engine must be turned off, the corrective action made, and the line pressure reading rechecked. If the fail-safe has not canceled out, look for internal causes. When the fail-safe mechanism triggers, it is telling you that something is wrong with the TV mechanical movement—that it is disconnected or greatly delayed.

- Pay close attention to pressure response time. The pressure reading may meet specifications, but how much time did it take to get there? Response time to meet specifications should be less than one second. Problems in this area are usually very definite and noticeable. The technician will know without question that the pressure buildup is taking too much time.

The technician needs to establish that the pump circuit is right if all the required pressure readings are to be meaningful.

- Clutch and band circuits are usually metered before the apply oil reaches the clutch or servo unit. Where the mainline pressure test is supposed to be reading the circuit, it will tell you what is happening up to the orifice but nothing about the circuit from the orifice to the clutch or servo unit. For example, in the D range, the forward clutch circuit is introduced into the mainline pressure reading (Figure 12–63). A normal reading in this case only identifies that a good pressure head is feeding the forward clutch circuit up to the orifice. It does not pick up any pressure loss from piston seals, oil sealing rings, or broken piston cylinder welds beyond the orifice. The orifice makes it possible for the pump to maintain its pressure head. A shift circuit orifice is typically located in the valve body separator plate in series with the shift valve feed oil, at the clutch piston cylinder feed port, or at the oil ring groove feed port to clutch drum (Figures 12–64 and 12–65).
- Always start your pressure test sequence with an isolated pump circuit test. This takes place in P and N when the line pressure is cut off from the clutch/band circuits at the manual valve. The possibility of auxiliary circuit leaks affecting pump-regulated output is limited.

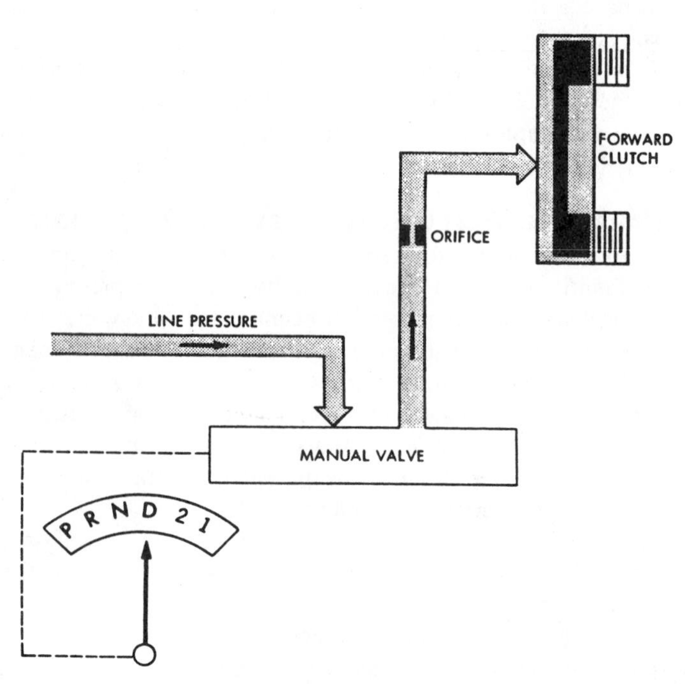

FIGURE 12–63 Basic forward clutch circuit.

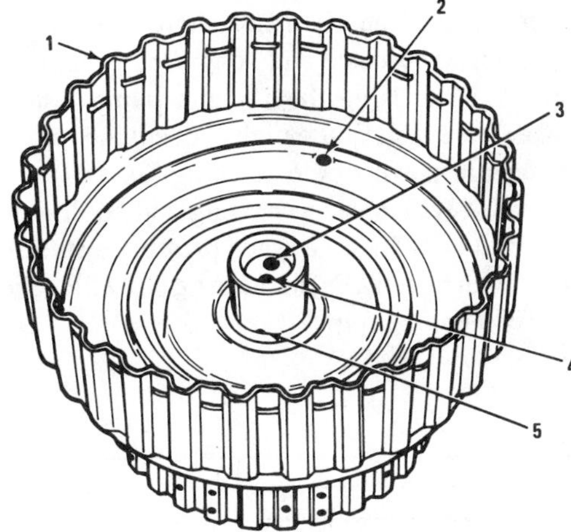

FIGURE 12–64 Clutch metered orifice feed located in the cylinder bore (#5). (Courtesy of General Motors Corporation, Service Technology Group)

If the line pressure is too low or too high, the diagnosis and fix must be made at this time.

- If two or more readings are low or two or more readings are high, they share a common problem. The technician is not dealing with two or more problems—only one solution is needed.
- As a valuable assist to any pressure testing diagnosis, the technician needs to be able to understand and work with the transmission hydraulic diagrams. It is nice to be able to take the pressure readings and do the diagnostic homework with diagram assistance.
- Check the oil pressure when the malfunction occurs. This could involve cold pressure test priority over the usual hot test. For example, the transmission might have an annoying late 1-2 shift, or the entire shift schedule may be late only when cold. The garage shifts N to D and N to R also may be harsh. A sticking TV valve can be the cause of both high TV pressure and line pressure. This will show up on the gauge reading as high line pressure when cold.

Several examples will be used to demonstrate a variety of pressure test procedures and interpretations. Although the examples primarily center on some basic three-speed units, the proedures and interpretations carry over to other three-speed and four-speed units. Before pressure testing, it is important that the prechecks have been completed.

- Transmission is at operating temperature.
- Fluid level is right.
- Where applicable, TV cable/linkage moves freely and is in adjustment.

FIGURE 12–65 Clutch metered orifice feed located in ring groove.

- Where applicable, vacuum flow to modulator and modulator diaphragm checkout is right.
- Outside manual linkage moves freely and is in adjustment.
- Engine is tuned and diagnostic codes are eliminated.

Where applicable, be sure to follow the manufacturer's total running-time limitation for a single test run or combination of test runs.

CAUTION: The brakes must be applied during test runs.

Line Pressure Diagnosis: Ford A4LD and Cruise-O-Matic Series. The test procedure is applicable to the entire three-speed Cruise-O-Matic series: C-6, C-5, C-4, C-3, and FMX, plus the four-speed A4LD. This automatic transmission group uses a vacuum modulator for TV control with the regulated TV oil used for shift delay scheduling and as a boost oil at the pressure regulator valve. The line pressure regulation is basically controlled as follows:

- Spring tension plus TV oil boost in P-N-OD-D-2-1 or P-N-D-2-1.
- Spring tension plus TV oil boost plus reverse mainline oil boost in R.
- At a specified vehicle speed and throttle opening in OD-D-2-1, the governor pressure switches a cutback valve and partially eliminates the TV oil boost. This results in a line pressure drop.

From a mainline pressure tap, the pump, TV, and governor circuits can be analyzed (Figure 12–66). Pressure test charts (Charts 12–4 and 12–5 on pages 298 and 299) give the pressure test procedure and representative specification data for a C-6 and A4LD transmission. Note that the engine is used as the vacuum source. The drive wheels must be held stationary, with the car brakes applied and the car weight on the wheels. Step on the accelerator as much as needed to achieve the necessary vacuum setting. Operation below 3 in hg (20.2 kPa) of vacuum is a wide open throttle stall condition. Therefore, each test run

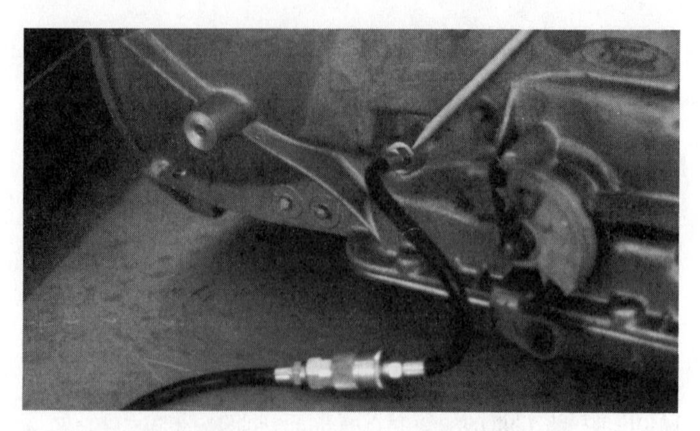

FIGURE 12–66 Mainline pressure tap location—Ford.

to a maximum of five seconds followed by a cooling cycle. To cool the transmission, place the selector in N and run the engine at 1000 rpm for at least thirty seconds.

As an alternative method for setting the required vacuum levels, tee a hand-operated vacuum pump tester into the transmission tester vacuum line. By using a long length of hose, the vacuum level can be adjusted and controlled from the driver's seat (Figure 12–67). Using the vacuum pump tester avoids running the pressure tests at the high stall engine rpm required to pull the vacuum down to the 10 in hg (34 kPa) or 3 in hg (10.2 kPa) levels. Transmission heating and cooling time between tests is reduced. The idle speed pressure tests remain the same. Runs at 10 in hg (34 kPa) and 3 in hg (10.2 kPa), however, are made at 1000 engine rpm.

			Manifold Vacuum						
			Idle						
Trans. Type	Vacuum Diaphragm Type and Ident.	Range	15" & Above	14"	13"	12"	11"	10"	W.O.T. Thru Detent
C6	SAD	D, 2, 1	53-88	74-94	80-100	86-106	92-112	98-118	151-180
	1 Black	R	66-138	116-147	125-156	135-166	144-175	153-184	236-295
	Stripe	P, N	53-88	74-94	80-100	86-106	92-112		
C6	SAD	D, 2, 1	53-75	62-82	69-89	76-96	83-102	89-109	150-180
	1 Purple	R	66-118	97-128	108-139	119-149	129-160	140-171	235-295
	Stripe	P, N	53-75	62-82	69-89	76-96	83-102		
C6	SAD	D, 2, 1	53-60	53-60	53-72	61-80	69-88	77-97	149-180
	1 Green	R	66-88	70-100	82-113	95-126	108-138	120-151	233-295
	Stripe	P, N	53-60	53-60	53-72	61-80	69-88		
C6	HAD	D, 2, 1	53-75	62-82	69-89	76-96	83-102	89-109	150-180
	Plain Sea	R	66-118	97-128	108-139	119-149	129-160	140-171	235-295
	Level	P, N	53-75	62-82	69-89	76-96	83-102		
	Bar = 29.5								

*Line Pressure Specifications (Control PSI) With Governor Pressure at Zero (Oil Temperature — 150°-200°F)
SAD: Single Area Diaphragm
S-SAD: Extra Large Single Area Diaphragm
HAD: High Altitude Diaphragm
S-HAD: Extra Large High Altitude Diaphragm
Using absolute barometric pressures:
(1) Determine the difference between the baseline, absolute barometric pressure of 29.25 and the absolute barometric pressure for the area.
(2) For the HAD device, multiply the result of (1) by 3. (C4) or by 3.62 (C6). For the S-HAD device, multiply the result of (1) by 4. (C4).
(3) For an absolute barometric pressure greater than the baseline barometric pressure, add the result of (2) to the given "D" and "P, N" specifications for the 29.75 to 28.75 barometric pressure range.
For an absolute barometric pressure less than the baseline barometric pressure, subtract the result of (2) from the given "D" and "P, N" specifications for the 29.75 to 28.75 barometric pressure range.

Pressure Test

Engine RPM	Manifold Vacuum In-Hg	Throttle	Range	PSI Record Actual	PSI Record Spec.
Idle	Above 15	Closed	P		
			N		
			D		
			2		
			1		
			R		
As Required	10	As Required	D,2,1		
As Required	Below 3	Wide Open	D		
			2		
			1		
			R		

Reprinted with permission of Ford Motor Co.

CHART 12–4 C-6, typical line pressure specifications.* (Reprinted with the permission of Ford Motor Company)

Transmission Type	Transmission Model	Range	Idle		WOT Stall through Detent
			15 in. and Above	10 in.	
A4LD	87GT-CAA	(1)$_{D}$,D,2,1 R P,N	50–70 60–80 50–70	92–112 128–148	205–235 278–314
A4LD	87GT-DAA-FAA HAA-KAA-MAA NAA	(2)$_{D}$,D,2,1	57–78 67–105 57–78	114–134 157–177	205–235 282-316
A4LD	87GT-FAA-KAA NAA	$_{D}$,D,2,1 R P,N	57–67 67–77 57–67	90–110 124–144	180–210 247–280
A4LD	87GT-ABA	(1)$_{D}$,D,2,1 R P,N	50–70 70–109 50–70	92–113 158–178	167–195 282–316
A4LD	87GT-AAA	(2)$_{D}$,D,2,1 R P,N	57–77 62–82 57–77	88–108 122–142	200–224 270–300
A4LD	87GT-BAA	D,$_{D}$,2,1 R P,N	57–77 62–82 57–77	79–99 108–128	190–210 260–280

Note: 29.0-30.0 in. Hg absolute barometric pressure (at sea level).
24.0-25.0 in. Hg absolute barometric pressure (at 5000 ft.).

Pressure Test

Engine rpm	Manifold Vacuum (in. Hg)	Throttle	Range	PSI	
				Record Actual	Record Spec.
Idle	Above 15	Closed	P N OD D 2 1 R		
As required	10	As required	OD, D,2,1		
As required	Below 3	Wide open	OD D 2 1 R		

CHART 12–5 A4LD typical line pressures. (Reprinted with the permission of Ford Motor Company)

Once accurate test results are obtained, the test data can be analyzed to identify the problem area and to verify the parts of the hydraulic system that are meeting specifications and working right. When test results indicate an abnormal pressure, refer to the manufacturer's pressure diagnosis chart (Chart 12–6, page 300). Chart interpretation and location of the problem still depend on your knowledge of hydraulic system operation and some commonsense deductions.

The following pressure test tips taken from the Ford service manual should help in your diagnosis.

Keep in mind that clutch and servo leakage may or may not show up on the control pressure test. This is because (1) the pump has a high output volume and the leak may not be severe enough to cause a pressure drop, and (2) orifices between the pump and pressure chamber may maintain pressure at the source, even with a leak downstream. Pressure loss caused by a less-than-major leak is more likely to show up at idle than at WOT, where the pump is delivering full volume.

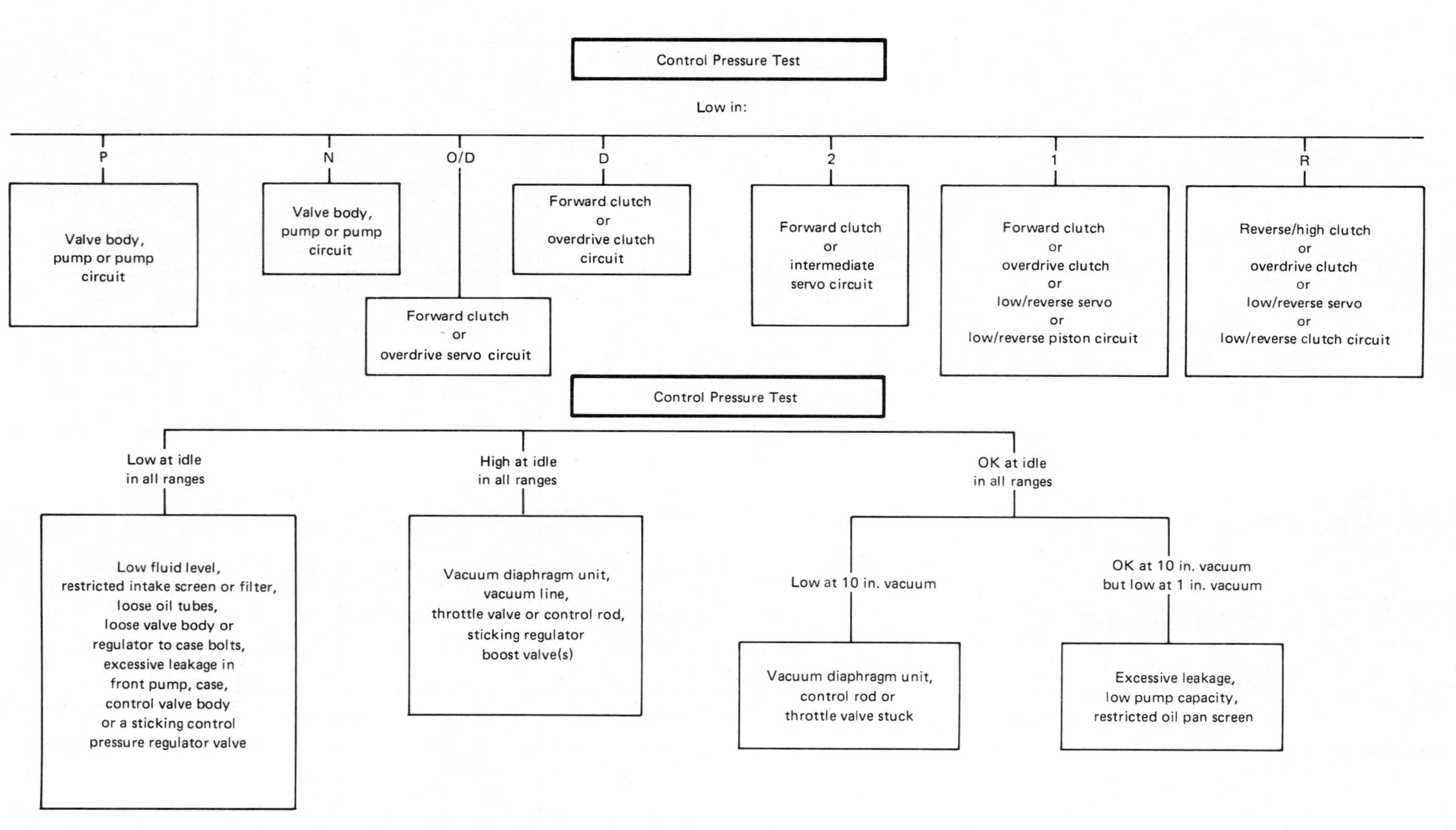

CHART 12–6 Combined troubleshooting chart: A4LD, C-6, C-4, and C-3. (Reprinted with the permission of Ford Motor Company)

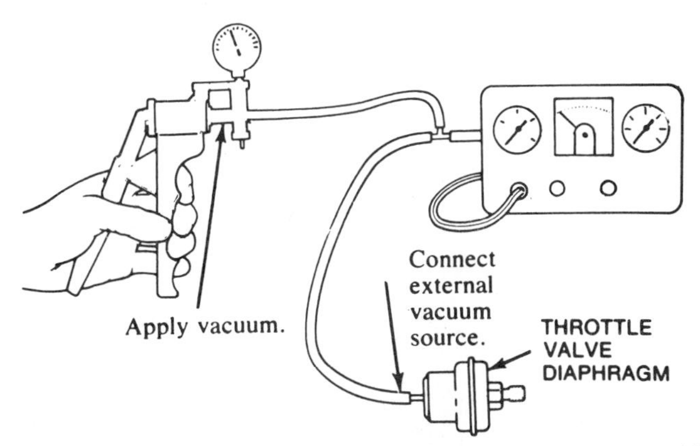

FIGURE 12–67 Mainline pressure test using vacuum pump to avoid high engine rpms. (Reprinted with the permission of Ford Motor Company)

To further isolate leakage in a clutch or servo circuit, it may be necessary to remove the oil pan and valve body. Then, perform hydraulic circuit air pressure tests (Figure 12–68).

Because the hydraulic systems of many automatic transmissions are much alike, a general analysis of the information can be applied to other transmissions. When engine idle is specified, this test is performed at curb-idle rpm. Tests at idle produce the biggest stress on the pump circuit for delivering volume and operating pressures. Therefore, tests done at engine idle or low-rpm conditions are excellent for analyzing the pump circuit for pump wear or circuit leaks. Some procedures, however, do not include the idle rpm test.

Beginning or existing pump circuit problems can be detected with an isolation test in P and N. The manual valve blocks the pump output to all clutch/band circuits and the governor circuit (Figure 12–69). The throttle circuit must be canceled out to eliminate any possibility of a TV boost at the regulator valve. This is done by drawing a vacuum to 18-20 in hg (61-69 kPa). With TV pressure now at zero psi, the pump is strictly working against the fixed spring tension of the regulating valve. Pressure readings should be at the low side of the specifications as determined by the spring (see Charts 12–4 and 12–5).

If the pressure readings in P and N are normal, the following is true about the hydraulic system:

- The suction side of the pump circuit is tight.
- The pump gears and assembly casing are not worn or defective.
- The pressure delivery side of the circuit is tight.
- The pressure regulator valve and boost valves are working.

It is important in pressure testing to establish that the pump system has normal capabilities. If the pump system is not right, other systems and circuits will not be right and can be falsely diagnosed as the problem area. If the P and N pressure readings are low, the other ranges also can be expected to read low. Refer to the diagnosis chart (Chart 12–6, page 300) for suggested problem cause in the pump circuit. Be sure the filter is not restricted and retorque the valve body to the case bolts to check for looseness. Should it be necessary to pull the transmission for further investigation, retorque the pump to the case bolts to check for looseness. All parts of the pump system must be treated as potential suspect areas. Be careful to approach the problem by starting with the easy and accessible checkpoints first.

Do not be fooled if the pressure readings at the 10 in hg (34 kPa) and 3 in hg (10.2 kPa) vacuum levels are normal. The higher engine or pump rpm can sometimes overcome the pump circuit pressure deficiency indicated at idle. Should the P and N idle pressures read zero or almost zero, the most likely problem area is the pressure regulator valve stuck in the exhaust position or a sheared pump drive. A worn pump always develops a pressure head, even though it cannot meet specification requirements. Low pressures cannot furnish the force to hold a band or a clutch, and the transmission slips. Obviously, at zero psi, the vehicle fails to move.

Idle pressure tests in P and N can be used as a checkout of the throttle system regulated output. Because the pressure boost valve is sensitive to throttle pressure, it makes sense that a mainline pressure reading isolated from the clutch and band circuits can reflect throttle system output. If test pressure is high at idle in P and N, the other ranges can be expected to read high at idle because the throttle system has not canceled out.

Consult the diagnosis chart (Chart 12–6) for suggested problem causes in the throttle circuit. Do not overlook an improper vacuum modulator application or calibrated modulator spring tension. Do not let the modulator fool you. What you are going to find as the fault is unknown, but you must hunt until the cause is found. Keep in mind that the throttle system includes everything from the engine through the internal valving. Besides the throttle valve, all related items, such as the TV boost valve, must be considered. At least you know that the transmission does not need to be removed and disassembled for inspection. This is an in-the-car fix. The modulator diaphragm should have been checked for leakage before the technician has gone through all this trouble.

It is also important to verify at what point the throttle system cuts in. Gradually drop the vacuum until the mainline pressure starts to rise from the throttle boost effect. The throttle system cut-in point should occur before the vacuum drops below 13 in hg (44 kPa). Where applicable, the cut-in point can be adjusted by changing the modulator spring load setting (Figure 12–70). If the adjustment cannot be attained, the technician should consider installing a new modulator. Adjustable modulators are available as an aftermarket item.

Further checkout of the throttle system continues with the pressure rise checks.

1. Pressure Rise Check at 10 in hg (34 kPa). If the pressure does not increase as the vacuum drops to 10 in, then the push rod to the throttle valve is possibly missing. This means that the spring force cannot be relayed to the throttle valve, and TV pressure is zero psi. Sometimes the TV pressure relief valve unknowingly gets lost during a fluid and filter change and gives the same results.

If the pressure increases but remains low, a modulator adjustment might correct the problem. Do not expect the adjust-

FIGURE 12–68 Air testing hydraulic circuit with test plate—Ford AOD.

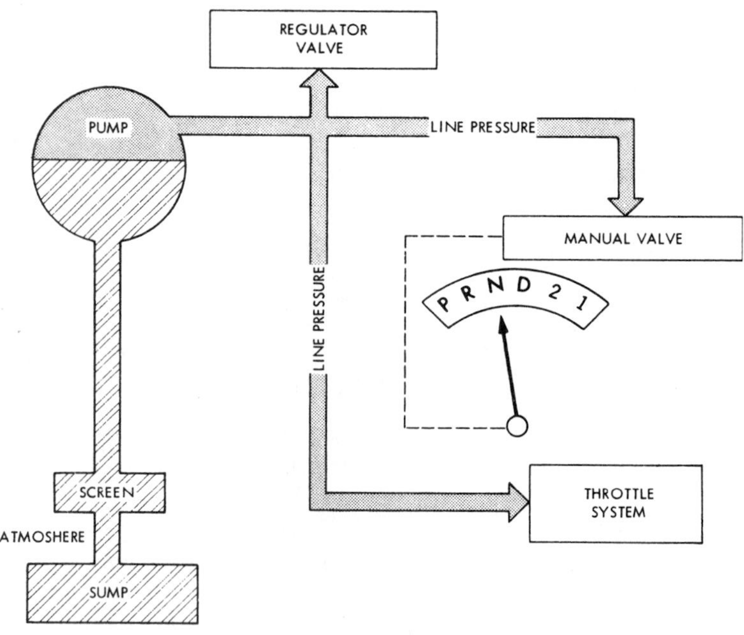

FIGURE 12–69 Basic neutral circuit.

CHANGING DIAPHRAGM PRESSURE SETTING

- Adjust ONLY if all pressures are high or low.
- Do not set below specified pressure for shift feel.

SCREW-IN DIAPHRAGM

- In to increase; out to decrease (A).
- One turn = 2-3 psi.

PUSH-IN DIAPHRAGM

- Change selective rod (B).
- Longer to increase; shorter to decrease.
- Each rod increment = 5-6 psi.

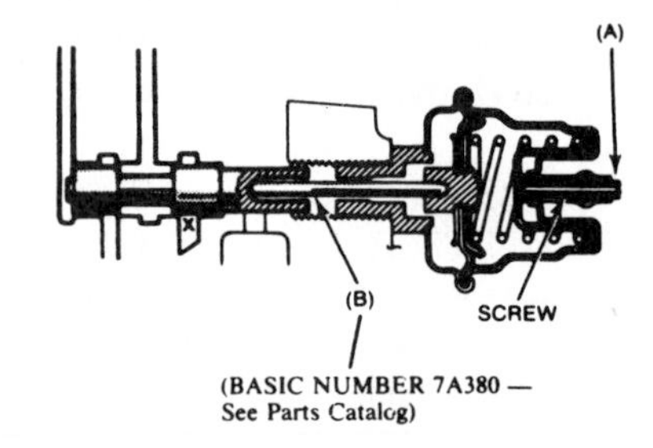

FIGURE 12–70 Vacuum modulator adjustment procedure. (Reprinted with the permission of Ford Motor Company)

ment to compensate for major pressure loss. One full turn of the screw only changes the pressure 2 to 3 psi (14 to 20 kPa). Different push rod lengths for minor adjustments are also available through the dealer. The modulator cut-in point, however, must remain correct—otherwise, a replacement modulator unit must be considered.

With a good cut-in point and proper operating pressure at 10 in hg (34 kPa), it is reasonable to believe that the pump and throttle systems are both right. The idea here is to test the midrange operation of the modulator and throttle system for accuracy. Should the pressure not meet specifications until 8 in hg (27 kPa), it is cause for a shift schedule problem and friction element failure under heavy throttle.

2. Pressure Rise Check at 1 in hg (3.4 kPa). Correct pressures at 10 in hg (34 kPa) usually mean good pressures at 1 in. Should the pressures prove to be low, a restricted suction screen is a high-suspect item, or possibly excessive leakage in the pump circuit resulting in low pump capacity. The pump is not worn out. The extra-high boost pressures can open up hairline fractures in the pump or even blow past defective or loose gasket areas: for example, the pump to case gasket.

To verify the condition, run the engine at 1000 rpm in P and N. Expose the modulator diaphragm to atmosphere, and compare readings to the OD, D, 2 and 1 ranges. Throttle valve system problems are the most frequent cause of delayed and harsh shifts and unbalanced pump output. Pressure testing is the only scientific method of determining if this system is working correctly.

Governor line pressure loss consideration applies only to governor circuits with unmetered line oil feed. Low pressures at idle in the D, 2, and 1 ranges (only) can be caused by excessive leakage in the governor circuit. In D, 2, and 1, the manual valve routes mainline pump pressure to the governor (Figure 12–71). Severely worn governor rings, stuck governor

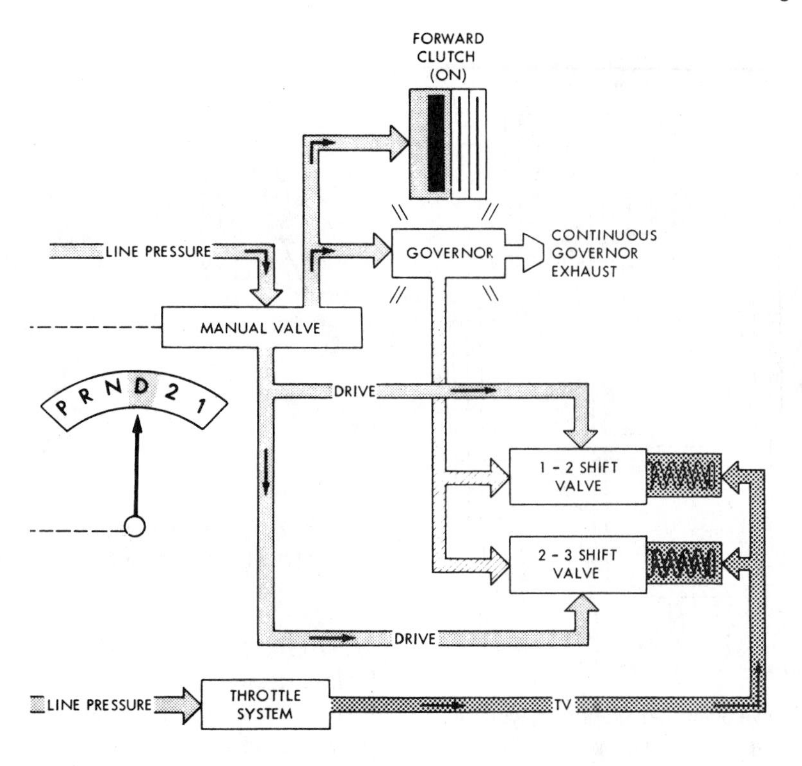

FIGURE 12-71 Basic drive circuit.

valving, loose governor body bolts, or loose valve body to case bolts represent possible problem areas. These problems are accompanied by a shift complaint.

Governor Testing: A4LD, C-6, C-5, C-4, and C-3. In this transmission group, the governor circuit is not equipped with a governor pressure tap. Governor operation, however, can be checked from the mainline pressure reading. A branch of the throttle pressure circuit to the pressure regulator boost valve is routed through a cutback control valve. At the cutback valve, the throttle pressure holds the valve up against governor pressure and is relayed to the pressure boost valve (Figure 12–72). Depending on the aggressiveness of the throttle foot, the governor pressure shifts the cutback control valve at some point between 10 and 15 mph (16 to 18 km/h). When the cutback control valve is shifted, throttle pressure is partially cut off from the pressure boost valve, and the mainline pressure drops (Figure 12–73). This pressure drop can be identified on the pressure gauge, and therefore, the governor system operation can be tested in the shop off the mainline pressure tap.

To prepare for the test:

1. Hook up the transmission tester or equivalent instrumentation as described earlier under pressure testing.
2. Isolate the engine vacuum from the modulator and provide an external vacuum control (Figure 12–74).
3. Support the rear axle on a hoist or floor stands to clear the drive wheels off the ground. Drive wheels must be free to turn.

To run the test:

1. Never exceed 60 mph (96 km/h).
2. Place the selector in the 2 range and proceed to test the governor system under moderate and heavy loads as out-

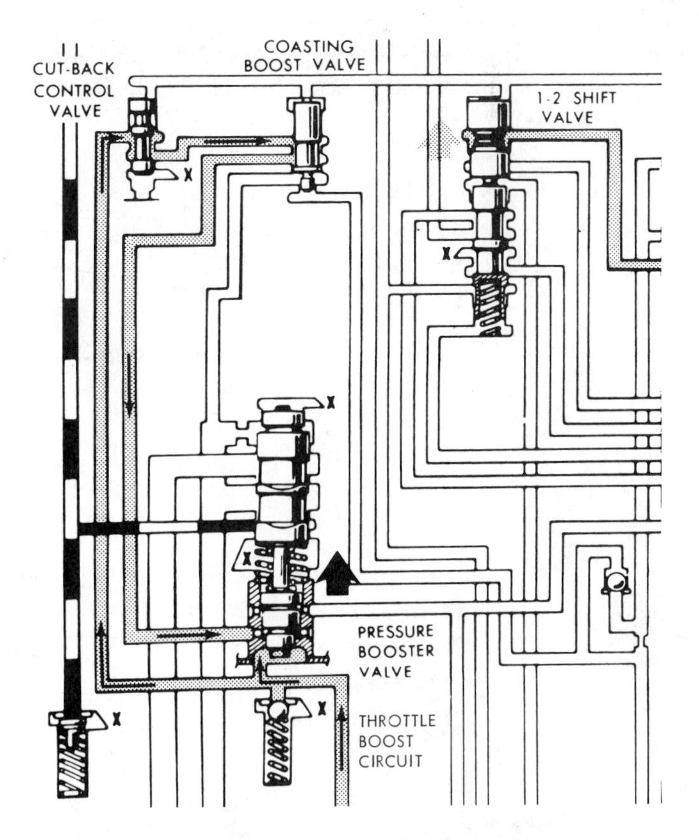

FIGURE 12–72 Throttle boost circuit. (Reprinted with the permission of Ford Motor Company)

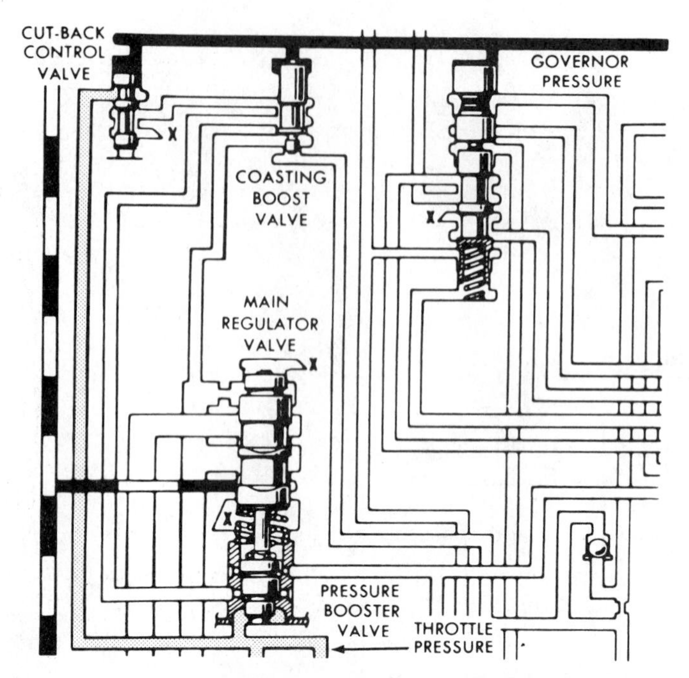

FIGURE 12–73 Cutback control valve operation. (Reprinted with the permission of Ford Motor Company)

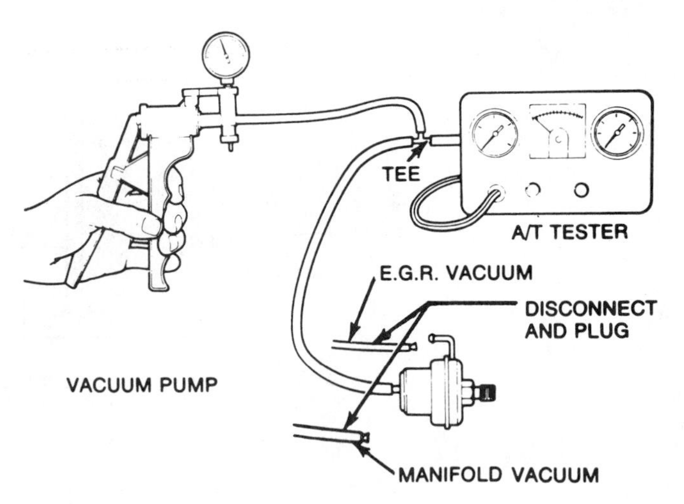

FIGURE 12–74 External control of vacuum modulator. (Reprinted with the permission of Ford Motor Company)

lined in Figure 12–75. (The A4LD selector is placed in the OD or D range.)

If the cutback speeds check out okay, the governor system is right.

Line Pressure Diagnosis: General Motors 3T40 (125-C) and 4L60 (700-R4). Line pressure checks to a 3T40 (125-C) or a 4L60 (700-R4) unit are valuable for diagnosing numerous hydraulic circuits. The units share many similar fundamental hydraulic principals of operation. Both use a TV cable to provide a throttle pressure to delay upshifts, force downshifts, adjust shift quality, and boost line pressure under load conditions. In addition, mainline boost is available to the pressure regulator valve from the manual valve in reverse and the manual 2 and low forward gear ranges.

The line pressure checks and basic hydraulic circuit diagnosis of a 3T40 (125-C) are emphasized at this point to illustrate diagnostic principles. The objective is to show how line pressure readings can be used to determine what part of the hydraulics is working right and then isolate the problem area. Before

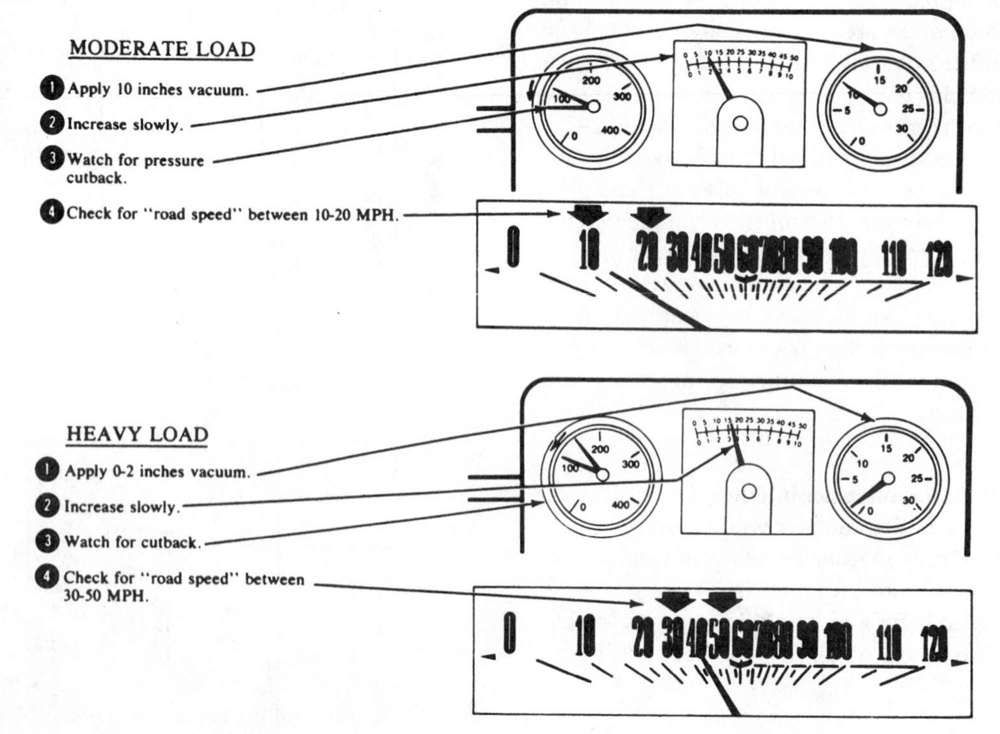

FIGURE 12–75 C-3, C-4, C-5, C-6, and A4LD governor test. (Reprinted with the permission of Ford Motor Company)

any pressure diagnosis starts, the technician should become familiar with the basic line pressure and throttle valve controls of the 3T40 (125-C).

1. The TV cable on the 3T40 (125-C) controls line pressure, shift points, shift feel, part throttle, and detent downshifts.
2. Maximum throttle pressure is limited to 90 psi (620 kPa) by a shift TV valve.
3. The mainline feed oil to the throttle system in P and L is completely cut off at the manual valve. Throttle oil boost at the regulator valve cannot be a factor.
4. A cable disconnect or extended cable adjustment puts the TV oil in high mode.
5. At hot idle, the N pressure should be equal to, or no more than, 10 psi (69 kPa) above the P pressure, verifying that the cable adjustment is not excessively advanced at closed throttle.
6. Pressure regulation in each of the operating ranges is based on the following:

P	spring tension only
R	spring tension plus regulated reverse boost oil that is dependent in part on TV oil
N	spring tension plus TV oil
D	spring tension plus TV oil
2	spring tension plus regulated intermediated boost oil at light throttle. TV oil replaces intermediate boost oil at medium and heavy throttle.
1	spring tension plus regulated intermediate boost oil. There is no TV pressure in Manual 1. At closed and light throttle, intermediate and low range pressure should be the same.

7. There is no governor cutback on the line oil, but a tap for a direct governor pressure reading is available. (Governor circuit and shift problems are discussed later in this chapter.)
8. There is only one line pressure tap.

Now let us look at several line pressure checks and discuss what they might mean. Chart 12–7, page 306 shows normal test readings at the minimum and maximum TV cable position. What does it show the technician?

- The isolated pump circuit test was run in park. It proved that the oil pump is working properly, the regulator valve is working, the oil filter is not plugged, and the suction and pressure sides of the pump circuit are not leaking. These circuits are shown in Figure 12–76, the hydraulic schematic for park range. If the pump circuit is right, it is the basis for normal readings for all the other pressure readings. It can then be further identified that:
- The oil pressure boost valves are working right in manual 2, manual low, and R.
- The TV system is working correctly.
- Everything in the hydraulic system is working properly up to the shift valves or metered circuits such as the forward clutch.

Again, it is important to establish what the pump circuit is doing first. Keep in mind that two or more low readings or two or more high readings usually share a common fault. High or low readings in N with normal line pressure in P indicates a throttle system problem with this transmission.

When a pressure problem is identified, it is helpful to refer to hydraulic schematics to identify which hydraulic components are used in a specific gear and range position. Knowing that certain devices worked properly in a gear and range position, the assumption can be made that they function accurately in the problem situation. Recognizing the devices that are unique to each gear and range position and eliminating known good components helps to pinpoint the cause of the problem.

Causes of Low Oil Pressure—3T40 (125-C).

1. Low fluid level
2. TV system (pressure low in neutral, drive, and low, and pressure low to normal in intermediate and reverse):
   a. TV cable misadjusted or sticking
   b. TV linkage binding
   c. Throttle valve stuck
   d. Shift TV valve stuck
3. Oil filter plugged
4. Oil filter O-ring seal leaking or damaged
5. Valve body and pump assembly bolts loose
6. Valve body related problems:
   a. Check ball #5 or #6 missing or mislocated
   b. Valve stuck or damaged (TV valve and plunger, shift TV valve, pressure regulator valve, TV boost valve, or pressure relief valve)
   c. 1-2 accumulator piston and/or seal leaking or missing
   d. Internal leaks
7. (Low only) Low blow-off valve damaged, or #4 check ball mispositioned or missing
8. (Reverse only) Low-reverse cup plug seal between case and clutch housing missing or leaking, or O-ring and/or washer on feed tube to the seal missing or leaking
9. Pump vane broken, seals cut or missing
10. Intermediate oil passages to pressure regulator blocked
11. Driven sprocket support to case cover leaks

Causes of High Oil Pressure—3T40 (125-C).

1. TV system (pressure high in neutral, and drive, and pressure normal to high in intermediate and reverse):
   a. TV cable misadjusted, sticking or broken
   b. TV linkage binding
   c. Throttle valve stuck
   d. Shift TV valve stuck
   e. TV lifter bent or damaged
2. Valve body and pump assembly:
   a. Stuck valve (TV valve and plunger, shift TV valve, pressure regulator valve, or TV boost valve)
   b. Pump slide stuck
3. (Low only) Low blow-off valve stuck closed
4. Internal pump or case cover leaks

HYDRA-MATIC 3T40 LINE PRESSURE CHECK PROCEDURE

1. Check fluid. See "3T40 Transaxle Fluid Checking Procedure" in this section.
2. Check and adjust Throttle Valve (TV) cable and manual linkage. Refer to Section 7A.
3. Remove line pressure plug and install J 21867 pressure gage.
4. Start the engine, set the parking brake and raise engine speed to 1000 RPM.
5. Check transaxle line pressure:
   - Minimum Line Pressure - no changes to normal TV cable travel.
   - Full Line Pressure - hold the TV cable to full travel.
   - Note pressure readings in all gear ranges for Minimum and Full Line Pressure and compare with information in the chart below.
6. Use the "Condition Diagnosis Charts" in this Section to diagnose line pressure readings that are too low or too high.

NOTICE: Total test running time should not be longer than 2 minutes or transaxle damage could occur.

CAUTION: Brakes must be applied at all times to prevent unexpected vehicle motion.

1995 HYDRA-MATIC 3T40 TRANSAXLE

	RANGE	MODELS	kPa	PSI
	P	KDC, SWC	396-436	57-63
		AJC, AKC, CRC, HTC, JCC, LAC, LCC	459-506	66-73
MINIMUM	R	SWC	673-764	98-11
LINE PRESSURE		AJC, AKC, HTC, JCC, KDC, LAC, LCC	760-885	110-128
		CRC	882-970	128-141
NORMAL TV TRAVEL	N, D	KDC, SWC	396-436	57-63
1000 RPM		AJC, AKC, CRC, HTC, JCC, LAC, LCC	459-506	66-73
	D2, D1	KDC, SWC	827-910	120-132
		AJC, AKC, CRC, HTC, JCC, LAC, LCC	960-1055	140-153
	P	KDC, SWC	396-436	57-63
		AJC, AKC, CRC, HTC, JCC, LAC, LCC	459-506	66-73
FULL	R	SWC	1378-1611	200-233
LINE PRESSURE		AJC, AKC, CRC, JCC, KDC, LAC, LCC	1507-1816	216-263
		HTC	1663-1949	241-282
	N, D	CRC, KDC, SWC	752-920	107-133
FULL TV TRAVEL		AJC, AKC, JCC, LAC, LCC	914-1042	132-151
1000 RPM		HTC	979-1123	142-163
	D2, D1	KDC, SWC	827-910	120-132
		AJC, AKC, CRC, HTC, JCC, LAC, LCC	960-1055	140-153

CHART 12–7 3T40 (125-C) line pressure check procedure. (Courtesy of General Motors Corporation, Service Technology Group)

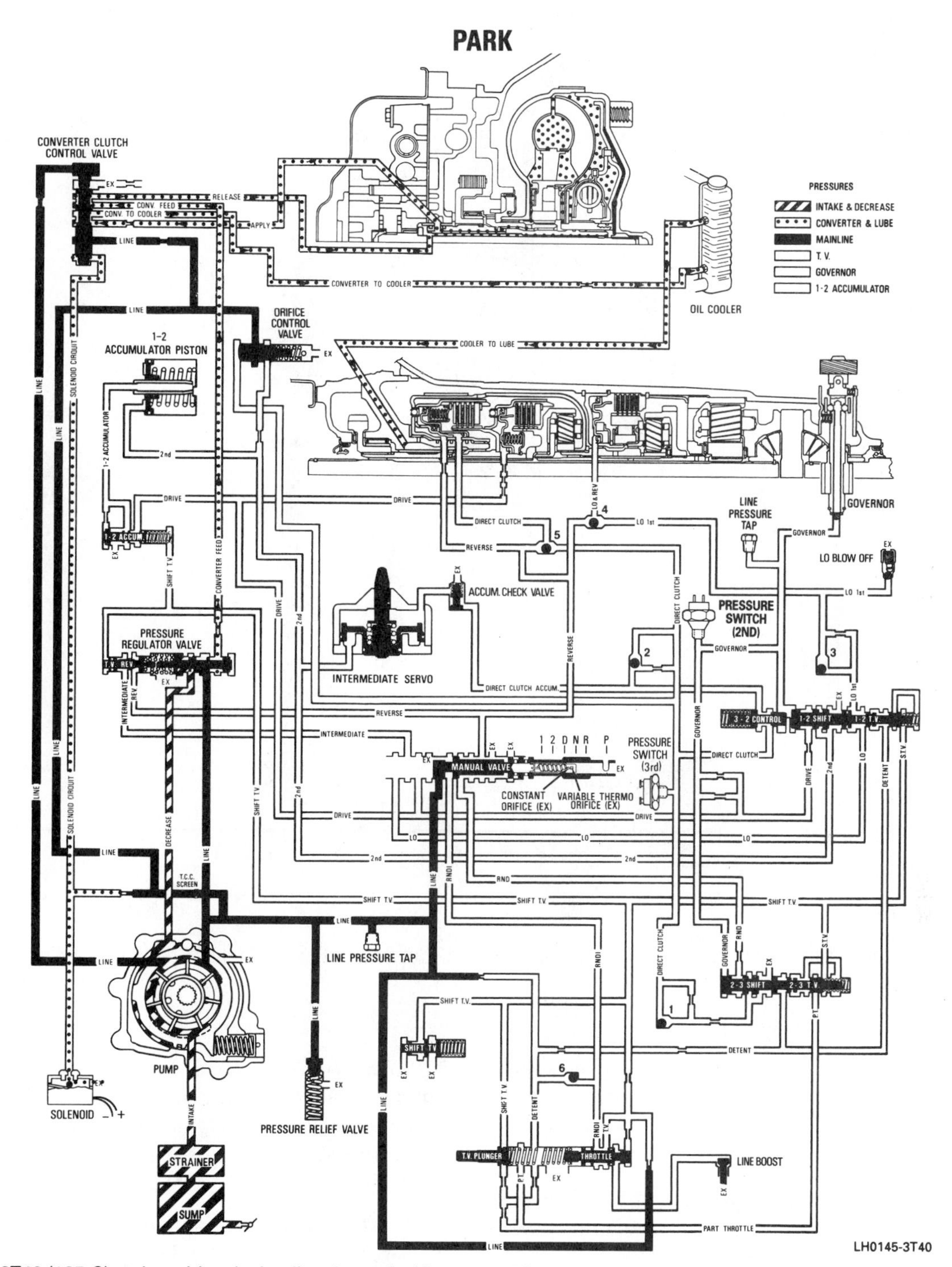

FIGURE 12–76 3T40 (125-C) park position, hydraulic schematic. (Courtesy of General Motors Corporation, Service Technology Group)

Line Pressure Diagnosis: General Motors 4T60 (440-T4) and 4T60-E The 4T60 (440-T4) and 4T60-E line pressures are regulated by the vacuum modulator as shown in Figure 12–77. The manual valve positioning provides pressure boost in park, reverse, neutral, and low. While shift points on a 4T60 (440-T4) are controlled by a TV cable, on a 4T60-E, they are controlled by the PCM and shift solenoids. It should be noted that modulator oil is also routed to the 1-2 and 2-3 accumulator valves to provide accumulator regulation in response to engine manifold vacuum.

Operation of the vacuum gauge/pump is recommended to realize minimum and maximum line pressures for each of the gear ranges. Minimum line pressure is caused by drawing 18 in hg (61 kPa) vacuum to the modulator. Maximum line pressure is achieved by providing 0 in hg (0 kPa) of vacuum.

Line pressure checks to a 4T60-E can be made according to the procedural steps listed in Chart 12–8, page 309. A line pressure check procedure chart for a 4T60 (440-T4) follows a similar format.

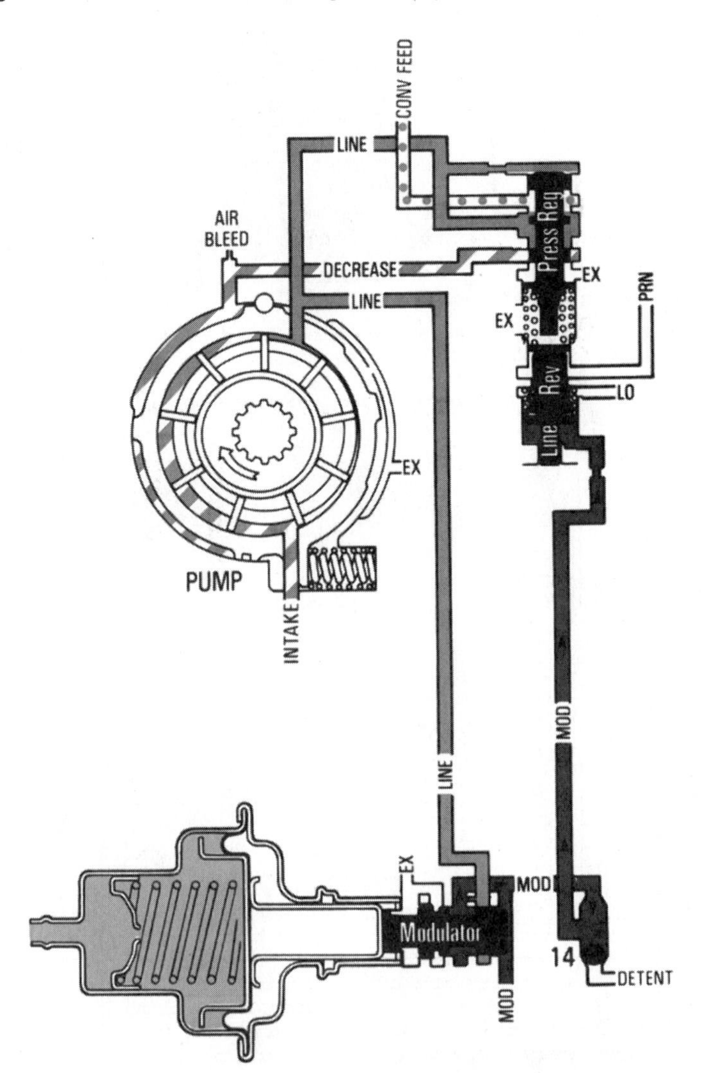

FIGURE 12–77 Modulator pressure directed to line pressure regulator valve. (Courtesy of General Motors Corporation, Service Technology Group)

This information helps isolate possible problem devices when trying to identify the causal factor in a gear range with too little or too much pressure.

Low or High Oil Pressure Related Components and Causes.

1. Fluid level: Too low or too high.
2. Vacuum line: Leaking, pinched, disconnected, or cut.
3. Modulator: Leaking, damaged diaphragm, or bent unit.
4. Modulator valve: Nicked, scored, or stuck.
5. Oil pump assembly: Stuck slide, leaking seals, or damaged vanes or pump drive shaft.
6. Pressure regulator valve and spring: Nicked, scored, stuck, or damaged.
7. Pressure relief valve: Damaged spring or ball missing.

The 4T60 (440-T4) uses a governor assembly and provides a pressure tap to check governor pressure. Governor circuit and shift problems are discussed later in this chapter.

When analyzing hydraulic circuit problems, work closely with hydraulic schematics and diagnostic charts available in shop manuals and training materials. With the numerous devices involved, any help available to be able to make an accurate diagnosis and repair should be considered.

Line Pressure Diagnosis: General Motors 4L60-E and 4L80-E. Unlike previously discussed GM transmissions that use a TV cable or vacuum modulator to adjust line pressure for engine load, the 4L60-E and 4L80-E use a **pressure control solenoid (PCS)**. The solenoid provides a boost oil pressure, referred to as a *torque signal,* to the mainline pressure regulator valve. The reverse oil circuit also provides a boost to mainline when the manual valve is positioned in the R range.

- **Pressure control solenoid (PCS):** a specially designed solenoid containing a spool valve and spring assembly to control fluid mainline pressure. A variable current flow, controlled by the ECM/PCM, varies the internal force of the solenoid on the spool valve and resulting mainline pressure. General Motors for-

HYDRA-MATIC 4T60-E LINE PRESSURE CHECK PROCEDURE

1. Check fluid. See "4T60-E Transaxle Fluid Checking Procedure" in this section.
2. Remove line pressure plug and install J 21867 pressure gage (A).
3. Disconnect vacuum line from transaxle modulator and connect J 23738 vacuum gage/pump (B) to engine vacuum line.
4. Start the engine, set the parking brake and note vacuum gage:
   - At sea level, engine vacuum should be at least 61 kPa (18" Hg.). Engine vacuum will drop about 3.5 kPa (1" Hg.) for every increase in altitude of 305 M (1000 ft.).
   - If engine vacuum is low, refer to Section 6A/6E for diagnosis.
5. Disconnect J 23738 vacuum gage/pump (B) from engine vacuum line. Connect J 23738 vacuum gage/pump to transaxle modulator. Install a plug in engine vacuum line to prevent a vacuum leak.
6. Raise engine speed to 1250 RPM.
7. Check transaxle line pressure:
   - Minimum Line Pressure - provide 61 kPa (18" Hg.) vacuum to modulator with J 23738 gage/pump (A).
   - Full Line Pressure - provide 0 kPa (0" Hg.) vacuum to modulator with J 23738 gage/pump (A).
   - Note pressure readings in all gear ranges for Minimum and Full Line Pressure and compare with information in the chart below.
8. Use the "Condition Diagnosis Charts" in this Section to diagnose line pressure readings that are too low or too high.
9. Reconnect engine vacuum line to transaxle modulator. Remove J 21867 pressure gage (B) with the engine off and install line pressure plug.

(B) J 23738
(A) J 21867

(A) ATTACH PRESSURE GAGE
(B) HAND OPERATED VACUUM PUMP

NOTICE: Total test running time should not be longer than 2 minutes or transaxle damage could occur.

CAUTION: Brakes must be applied at all times to prevent unexpected vehicle motion.

1995 HYDRA-MATIC 4T60-E TRANSAXLE

	RANGE	MODELS	kPa	PSI
MINIMUM LINE PRESSURE 1250 RPM 61 kPa (18" Hg) VACUUM	D4, D3, D2	ACW, ASW, BKW, BXW, CAW, KUW, PMW, YMW, YZW	422-475	61-69
		AJW, ATW, BFW, BLW, PAW, PBW, PCW, WFW, YDW, YNW	512-592	74-86
		AFW	512-596	74-86
	D1	AFW, ATW, BLW, PBW	921-1333	134-193
		ACW, ASW, BKW, BXW, CAW, KUW, PMW, YMW, YZW	998-1276	145-185
		AJW, BFW, PAW, PCW, WFW, YDW, YNW	1005-1289	146-187
	P, R, N	ACW, ASW, BKW, BXW, CAW	422-475	61-69
		KUW, PMW, YMW, YZW	423-536	61-78
		ATW, BLW	460-666	67-97
		AFW, PBW	512-666	74-97
		AJW, BFW, PAW, PCW, WFW, YDW, YNW	542-696	79-101
FULL LINE PRESSURE 1250 RPM 0 kPa (0" Hg) VACUUM	D4, D3, D2	AFW, ATW, BLW, PBW	1148-1400	166-203
		ACW, ASW, BKW, BXW, CAW, KUW, PMW, YMW, YZW	1150-1390	167-202
		AJW, BFW, PAW, PCW, WFW, YDW, YNW	1153-1400	167-203
	D1	AFW, ATW, BLW, PBW	921-1333	134-193
		ACW, ASW, BKW, BXW, CAW, KUW, PMW, YMW, YZW	998-1276	145-185
		AJW, BFW, PAW, PCW, WFW, YDW, YNW	1005-1289	146-187
	P, R, N	AJW, BFW, PAW, PCW, WFW, YDW, YNW	1540-1869	223-271
		ACW, ASW, BKW, BXW, CAW, KUW, PMW, YMW, YZW	1570-1898	228-275
		AFW, ATW, BLW, PBW	1774-2164	257-314

CHART 12–8 4T60-E line pressure check procedure. (Courtesy of General Motors Corporation, Service Technology Group)

merly identified this device as a Force Motor, and Ford as an EPC (electronic pressure control) solenoid.

The PCS is energized by a flow of current controlled by the PCM or TCM. The amount of current through the solenoid's coil determines the amount of oil allowed to pass through the solenoid to act upon the pressure regulator valve (Figure 12–78). As current is decreased to 0 amps, line pressure increases. As current is increased to 1.1 amps at 4.5 volts, line pressure decreases.

Line pressure has a direct relationship to shift quality. If line pressure is too high, harsh shifts are the result. Conversely, if line pressure is too low, shifts are soft and sluggish. If a problem with the pressure control solenoid is detected by the PCM, a code 73 is set. Always check for DTCs (diagnostic trouble codes) and make repairs before beginning line pressure checks.

In these transmissions, a mainline pressure range for drive, park, and neutral is provided for each transmission, as well as a pressure range for reverse (Chart 12–9).

To evaluate the operation of the PCS (pressure control solenoid) and its effect on mainline pressure, a scan tool must be used. As shown in Chart 12–9, with increased current flow to the solenoid with the scanner, line pressure lessens. A scan tool, such as a Tech 1, can perform this test only with the vehicle stopped and in either park or neutral. This prevents the clutches used in drive and reverse ranges from being damaged by being applied with extremely low or high pressures. Normally occurring line pressure readings for reverse and the forward gears can be checked on a diagnostic road test or on a hoist.

"Condition and Cause" charts supplied in shop manuals illustrate the numerous items to inspect when experiencing transmission problems. Also, possible causes of the conditions are listed. Chart 12–10, page 312 shows items that relate to high and low line pressure. In addition to the typical hydraulic devices, the PCM and pressure control solenoid are listed.

Additional Electronically Controlled Transmissions/Transaxles—Line Pressure Diagnosis

Electronically controlled transmissions and transaxles vary as to the amount of PCM control over them. As shown in the General Motors group, line pressure can be controlled either by engine vacuum or by the PCM with a pressure control solenoid. Most of the transmissions and transaxles recently or currently being designed by the manufacturers use electronic pressure control solenoids. Additional coverage of line pressure checks and diagnostic procedures specific to electronically controlled transmissions is presented in Chapter 14.

Pressure Rise Checks: TV or Modulator-controlled Transmissions/Transaxles

Line pressure checks always include a minimum TV line pressure check and full TV line pressure check. If these line pressures are normal, does it mean that the TV system is right? Not necessarily. The technician needs to know what the throttle boost is doing between closed throttle and full throttle. If the throttle boost on the line pressure is lagging behind the gas pedal movement, it is cause for concern. Heavy-duty throttle operation is bound to cause repetitive clutch failures. The throttle boost response to pedal movement between closed and full throttle is important, and the linkage/cable system can be checked as follows:

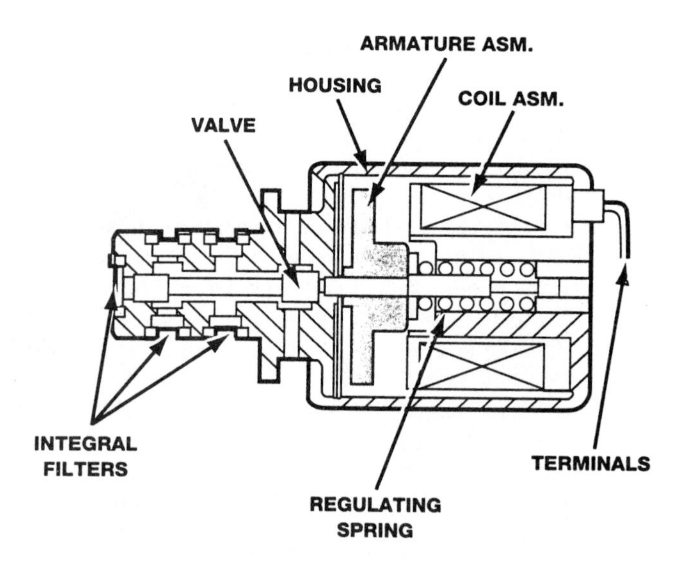

FIGURE 12–78 Pressure control solenoid (PCS), cutaway view. (Courtesy of General Motors Corporation, Service Technology Group)

- With engine at hot idle, put the range selector in D.
- Slowly press on the gas pedal and watch for the pressure rise starting point. It should begin within 1/8 in (3 mm) travel.
- Continue with the pedal movement to 50% throttle and observe that the throttle boost shows a steady progression.
- Continue to push the throttle another 25 to 50%. The line pressure should go full boost. The line pressure should always reach full boost before the pedal reaches full travel.

If full boost is not realized, identify the cause. Sometimes the throttle lever at the end of the throttle shaft is loose and does not respond immediately to pedal travel (Figure 12–79). On vacuum-controlled throttle systems, the point at which the vacuum modulator starts to activate the throttle boost should be observed. Look for a steady pressure rise as the pedal movement draws the vacuum supply down to zero. Ford specifications usually give a midrange check at 10 in hg (34 kPa).

Unfortunately, specifications for modulator rise evaluation are not generally given. As a guideline, pressure rise should usually begin when the vacuum drops 3 in hg (10 kPa). The technician needs to depend, in most cases, on personal experience. If the pressure line boost has not occurred by 11 in hg (37 kPa), it definitely means trouble.

For technician review, the 4T60 (440-T4) has published specifications for modulator system evaluation (Chart 12–11, page 313). Notice the differences in the pressure rise cut-in and rate of pressure rise between the models. Pressure rise is critical and can eliminate complaints in the area of shift quality.

4L80-E AND 4L60-E LINE PRESSURE CHECK PROCEDURE

Line pressures are calibrated for two sets of gear ranges — Drive-Park-Neutral, and Reverse. This allows the transmission line pressure to be appropriate for different pressure needs in different gear ranges:

Gear Range	4L80-E Line Pressure Range	4L60-E Line Pressure Range
Drive, Park, or Neutral	35 - 171 PSI	55 - 189 PSI
Reverse	67 - 324 PSI	64 - 324 PSI

Before performing a line pressure check, verify that the pressure control solenoid is receiving the correct electrical signal from the vehicle computer:

1. Install a scan tool.
2. Start the engine and set parking brake.
3. Check for a stored pressure control solenoid malfunction code and other malfunction codes.
4. Repair vehicle if necessary.

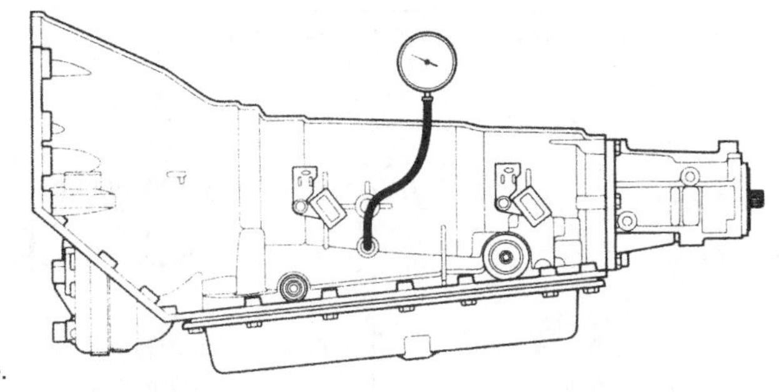

Inspect:

- Fluid level (see Section 7A)
- Manual linkage

Install or connect

- TECH 1 Scan tool
- Oil pressure gage at line pressure tap

5. Put gear selector in Park and set the parking brake.
6. Start the engine and allow it to warm up at idle.
7. Access the "override pressure control solenoid" test on the TECH 1 scan tool.
8. Increase PRESSURE CONTROL SOLENOID CURRENT in 0.1 amp. increments and read the corresponding line pressure on the pressure gage. (Allow pressure to stabilize for 5 seconds after each current change.)
9. Compare data to the Drive-Park-Neutral line pressure chart below.

Line pressure will pulse either high or low every ten seconds to keep the pressure control solenoid plunger free. This is normal and will not harm the transmission.

NOTICE: Total test running time should not exceed 2 minutes, or transmission damage may occur. Increasing the engine speed above idle without vehicle movement (such as holding the brake) in a forward or reverse gear causes transmission stall. Continued operation in the stall condition can result in transmission overheat, malfunction or fluid expulsion.

CAUTION: Brakes must be applied at all times to prevent unexpected vehicle motion.

If pressure readings differ greatly from the line pressure chart, refer to the Diagnosis Charts contained in this section.

The TECH 1 scan tool is only able to control the pressure control solenoid in Park and Neutral with the vehicle stopped at idle. This protects the clutches from extremely high or low pressures in Drive or Reverse ranges.

Pressure Control Solenoid (Amp.)	4L80-E Line Pressure (PSI)	4L60-E Line Pressure (PSI)
0.02	157 - 177	170 - 190
0.10	151 - 176	165 - 185
0.20	140 - 172	165 - 180
0.30	137 - 162	155 - 175
0.40	121 - 147	148 - 168
0.50	102 - 131	140 - 160
0.60	88 - 113	130 - 145
0.70	63 - 93	110 - 130
0.80	43 - 73	90 - 115
0.90	37 - 61	65 - 90
0.98	35 - 55	55 - 65

Pressures at 1500 RPM and 66°C (150°F). Line pressure drops as temperature increases.

CHART 12–9 4L60-E and 4L80-E line pressure check procedure. (Courtesy of General Motors Corporation, Service Technology Group)

CONDITION	INSPECT COMPONENT	FOR CAUSE
HIGH LINE PRESSURE	• Pressure Regulator Valve (231)	– Stuck at high torque signal due to undersized bore or sediment.
	• Reverse Boost Valve (228)	– Stuck at high torque signal due to undersized bore or sediment.
	• Retainer Pin (211)	– Broken.
	• Orificed Plug (210)	– Blocked.
	• Pressure Control Solenoid (320)	– Failed "off" – Loose connector.
	• PCM	– Loose connector.
	• Possible Codes ** – 73 Pressure Control Solenoid Current	
LOW LINE PRESSURE	• Pump (203)	– Cross channel air leak at body to cover, or body to case gasket.
	• Pressure Regulator Valve (231)	– Stuck at low torque signal due to undersized bore or sediment.
	• Reverse Boost Valve (228)	– Stuck at low torque signal due to undersized bore or sediment.
	• Pump Valve Bores	– Excessive valve clearance due to wear.
	• Spring (230)	– Broken.
	• Retainer Pin (211)	– Broken.
	• Valve Body (301)	– Cross channel leaks. – Cross valve land leaks.
	• Gasket/Spacer Plate	– Damaged or missing.
	• Pressure Control Solenoid (320)	– Stuck "on". – Broken clip causing leakage. – Pinched wire to ground. – Screen missing.
	• PCM	– Failed.
	• Possible Codes ** – 73 Pressure Control Solenoid Current	

ALL ILLUSTRATION NUMBERS REFERENCE HYDRA-MATIC 4L80-E UNIT REPAIR SECTION.

RH0020-4L80-E

CHART 12–10 High and low pressure conditions and causes—4L80-E. (Courtesy of General Motors Corporation, Service Technology Group)

FIGURE 12–79 Throttle lever inspection point.

VACUUM @ MOD (IN/Hg)	LINE PRESSURES					
	Models: BN, BS, BU, CP, CW, HT		Models: BA, BC, BX		Models: OB, OY	
	*KPA	*PSI	*KPA	*PSI	*KPA	*PSI
0	1145	166	1180	171	1140	165
2	1020	148	1055	153	1140	165
4	895	130	930	135	1140	165
6	770	112	805	117	1000	145
8	645	94	680	99	865	125
10	520	75	555	80	725	105
12	450	65	485	70	590	86
14	450	65	485	70	485	70
*(+ or - 35 KPA)/(+ or - 5 PSI)						

CHART 12–11 Vacuum modulator system evaluation chart—4T60 (440-T4).

In linkage/cable-operated TV systems, it might be necessary to observe the TV plunger for correct movement even though external adjustment and movement are observed as okay. A typical system observation for Hydra-Matic 200C, 200-4R, 325-4L, 3T40, 4L60, and 4T60 transmissions is shown in Figure 12–80.

Sometimes a high line pressure problem is difficult to isolate, and the technician is trying to make an educated guess about whether to go for the pump or the valve body. The following technique can be used for a positive diagnostic decision when trying to solve an elusive high line pressure problem in a Hydra-Matic 200C, 200-4R, 325-4L, 3T40, or 4L60.

- Remove the TV bracket, TV bushing and plunger, and TV spring. This will allow the TV oil to drain directly to the sump and eliminate the TV oil from the shift valves and pressure regulator valve boost.
- Replace the bracket bolts and oil pan, and retest.

If the shift points drop and pump pressure is back to normal, it is a TV valve body problem. Should the line pressure remain high, it is in the pump (slide type) or pressure regulator valve area.

GOVERNOR CIRCUIT AND SHIFT PROBLEMS

In this section, we add to what has been discussed regarding governor circuit testing—mainly the road test shift schedule and line pressure drop. Governor problems show up as no shifting, late shifting, early shifting, or missing gear starts. Some transmissions provide a governor pressure tap (Figure 12–81), which allows an excellent check on what the governor is doing in the primary and secondary stages of operation. For accurate test results on any governor testing, however, it is important first to check out the line pressure and the line pressure rise.

The governor is the speed sensor and is responsible for shifting the valves. When a gear is missing, however, how does the technician know whether the shift valve is moving? Does it really open? Is the valve stuck? Is there a misassembled clutch pack or stripped shaft splines? Is governor oil failing to open the valve? An incorrect separator plate or gasket, or a combination of the two, can cause a governor circuit problem. If a gear is missing, especially top gear, try the following diagnostic procedure:

- Remove the valve body and air check the clutch circuit. If the air check is okay, proceed as follows.
- Upshift the valve in question manually. Use a spacer in the bore.
- Torque-tighten the valve body to the case and check the results.
- If there is no gear, the probable cause is a misassembled clutch, broken welds, or stripped splines. Should the shift happen, suspect the governor circuit. A stuck shift valve should have been determined when the valve body was removed.

If the transmission consistently has a second-gear start, the concern is usually governor pressure versus stuck valve as the cause and can be determined as follows:

- If a governor pressure tap is available, the governor reading should not exceed 3 psi (20 kPa) with the vehicle stopped.
- If a pressure tap is not provided, remove the governor.
- Should second-gear starts continue with the governor pressure at 3 psi (20 kPa) or less or with the governor removed, the shift valve is stuck. Should the governor have a pressure tap, a stuck shift valve can sometimes be freed by using sharp blasts of air through the pressure port.

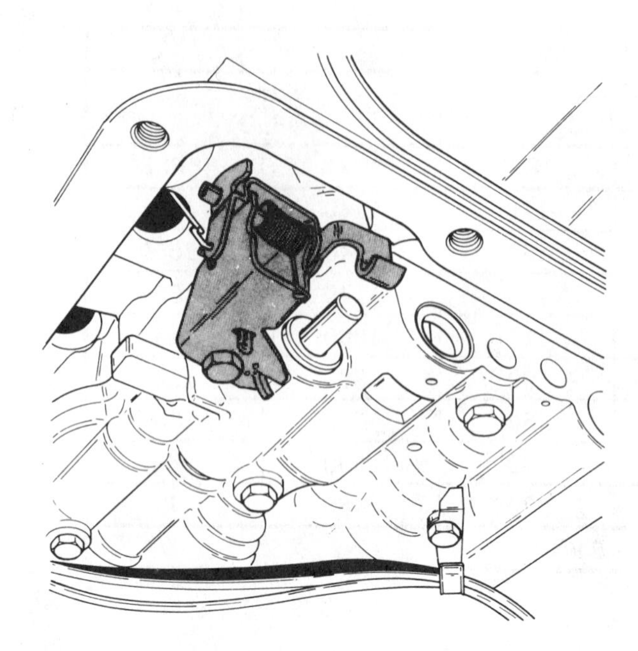

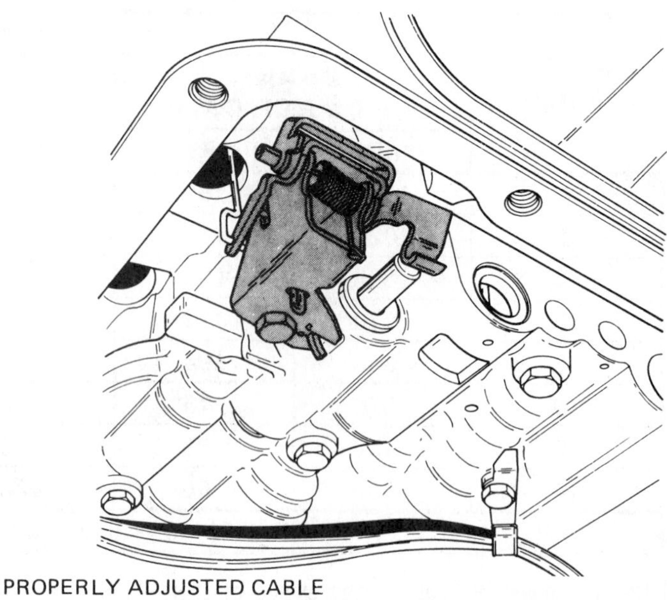

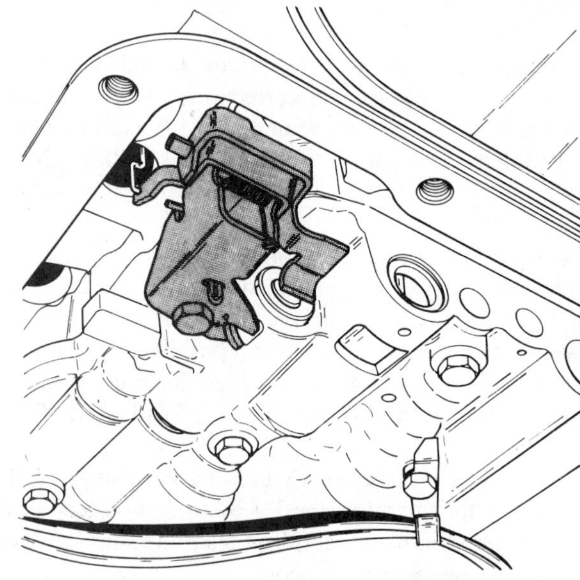

FIGURE 12–80 Throttle linkage positions, incorrect and correct. (Courtesy of General Motors Corporation, Service Technology Group)

To help locate governor circuit problem areas, some additional transmission shift behavior patterns can be observed and used as clues for where to look.

- Normal shift pattern when cold—late hot shifting: indicates a governor oil leak out of the circuit.
- Normal shift pattern when cold—early hot shifting or wrong gear starts: indicates an oil leak into the regulated side of the governor circuit.

Do not be fooled by the appearance of a governor. A governor may look good and pass your inspection but not produce desired results when put into action. Substitute a known good governor and check the results, especially when the top gear is missing.

The check ball-style governor lends itself to a self-check on a no-fourth-gear problem. The governor feed hole is blocked below the shaft sealing ring and the transmission retested (Figure 12–82). Governor circuit oil at the shift valves is now equal to line pressure. If the governor is causing the problem, there should be a fourth-gear start in OD range. To verify a fourth-gear start, run the transmission with the drive wheels off the ground at approximately 40 mph (64 km/h). Drop the selector into manual 3 and feel for the braking. This same procedure can be adapted for a three-speed unit.

GOVERNOR PRESSURE
REPRESENTATIVE 4T60 SCHEDULE

MPH (KM/H)	kPa	PSI
0 (0)	17-31	2.5-4.5
10 (16)	44-59	6.5-8.5
20 (32)	96-110	14-16
30 (48)	200-214	29-31
40 (64)	303-317	44-46
50 (80)	400-434	58-63

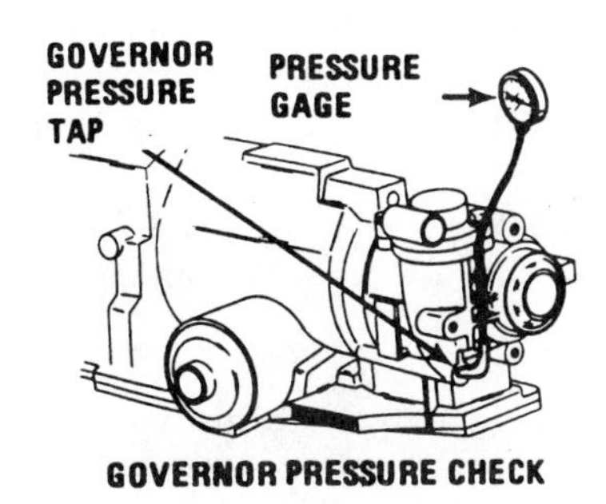

FIGURE 12–81 4T60 (440-T4) governor pressure check. (Courtesy of General Motors Corporation, Service Technology Group)

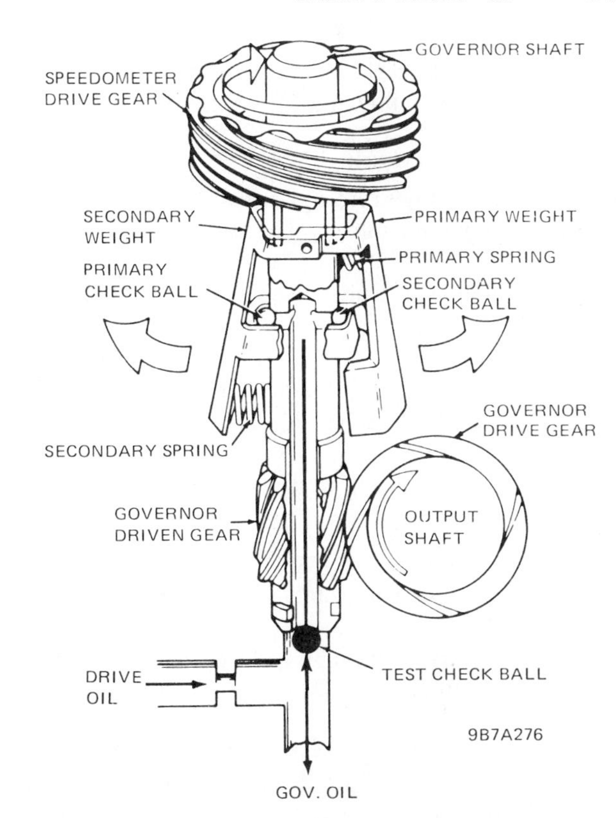

FIGURE 12–82 Test check ball placed beneath the governor. (Courtesy of General Motors Corporation, Service Technology Group)

❑ SECTION 5 REVIEW

Key Terms/Acronyms

Pressure control solenoid (PCS).

Questions

Checking the Modulator Vacuum Supply and System Reliability for Shift Timing and Pressure Control

1. When checking for the effective vacuum on the modulator, the technician should use a vacuum gauge at the engine manifold. ___T/___F
2. Evidence of fuel at the modulator diaphragm means that the modulator should be replaced. ___T/___F
3. A leaky vacuum accessory circuit can affect the modulator vacuum signal. ___T/___F
4. A modulator application using too large a diaphragm diameter causes late and harsh shifting. ___T/___F

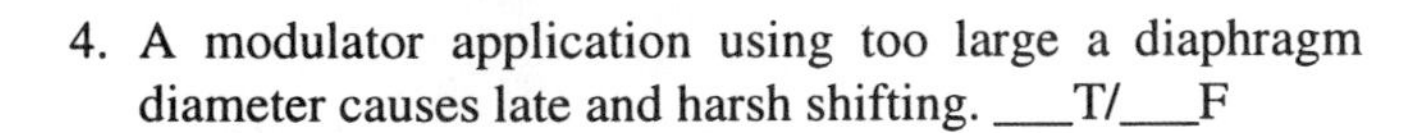

5. If the modulator spring weight or tension is too weak, the transmission shifts early and slips. ___T/___F
6. An iced modulator typically causes a maximum delayed shift pattern. ___T/___F
7. An unwanted restriction in the vacuum line typically causes a delayed shift pattern. ___T/___F
8. A disconnected vacuum line at the modulator causes maximum delayed shifts. ___T/___F
9. An unwanted restriction in the vacuum line causes transmission slippage with medium and heavy throttle performance. ___T/___F
10. Modulator operation has a direct bearing on the transmission shift schedule and quality. ___T/___F
11. Modulator operation has a direct bearing on the transmission operating pressure. ___T/___F

Pressure Testing

1. The pressure test sequence starts with an isolated pump circuit test. ___T/___F
2. If the isolated pump pressure is low, all the other operating pressures will be low. ___T/___F

3. If line pressure checks meet specifications at the minimum and maximum limits, the TV system is considered right. ___T/___F

4. Two or more high readings or two or more low readings share a common fault. ___T/___F

5. High mainline pressure can be caused by a plugged filter. ___T/___F

6. A disconnected vacuum modulator produces low mainline pressures and soft (sluggish) shifts. ___T/___F

_____ 7. If the isolated pump pressure is low, it could be caused by:

A. worn pump gears
B. loose valve body to case bolts
C. missing filter gasket or O-ring
D. defective pump body to case gasket
E. all of the above

Pressure Testing—General Motors

_____ 1. Which method is used to produce line pressure boost?

A. TV cable
B. Vacuum modulator
C. Pressure control solenoid
D. A and B
E. All of the above

2. If the TV cable becomes disconnected on a 3T40 (125-C), low line pressures are the result. ___T/___F

3. (3T40/125-C) The manual valve allows a direct boost to the pressure regulator valve in reverse, manual 2, and manual low. ___T/___F

4. (3T40/125-C) High or low pressures in N but normal pressures in park may indicated a problem with the TV circuit. ___T/___F

_____ 5. (3T40/125-C) Technician I says the TV cable controls mainline pressure. Technician II says the TV cable controls shift timing. Who is right?

A. I only C. both I and II
B. II only D. neither I nor II

_____ 6. (4T60/440-T4) Technician I says the TV cable controls mainline pressure. Technician II says the TV cable controls shift timing. Who is right?

A. I only C. both I and II
B. II only D. neither I nor II

_____ 7. (4T60/440-T4) To achieve maximum line pressure, Technician I says draw 18 in hg (61 kPa) of vacuum to the modulator. Technician II says the TV cable must be pulled to its maximum. Who is right?

A. I only C. both I and II
B. II only D. neither I nor II

_____ 8. (4T60-E) Technician I says the vacuum modulator controls mainline pressure. Technician II says the TV cable controls shift timing. Who is right?

A. I only C. both I and II
B. II only D. neither I nor II

_____ 9. (4L60-E, 4L80-E) Technician I says to increase line pressure, more current is fed to the pressure control solenoid. Technician II says the pressure control solenoid sends a "torque signal" oil to the pressure regulator valve. Who is right?

A. I only C. both I and II
B. II only D. neither I nor II

_____ 10. (4L60-E, 4L80-E) Technician I says check for computer codes prior to line pressure tests. Technician II says a scan tool can be used to control reverse oil pressure. Who is right?

A. I only C. both I and II
B. II only D. neither I nor II

_____ 11. (4L60-E, 4L80-E) Regarding line pressure boost in reverse, Technician I says a boost is caused by a manual valve circuit. Technician II says the pressure control solenoid provides boost. Who is right?

A. I only C. both I and II
B. II only D. neither I nor II

Section 6 Review of Diagnostic Procedures

CHALLENGE YOUR KNOWLEDGE

Define the following key terms/acronyms:

Throttle positions: minimum throttle, light throttle, medium throttle, heavy throttle, wide open throttle (WOT), full throttle detent downshift, zero-throttle coast down, and engine braking; shift conditions: bump, chuggle, delayed (late or extended), double bump (double feel), early, end bump (end feel or slip bump), firm, flare (slipping), garage shift, harsh (rough), hunting (busyness), initial feel, late, mushy, shudder, slipping, soft, surge, and tie-up; thermostatic element, malfunction indicator lamp (MIL); diagnostic trouble codes (DTCs); stall test; garage shift feel; shift busyness; deceleration bump; O_2 sensor, PROM (programmable read-only memory); brinneling; spalling; and pressure control solenoid (PCS).

List the following:

Seven failure conditions; and the seven-step diagnostic procedure.

Summarize the following procedures:

Each of the seven steps of the diagnosis: preliminary checks (steps 1–4), stall test, road test, and pressure test; manual linkage adjustment; TV cable/linkage adjustment; ECM/PCM diagnostic trouble code (DTC) identification; and mainline pressure test for a transmission controlled by TV cable/linkage, vacuum modulator, and pressure control solenoid (PCS).

Interpret and demonstrate application of the following:

Stall test results; shift-speed chart; road test summary report; clutch/band apply chart; and fluid pressure test chart.

Describe the techniques needed to analyze the following:

Torque converter clutch operation; mechanical failures; noise concerns; erratic shift problems; governor operation; and vacuum modulator concerns.

Describe the procedures needed to examine/test the following:

Fluid level; fluid condition; garage shift feel; clutches and bands; and torque converter clutch.

Identify the warning associated with:

Performing a stall test.

SUMMARY OF DIAGNOSTIC PROCEDURES

Efficient diagnosis of automatic transmission concerns is accomplished by following a systematic process. The process is essential and begins with identifying the customer's concern. If a problem exists, check the fluid level and condition. If those are good, continue with the diagnostic steps. During the diagnosis, if a problem is located, correct it and then check to see if the situation is resolved.

From the driver's seat, check the manual shift linkage for correct operation. If it is okay, make visual checks under the hood. Inspect wiring and vacuum connections, and check accelerator linkage for WOT.

Next, check for diagnostic trouble codes from the PCM/TCM. Pursue any that are present. Many transmission malfunctions are actually engine issues. Trouble codes provide a means of identifying failures.

Many manufacturers suggest performing a stall test. This procedure can assist in determining if the engine, torque converter, and certain clutches and bands are working properly.

When performing a complete road test, record your findings. Clues gathered from a road test should lead you in the right direction.

Prior to making an internal repair that requires transmission removal, determine fluid pressures. Procedures and specifications vary, but with hydraulically controlled transmissions, minimum and maximum readings are typically required.

After gathering the right information, a technician can make a wise decision concerning the repair to the vehicle that is needed to solve the problem. Logical and complete diagnosis leads to quality repairs, greater profits, and improved customer satisfaction.

13 Torque Converter Clutch Diagnosis

OBJECTIVE:

The objective of this unit is for you to become proficient with the terms and concepts contained in this chapter. Completion of this objective is essential to becoming a successful automatic transmission technician.

CHALLENGE YOUR KNOWLEDGE

Define the following key terms/acronym:

Open circuit, short circuit, high resistance, and high-impedance DVOM (digital volt/ohmmeter).

List the following:

Three wiring circuit conditions that cause TCC malfunctions; TCC circuit locations where a 12-volt test lamp can be used; two oil pressure switch designs; applications of an ohmmeter to check TCC circuitry; four circuit locations where an "open" condition may exist; and four causes for high resistance.

Describe the following:

Usage of a scanner to evaluate TCC electrical input sources and TCC operation; the rationale of using a high-impedance multimeter for checking electronic circuits; and the relationship between cooler line pressure fluctuations and TCC operation.

Identify the cautions associated with the following:

Jumper wire usage.

INTRODUCTION

Torque converter clutch hydraulic and electronic circuit control basics were presented in Chapter 10. Chapter 12 discussed converter clutch problem conditions. The objective here is to highlight some diagnostic concepts that can help the technician locate the problem. For detailed diagnostics and specifics of a particular system, consult the manufacturer's applicable wiring and hydraulic diagrams and diagnostic guide. Remember, the engine electronic control module *must* be cleared of any diagnostic codes. It is possible for a car to be driven with the malfunction indicator lamp (MIL)/check engine light ON. The computer-controlled engine management system simply covers up the problem, and the engine continues to run at what might appear normal to the customer. This nonideal condition, however, can affect transmission and TCC operation.

The technician must activate the code display. This procedure varies between the manufacturers, so we will use one particular system for discussion. On OBD I General Motors vehicles, a data link connector (DLC/ALDL) is located under the instrument panel (Figure 13–1). Connect terminal B to ground terminal A with a paper clip or metal diagnostic key and

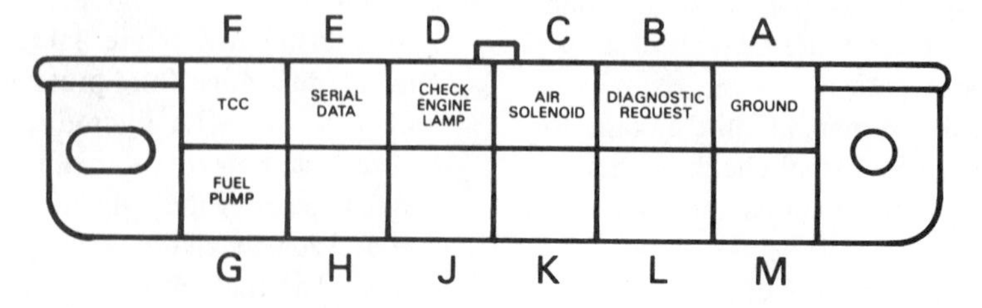

FIGURE 13–1 GM data link connector (DLC/ALDL) - OBD I. (Courtesy of General Motors Corporation, Service Technology Group)

observe that the MIL/check engine light flashes (Figure 13–2). The code will indicate a fault in a given circuit but will not identify the exact fault. Code 12 identifies that all circuits are clear of fault. Code 14 indicates a fault in the coolant sensor circuit.

The use of a scan diagnostic instrument is an effective and quick method of checking sensors and switches that are inputs to the ECM/PCM (Figure 13–3). Two simple connections are all that are required to gain access to the troubleshooting information. A power cord plugs into the cigarette lighter and the other cord into the DLC/ALDL.

The scan tool displays any trouble codes stored in the engine electronic module memory. It can also display what the control module is actually seeing as a signal from the sensor circuits. It displays engine coolant temperature (ECT), intake air temperature (IAT), engine rpm, vehicle speed in mph or km/h, throttle position opening, and so on. The sensor readings that are available vary, based on vehicle model and year, engine, transmission, and other criteria. Scanning conveniently allows the technician to check ECM/PCM system sensors while in the vehicle (Figure 13–4) and during a road test.

The TCC system can even be identified as working electrically. To determine if the clutch engaged, watch the engine rpm on the scan tool. The engine rpm should drop 200 to 300 rpm. In many vehicles, such as those using a 4L80-E, the amount of torque converter slippage as well TCC duty cycle can be shown on screen. The diagnostic system also features a checkout on the transmission switches and overdrive parameters that dictate overdrive engagement. A typical scan tool readout concept is illustrated in Figure 13–5.

There are several good scan testers on the market. Complete versatility of each scan tool depends on the technician's ability to understand what the electronic engine and transmission management system is all about.

Figure 13–6 shows the basic sensors that have an effect on TCC operation.

ELECTRICAL/ELECTRONIC CIRCUIT TROUBLESHOOTING

Before any attempt is made to troubleshoot an electrical/electronic circuit, the technician must be familiar with the circuitry: What is it supposed to do? What components and connectors make up the circuit? How is it routed in the vehicle? How does it work? The technician needs to know how electricity behaves in series and parallel circuits, and the relationships between voltage, resistance (ohms), and current flow (amperage). These basics are covered in Chapter 9, with an introduction to computer-controlled circuits.

The days of fumbling your way through an electrical/electronic circuit diagnosis are gone—no more experimental hot-wire spark tests. Even the voltage power must be correct. Electronic circuits work on very low voltage, usually 0 to 5 V or 0 to 8 V, and carry very low amperage in the milliampere range. "Zapping" an electronic circuit with 12 volts is sudden death for that circuit.

Circuit Troubles

When approaching an electrical problem for diagnosis, it is necessary to define the type of problem. There are three conditions that cause an inoperative circuit:

- Open circuit
- Short circuit
- High resistance

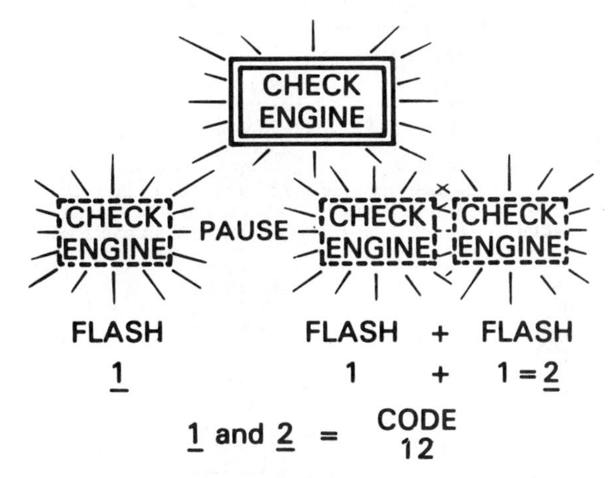

FIGURE 13–2 Check engine light flashes to display codes. (Courtesy of General Motors Corporation, Service Technology Group)

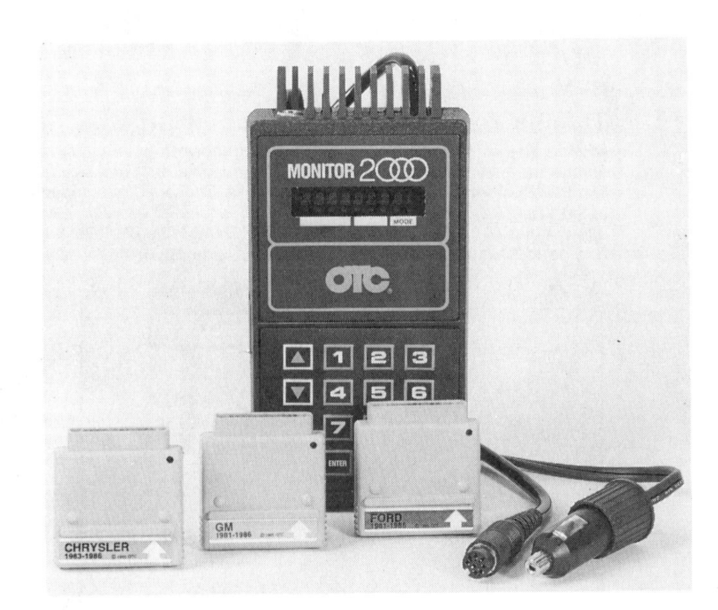

FIGURE 13–3 Scan diagnostic instrument.

FIGURE 13–4 Scan tool connected to vehicle.

Open Circuit. When there is a complete break or interruption of the circuit, it is identified as an **open circuit**. The current flow does not have a complete path from and back to the power source (battery). The following are examples of open circuits:

- Blown fuses
- Broken wires, especially at the connectors or in the connector plugs
- Internally open components, such as switches and bulbs
- Extremely high resistances, which often cause the same symptoms as an open circuit. Common examples are loose or corroded connections.

❖ **Open circuit:** A break or lack of contact in an electrical circuit, either intentional (switch) or unintentional (bad connection or broken wire).

MODE 9

RPM (Left Display) | 0-6,000 |

Throttle Position Sensor (Right Display) | Volts, 0-5 |

RPM: The RPM and throttle position are shown together. This will enable the user to monitor throttle position at different RPM's.
Throttle Position Sensor: The Throttle Position Sensor (TPS) is mounted in the carburetor body or on the air horn of EFI equipped vehicles. It emits a low voltage (usually less than 1 volt) when the throttle blade is closed and up to 5 volts when the throttle blade is wide open.

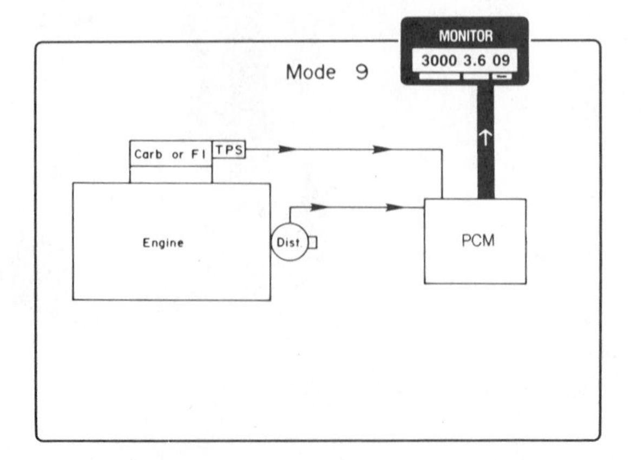

MODE 12

Overdrive Disable (Left Display) | On/Off |

Overdrive Request (Right Display) | On/Off |

Overdrive Disable: When the Throttle Position Sensor is above a specified limit, the ECM will disengage the overdrive portion. When the overdrive is engaged, the tester will show an ON display. When the overdrive is disengaged, the tester will show an OFF display.
Overdrive Request: The display indicates if TPS, coolant, gear selector, and other inputs have been received and evaluated by the ECM. ON indicates that all inputs are in the range where the overdrive can operate. OFF indicates that one or more of the inputs are not within the operating range to engage the overdrive.

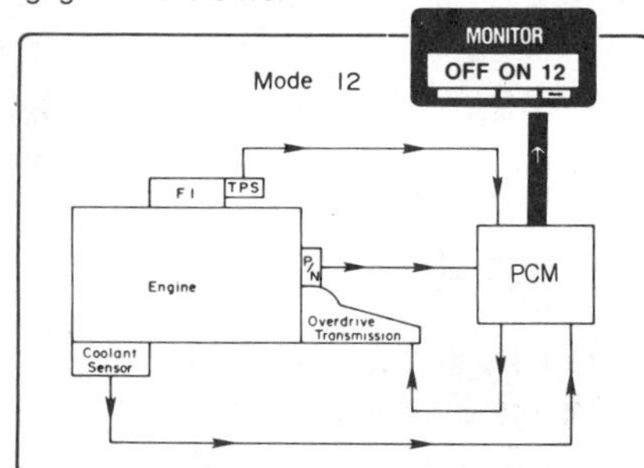

FIGURE 13–5 Monitor scan tool readout examples. (Courtesy of OTC Division, SPX Corp.)

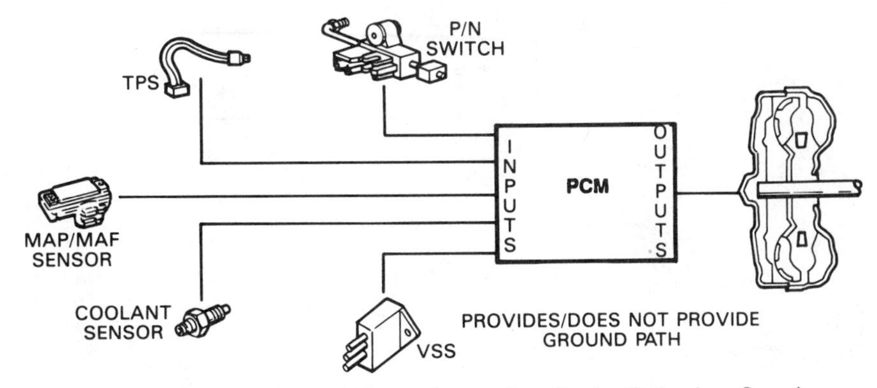

FIGURE 13–6 Basic TCC input sensors. (Courtesy of General Motors Corporation, Service Technology Group)

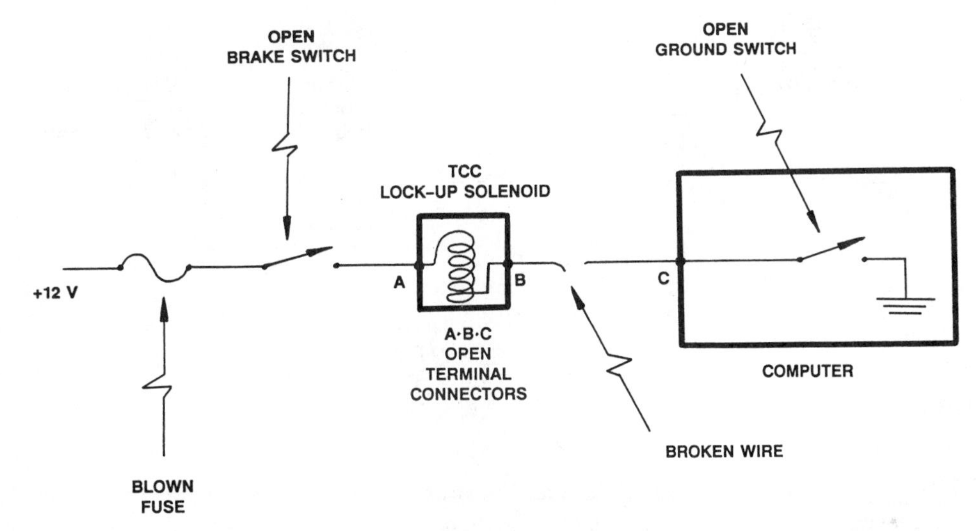

FIGURE 13–7 Open circuit possibilities; simplified TCC electrical circuit.

Possible points of open circuit problems are shown in a simplified TCC electrical diagram (Figure 13–7).

Short Circuit. A circuit that is accidentally completed in an electrical path for which it was not intended is referred to as a **short circuit**. A short circuit can also be defined as an accidental copper-to-ground or copper-to-copper contact. All short circuits do not cause blown fuses or generate smoke signals—it depends on where the circuit is shorted. It is important to relate the symptoms of the short circuit to its probable location. Figure 13–8 illustrates three shorted conductor-to-ground locations and shows the effect on the circuit.

- A short between the power source and load blows the fuse. If the circuit is not fused, the conductor wire overheats, smokes the insulation, and may even melt in two.
- If the short is between the load and circuit ground side switch, the load is constantly ON.
- Should the short be located between all the working components and ground, there is no effect on the circuit and, in all probability, a "no problem" condition.

❖ **Short circuit:** A circuit that is accidentally completed in an electrical path for which it was not intended.

Some grounded short-circuit effects in a TCC electrical circuit are illustrated in a simplified diagram in Figure 13–9. A conductor-to-conductor short-circuit effect is illustrated in Figure 13–10. A parallel path is formed between the shorted circuits caused by the unintentional touching of the wires.

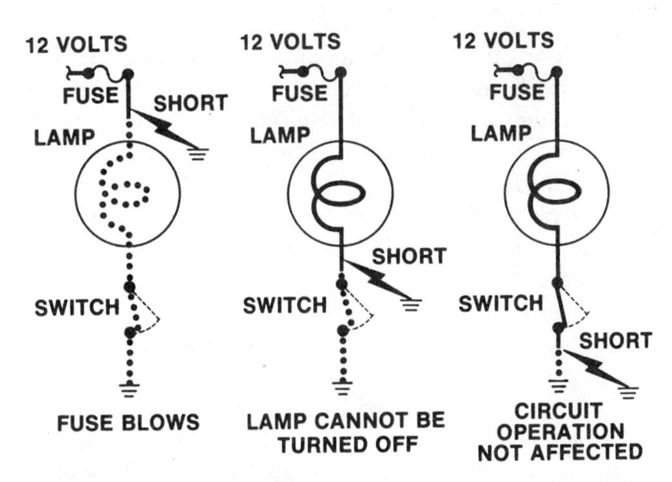

FIGURE 13–8 Short-circuit effects with switch located between load and ground. (Reprinted with the permission of Ford Motor Company)

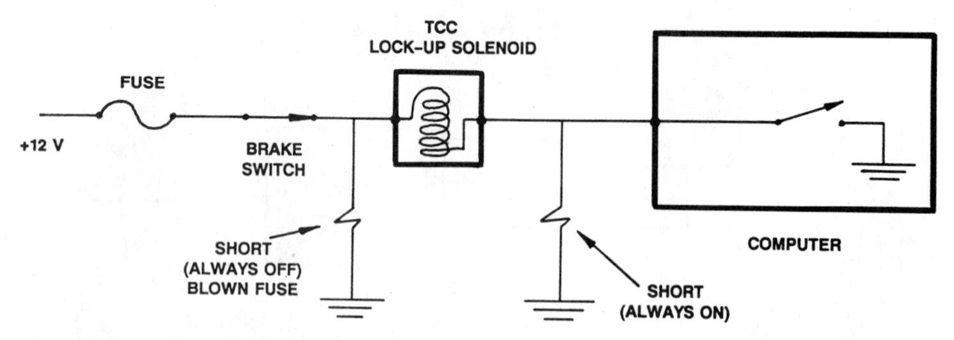

FIGURE 13–9 Short-circuit effects; simplified TCC electrical circuit.

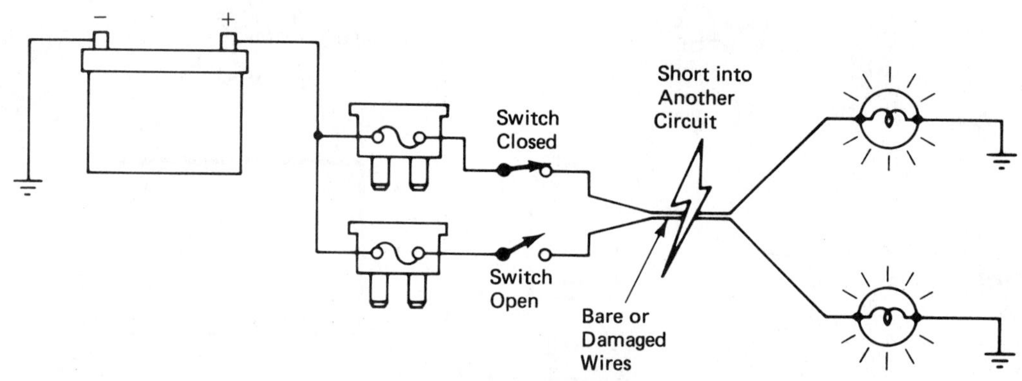

FIGURE 13–10 Conductor-to-conductor short circuit.

High Resistance. There is always a normal amount of design resistance in a circuit. A **high resistance**, in terms of a problem, is a resistance in any part of a circuit higher than normal. High resistances are caused by corrosion, connector looseness, improper wire end termination, or inadequate contact area at terminals, connectors, and grounds. High resistances even occur within the circuit components.

❖ **High resistance:** Often refers to a circuit where there is an excessive amount of opposition to normal current flow.

A high resistance creates an unwanted or extra load in the circuit, which results in a voltage or power loss needed to operate the circuit load. Lights will be dim, motors will run slower, and so on. Extreme high resistance acts like an open circuit and actually prevents the circuit load from operating. Loose or corroded grounds are frequent problem areas and are especially disruptive to circuits, since they are designed for zero resistance.

Resistance can be expressed in terms of "voltage drop" or voltage lost. In any electrical circuit, all the applied voltage is used by the time it goes from the battery positive post, all the way through the circuit, and back to the negative post. If the voltage at the battery is 12.5 V, all 12.5 V is used somewhere in the circuit (Figure 13–11). It would be ideal to have all the applied battery voltage available for use by the circuit load (Figure 13–12), but this ideal is never reached because there is always normal built-in circuit resistance that uses some of the voltage before it actually gets to the load. This normal resistance occurs at wire connections and switch contacts and is a result of the internal resistance of the circuit wire (Figure 13–13). What voltage is left over is used by the circuit load. In all instances, voltage drop is the result of resistance or load.

When an electrical circuit exceeds its normal resistance, more than normal voltage is lost on its way to the working component. Without adequate applied voltage, the working component malfunctions. Shown in Figure 13–14, a poor con-

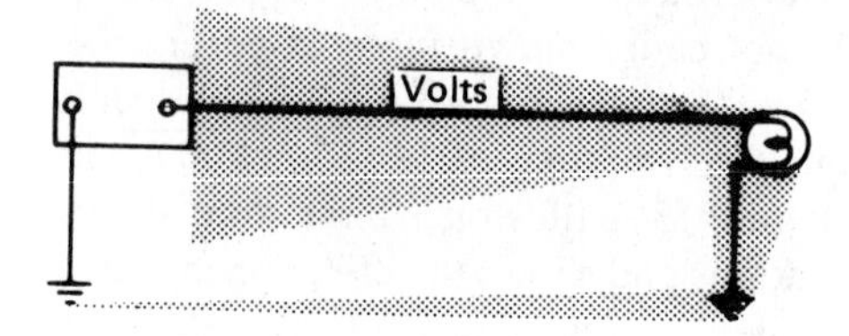

FIGURE 13–11 All of supplied voltage is used somewhere in the circuit.

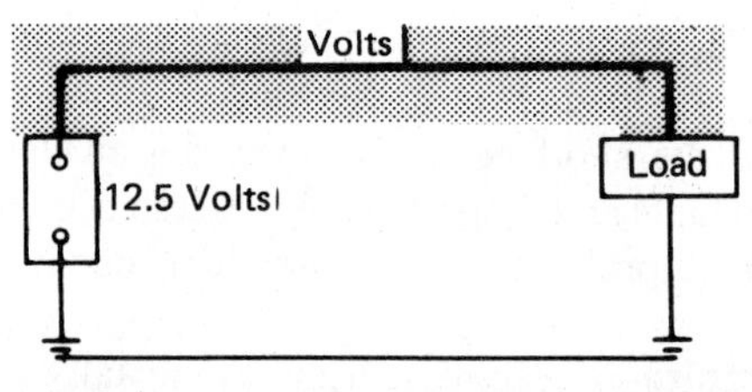

FIGURE 13–12 Ideally, all voltage would be available to circuit load.

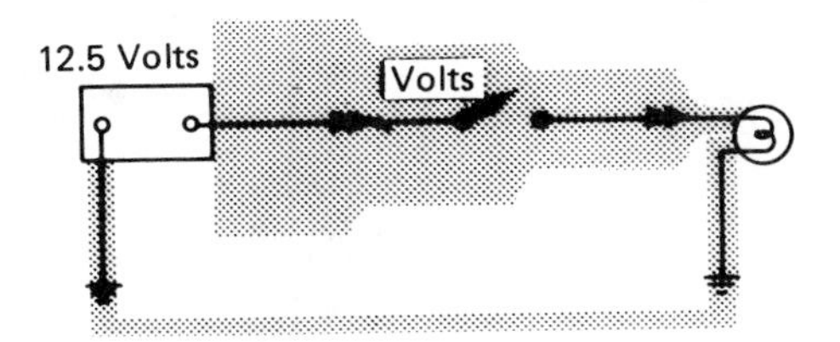

FIGURE 13–13 Normal circuit resistance uses some voltage.

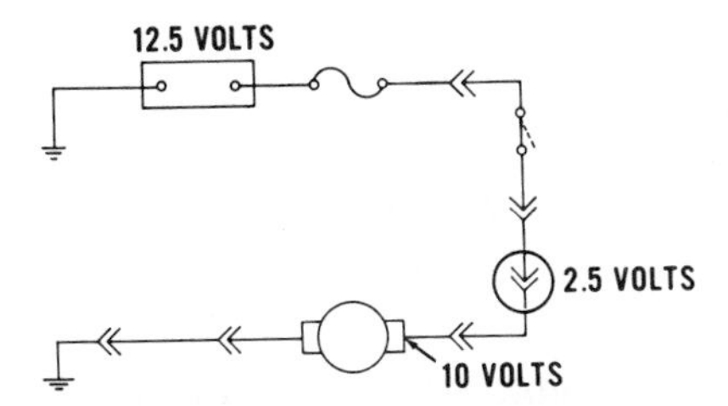

FIGURE 13–14 Poor connection drops 2.5 V. (Reprinted with the permission of Ford Motor Company)

nection drops or uses 2.5 V of the battery power before it gets to the motor. This means that the applied voltage to the motor is 10 V. In this situation, the motor lugs and operates at a speed slower than normal.

The best way to test for voltage drop is to connect a voltmeter in parallel with the circuit to the operating component. On the power feed side of the circuit, connect the positive lead of the voltmeter to the positive post of the battery and the negative lead into the circuit wire at the selected test point. The voltmeter in Figure 13–15 is connected to show the total amount of voltage loss on the power feed side of the circuit.

For the ground side, connect the negative lead of the voltmeter to the negative post of the battery and the positive lead into the circuit test point. The voltmeter in Figure 13–16 is connected to show the total amount of voltage loss on the ground side of the circuit. In essence, you are always reading the voltage drop between the leads of the voltmeter. For lighting and accessory circuits, the combined voltage drop for the ground and hot sides of the circuit load should not exceed 1.3 V. As a standard rule, the ground side of the circuit should not exceed 0.05 V, and preferably should be zero. The most accurate and fastest method to use for testing ground resistance is the voltage drop test. Use an analog or digital voltmeter for the following procedures.

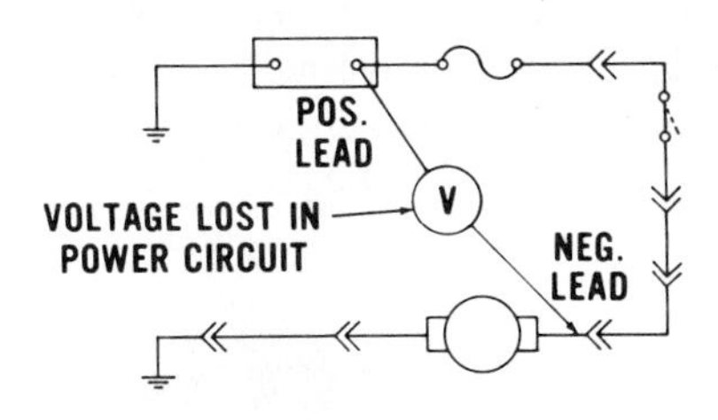

FIGURE 13–15 Voltmeter shows total loss on power feed side of circuit. (Reprinted with the permission of Ford Motor Company)

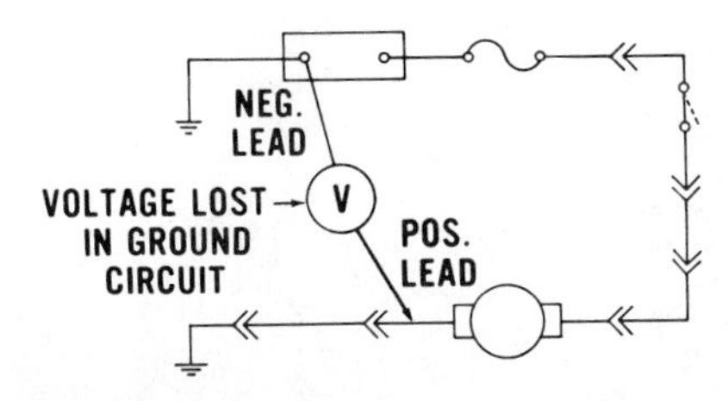

FIGURE 13–16 Voltmeter shows total loss on ground side of circuit. (Reprinted with the permission of Ford Motor Company)

Resistance Test at Battery Posts and Cable Connectors.

1. Turn on the headlights and heater-A/C blower motor to provide a battery load.
2. Connect or touch the positive meter lead to the positive battery post and the negative meter lead to the cable connector (Figure 13–17).
3. Read the voltage drop on the meter. Be sure to use the low-voltage scale.
4. Repeat the same procedure for the negative post and connector. Connect the negative meter lead to the negative battery post and the positive meter lead to the cable connector.

The voltage drops should preferably read zero, but not more than 0.05 V. If resistance is too high, remove the cables and clean the connectors and posts. Reattach and coat the connections with white lube or petroleum jelly.

Battery-to-Sheet Metal Resistance Test.

1. Turn on the headlights and heater-A/C blower motor to provide a battery load.
2. Connect the voltmeter negative lead to the battery negative post.
3. Take the voltmeter positive lead and firmly probe the sheet metal in several different places (Figure 13–18).
4. Read the voltage drop on the meter.

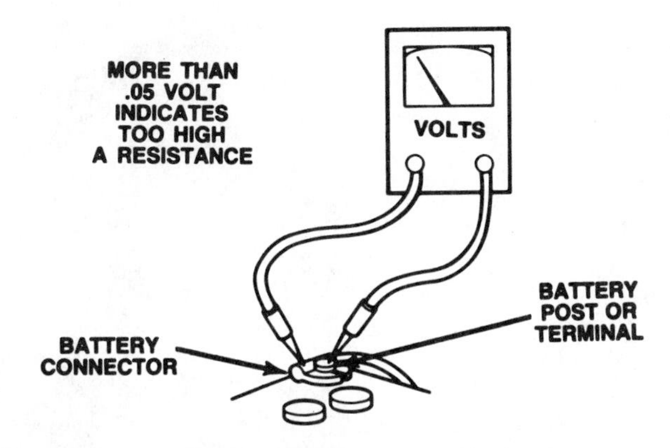

FIGURE 13–17 Checking for high resistance at the battery post. (Reprinted with the permission of Ford Motor Company)

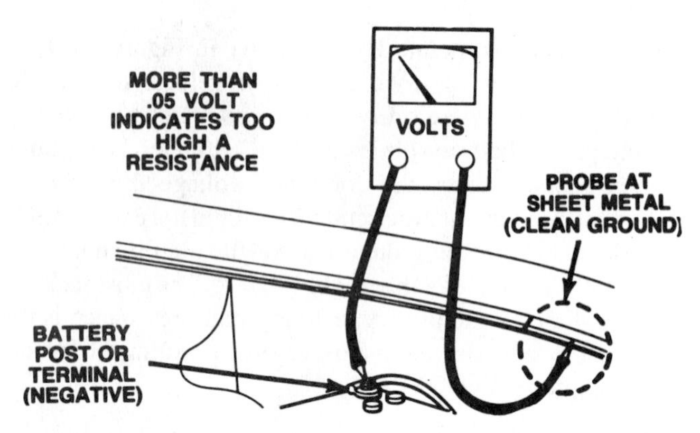

FIGURE 13–18 Checking for high resistance between negative battery post and body sheet metal. (Reprinted with the permission of Ford Motor Company)

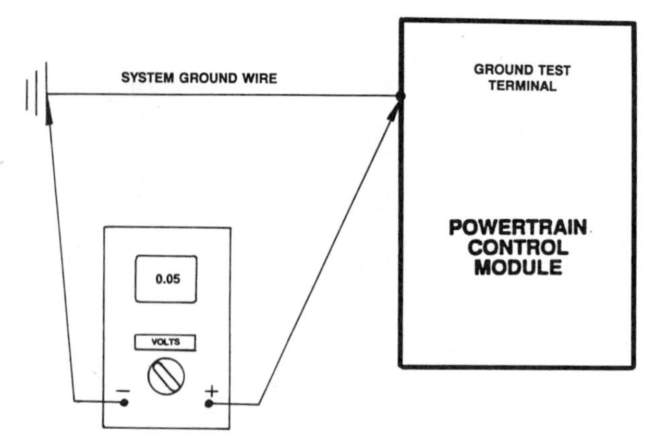

FIGURE 13–19 Test resistance at system ground wire.

The voltage drop should preferably read zero, but not more than 0.05 V. If the resistance is too high, check for resistance at the negative cable and ground strap connections. It is not unusual to find ground straps missing. If a voltmeter is not available, a 12-gauge jumper wire can be used for the test to verify the ground condition.

Computer-to-Engine Block Resistance Test. One of the most important wires on an engine computer control module is the system ground wire (Figure 13–19).

1. Turn on the ignition switch.
2. Connect the voltmeter positive lead to the module ground test terminal.
3. Take the voltmeter negative lead and firmly probe the engine block.
4. Read the voltage drop at the meter.

The voltage drop should preferably read zero, but no more than 0.05 V. As an alternative and quicker method, the ground resistance can sometimes be measured by removing the TP (throttle position) sensor connector at the throttle body and connecting the voltmeter leads to the TP sensor ground wire and engine block.

Oil Pressure Switches

It is not uncommon for TCC electronic circuitry to incorporate oil pressure switches for the necessary lockup control to coordinate engine-transmission performance (Figures 13–20 and 13–21). Pressure switches use hydraulic pressure acting on a diaphragm to change contact point position (Figure 13–22). The switch contacts are arranged to act as grounding contacts or insulated series contacts. The switches are also designed with either one or two terminals (Figure 13–23). A single terminal switch provides a circuit ground, while a two-terminal switch acts as a series connector in the circuit (Figure 13–24). The grounding switches are nothing more than signal switches that tell the computer the state of transmission operation. Switches are designed to be either normally open (NO) or normally closed (NC).

NO: The switch will not complete the electrical contact until acted upon by hydraulic pressure.

NC: The switch will not break the electrical contact until acted upon by hydraulic pressure.

FIGURE 13–20 Oil pressure switches—GM 4T60 (440-T4).

FIGURE 13–21 Oil pressure switches—Ford AXOD.

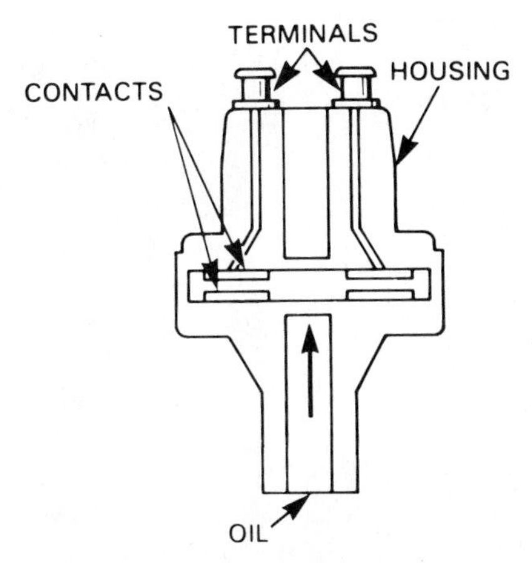

FIGURE 13–22 Hydraulic pressure closes contact points.

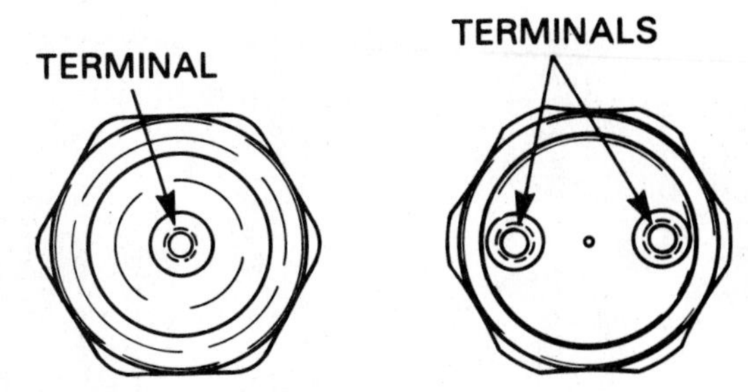

FIGURE 13–23 Switches use one or two terminals.

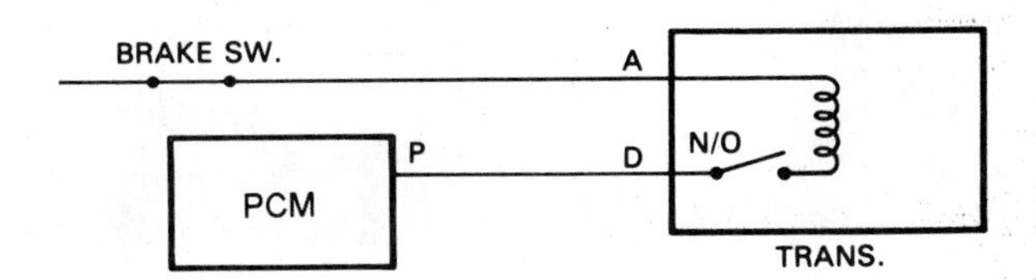

A PCM CONTROL W/SERIES SWITCH

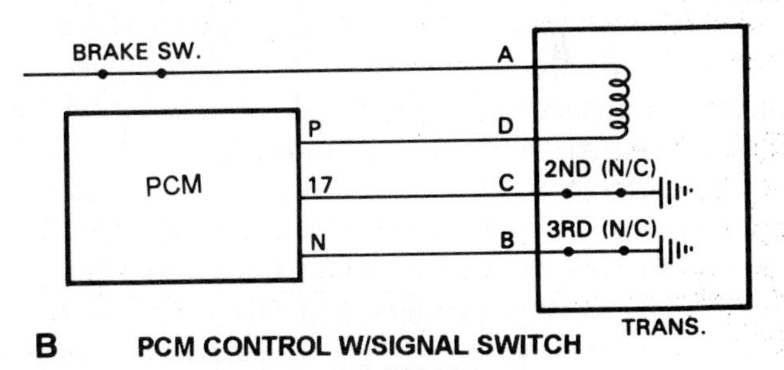

B PCM CONTROL W/SIGNAL SWITCH

FIGURE 13–24 (a) Two-terminal switch acts in series to circuit; (b) Single terminal switches provide circuit grounds.

Featured in Chart 13–1 is the procedure used in diagnosing 4L60 (700-R4) switches. This procedure is typical for all automatic transmissions. Be sure to consult switch test specifications for the applicable transmission. To avoid switch damage, never use unregulated air.

Switches and TCC solenoids can be tested with the "Smartester" (Figures 13–25 and 13–26). Because it uses regulated air pressure and works with realistic operating voltage and amperage load, it tests switches and diodes more accurately. Also, it can detect intermittent TCC solenoid behavior. Leaky switches that might cause late shifts or pressure loss in a clutch circuit can also be detected. It is always a good policy to check out any new replacement switch or TCC solenoid. "New" does not guarantee it will work.

OIL PRESSURE SWITCH CHECKING PROCEDURE

TOOL REQUIRED: AIR LINE ADJUSTABLE TO 60 PSI & BATTERY POWERED TEST LIGHT

Switch Type	Part No.	Step: Action	Result
Single Terminal Normally Open	8642473 Silver Housing	1: Connect test light leads to the terminal and switch body; air pressure < 10 psi 2: Apply air pressure to the switch; air pressure > 30 psi	1: Light off 2: Light on
Single Terminal Normally Closed	8634475 8642569 Black Housing	1: Connect test light leads to the terminal and switch body; air pressure < 30 psi 2: Apply air pressure to the switch; air pressure > 55 psi	1: Light on 2: Light off
Double Terminal Normally Open	8643710 Silver Housing	1: Connect the test light leads to the switch terminals; air pressure < 10 psi 2: Apply air pressure to the switch; air pressure > 30 psi	1: Light off 2: Light on
Double Terminal Normally Closed	8642346 Black Housing	1: Connect the test light leads to the switch terminals; air pressure < 7 psi 2: Apply air pressure to the switch; air pressure > 14 psi	1: Light on 2: Light off

IMPORTANT: For double terminal switches, it is also necessary to check for internal shorts on the switch. To do this, check each terminal one at a time by connecting the test light to the terminal and the metal body both with and without air pressure. The light should not come on during this test.

CHART 13–1 Oil pressure switch checking procedure; typical 4L60 (700-R4) specifications. (Courtesy of General Motors Corporation, Service Technology Group)

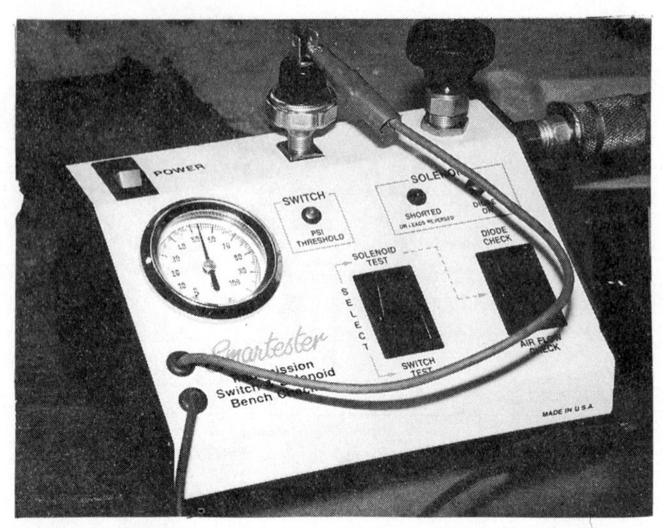

FIGURE 13–25 Smartester switch and solenoid checker; pressure switch being tested.

Test Meters and Test Lights

When working with electricity/electronics, the technician must be aware that all test meters and test lights are not alike. Conventional analog (needle) test meters and test lights cannot be used on computer circuitry or sensitive electronic circuitry anywhere on the vehicle. You are dealing with low voltage and milliampere circuitry. Conventional test meters and test lights require more current to operate—more than what can be handled by the computer circuit. If immediate damage does not result, the electronic components may suffer structural weakness, which results in eventual early failure or erratic operation: "Now it works, now it doesn't." It may take thirty days or more, but it will happen.

For electronic testing, the technician must use instrumentation capable of limiting circuit current flow to safe milliampere levels. For electronic testing, a **high-impedance DVOM** (digital volt/ohmmeter) gives excellent results. It has combined voltmeter, ammeter, and ohmmeter capabilities, plus a diode test feature. A high-impedance digital multimeter capable of AC or DC readings is featured in Figure 13–27.

❖ **High-impedance DVOM (digital volt/ohmmeter):** This styled device provides a built-in resistance value and is capable of limiting circuit current flow to safe milliampere levels.

Use of a conventional test light will also result in electronic circuit damage. Electronic circuit-powered test lights use a light-emitting diode (LED) for a test light in series with a 1/2 W, 470- to 580-ohm resistor. These lights can be used for both conventional and electronic circuitry. Keep in mind the limitations of a test light: it will tell you if you have power or if you do not have power, but it cannot tell you how much power.

Jumper Wires

Jumper wires come under the category of test equipment. They are simple lengths of wire that are used to temporarily bypass, or "jump," sections of a circuit where an open is suspected. The jumper wires are usually fabricated by the technician, for example, they may consist of a length of wire with an alligator clip at each end or a customized wire with different styles of terminal ends for special test hookups. Terminal ends must be crimped and soldered with 60:40 rosin-core solder.

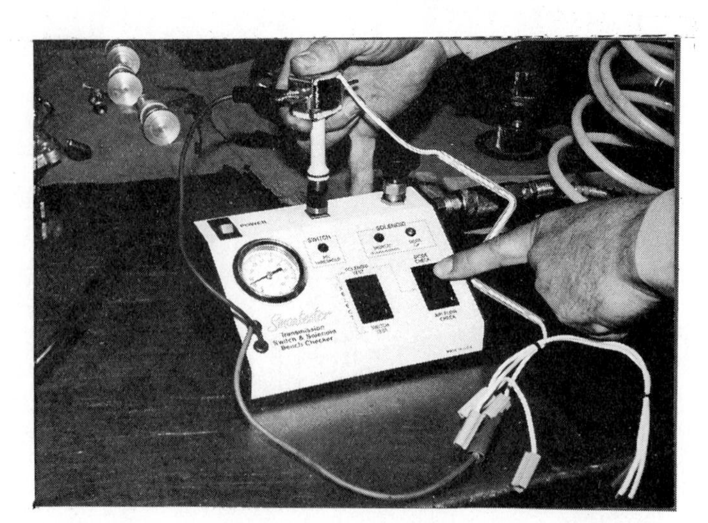
FIGURE 13–26 Smartester in use; TCC solenoid being tested.

CAUTIONS: Never use jumpers made from wire with a smaller gauge than is used in the test circuit. The small-gauge wire adds unwanted circuit resistance that turns to heat and may cause the wire to melt.

Never use a jumper to bypass the circuit high-resistance load. In effect, this creates a direct short to ground and irreversible electronic damage, including possible fire.

FIGURE 13–27 High-impedance digital volt/ohmmeter (DVOM).

CONVERTER CLUTCH DIAGNOSIS PROCEDURES

Chapter 12 discussed how to verify the TCC engagement on a road test. If there is an engagement problem, the concern at this time is to determine if the problem is electrical, hydraulic, or mechanical. Electrical and hydraulic problems can be diagnosed several different ways, limited only by the imagination of the technician. To illustrate diagnostic techniques, the discussion is limited to some of the approaches used with General Motors automatic transmissions. Many of the concepts are applicable directly or indirectly to all TCC-equipped automatic transmissions. It is a matter of adapting test techniques to the product. The technician must always be responsible for understanding the particular TCC system with emphasis on the electrical/electronic circuitry for specific applications.

TCC Always Applied—GM 3T40

1. Place the transmission selector in park and set the parking brake.
2. Start the engine.
3. Depress the brake pedal.
4. Place the transmission selector in drive.
5. If the engine stalls, the likely problem area is the TCC hydraulics.
6. If the engine stalls after the brake pedal is released, the likely problem area is the TCC electrical system.

No TCC Operation: Jumper Wire Testing

Attach a fused jumper harness to the transmission connector with a bypass 12-V source. Be sure to follow the manufacturer's connector codes. Accelerate vehicle to a steady 50 mph (81 km/h) in drive or manual (3) for overdrive transmissions, and plug and unplug the TCC test harness at the cigar lighter. Note whether there is TCC engagement.

- If the TCC engages, the transmission TCC electrical/hydraulic control circuit is working, and the converter clutch is functioning mechanically. Further electrical diagnosis is needed.
- If the TCC does not engage, further internal/hydraulic circuit control testing is necessary.

No TCC Operation: Cooler Line Pressure Test

Remove the cooler line from the radiator and couple a 100-psi (690-kPa) pressure gauge into the line with a T-fitting (Figure 13–28). This can be done on either the inlet or outlet line provided the cooler is not plugged. If cooler line pressure is going to be measured, the hookup must be on the inlet line. With the drive wheels supported off the ground, place the transmission selector in drive and accelerate to a steady 50 mph (81 km/h). If the lockup valve moves, you will see either a quick drop of 5 to 10 psi (35 to 69 kPa) or an increase in gauge pressure, depending on the converter circuitry. This should occur at the road speed cut-in of the TCC.

WARNING: Whenever operating a raised vehicle with the drive wheels turning, make certain that the vehicle is secure. Also, be cautious near turning driveline components.

- If you see this fluctuation, it verifies in one step that the "no engagement" problem is not in the computer, solenoid, switches, connectors, or the converter clutch/lockup valve. The likely problem area is confined to the converter, the input shaft O-ring, or the front pump.
- Should the pressure gauge fail to show converter clutch/lockup valve movement, it is an electrical/hydraulic circuit control problem. If the gauge pressure picks up a very minor needle flick, suspect an incorrectly installed converter clutch/lockup valve.

TCC GENERAL ELECTRICAL DIAGNOSIS

The following discussion represents examples of general procedures that can apply to most electrical/electronically controlled TCC systems. Where more specific information is needed, refer to the manufacturer's service manual or equivalent. Figure 13–29 illustrates a typical TCC electrical schematic.

Power Check and Brake Switch Diagnosis

To verify electrical system operation, start with the hot side of the circuit and test for power at the transmission connector.

- Disconnect the harness at transmission.
- Turn the ignition on, and connect a test light or voltmeter between the 12-V connector terminal and ground. Terminals and wires are all coded for wire identification by the

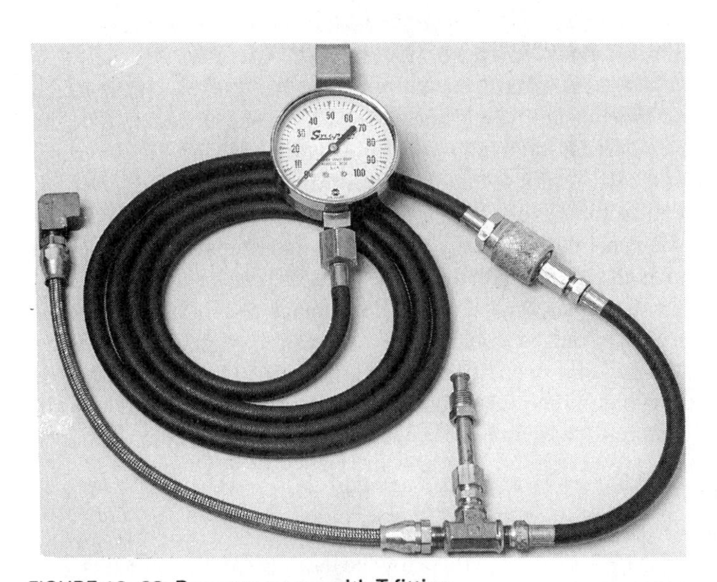

FIGURE 13–28 Pressure gauge with T-fitting.

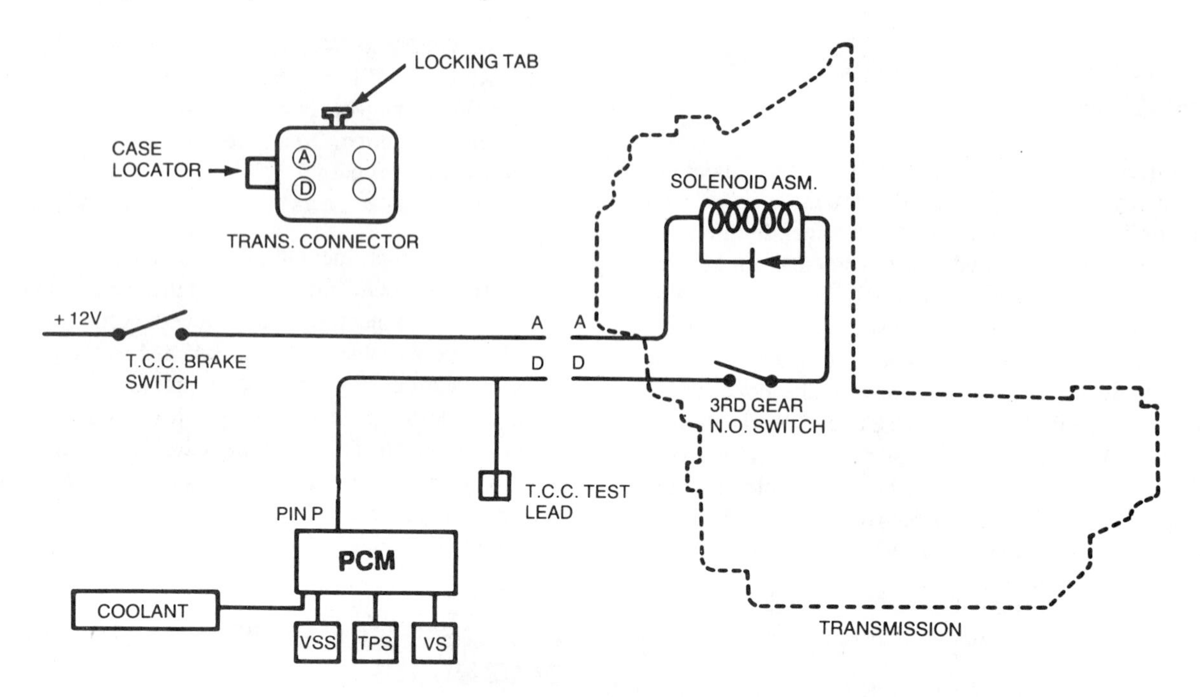

FIGURE 13–29 Simplified 3T40 (125-C) TCC solenoid circuit. (Courtesy of General Motors Corporation, Service Technology Group)

manufacturers. Terminal A is usually the 12-V hot connector to the TCC solenoid, while terminal D is often the return to ground.

- If the test light is ON or the voltmeter reads 12 V, proceed to depress the brake pedal momentarily. If the light stays on, the circuit has a faulty brake switch or adjustment. Otherwise, the problem is in the transmission. Test and correct internal circuitry as necessary.
- Should the test light or voltmeter fail to show evidence of power, verify the circuit fuse. Sometimes a blown fuse leaves only a hairline separation that is not normally visible. Use an ohmmeter to test fuses that look good.
- If the fuse tests OK, check the brake switch adjustment and continuity with a test light.

Testing TCC Electrical Circuit with DLC/ALDL Test Terminal: General Motors Computer Command Control

Starting in 1982, the TCC test point is located in the DLC/ALDL connector on all cars equipped with 3C electronic control—terminal F in the upper left corner of the DLC/ALDL connector (Figure 13–1). The DLC/ALDL location is under the dash near the steering column. Terminal F taps into the solenoid ground side return to the ECM/PCM and can be effectively used for diagnosis. Install a high-impedance test light between terminal F and ground. When road testing, watch for the light. When it comes on, it verifies that the ECM/PCM has completed the solenoid circuit. Observe the test light closely for an occasional blink, which indicates an intermittent circuit.

NOTE: On some GM transmission models, the light may be on and go off when the clutch comes on.

As an option, a jumper can be used between terminals F and D. The MIL/check engine light takes the place of the test light. Terminal D is MIL/check engine light ground. Should the MIL/check engine light fail to respond during the test, be sure to test for a defective light. Place the test light between terminals D and B (+5 V). If a code readout appears at the test light, the MIL/check engine bulb is defective or missing. A typical electrical test procedure using the DLC/ALDL is illustrated in Chart 13–2.

Ohmmeter Testing: Internal Electrical Diagnosis

All automatic transmission TCC internal circuitry can be checked with an ohmmeter. The technician needs to be familiar with the specific circuitry for the transmission application and, therefore, should refer to the manufacturer's service manual or equivalent as necessary. The following examples of checking internal TCC circuitry in General Motors Hydra-Matic applications are typical but are used here for illustrative purposes only. With some imagination, the techniques can be applied for Chrysler, Ford, and the imports. Follow service manual test recommendations.

1. Use an analog meter. The high-impedance DVOM applied voltage is too low and will not forward bias the solenoid diode. A good diode will test bad. Since the transmission harness is disconnected, the analog meter will not be destructive to the PCM/ECM electronics. The purpose of the diode is to localize or isolate the induced high-voltage spike when the solenoid circuit opens and prevent computer damage. The solenoid will function normally with an open diode.

2. Disconnect the harness at the transmission and connect the ohmmeter leads as shown in the diagrams. For in-the-car

Mechanical checks such as linkage, oil level, etc., should be performed prior to using this chart. Also, check for a Code 24. If present, see Chart 24 of the Service Manual.

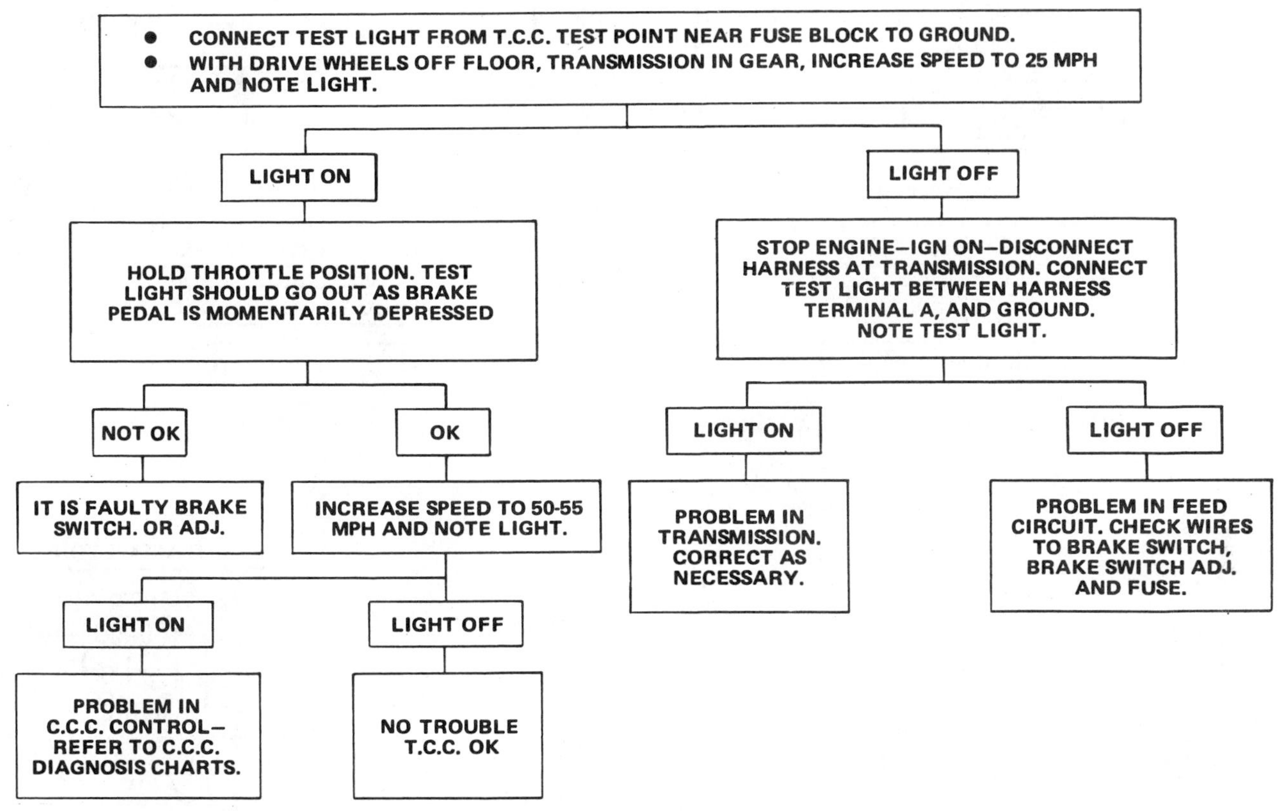

CHART 13–2 Transmission converter clutch (TCC) electrical diagnosis; cars equipped with computer command control (CCC). (Courtesy of General Motors Corporation, Service Technology Group)

testing, provide yourself with a harness jumper from a salvage car for easy meter hookup (Figure 13–30).

A TCC solenoid and diode test summary is given in Chart 13–3, page 330. The following GM transmission/transaxle internal TCC wiring circuits are representative only. There are usually variations of the TCC wiring circuit within each transmission product category. Consult the manufacturer's service manual or equivalent for exact service matchup. The diagrams are used to show diagnostic techniques that can be applied to other TCC wiring.

CCC Electronic—Typical 3T40 (Figure 13–31).

1. With the engine stopped, check for an open third-gear switch. Connect the meter leads to A and D terminals and then switch the leads. The meter should read infinity in both directions.
2. Engine running, check the continuity of the solenoid, diode, and third-gear switch. Connect the positive lead to A, the negative to D.
3. With the drive wheels off the ground and the transaxle in drive, run the vehicle until the third-gear switch closes. The meter reading should be in the range of 20 to 40 ohms.
4. Reverse the meter leads on A and D. Resistance should read in the range of 2 to 15 ohms.
5. Should the meter read infinity in steps 3 and 4, suspect the third-gear switch as the probable cause of an open circuit—it fails to close.

Non-CCC Model—Typical 4L60 (Figure 13–32).

1. With the engine stopped, check for continuity of the solenoid, diode, and 4-3 switch. Connect the positive lead to

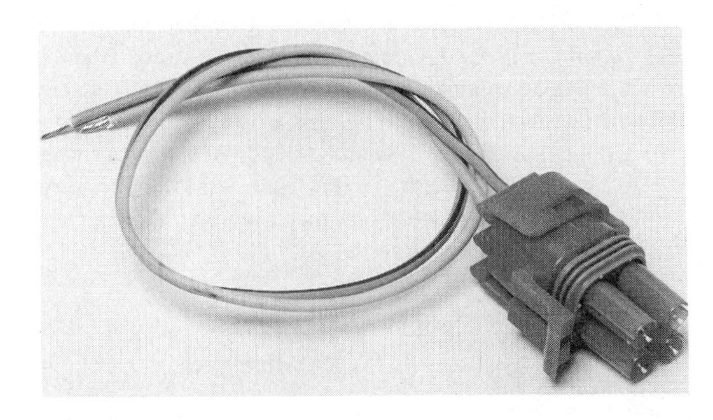

FIGURE 13–30 Harness connector jumper.

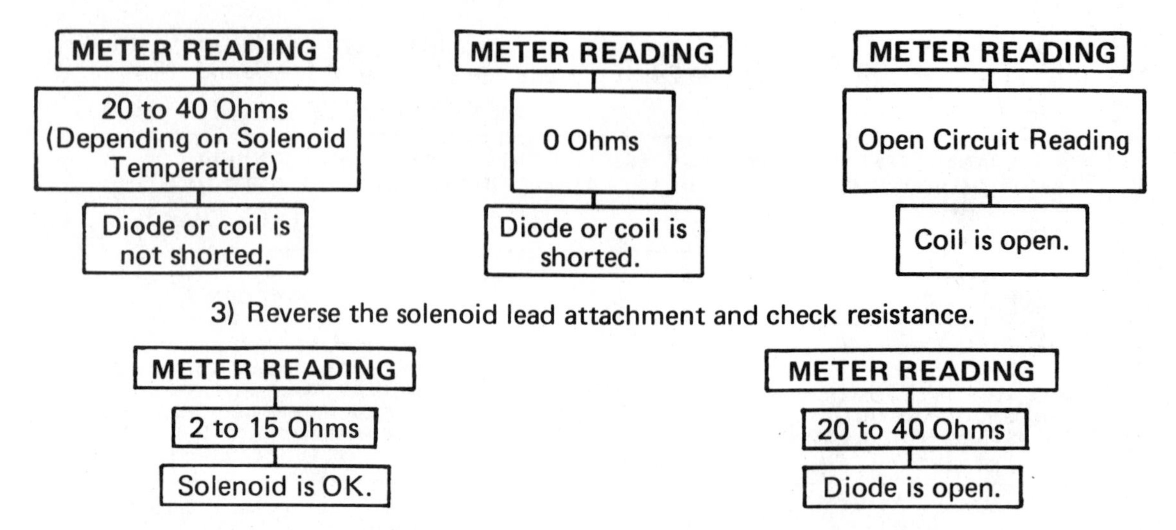

CHART 13–3 TCC solenoid and diode tests. (Courtesy of General Motors Corporation, Service Technology Group)

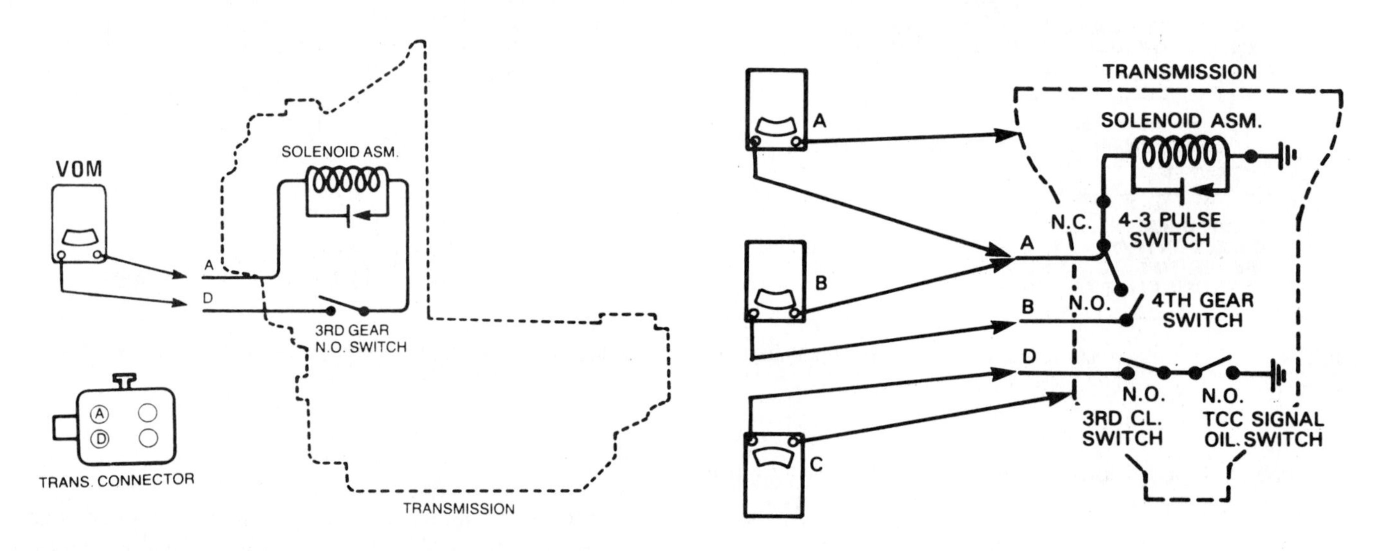

FIGURE 13–31 Typical 3T40 (125-C) TCC solenoid and switch circuit test. (Courtesy of General Motors Corporation, Service Technology Group)

FIGURE 13–32 Typical 4L60 (700-R4) TCC solenoid and switch tests. (Courtesy of General Motors Corporation, Service Technology Group)

A, the negative to ground. Reverse the leads and check the diode.

2. A fourth clutch switch check requires the rear wheels to be off the ground. Connect the meter to A and B and run the engine to make a 3-4 shift. The meter reading should change from infinity to near zero.
3. Connect the meter to D and ground. With the rear wheels off the ground, run the engine to close the third clutch and TCC signal switches. The meter reading should change from infinity to near zero.

TCC Electronic—Typical 4T60

Figure 13–33 shows the TCC internal wiring diagram for several popular models of the 4T60 (440-T4). Without further explanation and with a little imagination, the technician should be able to use the ohmmeter for internal circuit testing.

Solenoid Testing

Even though the solenoid internal circuitry passes the electrical tests, the solenoid needs to be tested mechanically. Where the solenoid is accessible, remove the oil pan and proceed as follows:

1. Remove the TCC solenoid without disconnecting wires.
2. Use a cigar lighter–powered jumper harness at transmission connector to supply power to the solenoid circuit (shown previously).
3. Plug and unplug the connector while blowing regulated air [5 to 10 psi (34.5 to 69 kPa)] into the solenoid. If the solenoid is cooled down from exposure to hot fluid, you can use your mouth to blow air.

NOTE: On transmissions/transaxles with a third- or fourth-gear switch, it will be necessary to use a jumper to bypass the switch.

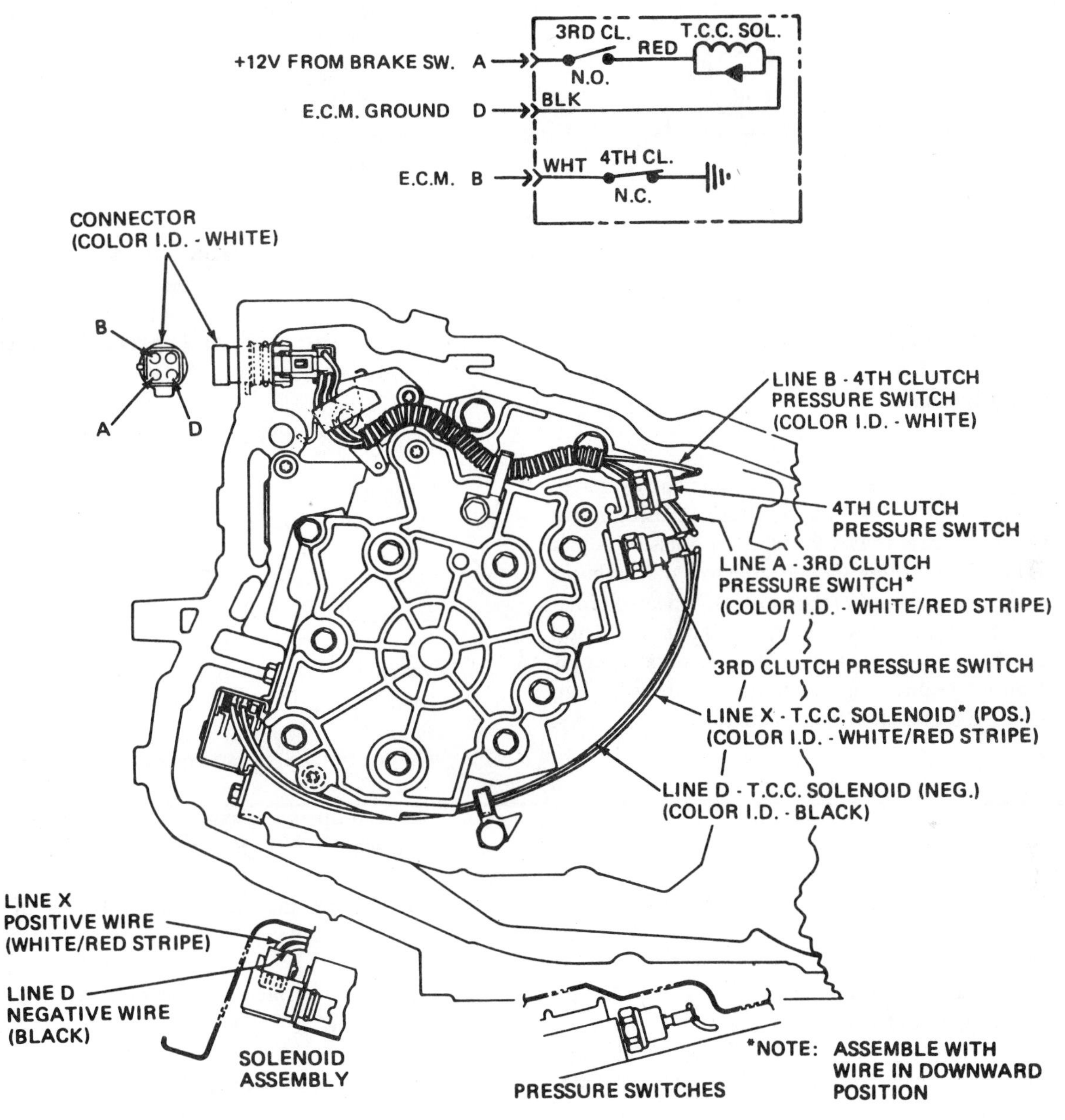

FIGURE 13–33 4T60 (440-T4) TCC solenoid and switch locations and circuit schematic. (Courtesy of General Motors Corporation, Service Technology Group)

This test actually verifies if the solenoid is functioning both mechanically and electrically. Removing the solenoid also gives the opportunity to check for a loose solenoid mounting or damaged O-ring seal.

Sometimes a TCC solenoid sticks closed once the transmission fluid reaches operating temperature. This can result in failure of the converter clutch to release. On closed-throttle braking, the engine is lugged almost to a stall just before the low speed downshift, high to intermediate. When checking the solenoid under cold or lukewarm temperature, the check valve action usually tests normal. For a thorough checkout, however, soak the solenoid in water close to the boiling point and retest. The check valve will trigger closed and remain stuck in the closed position.

Be careful when bench testing diode-equipped solenoids with an automotive battery. The internal diode will be instantly destroyed if the battery polarity is not properly connected to the solenoid leads. An analog ohmmeter is a suggested tool to prevent damage to the diode.

CAUTION: When bench testing solenoids with an automotive battery, use care that the polarity hook-up is correct. If not, diode damage will occur.

As a final note, be certain that the converter lockup is functioning correctly. If not, fuel economy is affected, and

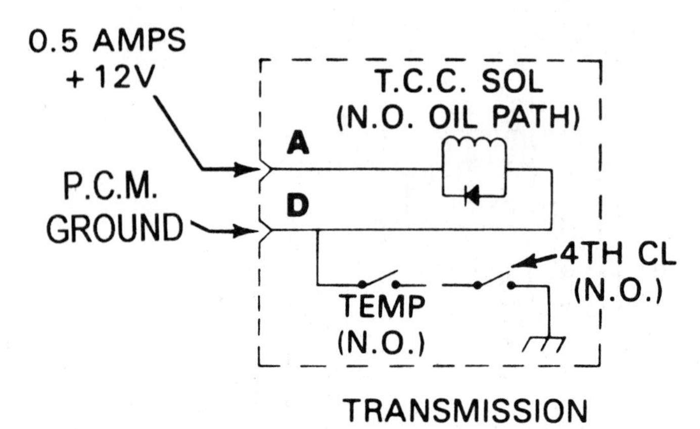

FIGURE 13–34 Oil temperature switch added to TCC circuit for transmission protection. (Courtesy of General Motors Corporation, Service Technology Group)

failure to engage will be a disaster for the automatic overdrive transmission. In overdrive, the converter turbine speed is approximately thirty percent slower than the transmission output shaft speed. This does not allow the converter to rest. It is almost always in torque phase and generating excessive heat. Normal cooling cannot handle the heat factor, and the transmission self-destructs. In case of a computer failure, some systems have added an oil temperature switch as a safety backup (Figure 13–34). TCC operation is essential in overdrive gears.

SUMMARY

Late model torque converter clutch systems are typically controlled by the PCM/TCM. Through the use of electrical input signals, the module can determine when the clutch should be applied and released.

When diagnosing converter clutch problems, scan the PCM/TCM for diagnostic trouble codes (DTCs). If any exist, identify the cause of the DTC and repair it. Remember that engine problems may have a definite relationship to converter clutch operation.

Many failures can be related to electrical circuits, including open, shorted, or grounded wiring conditions. In addition, high resistance in a circuit causes problems. Conventional electrical test meters can be used to perform most circuit checks.

Data link connectors, such as an ALDL, can be used to check electrical operation of the converter clutch control system. If the cause of the problem appears to be a pressure switch or solenoid located in the transmission, remove the item and bench test it. If these items are in good working order, an internal transmission hydraulic problem may be the cause of a clutch operation malfunction.

Because of the wide variety of electrical and hydraulic torque converter clutch designs, the appropriate service manual should always be used to efficiently diagnose converter clutch problems. A properly operating torque converter clutch system is able to enhance fuel economy, lower vehicle emissions, and assist in maintaining proper transmission fluid temperature.

❑ REVIEW

Key Terms/Acronym

Open circuit, short circuit, high resistance, and high-impedance DVOM (digital volt/ohmmeter).

Questions

TCC Electrical/Electronic Circuit Troubleshooting

1. A blown fuse will cause an open circuit. ___T/___F
2. If a short is located between the load and a ground-side circuit switch, the circuit fuse will blow. ___T/___F
3. A copper-to-copper short usually results in a blown fuse. ___T/___F
4. Every electrical circuit has a normal amount of design resistance. ___T/___F
5. Voltage drop is a measurement of resistance. ___T/___F
6. Voltage drop is always read between the leads of the voltmeter. ___T/___F
7. Voltage drop identifies power losses in the circuit. ___T/___F
8. To measure resistance with an ohmmeter, the circuit must be ON.___T/___F
9. To measure resistance on the hot side of the circuit with a voltmeter, the circuit must be OFF. ___T/___F
10. The voltage drop for ground resistance should not exceed 0.5 V. ___T/___F
11. A grounding hydraulic switch usually has two terminals. ___T/___F
12. "NO" means normally open. ___T/___F
13. Controlled regulated air must be used when testing oil pressure switches. ___T/___F
14. Always check new oil pressure switches for accuracy. ___T/___F
15. Conventional analog meters and test lights are compatible with electronic circuit testing. ___T/___F

16. Observing cooler line pressure is a valid means of testing the TCC electro/hydraulic control. ___T/___F

17. A high-impedance digital ohmmeter must be used when testing a diode-equipped TCC solenoid. ___T/___F

18. (On GM applications) Terminal B at the transmission/ transaxle connector is the TCC solenoid 12-V power terminal. ___T/___F

19. If the power test light fails to turn on at the transmission/ transaxle connector, check out the brake switch next. ___T/___F

20. Even if a fuse looks okay, it is still good practice to check it with an ohmmeter. ___T/___F

21. (GM DLC/ALDL) To test for a completed solenoid circuit, connect a high-impedance test light between the F and B terminals. ___T/___F

22. (GM DLC/ALDL) In place of a high-impedance test light, the solenoid circuit can be tested using the check engine light. ___T/___F

23. (GM DLC/ALDL) To test the check engine light for burnout, place a test light between terminals D and B. ___T/___F

24. When the TCC solenoid is diode-equipped, the ohmmeter should read 20 to 40 ohms (ohmmeter black lead connected to solenoid positive lead). ___T/___F

25. If a TCC solenoid checks out electrically, the technician can conclude that it is also mechanically working correctly. ___T/___F

26. A no-fourth-gear converter lockup may damage the transmission. ___T/___F

27. If the TCC fails to engage, the technician should first troubleshoot the hydraulic side of the circuit. ___T/___F

14 Electronic Shift Diagnosis

OBJECTIVE:

The objective of this unit is for you to become proficient with the terms, concepts, and procedures contained in this chapter. Completion of this objective is essential to becoming a successful automatic transmission technician.

CHALLENGE YOUR KNOWLEDGE

Define the following key terms/acronyms:

Hard codes, soft codes, fail-safe/backup control, limp-in mode, OBD II (On-Board Diagnostics Generation Two), electrostatic discharge (ESD), and transmission fluid temperature (TFT) sensor.

List the following:

Ten problem areas of electronic shift concerns; the seven steps of strategy-based diagnosis; three basic malfunctions that cause trouble codes to be set; and five elements of OBD II.

Describe the following:

Common methods to retrieve diagnostic trouble codes; the difference between hard and soft codes; the logical order to follow when multiple codes are present; a method to prevent electrostatic discharge damage to electronic components and why this technique is necessary; the two methods to evaluate transmission electrical devices; and how to use service manual information to understand and locate the cause of diagnostic trouble codes.

INTRODUCTION

Efficient diagnosis of electronically controlled transmissions requires following a systematic process. This procedure includes many of the diagnosing principles emphasized in Chapters 12 and 13. As with hydraulically controlled transmissions, following a logical sequence helps identify the cause of a problem in a productive manner.

This chapter emphasizes the use of diagnostic trouble codes to recognize and locate problem circuits. Codes that relate to transmission malfunctions can be caused or detected by engine and transmission input devices, as well as transmission output devices. Though codes can be generated by failed electronic input and output switches and sensors, faulty internal hydraulic and mechanical components may be the actual cause. This necessitates consideration of the many operating characteristics of electronically controlled transmissions.

To be effective at pinpointing the cause of a problem, a technician must use a variety of resources and tools. Use of diagnosis charts and wiring circuit schematics found in shop manuals is extremely helpful and will lead the technician in the right direction. With this information, scanners, DVOMs, and special transmission testers are recommended to determine whether electronic switches, sensors, and their circuits are in good working order. If they are functioning properly, they may be detecting internal transmission problems.

Correctly analyzing an electronically controlled transmission concern can be a rewarding experience. Though there are many varieties of transmissions, each with a host of possible trouble codes, every manufacturer provides tools and information to simplify the process. Following troubleshooting charts leads to accurate diagnosis. Correctly identifying the cause of a problem is vital to effectively servicing the transmission.

FUNDAMENTAL APPROACH TO DIAGNOSING ELECTRONICALLY CONTROLLED TRANSMISSIONS/TRANSAXLES

Modern electronically controlled transmissions respond to numerous PCM input sensors and switches. These are located in the engine compartment, on and within the transmission and drivetrain, and occasionally within the driver's reach. Output actuators controlled by the PCM perform the work, allowing clutches and bands to be activated. Though many people do not understand how these input and output devices function, their operation provides the smooth, responsive shift action desired by today's drivers (Figure 14–1).

The sensors and switches are often taken for granted until a malfunction is noticed. Unfortunately, the cause of a problem is frequently directed at these items because they are not understood. Assumptions are made and good components wrongfully replaced. Time, effort, and money is wasted in this type of careless practice which can easily be avoided. Fixing a problem right the first time requires following systematic steps of procedure.

Figure 14–2 is a problem-solving flow chart representing "strategy-based diagnostics." Many of these steps were highlighted in Chapter 12. Each step should be completed before moving onto the next. Following is a brief summary of the steps:

1. **Verify complaint.** Determine if the customer's concern represents a valid problem or not. Gather as much information about the concern as possible. A road test with the customer can identify the concern and help eliminate misunderstandings between the customer and service facility.

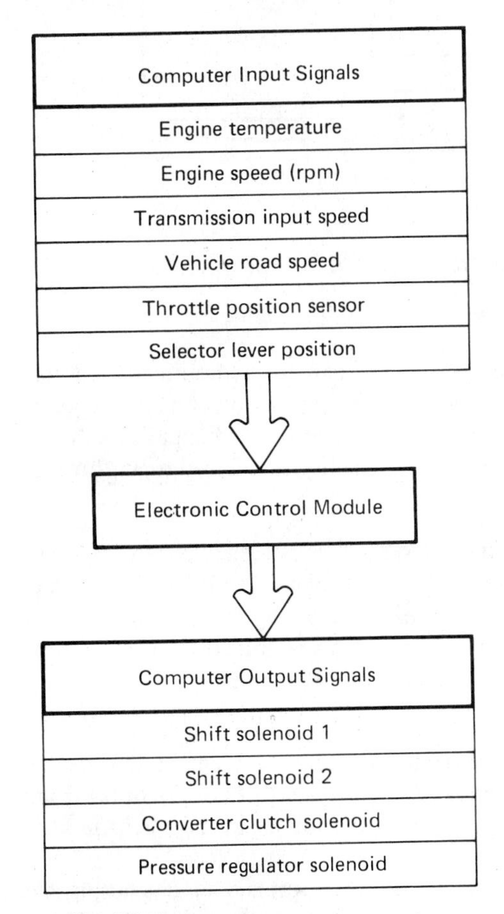

FIGURE 14–1 Simplified electronic control concept.

2. **Perform preliminary checks.** Visual inspections of the fluid level and condition, shift linkage, cables, wiring, vacuum hoses, and underhood components may indicate the cause of a problem. Also, check engine performance, and use a scanner to determine if any diagnostic trouble codes (DTCs) are present.
3. **Check bulletins and troubleshooting hints.** From the information already gathered, use bulletins and other resource materials to locate symptom diagnosis and repair strategies.
4. **Perform service manual diagnostic checks.** Work closely with the factory shop manual (or equivalent resource) for direct guidance with the data you now have.
5. **Diagnose Path Options.**
   - 5A. **Hard code.** Any DTCs present must be pursued next. **Hard codes** represent the detection of a problem in a specific circuit while testing.
   - 5B. **No code.** Symptom charts should be used to help locate probable causes when no DTCs are present.
   - 5C. **No matching symptom in service manual.** Analyze the operation of the unit and refer to wiring diagrams, hydraulic schematics, and solenoid/clutch/band application charts.

     The better one understands the principles of operation of electronically controlled transmissions, the easier it is to identify the cause of a puzzling situation. Use a process of elimination. Know what is working correctly to recognize what might not be operating right.
   - 5D. **Intermittent.** This category of malfunctions can be quite challenging—difficult to verify and diagnose with confidence. Use "intermittent diagnostic" helps when available.

     Often loose wiring connectors, sticking relays and solenoids, poor grounds, and add-on aftermarket equipment/accessories can influence intermittent malfunctions. The situation may cause a DTC to be set into the PCM's memory, but the problem is not present at the time of testing. These types of DTCs are referred to as **soft codes**. Check the service history of the vehicle for clues.
   - 5E. **Operating as Designed.** Trying to fix a normal condition is a worthless venture. Communicate to the customer that the transmission is operating properly and a repair is not needed.
6. **Reexamine the complaint.** If the cause of a problem cannot be found or isolated, return to the beginning of the process.
7. **Repair and verify the fix.** After fixing the vehicle and clearing any codes, perform a thorough road test and validate that the repair fixed the problem.

❖ **Hard codes:** DTCs that are present at the time of testing; also called continuous or current codes.

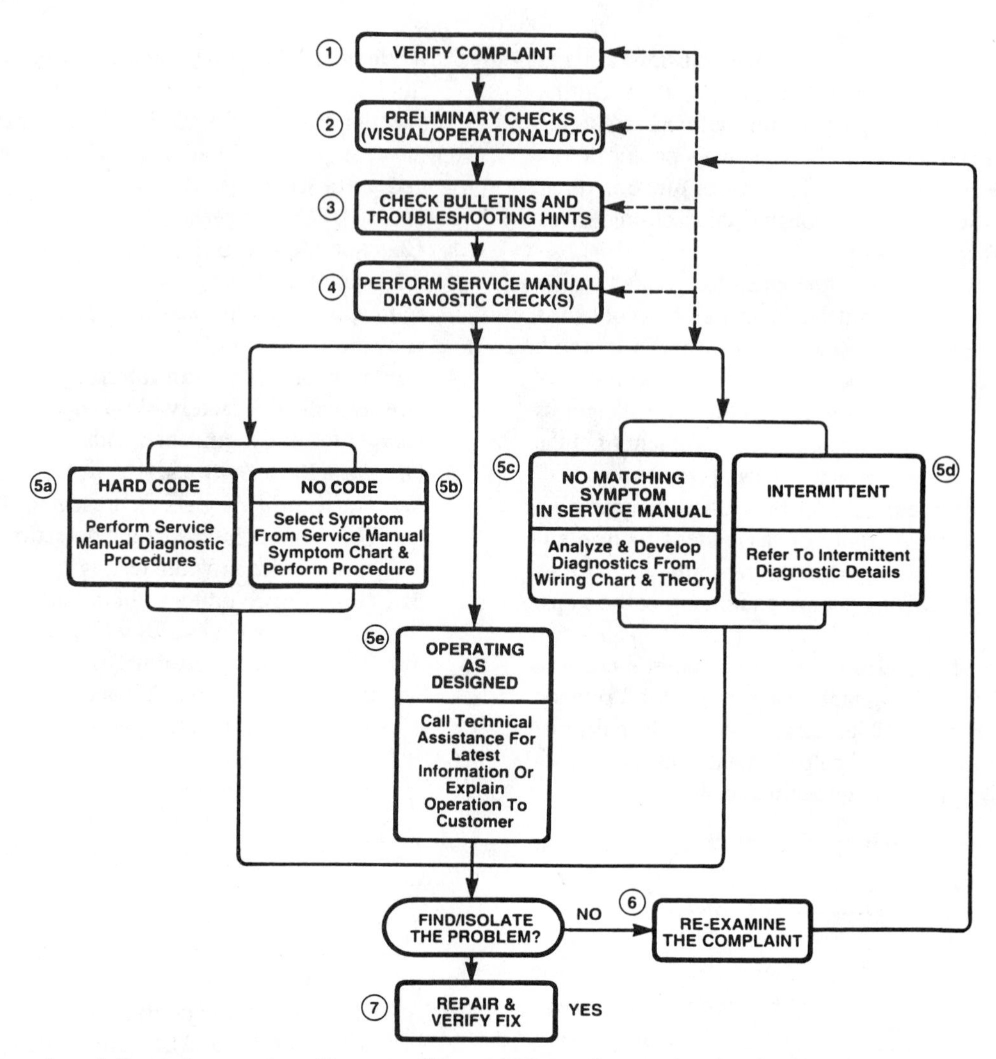

FIGURE 14–2 Strategy-based diagnosis procedure. (Courtesy of General Motors Corporation, Service Technology Group)

❖ **Soft codes:** DTCs that have been set into the PCM memory but are not present at the time of testing; often referred to as history or intermittent codes.

Recognizing the transmission devices that are inputs to the PCM, and those that are outputs controlled by the PCM, will aid in your understanding of the transmission operation. When these devices fail, many PCMs substitute the sensor reading with a default value or the signal of another sensor. The substitute value is used to provide a temporary **"fail-safe/backup control"** signal to allow the vehicle to be driven (Table 14–1).

❖ **Fail-safe/backup control:** A substitute value used by the PCM/TCM to replace a faulty signal from an input sensor. The temporary value allows the vehicle to continue to be operated.

At times, certain codes or combinations of codes will shut down the transmission electronic outputs. This typically produces a situation where only reverse and one forward gear in the overdrive range can be hydraulically energized. Some transmissions may provide another forward gear in certain manual ranges. This condition, which allows the vehicle to be driven to a service facility, is referred to as **"limp-in mode."**

❖ **Limp-in mode:** Electrical shutdown of the transmission/transaxle output solenoids, allowing only forward and reverse gears that are hydraulically energized by the manual valve. This permits the vehicle to be driven to a service facility for repair.

To analyze transmission problems, a few special tools and testers are needed (Figure 14–3). Some of these are essential, while others make the job easier and help produce efficient results (Figure 14–4).

Following a logical approach to diagnosing electronically controlled transmission concerns is important. Work intelligently and employ all the resources and tools available to locate the actual cause. Be patient as you tackle new types of problems. Realize that as you gain experience, you will become more efficient.

FAIL-SAFE/BACKUP FUNCTION (Limp-In Mode) TABLE 110001597

When the PCM detects the following malfunction(s), the PCM carries out fail safe/back-up control. In addition, the corresponding DTC number(s) can be identified by using the scan tool.

Diagnostic item	Control contents during malfunction
No cam signal at PCM	The PCM uses the crankshaft position sensor signal only to control fuel injection timing, etc. (Accordingly, normal sequential multiport fuel injection may not be carried out.)
Throttle position sensor voltage low	The PCM uses the value calculated from the MAP sensor signal instead of the throttle valve opening angle (voltage).
Throttle position sensor voltage high	The PCM uses the value calculated from the MAP sensor signal instead of the throttle valve opening angle (voltage).
Engine coolant temperature (ECT) sensor voltage too low	• The PCM uses the default value [°C (°F)] as the engine coolant temperature. • The PCM operates the radiator fan. • The PCM carries out open loop control.
Engine coolant temperature (ECT) sensor voltage too high	• The PCM uses the default value [°C (°F)] as the engine coolant temperature. • The PCM operates the radiator fan. • The PCM carries out open loop control.
No vehicle speed sensor signal	The PCM controls the engine as if the vehicle speed were 0 mph.
MAP sensor voltage too low	The PCM uses the value calculated from the throttle position sensor and the engine speed signals instead of the MAP value (mV).
MAP sensor voltage too high	The PCM uses the value calculated from the throttle position sensor and the engine speed signals instead of the MAP value (mV).
No change in MAP from start to run	The PCM uses the value calculated from the throttle position sensor and the engine speed signals instead of the MAP value (mV).
Intake air temperature sensor voltage low	The PCM uses the engine coolant temperature instead of the intake air temperature.
Intake air temperature sensor voltage high	The PCM uses the engine coolant temperature instead of the intake air temperature.
Throttle position sensor voltage does not agree with MAP	The PCM uses the value calculated from the MAP sensor signal instead of the throttle valve opening angle (voltage).
Timing belt skipped 1 tooth or more	The PCM uses the crankshaft position sensor signal only to control fuel injection timing, etc. (Accordingly, normal sequential multiport fuel injection may not be carried out.)
Intermittent loss of camshaft position or crankshaft position	The PCM uses the crankshaft position sensor signal only to control fuel injection timing, etc. (Accordingly, normal sequential multiport fuel injection may not be carried out.)

TABLE 14–1 Fail-safe/backup functions—41TE (A-604). (Courtesy of Chrysler Corp.)

ON-BOARD DIAGNOSTICS

Since the early 1980s, engine, transmission, and powertrain control modules (ECMs/TCMs/PCMs) have had the ability to test input and output circuits against assorted parameters. When outside of these boundaries, diagnostic trouble codes are set. They are designed to assist the technician to locate a problem cause. Many codes automatically illuminate the check engine or malfunction indicator lamp (MIL) to get the driver's attention when a problem exists.

Which sensors and switches apply to transmission operation, which ones are self-testing, how often their signals are checked, and how the PCM responds to such conditions vary dramatically. The changing level of sophistication through the evolutionary years of electronically controlled transmissions is quite apparent. Manufacturers' product lines are becoming somewhat standardized in how their PCMs integrate with all the electronic components, but differences still exist.

As will be discussed later in this chapter, the approaches used to deal with DTCs have similarities between the manufacturers. But when it comes to diagnosing a specific concern, product information and guidelines must be followed. Without them, diagnosis often becomes a guessing game.

Diagnostic Trouble Codes (DTCs)

DTCs generated by the PCM tell the technician that a malfunction in a particular electronic/electrical circuit has been identified. They provide a direction for the diagnostician to follow. Numerical codes and their meanings are different for each manufacturer group. Even within a manufacturer's product lines, code meanings can be different from one year to another. But in all vehicles, DTCs can be set by three different situations:

1. There is no signal from an input or output device (sensor or switch).

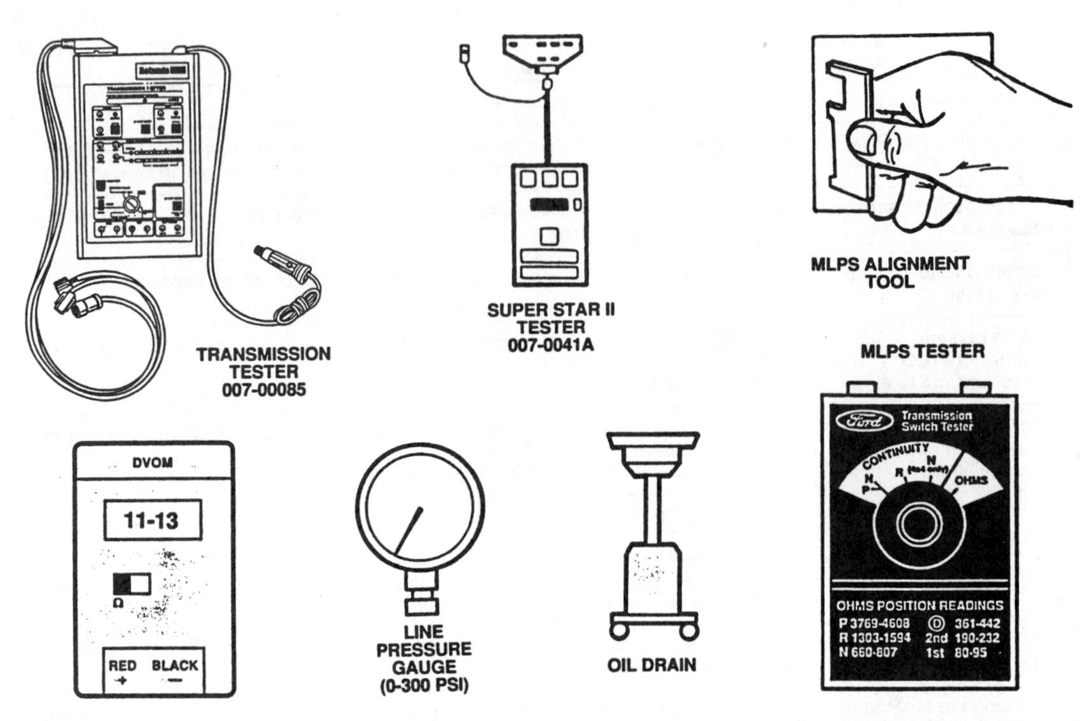

FIGURE 14–3 Typical transmission/transaxle diagnostic tools and testers. (Reprinted with the permission of Ford Motor Company)

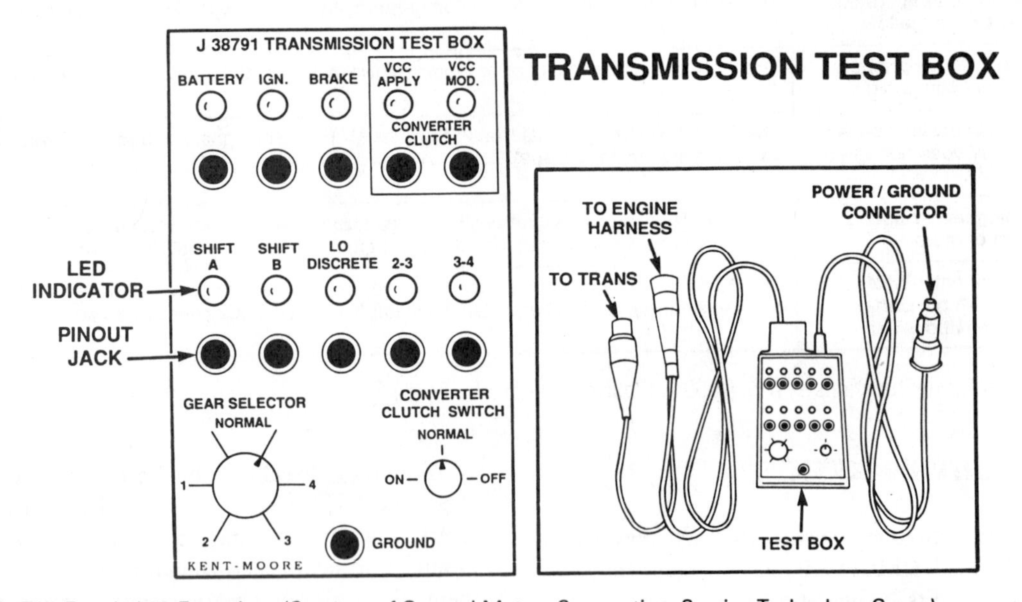

FIGURE 14–4 GM 4T60-E and 4L60-E test box. (Courtesy of General Motors Corporation, Service Technology Group)

2. The signal from a device is not within the normal operating range for a given vehicle operating condition.
3. A circuit self-test identifies that the sensor/switch circuit is not functioning as designed.

Codes are set into the PCM's memory for a certain number of drive cycles and can be accessed through different means. Assumptions are too often made, and careless technicians wrongly identify the sensor or switch in a problem circuit as being the cause for setting a DTC. This leads to unnecessary parts replacement, wasted time and effort, and a cure that still needs to be located.

Use codes to your advantage when diagnosing. The type of code and its relationship to other codes and the diagnosis procedure must be considered. Though a sensor or switch can indeed be faulty, other components may have failed and set the code, including loose electrical connectors, poor grounds, open or shorted wiring circuits, high or low hydraulic circuit pressures, certain fluid levels, and failed mechanical components. Essentially, trouble codes indicate that an abnormal condition occurred or exists, causing the engine and/or transmission not to function properly. At times, a circuit component or device malfunctioning may set more than one code.

When pursuing a transmission concern and multiple codes exist, follow up on input codes from the engine first. Though an engine may appear to be running fine, remember that many PCMs have the ability to substitute faulty values with default signals and improve an engine's performance. On the other hand, the transmission may be responding to errant signals and function poorly. It is necessary to have the engine operating properly if the transmission is to respond correctly.

Once engine codes are dealt with, the next step is typically to seek out the cause for transmission input codes, and then any output codes. Test the circuits and devices that most likely set the DTCs and are most accessible. Again, work closely with the appropriate shop manual. With experience, the technician will recognize certain situations and be able to locate and repair many problems with confidence. Be sure to follow a logical process, as described earlier in this chapter.

Hard Codes. A DTC that is current at the time of testing is referred to as a "hard code." Erased from the memory, it returns immediately. Hard codes are often called "continuous codes" or "current codes." Since they are present at the time of diagnosing, they are pursued first.

Soft Codes. A DTC set into the PCM memory, but not present at the time of testing, is referred to as a "soft code." A code of this nature is also called a "history code" or an "intermittent code." Since the condition that set the soft code is not currently present, diagnosis is more challenging. Work patiently and observantly as you strive to identify the cause of a soft code.

CODE RETRIEVAL

A variety of different methods and tools can be used to retrieve DTCs. These include a simple jumper wire, analog (needle-type) volt/ohmmeter, in-dash electronic displays, handheld scanners, and full-size computer-operated service bay diagnostic units.

Service bay diagnostic units, shown in Figures 14–5 and 14–6, provide various diagnostic and repair information. They are able to access codes and do all the functions of handheld scan tools, plus a lot more. The computer-based diagnostic units come in a variety of models and features. Most include a full-size computer monitor, printer, CD drive unit for service bulletins, and shop manuals information. Some provide a modem for direct phone line access to manufacturers' troubleshooting services. Though rather expensive, they are an excellent tool for the trained technician to use.

Each manufacturer recommends procedures and tools that can be used to retrieve codes in their vehicles. In addition, many independent companies produce and sell testing equipment. Whichever tool and/or procedure is used, follow the instructions for the equipment carefully. These instructions are usually found in the appropriate shop manual or reference. To help you keep track of the codes present in memory during the diagnostic process, write them down.

General Motors

Manual Reading Codes. By using a jumper wire or paper clip, jump terminal A to terminal B at the DLC/ALDL (Figure 14–7). Turn the ignition to the run position and the MIL (service engines soon/check engine light) will flash. The first set of

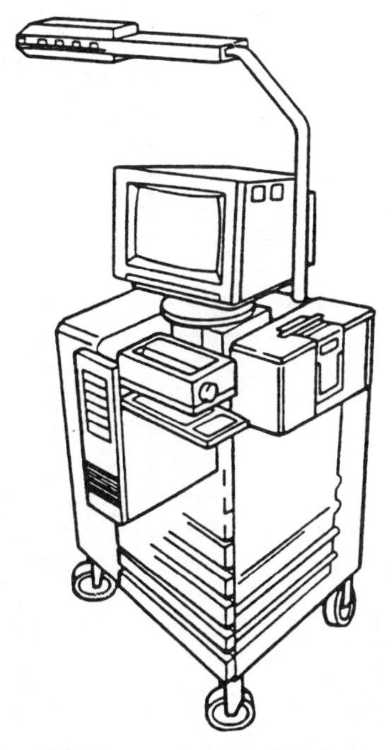

FIGURE 14–5 Ford SBDS—service bay diagnostic system. (Reprinted with the permission of Ford Motor Company)

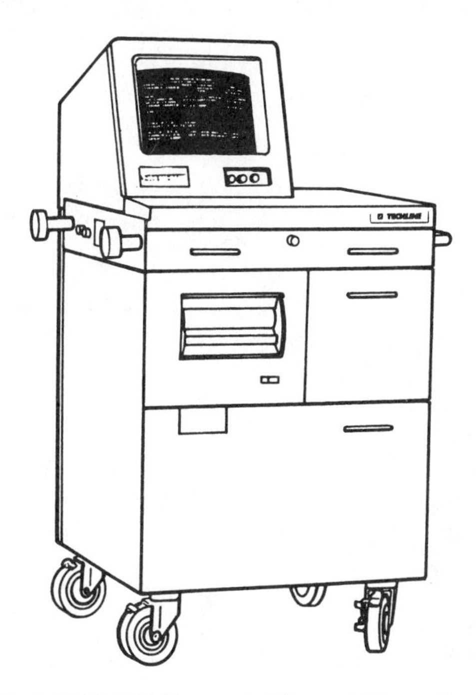

FIGURE 14–6 GM T-100. (Courtesy of General Motors Corporation, Service Technology Group)

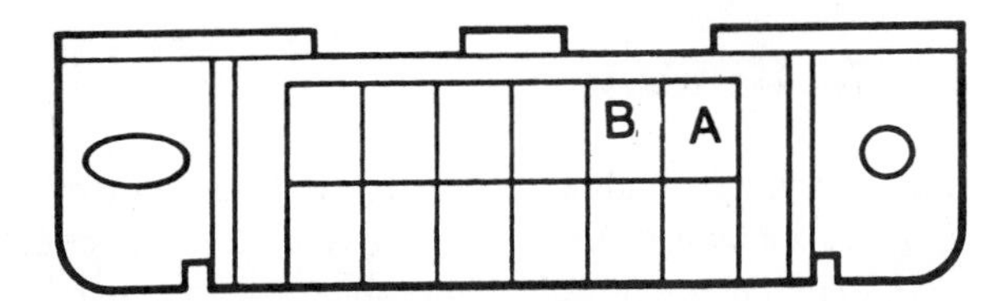

FIGURE 14–7 GM DLC/ALDL (OBD I). (Courtesy of General Motors Corporation, Service Technology Group)

flashes represents the digit in the "tenths" position, and the second set of flashes represents the "ones" position. Three code 12s (each one indicated by one flash, a pause, followed by two flashes) mean that the PCM's self-diagnostic function is operating. Codes will output in order from lowest to highest. When all codes in memory have been retrieved, four code 12s will indicate that the readout is complete. Turn the key off and remove the jumper.

To clear the codes from the PCM memory, a specific fuse must be removed from the fuse block. An alternate, but less preferred, method is to disconnect the battery. This procedure clears electronic radio, clock, temperature control, and other preset component memory.

Tech 1. The Tech 1 handheld scanner is a bidirectional tool/tester (Figure 14–8). When connected to the PCM's diagnostic link connector (DLC/ALDL), it can receive and send information and instructions. The use of a scan tool is suggested because it reduces errors and is an efficient diagnostic instrument.

With the appropriate cartridge inserted into the Tech 1, make the connection to the ALDL/DLC (data link connector) and insert the power plug into the cigar lighter. Turn the key to the ON position, and input the correct vehicle and engine information requested on the display screen. A variety of functions is available to the technician. A menu-type system allows the operator to choose the desired selection. Pressing button F2 while viewing the main menu presents DTCs.

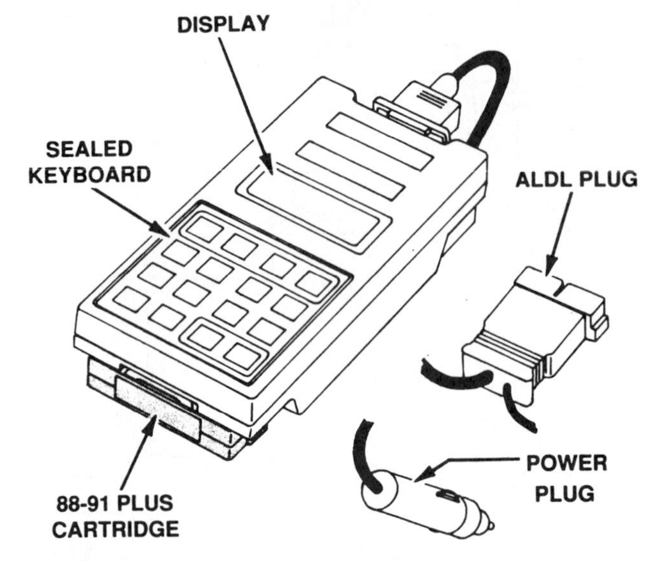

FIGURE 14–8 Tech 1 scan tool. (Courtesy of General Motors Corporation, Service Technology Group)

From the Tech 1, many input and output sensor and switch signals can be read with the engine off (key ON) or running. The technician is also able to control output devices such as TCC, pressure control, and shift solenoids. This feature can be extremely helpful in the diagnostic procedure. With the engine running, signals can be observed from both input and output devices (Chart 14–1). Also, codes can be cleared through the Tech 1.

Tech 2. The recently introduced Tech 2 handheld scan tool is considered an essential item for servicing late model GM vehicles designed with OBD II systems (Figure 14–9). In addition to a large screen to display multiple lines of data, the Tech 2 has graphing capabilities and numerous other desirable features.

In-Dash Displays. When equipped, vehicle codes can be read from driver information centers (Figure 14–10) and fuel data

TECH 1 PARAMETER	IN PARK AT IDLE
ENGINE SPEED	550-750 RPM
TRANS OUTPUT SPD	0 RPM (1)
ENG COOL TEMP	85°C - 105°C (5)
TRANS FLUID TEMP	30°C - 130°C (5)
THROT POSITION	.4 - .6 V
THROTTLE ANGLE	0%
A/B/C RNG	OFF ON OFF
TRANS RANGE SW.	Park/Neutral
COMMANDED GEAR	1
ADAPTABLE SHIFT	NO (3)
1-2 SOL / 2-3 SOL	ON ON
CTR FDBK 1/2 2/3	ON ON (4)
3-2 CONTROL SOL	0%
3-2 CONTROL FDBK	OFF
HOT MODE	NO
TCC SLIP RPM	600-750
TCC SOLENOID	OFF
CTR FDBK TCC	OFF (4)
DESIRED PCS	1.01
ACTUAL PCS	1.01
PCS DUTY CYCLE	40 - 60%
VEHICLE SPEED	0
4WD LOW SWITCH	OFF
CRUISE ENGAGED	OFF
TCC BRAKE SWITCH	Released
KICKDOWN ENABLED	NO
1-2 SHIFT TIME	.3 - 1.35 Sec. (2)(3)
1-2 SHIFT ERROR	0 - 6.25 Sec. (2)(3)
2-3 SHIFT TIME	.3 - 1.35 Sec. (2)(3)
CURR. ADAPT CELL	N/A < 25% TP
TRANS CALIB ID	Internal ID Only
SYSTEM VOLTS	13.6
PRESS ADAPT 25% TP	−5 - 10
PRESS ADPT 40% TP	−5 - 10
PRESS ADPT 70% TP	−5 - 10
CURR. ADAPT PSI	-5 - 10

CHART 14–1 Typical Tech 1 data. (Courtesy of General Motors Corporation, Service Technology Group)

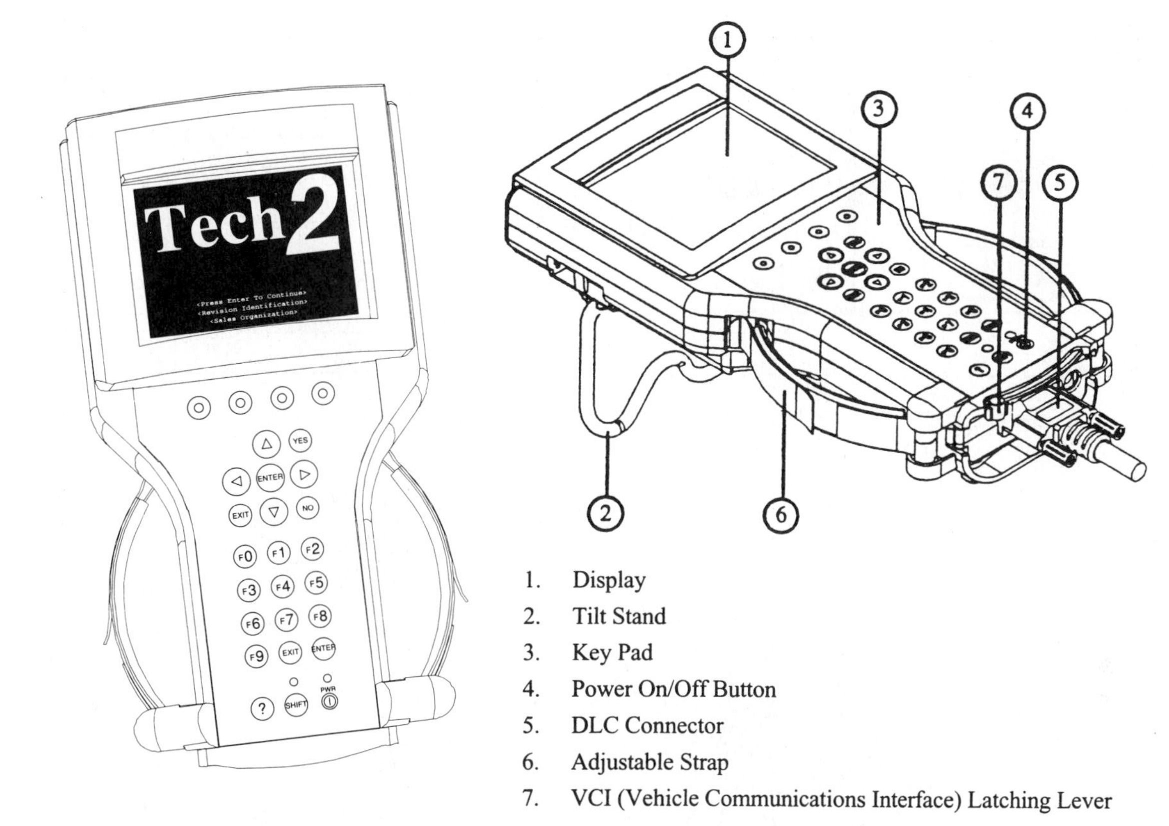

FIGURE 14–9 Tech 2 design and features. (Courtesy of General Motors Corporation, Service Technology Group)

center and climate control panels (Figure 14–11). By simultaneously pressing the ECC OFF and WARMER buttons, you enter the diagnostic functions. After five seconds, the codes will appear. ECM/PCM codes are prefixed with an "E." These are followed by BCM (body-chassis module) and SIR (supplemental inflatable restraint) codes. If no ECM/PCM codes are present, the message will read: "NO ECM CODES."

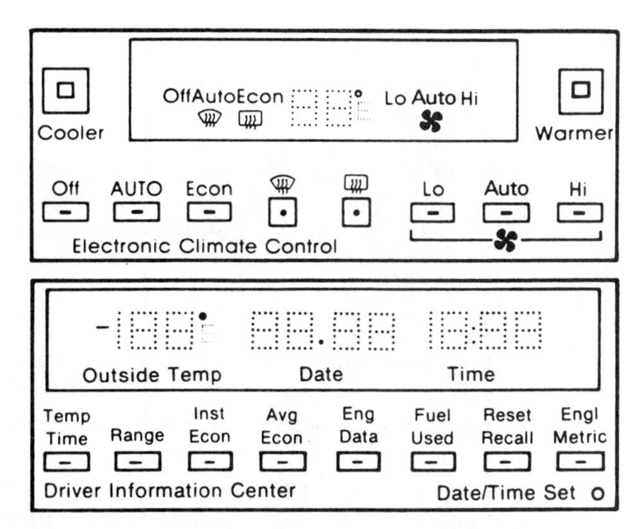

FIGURE 14–10 Driver information center. (Courtesy of General Motors Corporation, Service Technology Group)

To proceed to different levels, press and release the appropriate button. By pressing "OFF," you return to the next selection in the previous level. Press "RESET" to exit diagnostics.

Chrysler

Manual Readout. Certain Chrysler vehicles allow code retrieval by observing the flashing of the MIL. With these vehicles, cycle the key ON and OFF twice, and then turn the key to the ON position once again. Record the sequence of MIL flashes. Refer to the service manual for code descriptions and diagnosis/repair strategies.

Manually reading codes by counting the sequence of MIL flashes is not the preferred method. Rather, use a scan tool to promote accuracy.

DRB (Digital Readout Box) II and III. Chrysler 41TE (A-604) and 42LE transaxles are controlled by a TCM that interfaces with the engine's control module. The diagnostic connector is located in the C^2D circuit that connects the two modules and is positioned under the instrument panel. The DRB II and the recently introduced DRB III are the diagnostic scan tools recommended by Chrysler (Figure 14–12).

With the correct cable connections made and the appropriate software cartridge inserted, the technician can begin. Turn the ignition ON. With the keypad, identify the vehicle model year and the system to be tested. A green LED will illuminate on the DRB II tester if the self-test of the C^2D circuit passes. Other-

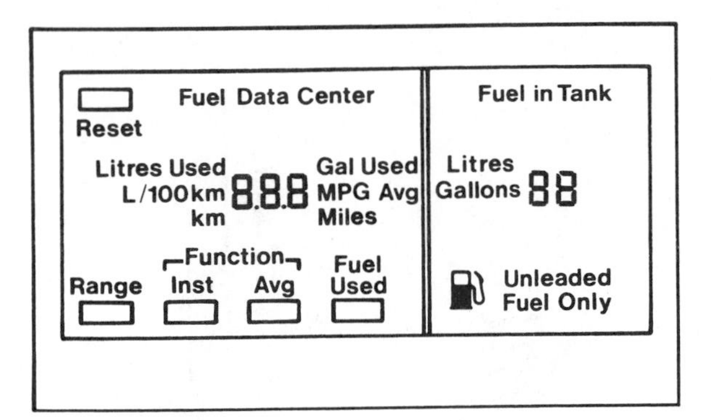

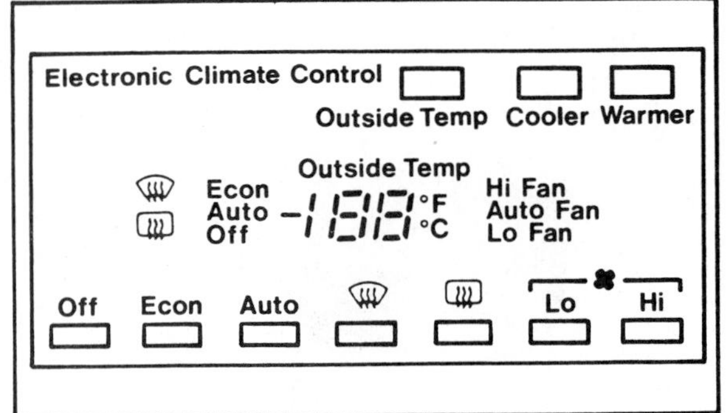

FIGURE 14–11 Fuel data center and climate control panel. (Courtesy of General Motors Corporation, Service Technology Group)

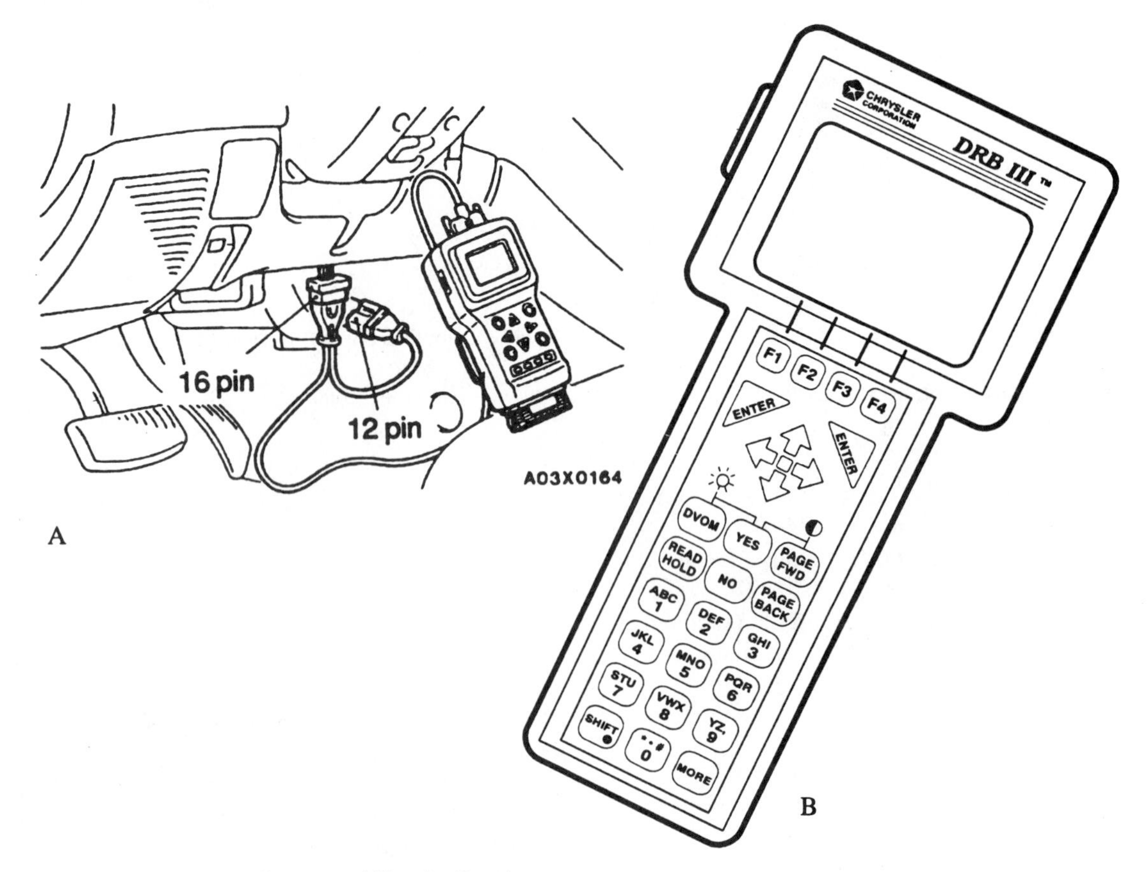

FIGURE 14–12 (a) DRB II; (b) DRB III. (Courtesy of Chrysler Corp.)

wise, a red LED will be lit. After the system has passed this preliminary check, stored trouble codes will be displayed.

The DRB II scan tool is able to enter into five different test systems. From the tester, the operator can perform the following tests:

1. *Diagnostic test:* Displays and erases codes in the TCM's memory.
2. *Circuit actuation test:* Tests output circuits and actuators by turning them on and off.
3. *Switch test:* Determines if switch input signals are being received by the controller.
4. *Sensor test:* Determines if sensor input signals are being received by the controller.
5. *Engine running test:* Displays sensor and switch signal readings while the engine is running. Some output devices can be controlled by the DRB II, such as shift solenoids.

Ford

Manual Readout. Codes can be manually read by grounding two terminals at the self-test output connecter (DLC). The MIL will flash in a sequence similar to the method used by GM. Another method is to observe the pulses on the needle of an analog volt/ohmmeter grounded through the connector. These methods are described in detail in the shop manual but are not recommended, since scan tools are readily available.

Super Star II Tester. The Super Star II Tester, shown in Figure 14–13, is an improved version of Ford's original scan tool, the STAR (Self-Test Automatic Readout) tester. Hooked up to the self-test output connector/DLC, this device is capable of displaying codes. Ford engine codes are displayed with two digit numbers and transmission codes with three digits.

The Super Star II Tester is used to perform these three "quick tests":

1. *Key on, engine off (KOEO):* Displays hard and soft codes. A code 11 or 111 indicates that no trouble codes are currently present.
2. *Key on, engine running (KOER):* With the engine running, the technician must perform certain tasks ("goose" the accelerator pedal, turn the steering wheel, apply the brakes, and so on). The engine performs self-checks to the operation of the various sensors and switches. Problems detected will set a code. A code 11 or 111 indicates that no trouble codes are present.
3. *Wiggle test:* This test can be quite helpful in locating the cause of an intermittent concern. While the technician moves wire, harnesses, and connectors, or taps or wiggles components, if an intermittent condition is detected, the tester will record a code and emit a noise.

To perform "quick tests," refer to the appropriate Ford powertrain control/emission diagnosis manual for the exact steps of procedure. There are variations depending on year, model, and engine.

New Generation Star (NGS) Tester. The NGS provides a great deal more than just code readout (Figure 14–14). From the display screen, multiple input and output signal readings can be obtained. In addition, in select vehicles, certain output devices can be controlled by the operator.

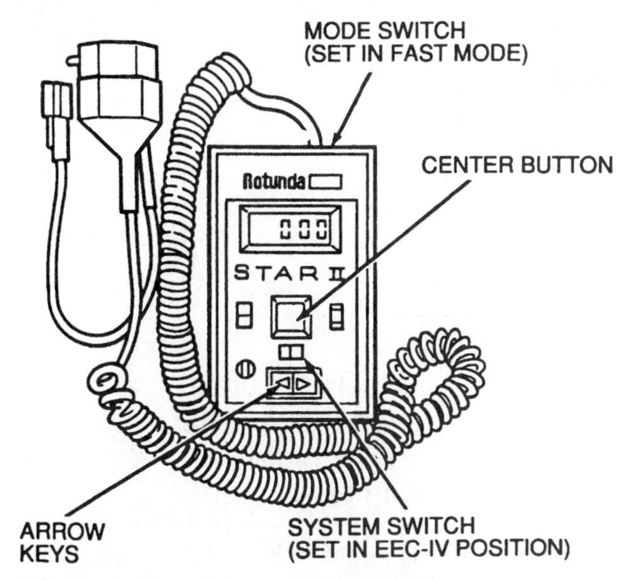

FIGURE 14–13 Super Star II tester. (Reprinted with the permission of Ford Motor Company)

OBD II (ON-BOARD DIAGNOSTICS GENERATION TWO)

OBD II refers to a federal law, requiring automobile manufacturers' compliance beginning with the 1996 model year, that forced automobile makers to reduce vehicle emissions and increase the level of sophistication of the on-board computer monitoring of vehicle emission devices. The vehicle must actively perform diagnostic tests on systems that relate to the level of emissions. If a system fails or deteriorates and allows the level of emissions to exceed one and one-half times the allowable standard, the MIL illuminates, even if the vehicle appears to be operating normally (Figure 14–15).

❖ **OBD II (On-Board Diagnostics Generation Two):** Refers to the federal law mandating tighter control of 1996 and newer vehicle emissions, active monitoring of related devices, and standardization of terminology, data link connectors, and other technician concerns.

There are many standards and requirements of OBD II that directly affect the transmission technician. These include a new DTC numbering system (Figure 14–16), additional DTCs, the configuration of the sixteen-pin data link connector (DLC) (Figure 14–17), standard location of the DLC (Figure 14–18), and many PCM/TCM-related component name changes. These conditions have stimulated uniformity in the automobile industry and should prove to be quite helpful to the diagnostician.

Since the transmission is an integral factor in the level of vehicle emissions, many transmission failures can cause the MIL to light. Shown in Figure 14–19 is a representation of a GM Tech 1 display format for an OBD II DTC. Due to the new code format and numbering system, manual retrieval of DTCs is not possible.

The shop manual will guide the technician to an understanding of what conditions may have set the code. Further

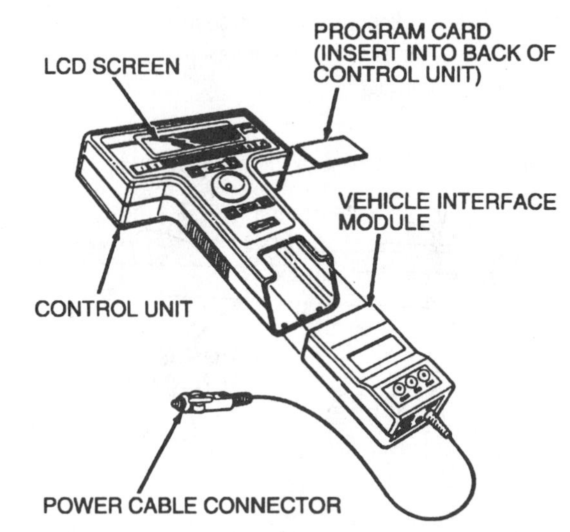

FIGURE 14–14 New Generation Star (NGS) tester. (Reprinted with the permission of Ford Motor Company)

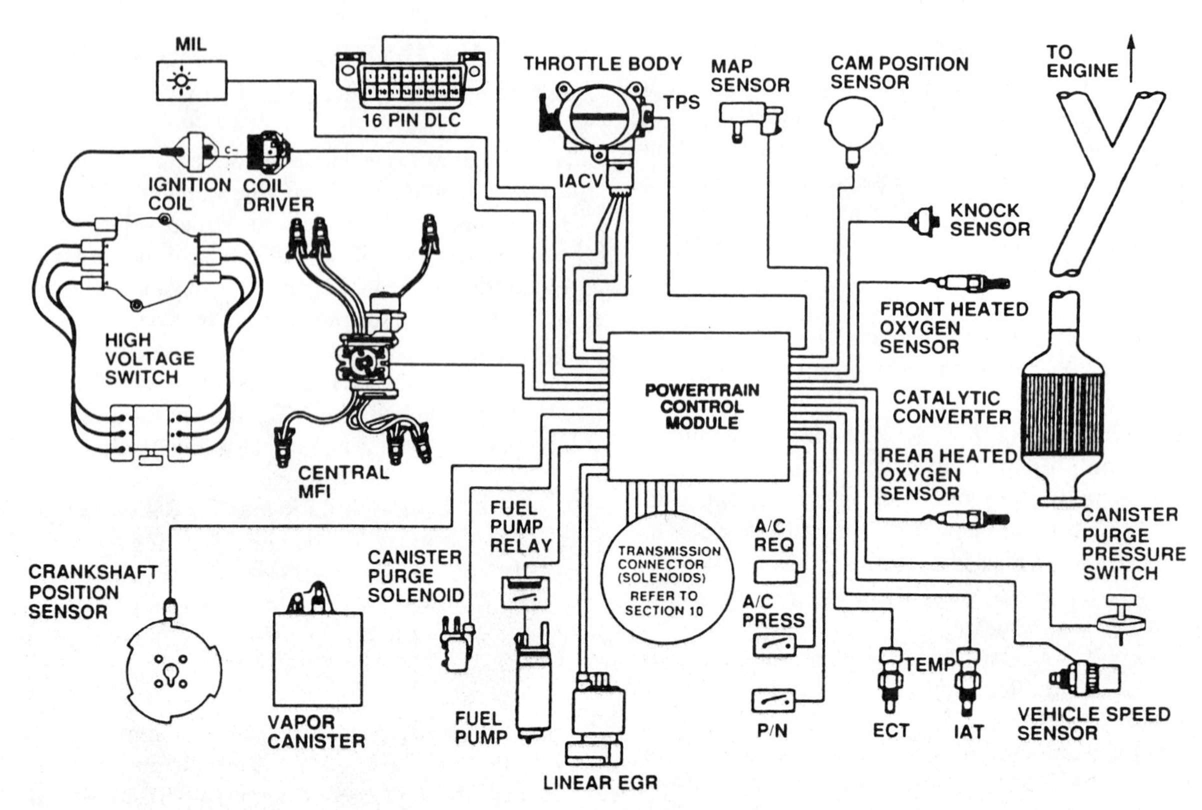

FIGURE 14–15 Common OBD II input and output devices. (Courtesy of General Motors Corporation)

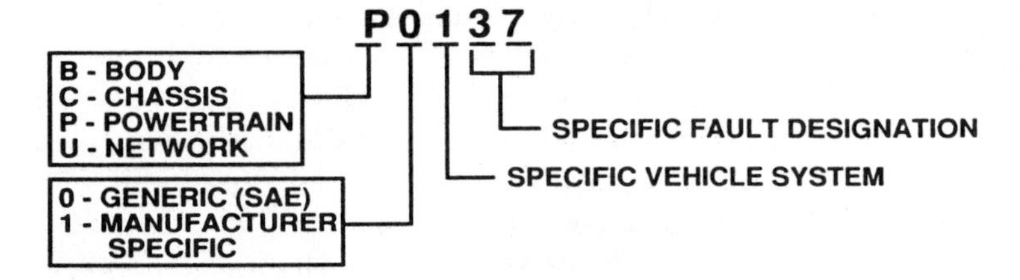

FIGURE 14–16 OBD II diagnostic trouble code (DTC) format. Transmission "Specific Vehicle System" codes are represented by a #7 or #8. (Courtesy of General Motors Corporation)

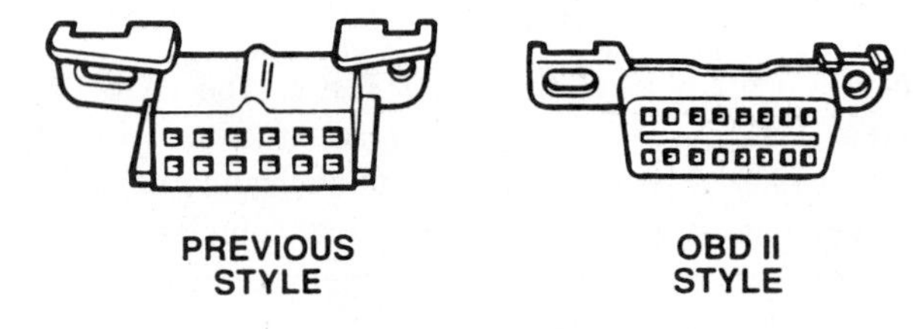

FIGURE 14–17 Data link connectors (DLCs)—original GM ALDL (OBD I) and OBD II styles. (Courtesy of General Motors Corporation)

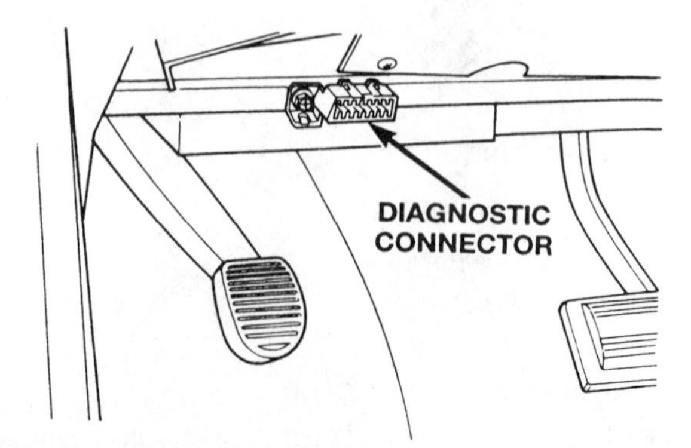

FIGURE 14–18 OBD II DLC location. (Courtesy of Chrysler Corp.)

DTC	Pxxxx
Last Test Failed	
Failed Since Clear	
History DTC	
MIL Requested	

Tech 1 Display

FIGURE 14–19 Tech 1 OBD II display. (Courtesy of General Motors Corporation) study and awareness of how to diagnose and repair vehicles equipped with OBD II technology is encouraged.

TROUBLESHOOTING PRINCIPLES

The poor operating condition of certain components tends to have a strong influence on the proper functioning of the transmission electrical and electronic circuitry. Below is a list of common causes of electronically controlled transmission concerns. This is followed by lists of troubleshooting tips and jump starting recommendations.

Electrical Problems

- Battery voltage—minimum 11.5 V.
- Fuse.
- Misalignment of terminals at connections.
- Defective throttle position sensor.
- Defective speed sensor.
- Defective solenoids.
- Pulse generator wires crossed.
- Solenoid wires crossed.
- Improper accessory installation—CB or cellular phone.
- Intermittent continuity of circuit wire within the connector.

Troubleshooting Tips

1. To function, the computer must have a minimum power supply of 11.5 V.
2. Fuses that "blow" immediately indicate a copper-to-ground short circuit before the load device in the power supply. Do not overlook crossed wiring on pressure switches due to human error or an incorrect wiring color code.
3. Never replace a fuse with a higher rating than specified.
4. Always use a high-impedance test light. A standard circuit-powered 12-V test light will allow too much current to flow in the circuit, resulting in instant or delayed failure of electronic components. Self-powered test lights must use a high-impedance bulb with no more than a 1.5-V battery.
5. Use a high-impedance (minimum ten megohm) digital DVOM for circuit testing. An analog (needle-style) volt/ohmmeter will cause electronic component damage. Again, too much current will flow in the circuit.
6. If a circuit short is suspected to be the cause of a computer failure, use an ohmmeter to check the continuity of the output wires to ground.
7. Voltage spikes from electric welding can cause computer failure. Always disconnect and remove the computer from the vehicle before welding. Disconnecting the battery ground will not protect the computer.
8. Before replacing the computer, be sure to follow the manufacturer's procedure for checking out the solenoid isolation diodes. These are used in various circuit controls other than the transmission, such as the air conditioner compressor magnetic clutch circuit.
9. Because of the low circuit current flow, always check the ground circuitry using a voltage drop test. A good ground will read zero to 0.05 V. Marginal grounds will read 0.5 V plus. An open ground will read 12 V. The ground is critical to the operation of the computer and is, surprisingly, the fault in a significant number of computer malfunctions.
10. To prevent possible damage from voltage spikes, the ignition must be turned off when:
    - jump starting.
    - battery charging.
    - disconnecting or connecting the computer or computer circuit components.
11. Do not probe the wires with a pointed instrument under any circumstances. This opens the wire insulation, and eventually outside contaminants start a corroding (resistance) effect. Minor resistance greatly affects milliampere circuitry. You will need to make connector jumper wires compatible to break into the circuit.
12. Wire terminal ends must be both crimped and soldered, using 60:40 rosin-core solder. Follow the manufacturer's procedure for repairing weatherproof connectors.
13. Connectors are a major area of concern, especially on intermittent problems. Wiggle circuit connector halves and wires back and forth, and watch your multimeter. If the voltage readings change, separate the connector halves and look for misaligned terminals and corrosion. Should the terminals look good, give your attention to the wire and terminal connection in the connector. Repair or replace as necessary.
14. Two supply items are available at your local radio-TV supply store:
    - tuner spray to clean connector terminals.
    - connector lubricant to prevent terminal corrosion.
15. Look for crossed wiring to solenoids and pressure switches. Sometimes wire color codes are not true to the wiring diagram.
16. Check for improper accessory installation, especially a CB or cellular phone. Hookups within 18 in (483 mm) may cause the computer operation and transmission to act up. Accessory hookups cannot be allowed to interrupt the power supply to the computer or cause magnetic interference.
17. Computer circuit wiring cannot run parallel with a high-current-carrying wire. The magnetic induction effect will disrupt computer operation. The wires can cross each other at right angles but never run parallel. Computer wiring also must be kept out of the influence of the ignition secondary. It is not unusual to solve a computer problem by simply correcting the wire routing of a sensor to its original placement. The electromagnetic environment of the vehicle, especially under the hood, is very hostile to electronics. Sensor wire routing is critical.
18. Whenever the battery is disconnected or removed on a computer-controlled car, the memory for radio station selection, seat position, climate control setting, and so on is erased. You may record the settings before battery removal, and then reprogram these before returning the

vehicle to the customer, although a better method is to plug a transistorized battery-powered "memory keep-alive" device into the cigar lighter prior to battery removal.

Jump Starting

1. The ignition switch and all electrical devices, including the dome light and radio, must be OFF before connecting or disconnecting the jumper cables. This avoids damage from voltage spikes.
2. Cable polarity must be correct. Reverse polarity, even for a brief moment, will cause component damage.
3. There must be absolutely no 24-V jumps.
4. Once the "dead" car is running, remove both jumper connections before turning on any electrical circuits.

Electrostatic Discharge (ESD)

The human body is a great generator of electrostatic voltage. As shown in the data (Chart 14–2), voltages can be generated as high as 35,000 V, depending on atmospheric conditions. When an **electrostatic discharge (ESD)** of high voltage occurs, it is an obvious threat to the delicate on-board automotive electronics used in engine management controls, instrument panels, and radios. The technician needs only to touch the terminal ends of an electronic control device to invite trouble. What is not usually known is that the human body never feels a static discharge below 3000 V and that ESD levels much lower than 3000 V are very capable of destroying or structurally weakening electronic components. If you cannot feel or hear the ESD, it is hard to be convinced that damage has occurred.

❖ **Electrostatic discharge (ESD):** An unwanted, high-voltage electrical current released by an individual who has taken on a static charge of electricity. Electronic components can be easily damaged by ESD.

Electrostatic discharge (ESD) often does not result in the immediate destruction of electronic components, but it does have a weakening effect. This results in eventual early failure or erratic operation: "Now it works, now it doesn't." It is obvious that the technician must learn some new and special service procedures when working with automotive electronics.

Static Grounding

The technician should be grounded to safely drain off any static buildup in the human body. Similarly, a grounded conductive mat is needed as a safe place on which to set down sensitive electronic parts, rather than the carpeting or seats which can carry thousands of static volts. Synthetic clothing materials worn by the technician can have the same effect. The working area and technician must be statically neutralized, whether working with electronics in the car or at the bench.

Featured in Figure 14–20 is a static protection kit. Essentially, the technician and conductive mat share a common ground connected to the car body. Some technicians may statically neutralize themselves by touching the metal bench or car body before working on sensitive electronics, but they should remember that there is still a failure to prevent regeneration of a static body charge. Any foot or arm movement, or body movement on a seat cover, quickly regenerates a charge at destructive levels.

	Electrostatic Voltages	
Means of Static Generation	10 to 20 Percent Relative Humidity	65 to 90 Percent Relative Humidity
Walking across carpet	35,000	1,500
Walking over vinyl floor	12,000	250
Worker at bench	6,000	100
Vinyl envelopes for work instructions	7,000	600
Common poly bag picked up from bench	20,000	1,200
Work chair padded with polyurethane foam	18,000	1,500

CHART 14–2 Electrostatic voltages. (Courtesy of 3M Electrical Specialities Division)

SERVICING DIAGNOSTIC TROUBLE CODES

Transmission malfunctions that occur and are associated with diagnostic trouble codes (DTCs) can be typically solved through the use of service manuals. By following the suggested methods in the manuals in an orderly manner, the technician usually can determine efficiently the cause of the problem. Avoid skipping steps of procedure with which you are unfamiliar just because they will slow you down. Complete each step before moving on to the next. Disregard of diagnostic recommendations increases the chance that your final determination will be incorrect—if you reach a conclusion. A careless approach may lead to wasted time and effort. Rather, stay on course.

As mentioned earlier, there needs to be a logical approach to tracing codes that are generated. Remember to pursue engine input codes, then transmission input codes, and lastly transmission output codes. Most manufacturers recommend starting with the lowest numerical engine DTC first. Also, follow up on hard codes before soft codes, since they are actively present and easier to isolate. Before clearing codes from memory to determine if they will appear again, record them on paper for reference.

To demonstrate the use of various service manual diagnostic charts and references, the next part of this chapter is divided into three units covering engine input codes, transmission input codes, and transmission output codes. Each unit highlights a different DTC and a different manufacturer's shop manual diagnostic references. This will illustrate the variety of resources available to technicians.

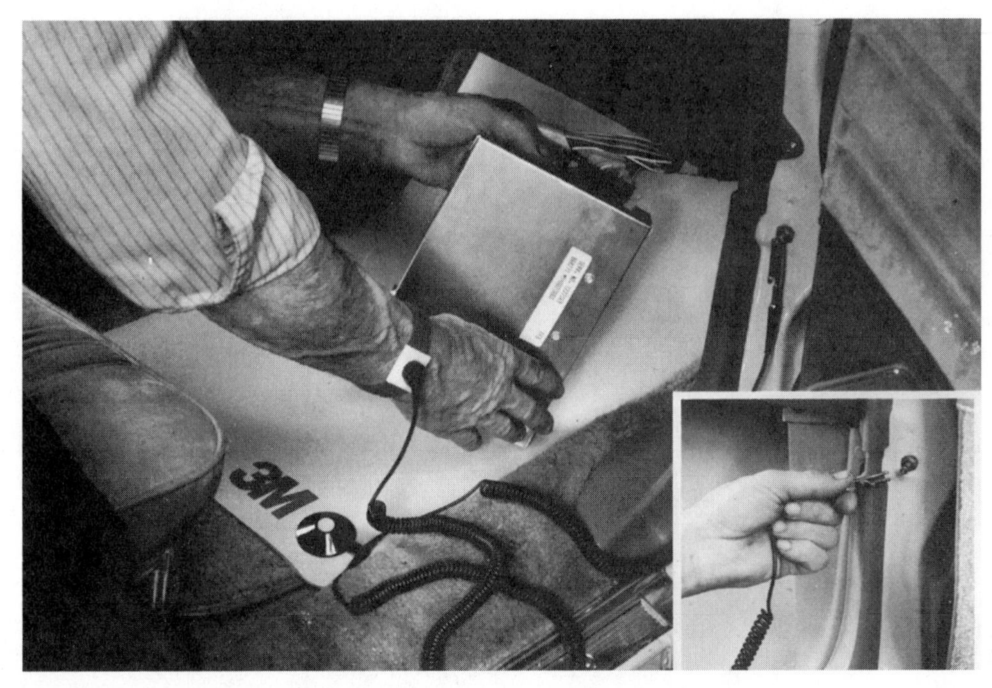

FIGURE 14–20 Static protection kit. (Courtesy of 3M Electrical Specialities Division)

Engine Input DTC Diagnosis

GM 4L60-E (1994 C/K Series Pickup Truck). Transmission symptoms described by the customer and verified by the technician include no TCC application, fixed shift points, harsh engagements, and an illuminated MIL. In addition, the engine runs rough.

After checking the fluid level and condition and making a visual inspection of mechanical and electrical components, connect a scan tool to the DLC. In this example, a code 21 is present. Referring to Chart 14–3, page 348, Diagnostic Trouble Codes and Default Actions (a partial listing), review code 21. The PCM senses a voltage reading from the TP sensor circuit that is too high—greater than 4.88 V. The vehicle symptoms are listed in the Default Actions column. From the scan tool, TP sensor voltage can be read. The tool shows that it is, indeed, over 4.88 V, regardless of throttle plate angle.

Making an assumption that the TP sensor is faulty and replacing it would be a poor choice. Rather, use the diagnostic help available in the Driveability and Emissions section (3A) in the shop manual. There, as shown in Figures 14–21A and 14–21B, you will find a wiring diagram and troubleshooting chart to assist you in locating the cause of a DTC 21.

Follow the steps in Figure 14–21B.

Step 1: The scan tool reads greater than 4.88 V at all throttle plate positions.

Step 2: By disconnecting the TP sensor, the scan tool now reads less than .2 V. Though the voltage has dropped, do not condemn the TP sensor as being faulty.

Step 3: Probe the sensor ground terminal at the detached connector with a test light connected to battery voltage. Since the test light remains off, inspect and locate the open ground circuit. If a broken #452 BLK wire or faulty terminal end is located, repair or replace as needed. (If not found, perhaps the PCM is not functioning properly).

After repairing the open condition, reconnect the TP sensor and read the voltage on the scan tool. It should now sweep between 0.5 V at 0% opening and 4 V at wide open throttle. Clear the code from the memory. Road test the vehicle to verify that the engine and transmission operate properly and that the MIL remains off.

Determining the cause of an engine/transmission-related DTC can be a manageable process. Locate the proper reference sources and follow the procedures described.

Transmission/Transaxle Input DTC

Chrysler 41TE (A-604) (1996 Plymouth Voyager). Transaxle symptoms described by the customer and verified by the technician include: only park, neutral, reverse, and second gear are available; there are no upshifts or downshifts; although the vehicle lacked performance, it was able to be driven to the service facility; and the MIL is illuminated.

These symptoms signify a "limp-in" condition, where electrical power to the transaxle is shut down at the relay, which helps prevent additional internal transaxle deterioration and unsafe vehicle operation. Numerous transaxle-related problems and codes will put the unit into this mode, including

DIAGNOSTIC TROUBLE CODES AND DEFAULT ACTIONS

Refer to the Driveability and Emissions Electrical Diagnosis Manual

TROUBLE CODE	CODE PARAMETERS	DEFAULT ACTION
14 Engine Coolant Temp Sensor Circuit (High)	Engine coolant temp over 151°C (306°F) for 1/2 second.	• TCC apply cold.
15 Engine Coolant Temp Sensor Circuit (Low)	Engine coolant temp less than -40°C (-40°F) for 1/2 second.	• TCC apply cold.
21 Throttle Position Sensor Circuit (High)	TP voltage greater than 4.88 volts for four seconds.	• No TCC. • Fixed shift points. • Harsh shifts. • Maximum line pressure. • No fourth gear in hot mode.
22 Throttle Position Sensor Circuit (Low)	With engine running, TP voltage less than .06 volts for four seconds. (Diesel is less than .16 volts.)	• No TCC. • Fixed shift points. • Harsh shifts. • Maximum line pressure. • No fourth gear in hot mode.
24 Vehicle Speed Sensor Signal Low	In Drive or Reverse with engine speed greater than 3000 rpm, output speed is less than 250 rpm for three seconds. (MAP is 100-255 kPa, TP is 10-100%)	• Maximum line pressure. • Second gear only.
28 Fluid Pressure Switch Assembly Fault	PCM detects one of two "invalid" combinations of PSM signals for five seconds.	• No TCC. • Harsh shifts. • No fourth gear in hot mode. • No 2nd gear in D2.
37 Brake Switch Stuck "ON"	With brake on, vehicle speed is 5-20 mph for six seconds, then vehicle speed is >20 mph for six seconds. After vehicle speed is < 5 mph. This must happen seven times.	• No TCC. • No fourth gear in hot mode.
38 Brake Switch Stuck "OFF"	With brake off, vehicle speed is >20 mph for six seconds, then vehicle speed is 5-20 mph for six seconds. After vehicle speed is < 5 mph. This must happen seven times.	• No TCC. • No fourth gear in hot mode.
52 Long System Voltage High	Generator voltage is greater than 16 volts for 109 minutes.	• No TCC. • Maximum line pressure. • Third gear only.
53 System Voltage High	Generator voltage is greater than 19.5 volts for two seconds.	• No TCC. • Maximum line pressure. • Third gear only.
58 Transmission Fluid Temp Sensor Circuit (High)	Transmission fluid temperature is greater than 151°C (306°F) for one second.	• No default action.
59 Transmission Fluid Temp Sensor Circuit (Low)	Transmission fluid temperature is below -40°C (-40°F) for one second.	• No default action.

RH0026-4L60- E

CHART 14–3 4L60-E DTCs and default action. (Courtesy of General Motors Corporation, Service Technology Group)

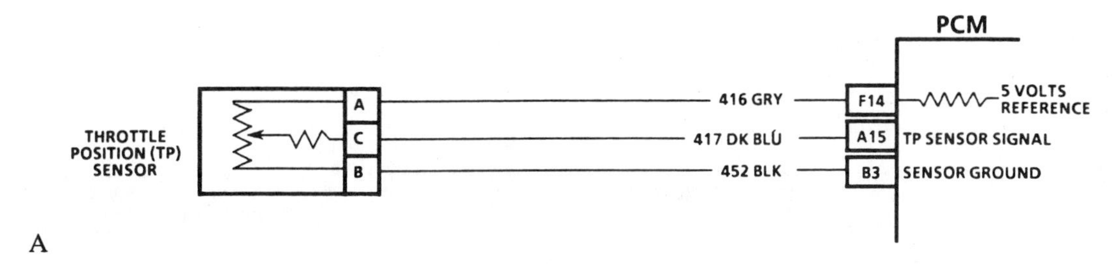

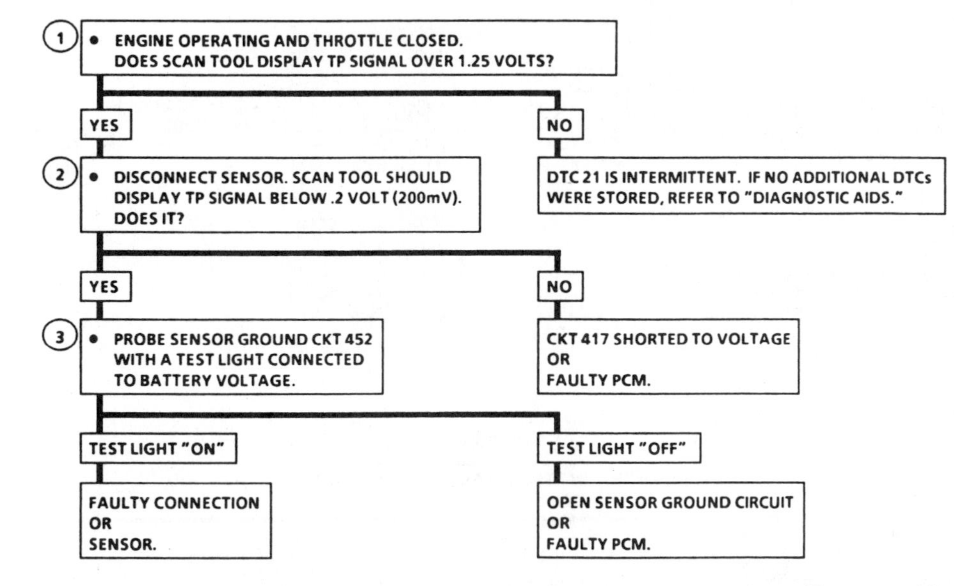

FIGURE 14–21 (A) DTC 21 throttle position (TP) sensor circuit. (B) DTC 21 troubleshooting procedures. (Courtesy of General Motors Corporation, Service Technology Group)

hydraulic, mechanical, and electrical problems. Hydraulic circuits fed by the manual valve position allow the operation of the transaxle, as described.

Figure 14–22 shows a Chrysler flow chart for diagnostic troubleshooting. Chrysler recommends checking for diagnostic trouble codes, using the powertrain diagnostic test procedure manual, before attempting any repairs.

Before taking a road test, check the fluid level and condition and the gearshift cable adjustment. If a problem continues after the prechecks and repairs are completed, hydraulic pressure checks should be performed.

Most likely, a PCM code 45 (OBD II generic code P0700) would appear. Reviewing the service manual, Group 25—"Emission Control Systems," the code states that an automatic transmission DTC is present in the TCM. It directs the technician to Group 21.

Group 21 is titled "Transaxle" and includes transaxle diagnostic and repair procedures. By requesting TCM fault codes through the scan tool, a code 21 appears. Chart 14–4, page 350, shows the possible causes for setting a code 21 when the overdrive clutch pressure is too low. It should be apparent that one must not jump to conclusions in the diagnostic procedure. Rather, follow a logical process.

If more than one TCM code is displayed, a matrix style chart showing common causes can be beneficial (Chart 14–5, page 351). Become familiar with the description and background of code 21 (Chart 14–6, page 352).

Figure 14–23 shows the electrical circuit layout for the PCM and the solenoid and pressure switch assembly. Notice that TCM pin #9 reads whether or not the OD pressure switch is hydraulically closed. Located between the TCM and the solenoid and pressure switch assembly is the eight-way in-line connector. Many pinpoint circuit tests can be made at this junction, located at the solenoid and pressure switch assembly. The assembly is mounted externally on the transaxle.

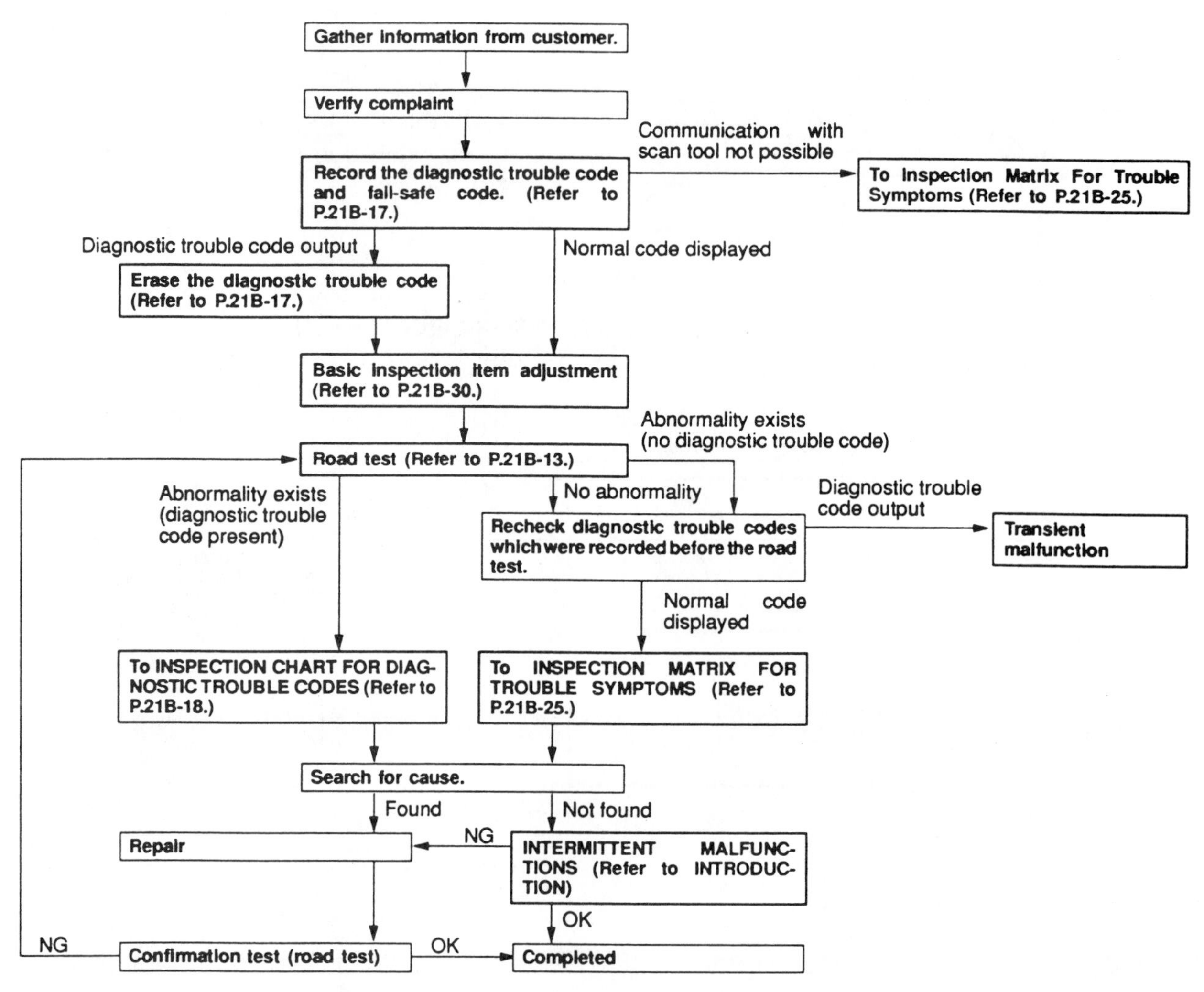

FIGURE 14–22 Chrysler diagnostic troubleshooting procedures. (Courtesy of Chrysler Corp.)

CONDITION	POSSIBLE CAUSES	CORRECTION
DTC 21- OD CLUTCH PRESSURE TOO LOW	1. Faulty cooling system.	1. Check and repair cooling system.
	2. Torque converter clutch failure.	2. Repair torque converter clutch circuit.
	3. Internal solenoid leak.	3. Check solenoid clutch pack.
	4. Pressures too high.	4. Check fluid pressure at ports.
	5. Valve body leakage.	5. Check valve body surface for warpage.
	6. Regulator valve.	6. Check regulator valve for sticking.
	7. Sticking valves.	7. Check solenoid pack and valve body for sticking valves.
	8. Plugged filter.	8. Replace trans. oil filter.
	9. Worn accumulator seal rings.	9. Replace accumulator seal rings.
	10. Failed clutch seals.	10. Replace clutch seals.
	11. Failed OD clutch.	11. Repair OD clutch.
	12. Worn oil pump.	12. Replace oil pump.
	13. Worn reaction shaft support seal rings.	13. Replace reaction shaft support seal rings.
	14. High fluid level.	14. Drain fluid to proper level.
	15. Low fluid level.	15. Fill fluid to proper level.

CHART 14–4 DTC 21 possible causes and corrections. (Courtesy of Chrysler Corp.)

INSPECTION MATRIX FOR DIAGNOSTIC TROUBLE CODE (Internal transaxle problem)

110001736

Code No.	Condition	Code No.	Condition
21	OD clutch – pressure too low	38	Partial lockup control out of range
22	2/4 clutch – pressure too low	46	UD clutch – not lowering pressure
23	OD & 2/4 clutch – pressure too low	47	Solenoid switch valve stuck in the LR position
24	LR clutch – pressure too low	50	Speed ratio default in Reverse
25	OD & LR clutch – pressure too low	51	Speed ratio default in 1st
26	2/4 & LR clutch – pressure too low	52	Speed ratio default in 2nd
27	OD, 2/4 & LR clutch – pressure too low	53	Speed ratio default in 3rd
31	OD clutch pressure switch response failure	54	Speed ratio default in 4th
32	2/4 clutch pressure switch response failure	60	Inadequate LR element volume
33	OD & 2/4 clutch pressure switch response failure	61	Inadequate 2/4 element volume
37	Solenoid switch valve stuck in the LU position	62	Inadequate OD element volume

Probable cause	Code No.	21	22	23	24	25	26	27	31	32	33	37	38	46	47	50	51	52	53	54	60	61	62
Low fluid level		X	X	X	X	X	X	X	X	X	X					X	X	X	X	X			
Aerated fluid (High fluid level)		X	X	X	X	X	X	X	X	X	X			X									
Worn or damaged reaction shaft support sealing		X														X	X	X	X	X			
Worn or damaged input shaft sealing													X			X	X	X	X	X			
Worn pump		X	X	X	X	X	X	X	X	X	X		X			X	X	X	X	X			
Damaged or failed clutch(es)	Underdrive clutch													X			X	X	X				
	Overdrive clutch	X							X										X	X			X
	Reverse clutch															X							
	2/4 clutch		X							X								X		X		X	
	Low/ Reverse clutch				X											X	X				X		
Damaged clutch seal		X	X	X	X	X	X	X						X		X	X	X	X	X	X	X	X
Worn or damaged accumulator sealing		X	X	X	X	X	X	X						X		X	X	X	X	X	X	X	X
Plugged filter		X	X	X	X	X	X	X			X					X	X						
Stuck/sticky valves		X	X	X	X	X	X	X				X	X		X	X	X	X	X	X			
Solenoid switch valve												X		X	X								
Lock-up switch valve													X										
Torque converter control valve													X										
Regulator valve		X	X	X	X	X	X	X						X		X	X	X	X	X	X	X	X
Valve body leakage		X	X	X	X	X	X	X	X	X	X	X	X	X	X	X	X	X	X	X			
Pressures too high		X	X	X	X	X	X	X					X	X		X	X	X	X	X			
Internal solenoid leak		X	X	X	X	X	X	X	X	X	X	X	X	X	X	X	X	X	X	X			

CHART 14–5 41TE (A-604)/42LE DTC matrix chart. (Courtesy of Chrysler Corp.)

DIAGNOSTIC TROUBLE CODE 21-27

DIAGNOSTIC TROUBLE CODE:	21-27 Pressure Switch Circuits Code 21 OD Pressure Switch Circuit Code 22 2/4 Pressure Switch Circuit Code 23 2/4-OD Pressure Switch Circuit Code 24 LR Pressure Switch Circuit Code 25 LR-OD Pressure Switch Circuit Code 26 LR-2/4 Pressure Switch Circuit Code 27 All Pressure Switch Circuits
BACKGROUND:	The transmission system uses three pressure switches to monitor the fluid pressure in the LR, 2/4, and OD elements. The pressure switches are continuously checked for the correct states in each gear as indicated below: Normal Pressure Switch States (see table below) O = Switch is open C = Switch is closed When a pressure switch mismatch is detected, the solenoid circuits are tested for continuity. If that test fails, solenoid circuits are blamed for the pressure switches mismatch. Otherwise the appropriate pressure switch code is set.
WHEN CHECKED:	Every 0.007 second.
ARMING CONDITIONS:	(1) More than 2.0 seconds since start-up. (2) No loss of transaxle oil pump prime. (3) Engine speed greater than 500 rpm. (4) No shift in progress. (5) Pressure switch mask inconsistent with the normal pressure switch state table. Use DRB II State Input/Output display.
CONDITIONS:	Pressure switch error count must equal 255.
SET TIME:	For hard faults when super cold = 3.3 seconds For hard faults when cold = 2.2 seconds For hard faults when warm = 1.4 seconds For hard faults when hot = 0.6 second (Temperature description based off of DRB II transaxle state display)
EFFECT:	Transmission limp-in.
POSSIBLE CAUSES:	Low/high fluid level in transmission. Short/open in LR Pressure Switch circuit, 2/4 Pressure Switch circuit, or OD Pressure Switch circuit. Solenoid pack internal problem. Internal transmission problem. 40-way connector problem (cavities 9, 47, and 50). Internal Transmission Control Module failure.

Normal Pressure Switch States

GEAR	LR	2/4	OD
R	O	O	O
N	C	O	O
1ST	C	O	O
2ND	O	C	O
3RD	O	O	C
4TH	O	C	C

O = Switch is open
C = Switch is closed

9321-310

CHART 14–6 41TE (A-604)/42LE DTC 21-27 description. (Courtesy of Chrysler Corp.)

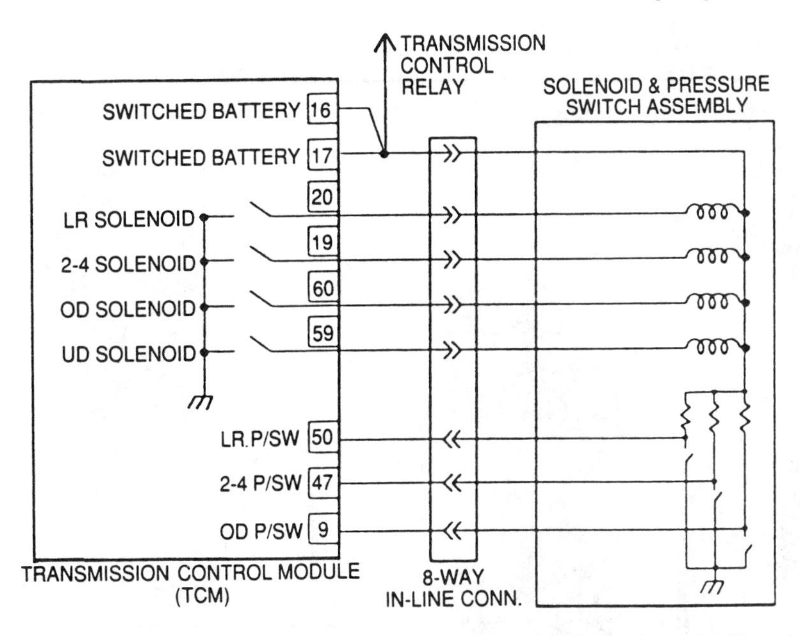

FIGURE 14–23 41TE (A-604)/42LE TCM and solenoid and pressure switch assembly electrical schematic. (Courtesy of Chrysler Corp.)

The powertrain diagnostic test procedure manual will lead the technician through specific electrical circuit tests. For a code 21, perform the necessary tests shown in Figures 14–24A and 14–24B, and in Figures 14–25A and 14–25B. Electrical open and grounded circuits can be located between the solenoid/pressure switch assembly connector, the relay, and the TCM. Also, the malfunctioning of the TCM, and the solenoid/pressure switch assembly as it relates to this circuit, can be determined. Executing the tests should pinpoint the cause of the problem.

If all of the wiring and electrical/electronic devices check out good, the problem is within the transaxle. The cause is either a hydraulic or a mechanical problem. At this point, refer back to the shop manual, Group 21, and conduct conventional transmission/transaxle diagnosis.

Transmission Output DTC

Ford E4OD (1995 Econoline Van). Transmission symptoms described by the customer and verified by the technician are as follows. Park, neutral, and reverse are normal. In overdrive range, first gear is available, with a late upshift to fourth gear that appears when a shift into third should take place. In manual 2 range, only second gear is present, with engine braking, while in manual 1 range, first gear is available. On a high-speed drop from overdrive range to manual 1, second gear engine braking is apparent, and then a downshift to first. The transmission control indicator lamp (TCIL) is flashing. The TCIL is located on the shift lever and next to the transmission control switch (TCS), which is used to cancel overdrive (fourth gear) (Figure 14–26).

Shown in Figure 14–27, from section 07-01A of the service manual, are the numerous electrical/electronic PCM input and output devices that relate to transmission operation. To pinpoint the cause of a transmission problem, Ford suggests using the diagnostic flow chart in Figure 14–28.

After performing the preliminary checks (know and understand the customer concern, check fluid level and condition, verify the concern, and so on), check the PCM for codes. If using a Super Star Tester II, a hard code 617 would be present. A four-digit code (P0781) can be accessed with a New Generation Star (NGS) tester.

Reviewing the description for code P0781/617, shown in Chart 14–7, page 360, the problem could be with a shift solenoid or internal transmission components. Chart 14–8, page 361, lists the electrical and hydraulic/mechanical components that could be at fault, and the action to take. Always pursue the electrical possibilities first. Refer to the solenoid operation chart (Chart 14–9, page 361).

The shift solenoids can malfunction in either the ON or OFF position and cause different situations, as shown in Chart 14–10, page 362. The actual gears obtained in the forward ranges match the symptoms when shift solenoid 2 is always OFF. Shift solenoid 2 and its circuitry is suspected at this time, but avoid jumping to conclusions or skipping steps of procedure.

Shift solenoid 2 is part of the transmission solenoid body, shown in Figure 14–29. The unit contains both shift solenoids, the pressure control solenoid (PCS)/electronic pressure control (EPC) solenoid, and the TCC solenoid. When replacing an expensive component like this, you want your diagnosis and repair to be accurate.

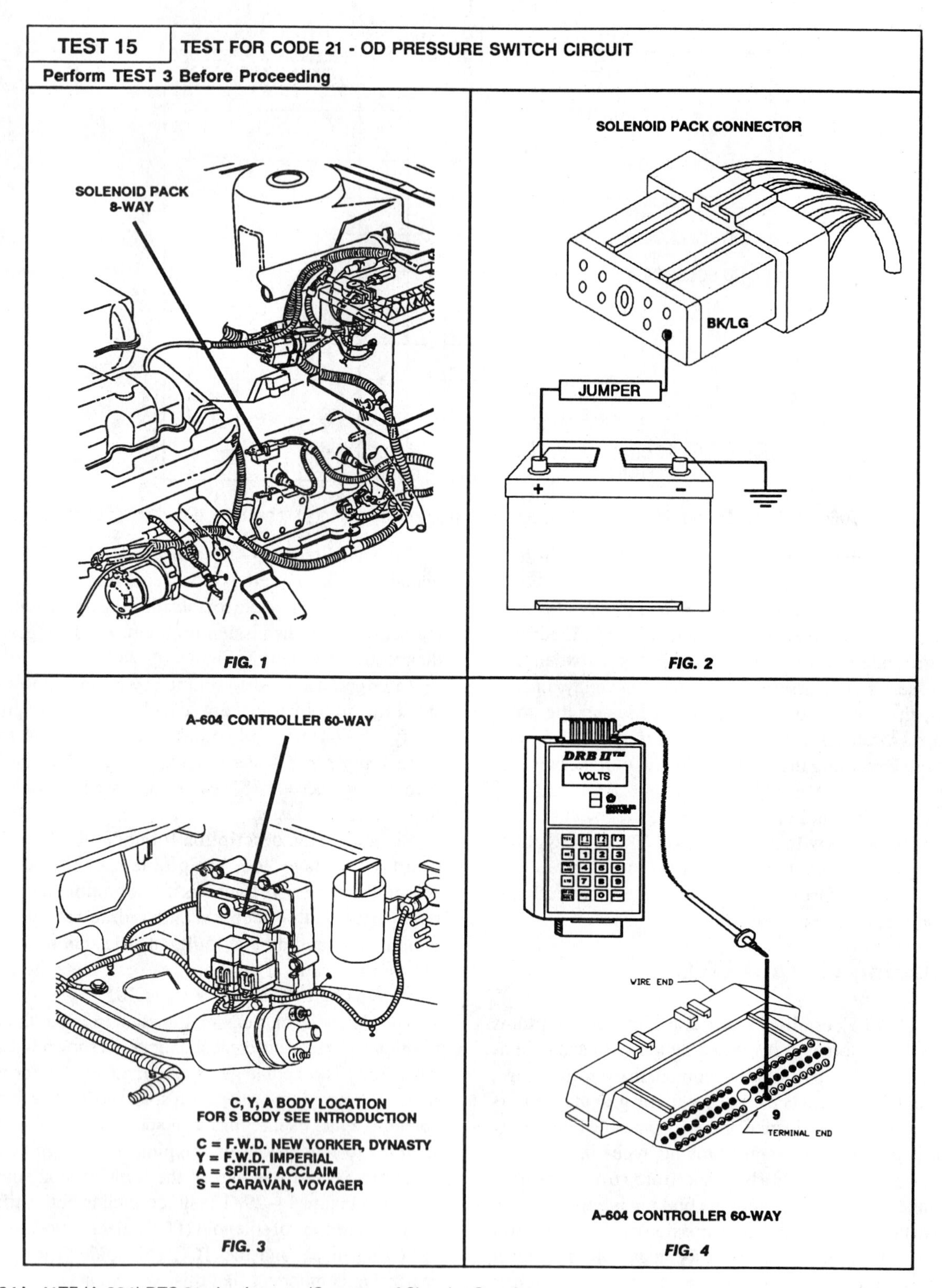

FIGURE 14–24A 41TE (A-604) DTC 21 circuit tests. (Courtesy of Chrysler Corp.)

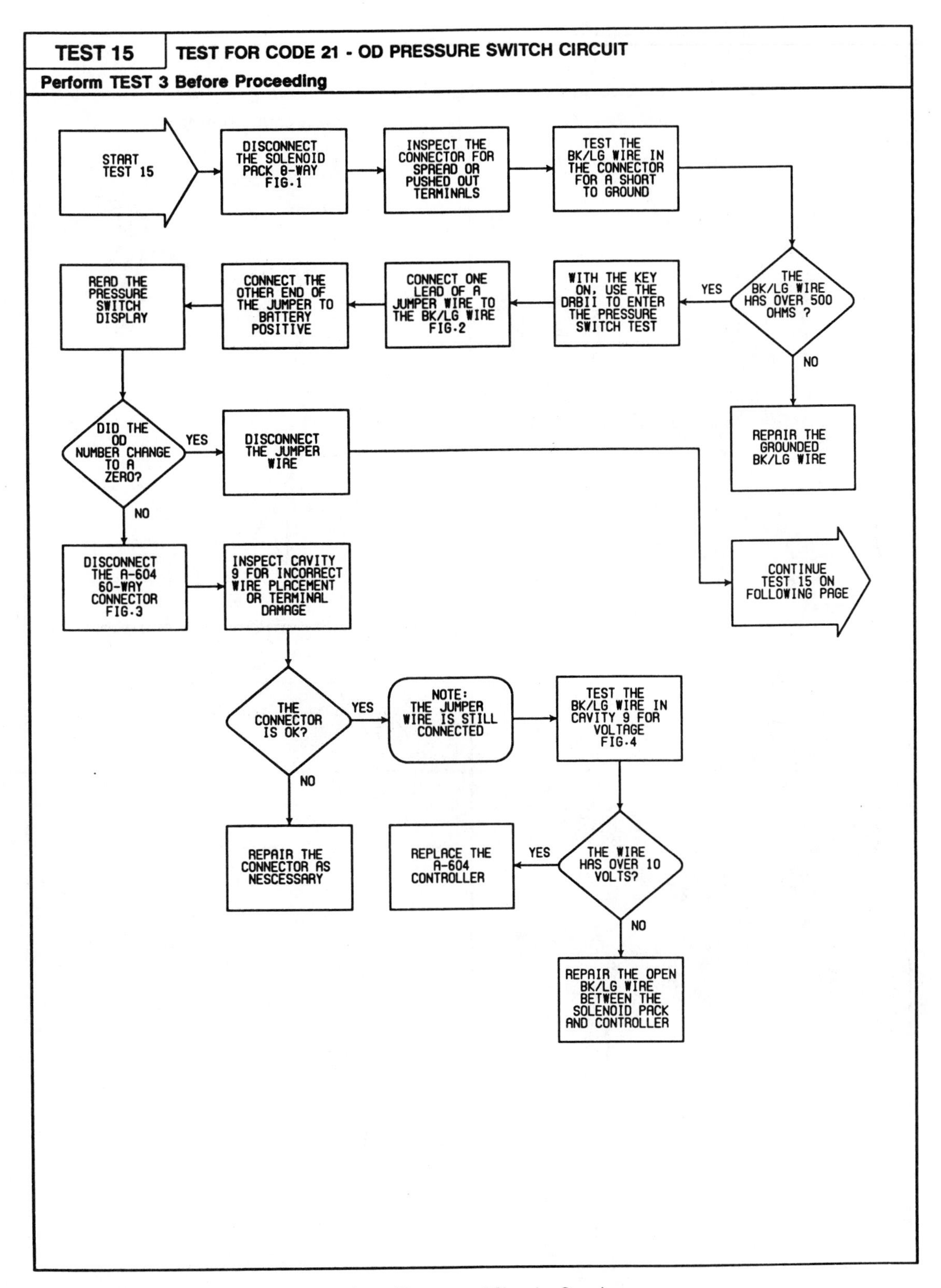

FIGURE 14–24B 41TE (A-604) DTC 21 circuit test flow chart. (Courtesy of Chrysler Corp.)

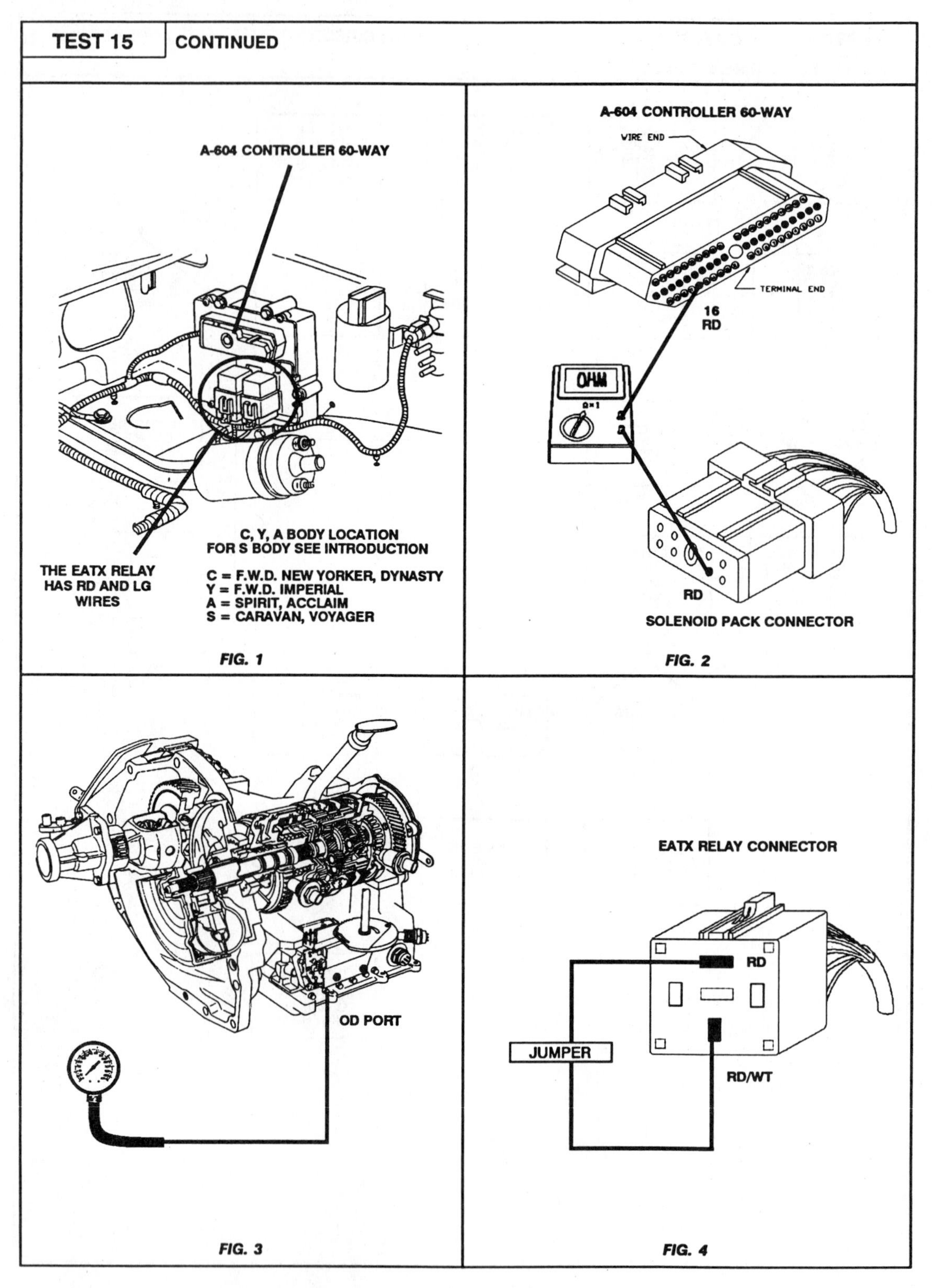

FIGURE 14–25A 41TE (A-604) DTC 21 circuit tests continued. (Courtesy of Chrysler Corp.)

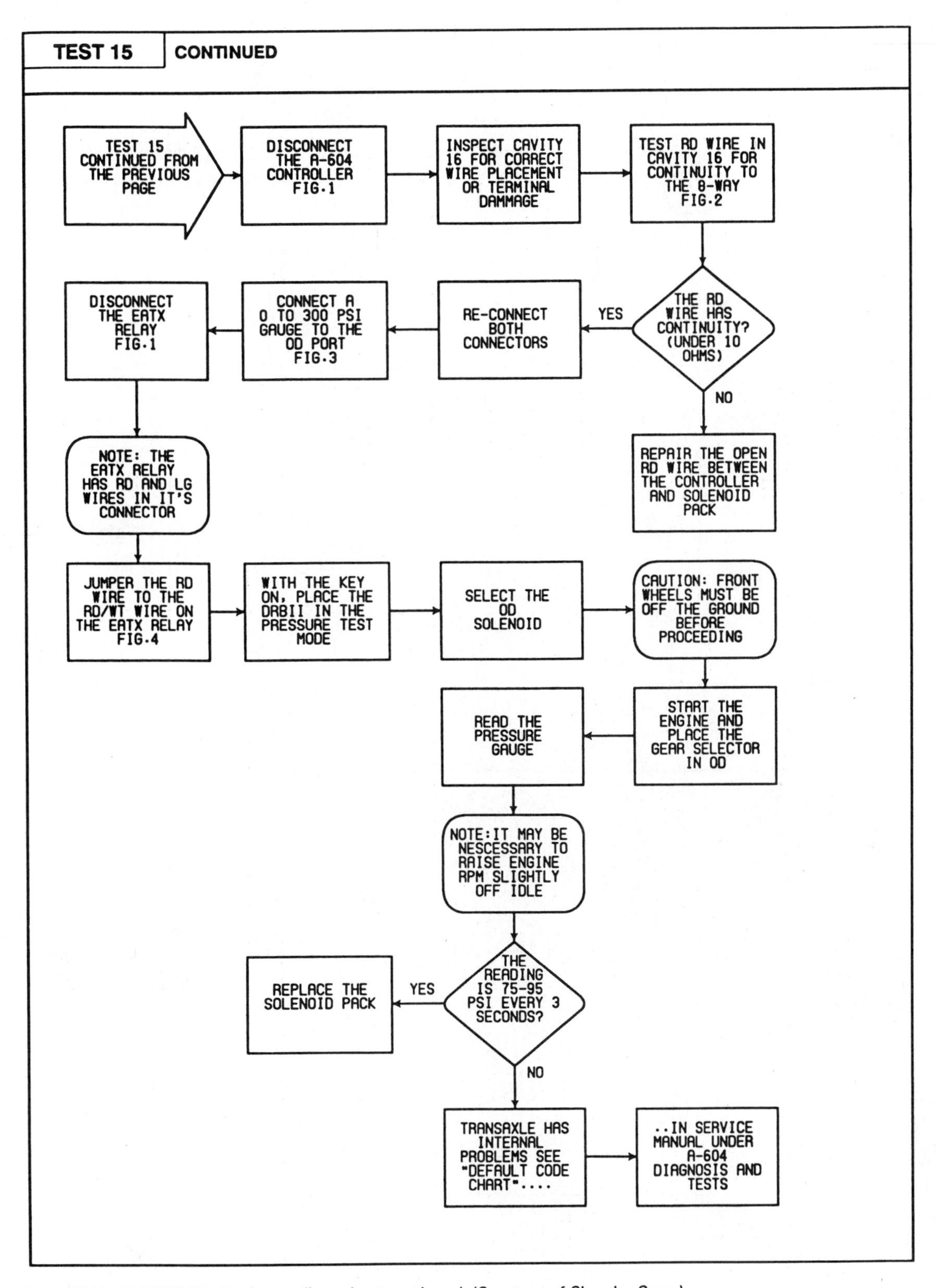

FIGURE 14–25B 41TE (A-604) DTC 21 circuit test flow chart continued. (Courtesy of Chrysler Corp.)

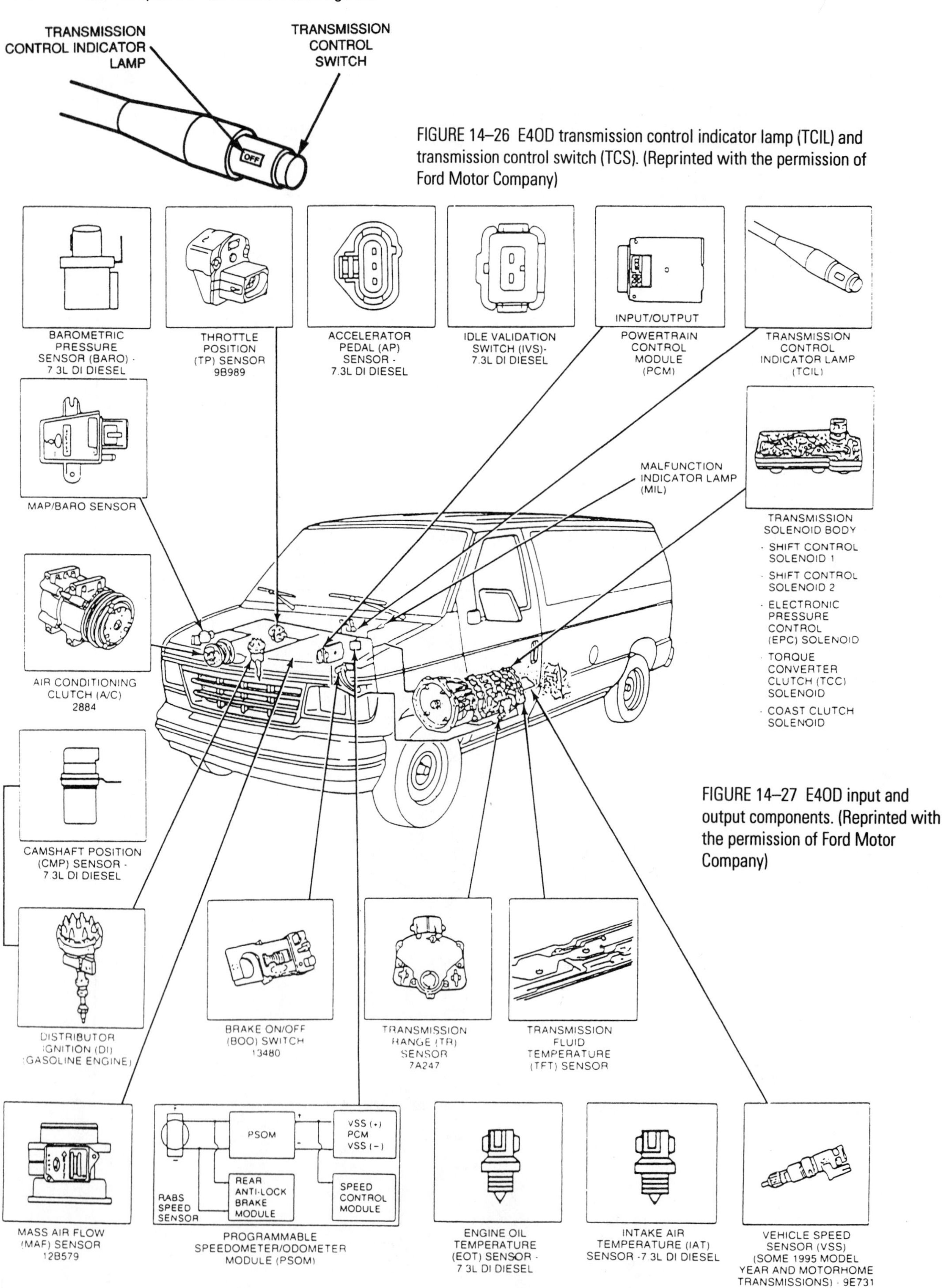

FIGURE 14–26 E4OD transmission control indicator lamp (TCIL) and transmission control switch (TCS). (Reprinted with the permission of Ford Motor Company)

FIGURE 14–27 E4OD input and output components. (Reprinted with the permission of Ford Motor Company)

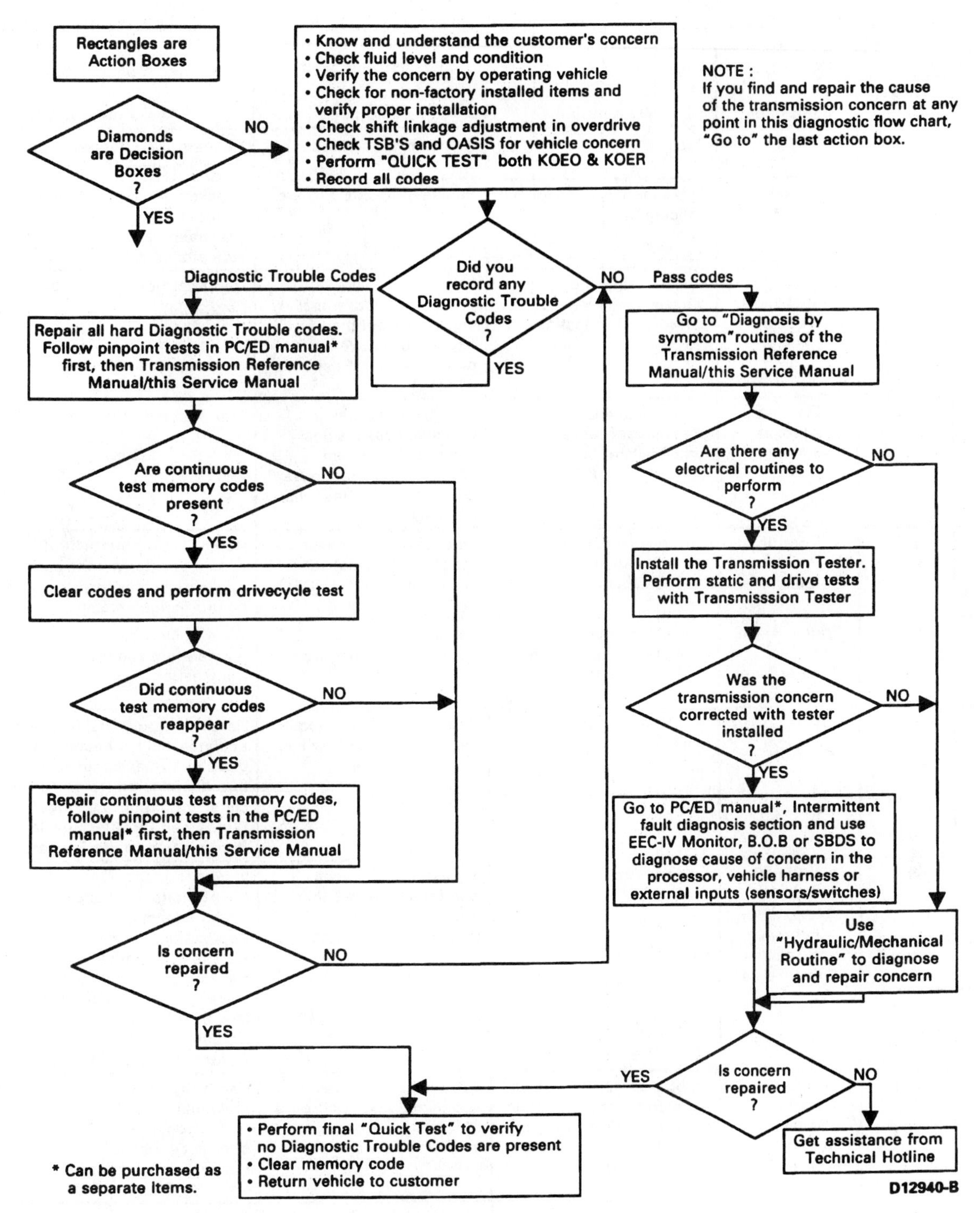

FIGURE 14–28 Diagnosis flow chart. (Reprinted with the permission of Ford Motor Company)

Diagnostic Trouble Code					
Four Digit	**Three Digit**	**Component**	**Description**	**Condition**	**Symptoms/Actions**
P1111	111	System	Pass	No concern detected.	Concern not detected by powertrain control module. [ab]
P0340 P0341 P0344	211	Crankshaft position sensor	Crankshaft position sensor concern.	Engine RPM circuit failure.	Crankshaft position sensor failure / engine will stall or not run.) May flash transmission control indicator lamp.[b]
—	126	Manifold absolute pressure sensor	Manifold absolute pressure sensor out of On-Board Diagnostics range.	Manifold absolute pressure sensor signal higher or lower than expected or no response during Dynamic Response (Goose) Test.	Rerun On-Board Diagnostics.
—	128	Manifold absolute pressure sensor	Manifold absolute pressure sensor vacuum circuit failure.	Manifold absolute pressure sensor signal higher or lower than expected or no response during Dynamic Response (Goose) Test.	Firm shift fell, late shifts at altitude.[b]
—	121	Throttle Position Sensor	Throttle position sensor out of On-Board Diagnostics range.	Throttle position sensor (gasoline engines) not at idle position during KOEO.	Rerun at appropriate throttle position sensor position per the engine application. May flash transmission control indicator lamp.
P1711	636	Transmission Fluid Temperature Sensor	Transmission fluid temperature sensor out of On-Board Diagnostics range.	Transmission not at operating temperature during On-Board Diagnostics.	Warm vehicle to normal operating temperature and rerun On-Board Diagnostics.
P0500	452	Vehicle Speed Sensor	Insufficient vehicle speed input.	Powertrain control module detected a loss of vehicle speed signal during operation.	Harsh engagements, firm shift feel, abnormal shift schedule, unexpected downshifts may occur at closed throttle, abnormal torque converter clutch operation or engages only at Wide Open Throttle. May flash transmission control indicator lamp.[a]
P0781[c]	617[c]	Shift Solenoid 1, Shift Solenoid 2 or Internal Transmission Components	1-2 shift error.	Engine RPM drop not detected when 1-2 shift was commanded by powertrain control module.	Improper gear selection depending on failure mode and transmission range selector lever position. Refer to Shift Application Chart. Shift errors may also be due to other internal transmission concerns such as stuck valves or damaged friction material. May flash transmission control indicator lamp. **Refer to Pinpoint Test A.**
P0123	123	Throttle Position Sensor or accelerator pedal position sensor	Throttle position sensor or accelerator pedal position sensor circuit above maximum voltage, (short to vehicle power).	Voltage above or below specification for On-Board Diagnostics or during normal vehicle operation.	Harsh engagements, firm shift feel, abnormal shift schedule, abnormal torque converter clutch operation or does not engage. May flash transmission control indicator lamp.[b]
P0713	637	Transmission Fluid Temperature Sensor	-40°C (-40°F) indicated, transmission fluid temperature sensor circuit open.	Voltage drop across transmission fluid temperature sensor exceeds scale set for temperature -40°C (-40°F).	Torque converter clutch and stabilized shift schedule may be enabled sooner after cold start. **Refer to Pinpoint Test B.**

CHART 14–7 DTC description chart. (Reprinted with the permission of Ford Motor Company)

SHIFT CONCERNS: NO 1-2 SHIFT (AUTOMATIC)

Possible Component	Reference/Action
220 — ELECTRICAL ROUTINE	
Powertrain Control System • Electrical inputs/outputs, vehicle wiring harnesses, powertrain control module, throttle position sensor, vehicle speed sensor, shift solenoid 1, shift solenoid 2	• Run On-Board Diagnostics. Refer to Powertrain Control/Emissions Diagnosis Manual[a] for diagnosis. Perform Service Manual Pinpoint Test "A" using the Transmission Tester 007-0085C as outlined in this section. Service as required. Clear codes, road test and rerun On-Board Diagnostics.
320 — HYDRAULIC/MECHANICAL ROUTINE	
Shift Linkage (Internal/External) or Cable • Damage, misadjusted • Transmission Range (TR) sensor damaged, misadjusted	• Inspect for damage. Service as required. Adjust linkage as outlined. After servicing linkage, verify that the Transmission Range (TR) sensor is properly adjusted. Refer to Section 07-14A for Transmission Range (TR) sensor adjustment.
Main Controls • Bolts not tightened to specification • Gaskets damaged, misaligned • Shift solenoid 2 malfunction • D2 valve, 1-2 shift valve, 1-2 manual transition valve, intermediate clutch accumulator regulator valves, springs — stuck, damaged, missing or misassembled • Air bleed for shift solenoid 2 circuit damaged or missing • Wrong parts used in rebuild	• Retighten bolts to specification. • Inspect for damage and replace. • Refer to Electrical Routine No. 220. • Inspect for damage. Service as required. • Inspect for damage. Replace case. • Verify that proper parts were used.
Intermediate Clutch Assembly • Assembly • Seals or piston damaged • Friction elements worn, missing, damaged, misassembled • Ball check stuck/missing • Feedbolt torque incorrect, leaks, missing • Cylinder assembly outer diameter/case bore damaged, leaking	• Air check clutch assembly as outlined in this section. • Inspect for damage. Service as required. • Inspect for damage. Service as required. • Inspect for damage. Service as required. • Inspect and retighten bolts as required. • Inspect for damage. Service as required.
Intermediate One-Way Clutch Assembly • Cage/sprags damaged, improperly assembled on inner race • Improper components used in rebuild	• Inspect for damage. Service as required. • Verify that proper components are used.

a May be purchased as a separate item.

CHART 14–8 No 1-2 shift possible causes—E40D. (Reprinted with the permission of Ford Motor Company)

Gearshift Selector Lever Position	Powertrain Control Module Commanded Gear	Shift Control Solenoid 1	Shift Control Solenoid 2	Torque Converter Clutch Solenoid	Coast Clutch Solenoid
Park	First	ON	OFF	OFF	OFF
Reverse	First	ON	OFF	OFF	OFF
Neutral	First	ON	OFF	OFF	OFF
Ⓓ	First	ON	OFF	[a]	[a]
Ⓓ	Second	ON	ON	[a]	[a]
Ⓓ	Third	OFF	ON	[a]	[a]
Ⓓ	Fourth	OFF	OFF	[a]	OFF
Ⓓ Cancel	First Through Third Gear Only, Shift Solenoid 1, Shift Solenoid 2, and Torque Converter Clutch, Same as Overdrive, Coast Clutch Solenoid Always On.				
Manual 2	Second	OFF	OFF	[a]	ON
Manual 1	Second	OFF	OFF	OFF	ON
Manual 1	First	ON	OFF	OFF	ON

CHART 14–9 E40D solenoid operation. (Reprinted with the permission of Ford Motor Company)

Shift Solenoid Failure Mode Charts

Shift Control Solenoid Failure "Always ON"

Failed ON due to powertrain control module and / or vehicle wiring concerns; solenoid electrically or hydraulically stuck ON.

Shift Control Solenoid Failure "Always OFF"

Failed OFF due to powertrain control module and / or vehicle wiring concerns; solenoid electrically or hydraulically stuck OFF.

Shift Solenoid 1 ALWAYS OFF

Powertrain Control Module Gear Commanded	Position Selected: Overdrive	Position Selected: 2	Position Selected: 1
	Actual Gear Obtained		
1	4	2	1
2	3	2	2
3	3	2	2
4	4	2	2

Shift Solenoid 1 ALWAYS ON

Powertrain Control Module Gear Commanded	Position Selected: Overdrive	Position Selected: 2	Position Selected: 1
	Actual Gear Obtained		
1	1	2	1
2	2	2	1
3	2	2	1
4	1	2	1

Shift Solenoid 2 ALWAYS OFF

Powertrain Control Module Gear Commanded	Position Selected: Overdrive	Position Selected: 2	Position Selected: 1
	Actual Gear Obtained		
1	1	2	1
2	1	2	1
3	4	2	2
4	4	2	2

Shift Solenoid 2 ALWAYS ON

Powertrain Control Module Gear Commanded	Position Selected: Overdrive	Position Selected: 2	Position Selected: 1
	Actual Gear Obtained		
1	2	1	1
2	2	2	1
3	3	2	2
4	3	2	2

CHART 14–10 Solenoid malfunctions. (Reprinted with the permission of Ford Motor Company)

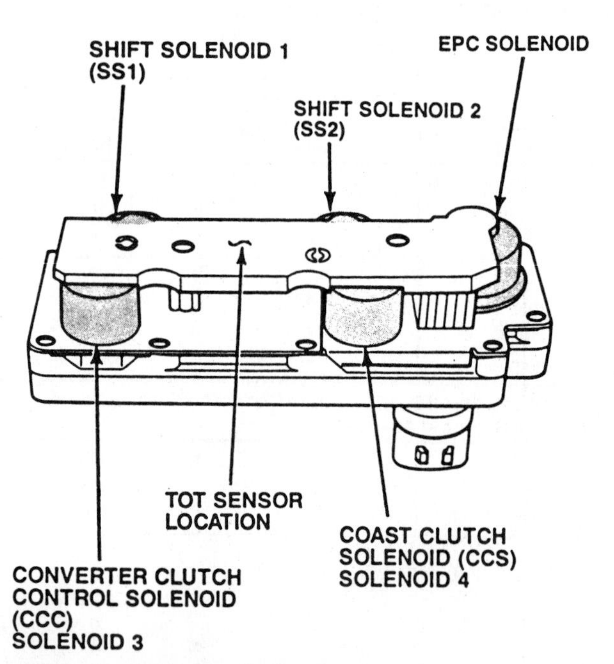

FIGURE 14–29 E4OD solenoid body. (Reprinted with the permission of Ford Motor Company)

Follow the pinpoint tests to determine the cause. After completion of step A1, with a YES result, move to step A2 (Chart 14–11). When identifying and then checking the proper test pin on the transmission vehicle harness connector (Figure 14–30), a voltage change to shift solenoid 2 is observed.

If directed to proceed to step A3 and then A4 (Chart 14–12, page 364), a Rotunda EEC-IV 60-Pin Breakout Box provides the junction points to check circuit voltage and resistance values. Shown in Figure 14–31 is a DVOM checking resistance at the breakout box pin jacks.

To accomplish steps A5 through A8 (Charts 14–13 and 14–14, pages 365 and 367), Ford recommends using the transmission tester shown in Figure 14–32. A variety of overlays (Figure 14–33) and wiring harness cables with connectors adapt the tester to most of Ford's electronically controlled transmissions and transaxles. A competent technician using the tester can perform numerous tests to a transmission. Figure 14–34 illustrates the E4OD transmission overlay, the devices that can be checked, their circuit jacks, and LED lamp locations.

WARNING: Ford's transmission testers and many other testers, including ones made by a technician, provide direct control of the solenoids. Careless selection of gears at improper speeds may damage internal components and cause loss of control of the vehicle.

PINPOINT TEST A: SHIFT SOLENOIDS

	TEST STEP	RESULT ▶	ACTION TO TAKE
A1	E4OD ELECTRONIC DIAGNOSTICS		
	• Check to make sure the transmission harness connector is fully seated, terminals are fully engaged in connector and in good condition before proceeding. • Enter Output State Diagnostic Test Mode (DTM). Refer to Special Test Modes in this section. • Connect Star tester. • Perform KOEO test until continuous memory DTC have been displayed. • Depress throttle to Wide Open Throttle (WOT) and release. • **Does vehicle enter Output State DTM?**	Yes ▶ No ▶	REMAIN in Output State DTM. GO to **A2**. DEPRESS throttle and release. If vehicle did not enter Output State DTM, REFER to Powertrain Control / Emission Diagnosis Manual.[9]
A2	CHECK ELECTRICAL SIGNAL OPERATION		
	CAUTION: Remove heat shield from transmission before removing connector. Remove Solenoid Body connector by pushing on center tab and pulling on wiring harness. Do not attempt to pry tab with a screwdriver. Reinstall heat shield after service. NOTE: Refer to the schematic and charts preceding this Pinpoint Test. • Disconnect transmission connector. • Using a mirror, inspect both ends of connector for damaged or pushed out pins, corrosion, loose wires, and missing or damaged seals. • Connect a volt-ohmmeter positive lead to vehicle power (VPWR) circuit and negative test lead to solenoid circuit of transmission vehicle harness connector. • Place volt-ohmmeter on 20 volt scale. • While observing volt-ohmmeter, depress and release throttle to cycle solenoid output On and Off. • **Does the suspected solenoid output voltage change at least 0.5V?**	Yes ▶ No ▶	GO to **A5**. GO to **A3**.

CHART 14–11 Shift solenoid pinpoint tests—steps A1 and A2. (Reprinted with the permission of Ford Motor Company)

TRANSMISSION CONNECTOR PIN ASSIGNMENTS

12-WAY CONNECTOR PIN	DESCRIPTION	EEC-IV 60-WAY CONNECTOR PIN		
		GAS	DIESEL	4.7L (CALIF) 5.0L 5.8L (CALIF)
1	VEHICLE POWER (VPWR)	37. 57	71, 97	37. 57
2	SHIFT SOLENOID 2	19	1	52
3	SHIFT SOLENOID 1	52	27	51
4	TORQUE CONVERTOR CLUTCH	53	28	53
5	COAST CLUTCH SOLENOID	55	28	55
6		-	-	-
7	TRANSMISSION OIL TEMPERATURE SENSOR	42	37	49
8	SIGNAL RETURN (SIG RTN)	46	91	46
9		-	-	-
10		-	-	-
11	ELECTRONIC PRESSURE CONTROL	38	91	38
12	ELECTRONIC PRESSURE CONTROL POWER	37.57	71. 97	37.57

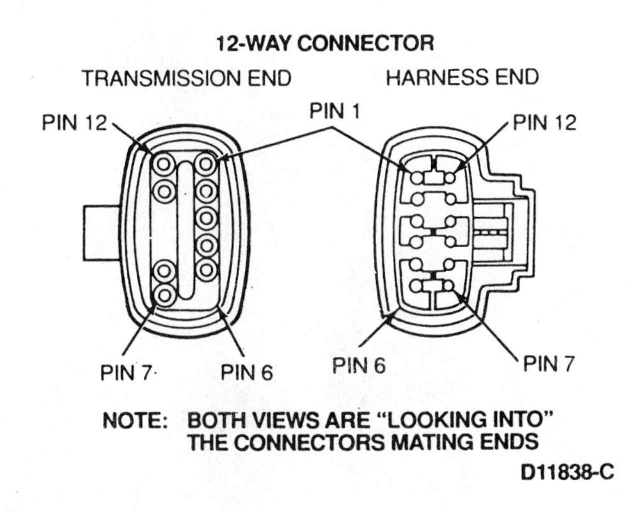

FIGURE 14–30 E4OD connector pin identification. (Reprinted with the permission of Ford Motor Company)

PINPOINT TEST A: SHIFT SOLENOIDS (Continued)

	TEST STEP	RESULT ▶	ACTION TO TAKE
A3	CHECK CONTINUITY OF SOLENOID SIGNAL AND VEHICLE POWER HARNESS CIRCUITS		
	• Turn ignition switch off. • Make sure harness transmission connector is disconnected. • Disconnect powertrain control module. Inspect for damaged or pushed out pins, corrosion, or loose wires. • Install Rotunda EEC-IV 60-Pin Breakout Box 007-00033, leave powertrain control module disconnected. NOTE: Refer to the schematic and charts preceding this Pinpoint Test. • Measure resistance between powertrain control module signal test pin 52 or 19 (51, 52, — 4.9L Calif, 5.8L Calif, 5.0L) (1,27-Diesel) at Breakout Box and signal pin at transmission harness connector. • Measure resistance between powertrain control module signal test pins 37 and 57, 71, 97 (Diesel) at Breakout Box and signal pin at transmission harness connector. • **Is each resistance less than 5.0 ohms?**	Yes ▶ No ▶	GO to **A4**. SERVICE open circuit(s). REMOVE Breakout Box. RECONNECT all components. REPEAT Quick Test.
A4	CHECK SOLENOID HARNESS FOR SHORTS TO POWER AND GROUND		
	• Make sure Rotunda EEC-IV 60-Pin Breakout Box 007-00033 is installed and powertrain control module is disconnected. • Ensure transmission harness connector is disconnected. NOTE: Refer to the schematic and charts preceding this Pinpoint Test. • Measure resistance between powertrain control module signal output pin and test pin 37 / 57 (71, 97 Diesel) at Breakout Box. • Measure resistance between test pins 40 / 60 77, 103 and 46, 91 (Diesel) at Breakout Box and chassis ground. • **Is each resistance greater than 10,000 ohms?**	Yes ▶ No ▶	GO to **A5**. SERVICE short circuit. REMOVE Breakout Box. RECONNECT all components. REPEAT Quick Test.

CHART 14–12 Shift solenoid pinpoint tests—steps A3 and A4. (Reprinted with the permission of Ford Motor Company)

FIGURE 14–31 Rotunda DVOM and 60-Pin Breakout Box.

	TEST STEP	RESULT	ACTION TO TAKE
A5	SOLENOID FUNCTIONAL TEST		
	• Ensure that the vehicle harness is disconnected at the transmission connector. • Install Transmission Tester 007-0085C. • Using tests outlined under Tester Instructions, perform the Solenoid Voltage Test. NOTE: LED will turn "**GREEN**" when solenoid activates and turn "**OFF**" when deactivated. LED will turn "**RED**" if activated solenoid is shorted to B+. LED will remain "**OFF**" if an activated solenoid is shorted to ground or no continuity. • **Do solenoids activate? (LED GREEN)**	Yes ▶ No ▶	GO to **A6**. GO to **A7**.
A6	TRANSMISSION DRIVE TEST		
	• Perform Transmission Solenoid Cycling Test Procedures as outlined under Transmission Tester Instructions in this section. • **Does vehicle upshift when commanded by the tester?**	Yes ▶ No ▶	ERASE all codes and PERFORM Drive Cycle Test in this section. REPEAT Quick Test. GO to **A7**.

CHART 14–13 Shift solenoid pinpoint tests—steps A5 and A6. (Reprinted with the permission of Ford Motor Company)

The E4OD overlay and the E4OD tester harness make the unit extremely valuable for accurately completing the pinpoint tests. The tester's shift solenoid LEDs will be lit green (good), lit red (shorted to B+), or remain off (open or shorted to ground circuit) during step A5. In step A6, the technician can command upshifts and downshifts with the tester. A volt/ohmmeter can be used to read circuit resistance and check for shorts to ground in steps A7 and A8, respectively (Chart 14–14).

Upon pinpointing the cause, repair or replace the component as needed. Record the codes, erase them, and then run a quick test with the scan tool. If by chance the problem was not located, refer to the "Diagnosis by Symptom" section in the shop manual for additional instructions.

FIGURE 14–32 Rotunda transmission tester.

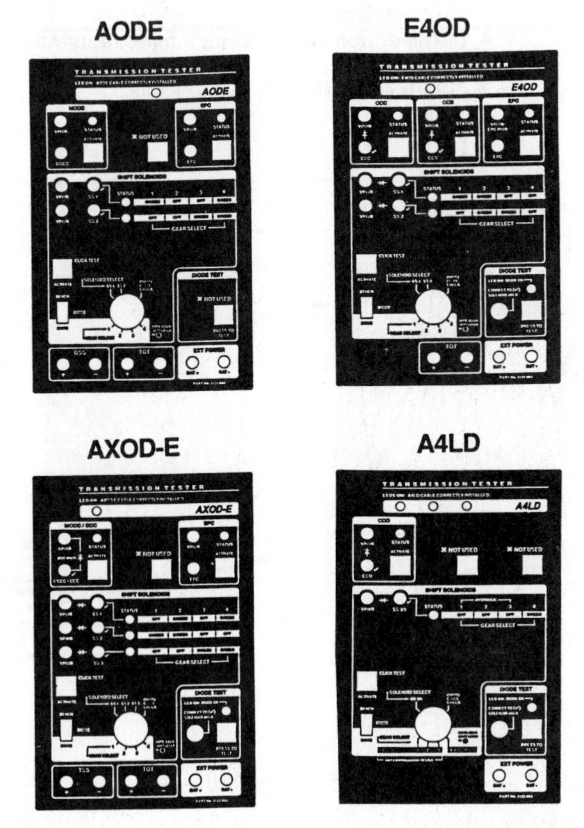

FIGURE 14–33 Overlays for transmission tester. (Reprinted with the permission of Ford Motor Company)

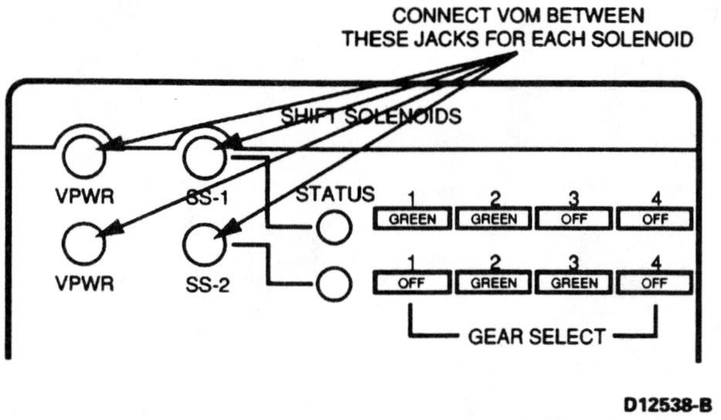

	TEST STEP	RESULT ▶	ACTION TO TAKE
A7	CHECK RESISTANCE OF SOLENOID		
	• Bench / Drive switch in BENCH mode. • Rotate gear selector switch to OHMS CHECK position. • Connect a volt-ohmmeter negative lead to the shift solenoid 1 jack and positive lead to vehicle power jack on tester. This is to test shift solenoid 1. • Record resistance. • Connect a volt-ohmmeter negative lead to the shift solenoid 2 jack and positive lead to the vehicle power jack on tester. This is to test shift solenoid 2. • **Are shift solenoids resistance between 20 and 30 ohms?**	Yes ▶ No ▶	GO to **A8**. REPLACE solenoid body assembly. RECORD and ERASE codes. REPEAT Quick Test.
A8	CHECK SOLENOID FOR SHORT TO GROUND		
	• Check for continuity between BAT(-) (engine ground) and appropriate jack with a digital volt-ohmmeter or other low current tester (less than 200 milliamps). • Connection should show infinite resistance (no continuity). • **Is there continuity?**	Yes ▶ No ▶	REPLACE solenoid body assembly. RECORD and ERASE codes. REPEAT Quick Test. GO to Diagnosis by Symptom Section.

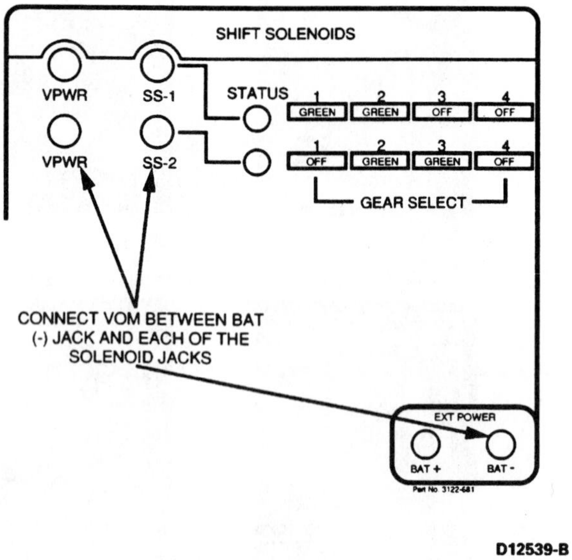

CHART 14–14 Shift solenoid pinpoint tests—steps A7 and A8. (Reprinted with the permission of Ford Motor Company)

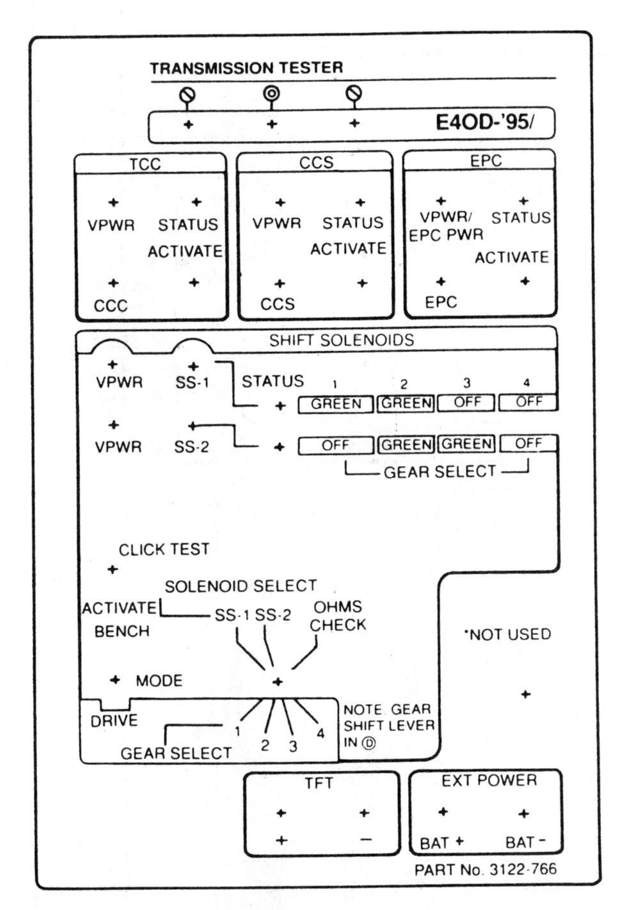

FIGURE 14–34 E4OD transmission tester pinpoint tests. (Reprinted with the permission of Ford Motor Company)

OPTIONAL TESTING METHODS

Though the transmission testers available from manufacturers and the aftermarket are efficient tools to use for diagnosing and testing solenoids, other methods are available. Practical means can be used for diagnostic work, often with relatively inexpensive tools.

Making accurate pinpoint tests to internal transmission solenoid, as well as to pressure switches and sensors, requires a thorough understanding of the unit's electrical devices and their functions. Service manuals provide the necessary information such as component locations, description of their operation, diagnostic charts, wiring diagrams of the circuits, and connector pin identification (Figure 14–35). Knowing how the devices function properly will help you correctly recognize when one does not function properly.

The techniques for solenoid diagnosis are some of the same basic approaches used in TCC electrical/electronic trouble-shooting (Chapter 13). Some transmissions are only semielectronic—the TCC engagement and 3-4 overdrive shift are solenoid-operated and controlled by the on-board ECM/PCM. In full electronic control, the transmission electronics are also interfaced with the PCM/TCM. The TCC engagement and all shifts are transmission computer-controlled. This discussion will be limited to basic trouble-shooting tips that can be adapted or modified for the transmission product and model.

Before performing pinpoint tests, determine if the solenoids are case-grounded or computer-grounded, and their ON/OFF sequence for all operating ranges and gears. Solenoid diagnosis is similar to clutch/band failure analysis.

When diagnosing a problem on an electronically controlled transmission, it is important to separate electronic problems from mechanical and hydraulic problems. It is not unusual to get a complaint of high gear or overdrive takeoffs, or a light tie-up in one or more forward gears. A recommended approach is to isolate the electronic control system from the transmission and start with the shift solenoids.

Strategies

Depending on circuit design, different strategies can be used to operate the solenoids manually in place of the computer. One approach that can be used effectively on some electronic transmissions begins by disconnecting the wire harness at the transmission. Substitute the vehicle harness with a test harness and switch box that enables you to have manual control over the solenoids (Figure 14–36).

The test unit can easily be made. From a salvage vehicle, retrieve the transmission harness wiring and connector. Using a cigar lighter adapter, connectors, and switches from a radio-TV parts supply store, assemble your own tester. It is now a matter of manually coordinating the test switches with the ON/OFF settings of the solenoid valves and the selected gear. Chart 14–15, page 369, shows an ON/OFF solenoid valve and gear sequence pattern that might be found in a typical four-speed electronic transmission. Notice the effects of malfunctioning solenoid combinations on the gear selection.

With the drive wheels off the ground, determine that the transmission has all four gears. You can verify shift changes by watching the increase or decrease in the tachometer rpm or the dash speedometer. If all four gears are present, test each gear for slipping in each gear's operating range. Use the stall test to check for holding ability. Do not forget to pause between each test run and fast idle in neutral for thirty seconds to cool the converter. If the transmission does not slip and has all four gears, the problem area is in the electronics.

Sometimes a solenoid functions at idle or light throttle but leaks under heavy throttle when the hydraulic pressures are higher. Use the same test hookup switch, the S_2 solenoid ON, with the drive wheels off the ground. Set the engine rpm at a fast idle speed of approximately 1200 rpm, and increase line pressure if possible. Line pressure regulation methods in electronic shift control transmissions vary. Some use a vacuum modulator, some a TV cable, and the most sophisticated use electronic pressure control solenoids. Always follow the manufacturer's recommendations regarding line pressure increases while testing transmissions.

The transmission should remain in second gear. If it downshifts to first gear, the S_2 solenoid cannot hold pressure and is defective. A shift into third gear indicates that pressure is not holding in the S_1 solenoid. Leaking solenoids should be the high-suspect area when the transmission shifts okay under

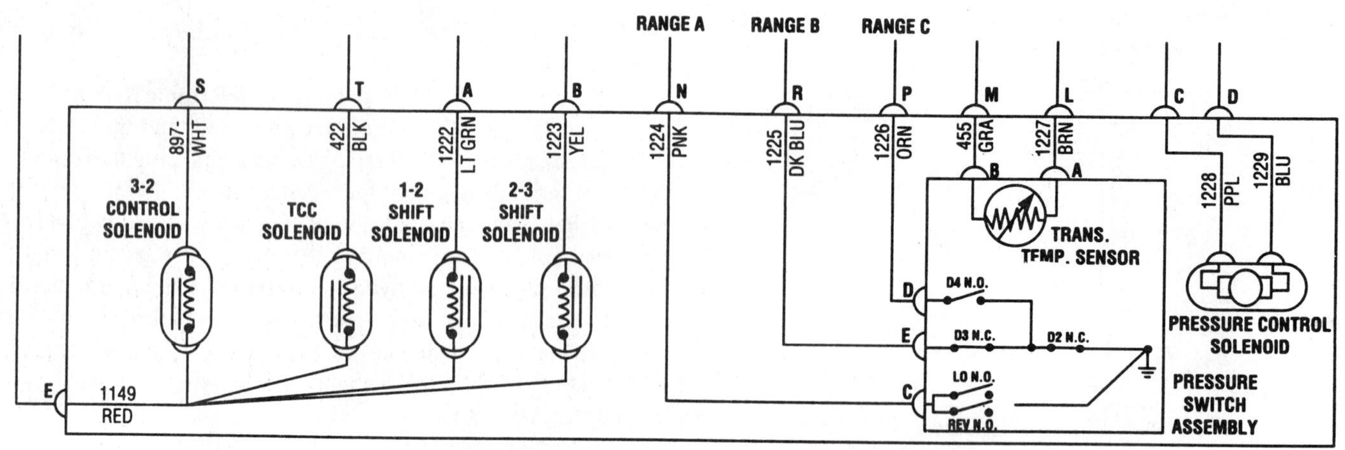

N.C. = NORMALLY CLOSED SWITCH
N.O. = NORMALLY OPEN SWITCH

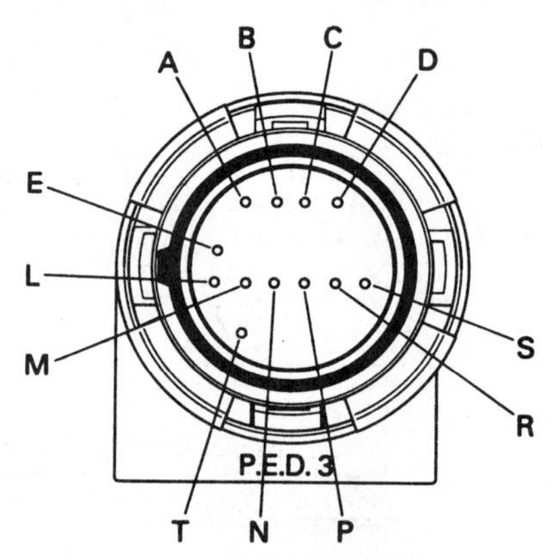

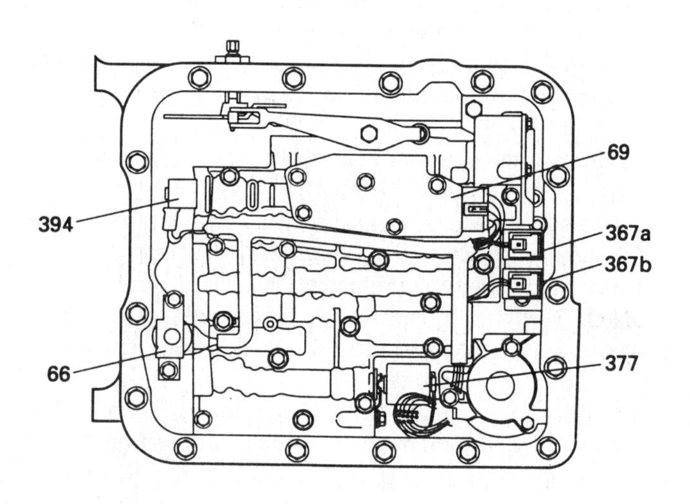

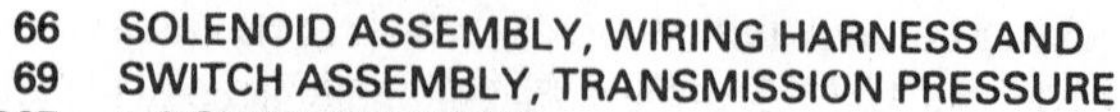

66 SOLENOID ASSEMBLY, WIRING HARNESS AND
69 SWITCH ASSEMBLY, TRANSMISSION PRESSURE
367a 1-2 SHIFT SOLENOID
367b 2-3 SHIFT SOLENOID
377 PRESSURE CONTROL SOLENOID
394 3-2 CONTROL SOLENOID

CAVITY	FUNCTION
A	1-2 SHIFT SOLENOID (LOW)
B	2-3 SHIFT SOLENOID (LOW)
C	PRESSURE CONTROL SOLENOID (HIGH)
D	PRESSURE CONTROL SOLENOID (LOW)
E	BOTH SHIFT SOLENOIDS, TCC SOLENOID, AND 3-2 CONTROL SOLENOID (HIGH)
L	TRANSMISSION FLUID TEMPERATURE (HIGH)
M	TRANSMISSION FLUID TEMPERATURE (LOW)
N	RANGE SIGNAL "A"
P	RANGE SIGNAL "C"
R	RANGE SIGNAL "B"
S	3-2 CONTROL SOLENOID (LOW)
T	TCC SOLENOID (LOW)

FIGURE 14–35 4L60-E electrical diagram and device location. (Courtesy of General Motors Corporation, Service Technology Group)

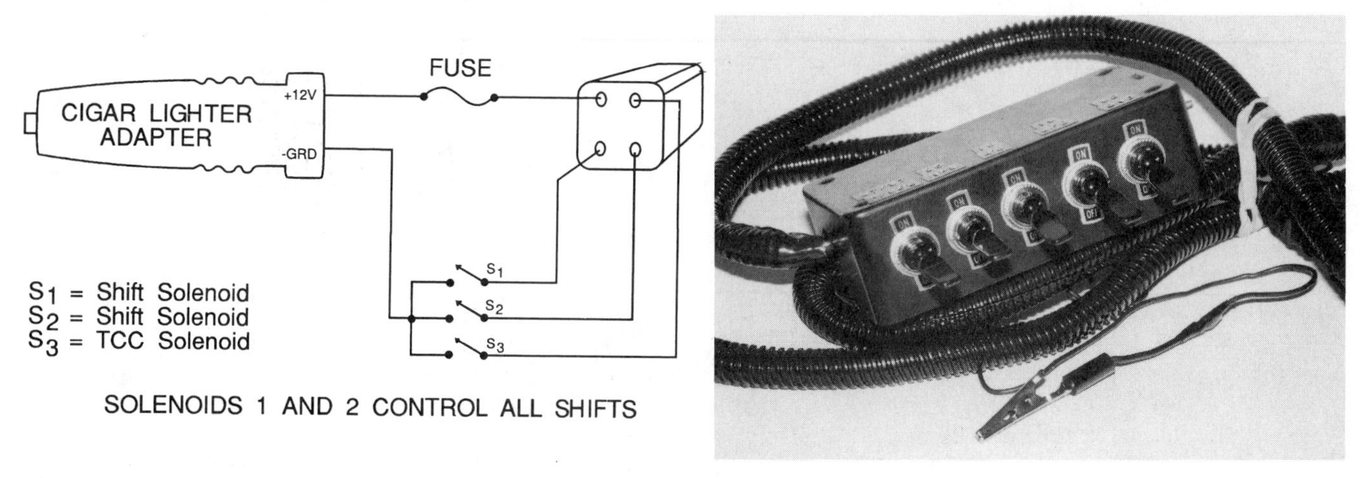

FIGURE 14–36 Manual control test harness for computer-grounded solenoids.

	Normal			*Solenoid No. 1 Malfunctioning*			*Solenoid 2 Malfunctioning*			*Both Solenoids Malfunctioning*
	Solenoid valve			*Solenoid valve*			*Solenoid valve*			
Range	*SS 1*	*SS 2*	*Gear*	*SS 1*	*SS 2*	*Gear*	*SS 1*	*SS 2*	*Gear*	*Gear*
OD	ON	OFF	1st	x	OFF	2nd	ON	x	OD	3rd
	OFF	OFF	2nd	x	OFF	1st	OFF	x	3rd	OD
	OFF	ON	3rd	x	ON	OD	OFF	x	2nd	1st
	ON	ON	OD	x	ON	3rd	ON	x	1st	2nd
D	ON	OFF	1st	x	OFF	2nd	ON	x	OD	3rd
	OFF	OFF	2nd	x	OFF	1st	OFF	x	3rd	OD
	OFF	ON	3rd	x	ON	OD	OFF	x	2nd	1st
L	ON	OFF	1st	x	OFF	2nd	ON	x	OD	3rd
	OFF	OFF	2nd	x	OFF	1st	OFF	x	3rd	OD
P-N	ON	OFF		No Effect						
R	ON	OFF	R	No Effect						

CHART 14–15 Representative solenoid valve operation and gear sequence pattern with effects of malfunctioning solenoids.

light throttle but takes off in third or fourth gear on heavy acceleration.

Solenoids can be bench tested or tested in the car with a DVOM for an open, short, or ground. Resistance values are usually given by the manufacturer for both cold and hot readings. A solenoid may test okay electrically, but remember it must also test out mechanically and hydraulically. The check valve may fail to seat, or the porting can be plugged. When the solenoid is energized, the magnetic field removes microparticles of metal flowing in the fluid. The particles accumulate in the solenoid housing and prevent proper plunger or ball check seating, or plug the hydraulic porting. These effects can cause an erratic shift pattern such as no shift or binding shifts.

Shift solenoids or any control solenoid can be checked for circuit resistance and shorts to ground. This is done as an in-the-car diagnostic operation and without pulling a pan or side cover. These checks should be made first before getting involved with any mechanical checks on the solenoids.

As an example, let us look at the Ford E4OD transmission case connector and pin identification (Figure 14–37). Five valve body solenoids and one sensor are used and can be circuit checked with a DVOM. For easy test access to the pin connector, it is a good practice to use a test harness (Figure 14–37).

Digital ohmmeter hookups should produce the following results:

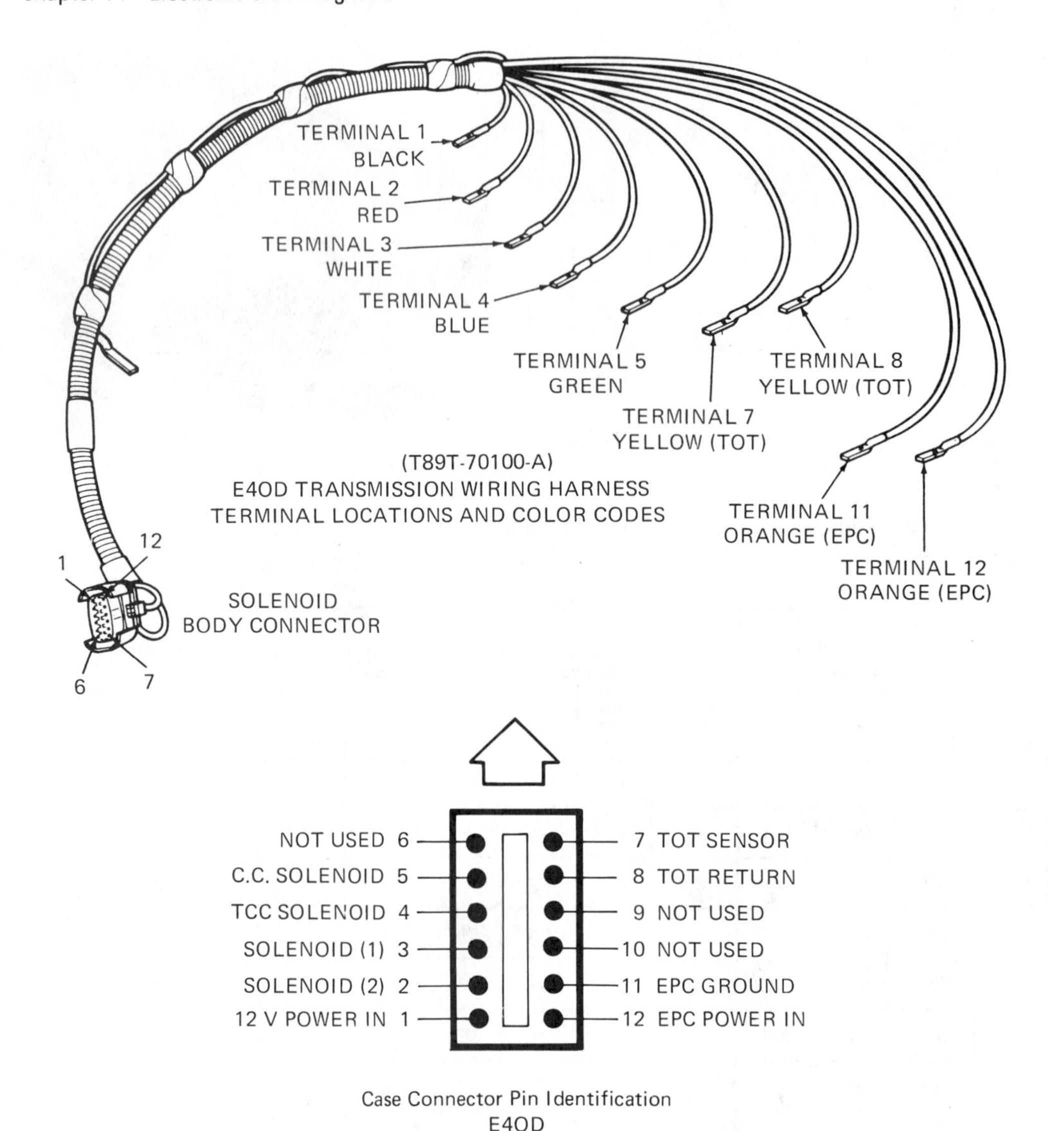

FIGURE 14–37 E4OD test harness and case connector identification.

- Shift solenoid 1—pins 1 and 30 (20 to 30 ohms)
- Shift solenoid 2—pins 1 and 2 (20 to 30 ohms)
- Coast clutch solenoid—pins 1 and 5 (20 to 30 ohms)
- Pressure control solenoid (PCS)/EPC solenoid—pins 11 and 12 (4.0 to 6.5 ohms)

To check for a short ground on all of the above, individually connect pins 1, 2, 3, 4, 5, 11, and 12 with the test ohmmeter to engine ground.

Some solenoids can be disassembled and cleaned and put back into service, while others cannot be reused and must be replaced. To verify if the solenoid can mechanically seat the check valve and allow a free flow of fluid through the hydraulic porting, it can be air checked. Remember that a solenoid can be normally open or normally closed. Using a normally closed solenoid as an example, apply 70 psi (483 kPa) of regulated air to the metered orifice check valve port at the solenoid valve.

- Deenergized: No air should flow through the exhaust porting (see Figure 14–38).
- Energized: A free flow of air should pass through the exhaust porting.

For a normally open solenoid, the test results are reversed.

To avoid being fooled when working with solenoids, it is extremely important to understand that shift solenoids can be either normally closed or normally open when deenergized. Even solenoids used in the same transmission can be different—some may be normally closed and some may be normally open when deenergized. Product knowledge is essential regarding the operation of each solenoid.

Transmission Fluid Temperature (TFT) Sensor

To check the resistance of a **transmission fluid temperature (TFT) sensor** (originally called a TOT/transmission oil temperature sensor) in an E4OD, connect an ohmmeter to pins 7 and 8 (Figure 14–37). Note that resistance specifications vary with transmission temperature (Chart 14–16).

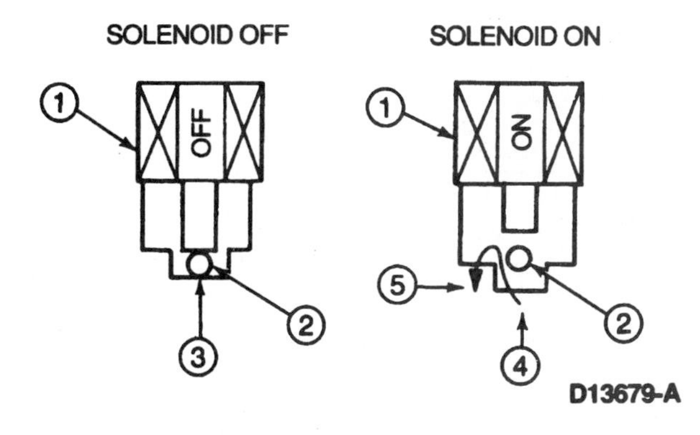

Item	Part Number	Description
1	—	Solenoid (Part of Transmission Solenoid Body)
2	—	Check Ball (Part of Solenoid)
3	—	Feed (Blocker)
4	—	Feed
5	—	Output (To Shift Valves, etc.)

FIGURE 14–38 E4OD shift solenoid operation and components. (Reprinted with the permission of Ford Motor Company)

❖ **Transmission fluid temperature (TFT) sensor:** An input device to the ECM/PCM that senses the fluid temperature and provides a resistance value. This sensor, originally called a TOT/transmission oil temperature sensor, operates on the thermistor principle.

As a guideline, carefully touch the transmission oil pan opposite the exhaust. If the pan temperature is warm but not hot to the touch, use the 41 to 70°C (105 to 158°F) specification range.

Temperature		Resistance
°C	°F	Ohms (K)
(-)19-(-)1	(-)3-31	967K-248K
(-)40-(-)20	(-)40-(-)4	248K-100K
0-20	32-68	100K-37K
21-40	69-104	37K-16K
41-70	105-158	16K-5K
71-90	159-194	5K-2.7K
91-110	195-230	2.7K-1.5K
111-130	231-266	1.5K-0.8K
131-150	267-302	0.8K-0.54K

CHART 14–16 Transmission fluid temperature (TFT) sensor resistance values. (Reprinted with the permission of Ford Motor Company)

SUMMARY

As with hydraulically controlled transmissions, when diagnosing electronically controlled units, a systematic process must be followed. This will lead the technician to the cause of the problem efficiently.

Electronically controlled transmission operation is dependent upon the commands from the PCM/TCM. The PCM's/TCM's decision-making process to activate the output devices, such as shift, TCC, and electronic pressure control solenoids, is based upon the input signals it receives. Different transmissions use a variety of input devices that typically can be categorized into three areas: signals sent from sensors and switches in the engine compartment, from the driver, and from the transmission itself.

Transmission malfunctions can be related to electrical/electronic, hydraulic, and mechanical problems. Begin your diagnosis by checking the fluid level and condition and inspecting wiring and vacuum connections under the hood. If these appear good, use a scan tool to retrieve any diagnostics trouble codes (DTCs) that may be present. The codes will be set if an electrical, hydraulic, or mechanical failure is present. Codes present at the time of testing are called hard codes. Those in memory are referred to as soft codes. The DTCs help identify the source of the problem and must be pursued next.

When multiple codes are detected, there is a definite order to follow. Typically the technician works with hard codes prior to soft codes and with lower-numbered codes before higher-numbered codes. Since the engine provides the power to the transmission, engine input codes should be diagnosed first. Next, transmission input codes, followed by transmission output codes, are fixed. Work closely with the service manual to pinpoint the cause of the malfunction.

The OBD II standards are now a part of the modern vehicle. To properly diagnose and service today's powertrains, technicians need to understand the increasingly sophisticated technology used in late model vehicles.

❑ REVIEW

Key Terms/Acronyms

Hard codes, soft codes, fail-safe/backup control, limp-in mode, OBD II (On-Board Diagnostics Generation Two), electrostatic discharge (ESD), and transmission fluid temperature (TFT) sensor.

Questions

Fundamental Approach

1. Number in correct order the basic steps of strategy-based diagnosis.

_____ Check bulletins and troubleshooting hints.

_____ Perform preliminary checks.

_____ Perform service manual diagnostic checks.

_____ Reexamine the complaint.

_____ Repair and verify the fix.

_____ Select diagnosis path.

_____ Verify complaint.

_____ 2. Technician I says hard codes are tougher to fix because the problem is not present during testing. Technician II states that soft codes are intermittent and are sometimes referred to as history codes. Who is right?

A. I only C. both I and II
B. II only D. neither I nor II

_____ 3. Technician I says that "fail-safe/backup control" refers to the PCM substituting faulty input signals with temporary values. Technician II states that "limp-in mode" is present when the transmission output solenoids are electrically deenergized, allowing limited gear engagement. Who is right?

A. I only C. both I and II
B. II only D. neither I nor II

On-Board Diagnostics

1. List the three situations that cause DTCs to be set.

____

____

____

2. Number in correct sequence the order in which to pursue multiple codes.

_____ Engine input DTCs.

_____ Transmission/transaxle input DTCs.

_____ Transmission/transaxle output DTCs.

3. The technician should pursue soft codes prior to determining the cause of hard codes. ___T/___F

Code Retrieval

1. List three methods to retrieve codes.

____

____

____

2. List five tests that a scan tool, like Chrysler's DRB II, can perform.

____

____

____

____

____

3. A bidirectional scanner, like GM's Tech I, can display and send information. ___T/___F

4. To erase codes, the best method is to disconnect the battery. ___T/___F

5. Ford's Super Star II tester displays codes and sensor input signal readings and provides control over output actuators. ___T/___F

6. Ford's NGS tester displays codes and sensor input signal readings and provides control over output actuators. ___T/___F

7. A "wiggle test" is helpful in checking wiring and connectors while determining the cause of intermittent problems. ___T/___F

OBD II

1. List five requirements and standards of OBD II that affect the transmission technician.

____

____

____

____

____

Troubleshooting Principles

1. For proper computer-controlled transmission operation, the battery should have a minimum charge of 11.5 V. ___T/___F

2. Probing through wire insulation to check circuit conditions is an acceptable practice. ___T/___F

3. Magnetic induction can occur due to improper wire routing. ___T/___F

4. An electrostatic discharge from a human body can be as great as 35,000 V. ___T/___F

5. A static discharge of less than 3000 V is not felt by the human body but is capable of damaging electronic components. ___T/___F

6. To prevent static discharge, a technician should:

____

Servicing Diagnostic Trouble Codes (DTCs)

1. Usually, higher-numbered codes should be pursued before lower-numbered codes. ___T/___F

2. A DTC relating to a throttle position (TP) signal voltage that is too high means that the TP sensor is faulty and needs to be replaced. ___T/___F

3. List three categories of problems that will put a vehicle into "limp-in mode."

____

____

____

4. From the answers in question 3, which should be investigated first (after preliminary checks)?

____

Optional Testing Methods

1. To test the shift solenoids, the technician needs to be manually controlling them. ___T/___F

2. To control the solenoids, the technician must unplug the transmission harness and substitute a combination test harness and powered switch box. ___T/___F

3. To test the shift solenoids for an open, short, or ground, the solenoids must be removed and bench tested. ___T/___F

4. To bench test the shift solenoid check valve for holding, regulated air at 50 psi (345 kPa) is used. ___T/___F

5. TFT resistance values vary with fluid temperature. ___T/___F

15 Transmission/Transaxle Removal and Installation

OBJECTIVE:

The objective of this unit is for you to become proficient with the terms, concepts, and procedures contained in this chapter. Completion of this objective is essential to becoming a successful automatic transmission technician.

CHALLENGE YOUR KNOWLEDGE

Define the following key term:

Electrolysis.

List the following:

Five undervehicle suspension and drivetrain items to be inspected during R&R (removal and replacement); four inspections performed to a flexplate; and the amount of oil flowing through typical cooler systems.

Describe the following:

Correction method for crankshaft/converter pilot wear concerns; cooler system flushing procedures; and cooler system oil flow volume testing.

Identify the warnings associated with:

The need for eye and foot protection when removing or installing transmissions and transaxles; the battery ground cable; and the torque converter secured inside the converter housing during R&R.

INTRODUCTION

Automatic transmission/transaxle removal and replacement (R&R) is an art that requires special technician skills and is as important as the transmission rebuild itself. Carelessness, especially during installation, can destroy a perfectly good rebuild unit. Although the dramatic change to front-wheel drive has provided desirable engineering advantages, transaxle rebuilding and R&R has become more time-consuming and complex. Transaxle R&R requiring the removal of a subframe can require six hours of labor. It is obvious that a mistake in the rebuild or R&R process requiring the transaxle to be removed again is extremely costly.

Removal and replacement is more than a nuts-and-bolts operation. Clamping the transmission to the engine block and the converter to the flywheel must be coordinated with a definite procedure. Failure to follow the recommended procedure can result in vibration or damage to the transmission pump or converter. Installation is concluded with TV and manual linkage adjustments, electrical/vacuum line hookups where applicable, fluid fill, engine cold and hot idle check, and a checkout on the auxiliary cooling fan operation where applicable.

Other service requirements of an R&R technician include a variety of tasks, such as:

- Cross and yoke/CV joint replacement
- Evaluation of subframe and suspension damage in transaxle applications
- Cooler line repair
- Auxiliary cooler installation
- Cooler system flushing and fluid volume testing
- Location of external fluid leaks
- Installation of pressure gauges
- In-the-car band adjustments
- In-the-car R&R of subassemblies such as the valve body and governor

WARNING: Because of the various hazards that exist under the hood and under the vehicle, the technician should always wear safety glasses and safety shoes.

TRANSMISSION/TRANSAXLE REMOVAL

Undercar Inspection

Before actual removal, the technician needs to make an undercar hoist inspection of the drivetrain components. Consider this an opportunity to sell more needed service and protect yourself against drivetrain disorders that are transmission-related. These can include items such as worn universal joints, damaged drive shaft/half-shafts, and loose motor mounts. Even note the condition of the tires on the drive wheels. Once the transmission has been repaired, customers will be quick to point to the technician if there is any vibration, clunk, or steering problem.

Front-Wheel Drive. On front-wheel drive units, inspect the CV joints for torn, collapsed, cracked, or rotted boots. If external lube leakage is not visible, squeeze the boots and listen for escaping air. Once the boots open, loss of lubricant plus dirt and water contamination quickly accelerate joint wear. Boot replacement must be accompanied by a complete disassembly, cleaning, and inspection of the joint. Be sure to use the proper CV joint lubricant and the proper clamping tool for the boot straps during assembly. Do not forget to "burp" the boot.

The front-drive system also should be inspected carefully for:

- Worn, loose, or damaged control arm bushings
- Worn, loose, or damaged sway bar links and mounting bushings
- Front strut upper mount worn, loose, or damaged
- Rack and pinion mount bushings worn, loose, or damaged

Front-drive inspection should include a close look at the subframe for road or collision damage. The subframe geometry must be correct to ensure proper positioning of the transaxle and engine. Before the transaxle is pulled, the vehicle may need to see a frame shop.

Rear-Wheel Drive. Check the drive shaft cross and yoke universal joints in RWD systems as follows:

- Make sure the joint angles are in their normal position. If the rear axle and wheels stay in their dropped position on the hoist, worn universal joints will test out tight. Adjustable high-style underhoist support stands will be needed to position the axle in its normal position.
- Check the joints for looseness by pushing up and down and twisting the drive shaft back and forth. Rotate the drive shaft one-fourth turn and repeat.
- Look for outward signs of rust or grease loss.
- Once the drive shaft is removed, countercheck the joint for free movement. A stiff joint or binding movement indicates needle and cross-shaft trunnion failure.
- Drive shaft double Cardon joints use two cross and yoke joints closely coupled together (Figure 15–1). The joints are centered on a spring-loaded ball stud and socket to maintain the relative position of the two joint units. When the drive shaft is removed, test the movement of the assembly. It should twist around smoothly and strongly resist any effort to straighten at the joint. The ball stud and socket does wear, and the centering spring is known to fracture. Ball and seat wear causes short-duration roughness on heavy acceleration.

Removal Highlights

The technician soon learns the difference in the complexity and difficulty level between the R&R of a RWD unit and a FWD unit. Front-wheel drive transaxle R&R can be very time-consuming, as already mentioned, requiring as much as six hours where the subframe needs to be removed. On some installations, the technician may prefer to remove the transaxle and engine together after pulling the half-shafts. The R&R time for most rear-wheel drive units should not exceed one and one-half hours. When removing the transmission, the following guidelines are important to follow:

1. As a safety precaution, remove the battery ground cable.

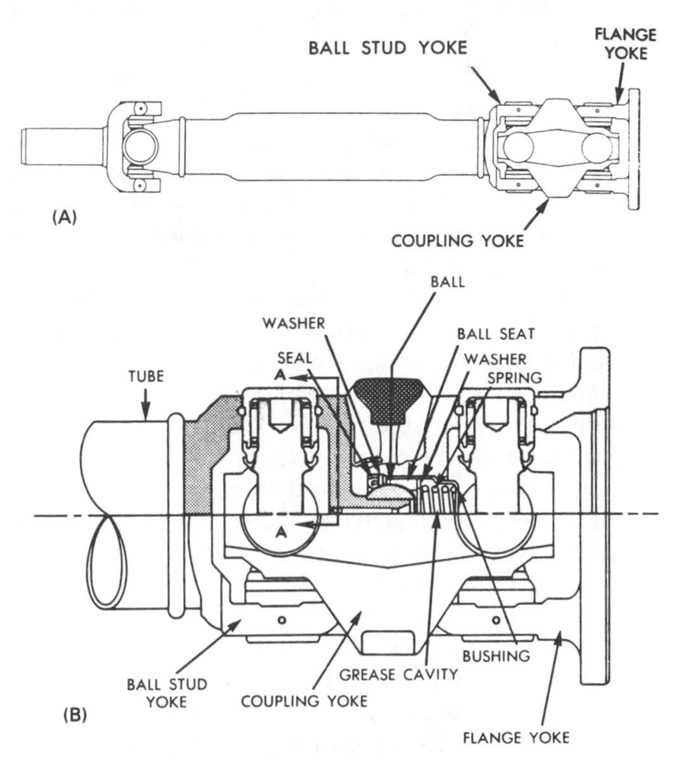

FIGURE 15–1 (a) Type 1-2 driveshaft; (b) typical rear double Cardon universal joint. (Courtesy of General Motors Corporation, Service Technology Group)

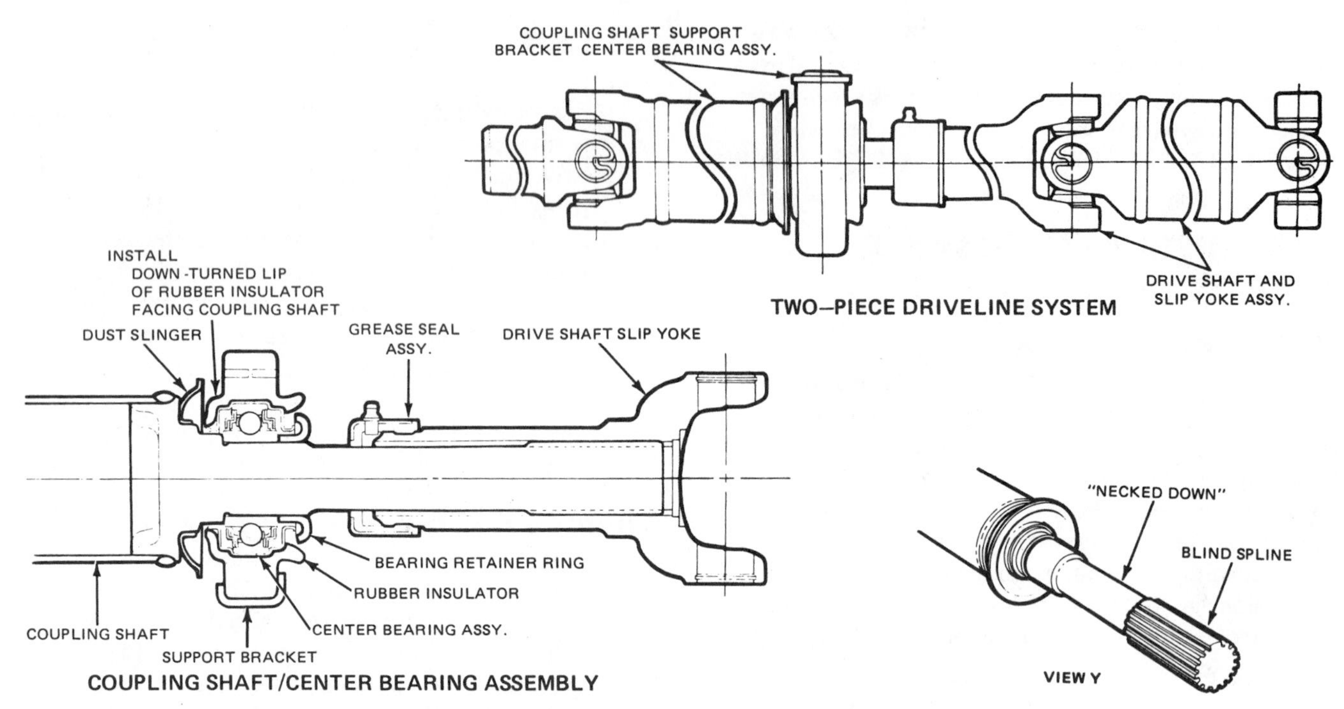

FIGURE 15–2 Drive shaft yokes correctly phased. (Reprinted with the permission of Ford Motor Company)

WARNING: Removal of the battery ground cable is a good, safe practice. It reduces the possibility of creating unwanted sparks and short circuits during removal and installation procedures, by eliminating the normal electrical path back to the battery.

2. To avoid a fluid mess, lower the transmission pan and drain the fluid. Loosen the front or rear pan bolts and then remove all the remaining bolts. The oil pan can simply be pried loose and tilted over a drain container without losing control of the fluid. Some technicians use the plastic cup plugs used in shipping to seal the transmission outlets and bench drain the fluid.
3. RWD applications. Scribe a mark or use a light paint spray from an aerosol can for marking the drive shaft and companion flange for correct reassembly. This should be standard practice whenever removing the drive shaft to ensure that drive shaft runout is not disturbed. Otherwise, an unwanted driveline vibration may result. Where slip yoke assemblies are splined to the drive shaft, they should preferably be kept in place to ensure correct yoke phasing of the shaft sections. Incorrect yoke phasing causes vibrations. Some splined sections use a blind spline and the sections can fit together only one way for correct yoke phasing. Do not always depend on this feature, however—make it a practice to mark the splined joint for realignment. As a general rule, the yokes on the end of a drive shaft or drive shaft section should be in the same plane (Figure 15–2).
4. FWD applications. Loosen the hub nut and wheel nuts with the vehicle on the floor and brakes applied (Figure 15–3).
5. FWD applications. In most cases, the outer CV joint splined shaft can be separated from the hub by holding the CV housing and moving the hub (knuckle) assembly away (Figure 15–4). If the splined joint refuses to budge with reasonable effort, use a hub remover (Figure 15–5). Never use a hammer to separate the outboard CV joint stub shaft from the hub—damage to the stub threads, wheel bearing, front brake, or CV joint is inevitable.
6. To minimize removal time, start by disconnecting the accessible under-the-hood items before lifting the vehicle. At this time, it might be easier to remove the converter housing upper bolts (Figure 15–6). On FWD applications, now is the time to install the engine support fixture (Figure 15–7).

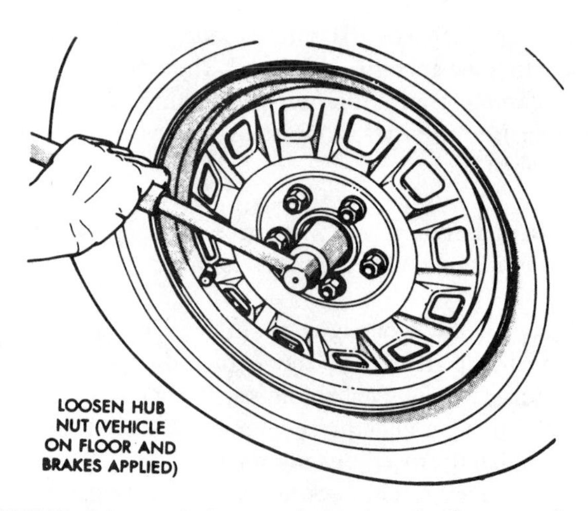

FIGURE 15–3 Loosen hub nut and wheels nuts. (Courtesy of Chrysler Corp.)

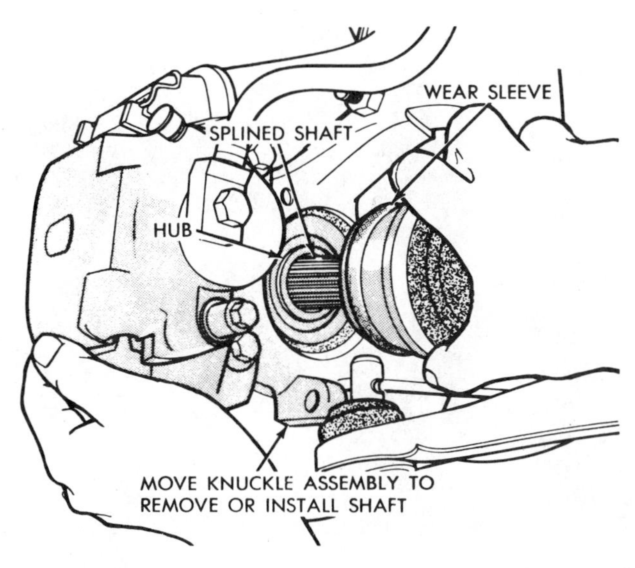

FIGURE 15–4 Separate outer CV joint shaft from hub. (Courtesy of Chrysler Corp.)

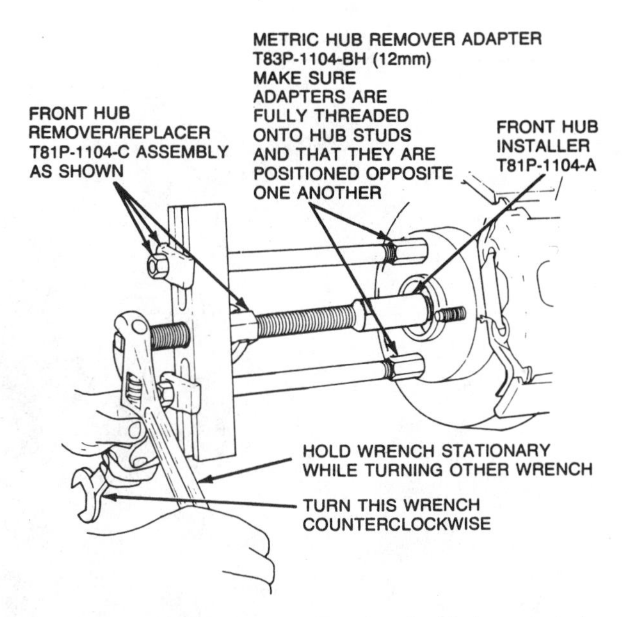

FIGURE 15–5 Front hub remover. (Reprinted with the permission of Ford Motor Company)

7. Note any items that might appear to interfere with clearances and get damaged when the engine tilts from the weight of the transmission. On RWD applications, some distributor installations close to the firewall are in jeopardy—remove the distributor cap or place a wood block between the engine block and firewall. Even an engine cooling fan may want to tilt into the radiator or shroud.
8. Study the exact routing of cables, vacuum lines, and electrical wiring. Routing must be duplicated for installation.
9. Always use a transmission jack for R&R. It provides maximum security and necessary tilting action and holding position to maneuver the transmission in tight quarters (Figures 15–8 and 15–9).
10. Some items, such as cooler line fittings and bell housing bolts, are more accessible when the crossmember is removed and transmission lowered.
11. When disconnecting the cooler lines, hold the case fitting with a wrench. This prevents the case and line fittings from turning together and twisting the line. A twisted line restricts normal cooler and lubrication fluid flow.
12. Mark the flywheel (flexplate) and converter so they can be installed in the same position for balance purposes.
13. Remove the converter-to-flywheel attaching bolts (or nuts) before removing all the converter housing bolts. Otherwise, the weight of the transmission can distort the flywheel or result in transmission pump damage. When possible, it is preferable to rotate the engine from the front instead of rotating at the flywheel to reach the converter attaching bolts/nuts through the access opening (Figure 15–10).
14. The transmission and converter are removed together as a unit.

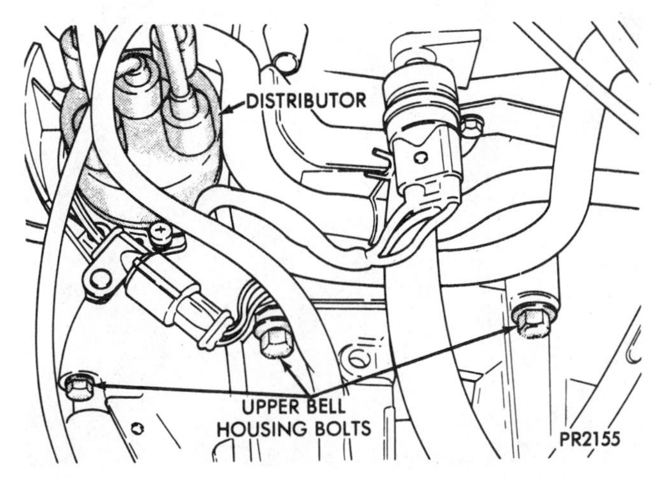

FIGURE 15–6 Remove converter housing upper bolts. (Courtesy of Chrysler Corp.)

FIGURE 15–7 Install engine support fixture.

FIGURE 15–8 Always use a transmission jack.

WARNING: Torque converters are heavy and easily slip off the turbine and stator support shafts, and out of the converter housing. To avoid personal injury and damage to the converter, slightly tip the transmission up at the converter end. You may want to design a hold-in clamp that can bolt onto the opening of the transmission converter housing or secure the converter in position with mechanic's wire (Figure 15–11).

Flywheel/Flexplate Inspection

Once the transmission is removed, make an immediate inspection of the flexplate (Figure 15–12). Areas of concern are:

- Flexplate cracks (Figure 15–13). The cracks are not always obvious, and therefore, critical visual inspection is necessary for the not-so-obvious condition. As a diagnostic check, tap the flexplate lightly with a hammer. A solid flexplate produces a definite ringing sound versus a short ring or no ring from a cracked flexplate.
- Starter ring gear tooth wear. The engine always stops the flexplate in a definite pattern: one of two different positions for four-cylinder, one of three positions for six-cylinder, and one of four positions for eight-cylinder. These are the starter engagement positions where the tooth wear occurs. As a general rule, if any of the teeth are worn back by one-third, the flexplate needs to be replaced. The same evaluation holds true for torque converters with the ring gear attached. Do not forget to examine the starter drive pinion gear.
- Wear on the converter contact pads and bolt/stud hole elongation. This is usually caused by a previous loose

A

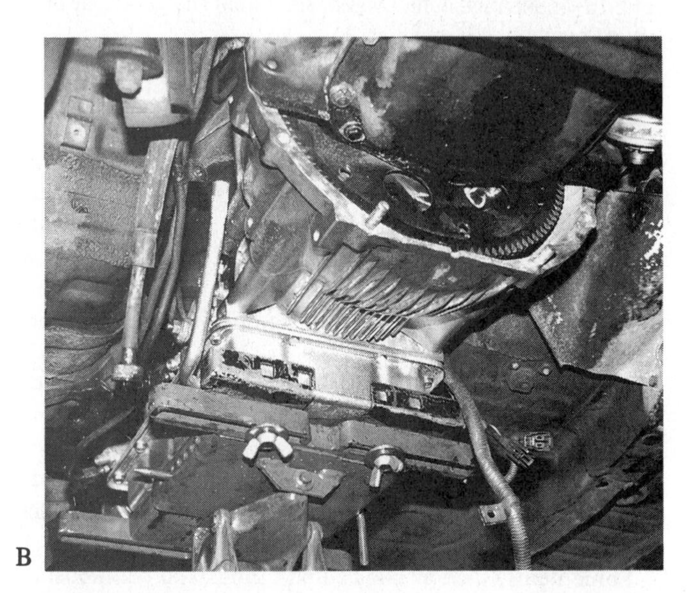

B

C

FIGURE 15–9 Transmission jack provides safe maneuvering in tight quarters.

FIGURE 15–10 Rotate the engine from the front to reach the converter attaching bolts/nuts.

FIGURE 15–11 Mechanic's wire used to prevent converter from falling out.

FIGURE 15–12 Typical flexplates.

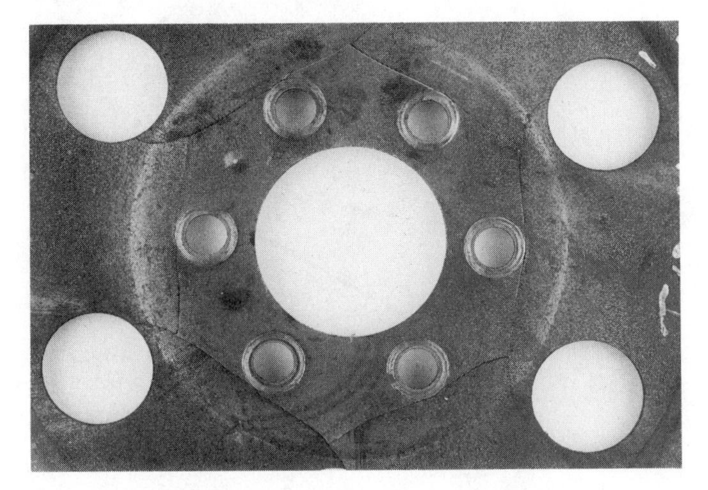

FIGURE 15–13 Flexplate cracks.

flexplate and calls for replacement. This condition can pull the converter off center and stress the flexplate.
- Excessive runout. This should be suspected whenever converter hub and transmission pump bushing wear is heavy and/or transmission pump damage has occurred. Included on the list of probable runout causes are converter hub breakage, hub weld cracking, and vibration.

Flexplate runout is taken with a dial indicator (Figure 15–14). You may want to use a magnetic base dial indicator with a lockup flex arm for this operation. If total runout exceeds 0.010 in (0.254 mm), replace the flexplate. As the flexplate is rotated, be sure to keep the crankshaft endplay eliminated so that it will not appear as a factor in the runout reading.

TRANSMISSION/TRANSAXLE INSTALLATION

Installation of the transmission converter unit basically requires reversing the removal procedure. There are concerns in following several new procedures and checks to ensure the proper centering of the converter and housing to the engine.

Preinstallation

- Make sure both transmission dowel pins are in the engine block and protruding far enough to hold the transmission in alignment.
- Check the rear of the engine block for leaking core plugs.
- Clean and examine the converter hub pilot hole in the end of the crankshaft. The pilot hole will wear in one place only. If pilot wear is present, turn the crankshaft until the wear area is on top and mark the bottom of the flexplate (flywheel). Generally, converter crankshaft pilot wear and converter pilot wear occur at their common point of contact established from previous installation. There is always some movement allowed by the flexplate. Converter pilot wear also must be located toward the top. The objective of locating pilot wear toward the top is to enable the converter to properly center itself on new piloting surfaces while clamping the converter to the flexplate. When installing a new or reconditioned converter, crankshaft pilot hole wear must still be located toward the top.
- Always coat the crankshaft pilot hole with multipurpose grease.
- When installing the converter in the transmission, be careful not to force it into position. Damage to the seal, bushing, or internal pump area must be avoided.

The converter must align itself with the stator shaft splines, plus a pump drive gear where the hub is required to drive the pump. To achieve this alignment and full engagement in the transmission, hold the converter by the pilot. While lightly pushing forward, rotate the converter with your free hand (Figure 15–15). This allows the converter to make the necessary alignments for full engagement.

If in doubt whether full engagement has taken place, a measurement can be made from the transmission converter housing face to the converter front cover lugs/pads (Figure 15–16). This dimension varies between product transmissions. If the converter is not fully engaged, it will butt against the flexplate and prevent normal mating of the converter to flexplate and transmission to engine block during installation. Because the flexplate will flex, the installation can be forced,

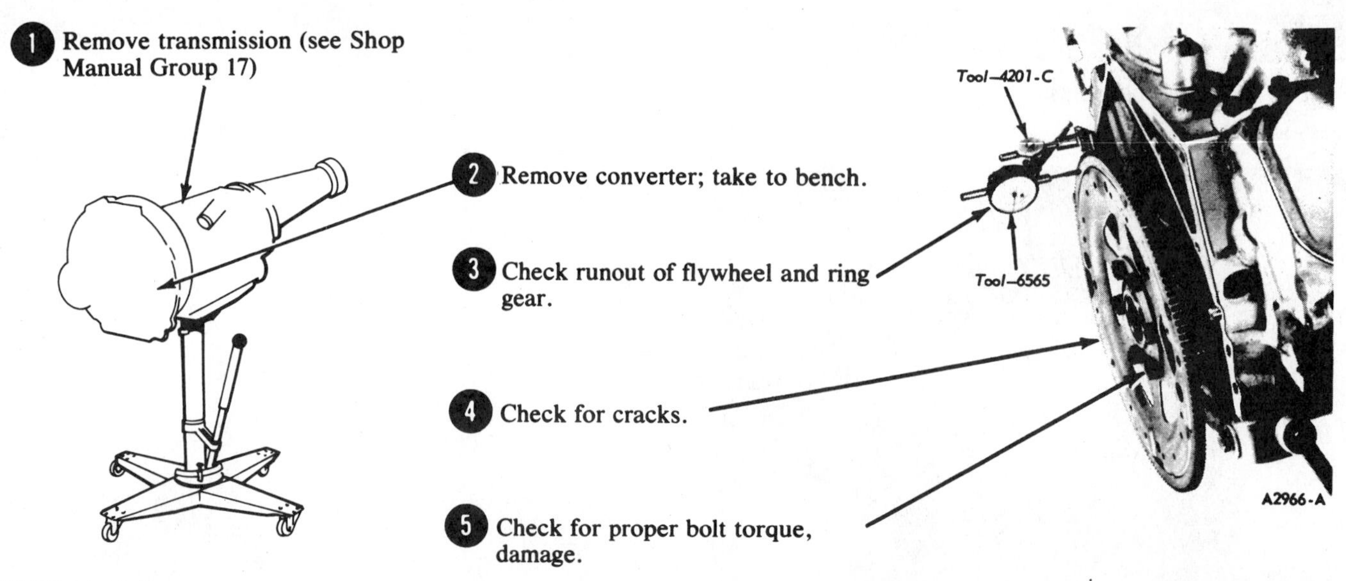

FIGURE 15–14 Remove converter; check for excessive flexplate/ring gear runout. (Reprinted with the permission of Ford Motor Company)

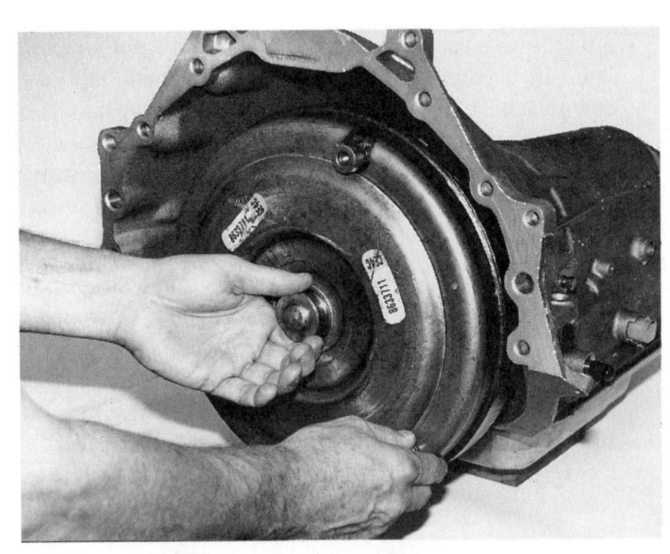

FIGURE 15–15 When installing the converter, support the pilot and rotate the converter to align shaft splines and pump.

but not without serious damage that can include the destruction of the transmission pump. Figure 15–17 shows what can happen when the converter drive hub wedges against the pump drive gear, even before the engine is started (Ford application).

Installation Highlights

- Full converter engagement must be maintained during installation.
- Move the transmission into position as close to the engine as possible.
- Rotate the converter to match the flexplate, attaching holes and converter/flexplate alignment marks made during removal. Chrysler rear-drive converters have a staggered bolt pattern. Therefore, the flexplate holes are offset and will match up only one way. The converter and flexplate are marked for correct matchup (Figure 15–18).
- Adjust the jack tilt to get an equal spacing between the engine block and transmission, along with alignment to the engine block dowel pins.
- While bolting the transmission to the engine, check the converter to make sure it is free to rotate. There must be clearance between the flexplate and converter at all times. Checking for rotation indicates that the converter is still fully engaged in the transmission and the converter pilot hub is not binding in the crankshaft pilot hole. Proper bolting is very critical to reestablishing engine-transmission-converter alignment (Figure 15–19).
- The converter always installs into the transmission in a position that is deeper than its normal operating position. When the transmission is bolted, there should be an approximate 1/8- to 3/16-in (3.175- to 4.763-mm) clearance between the flexplate and the converter pads. The converter should slide forward and into position. Hand-start the converter bolts/nuts. Starting with the pilot wear positioned to the top, lightly snug the bolts/nuts, and then full-torque to

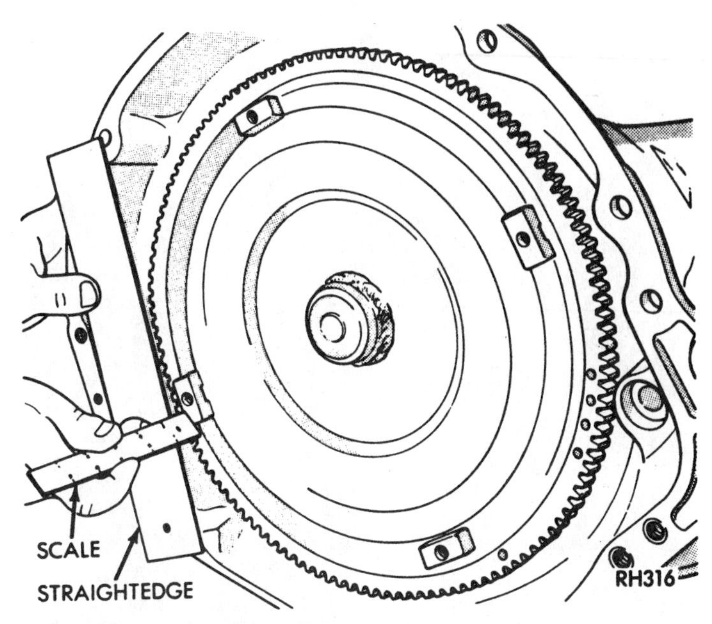

FIGURE 15–16 Measure to check for full converter engagement. (Courtesy of Chrysler Corp.)

specifications, again starting with the pilot wear at the top. Absolutely do not bring each bolt/nut to full torque one at a time without first "fussing" with a light pretorque all the way around. Following correct procedure will ensure proper converter piloting and alignment.

- Do not attempt to mix metric and USS (English) threads when clamping the converter. A metric converter bolt will not thread into a standard thread converter lug. A standard converter bolt will thread into a metric converter lug, but

FIGURE 15–17 Damage to pump from forced installation.

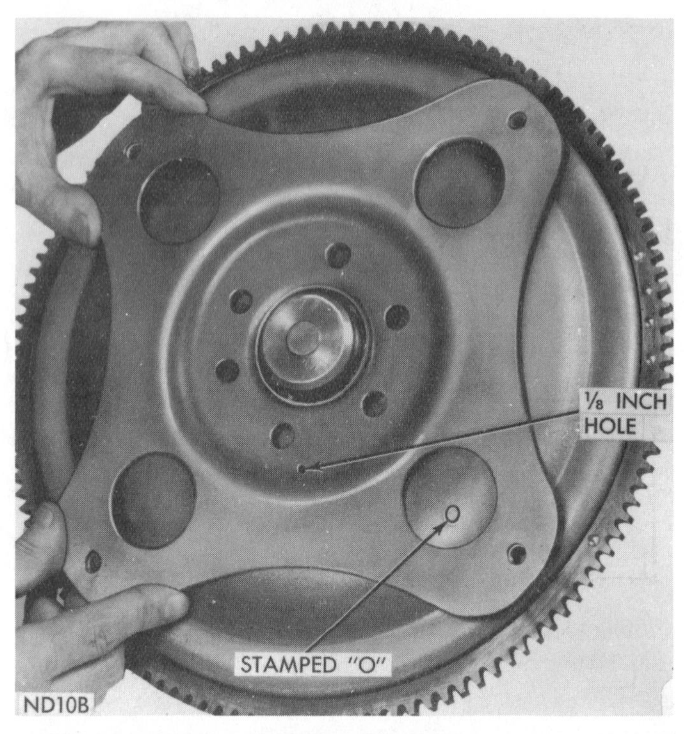

FIGURE 15–18 Chrysler converters and flexplates are marked for correct alignment. (Courtesy of Chrysler Corp.)

not without eventually stressing out. Converter attaching bolts are special bolts using extradimensional bolt heads and must have a grade 5 rating or better. Dimensional bolt length is also important. Longer-than-specified bolts will bottom out on the converter cover and dimple the cover in the exact area of the converter clutch contacting surface. This will cause clutch shudder and early failure.

- Be sure that the engine-transmission ground strap to the car body is in place and free of corrosion. Corroded or missing ground straps must be replaced. Otherwise, an electrolysis action sets up between the ATF and rotating transmission parts, causing a fusion or welding-like corrosion that destroys support bushings and roller thrust surfaces. Another item that may suffer is the manual control cable.

❖ **Electrolysis:** A surface etching or bonding of current-conducting transmission components that may occur when grounding straps are missing or in poor condition.

- Route and anchor control cables, vacuum lines, cooler lines, and wiring to original factory location.
- If the transmission has a vacuum modulator, replace the steel tube line if it is pinched, kinked, or twisted. Replacement of the rubber nipple tube connectors is preferable, as

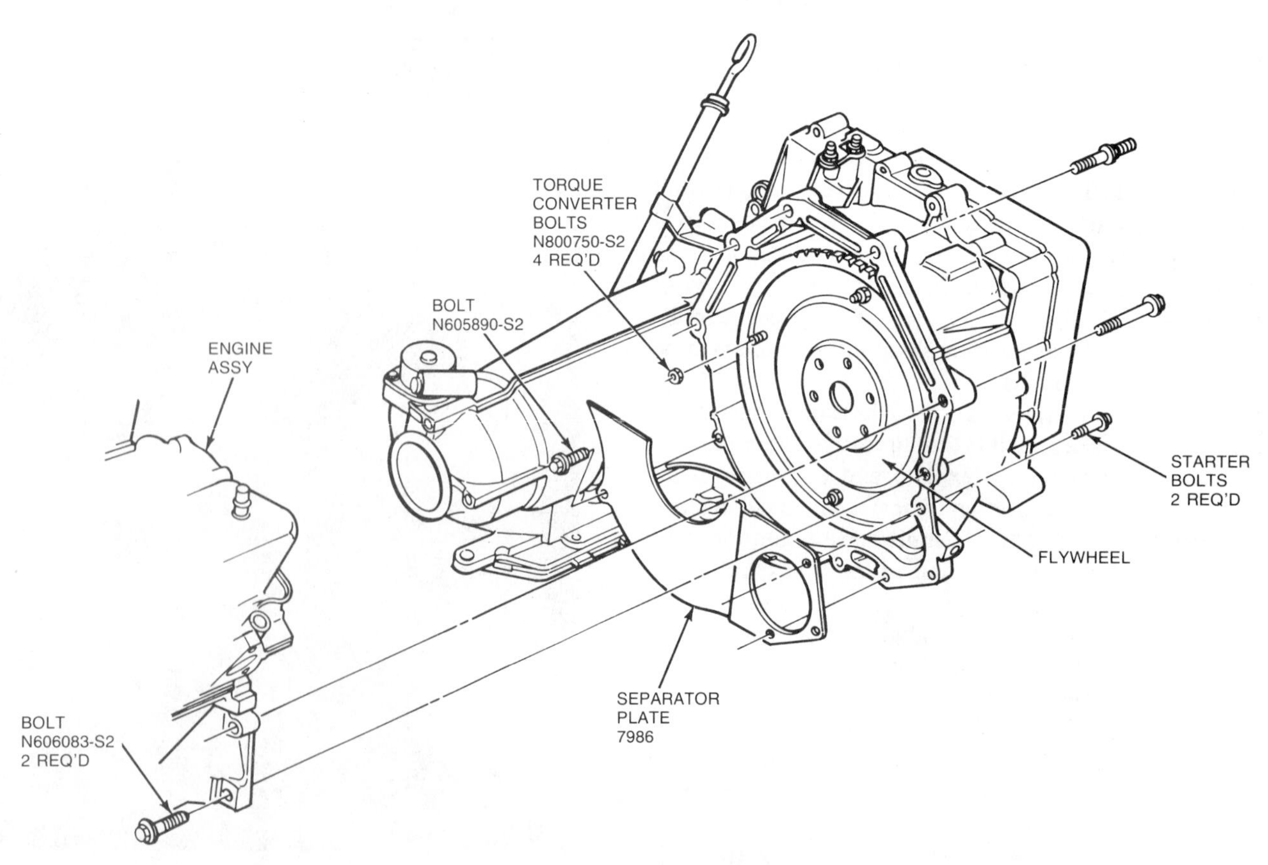

FIGURE 15–19 Proper bolting is critical. (Reprinted with the permission of Ford Motor Company)

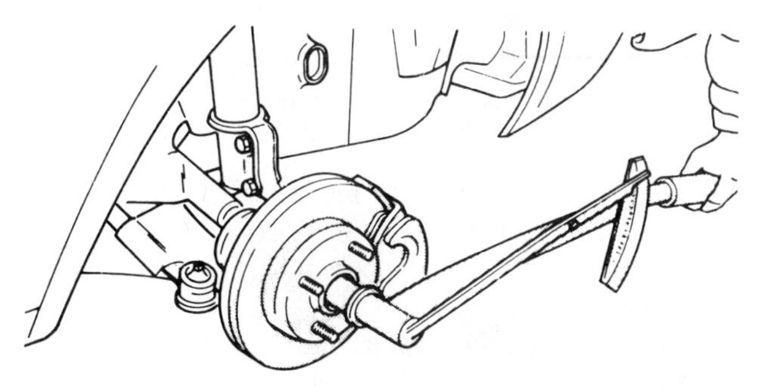
FIGURE 15–20 Torque tighten hub nut. (Courtesy of Chrysler Corp.)

they get brittle and crack from heat exposure. Do not use straight nipple hose where a ninety-degree molded nipple is required. The straight hose crimps and restricts the vacuum signal.

- Replace all missing bolts or minor exterior items related to the installation. This makes the job look professional and helps eliminate future complaints.
- On Chrysler FWD applications, be sure that the engine/transaxle assembly is properly centered. This can be verified by measuring the drive shaft lengths. Slotted engine mounts allow for a centering adjustment—refer to the service manual. Engine centering can be disturbed during engine or transaxle R&R or after vehicle frame damage and repair. Out-of-dimension drive shaft lengths can result in weird noises, repeated differential side bearing failure, or axle "popout."
- Some FWD applications require a special tool to pull the outer axle spline through the wheel bearing. Do not try to hammer it through. The axle hub nut is usually self-locking and should not be reused. It must be torqued to specifications (Figure 15–20). Note that this is done with the brakes applied. Do not use any power or impact tools.
- Fill the transmission with four quarts of the correct fluid. Start the engine and bring the fluid level to the add or cold mark on the dipstick, using proper level check method.(See

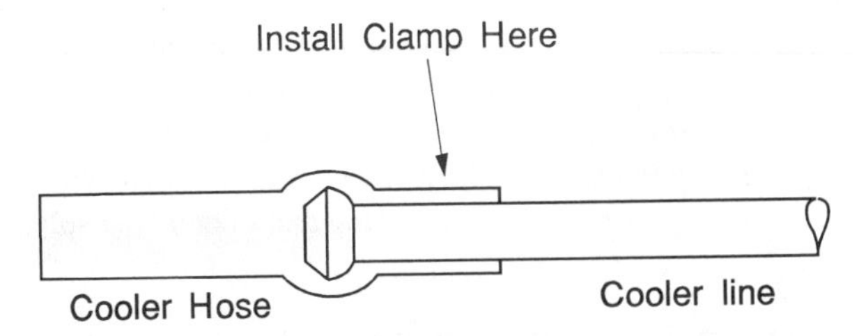

FIGURE 15–21 Location of cooler hose clamp.

section 2 in Chapter 12, "Diagnostic Step 2: Check the Fluid Level and Condition.").

- Road test and make any necessary external linkage adjustments for proper shift feel and kickdown performance.
- Make a final check for external fluid leaks and proper fluid level.
- To protect against coolant and ATF overheating, test the operation of the electric fan motor where applicable. On air-conditioner-equipped vehicles, the radiator fan motor must be ON whenever the air conditioner is turned ON.

Cooler Line Installation/Repair

Do not overlook the cooler lines during installation:

- Cooler lines that are twisted, have sharp bends/kinks, or show any sign of abrasion wear/corrosion should be replaced. Line flares should not be crushed or distorted.
- Brass compression fittings should be held to one per line.
- Always fabricate replacement tubing with the same-size steel tubing as the original line. Do not use copper or aluminum lines.
- For flared fittings, double-flare the ends of the steel tubing for extra strength and a positive seal. Where flexible reinforced hose is used as a line coupling, the hose needs to fit over a double flare at the end of the cooler line. A single flare eventually cuts through the hose. Locate the hose clamp as shown in Figure 15–21.
- Featured in Figure 15–22 is a Ford cooler line push-style connector fitting. To remove the line, a special cooler line disconnect tool is needed (Figure 15–23). The cooler line

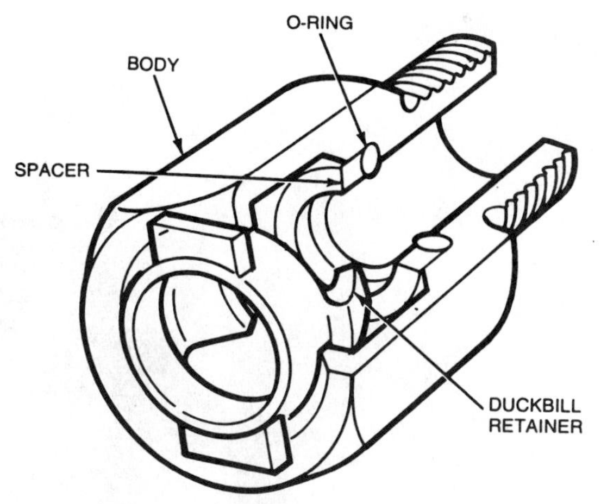

FIGURE 15–22 Push-style cooler line connector fitting. (Reprinted with the permission of Ford Motor Company)

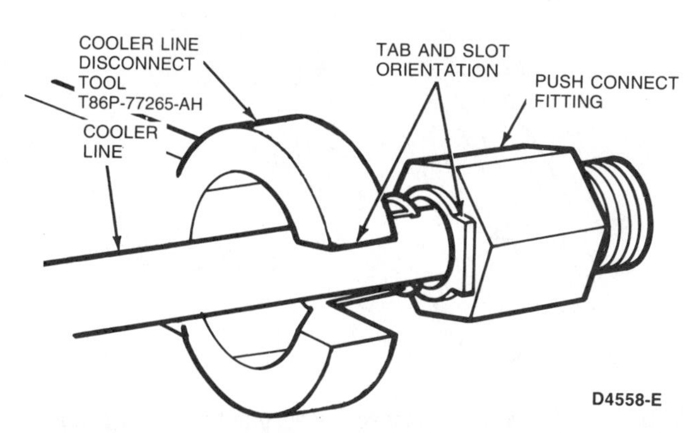

FIGURE 15–23 Disconnect tool required for push-style connector. (Reprinted with the permission of Ford Motor Company)

easily connects by pushing the line into the connector. As you do this, listen for a click. This means that the retainer engaged the tube bead. Pull back on the line to check for full engagement. Should leakage occur, it will require a new fitting or combination fitting and cooler line splice.

COOLER SYSTEM FLUSHING

Much is said about the contribution of downsized radiators and road load conditions to ATF heat buildup and eventual transmission failure. Although an add-on transmission cooler is a priority choice as a solution to the problem, it might not do the job. Why? A restricted cooler flow could be the real problem. Restricted cooler flow is the second most common reason for transmission overheating. It starves the transmission lubrication system.

The transmission cooler element in the radiator with its honeycomb cavities provides a great hideaway for contaminants cycling out of the converter. Even auxiliary coolers are affected. The contaminants come primarily from internal converter metallic wear/damage or loose lockup friction material. Transmission pump damage also can be a contributing factor. Any foreign matter in the system has the potential to circulate back to the transmission, where microparticles can slip through the filter and cause a malfunction in items such as the governor and the TV and line pressure regulator valves.

Contaminants are a major cause of recurring troubles in automatic transmissions. How many times have you heard of technicians cleaning a valve body or governor several times after an overhaul, or experiencing an early transmission overhaul failure due to overheating? For protection, the technician should make it a practice to flow-test and flush the cooler system.

Depending on the type of equipment available in the shop, several different techniques can be used. The best results are achieved with a flusher designed specifically for the purpose (Figure 15–24). These are portable units that in most cases pump a mineral spirit or biodegradable solvent through the system. Reverse flush as follows:

- Disconnect the cooler inlet line at the radiator, carefully remove any debris that may have collected at the inlet, and then reconnect. Always use a backup wrench to prevent damage.
- Start by flushing the cooler and lines in reverse to the normal flow.
- Reconnect the flusher, and flush the system in the direction of normal flow.
- Flush for a minimum of fifteen minutes in both directions.
- Purge the solvent from the system, using an air supply at no more than 25 psi (172 kPa). Continue the process until moisture is no longer seen leaving the discharge line.
- Conclude with a final flush, using ATF to remove any remaining solvent in the system.

Reverse flushing is not guaranteed to clear the cooling system of unwanted debris, nor does it open pinched/restricted cooler lines. After the flushing process, a system volume delivery test should be performed. A Kent-Moore oil cooler and line flusher utilizes the flow of tap water combined with special flushing fluid and regulated air pressure (Figure 15–25). The flushing fluid is a biodegradable concentrated blend of detergents and silicates. The equipment has its own unique operation, with the flushing procedure maintaining the reverse flush process.

In the absence of special equipment, a clean shop solvent from a tank equipped with a pump can be used for the flushing operation (Figure 15–26). Another method is to use a suction/pump gun (Figure 15–27). With this technique, fill the cooler with solvent. Reload the gun and proceed to work the gun handle back and forth to surge the solvent for the best cleaning effect. Continue the process using as many fresh loads of solvent as necessary to clear the system. Reverse flushing is still in order.

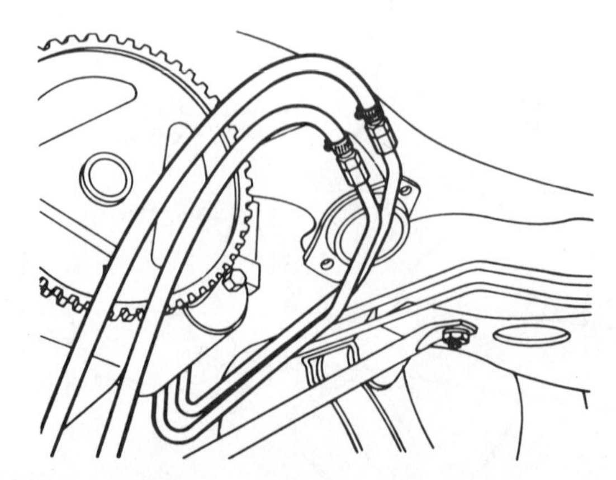

FIGURE 15–24 Converter flush machine adapter hoses attached to the cooler system. (Courtesy of General Motors Corporation, Service Technology Group)

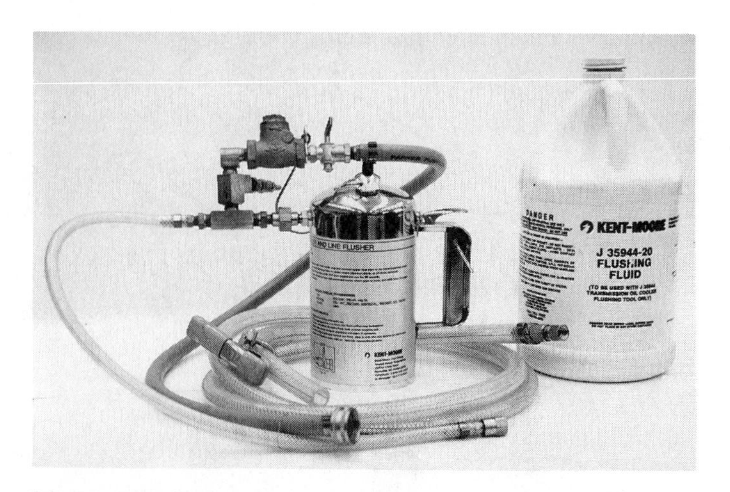

FIGURE 15–25 Special flushing equipment uses concentrated solvent and tap water.

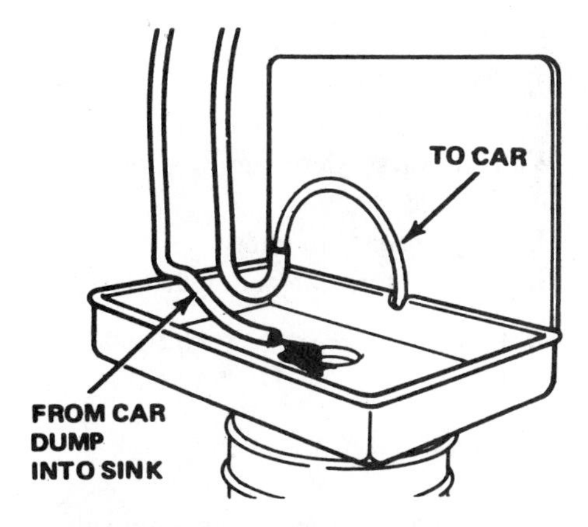

FIGURE 15–26 Solvent tank and pump for flushing. (Courtesy of General Motors Corporation, Service Technology Group)

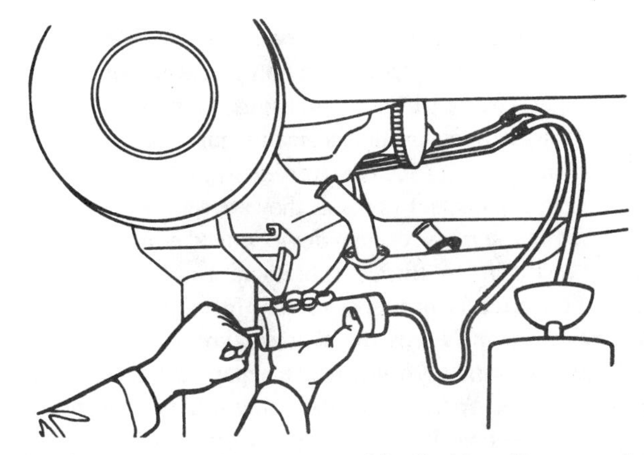

FIGURE 15–27 Suction/pump gun used for flushing. (Courtesy of General Motors Corporation, Service Technology Group)

Volume Testing

After reverse flushing, the reliability of the cooler system to flow transmission fluid should be checked with a volume test. The procedure is as follows:

- Add one extra quart or liter of ATF to the transmission.
- Disconnect the cooler return line at the transmission, and arrange for a container to collect the fluid discharge.
- Run the engine at test rpm with the transmission in park (TorqueFlite 900 and 700 RWD series must be in neutral). Follow these guidelines:

1. Chrysler automatic transmissions must deliver a continuous flow of 1 q (0.956 L) in 20 seconds or less (engine at curb idle).
2. General Motors automatic transmissions must deliver a continuous flow of 2 q (1.90 L) in 30 seconds or less (engine at curb idle).
3. Ford simply requires that the technician observe a liberal flow of fluid. This generally falls in the range of 2.25 q (2.2 L) in 30 seconds or less (engine at 1000 rpm).

CAUTION: Limit testing to thirty seconds to avoid pumping the transmission dry.

If the fluid flow does not meet specifications, the technician will need to determine if the problem is a plugged radiator cooler, pinched lines, or restricted add-on cooler. Reverse flushing does not always clear a plugged cooler element. Transmission radiator coolers requiring replacement should be replaced professionally at a radiator shop.

A less time-consuming technique that can be used for checking a transmission cooler system restriction involves the use of a pressure gauge. Install the gauge in the pressure side of a cooler system flusher. The technician must first establish a data base for each product transmission. How much pressure does it take for the pump on the flushing machine to move solvent through a normally open cooling system disconnected from the transmission? If 25 psi (173 kPa) is established as the norm for a product system, an actual test reading showing 40 psi (276 kPa) indicates a restricted system. There is no need to hook up the system to the transmission or to have the engine running.

As a preventative maintenance item and to protect the long-term reliability of an overhaul/rebuild, a transmission filter should be added into the cooler return line (Figure 15–28). Reverse flushing never eliminates every particle of debris. An in-line filter helps extract ferrous and nonferrous particles from the fluid if some remains after service/overhaul procedures, and as the transmission wears in the future.

SUMMARY

Removing and replacing a transmission or transaxle can be a rather involved process. Because of the complexity of today's vehicles, careful attention must be given to the components that must be detached and returned to their original positions.

Begin the removal procedure with an undervehicle inspection. Pay special attention to components that affect transmission operation.

Work safely, and be aware of the hazards surrounding you. For your protection, wear safety glasses and safety shoes. To

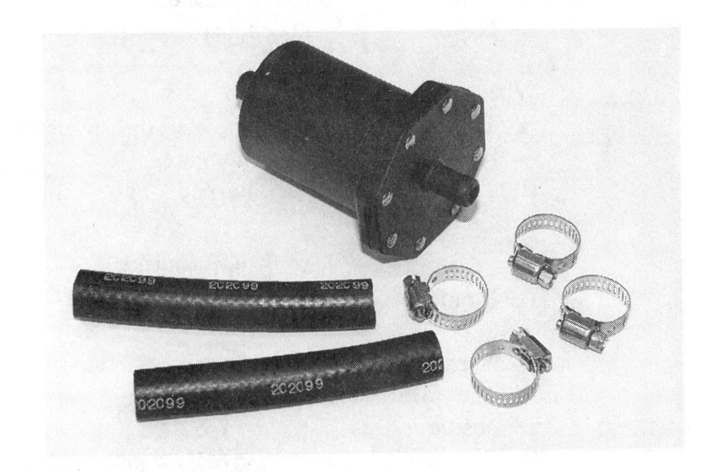

FIGURE 15–28 In-line filter.

prevent electrical sparks, disconnect the negative battery cable. When removing or replacing the transmission, use a transmission jack, and make sure that the engine is properly supported.

Refer to the appropriate service manual for the exact steps of procedure in the R&R of the transmission. Index components and thoroughly inspect items as they are removed or exposed.

During reassembly, certain devices need extra care. Always tighten fasteners to torque specifications. Be sure that the torque converter is properly seated. In addition, service the cooler system, and check that the flow volume is adequate.

A quality transmission service repair is important to the proper and safe operation of the vehicle. Work carefully, and do the best job possible.

❑ REVIEW

Key Term

Electrolysis.

Questions

Undercar Inspection

____ 1. Before removal of a FWD transaxle unit, carefully inspect for:

A. subframe collision damage.
B. worn rack and pinion mount bushings.
C. loose motor mounts.
D. all of the above.

____ 2. When inspecting front-wheel drive CV joints:
I. A collapsed boot is considered a normal condition.
II. Squeeze the boots and listen for escaping air.

A. I only C. both I and II
B. II only D. neither I nor II

____ 3. When replacing a CV boot:
I. The CV joint must be completely disassembled for cleaning and inspection.
II. The boot must be "burped" before final clamping.

A. I only C. both I and II
B. II only D. neither I nor II

____ 4. To check cross and yoke U-joints:
I. The joint angles must be in their normal position.
II. With the drive shaft removed, the joints should have a stiff movement.

A. I only C. both I and II
B. II only D. neither I nor II

____ 5. With the drive shaft removed, a double Cardon joint should:
I. twist around smoothly.
II. strongly resist any straightening effort.

A. I only C. both I and II
B. II only D. neither I nor II

Removal Highlights

1. The battery ground cable should be removed. ___T/___F

2. (RWD) The drive shaft must be marked for original position. ___T/___F

3. (RWD) The splined sections of the drive shaft must be correctly joined for yoke phasing. ___T/___F

4. (FWD) The use of a heavy-duty hammer to separate the outboard CV joint stub shaft from the hub is an acceptable practice. ___T/___F

5. The usual practice is to remove the converter housing bolts before removing the converter-to-flexplate fasteners. ___T/___F

Flexplate Inspection

1. A cracked flexplate responds to a hammer tap with a definite ringing sound. ___T/___F

2. Wear on the converter contact pads calls for a flexplate replacement. ___T/___F

3. Suspect excessive flexplate runout when the transmission pump has internal damage. ___T/___F

4. Suspect excessive flexplate runout when the converter hub is fractured. ___T/___F

5. Flexplate runout should not exceed 0.010 in (0.254 mm). ___T/___F

Installation

1. The converter housing-to-engine alignment is a function of the housing bolts. ___T/___F

2. Always coat the crankshaft pilot hole with multipurpose grease. ___T/___F

3. If the converter does not freely move forward against the flexplate, the pilot hub may be binding in the crankshaft pilot hole. ___T/___F

4. Converter attaching bolts should have a minimum grade rating of 3. ___T/___F

5. A missing engine-transmission ground strap may cause electrolysis conditions and damage transmission components. ___T/___F

6. Incorrect routing of the transmission wiring could upset the transmission computer operations. ___T/___F

7. (FWD) It is an acceptable practice to drive the outer axle spline into place with a heavy-duty hammer. ___T/___F

8. Before starting the engine, fill the transmission to full capacity. ___T/___F

_____ 9. If the converter butts against the flexplate, it could be caused by:
   I. improper pump gear installation.
   II. converter drive hub failure to engage the pump drive gear.

   A. I only
   B. II only
   C. both I and II
   D. neither I nor II

_____ 10. (FWD) The axle hub nut:
   I. should not be reused.
   II. can correctly be torqued with an impact tool.

   A. I only
   B. II only
   C. both I and II
   D. neither I nor II

Cooler System Flushing

1. ATF contaminants are a major cause of recurring troubles. ___T/___F

2. Before flushing starts, disconnect the inlet line at the cooler and remove contaminants. ___T/___F

3. Start flushing the system in the normal flow direction. ___T/___F

4. The system should be flushed in each direction for a minimum of five minutes. ___T/___F

5. When purging the solvent from the system, use an air supply at no more than 75 psi (518 kPa). ___T/___F

6. Reverse flushing is guaranteed to purge the cooling system of unwanted junk. ___T/___F

Volume Testing

1. The volume is measured from the cooler return at the transmission. ___T/___F

2. An intermittent flow is considered normal. ___T/___F

3. The maximum time limit for running the volume test is thirty seconds. ___T/___F

4. It is advisable to add 1 q (0.956 L) of ATF to the transmission before testing. ___T/___F

5. A cooling system restriction can be detected with the use of a pressure gauge and system flusher. ___T/___F

16 Torque Converter Evaluation

OBJECTIVE:

The objective of this unit is for you to become proficient with the terms, concepts, and procedures contained in this chapter. Completion of this objective is essential to becoming a successful automatic transmission technician.

CHALLENGE YOUR KNOWLEDGE

Define the following key term:

Endplay.

List the following:

Ten reasons to replace a torque converter; two ways to check bushing to converter support clearance; and three methods to seal a converter drain hole drilled by a technician.

Describe the following:

The method to check for internal converter interference problems; a method to bench test a stator one-way clutch; and the procedure used to flush a converter.

INTRODUCTION

Converter service, as well as cooler circuit service, must be accomplished as a part of the overhaul process. These are often neglected and, when overlooked, can lead to a quick repeat failure of the transmission.

Contaminants not removed from the converter or cooling system will cycle back into the transmission through the lube circuit, attacking bushing and thrust washer surfaces. If the cooler element or lines are restricted, overheating in the converter and damage to the transmission occur.

Modern converters are welded assemblies and therefore cannot be disassembled in field service for cleaning and inspection. There are field bench check procedures that can be used by the technician to evaluate and determine converter serviceability.

The converter is a high-cost item that often is needlessly replaced. Studies have shown that over fifty percent of replaced converters are not defective and are serviceable. On the other hand, failure to detect a defective converter can result in an early breakdown of a perfectly good overhaul/rebuild. It should be obvious that it is important to understand the various conditions that determine whether the converter can be put back into service or must be replaced. The following guidelines can be used as a reference.

CONVERTER INSPECTION AND REPLACEMENT GUIDELINES

If one of the following conditions exist during bench inspection, the converter should be replaced.

1. The transmission pump is badly damaged, resulting in metallic particles entering the converter. The metallic content can never be flushed out one hundred percent. Although immediate internal wear may not always be apparent, the long-term reliability of the unit is questionable.
2. Internal converter failure, such as worn thrust washers/bearings and thrust surfaces, or interference between the member elements. Internal wear failures are usually associated with "aluminized" oil from the converter sampling.
3. The sample converter ATF has the color of a strawberry milkshake. This indicates that engine coolant has contaminated the fluid. Proceed as follows:

   - Nonlockup converters should be flushed using proper procedure and put back into service.
   - Converters with lockup clutches should be replaced.

NOTE: The converter does not need to be replaced if the sample ATF simply has an odor or dark discoloration.

FIGURE 16–1 Damaged converter hub.

FIGURE 16–2 Minor fretting wear.

4. Stator roller clutch failure. It is either frozen or freewheels in both directions.
5. A scored or damaged hub could cause a repeat front pump seal or bushing failure (Figure 16–1). Minor fretting wear on the end of the hub is an acceptable condition and is not a reason to replace the converter (Figure 16–2).

   Minor scuff marks and grooving are acceptable but should be hand-polished with crocus cloth or 600 wet/dry abrasive paper (Figure 16–3). Polishing marks should be in a circular pattern around the hub. Angularity can form leads that could cause leaks. Extensive polishing should not be necessary. Otherwise, reject the converter. The OD (outside diameter) of the hub should not be reduced by more than 0.001 in (0.025 mm). Flat spots cannot be removed by polishing.
6. External leakage, such as cracks at the hub weld area (Figure 16–4).
7. The drive studs on the converter cover are loose or have damaged pilot shoulders or stripped threads (Figure 16–5). These studs mate the converter to the crankshaft drive plate to (1) drive the converter and (2) pilot the converter to run true with the crankshaft. Any misalignment of the converter with the engine and transmission would run the drive hub off center to the pump and wipe out the bushing and pump gears.

   Converters using threaded pads need to be inspected for broken welds, stripped threads, and worn pad surfaces (Figure 16–6). Stripped or crossed threads are usually corrected with a helicoil installation. Worn studs or pads usually require a combined converter and flexplate replacement to maintain converter alignment.
8. Look for a broken or damaged converter pilot (Figure 16–7). Pilot damage prevents proper fit of the converter into the crankshaft bore and results in converter misalignment. Torque converters not properly piloted into the crankshaft can also cause the converter drive hub to bottom out on the pump drive gear. This condition can cause the hub wear pattern shown in Figure 16–8. When improper

FIGURE 16–3 Hand-polish minor scuffs and grooves.

FIGURE 16–4 Cracked converter hub weld.

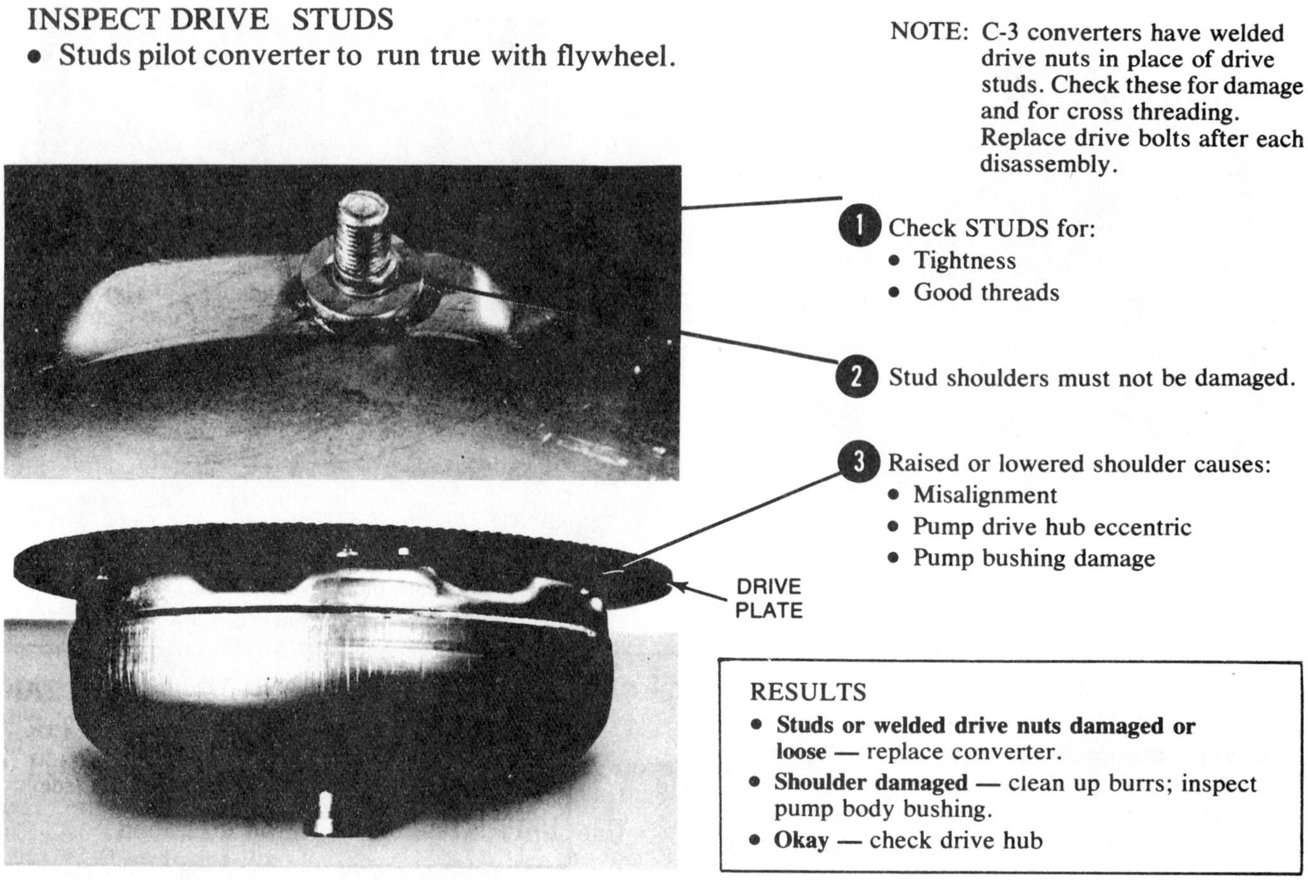

INSPECT DRIVE STUDS

- Studs pilot converter to run true with flywheel.

NOTE: C-3 converters have welded drive nuts in place of drive studs. Check these for damage and for cross threading. Replace drive bolts after each disassembly.

1 Check STUDS for:
- Tightness
- Good threads

2 Stud shoulders must not be damaged.

3 Raised or lowered shoulder causes:
- Misalignment
- Pump drive hub eccentric
- Pump bushing damage

DRIVE PLATE

RESULTS
- **Studs or welded drive nuts damaged or loose** — replace converter.
- **Shoulder damaged** — clean up burrs; inspect pump body bushing.
- **Okay** — check drive hub

FIGURE 16–5 Inspect drive studs. (Reprinted with the permission of Ford Motor Company)

piloting is evident, the transmission pump assembly usually has excessive wear damage.

9. The converter is blue. This was caused by an overheating condition due to abuse, overloading/overworking, faulty stator/reactor, and/or a restricted/inadequate cooler system.
10. The converter is dropped on the floor.

INTERNAL CONDITION

To determine the internal condition of the converter, some simple bench checks can be performed by the technician to evaluate thrust wear, stator one-way roller clutch action, and lockup clutch holding torque. For discussion purposes, a converter style with a pump drive hub is featured in steps 1 through 6. The converter can be checked for wear by several methods.

1. Hold the converter in a vertical position and observe the internal parts through the hub (Figure 16–9). If a thrust washer is out of place, excessive **endplay** exists, and the converter must be replaced. If a thrust washer is not seen as out of place, it cannot be assumed that the endplay is not excessive. While observing, make it a practice to rotate the converter slowly and listen for broken or loose parts.

❖ **Endplay:** The clearance/gap between two components that allows for expansion of the parts as they warm up, to prevent binding and to allow space for lubrication.

2. The preferred method for checking converter internal endplay is to use a special endplay tool equipped with a T-handle and dial indicator (Figure 16–10). When the tool

FIGURE 16–6 Inspect converters using threaded pads.

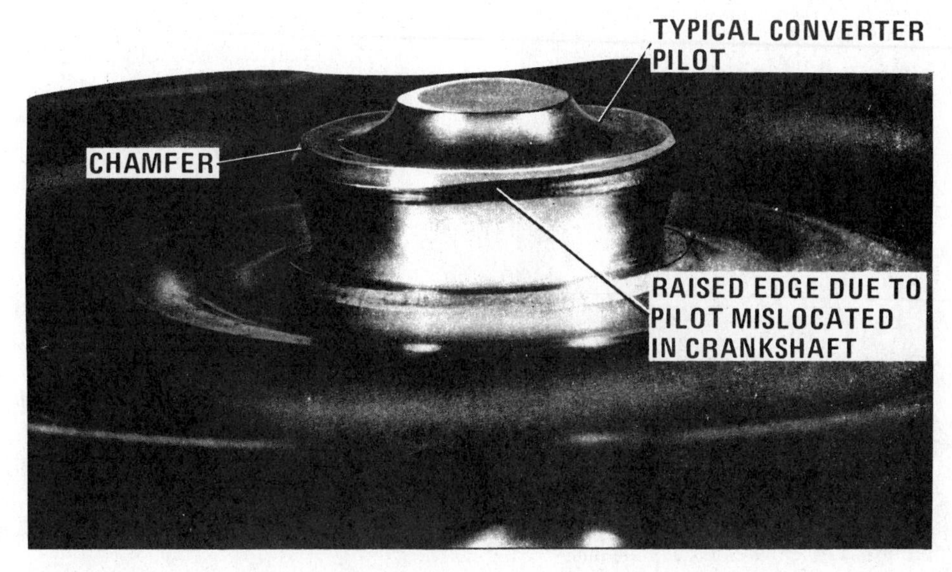

FIGURE 16–7 Typical torque converter pilot. (Courtesy of General Motors Corporation, Service Technology Group)

is installed, it is secured to the turbine splined hub. Refer to the shop manual for procedure and specifications. In most cases, the endplay should not exceed 0.050 in (1.27 mm), but this is no longer an absolute.

Many of the new converters with a lockup clutch are spring-loaded toward the hub, and the use of pliers as an alternate endplay check is no longer valid. Special endplay tooling must be used.

3. In non–spring-loaded converters, snap-ring pliers can be used in the absence of special tooling to measure endplay. Extend the pliers into the hub opening and clutch the stator hub splines (Figure 16–11). A firm up-and-down movement should not exceed 1/16 in (1.59 mm).
4. Internal interference inside the converter should be checked as follows:
   A. Stator-to-Turbine and Turbine-to-Cover Interference
      (1) Position the converter face down on the bench.
      (2) Insert the transmission pump and input shaft into the converter and engage the splines of the stator and turbine (Figure 16–12).
      (3) Hold the pump and converter and rotate the input shaft in both directions (Figure 16–13). Any binding or scraping indicates internal wear and interference.
   B. Stator-to-Impeller Interference
      (1) Invert the pump and converter on the bench as shown in Figure 16–14. With the converter in this

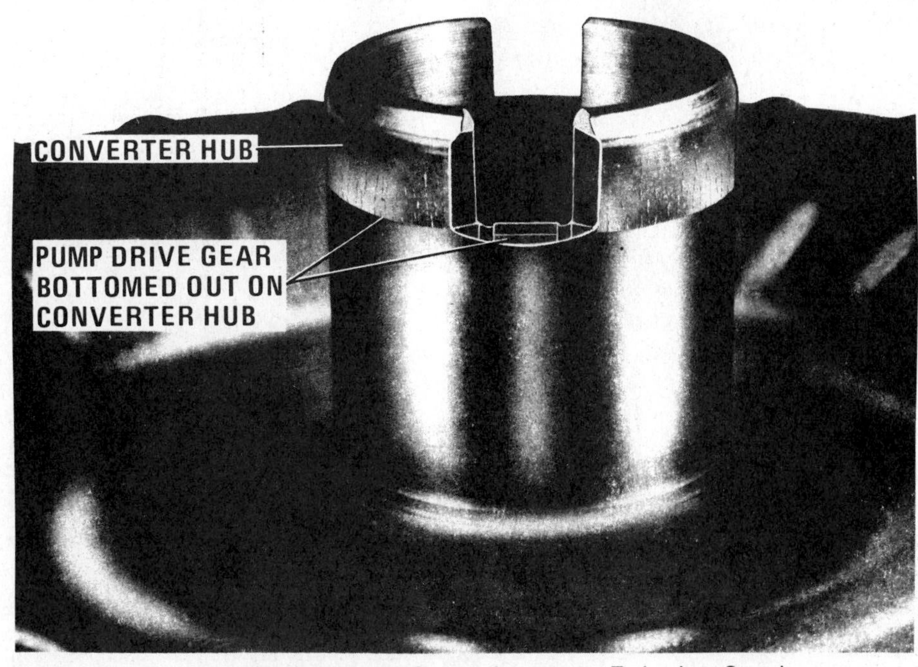

FIGURE 16–8 Torque converter hub. (Courtesy of General Motors Corporation, Service Technology Group)

FIGURE 16–9 Inspect internal converter parts through the hub.

position, the stator will drop against the impeller thrust washer.

(2) Hold the pump stationary and rotate the converter in both directions. Check for any binding or scraping, which would indicate a worn thrust washer or thrust surface.

5. To test the action of the stator one-way clutch, a dummy stator shaft can be inserted in the converter to engage the one-way clutch race (Figure 16–15). The inertia of a counterclockwise snap action should produce a freewheeling sensation, and a clockwise snap action a lockup sensation (Figure 16–16). A freewheeling action or lockup action in both directions indicates a faulty one-way clutch. For best results, the converter should be full of fluid.

Snap-ring pliers can be used to hold the splines of the one-way clutch race in place of the stator shaft. The same

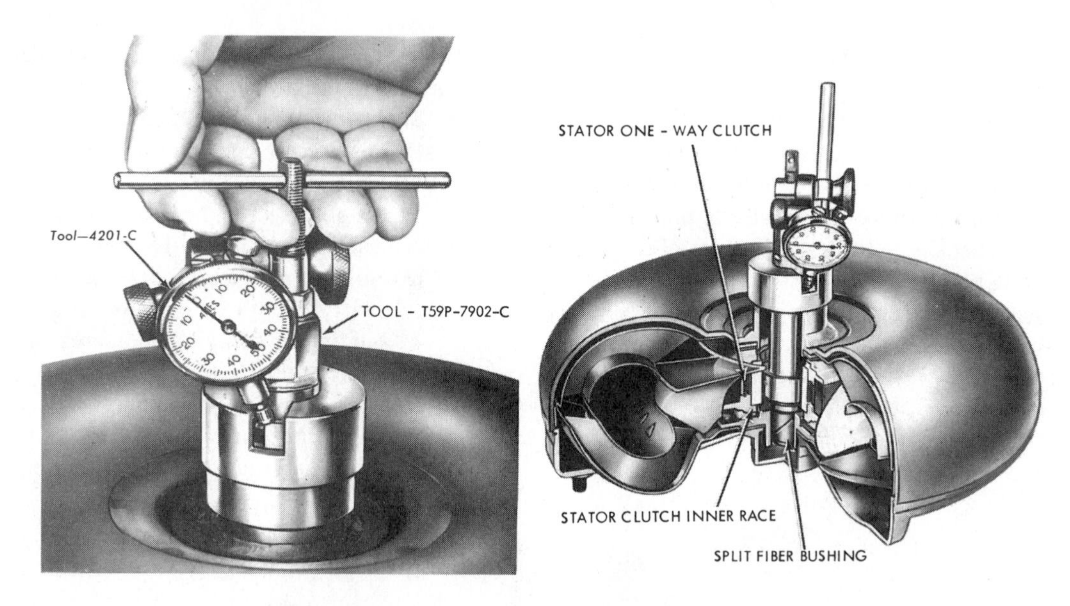

FIGURE 16–10 Check for internal converter endplay. (Reprinted with the permission of Ford Motor Company)

FIGURE 16–11 Snap-ring pliers can be used to check endplay in non–spring-loaded converters. (Top photo: converter split to show the endplay check procedure.)

FIGURE 16–12 Insert pump and input shaft into the converter.

FIGURE 16–13 Rotate input shaft in both directions.

rotational snap action will check the one-way clutch. An alternative method for checking the stator roller clutch calls for inserting a finger into the splined inner race and trying to turn the race in both directions.

6. Lockup clutches can be tested on a bench lockup tester for torque holding power and damper spring action (Figure 16–17).

   In place of a spring-style clutch damper assembly, some GM-designed lockup converters use a viscous silicone filled clutch damper assembly. Lockup is never complete, but ninety-eight percent efficiency is achieved. A viscous clutch plate provides extra smooth clutch engagement, while reducing chuggle problems. When tested on a lockup bench tester, the converter viscous assembly offers a very stiff but yielding resistance to the torque wrench. Silicone can leak from the damper housing and prevent converter lockup.

   A converter that fails to lock up will continually multiply torque in overdrive gear, resulting in an overheated transmission and destruction of a good overhaul/rebuild.

7. In transaxle applications where the converter hub is not used to drive the pump, the hub contains an inside support bushing that needs evaluation. If the bushing and drive support have been working in a clean fluid environment, there should be almost no evidence of wear (Figure 16–18).

   To prevent repeated front seal leaks, it is important that the clearance between the bushing and drive support not exceed 0.004 in (0.10 mm). This can be determined by

FIGURE 16–14 Invert pump and converter to allow stator to drop against impeller thrust washer.

FIGURE 16–15 Check action of stator one-way clutch with a dummy shaft. (Stator removed from a converter to illustrate the procedure.)

FIGURE 16–16 Rotate the shaft to check the stator one-way clutch.

measuring the inside bushing and outside drive support dimensions. A dial or digital vernier caliper, or combination telescoping gauge and micrometer, can be used (Figures 16–19 and 16–20).

If precision measuring tools are not available, use an ordinary strip of paper and overlap the bushing and hub (Figure 16–21). If the drive support does not fit the bushing, the clearance is okay. The nominal thickness of ordinary paper is 0.003 to 0.004 in (0.07 to 0.10 mm). The original factory bushing has a precision close-tolerance fit. If wear is not evident in this area, do not replace the bushing as a service practice.

8. If the converter drives an oil pump drive shaft, insert the shaft into the converter and check for engagement. There should be no binding in the spline engagement or excessive play when twisting the shaft.

FIGURE 16–18 Inspect inside support bushing.

FIGURE 16–17 Converter lockup clutch tester.

Most converters will pass bench checks and are reusable. It is preferred to flush the converter before putting it back into service. If flushing equipment is not available, use a suction gun or electric motor pump equipment used with the lockup bench tester to evacuate the converter fluid (for converters without drain plugs).

FIGURE 16–19 Use a vernier caliper to evaluate bushing-to-drive support clearance.

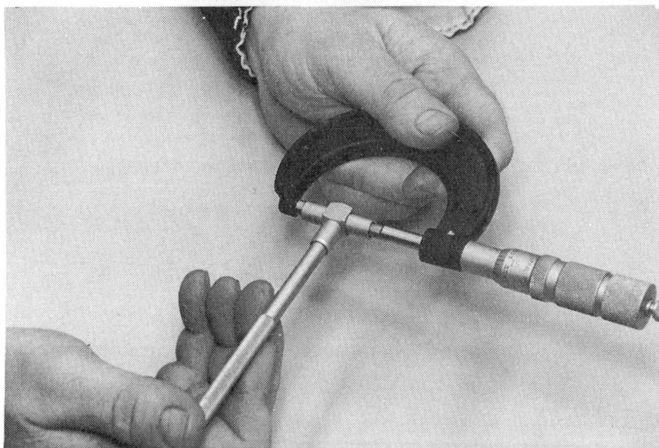

FIGURE 16–20 Use a telescoping gauge and micrometer to evaluate bushing-to-drive support clearance.

FIGURE 16–21 Use paper strip to evaluate bushing-to-drive support clearance.

Converters without drain plugs can be drilled with a 1/8-in (3.57-mm) hole between the top end of the impeller fin dimples (Figure 16–22). (Hint: Tip the converter when drilling, and drill from the bottom. This will allow the metal chips to wash away as the hole is made and fluid leaks out.) For converters without fin dimples, never drill an air bleed hole anywhere into the converter.

Once the converter is completely drained, it is ready for mounting in a converter flusher (Figure 16–23). This piece of equipment slowly rotates the converter while a cleaning solvent is cycled in and out of the converter to purge the old oil and contaminants. The machine automatically adds timed blasts of compressed air to the solvent for agitation as it enters the converter.

Ideally, the flushing machine screens the debris flushed from the converter so the technician can determine the origin and extent of contamination. For thorough cleaning, the machine flushing operation should continue for a minimum of twenty minutes. After flushing, drain the cleaning solvent, fill the converter with ATF, and drain again. This removes the residual solvent and avoids possible internal seal damage in the converter-transmission.

Where the drain hole was drilled, coat a 1/8-in (3.57-mm) closed-end pop rivet with Loctite and install. This is a special

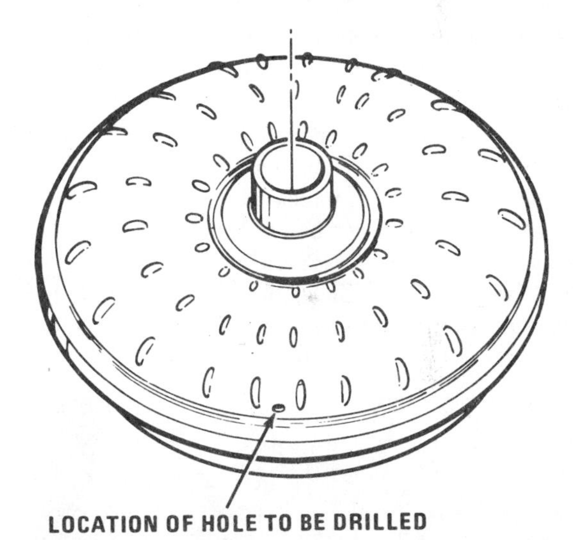

FIGURE 16–22 Drill 1/8-in air bleed to evacuate converter fluid. (Courtesy of General Motors Corporation, Service Technology Group)

FIGURE 16–23 Converter mounted in flusher.

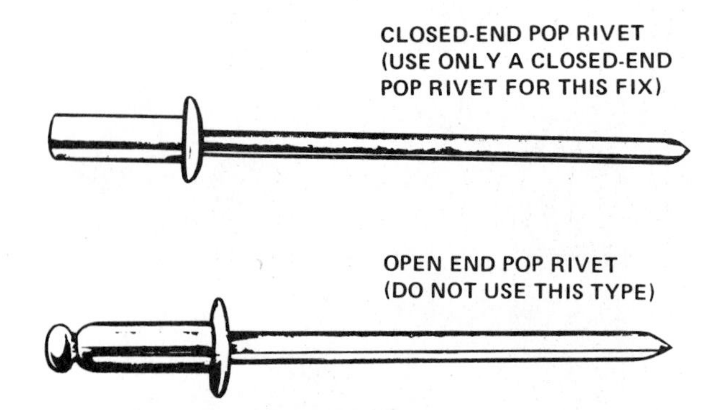

FIGURE 16–24 Use a closed-end pop rivet to close technician-drilled hole. (Courtesy of General Motors Corporation, Service Technology Group)

FIGURE 16–25 Pop rivet installed. A 1/8-in pipe plug installation is an alternative.

rivet that can be purchased at jobber outlets (Figure 16–24). The completed rivet installation is shown in Figure 16–25.

Also shown in Figure 16–25, as an example, is a drain hole drilled and tapped to receive a 1/4-in pipe plug (coated with Loctite). The pipe plug can be removed and replaced when performing future draining procedures. A drawback must be noted before drilling the larger-sized hole. In doing so, the possibility of nicking an impeller fin and introducing metal chips into the converter is increased.

A MIG welder also can be used to close a drain hole and produces excellent results (Figure 16–26).

As a final reliability check, the converter should be tested for leakage. The converter is submerged in a water tank and pressurized with a minimum of 80 psi (552 kPa). In Figure 16–27, the technician has clamped a radiator hose to the converter hub and uses a rubber check ball with a steel tube insert on the top end for pressurizing. If the converter has a leak area,

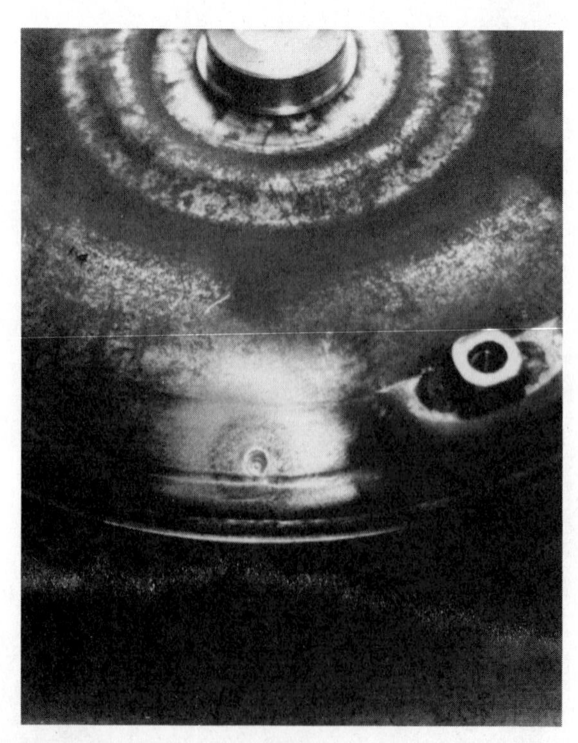

FIGURE 16–26 Converter bleed hole sealed with a MIG weld (non-converter clutch equipped unit).

FIGURE 16–27 Check the converter for leaks in water tank.

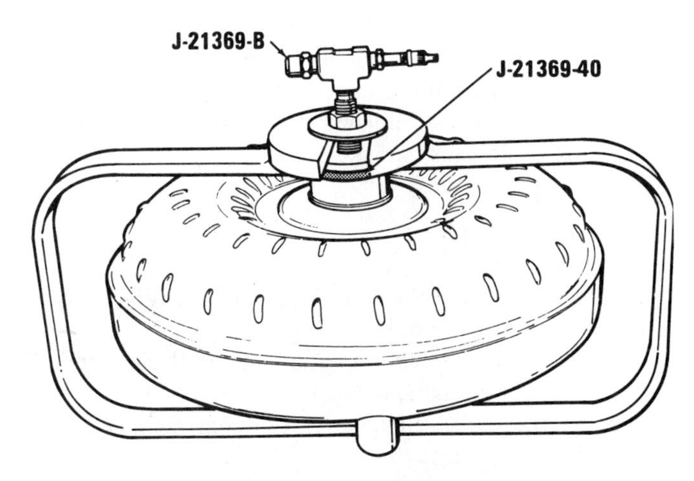

FIGURE 16–28 Torque converter pressurization kit. (Courtesy of General Motors Corporation, Service Technology Group)

it is easily identified by air bubbles in the water. Special torque converter pressurization kits are available (Figure 16–28).

Do not attempt to clean converters with solvent and hand agitation. This method never adequately flushes the trapped converter materials and usually makes the situation worse.

REPLACEMENT CONVERTERS

Just because a converter fits into place does not mean that it is the correct one or that it is a quality unit. It is good practice to perform the inspections listed earlier to validate the integrity of the converter you are about to install.

Torque Converter Identification

With today's wide range of engines and vehicles, a typical product transmission requires multiple converter usages. A converter with the wrong stall speed will definitely affect the vehicle acceleration and fuel economy. If the transmission and engine are computer-controlled, this adds to the confusion.

Converters are typically identified with a sticker or stamped number code on the converter. Make sure that the replacement converter matches the code. As an example, the Hydra-Matic 4T60 (THM 440-T4) converter in Figure 16–6 has a sticker code (FG2). The code can then be matched to the following data.

GM Sticker Codes

1. First digit = type of transmission
   F = 3T40 (123-C) or 4T60 (THM 440-T4)
   S = 3L30 (THM-180)
   H = THM-200 or 4L60 (THM 700-R4)
2. Second digit = K factor (stall)
   A = 240 K (2795) H = 128 K (1525)
   B = 220 K (2560) J = 177 K (2060)
   C = 205 K (2385) K = 237 K (2760)
   D = 180 K (2095) L = 130 K (1630)
   E = 160 K (1865) M = 131 K (1630)
   F = 148 K (1750) Y = 122 K (1375)
   G = 160 K (1865) Z = 203 K (2375)
3. Third digit = damper spring size
   1 = light tension
   2 = light tension
   3 = medium tension
   4 = medium tension
   5 = coast relief valves (diesel)
   6 = strong tension
   7 = strong tension
   8 = viscous
4. Fourth digit = converter-to-engine marriage (not always given) (bolt pattern or pad design)
   A = foreign bolt circle
   B = domestic bolt circle
   C = viscous pad design

If the sticker code has fallen off or cannot be read, the technician can read the impeller fin dimple angle, if externally visible. This system will identify three groups of stall speeds within a range of 1 to 300 rpm.

- High stall. Dimple angle definitely points away from converter rotation.
- Medium stall. Dimple angle is close to converter center line.
- Low stall. Dimple angle points in direction of converter rotation.

NOTE: The visual system does not identify converter stall torque ranges.

Converter Balancing

Converter balancing is another concern. Be sure that the replacement converter is balanced when reconditioned by the rebuilding source. Being neutral-balanced, its position on the flexplate should not induce or offset any powertrain vibrations.

When reusing the original converter, it should be reinstalled in its initial factory location to the flexplate. Many manufacturers position a slightly out-of-balance torque converter in a setting with the flexplate and crankshaft that will counteract and reduce engine vibrations.

Ford ATX Converter Replacement

The Ford ATX used three different torque converter designs.

- Original planetary splitter gear converter
- CLC (centrifugal locking clutch)
- FLC (fluid link converter—conventional three-element design), used since 1987. This is the recommended replacement for the two earlier converter styles as well.

When an FLC converter is returned as a replacement core, be sure to remove the adaptor sleeve (part number E7SZ-7A878-A) that splines the two input shafts together to the turbine. The replacement converter will not contain the adapter sleeve. If it is not installed, the vehicle will not move.

SUMMARY

Torque converters and cooler lines are too often neglected during a transmission overhaul. At times, they are not properly evaluated and/or serviced, while at other times, converters are needlessly replaced, and the cooler circuit may or may not be serviced. A complete and long-lasting transmission rebuild includes determining if the converter is reusable and, if so, servicing it properly along with the cooler lines.

Evaluating the torque converter includes making a variety of checks. Visually inspect the threaded fasteners for damage, the fluid condition for metal particles and water contamination, the cover for blueing, and the hub for wear. Internal converter checks include determining if the endplay of the components is within specifications, that splined shafts fit correctly, that the components do not interfere with one another, and that the clutch apply plate functions properly.

When a converter is considered to be reusable, the converter and cooler lines must be serviced by flushing out the old fluid. This will ensure that a freshly overhauled transmission will not be contaminated by the original fluid.

From the outside, a replacement converter may look the same as the original. Be careful not to use the wrong one, as vehicle performance and fuel economy will be changed. It is a good practice to thoroughly inspect and check over a replacement torque converter to guarantee its quality. And always remember to service the cooler system.

❑ REVIEW

Key Term

Endplay.

Questions

____ 1. The converter should be replaced:
   I. when the converter fluid sample is "metallic."
   II. when the converter fluid sample has dark discoloration.

   A. I only C. both I and II
   B. II only D. neither I nor II

____ 2. When hand-polishing the converter hub:
   I. the use of emery cloth is acceptable.
   II. polishing marks should form a crosshatch or angular pattern.

   A. I only C. both I and II
   B. II only D. neither I nor II

____ 3. The converter should be replaced:
   I. when it is dropped on the floor.
   II. when a threaded pad is found with stripped threads.

   A. I only C. both I and II
   B. II only D. neither I nor II

____ 4. If the sample ATF fluid shows engine coolant contamination:
   I. nonlockup converters can be flushed and put back into service.
   II. lockup clutch equipped converters should be replaced.

   A. I only C. both I and II
   B. II only D. neither I nor II

____ 5. The internal endplay of a converter should generally not exceed:

   A. 0.020 in (0.50 mm)
   B. 0.030 in (0.75 mm)
   C. 0.050 in (1.30 mm)
   D. 0.070 in (1.85 mm)

____ 6. The stator one-way clutch action can be tested:

   A. with a dummy stator shaft.
   B. with snap-ring pliers.
   C. with your finger.
   D. all of the above.

____ 7. The converter support bushing-to-hub clearance should not exceed:

   A. 0.004 in (0.10 mm)
   B. 0.006 in (0.15 mm)
   C. 0.010 in (0.25 mm)
   D. 0.012 in (0.30 mm)

____ 8. Converter support bushing-to-hub clearance can be checked by using:
   I. a paper strip.
   II. vernier calipers.

   A. I only C. both I and II
   B. II only D. neither I nor II

____ 9. When drilling an air bleed in a converter for flushing:
   I. the hole should be drilled on the top side of the converter cover.
   II. the hole size should be 1/4 in (6.35 mm).

   A. I only C. both I and II
   B. II only D. neither I nor II

_____ 10. To seal the bleeder hole:

I. treat a closed-end pop rivet with Loctite and install.

II. a MIG weld can be used.

A. I only
B. II only
C. both I and II
D. neither I nor II

_____ 11. A replacement converter with the wrong stall speed will:

A. confuse the engine-transmission computer control.
B. affect vehicle acceleration.
C. affect vehicle fuel economy.
D. all of the above.

12. After flushing, the converter should be filled with ATF and drained. ___T/___F

13. Flushing converters with solvent and manual agitation is not recommended. ___T/___F

14. A high-stall-speed converter will have the impeller dimple angle pointing in the direction of converter rotation. ___T/___F

17 Transmission/Transaxle Disassembly

OBJECTIVE:

The objective of this unit is for you to become proficient with the terms, concepts, and procedures contained in this chapter. Completion of this objective is essential to becoming a successful automatic transmission technician.

CHALLENGE YOUR KNOWLEDGE

List the following:

Three problems that occur from excessive endplay; three methods used to adjust endplay; and two means used to adjust servo pin travel.

Describe the following:

How to measure drive chain/link wear; the purpose of a colored drive chain, guide link; and the reason why accumulator springs should be labeled.

INTRODUCTION

The new generation of automatic transmissions are quite sophisticated, not only in operation but also in the complexity of overhauling/rebuilding. Although disassembly and reassembly techniques follow similar patterns, each product transmission features its own special requirements, requiring special tools for endplay checks and disassembly/assembly of component units. Do not try to work on the new breed of transmissions without using the required special tools. Performing an overhaul/rebuild without a special tool or entirely ignoring recommended procedures can damage a replacement part or cause a costly comeback. Special tools make the job easier, save time, and make it possible to do the job right the first time around. In addition, the technician must be knowledgeable about the mandatory part changes that need to be made in each automatic transmission product to avoid repeat failure and to improve performance.

Before disassembling a unit, thoroughly clean the exterior using either a steam cleaner or mineral spirits. A clean outside condition avoids contaminating the work area and makes handling and disassembly of the transmission easier. As you are aware, when working with hydraulic units, cleanliness cannot be overemphasized. A little dirt or debris left inside an overhauled/rebuilt unit can undo many hours of otherwise careful work. Therefore, keep your work bench, tools, and parts clean at all times.

There are highly polished and precision mating surfaces in the transmission, and careful handling of parts is required. Avoid using tools or techniques that will cause nicks, scores, or distortion of parts. For example, do not groove or score the inside of a front pump seal housing by carelessly extracting the oil seal with a chisel or screwdriver. Use proper tru-arch and snap-ring pliers to remove and reinstall snap rings without distorting them. Avoid nicking ring grooves and their sealing rings.

In line with the industry commitment to convert to the metric system, current transmission and transaxle fasteners, dimensions, and specifications are based on metric measurements. Older domestic transmissions primarily used SAE fasteners. Some units, however, used a combination of SAE and metric bolts. The manufacturers' service manuals will identify the types and dimensions of the fasteners used in a specific unit. Metric and SAE fasteners should not be interchanged or substituted.

REMOVAL OF UNIT ASSEMBLIES

There are product differences between transmissions and, therefore, variations in disassembly requirements. The manufacturer's service manual or equivalent should be consulted for each transmission. Rather than attempting to duplicate a detailed disassembly procedure, this text highlights typical service techniques.

Disassembly often involves mounting the transmission in a bench holding fixture (Figure 17–1) and following the recommended sequence for minor parts and subassembly removal. Repair holding fixtures are available for many

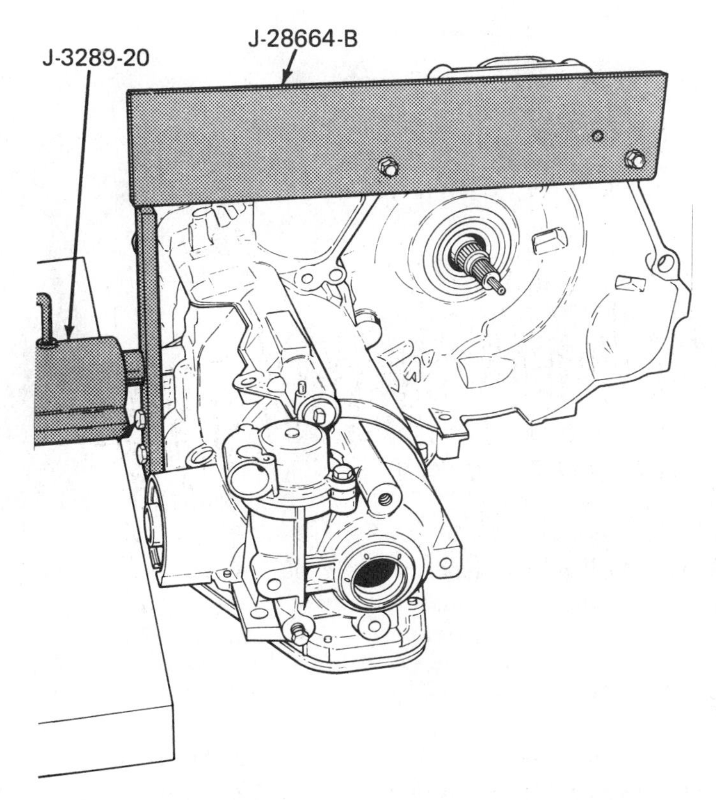

FIGURE 17–1 4T60 (440-T4) in holding fixture. (Courtesy of General Motors Corporation, Service Technology Group)

transmissions and offer the advantage of working convenience to save repair time. The goal of overhauling/rebuilding is to be efficient. Speed, reliability, and profit are important factors.

The following sequence uses examples from a variety of transmissions to illustrate some of the special disassembly requirements the technician will encounter or perform as a matter of good practice.

CONVERTER REMOVAL

Converter removal is usually a simple matter of rotating the converter and sliding it away from the case converter housing. Keep the converter hub facing up to avoid a mess from fluid spill. Once removed, cap the opening to prevent debris from entering the converter.

There are times when internal damage prevents normal removal of the converter and makes it difficult to budge. In such instances, drill a hole through the center of the converter front hub to accommodate a 3/8- or 5/16-inch standard bottoming tap. After tapping the hole, use a long bolt to push against the turbine shaft to remove the converter. On applications where a pump shaft splines into the converter cover, use a heavy-duty slide hammer with the proper threaded adaptor.

ENDPLAY CHECKS

When stripping a transmission case, a major concern is measuring the endplay between subassemblies. A transmission may require only one endplay check, or it may require several. During acceleration and deceleration, the clutch and planetary units move back and forth in the case. This action produces the following effects:

- Oil rings are scuffed in the drum bores, which shortens the ring life. This imposes a high load against the ring groove lands, resulting in a combination of ring and groove wear.
- The slip wear between shaft splines produces metallic micro particles that contaminate the ATF. The spline wear also contributes to N-D and N-R engagement clunk.
- The needle-style thrust bearings, although having excellent antifriction and load-carrying qualities, cannot hold up against constant and heavy shock loads from internal components, especially in the planetary system. The needles are brittle, and when they break, it often causes severe damage to the planetary gears, shaft bushings, and other thrust surface areas.

Taking endplay measurements during disassembly is important. An excessive endplay measurement identifies a problem

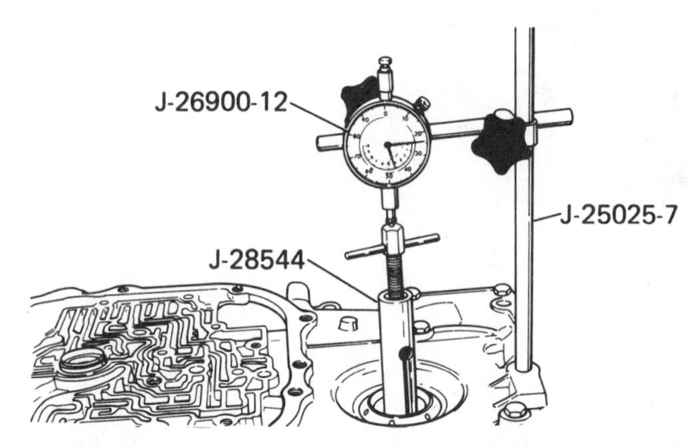

FIGURE 17–2 Input shaft to case cover endplay—3T40 (125-C). (Courtesy of General Motors Corporation, Service Technology Group)

area. If the endplay is on the high end of specifications, the technician can plan to adjust to the low limits during reassembly. Some specifications, especially input shaft or front unit endplay, may have a wide spread. Technicians should always adjust the endplay to the low end, realizing that the gap will increase due to normal wear over the life of the transmission.

The following illustrations demonstrate some of the special tool setups and transmission endplay requirements that may be encountered. Refer to the manufacturer's service manual for a complete list of procedures. Endplay adjustments are typically made with selective (1) thrust washers, (2) spacers, or (3) snap rings.

INPUT SHAFT SELECTIVE SNAP RING (621)

THICKNESS		IDENTIFICATION COLOR
1.83-1.93mm	(0.071″-0.076″)	WHITE
2.03-2.31mm	(0.078″-0.084″)	BLUE
2.23-2.33mm	(0.088″-0.092″)	RED
2.43-2.53mm	(0.095″-0.099″)	YELLOW
2.63-2.73mm	(0.103″-0.107″)	GREEN

FIGURE 17–3 Input shaft selective snap ring location and selection chart—GM 3T40 (125-C). (Courtesy of General Motors Corporation, Service Technology Group)

GM Hydra-Matic 3T40 (THM 125C).

1. Input shaft to case cover endplay (Figure 17–2). Endplay controlled by selective snap ring on the input shaft (Figure 17–3).
2. Reaction sun gear to input drum endplay (Figure 17–4). Endplay controlled by selective snap ring on reaction sun gear (Figure 17–5).
3. Low roller clutch race endplay (Figure 17–6). Endplay controlled by selective spacer on top of roller clutch (Figure 17–7).
4. Final drive to case endplay (Figure 17–8). Endplay controlled by selective thrust washer on final drive (Figure 17–9).

GM Hydra-Matic 4L60-E/4L60 (THM 700-R4).

1. Transmission input endplay check (Figure 17–10). Endplay controlled by selective washer located between the input housing (Figure 17–11). Only mandatory endplay check during disassembly.

Ford ATX (Domestic Design).

1. Transaxle endplay check (Figure 17–12). Endplay controlled by #12 selective thrust washer located on pump clutch support (Figure 17–13). Only required endplay

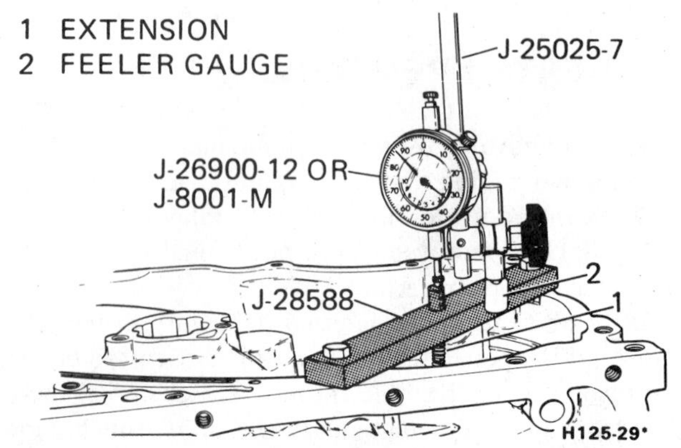

FIGURE 17–4 Sun gear/input drum endplay—3T40 (125-C). (Courtesy of General Motors Corporation, Service Technology

REACTION SUN GEAR
TO INPUT DRUM
SELECTIVE SNAP RING
(644)

Thickness	Identification/Color
2.27 - 2.37mm (0.089" - 0.093")	Pink
2.44 - 2.54mm (0.096" - 0.100")	Brown
2.61 - 2.71mm (0.103" - 0.107")	Lt. Blue
2.78 - 2.88mm (0.109" - 0.113")	White
2.95 - 3.05mm (0.116" - 0.120")	Yellow
3.12 - 3.22mm (0.123" - 0.127")	Lt. Green
3.29 - 3.39mm (0.129" - 0.133")	Orange
3.46 - 3.56mm (0.136" - 0.140")	No Color

FIGURE 17–5 Selective snap ring location and selection chart—GM 3T40 (125-C). (Courtesy of General Motors Corporation, Service Technology Group)

check—omitted during disassembly, but mandatory during assembly.

Ford AXOD/AX4S (AXODE).

1. Final drive to case endplay (Figure 17–14). Endplay controlled by selective thrust washer (#18) on the final drive carrier (Figure 17–15).
2. Sprocket support endplay (Figure 17–16). Endplay requires the combined selection of #5 and #8 thrust washers located on the sprocket support (Figure 17–17). Endplay checks omitted during disassembly, but mandatory during assembly.

Chrysler 30RH (A-900) Series and 31TH (A-400/670) Series.

1. Input shaft endplay check (Figure 17–18). Endplay controlled by selective thrust washer on end of output shaft

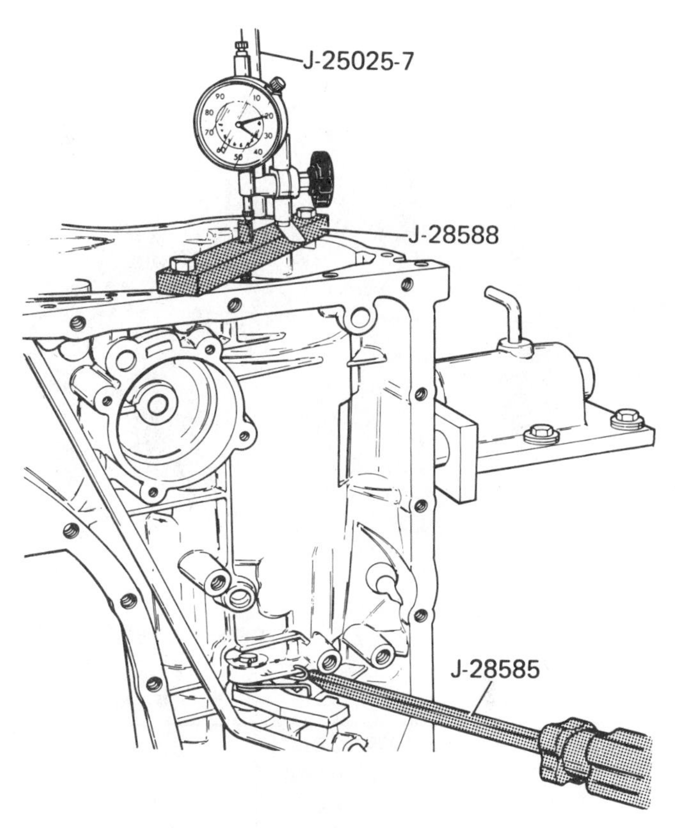

FIGURE 17–6 Low roller clutch race endplay—3T40 (125-C). (Courtesy of General Motors Corporation, Service Technology Group)

(Figure 17–19). Only mandatory endplay check during disassembly.

Chrysler 36RH (A-727) and 40RH (A-500) Series.

1. Input shaft endplay check (Figure 17–20). Endplay controlled by selective thrust washer located on pump assembly reaction shaft support (Figure 17–21).

SERVO PIN TRAVEL CHECK

A servo pin travel check measures the distance a servo pin/piston travels from its fully released position to the band apply position. The stroke travel is controlled by a select apply pin and, in effect, adjusts the band. The select pin takes the place of an adjusting screw.

This check may be taken during disassembly, but it must be done during assembly. Figures 17–22 to 17–25 demonstrate some of the special tool setups needed for checking servo travel. Refer to the manufacturer's service manual for a complete list of procedures.

REVERSE CLUTCH HOUSING TO LO RACE SELECTIVE SPACER (660)

THICKNESS		IDENTIFICATION
1.00-1.10mm	(0.039"-0.043")	1
1.42-1.52mm	(0.056"-0.060")	2
1.84-1.94mm	(0.072"-0.076")	3
2.26-2.36mm	(0.089"-0.093")	4
2.68-2.78mm	(0.105"-0.109")	5
3.10-3.20mm	(0.122"-0.126")	6

FIGURE 17–7 Low race selective spacer location and selection chart. (Courtesy of General Motors Corporation, Service Technology Group)

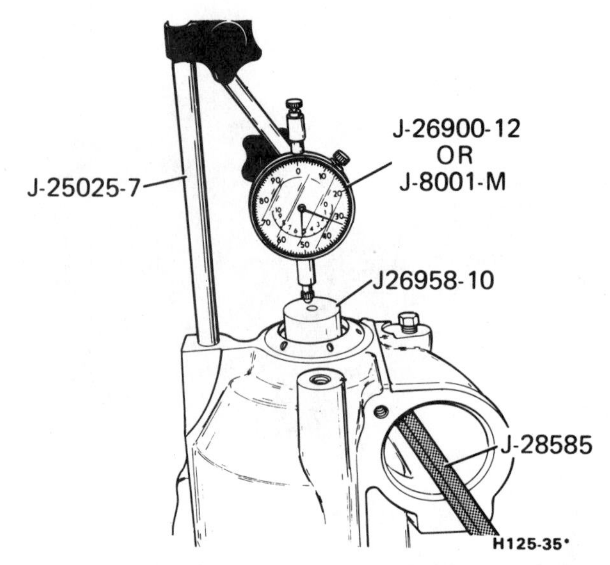

FIGURE 17–8 Final drive to case endplay measurement—3T40 (125-C); similar procedure for a 4T60 (440-T4). (Courtesy of General Motors Corporation, Service Technology Group)

FINAL DRIVE TO CASE END PLAY
SELECTIVE THRUST WASHER (680)

THICKNESS		IDENTIFICATION NO./COLOR
1.40 - 1.50mm	(0.055" - 0.059")	0/Orange
1.50 - 1.60mm	(0.059" - 0.062")	1/White
1.60 - 1.70mm	(0.062" - 0.066")	2/Blue
1.70 - 1.80mm	(0.066" - 0.070")	3/Pink
1.80 - 1.90mm	(0.070" - 0.074")	4/Brown
1.90 - 2.00mm	(0.074" - 0.078")	5/Green
2.00 - 2.10mm	(0.078" - 0.082")	6/Black
2.10 - 2.20mm	(0.082" - 0.086")	7/Purple
2.20 - 2.30mm	(0.086" - 0.091")	8/Purple & White
2.30 - 2.40mm	(0.091" - 0.095")	9/Purple & Blue

FIGURE 17–9 Final drive thrust washer and selection chart—GM 3T40 (125-C). (Courtesy of General Motors Corporation, Service Technology Group)

GENERAL DISASSEMBLY PRACTICES

As unit assemblies are removed, organize them on the bench in the exact order of removal. Special attention should be given to the location of thrust washers, spacers, snap rings, valve body/case cover bolt sizes, and check ball locations. Note the size and location of bolts.

Use handtools for removing fasteners throughout the transmission. Avoid power tools, which eliminate any sensitivity to loose fasteners that may be evidence of why a unit or component failed. By using hand tools, undertorqued bolts in critical areas such as the valve body, channel plate, case cover, servo cover, and pump can be detected.

On chain or drive link assemblies, check the chain assembly for possible wear. Grasp the chain midway between the sprockets and stretch it to its fully extended position, away from the center. Select a reference point on the transaxle case, and measure the low end of the extension with a steel rule (Figure 17–26). Then push the chain to its full high end, toward the center. Using the same reference point, measure the high spot (Figure 17–27). If the measured difference between the two

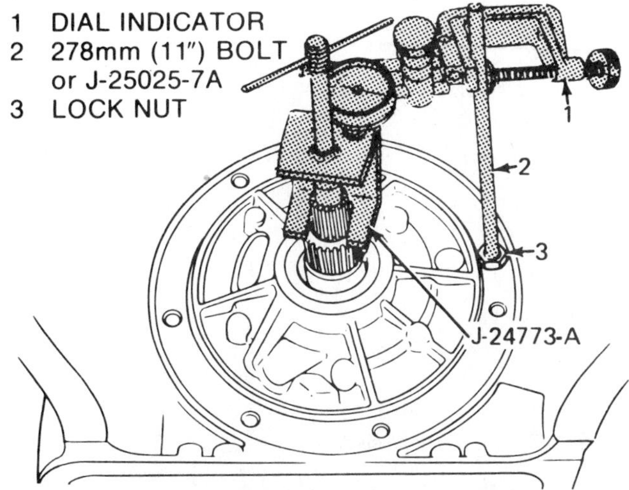

FIGURE 17–10 GM 4L60-E/4L60 (700-R4) transmission input endplay; similar procedure for 200-4R and 200-C transmissions. (Courtesy of General Motors Corporation, Service Technology Group)

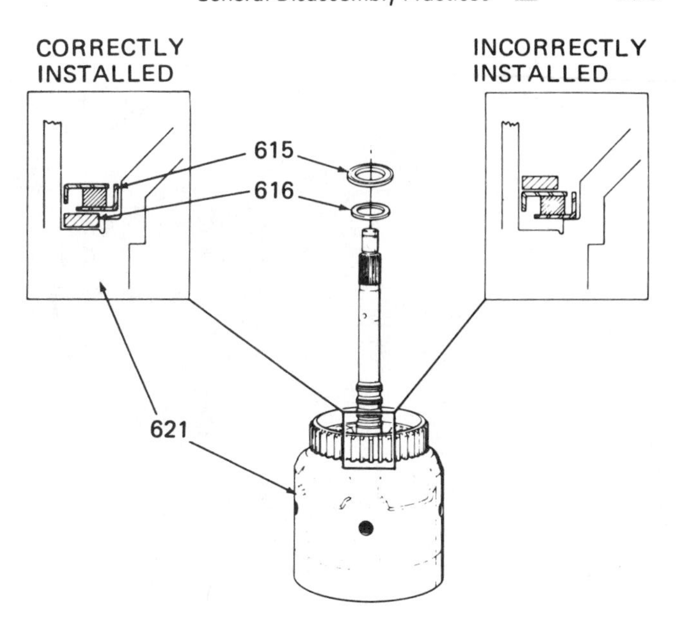

ILL. NO.	DESCRIPTION
615	BEARING ASSEMBLY, STATOR SHAFT/ SELECTIVE WASHER
616	WASHER, THRUST (SELECTIVE)
621	HOUSING & SHAFT ASSEMBLY, INPUT

TRANSMISSION END PLAY WASHER SELECTION CHART

WASHER THICKNESS		I.D.
1.87 - 1.97 mm	(.074" - .078")	67
2.04 - 2.14 mm	(.080" - .084")	68
2.21 - 2.31 mm	(.087" - .091")	69
2.38 - 2.48 mm	(.094" - .098")	70
2.55 - 2.65 mm	(.100" - .104")	71
2.72 - 2.82 mm	(.107" - .111")	72
2.87 - 2.99 mm	(.113" - .118")	73
3.06 - 3.16 mm	(.120" - .124")	74

FIGURE 17–11 Selective washer location and selection chart—4L60-E/4L60 (700-R4). (Courtesy of General Motors Corporation, Service Technology Group)

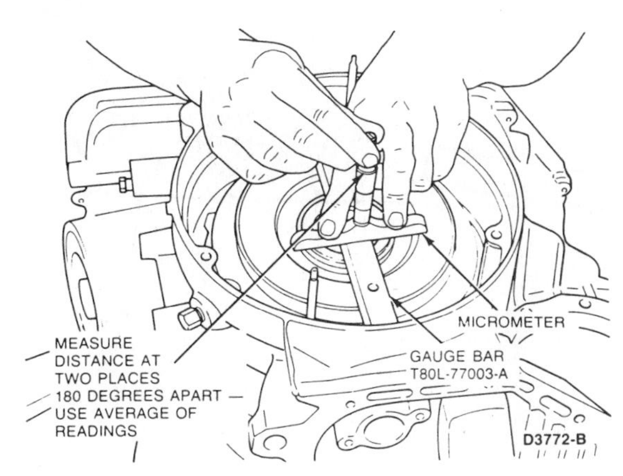

THRUST WASHER SELECTION CHART

For This Reading	Steel/Bronze-Type Washer Part ID	Nylon Washer (Snap-On)-Type Color ID
2.00-1.77mm (0.079-0.070 inch)	AA	Black
2.20-2.00mm (0.087-0.079 inch)	BA	Natural
2.41-2.20mm (0.095-0.087 inch)	CA	Blue
1.77-1.46mm (0.070-0.057 inch)	EA	Green

THRUST WASHER THICKNESS CHART

Thickness		Steel/Bronze ID	Nylon ID
MM	Inch		
1.40-1.45	(.055-.057)	AA	Black
1.60-1.65	(.063-.065)	BA	Natural
1.80-1.85	(.071-.073)	CA	Blue
1.15-1.20	(.045-.047)	EA	Green

FIGURE 17–12 Transmission endplay check and thrust washer selection charts—ATX (domestic design). (Reprinted with the permission of Ford Motor Company)

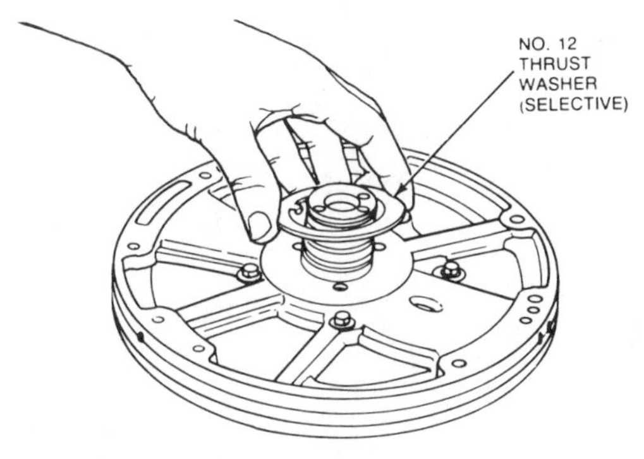

FIGURE 17–13 #12 selective thrust washer located on pump—ATX (domestic design). (Reprinted with the permission of Ford Motor Company)

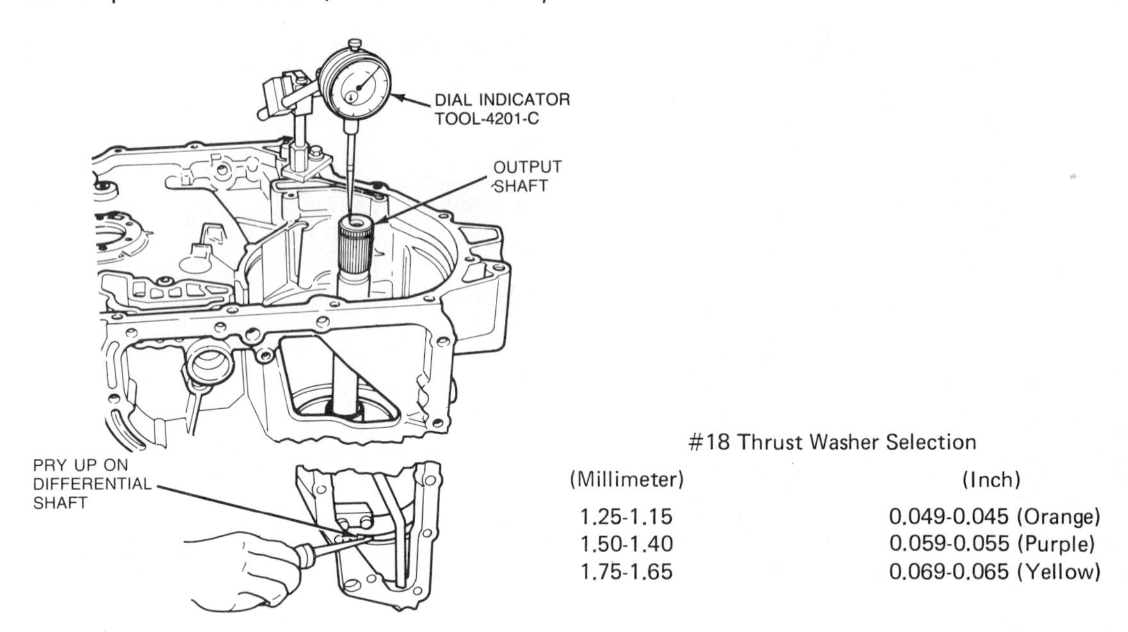

#18 Thrust Washer Selection

(Millimeter)	(Inch)
1.25-1.15	0.049-0.045 (Orange)
1.50-1.40	0.059-0.055 (Purple)
1.75-1.65	0.069-0.065 (Yellow)

FIGURE 17–14 Final drive to case endplay—AXOD/AX4S (AXODE). (Reprinted with the permission of Ford Motor Company)

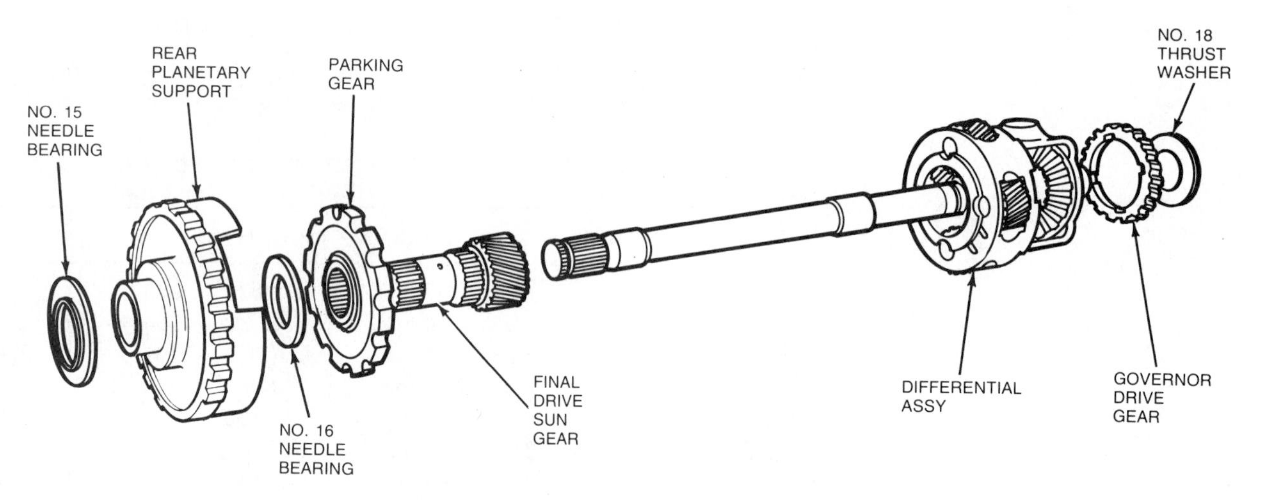

FIGURE 17–15 #18 final drive selective thrust washer location— AXOD/AX4S (AXODE). (Reprinted with the permission of Ford Motor Company)

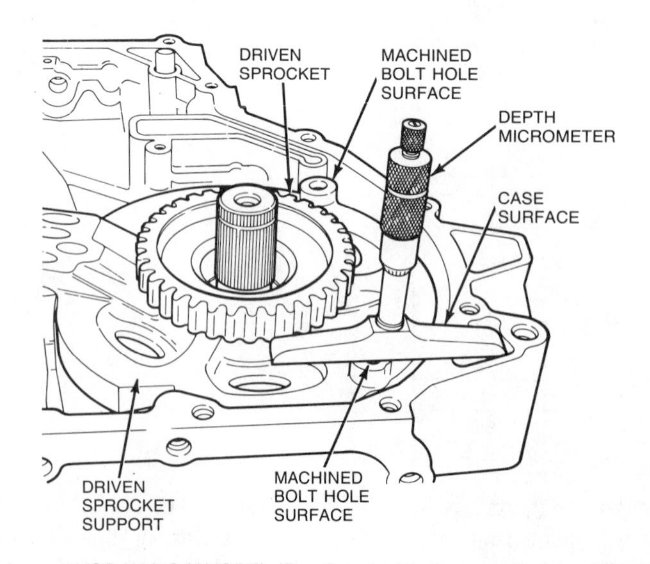

FIGURE 17–16 Sprocket support endplay—AXOD/AX4S (AXODE). (Reprinted with the permission of Ford Motor Company)

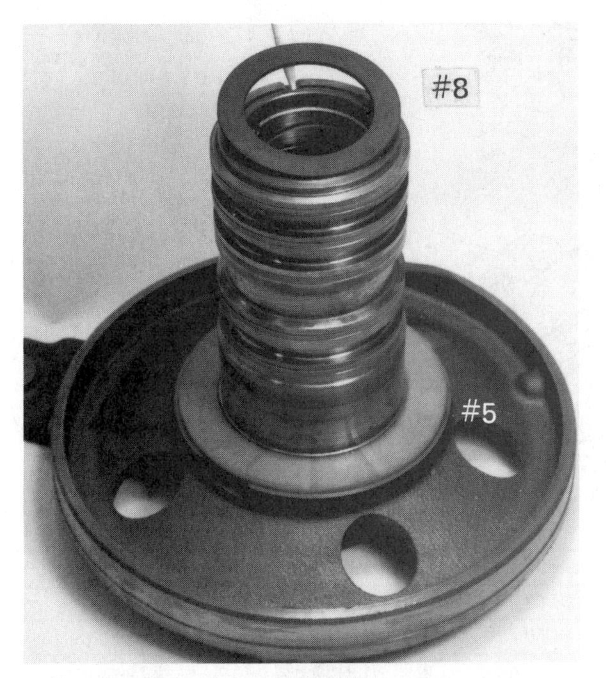

NO. 5 THRUST WASHER SELECTION

Thrust Washer Thickness		Color
mm	Inches	
2.28-2.18	0.090-0.086	Green
2.53-2.43	0.099-0.095	Black
2.77-2.67	0.109-0.105	Natural
3.02-2.92	0.118-0.115	Red

NO. 8 THRUST WASHER SELECTION

Thrust Washer Thickness		Color
mm	Inches	
1.53-1.43	0.060-0.056	Natural
1.78-1.68	0.070-0.066	Dark Green
2.02-1.92	0.079-0.075	Light Blue
2.27-2.17	0.089-0.085	Red

FIGURE 17–17 #5 and #8 selective thrust washer location and selection charts—AXOD/AX4S (AXODE). (Reprinted with the permission of Ford Motor Company)

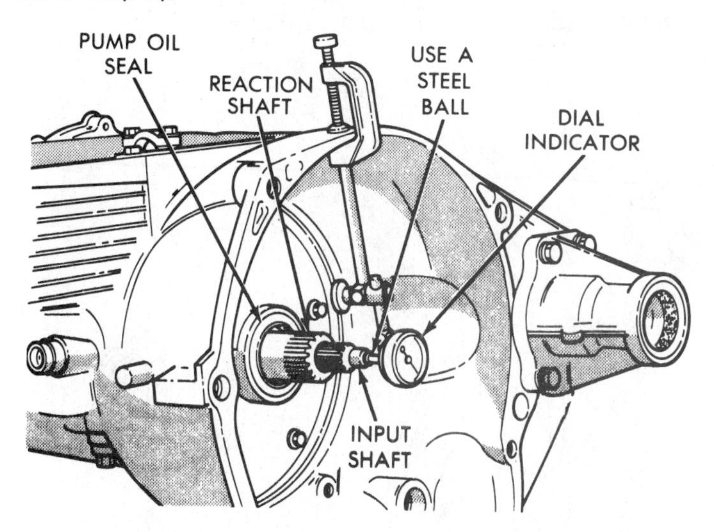

FIGURE 17–18 Measure input shaft endplay—31TH (A-400/670) series. (Courtesy of Chrysler Corp.)

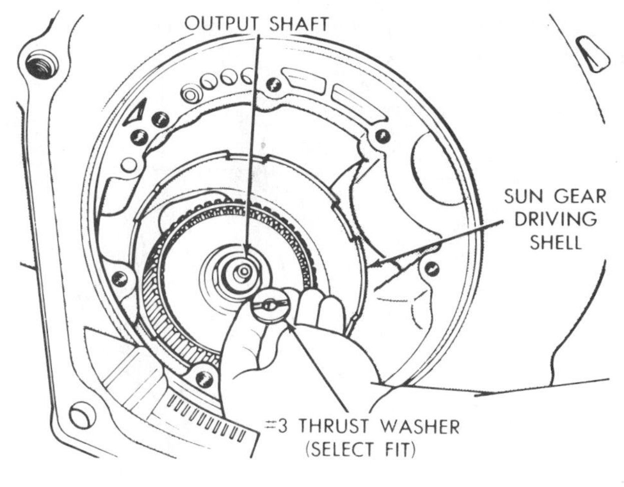

#3 Select Washer

(Millimeter)	(Inch)
1.98-2.03	.077-.080
2.15-2.22	.085-.087
2.34-2.41	.092-.095

FIGURE 17–19 #3 thrust washer—31TH (A-400/670) series. (Courtesy of Chrysler Corp.)

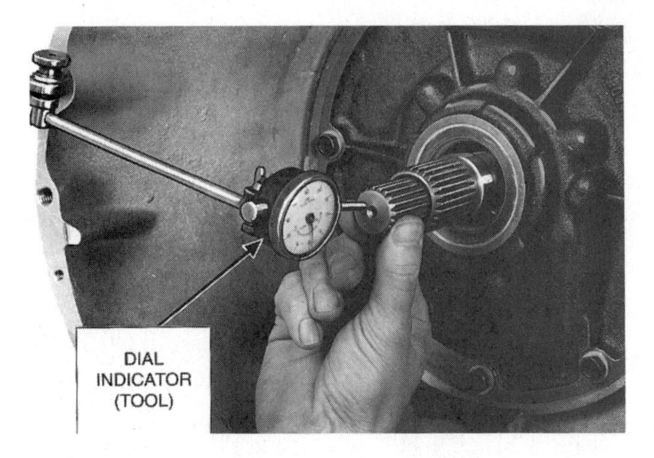

FIGURE 17–20 30RH (A-900 series) transmission input shaft endplay check. (Courtesy of Chrysler Corp.)

#1 Select Washer

(Millimeter)	(Inch)
1.55-1.57	.061-.063 (Natural)
2.13-2.18	.084-.086 (Red)
2.59-2.64	.102-.104 (Yellow)

FIGURE 17–21 #1 select thrust washer located on back of pump—30RH (A-900 series).

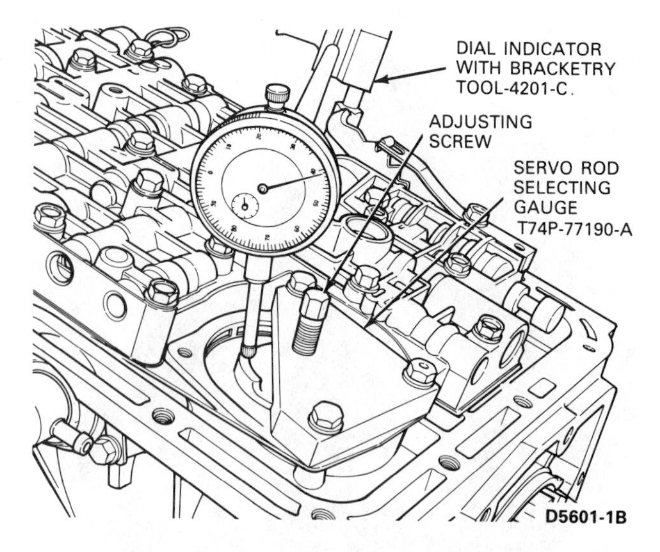

Length — mm	Length — Inches	I.D.
54/53 mm	2.112/2.085	1 Groove
51/50 mm	2.014/1.986	No Groove
49/48 mm	1.915/1.888	2 Grooves

FIGURE 17–22 A4LD reverse servo pin selection. (Reprinted with the permission of Ford Motor Company)

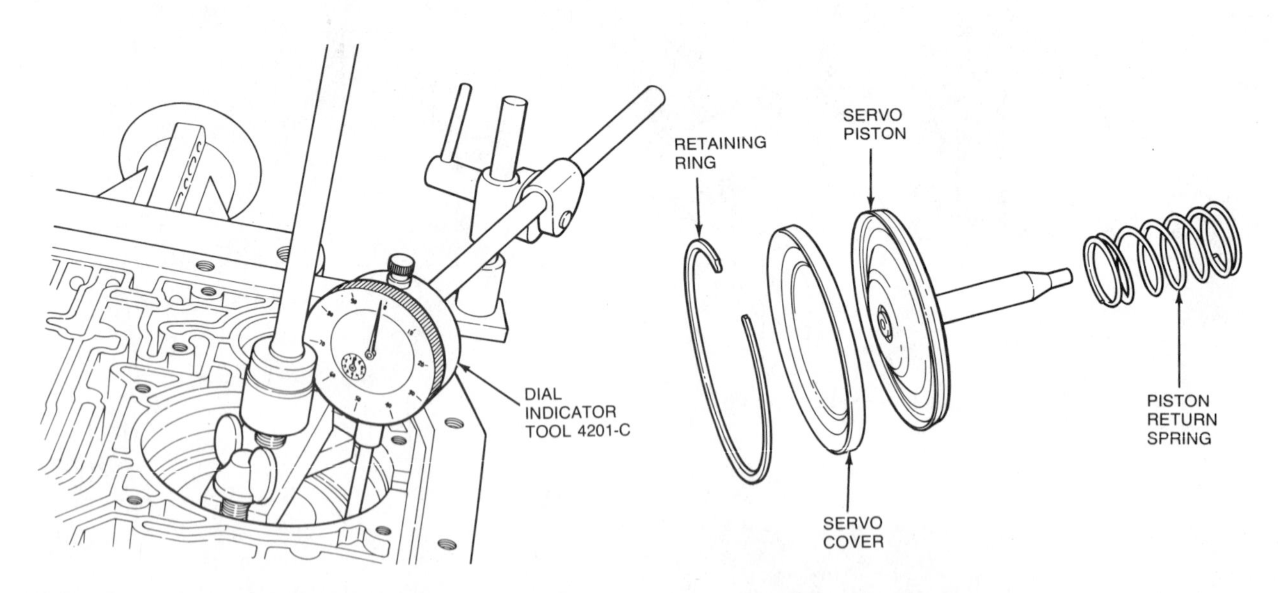

FIGURE 17–23 AOD low/reverse servo pin selection; apply pin available in three lengths. (Reprinted with the permission of Ford Motor Company)

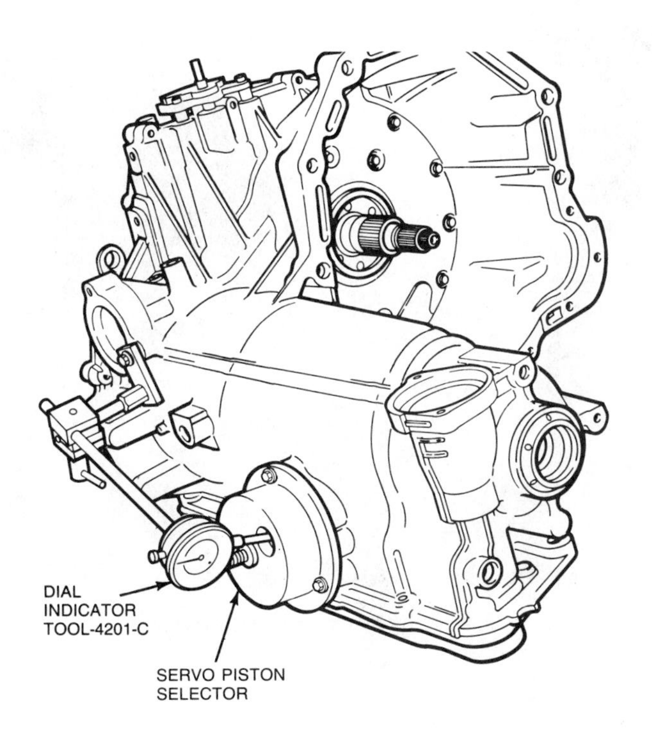

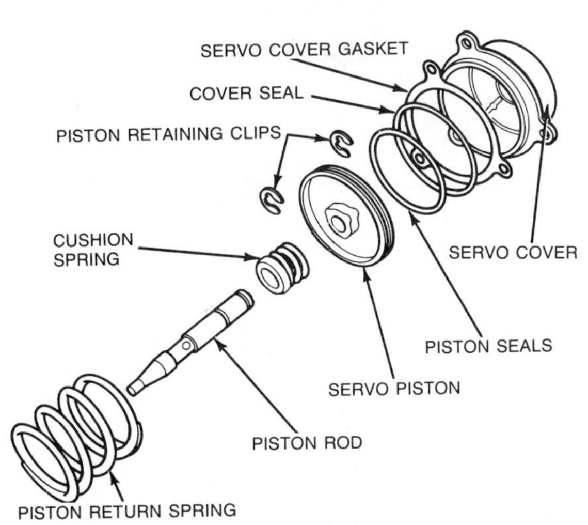

FIGURE 17–24 AXOD, AX4S (AXODE) low/intermediate servo pin (rod) selection; apply pin available in three lengths. (Reprinted with the permission of Ford Motor Company)

readings exceeds specifications, the chain should be replaced (Figure 17–28). The typical maximum specification is 1 1/16 in (27.0 mm).

A chain or drive link assembly needs to be reinstalled in its original upright position. If installed upside down, a new wear pattern will develop and noise will be generated. General Motors units use a colored guide link to identify the up side (Figure 17–29). On the Ford AXOD, a yellow chain mark is sometimes but not always used to identify the down side.

Pay careful attention to valve body wire routing, fasteners, and connectors (Figure 17–30). It is easy to cross wire connectors. The service manual wiring code does not always agree

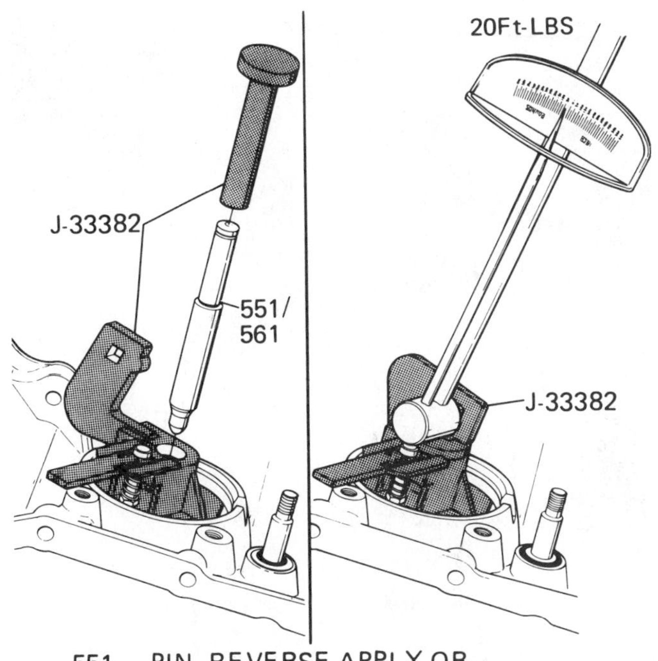

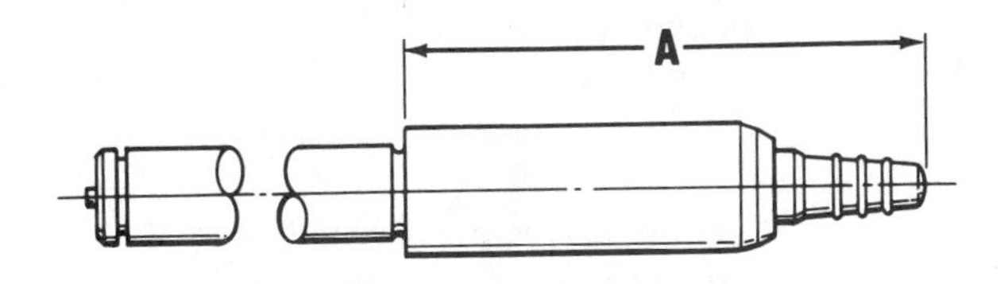

REVERSE BAND APPLY PIN

IDENTIFICATION	DIMENSION A
2 WIDE BANDS	70.86-71.01mm (2.789"-2.795")
3 GROOVES & WIDE BAND	71.91-72.06mm (2.831"-2.837")
2 GROOVES & WIDE BAND	72.96-73.11mm (2.872"-2.878")
1 GROOVE & WIDE BAND	74.01-74.16mm (2.913"-2.919")
NO GROOVE	75.03-75.18mm (2.953"-2.959")
1 GROOVE	76.08-76.23mm (2.995"-3.001")
2 GROOVE	77.13-77.28mm (3.036"-3.042")
3 GROOVE	78.18-78.33mm (3.077"-3.083")
4 GROOVE	79.20-79.35mm (3.118"-3.124")

1-2 BAND APPLY PIN

IDENTIFICATION	DIMENSION A
1 RING & WIDE BAND	56.24-56.39mm (2.214"-2.220")
1 RING	57.23-57.38mm (2.253"-2.259")
2 RINGS	58.27-58.42mm (2.294"-2.299")
3 RINGS	59.31-59.46mm (2.335"-2.340")
WIDE BAND	60.34-60.49mm (2.375"-2.381")
2 RINGS & WIDE BAND	61.34-61.49mm (2.414"-2.420")

FIGURE 17–25 4T60 (440-T4) reverse and 1-2 bands servo pin selection method and charts. (Courtesy of General Motors Corporation, Service Technology Group)

FIGURE 17–26 Measure low end of chain extension.

FIGURE 17–27 Measure high end of chain extension.

with the actual wire code. Some technicians play it safe and take a Polaroid picture for reference.

Valve body wire connectors can be difficult to remove. Avoid pulling or jerking on connector wires. Use the technique shown in Figure 17–30.

Some pump assemblies require special tooling for removal from the case—either a slide hammer or a puller arrangement. The slide hammer is used to pull pumps on all Chrysler automatic transmissions, Ford AOD and ATX, and GM 250-C, 350-C, and 3L80 (400) (Figure 17–31). The puller arrangement in Figure 17–32 is required for GM 200C, 200-4R, and 4L60-E/4L60 (700-R4) transmissions. Before the pumps are pulled in most modern RWD transmissions, the converter clutch solenoid and/or transmission filter must be removed.

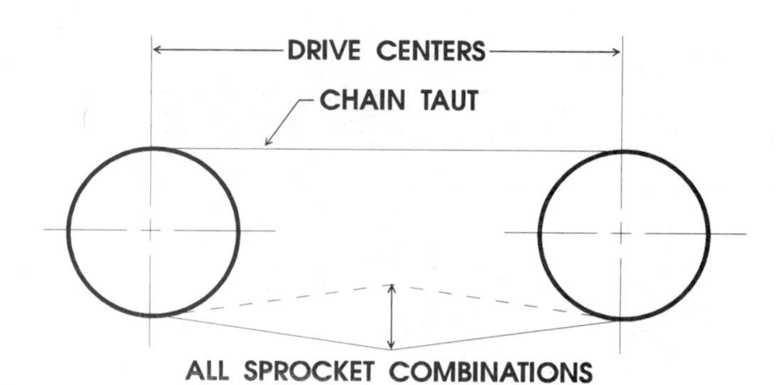

FIGURE 17–28 Replace chain if difference between the two measurements exceeds specifications.

FIGURE 17–29 Colored guide link identifies chain position for reassembly.

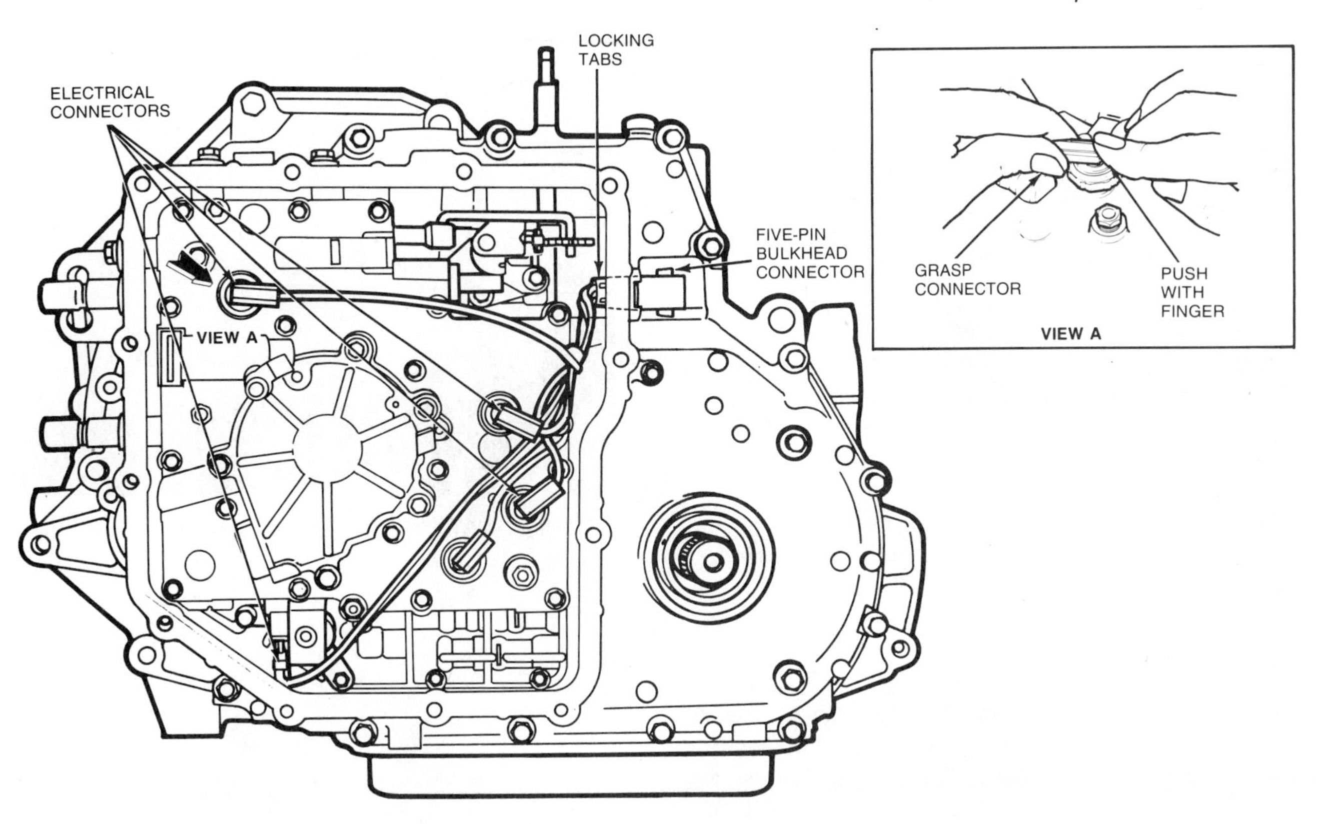

FIGURE 17–30 Valve body wire routing, fasteners, and connectors—AXOD. (Reprinted with the permission of Ford Motor Company)

FIGURE 17–31 RWD transmission front pump removal. (Courtesy of Chrysler Corp.)

FIGURE 17–32 Front pump puller arrangement—GM 4L60-E/4L60 (700-R4); similar setup for 200-4R and 200-C transmissions.

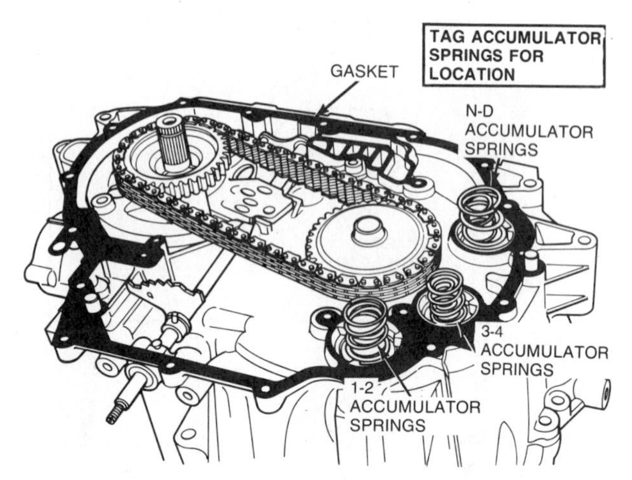

FIGURE 17–33 Accumulator springs should be tagged to identify location—AXOD. (Reprinted with the permission of Ford Motor Company)

FIGURE 17–34 3T40 (125C) case cover mounted thermostatic element.

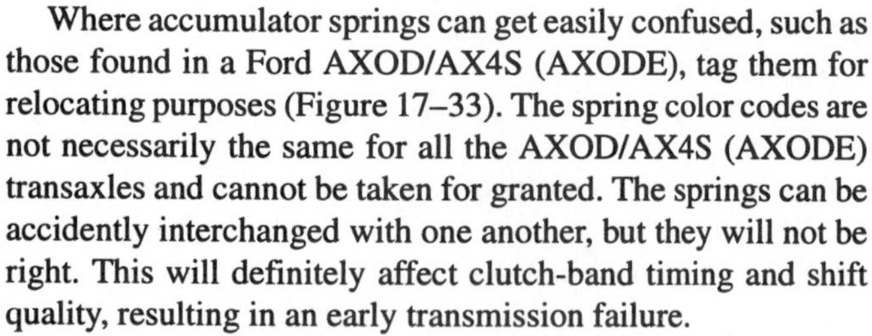

Where accumulator springs can get easily confused, such as those found in a Ford AXOD/AX4S (AXODE), tag them for relocating purposes (Figure 17–33). The spring color codes are not necessarily the same for all the AXOD/AX4S (AXODE) transaxles and cannot be taken for granted. The springs can be accidently interchanged with one another, but they will not be right. This will definitely affect clutch-band timing and shift quality, resulting in an early transmission failure.

Where a thermostatic element is used for controlling the fluid levels in transaxles, avoid abusive handling that will distort its normal configuration (Figure 17–34). Do not try to straighten one out. Replacement is the only cure, following an exact procedure that involves correct height spacing.

Avoid distorting the normal shape of a band during removal by extreme stretching or twisting. Otherwise, it may need to be replaced. To maintain its correct shape, fix the band with a paper clip when it is removed (Figure 17–35).

FIGURE 17–35 Fix the band with a paper clip when it is removed.

SUMMARY

Careful disassembly and reassembly practices will help guarantee a quality service repair. It is important to follow manufacturer service procedures, perform the required endplay measurements, and use special tools. By doing so, worn and damaged components can be identified, adjustments made, and transmission components accurately reassembled.

An organized approach to disassembling and reassembling procedures is helpful. The bench and work area where the technician performs the tearing down and rebuilding should be clean and uncluttered. Good work practices enhance the rebuilding process while maintaining personal safety.

❑ REVIEW

Questions

____ 1. Excessive internal endplay will cause:

I. a combination of ring and groove wear.
II. damage to needle-style thrust washers.

A. I only   C. both I and II
B. II only   D. neither I nor II

____ 2. Endplay adjustments are typically made with selective:

A. spacers.

B. thrust washers.
C. snap rings.
D. two of the choices.
E. all of the choices.

____ 3. What is true about disassembly/assembly?
I. Transmission units normally require an input shaft or front endplay check.
II. The use of power tools is recommended for time efficiency.

A. I only
B. II only
C. both I and II
D. neither I nor II

____ 4. Which one of the following units uses a selective snap ring to control input shaft endplay?

A. Hydra-Matic 3T40
B. Hydra-Matic 4T60
C. Ford AXOD
D. None of the above

____ 5. On chain or drive link assemblies:
I. measured chain slack should not exceed 3/4 in (19 mm).
II. the chain must be kept in the same upright position for installation.

A. I only
B. II only
C. both I and II
D. neither I nor II

____ 6. What is true about disassembly/assembly?
I. Servo travel checks are mandatory.
II. Valve body wiring connectors are removed by pulling on the connector wires.

A. I only
B. II only
C. both I and II
D. neither I nor II

____ 7. Which of the following transmissions requires a slide hammer to remove the pump assembly?

A. All Chrysler units
B. Ford ATX
C. Hydra-Matic 3L80
D. Two of the choices
E. All of the choices

8. When accumulators have springs that appear to be similar, except for their color, the springs are interchangeable. ___T/___F

9. A thermostatic element used for controlling the fluid levels in a transaxle must be replaced when distorted. ___T/___F

10. It is permissible to stretch a friction band to its maximum limit for inspection purposes. ___T/___F

18 Transmission/Transaxle Overhaul Practices

OBJECTIVE:

The objective of this unit is for you to become proficient with the terms, concepts, and procedures contained in this chapter. Completion of this objective is essential to becoming a successful automatic transmission technician.

CHALLENGE YOUR KNOWLEDGE

List the following:

Eight categories of parts/components typically discarded and replaced during an overhaul; nine categories of parts/components typically renewed on a "replace as needed" basis; and two assembly lubes.

Describe the following procedures:

Cleaning parts/components; removing gasket material; fitting metal and O-ring seals; installation techniques for solid Teflon rings; installing one-way clutches accurately; filter and screen servicing; inspection and service of bands and drums.

Identify:

Three safety warnings that are associated with overhaul practices.

INTRODUCTION

During an overhaul, it is standard practice to discard certain parts and replace them with new ones. Other parts are replaced according to need when they are inspected and evaluated to be worn excessively or damaged. This chapter presents general guidelines and practices for making these determinations and overhauling transmission components.

TYPICAL DISCARD AND REPLACEMENT PRACTICES

Discard and replace	Replace as needed
Gaskets	Gears
O-rings	Bands
Oil seals	Bushings and thrust washers
Metal/Teflon sealing rings	Pump
Clutch steel plates	Converter
Clutch friction plates	Governor
Modulator	Clutch housings
Fluid intake filter	One-way clutches (sprag and roller units)
	Minor parts

A typical overhaul kit is shown in Figure 18–1. It contains all the necessary replacement gaskets, seals, and clutch plates to recondition the subassemblies and reseal the transmission case. The vacuum modulator, bands, bushings, thrust washers, and fluid filter are typically not included in overhaul kits. The parts furnished in an overhaul kit must be checked for proper sizing and fit, and prepared for installation. Each metal sealing ring, rubber seal, and gasket should be given a final quality control check to make sure that it is going to work. This is a critical part of the transmission overhaul, and serious attention must be given to the preparation and installation of these items.

GENERAL GUIDELINES

WARNINGS:

- **Wear safety glasses and safety shoes.**
- **Keep work area and tools organized and clean.**
- **Use special tools or equivalent as required for special disassembly/assembly requirements. These tools save time and prevent unnecessary damage to parts. In some cases, there is also a safety factor, especially where clutch spring forces are involved. Figure 18–2 shows a heavy-duty press operated spring tool that is used to**

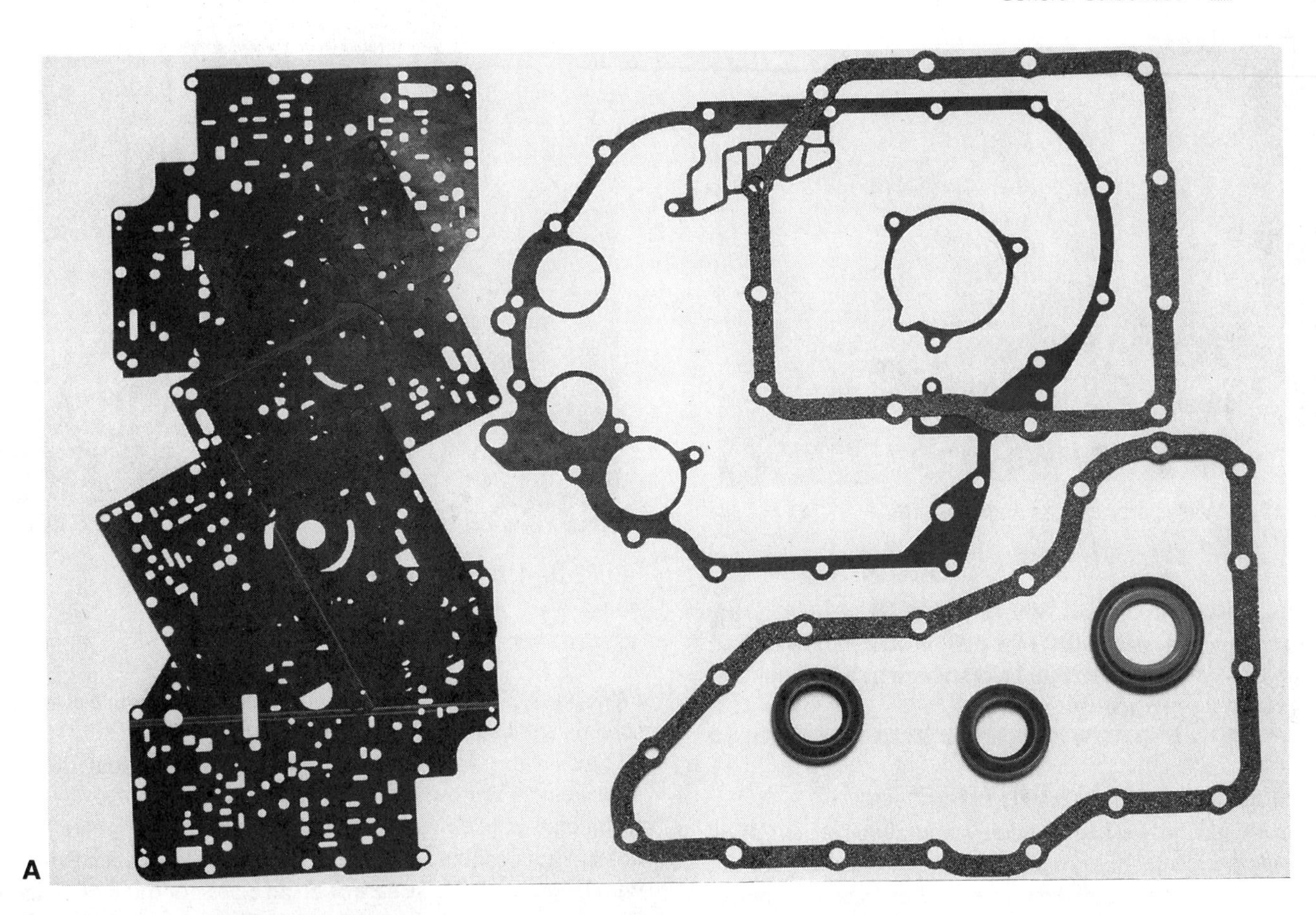

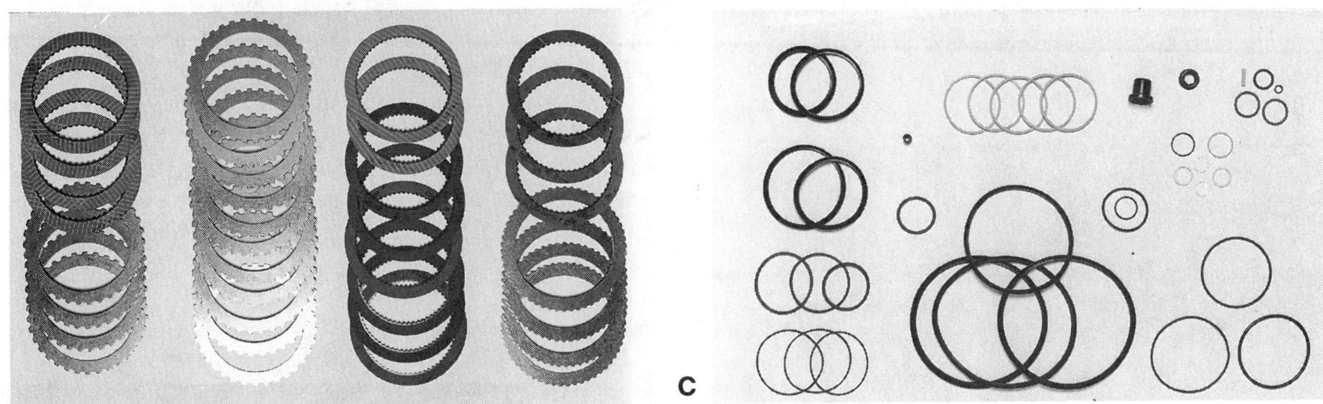

FIGURE 18–1 Ford AXOD: (a) gaskets, (b) clutch plates, and (c) seals.

FIGURE 18–2 Chrysler 40RH (A-500)/46RH (A-518) direct clutch hub spring being compressed for removal/installation; spring has 800 lb (3558 N) of force.

FIGURE 18–3 Wash parts in clean mineral spirits.

FIGURE 18–4 Dry parts with compressed air.

compress and control 800 lb (3558 N) of hub spring force of a Chrysler 40RH (A-500)/46RH (A-518) direct clutch. A substandard tool arrangement could be dangerous to your health.

Parts Cleaning and Inspection

Parts should be washed in clean mineral spirits and dried with compressed air (Figures 18–3 and 18–4). As the parts are cleaned, they can be organized on a parts tree to help eliminate bench clutter (Figure 18–5). Shop towels or rags should not be used to wipe parts after cleaning because of the lint that adheres to the parts. This lint accumulation has been known to clog filter screens and cause valve seizing.

During the drying process, be sure to blow out the fluid passages to remove any obstructions. In Figure 18–6, the technician is blowing out and checking the output shaft lube holes. Small orificed passages may need to be checked with tag wire (Figure 18–7). Be sure to do an extra good job of cleaning the case. After the mineral spirits bath, use hot water to flush the case, and then air dry it (Figure 18–8). The case passages all must be blown out.

Once parts are cleaned and air dried, they should be inspected to determine their condition for reuse or replacement. Figure 18–9 shows a technician inspecting planet pinion gears for loose or worn pins and chipped or worn gears. Give extra attention to bushings and thrust washers and their contact surfaces.

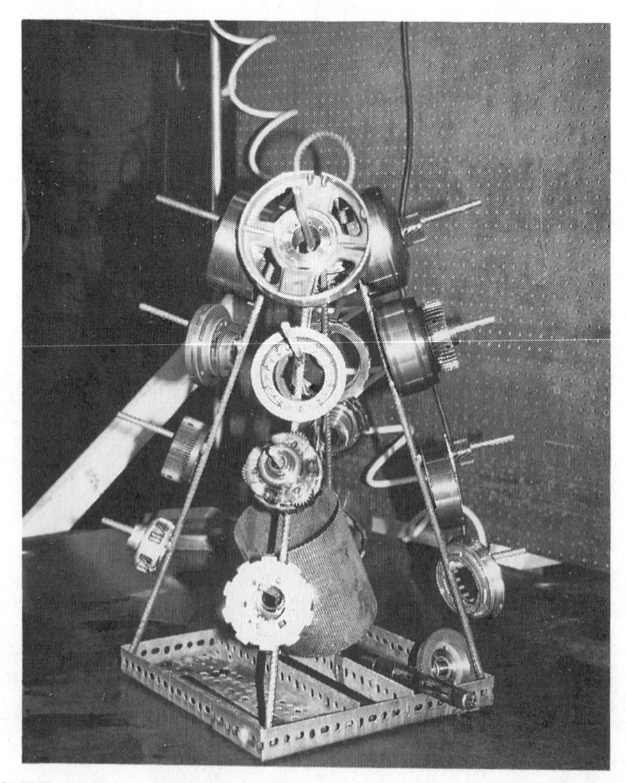

FIGURE 18–5 A parts tree reduces clutter.

FIGURE 18–6 Blow out fluid passages.

FIGURE 18–7 Check small orificed passages with tag wire.

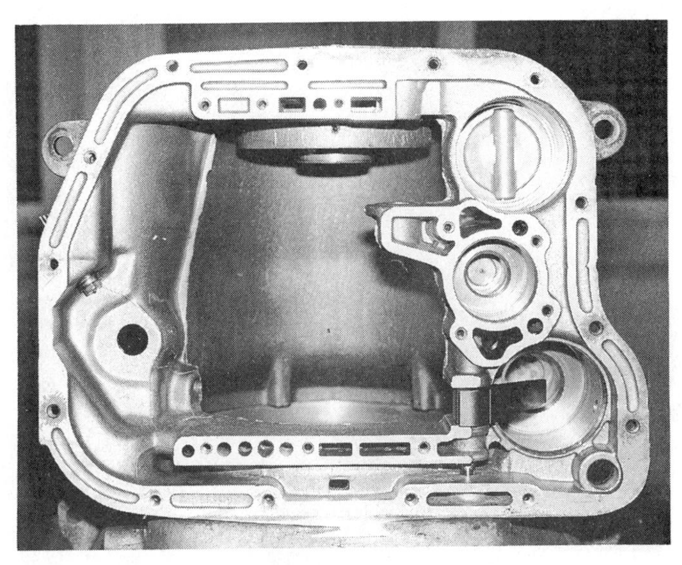
FIGURE 18–8 Clean case thoroughly.

Needle-type thrust bearings need to be observed closely for pitting, flaking, and brinneling (Figure 18–10). Where needle-type thrust bearings assemblies are nonseparable, rotate the washer many times over in the palm of the hand with finger and thumb pressure to feel for roughness or catching (Figure 18–11). Some needle-type thrust bearing assemblies in planet carrier assemblies are nonremovable. Use a sun gear, pipe extension, or long bushing of appropriate diameter to reach in and rotate the bearing assembly (Figure 18–12).

Gasket material can be very difficult to remove. To avoid damage to the aluminum contacting surface, spray the gasket with Permatex gasket remover and scrape it off, preferably with a plastic scraper. The use of a metal scraper can cause aluminum surface damage and create a potential leak area.

All paper gaskets are installed dry. Do not use any gasket cement or sealers.

If tapped aluminum threads are stripped/damaged, they should be made serviceable again with the use of a Heli-Coil or equivalent (Figure 18–13). This should be done before any assembly work begins.

FIGURE 18–9 Technician inspects planet pinion gears.

All fasteners must be torque-tightened to specifications with a torque wrench. An impact wrench will not do. Improper torquing is a leading cause of valve seizure, oil leaks, and parts damage.

Preliminary Steps to Rebuilding Subassemblies

For assembly work, use TransJel or Vaseline to prelubricate and hold in place needle-type thrust bearings, thrust washers, sealing rings, and rubber seals. Figure 18–14 shows the use of TransJel to hold thrust washers in place. Do not use white lube or grease, as their high melting points make them incompatible with transmission fluid. White lube and grease can get caught in the fluid stream and clog filter screens or small metered orifices.

Presoak new friction plates for thirty minutes in transmission fluid before installation (Figure 18–15). If new bands are installed, they also must be soaked. (Refer to the section entitled "Clutch Service" in Chapter 19.)

FIGURE 18–10 Closely inspect needle-type thrust bearings for pitting, flaking, or brinneling.

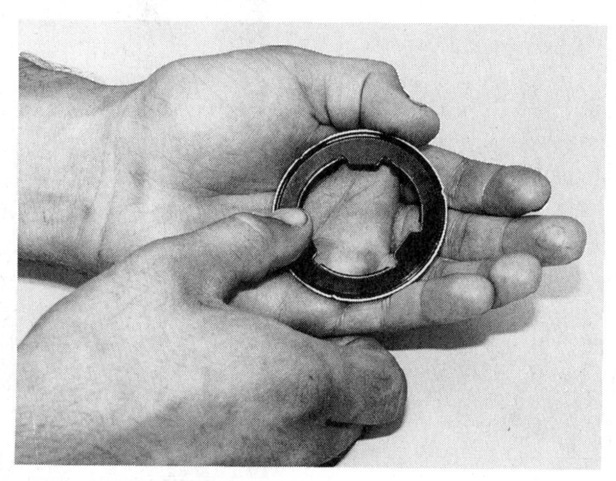

FIGURE 18–11 Rotate enclosed needle-type thrust bearing for roughness.

FIGURE 18–12 Testing AOD nonremovable needle-type thrust bearing located inside carrier assembly.

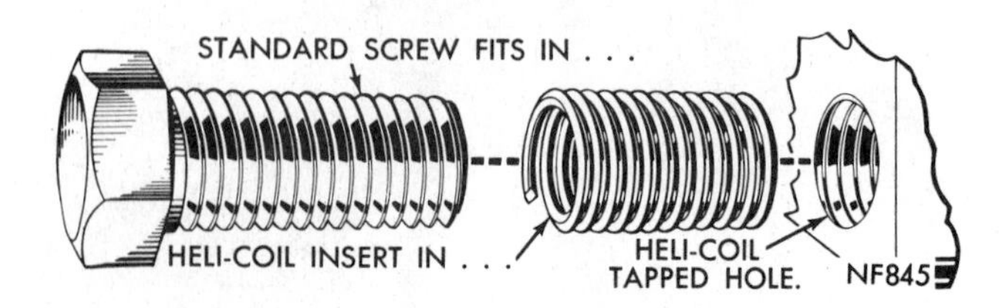

FIGURE 18–13 Damaged aluminum threads made serviceable with Heli-Coil. (Courtesy of Chrysler Corp.)

Seals

Metal Oil Rings. Check the fit of the metal oil rings in bore and grooves. Inside the bore (Figure 18–16), the ring must have good tension and conform to the bore diameter. Test the ring with your fingers for any side motion. Side motion in the ring cannot be tolerated. The ring must be in complete contact with its bore. A nonconforming ring to bore contact does not necessarily mean a bad ring—the bore itself may not be truly round. The ring and bore contact can be checked with a light source.

The ring gap should not exceed 0.012 in. If an oversized ring is installed, it will break when slipped into the bore, or the gap ends will butt. Undersized rings will actually fall through the bore. Nonconforming rings will cause leakage between ring and bore.

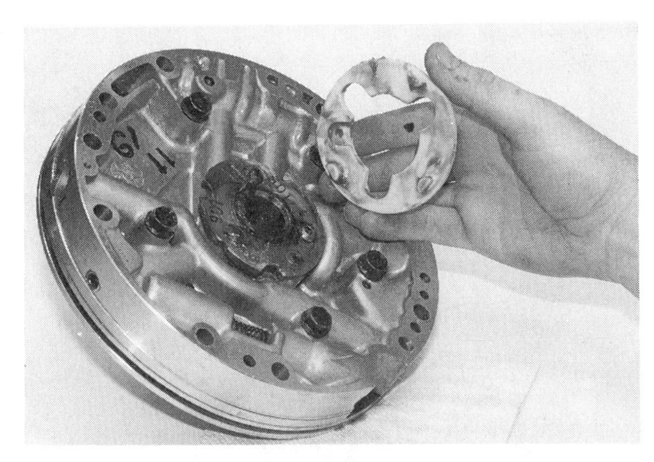

FIGURE 18–14 Use TransJel transmission assembly lube to hold thrust washers in place.

To check for ring and groove fit, place the ring in the groove as illustrated in Figure 18–17. Ring and groove clearance should not exceed 0.005 in (.127 mm). With the ring in place, be sure to check the grooves for step and taper wear and ridging (Figure 18–18). When the ring is installed, make sure that it rotates freely in the groove. Push the ring against the sides of the groove to check for complete contact around the groove. If

FIGURE 18–15 Presoak new friction plates (and bands).

the groove is not machined straight, the ring will not conform. Nicks or minor damage at the groove that bind the ring can be removed with a file. A good groove should have well-defined sharp corners on the top and bottom.

Ring gap and groove clearance measurements normally are not required. These specifications can be used for making a quick visual judgment about whether the ring is right for the bore and groove or whether wear is the problem. If there is a ring to bore or ring to groove problem, the clearances will greatly exceed specifications, and measurements generally will not be necessary. Ring to bore and ring to groove conditions are too often taken for granted and overlooked. Be sure to make this a regular part of your overhaul practice and avoid a repeat clutch failure. Every new ring should be quick checked in the bore and groove.

Rubber Seals. Check the fit of the rubber seals if you are in doubt about their dimensional integrity or need to decide which seal size to use from a seal package. This is especially critical on clutch and servo piston applications.

O-Ring Seals. Round or square-cut O-ring seals must meet criteria for (1) correct diameter and (2) correct thickness. Standard manufacturing tolerances require that the O-ring seal diameter must

FIGURE 18–16 Sizing ring and gap in Ford C-6 governor sleeve bore.

FIGURE 18–17 Check for ring and groove fit.

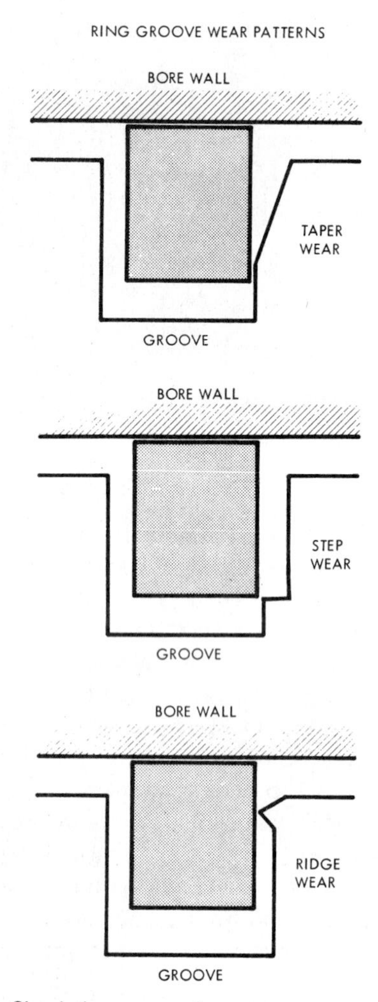

FIGURE 18–18 Check the grooves for wear.

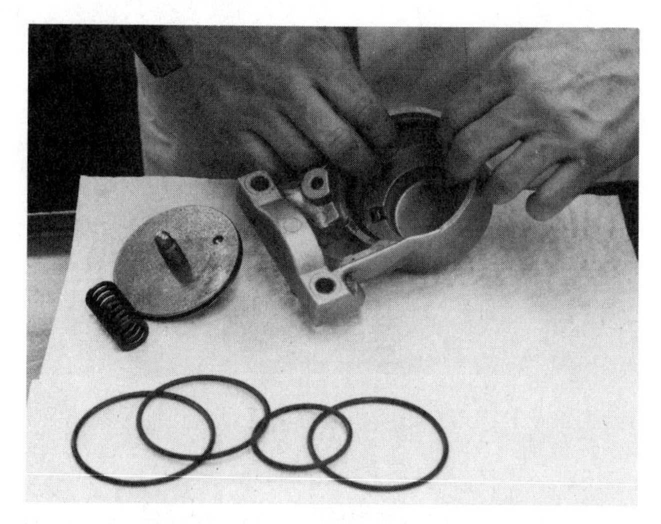

FIGURE 18–19 Check O-ring seal diameter by placing in bore.

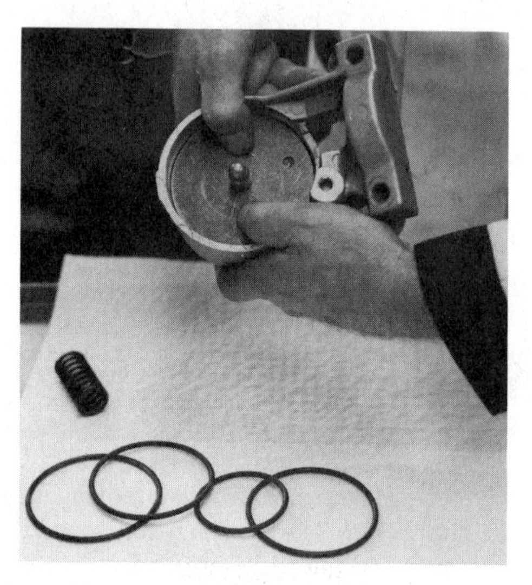

FIGURE 18–20 Check O-ring seal thickness by checking the drag in the assembly unit.

be within ±3% of the bore size. Check this dimension by placing the seal in its bore (Figure 18–19). The seal should have a slight drag without losing its configuration or fit loosely within a 3% limit. For example, a 2-1/2-in (63.5-mm) servo piston bore requires that the seal clearance be no more than approximately 5/64 in (2 mm). Because you are working with a circle, the maximum clearance to the bore on all sides of the seal is one-half of 5/64 in, or slightly more than 1/32 in (0.8 mm).

Although O-ring clearance measurements are not required during overhaul, the figures in the preceding paragraph can give you a basis for quick checking a seal for correct diameter. A correct diameter seal should not fit loosely on the piston, nor should it need to be overstretched to be put in place.

The O-ring seal thickness determines the amount of squeeze or compression on the rubber when it is installed and is related to the ability of the rubber to seal. Seal thickness in the servo piston can be checked by examining the drag in the assembly unit (Figure 18–20). With the seal installed on the piston and lubricated, round seals should have a firm drag, while square seals give only a light drag. Round seals have a tendency to return to their normal set and get looser and, therefore, require the heavy drag.

There are usually two or more seals involved with a piston installation, which makes it necessary to test and remove each seal one at a time so they can be checked individually.

A seal with good thickness but a small diameter sometimes gets wrongfully overstretched into place in the piston groove. The seal sticks out of groove and looks like it will have a good drag. The problem is that the overstretched seal wants to return to its original size, and what was a good drag eventually is nonexistent. The so-called good seal becomes a leaker. O-rings will shrink 0.001 in for every one percent they are stretched.

Lip Seals. These seals can be individually checked with the clutch piston in the drum cylinder. Because these seals depend on hydraulic pressure for sealing, they will provide only a very light drag. The seals need to be lubricated for check out. Start with the inner seal by holding the drum upside down and letting the piston free-fall against your hand. Work the piston up and down in the cylinder and feel for a light drag (Figure 18–21). If there is an absence of drag, the seal is undersized.

FIGURE 18–21 Checking TorqueFlite rear clutch piston drag.

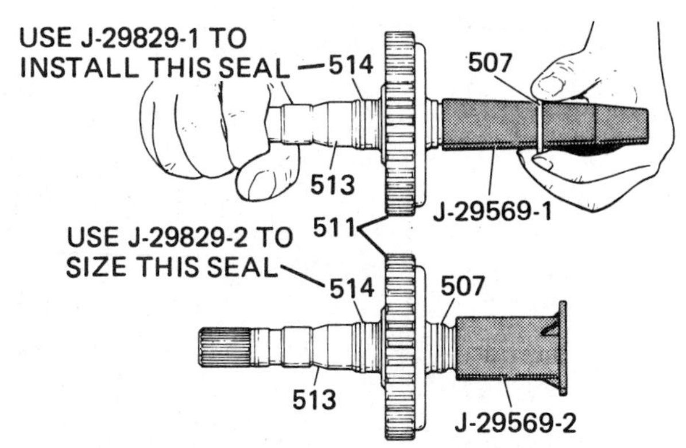

507 RING, OIL SEAL (TURBINE SHAFT/SUPPORT)
511 SPROCKET, DRIVE
513 SHAFT, TURBINE
514 SEAL, "O" RING (TURBINE SHAFT/ TURBINE HUB)

FIGURE 18–22 Installing (expanding and sizing) solid Teflon rings on turbine shaft—GM 3T40 (125-C), 4T60 (440-T4), and 4T60-E procedure. (Courtesy of General Motors Corporation, Service Technology Group)

Leave the inner seal in place and install the outer seal on the piston. Using the same procedure with both seals, a definite increase in a seal drag should be felt. The piston may even stay in the cylinder and not free-fall. When the piston uses a center or third seal, each seal should be checked one at a time.

A good fitting lip seal has a slight stretch when installed in its groove. If a lip seal requires excessive stretching, the sealing lip does not correctly contact the bore. A very tight stretch can prevent the lip from contacting the bore entirely. A loose groove fit will cause the lip end to roll and wedge between the piston and cylinder bore when the piston strokes.

Teflon Sealing Rings. When handling Teflon sealing rings, remember that the material is soft and cannot afford any nicks, scratches, or other deformities. Careless handling can cause rings not to seal properly.

Solid or continuous Teflon rings need special installation tools. Without a tool to assist, the seals can be easily distorted and overstretched. The seals need to be slightly stretched/expanded to fit over the shaft. Figure 18–22 (top illustration) shows the installation technique for the solid Teflon rings used on the Hydra-Matic 4T60 (440-T4), 4T60-E, and 3T40 (125-C) turbine shaft. The tool sleeve is coated with TransJel/Vaseline so that the rings can slide into place easily. Once installed, the rings need to be sized/compressed (Figure 18–22, bottom illustration). In Figure 18–23, the technician is sizing the solid Teflon rings on the input housing of a 4T60 (440-T4).

FIGURE 18–23 Sizing solid Teflon rings on 4T60 (440-T4) input housing.

One-way Clutches

When installing one-way sprag clutch units, follow the exact manufacturer's instructions for installation. Many sprag cage assemblies can be easily installed backward, which results in the wrong freewheeling action. In Figure 18–24, the intermediate sprag assembly of a Hydra-Matic 3L80 (400) is shown in its outer race. One side of the sprag assembly has a shoulder that must face down when mounted onto the direct clutch housing in order to give the correct freewheeling direction (Figure 18–25). If the sprag cage is installed backward, the transmission will start in first gear, miss second gear, and lock up in third gear. The engine will lug down until the transmission downshifts back to first gear. This hazardous cycle then repeats itself: 1-3-1-3, and so on.

One-way roller clutches usually pose no special installation problem. In some settings, however, they can be installed improperly. Correct and incorrect assembly installations are shown for a THM-350 intermediate roller clutch (Figure 18–26). Note how the cage and cam configurations are matched and mismatched in the two photographs. An incorrect installation gives a freewheeling action in both directions, and the roller fails to hold.

Sprag and roller clutch functional checks should always be made to verify the correct freewheeling and holding action (Figure 18–27). When inspecting a sprag assembly for wear, closely examine the leading edge of the sprag. If present, the wear mark will run along the edge. If it is less than 1/32 in (0.80 mm), it is acceptable. If the wear is greater than that, the sprag is close to the failure point and should be replaced. Be very critical when inspecting sprag and roller clutch assemblies for wear. They can work fine at room temperature but fail on a cold day or under heavy acceleration.

Filters/Screens

Be sure to put back in place all the small filter screens that are usually tucked away in holes and passages. The fluid passing through these filters has already gone through the main filter, but the main filter cannot do it all. The small filters are backup filters to protect valves, governors, and electric sole-

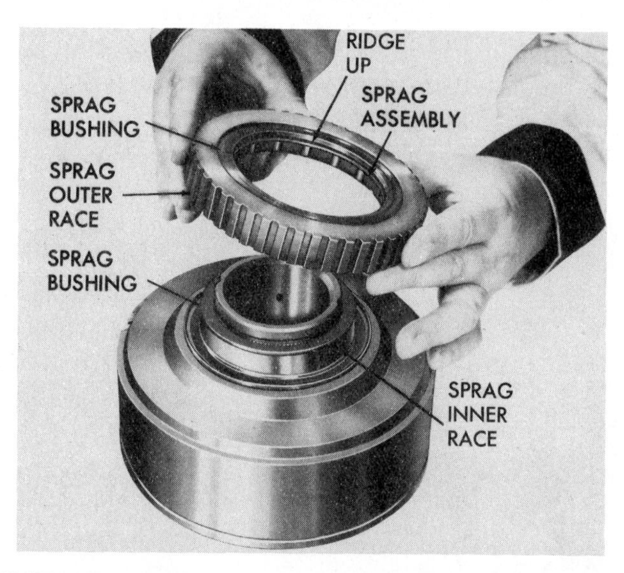

FIGURE 18–24 Intermediate sprag assembly—Hydra-Matic 3L80 (400). (Courtesy of General Motors Corporation, Service Technology Group)

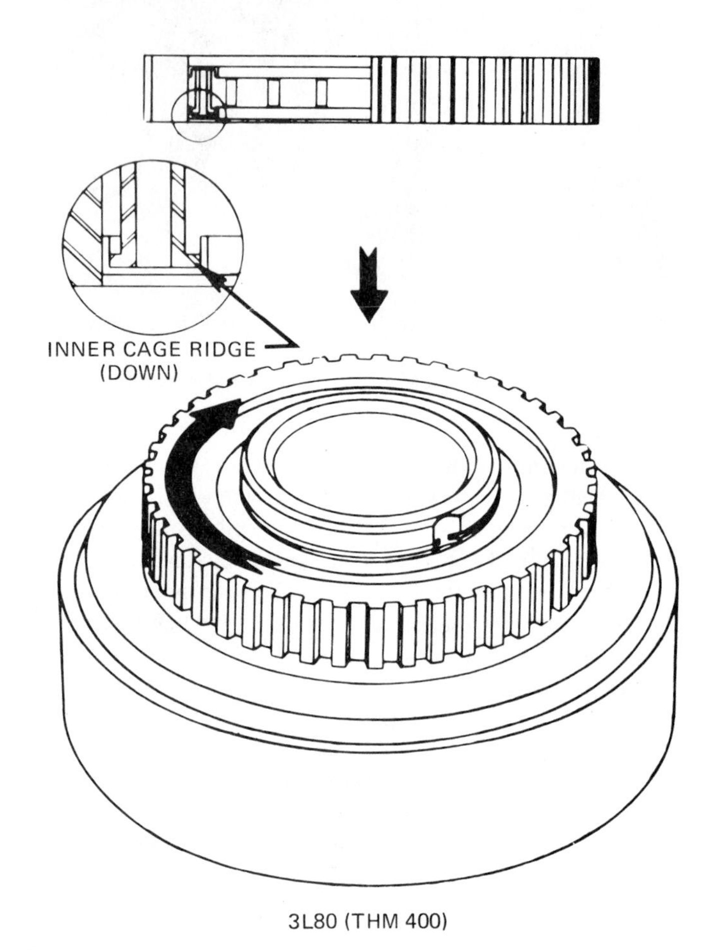

FIGURE 18–25 Hydra-Matic 3L80 (400) sprag assembly shoulder must face down. (Courtesy of General Motors Corporation, Service Technology Group)

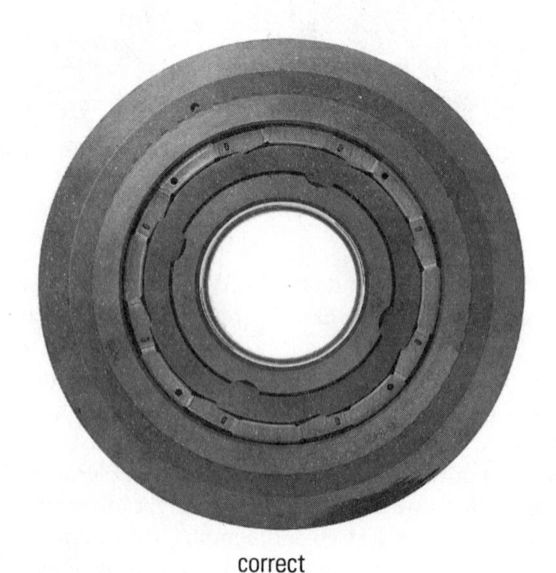

FIGURE 18–26 Hydra-Matic roller clutch installations.

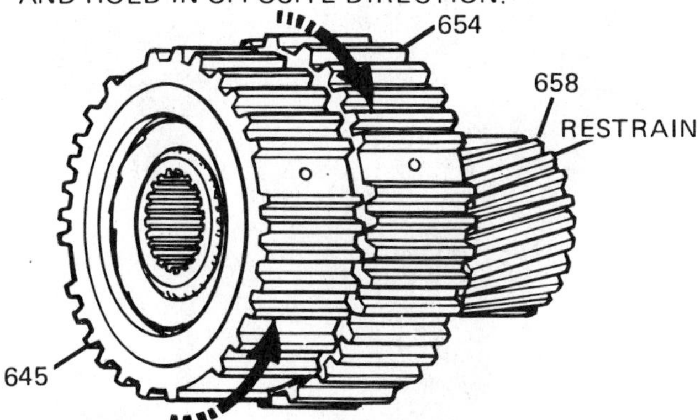

FIGURE 18–27 GM 4T60 (440-T4). (Courtesy of General Motors Corporation, Service Technology Group)

FIGURE 18–28 Pump filter screen—GM 4L60.

noids from foreign material not caught by the main filter. It is important for these small filters to trap material from the ATF that could cause valves to stick or have an abrasive effect.

The practice of throwing these screens away is not recommended. Be careful not to clean the small nylon microfilters with a highly toxic cleaner. The nylon mesh will swell and seal up.

Several small filter installations are shown in Figures 18–28 to 18–31. Normally, the governor screen in the Hydra-Matic 4T60 (440-T4) and 3T40 (125C) is not removed unless the shaft sleeve is scored or the screen is damaged. The governor screen can be back-flushed with the governor and the governor pressure tap plug removed. For a thorough back-flush, also remove the valve body (not the governor feed pipe) on the 3T40. On the 4T60, remove the governor oil pipe retainer. On an overhaul/rebuild, it is necessary to remove the governor pressure plug only when cleaning the case.

The technician should know that the 3T40 and 4T60 both use check ball–style governors. Therefore, a plugged filter screen may not be the cause of a "no upshift" condition. Since the screen is often part of the governor shaft sleeve, it is not a throwaway item. If the pump screen is left out in the 4L60 (700-R4) (Figure 18–28), a significant drop in line pressure occurs. The rubber O-ring on the filter is needed to seal the pump line pressure.

FIGURE 18–29 Pump filter screen—TorqueFlite.

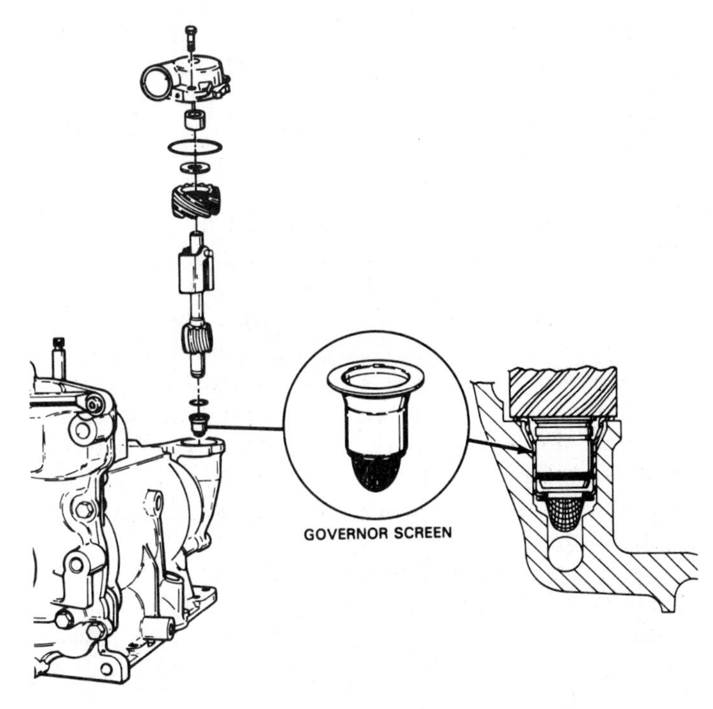

FIGURE 18–30 Governor filter screen—4T60 (440-T4)/3T40 (125-C). (Courtesy of General Motors Corporation, Service Technology Group)

FIGURE 18–31 Bypass solenoid filter screen—Ford AXOD.

Bands and Drums

Paper-based friction bands that are burned or flaked, have crack marks, or are heavily discolored with black should be replaced. A black discolored band has picked up a heavy carbon surface from the cast iron drum. The drum probably has a glazed surface to complement the band condition. If this combination is reused, long drawn-out 1-2 shifts or 1-3 skip shifts will be built back into the transmission.

When new flex bands are used, the drum surface should be deglazed with a medium-coarse 180-grit emery cloth to break in the friction surfaces. Always work in drum rotation (Figure 18–32). Working back and forth across the drum causes unwanted band wear. In some special situations, the drum needs to be deglazed by working the emery cloth across the drum rotation. The usual deglazing process with drum rotation results temporarily in minor rough gear engagement. Examples of minor rough engagement include:

- 4T60: rough reverse (second clutch drum)
- 4L60: rough second (reverse input drum)

On smooth, hard-surfaced drums, such as the RWD/FWD TorqueFlite front drum, deglazing is not required. This type of surface is engineered with the band to give the desired 1-2 shift feel and must not be upset. Hard-surfaced drums are easily identified when the medium-coarse grit has very little effect on deglazing the drum surface. Where the drum surface has been scored by metal-to-metal contact with the band, the drum must be replaced.

Do not overstretch a flex band during inspection. If overstretched, the band loses its normal configuration and causes surface cracking of the friction facing. It cannot be used. Keep the band ends together with a paper clip until ready for installation.

FIGURE 18–32 Deglaze drum surface in rotational direction.

SUMMARY

The process of overhauling a transmission includes many different steps of procedure. During the overhaul, certain items are always replaced, while the remainder are inspected and checked to determine if they are reusable. When working in a shop setting, always wear safety glasses, and be especially cautious as snap rings are removed on spring-loaded components.

To be able to accurately inspect components, clean them in minerals spirits and dry them with compressed air. Check for unusual wear and that lubrication passages are clear of debris. Avoid using shop rags or paper towels to wipe parts, as fibers left behind may jam valves or plug filters.

Prepare components for reassembly. Use TransJel or Vaseline to hold thrust washers and bearings in place. Soak clutch friction plates and bands in fresh ATF for a minimum of thirty minutes. Prior to installing seals, check their sizing to make sure they will seat and seal properly. Wet seals with ATF, and use seal installers when appropriate. It is a good practice to use a low air pressure to air test clutch devices when secured in their housings. It is also recommended that all components be wet with ATF when installing them.

Certain components often are not given a careful enough inspection and need to be installed properly for correct operation. These include one-way clutches, filters and screens, governors, and housing surfaces for band applications.

When rebuilding a transmission, spend the time and effort needed to do a complete, accurate job. The reliability of an overhauled unit is dependent upon the amount of attention given to the internal components.

REVIEW

Questions

1. Rubber seals should be prelubricated with a petroleum jelly. ___T/___F
2. Once the parts are flushed clean, they should be wiped dry immediately with shop towels. ___T/___F
3. A metal scraper is preferred for removal of gasket material on aluminum surfaces. ___T/___F
4. Gaskets typically are treated with aviation Permatex before installation. ___T/___F
5. Nonseparable needle-type thrust bearings are evaluated with a rotating finger and thumb pressure. ___T/___F
6. Improper torquing is a leading cause of valve seizure, oil leaks, and parts damage. ___T/___F
7. Petroleum jelly is a legitimate lubricant for prelubing clutch plates. ___T/___F
8. It is an acceptable practice to hold thrust washers in place with white lube. ___T/___F
9. Ring and groove clearance for metal rings should not exceed 0.010 in (0.254 mm). ___T/___F
10. It is an acceptable practice to remove ring groove nicks with a fine-cut file. ___T/___F
11. A lip seal should produce a slight drag in the cylinder bore. ___T/___F
12. If the O-ring rubber seal sticks out of the groove, it can be said that it is dimensionally correct. ___T/___F
13. A good fitting O-ring seal takes a heavy stretch to fit it into place. ___T/___F
14. When a roller clutch is installed backward, it may freewheel in both directions. ___T/___F
15. When a sprag is installed backward, it will freewheel in both directions. ___T/___F
16. To avoid overstretching and distorting solid Teflon seals, they are installed with a special tool. ___T/___F
17. Once installed, solid Teflon seals must be sized. ___T/___F
18. It is a good practice to deglaze clutch drum surfaces where flex bands are used. ___T/___F
19. It is a good practice to keep the flex band ends together with a paper clip. ___T/___F

19

Subassembly Reconditioning

OBJECTIVE:

The objective of this unit is for you to become proficient with the terms, concepts, and procedures contained in this chapter. Completion of this objective is essential to becoming a successful automatic transmission technician.

CHALLENGE YOUR KNOWLEDGE

Define the following key terms:

Lubrite coating and anodized.

List the following:

Ten inspections performed to a typical IX gear pump; eight inspections performed to a typical clutch housing; three inspections performed when evaluating clutch fiber plates and steel plates for possible reuse; two manufacturer-designed methods for adjusting clutch plate clearance; four purposes of air checking a rebuilt clutch housing; seven tips for reassembling valve bodies; and typical bushing clearance.

Describe the following:

Why pump gears, rotors, and vanes must be installed properly; clutch pack endplay measurement and adjustment procedures; carrier pinion endplay specification guidelines and measurement procedures; valve body disassembly techniques; inspection points for each of the three types of governors; and bushing removal and installation procedures.

INTRODUCTION

This chapter highlights the servicing of subassemblies such as pumps, clutch housings, valve bodies, and governors. Planetary carrier service and bushing replacement procedures are also included. Reconditioning involves the disassembly of subassemblies, cleaning of components, evaluation of parts for wear or damage, replacement of necessary items, and correct reassembly of the units.

PUMP SERVICE

The pumps used in automatic transmissions are highly reliable and give long life with very little wear. They seldom need to be replaced. When conditions of very excessive wear and damage are noticed (normally the fault of the pump), replace the pump. To determine whether the pump can be serviced and reused, it first must be separated, and all of its parts must be cleaned with mineral spirits and then air dried. In most cases, a close visual inspection picks up excessively worn parts. Some pump assemblies include the pressure regulator and converter clutch valves (Figure 19–1), which are included in pump reconditioning.

Pump Inspection Guidelines

IX gear, IX rotor, and variable displacement/capacity vane pumps are found in both automatic transmission (RWD) and transaxle (FWD) applications (refer to Chapter 6). The reconditioning guidelines all follow similar patterns.

Because visual inspection is very critical in evaluating pump wear, it is advantageous for the technician to know where to look. Knowledge of how the pump works can be put to use in locating logical wear points. For illustrative purposes, an IX gear pump is examined for gear and pocket wear (Figure 19–2). Figure 19–3 identifies the suction (intake) and pressure (output) pockets of a gear pump. Note that the pressure side has the small pocket. The fluid in the pressure pocket exerts a side thrust (1) on the drive gear and converter drive hub toward the suction pocket, and (2) on the driven internal gear opposite the suction pocket. These forces result in the following wear patterns:

- The side thrust on the drive gear and converter hub puts constant pressure on one side of the pump body bushing.

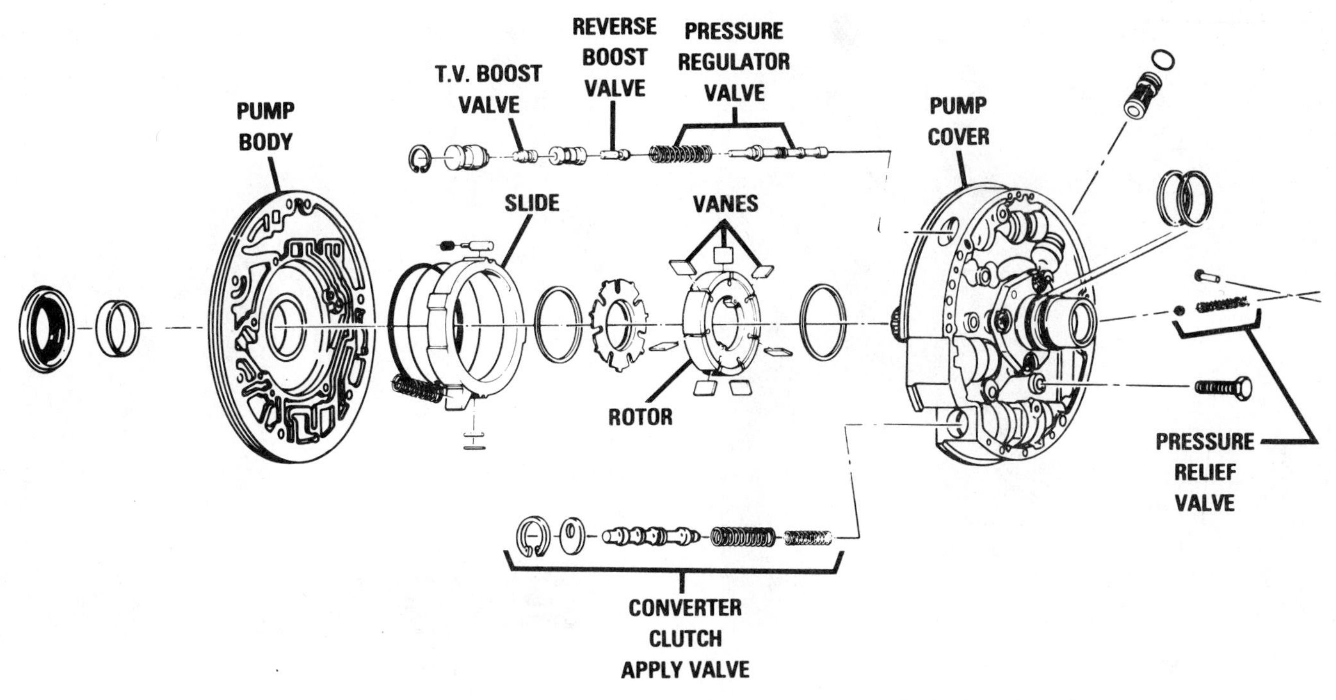

FIGURE 19–1 4L60 (700-R4) pump assembly. (Courtesy of General Motors Corporation, Service Technology Group)

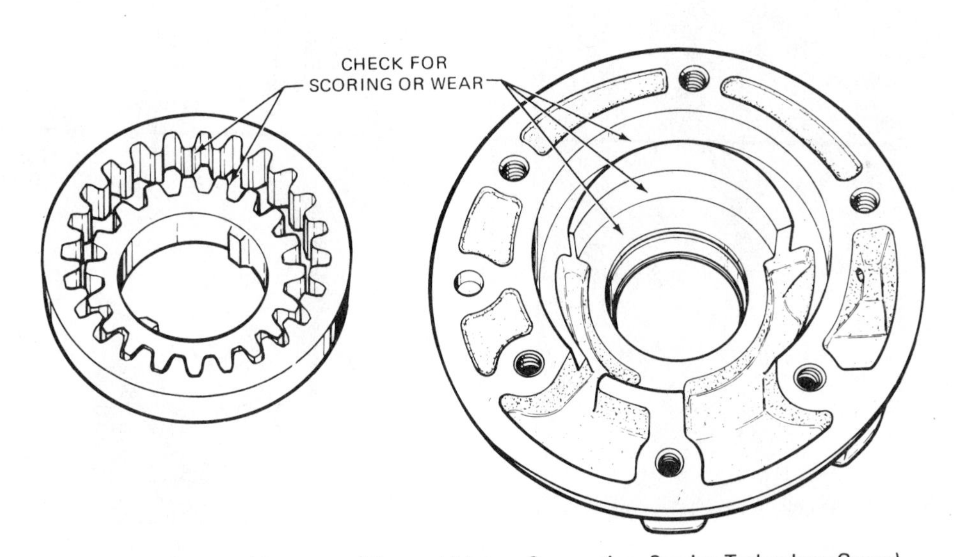

FIGURE 19–2 Oil pump pocket and gear wear. (Courtesy of General Motors Corporation, Service Technology Group)

This wear area is located on the suction/intake side (Figure 19–4).

- Should the body bushing wear be excessive, the drive gear will undercut the crescent or divider. This damaged area is located at the leading edge of the divider on the suction/intake side (Figure 19–5). This type of wear produces fine metallic particles that can severely damage the pump and contaminate the converter, cooling, and lube circuits. Minor scuffing in this area can be tolerated, provided the body bushing is replaced.
- The side thrust on the internal gear produces a wear area in the body on the pressure/output side (Figure 19–6).
- Should the internal gear to pump body wear be excessive, the internal gear will undercut the divider in the area of the leading edge on the suction/intake side (Figure 19–7).

The following items also need special attention during inspection of the pump assembly:

FIGURE 19–3 Pressure pockets of an IX gear pump.

FIGURE 19–4 Wear occurs on one side of the pump body bushing.

FIGURE 19–5 Inspect for damage at leading edge of divider.

FIGURE 19–6 Wear area in the pump body, pressure side.

- Examine the land area between the pressure and suction pockets. The location depends on pump design (Figure 19–8). If the area is grooved or undercut, part of the pump pressure head is lost back to the suction side. Minor scuffing can be polished with crocus cloth.
- Inspect the gears for cracks, scoring, or galling. A cracked drive member usually is an indication that the transmission was bolted to the engine with the converter drive hub not engaged in the drive member (Figure 19–9).
- A production practice is to apply a black **lubrite coating** on the pump gears for initial break-in. It is normal for the lubrite coating to wear off unevenly. This is not a reason to replace the pump body assembly.

❖ **Lubrite coating:** A black-colored coating applied to pump components that allows contact surfaces to wear in smoothly. It is normal for the coating to disappear on the contact surfaces as the components are in service.

- Inspect the pump body and pump cover for scoring or galling (Figure 19–10).
- An item usually neglected is the internal pump bushings (Figure 19–11). Although they seldom need replacement, they must be inspected very closely for being out-of-round. Excessively worn stator shaft bushings can affect the converter, cooling, and lubrication circuits and may cause fluid foaming and fluid loss out the filler tube.

FIGURE 19–7 Look for internal gear to pump body wear.

- ❍ Give attention to the stator shaft splines (Figure 19–8). Worn, stripped, or wave-like splines call for a pump cover replacement. Diesel applications are noted for causing spline damage.
- ❍ Inspect oil ring grooves and clutch bushing supports on the oil delivery/clutch support sleeve for groove damage or wear (Figure 19–12).
- ❍ Check the pump body bushing for tightness to the bushing bore hole. If the bushing is loose and spins, be sure that the replacement bushing fits tightly. Otherwise, the pump body will need to be replaced.
- ❍ A severely damaged pump is usually an indication of an external problem such as converter/flex plate alignment or faulty transmission installation practice. If this is overlooked, your overhaul will soon fail.

FIGURE 19–8 Examine the land area between inlet and outlet ports (bottom arrow) and the stator shaft splines (top arrow).

- ❍ Figure 19–1 shows a Hydra-Matic 4L60 (700-R4) pump assembly. When working with this style of pump unit, be sure to remove the valving and filter screens for cleaning and inspection. Not only does this clean out the valve bore assembly and check the valves for free movement, but it makes it possible to flush out all the pump cover oil passages. Otherwise, all foreign material cannot be flushed out. The valving might be free in the bore on bench assembly, only to get hung up when the transmission hits the road.

As a pump-related matter on the 4L60 (700-R4), be sure to clean and check the operation of the input shaft check ball/retainer assembly (Figure 19–13). The check ball is keyed to the control of the torque converter clutch lockup and unlock circuitry. During the unlock mode, the torque converter fluid is

FIGURE 19–9 Inspect gears for cracks, scoring, or galling.

FIGURE 19–10 Inspect the pump cover for scoring or galling.

FIGURE 19–11 (a) Front support bushing for turbine/input shaft, located inside stator shaft; (b) rear support bushing for turbine/input shaft, located inside the pump cover's oil delivery/clutch support sleeve.

FIGURE 19–12 Inspect oil ring grooves and clutch housing bushing supports.

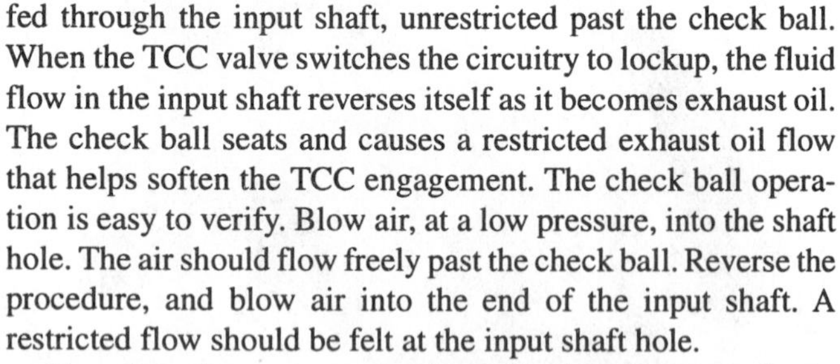

FIGURE 19–13 Turbine/input shaft check ball and retainer assembly.

fed through the input shaft, unrestricted past the check ball. When the TCC valve switches the circuitry to lockup, the fluid flow in the input shaft reverses itself as it becomes exhaust oil. The check ball seats and causes a restricted exhaust oil flow that helps soften the TCC engagement. The check ball operation is easy to verify. Blow air, at a low pressure, into the shaft hole. The air should flow freely past the check ball. Reverse the procedure, and blow air into the end of the input shaft. A restricted flow should be felt at the input shaft hole.

Complete the pump evaluation with the clearance checks outlined in the manufacturer's service manual. Pumps normally have a maximum end clearance limit of 0.0025 in (0.058 mm). This is illustrated in Figures 19–14 and 19–15. As a substitute, use plasti-gauge over the slide gear area, and torque-tighten the pump assembly together.

Pump Assembly Guidelines

To prepare a pump for assembly, begin with thoroughly cleaned and dried assembly parts. Replace the pump bushing

FIGURE 19–14 Check pump clearances (IX gear pump). (Courtesy of General Motors Corporation, Service Technology Group)

FIGURE 19–15 Measure pump clearances (variable displacement/capacity vane pump).

FIGURE 19–16 Check the fit of the seal on the converter hub.

as necessary. To avoid blow out of the pump seal or leakage past the seal, converter hub to bushing clearance should not exceed 0.004 in (0.10 mm).

Before installation of the pump seal/converter hub seal, check the fit of the seal on the hub (Figure 19–16). Lightly lubricate the seal and hub and, with a rotational motion, feel for the lip hub tension. A seal that slides down the shaft with practically no resistance is worn and must be replaced.

To install the pump seal, treat the metal casing with a sealer. Most metal casings come already prepared with a sealer, so be on the conservative side during application. Use caution to avoid letting the sealing compound close the oil drain-back hole between the seal and bushing. It is also a good practice to pack the inside of the seal cavity with TransJel or Vaseline to retain the seal spring. During installation, the hammer shocks on the seal driver can pop the spring out of place, which may go unnoticed. Without the spring tension on the lip, the seal will leak.

To install the seal, the use of a seal driver is recommended (Figure 1–17). In the absence of a seal driver, the weight of a heavy hammer carefully striking the outside shoulder of the seal can be used (Figure 19–18). Some pump seals are required to be staked or fit with a retainer ring to prevent seal blowout (Figures 19–19 and 19–20).

Dip the drive and driven gears (or lobed rotors) in transmission fluid and install into the pump body. Gear-type pumps must have the pump gears installed in the same position in which they were removed for two reasons: (1) the gears have established a wear pattern that cannot be upset without causing pump gear noise, and (2) the drive gear must be installed with the drive tangs or drive slots in the correct relationship to the converter drive hub. Incorrect drive gear installation will wedge the flex plate, converter, and drive gear when the transmission is bolted

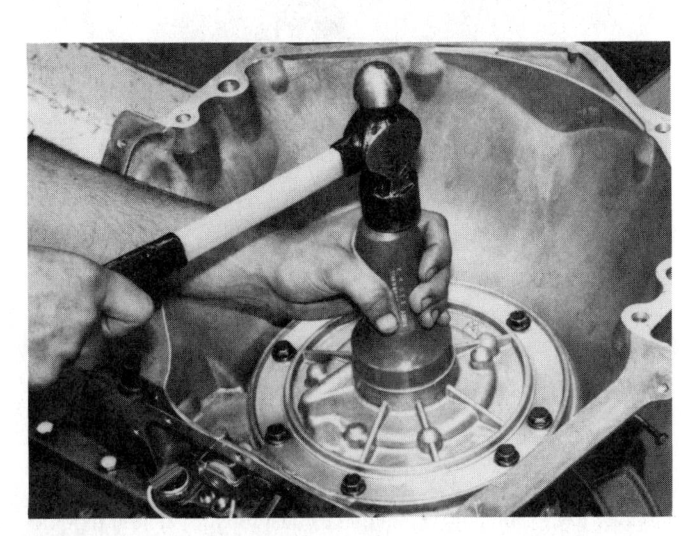

FIGURE 19–17 Seal driver used to install pump seal.

FIGURE 19–18 Strike outside shoulder of pump seal with a hammer.

FIGURE 19–19 A4LD pump seal is staked.

FIGURE 19–20 4L60 (700-R4) pump seal retainer.

to the engine. Refer to the manufacturer's directions for correct installation.

The current practice is to provide production marks in the gear faces to identify correct installation position. Shown in Figure 19–21 are the gear marks provided for a Hydra-Matic pump assembly. Note that the drive gear is positioned with the converter drive tangs facing up. The driven gear mark may face up or down, as required by the product manufacturer. In the absence of production marks, the gears can be installed by matching the wear pattern of the teeth. The gear teeth show wear on one side only in relation to rotation.

When dealing with lobed rotating members of an IX rotor pump, follow practices similar to those extended to the gears of an IX gear pump.

When a pump unit is designed with a pump cover and pump body, the two halves must be bolted together on center. If this is not done, the pump assembly may not fit into the transmission case. With the pump bolts in place finger-tight, use a special band for the centering operation, and then torque-tighten the bolts (Figure 19–22).

Do not install any valving into the pump until the pump cover bolts are torque-tightened. This ensures that the valves are free to move and will be unaffected by the bolt torque.

When the new pump body to case O-ring seal is installed, check it for correct sizing. You should feel a slight lip edge extending from the seal groove (Figure 19–23). If the lip edge is not felt, it means that the O-ring does not extend out of the groove and is undersized.

FIGURE 19–21 IX pump gears with identification marks.

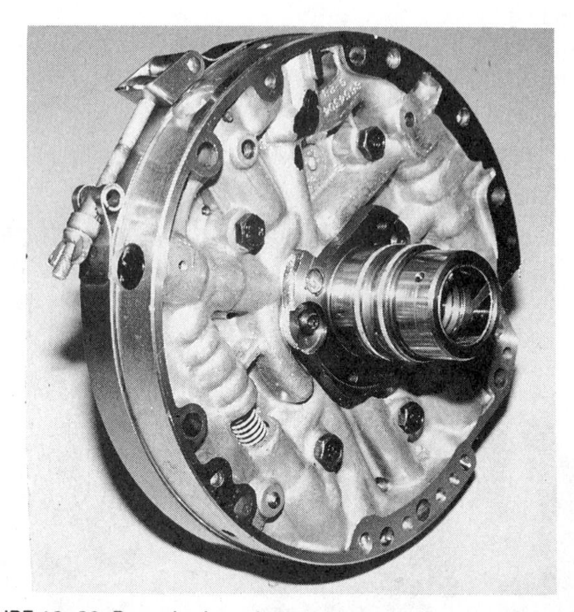

FIGURE 19–22 Pump body and pump cover must be bolted together on center; use of a special strap is recommended.

FIGURE 19–23 Feel for a slight lip on pump body to case O-ring.

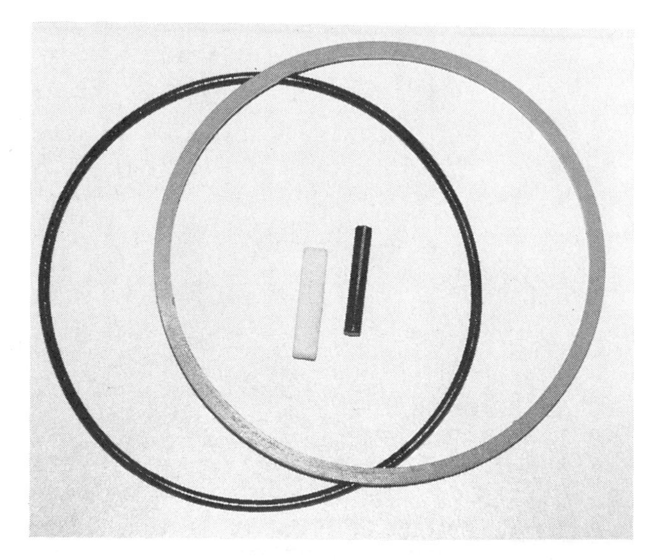

FIGURE 19–24 Variable displacement/capacity vane pump seals.

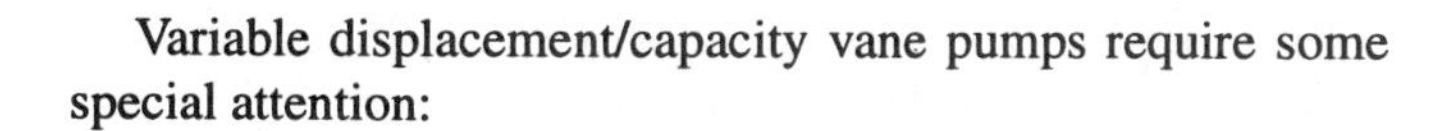

Variable displacement/capacity vane pumps require some special attention:

1. A seal kit is shown in Figure 19–24.
2. The pump rotor must be installed facing the proper direction. Refer to the product manufacturer's service directions.
3. The vanes must be installed in their original wear pattern, not necessarily in original rotor slots. When examining the vanes, look for the wear edge that works against the top and bottom rings and the wear side that faces rotor rotation (Figure 19–25).
4. Depending on the product manufacturer, serviceable parts in addition to the seal kit may not be available.

Hydra-Matic permits replacement of the pump rotor and slide if they show excessive wear or damage. If either the pump body pocket or pump body cover are scored, do not bother to service—replace the pump. The pump slide and rotor have select fits, and therefore, the thickness of the replacements must match the original parts measurements (Chart 19–1, page 434). Incorrect selection can result in pump damage or low line pressure. Make it a practice to run a fine hone over both sides of the replacement rotor or slide to remove any burrs.

A

B

FIGURE 19–25 (a) Vane edge wear faces guide rings; (b) vane surface wear pattern.

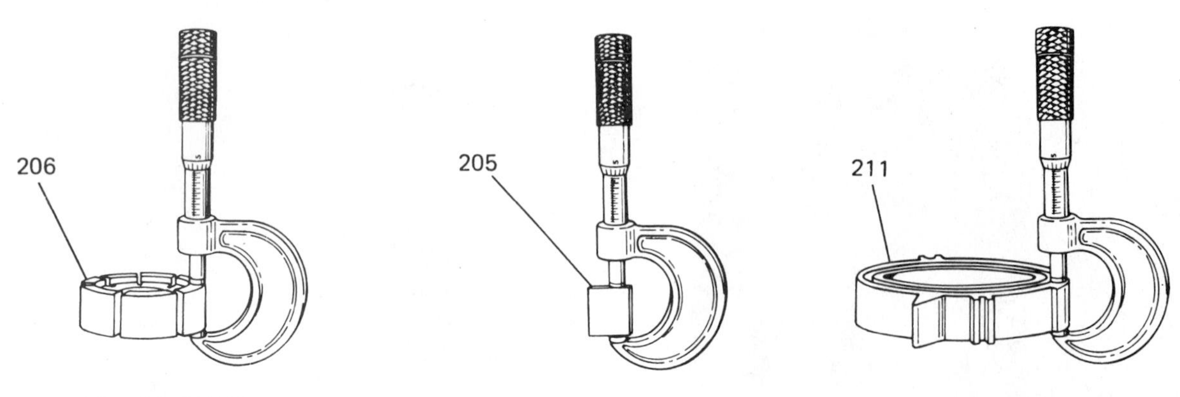

205 VANE, OIL PUMP
206 ROTOR, OIL PUMP
211 SLIDE, OIL PUMP

ROTOR SELECTION		VANE SELECTION		SLIDE SELECTION	
THICKNESS (mm)	THICKNESS (in.)	THICKNESS (mm)	THICKNESS (in.)	THICKNESS (mm)	THICKNESS (in.)
17.593 - 17.963	.7068 - .7072	17.943 - 17.961	.7064 - .7071	17.983 - 17.993	.7080 - .7084
17.963 - 17.973	.7072 - .7076	17.961 - 17.979	.7071 - .7078	17.993 - 18.003	.7084 - .7088
17.973 - 17.983	.7076 - .7080	17.979 - 17.997	.7078 - .7085	18.003 - 18.013	.7088 - .7092
17.983 - 17.993	.7080 - .7084			18.013 - 18.023	.7092 - .7096
17.993 - 18.003	.7084 - .7088				

CHART 19–1 Representative vane rotor and slide selection chart. (Courtesy of General Motors Corporation, Service Technology Group)

CLUTCH SERVICE

Clutch servicing requires some special attention and know-how. When a clutch unit is disassembled, it is a good practice to keep the clutch plates and related assembly parts in order. This provides guidance for the assembly sequence and is especially important when working with input housings that may contain two or three clutch units. There are parts that look similar but cannot be interchanged. What you have taken apart may not always match the pictorial view in a service manual.

Even keep the clutch plates, which will be discarded later, in the lineup. A typical clutch assembly is shown in Figure 19–26.

Clutch Unit Inspection and Reconditioning Guidelines

To prepare for inspection, clean all metal parts in mineral spirits and air dry them. Do not rinse or clean the composition plates if they are to be put back into service. The plate material must retain its oil soak.

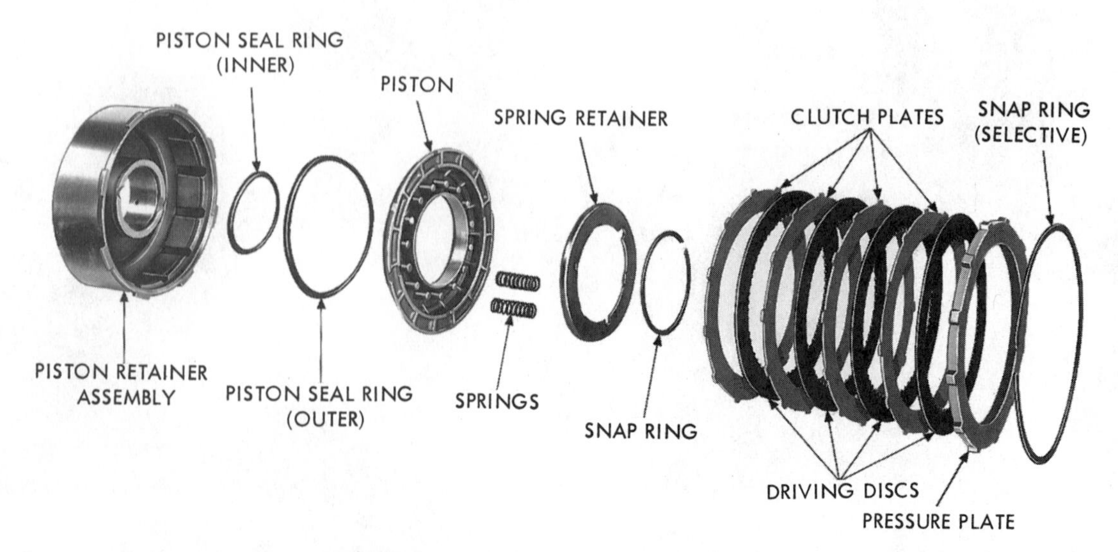

FIGURE 19–26 Front clutch assembly—TorqueFlite 36RH (A-727). (Courtesy of Chrysler Corp.)

FIGURE 19–27 Inspect oil ring grooves.

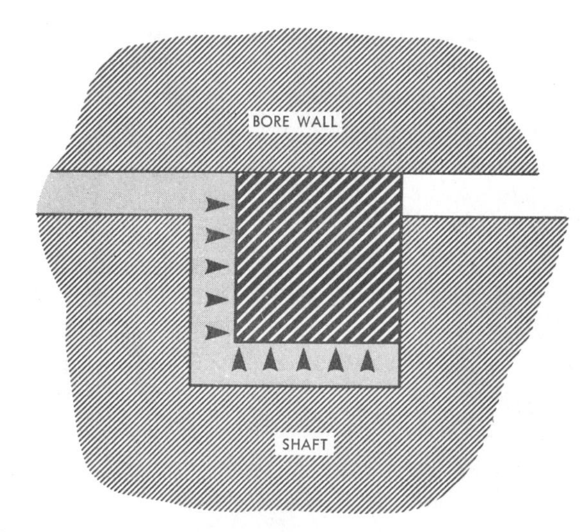

FIGURE 19–28 Metal rings depend on clutch apply oil to seal.

Inspect the drum housing bushing support for wear or scoring. The bushing typically is a serviceable part and can be replaced.

Inspect the bore and grooves in which the oil rings fit (Figure 19–27). Metal rings depend on the clutch apply oil to seal against the ring groove and the bore wall as illustrated in Figure 19–28. In a rotating drum application, the fluid force seats the ring tightly against the drum bore, and the ring rotates with the bore and drum. As can be observed in Figure 19–28, a sealing action also takes place between one side of the groove and the ring while the ring spins in the groove.

Contaminants in the fluid stream, especially metallic particles, can cause ring and groove wear. Excessive side clearance results in failure of the ring to seal (Figure 19–29). Also, the ring contact to the drum bore can develop a wear pattern. During vehicle acceleration and deceleration, the drum unit moves back and forth as allowed by the normal internal transmission/transaxle endplay. The polished ring spins in the bore and causes ring grooving and excessive leaking (Figure 19–30). A polished ring is compared to a new ring in Figure 19–31. Teflon rings are designed to perform in a manner similar to the metal ring. Hydraulic pressure underneath the ring forces it to expand and fit against the bore and ring groove. Even though there is less wear with a Teflon ring than a metal ring, bore and ring groove wear damage can occur.

In most cases, the oil ring bore wall shows no significant wear except for polish marks. To ensure that new metal rings will rotate with the drum, the polish surface should be removed from the bore wall with medium-coarse emery or sandpaper grit. For this hand operation, a piece of sponge can be used to work the emery or sandpaper in the bore (Figure 19–32). Glass bead blasting has also given good results.

A rotating clutch unit often has a check ball in either the drum housing or clutch piston (Figure 19–33). Clutch housing and piston check ball operation was discussed in Chapter 7. Where a piston has a dual apply chamber, the extra chamber may have an additional check ball in the piston (Figure 19–34).

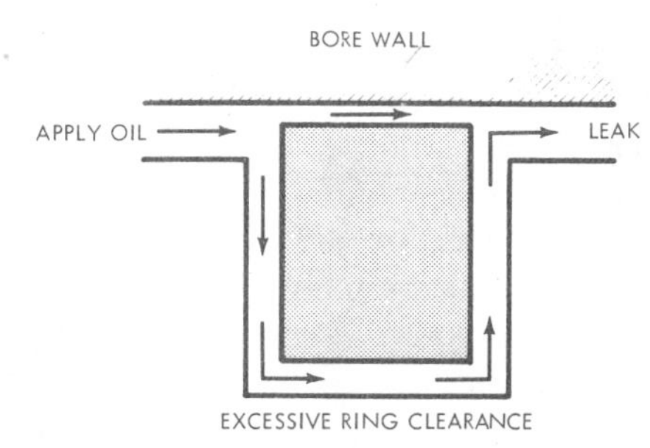

FIGURE 19–29 Excessive side clearance from wear produces poor sealing.

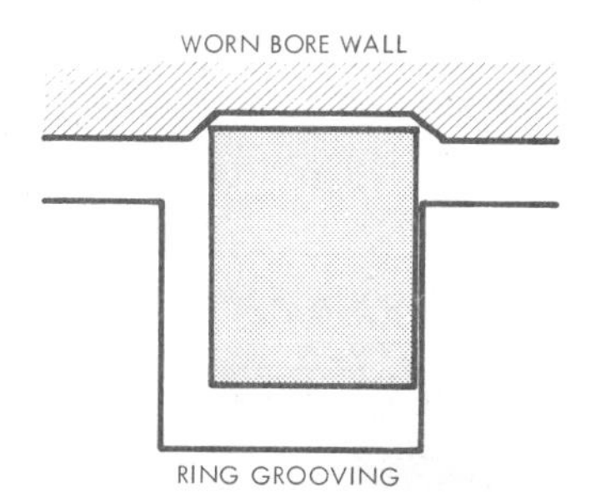

FIGURE 19–30 Ring grooving in clutch housing causes excessive leaking.

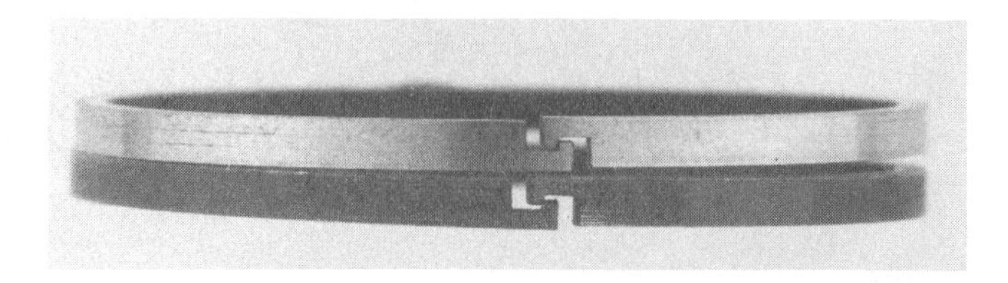

FIGURE 19–31 Top: Polished ring. Bottom: New ring.

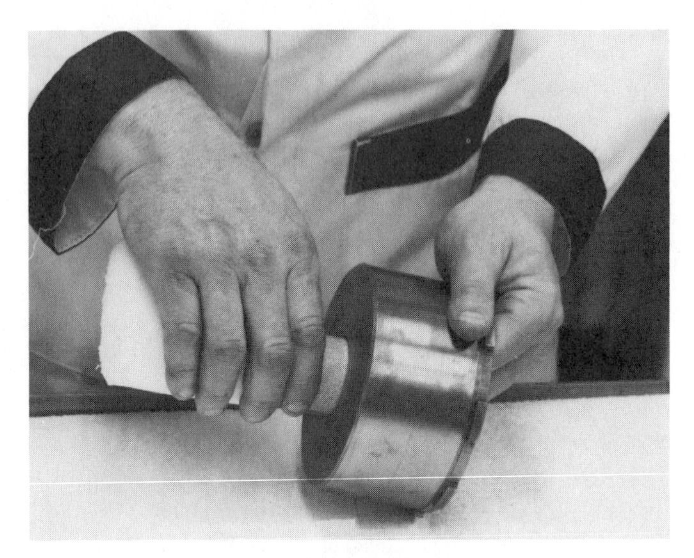

FIGURE 19–32 Use a sponge to work emery paper inside oil ring bore.

The check ball must be free to rattle. If the check ball area is sealed closed by severe varnish or metallic contamination, it must be cleaned out to allow free action of the check ball. Without proper check ball action, the clutch plates will fail again. Depending on location, the check ball can be pretested for leakage before clutch assembly by filling the clutch piston bore or piston cavity with fresh solvent. There should be no evidence of leakage.

FIGURE 19–33 Rotating clutch unit with check ball.

A rotating clutch housing that is not designed with a check ball may have a small fluid bleed hole in the housing or piston. Inspect bleed holes for unwanted blockage.

Inspect the piston bore for score marks. Light scores usually can be removed with crocus cloth. This leaves a polished bore surface. Emery cloth or abrasive paper produces a scratched surface finish that will quickly wear the piston seals.

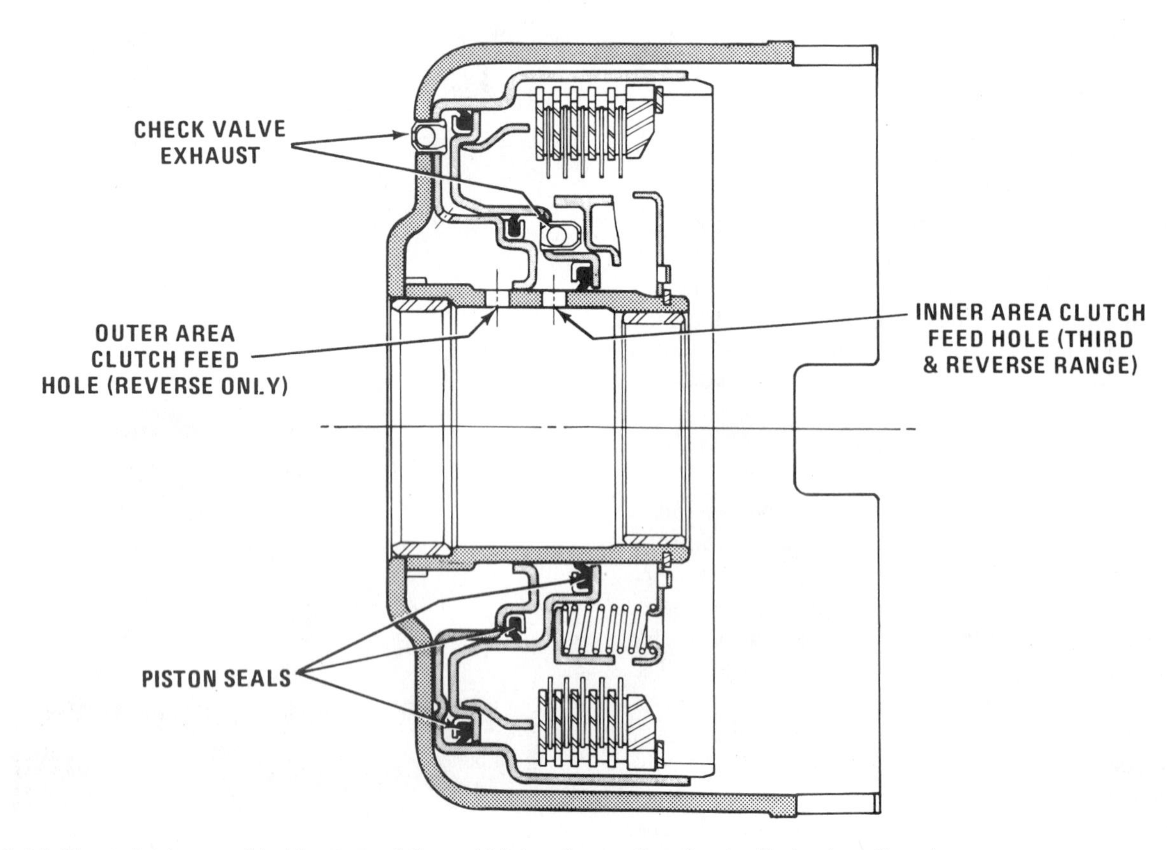

FIGURE 19–34 Direct clutch assembly. (Courtesy of General Motors Corporation, Service Technology Group)

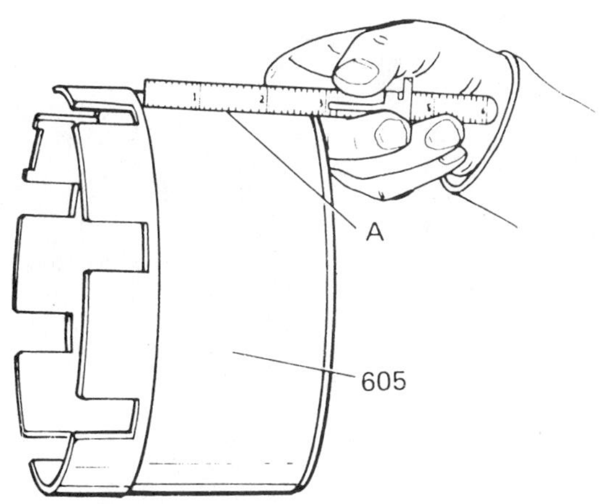

FIGURE 19–35 Checking reverse input housing for dishing—4L60 (700-R4). (Courtesy of General Motors Corporation, Service Technology Group)

Check the clutch drum band surface for dishing with a steel straightedge (Figure 19–35). This is extremely important in GM 4L60-E, 4L60 (700-R4) and Ford A4LD transmissions.

Inspect the piston return springs for distorted or collapsed coils. If the springs show visual evidence of being affected by clutch overheating, they must be replaced. To ensure correct piston movement in the cylinder bore, the entire spring assembly should be replaced. The piston return springs are an important factor in the clutch apply and release timing, especially in transmission shifts with a band.

If a single-disc (diaphragm or belleville) spring is used, look for finger wear, distortion, and hairline cracks (Figure 19–36). Occasionally, a weak single-disc spring may fold over and give the appearance that it was installed backward. Rather than reusing it, replace it.

Inspect the clutch hub and housing for spline and groove wear caused by the clutch plates (Figure 19–37).

Inspect the friction and steel clutch plates for possible reuse. When the plates are obviously worn, burned, warped, or cone-shaped, replacement is the only answer (Figure 19–38). Overhaul kits contain replacement plates for all the clutch units. Therefore, it is a common practice to recondition the clutch units with all new plates regardless of the reuse value of the original plates. New plates are good insurance for positive clutch action and a long-term overhaul.

When necessary, reuse of clutch plates is a legitimate service practice, but it does call for a careful evaluation of the plates. It

FIGURE 19–36 Examine disc spring for wear or damage.

FIGURE 19–37 Inspect clutch hub and housing.

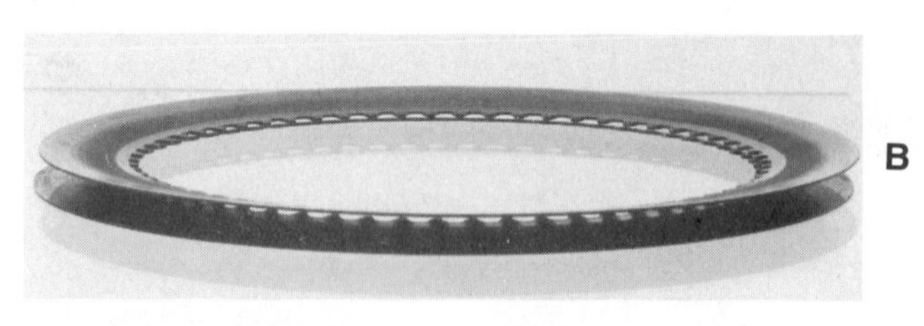

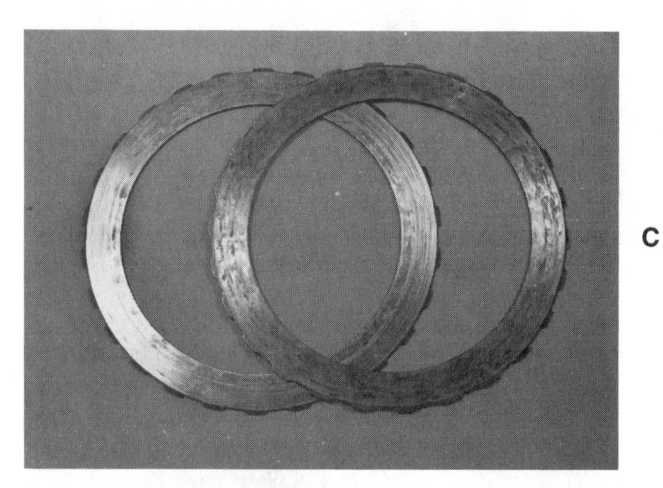

FIGURE 19–38 Reject clutch plates if you identify (a) worn and pitted friction plates, (b) warped friction plates, or (c) worn steel plates.

can become very time-consuming when the steel plates need reconditioning.

Reuse of Friction Plates. When operating in a proper oil and temperature environment, friction plates can give thousands of miles of service without showing significant wear. Even the printed trade numbers and letters may remain visible. Friction plates become oil and temperature darkened but can be reused if (1) the plates are flat, and not warped or cone-shaped; (2) the fiber material is firm and not pitted, flaked, glazed, or loose; and (3) when the plate facing is squeezed between the thumb and fingers, oil seeps out of the fiber. A glazed lining sometimes looks good but will not give up oil when squeezed. To reuse friction plates in a clutch unit, they should all be acceptable. Do not fuss with partial replacement. Decide whether they can all be reused, and if not, replace all of them.

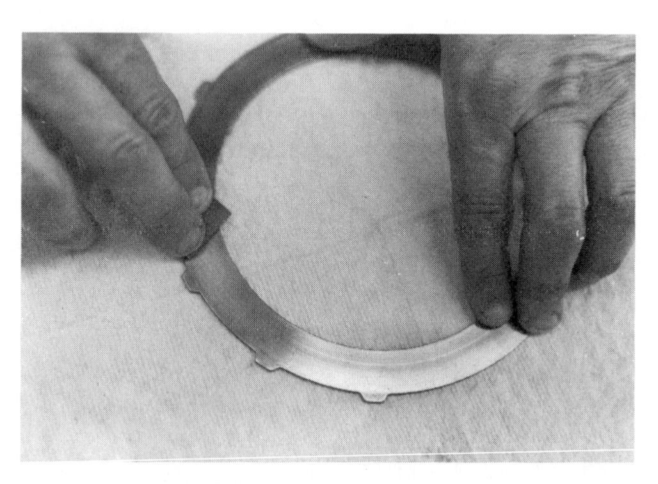

FIGURE 19–39 Recondition steel plates with abrasive paper.

Reuse of Steel Plates. If the friction plates pass inspection, the steel plates are usually salvageable. The only requirement is to replace the plates in the exact order of removal, which keeps the mating contacting surfaces matched for the right action with the friction plates. Steel plates can be reused if (1) they are flat, and not warped or cone-shaped; (2) they show no evidence of wear, scuffing, or grooving; and (3) they suffer from only minor hot or oxidation spots. Some clutch packs use a waved steel plate on top of the piston to cushion the clutch apply. Do not reject this waved plate because of warpage.

To recondition steel plates, use a medium-coarse emery or abrasive paper to lightly work the surface until all of the coating is removed (Figure 19–39). This provides a finish similar to that of new steel plates and ensures proper application and break-in of the clutch pack. The polished surface of used steel plates matched with new friction plates causes friction material glazing and drawn-out, lazy shifts. The clutch soon fails completely. Steel plate reconditioning is not recommended for the front (direct) clutch in TorqueFlite transmissions.

A study by Borg-Warner was made on the use of grit blasting, sand blasting, and glass bead blasting as salvage techniques on steel plates. Severe abrasion wear on the clutch pack resulted from the resurfacing when using the grit blast and sand blast. These two treatments are not recommended. The refinished surface produced by glass bead blasting at 90 psi gave satisfactory wear and friction results. Some transmission shops are already using this technique with excellent results. Glass beading does not remove any metal. It only cleans and reconditions the surface. It is critical that the plates be cleaned after the beading operation.

Clutch Unit Assembly Guidelines

The technician needs to address concerns such as clutch piston height, clutch plate thickness, required number of clutch plates, plate clearance, and mechanically correct assembly. When reassembling a clutch unit, it is important that the assem-

FIGURE 19–40 Inner clutch piston seal location.

FIGURE 19–41 Outer clutch piston seal location.

bly parts and service specifications be matched with the transmission/transaxle model code and engine application. This information is usually provided in the manufacturer's service manual.

Lip seals used on the piston must have the lip facing the pressure side. The inner and outer seals are always installed with the lips facing down into the piston cylinder (Figures 19–40 and 19–41). If the piston is designed with two apply chambers, a center seal is used on the cylinder sleeve hub. In a GM direct clutch housing, the lip must face up or away from the piston cylinder (Figure 19–42). The three-seal arrangement is shown in a sectional view in Figure 19–34.

The inside seal is sometimes located on the cylinder sleeve hub (Figure 19–43). Failure to replace the seal, especially if the clutch pack was burned, usually results in another clutch failure. Seals exposed to the heat generated in a burned clutch pack lose their elasticity, becoming hard and brittle, and no longer seal properly.

Prelubricate the seals with transmission fluid, TransJel, Vaseline, or a silicone wax. Silicone wax (Door Ease) works especially well on square-cut piston seals. The piston usually slips into the cylinder with light hand pressure. When using this method, the piston cylinder walls must be dry and free of transmission fluid (Figure 19–44).

When installing a piston with lip seals, it is sometimes necessary to use a smooth 0.010-in (0.254-mm) feeler gauge to work the lip edge into the cylinder (Figure 19–45). Be careful not to cut the seal, and do not force the installation. Otherwise, the seal lips will fold over and wedge. When the piston is ready, it should drop in the cylinder with ease. Use special seal protectors when recommended (Figure 19–46). This helps to prevent seal damage and makes piston installation easy. A piano-wire loop fixed in a holder is an excellent tool for helping install the piston and lip seal (Figure 19–47).

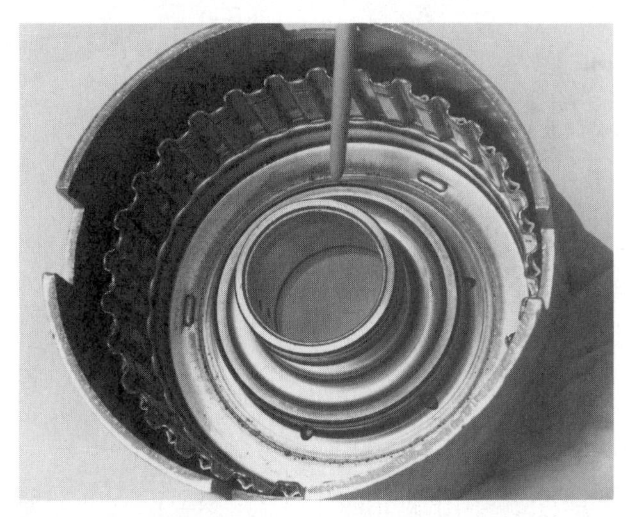

FIGURE 19–42 Direct clutch housing center piston seal location—GM 3T40 (125-C), 200, 350-C, and 200-R4.

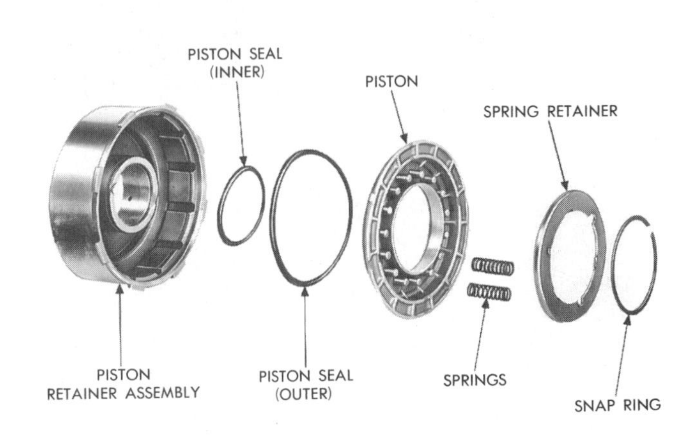

FIGURE 19–43 Front clutch housing assembly. (Courtesy of Chrysler Corp.)

Always install the proper number of piston return springs in exactly the required positions. Figure 19–48 shows a Ford C-6 ten-spring-design reverse-high clutch and a Chrysler TorqueFlite 36RH (A-727) thirteen-spring-design front clutch.

New friction plates must be presoaked in automatic transmission fluid for a minimum of thirty minutes and matched with new or reconditioned steel plates. The friction plate material

FIGURE 19–44 Prelubricate seals with silicone wax.

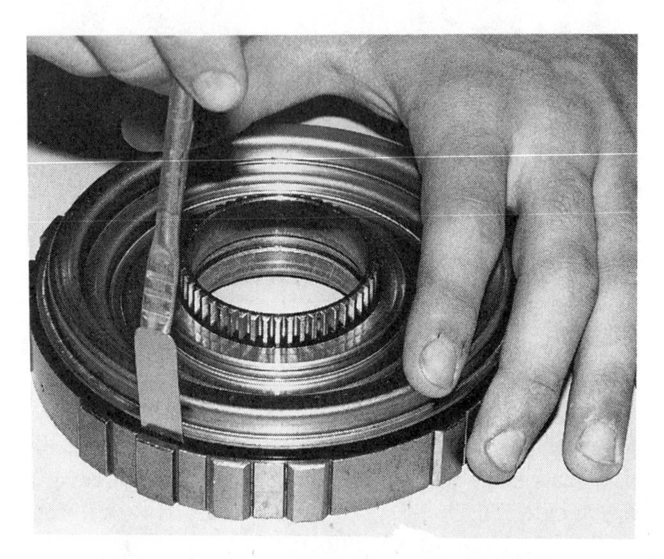

FIGURE 19–45 Use a seal installer or feeler gauge to work seal lip into clutch housing.

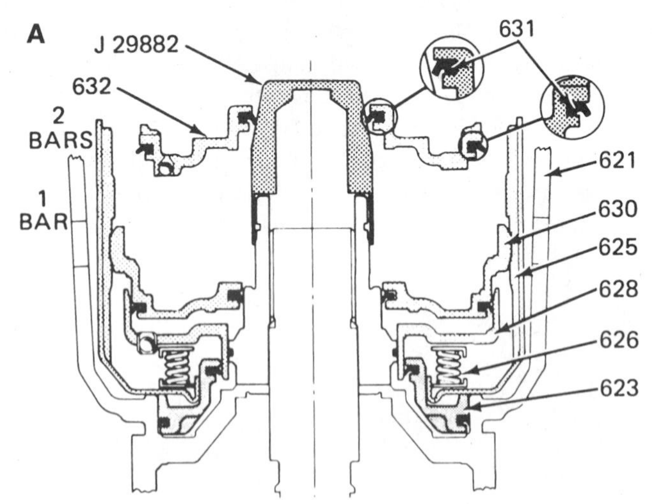

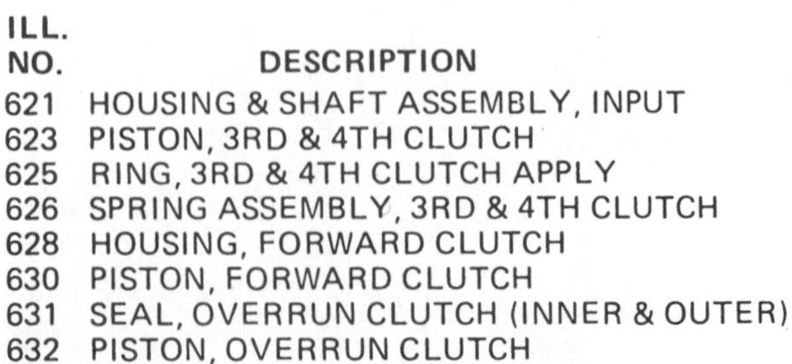

ILL.
NO. DESCRIPTION
621 HOUSING & SHAFT ASSEMBLY, INPUT
623 PISTON, 3RD & 4TH CLUTCH
625 RING, 3RD & 4TH CLUTCH APPLY
626 SPRING ASSEMBLY, 3RD & 4TH CLUTCH
628 HOUSING, FORWARD CLUTCH
630 PISTON, FORWARD CLUTCH
631 SEAL, OVERRUN CLUTCH (INNER & OUTER)
632 PISTON, OVERRUN CLUTCH

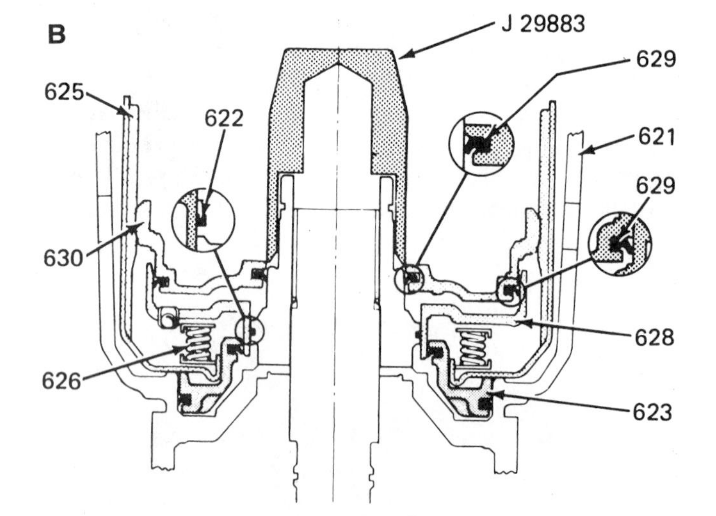

621 HOUSING & SHAFT ASSEMBLY, INPUT
622 SEAL, O-RING (INPUT/FWD. HSG.)
623 PISTON, 3RD & 4TH CLUTCH
625 RING, 3RD & 4TH CLUTCH APPLY
626 SPRING ASSEMBLY, 3RD & 4TH CLUTCH
628 HOUSING, FORWARD CLUTCH
629 SEALS, FWD. CLUTCH (INNER & OUTER)
630 PISTON, FORWARD CLUTCH

FIGURE 19–46 4L60 (700-R4): (a) installation of forward clutch piston using seal protector; (b) installation of overrun clutch piston using seal protector. (Courtesy of General Motors Corporation, Service Technology Group)

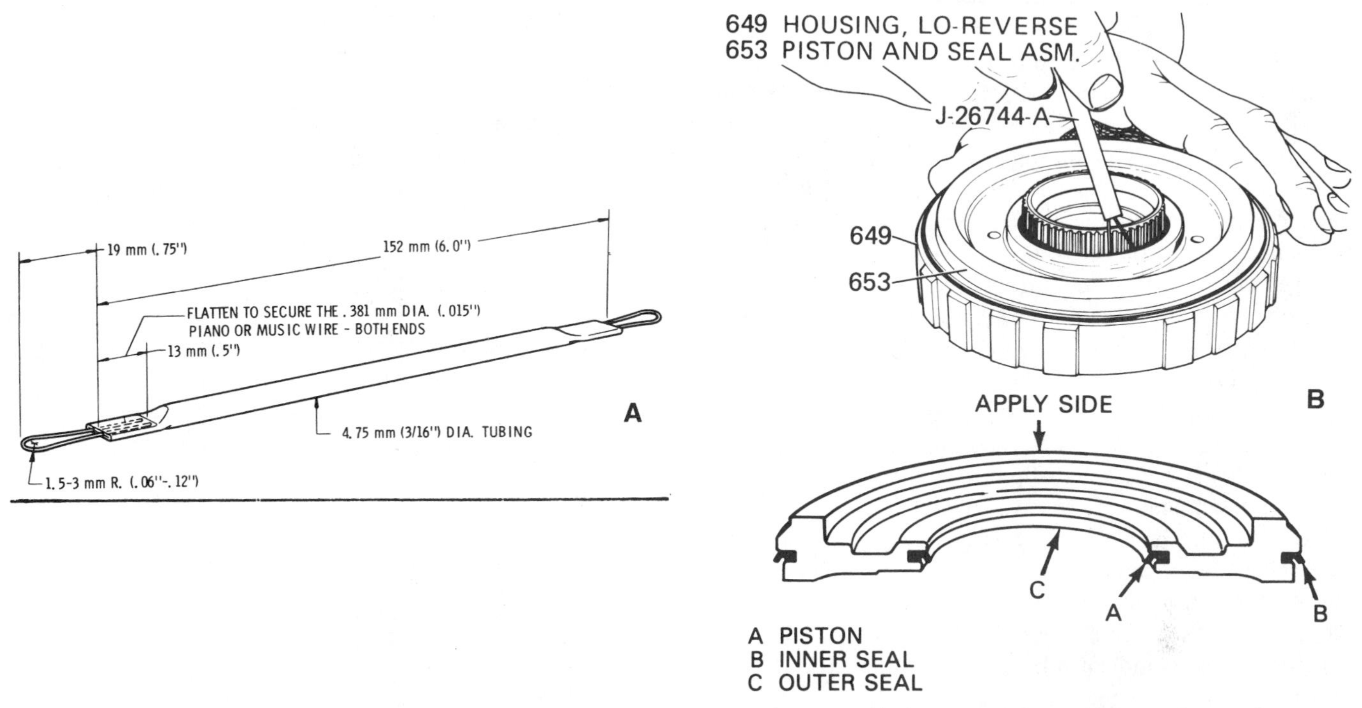

FIGURE 19–47 (a) Wire loop seal installer; (b) installing low and reverse piston—3T40 (125-C). (Courtesy of General Motors Corporation, Service Technology Group)

must be saturated with oil to prevent glazing or burning during break-in. It requires 10 to 15 mi (16 to 24 km) of driving before the oil mist surrounding the clutch pack can fully soak the friction plates. When the clutch is applied, the oil squeezed out of the friction lining carries away the heat. Friction plates applied dry or with limited prelubrication can fail within a few miles after an overhaul, especially if abusive shifts are the customer's immediate criterion for testing. Some of the new packaged friction plates come prelubricated, but it is still a good practice to soak all friction plates.

Be sure that the correct number of friction and steel plates are installed. There may be a difference in the friction and steel plate thickness between the forward and direct clutch units. As a rule, the thinner plates, whether friction or steel, go in the

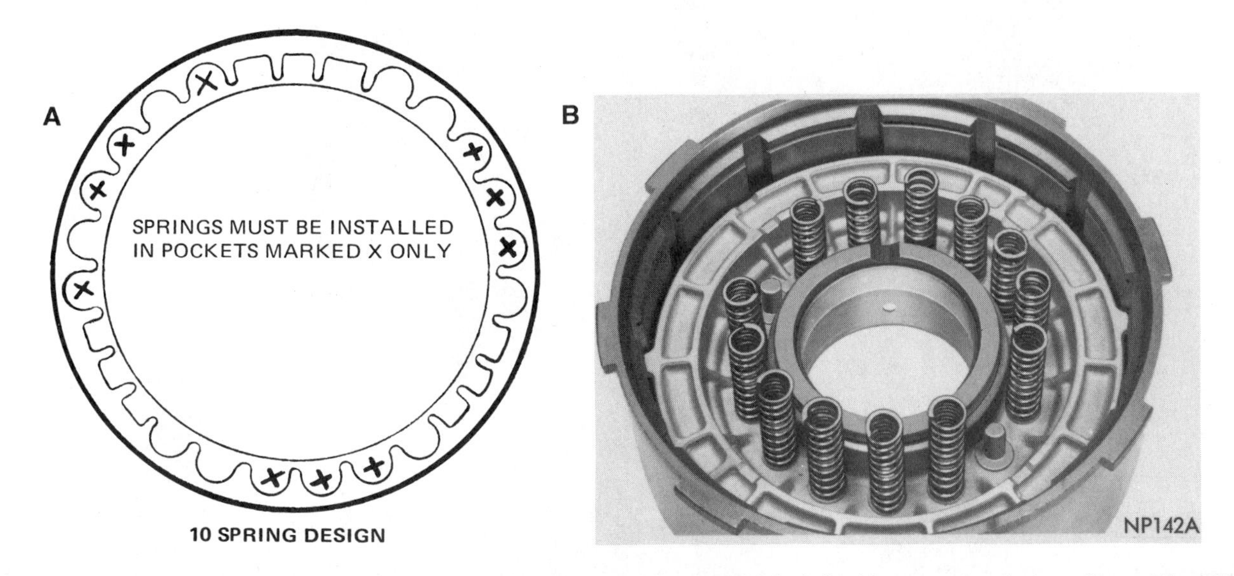

FIGURE 19–48 (a) Ten-spring design—Ford C-6 (Courtesy of Ford Parts and Service Division); (b) thirteen-spring design—TorqueFlite 36RH (A-727). (Courtesy of Chrysler Corp.)

forward clutch. The manufacturer normally provides the necessary clutch plate usage information for the clutch units. An example is given in Chart 19–2. Note that the forward clutch steel flat plates are thinner than the direct clutch steel flat plates. They do interchange, but the clutch pack clearances will never be right, and the clutches are doomed to failure. It does not happen very often, but clutch plate thickness can get changed from the previous model year. Always check for a possible change in order to stay out of trouble.

Each clutch assembly must be checked for plate clearance. Correct clearance cannot be established if (1) the clutch plate count is not right, (2) the friction or steel plate thickness is not right, (3) the apply ring width on steel clutch piston applications is not right (refer to Chart 19–2 and Figure 19–49), or (4) the clutch is not mechanically assembled properly.

Some clutch units do not have clearance specification provided. If the clutch unit has been assembled properly, it should have acceptable clearance. It is a good practice to observe if the clutch pack is too tight or too loose, which would indicate an error. A tight clearance is sensed by a heavy clutch pack drag, while looseness is indicated by a gap of 5/32 to 1/4 in (4 to 6 mm) or more between the clutch pack and pressure plate. In the absence of specifications, allow a minimum 0.010 in (0.25-mm) clearance for each friction plate.

When the service requirement calls for measuring the clutch pack clearance, the check is usually made with a feeler gauge between the snap ring and pressure plate, or between the

3T40 (125C) CLUTCH PLATE AND APPLY RING USAGE CHART

CLUTCH	FLAT STEEL PLATE		COMP. FACED PLATE	WAVED PLATE		APPLY RING	
DIRECT	No.	Thick-ness	No.	No.	Thick-ness	I.D.	Thick-ness
CDC,CJC,CPC,CTC CUC,CXC,HLC, HRC,PMC,PPC	5	2.3mm (0.09")	5	–	––	7	19.0mm (0.75")
JFC,JKC,JXC	3	2.3mm (0.09")	3	–	––	2	27.4mm (1.08")
ALL OTHERS	4	2.3mm (0.09")	4	–	––	1	23.1mm (0.91")
FORWARD ALL	3	1.9mm (0.07")	4	2	1.25mm (0.05")	–	––
LO & REVERSE ALL	4	2.2mm (0.09")	5	2	1.94mm (0.08")	–	––

The direct and forward clutch flat steel clutch plates and the forward clutch waved steel plate should be identified by their thickness.

The direct and forward production installed composition-faced clutch plates must not be interchanged. For service, direct and forward clutch use the same compositioned-faced plates.

The forward clutch backing plate is selective. Refer to the Forward Clutch End Play Chart.

Measure the width of the clutch apply ring for positive identification.

Chart 19–2 3T40 (125-C) clutch plate and apply ring usage chart. (Courtesy of General Motors Corporation, Service Technology Group)

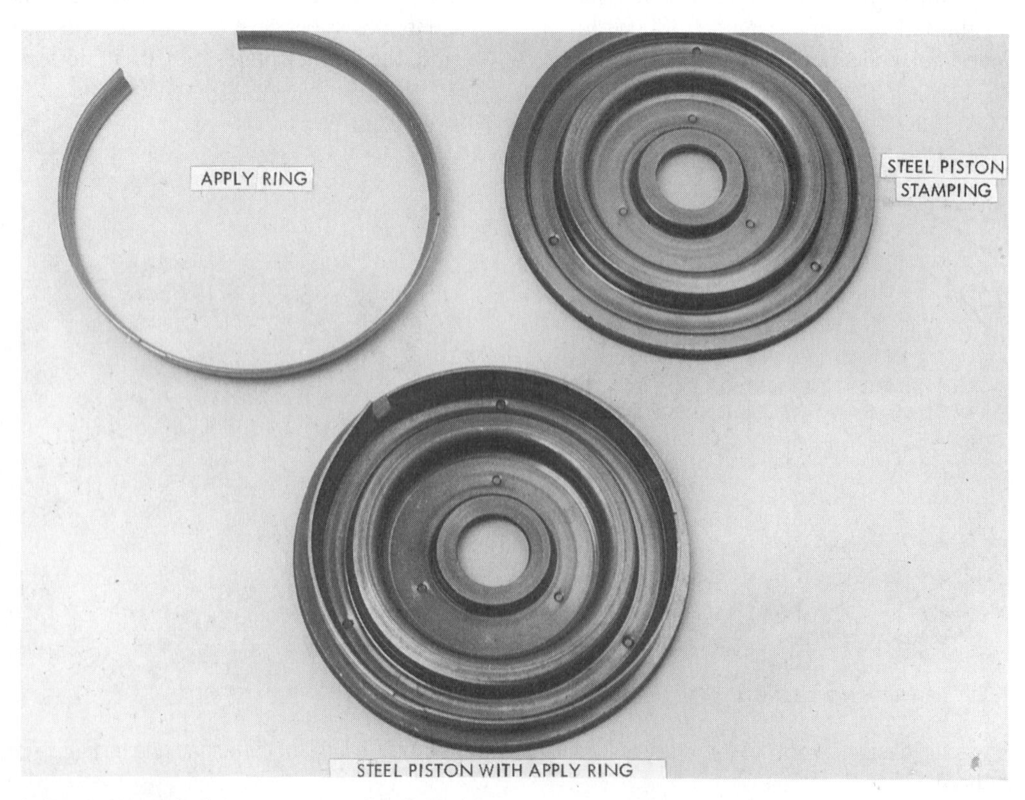

FIGURE 19–49 Steel piston and apply ring, separate and assembled.

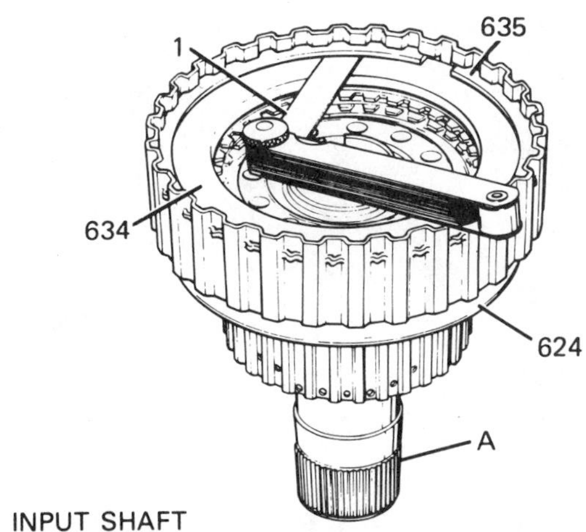

A INPUT SHAFT
1 FEELER GAGE 1.0-1.5mm (.04"-.06")
624 HOUSING ASSEMBLY, FORWARD CLUTCH
634 PLATE, FORWARD CL. BACKING (SELECTIVE)
635 RING, SNAP

BACKING PLATE THICKNESS		IDENTIFICATION CODE	
MM	Inches	Steel	Powdered Metal
5.0 - 4.9	.197 - .191	A	6
4.5 - 4.3	.175 - .170	B	7
3.9 - 3.8	.154 - .148	C	8
3.3 - 3.2	.132 - .126	D	9

FIGURE 19–50 Measuring 3T40 (125-C) forward clutch plate clearance; selective backing plate chart. (Courtesy of General Motors Corporation, Service Technology Group)

pressure plate and top of the clutch pack (Figures 19–50 to 19–52). Another technique to obtain a reading is to use a dial indicator with the plunger positioned on top of the pressure plate. Apply air, 35 psi (242 kPa) for most clutch assemblies, to the housing while mounted on its support (Figure 19–53). In place of the air, the technician can lift the pressure plate with a shop-manufactured hook and rod from a welding rod. Depending on the transmission, the clearance is adjusted with a selective snap ring or pressure/backing plate.

FIGURE 19–52 Measuring clutch plate clearance between pressure plate and clutch pack.

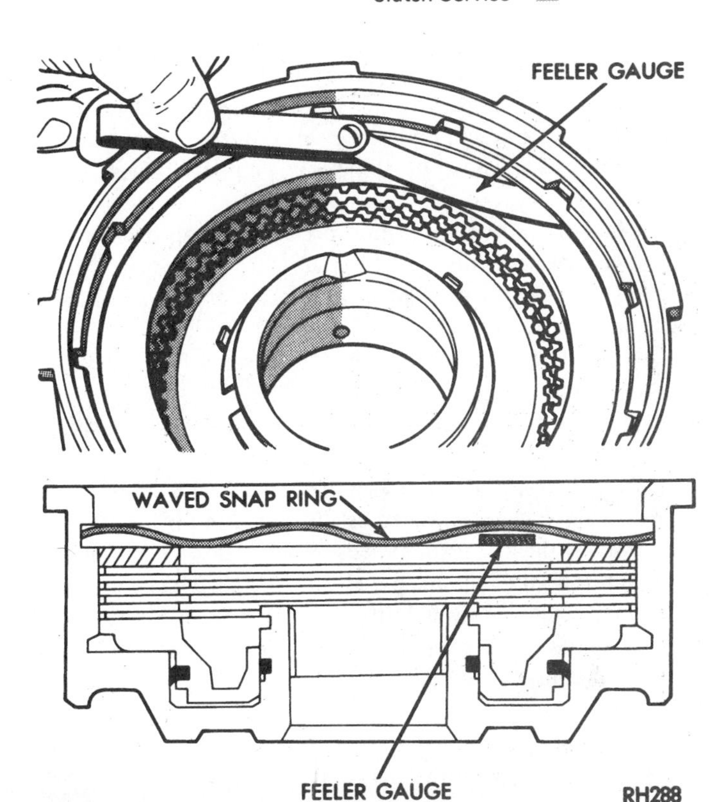

FIGURE 19–51 Measuring front clutch plate clearance, designed with a waved snap ring. (Courtesy of Chrysler Corp.)

Always set the clutch pack clearance to the minimum side of the given specifications, but never adjust it below the minimum. Adjusting the clearance to the low end shortens the piston stroke for longer seal life and crisper shift quality.

Many clearance tolerances are large, and some technicians feel that this can pose problems. For example, suppose the forward clutch pack clearance on a particular transmission calls

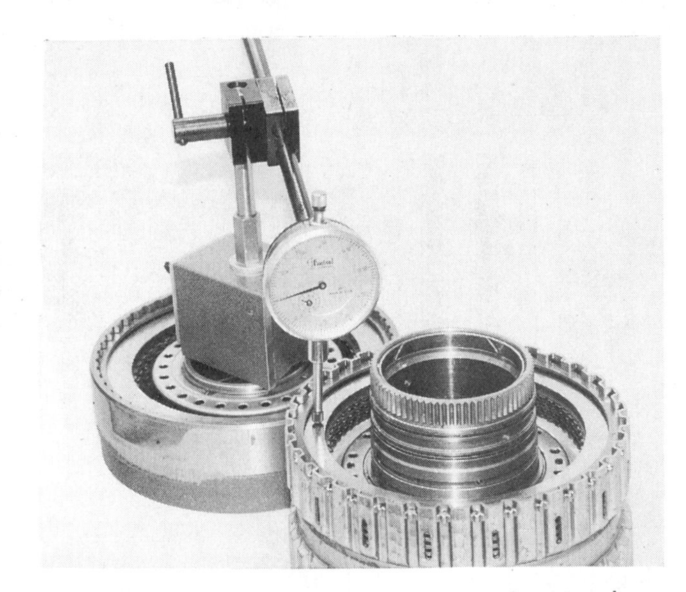

FIGURE 19–53 Magnetic base dial indicator setup for measuring clutch plate clearance.

for 0.010 to 0.080 in (0.25 to 2 mm), but selective pressure plates or snap rings may not provide the desired minimum specification the technician hopes to produce. A custom adjustment can be made by one of the following techniques:

1. Add an extra composition plate and install back to back with another composition plate.
2. Add an extra steel plate and install back to back with another steel plate.
3. Use extra thick steel plates as necessary for replacement when available.
4. Use one or two composition plates shaved on one side.

The friction plates are easy to shave in the shop (Figure 19–54) and can be arranged as illustrated in Figure 19–55 to achieve the desired results. Because the plates are mounted on the same splined hub, the metal sides can face one another.

Finally, check the reliability of the reconditioned clutch with an air check (Figure 19–56). This ensures that (1) the piston seals were not damaged or installed incorrectly during assem-

FIGURE 19–54 Shave composition plates for custom adjustment.

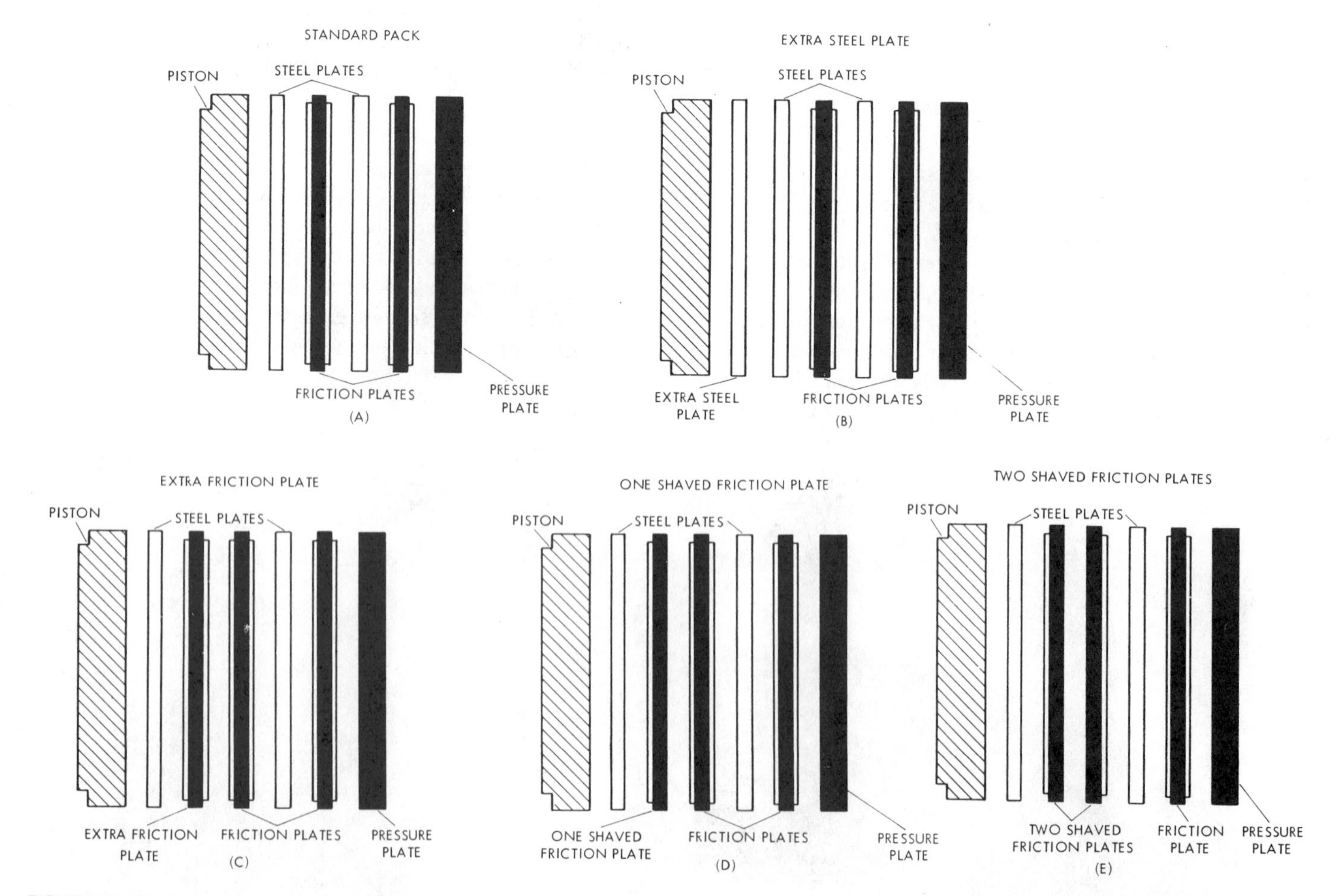

FIGURE 19–55 Sample custom plate arrangements.

bly, (2) the oil rings are right, (3) the check balls are seating, and (4) mechanically, the piston will apply and release. When air checking some clutch units, it is necessary to block certain holes while feeding line air pressure into others. In Figure 19–57, the technician is required to block the clutch feed hole while charging the clutch cylinder with apply air through the check ball. When the air nozzle is removed and with the feed hole still blocked by the finger, the clutch should remain applied. This is an excellent test on the check ball to see if it leaks. If the check ball leaks, the clutch cannot maintain apply pressure and will fail again.

When possible, check the clutch units on the pump assembly or sprocket support (Figure 19–58). This gives an added check to ensure that the clutch oil circuits are open and all the cup/ball plugs are in place. Restricted circuits can be detected by giving attention to the response time of the clutch apply. On pump assemblies with a cover to body plate, it is advisable to torque several nuts, bolts, and flat washers in the pump to case bolt holes (Figure 19–59). This helps to keep the oil passages tight and gives a better air check.

For effective air checking, the line pressure needs to be regulated down to 35 psi (242 kPa) (Figure 19–60). Higher pressure overcomes excessive sealing leaks and gives a false indication that the clutch unit and circuit are okay. Normal air leakage past oil rings and bushings will be audible during the test. By checking good clutch units, the technician can establish what is normal, and then sense the aggressiveness of the pistons applying and detect which ones are satisfactory. Even though a faulty clutch pack may apply and release, the amount of air leakage is pronounced, and the clutch piston action is sluggish.

The air check is especially useful in checking an input housing for misassembly where two to three clutch units are stacked into the housing. Following is a discussion concerning problems with a Hydra-Matic 4T60 (440-T4) input housing and shaft assembly, featured in Figures 19–61 and 19–62.

FIGURE 19–56 Air check reconditioned clutch assembly.

- *Third clutch snap ring (642) to the back plate missing.* The third clutch overstrokes and applies both the third clutch and the input clutch. This will cause a power flow lockup in fourth overdrive when the input clutch should be released.
- *Third clutch snap ring (642) and input clutch snap rings (649) are reversed.* The wider (thicker) input clutch snap ring will interfere with the input clutch piston stroke and prevent clutch engagement. The transaxle will not have forward or reverse gears.
- *Input clutch apply plate (646) is reversed.* The undercut side of the plate must face the snap ring. Otherwise, the

FIGURE 19–57 Block clutch feed hole while air checking.

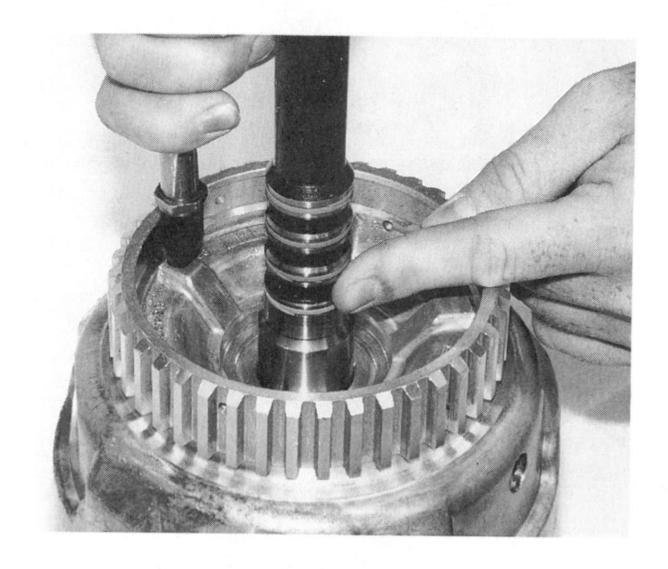

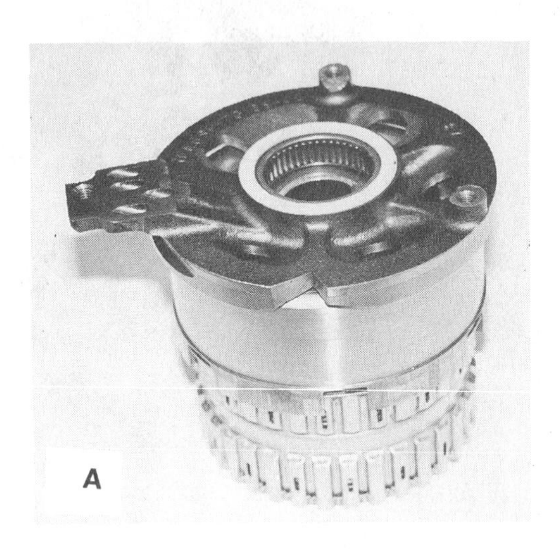

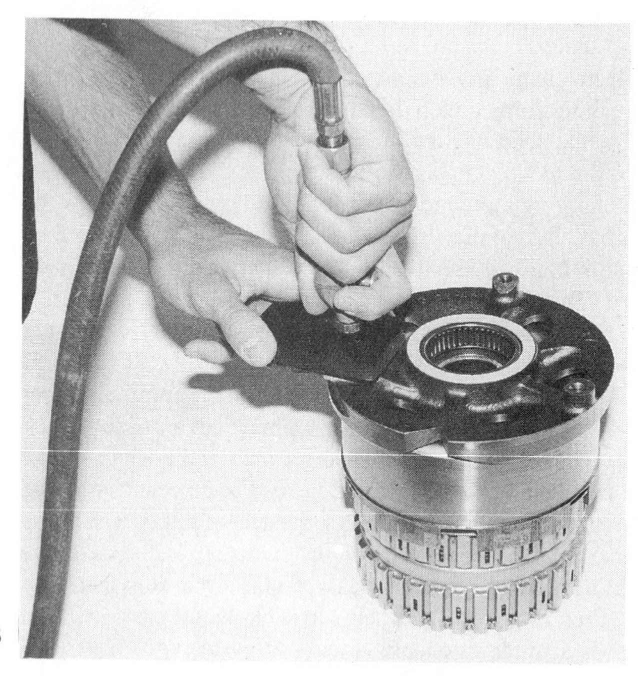

FIGURE 19–58 AXOD and AX4S (AXODE): (a) clutch units mounted on sprocket support, and (b) technician using rubber block with air feed hole to test clutch units.

FIGURE 19–59 Fasten several pump to case bolt holes before air check.

FIGURE 19–60 Air check clutch units with regulated air pressure; 35 psi recommended for most assemblies.

FIGURE 19–61 Input clutch housing—4T60 (440-T4).

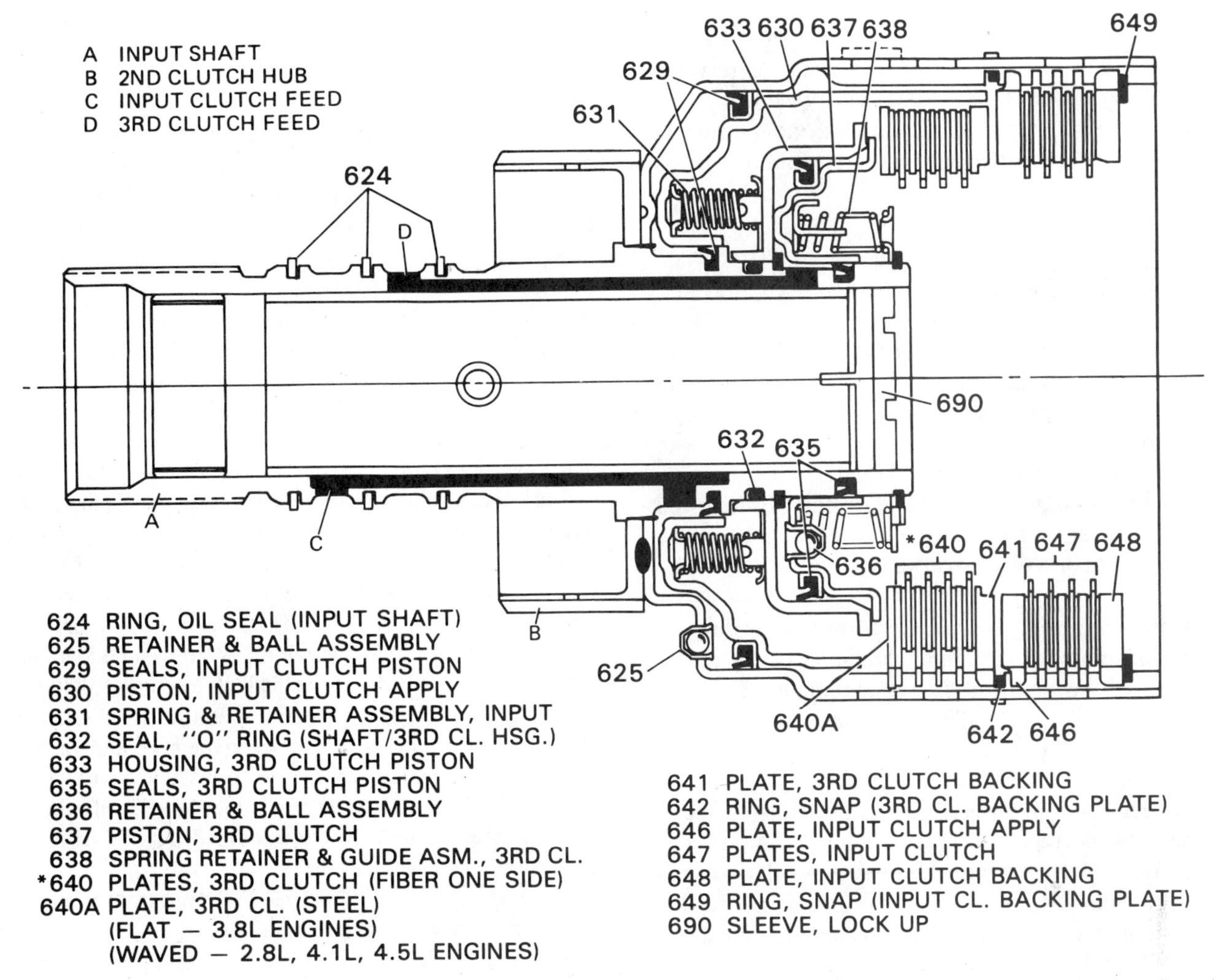

FIGURE 19–62 Input clutch housing cross-section—4T60 (440-T4). (Courtesy of General Motors Corporation, Service Technology Group)

input clutch clearance is greatly reduced. This usually causes a heavy clutch plate drag and a "bind-up" condition in fourth gear.

PLANETARY CARRIER SERVICE

When examining the planetary units, pay attention to planet carriers. Pinion gear wear can cause objectionable whining noise when the planetary power flow is functioning in reduction or overdrive. Make it a practice to closely inspect the pinion gears for gear damage and loose or worn pins. Roll the gears over and rock the gears by hand to assist in gear or pin evaluation. Do not use shop air to spin-up the gears. Pinion gear endplay is another factor that can cause excessive gear noise.

When inspecting the carrier unit, be sure to measure the planet pinion gear endplay with a feeler gauge and compare it to the manufacturer's specifications (Figure 19–63). In the absence of specifications, use 0.009 to 0.024 in (0.24 to 0.64 mm). In most cases, excessive endplay is obvious.

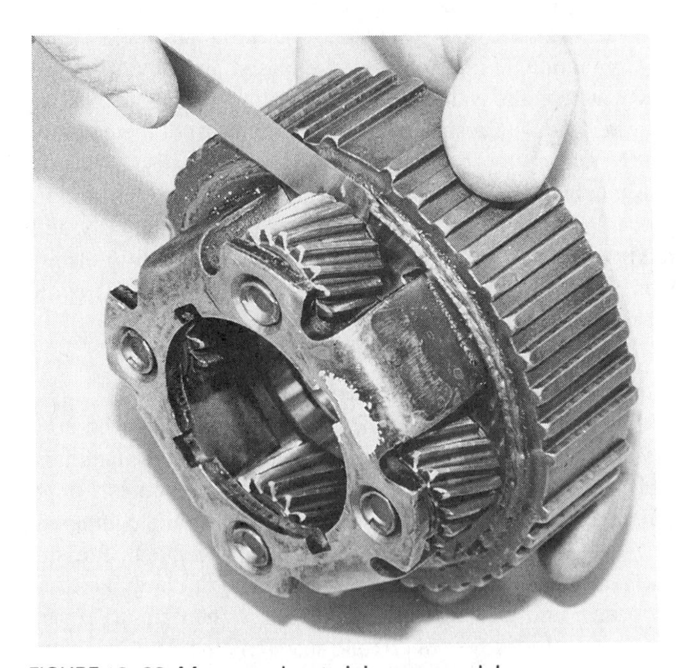

FIGURE 19–63 Measure planet pinion gear endplay.

NOTE: Some carriers are rebuildable, while others are not. Rebuildable carriers with too much endplay may be serviced by disassembling and installing new steel or bronze thrust washers. For carriers that cannot be taken apart, the aftermarket provides shims that can be locked and tabbed into location beneath the pinion gears.

VALVE BODY SERVICE

The time to eliminate stuck valves or the potential for valve sticking is when the transmission is still on the bench. Cleaning the valve body to ensure trouble-free valve movement is an important part of the overhaul/rebuild, demanding extra clean handling and attention to detail. While disassembling, cleaning, and reassembling, an organized procedure needs to be followed to produce dependable results. This section highlights service to cast iron valve bodies using steel valves.

NOTE: Valve body service procedures vary, depending on whether the assembly is cast iron or aluminum. Many manufacturers currently recommend not servicing their aluminum valve body assemblies, but rather, replacing them if faulty. Follow service manual procedures.

Valve bodies that are severely contaminated with metal particles usually are replaced. Reconditioning a metal-contaminated unit is often an impossible task, particularly if it is aluminum. It is a time-consuming and not very cost-effective procedure that may not produce reliable results. You are never sure that it will be one hundred percent effective.

Valve bodies that are contaminated with minor deposits of metal particles or severe deposits of sludge or varnish can be overhauled with reliable results. Some valve bodies may look like a disaster but still be serviceable. The best practice is to always overhaul the valve body, even when the valve body looks clean.

Contaminants are not always exposed and obvious, especially where the valves are riding in a valve bore bushing (Figure 19–64). Fresh ATF can flush out most of the contaminants, but any loose particles remaining can cause a valve to jam in a bushing. This does not happen often, though, and the process of properly overhauling a valve body assembly definitely improves the reliability of the freshly rebuilt transmission. The following guidelines cover general shop practices and service tips for the successful overhaul.

Valve Body Overhaul Guidelines

For a successful overhaul, the valve body must be completely disassembled to properly clean out the contaminants and ensure that all the valves move freely in their bores. For job efficiency, clean and organize your bench area and equipment.

For cleaning, use a fresh mineral spirits solvent. Pressure washers with water-based detergents spray a combination of soap and fine particles at the castings. When air drying the castings, a sludge film of dirt and soap remains. This will cause heavy valve dragging and/or sticking valves during reassembly. The valve castings need to be flushed in clean solvent to remove the film. Cleaning the valve body castings and valves the old-fashioned way produces affordable and excellent results.

When cleaning valve body parts, do not soak any rubber check balls in mineral spirits or other solvents. The solvent penetrates and distorts the rubber and diminishes the sealing capability of the check ball. Rubber check balls can get worn and still look as though they are correct. If there appears to be one smaller-sized ball in the group, it probably has been worn down in size and needs to be replaced. Remember to inspect the separator plate for damage at the check ball seats.

Presoak the valve body in mineral spirits and clean the exterior with a brush (Figure 19–65). Using the shop manual, disassemble the valve body over a large parts tray filled with mineral spirits. The tray can catch the parts as they are removed. Otherwise, the springs, check balls, and valves can easily roll off the bench and get lost (Figure 19–66). These parts are not normally replaceable items. Be sure to remove rubber check balls from the solvent. Valves that are severely coated with varnish need to be sprayed with carburetor cleaner to free them from their bores (Figure 19–67).

Valves that are stuck in their bores from embedded metal or scratches are sometimes very difficult to remove, but it can be done. Scratches are caused by fine metallic particles in the fluid stream that work their way between the bore and valve. Each scratch on the valve or the bore forms a groove with flared edges, which results in wedging the valve in the bore. The sticking effect of scratches and embedded metal on the valves is illustrated in Figure 19–68.

In most cases, wedged valves can be removed from the bore by carefully prying the spool valves with a miniature screwdriver blade or dental pick (Figure 19–69). Using sharp, light taps with a plastic hammer on the end of the bore casting also may be successful (Figure 19–70). Should these two techniques fail, the following shock treatment can be used to overcome the interference of the flared grooves and embedded particles:

1. Place a clean shop towel on a flat bench surface. With the flat or passage side facing down, slam the valve body on the flat surface several times (Figure 19–71).
2. After the bench slam operation, place a screwdriver blade in the grooved section of the valve. Rap the screwdriver from several positions, top and sides of the valve, with an open-end wrench. The use of a hammer in place of an open-end wrench may produce too much force and bend the valve. Figure 19–72 illustrates the effect of the procedure on the flared edges and metal particles.

Keep each valve bore assembly separated and aligned (Figure 19–73). The photograph shows the valve trains on a paper towel. A better method of keeping the valve trains in order and lint-free is to place them in the grating of an old refrigerator shelf rack. Using a wire-type rack will keep the valves separated and prevent them from rolling away. Use the manufacturer's pictorial view in the shop manual as a guideline. This practice prevents accidental interchange of look-alike springs and valves. Properly sorting and organizing the parts also saves time during reassembly.

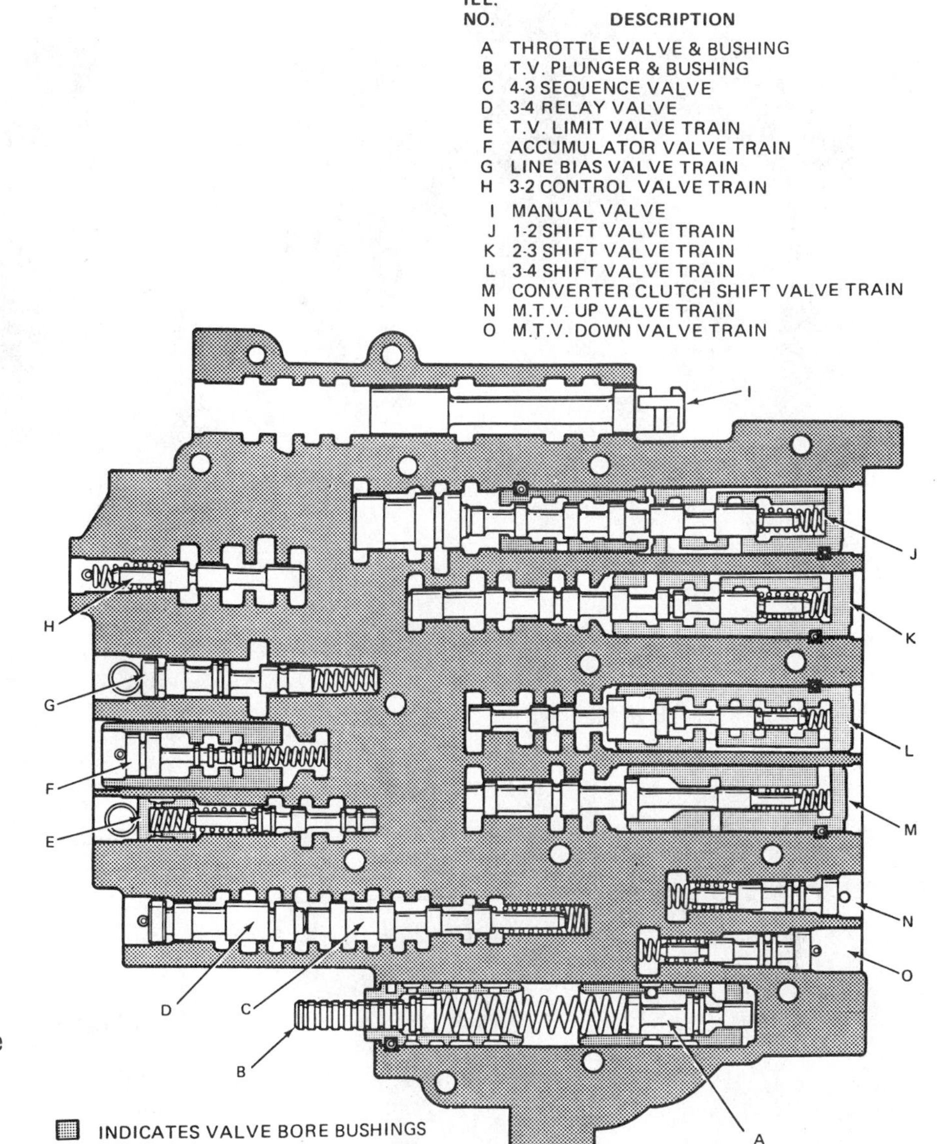

FIGURE 19–64 Valve body assembly with valve bore bushings—typical 4L60 (700-R4). (Courtesy of General Motors Corporation, Service Technology Group)

FIGURE 19–65 Soak the valve body in mineral spirits and clean with a brush.

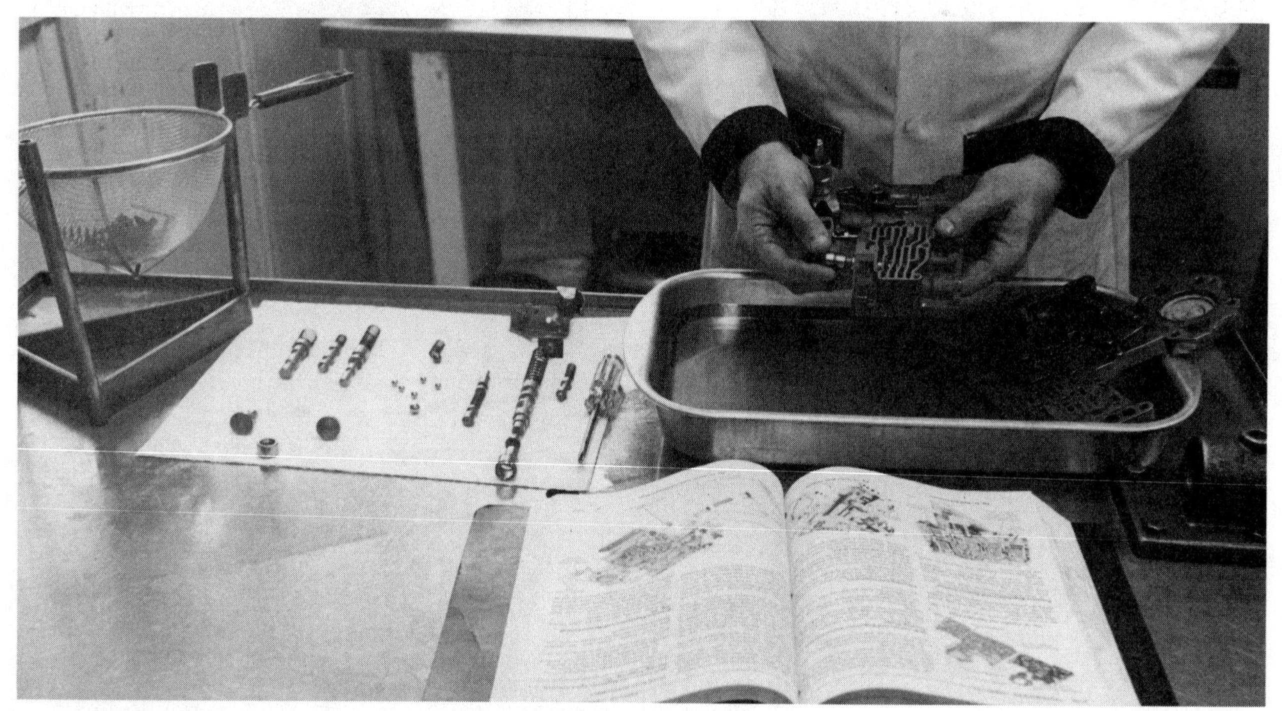

FIGURE 19–66 Disassemble over a tray.

FIGURE 19–67 Spray varnished valves with carburetor cleaner.

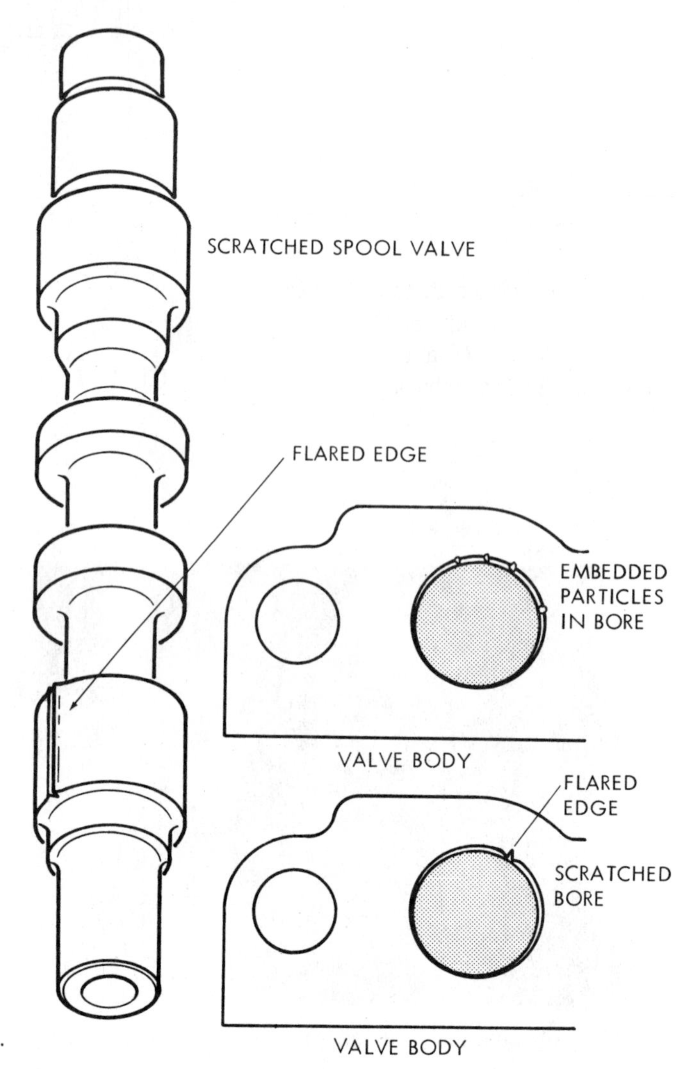

FIGURE 19–68 Sticking effect of scratches and embedded particles.

FIGURE 19–69 (a) TorqueFlite valve body; (b) 4T60 (440-T4) valve body.

FIGURE 19–70 Light taps with a plastic hammer may loosen a wedged valve.

FIGURE 19–71 Strike valve body on a towel-covered flat surface to loosen valve.

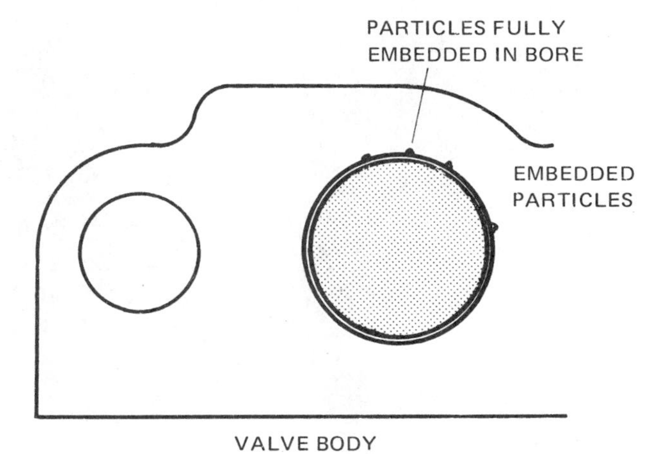

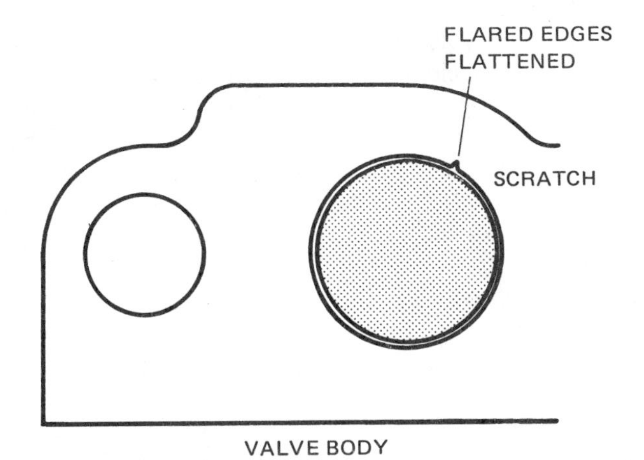

FIGURE 19–72 Impact effect on valve body bore to remove embedded particles or scratches.

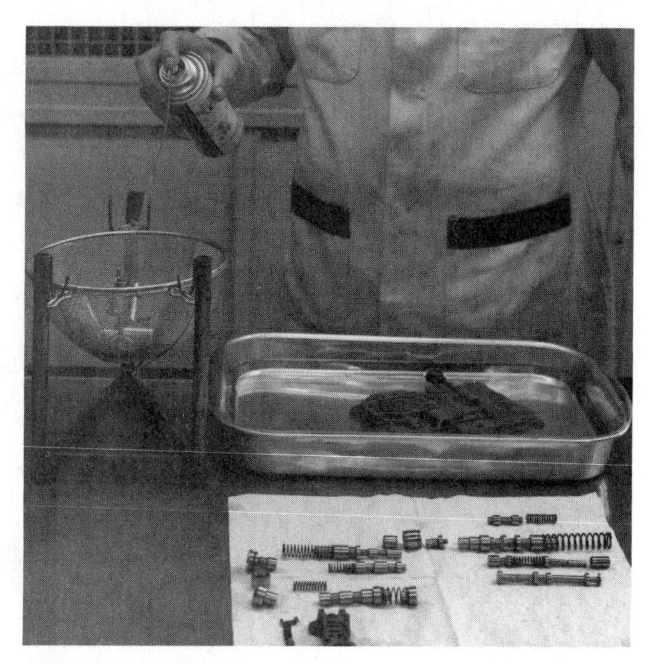

FIGURE 19–73 Keep the valve train components aligned. Flush with carburetor spray, followed by mineral spirits. Wet valves with ATF prior to installation.

The valve body castings and valve train assembly parts normally can be flushed and cleaned with carburetor spray, followed by a clean mineral spirit flush and air drying (Figure 19–73).

For final reassembly preparation, test the valves that might appear to be a problem by checking for free movement in their bores under dry conditions. Simply insert each suspect valve in its bore. The valve should move freely in its bore as the valve body is tipped back and forth (Figure 19–74).

Valve bores and valves that have scratches or embedded metal particles can be reconditioned to size in several ways. For many bore diameters, a rolled piece of crocus cloth can be inserted and rotated. By rotating in a direction that unwinds the roll, the crocus cloth hugs the bore and polishes it to size (Figure 19–75). For small bores, the crocus cloth can be rigged on the end of a wooden dowel.

Crocus cloth also can be used to recondition a valve. Place a sheet of crocus cloth on a flat plate surface, such as automobile safety plate glass. Rotate the valve as you scrub it across the cloth. Valve nicks and scratches also can be dressed down with a fine honing stone. These techniques are illustrated in Figures 19–76 and 19–77. Do not round off the sharp edges of the valve ends. The razor sharp edges are needed to cut away any contaminants in the bore that might seize the valve. A rounded edge forms a pocket for the contaminants to gather and the valve easily seizes.

Another technique that can produce good results is the use of fine car body polishing compound mixed with transmission fluid and graphite, such as Dri-Slide. Apply a thin coat of the mix on the valve lands, and insert the valve in the bore (Figure 19–78). Be careful not to use excessive force.

FIGURE 19–74 Check valves for free movement.

If the valve is jammed, it is probably caused by a heavy coating of the compound mix. Correct the problem, and start over. Once in the bore, work the valve back and forth and rotate it. In small bores, a miniature screwdriver and wooden dowel can be used to stroke the valve. Large valves usually pose no problem, and in some cases, a rubber hose can fit on the valve end to assist the operation (Figure 19–79).

The new **anodized** aluminum valves in aluminum bodies have closer bore tolerances, since like metals expand evenly. The anodized surface is hard and can cut foreign material. It is extremely important that the anodized surface not get removed. Therefore, avoid using a car body polishing or household cleanser compound to recondition a stuck valve in the bore. To remove a valve nick or burr, use the Arkansas stone on the problem area only.

❖ **Anodized:** A special coating applied to the surface of aluminum valves for extended service life.

FIGURE 19–75 Rotate rolled crocus cloth to polish valve bores (cast iron valve body).

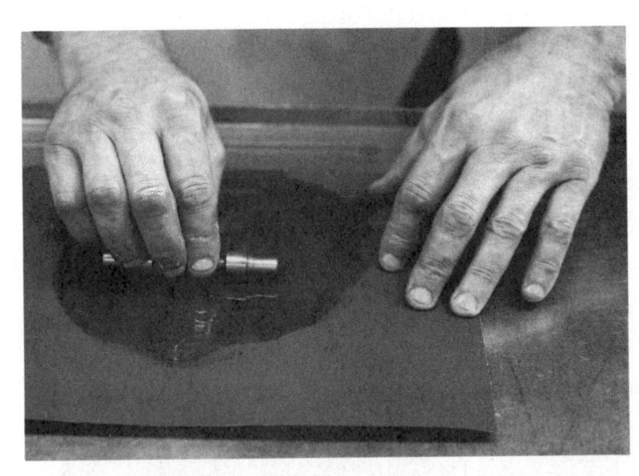

FIGURE 19–76 Rotate steel valve on flat plate surface covered with crocus cloth.

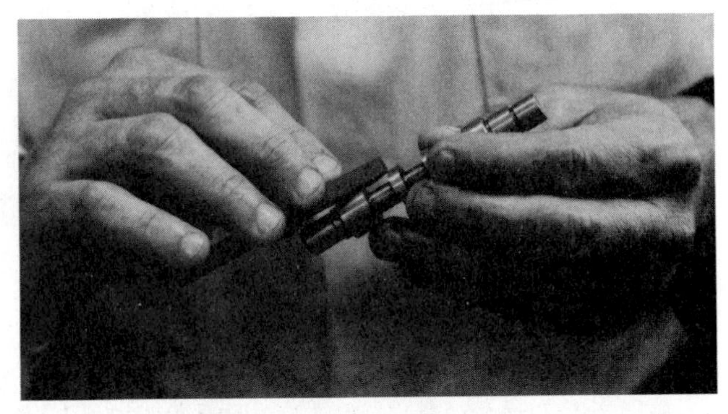

FIGURE 19–77 Use the fine honing stone to smooth steel valve nicks and scratches.

Valves are easily distorted if they drop to the shop floor. If a dropped valve drags or binds in its bore, it must be replaced. It is almost impossible to put the valve back into proper alignment. Since individual valves are not normally available as a parts distribution item, the technician now has a problem.

If reasonable effort does not produce a free-moving valve, the valve body needs to be replaced. Creating excessive clearance between the valve lands and bore will allow penetration by ATF contaminants. The valve will seize, even though it was tested and moved freely in its bore. Be sure to flush and clean the casting bore and valve immediately. Air dry the bore, and test the free fall of the valve.

Always flush the valve and valve body with cold solvent and air dry after removing scratches and embedded particles. Never remove scratches or embedded particles with emery cloth or abrasive paper. This only produces more scratches and adds to the problem you are trying to eliminate. The valves will seize once the vehicle hits the road.

Once the valve body and valve bore assemblies have been evaluated and prepared, the valve body unit is ready for assembly. Observe the following tips:

1. Prelubricate the valves in transmission fluid.
2. Each valve should slide into the bore freely and easily without dragging or binding (Figure 19–80). When necessary, spring-loaded valves can be pushed in their travel range with a wooden dowel and locked into place with a miniature screwdriver. The smaller valves might need some alignment help to slip into place. Never force any valve.
3. On valve body bores requiring end plugs, be sure to install the plugs with the countersink hole facing toward the outside, smooth side in (Figure 19–81). Valve springs are known to catch the spring end in the hole. This affects valve train operation. The valve body bores and valving do not

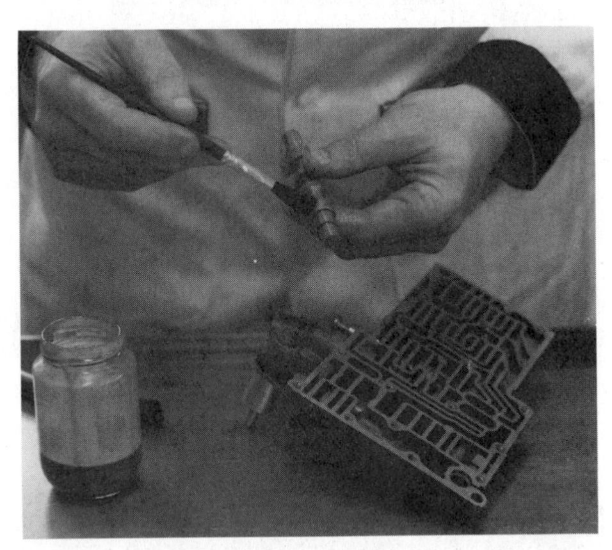

FIGURE 19–78 Use a fine coat of car body polishing compound, transmission fluid, and graphite to polish valve in the bore (cast iron valve body).

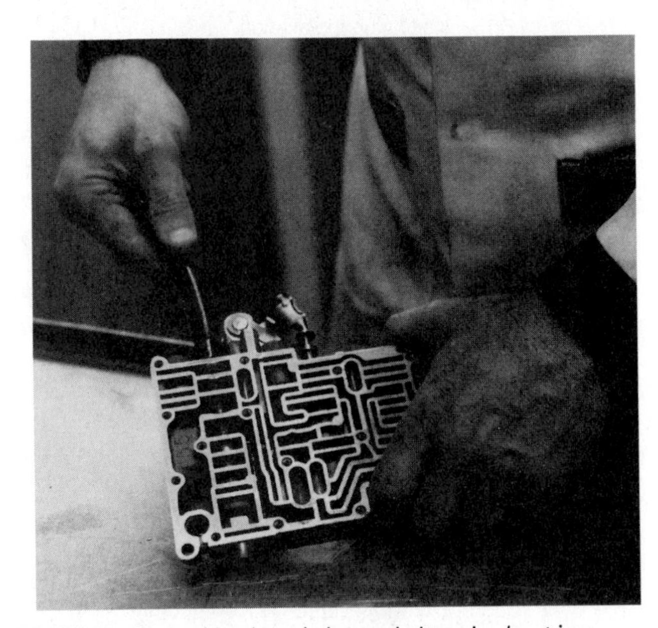

FIGURE 19–79 A rubber hose helps work the valve (cast iron valve body.)

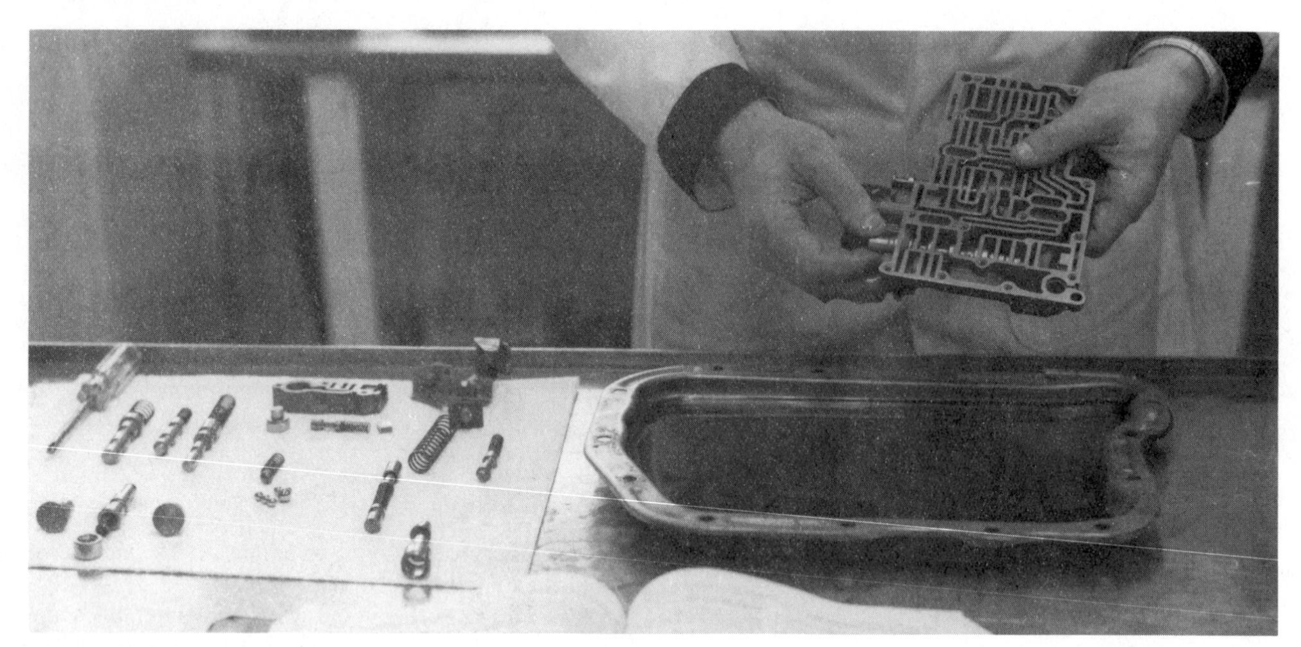

FIGURE 19–80 Each valve should slide in easily.

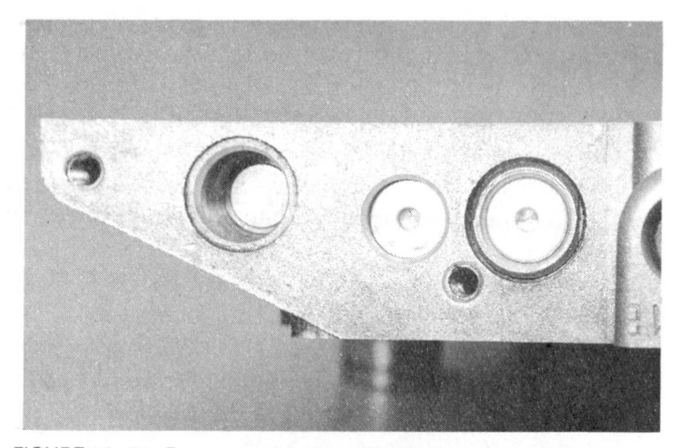

FIGURE 19–81 Proper end plug installation.

significantly change dimensionally with use, nor is there a significant change in the valve spring tension. The main concern with springs is that they show no signs of collapse or distortion. If the valve body was good when the vehicle was new, it should operate correctly when properly overhauled.

- Installing shift kits to modify shift quality for longer transmission life or to cure inherent erratic shifting problems is an acceptable practice. Many technicians install the kits as a matter of practice on every overhaul/rebuild. The kits are especially useful in curing soft or slipping shifts that are detrimental to long-term transmission life. The shift feel is made firmer, but not objectionably overaggressive.
- In some special situations, a service bulletin may require the technician to verify the spring weight of a valve to correct an erratic performance problem. This is measured on a kitchen scale, as illustrated in Figure 19–82. From a free-standing height, gently compress

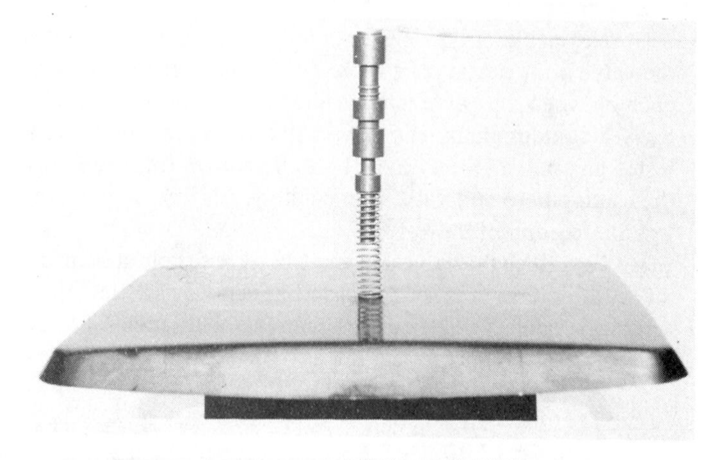

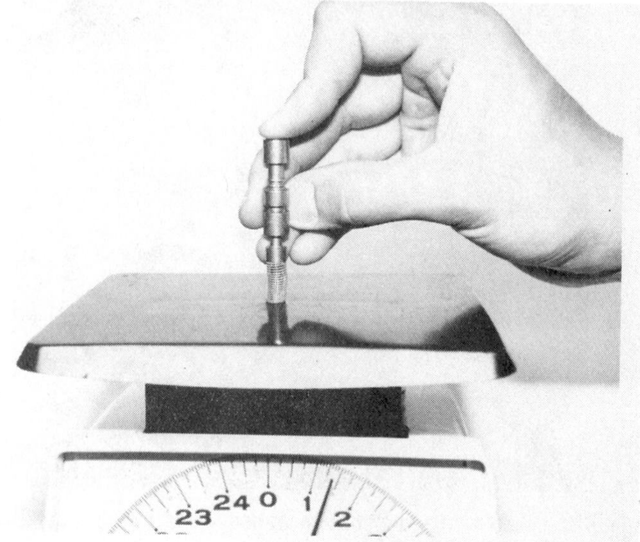

FIGURE 19–82 Verifying valve spring weight.

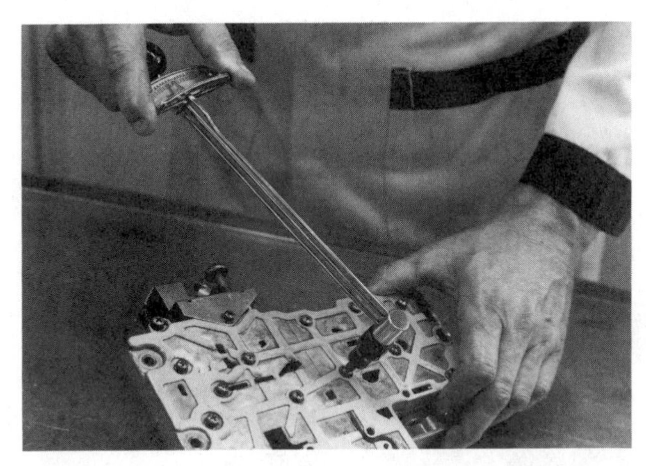

FIGURE 19–83 Torque-tighten valve body bolts and fasteners.

the spring until the valve stem contacts the scale. If the weight does not meet specifications, replace the spring with the required calibration.

4. Check balls should be held in place with TransJel/petrolatum. Remember that check ball locations are not always the same from year to year. If in doubt, be sure to consult the appropriate service technician information for correct location.
5. On two-piece valve bodies, it is advisable to bolt together completely and torque-tighten the upper and lower bodies before installing the valves. However, some valves need to be installed before bolting. This procedure helps ensure that the valve body will not be distorted by bolt tightening, which can cause valve seisure.
6. Always torque-tighten the valve body bolts and fasteners to avoid valve seizure or unwanted circuit leaks (Figure 19–83). Refer to the manufacturer's shop manual for specifications. Some valve body bolts require only 35 in-lb (4 N-m), just slightly over finger-tight. Do not exceed the specification limits. To avoid stripping threads, be sure to install all the valve body bolts or screws before torque-tightening. This ensures alignment of the holes between the valve body halves and separator plate.
7. There are often several sets of separator plate gaskets in an overhaul kit. In many applications, the correct gasket should match the separator plate configuration, and all plate holes should have a matched opening through the gasket (Figure 19–84). When possible, use the original gaskets to help select the correct one. Like check ball locations, valve body gaskets are not always the same from year to year, so the technician needs to be careful. The gaskets are installed dry.

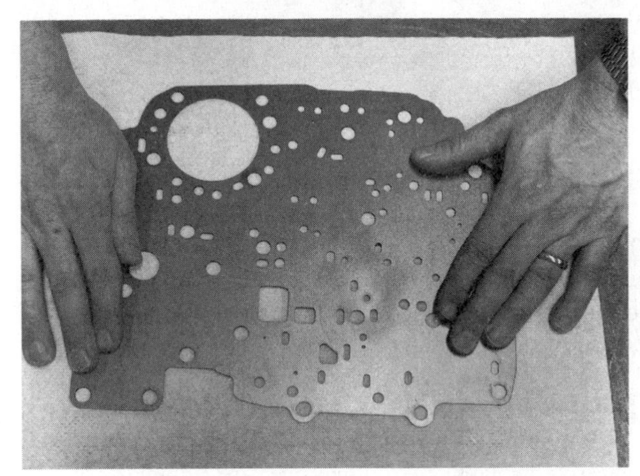

FIGURE 19–84 Match separator plate holes to gasket holes.

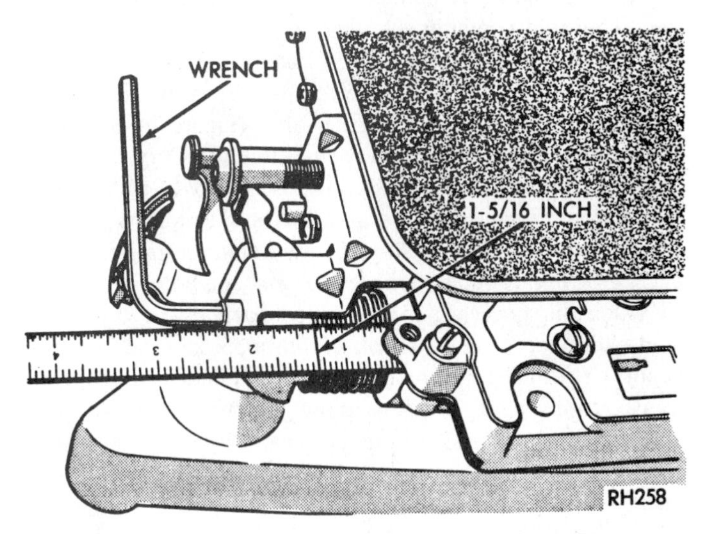

FIGURE 19–85 Line pressure adjustment check—TorqueFlite 30RH/36RH (A-904/A-727). (Courtesy of Chrysler Corp.)

Some valve bodies require external adjustments, and these should always be checked for accuracy. Figures 19–85 and 19–86 show the line pressure and throttle pressure adjustment checks on a Chrysler TorqueFlite valve body. Most service manuals provide detailed exploded views of the valve body to help the technician identify the location and position of the valves, springs, and check balls. However, these illustrations sometimes include errors and may not show the exact valve body assembly.

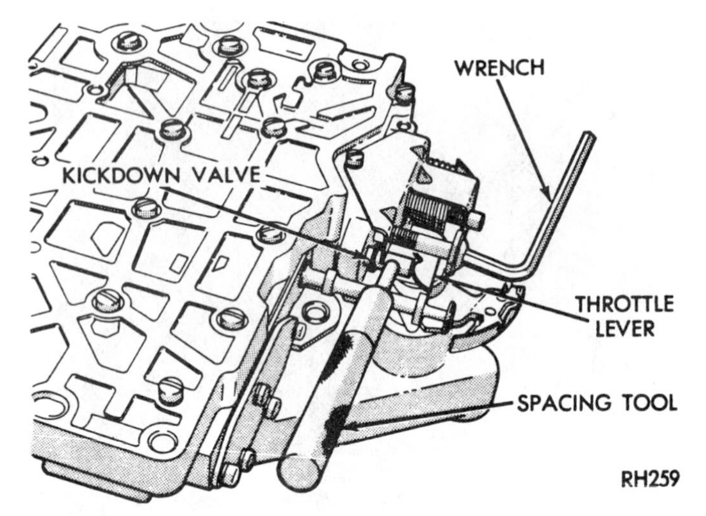

FIGURE 19–86 Throttle pressure adjustment check—TorqueFlite 30RH/36RH (A-904/A-727). (Courtesy of Chrysler Corp.)

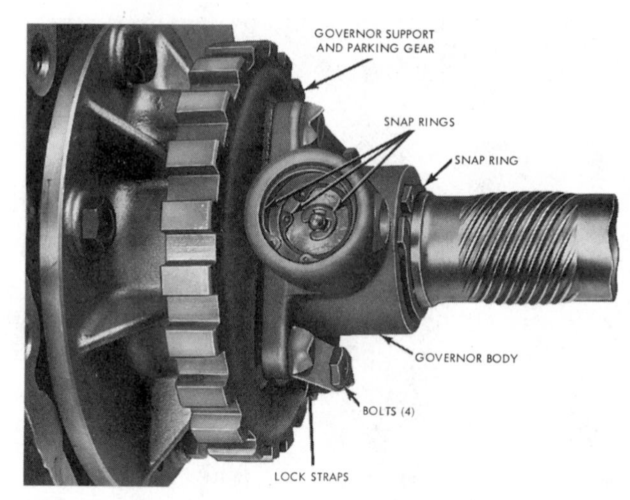

FIGURE 19–87 Governor assembly. (Courtesy of Chrysler Corp.)

GOVERNOR SERVICE

The governor requires attention during overhaul and is another service item often overlooked. Spending a few minutes to check out the working condition of the governor system will avoid those erratic or no-shift problems after overhaul.

Output Shaft-Mounted Governors

With output shaft-mounted governors, disassemble, clean, and assemble the unit with the same care required of the valve body. A typical unit is illustrated in Figures 19–87 and 19–88.

In this type of output shaft-mounted design, be sure to separate the inner and outer weights for cleaning and inspection. The fluid contaminants tend to collect in this area and cause improper secondary governor action.

The governor body must be removed from its support or distributor to check for a clogged filter (Figure 19–88). The filter is located on the inlet side of the circuit on Chrysler, Ford, and GM Hydra-Matic transmissions. Be sure to inspect governor support oil ring grooves and ring bore condition (Figure 19–89). New metal rings should be checked for groove clearance, groove-free rotation, bore ring gap, and bore fit.

When the governor is positioned back in the transmission/transaxle, it can be checked with a 35 psi (242 kPa) air supply or less for proper action. On all Chrysler and Ford RWD transmissions, apply pressure to the governor out passage located in the transmission case. For best test results, cover the governor case inlet passage with your finger (Figures 19–90 and 19–91).

- *Ford.* Listen for a buzzing noise similar to that of a model airplane engine. This means that the governor valve is properly "hunting" or oscillating.
- *Chrysler.* The governor must be positioned with the weights working against the valve. A fast-acting audible "thud" should be heard. On occasion, however, the valve will oscillate in a low-keyed tone.

As a countercheck on both Ford and Chrysler governors, apply air to the governor case inlet passage. With air charging the circuit, cover the outlet passage temporarily with your finger. The governor valve should snap closed and stay in this position. With some imagination, these same governor air checks can be made on Chrysler FWD TorqueFlite units. The governor is mounted on the transfer shaft (Figure 19–92).

Case-Mounted Governors

Case mounted governors use either a valve-style or a check ball–style assembly. Both styles are found in FWD and RWD applications (Figures 19–93 and 19–94). The valve style is used in GM 3L80 (400), 4L60 (700-R4), THM 350-C, THM 250-C, and Ford ATX units. Check ball styles are used in GM 3T40 (125C), 4T60 (440-T4), THM 325-4L, THM 200-C, and Ford AXOD units. To check these governor assemblies, flush in mineral spirits, air dry, and make the following inspections.

Valve Style.

- Inspect governor sleeve finish for scoring.
- Inspect governor-driven gear for damage and looseness.

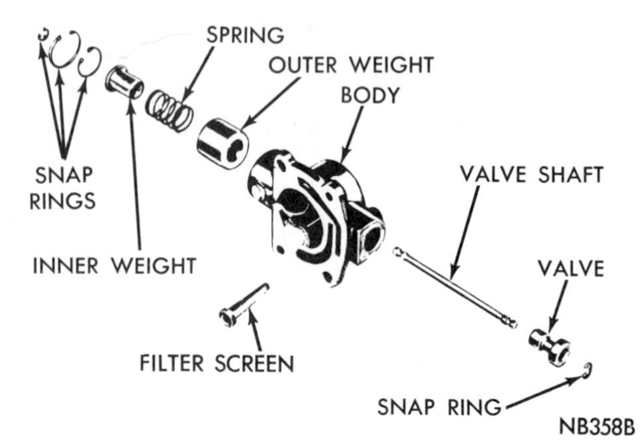

FIGURE 19–88 Check governor body for clogged filter. (Courtesy of Chrysler Corp.)

FIGURE 19–89 Ford A4LD: (a) governor, and (b) governor support and ring bore.

FIGURE 19–90 Air checking governor—TorqueFlite RWD three-speed transmission.

FIGURE 19–91 Air checking governor—Ford RWD three-speed transmission.

FIGURE 19–92 Transfer shaft-mounted governor—TorqueFlite 31TH (A-400 series and A-670) FWD three-speed transaxles.

FIGURE 19–93 Case-mounted governor assembly—GM RWD three-speed transmission.

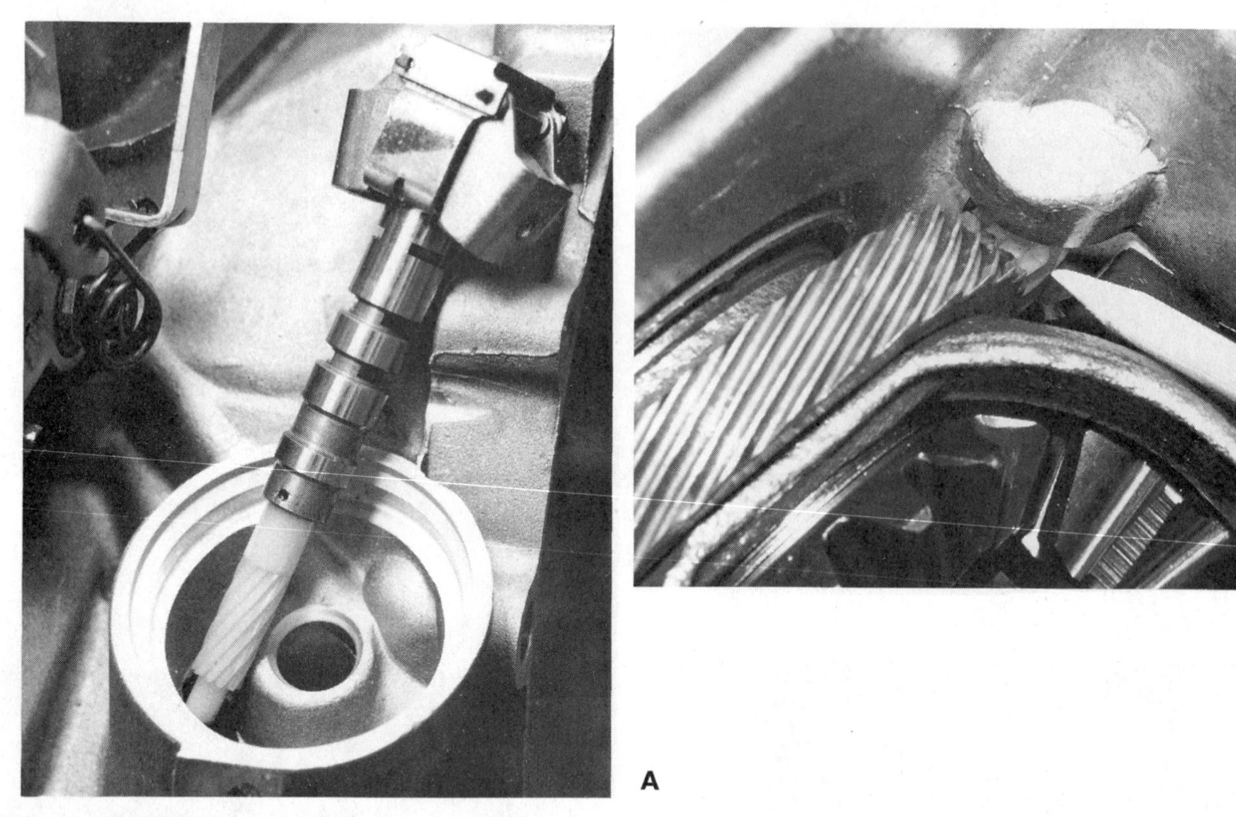

FIGURE 19–94 Ford ATX (domestic design): (a) case-mounted valve-style governor, and (b) governor driven from speedometer drive gear mounted on differential case.

- Inspect governor weight springs for distortion.
- Check governor weights for free operation in their retainers.
- Check the free movement of the governor valve by moving the weights in and out (Figure 19–95).
- It is important to check the valve inlet and exhaust port openings. These openings should be at least 0.020 in (0.50 mm). If they are right, the gaps are usually large enough for the technician to make a good judgment without measuring. Should a gap be questionable, use a 0.020-in (0.50-mm) feeler gauge as a go/no-go measurement. To measure the inlet port, the weights are extended completely outward. The exhaust system port opening is checked with the weights completely held in (Figure 19–96). If the gap openings are not correct, replace the governor or carefully bend the weight tangs equally.
- Check the governor for free rotation in the transmission/transaxle case bore. If the governor sleeve was found to be scored on an earlier inspection, the case bore must be closely examined for scoring.

If the case bore is scored, it can be reamed out and sleeved back to size with a service bushing (see the following section entitled: "Bushing Replacement"). As another choice, the bore can be lightly polished by hand with a brake wheel cylinder hone (Figure 19–97) or silicone honing brush (Figure 19–98). The edges of the honing stones must be rounded off to avoid biting into the aluminum. Oil the stones and turn the hone by hand with a tap handle about ten turns. Both ends of the bore must be conditioned separately and equally. This means that the inner and outer halves are polished with the same amount of turns. Avoid moving the hone in and out when turning. With a silicone brush, use short, rotating, in-and-out strokes.

After the honing operation, flush and air dry the case. It is important that the shaft lands all measure at no less

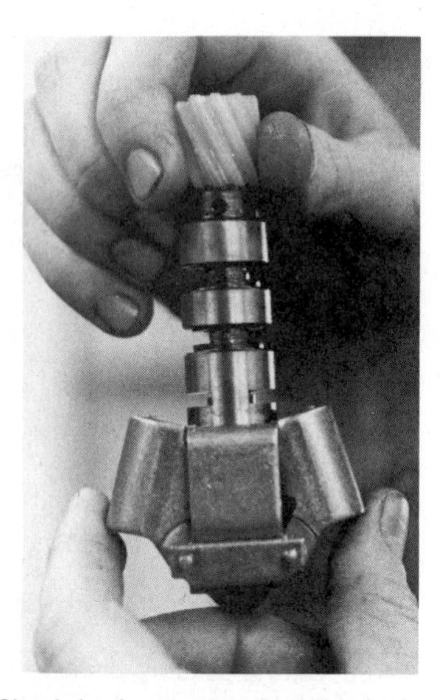

FIGURE 19–95 Check for free movement of governor valve.

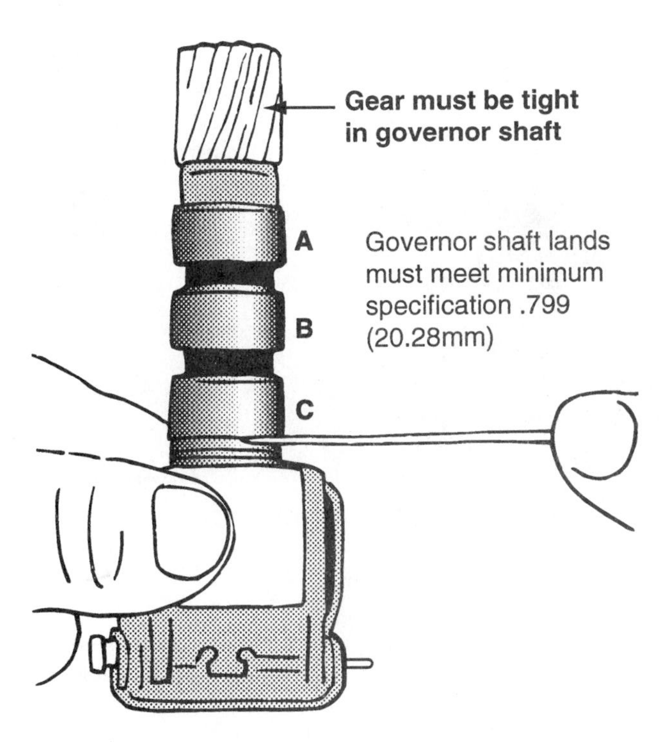

FIGURE 19–96 Measuring governor exhaust circuit port. (Courtesy of General Motors Corporation, Service Technology Group)

than 0.799 in (20.28 mm). Otherwise, replace the governor. For reliability, be sure to check the new governor for undersize. This dimension is valid for the General Motors THM 300, THM 250, THM 350, 3L80 (400), and 4L60 (700-R4) transmissions.

With a governor, perform the following leak check to test for excessive clearance between the governor shaft lands and case bore:

1. Place the transmission case on the bench with the valve body mounting surface up.

FIGURE 19–98 Silicone honing brush used to polish case bore.

FIGURE 19–97 Brake wheel cylinder hone used to polish case-governor bore.

2. Install the governor and fill the governor case passages with solvent to the level shown in Figure 19–99.
3. Rotate the governor to remove air bubbles in the circuit, and then refill the passages back to level.
4. With the governor at rest, time the leakdown for thirty seconds. If the solvent level falls below point A (Figure 19–99), the sealing tolerance between the governor shaft and the case bore is excessive. This means that the governor case bore can be serviced with a bushing (GM only).

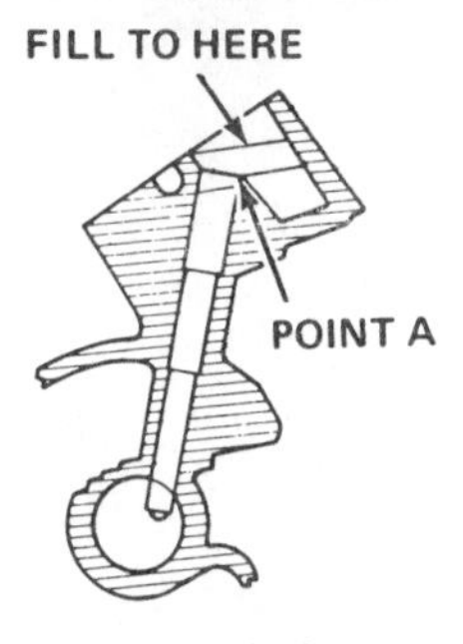

FIGURE 19–99 Fill governor case shaft passages with solvent.

Examine the driven gear for looseness. It must be tight to the governor shaft (Figure 19–96). The gear actually plugs the end of the governor valve bore. Any leakage around the gear will affect governor valve balancing and regulated output. To air check the governor action in the case bore, use the same procedures as those used for output shaft-mounted governors. Governor weights must be in the vertical position (Figure 19–100). For a good oscillating action, place your finger on the inlet passage while applying 35 psi (245 kPa) of air to the outlet.

Many case-mounted governors can be completely disassembled and serviced if necessary. Small parts replacements are usually available, including the nylon driven gear. Consult the manufacturer's service manual for procedures. The pins can be replaced using number 6 finishing nails.

Check Ball Style. On check ball–style governors, the balls and seats can get worn and distorted, failing to seal. To check the balls and seats for leakage, position the governor vertically with the weights at the bottom. Fill the governor shaft with mineral spirit solvent and observe for leakage past the check balls (Figure 19–101). If any leakage is observed, replace the governor. Be sure to examine the governor seal bore and replace the shaft seal.

Check the weights for free movement and clearances (Figure 19–102). There must be clearance:

- between the governor weight tangs and the slot end of the governor-driven gear
- between the weights at point A
- between the secondary weight and governor-driven gear at point B

FIGURE 19–100 Air check governor action.

If any binding or contact takes place in these areas, carefully bend the arms with a small screwdriver to create a clearance.

BUSHING REPLACEMENT

The transmission bushings align and support shafts, gears, and clutch drums. They also act as restrictors for the converter and lubrication circuits that are routed between inner and outer shaft members. Worn bushings replaced during overhaul can reduce gear noise, sealing and ring groove wear, converter drain-back, and pressure losses in the converter/lubrication circuits.

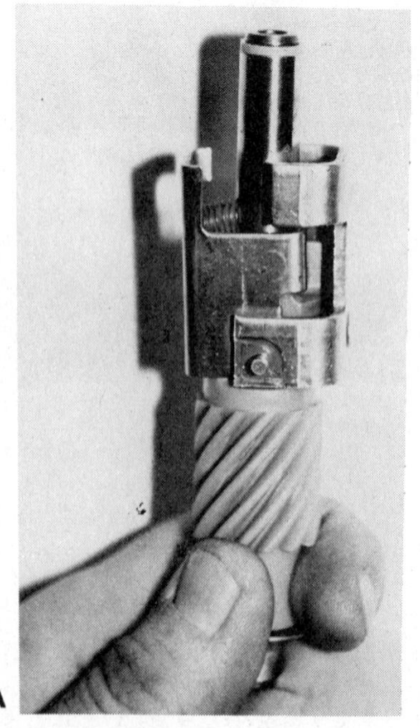

FIGURE 19–101 (a) With governor shaft loaded with mineral spirits, technician observes for check ball leakage. (b) Replace shaft seal and evaluate condition of shaft.

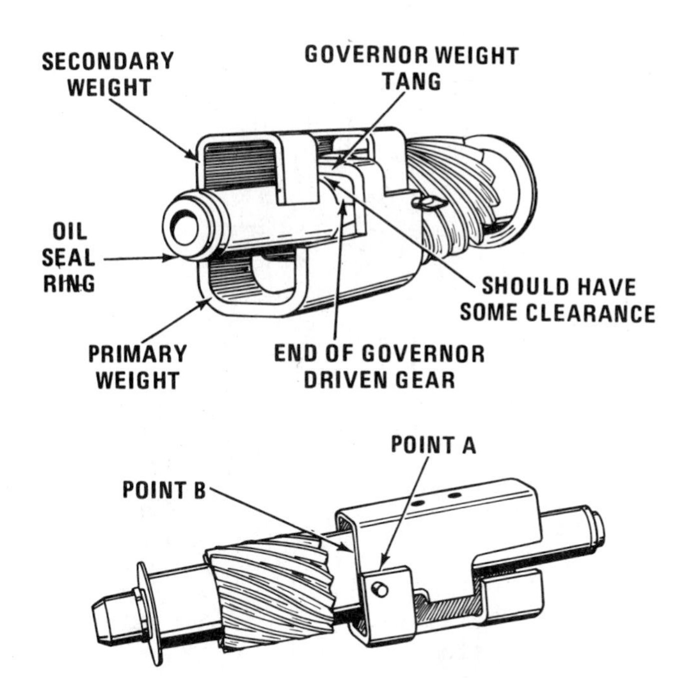

FIGURE 19–102 Check weights for free movement and clearances. (Courtesy of General Motors Corporation, Service Technology Group)

FIGURE 19–103 Service bushing kit.

Service bushings are available as a dealer or jobber item. They can be purchased individually or in a kit package (Figure 19–103). The replacement bushings are precision made and do not require boring or reaming after installation.

On TorqueFlite transmissions, the front stator shaft bushing is eliminated in favor of supporting the input shaft and turbine with a bushing installation in the converter cover. With any TorqueFlite rear stator shaft and/or rear clutch ring and groove wear, you should take into consideration the condition of the converter bushing with a good bench analysis.

During cleaning and inspection of the transmission, most bushings will show little wear and do not need replacement. It is usually obvious when a bushing needs replacement. A visual inspection will easily pick up a galled, scored, or excessively worn condition. Where bushing wear is not obvious but is suspected, fit the mating part to the bushing with a paper strip to check the clearance (Figure 19–104). The paper strip should fit snugly. If a paper strip is folded and the doubled layered paper strip fits, a problem definitely exists. Paper thickness measures from 0.0025 to 0.003 in (0.063 to 0.075 mm). The maximum bushing clearance should not exceed 0.004 in (0.10 mm).

Although extensive factory tooling is available for bushing removal and installation, bushing service can be done with some simple tooling and your own innovations. Most bushings can be split and removed with a bushing cutter chisel and hammer (Figure 19–105). A variety of cutter chisels are shown in Figure 19–106. Be sure to select the proper chisel for the job, and avoid using a hammer that is too light. Let the weight of a heavy hammer do the work. Because split-type bushings are usually used, drive the chisel on the seam of the bushing split.

On extra-small bushings that are located in dead-end bores, an N/C (National Course) thread tap can do the job. A 9/16 N/C tap can be used to remove the output shaft bushing in some Hydra-Matic RWD transmissions. To prevent breaking the tap, a large bearing ball must be placed in the bottom of the bushing bore.

FIGURE 19–104 Use paper strip to check bushing wear.

FIGURE 19–105 Use of a bushing cutter chisel.

FIGURE 19–106 Bushing (and seal) cutter chisels.

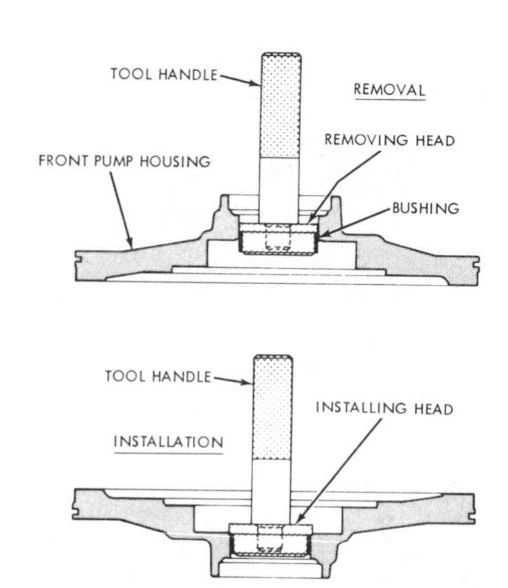

FIGURE 19–107 Top: Pump bushing removal. Bottom: installation. (Courtesy of Chrysler Corp.)

Replacing a bushing requires an exact technique if the job is to be done properly. You will definitely need a bushing installer adapted for the bushing size to drive or press the bushing into place. Otherwise, the bushing will be damaged or misaligned. A proper bushing installation using adapter heads for removal and installation is illustrated in Figure 19–107. Some bushings must be staked, as required by the manufacturer (Figure 19–108).

Where case-mounted governors are used in the GM3L80 (400), 4L60 (700-4R), 350-C, and 250-C transmissions, the governor bore occasionally gets worn. To salvage the case, special factory tooling is available that permits reaming the bore oversized to accept a governor bushing sleeve. Be sure to align the bushing with the governor bore oil holes when driving the bushing into place.

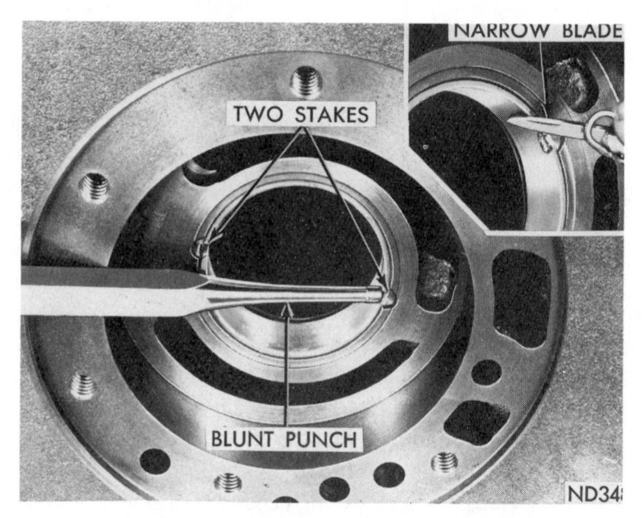

FIGURE 19–108 Staked bushing. (Courtesy of Chrysler Corp.)

SUMMARY

Subassembly reconditioning refers to disassembling, inspecting, and reassembling components such as pumps, clutch housings, planetary carriers, valve bodies, governors, and bushings. Special techniques are involved in properly servicing these units.

Pumps vary in design and construction, but certain inspection and service procedures are quite standard. Always inspect shafts and splines, the pumping elements and the housing in which they revolve, bushings, and valves for unusual wear. Replace these items as needed. All internal and external pump seals are to be renewed. It is important to wet pumping mechanisms and valves with ATF when installing these items. When required, use special tools for correct reassembly of the pump assembly.

Many clutch housings contain multiple pistons and clutch packs that can easily be misassembled by a careless technician. Work closely with a service manual to identify the exact teardown and rebuild of these units. Inspect the housings, bushings, return springs, pistons, check ball capsules, and sealing surfaces for wear and damage. When reassembled, air test the units and check the clutch pack clearance.

Planetary carriers transmit a great amount of engine torque and will wear out. Wear and damage can occur to the gears, splines, pinion bearings, thrust washers, and thrust bearings. Some carriers are serviceable, while others must be replaced when they are determined to be bad.

Follow the manufacturer's recommendations when servicing valve bodies. When disassembling these complicated devices, work on a clean bench top, keep the valve assemblies organized, and clean the pieces in fresh mineral spirits and air dry them. When reassembling, dampen all valves with ATF. When reinstalled, they must be able to move freely in their bores.

Clean governors, inspect governor bushings, and check their operation. Fluid leaks and low governor pressures have a direct affect on shift timing.

If support bushing clearances are too great, replace the bushing, and shafts as needed. Special bushing drivers are available to seat bushings properly in their locations.

All internal components must be in good working order to guarantee that the rebuilt transmission will provide dependable service. When servicing the assemblies discussed, completely inspect them for wear, and repair as needed.

❑ REVIEW

Key Terms

Lubrite coating and anodized.

Questions

Pump Service

____ 1. In IX gear pump applications:
I. the driven gear may be installed with either side facing down.
II. drive gears designed with drive tangs must be installed with the tangs facing down.

A. I only C. both I and II
B. II only D. neither I nor II

____ 2. Pump end clearance checks:
I. are normally measured with a straightedge and feeler gauge.
II. should not exceed 0.004 in (0.102 mm).

A. I only C. both I and II
B. II only D. neither I nor II

____ 3. The pump body bushing to converter hub clearance should not exceed:

A. 0.002 in (0.051 mm).
B. 0.004 in (0.10 mm).
C. 0.008 in (0.203 mm).
D. 0.012 in (0.305 mm).

____ 4. A severely damaged pump can be caused by:
I. a converter/flex plate alignment problem.
II. faulty transmission installation practice.

A. I only C. both I and II
B. II only D. neither I nor II

5. In a variable-capacity pump, the wear side of the vanes should be reversed when reinstalled in the rotor slots. ___T/___F

6. The pump/converter hub seal should be packed with petroleum jelly. ___T/___F

7. Variable-capacity pumps must be serviced with new sealing kits. ___T/___F

8. Pump assembly valving should be installed after the pump body and cover bolts are torque-tightened. ___T/___F

Clutch Service

1. Clutch drum bushings are a nonserviceable part item. ___T/___F

2. Emery cloth is an excellent material for polishing clutch cylinder bores. ___T/___F

3. If the piston return springs show signs of overheating, they must be replaced. ___T/___F

4. Failure of a rotating clutch check ball to seat will cause a repeat clutch failure. ___T/___F

5. If a rotating clutch is assembled with a plugged check ball area, the clutch will fail. ___T/___F

6. Lip seals used on the piston must have the lip facing the pressure apply side. ___T/___F

7. Friction plates should be presoaked in ATF for thirty minutes before installation. ___T/___F

8. What are the two usual production provisions for adjusting clutch plate clearance?
(a) ______________________________
(b) ______________________________

9. It is preferred to keep the clutch plate clearance on the high side. ___T/___F

10. Clutch plate clearance can be measured effectively with a dial indicator. ___T/___F

11. When air checking clutch units, a minimum pressure of 75 psi (518 kPa) is recommended. ___T/___F

Valve Body and Governor Service

1. It is a good practice to flush and clean a valve body assembly with fresh mineral spirits solvent. ___T/___F

2. Rubber check balls should not be soaked in mineral spirits. ___T/___F

3. To free stuck valves, it is an acceptable practice to tap the side of the valve body with a plastic hammer. ___T/___F

4. Emery cloth is an excellent material for reconditioning a stuck valve and its bore. ___T/___F

5. When reconditioning a stuck valve, the spool edging must be kept razor sharp. ___T/___F

6. The use of carburetor spray for flushing and cleaning valves and the valve body should be avoided. ___T/___F

7. In most cases, the separator or spacer plate gasket is right when all the plate holes have an opening through the gasket. ___T/___F

8. It is an acceptable practice to use white lube to hold the check balls in place. ___T/___F

9. It is a good practice to omit the governor filter screen. ___T/___F

10. On valve-style, case-mounted governors, mineral spirits can be used to perform a case to shaft leakdown test. ___T/___F

11. On check ball-style, case-mounted governors, mineral spirits can be used to perform a check ball leakdown test. ___T/___F

12. When air checking for correct governor action in the case, either a "thud" or an oscillating noise should be heard. ___T/___F

20 Transmission/Transaxle Assembly Practices

OBJECTIVE:

The objective of this unit is for you to become proficient with the terms, concepts, and procedures contained in this chapter. Completion of this objective is essential to becoming a successful automatic transmission technician.

CHALLENGE YOUR KNOWLEDGE

Define the following key terms/acronym:

Variable orifice thermal valve and RTV.

List the following:

Transaxle differential inspection concerns; three selective fit items used to control endplay clearances; and two methods used to adjust a band.

Describe the following:

Techniques used to identify correct installation of needle-type thrust bearings; two methods used to test clutch and servo hydraulic circuits; and the importance of identifying valve body, spacer plate and case gaskets, as well as check ball locations.

INTRODUCTION

Once the subassemblies and related parts have been evaluated and prepared, they are ready to be installed into the transmission/transaxle case. Basically, this requires reversing the disassembly procedure. Again, use the manufacturer's service manual or equivalent as a reference.

A main concern during assembly is to correctly mate the subassemblies to the related parts. The transmission/transaxle should go together without excessive force. If parts do not assemble freely, find the cause and correct the trouble before proceeding. Be familiar with the applicable update service bulletins and mandatory parts changes to avoid repeat failures or performance concerns.

ASSEMBLY GUIDELINES

Selected illustrations are used to show various key points in general assembly procedures that apply to many automatic transmissions and transaxles.

Transaxle Differential

- On transaxle designs with a helical ring and pinion/differential assembly, it will not be necessary to readjust side bearing preload or endplay provided that there were no visual signs of bearing/cup failure. The extension housing or bearing retainers will require replacement seals (Figure 20–1). Should any of the differential side bearing and cup assemblies require replacement or if the transaxle case is replaced, bearing preload and turning torque must be checked and readjusted (Figure 20–2). (Refer to the manufacturer's service manual for the exact steps of procedure.)

NOTE: Turning torque measurement must be made without interference from another gear intermesh.

Inspect the differential gear assembly (Figur 20–3). Give close attention to:

1. Side gear thrust washer and case wear
2. Pinion gear thrust washer and shaft wear
3. Pinion shaft to case wear

The technician needs to determine whether any driveline engagement or acceleration-deceleration "clunk" concerns are related to the final drive-differential unit. In some installations, differential side gear endplay is adjustable with a select side gear thrust washer (Figure 20–4).

• On the Chrysler TorqueFlite three-speed FWD series, the ring and pinion gears must have the same identification mark. Otherwise, there may be a mismatch in tooth count and/or helical angle. In Figure 20–1, the ring gear has a single line around its circumference, formed by notches in the gear teeth. The ring and pinion gears come marked with one line, two lines, or no lines. Always make sure that the tooth count is the same

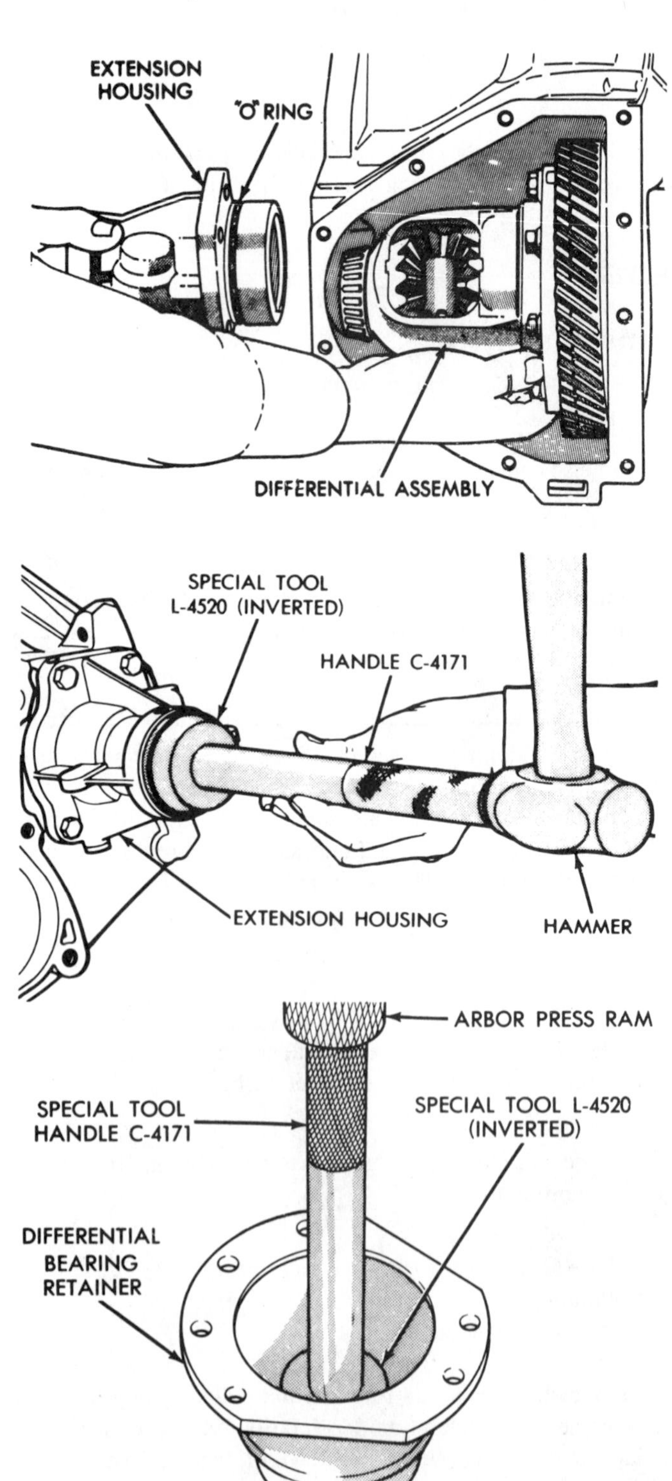

FIGURE 20–1 Final drive seal/bushing service—Chrysler three-speed FWD series. (Courtesy of Chrysler Corp.)

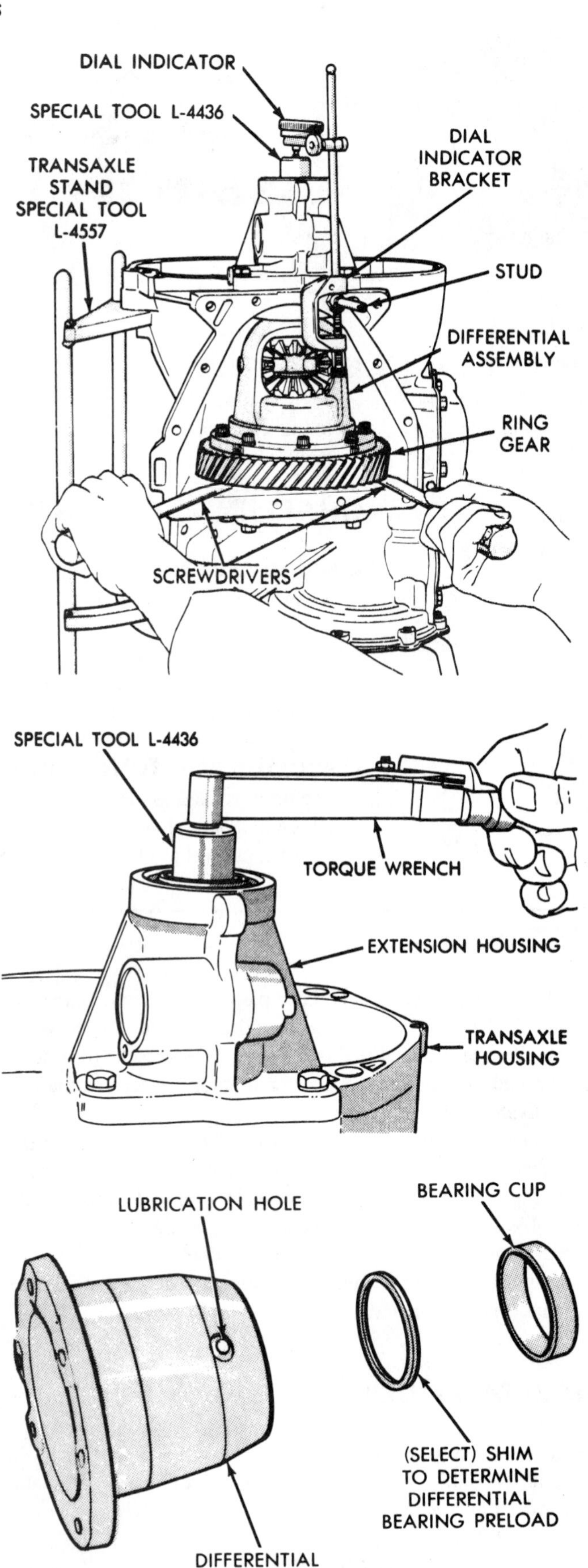

FIGURE 20–2 Checking differential endplay and bearing preload. (Courtesy of Chrysler Corp.)

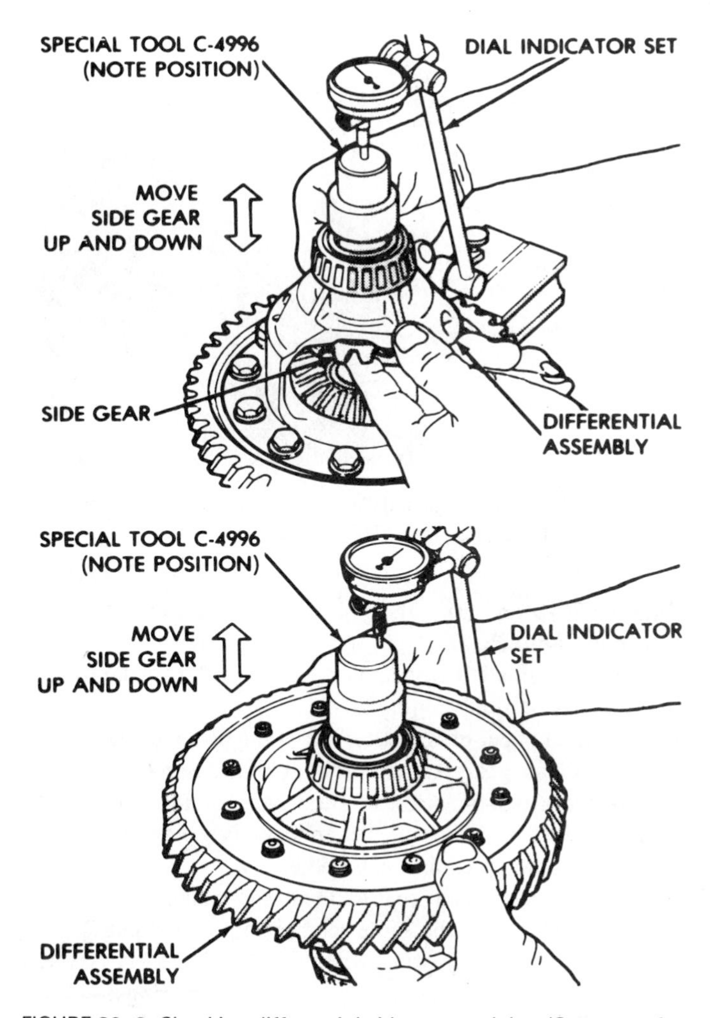

FIGURE 20–3 Checking differential side gear endplay. (Courtesy of Chrysler Corp.)

if replacing a ring or pinion gear. The helical ring and pinion gears do not need to be replaced as a matched set.

- When working with tapered roller bearings, it is important to maintain correct endplay and preload/turning torque to avoid premature bearing failures. Turning torque must always be checked without interference from another gear and shaft intermesh. In Figure 20–5, the output shaft bearing turning torque is being checked on a TorqueFlite FWD three-speed unit. The turning torque is adjustable with a select shim (Figure 20–6).

Not all tapered roller bearings have a preload turning torque when bench checked. The transfer shaft endplay is taken with a dial indicator (Figure 20–7). The endplay clearance allows for heat expansion of the parts, and for correct fit once warmed up to operating temperatures. Select shims are usually used to control bearing endplay (and preload turning torque). It is not necessary to go through a bearing adjustment procedure on reassembly unless the following are replaced: (1) transaxle case, (2) bearings, and (3) bearing shaft or related shaft assembly parts.

Do not be alarmed when checking used bearings. They lose fifty percent of their turning torque and offer almost no turning resistance. Should they require adjustment, the turning torque

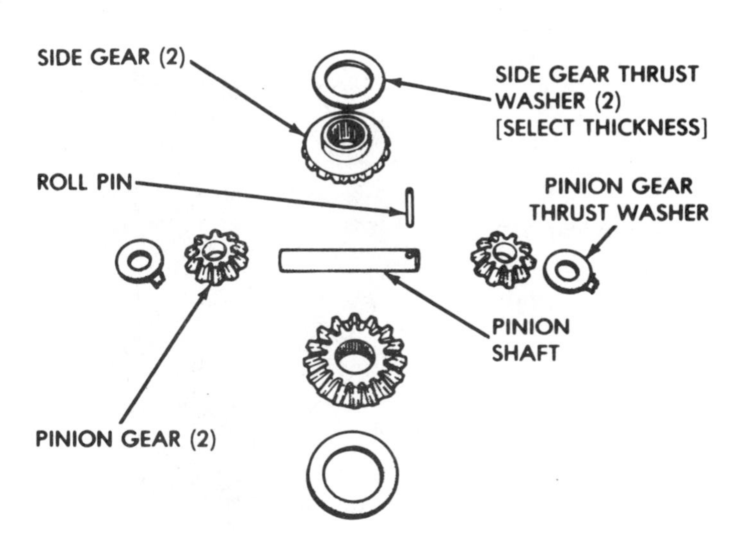

FIGURE 20–4 Differential gears with side gear selective washer. (Courtesy of Chrysler Corp.)

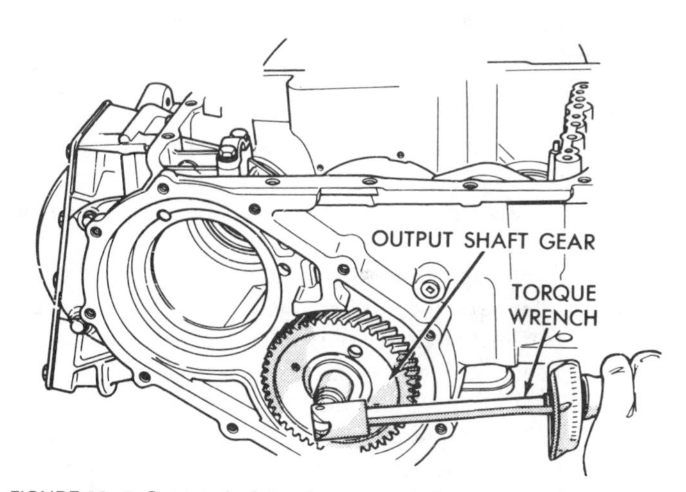

FIGURE 20–5 Output shaft turning torque being measured. (Courtesy of Chrysler Corp.)

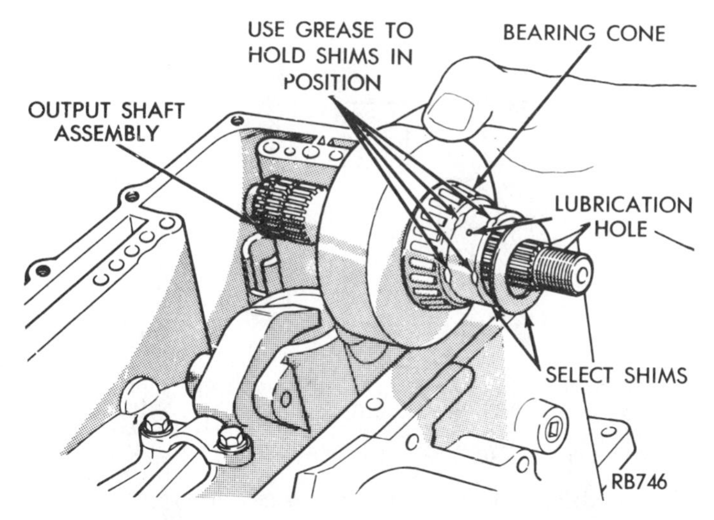

FIGURE 20–6 Select shim location. (Courtesy of Chrysler Corp.)

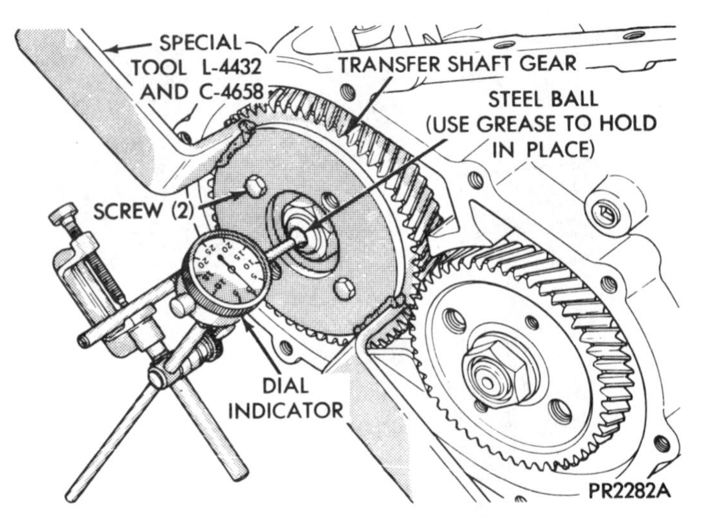

FIGURE 20–7 Checking transfer shaft endplay. (Courtesy of Chrysler Corp.)

should be set up to half of new bearing specifications. Refer to the individual manufacturer's service manual for exact adjustment procedures.

Transmission/Transaxle

- Be sure to replace the manual shaft case seal and throttle shaft seal.
- Presoak new bands and clutch discs for thirty minutes in transmission fluid.
- Careful attention must be given to the placement of thrust washers. They should be tacked into place with a petroleum jelly or TransJel on one side only (Figure 20–8). When applicable, treat the side of the thrust washer that has tangs or notches. Avoid using white lube or other greases. A thrust washer incorrectly positioned or one that has dropped out of place will cause assemblies to bind and may prevent proper snap-ring installations. Also, endplay clearances will be dramatically affected. Make sure that needle-type thrust bearings get installed correctly. Otherwise, they will be subject to rapid wear and failure. The ninety-degree angle of the bearing race must always match the ninety-degree angle of the mating part

FIGURE 20–8 TransJel transmission assembly lube applied to tanged side of thrust washer.

(Figures 20–9 and 20–10). Typical installation directions are shown in Figure 20–9.

- Prelubricate all O-ring and case seals with petroleum jelly or transmission fluid (Figure 20–11).
- Install all paper gaskets dry. Avoid the use of sealers.
- Prelubricate gears, sprag or roller clutch assemblies, shafts, and bushings with transmission fluid. Most parts can be dipped in fluid before installation. The tendency is to neglect lubrication and put the transmission together dry.
- On scarf-cut Teflon seal applications, make sure that the ends are matched in the same relationship as the cut (Figure 20–12). Otherwise, the ring will fail to seal. Teflon is a soft material, and care must be taken not to nick the rings.
- Install all the miniature filter screens. These filters are not throwaway items.
- Make all the required clearance checks during the buildup and installation of the planetary and clutch units into the case.

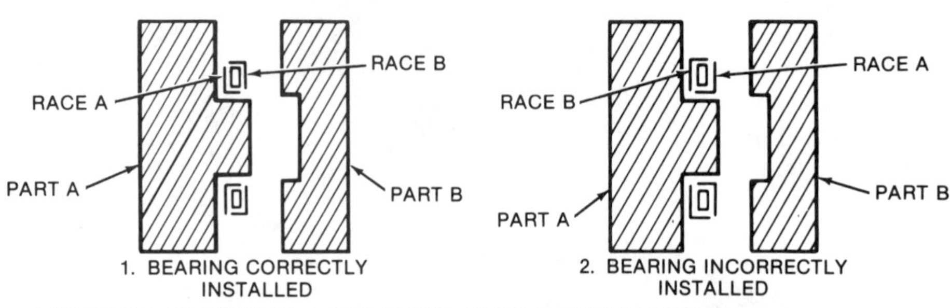

FIGURE 20–9 Needle-type thrust bearing installation concept. (Courtesy of General Motors Corporation, Service Technology Group)

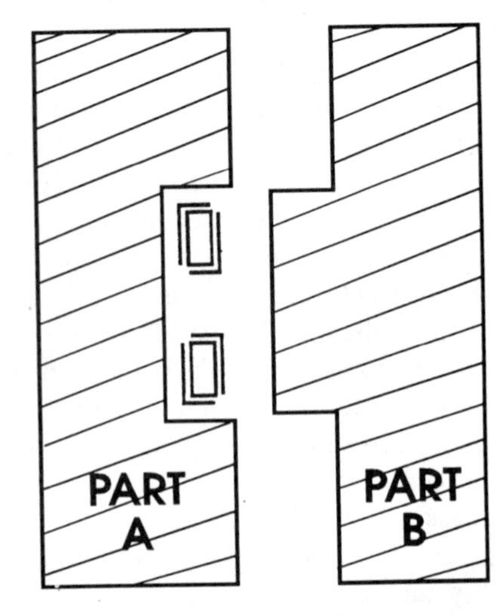

FIGURE 20–10 Thrust bearing outer race matched with counter bore in "Part A" for correct installation.

FIGURE 20–11 (a) Prelubricated servo piston seal; (b) prelubricated servo cover seal.

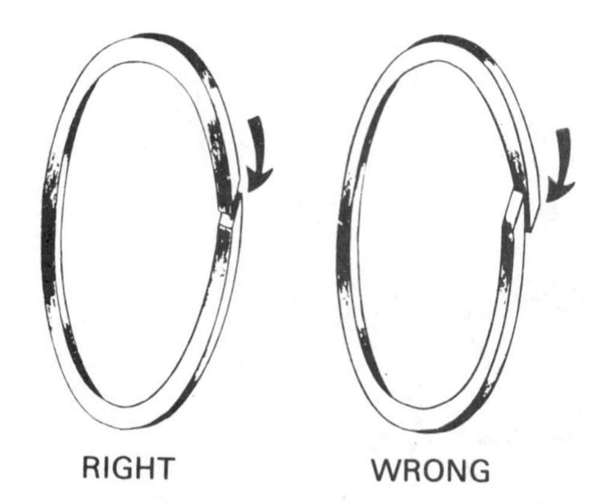

FIGURE 20–12 Scarf-cut Teflon sealing ring installation. (Courtesy of General Motors Corporation, Service Technology Group)

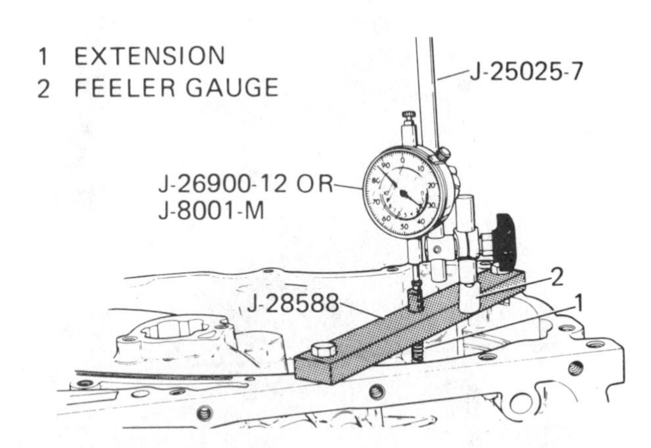

FIGURE 20–13 Measuring selective snap-ring endplay (sun gear/input drum)—3T40 (125-C). (Courtesy of General Motors Corporation, Service Technology Group)

(Refer to Chapter 17, "Transmission/Transaxle Disassembly," Figures 17–2 to 17–21). Clearance/endplay checks may or may not require special tooling (Figures 20–13 and 20–14). Clearances and endplays are adjustable by either a (1) selective thrust washer, (2) selective spacer, or (3) selective snap ring. If the clearance or endplay check is on the high side of the specification range, make it a practice to adjust it to the low end.

- A critical assembly step is the installation of the input clutch units (Figure 20–15). The friction discs for each clutch must all be indexed on their splined hub. If the discs are not all indexed, the clutch units will not rest fully in the case and ride high. When the pump or sprocket support (FWD) is bolted into place, it will wedge against the clutch units and bind the misindexed discs. Front endplay is nonexistent, and the input shaft will not rotate.

For a typical three-speed unit, index the direct clutch discs on the forward hub, and install the combined units into the case (Figure 20–15). The difficult part is indexing the forward clutch discs onto the front internal gear clutch hub. While this is taking place, the direct clutch drum lugs must engage the drive shell without disturbing the direct clutch friction disc indexing. Manipulate the clutch units into position until forward clutch contact is made with the internal gear clutch hub. It is now a matter of rotating the input shaft back and forth, using a light rocking action.

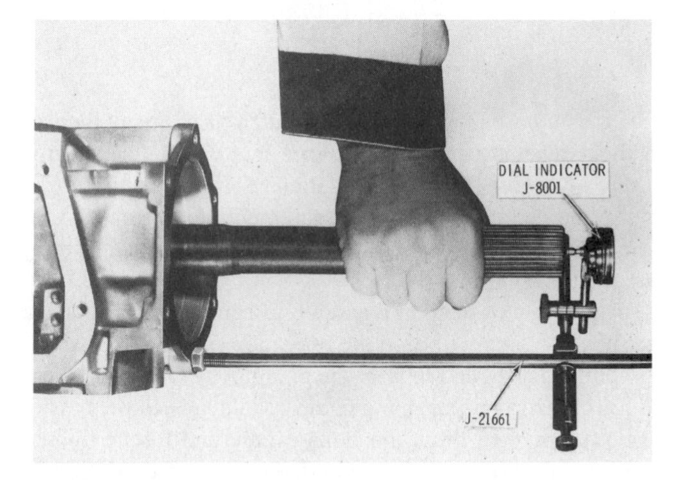

FIGURE 20–14 Checking output shaft endplay—3L80 (400). (Courtesy of General Motors Corporation, Service Technology Group)

FIGURE 20–15 Installing input clutch units—TorqueFlite 36RH (A-727).

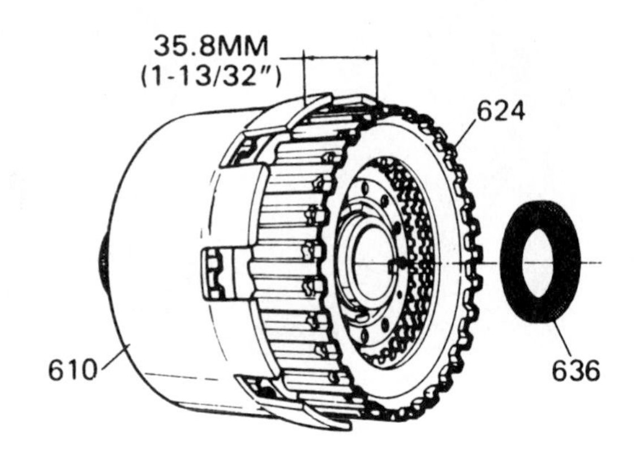

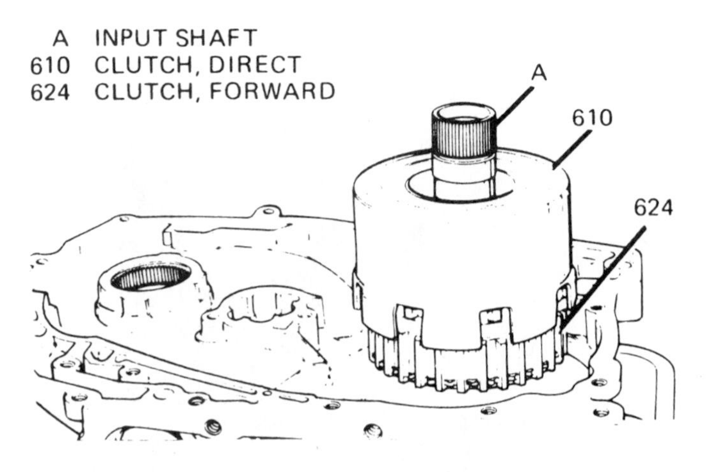

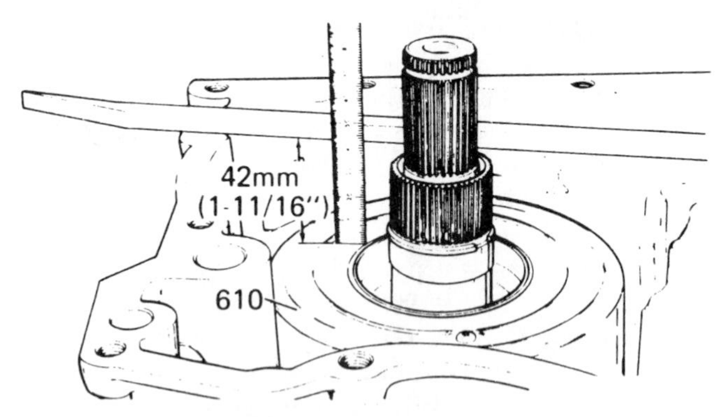

FIGURE 20–16 Installing input clutch units—3T40 (125-C). (Courtesy of General Motors Corporation, Service Technology Group)

If the forward clutch and direct clutch discs are all indexed on their hub, proper case clearance is attained for pump or sprocket support (FWD) installation. Some transmission assembly procedures include measurements to ensure that the clutch discs are all indexed and the clutch units are at the proper case depth (Figure 20–16).

- Installation of the pump assembly requires extra care. The use of guide pins will ease the operation. Be sure to install a new pump gasket and to treat the pump body O-ring seal with a lubricant. To avoid stripping pump bolt threads, loosely install all bolts and then draw the bolts evenly and torque-tighten. Check to make sure that the input shaft rotates freely during the tightening process. If the input shaft binds up, it means that the clutch discs are not all indexed to their hub or a thrust washer is out of place.

On some transmissions, such as the Ford A4LD, the pump bolts onto a separator plate, and the converter housing is not integral with the transmission case [Figure 20–17(a)]. The pump must be center aligned on the converter housing during installation [Figure 20–17(b)]. This prevents seal leakage, pump breakage, and bushing failure.

- When installing a servo assembly in the servo bore, do not get careless. A 4L60-E/4L60 (700-R4) 2-4 servo assembly is illustrated in Figure 20–18. Failure to lube the seals can cause the O-rings to twist in the groove or get pinched in the bore. It

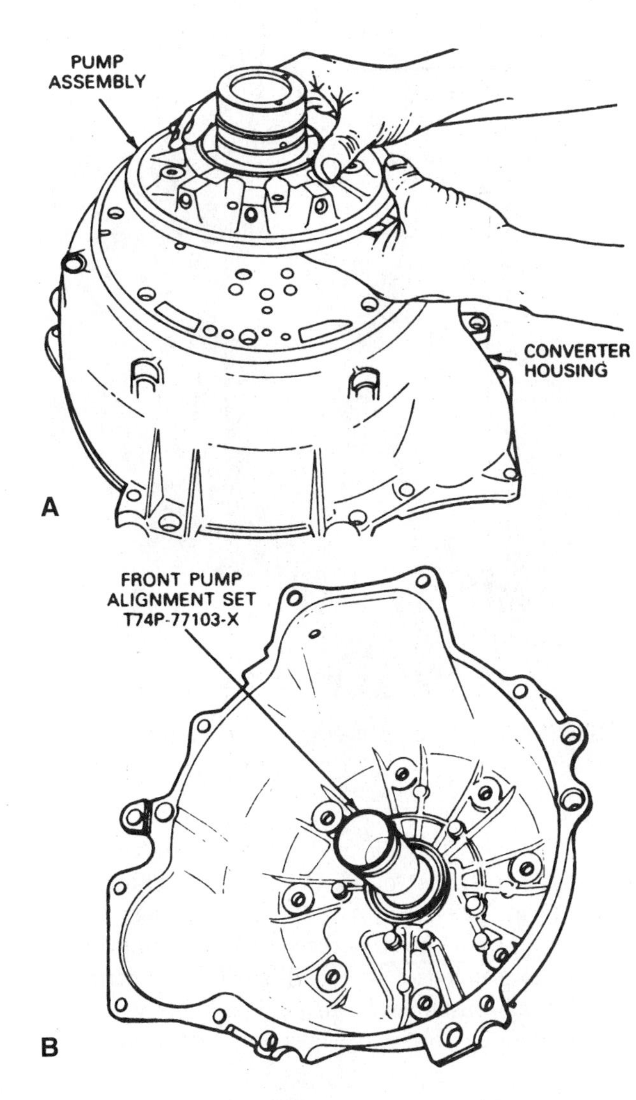

FIGURE 20–17 Ford A4LD: (a) pump housing unit; and (b) pump housing alignment tool. (Reprinted with permission of the Ford Motor Company)

FIGURE 20–18 2-4 servo assembly—GM 4L60-E/4L60 (700-R4).

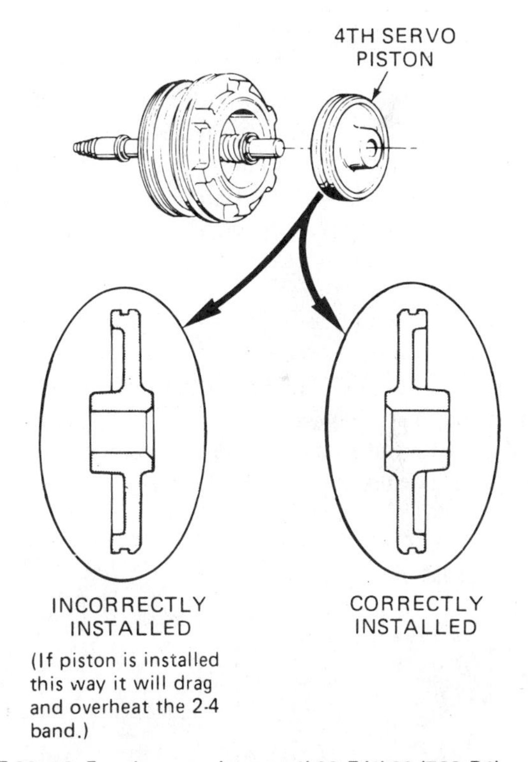

FIGURE 20–19 Fourth servo piston—4L60-E/4L60 (700-R4). (Courtesy of General Motors Corporation, Service Technology Group)

is also easy to misassemble. Figure 20–19 shows the incorrect and correct installation of the fourth servo piston.

- On band adjustments employing an adjustment screw, the adjustment specification must match the engine displacement and transmission model. A torque wrench or special torque-sensitive tool is used for preloading the band before adjusting the band clearance (Figure 20–20).
- Most band adjustments are now made with a select servo apply pin. The select apply pin controls the servo stroke travel and, in effect, takes the place of an adjusting screw. Because it takes special tooling and "fuss," this adjustment check is often bypassed. Even if the same band is used over again, the servo travel is a mandatory check.

In the absence of special servo adjustment tooling, an old servo cover can be drilled to accommodate a 5/16-in (7.94-mm) Phillips screwdriver. Manually stroke the servo piston to apply the band and measure the stroke travel (Figure 20–21). Stroke travel should measure 1/18 to 3/16 in (3.18 to 4.763 mm). The band adjustment is critical. It plays an important part in the timing of transitional shifts between a clutch releasing and a band applying, and when a band is releasing and a clutch is applying. The band might apply tightly around the drum during a servo air check, but that is not good enough.

- Once assembled into the case, run a reliability air check on the clutches, servo, and governor action using regulated air at 35 psi (245 kPa). The TorqueFlite 30RH/36RH (A-900/A-725) series case feed holes for these circuits are illustrated in Figure 20–22. The technician in Figure 20–23 is running a check on the rear clutch apply circuit. You should be able to hear and feel the clutch apply and release. The front clutch, governor, and the front and rear servo operation can all be air tested for operation

FIGURE 20–20 Adjusting band with torque wrench, overdrive band—Ford A4LD.

FIGURE 20–21 Servo travel check, using test servo cover with hole in center.

The 4T60 (440-T4) driven sprocket support clutch passages are identified in Figure 20–24. In Figure 20–25, an air check is being performed on the fourth clutch. Because the fourth clutch plates are not yet installed, the apply air should be limited to 10 psi (69 kPa). Otherwise, the piston will be blown out of its bore. The 1-2 servo apply circuit can be checked with the pipe in place (Figure 20–26).

To air check clutch circuits when the upper valve body channels are part of the case, it may be necessary to use a different technique. One method is to form a seal around the feed hole with putty (Figure 20–27). Remove putty completely when done with the air test. A rubber ball with a metal tube insert can then be used for the air check (Figure 20–28). Some manufacturers and equipment suppliers provide a special case test plate for the servo, accumulator, and clutch circuit air checks (Figure 20–29).

Accurately air checking the circuits and units in the transmission takes practice. There will always be a certain amount of air leakage past the seal rings and bushings, which you must determine to be normal or excessive. The more checks you

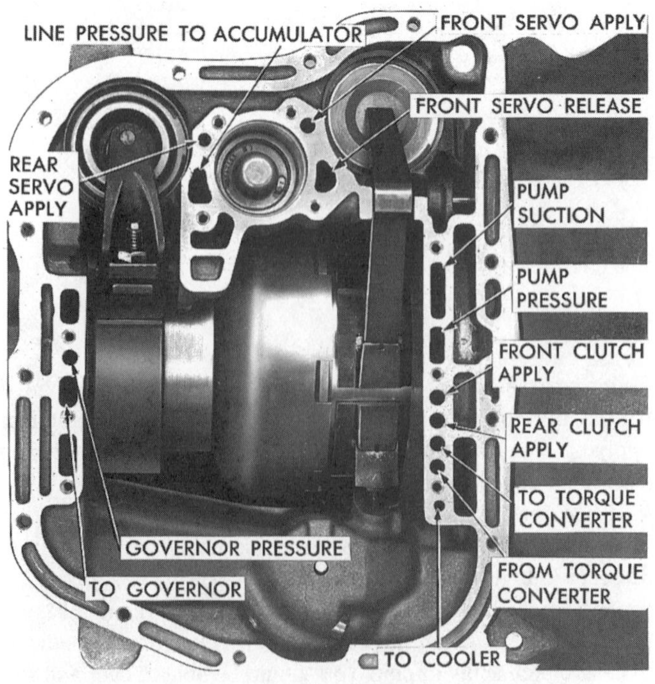

FIGURE 20–22 Oil circuit feed hole identification. (Courtesy of Chrysler Corp.)

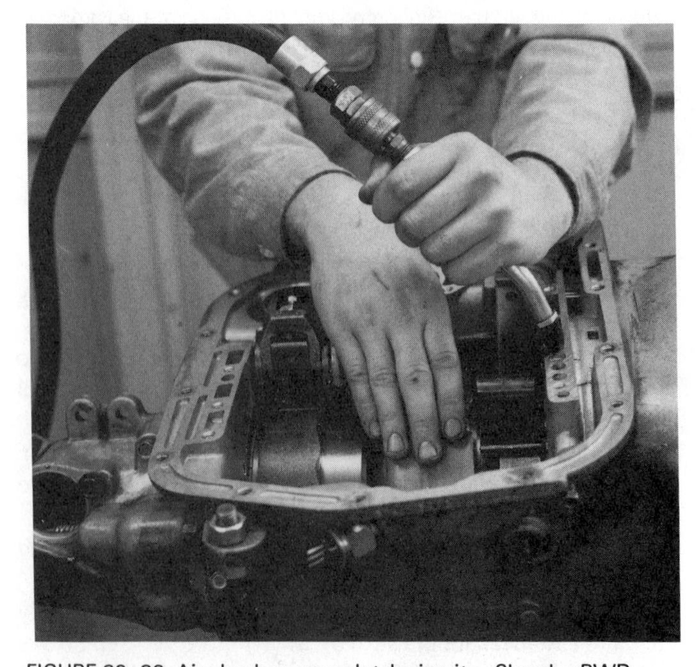

FIGURE 20–23 Air check on rear clutch circuit—Chrysler RWD three-speed transmission.

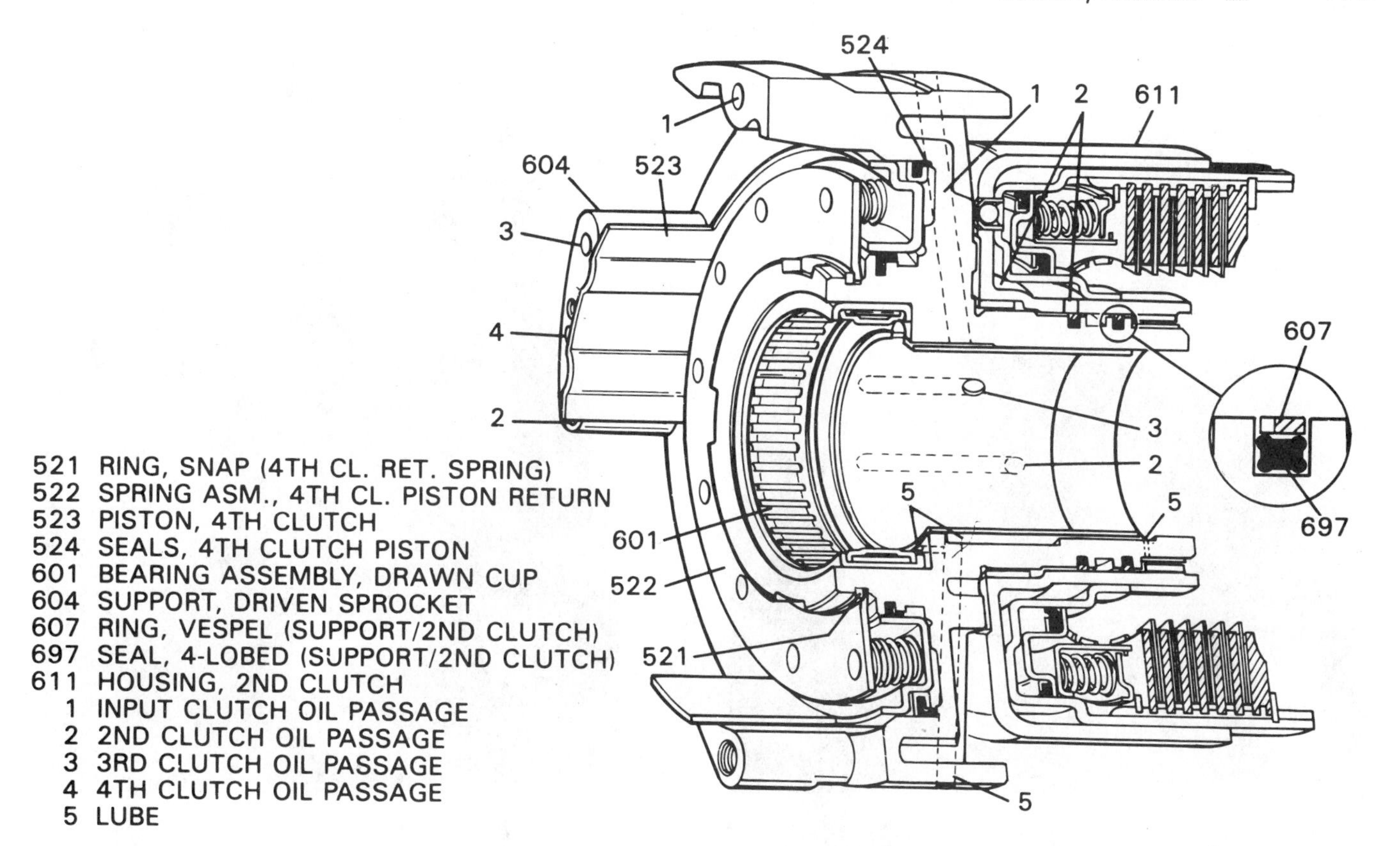

FIGURE 20–24 4T60 (440-T4) driven sprocket support hydraulic circuit passages. (Courtesy of General Motors Corporation, Service Technology Group)

make, the more proficient you can become in the art of air checking in the case.

A Hydraulic Circuit Analyzer, produced by Answermatic, Inc., uses case-mounted test plates with various line fittings (Figure 20–30). Individual clutch/servo circuits can be charged with oil. The oil pressure is measured at the circuit and compared to the specifications provided. If the charge oil does not read within 10 psi (69 kPa) of the specified pressure, excessive circuit leakage is indicated and must be corrected. This test can be performed on the bench or in the car, before or after a rebuild.

- It may be necessary to remove certain distribution tubes that depend on a case press fit for sealing (Figure 20–31). When installing, lightly tap the tubes until fully sealed. Apply Loctite or equivalent around the tube to case surface.

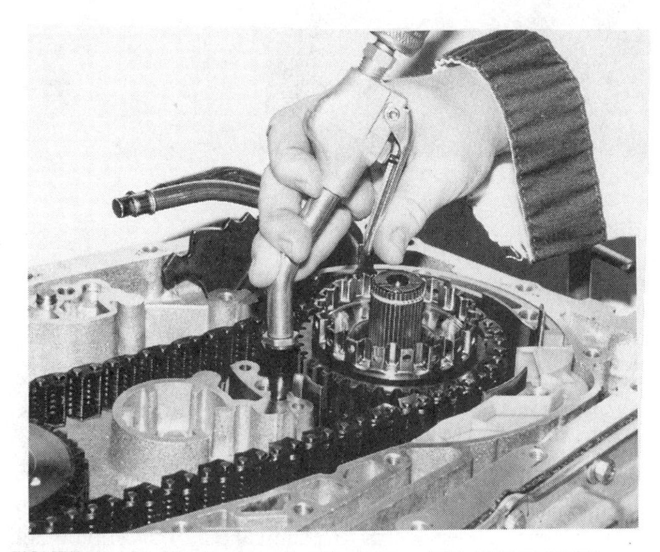

FIGURE 20–25 Air checking fourth clutch—GM 4T60 (440-T4).

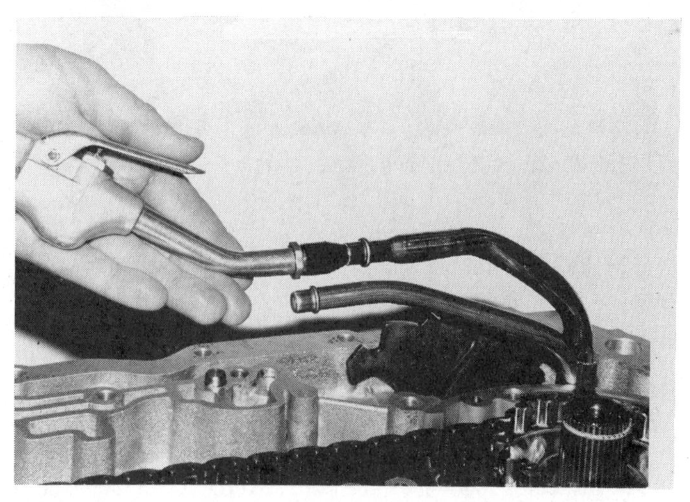

FIGURE 20–26 Air checking 1-2 servo apply circuit—GM 4T60 (440-T4).

FIGURE 20–27 Preparing circuit for air testing.

FIGURE 20–28 Air testing hydraulic circuit with rubber ball.

FIGURE 20–29 Ford AOD and air test plate.

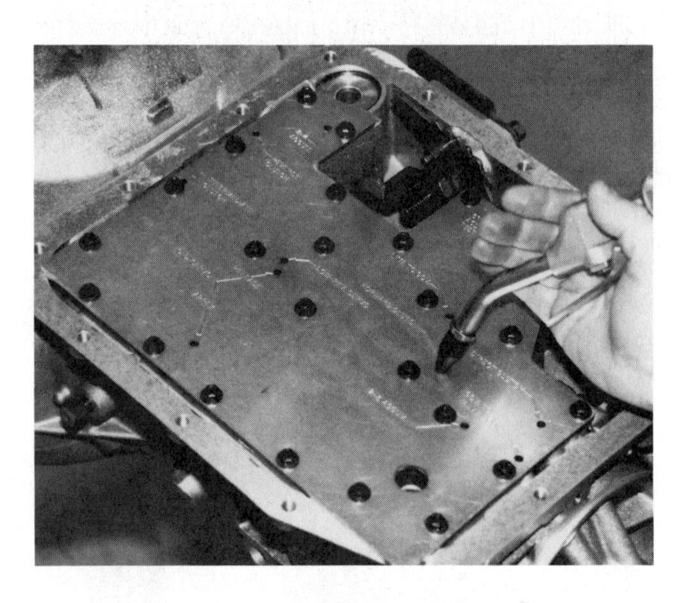

• Make sure that the valve body gasket and separator plate are not mismatched. With the gasket aligned to the separator plate, hold the gasket against the light. The separator plate should have no holes blocked. Some valve body gaskets will need to be sequenced with the separator plate and are usually marked (Figure 20–32). Occasionally, new gaskets and separator plates are introduced. In Figure 20–33, three sets of valve body gaskets are shown for the Ford AXOD. As another example, Ford introduced new valve body gaskets for the 1989 AOD. They will retrofit all previous models, but do not use 1988 or earlier gaskets on a 1989 or newer models. The earlier gaskets are minus a few holes, and if these gaskets are used, second gear is blocked out on a 1989 or newer AOD model. Expect a shift complaint of 1-3-4. The technician must be alert to valve body gasket and separator plate changes on a year-to-year basis.

• Valve body and case/channel plate check ball locations are very important. These parts definitely must all be there and be in the right places. The technician who is unfamiliar with the unit and does not have access to technical pictorials should not trust memory but should take a Polaroid picture during disas-

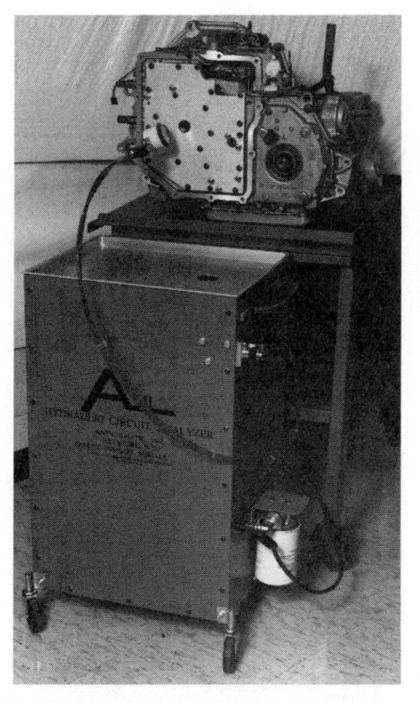

FIGURE 20–30 Hydraulic Circuit Analyzer; used to check clutch, servo, and accumulator circuits for leakage with oil pressure, gauge unit, and test plate—Ford AX4S (AXODE).

Tighten park rod abutment bolts to 27-30 N·m (20-22 lb-ft).

Tighten reverse drum 6mm Allen head anchor bolt to 10-12 N·m (7.5-9 lb-ft) and 19mm locknut to 34-47 N·m (25-35 lb-ft).

Install tubes in position and tap lightly until fully seated. Apply Threadlock 262, E2FZ-19554-B or equivalent around tube-to-case surface.

Install tube retaining brackets.

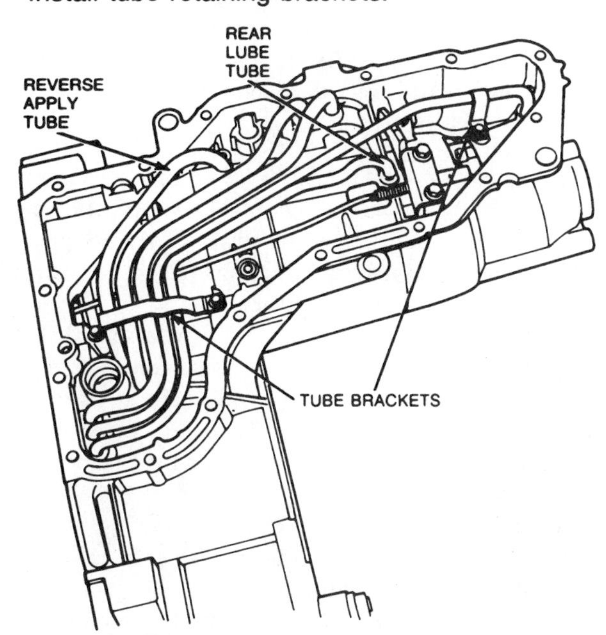

FIGURE 20–31 AXOD oil tubes. (Reprinted with permission of the Ford Motor Company)

sembly when the check balls are exposed. Typical check ball locations are shown for the Ford AXOD in Figure 20–34 (illustrative purposes only).

The technician must be alert to check ball location changes on a year-to-year basis. In Figure 20–35, valve body check ball locations are compared between Hydra-Matic 4L60 (700-R4) models with and without an auxiliary valve body. Note that check ball number 5 is eliminated from its bathtub-shaped cavity in auxiliary valve body applications. Installation of forward clutch number 12 check ball in the valve body or failure to install number 12 check ball in the auxiliary valve body can cause a variety of problems:

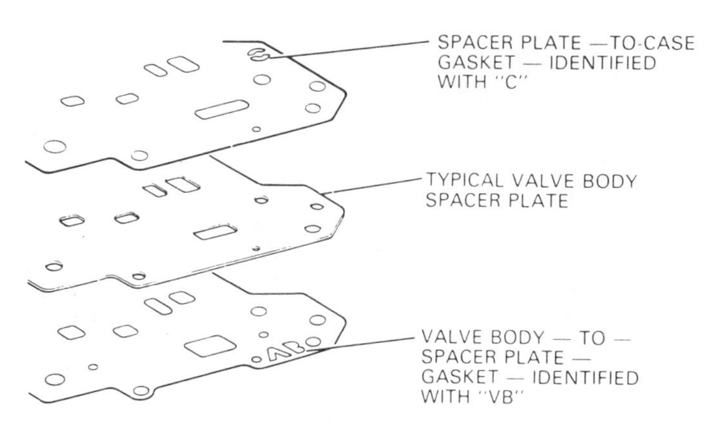

FIGURE 20–32 Typical Hydra-Matic spacer plate gasket identification. (Courtesy of General Motors Corporation, Service Technology Group)

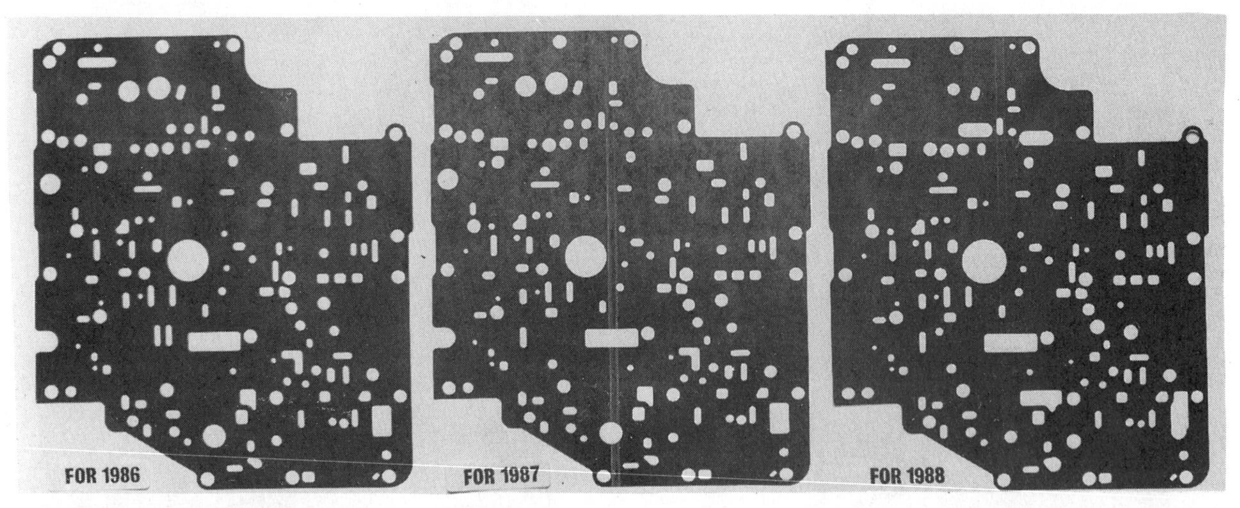

FIGURE 20–33 Three different Ford AXOD valve body gaskets.

1. delayed drive engagement
2. harsh drive engagement
3. slipping on hard acceleration
4. burned forward clutch
5. no upshift
6. 1-2 tie-up, burned low and reverse clutch pack, burned band

To help in check ball location, examine the separator plate holes over the bathtub-shaped cavities. The presence of two holes over the tub area usually calls for a check ball. Check balls can be difficult to keep in place during installation. Use a petroleum jelly to keep them fixed in their cavities.

In the servo bore of Hydra-Matic 4L60-E, 4L60 (700-R4), 200-4R, and 200-C transmissions, it is important to run a leak check on the retainer and ball assembly. The tube-style assembly is located in the release side of the servo acting as an accumulator (Figure 20–36). If it is leaking or missing, it will mean a repeat 3-4 clutch failure in the 4L60-E/4L60 or a direct clutch failure in the 200-4R and 200C.

The check ball sealing is checked with the servo assembly in place. Fill the servo accumulator cavity with mineral spirits and watch for leakage inside the case (Figure 20–37). If any leakage occurs, replace the retainer and ball assembly. Refer to the service manual.

- In the new generation of transmissions, it is common to find a **variable orifice thermal valve** located on the separator/spacer plate (Figure 20–38). The purpose is to provide a variable feed rate of apply oil in a clutch or servo circuit, depending on oil temperature. At 0°F (-18°C) the thermal orifice is wide open, and two flow paths charge the circuit to compensate for cold oil viscosity. As the oil gets warm, the thermal valve gradually shuts down one of the flow paths. At 70oF (21oC), the thermal valve completely closes, and one charge path only feeds the apply circuit.

❖ **Variable orifice thermal valve:** Temperature-sensitive hydraulic oil control device that adjusts the size of a circuit path opening. By altering the size of the opening, the oil flow rate is adapted for cold to hot oil viscosity changes.

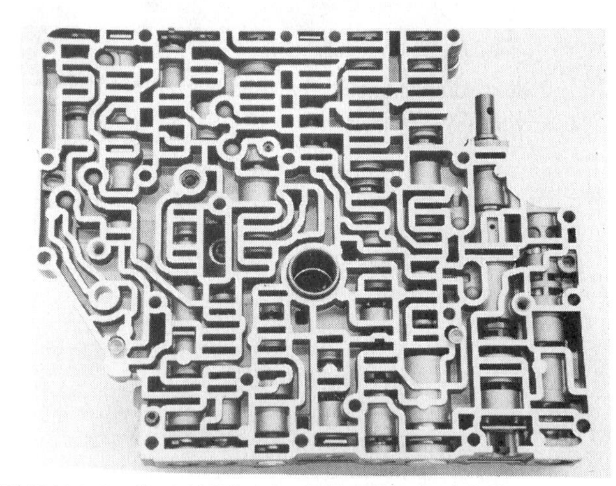

FIGURE 20–34 Ford AXOD valve body with check balls.

It is extremely important that these thermal valves are not damaged and are responsive to heat. The valve action can be tested with a heat gun or match. These valves are generally keyed to reverse and forward clutch engagement and can affect the timing and feel of the engagement.

- On FWD units with chain covers, it is essential that the different bolt sizes are located properly and torqued in sequence (Figure 20–39). The valve body and chain cover must be clamped correctly to avoid unwanted circuit leaks. After installing the chain cover, the input shaft should have some endplay and rotate freely.
- Remember that the leading cause of sticking valves and internal leakage is improper torquing. Tighter is definitely not better. A standard M.6 bolt torqued to 10 to 12 ft-lb (14 to 16 N-m) will produce an 1800-lb (9200-N) clamping force on the bolt surface area of the valve body, gasket, and case. Some valve body bolt torques are as low as 35 to 60 in-lb (5 to 7 N-m).

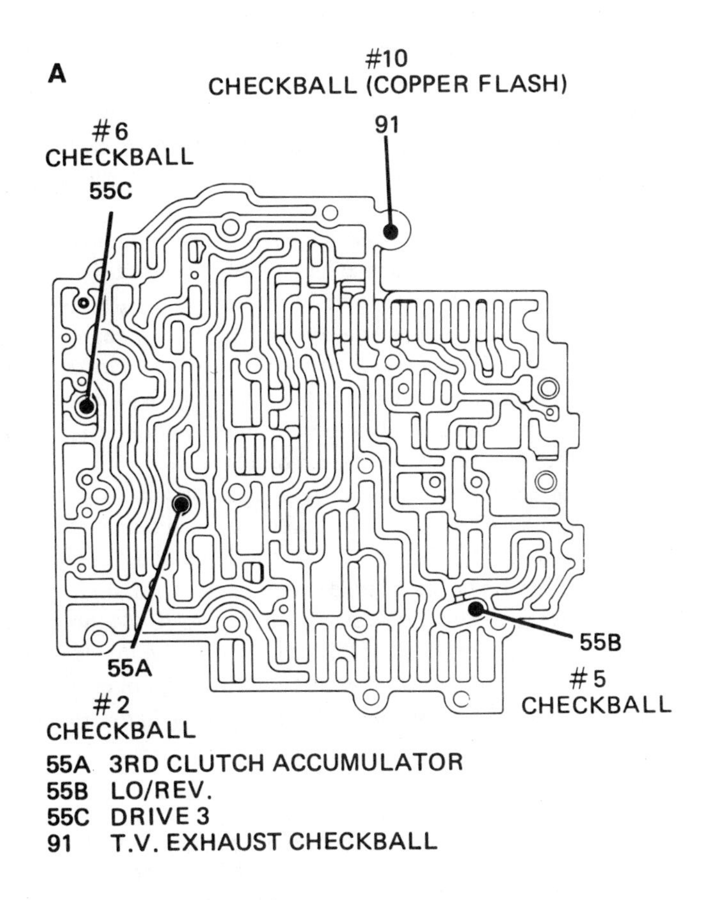

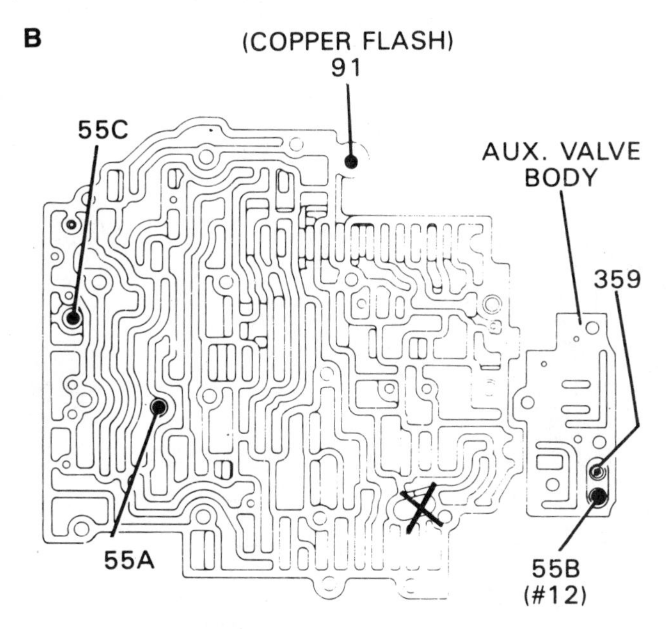

55A #2 CHECKBALL (3RD CLUTCH ACCUM.)
55B #12 CHECKBALL (FORWARD CLUTCH)
55C #6 CHECKBALL (DRIVE 3)
91 #10 CHECKBALL (T.V. EXHAUST)
359 CUP PLUG – ORIFICE

FIGURE 20–35 4L60 (700-R4): (a) without auxiliary valve body; (b) with auxiliary valve body. (Courtesy of General Motors Corporation, Service Technology Group)

FIGURE 20–36 2-4 servo bore—GM 4L60-E/4L60 (700-R4).

FIGURE 20–37 GM 4L60-E/4L60 (700-R4): (a) filling servo bore with ATF (2-4 servo assembly in place); (b) inspecting for servo check ball leakage.

FIGURE 20–38 Variable orifice thermal valve located on valve body spacer plate.

To install the valve body properly, loosely install all the attaching bolts. If a torque-tightening sequence is not provided in the service manual, start at the center and use a spiral or circular pattern to the outside. Torque-tightening is illustrated in Figures 20–40 and 20–41. Note that an actual torque-tightening sequence is provided for the Ford AXOD, following a center to outside pattern. When torquing, keep all the torque readings either to the high or the low side of the specification range. In some cases, after torquing, the shift valve movement can be checked with air through the governor output circuit (Figure 20–42).

- Do not forget to replace the magnet in the pan bottom (Figure 20–43). This is added insurance for collecting the dust-like metal particles out of the fluid that would normally pass through the filter. The oil filter cannot do a perfect job. Otherwise, it would plug in a hurry. The filter and magnet are important to keeping the oil clean and ensuring trouble-free valving. Make sure the magnet fits over the pan dimple so that it does not interfere and damage the filter when the pan is bolted up.
- It is not uncommon to eliminate the pan or side cover paper gasket in favor of **RTV.** It is an effective gasket sealant if the oil pan or side cover flange is flat or has depressed ribs (Figure 20–44). If the flange is designed with a raised rib, a gasket is necessary (Figure 20–45). Referring to Figure 20–44, note that a 1/16-in (0.40-mm) bead of RTV must be on the inside of the bolt holes. The bead must be continuous and unbroken. Otherwise, a leak path is formed. Avoid applying an excessive amount of RTV.

Start remaining chain cover bolts and tighten 10mm bolts to 27-33 N·m (20-26 lb-ft). Tighten 8mm bolt to 9-12 N·m (7-9 lb-ft). Tighten 13mm bolt to 34-48 N·m (25-35 lb-ft). Tighten bolts in sequence shown.

NOTE: After installing chain cover, input shaft should have some end play and should rotate freely. If it will not rotate freely, remove chain cover and inspect cast iron seal for damage.

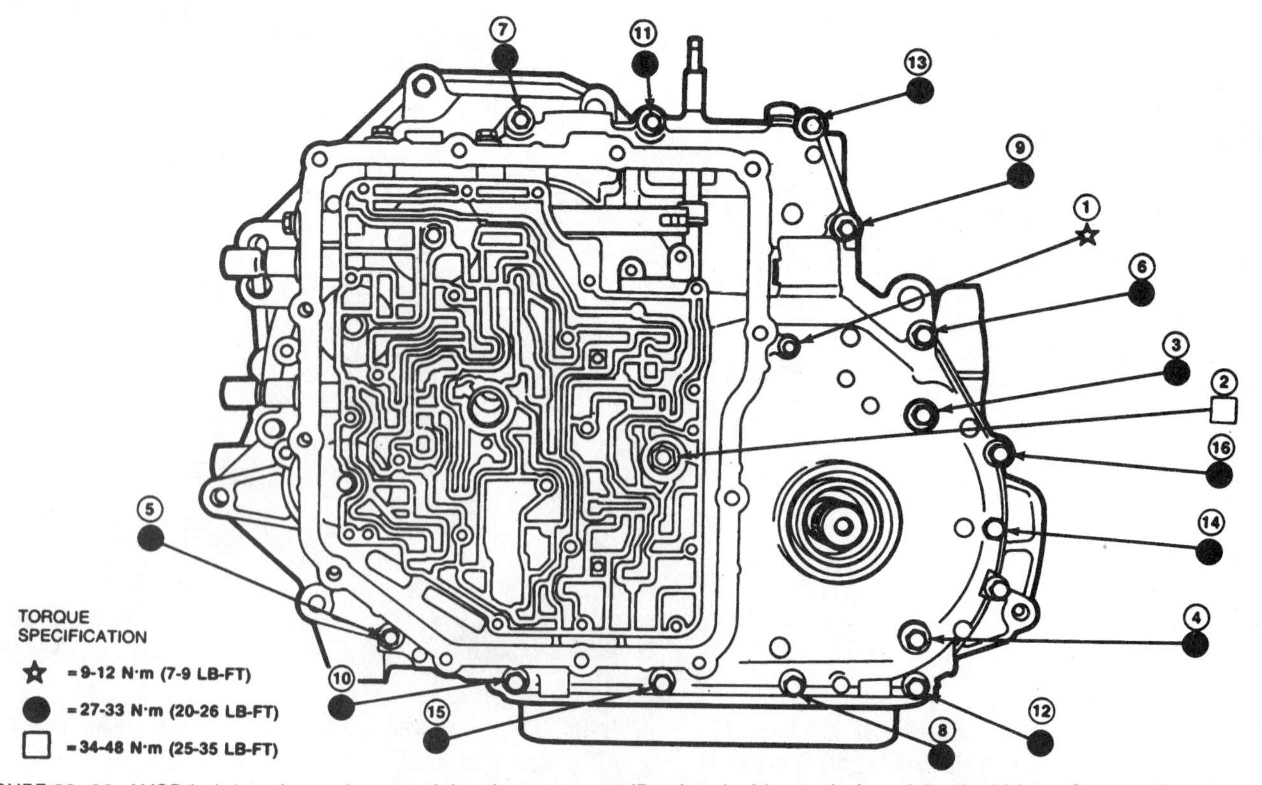

FIGURE 20–39 AXOD bolt location and torque-tightening sequence. (Reprinted with permission of the Ford Motor Company)

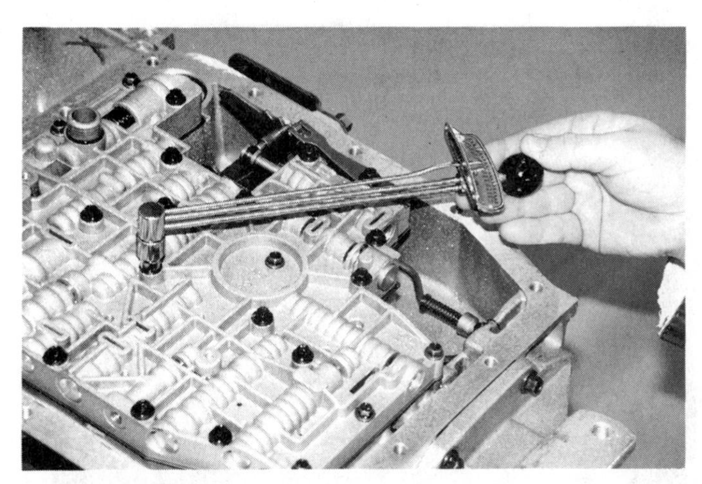

FIGURE 20–40 Ford AOD valve body, torque-tightening fasteners.

❖ **RTV:** A gasket making compound that cures as it is exposed to the atmosphere. It is used between surfaces that are not perfectly machined to one another, leaving a slight gap that the RTV fills and in which it hardens. The letters RTV represent room temperature vulcanizing.

The case and oil pan flanges must be dry and free of any oil film for a leakproof seal to form. Do not use a petroleum-based solvent for cleaning. It leaves an oily film to which RTV will not stick. Carburetor spray is an effective cleaner. Install the pan while the sealant is wet. Do not wait for it to skin over. Special pan bolts are required (Figure 20–46). If the bolt head has incurred a reverse dished flange, it must be replaced or else sealing may not take place. Whether using RTV sealant or a paper gasket, the pan bolt holes must be flat with the flange. If they are dished from previous bolt installation, use the lip edge of the bench and a heavy hammer to recondition (Figure 20–47).

NOTE: Many late model transmissions use reusable pan gaskets. These rigid rubber-based gaskets should be inspected prior to reinstallation. Refer to the service manual to determine if they should be reused.

- The governor cover flange on the THM-250C, THM-350C, 4L60 (700-R4), and 3L80 (400) should be treated with Loctite (Figure 20–48). During acceleration and deceleration, a fore-and-aft hammering action of the governor on the cover can loosen the cover and push it out of the governor hole. The transmission will lose governor pressure and downshift to first gear regardless of vehicle speed.

Install 22 valve body bolts and tighten in sequence to 9-12 N·m (7-9 lb-ft).

NOTE: Install three short bolts where indicated.

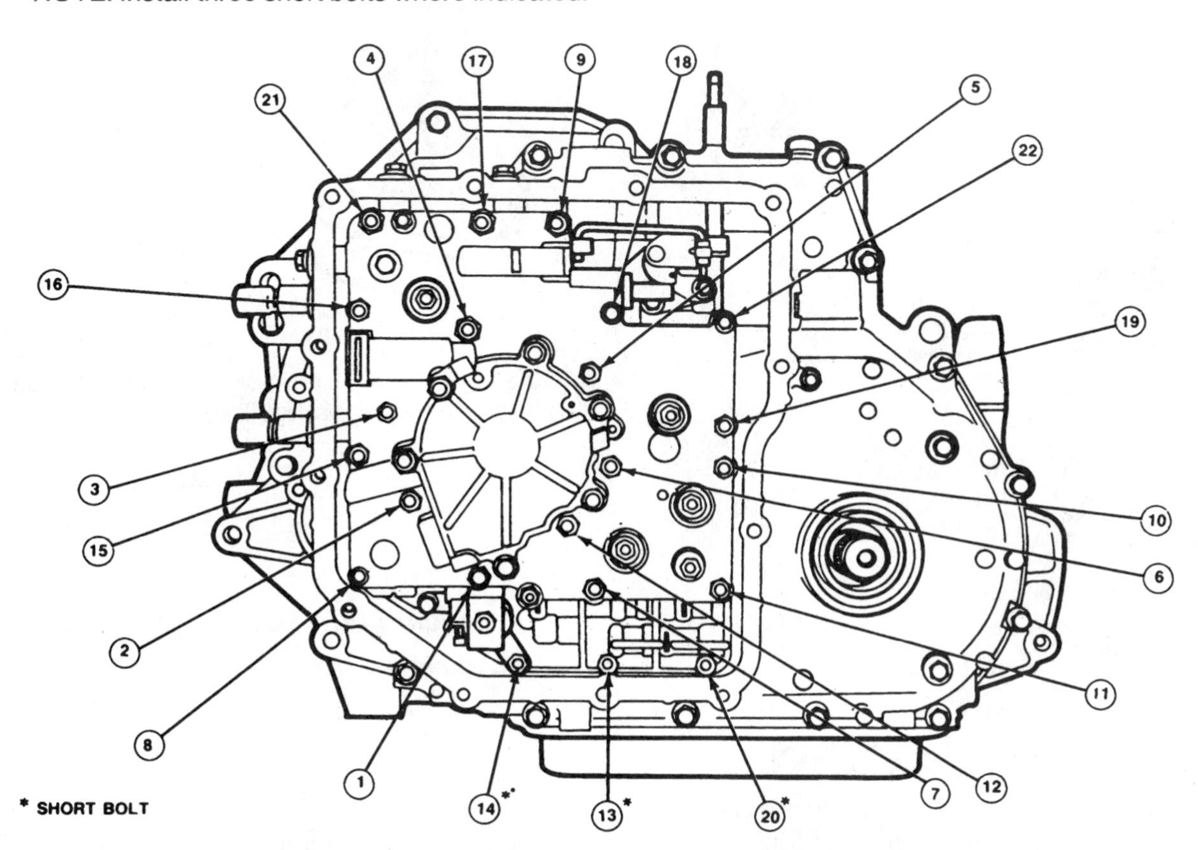

FIGURE 20–41 Ford AXOD valve body, torque-tightening fasteners. (Reprinted with permission of the Ford Motor Company)

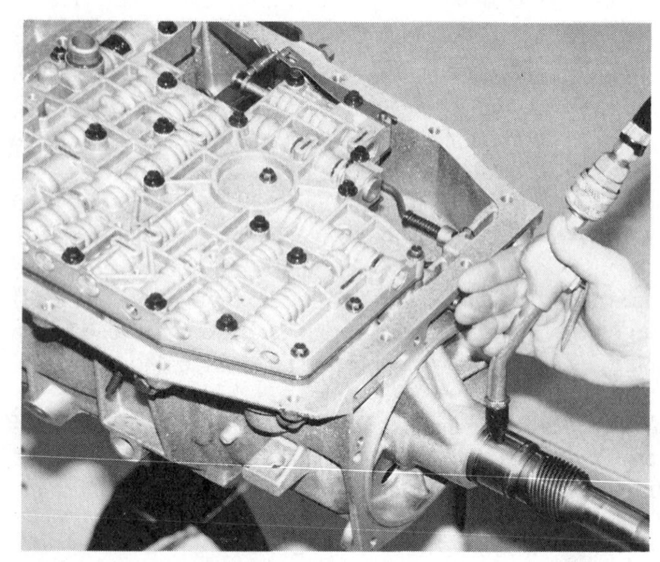

FIGURE 20–42 Air checking shift valve movement from governor port—Ford AOD.

FIGURE 20–43 Magnet positioned in bottom pan.

- Align and adjust the neutral safety/start switch when applicable (Figure 20–49).
- If the transmission is equipped with a vacuum modulator, test it before installing it.

After installing the transmission in the car, always perform a complete road test. Make a final check on the fluid level and inspect for fluid leaks. If the radiator system is equipped with an auxiliary cooler fan, make sure that it is working.

SUMMARY

When assembling the many internal transmission components, care must be taken to avoid making mistakes. Use a service

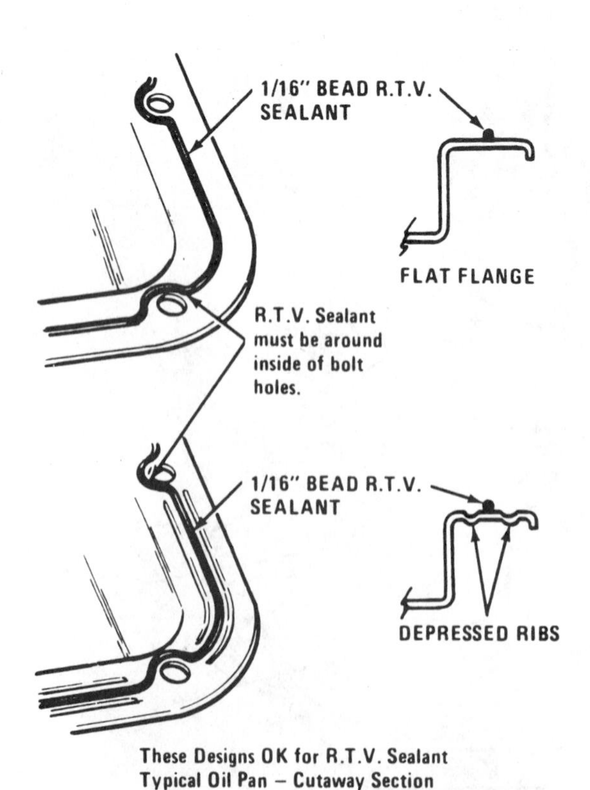

FIGURE 20–44 Bottom pan, flange-style and depressed ribs designs. (Courtesy of General Motors Corporation, Service Technology Group)

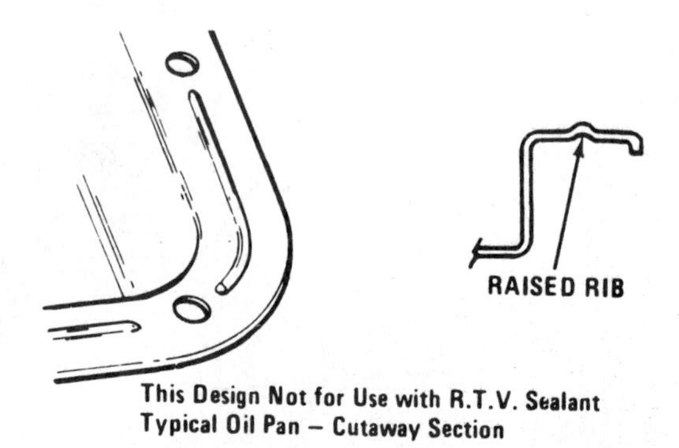

FIGURE 20–45 Raised rib pan design. (Courtesy of General Motors Corporation, Service Technology Group)

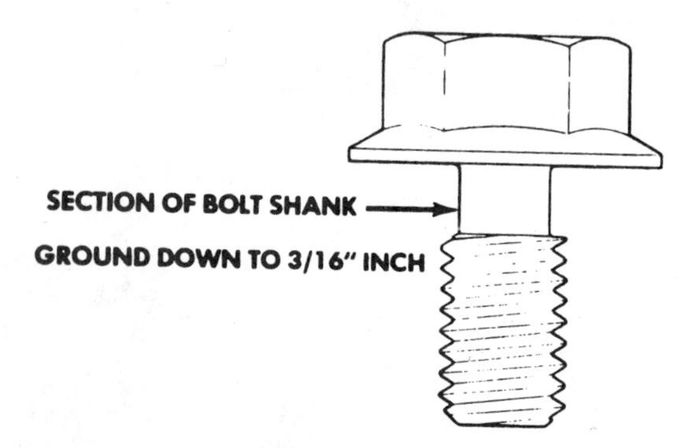

FIGURE 20–46 Special oil pan bolts. (Courtesy of General Motors Corporation, Service Technology Group)

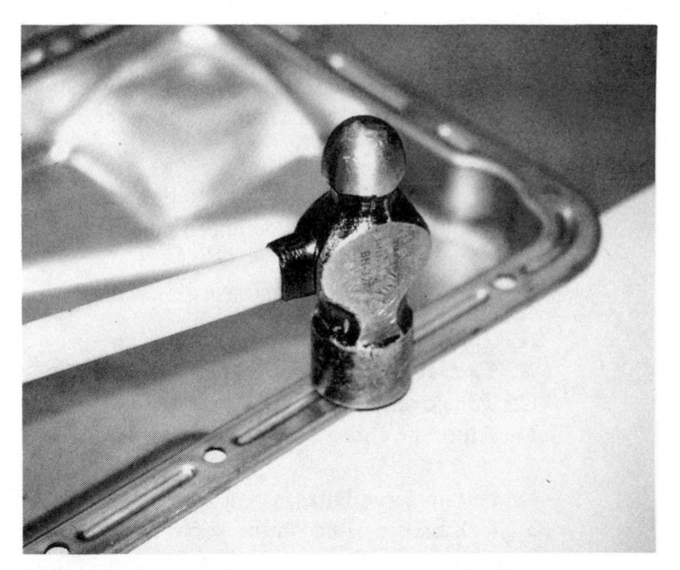
FIGURE 20–47 Correcting oil pan flange distortion.

FIGURE 20–48 Governor cover installed; cover bore flange treated with Loctite.

manual as a guide to establish an order for reassembly and recognize critical checks. If at any time components are not seated fully or do not turn freely, stop and correct the cause of the problem. In addition, to prevent future problems, perform any critical transmission updates and modifications described in service bulletins.

Any replacement of components that is called for must not alter the performance of the transmission or transaxle. This is especially important when renewing final drive units in transaxles. Since a variety of gear combinations are available, they must match the original setup.

When assembling internal items, always prepare and position them accurately. Devices such as thrust washers, thrust bearings, seals, and gaskets must be properly installed. Fiber clutch plates and bands must be soaked in ATF for a minimum of thirty minutes. Also, prelubricate gears, shafts, bushings, and one-way clutches with ATF. Carelessly ignoring the importance of these concerns will jeopardize the quality of the rebuilt transmission.

While rebuilding the unit, perform measurement checks as required. Endplay dimensions, band/servo pin travel lengths, and clutch pack clearances can be easily adjusted.

Complete your overhaul by making sure that hydraulic circuits, electrical systems, and mechanical devices function as designed. Typical shop tools and special testers help to make this task easier.

After installing the unit into the vehicle, perform a complete road test to verify that the transmission is performing properly. Afterward, make one final check of the fluid level and adjust if needed. The service life of the transmission, as well as the reputation of you and your shop, will soon be determined by the customer.

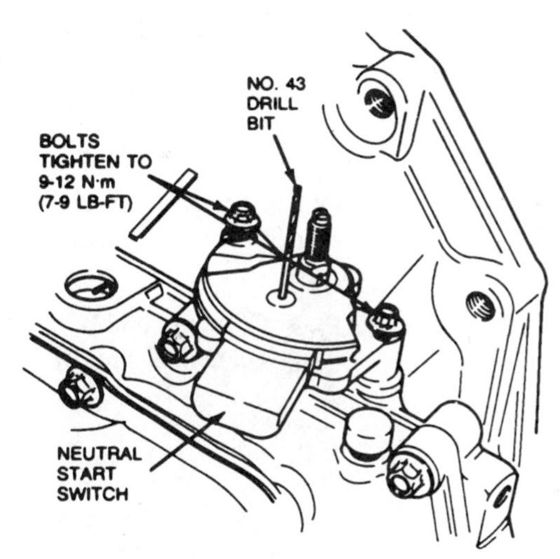

FIGURE 20–49 Neutral safety/start switch—AXOD. (Reprinted with permission of the Ford Motor Company)

❑ REVIEW

Key Terms/Acronym

Variable orifice thermal valve and RTV.

Questions

____ 1. What is true when assembling a transaxle differential and helical final drive?

I. Reused side bearings must be adjusted for preload.

II. Replacement helical ring and pinion gears come as a matched set.

A. I only C. both I and II
B. II only D. neither I nor II

____ 2. What is true when assembling a transaxle with taper roller bearing applications?
I. Turning torque is adjusted with selective shims or spacers.
II. Used bearing turning torque must be adjusted to new bearing specifications.

A. I only C. both I and II
B. II only D. neither I nor II

3. What are three points of wear concern when inspecting the differential gear assembly on a transaxle unit?

(a)________________________________

(b)________________________________

(c)________________________________

____ 4. Technician I says, "Nonseparable needle-type thrust bearings can be installed correctly with either side facing up or down." Technician II says, "Prelubricate all O-rings and case seals with a petroleum jelly." Who is right?

A. I only C. both I and II
B. II only D. neither I nor II

____ 5. What is true about Teflon seal rings?
I. Mismatched scarf-cut ends will cause sealing failure.
II. Teflon is a soft material, and care must be taken not to nick the rings.

A. I only C. both I and II
B. II only D. neither I nor II

____ 6. If the input or turbine shaft fails to turn after the pump or a sprocket support is bolted into place, the cause may be:
I. a mislocated thrust washer.
II. an input clutch unit that is not fully indexed to its hub.

A. I only C. both I and II
B. II only D. neither I nor II

____ 7. On servo units not equipped with a band-adjusting screw:
I. band clearance is adjusted with a select fit apply pin.
II. the clearance check on a reused band is not necessary.

A. I only C. both I and II
B. II only D. neither I nor II

____ 8. Occasionally, the technician may observe:
I. valve body gasket and separator plate changes.
II. valve body and case/channel plate check ball location changes.

A. I only C. both I and II
B. II only D. neither I nor II

____ 9. Which is true about in-the-case air checks?

A. The clutches can be tested.
B. The servos can be tested.
C. The governor can be tested.
D. All of the above.

____ 10. On certain servo bores, when a tube retainer with a check ball inserted on the servo release side is used:
I. a leaky check ball will cause a repeat band failure.
II. the technician should check for leakage by filling the servo release cavity with mineral spirits.

A. I only C. both I and II
B. II only D. neither I nor II

____ 11. When torquing valve body bolts:
I. keep all the torque readings to either the high or low side of specifications.
II. start at the center and use a spiral pattern to the outside if torque sequencing is not given.

A. I only C. both I and II
B. II only D. neither I nor II

____ 12. Technician I says, "Usually, the presence of two separator plate holes over a bathtub-shaped cavity calls for a check ball." Technician II says, "Check ball location is always the same from year to year." Who is right?

A. I only C. both I and II
B. II only D. neither I nor II

____ 13. RTV sealant:
I. cannot be used on flanges with raised ribs.
II. must form a dry skin on the flange before installation.

A. I only C. both I and II
B. II only D. neither I nor II

21 External Fluid Leaks

OBJECTIVE:

The objective of this unit is for you to become proficient with the terms, concepts, and procedures contained in this chapter. Completion of this objective is essential to becoming a successful automatic transmission technician.

CHALLENGE YOUR KNOWLEDGE

List the following:

Four causes of gasket leaks; six locations of external seals; eight causes of seal leaks; and five possible leak sources within a transmission converter housing.

Describe the steps of procedure when using the following leak source detection methods:

Visual inspection; fluorescent dye and black light; and powder spray.

INTRODUCTION

Locating and repairing external fluid leaks is another type of service required for automatic transmissions/transaxles. Most fluid leaks can be easily located and repaired, although some may be tricky and difficult to pinpoint.

Underbody air currents sometimes blow the fluid around and make the job of locating the leak more difficult. Figure 21–1 illustrates a leak at a transmission cooler line fitting, often a tricky problem to identify. Note how the fluid leak leads to the head of the pan and flows off at the rear. Air currents also pick off the fluid and work it against the transmission extension housing. Be careful with this type of leak. Do not replace an extension housing seal or pan gasket, only to find out later that the solution to the problem was to tighten or repair a cooler line fitting.

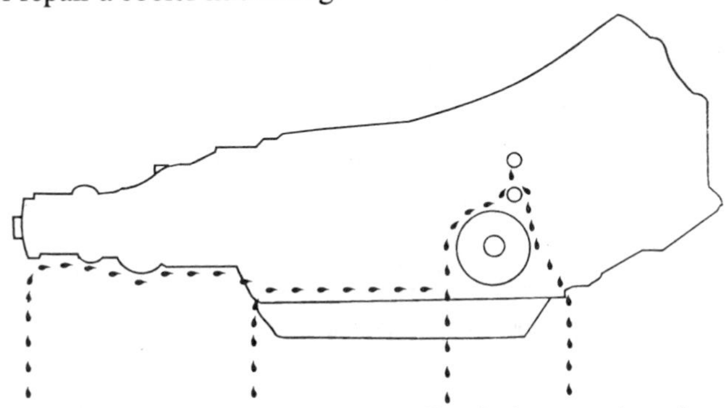

FIGURE 21–1 Underbody air currents affect leak source detection.

What can also complicate a leak detection problem is engine oil or a combination of transmission/transaxle fluid and engine oil dripping from the transmission/transaxle area. Engine oil leaks from the rear crankshaft seal or rocker arm covers and oil galley plug leaks from the rear of the engine block are some of the possible sources. The best policy on external leaks is to take a few extra minutes and examine all possibilities visually. Figure 21–2 shows the possible leak sources on a GM Hydra-Matic 4L60 (700-R4).

LEAK DETECTION

Visual—Preliminary Checks

- ❍ Check the transmission/transaxle fluid level. A high fluid level can cause external leaks.
- ❍ Identify the fluid source. Is the fluid engine oil, actual transmission/transaxle ATF, or even possibly power steering fluid?
- ❍ If a quick visual inspection does not locate the source, take the vehicle on the road and bring the ATF operating temperature to normal. Park the vehicle and place a large sheet of paper or cardboard under the transmission/transaxle area. Wait ten minutes, and then examine the paper/cardboard for drips. This is useful in locating the approximate leak area. Keep in mind that the leak area is not necessarily the leak source. Some leaks do not occur until the ATF trickles down into the pan, after the engine is OFF and has

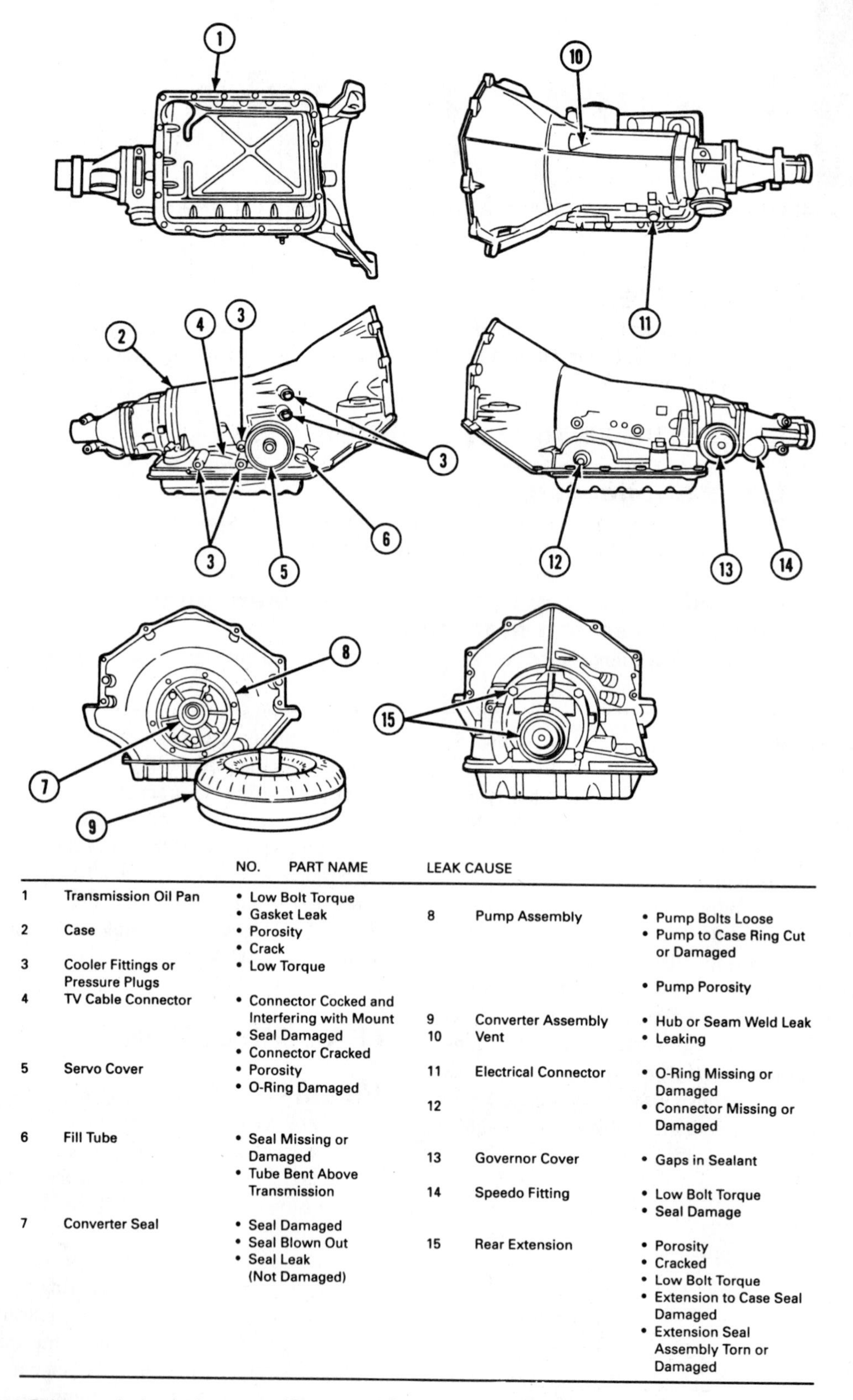

NO.	PART NAME	LEAK CAUSE
1	Transmission Oil Pan	• Low Bolt Torque • Gasket Leak
2	Case	• Porosity • Crack
3	Cooler Fittings or Pressure Plugs	• Low Torque
4	TV Cable Connector	• Connector Cocked and Interfering with Mount • Seal Damaged • Connector Cracked
5	Servo Cover	• Porosity • O-Ring Damaged
6	Fill Tube	• Seal Missing or Damaged • Tube Bent Above Transmission
7	Converter Seal	• Seal Damaged • Seal Blown Out • Seal Leak (Not Damaged)
8	Pump Assembly	• Pump Bolts Loose • Pump to Case Ring Cut or Damaged • Pump Porosity
9	Converter Assembly	• Hub or Seam Weld Leak
10	Vent	• Leaking
11	Electrical Connector	• O-Ring Missing or Damaged
12		• Connector Missing or Damaged
13	Governor Cover	• Gaps in Sealant
14	Speedo Fitting	• Low Bolt Torque • Seal Damage
15	Rear Extension	• Porosity • Cracked • Low Bolt Torque • Extension to Case Seal Damaged • Extension Seal Assembly Torn or Damaged

FIGURE 21–2 4L60 (700-R4) transmission leak sources. (Courtesy of General Motors Corporation, Service Technology Group)

FIGURE 21–3 Spray powder or florescent dye can be used for locating fluid leak sources.

built up a high sump level. It takes a few minutes with the vehicle at rest to get these leaks working.

- On the hoist, visually inspect the suspect area and component. In hard-to-inspect locations, use of a mirror is helpful.
- For a suspected servo cover leak, temporarily run the transmission/transaxle on the hoist in the appropriate operating range to activate the intermediate, overdrive, or reverse servo where applicable.

CAUTION: Follow the manufacturer's recommendation for operating the powertrain on the hoist. Basically, the drive suspension must be supported at or near normal height.

FIGURE 21–4 Black light picks up fluorescent dye trace from fill tube and servo cover.

If the leak remains a mystery, use either a black light and fluorescent dye or a powder spray for leak detection (Figure 21–3).

Fluorescent Dye and Black Light

Adding a fluorescent dye to the oil and using a black light may be necessary when locating certain leak sources. This is the favored technique for tough leaks. It does an excellent job.

- Add one ounce of fluorescent dye to the ATF.
- Operate the vehicle on the road and thoroughly warm up the ATF to operating temperature.
- Direct the black light (ultraviolet lamp) at the suspect area (Figure 21–4). The dyed fluid, which appears as a bright red-orange, should lead to the leak source.

Powder Spray

For this method, you must thoroughly clean the suspect area, preferably with a spray solvent. Then dry the area.

- After surface cleaning and drying of the suspect area, apply several coats of an aerosol spray powder to the area. A foot powder can be used.
- Operate the vehicle on the road and warm up the ATF thoroughly.
- The ATF leak will leave an oil path on the powder surface, leading to the leak source (Figure 21–5).

LEAK REPAIR

Most leak repairs involve replacement of a gasket or seal, which is simple enough. The most important part of replacing a seal or gasket, however, is determining what caused the seal or gasket to leak. By identifying and repairing the cause, a long-term, quality repair can be expected.

FIGURE 21–5 Powder trace picks up oil leak at servo cover.

Gaskets

Common gasket leak causes are as follows:

- Improperly installed gasket.
- Damaged gasket.
- Improperly torqued fasteners.
- Improper sealant used or no sealant used (where applicable). Some fasteners are required to have a sealer treatment or a washer gasket on the bolt head.

Areas of concern that may not get proper attention are the following:

- Bent or warped flanges in the oil pan or side cover (FWD).
- A cracked or porous component casting: a servo cover, pump body, converter weld, or the transmission/transaxle case.
- Where a gasket is required to seal against a pressure head and there appears to be a blowout of the gasket, check the transmission/transaxle line pressures.
- The sealing surface may be damaged with deep scratches. Carelessness in removing old gasket material from an aluminum surface with a metal scraper is the usual cause.

Seals

Most seal leaks are due to the following factors:

- A weak/worn lip seal used at the converter, the manual and throttle shafts, or the extension housing/output.
- Missing, pinched, twisted, or torn O-ring seals, such as those found on the servo cover, filler tube, electrical connector, and/or speedometer adapter. Do not overlook the speedometer inner seal in the cable housing (Figure 21–6).
- Improper installation of lip seal—the lip faces in the wrong direction. Some seals are designed with flutes that help push fluid back into the bore. Reversing the seal allows the flutes to push out the fluid.
- Garter spring out of place or missing from inside the lip seal.

Areas of concern that may not get attention are the following:

- Damaged seal bore—nicks or deep scratches.
- Grooved or scuffed shaft surface, manual shaft, converter hub, or propeller shaft splined slip yoke/CV joint.
- Cracked component casting.
- Worn or loose shaft support bushings or bearings, preventing the shaft from running on center.
- Plugged case air vent.
- Drain-back circuit plugged or restricted—a special concern for some converter seal leaks.
- Excessive bushing to shaft clearance. Excess fluid escapes past the bushing and cannot be handled adequately by the drain-back circuit. The fluid pocket behind the seal gets pressurized and pushes out past the seal. It may even blow out the seal casing.

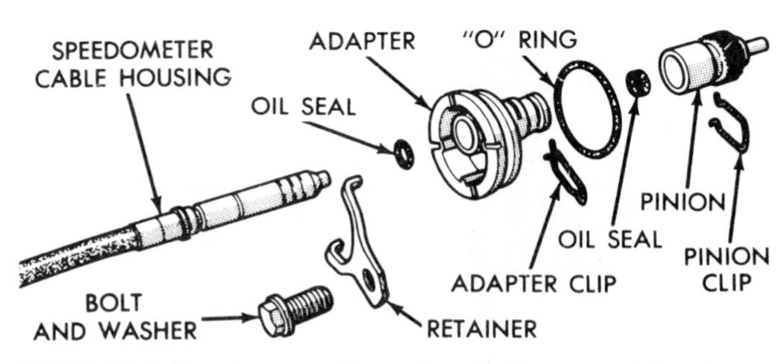

FIGURE 21–6 Speedometer drive and seals. (Courtesy of Chrysler Corp.)

CONVERTER HOUSING AREA LEAKS

This area makes up a significant portion of the leak sources. Typical converter housing area leaks are shown in Figure 21–7. Note the various paths the leaks follow to the bottom of the housing:

- Oil pump (converter) seal or cracked converter hub weld
- Oil pump body
- Oil pump to case bolt
- Oil pump to case gasket/body O-ring seal
- Crankshaft seal

Although the pump (converter) seal is high on the priority list, it is good practice to treat all the possible pump leak areas for long-term reliability. Pump to case bolts usually require seals, and in some instances, certain bolts require thread sealer. Shown in Figures 21–8 to 21–17 are typical resealing operations performed on a 4L60 (700-R4) pump. Do not forget to evaluate the converter hub bushing and converter hub surface condition. If for any reason you suspect that the converter has a cracked weld, perform an air check.

If the converter seal was blown out of its bore, review the technical bulletins. There may be a special problem with the drain-back circuit or converter bushing clearance that needs attention to prevent a repeat failure. It is also advisable to check the transmission/transaxle for high line pressures. Where applicable, be sure to clamp (with a retainer) or stake the converter seal in place. The GM 4L60 (700-R4) converter hub seal should be held in place with a retainer (Figure 21–17). Shown in Figure 21–18 is a Ford A4LD converter hub seal staked into position.

When resealing the converter area housing, it is considered good practice to replace all the external transmission seals to achieve long-term, leakproof mileage. It is embarrassing to seal the converter area and three months later have the manual shaft or extension housing seal leak. At the same time, include some preventative maintenance. Drop the pan for inspection, change the filter and ATF, and adjust the bands where applicable. Make sure that the throttle system and the road test verifies correct shift performance. Before returning the vehicle to the customer, verify that the original oil leak problem is fixed.

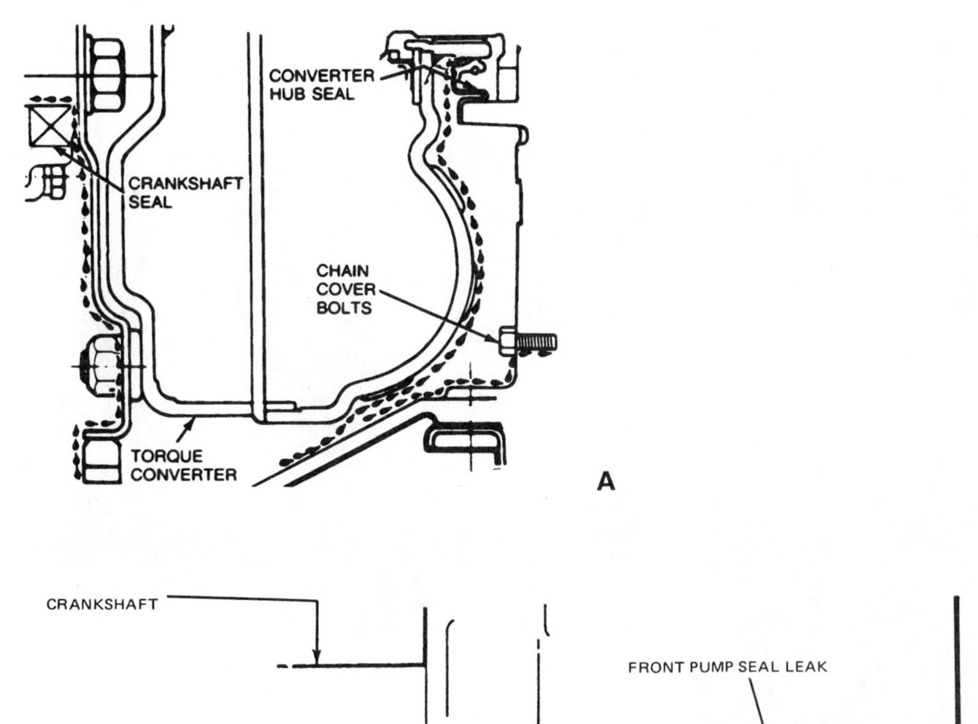

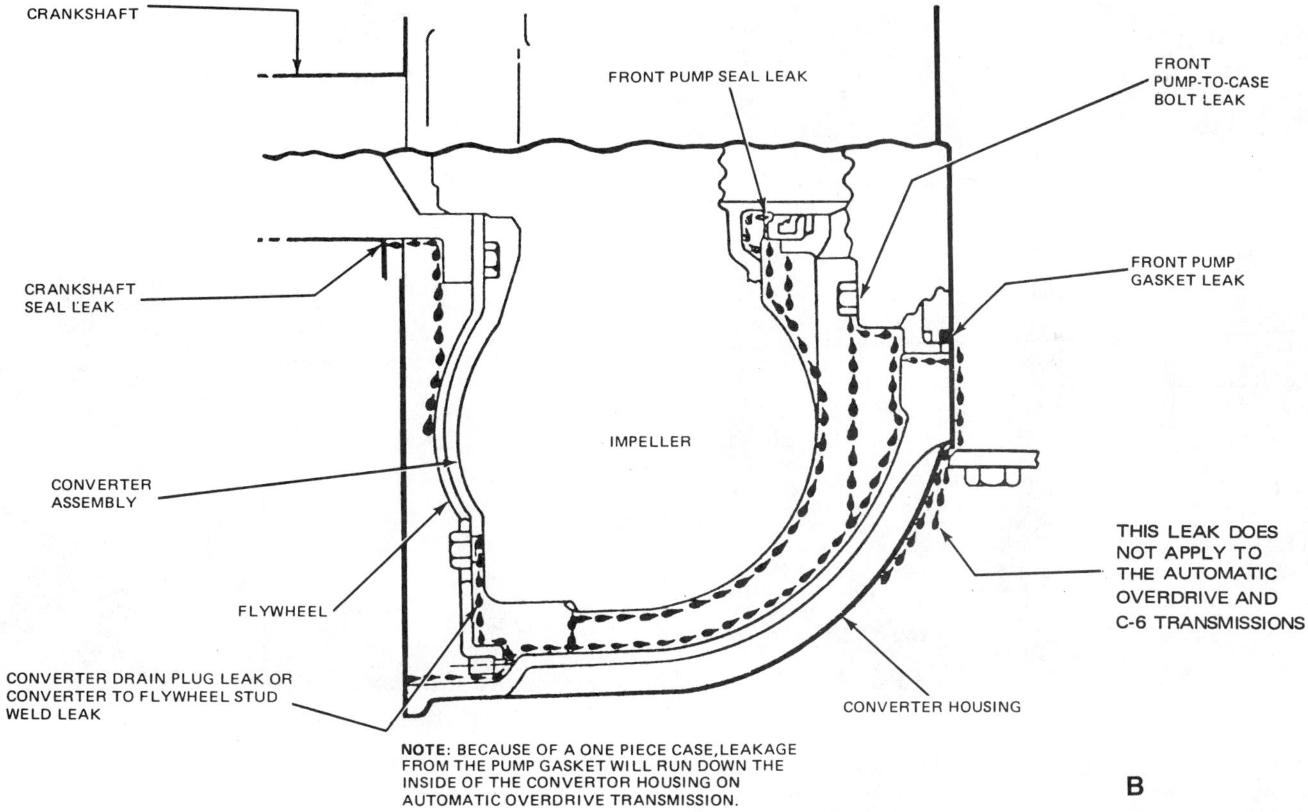

FIGURE 21–7 Converter area leaks: (a) typical FWD; (b) typical RWD. (Reprinted with the permission of Ford Motor Company)

FIGURE 21–8 Remove TCC solenoid and transmission filter.

FIGURE 21–9 Carefully remove seal with hammer and chisel.

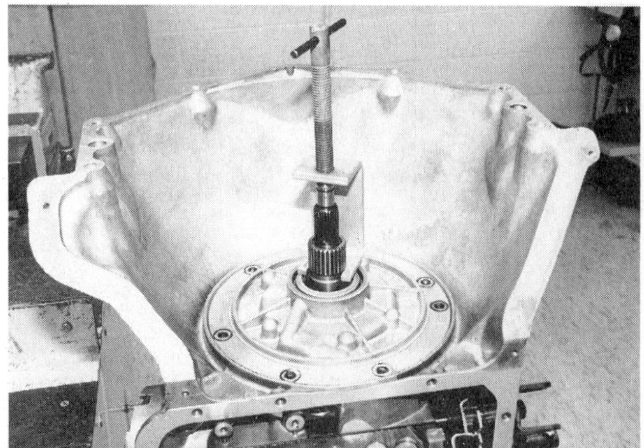

FIGURE 21–10 Remove pump with special puller.

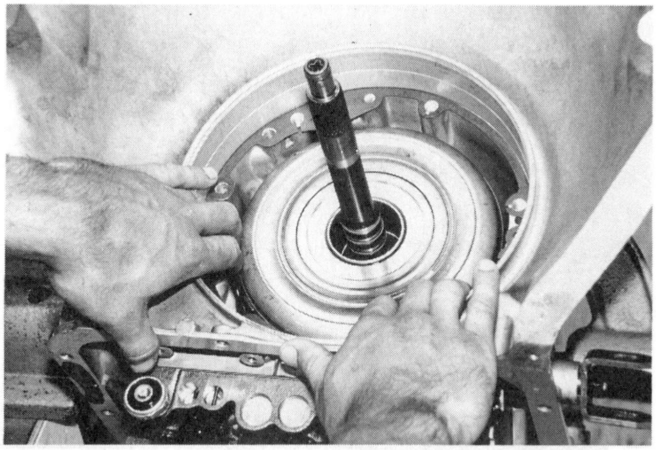

FIGURE 21–11 Installation of new pump to case gasket.

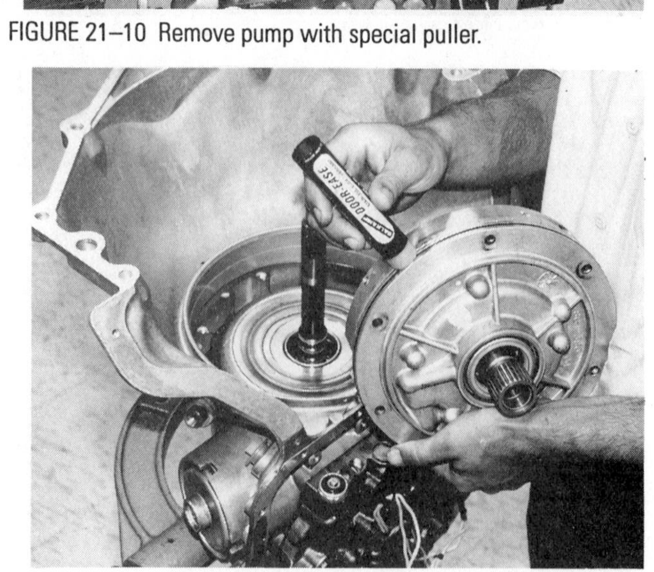

FIGURE 21–12 Treat new pump body O-ring seal with Door-Ease, TransJel, or Vaseline.

FIGURE 21–13 Visually align pump bolt holes to case threads and let the pump rest on its body O-ring seal. (Guide pins are helpful for pump to case alignment.)

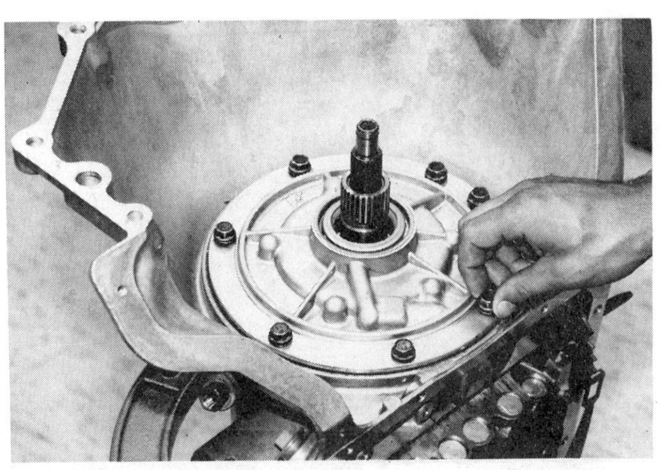

FIGURE 21–14 Use new bolt seals and hand-thread pump to case bolts. The bolts act as guide pins.

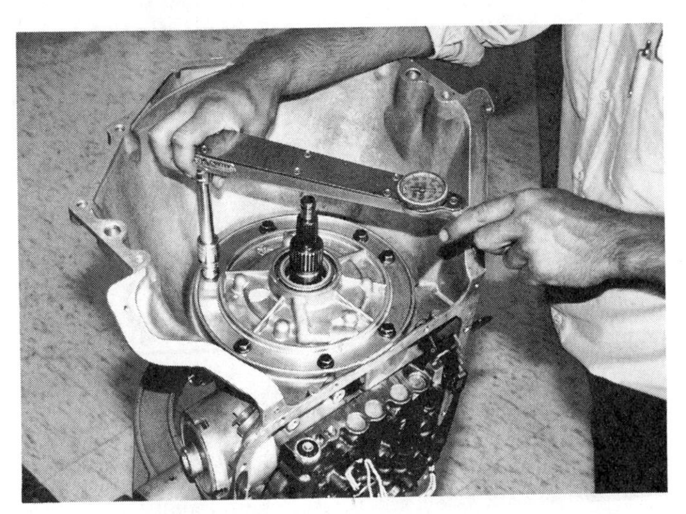

FIGURE 21–15 Torque-tighten pump body to case bolts. Check input shaft for endplay and free rotation.

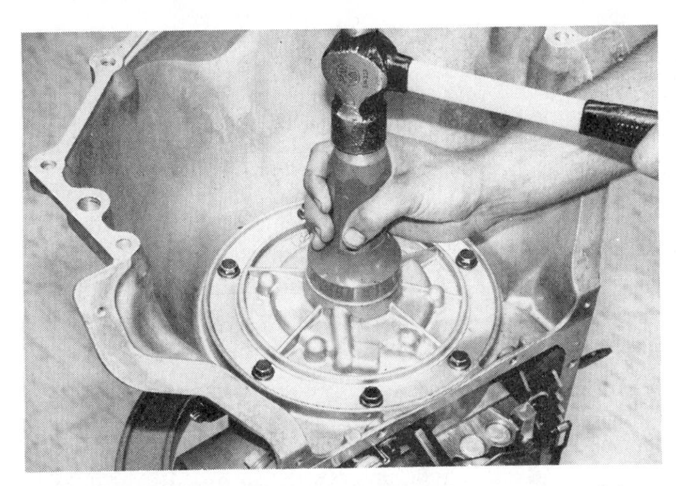

FIGURE 21–16 When installing the pump body converter seal, be sure to pack the garter spring with Vaseline or TransJel. Treat the metal seal casing with a thin coat of sealer.

FIGURE 21–17 GM 4L60 (700-R4) pump seal retainer.

FIGURE 21–18 Ford A4LD staked converter hub seal.

SUMMARY

Identifying the source and cause of a fluid leak can sometimes be challenging. First determine whether or not the fluid is from the transmission. Examine its color and check fluid levels. Next, identify the general area of the source. After pinpointing the source of the leak, perform the necessary repair. Add fluid if necessary, and check to be sure that the repair has fixed the problem.

A number of different methods can be used to help pinpoint the source of a fluid leak, including visual inspection, fluorescent dye and black light, and powder spray. After locating the source, attempt to determine what may have caused the item to leak. Identifying the cause as well as the source will help prevent a recurring leakage concern.

At times it may be difficult to pinpoint the source, especially in areas where it is hard to see. Leaks from inside a RWD converter housing fit this category. Since there are numerous possibilities, it is recommended that all seals and gaskets in the pump and converter area be replaced. Also, pressure test the converter and inspect the converter hub.

Fluid leaks can be a nuisance to a vehicle owner and detrimental to transmission operation. Diagnose the source and cause, make a good permanent repair, add fluid as needed, and verify that the problem is fixed.

❑ REVIEW

Questions

____ 1. To locate the leak source:
I. a fluorescent ATF dye and black light can be used.
II. an aerosol foot powder spray on the suspect area can be used.

A. I only   C. both I and II
B. II only   D. neither I nor II

____ 2. Which of the following is an ATF leak concern?
A. high fluid level
B. worn bushings
C. high line pressures
D. all of the above

____ 3. When looking for the leak, it is a good practice:
I. to check the fluid level and add ATF as necessary.
II. to run the vehicle on the road and bring the ATF up to operating temperature.

A. I only   C. both I and II
B. II only   D. neither I nor II

____ 4. A converter seal blown out of its bore may be caused by:
I. a restricted drain-back circuit.
II. excessive converter bushing clearance.

A. I only   C. both I and II
B. II only   D. neither I nor II

____ 5. Repair of a seal leak in the front pump area should include:
I. pulling the pump and replacing the case gasket and body O-ring.
II. replacing only the seal or gasket that is leaking.

A. I only   C. both I and II
B. II only   D. neither I nor II

6. List four common causes of gasket leaks:

(a)____________________

(b)____________________

(c)____________________

(d)____________________

7. List four common causes of seal leaks:

(a)____________________

(b)____________________

(c)____________________

(d)____________________

22 Chrysler Automatic Transmission/Transaxle Summaries

INTRODUCTION

This chapter is a summary study of Chrysler automatic transmissions/transaxles featured in the late 1980s and introduced in the early- and mid-1990s. In brief, the discussion highlights the historical background, transmission description, operating characteristics, mechanical powerflow summaries, and general support data. Chrysler and other manufacturers have a system of transmission application generally based on engine displacement, transmission availability, vehicle usage, and vehicle weight. The engine displacements, vehicle models, and transmission applications mentioned in the chapter are to be treated as typical only for the time of publication. For precise product application, the manufacturer's service or parts manuals should be consulted.

It will be to the technician's advantage to have a reasonable mastery of planetary gear fundamentals and operating concepts of Simpson and Ravigneaux geartrains. Many of the geartrains in the transmissions feature Simpson or Ravigneaux setups or modified versions already presented in detail in Chapters 3 and 4. Technicians should refer to the torque converter lockup design and basic hydraulic/electronic controls featured in Chapters 2 and 10 and to the fundamentals of hydraulic and electronic shift controls featured in Chapters 7 and 11.

Until recently, Chrysler Corporation has referred to their automatic transmission family under the TorqueFlite trade name. Typical reference of these TorqueFlite units included the familiar RWD A-900, A-727, and A-500 series, and the FWD A-400 and A-600 series. Evolving car and truck product designs for the 1990s demanded a wide variety of transmission adaptations of the TorqueFlite line.

New innovations to the product line included the introduction of four-speed transmissions and electronic shift control. The RWD A-500 overdrive series uses a split hydraulic/electronic control. An overdrive planetary section was added to the rear or output end of the basic three-speed TorqueFlite. A four-speed electronic shift transaxle, the A-604 (41TE), was first produced for the 1989 model year, followed by the 42LE for the 1993 LH model cars (Vision, Intrepid, and Concorde).

To more clearly define the variety of transmission applications, a new designation code system was adopted starting with production year 1992.

CHRYSLER PRODUCT DESIGNATION SYSTEM

1st digit:	Number of forward speeds.
2nd digit:	Relative torque capacity (0 = light).
3rd digit:	R = Rear drive; T = Transverse; L = Longitudinal; and A = All-wheel drive.
4th digit:	H or E = Hydraulic or electronic control.

Previous Designations	New Designations
A-404	30TH
A-413 Turbo/A-670	31TH
A-604 Light	40TE
A-604	41TE
A-606	42LE
A-904	30RH
A-999	31RH
A-727	36RH
A-727HD (Diesel)	37RH
A-500 (A-904 body)	40RH
A-500 (A-999 body)	42RH
A-500ES (Electronic shift)	42RE
A-500ES (1996 and later)	44RE
A-518	46RH
A-518ES	46RE
A-618	47RH

CHRYSLER RWD THREE-SPEED SERIES: 30RH/31RH/32RH (A-904 series) and 36RH/37RH (A-727 series)

This series of automatic transmissions using the new designation code system is a continuation of the original TorqueFlite A-904/A-727 series used by Chrysler for their RWD passenger car and truck applications. The A-904 was introduced in 1960 and used with Chrysler's slant six engine. In production year 1962, the A-727 replaced the cast iron TorqueFlite for high-torque, heavy-duty engine applications. Through the years, both transmission series have been adapted to handle a wide range of Chrysler engine and vehicle product lines.

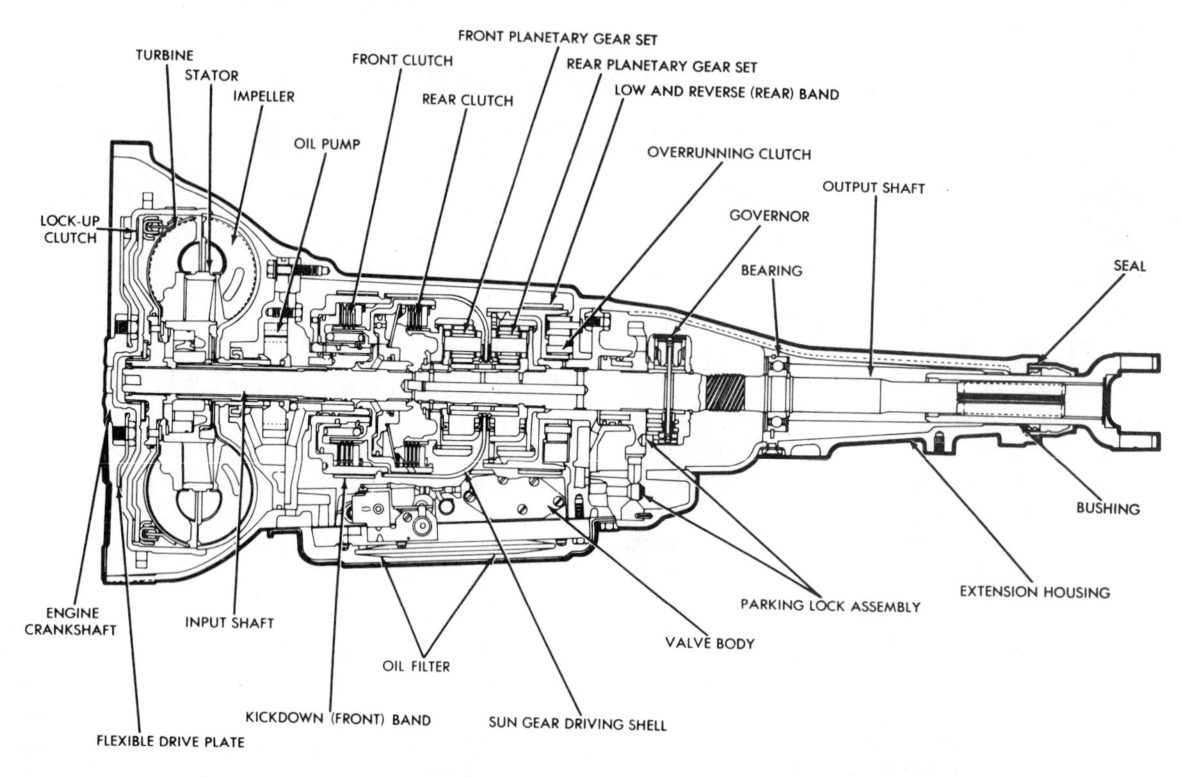

FIGURE 22–1 30RH (A-904), 31RH (A-998), and 32RH (A-999) sectional view. (Courtesy of Chrysler Corp.)

Three-speed RWD automatic transmissions are still in current production. The 1997 Jeep Wrangler 2.5-L engine applications may be equipped with a 30RH (A-904), and the 4.0-L engine applications with a 32RH. Operation of the torque converter clutch is controlled by the PCM.

The A-904 and A-727 pioneered the use of the one-piece integrated aluminum case. The converter housing and transmission case are molded as a single casting with a bolt-on extension housing.

Figures 22–1 and 22–2 show sectional views of the assemblies. Basically, they differ in physical size and torque handling capacity. As an aid to in-vehicle identification, transmission pan size and configuration can be compared (Figure 22–3).

The transmissions use a three-element torque converter with a lockup clutch, introduced in the 1978 model year. The lockup circuitry featured an all-hydraulic control system. For 1986 and 1987 production, the systems converted to electronic control, identified by an electrical connector threaded into the left rear side of the case. On heavy-duty diesel and V-10 engine applications, the 36RH and 37RH do not use a lockup converter.

The transmissions use a three-speed Simpson compound planetary geartrain controlled by:

- Two multiple-disc drive clutches.
- Two bands and servo assemblies.
- Low roller reaction clutch.

Operating range selection is through a six-position quadrant (P-R-N-D-2-1). Figure 22–4 represents a simplified schematic view of the geartrain and control elements. A powerflow clutch/band application and gear ratio summary is provided in Chart 22–1, page 501.

For better first and second gear performance, the 900 series geartrain was modified into a wide-ratio gearset beginning with the 1980 model year. Basically, the wide-ratio feature is achieved by a change in the sizing and ratio of the front planetary. The front planetary unit is larger, with a noticeable increase in the sun gear size. The rear planetary unit remains unchanged. The first-gear ratio is now 2.74:1 and the second-gear ratio 1.54:1, compared to the previous 2.45 and 1.45 ratios.

Though RWD passenger cars were no longer produced beginning with 1990 model year, this automatic transmission group continued to be built. They are found in the truck and van/wagon lineup: 1500, 2500, and 3500 models.

- 30RH is used with 2.5-L engines.
- 32RH is used with 3.9-, 4.0-, and 5.2-L engines.
- 36RH is used with HD 5.2- and 5.9-L engines.
- 37RH is used with diesel and V-10 engines.

Starting with production year 1995, all BR model trucks use four-speed automatic transmissions rather than three-speed units. The three-speed units are still available for Jeeps and Ram vans.

External Transmission Controls and Features

- *Manual range selection:* linkage operated.
- *Throttle system control:* linkage operated.

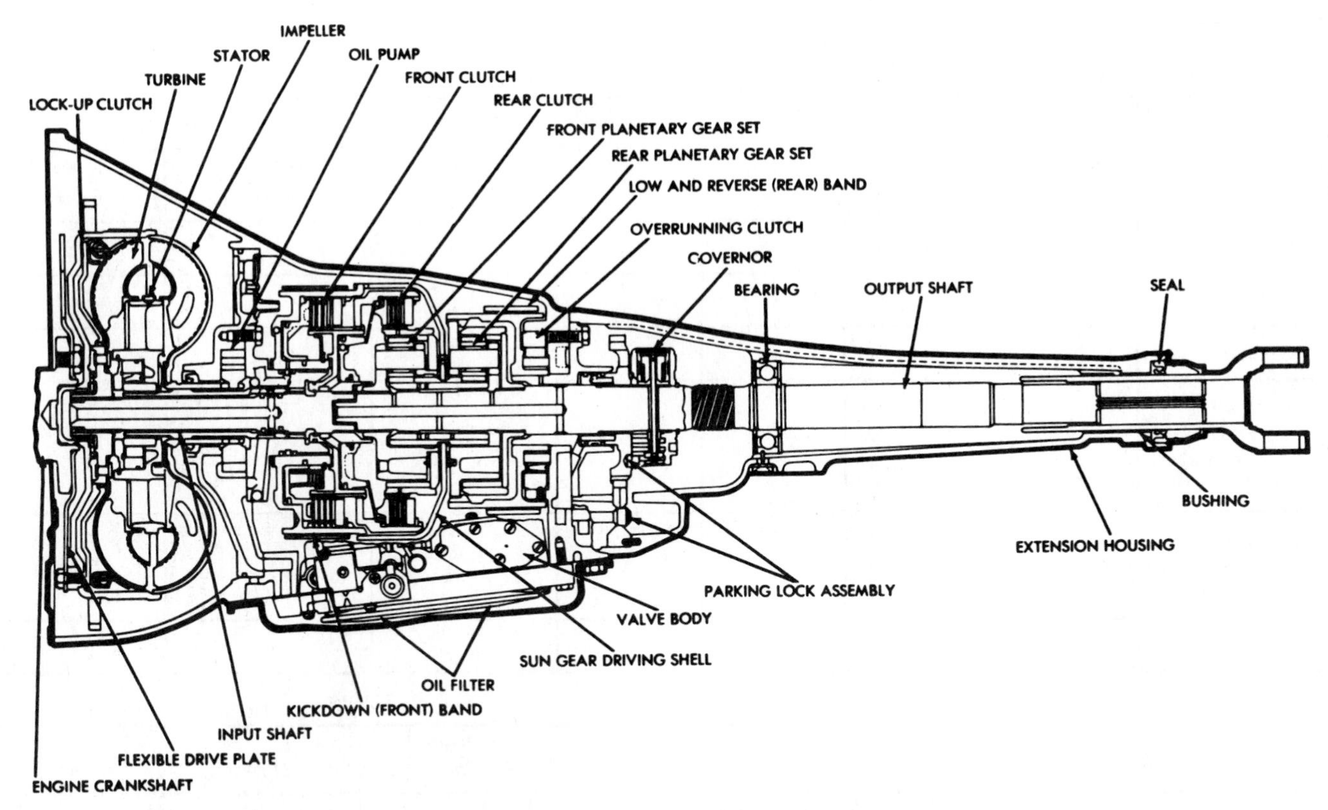

FIGURE 22–2 36RH (A-727) and 37RH (A-727HD) cutaway view. (Courtesy of Chrysler Corp.)

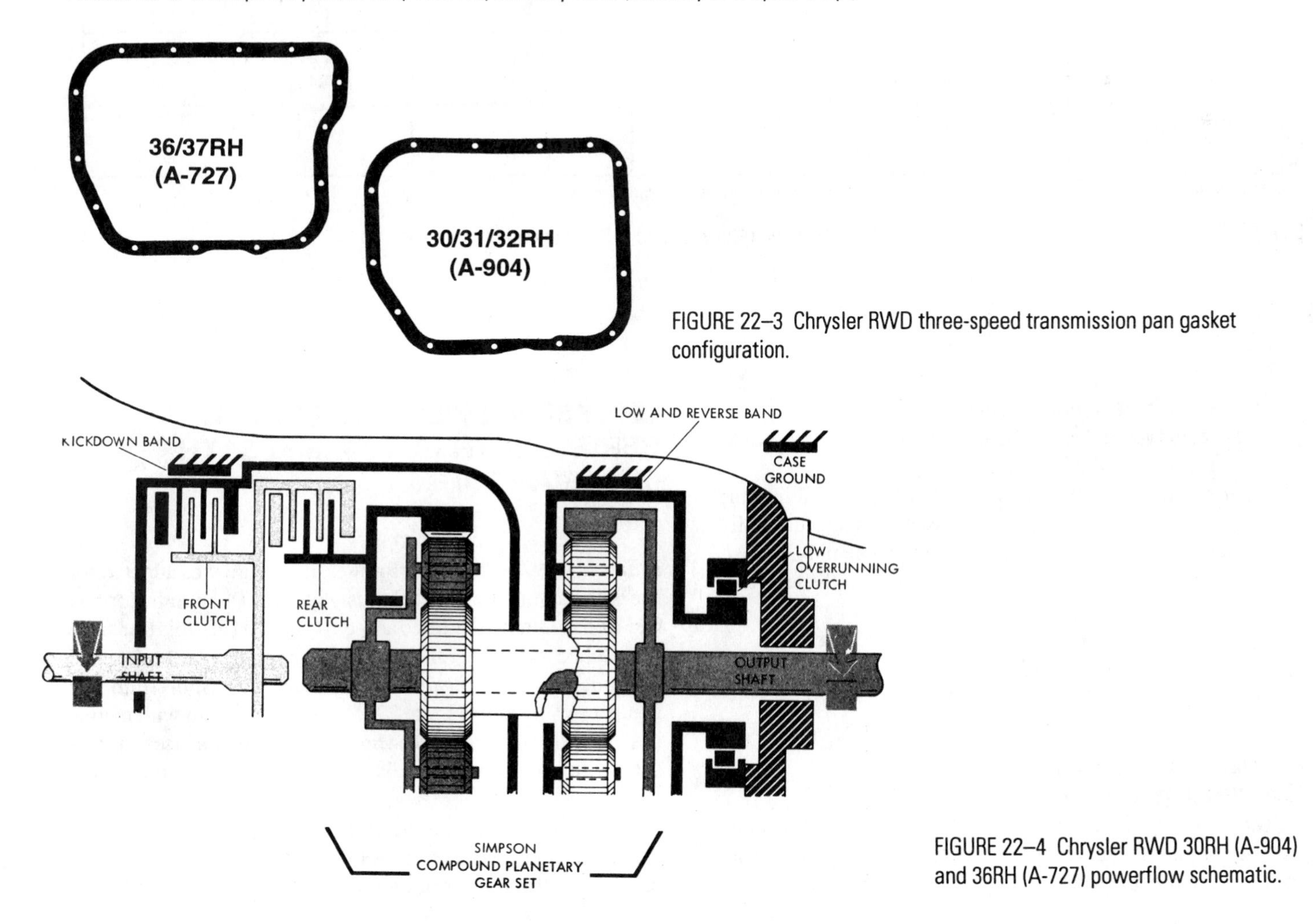

FIGURE 22–3 Chrysler RWD three-speed transmission pan gasket configuration.

FIGURE 22–4 Chrysler RWD 30RH (A-904) and 36RH (A-727) powerflow schematic.

Range Reference and Clutch/Band Apply Summary

Lever Position	Start Safety	Parking Sprag	Clutches: Front	Clutches: Rear	Clutches: Over-Running	Bands: (Kickdown) Front	Bands: (Low-Rev.) Rear
P—Park	X	X					
R—Reverse			X				X
N—Neutral	X						
D—Drive							
First				X	X		
Second				X		X	
Direct			X	X			
2—Second							
First				X	X		
Second				X		X	
1—Low (First)				X	X		X

Transmission	Model Year	Gear Ratio: 1 or Drive Breakaway	2 or Drive Second	Direct	Reverse
Auto Transaxle (FWD)	78–80 81–Current	2.48:1 2.69:1	1.48:1 1.55:1	1:1 1:1	2.10:1 2.10:1
904 Family (RWD)	60–79 *80–Current	2.45:1 2.74:1	1.45:1 1.54:1	1:1 1:1	2.22:1 2.22:1
A727 Family (RWD)	62–Current No Change	2.45:1	1.45:1	1:1	2.22:1

*Some 80 & 81 model year A904 family transmissions have the previous style gear ratios

CHART 22–1 Chrysler RWD and FWD three-speed series range reference and clutch/band apply summary.

- Kickdown system control: throttle and kickdown valves are located in tandem in the same valve body bore. Kickdown works from the TV linkage.
- *Neutral safety and backup light switch:* threaded into the left rear side of the case. The switch is mechanically operated internally by the manual lever assembly.
- *Kickdown band adjusting screw:* located on the left front side of the case.
- *Low/reverse band adjusting screw:* located internally.
- *Pressure test plugs:* (1) line pressure, (2) front servo release, (3) rear servo apply, and (4) governor.

Powerflow Summary and Operating Characteristics

The operating characteristics of RWD three-speed transmissions are similar to their front-wheel drive counterparts and are combined in the following section.

CHRYSLER FWD THREE-SPEED SERIES: 30TH (A-404) and 31TH (A-413/A-670)

In January 1978, the RWD TorqueFlite design was adapted for use in the automatic transaxle design for the Plymouth Horizon and Dodge Omni. Identified as the A-404 (30TH), this transaxle is matched with the 1.8-L engine. It combines a three-speed transmission and a final drive/differential assembly into one compact FWD unit (Figure 22–5). The powertrain with engine is a transverse (east-west) system. The transmission uses a three-speed Simpson compound planetary geartrain controlled by:

- Two multiple-disc drive clutches.
- Two bands and servo assemblies.
- Low roller reaction clutch.

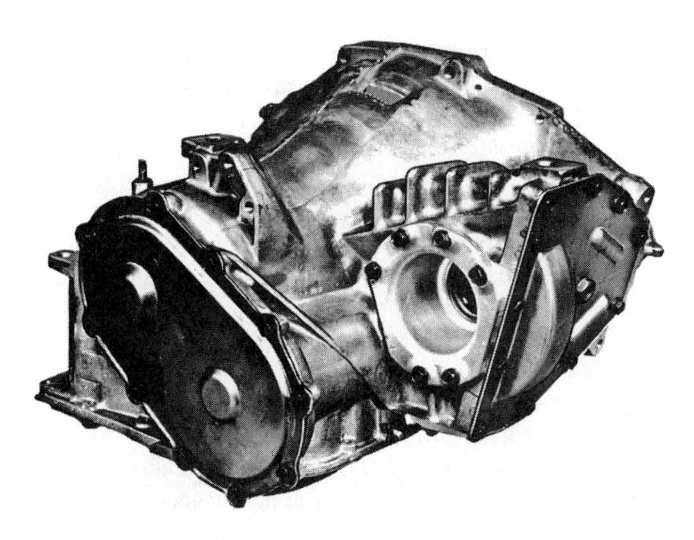

FIGURE 22–5 30TH (A-404) case assembly. (Courtesy of Chrysler Corp.)

Operating range selection is through a six-position quadrant lever (P-R-N-D-2-1). Referring to Figures 22–6 and 22–7, the geartrain output is relayed through a pair of helical gears to the transfer shaft. A helical pinion gear at the inner end of the transfer shaft drives the ring gear. Two final drive ratios are used: 3.48:1 federal (forty-nine states) and 3.67:1 (California).

Different variations of the transaxle design were developed for the 1980s and 1990s to create a versatile lineup for use behind new engine designs. Higher torque capacity versions were built with the same basic construction and control systems. Differences in the bell housing configurations prevent interchanging the transaxles. The A-413 (31TH) is matched with the 2.2-L and 2.5-L engines, and the A-470 with the 2.6-L engine. The A-413 (31TH) uses three final drive ratios (2.78, 3.02, and 3.22:1) with only one ratio (3.22:1) utilized in the A-470.

This FWD group originally used nonlockup converters. For production year 1987, these transaxles were equipped with

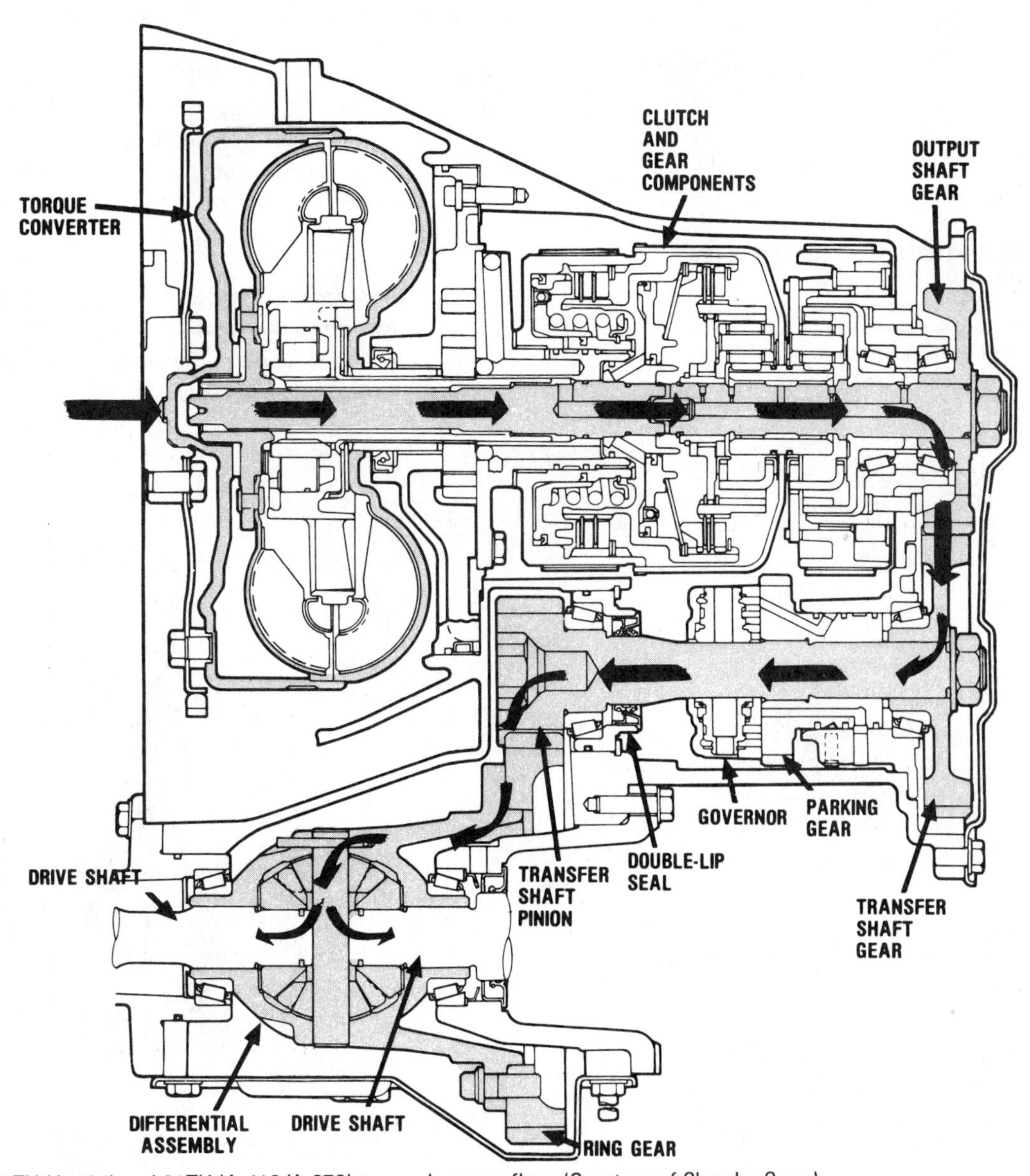

FIGURE 22–6 30TH (A-404) and 31TH (A-413/A-670) transaxle powerflow. (Courtesy of Chrysler Corp.)

FIGURE 22–7 Chrysler 30TH (A-404) transaxle transfer and final drive gears.

electronically controlled lockup, except for the 2.2-L and 2.5-L engine applications in mini-vans and cars equipped with turbo-charged engines.

The 31TH automatic transaxle continues to be produced and is used in 1996 Chrysler mini-vans and Neons.

External Transmission Controls and Features

- *Manual range selection:* linkage or cable operated.
- *Throttle system control:* cable operated.
- *Kickdown system control:* throttle and kickdown valves are located in tandem in the same valve body bore. Kickdown works from the TV cable.
- *Electronic lockup:* solenoid electrical connector on the left front side of the case is internally attached to the valve body.
- *Neutral safety and backup light switch:* threaded into the lower left rear side of the case. The switch is mechanically operated internally by the manual lever assembly.
- *Kickdown band adjusting screw:* located on the upper left side of the case.
- *Low/reverse band adjusting screw:* located internally.
- *Pressure test plugs (Figure 22–8):* (1) line pressure, (2) kickdown band apply, (3) kickdown release (front clutch), (4) kickdown apply at accumulator, and (5) governor.

Powerflow Summary and Operating Characteristics

Chrysler Three-speed FWD and RWD Transmissions. As an assist to understanding this section, the technician should study the clutch/band apply summary and range reference chart (Chart 22–1), which includes a complete review of the power-flow gear ratios. Details on TorqueFlite planetary powerflow are presented in Chapter 4.

N *Neutral.* The powerflow dead-ends at the front and rear drive clutches.

R *Reverse.* This is a function of the rear planetary unit (front clutch driving).

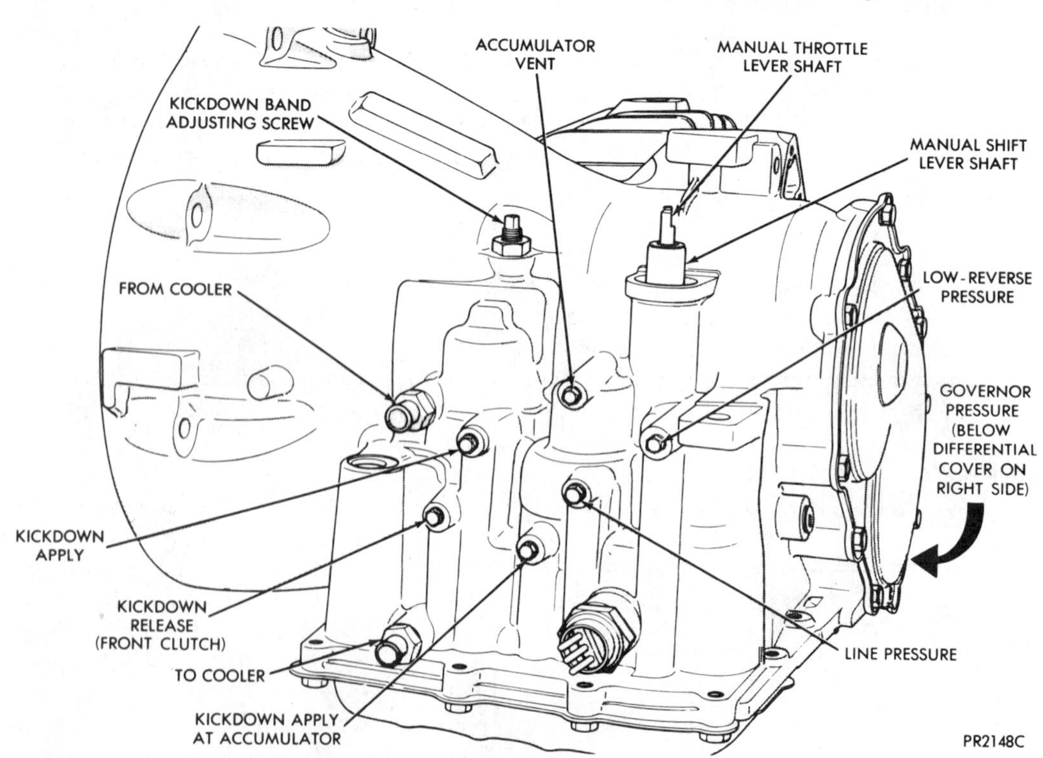

FIGURE 22–8 Chrysler 30TH (A-404) transaxle pressure test port locations. (Courtesy of Chrysler Corp.)

D *Drive.* This range is used for normal driving conditions and maximum economy. Full automatic shifting and three gear ratios are available.

1 *First gear (breakaway).* This is a combined function of the front and rear planetary units (rear clutch driving).

2 *Second gear.* This is a function of the front planetary unit (rear clutch driving).

3 *Third gear.* The front and rear drive clutches are applied and lock the planetary units for a 1:1 ratio. Converter clutch lockup follows after the shift into third and above a minimum preset vehicle speed. The minimum schedule falls in the approximate range of 30 to 40 mph (48 to 64 km/h).

For added acceleration, full-throttle kickdown shifts (3-1, 2-1, and 3-2) are available within the proper vehicle speed and engine torque range. As another performance feature, a 3-2 part-throttle downshift is usually incorporated in late model units for controlled acceleration during low-speed operation. The closed-throttle downshift pattern is 3-1.

2 *2 Range.* This provides a normal 1-2 shift pattern and keeps the transmission in second gear for extra pulling power or engine braking in hilly terrain. It can also be used for performance in congested traffic. The 2 range can be selected at any high gear vehicle speed, and the transmission will downshift to second. At closed throttle, the shift pattern is 2-1. A full-throttle kickdown shift 2-1 is available.

1 *1 Range.* First-gear operation is maintained for heavy pulling power and engine braking, especially for mountainous terrain. The 1 range can be selected at any vehicle speed range and will shift to second if the vehicle speed is too high. When the vehicle slows down to a safe speed (approximately 30 mph or 48 km/h), manual low engages, and the transmission will not upshift regardless of throttle opening.

CHRYSLER RWD FOUR-SPEED OVERDRIVE SERIES: 40RH, 42RH, 42RE, 46RH, 46RE, and 47RH

Originally referred to as the A-500 series, the RWD four-speed overdrive was introduced in production year 1989 as an option for all trucks and RWD vans. This was achieved by adding an additional planetary gearset to the output end of the existing A-998, A-999, and A-727 LoadFlite three-speed units. The overdrive assembly is placed into the extension housing, which was redesigned and extended 6 1/2 in (160 mm). Refer to Figures 22–9 (a and b) and 22–10. The 42RH is used with the 3.9-L V-6 engine in the 4x2 1500 truck models. For 5.2- and 5.9-L V-8 engine applications, the 46RH is used. The high torque 47RH is teamed with the 8.0-L V-10 and Cummins diesel engines.

Transmission Gear Ratios

Gear	42RH Wide Ratio	46RH/47RH
1st	2.74:1	2.45:1
2nd	1.54:1	1.45:1
3rd	1.00:1	1.00:1
4th	0.69:1	0.69:1
Reverse	2.21:1	2.21:1

The hydraulic control system for the three forward gears and reverse is basically identical to the three-speed RWD units with hydraulic shift control. The electronic controls for fourth gear (overdrive) and torque converter clutch lockup include an overdrive switch, overdrive control module, PCM (powertrain control module), and two solenoids (overdrive and converter

A

B

FIGURE 22–9 Chrysler RWD 46RH (A-518): (a) overdrive assembly located in extension housing; (b) rear view of main transmission body showing overdrive clutch position and intermediate shaft.

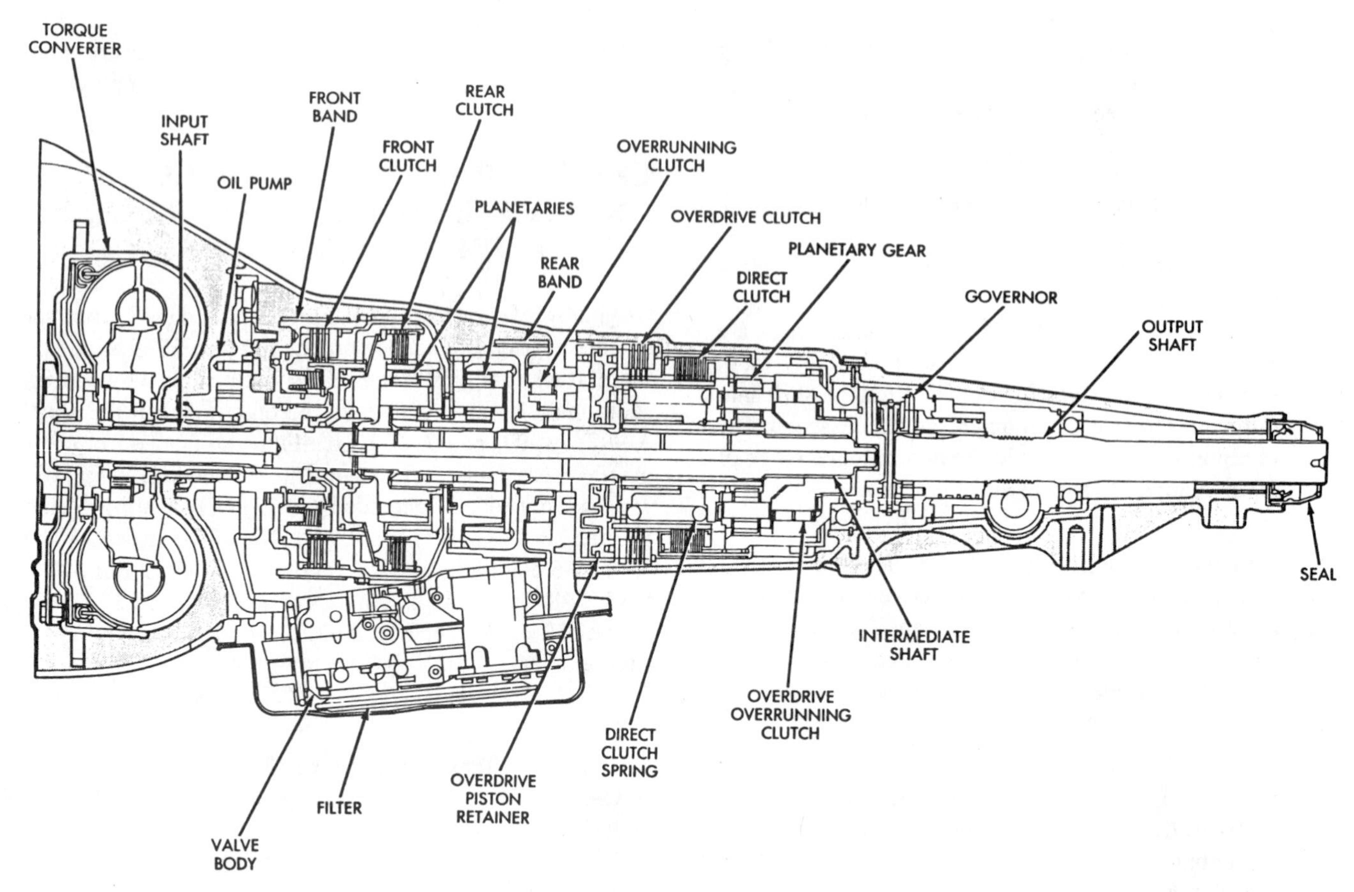

FIGURE 22–10 46RH (A-518) cutaway view. (Courtesy of Chrysler Corp.)

clutch). The valve body is modified to accommodate a new grouping of valves and solenoid mounts for the 3-4 shift and converter clutch control (see Figure 22–11).

The 1-2 and 2-3 shifts are controlled with hydraulic throttle and governor pressures. Fourth gear is electronically controlled and hydraulically activated. The overdrive switch, located on the dash, must be in the ON position. The PCM evaluates the

A

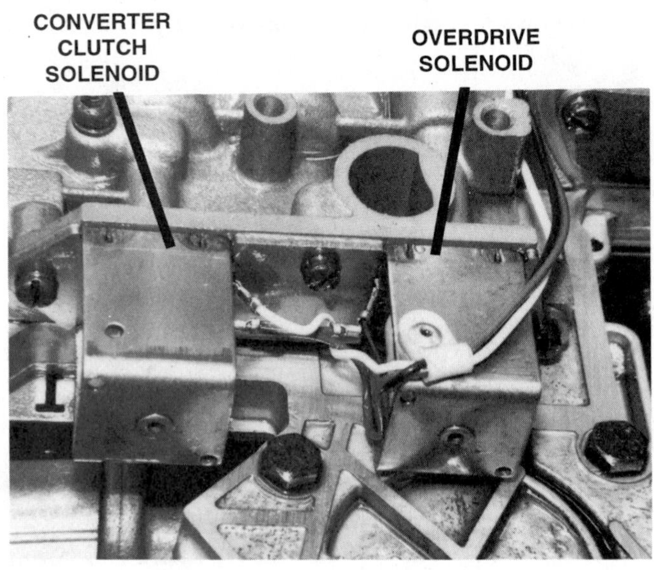

B

FIGURE 22–11 Chrysler 42RH (A-500): (a) valve body; (b) solenoids.

signals from input sensors and, when conditions are desirable, provides a ground for the solenoid circuit. The solenoid check ball closes a vent port and allows line pressure to overcome spring pressure and open the 3-4 shift valve.

Converter clutch engagement in fourth gear is controlled by the PCM. The energized lockup solenoid supplies line pressure to the converter clutch valve for lockup clutch apply.

The PCM allows a 3-4 shift and converter clutch apply when vehicle speed is above 25 mph and in third gear. The 3-4 shift timing is based on inputs from the following sensors and switch:

- Transmission fluid temperature (TFT)—minimum 60°F (140°C).
- Engine speed.
- Throttle position (TP).
- Manifold absolute pressure (MAP).
- Overdrive control switch position.

New for 1994 was the introduction of the 42RE for 4.0-L engine applications. The main feature is the elimination of the traditional centrifugal governor in favor of an electronic pulse width governor pressure control solenoid and sensor (Figure 22–12). The valve body transfer plate has been redesigned to accept a governor body with governor solenoid and sensor attachments. The converter clutch and overdrive solenoids are retained. A TCM (transmission control module) is added to the electronic control circuitry. The 1-2 and 2-3 shifts remain hydraulically activated.

The governor pressure output is programmed by the TCM to adjust the governor pressure curve and shift schedule under the following conditions:

- Transmission operation when the fluid temperature is at or below 30°F (–2°C). Fluid temperature thermistor is located on the underside of the converter clutch solenoid (Figure 22–13).
- Operation when fluid temperature is at or above 31°F (–1°C) during normal city or highway driving.
- Wide open throttle operation.
- Driving with the transfer case in 4WD low range.

Temperature readings from the thermistor are also used by the TCM to prevent converter clutch and overdrive clutch engagement when fluid temperature is below 30°F (–2°C). When fluid temperature exceeds 260°F (126°C), the TCM downshifts the transmission 4-3 and engages the converter clutch.

Additional electronics include a transmission speed sensor that monitors output shaft rotation. The input signals are processed by the TCM. Shared by both the TCM and PCM are input signals from the vehicle speed sensor (VSS) and throttle position (TP) sensor.

For 1996, the 42RE uses a single control module. The new PCM combines the functions of the TCM and PCM. Conversion to full electronic control is planned for 1997.

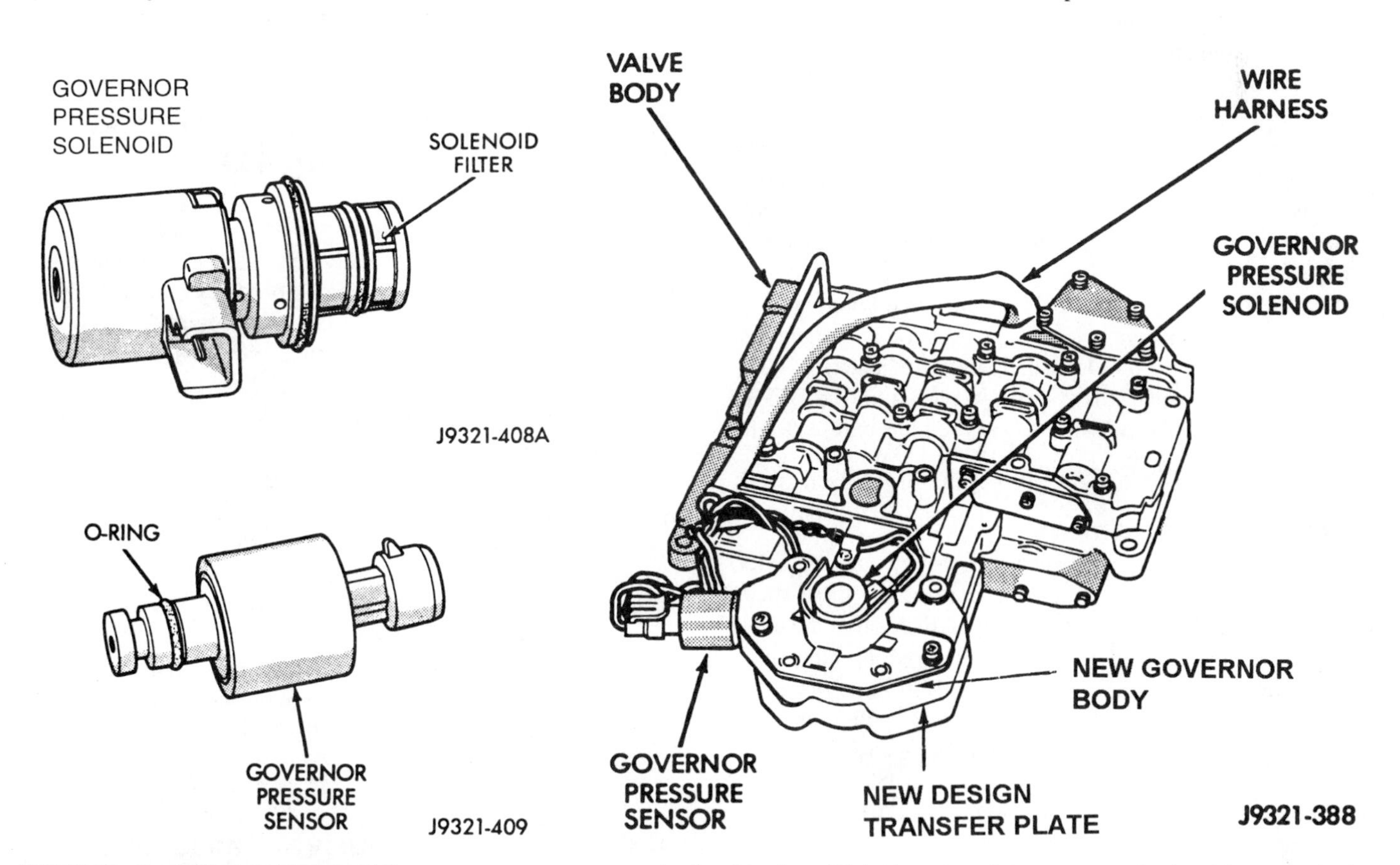

FIGURE 22–12 42RE (A-500ES): (Left) Governor pressure sensor and solenoid valves. (Right) Location of the sensor and solenoid on valve body. (Courtesy of Chrysler Corp.)

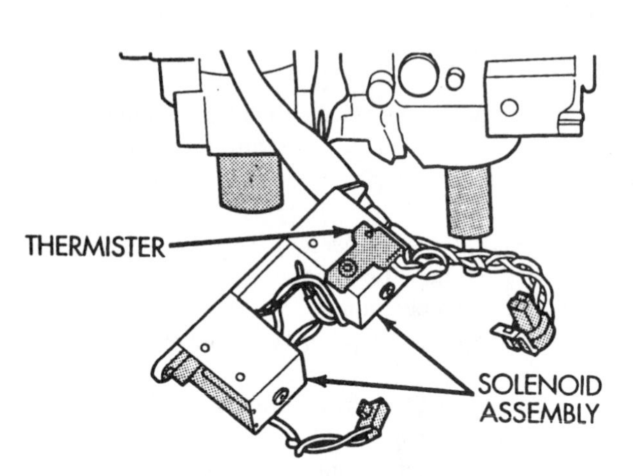

FIGURE 22–13 Temperature thermistor located on underside of converter clutch solenoid. (Courtesy of Chrysler Corp.)

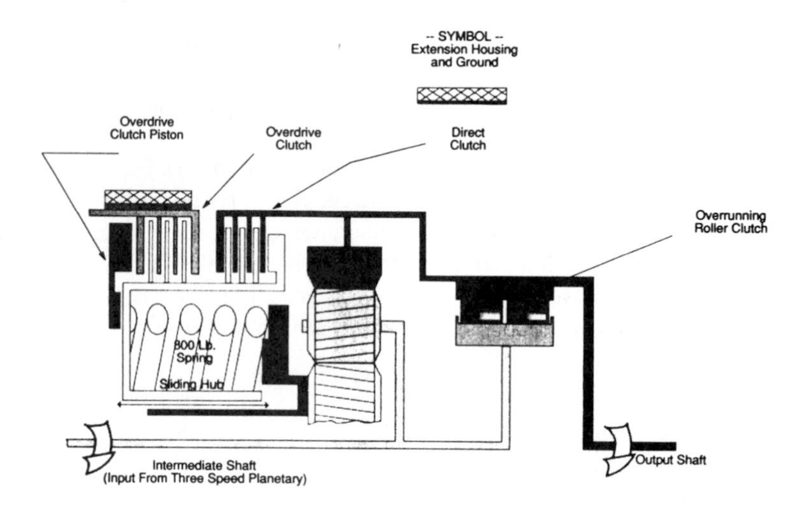

FIGURE 22–14 Chrysler 42RH (A-500) and 46RH (A-518) overdrive unit schematic.

Operating Characteristics

The operating range selection is through a six-position quadrant (P-R-N-D-2-1). The operating characteristics in manual 2 and 1 ranges and the three forward gears in the OD range remain comparable to those of the three-speed RWD LoadFlite/TorqueFlite transmissions. Overdrive automatically engages in the OD range when the overdrive selector switch is in the ON position. All minimum throttle 3-4 shifts occur between 25 and 28 mph (40 and 45 km/h), with 4-3 closed-throttle coast shifts set at 25 mph (40 km/h). When the throttle opening exceeds seventy percent, the 3-4 upshift is canceled regardless of vehicle speed. Prior to a kickdown 4-3 downshift, the lockup converter is released.

The driver has the option to cancel the overdrive feature by switching the dash-mounted overdrive switch to the OFF position. The transmission is then limited to first, second, and third gears in the OD range. Converter clutch lockup is available in third gear.

The overdrive switch circuit does not directly energize or de-energize the overdrive solenoid. Its function is to signal the TCM to either open or close the overdrive solenoid PCM circuit. Whenever the ignition key is turned from the OFF to the RUN position, the overdrive switch circuit is reset to the ON setting.

Powerflow Summary

Details of the overdrive planetary powerflow are presented in Chapter 4. In review, the output shaft of the three-speed LoadFlite becomes the intermediate shaft that drives the overdrive planet carrier. Total transmission output must pass through the overdrive assembly (Figure 22–14). In the direct mode, the 800-lb (3552-N) heavy-duty spring applies the direct clutch and locks the sun and annulus gears together. The intermediate (input) and output shafts turn at 1:1. In the overdrive mode, the overdrive clutch is applied, and the heavy-duty spring apply force is released from the direct clutch. The overdrive clutch holds the sun gear, and the planet carrier is driving. The effective ratio is 0.69:1.

Chart 22–2 summarizes the operating ranges and powerflow clutch/band applications. Note that the one-way roller clutch in the overdrive unit is also effective for first, second, and third gears. During the 3-4 shift, it keeps the connection between the intermediate shaft and output shaft tight. This prevents an engine flare condition from occurring while the direct clutch is released and the overdrive clutch is applied.

Pressure test port locations are identified in Figure 22–15.

CHRYSLER ELECTRONIC FOUR-SPEED FWD: 41TE (A-604) and 42LE (A-606)

The operation of this group of transaxles is controlled by a sophisticated electronic management system. The powertrain design is unique, void of any band or one-way clutch elements. The 41TE has an east-west/transverse orientation and was introduced in model year 1989 for use with 3.0-L and 3.3-L V-6 engine applications. Originally the 41TE was referred to as the Ultradrive A-604. In the 1995 Eagle Talon, the 41TE is referred to as the F4AC1.

For the 1993 model year, the 42LE (A-606) was put into production for Chrysler's LH car group (Concorde, Intrepid, and Vision). The 42LE was developed with a north-south longitudinal orientation to complement the 3.3-L and 3.5-L engines and vehicle design. This transaxle can handle high torque and high speed shifts.

Although the 42LE geartrain and most of the electronic circuit elements are similar to those of the 41TE, there are some significant differences. These will be highlighted in separate discussions of each transaxle.

41TE (A-604 Ultradrive) Features

Although the early trend in transmission electronification was to add electronics to already existing conventional auto-

Chrysler A-500/A-518
Range Reference and Clutch/Band Apply Summary Selector Pattern PRND21

Lever Position	A500 Over-Drive	Start Safety	Parking Sprag	Transmission						Overdrive		
				Clutches				Bands		Clutches		
				Front	Rear	O'Run	Lockup	K/D Front	Reverse/Rear	O/D	O'Running	Direct
P—Park		X	X									
R—Reverse	2.21			X					X			X
D—Drive												
First	2.74				X	X					X	X
Second	1.54				X			X			X	X
Third	1.00			X	X		X				X	X
O/D	.69			X	X		X			X		
2—Second												
First	2.74				X	X					X	X
Second	1.54				X			X			X	X
1—Low	2.74				X	X			X		X	X

1. Overdrive direct clutch and overdrive overrun clutch lock together the same components; intermediate (input) shaft to output shaft.
2. L/R band and trans overrun clutch hold the same component; rear planetary carrier.
3. Torque converter clutch lockup available in third and fourth; lockup shift determined by SMEC.

CHART 22–2 Chrysler RWD four-speed overdrive series range reference and clutch/band apply summary. (Courtesy of Chrysler Corp.)

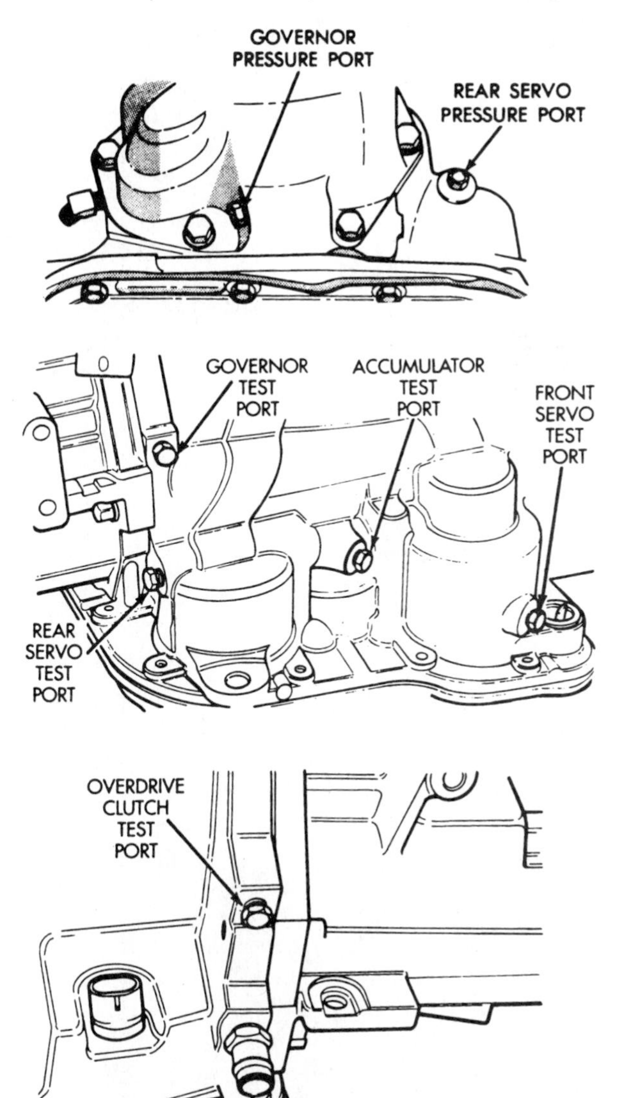

FIGURE 22–15 42RH/46RH/47RH pressure test port locations. (Courtesy of Chrysler Corp.)

matics, the 41TE (A-604) was truly an innovation. Externally, the case configuration looks similar to a typical 30TH (A-404) series automatic transaxle, with casting ribs added on the outside of the case differential section for extra strength (Figure 22–16). It weighs only 16 lb (7.2 kg) more and is 1/2 in (12.7 mm) longer. The distinguishing external feature is the solenoid assembly, located on the left side of the case.

Inside the 41TE (A-604), a new compact clutch and gear arrangement eliminates the use of bands and one-way clutches (Figure 22–17). Three multiple-disc drive clutches and two multiple-disc reaction clutches control the geartrain operation with electronic control precision. A three-element torque converter incorporates a lockup clutch.

This four-speed design, because of its simplicity, requires substantially fewer parts compared to other four-speed units. The traditional governor and throttle systems have been re-

FIGURE 22–16 Chrysler 41TE (A-604) case assembly.

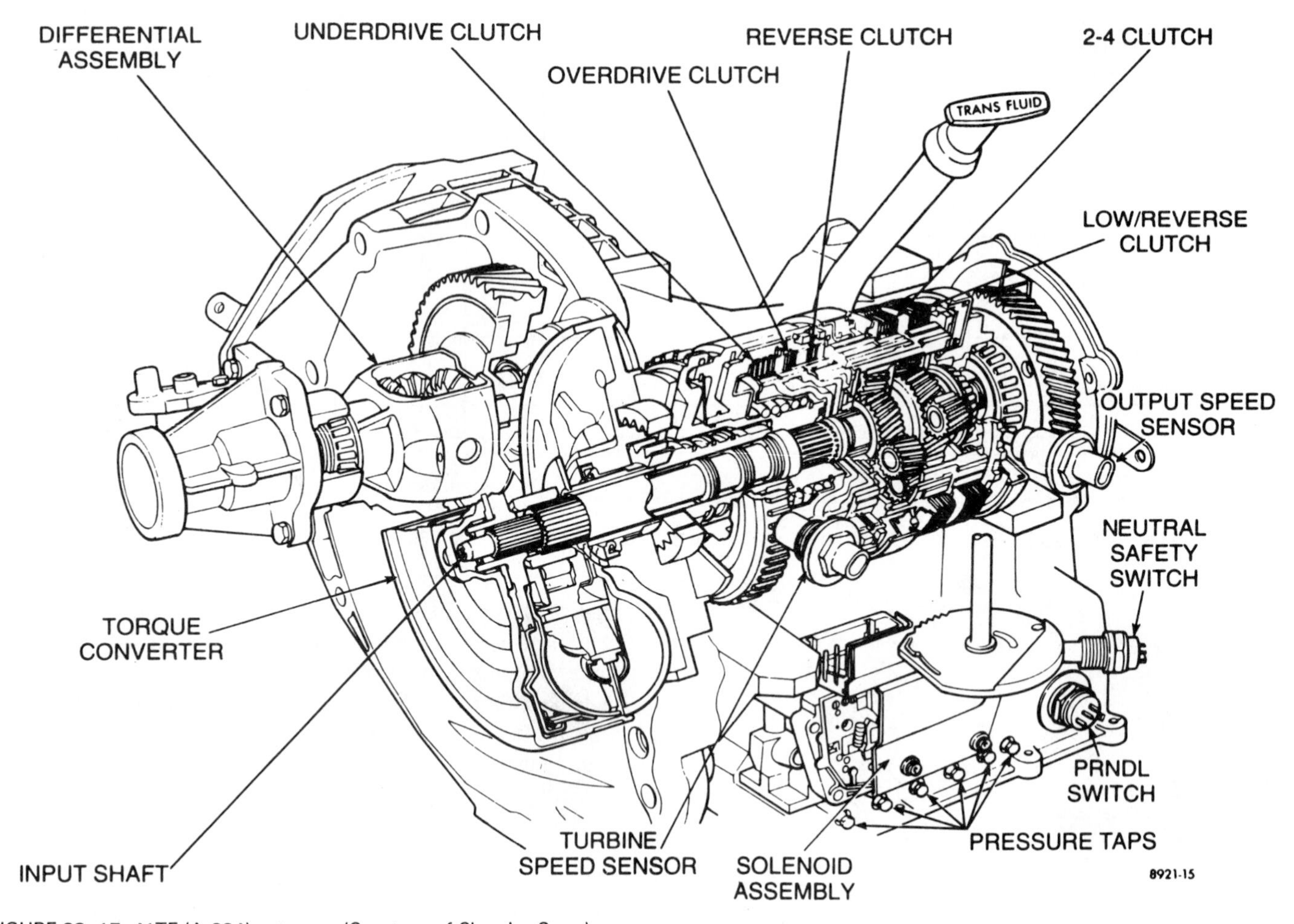

FIGURE 22–17 41TE (A-604) cutaway. (Courtesy of Chrysler Corp.)

placed by electronic speed sensors. Most of the hydraulic valve body has been eliminated, replaced by a solenoids assembly. The valve body is responsible only for controlling the mainline and torque converter pressures and directing pressurized fluid to the clutches, torque converter, and solenoid valves (Figures 22–18 to 22–20).

The Ultradrive is the first in the industry to use fully adaptive electronic controls. This provides the ultimate in transmission performance. Whether dealing with light throttle shifts or heavy throttle shifts, clutch engagement quality is smooth and consistent. Even the kickdown shifts are extremely smooth. It is impossible to squawk the tires or jar the driveline on N-to-D or N-to-R manual shifts.

For shift decisions and control of shift quality, a transmission control module (TCM) is used. A sixty-way electrical connector couples the control module with the transaxle and the engine control module. The engine control module and the TCM are linked together by a two-way communication data link system that shares information. This sophisticated multiplexed system, which is referred to as the CCD or C^2D, eliminates the use of a large number of wires that normally would be needed for sending information between various electronic control modules that share data. (Figure 22–21).

FIGURE 22–18 Chrysler 41TE (A-604) valve body.

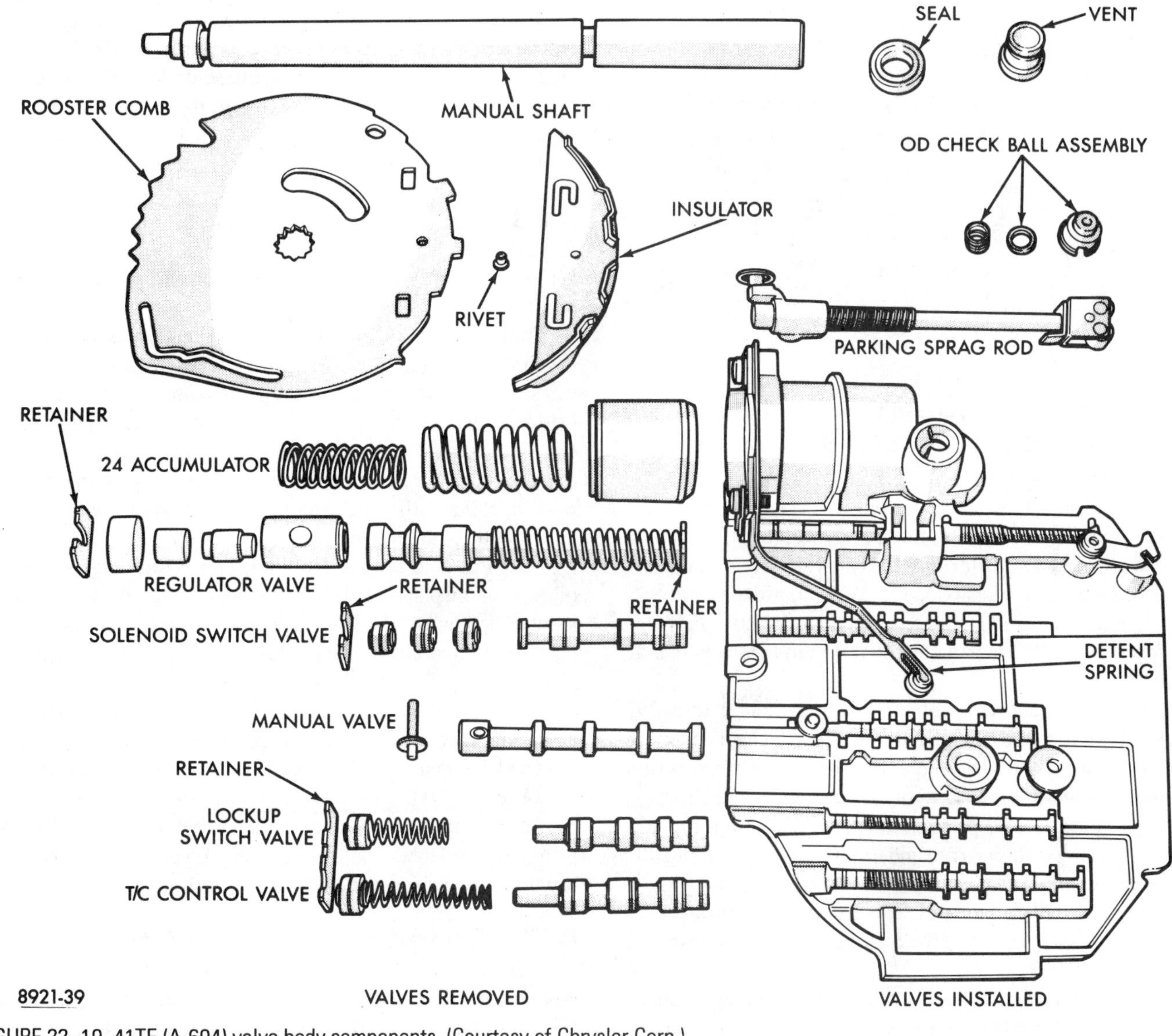

FIGURE 22–19 41TE (A-604) valve body components. (Courtesy of Chrysler Corp.)

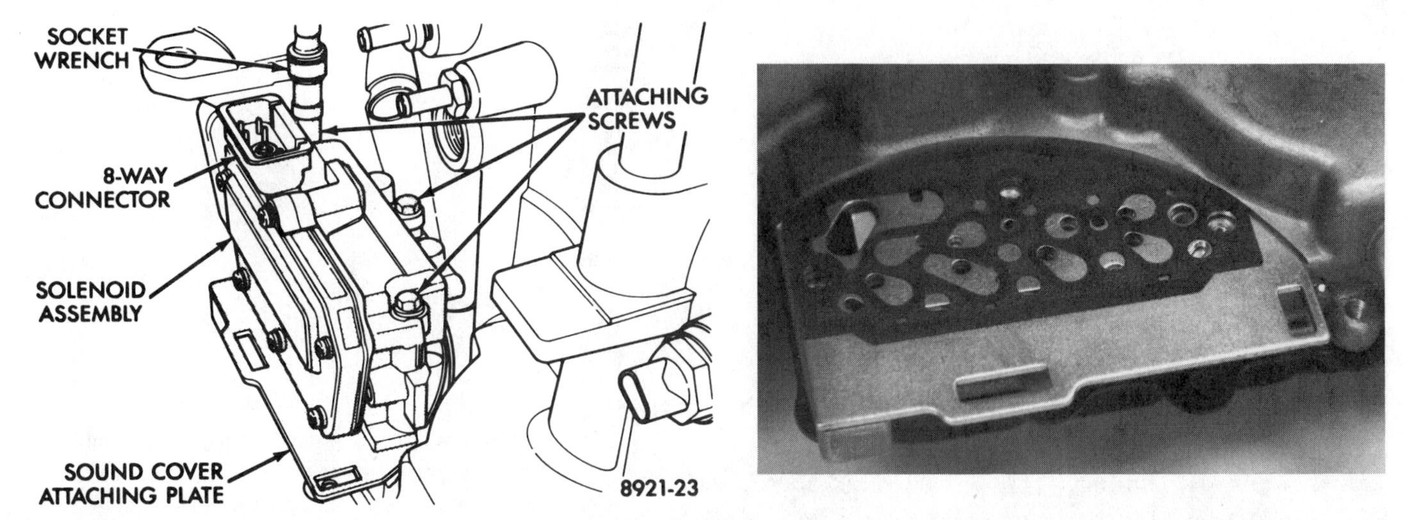

FIGURE 22–20 41TE (A-604) solenoid assembly mounts to transaxle case and integrates with valve body. (Courtesy of Chrysler Corp.)

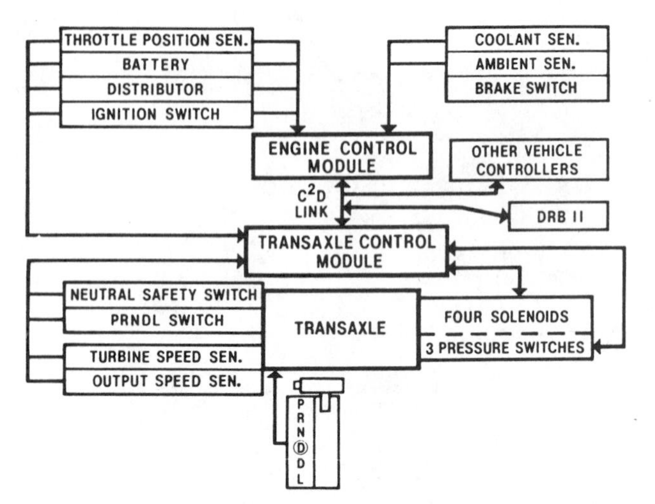

FIGURE 22–21 Simplified 41TE (A-604) multiplex control circuitry. (Courtesy of Chrysler Corp.)

As shown in Figure 22–21, the TCM receives inputs from numerous devices. This information is analyzed by the control module to operate the transaxle. The TCM controls four duty-cycle solenoids assembly-mounted to the valve body (Figures 22–22A to 22–22C). The solenoids are ground-side switched through the TCM.

The engine coolant and ambient air temperatures directly affect shift timing and the converter clutch. The converter lockup is inhibited until the engine reaches a normal temperature range. A unique feature of the converter electronic-hydraulic control system permits the converter clutch to partially apply until the engine rpm and turbine rpm are within 100 rpm of each other. At that point, full converter clutch engagement occurs, virtually unnoticed. This phasing is controlled by inputs from the tach sensor at the distributor and the transmission input sensor.

The controller is continuously evaluating the characteristics of the engine and the transaxle to maintain consistent shift quality performance. Based on input speed and output speed sensors, it monitors the release and apply rate of the clutch assemblies 140 times per second. Essentially, the controller is concerned with the exact time it is going to take to complete the shift with the prevailing engine torque.

To achieve the desired apply and release of the clutch assemblies, the TCM adjusts the pulse width of the duty-cycle solenoids. The release and fill rate for each clutch can be altered and relearned to meet the programmed standards to compensate for transmission and engine wear and personal driving characteristics.

The computer learning process, referred to as adaptive memory, may require up to ten shift cycles to adapt to a new transmission characteristic. When the computer loses its memory through a loss of battery power or a disconnect of the sixty-pin connector, it will require the necessary shift cycles for the computer to relearn the transaxle operating characteristics. Shift cycle/adaptive learning procedures are provided in service and training manuals.

The electronic control system has on-board diagnostic capabilities and displays fault code numbers to identify which circuit is malfunctioning. The DTCs (diagnostic trouble codes) can appear due to hydraulic, mechanical, and electrical problems. The Chrysler DRB II, DRB III, and other handheld testers are able to identify fault codes from the transmission computer. When you use a computer scanner for engine control diagnostics, only engine fault codes appear, not transmission fault codes. Once a diagnostic trouble code (DTC) is present, the transaxle may go into limp-in mode. In limp-in mode, only park, neutral, reverse, and second gears are available. The transaxle is protected from potentially hazardous operation that could lead to serious damage.

41TE Operating Characteristics

Neutral/Park. These two quadrant positions provide for engine start. The drive clutches are OFF, and there is no power

FIGURE 22–22A Disassembled solenoid assembly for viewing purposes only. Do not disassemble to service or to inspect. Assembly service requires a complete unit replacement.

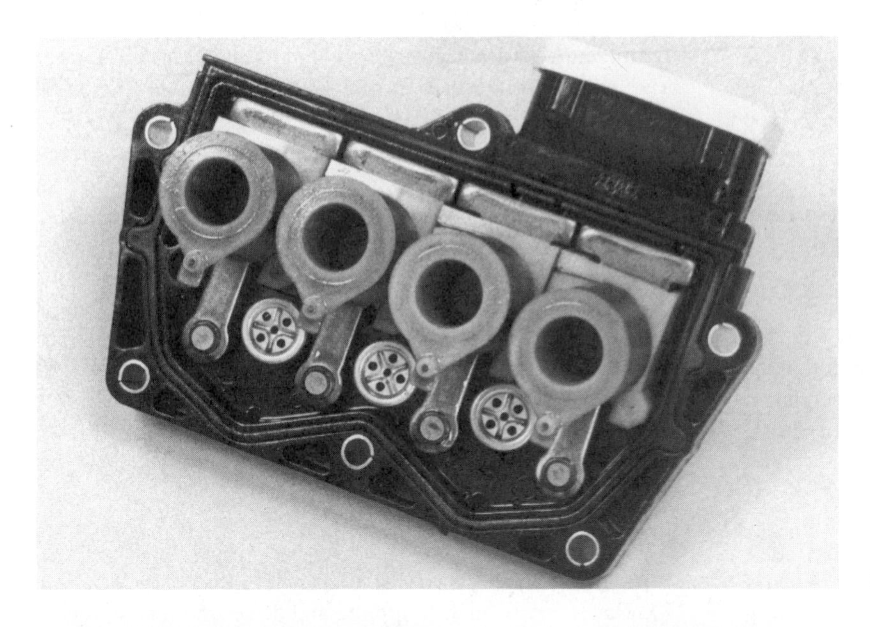

FIGURE 22–22B Solenoids and pressure switches.

FIGURE 22–22C Solenoid assembly pressure switch springs and resistors.

low through the geartrain. The low/reverse reaction clutch is applied in anticipation of transmission engagement. In the park position, an internal park rod assembly operated through the manual lever engages the park pawl into the park lugs on the rear carrier. This locks the output shaft to the transaxle case.

Overdrive. The transmission engages first gear and is ready for full automatic shifting, 1-2, 2-3, and 3-4. The shift points are predetermined by engine load and vehicle speed. Minimum throttle engagement of fourth gear occurs at approximately 30 mph (48 km/h). Once in fourth gear, the torque converter clutch locks up when load and speed conditions are right.

The driver can overrule the shift system by depressing the accelerator. Within the proper vehicle speed and engine torque range, WOT forced downshifts 4-3 and 4-2, as well as 3-2, 3-1, and 2-1 combinations, are available. For part-throttle acceleration, the transmission can downshift either 4-3 or 4-2. Closed-throttle coast downshifts are exceptionally hard to detect. The 4-3 shift occurs at 23 mph (37 km/h) and the 3-1 shift at 5 mph (8 km/h). Starting with 1992 production, the sequence is 4-3-2-1.

Drive. The D range is used for moderate to heavy-duty road load conditions such as hilly/mountainous driving or trailer pulling. The transaxle uses only first, second, and third gears, with the 2-3 minimum shift delayed to 40 mph (64 km/h). The 3-4 shift is inhibited. Within the proper vehicle speed and engine torque range, WOT forced downshifts 3-2 and 2-1 are available. In D range, the converter clutch is applied in third gear only, unless the engine coolant temperature rises above 240°F (116°C), in which case the converter clutch is also applied in second gear.

Manual Low. The L range provides extra pulling power and engine braking when handling steep grades. Normally, manual low permanently "locks in" first gear. Upshifts to second and third gears are programmed to provide maximum engine performance and overspeed protection. The 1-2 shift occurs at 38 mph (61 km/h) and the 2-3 shift at 70 mph (113 km/h). The torque converter clutch can be applied in third gear.

The downshift points occur at higher vehicle speeds to provide for maximum engine braking when descending steep grades. The 3-2 shift occurs at 68 mph (110 km/h) and the 2-1 at 35 mph (56 km/h).

Reverse. The R range should normally be engaged with the vehicle completely stopped. If the output sensor detects that the vehicle is moving forward at 8 mph (13 km/h) or greater, reverse engagement is blocked. The transaxle is put into neutral and is protected from potential damage.

41TE Powerflow Summary

The planetary geartrain is composed of a compact compound assembly integrating two planetary units. A transfer shaft relays the torque to a final drive unit designed with a 3.43:1 ratio. Controlling the geartrain are three input clutches and two reaction clutches (Figures 22–17 and 22–23).

The input clutches are contained in a single assembly and drive-selected components in the planetary geartrain (Figures 22–24 and 22–25).

- Underdrive clutch drives the rear sun gear.
- Overdrive clutch drives the front carrier assembly.
- Reverse clutch drives the front sun gear.

The OD/REV piston is unique in design, as it will stroke either forward or backward, giving a push or pull effect. In fourth gear/overdrive, it strokes forward and clamps the overdrive clutch. For reverse, it strokes backward and clamps the reverse clutch. Study Figures 22–17 and 22–23. The 2-3 reaction clutch holds the front sun gear, and the low/reverse reaction clutch holds the front carrier assembly. The output of the planetary geartrain always passes through the rear carrier assembly (Figure 22–26). Study the clutch apply summary and range reference chart (Chart 22–3, page 508).

Neutral/Park. None of the input clutches are applied, so no power is transmitted to the planetary geartrain. The low/reverse clutch is applied in preparation for the manual shift into OD or R.

First Gear. (Figure 22–27)

- Underdrive clutch is applied and drives the rear sun gear.
- Low/reverse reaction clutch is applied and holds the front carrier assembly.
- Low gear is a function of the rear planetary providing a reduction of 2.84:1.

Second Gear. (Figure 22–28)

- Underdrive clutch is applied and drives the rear sun gear.
- 2-4 clutch is applied and holds the front sun gear.
- Second gear is a combined function of the front and rear planetary units, providing a compound ratio of 1.57:1.

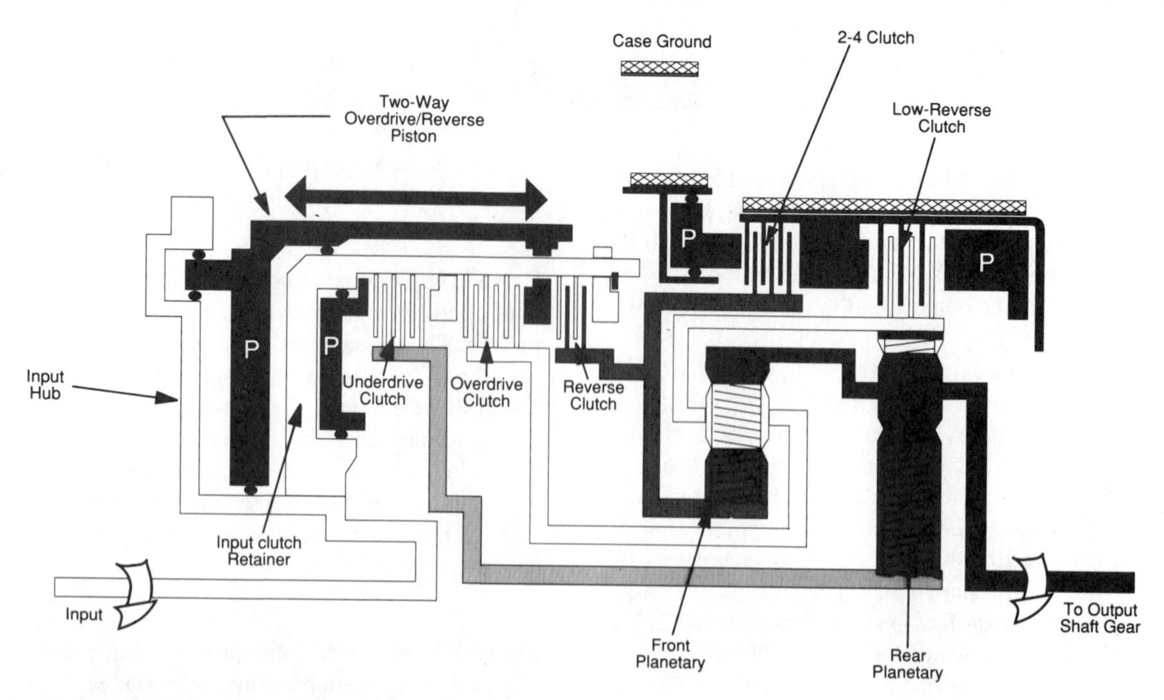

FIGURE 22–23 Chrysler 41TE (A-604) powerflow schematic.

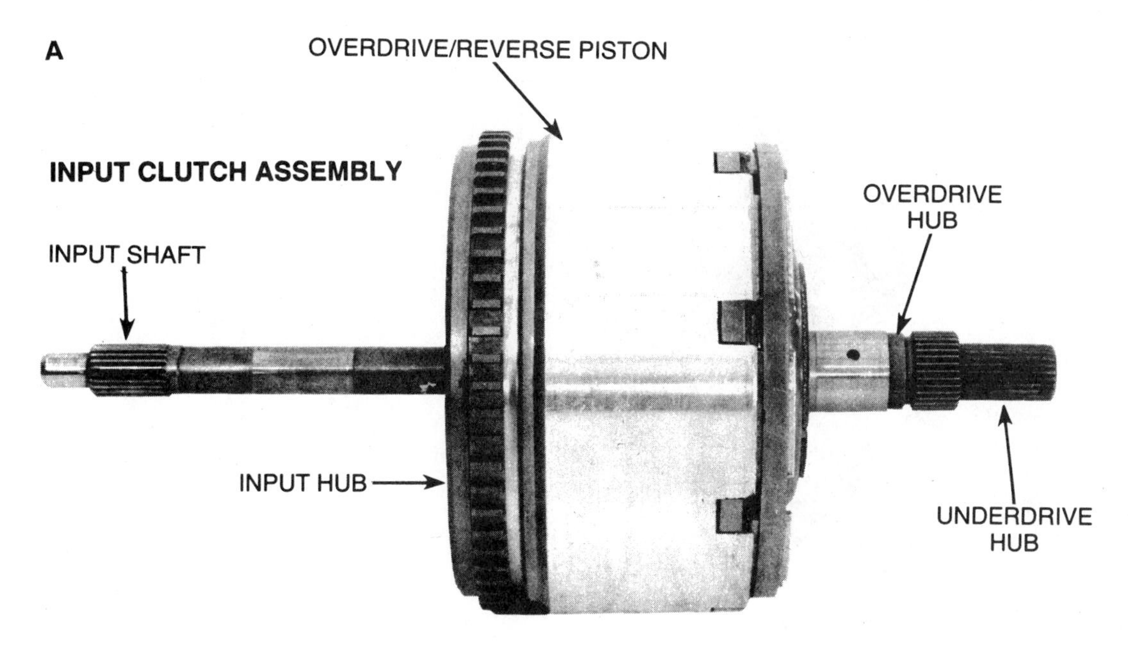

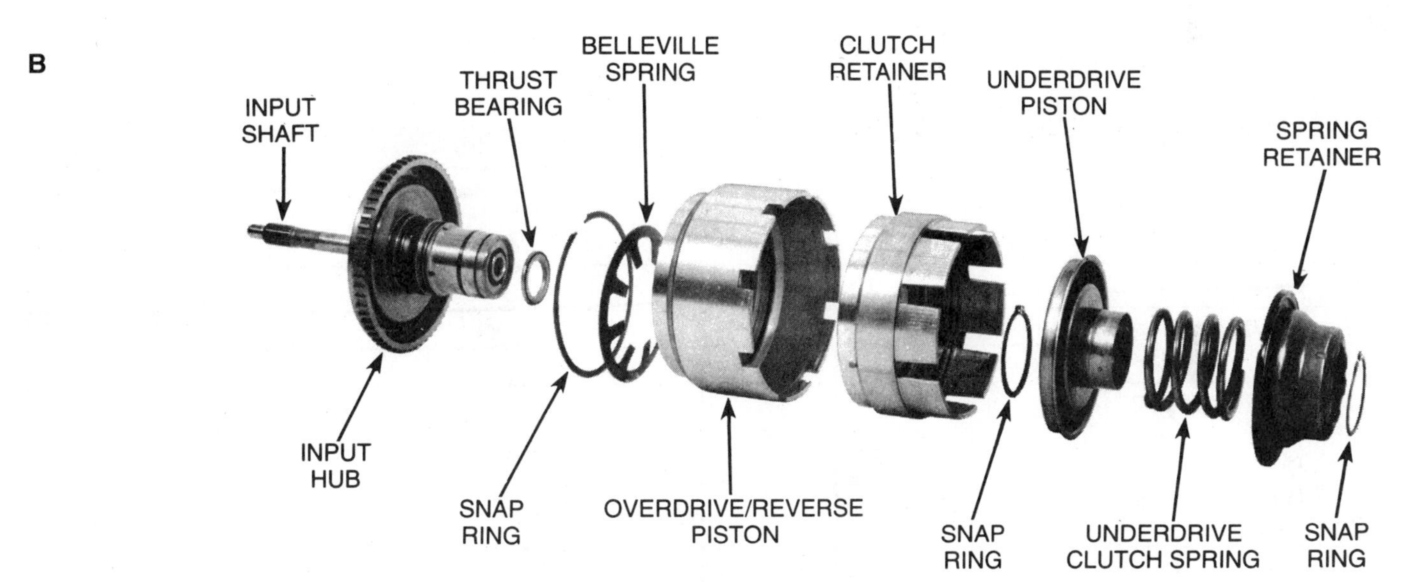

FIGURE 22–24 Chrysler 41TE (A-604): (a) input clutch assembly; (b) component identification. (Courtesy of Chrysler Corp.)

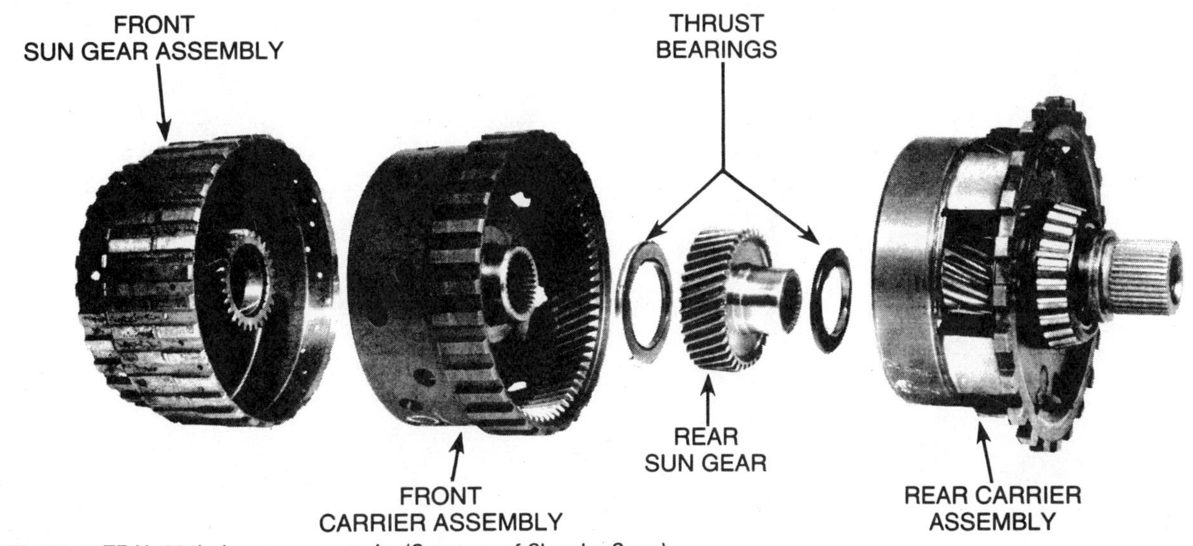

FIGURE 22–25 41TE (A-604) planetary geartrain. (Courtesy of Chrysler Corp.)

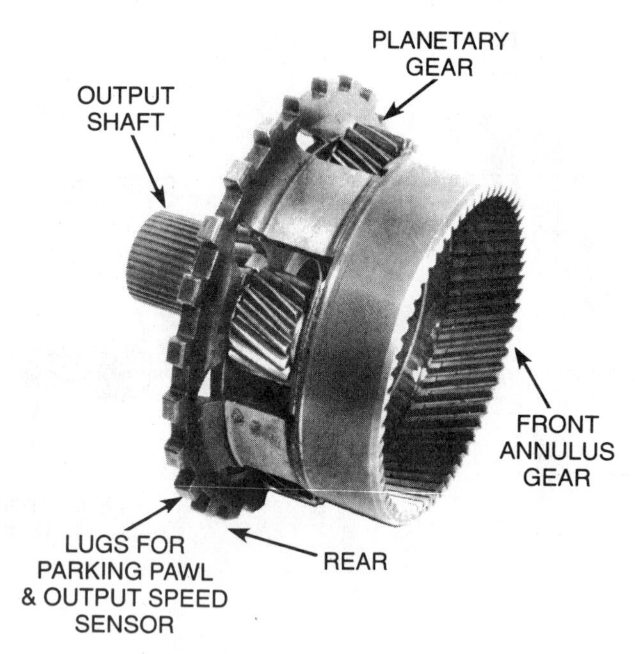

FIGURE 22–26 41TE (A-604) rear carrier assembly. (Courtesy of Chrysler Corp.)

Shift Lever Position	Start Safety	Park Sprag	Clutches: Underdrive	Clutches: Overdrive	Clutches: Reverse	Clutches: 2/4	Clutches: Low/ Reverse	Lock-Up
P — PARK	X	X					X	
R — REVERSE					X		X	
N — NEUTRAL	X						X	
OD — OVERDRIVE								
First			X				X	
Second			X			X		
Direct			X	X				
Overdrive				X		X		X
D — DRIVE*								
First		X					X	
Second			X			X		•X
Direct			X	X				X
L — LOW*								
First			X				X	
Second			X			X		
Direct			X	X				X

*Vehicle upshift and downshift speeds are increased when in these selector positions.
•Engages only when engine temperature rises to 240°F (116°C)

CHART 22–3 Chrysler 41TE (A-604) and 42LE (A-606) range reference and clutch/band apply summary. (Courtesy of Chrysler Corp.)

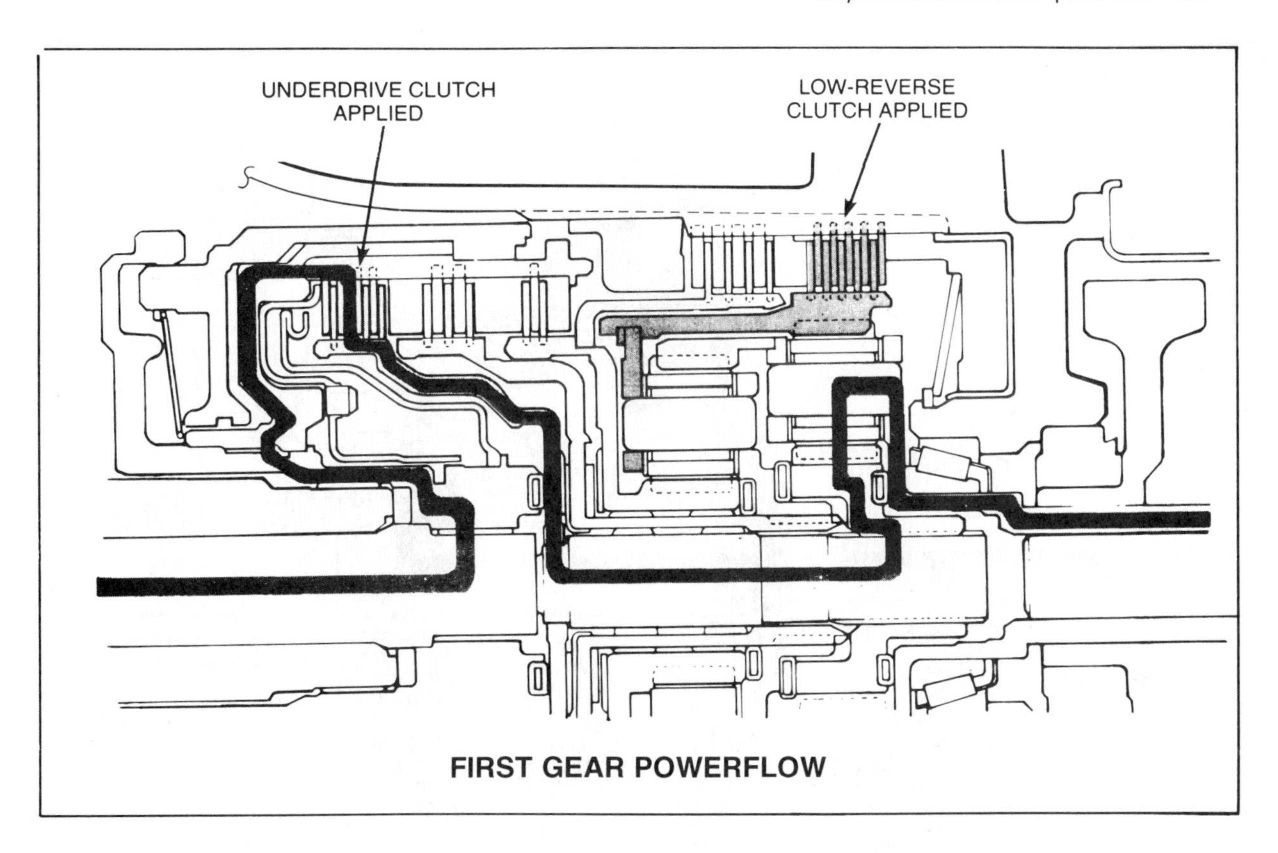

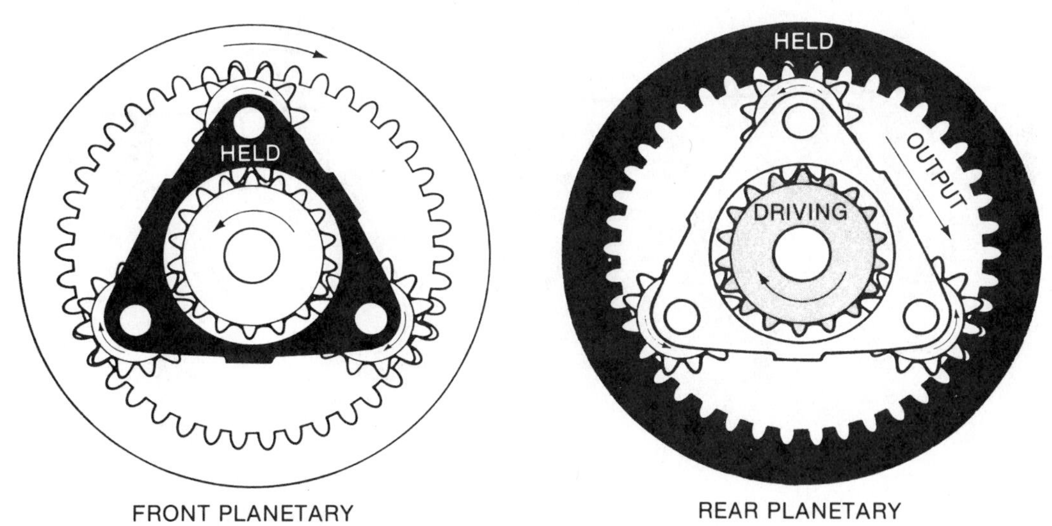

- REAR ANNULUS/FRONT CARRIER (HELD)
- REAR SUN GEAR DRIVING
- REAR SUN GEAR DRIVES THE PINIONS
- PINIONS WALK AROUND THE ANNULUS GEAR AND DRIVE THE REAR CARRIER TO PROVIDE THE OUTPUT TORQUE 2.84:1
- FRONT PLANETARY FREEWHEELS

FIGURE 22–27 Powerflow in low gear. (Courtesy of Chrysler Corp.)

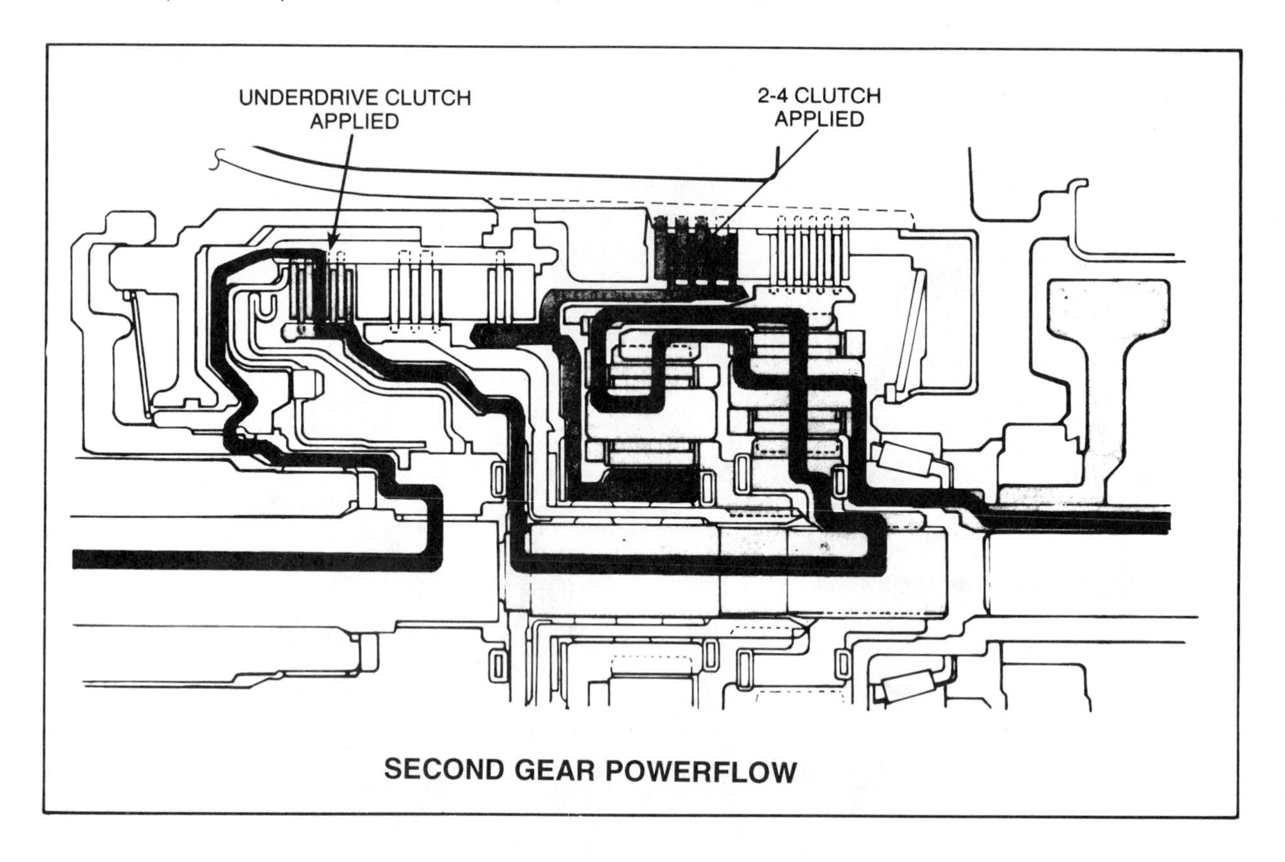

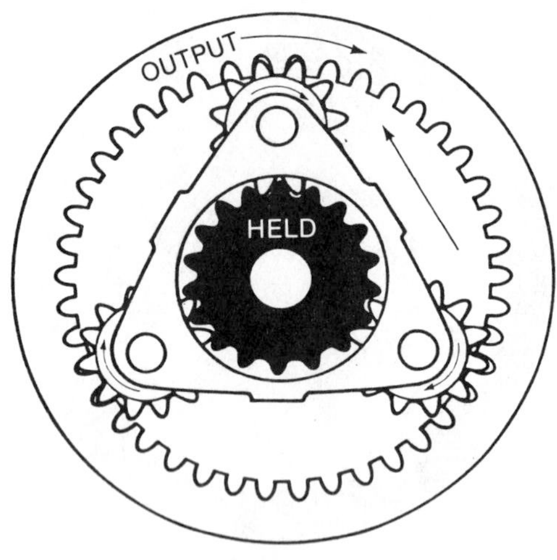

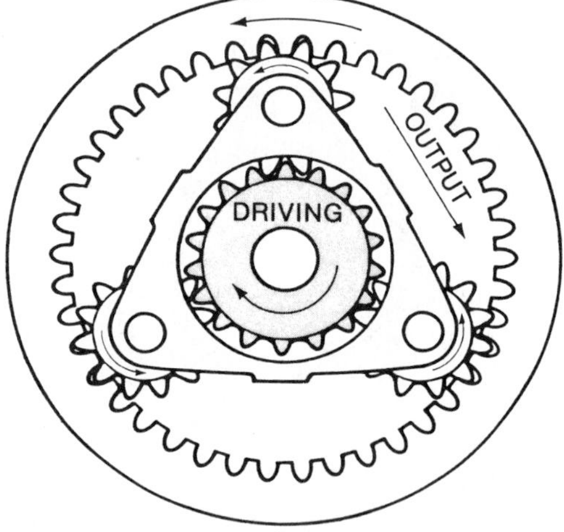

- FRONT SUN GEAR (HELD)
- REAR SUN GEAR DRIVING
- REAR SUN GEAR DRIVES THE PINIONS
- PINION ROTATION DRIVES THE REAR ANNULUS/FRONT CARRIER ASSEMBLY. THE POWER ROTATION ENTERS THE FRONT PLANETARY THROUGH THE FRONT CARRIER
- FRONT CARRIER ROTATION DRIVES THE PINIONS
- PINIONS WALK AROUND THE SUN GEAR AND DRIVE THE FRONT ANNULUS/REAR CARRIER TO PROVIDE THE OUTPUT TORQUE 1.57:1

FIGURE 22–28 Powerflow in second gear. (Courtesy of Chrysler Corp.)

Third Gear. (Figure 22–29)

- Underdrive clutch is applied and drives the rear sun gear.
- Overdrive clutch is applied and drives the front carrier/rear annulus assembly.
- The planetary gear units are locked together for a direct ratio of 1:1.

Fourth Gear/Overdrive. (Figure 22–30)

- Overdrive clutch is applied and drives the front carrier/rear annulus assembly.
- 2-4 clutch is applied and holds the front sun gear.
- Fourth gear is a function of the front planetary and provides an overdrive ratio of 0.69:1.

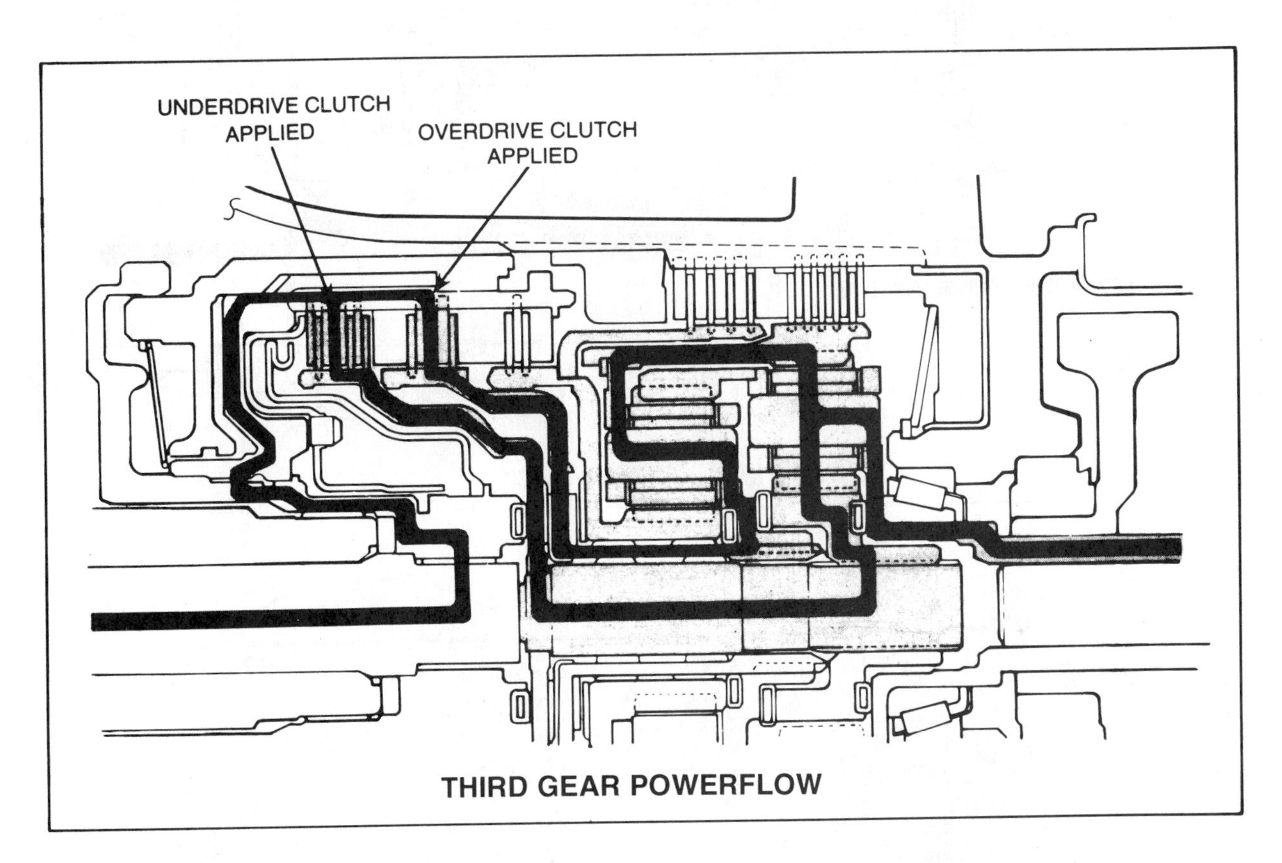

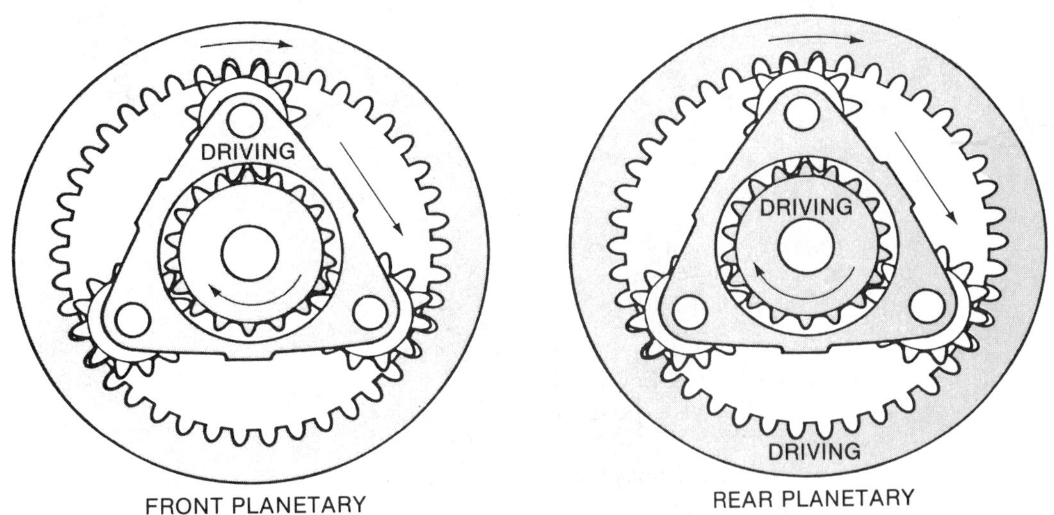

- REAR SUN GEAR AND FRONT CARRIER/REAR ANNULUS ARE CLUTCHED TOGETHER AND DRIVING
- FRONT AND REAR PLANETARY UNITS ARE LOCKED. DIRECT RATIO IS 1:1

FIGURE 22–29 Powerflow in third gear. (Courtesy of Chrysler Corp.)

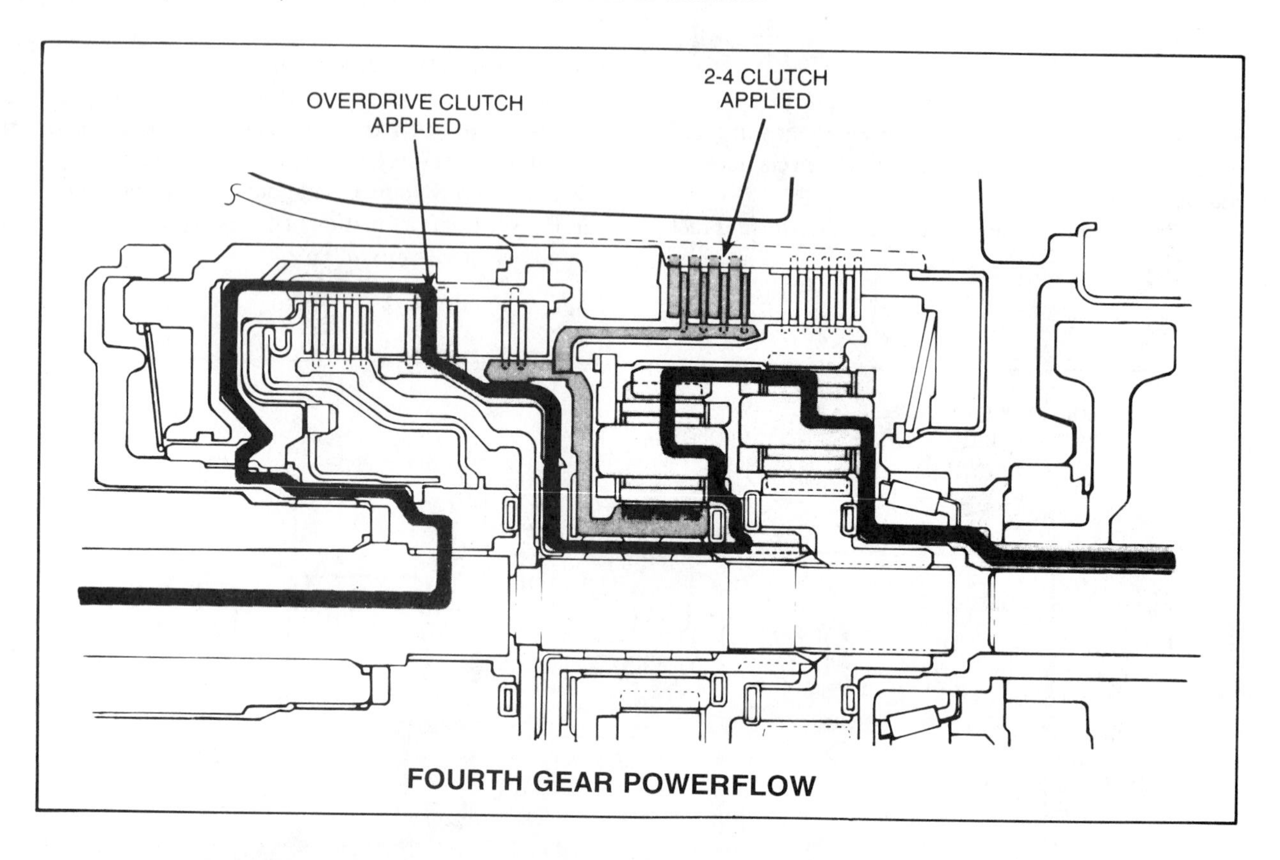

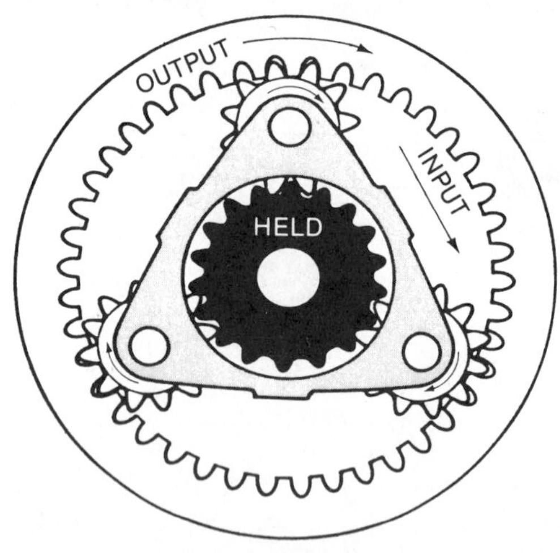

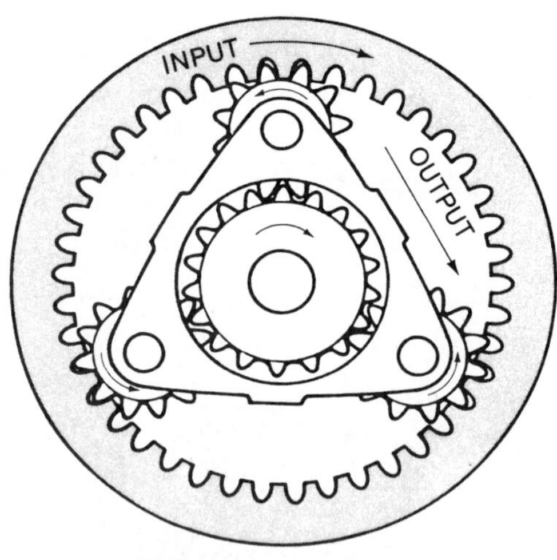

- FRONT SUN GEAR IS (HELD)
- FRONT CARRIER/REAR ANNULUS IS DRIVING
- CARRIER PINIONS ROTATE AND WALK AROUND THE SUN GEAR AND DRIVE THE FRONT ANNULUS/REAR CARRIER ASSEMBLY AT AN OVERDRIVE RATIO 0.69:1
- REAR PLANETARY FREEWHEELS

NOTE: FINAL DRIVE RATIO IS 3.42 TO 1. THE OVERALL TOP GEAR RATIO IN OVERDRIVE IS 2.36 TO 1.

FIGURE 22–30 Powerflow in fourth gear. (Courtesy of Chrysler Corp.)

Reverse. (Figure 22–31)

- Reverse clutch is applied and drives the front sun gear.
- Low/reverse reaction clutch is applied and holds the front carrier/rear annulus assembly.
- Reverse gear is a function of the front planetary, providing a reverse ratio of 2.21:1.

42LE (A-606) Features

The 42LE case assembly is featured in Figure 22–32. Although this transaxle configuration has a north-south/longitudinal powertrain orientation, it retains many of the design features of the geartrain and electronic control of the 41TE (A-604). The geartrain and friction elements basically remain

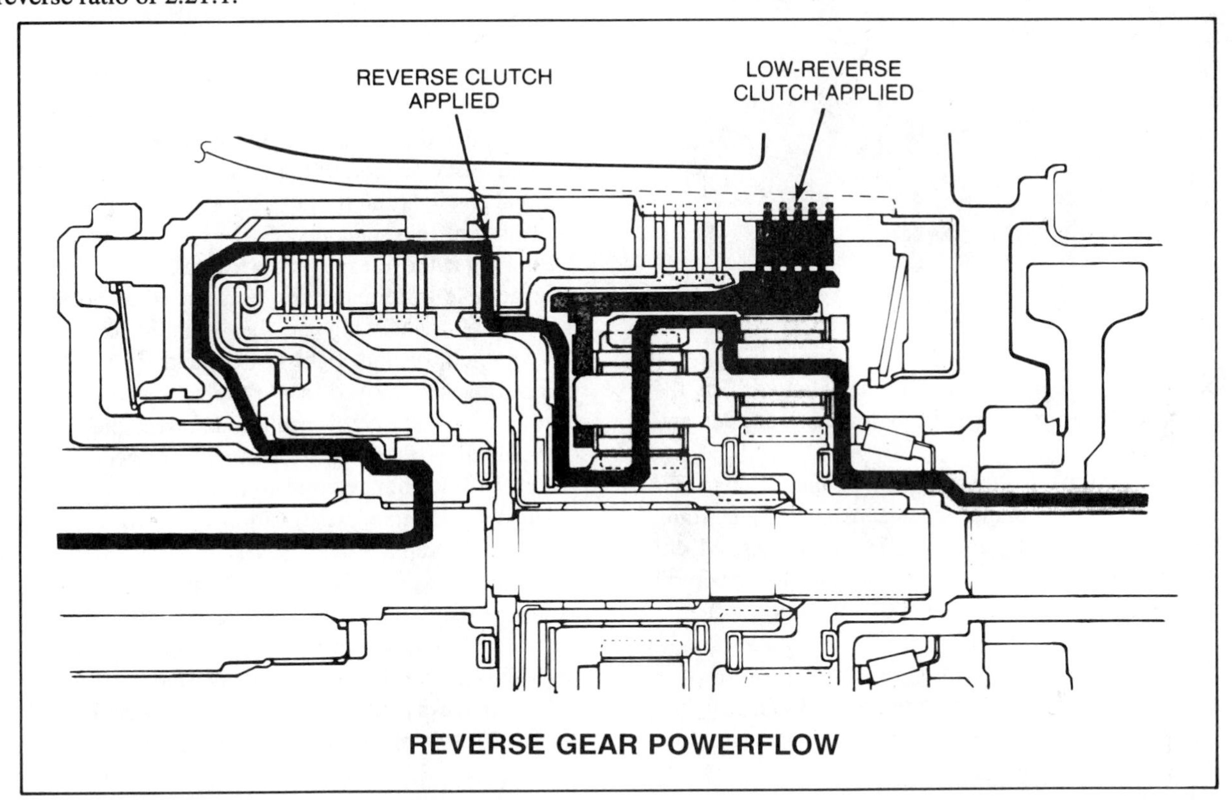

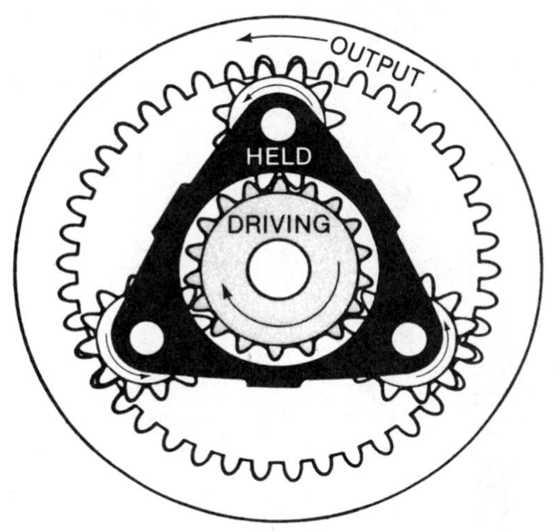

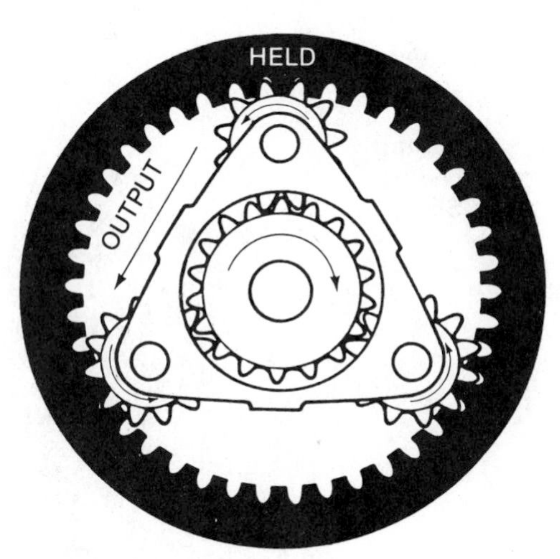

- FRONT CARRIER/REAR ANNULUS (HELD)
- FRONT SUN GEAR DRIVING
- FRONT SUN GEAR DRIVES THE PINIONS CCW
- PINIONS ACTING AS REVERSE IDLERS DRIVE THE FRONT ANNULUS/REAR CARRIER ASSEMBLY CCW TO PROVIDE THE OUTPUT TORQUE 2.21:1
- REAR PLANETARY FREEWHEELS

FIGURE 22–31 Powerflow in reverse gear. (Courtesy of Chrysler Corp.)

FIGURE 22–32 Chrysler 42LE case assembly.

the same, including identical gear ratios. Upgrades in the rotating parts were made to handle the increased torque demand of the 3.5-L engine.

Electronic control is also quite similar to that of the 41TE (A-604), with a few refinements. The geartrain and electronic controls are covered in the 41TE section in this chapter and will not be repeated here. This section highlights the structural features and characteristics that make this transaxle different from the 41TE.

The cutaway section view in Figure 22–33 identifies the main internal components. Referring to illustration highlights, the geartrain output torque drives a chain link for transfer to a hypoid pinion shaft and drive gear (Figure 22–34). The final drive geartrain includes the transfer shaft, ring gear with a top-mounted pinion drive gear, and differential assembly (Figures 22–35 and 22–36).

The hypoid ring and pinion final drive handles higher torque loads, runs quietly, and changes the final torque drive direction to an east-west orientation. The left side output shaft is permitted to cross under the input shaft (Figure 22–37). Gear backlash and side bearing preload adjustment is achieved with threaded bearing and backlash adjustment sleeves. Transfer shaft bearing preload and pinion depth is controlled by selective shims. The final drive section of the transaxle uses hypoid gear lube. Two seals on the pinion shaft keep the gear lube and automatic transmission fluid separated.

Looking at the left side of the transaxle case (Figure 22–38), some notable differences can be seen when compared to the 41TE (A-604). Missing are the solenoid assembly, neutral/safety switch, backup light switch, and PRNDL switch. The solenoid pack is moved inside the case and mounted on the valve body. The internal location dampens the noise from the solenoid valve operation and provides protection to the solenoid assembly from external contaminants.

The manual valve lever position sensor (MVLPS), shown in Figure 22–39, replaces the PRNDL, neutral/safety, and backup light switches used in the 41TE. The purpose of the MVLPS is to inform the TCM which selector range lever position the driver has chosen. Worth noting on the left side are the pressure tap provisions for diagnosing the hydraulics.

The eight-way solenoid pack connector and differential fill plug are located on the right side of the case.

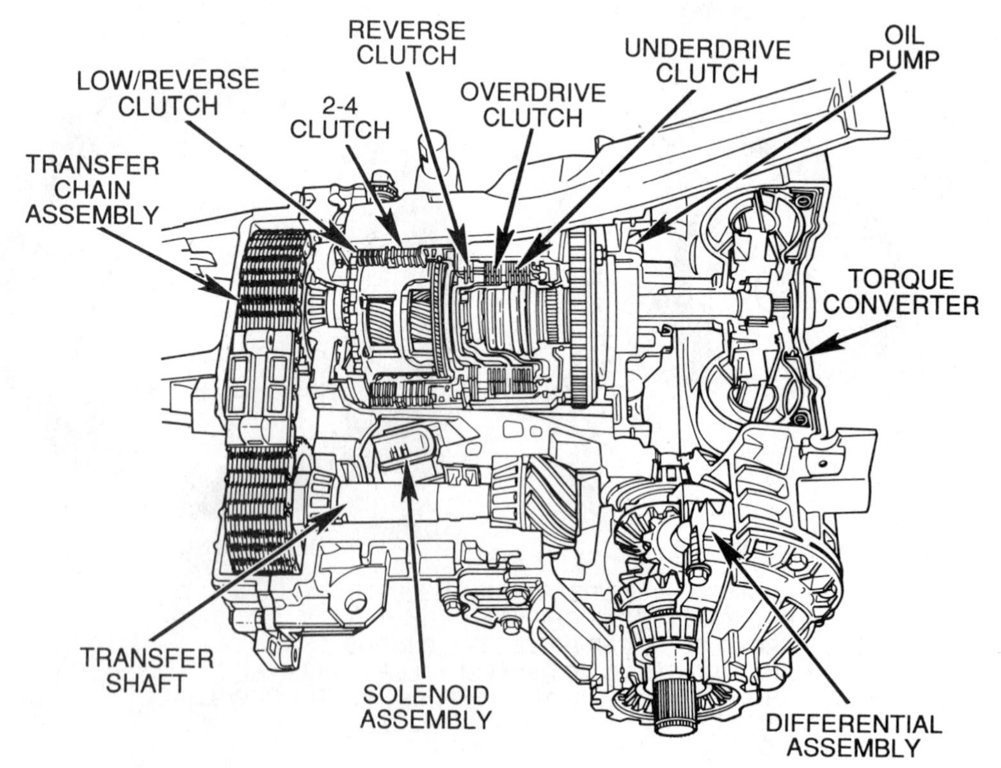

FIGURE 22–33 42LE sectional view. (Courtesy of Chrysler Corp.)

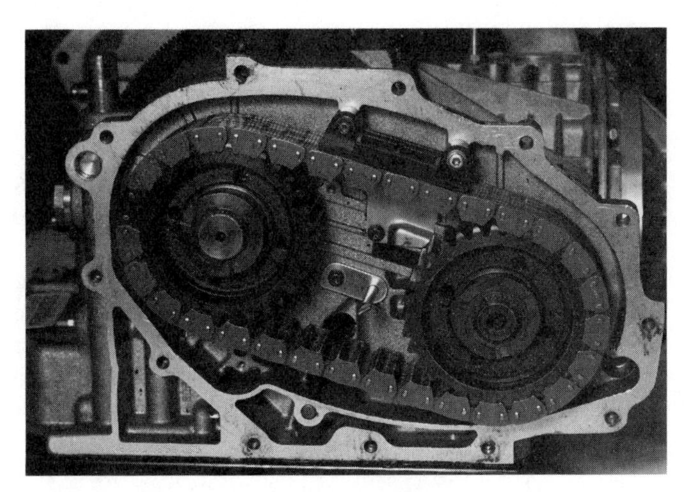

FIGURE 22–34 Chrysler 42LE transfer chain.

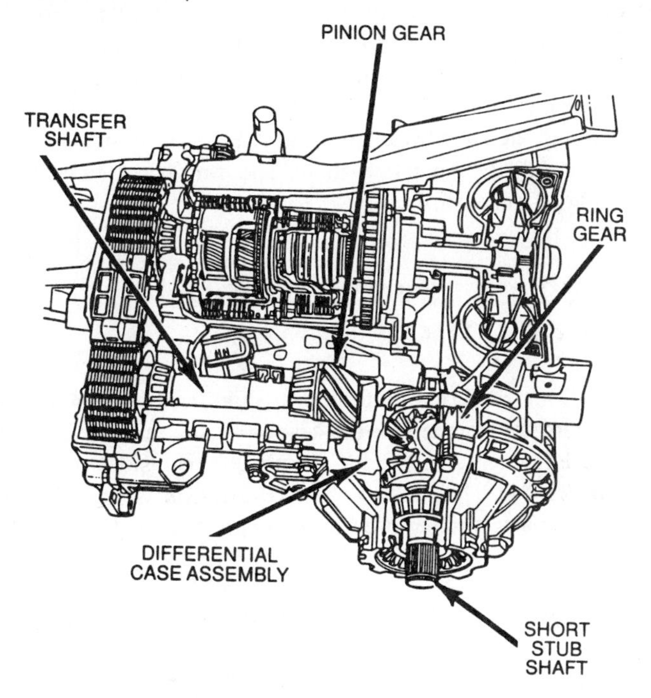

FIGURE 22–35 42LE transfer shaft and final drive components. (Courtesy of Chrysler Corp.)

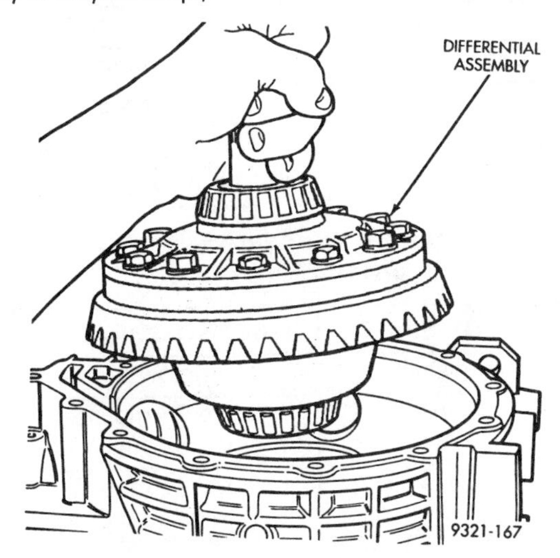

FIGURE 22–36 42LE differential assembly. (Courtesy of Chrysler Corp.)

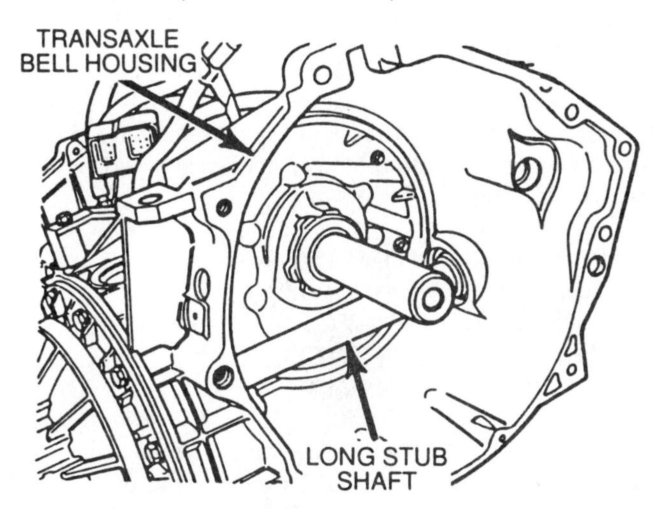

FIGURE 22–37 42LE left side output stub shaft crosses under input shaft. (Courtesy of Chrysler Corp.)

FIGURE 22–38 Chrysler 42LE left side case view.

FIGURE 22–39 Chrysler 42LE valve body and solenoid assembly.

42LE Operating Characteristics

These essentially are similar to the 41TE (A-604). A special feature is added to the TCM logic on transmission fluid temperature to determine transaxle operating characteristics. The temperatures are:

- Extreme cold: less than –15°F (–26°C).
- Super cold: less than 0°F (–18°C).
- Cold: less than 36°F (–2°C).
- Warm: less than 76°F (24°C).
- Hot: greater than 76°F (24°C).

The transaxle does not use a transmission fluid temperature (TFT) sensor to directly read the temperature of the fluid. Rather, the fluid temperature is calculated from the readings of various other input sensors, including engine coolant temperature, ambient temperature, vehicle speed, operating range, and ignition OFF time. Fluid temperature has a definite bearing on the TCM select shift schedule, N and R excluded.

In extreme cold conditions, all shifts are inhibited. The forward operating ranges are restricted to second-gear or limp-in mode operation. Once the temperature reaches the super cold range, the full complement of shifts is allowed. Super cold will inhibit 4-2 and 3-1 full throttle kickdowns. As the fluid temperature increases and phases into the warm and hot zones, the TCM tightens up on the shift schedule spread.

The EMCC (electronically modulated converter clutch) operation is based on engine temperature. When the coolant temperature reaches 151°F (66°C), EMCC operation is permitted.

42LE Powerflow Summary

Refer to the 41TE (A-604) powerflow summary, including the range reference and clutch apply summary chart (Chart 22–3, page 508).

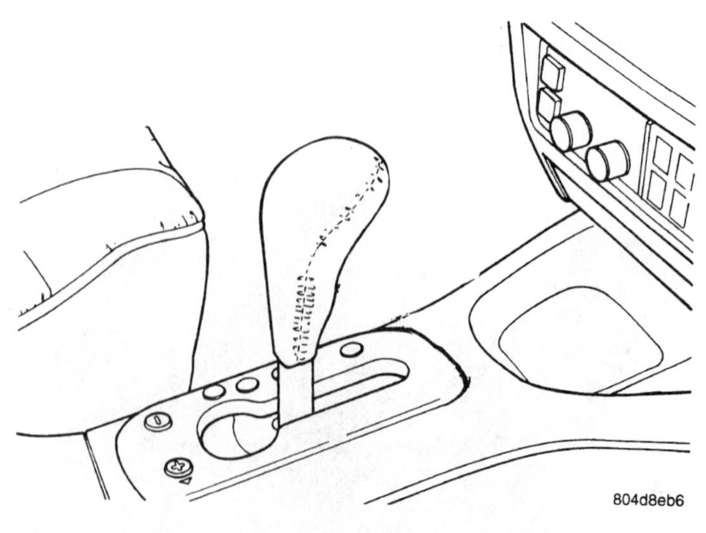

FIGURE 22–40 Autostick shift lever. Note the shape of the shift lever pattern. (Courtesy of Chrysler Corp.)

42LE Autostick

Available in the 1996 Eagle Vision TSI, the "Autostick" allows drivers to select either automatic or manual shifting. The shift lever quadrant pattern is P-R-N-OD-A/S. Referring to Figure 22–40, when the shift lever is moved into the A/S position, Autostick is activated, and side-to-side movement of the lever is possible.

When Autostick is chosen, the transaxle remains in the gear that it was in prior to selecting A/S. The driver now controls the upshifts and downshifts. Moving the lever to the right causes an upshift, and when the lever is moved to the left of center, a downshift occurs. The instrument cluster identifies to the driver which gear has been selected.

The Autostick logic program includes many special features. The vehicle can pull away from a stop in first, second, or third gear. Speed/cruise control is available in third and fourth gears but is deactivated if downshifted into second gear. For safety, durability, and driveability, some upshifts and downshifts are prevented or commanded automatically (Chart 22–4).

Moving the shift lever from the A/S position into the OD range cancels the Autostick mode. The transaxle then follows normal OD range shift timing.

CHRYSLER TRANSMISSION IDENTIFICATION

On RWD and FWD hydraulic control units, a series of numbers and letters are stamped on the case lip just above the oil pan flange: left front side for RWD and left rear side for FWD (Figure 22–41). The parts number is the transmission assembly reference parts number used for parts procurement. The build

Automatic shifts will occur under the following conditions

TYPE OF SHIFT	APPROXIMATE SPEED
4-3 coast downshift	13 mph
3-2 coast downshift	9 mph
2-1 coast downshift	5 mph
1-2 upshift	6300 engine rpm
2-3 upshift	6300 engine rpm
4-3 kickdown shift	13-31 mph w/sufficient throttle

Manual shifts are NOT permitted under the following conditions

TYPE OF SHIFT	APPROXIMATE SPEED
3-4 upshift	Below 15 mph
3-2 downshift	Above 74 mph @ closed throttle or 70 mph otherwise
2-1 downshift	Above 41 mph @ closed throttle or 38 mph otherwise

CHART 22–4 42LE Autostick overrides. (Courtesy of Chrysler Corp.)

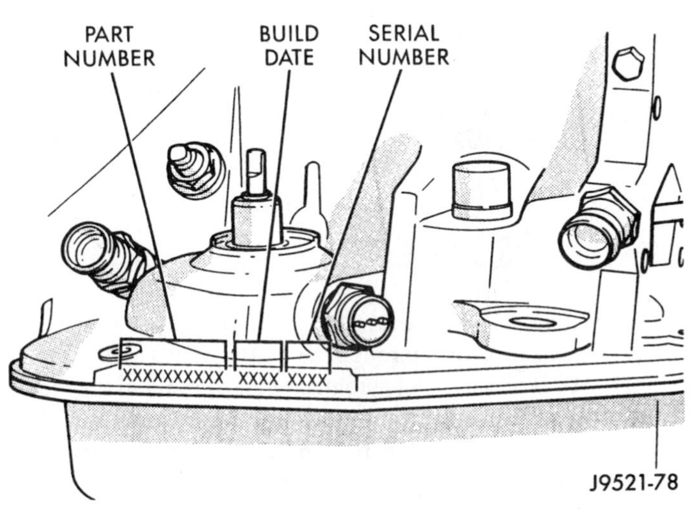

FIGURE 22–41 Transmission part and serial number location. (Courtesy of Chrysler Corp.)

FIGURE 22–42 Chrysler 41TE/42LE ID tag.

date is the warranty date code based on a 10,000-day calendar. The year and date of production can be identified by using the date code chart in the parts manual. The serial number represents the daily production number.

The 41TE (A-604) and 42LE (A-606) use identical ID tags (Figure 22–42). The 41TE ID tag is located on the bell housing, and the 42LE ID tag is located at the rear left of the transaxle case. The *K* means it was built in Kokomo, Indiana. For parts reference, use the three-digit number.

The three-speed FWD transaxle models are sometimes difficult to identify because of their look-alike features. The bell housing opening configurations and different starter locations offer a method for identification (Figure 22–43).

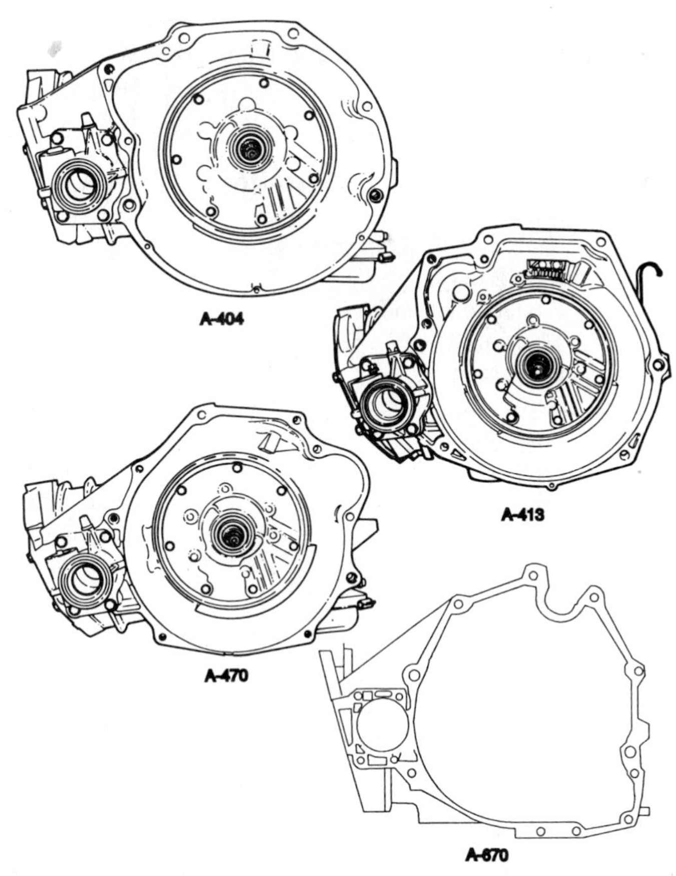

FIGURE 22–43 Three-speed transaxle bell housing configurations. (Courtesy of Chrysler Corp.)

SUMMARY

In the early- to mid-1990s, Chrysler refined their TCM electronically controlled 41TE (A-604) and 42LE automatic four-speed transaxles for their domestic car line. These transaxles will continue to be used in late model FWD vehicles.

Current attention is being focused on the evolution of transmissions that are hydraulically controlled and those that are hydraulically/electronically controlled. Beginning in 1996 models, the 42RE (A-500ES) four-speed hydraulically/electronically controlled automatic transmission powertrains use a single control module. This PCM combines the functions of the TCM and PCM/ECM. Presently, RWD automatic transmissions with full electronic shift control are being produced.

Chrysler continues to develop automatic transmissions and transaxles that will enhance vehicle fuel efficiency and emission levels and meet the needs of today's consumers.

23 Ford Automatic Transmission/Transaxle Summaries

INTRODUCTION

Ford produces a wide variety of automatic transmissions and transaxles to meet the requirements of their product line. Current production units reflect new engineering design concepts featuring advanced electronic control of four- and five-speed powertrain configurations. Highlighted in this chapter are Ford automatic transmissions and transaxles identified by the following code designations:

Front-wheel Drive—Automatic Transaxles

AX4S	AX4N	CD4E	ATX

Rear-wheel Drive—Automatic Transmissions

AODE	4R70W	C-6	E4OD
A4LD	4R44E	4R55E	5R55E

Code Translation

A	Automatic	O	Overdrive
C	Vehicle application size	S	Synchronized
	(CD4E only)	T	Transmission
C	Production decade	W	Wide-ratio
	(C-6 only)	X	Transaxle
C	Close ratio	4	Four-speed
D	Drive	5	Five-speed
E	Electronically controlled	44	440 ft-lb max.
F	Front-wheel drive		input torque*
H	Taurus SHO vehicle	55	550 ft-lb max.
HD	Heavy-duty		input torque*
L	Lockup	70	700 ft-lb max.
M	Manual		input torque*
N	Nonsynchronized		

* As measured at the torque converter output and transmission input shaft.

FORD FOUR-SPEED RWD TRANSMISSIONS: AOD, AODE, and 4R70W Series

This transmission series began with the 1980 model year when the hydraulically controlled AOD was introduced.

The AOD, which served as the forerunner to the AODE and 4R70W, was produced through the 1991 product year.

The AOD was the first automatic transmission in the industry to feature a built-in overdrive: fourth-gear ratio is 0.67:1. Engine rpm in overdrive is reduced by thirty-three percent to sustain a given road speed when compared to a conventional high gear ratio of 1:1.

For 1980, the transmission was standard equipment on the Lincoln Continental and Mark VI, and optional with the 5.0-L Thunderbird and Cougar XR-7, as well as the 5.0- and 5.8-L Ford LTD and Mercury Marquis. For 1981, it completely replaced the C-6 transmission for use in passenger cars. In updated applications, the AOD has been teamed with the 3.8- and 5.0-L engine, including the E and F series van/light truck line.

Figures 23–1 and 23–2 show cutaway and schematic views of the AOD automatic overdrive transmission. As can be observed, the AOD uses a Ravigneaux planetary geartrain with the ring gear as the output member. In third gear, the ratio is 1:1, with the input torque split between two shafts that drive the reverse sun gear and planet carrier. Overdrive is achieved by the addition of a band to hold the reverse sun gear while driving the planet carrier. For control of the geartrain, the AOD transmission uses three multiple-disc drive clutches, one multiple-disc reaction clutch, two bands, and two one-way roller clutches.

The transmission case is a one-piece aluminum casting with the converter housing and has a bolt-on extension housing. Bands require no scheduled maintenance adjustments during regular service.

The AOD uses a three-element torque converter with a design feature vital to overdrive: a split torque powerflow. Two input shafts provide both hydraulic and mechanical drive. A turbine tube shaft splines into the turbine hub and hydraulically drives the forward clutch cylinder. A solid direct-drive input shaft splines into a torsional damper assembly, which is part of the converter cover. The latter arrangement mechanically links the direct clutch to the engine crankshaft.

Input to the geartrain varies from one hundred percent hydraulic drive in first, second, and reverse gears to approximately forty percent hydraulic and sixty percent mechanical in third gear (split torque), and one hundred percent mechanical (lockup) in overdrive fourth gear. The split torque and lockup feature of the converter offers additional fuel economy.

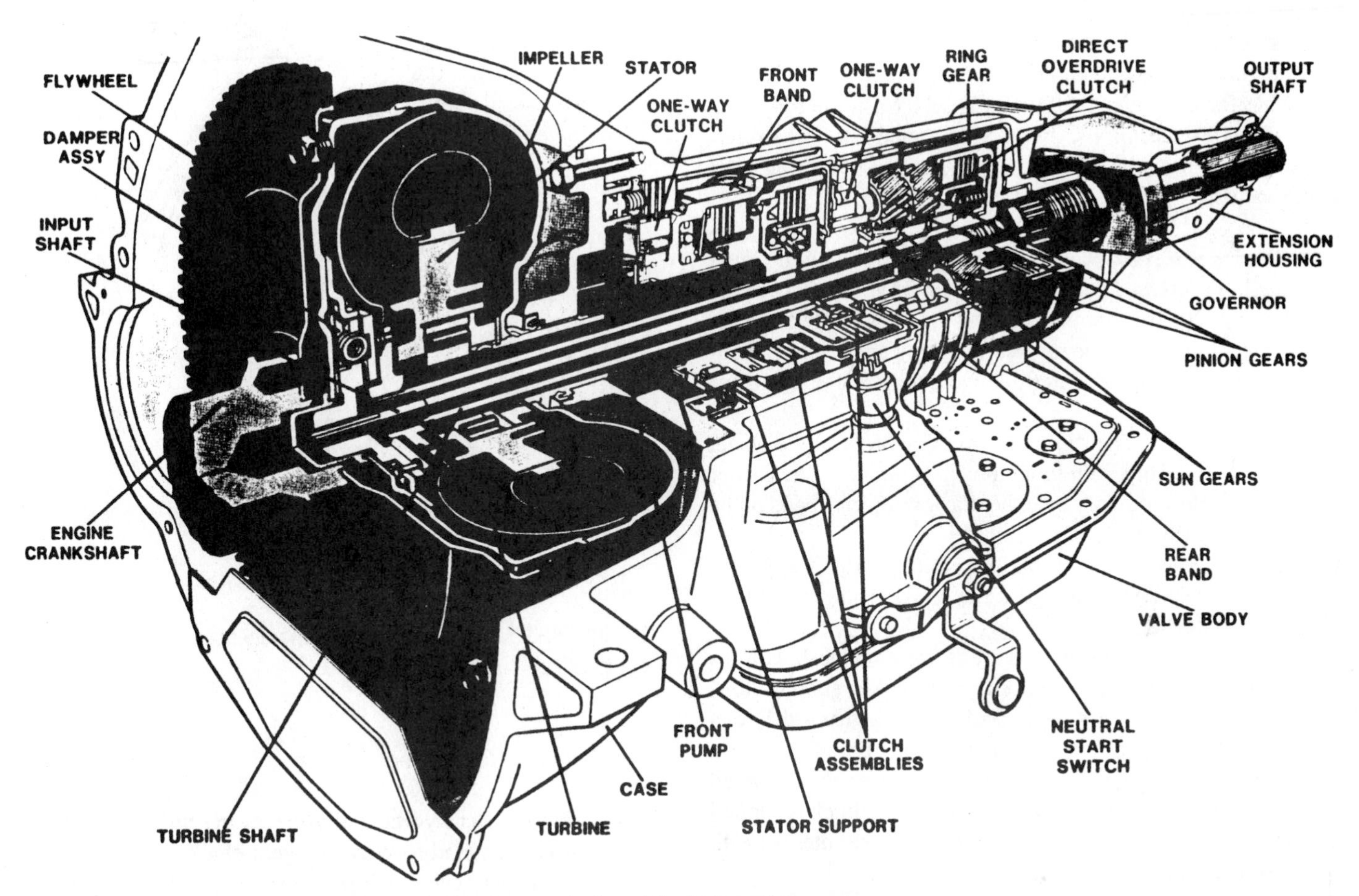

FIGURE 23–1 AOD cutaway view. (Reprinted with the permission of Ford Motor Company)

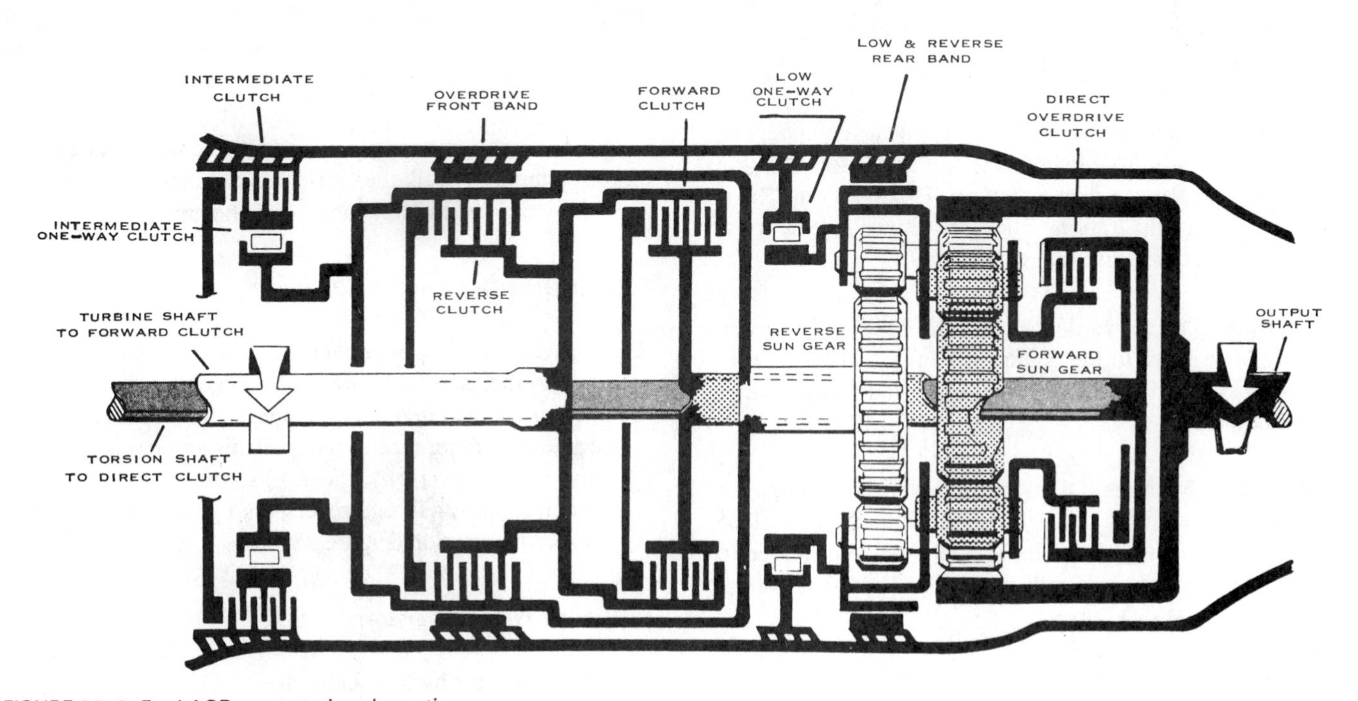

FIGURE 23–2 Ford AOD powertrain schematic.

Powerflow Summary

The geartrain design and clutch/band control is essentially the same for the AOD, AODE, and 4R70W. In the following powerflow discussion, use Figures 23–1 and 23–2 as a reference. Following is a description of the geartrain planetary components.

1. *Reverse sun gear.* This sun gear is integral with the input shell. It is driven by the turbine (tube) input shaft when the reverse clutch is applied. It turns independently from the forward sun gear. The reverse sun gear can be held stationary by applying the intermediate clutch or the overdrive band.
2. *Planet carrier.* The AOD transmission has a single planet carrier assembly with long and short pinions. The carrier can be held stationary by either the low-reverse band or the planetary (low) one-way clutch.
3. *Long pinions.* The long pinions in the carrier are in constant mesh with the ring gear, the reverse sun gear, and the short pinions.
4. *Short pinions.* Short pinions are in constant mesh with the forward sun gear and the long pinions. The short pinions in the gearset do not mesh with the ring gear—they can drive the ring gear only through the long pinions.
5. *Forward sun gear.* The forward sun gear is so named because it is driven whenever the forward clutch is applied in first, second, and third gears. The sun gear meshes with the short planet pinions.
6. *Ring gear.* The gearset assembly has only one ring gear, which meshes with the long planet pinions. It is splined to a flange on the output shaft. Thus, the output from this gearset is always through the ring gear.

With the relationship of the planetary gear components, clutches, and bands established, we are ready to show the powerflow in the various driving ranges. To simplify the Ravigneaux planetary illustrations, they are shown in rear-view perspective with clockwise input rotations.

Neutral/Park. The forward, reverse, and direct clutches are off. There is no input to the geartrain (Figure 23–2). In park, the low-reverse band is applied but does not cause any drive or lock condition.

Overdrive and 3 Range. OD is the normal driving range for maximum fuel economy. Full automatic shifting takes place with a fourth-gear overdrive. The manual 3 range is limited to first, second, and third gears, with powerflows that are identical to those used in the OD range.

First Gear—OD and 3 Range.

- *Forward clutch applied:* acts as the input clutch and locks the turbine shaft to the forward sun gear.
- *Low one-way clutch effective:* holds the planetary carrier against counterclockwise rotation (Figure 23–3).

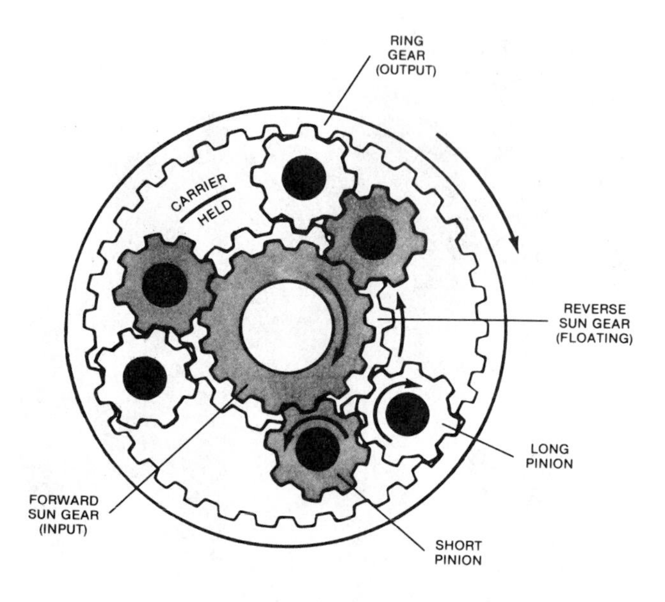

FIGURE 23–3 Ford AOD first-gear planetary operation (rear-view perspective).

First-Gear Powerflow.

- The converter hydraulically drives the turbine shaft, forward clutch, and forward sun gear clockwise.
- The sun gear drives the short pinions counterclockwise.
- The short pinions drive the long pinions clockwise.
- The long pinions drive the ring gear and output shaft clockwise at a reduced speed. First gear is a 2.4:1 reduction.

While driving the ring gear against the vehicle load, the long pinions attempt to walk the carrier counterclockwise. The low one-way clutch takes effect and holds the carrier.

Second Gear—OD and 3 Range.

- *Forward clutch applied:* acts as the input clutch and locks the turbine shaft to the forward sun gear.
- *Intermediate clutch applied:* locks the intermediate one-way clutch outer race to the transmission case.
- *Intermediate one-way clutch effective:* holds the reverse clutch drum, shell, and reverse sun gear against counterclockwise rotation (Figure 23–4).

Second-Gear Powerflow.

- The converter hydraulically drives the turbine shaft, forward clutch, and forward sun gear clockwise.
- The forward sun gear drives the short pinions counterclockwise.
- The long pinion gears are driven clockwise and walk clockwise around the stationary reverse sun gear.

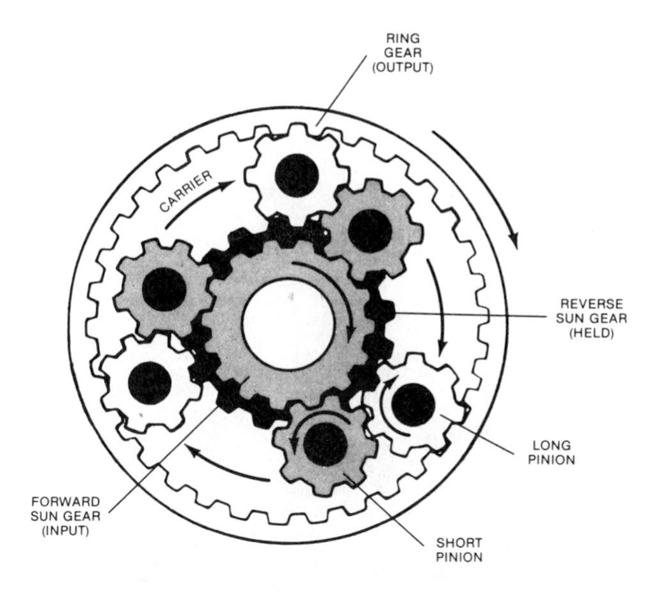

FIGURE 23–4 Ford AOD second-gear planetary operation (rear-view perspective).

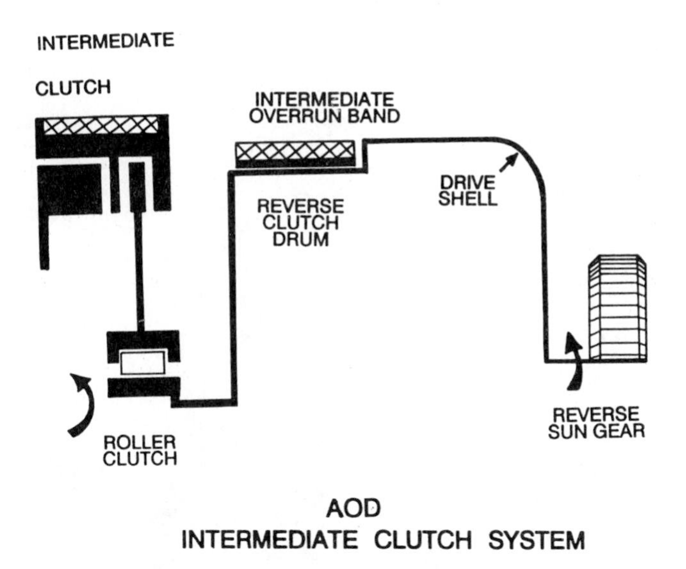

FIGURE 23–5 Ford AOD 1-2 shift sequence.

During first gear, the reverse sun gear and shell, and the reverse clutch drum, are turning counterclockwise. The intermediate one-way roller clutch remains ineffective. When the intermediate clutch is applied, it grounds the outer race of the roller clutch, resulting in a lockup. The reverse sun gear is held stationary against counterclockwise rotation (Figure 23–5).

- As the long pinions walk around the stationary reverse sun gear, they drive the ring gear and output shaft clockwise at a reduced speed. Second gear ratio is 1.47:1.
- The walking of the long pinions also drives the planet carrier in a clockwise direction. This releases the low roller clutch when second gear takes over the torque load.

Third Gear—OD and 3 Range.

- *Forward clutch applied:* acts as an input clutch and locks the tube-style turbine shaft to the forward sun gear.
- *Direct clutch applied:* acts as an input or drive clutch and locks the torsion shaft to the planet carrier (Figure 23–6).

Third-Gear Powerflow.

- The converter hydraulically drives the turbine shaft forward clutch and forward sun gear.
- The direct-drive torsion shaft, splined into the torsion damper and converter cover, mechanically drives the direct clutch and the planet carrier.
- With two members of the planetary unit driving at the same time and same speed, the pinion gears are trapped and cannot rotate on their centers. This locks together the entire gearset, and it rotates as a unit with 1:1 drive ratio.

Actually, there is some slip in the converter hydraulic drive, and therefore, an absolute direct drive cannot be attained. The planet pinions have minor rotation on their center shafts. For practical purposes, the planetary system is locked up. To minimize the slip, the engine torque input splits and takes a dual path through the converter, forty percent hydraulic and sixty percent mechanical (Figure 23–7). Because the planet carrier has a larger turning radius or lever arm than the forward sun gear, it will absorb more of the input torque.

- The intermediate clutch stays applied; however, with the planetary set lockup, the input to the roller clutch is clockwise. The intermediate roller clutch freewheels and is ineffective.

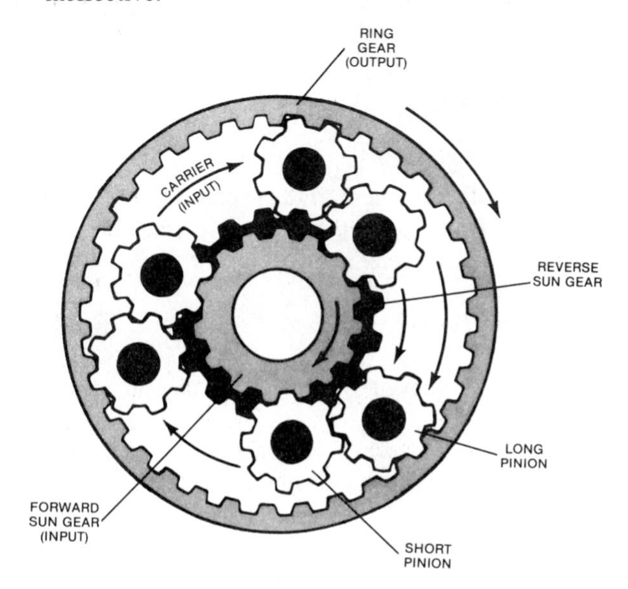

FIGURE 23–6 Ford AOD third-gear planetary operation (rear-view perspective).

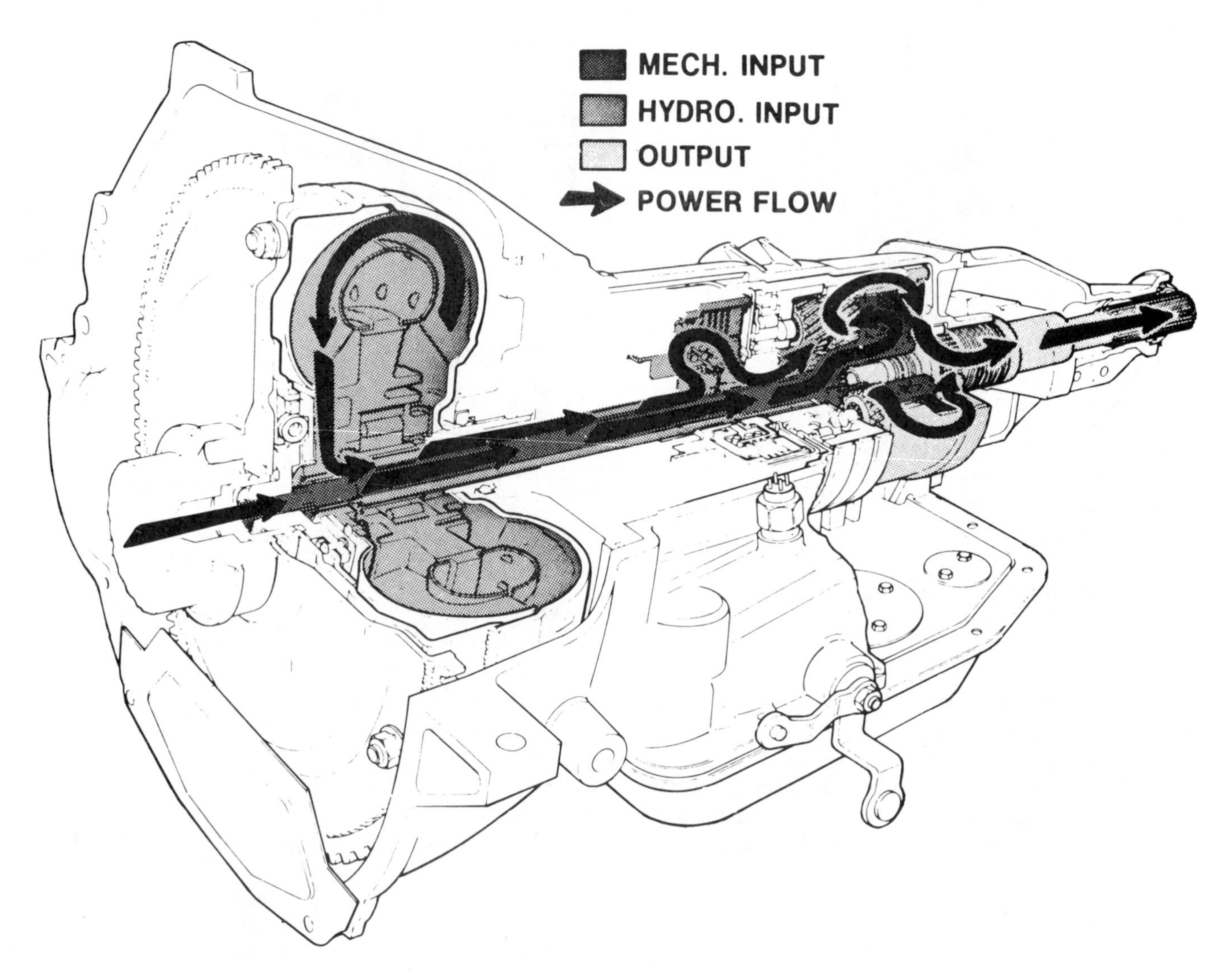

FIGURE 23–7 Ford AOD third-gear split-torque input (60% mechanical, 40% hydraulic). (Courtesy of SAE)

Fourth Gear—OD.

- *Direct clutch applied:* acts as the input or drive clutch and locks the solid torsion shaft to the planet carrier.
- *Overdrive band applied:* holds the reverse clutch drum, drive shell, and reverse sun gear stationary.

Fourth-Gear Powerflow.

- Overdrive is achieved by holding the sun gear and driving the planet carrier.
- The converter cover and torsional damper mechanically drive the direct-drive torsional shaft and direct clutch.
- The direct clutch drives the planet carrier assembly clockwise at engine speed.
- The long planet pinions walk around the stationary reverse sun gear in a clockwise direction.
- The ring gear and output shaft are driven by the long pinions at a faster speed. In overdrive, the gear ratio is 0.667:1. Because the input is one hundred percent mechanical drive, there is no converter slip (Figures 23–8 and 23–9).

Reverse.

- *Reverse clutch applied:* acts as the input clutch and locks the turbine shaft to the reverse sun gear.
- *Low/reverse band:* holds the planet carrier stationary.

Reverse-Gear Powerflow.

- The converter hydraulically drives the turbine shaft, reverse clutch, shell, and reverse sun gear clockwise.
- The reverse sun gear drives the long pinions counterclockwise.
- The long pinions drive the ring gear and output shaft counterclockwise. In effect, the long pinions are acting as reverse idler gears. The gear ratio is 2:1.
- The short pinions and forward sun gear turn but are not involved in the powerflow (Figure 23–10).

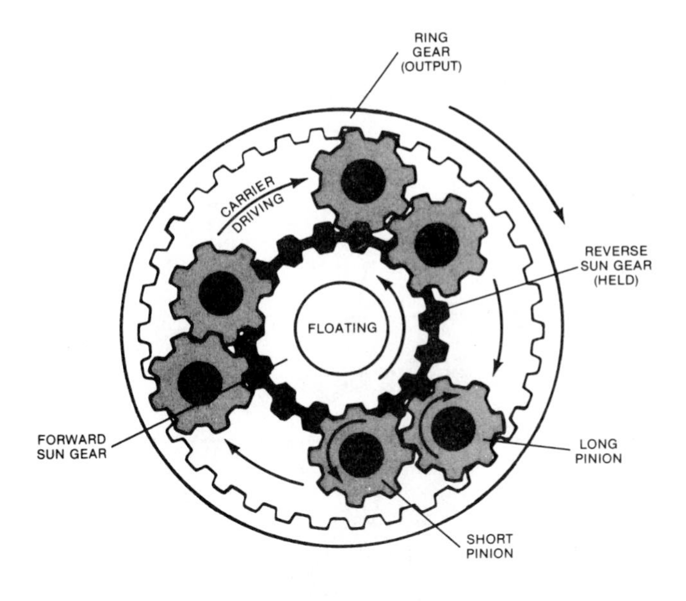

FIGURE 23–8 Ford AOD fourth-gear planetary operation (rear-view perspective).

External Transmission Controls and Features

- *Manual range selection:* linkage operated.
- *Throttle system control:* linkage or cable operated.
- *Forced downshift or kickdown control:* throttle plunger in tandem with throttle valve share the same valve body bore. TV plunger movement varies the spring force on the TV valve and also operates the kickdown system at WOT; linkage or cable operated with throttle system control.
- *Neutral safety and backup light switch:* threaded into the case above the control arms.
- *Pressure test plugs:* (1) line pressure tap on the left side of the case above the control arms, (2) TV limit pressure tap on right rear side of case, (3) forward clutch tap on right rear side of case, (4) direct clutch tap on right rear side of case (Figure 23–11).

Operating Characteristics

The shift selector offers six operating range positions. There are two different quadrant arrangements that alter the operating characteristics, as shown in Figure 23–12.

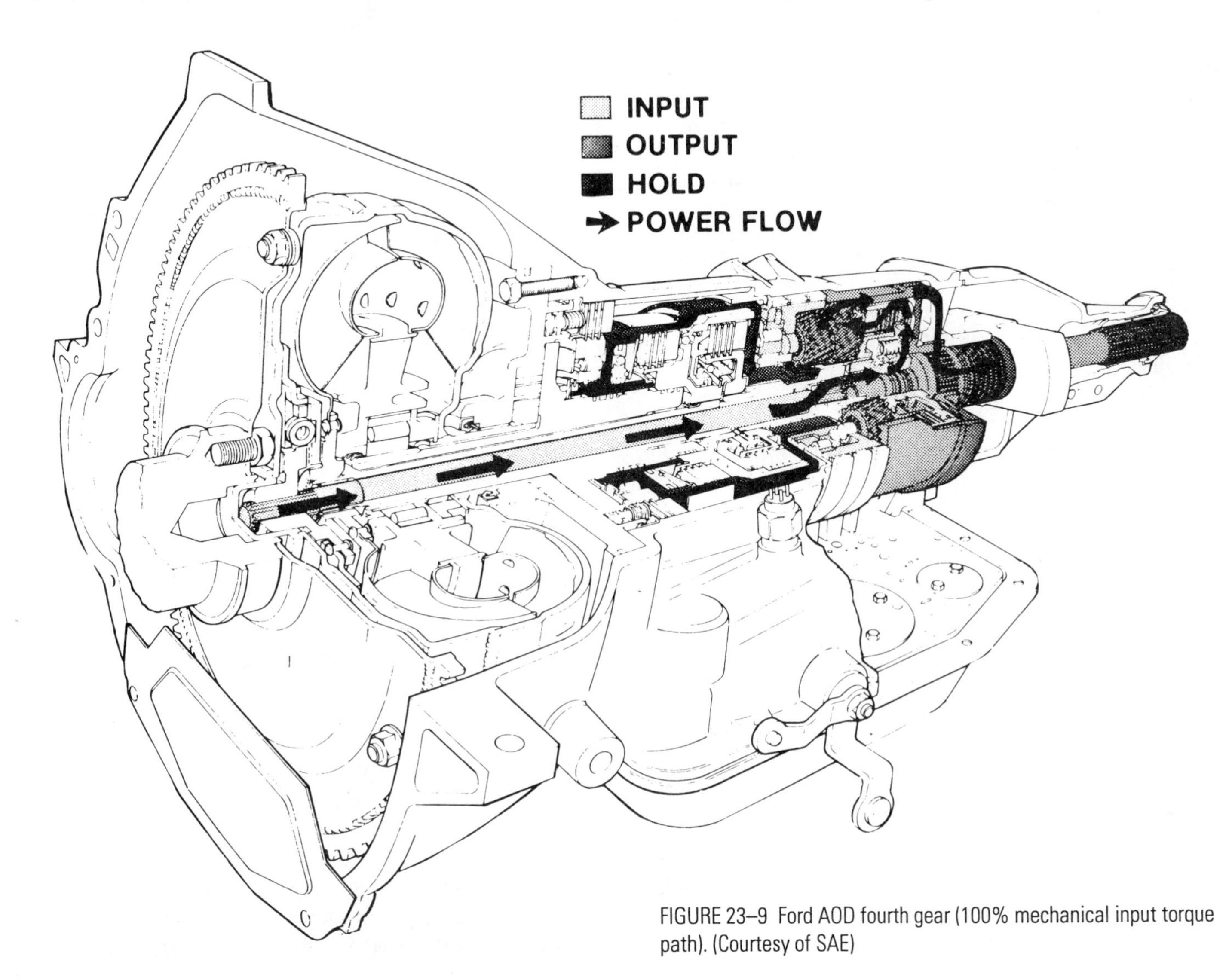

FIGURE 23–9 Ford AOD fourth gear (100% mechanical input torque path). (Courtesy of SAE)

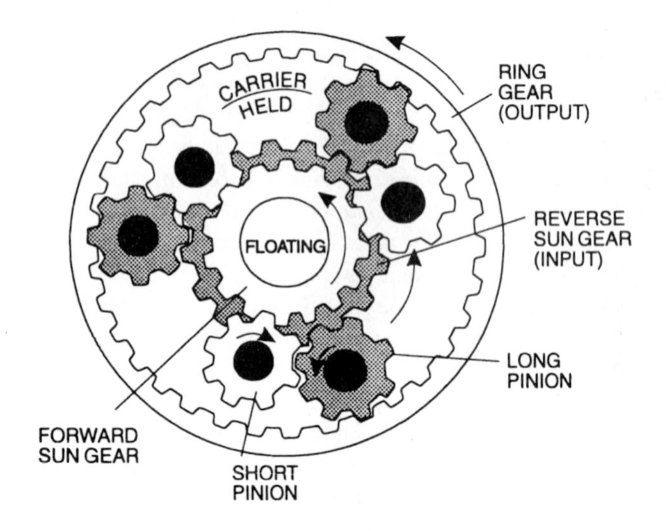

FIGURE 23–10 Ford AOD reverse-gear planetary operation (rear-view perspective).

Overdrive (1992 and Earlier). The transmission starts in first and schedules a normal 1-2 and 2-3 shift pattern, with the 3-4 overdrive shift delayed for road cruising. At minimum throttle effort, the 3-4 overdrive shift occurs at approximately 40 mph (65 km/h). The transmission will not shift into or remain in fourth gear under WOT. Full-throttle forced kickdown shifts (4-3, 4-2, 3-2, and 3-1) are available within the proper vehicle speed range for added performance. In OD, the transmission downshifts 4-3 when the accelerator is pushed for moderate or heavy acceleration. The closed-throttle downshift pattern shifts through all the gears: 4-3, 3-2, and 2-1.

Overdrive (1993 and Later). The OD position provides for the usual first through fourth gears and converter lockup. Overdrive may be canceled by pressing the transmission control switch (TCS) on the range selector lever. When it is pressed, the transmission functions as an automatic three-speed unit. If the driver does not reengage fourth gear, it automatically resets when the ignition is cycled OFF and ON.

Overdrive Lockout (1992 and Earlier). The objective of the D or 3 range is to lock out the fourth gear. It is used to achieve better performance when driving on hilly or mountainous roads. Additional engine braking is also realized. The same operating characteristics prevail as in OD except for the fourth-gear feature. The selector may be shifted from OD to D or from D to OD at any vehicle speed.

Intermediate (1993 and Later). Selection of the 2 range features a second-gear start and hold. Upshifts to third and fourth gears are inhibited. This is useful for engine braking when the TCS is in the OFF position and for start-ups on slippery road surfaces. Whenever 2 is selected with the vehicle moving, the transmission engages second gear regardless of vehicle speed.

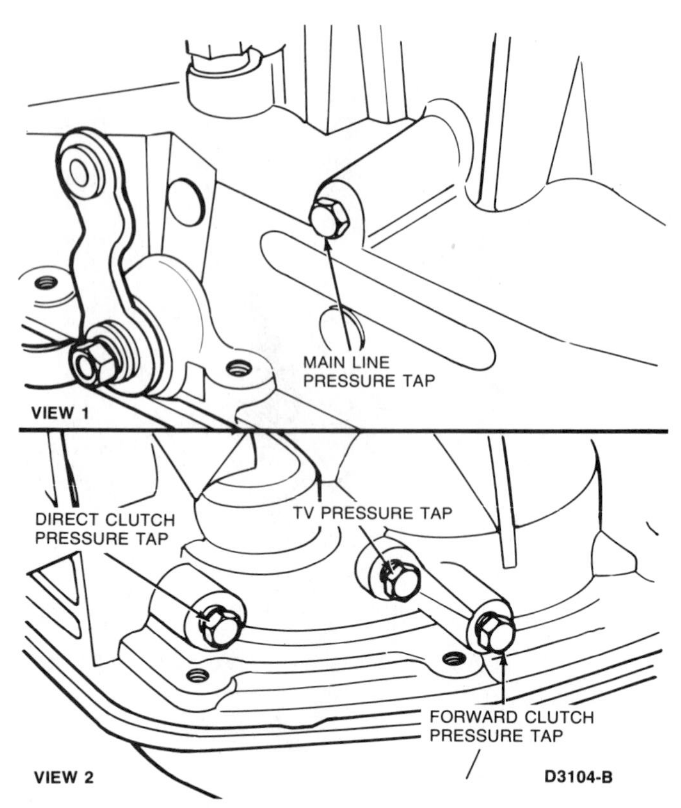

FIGURE 23–11 AOD pressure taps. (Reprinted with the permission of Ford Motor Company)

1992 AND EARLIER **1993 AND LATER**

FIGURE 23–12 Ford AOD selector pattern.

Manual Low. The objective is to stay in first gear for extra pulling power or maximum braking. The 1 range can be selected at any vehicle speed. If the vehicle speed is too high, however, second gear is engaged. When the vehicle slows down to a safe speed [approximately 25 mph (40 km/h)], it will downshift to first gear. Once first gear is engaged in manual low, the transmission stays in first, with the low-reverse band applied for braking action on coast. Upshifts to D and OD can be made manually for driver select shifting.

AODE (ELECTRONICALLY CONTROLLED) FEATURES

For 1992 full-size RWD cars and light-duty trucks/vans, the AOD was upgraded to feature expanded use of electronics and identified as the AODE. In addition to the electronic controls, other design features were incorporated into the new version.

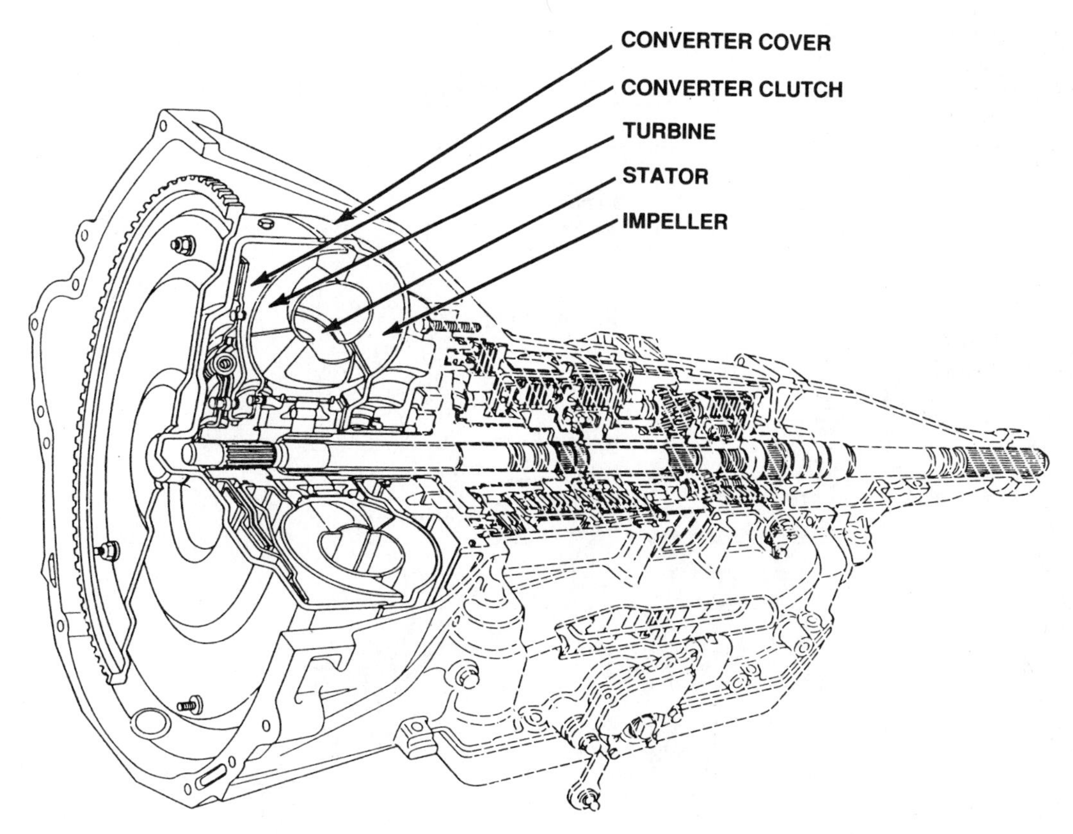

FIGURE 23–13 AODE: designed with a conventional torque converter clutch. (Reprinted with the permission of Ford Motor Company)

Torque Converter

The mechanical split torque–style converter is replaced with a traditional pressure plate–style torque converter lockup clutch (Figure 23–13). It can be applied in second, third, and fourth gears.

Case Pump

A new pump uses a permanent-mold cast aluminum body incorporating a gerotor (rotary lobe) pump design. This replaces the previous cast iron body using a gear and crescent pump design.

Geartrain Powerflow

The AODE geartrain uses the same planetary system and clutch/band control as the AOD (Chart 23–1, page 526). Even the gear ratios are the same. The significant difference is in the elimination of the torsion shaft that previously provided a direct mechanical connection between the direct clutch and engine crankshaft via the torque converter cover. The new design configuration has the converter turbine or converter lockup driving all three input clutches (forward, direct, and reverse). An intermediate (stub) shaft connects the direct clutch to the input/turbine shaft (Figure 23–14).

Electronics

Rather than hydraulic controls, electronic logic in the PCM is used to control shift scheduling, shift feel, and converter clutch operation. The PCM has complete management control of engine and transmission performance.

Computer transmission decisions are determined by a combination of input data provided by sensors and switches. These devices reflect output shaft speed, vehicle speed, transmission fluid temperature, manual lever position, and numerous other signals (Figure 23–15).

Using the full complement of input data, the PCM output drives four solenoids: MCCC (modulated converter clutch control), EPC (electronic pressure control), and two shift solenoids.

The MCCC solenoid provides a pulse-width duty cycle valve action that controls the converter lockup pressure. The solenoid works in conjunction with a converter bypass clutch control valve. This maximizes smooth lockup feel and smooth transitions between lock and unlock.

The PCM can command a duty cycle for either fully OFF, full clutch apply, or controlled slip (Figure 23–16). At zero duty cycle, the ball valve is closed, the clutch is OFF, and there is zero output pressure. At one hundred percent duty cycle, the ball valve is fully open, the clutch is fully applied, and there is full output pressure. For controlled slip, the PCM system rapidly cycles the solenoid ON and OFF at the required rate between zero and one hundred percent. The MCCC essentially controls the amount of solenoid-regulated valve pressure relayed to a converter bypass control valve.

The EPC solenoid is a variable force solenoid containing an integral spool valve and spring assembly. It replaces the cable/linkage-controlled TV system used in the AOD.

To control the EPC output, the PCM sends a variable current flow to the solenoid. This varies the internal force of the

FORD AOD, AODE, And 4R70W
RANGE REFERENCE AND CLUTCH/BAND APPLY SUMMARY

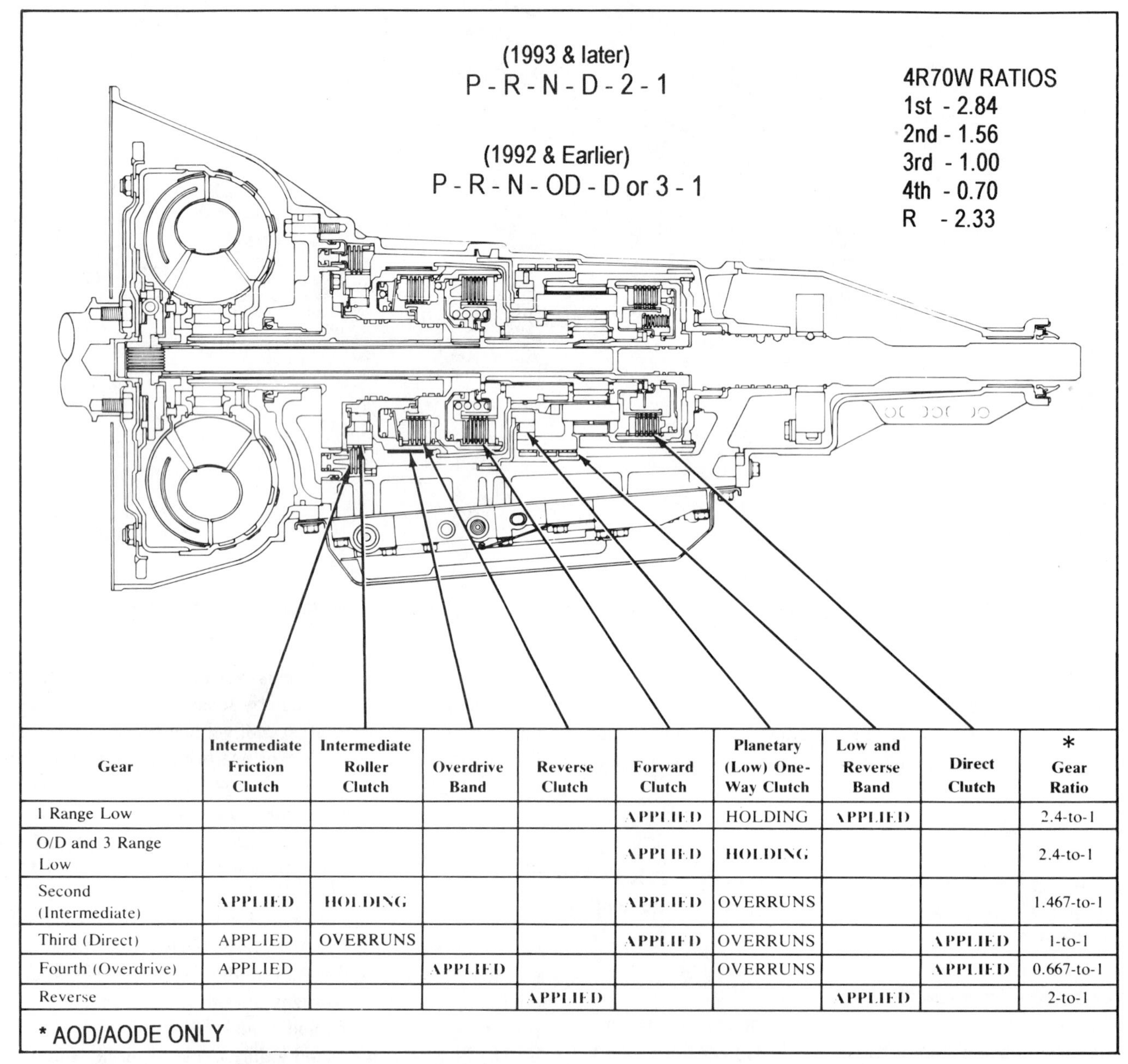

Gear	Intermediate Friction Clutch	Intermediate Roller Clutch	Overdrive Band	Reverse Clutch	Forward Clutch	Planetary (Low) One-Way Clutch	Low and Reverse Band	Direct Clutch	* Gear Ratio
1 Range Low					APPLIED	HOLDING	APPLIED		2.4-to-1
O/D and 3 Range Low					APPLIED	HOLDING			2.4-to-1
Second (Intermediate)	APPLIED	HOLDING			APPLIED	OVERRUNS			1.467-to-1
Third (Direct)	APPLIED	OVERRUNS			APPLIED	OVERRUNS		APPLIED	1-to-1
Fourth (Overdrive)	APPLIED		APPLIED			OVERRUNS		APPLIED	0.667-to-1
Reverse				APPLIED			APPLIED		2-to-1

* AOD/AODE ONLY

CHART 23–1 AOD, AODE, and 4R70W range reference and clutch/band apply summary and gear ratios. (Reprinted with the permission of Ford Motor Company)

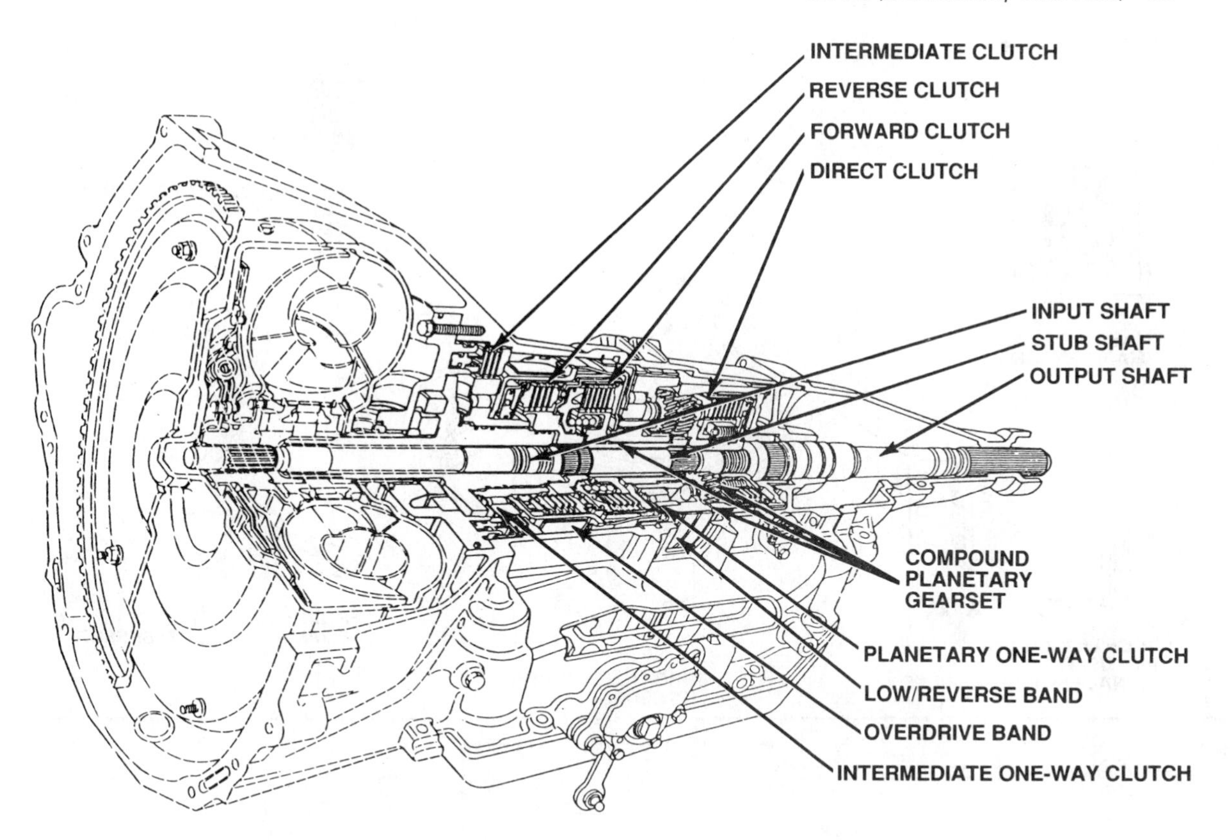

FIGURE 23–14 AODE: designed with an intermediate (stub) shaft, instead of a torsion shaft. (Reprinted with the permission of Ford Motor Company)

solenoid on the spool valve. With no current flow, the internal force at maximum holds the valve in the bore for maximum output, and the exhaust port is closed.

As current is allowed to flow, the internal solenoid force on the valve is reduced. The valve is pushed back by spring force for exhaust porting and variable output pressure regulation. Operation of the EPC is summarized in Figure 23–17.

The shift solenoids offer four possible ON/OFF combinations to control the fluid flow to the shift valves. Operation of the solenoids is illustrated in Figure 23–18. Note that the shift solenoid valves are normally closed (electrically OFF) and hydraulically open (electrically ON).

Chart 23–2, page 530, summarizes the operation of the four AODE solenoids. Figure 23–19 shows solenoid locations.

Should the AODE version lose all electrical signals from the PCM, the transmission will lock into second gear in all forward ranges. Reverse is available because it is a function of the manual valve.

4R70W (ELECTRONICALLY CONTROLLED)

The 4R70W is basically a heavy-duty version of the AODE. Initially introduced in 1993 for use in the Mark VIII, vehicle applications now include the Thunderbird, Cougar, Town Car, Crown Victoria, Grand Marquis, and E- and F-150 series trucks. For 1997, the 4R70W is used in the restyled F-150 series light-duty pickup trucks, the newly designed Ford Expedition, and in the Ford Explorer and Mecury Mountaineer sport-utility vehicles.

While the 4R70W retains many of the AODE design features, it has a number of unique features. Special wide-ratio gearing between first and second gears matches the performance curve of the new generation of engines. The wide ratio is achieved by changing the sizing and gear tooth count of the planetary components.

The new 4R70W ratios provide a sixteen percent increase in low- and reverse-gear output torque. Improved start-up acceleration is realized, as well as quieter operation in reverse.

Gear Ratio Comparisons

	1st	2nd	3rd	4th	Reverse
4R70W	2.84	1.56	1.00	.70	2.33
AOD/AODE	2.40	1.47	1.00	.67	2.00

A new torque converter is matched with the higher rpm range of the new engines. Wide open throttle shifts occur at 6000 rpm.

Other upgrades include changes to the intermediate one-way clutch, clutch plate materials, pump assembly, and clutch piston cylinder ball checks.

The six-position range selector pattern and operating characteristics are identical to the AODE (1993 and later).

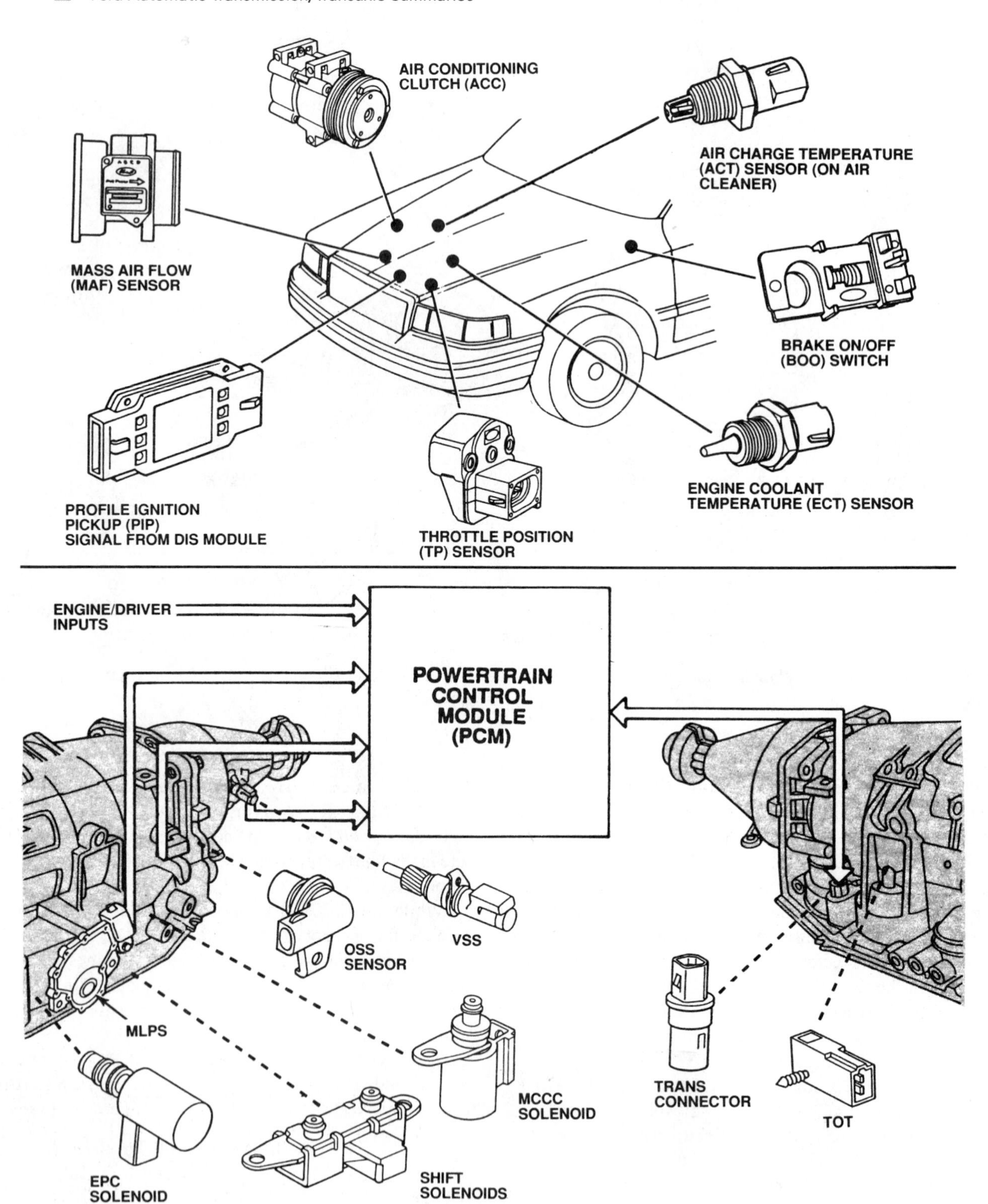

FIGURE 23–15 AODE-related sensors and solenoids. (Reprinted with the permission of Ford Motor Company)

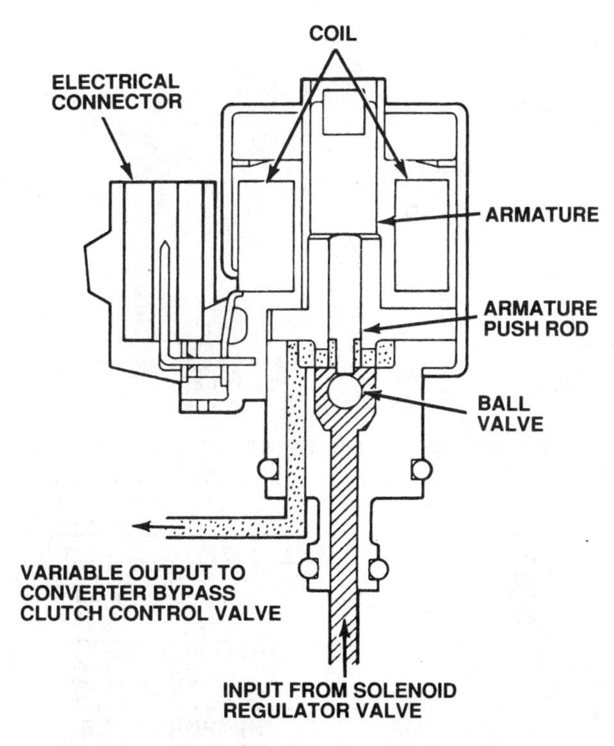

FIGURE 23–16 MCCC solenoid (AODE). (Reprinted with the permission of Ford Motor Company)

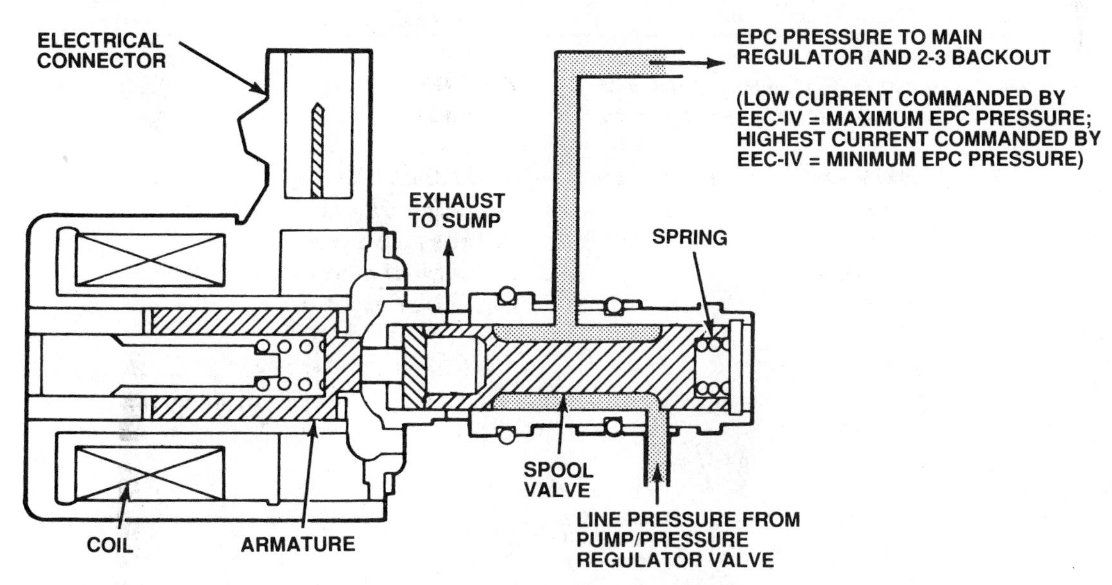

FIGURE 23–17 EPC solenoid (AODE). (Reprinted with the permission of Ford Motor Company)

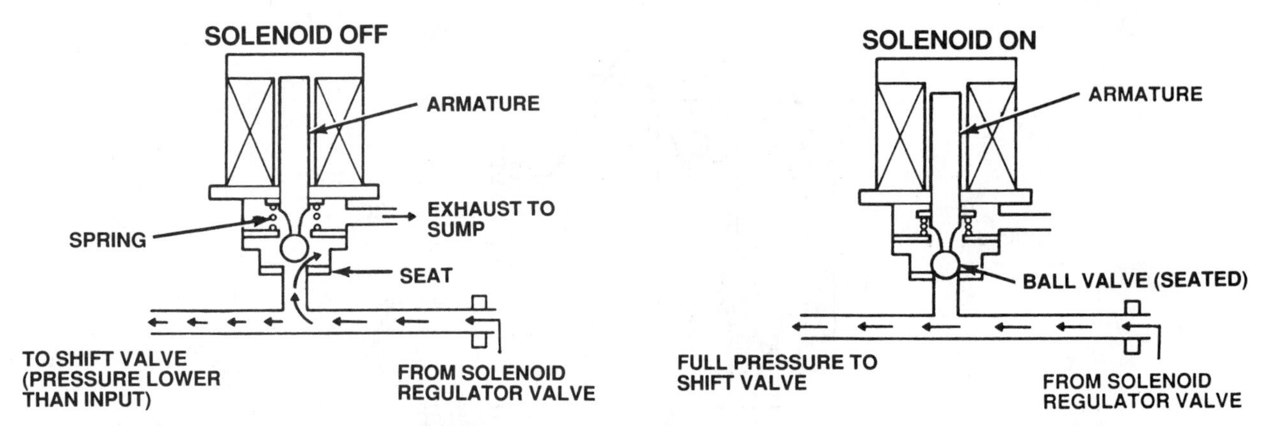

FIGURE 23–18 AODE shift solenoid operation. (Reprinted with the permission of Ford Motor Company)

	GEAR	SHIFT SOLENOIDS SS1	SS2	MODULATED CONVERTER CLUTCH CONTROL SOLENOID	EPC SOLENOID
Ⓓ RANGE	4	ON	ON	CONTROLLED BY EEC-IV STRATEGY	PRESSURE OUTPUT CONTROLLED BY EEC-IV STRATEGY IN ALL RANGES
	3	OFF	ON		
	2	OFF	OFF		
	1	ON	OFF	CLUTCH HYDRAULICALLY DISABLED	
D RANGE	3	OFF	ON	CONTROLLED BY EEC-IV STRATEGY	
	2	OFF	OFF		
	1	ON	OFF	CLUTCH HYDRAULICALLY DISABLED	
"1" RANGE	2	OFF	OFF	CONTROLLED BY EEC-IV STRATEGY	
	1	ON	OFF	CLUTCH HYDRAULICALLY DISABLED	
REVERSE, PARK, NEUTRAL	—	ON	OFF	CLUTCH HYDRAULICALLY DISABLED	

Courtesy of Ford Parts and Service Division

CHART 23–2 AODE solenoid operation chart. (Reprinted with the permission of Ford Motor Company)

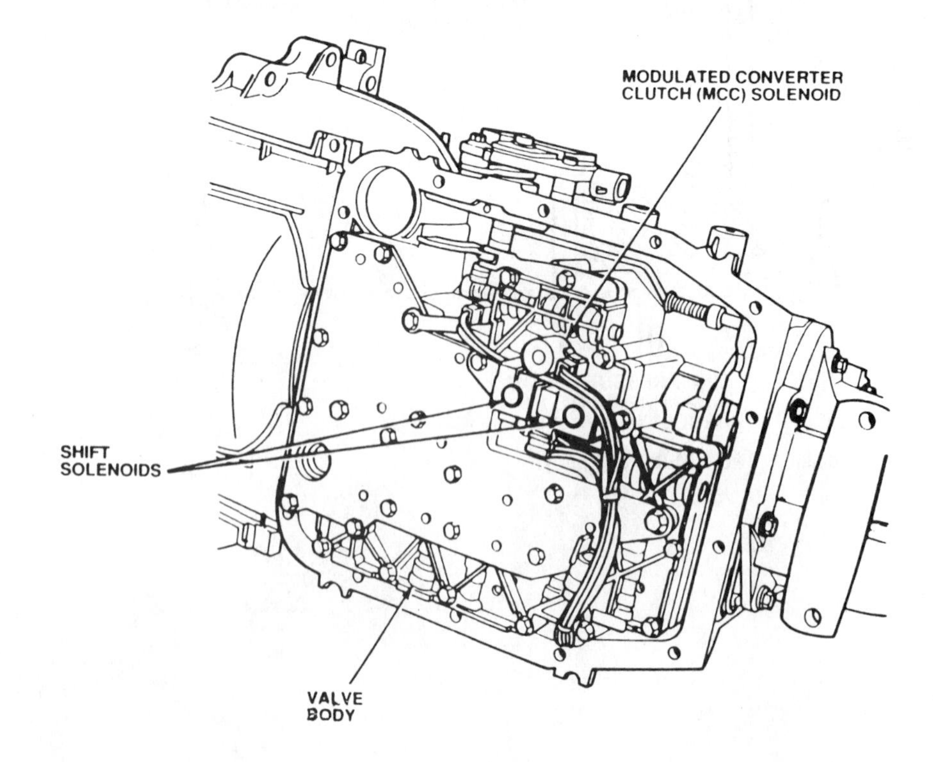

FIGURE 23–19 AODE solenoid locations. (Reprinted with the permission of Ford Motor Company)

A4LD, 4R44E/4R55E, AND 5R55E GROUP

Manufactured at the Ford transmission plant in Bordeaux, France, these four- and five-speed rear-wheel drive transmissions share a common geartrain layout. An input overdrive planetary unit is attached to a common Simpson three-speed planetary gearset. The transmission cases are essentially the same, featuring bolt-on converter and extension housings (Figures 23–20 and 23–21). The main differences occur in the style of transmission control: hydraulic/electronic control as compared to full electronic control.

All units use a three-element torque converter equipped with a clutch plate. For control of the geartrain powerflow, the transmissions use three multiple-disc clutches, three bands, and two one-way clutches. Refer to Chapter 4 for geartrain operation highlights.

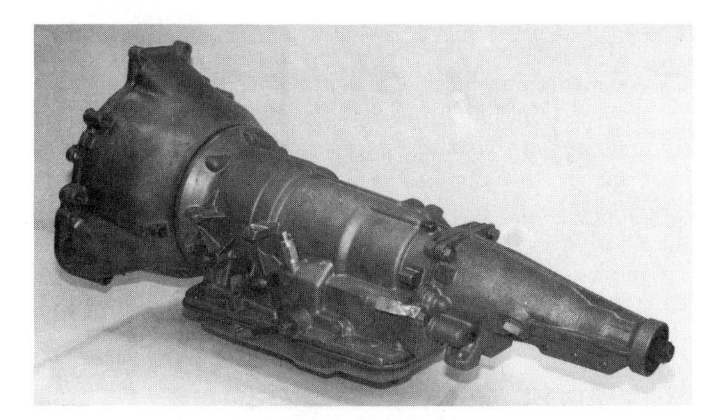

FIGURE 23–20 Ford A4LD left side case view.

A4LD TRANSMISSION

Built for light-duty applications, the A4LD was initially introduced for the 1985 production year in the Ranger and Bronco II product lines. In the following years, its use was expanded through 1994 in Mustangs, Thunderbirds, Aerostars, and Explorers. In 1995, production was limited to use in Aerostars.

The A4LD transmission features a converter lockup circuit control using combined transmission hydraulics and on-board (PCM) electronic controls. Normal converter clutch engagements and disengagements are scheduled hydraulically but can be overruled electronically.

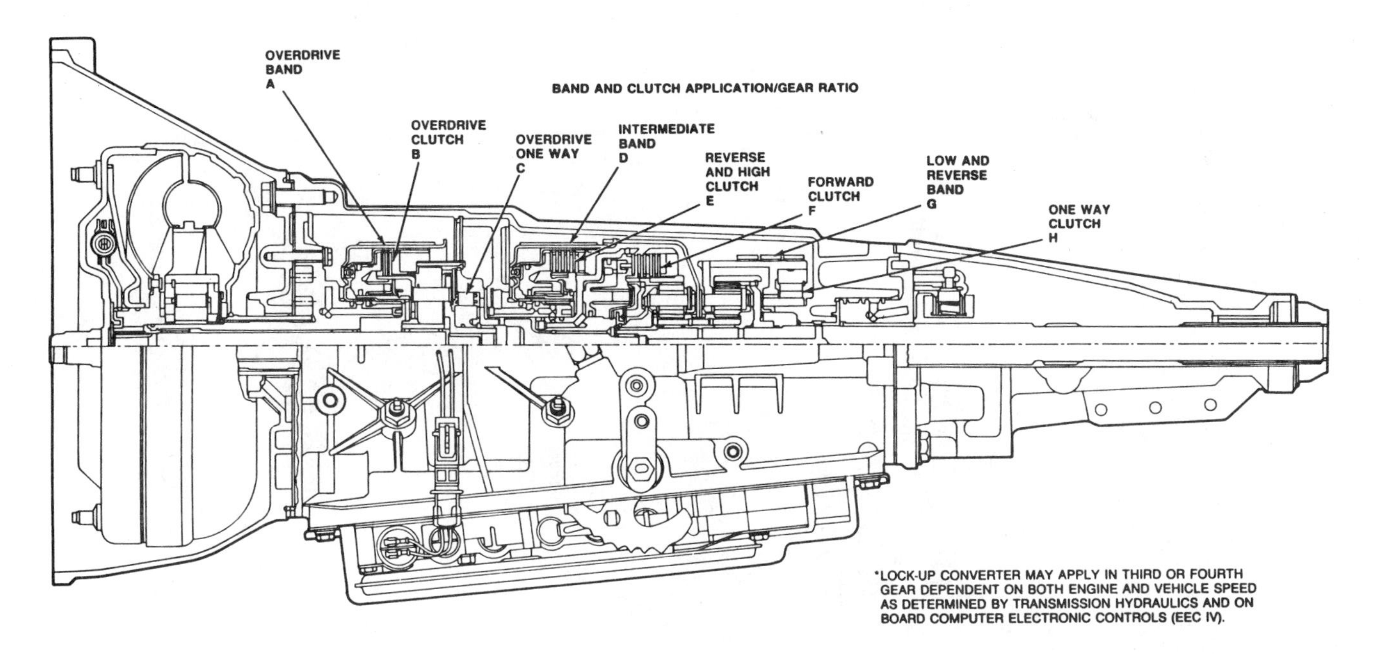

GEAR	OVERDRIVE BAND A	OVERDRIVE CLUTCH B	OVERDRIVE ONE WAY CLUTCH C	INTERMEDIATE BAND D	REVERSE AND HIGH CLUTCH E	FORWARD CLUTCH F	LOW AND REVERSE BAND G	ONE WAY CLUTCH H	GEAR RATIO
1 — MANUAL FIRST GEAR (LOW)		APPLIED	HOLDING			APPLIED	APPLIED	HOLDING	2.47:1
2 — MANUAL SECOND GEAR		APPLIED	HOLDING	APPLIED		APPLIED			1.47:1
D — DRIVE AUTO — 1ST GEAR		APPLIED	HOLDING			APPLIED		HOLDING	2.47:1
D — O/D AUTO — 1ST GEAR			HOLDING			APPLIED		HOLDING	2.47:1
D — DRIVE AUTO — 2ND GEAR		APPLIED	HOLDING	APPLIED		APPLIED			1.47:1
D — O/D AUTO — 2ND GEAR			HOLDING	APPLIED		APPLIED			1.47:1
D — DRIVE AUTO — 3RD GEAR		APPLIED	HOLDING		APPLIED	APPLIED			1.0:1
D — O/D AUTO — 3RD GEAR			HOLDING		APPLIED	APPLIED			1.0:1
D — OVERDRIVE AUTOMATIC FOURTH GEAR	APPLIED				APPLIED	APPLIED			0.75:1
REVERSE		APPLIED	HOLDING		APPLIED		APPLIED		2.1:1

FIGURE 23–21 (Top) A4LD clutch and band locations. (Bottom) Range reference and clutch/band apply summary. (Reprinted with the permission of Ford Motor Company)

The PCM controls an override solenoid mounted on the valve body and either inhibits lockup or allows it to occur. When the solenoid is de-energized, line pressure flows through the solenoid valving and acts as a shift inhibitor on the converter clutch valve group. The solenoid energizes when the PCM senses the appropriate engine and drive conditions for lockup to happen. The solenoid blocks the line pressure to the inhibitor circuit, and the hydraulic lockup valve group takes over. Refer to Chapter 10 for further explanation.

On A4LD units built prior to 1987, the 3-4 shift circuit is hydraulically controlled. To ensure that fourth gear will not engage unless conditions are right, the 3-4 shift circuit is combined with hydraulic and PCM electronic control in 1987 Thunderbird Turbo 2.3-L applications. This feature was added in 1988 to the A4LD units combined with 2.9-L engines in Rangers, Bronco IIs, and Aerostars. A 3-4 shift solenoid was added to the valve body next to the converter override solenoid.

The 3-4/4-3 shift solenoid is normally de-energized, permitting line pressure to flow through the solenoid valving and inhibit the 3-4 shift valve. When the PCM evaluates the 3-4 shift parameters, it energizes the solenoid, and the inhibitor oil is cut off and exhausted. This allows the hydraulic shift valve group to make the shift.

A malfunction in either solenoid circuit will turn on the "check engine" light and display an appropriate diagnostic trouble code (DTC). The technician should know that any malfunction of an engine sensor also can affect the solenoid circuits, and therefore, a DTC may not necessarily point to a solenoid circuit fault.

External Transmission Controls and Features

- *Manual range selection:* cable or linkage operated.
- *Throttle system control:* engine vacuum, modulator unit located on right side center of case.
- *Forced downshift system control:* linkage operated downshift lever located next to manual control lever.
- *Neutral safety and backup light switch:* threaded into case above manual control lever.
- *Intermediate band adjusting screw:* located on left side of case next to neutral safety and backup light switch.
- *Solenoid case connector:* left side of case.
- *Overdrive band adjusting screw:* located on left front side of case ahead of intermediate band adjusting screw.
- *Mainline pressure tap:* left side rear of manual control arm.

4R44E/4R55E TRANSMISSIONS

The 4R44E/4R55E are full electronically controlled versions of the A4LD. Introduced in the 1995 product line, the 4R44E is available in Rangers with 2.3-L and 3.0-L engines. The 4R55E is used with Ranger and Explorer 4.0-L engine applications.

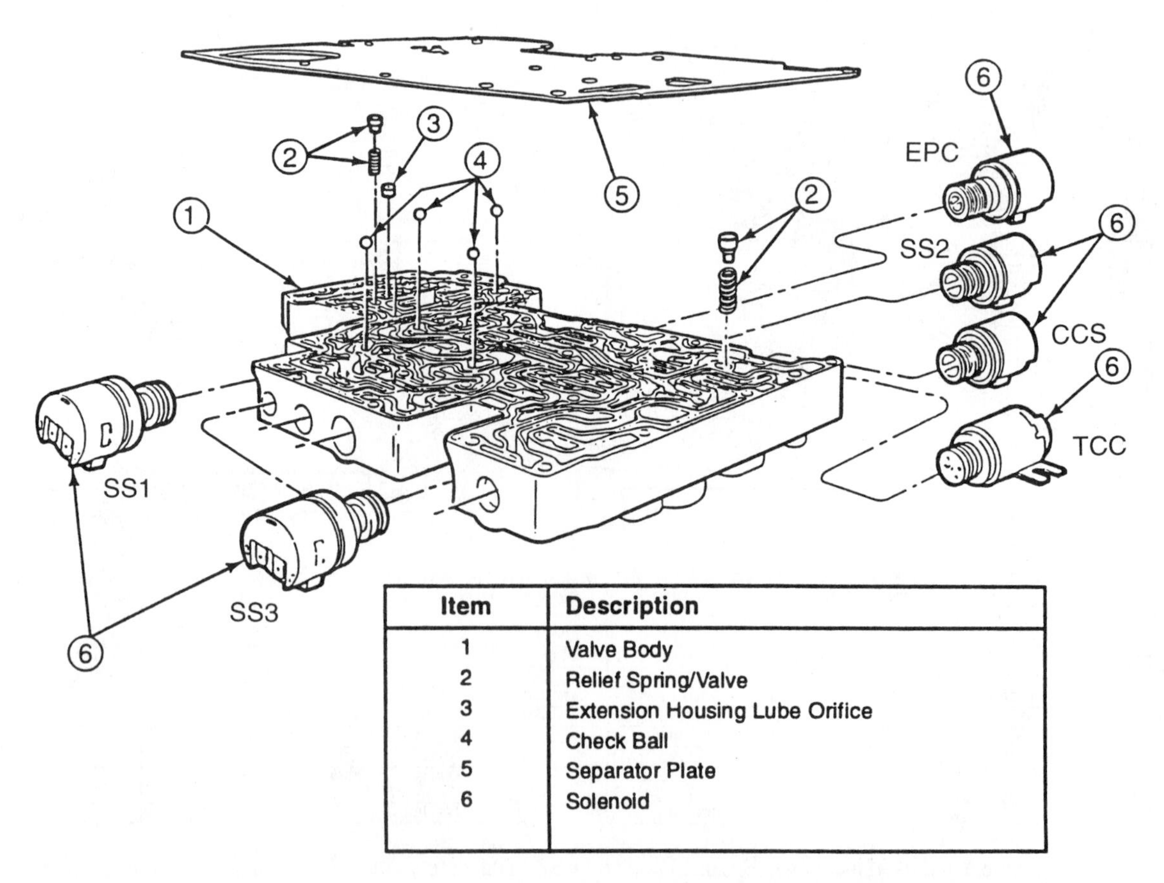

Item	Description
1	Valve Body
2	Relief Spring/Valve
3	Extension Housing Lube Orifice
4	Check Ball
5	Separator Plate
6	Solenoid

FIGURE 23–22 4R44E/4R55E solenoids mount into the valve body. (Reprinted with the permission of Ford Motor Company)

The electronic system is controlled by the PCM, which manages the transmission's six valve body mounted solenoids (Figure 23–22). Responding to the PCM output signals, the solenoids control the transmission's hydraulic circuits for shift timing and quality, as well as TCC application (Figure 23–23).

OPERATING CHARACTERISTICS—A4LD AND 4R44E/4R55E

The shift selector quadrant arrangements differ between the A4LD and the 4R44E/4R55E transmissions.

A4LD:	P-R-N-OD-D-2-1
4R44E/4R55E:	P-R-N-OD-2-1

Overdrive—A4LD. The transmission schedules a normal 1-2 and 2-3 shift pattern, with the 3-4 shift delayed for road cruising. The minimum shift into fourth gear can vary between 30–50 mph (48–80 km/h). A shift into fourth gear is not possible at WOT. The converter clutch engagement follows at approximately 5 mph (8 km/h) above the 3-4 shift point. During delayed shift schedules, it is possible for the converter clutch to engage after the shift into third. The converter clutch is prevented from engaging or is disengaged during the following conditions:

- Engine coolant below 128°F (54°C) or above 240°F (116°C).
- Application of brakes.
- Closed throttle.
- Heavy or WOT acceleration.
- Quick tip-ins or tip-outs.
- Prior to part throttle, forced shifts, or anytime more power is needed to maintain vehicle speed in the fourth or third gears.

Depending on the vehicle speed range and engine crankshaft torque reserve, WOT forced downshifts are available for added performance (4-3, 3-2, and 2-1). Part-throttle shifts and closed-throttle shifts are 4-3, 3-2, and 2-1.

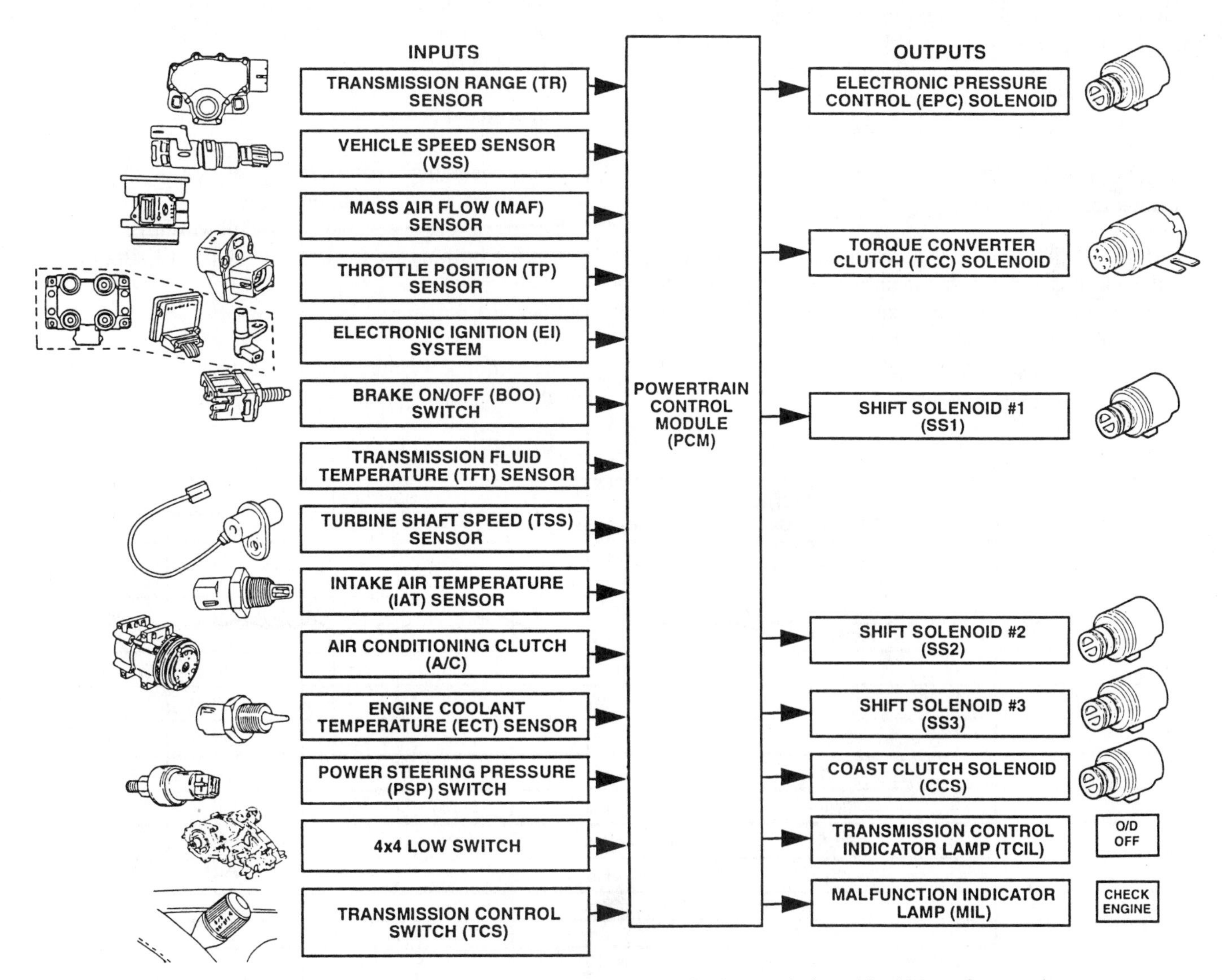

FIGURE 23–23 4R44E/4R55E electronic input and output devices. (Reprinted with the permission of Ford Motor Company)

Overdrive—4R44E/4R55E. The overdrive position provides for the usual first through fourth gears and converter lockup, similar to the A4LD. Overdrive, however, can be canceled by pressing the transmission control switch (TCS) located on the range selector lever. A transmission control indicator lamp (TCIL) illuminates on the instrument cluster to indicate that O/D fourth gear has been canceled.

Overdrive Lockout—A4LD. The D range locks out fourth-gear operation. The D position is used to achieve better performance on hilly/mountainous roads and when pulling heavy loads, or to provide engine braking in second and third gears.

Manual 2. This position provides a second-gear start and hold—upshifts and downshifts are inhibited. This feature allows for improved traction on slippery road surfaces. In addition, engine braking for second gear is realized. Whenever manual 2 is selected, second gear is engaged regardless of vehicle speed.

Manual 1. This position is intended for first-gear operation to provide extra pulling power and maximum engine braking. Manual 1 can be selected at any vehicle speed. If the vehicle speed is too high, second gear (engine braking) is engaged. When a safe speed is reached, a downshift to first occurs. No upshifts are allowed from first to second gear.

Reverse. For the 4R44E/4R55E units, the TCS must be in the off position for engine braking.

If the 4R44E/4R55E unit loses all electrical signals from the PCM, the transmission will start in third gear in OD range, and second gear in manual 2 and manual 1. Reverse is a function of the manual valve and is available.

Chart 23–3 summarizes the clutch/band and solenoid combinations for each operating range.

Overdrive Powerflow Summary

Detailed overdrive planetary powerflow is presented in Chapter 4. The 4R44E/4R55E overdrive gearset operates the same as the one used in the A4LD. The coast clutch function is identical to the overdrive clutch in the A4LD.

As a brief review, the converter output must always pass through the overdrive planetary to drive the main compound planetary gearset. During the first three forward gears and reverse, the overdrive planetary is in a direct-drive mode and does not affect the ratio output. In the overdrive mode, the overdrive band is applied and holds the sun gear. With the planet carrier driving, a 0.75:1 ratio is achieved.

5R55E TRANSMISSION

The 5R55E automatic transmission is an electronically controlled five-speed unit used in 1997 Aerostar and Ranger vehicles with 4.0-L engines. Its basic construction, geartrain powerflow, shift quadrant selector pattern, and electronic control system originate from the 4R44E/4R55E transmission design.

The hydraulic functions are directed by six electronic solenoids to control:

Engagement feel
Shift feel
Shift scheduling
Modulated TCC applications
Timing of torque demand and kickdown shifts
Engine braking (with O/D canceled) utilizing the coast clutch
Manual 1 timing

In addition to a torque converter utilizing a torque converter clutch (TCC) apply plate, the geartrain includes:

GEAR SELECTOR POSITION	POWERTRAIN CONTROL MODULE (PCM) GEAR COMMANDED	SOLENOIDS				
		ENGINE BRAKE	SS1	SS2	SS3	CCS
P	P	NE	On	Off	Off	Off
R	R	No	On	Off	Off	Off
R	R*	Yes	On	Off	Off	On
N	N	NE	On	Off	Off	Off
Ⓓ	1st	No	On	Off	Off	Off
Ⓓ	1st*	Yes	On	Off	Off	On
Ⓓ	2nd	No	On	On	Off	Off
Ⓓ	2nd*	Yes	On	On	Off	On
Ⓓ	3rd	No	Off	Off	Off	Off
Ⓓ	3rd*	Yes	Off	Off	Off	On
Ⓓ	4th	No	Off	Off	On	Off
MAN 2nd	2nd	Yes	On	On	Off	On
MAN 1st	1st	Yes	On	Off	Off	On

NE = No Effect * = Overdrive Cancelled – TCIL ON

CHART 23–3 4R44E/4R55E solenoid applications summary. (Reprinted with the permission of Ford Motor Company)

Three planetary gearsets
Three bands
Three multiple-disc clutches
Two one-way clutches

The names of most internal components of the 5R55E, as compared to the 4R44E/4R55E design, remain the same. Exceptions include identifying the 4R44E/4R55E overdrive planetary gearset, overdrive band, and overdrive one-way clutch as the front planetary gearset, front band, and front one-way clutch in the 5R55E.

A transmission control switch (TCS) is located on the selector lever. When the TCS is switched to the ON position, overdrive (fifth-gear) operation is canceled, the coast clutch is applied, and the transmission control indicator lamp (TCIL) will be illuminated.

The TCIL is located on the transmission range selector lever in the Aerostar and on the instrument panel in the Ranger. In addition to identifying whether the overdrive has been canceled by the TCS, the TCIL flashes if the powertrain control module (PCM) detects certain transmission-related sensor or solenoid malfunctions.

Operating Characteristics

The 5R55E transmission range selector lever has six positions: P-R-N-O/D-2-1.

Park. There is no powerflow through the transmission. The parking pawl is engaged to hold the output shaft and drive shaft stationary.

Reverse. A rearward vehicle direction is provided at a reduced gear ratio. When overdrive has been canceled by switching the TCS to the ON position, engine braking is provided by the application of the coast clutch.

Neutral. As in park position, there is no powerflow through the transmission. However, the parking pawl is not engaged as in the park position, and the vehicle wheels are able to rotate.

Overdrive. In the O/D position, the transmission upshifts or downshifts 1-2-3-4-5 automatically. When the overdrive has been canceled by turning the TCS on, overdrive (fifth gear) is canceled, and the coast clutch is applied to provide engine braking.

Manual 2. Selection of manual 2 provides a third-gear hold. This allows for maximum traction on slippery surfaces. Engine braking is always provided, whether the TCS is on or off.

Powerflow Summary

Engaging various combinations of the front planetary gearset (two ratios) and the rear Simpson compound planetary gearset (three forward ratios), the transmission design can provide six forward speeds. Based on performance and gear ratios available, five forward combinations and one reverse combination were selected. They are as follows:

Gear	Gear Ratio
1st	2.47:1
2nd	1.87:1
3rd	1.47:1
4th	1.00:1
5th	0.75:1
Reverse	2.00:1

When the ratios are compared to those of the 4R44E/4R55E transmissions, it is apparent that second gear is the additional gear provided.

Following is a listing of the applied clutches/bands for the five forward gears in the O/D range and for reverse gear.

First gear (O/D range): Front one-way clutch, forward clutch, and rear one-way clutch.
Second gear (O/D range): Front band, forward clutch, and rear one-way clutch.
Third gear (O/D range): Front one-way clutch, forward clutch, and intermediate band.
Fourth gear (O/D range): Front one-way clutch, forward clutch, and direct clutch.
Fifth gear (O/D range): Front band, forward clutch, and direct clutch.
Reverse gear (TCS off): Front one-way clutch, direct clutch, and low/reverse clutch.

The geartrain powerflow of the 5R55E is quite similar to the powerflow used in the 4R44E/4R55E design. The primary difference is second-gear operation.

Essentially, second gear is realized by engaging the front planetary gearset into the overdrive mode, while retaining the gear reduction in the Simpson compound planetary gearset from first-gear operation. A 1-2 upshift occurs due to the application of the front band. The 2-3 upshift requires that the front band be released and the intermediate band applied.

Electronic controls are responsible for keeping the 5R55E shift scheduling and shift quality accurate.

E4OD ELECTRONIC FOUR-SPEED OVERDRIVE TRANSMISSION (RWD/TRUCK)

In mid-production year 1989, Ford introduced its first electronically controlled automatic transmission for trucks. The E4OD is a four-speed automatic overdrive with a lockup converter and is used in Bronco, Econoline, F-Series, and F-Super Duty applications. It has been coupled to 5.8-, 7.3-, and 7.5-L engines. Recent applications include the 4.9- and 5.0-L engines. An external view of the transmission is shown in Figure 23–24. A reinforced aluminum case with a bolt-on extension housing is used. The E4OD is used in both two- and four-wheel drive configurations.

The E4OD replaces many applications of the C6 (three-speed). Components from the intermediate drum rearward are similar to the C6. Although the basic three-speed planetary gear

FIGURE 23–24 Ford E4OD left side case view.

assemblies look alike, the E4OD planetary gears have a higher contact ratio for quieter operation. Wide ratios are used in the forward gears.

	C6	E4OD
First	2.46	2.71
Second	1.46	1.538
Third	1.00	1.00
Fourth		0.712
Reverse	2.18	2.18

The main components that have been added up front are (1) an intermediate clutch and sprag assembly backed up by an overrun band, (2) an overdrive planetary assembly, and (3) a converter-equipped lockup clutch with a damper assembly. The main components that control the geartrain operation include six multiple-disc friction clutches, one band, two sprag one-way clutches, and a roller one-way clutch (Figure 23–28).

Electronic Control

The E4OD is electronically controlled by the PCM. Transmission shift feel and scheduling in the OD range and converter clutch operation are controlled by the PCM. The PCM receives and evaluates information from numerous sensors and switches and makes decisions that provide the best state of transmission operation for a given vehicle demand performance. The PCM achieves the objective by sending output signals to the solenoid body assembly containing five solenoids (Figure 23–25A).

Valve groups are still maintained in the valve body for switching the shift circuits and converter clutch circuit. An EPC (electronic pressure controller) is included in the solenoid assembly. A solenoid applications chart is shown in Figure 23–25B.

The electronic control also provides diagnostic capabilities by providing DTCs. Use a scan tool to identify the presence of DTCs. For the driver, the overdrive transmission control indicator lamp (TCIL) will flash during certain malfunction conditions that require immediate attention, such as a faulty EPC.

External Transmission Controls and Features

- *Manual range selection:* linkage operated.
- *Manual lever position sensor:* left side of case.
- *Transmission solenoid body case connector:* right side of case.
- *Vehicle speed sensor:* left side of extension housing.
- *Mainline pressure tap:* left front side of case ahead of manual control arm.

Operating Characteristics

The E4OD shift lever has six positions (Figure 23–26).

Overdrive (Normal). Full automatic shifting is provided. The transmission operates in a typical four-speed pattern. Depending on engine, axle ratios, and tire sizes, the minimum shift into third can vary between 25 and 30 mph (40 to 48 km/h). This is followed by a delayed minimum shift into fourth gear that can vary between 35 and 45 mph (56 to 74 km/h). A shift into fourth gear is not possible at WOT.

Depending on engine torque reserve, vehicle speed range, and road load, forced downshifts can occur in any sequence. If the conditions are right, it is possible to force a 4-1 downshift. Part-throttle and closed-throttle shifts are 4-3, 3-2, and 2-1. All upshift and downshift scheduling patterns are determined by the transmission calibration and the PCM.

Converter clutch engagement can occur in any gear, depending on the PCM strategy. The usual factors affecting the lockup and unlock modes are the same as those common to other lockup systems. E4OD converter clutch release logic is used for both upshifts and downshifts to minimize the perceived engine rpm change, which enhances shift quality. Like other systems, all downshifts are made with the converter clutch released.

Overdrive—With OD Canceled. An overdrive TCS (transmission control switch) allows the driver to lock out fourth gear (Figure 23–27). An amber TCIL (transmission control indicator lamp) illuminates when the TCS is switched ON (fourth gear canceled). This operating mode is used for trailer hauling, traveling in hilly/mountainous terrain, or when facing a strong headwind. It also provides engine braking for descending grades. Returning the system back to normal is a matter of pushing the switch button again. The TCS always resets to automatic overdrive (fourth-gear availability) when the ignition key is turned OFF.

Manual 2 and Manual 1. Manual 2 and manual 1 operating ranges are typical of the Ford family of automatic transmissions. One variation occurs in 2. Selection of 2 at higher vehicle speeds results in temporary third-gear rather than second-gear engagement. When a safe vehicle speed range is reached, second gear is locked in.

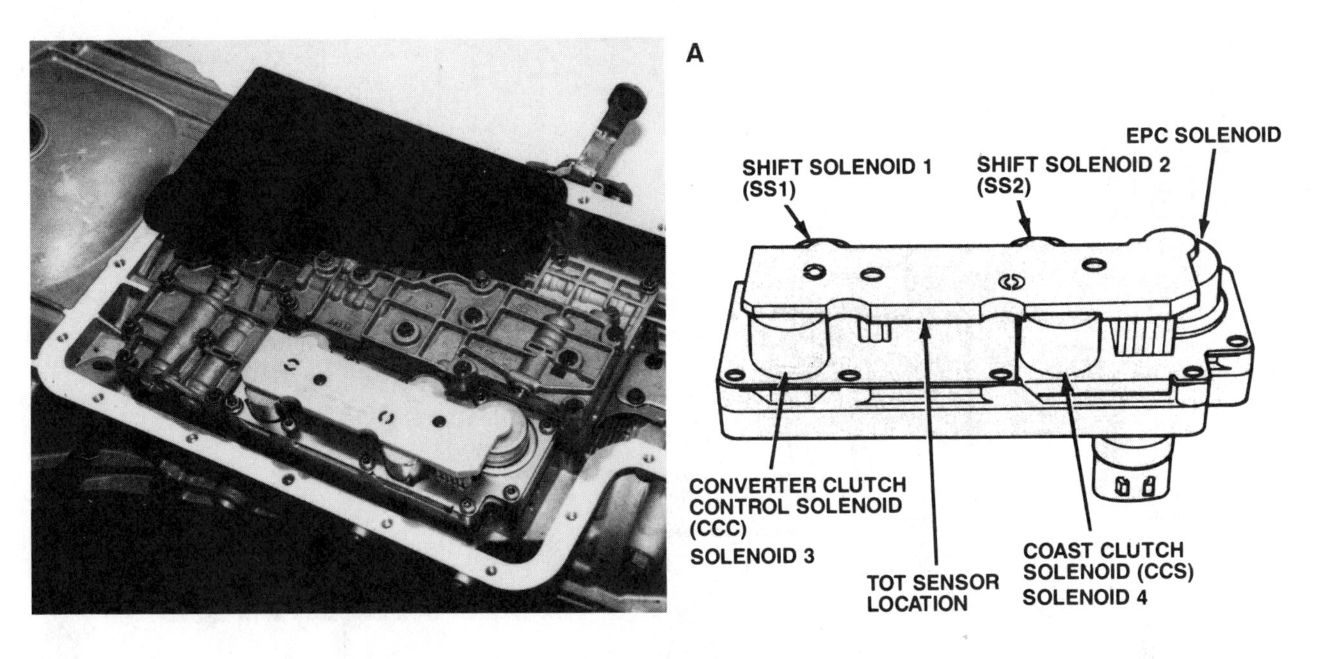

B

SOLENOID APPLICATION CHART

GEAR SELECTOR POSITION	ECA COMMANDED GEAR	SHIFT CONTROL			
		SOLENOID 1	SOLENOID 2	TORQUE CONVERTER CLUTCH SOLENOID 3	COAST CLUTCH SOLENOID 4
PARK	FIRST	ON	OFF	OFF	OFF
REVERSE	FIRST	ON	OFF	OFF	OFF
NEUTRAL	FIRST	ON	OFF	OFF	OFF
OD	FIRST	ON	OFF	*	OFF
OD	SECOND	ON	ON	*	OFF
OD	THIRD	OFF	ON	*	OFF
OD	FOURTH	OFF	OFF	*	OFF
OD CANCEL	FIRST THROUGH THIRD GEAR ONLY, S1, S2 AND TCC SAME AS OD, CCS ALWAYS ON.				
MANUAL 2	SECOND	OFF	OFF	*	ON
MANUAL 1	SECOND	OFF	OFF	OFF	ON
MANUAL 1	FIRST	ON	OFF	OFF	ON

*PCM CONTROLLED

FIGURE 23–25 (A) E4OD solenoid assembly mounted on valve body. (B) Solenoid applications chart. (Reprinted with the permission of Ford Motor Company)

Powerflow Summary

Using Figure 23–28 as a reference, the powerflow can be easily highlighted. With the overdrive planetary located up front, the converter output must always pass through the overdrive planetary to drive the main planetary system. During the first three gears and reverse, the overdrive planetary is in direct mode and does not affect the ratio output. The powerflows in neutral, the three forward speeds, and reverse are typical Simpson compound gearset patterns.

FIGURE 23–26 E4OD selector pattern. (Reprinted with the permission of Ford Motor Company)

Overdrive Unit—Direct Mode. In the direct mode, with the overdrive planet carrier driving, the sun gear wants to turn faster

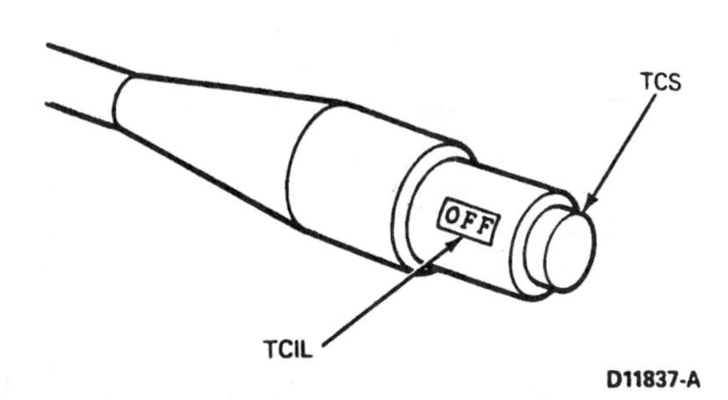

FIGURE 23–27 E4OD TCS (transmission control switch) and TCIL (transmission control indicator lamp) are located on the selector lever. (Reprinted with the permission of Ford Motor Company)

FORD E40D

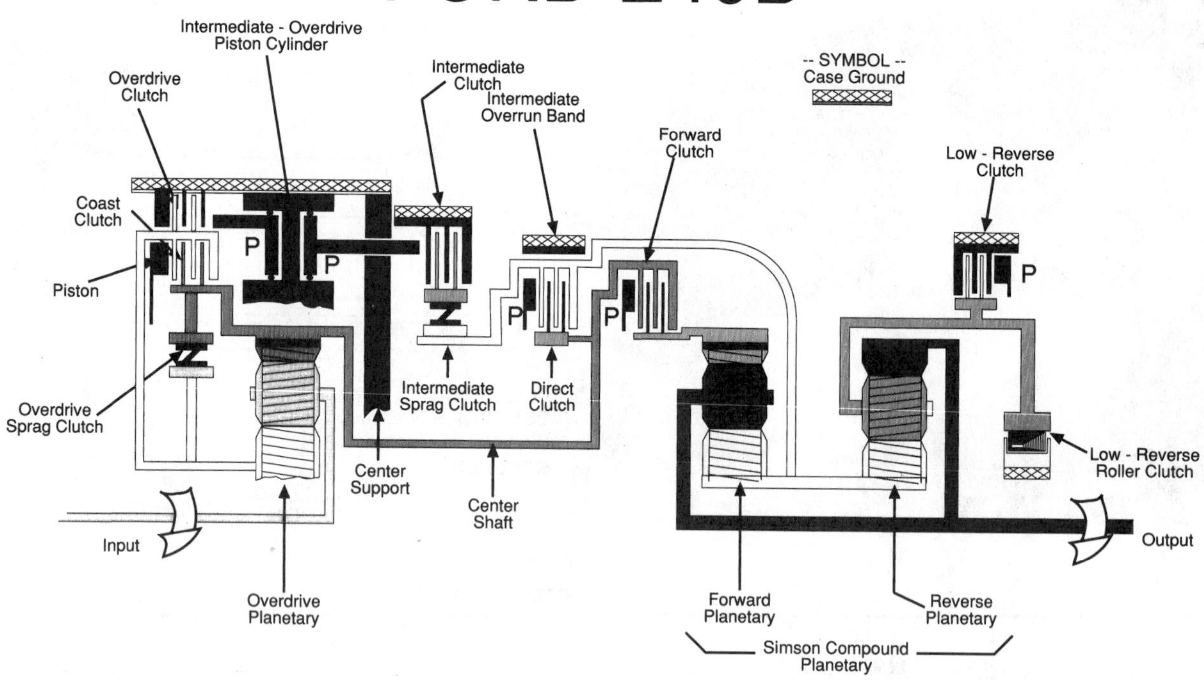

FIGURE 23–28 Ford E40D powertrain schematic.

than the ring gear. This results in a lockup of the one-way clutch between the sun gear and ring gear.

During coast or deceleration, the ring gear turns faster than the sun gear, and the one-way clutch overruns. The coast clutch is applied in reverse, manual 1, manual 2, and in OD range when OD range is canceled. This action provides engine braking through the overdrive gearset. Note that engine braking through the entire transmission is not achieved unless the three-speed section of the transmission also has engine braking capabilities.

Refer to Chapter 4 for additional information on overdrive units.

Overdrive Unit—Overdrive Mode. In the overdrive mode, the overdrive clutch is applied and holds the sun gear. With the planet carrier driving, a 0.712:1 ratio is achieved, and the one-way clutch overruns.

Intermediate System. The planetary clutch/band control incorporates an intermediate clutch and one-way sprag clutch used for second gear. During first-gear operation, the planetary sun gears turn counterclockwise (CCW). The CCW motion is transferred to the intermediate sprag assembly mounted on the front of the intermediate drum. The sprag clutch is ineffective because it lacks a case ground.

On the 1-2 shift, the sun gears continue with a CCW effort, and the intermediate clutch grounds the outer race. The sprag unit locks and holds the sun gear. On deceleration, the sprag unit freewheels. To achieve engine braking, the intermediate overrun band is applied in the manual ranges (Figure 23–29).

Chart 23–4 summarizes the clutch/band applications and operating ranges. Notice that the intermediate clutch remains applied in third and fourth gears. Since the intermediate drum is rotating clockwise, the sprag clutch overruns. On forced downshifts 4-2 or 3-2, the sprag clutch can take hold without waiting for the clutch to apply. The operating characteristics of the intermediate clutch and intermediate sprag simplify upshift and downshift timing.

ATX (AUTOMATIC TRANSAXLE/DOMESTIC DESIGN)

The ATX automatic transaxle is a transverse-mounted three-speed unit of metric design. It was introduced in production year 1981 as an option on the 1.6-L engine used in the Ford

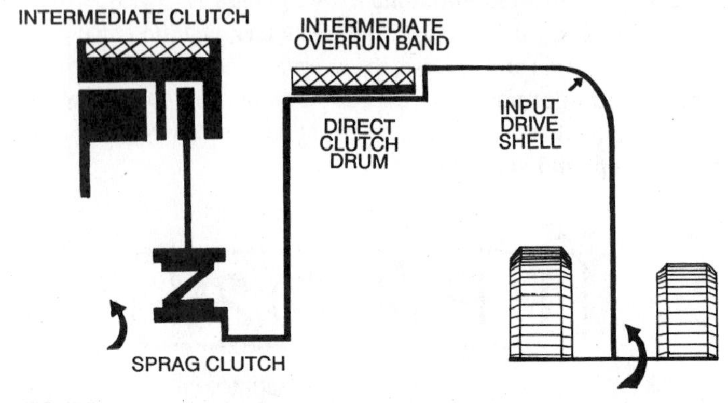

FIGURE 23–29 Ford E40D intermediate clutch system.

FORD E40D
OPERATING RANGE REFERENCE AND CLUTCH/BAND APPLY CHART

	OVERDRIVE UNIT			*SPEED PLANETARY UNIT*							
RANGE SELECTION	*SPRAG CLUTCH*	*COAST CLUTCH*	*O/D CLUTCH*	*FORWARD CLUTCH*	*DIRECT CLUTCH*	*INTER CLUTCH*	*INTER SPRAG CLUTCH*	*INTER OVERRUN BAND*	*REVERSE CLUTCH*	*L/R ROLLER CLUTCH*	*GEAR RATIO*
NEUTRAL/PARK	**HOLDS**										
REVERSE	**HOLDS**	**ON**			**ON**				**ON**		**2.18:1**
1st Gear-O/D	**HOLDS**	**(ON) LOCKOUT ONLY**		**ON**						**HOLDS**	**2.71:1**
2nd Gear-O/D	**HOLDS**	**(ON) LOCKOUT ONLY**		**ON**		**ON**	**HOLDS**				**1.538:1**
3rd Gear-O/D	**HOLDS**	**(ON) LOCKOUT ONLY**		**ON**	**ON**	**ON**					**1.00:1**
4th Gear-O/D			**ON**	**ON**	**ON**	**ON**					**.712:1**
2nd Gear-2	**HOLDS**	**ON**		**ON**		**ON**	**HOLDS**	**ON**			**1.58:1**
***3rd Gear-2 Deceleration**	**NOT HOLDING**	**ON**		**ON**	**ON**	**ON**					**1.00:1**
1st Gear-1	**HOLDS**	**ON**		**ON**					**ON**	**HOLDS**	**2.71:1**
***2nd Gear-1 Deceleration**	**NOT HOLDING**	**ON**		**ON**		**ON**	**NOT HOLDING**	**ON**			**1.538:1**

*(3rd Gear-2) and (2nd Gear-1) Temporarily Engaged WhenThe Operating Range Is Selected At An Excessive Vehicle Speed.

CHART 23–4 Ford E40D operating range reference and clutch/band apply chart. (Reprinted with the permission of Ford Motor Company)

Escort and Mercury Lynx. Additional application of the unit included a variety of adaptations with 1.9-, 2.3-, and 2.5-L four-cylinder engines used in Escort/Lynx, Tempo/Topaz, and Taurus/Sable models.

Front and rear external views are shown in Figures 23–30 and 23–31. The ATX uses a wide-ratio Ravigneaux planetary with a 2.8:1 first gear, 1.6:1 second gear, and a 1:1 third gear. The final drive is through a helical gearset and differential with 3.31, 3.26, 3.09, and 3.07 ratios available. The final drive ratio is dependent upon engine displacement, tire size, and converter style. For control of the geartrain, the ATX has three multiple-disc clutches, one band with an actuating servo, and a one-way roller drive clutch (Figure 23–32).

FIGURE 23–30 Ford ATX front view.

FIGURE 23–31 Ford ATX rear view.

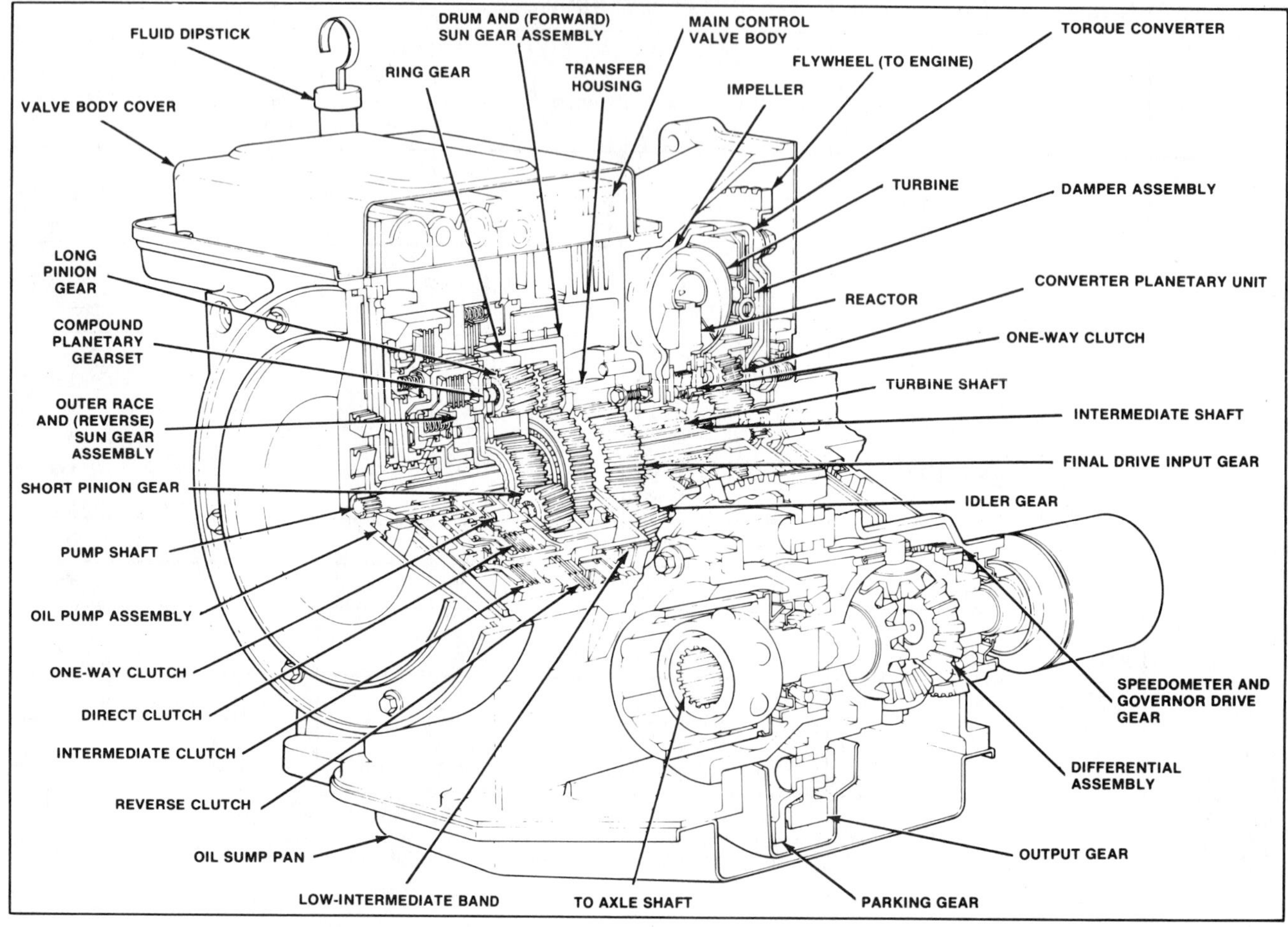

FIGURE 23–32 ATX sectional view. (Reprinted with the permission of Ford Motor Company)

In original design applications, a three-element torque converter uses a unique patented split torque concept that greatly reduces converter slippage. A planetary gearset is incorporated in the converter and acts as an engine torque splitter in second and third gears. Part of the torque is transmitted mechanically, as in manual transmissions, and the remainder by the fluid force on the turbine. In second gear, sixty-two percent of the torque is transmitted mechanically; in third gear, ninety-three percent is so transmitted. A torsional damper in the converter cover absorbs engine pulsations to the splitter gearset. In low gear and reverse, full torque is transmitted through the normal converter fluid drive (Figure 23–33).

In 1987, the planetary gearset–styled converter was replaced with two different units: the CLC (centrifugal locking clutch) and the FLC (fluid link converter). The FLC is a conventional three-element unit, open design. Eventually the FLC design was used exclusively, and the ATX was limited to the Tempo/Topaz car line.

External Transmission Controls and Features

- *Manual range selection:* linkage or cable operated lever arm.
- *Throttle system control:* linkage operated lever arm.
- *Forced downshifts/kickdown (KD) control:* linkage operated in common with the throttle system. The TV and KD valve group are in tandem in the same valve body bore and work from the same single external control.
- *Mainline pressure tap:* located on the left side top of the case and to the rear of the valve body cover.
- *Servo release pressure tap:* located forward of the mainline pressure tap next to the forward servo cover.

Operating Characteristics

The ATX shift selector has six positions: P, R, N, D, 2, and 1. The operational highlights for D, 2, and 1 are as follows.

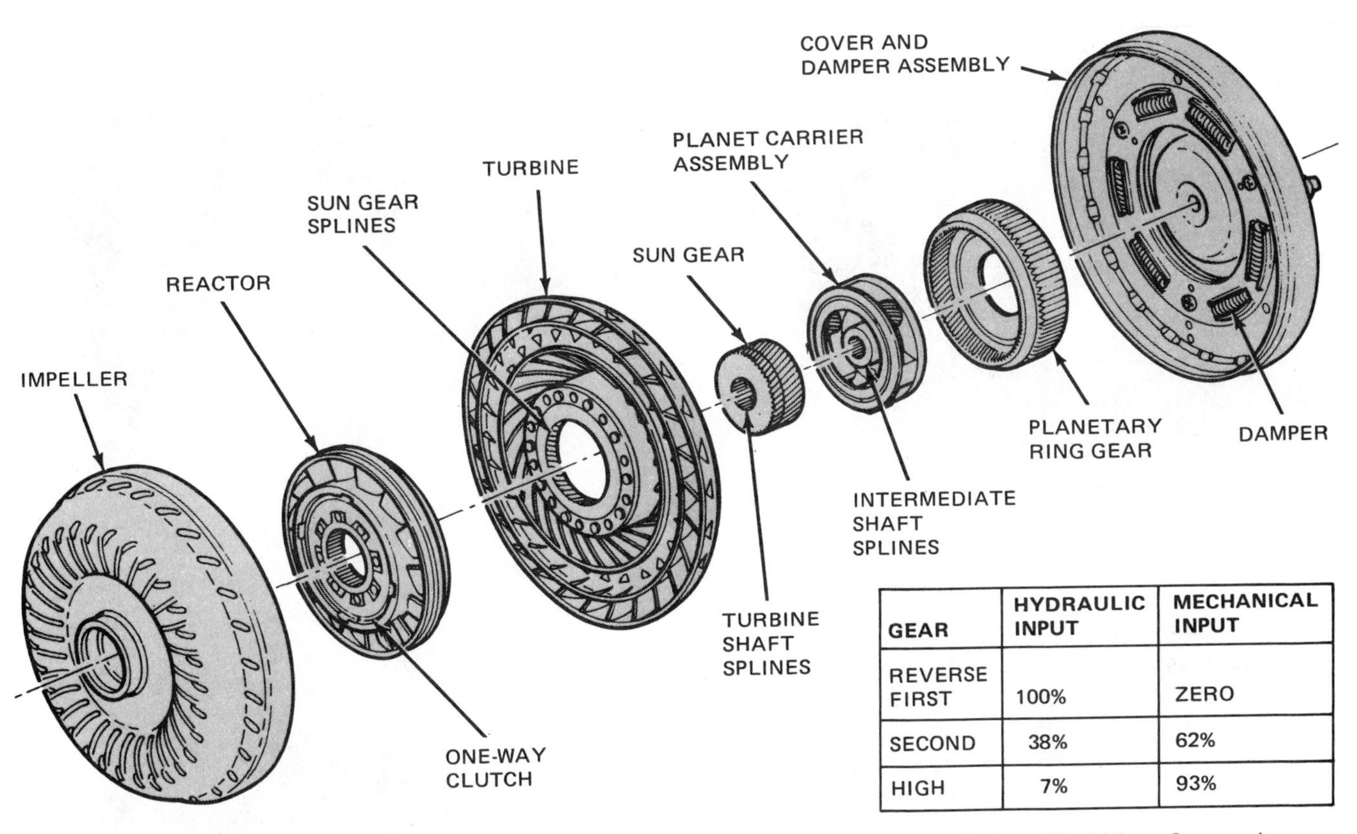

GEAR	HYDRAULIC INPUT	MECHANICAL INPUT
REVERSE FIRST	100%	ZERO
SECOND	38%	62%
HIGH	7%	93%

FIGURE 23–33 ATX converter assembly components with splitter planetary. (Reprinted with the permission of Ford Motor Company)

Drive. Selector position D is used for normal driving and maximum economy. Fully automatic shifting and three forward speeds are provided. The transmission automatically downshifts at closed throttle (coast) 3-2 and 2-1. When coasting in first gear, there is no engine braking action. For added performance, forced downshifts (kickdowns) 3-2, 3-1, and 2-1 are attained at wide open throttle through detent within the proper vehicle speed and engine torque range. A 3-2 part-throttle downshift feature offers controlled acceleration for low-speed operation in congested traffic.

Intermediate. In the 2 range, the transmission starts in first gear and automatically shifts to second. Third gear is locked out. There is no engine braking when coasting in first gear. A 2-1 forced downshift (kickdown) is available.

Manual 1. In the 1 range, the transmission is locked into low gear for extra pulling power or engine braking. If the selector is moved from D or 2 to 1 at an excessive speed, the transmission downshifts to second gear. When a safe road speed is attained, first gear engages.

Geartrain Powerflow Summary

The compound planetary gearset in the ATX is a three-speed Ravigneaux with reverse and neutral. The geartrain assembly components and holding members are illustrated in Figure 23–34. Although the ATX may use a planetary, CLC, or FLC styled converter, the FLC is used for purposes of explaining the geartrain powerflow. A splined insert is placed in the turbine element that permits both the intermediate and turbine shafts to be driven hydraulically.

Following is a brief summary of each planetary component and geartrain clutch/band member. Use Figures 23–32 and 23–34 as references.

Planetary Components

- *Planet carrier.* A single carrier assembly is used with three sets of dual pinions (long and short pinions) that are in

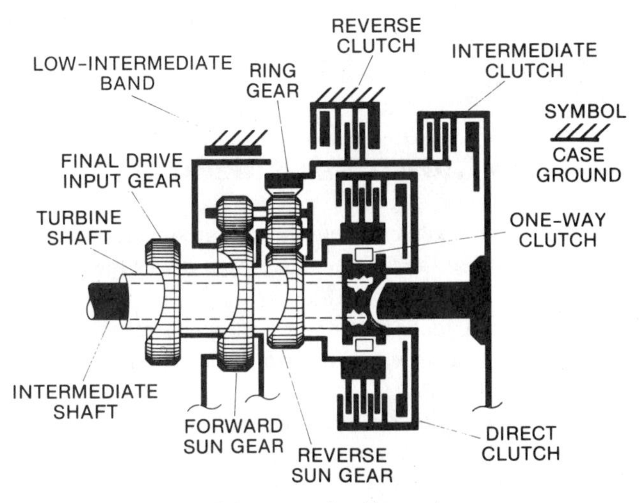

FIGURE 23–34 Ford ATX powertrain schematic.

FIGURE 23–35 Ford ATX carrier and final drive input gear.

FIGURE 23–36 Ford ATX assembled geartrain.

constant mesh. It is splined to the final drive input gear (Figure 23–35). Thus, the planet carrier is always the output member of the gearset.

- *Short pinions.* The short pinions in the carrier are in constant mesh with the long pinions and the reverse sun gear. Essentially, the short pinions act as idler gears between the reverse sun gear and long pinions.
- *Long pinions.* The long pinions in the carrier are in constant mesh with the short pinions, the forward sun gear, and the ring gear.
- *Reverse sun gear.* The reverse sun gear is integral with the one-way clutch outer race and direct clutch hub. It is driven by the turbine shaft when the one-way clutch is effective. The reverse sun gear also can be locked to the turbine shaft by applying the direct clutch. The reverse sun gear is in constant mesh with the planet carrier short pinions.
- *Drum and forward sun gear.* The sun gear meshes with the long planet pinions. The low-intermediate band holds the drum and forward sun gear stationary.
- *Ring gear.* The ring gear is integral with a single clutch hub for the intermediate and reverse clutches. The gear teeth mesh with the long planet pinions.

The gearset components are shown as a complete assembly in Figure 23–36.

FIGURE 23–37 Ford ATX turbine shaft and one-way roller drive clutch.

Geartrain Clutch/Band Members

- *One-way drive clutch:* locks the turbine shaft to the reverse sun gear when the turbine shaft input drives the sun gear (Figures 23–37 and 23–38).
- *Direct clutch:* locks the turbine shaft to the reverse sun gear to prevent the one-way clutch from freewheeling in manual low (Figure 23–38).
- *Intermediate clutch:* allows the intermediate shaft to engage and drive the ring gear (Figure 23–39).
- *Reverse clutch:* holds the ring gear stationary (Figures 23–32 and 23–34).
- *Low-intermediate band:* holds the forward sun gear stationary (Figures 23–32 and 23–34).

Geartrain Powerflow

With the assembled relationship of the planetary gearset and clutch/band components now established, we can review the powerflow in neutral, the three forward gears, and reverse. Reference to Figures 23–32 and 23–34 will support the discussion.

The geartrain planet carrier is always the output member and runs the final drive input gear (Figure 23–35). The final drive components are shown as they appear in the transfer gear housing in Figures 23–40 and 23–41. Note that the helical-cut

FIGURE 23–38 Ford ATX reverse sun gear and hub assembly.

FIGURE 23–40 Ford ATX final drive input gear and idler gear.

output gear corresponds to the ring gear in a conventional drive axle. The differential assembly is also identical to those used in other conventional axle differentials.

To simplify the Ravigneaux planetary illustrations, they are shown in rear-view perspective with front-view rotations.

Neutral/Park. The one-way clutch is the only holding member. Because the hydraulic turbine drive attempts to turn faster than the reverse sun gear, the one-way clutch locks the turbine shaft to the sun gear. There is powerflow input to the geartrain. The planetary gears spin-up but do not transmit power to the final drive because there is no reactionary member (Figure 23–42).

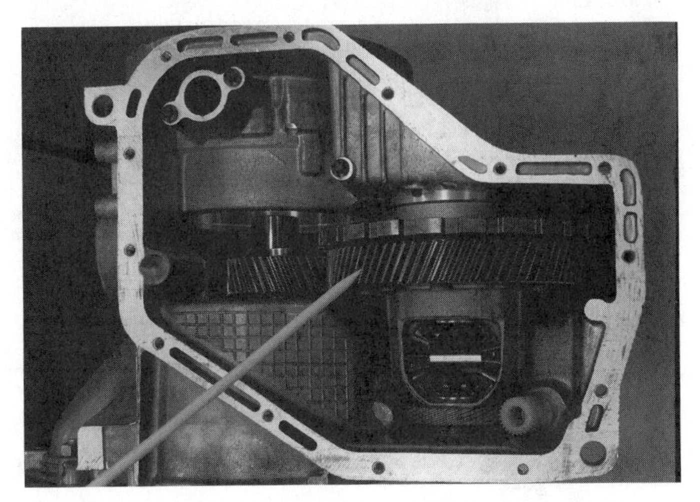

FIGURE 23–41 Ford ATX output gear and differential assembly.

First Gear. First-gear active clutch/band members:

1. *Low-intermediate band:* holds the forward sun gear stationary.
2. *One-way roller drive clutch:* locks the reverse sun gear to the turbine shaft input.

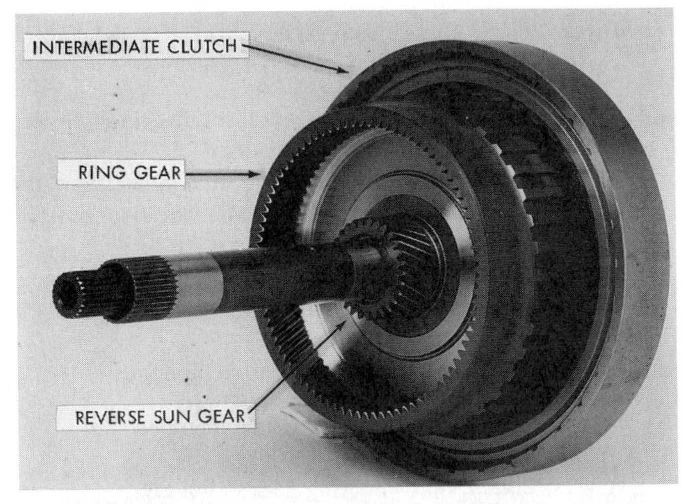

FIGURE 23–39 Ford ATX intermediate clutch splined to the ring gear.

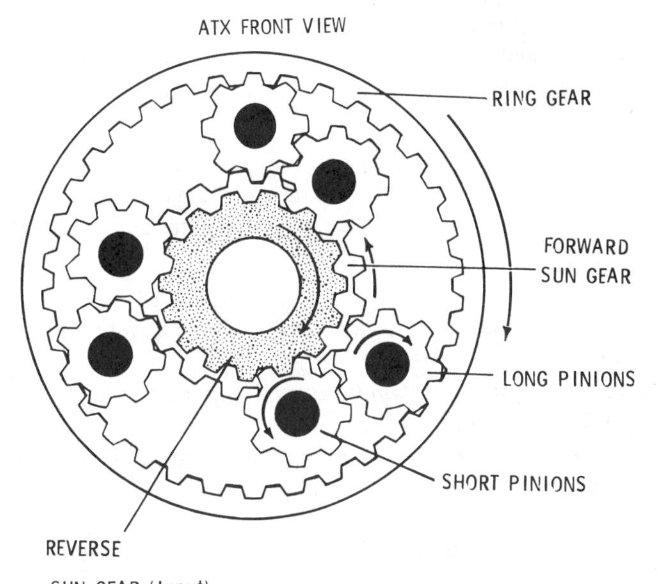

FIGURE 23–42 Ford ATX neutral and park planetary gear rotation.

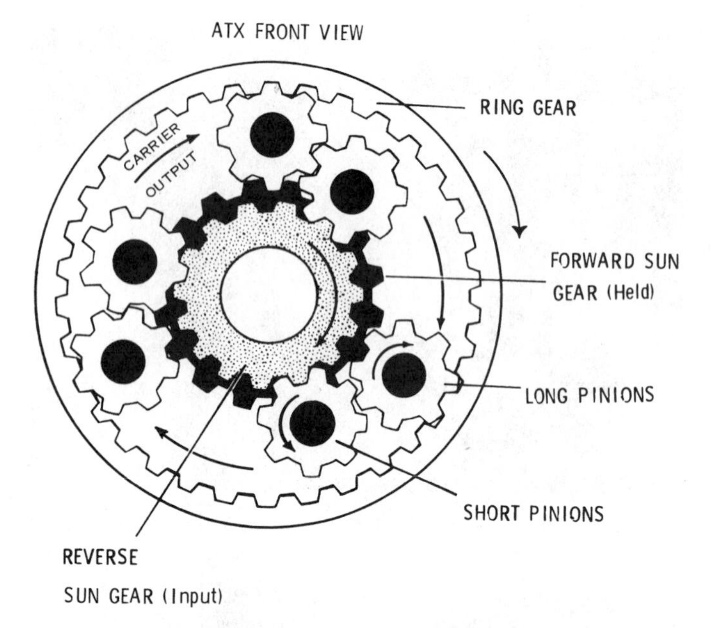

FIGURE 23–43 Ford ATX first-gear planetary rotation.

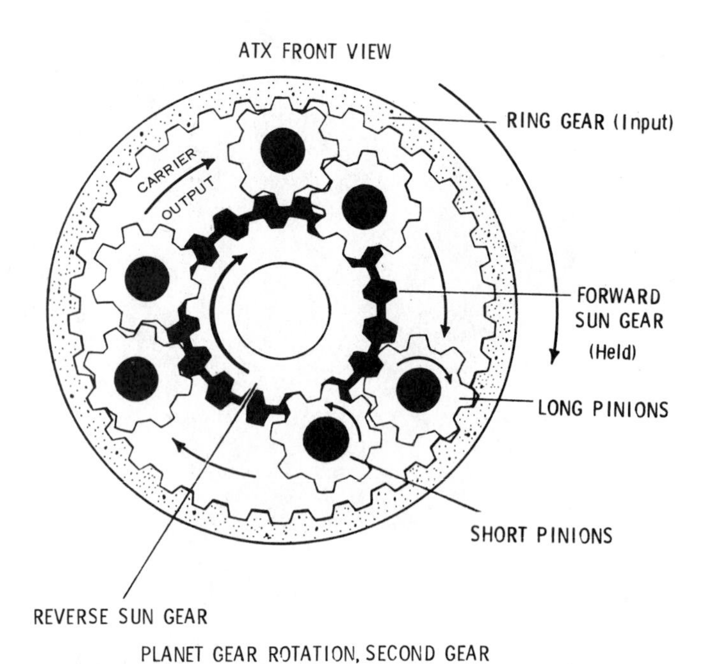

FIGURE 23–44 Ford ATX second-gear planetary rotation.

3. *Direct clutch:* applied in manual low (1) for engine braking.

- The converter turbine drives the turbine shaft and the reverse (input) sun gear.
- The turbine shaft attempts to turn faster than the reverse sun gear, and the one-way clutch locks the turbine shaft to the sun gear.
- The reverse sun gear rotating clockwise drives the short pinions counterclockwise.
- The short pinions drive the long pinions clockwise.
- The long pinions walk around the stationary forward sun gear and drive the planet carrier and final drive input gear clockwise (Figure 23–43).

First gear is a 2.79:1 reduction. At maximum converter torque, the available reduction in first gear can reach 5.6:1. The ring gear in mesh with the long pinions rotates clockwise but has no function in the powerflow.

Second Gear. Second-gear active clutch/band members:

1. *Low-intermediate band:* remains applied and holds the forward sun gear.
2. *Intermediate clutch:* locks the intermediate shaft to the gear ring.

- The hydraulic turbine torque is transmitted to the intermediate shaft.
- The intermediate clutch locks the ring gear to the intermediate shaft. The ring gear rotation causes the long pinions to turn clockwise.
- The long pinion gears walk around the stationary forward sun gear and drive the planet carrier and final drive input gear clockwise (Figure 23–44).

The second gear ratio is 1.61:1. The long pinions also drive the short pinions counterclockwise. The effect of the planet carrier rotation causes the short pinions to drive the reverse sun gear at an overdrive speed. The sun gear now turns faster than the turbine shaft, and the one-way clutch overruns. In second gear, the intermediate shaft provides the input torque to the geartrain. The turbine input shaft turns with the turbine but does not transmit any power.

Third Gear. Third-gear active clutch/band members:

1. *Intermediate clutch:* remains applied and locks the intermediate shaft to the ring gear.
2. *Direct clutch:* locks the reverse sun gear to the turbine shaft.

- The hydraulic turbine torque is transmitted through both the turbine and the intermediate shafts.
- The reverse sun gear driven by the turbine shaft and the ring gear driven by the intermediate shaft are also common input members to the geartrain, turning at turbine speed. This traps the long and short pinions between the ring gear and reverse sun gear, resulting in total lockup of the geartrain and a 1:1 ratio (Figure 23–45).
- The planet carrier turns the final drive input gear.

Reverse. Reverse gear active clutch/band members:

1. *Reverse clutch:* holds the ring gear stationary.
2. *Direct clutch:* locks the reverse sun gear to the turbine shaft.

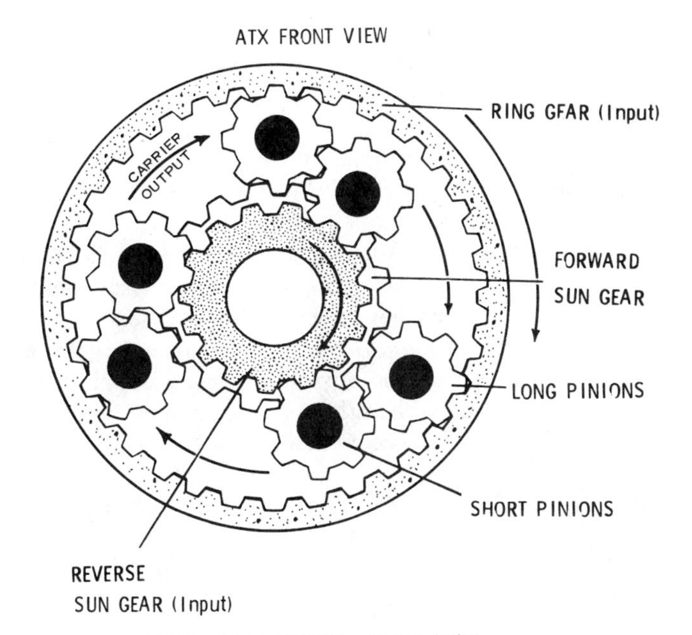

FIGURE 23–45 Ford ATX third-gear planetary rotation.

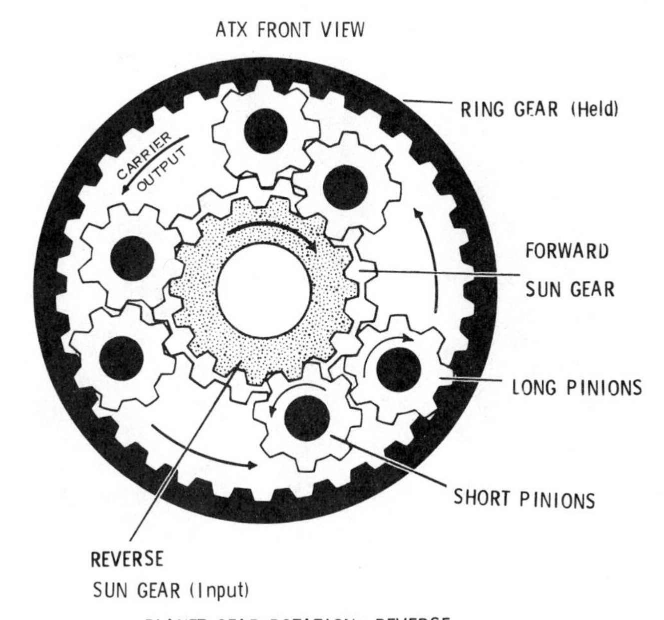

FIGURE 23–46 Ford ATX reverse-gear planetary rotation.

3. *One-way roller clutch*: locks the reverse sun gear to the turbine shaft input.

Reverse gear is hydraulically driven by the turbine.

- The hydraulic turbine drives the converter sun gear and turbine shaft clockwise.
- The turbine shaft attempts to turn faster than the reverse sun gear, and the one-way clutch locks the turbine shaft to the sun gear.
- The reverse sun gear rotating clockwise drives the short pinions counterclockwise.
- The short pinions turn the long pinions clockwise.
- The long pinions in mesh with the stationary ring gear walk around the inside of the ring gear and drive the planet carrier and final drive input gear CCW (Figure 23–46).

The reverse gear ratio is 1.97:1. At maximum converter torque, the available reduction can reach 4:1.

Chart 23–5 summarizes the operating ranges and clutch/band applications.

AX4S/AXODE AND AXOD TRANSAXLES

The AX4S and AXODE are the same transaxles featuring full electronic control. Ford has changed its transmission designations and the AXODE now is known as the AX4S. The original transaxle design was introduced as the AXOD in the 1986 Ford Taurus and Mercury Sable models with 3.0-L engines (Figure 23–47). Later applications included the 3.8-L engine. For 1991, the AXOD was upgraded to full electronic control. Vehicle applications remain with the Taurus/Sable and Windstar models.

The original design concept of the converter and geartrain inclusive of clutch/band control has remained constant. Sectional and schematic illustrations are shown in Figures 23–48

Selector Pattern P R N D 2 1

Selector Position	Gear	Int-Low Band	Direct Clutch	Intermediate Clutch	Reverse Clutch	One-Way Clutch	Gear Ratio
Park Neutral						Effective	------
Reverse			On		On	Effective	1.97-1
D	1st	On				Effective	2.79-1
	2nd	On		On			1.61-1
	3rd		On	On			1-1
2	1st	On				Effective	2.79-1
	2nd	On		On			1.61-1
1	1st	On	On			Effective	2.79-1

CHART 23–5 Ford ATX operating range and clutch/band apply summary.

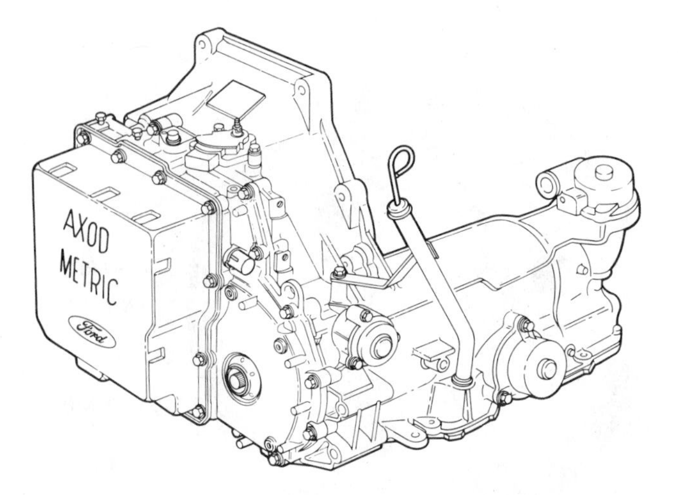

FIGURE 23–47 AXOD. (Reprinted with the permission of Ford Motor Company)

and 23–49. A wide-ratio compound planetary made up of two gearsets provide 2.77:1 first, 1.543:1 second, 1:1 third, and 0.694:1 fourth gears.

The geartrain is combined with a planetary final drive/differential assembly, which is located within the case and on the same centerline of the powerflow. Available final drive ratios are 3.19 for the 3.8-L engine and 3.337 for the 3.0-L engine. For control of the geartrain, four multiple-disc clutches, two bands, and two one-way drive clutches are used.

The three-element lockup torque converter drives the geartrain through a dual sprocket and drive link (chain) assembly. The number of teeth on the drive/driven sprockets are 38/35 (3.8 L) and 37/36 (3.0 L). The respective ratios are 0.921:1 and 0.973:1.

AXOD Hydraulic/Electronic Highlights

- *Oil pressure system:* uses a variable capacity rotary vane pump with a main regulator valve for line pressure control.

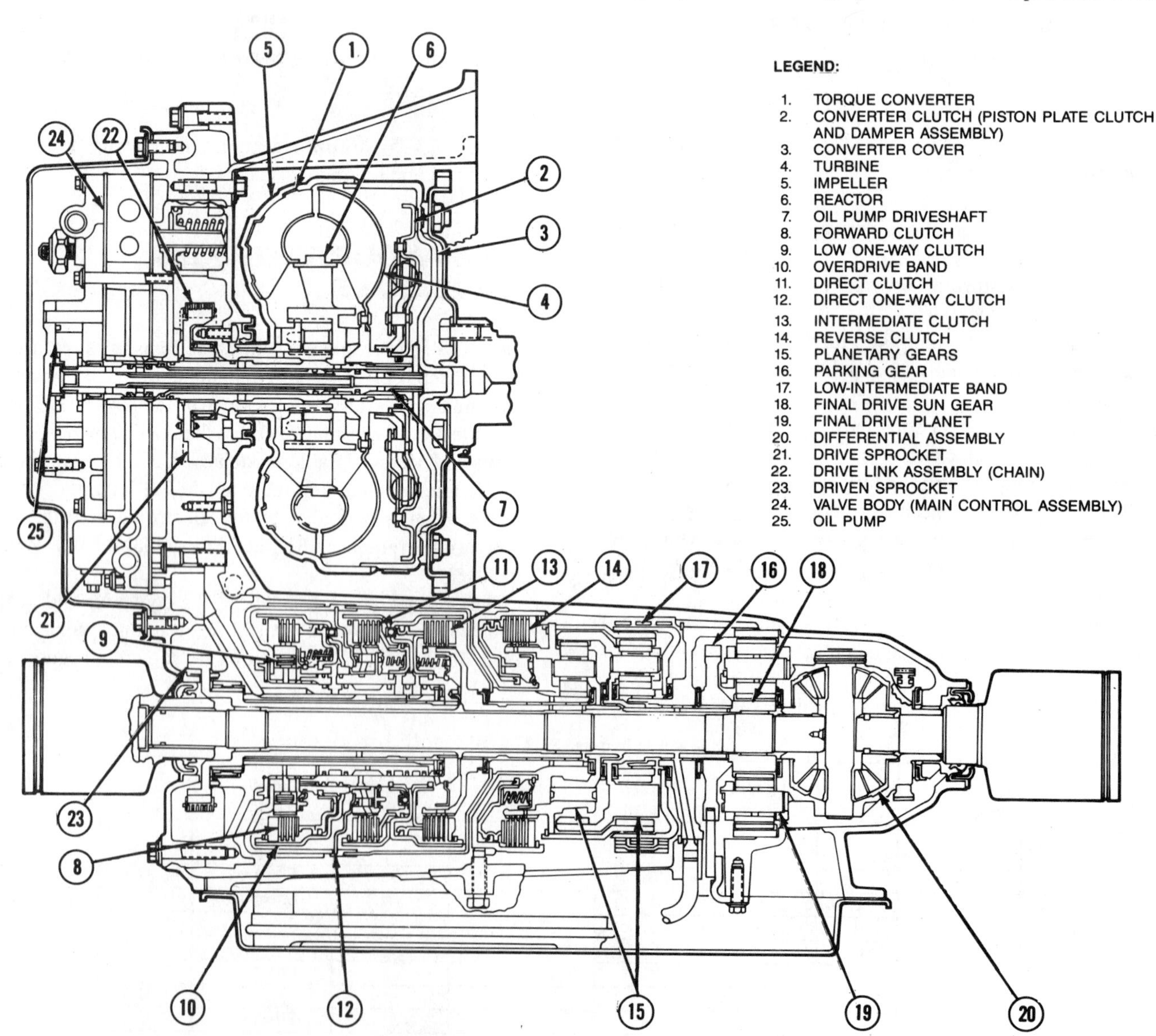

FIGURE 23–48 AXOD/AX4S (AXODE) sectional view. (Reprinted with the permission of Ford Motor Company)

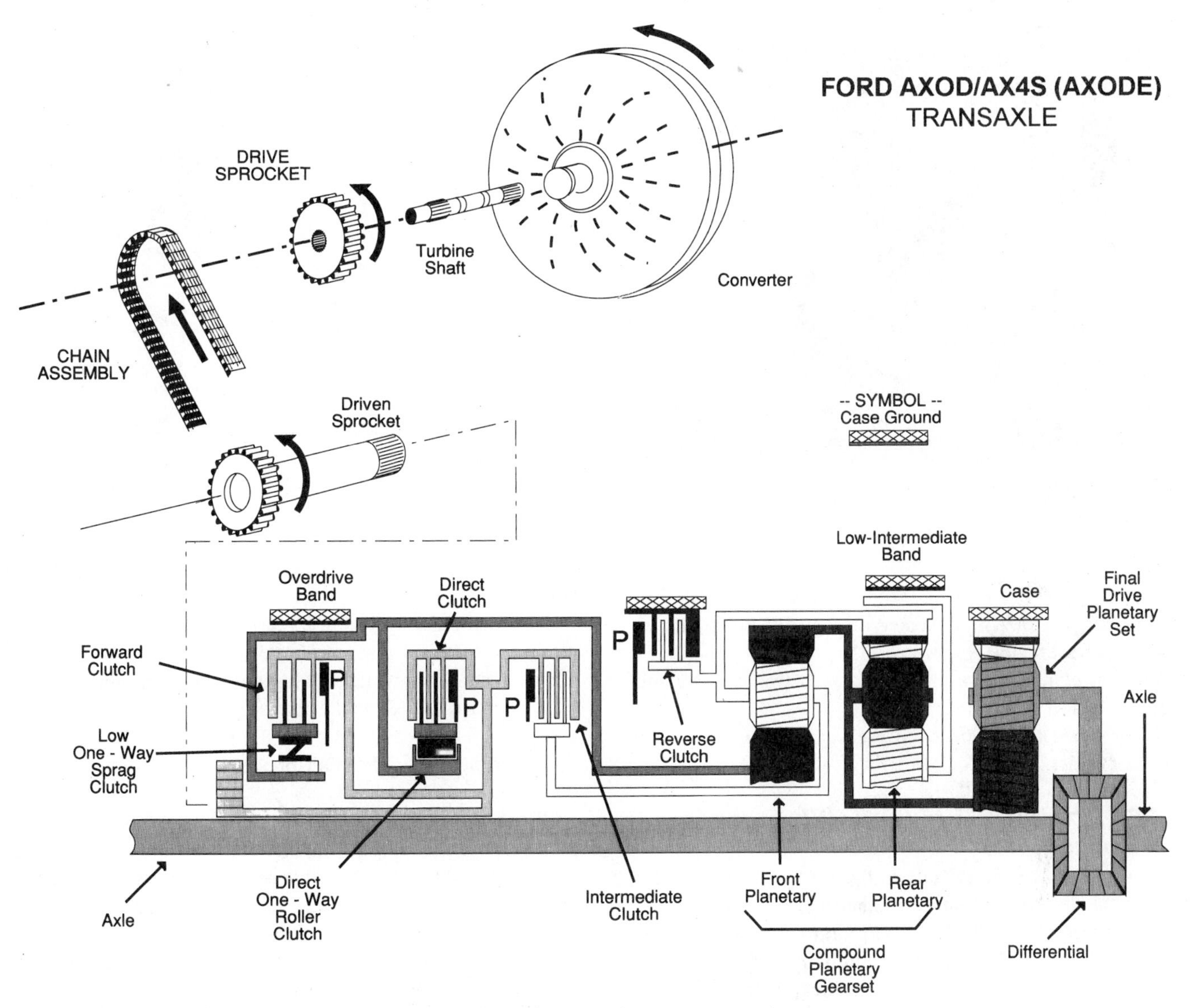

FIGURE 23–49 Ford AXOD/AX4S (AXODE) powertrain schematic.

- *Operating range selection:* preprogrammed by the manual valve.
- *Shift system:* features a conventional all hydraulic logic system using a check ball–style governor and mechanically operated TV system to control shift valve movement and shift scheduling.
- *Converter clutch lockup control system:* a combined hydraulic/electronic control system. A converter clutch solenoid circuit is electronically interfaced with the PCM. Converter clutch engagement and release are controlled by the solenoid and electronically scheduled by the PCM. The solenoid is ground-side switched through the PCM and is normally open (de-energized). When energized, the solenoid switch closes, resulting in a hydraulic movement of the converter clutch control valve and lockup engagement.

To manage the system, the PCM receives information from three pressure switches located on the valve body (Figure 23–50). They identify the operational status of the transaxle. All switches are normally open when no pressure is applied. A temperature switch is added for the 3.8 L. It provides for additional lockup operation to protect the transaxle from exceeding 275°F (132°C).

AXOD External Transmission Controls and Features

- *Manual range selection:* cable operated.
- *Throttle system control:* cable operated.
- *Forced downshift/kickdown control:* cable operated in common with the throttle system. The TV and KD valve groups are in tandem in the same valve body bore and work from the same single external control.
- *Neutral safety and backup light switch:* mounted over the manual control arm shaft.
- *Electronic control bulkhead connector:* located at the top right side of the chain cover.

FIGURE 23–50 AXOD pressure switches and bypass clutch solenoid location. (Reprinted with the permission of Ford Motor Company)

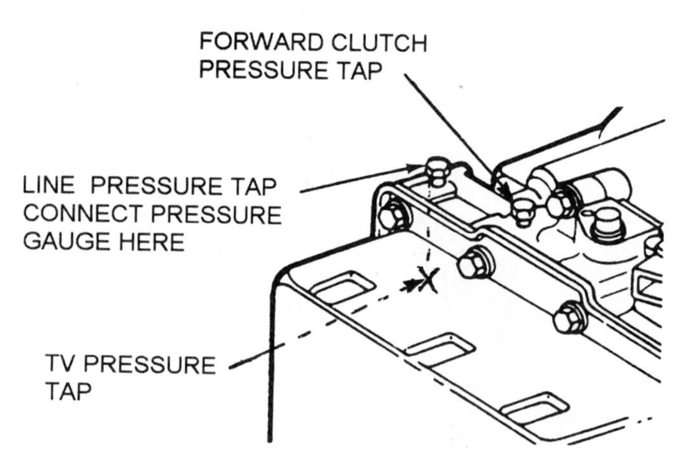

FIGURE 23–51 AXOD pressure taps. (Reprinted with the permission of Ford Motor Company)

- *Mainline pressure tap:* located on top of the chain cover toward the left side (Figure 23–51).
- *Forward clutch pressure tap:* located on top of the chain cover toward the center.
- *TV pressure tap:* located on the chain cover above the cooler outlet.

AX4S (AXODE) Hydraulic/Electronic Highlights

The upgrading of the AXOD to the AX4S (AXODE), designed with electronic controls, enhances the total engine/transaxle performance. The PCM is in command of shift scheduling, shift feel, and converter clutch operation. This is achieved through the use of five electronically managed solenoids mounted on the valve body (Figure 23–52). The added electronics is the major difference between the AX4S and the AXOD transaxles.

Electronic Controls (Input Devices). Illustrated in Figure 23–53 are the AX4S (AXODE) electronic input and output devices. Note that the PCM receives and processes input data from both engine and transaxle sensors in making management decisions. This is essential since the engine and transaxle must always work as a team. When the PCM is executing shifts, it simultaneously controls engine output. The transaxle input sensors and their locations are featured in Figure 23–54.

Electronic Controls (Output Devices). Output solenoid locations are shown in Figure 23–52. The five solenoids are controlled by the PCM.

Shift Solenoids. These solenoids are normally off and in the open position. The PCM solenoid shift control in the various operating ranges is summarized in Chart 23–6, page 551.

Lockup Solenoid. This solenoid controls the apply and release of the converter lockup clutch. Circuit function and operation is identical to the AXOD.

Modulated Lockup Solenoid. This solenoid is used on Continental vehicles. It provides a more sophisticated method of circuit control over the torque converter clutch apply oil (Figure 23–55).

The modulated lockup solenoid is a duty cycle (or variable pressure) solenoid and is normally OFF with the feed port closed. This blocks solenoid regulator oil from the converter clutch control valve, and the converter clutch is prevented from applying. When the PCM turns the solenoid ON and OFF, it will vary the solenoid output pressure to the converter clutch control valve.

When the PCM senses a need for more solenoid output pressure, it lengthens the time the solenoid stays ON (lengthens the pulse width). The variable solenoid pressure works the converter clutch control valve against spring tension (to the left). At maximum pulse width, the converter clutch control valve exhausts the release side converter clutch pressure, and the clutch fully applies. As the pulse width is shortened, the solenoid pressure is reduced at the converter clutch valve. This results in an increasing pressure head in the release side of the circuit—the release oil never truly exhausts. The release oil now opposes apply oil.

Depending on the pressure differential, the clutch stays applied but allows a controlled slip to accommodate engine torque demand. Depending on pulse width control by the PCM, the converter can be fully applied, fully released, or in a variable-controlled slip mode.

Electronic Pressure Control (EPC) Solenoid. The EPC is a variable force solenoid. It controls line pressure by providing a throttle pressure boost oil to the main pressure regulator valve (Figure 23–56). The PCM varies the current to the solenoid to produce the appropriate spool valve regulating action and torque signal oil. As current is increasingly applied, the boost oil to the main regulator valve drops. This results in a corresponding decrease in mainline pressure.

Geartrain Powerflow Summary

The geartrain powerflow is identical for the AXOD and AX4S (AXODE). Since some gear changes involve a clutch/band transition, the release and apply overlap must be

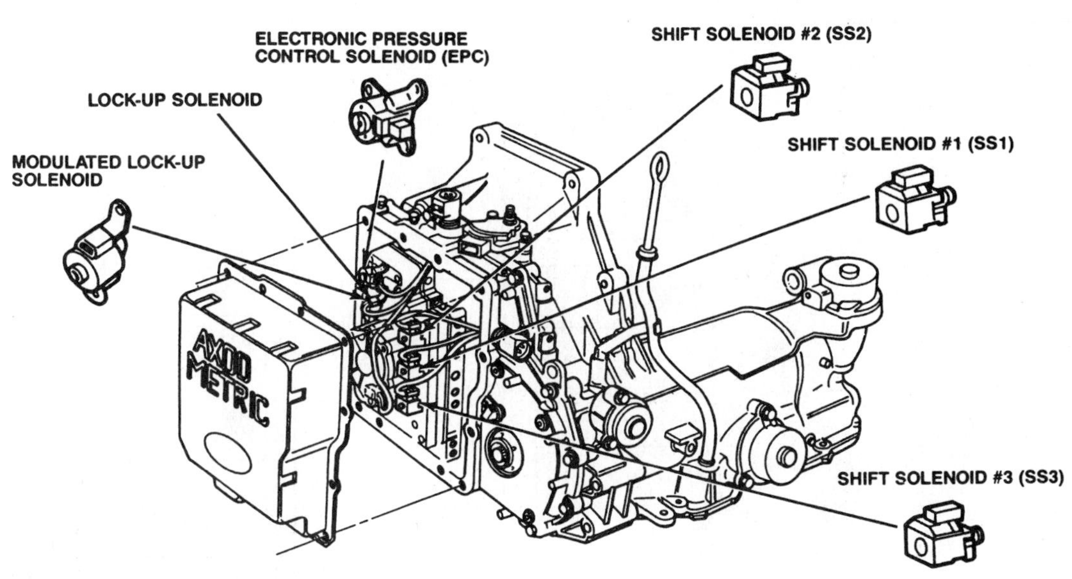

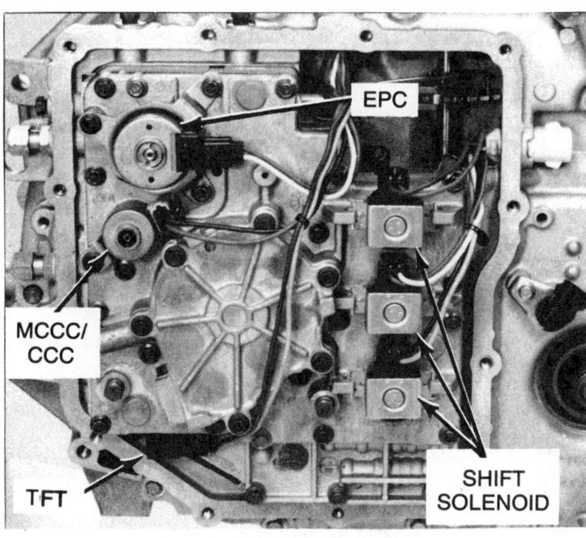

FIGURE 23–52 AX4S (AXODE) solenoids. (Reprinted with the permission of Ford Motor Company)

timed or synchronized. This distinguishes the AX4S geartrain from the AX4N geartrain, which is nonsynchronized. For technician reference, use the sectional and schematic illustrations shown in Figures 23–48 and 23–49.

Geartrain Clutch/Band Members

- *Low one-way drive clutch (sprag design):* transmits input drive torque from the forward clutch to the front sun gear in first gear and reverse.
- *Direct one-way drive clutch (roller design):* transmits input drive torque from the direct clutch to the front sun gear in third gear.
- *Forward clutch:* engages the input drive torque and drives the front sun gear through the direct one-way sprag clutch in first gear and reverse.
- *Direct clutch:* engages the input drive torque and drives the front sun gear through the direct one-way roller clutch in third gear.
- *Intermediate clutch:* engages the input drive torque and drives the front carrier/rear ring unit in second and third gears.
- *Reverse clutch:* reaction clutch that holds the front carrier/rear ring gear unit for reverse gear.
- *Low-intermediate band:* holds the rear sun gear stationary for first and second gears.
- *Overdrive band:* holds the front sun gear stationary for fourth-gear overdrive.

Geartrain Powerflow

The technician will notice that the planetary system uses a modified design and does not fall into the traditional Simpson configurations. Planetary principles of operation, however, remain the same and will apply when explaining the powerflow. In keeping with SAE standards, powerflow rotations are explained facing the planetary input front to rear. In the AXOD/AX4S (AXODE), the input is left-handed or counterclockwise.

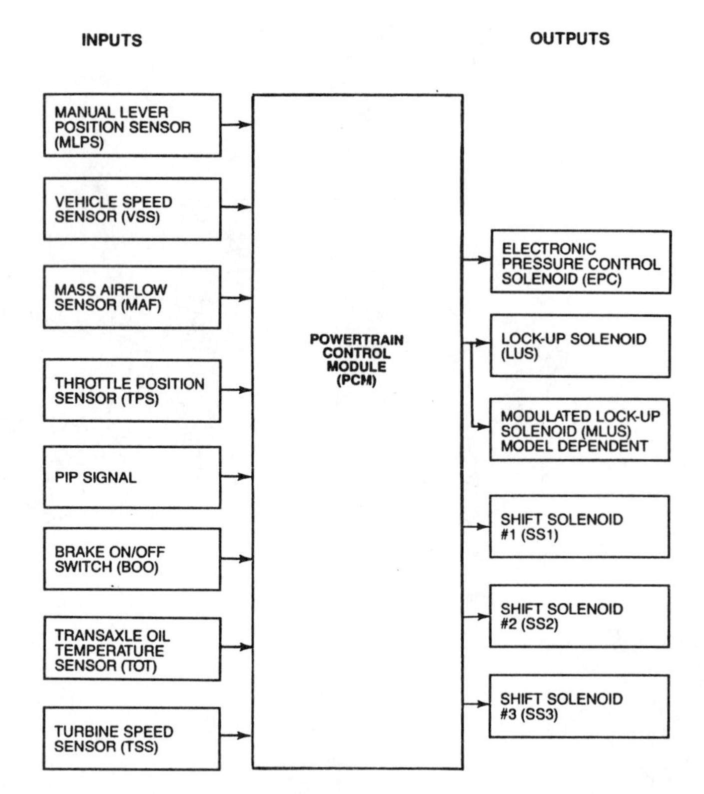

FIGURE 23–53 AX4S (AXODE) electronic inputs and outputs. (Reprinted with the permission of Ford Motor Company)

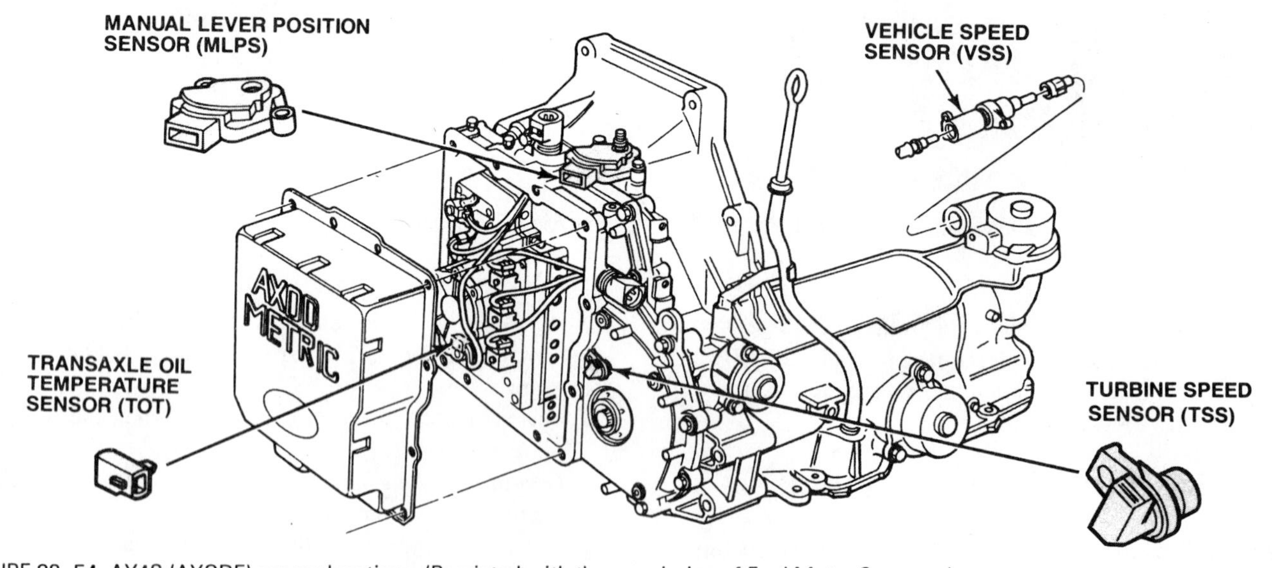

FIGURE 23–54 AX4S (AXODE) sensor locations. (Reprinted with the permission of Ford Motor Company)

SOLENOID APPLICATIONS CHART

PRNDL	GEAR	ENGINE BRAKING	SS1	SS2	SS3
Ⓓ	1	NO	OFF	ON	OFF
	2	YES	ON	ON	OFF
	3	NO	OFF	OFF	ON
	4	YES	ON	OFF	ON
D	1	NO	OFF	ON	OFF
—	2	YES	ON	ON	OFF
—	3	YES	OFF	OFF	OFF
	4	NOT ALLOWED BY STRATEGY			
1	1	YES	OFF	ON	OFF
	2	YES	OFF	OFF	OFF
	3	NOT ALLOWED BY HYDRAULICS			
	4	NOT ALLOWED BY HYDRAULICS			
R	R	NO	OFF	ON	OFF
P/N	P/N	NO	OFF	ON	OFF

CHART 23–6 AX4S (AXODE) solenoid applications chart. (Reprinted with the permission of Ford Motor Company)

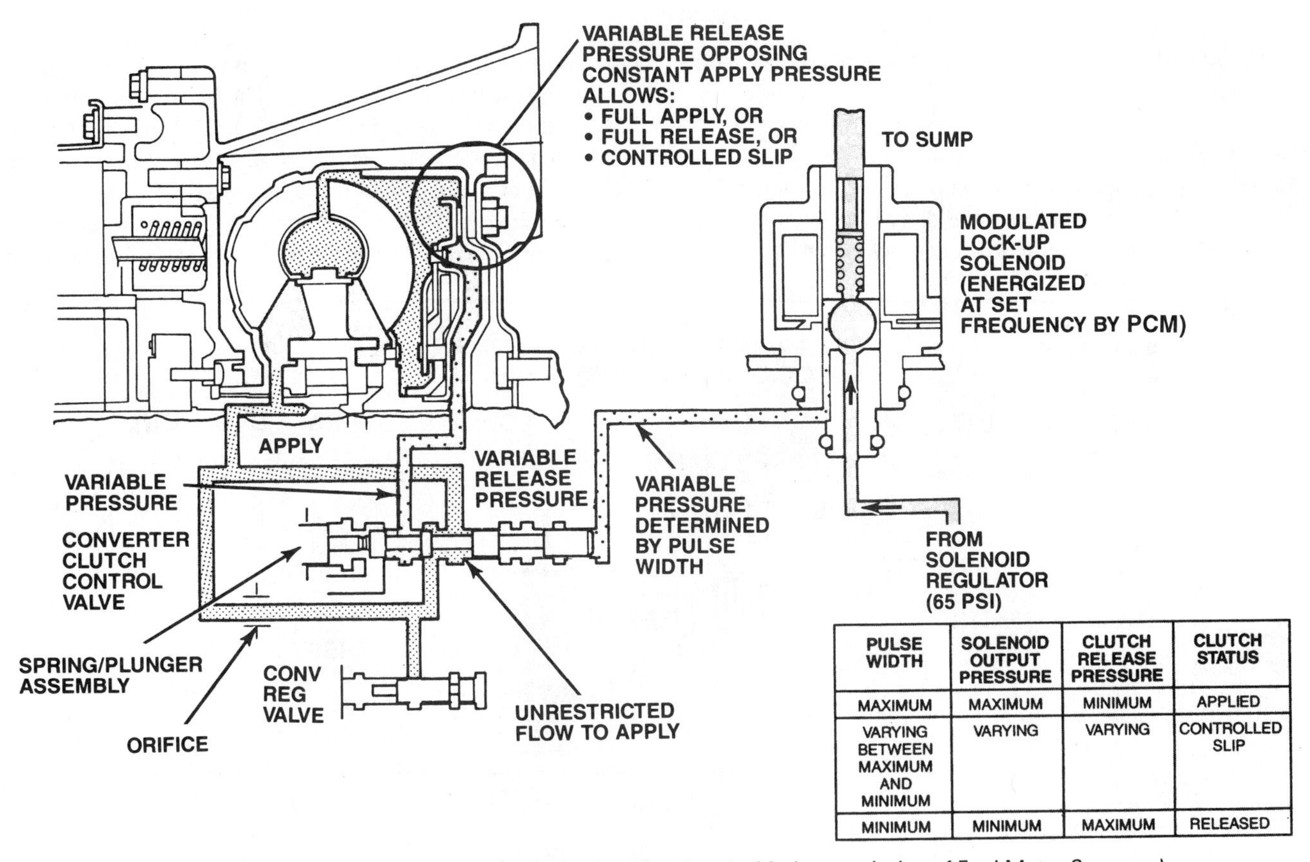

PULSE WIDTH	SOLENOID OUTPUT PRESSURE	CLUTCH RELEASE PRESSURE	CLUTCH STATUS
MAXIMUM	MAXIMUM	MINIMUM	APPLIED
VARYING BETWEEN MAXIMUM AND MINIMUM	VARYING	VARYING	CONTROLLED SLIP
MINIMUM	MINIMUM	MAXIMUM	RELEASED

FIGURE 23–55 AX4S (AXODE) modulated lockup solenoid operation. (Reprinted with the permission of Ford Motor Company)

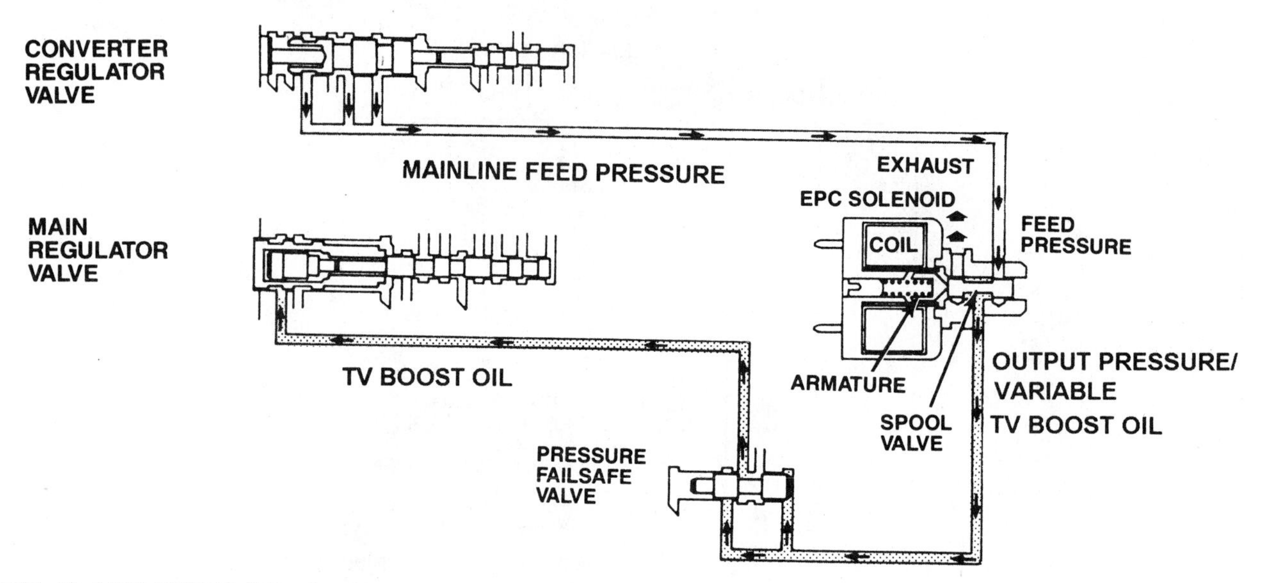

FIGURE 23–56 AX4S (AXODE): EPC solenoid operation. (Reprinted with the permission of Ford Motor Company)

Neutral/Park. The drive clutches are not applied, and therefore, there is no input to the planetary unit.

OD and D Ranges.

First Gear (Figure 23–57). With the range selector in OD or D, the forward clutch is applied and drives the front sun gear counterclockwise through the low one-way clutch. The CCW rotation of the front sun gear causes a clockwise rotational effort by the front carrier pinions. The front carrier pinions work against the vehicle weight or drive load through the front ring gear/rear carrier. The front ring gear, therefore, is reactionary to the front pinion rotation and induces a CCW pinion walking motion that transfers to the rear ring gear at a reduction. The rear ring gear now drives the rear carrier pinions CCW, which

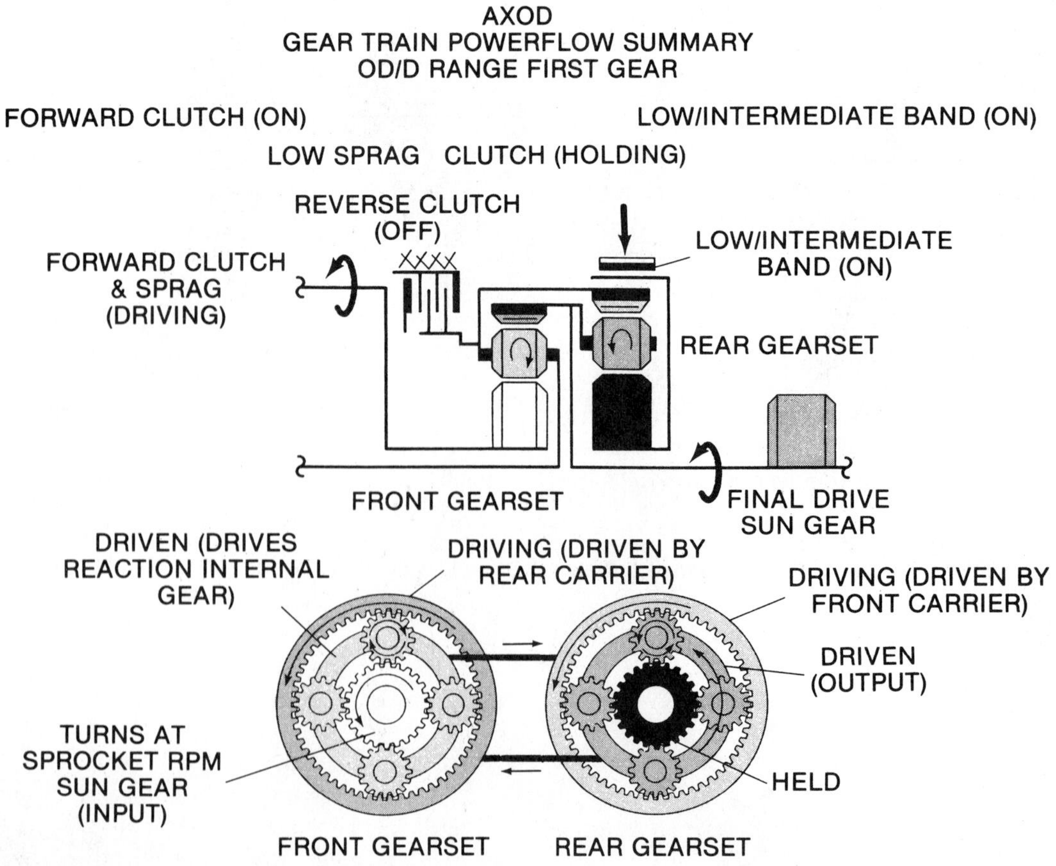

AXOD
GEAR TRAIN POWERFLOW SUMMARY
OD/D RANGE FIRST GEAR

FORWARD CLUTCH (ON)
LOW/INTERMEDIATE BAND (ON)
LOW SPRAG CLUTCH (HOLDING)
REVERSE CLUTCH
(OFF)
FORWARD CLUTCH
& SPRAG
(DRIVING)
LOW/INTERMEDIATE
BAND (ON)
REAR GEARSET
FRONT GEARSET
FINAL DRIVE
SUN GEAR
DRIVEN (DRIVES
REACTION INTERNAL
GEAR)
DRIVING (DRIVEN BY
REAR CARRIER)
DRIVING (DRIVEN BY
FRONT CARRIER)
DRIVEN
(OUTPUT)
TURNS AT
SPROCKET RPM
SUN GEAR
(INPUT)
HELD
FRONT GEARSET
REAR GEARSET

FIGURE 23–57 Ford AXOD/AX4S (AXODE) first-gear operation.

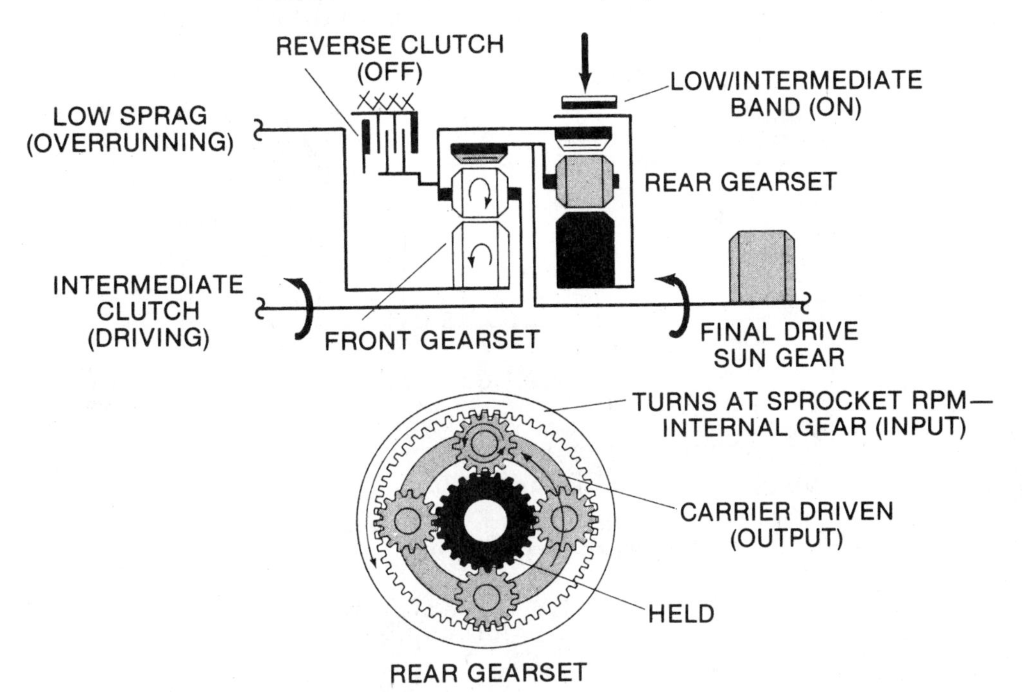

FIGURE 23–58 Ford AXOD/AX4S (AXODE) second-gear operation.

results in a CCW pinion walking motion as the rotating pinions work against the stationary rear sun gear (held by the low-intermediate band). The walking pinions drive the rear carrier/front ring gear at an output ratio of 2.77:1, which is passed on to the final drive sun gear. Both gearsets combine to provide the first-gear ratio.

Second Gear (Figure 23–58). The intermediate clutch is applied and drives the front planet carrier/rear ring gear unit counterclockwise. The rear ring gear drives the rear carrier pinions CCW, which results in a CCW pinion walking motion as the rotating pinions work against the stationary rear sun gear (held by the low-intermediate band). The walking pinions drive the rear carrier/front ring gear at an output gear ratio of 1.543:1, which is passed on to the final drive sun gear.

This is the same action that occurs in first gear except that the front carrier/rear ring unit is driven at input rpm rather than at a reduction. Although the front planetary does not contribute to second-gear ratio, the front carrier pinions react off the slower turning front ring gear and drive the sun gear at an overdrive counterclockwise speed. The inner race of the low sprag clutch turns faster than the outer race, and the sprag overruns.

Third Gear (Figure 23–59). With the range selector in OD or D, all three drive clutches are applied. The intermediate clutch remains applied from second gear and drives the front input carrier. The direct clutch is also applied and drives the input sun gear through the direct roller clutch. The intermediate and direct clutches essentially lock the front carrier and front sun gear to the input torque and rpm. The front carrier pinions are trapped and cannot rotate. The result is a direct-drive ratio of 1:1.

Lockup of the front planetary also locks the rear ring gear and rear carrier together, achieving a complete couple of the front and rear planetary units. During deceleration, the coast effect on the geartrain wants to drive the sun gear clockwise. This would overrun the direct roller clutch and put the geartrain in a neutral mode. The low sprag takes hold and keeps the front sun gear connected to the powerflow through the forward clutch. Engine braking is maintained.

Fourth Gear (OD Range Only) (Figure 23–60). The intermediate clutch is applied and drives the front carrier/rear ring gear counterclockwise while the overdrive band holds the front sun gear. The front carrier pinion gears are also rotating CCW on their centers, resulting in a walking motion as the pinions react to the stationary front sun gear. The pinion gear rotation and walking action drives the front ring gear/rear carrier unit at an output torque ratio of .694:1, which is passed on to the final drive sun gear.

Fourth-gear overdrive is a function of the front planetary. The direct-drive clutch, although applied, does not transmit any drive torque. The direct one-way clutch is overrunning because the inner race is stationary with the overdrive drum.

AXOD
GEAR TRAIN POWERFLOW SUMMARY
OD/D RANGE THIRD GEAR

FORWARD CLUTCH (ON) DIRECT CLUTCH (ON) INTERMEDIATE CLUTCH (ON)

LOW ONE-WAY SPRAG CLUTCH (HOLDING) ON DECELERATION
DIRECT ONE-WAY ROLLER CLUTCH (HOLDING) ON ACCELERATION

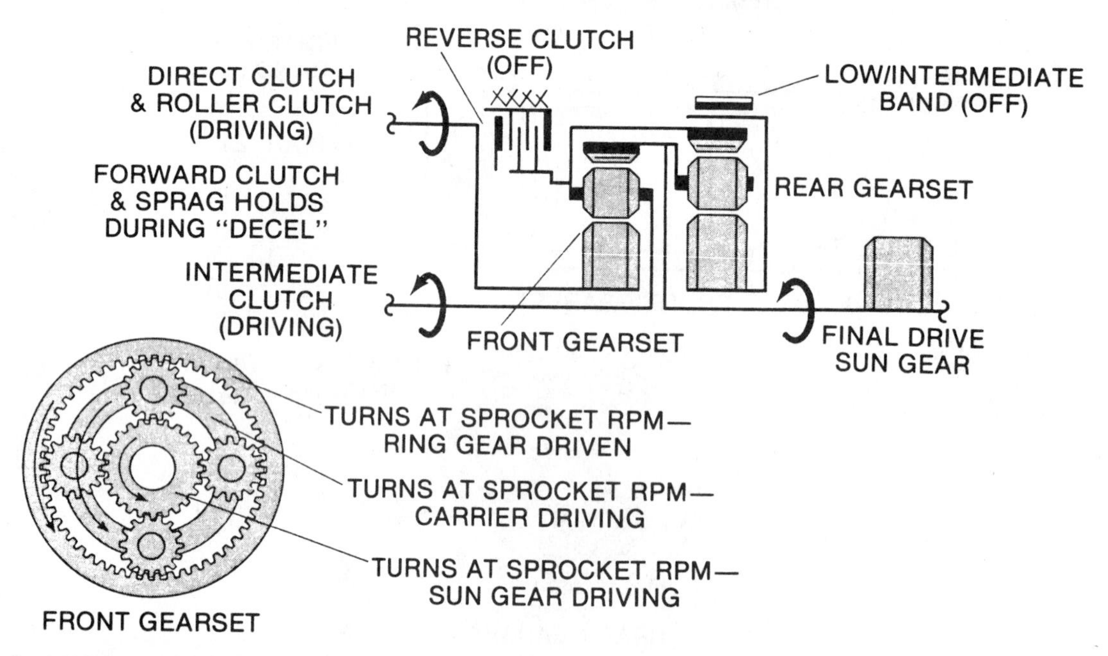

FIGURE 23–59 Ford AXOD/AX4S (AXODE) third-gear operation.

AXOD
GEAR TRAIN POWERFLOW SUMMARY
(OD) RANGE FOURTH GEAR

DIRECT CLUTCH (ON) OVERDRIVE BAND (ON) INTERMEDIATE CLUTCH (ON)

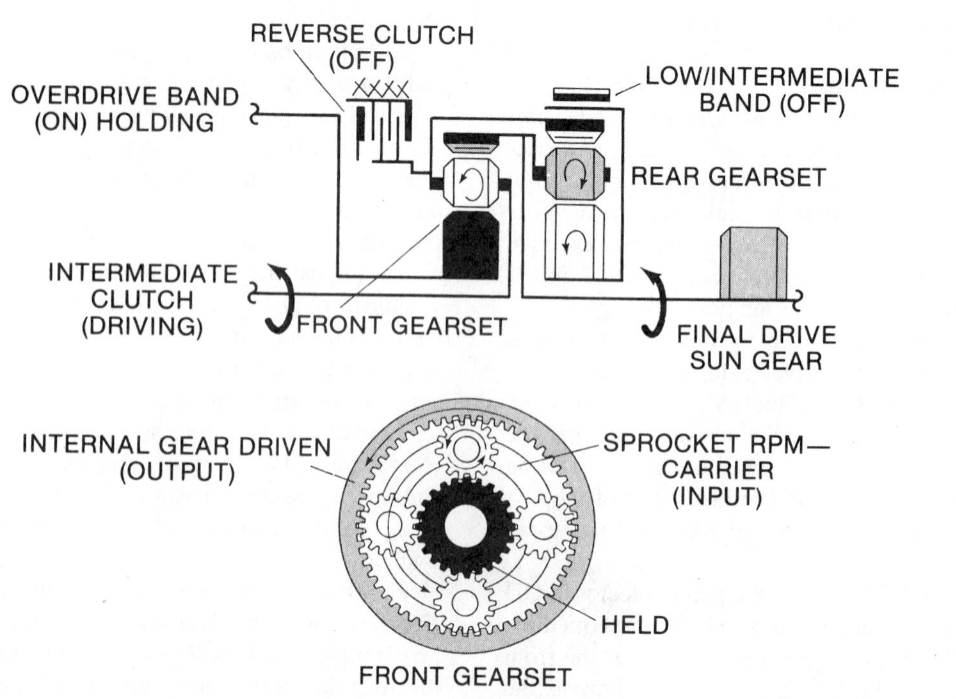

FIGURE 23–60 Ford AXOD/AX4S (AXODE) fourth-gear operation.

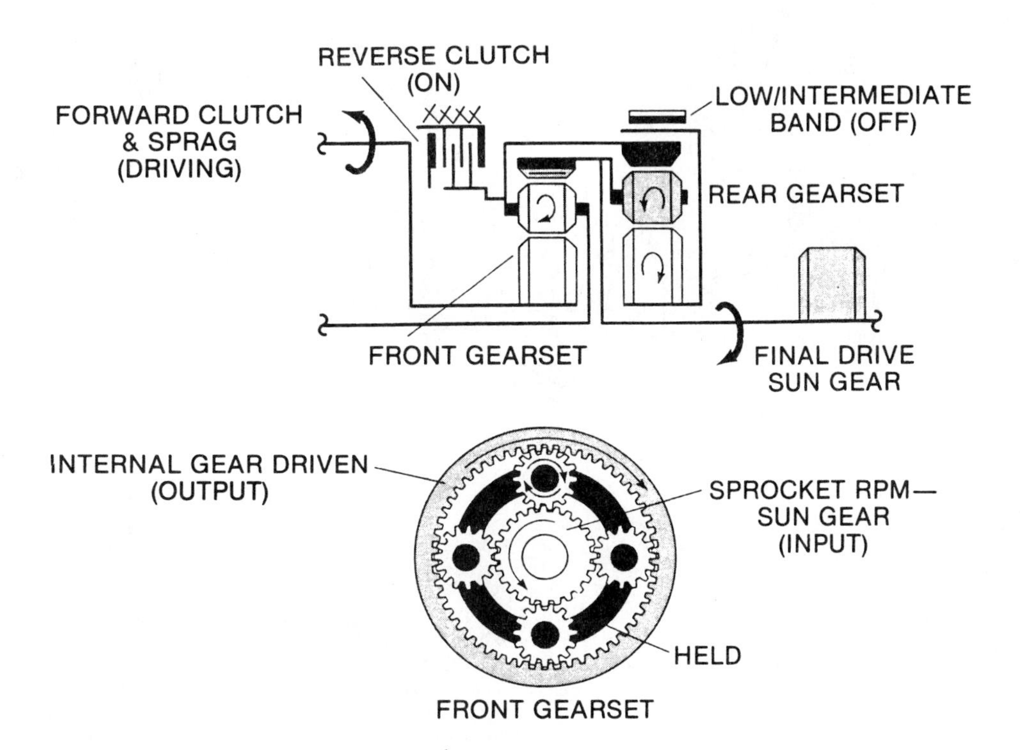

FIGURE 23–61 Ford AXOD/AX4S (AXODE) reverse-gear operation.

Reverse. The forward clutch is applied and drives the front sun gear counterclockwise through the low sprag clutch (Figure 23–61). The reverse clutch is applied and holds the front carrier/rear ring gear unit. With a stationary carrier, the CCW rotation of the front sun gear drives the carrier pinions and front ring gear/rear carrier unit at a reverse output gear ratio of 2.263:1, which is passed on to the final drive sun gear.

Manual Low. The powerflow in manual 1 is identical to the first-gear powerflow in the OD and D ranges. The engagement of the direct clutch is added to provide engine braking during coast (deceleration). Vehicle deceleration causes the front sun gear rotational effort to overrun the low one-way clutch; however, the direct one-way clutch with its opposite action acts as a backup and keeps the front sun gear locked to the powerflow through the direct clutch.

Final Drive/Differential. The AXOD uses a planetary final drive/differential assembly at the output end of the geartrain. Located to the rear of the case, it is positioned and supported on the centerline with the transmission geartrain powerflow (Figure 23–62). To achieve a reduction, the geartrain drives the final drive sun gear, the ring gear is lugged to the case, and the planet carrier (mounted to the differential case) is the output member.

Operating Characteristics

The six-position range selector pattern (P, R, N, OD, D, 1) and operating characteristics are identical for the AX4S/AXODE and AXOD transaxles. Fully automatic operation is provided in either the OD (overdrive) or D (overdrive lockout) positions. Manual select shifting is available by using the OD, D, and 1 positions. Park, reverse, and neutral are typical of other Ford automatic transmissions.

Overdrive. This is the normal range for most driving conditions. The transmission starts in first and will schedule a normal 1-2 and 2-3 shift pattern, with the 3-4 overdrive shift delayed for road cruising. At minimum throttle effort, the 3-4 shift can occur between approximately 32 and 43 mph (53 to 72 km/h). The transmission will not shift into or remain in fourth gear at WOT.

For added performance, WOT forced kickdown shifts (4-3, 4-2 and 3-2, 3-1, 2-1) are available within the proper vehicle speed range. For moderate controlled acceleration, the transmission provides a part-throttle 4-3 downshift. The closed-throttle downshift pattern is 4-3 and 3-1. The 3-2 coast downshift is inhibited. The converter clutch can be engaged in third or fourth gears by the PCM, provided that the engagement criteria have been satisfied.

Overdrive Lockout. In the D range, the transmission is fully automatic through third gear. The same operating characteristics prevail as in OD except for the overdrive feature. The converter clutch can be engaged in third gear by the PCM

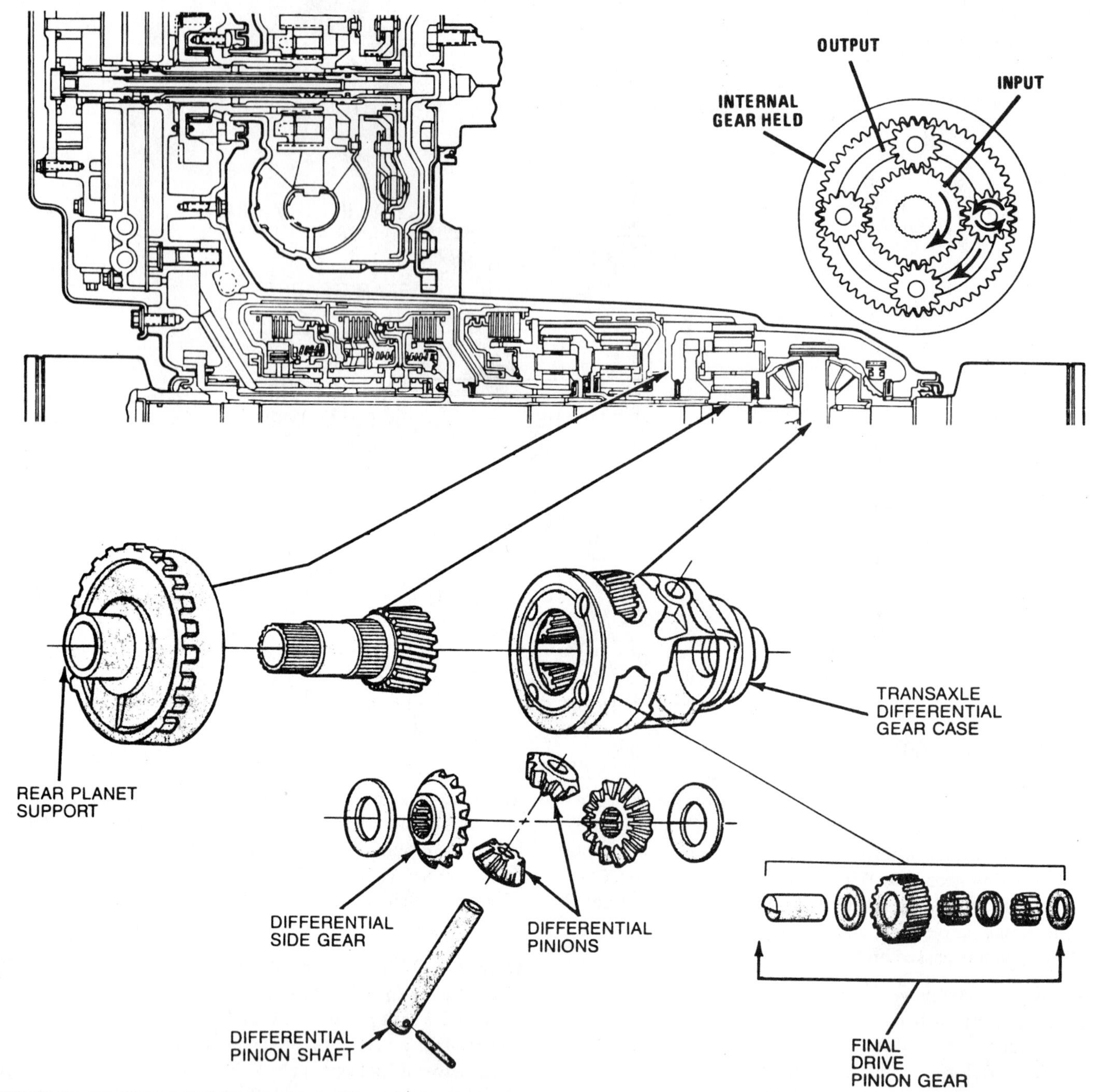

FIGURE 23–62 AXOD/AX4S (AXODE) final drive and differential assembly. (Reprinted with the permission of Ford Motor Company)

provided that the engagement criterion has been satisfied. This range should be used for towing, driving in strong headwinds, or operating in hilly/mountainous terrain.

Manual 1. The objective is to lock in first gear for maximum power and braking. If manual 1 is selected at an excessive vehicle speed, the transmission engages second gear until a safe speed is reached. When the vehicle slows to a safe speed [approximately 28 mph (47 km/h)], the transmission will downshift to first gear. Once first gear is engaged, it stays locked in. Converter clutch engagement is not available in first or second gears.

Figure 23–63 features a summary of the operating ranges and clutch/band combinations to control the geartrain.

AX4N TRANSAXLE

The AX4N is a fully automatic, four-speed, transverse-mounted FWD transaxle. It is electronically controlled and is equipped with a three-element clutching converter. Production began in January 1994, with this transaxle being available in Taurus and Sable models with the 3.0-L, two-valve engine. For 1995, it was installed in the new model Continental. Although

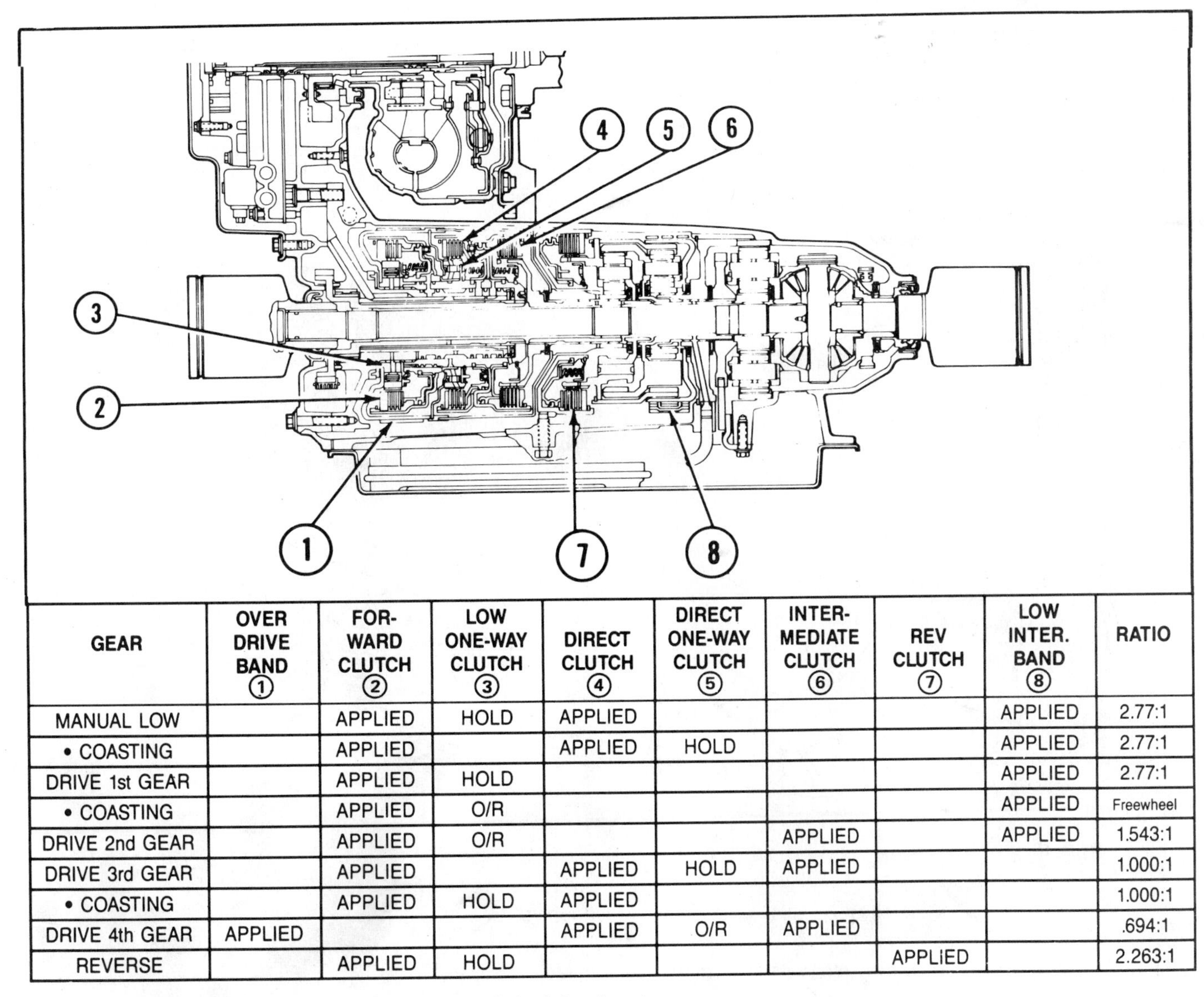

GEAR	OVER DRIVE BAND ①	FOR-WARD CLUTCH ②	LOW ONE-WAY CLUTCH ③	DIRECT CLUTCH ④	DIRECT ONE-WAY CLUTCH ⑤	INTER-MEDIATE CLUTCH ⑥	REV CLUTCH ⑦	LOW INTER. BAND ⑧	RATIO
MANUAL LOW		APPLIED	HOLD	APPLIED				APPLIED	2.77:1
• COASTING		APPLIED		APPLIED	HOLD			APPLIED	2.77:1
DRIVE 1st GEAR		APPLIED	HOLD					APPLIED	2.77:1
• COASTING		APPLIED	O/R					APPLIED	Freewheel
DRIVE 2nd GEAR		APPLIED	O/R			APPLIED		APPLIED	1.543:1
DRIVE 3rd GEAR		APPLIED		APPLIED	HOLD	APPLIED			1.000:1
• COASTING		APPLIED	HOLD	APPLIED					1.000:1
DRIVE 4th GEAR	APPLIED			APPLIED	O/R	APPLIED			.694:1
REVERSE		APPLIED	HOLD				APPLIED		2.263:1

FIGURE 23–63 AXOD/AX4S (AXODE) range reference and clutch/band apply summary.

similar to the AX4S (AXODE) transaxle, the AX4N is designed with higher speed and torque capacities and features nonsynchronized operation for smoother shifts (Figures 23–64 and 23–65).

Geartrain Design and Powerflow

The geartrain configuration and powerflow are identical to those of the AX4S (AXODE) (Figure 23–66). The converter chain drive ratio is .0973:1, with the transaxle gear ratios remaining the same: first gear, 2.77:1; second gear, 1.543:1; third gear, 1:1; fourth gear, 0.694:1; and reverse, 2.263:1. The combined drive chain and final drive ratio is 3.373:1.

The major design change in the geartrain eliminated the low-intermediate band used in the AX4S (AXODE) in favor of a low-intermediate clutch and intermediate one-way roller clutch for first- and second-gear operation. The operation of the low-intermediate roller clutch system eliminates synchronizing the release/apply transition during the 2-3 and 3-2 shifts,

AX4N AUTOMATIC TRANSAXLE

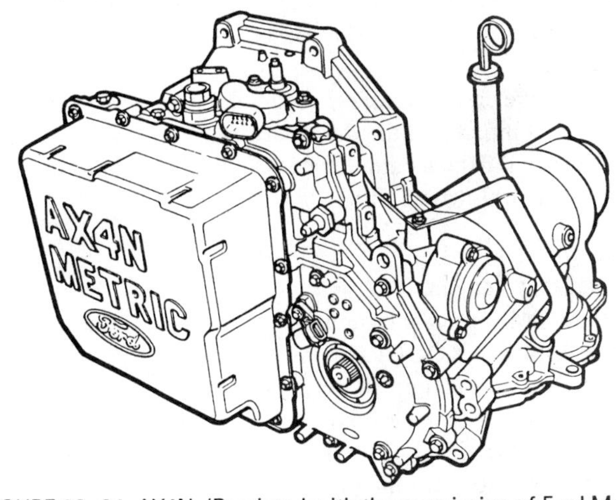

FIGURE 23–64 AX4N. (Reprinted with the permission of Ford Motor Company)

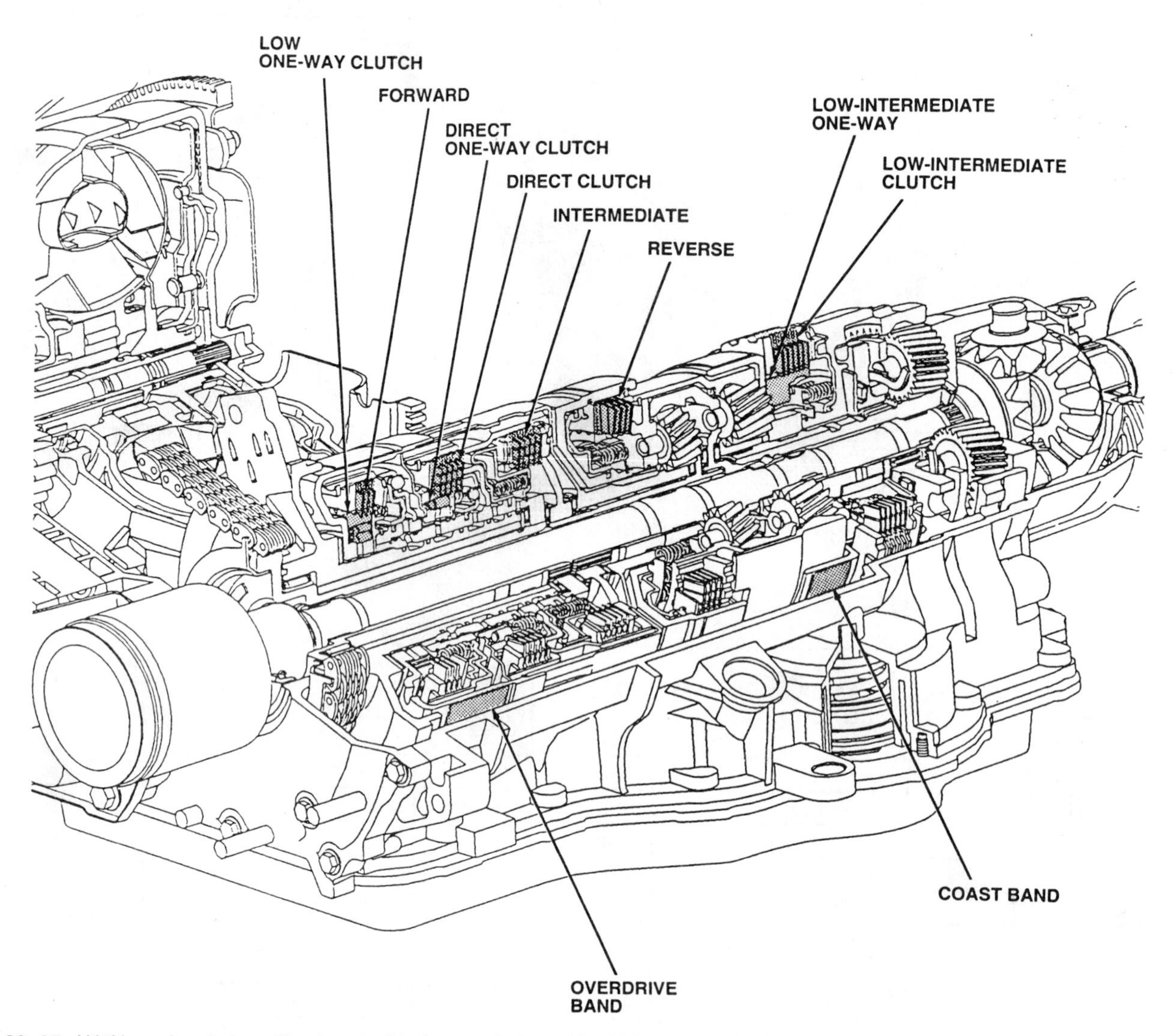

FIGURE 23–65 AX4N sectional view. (Reprinted with the permission of Ford Motor Company)

enhancing shift quality. This classifies the AX4N as a nonsynchronized transaxle, unlike the AX4S and AXOD units.

What formerly was the low-intermediate band is now called the coast band and is used for manual 1 engine braking in first and second gears. The coast band servo bore is located internally.

Within the case cover, four accumulator pistons and springs are used, compared to the three used in the AX4S (AXODE). An additional fifth accumulator is located in the bottom of the case for neutral-drive shifts.

Electronic Control

The PCM inputs and outputs are identical to those found in the AX4S (AXODE) systems. The shift, EPC, and TCC solenoids are mounted in new positions on the valve body (Figure 23–67). The shift solenoid order (1-2-3) is different from the AX4S order (2-1-3). The PCM retains its self-diagnostic ability to warn the driver of detected faults. Stored as DTCs, a technician can retrieve these warnings with the use of a scan tool.

Operating Characteristics

The six-position range selector pattern (P, R, N, OD, D, 1) and operating characteristics are basically identical to those of the AX4S (AXODE).

- *OD:* provides full automatic shifting, 1-2-3-4. Driver throttle demand can delay upshifts or force downshifts for added performance.
- *D:* provides full automatic shifting, 1-2-3; fourth gear is locked out. Coast braking occurs in third gear.
- *1:* provides first-gear operation, with engine braking capabilities and no upshifts to second gear. If the selector lever is dropped into manual 1 at a high vehicle speed, second gear engages with engine braking. Once a safe vehicle speed is attained, the transaxle downshifts and remains in first gear.

A loss of electronic control from the PCM will operate the AX4N in a fail-safe mode with the following features:

- Maximum line pressure in all positions.
- Functional P, R, and N.

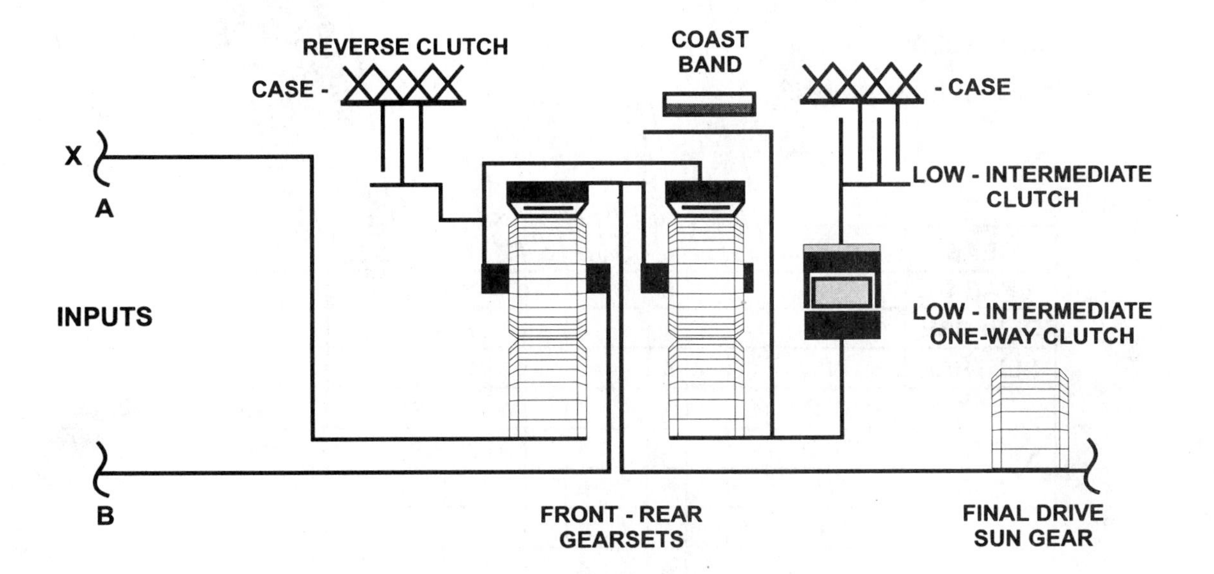

FIGURE 23–66 Ford AX4N powertrain schematic.

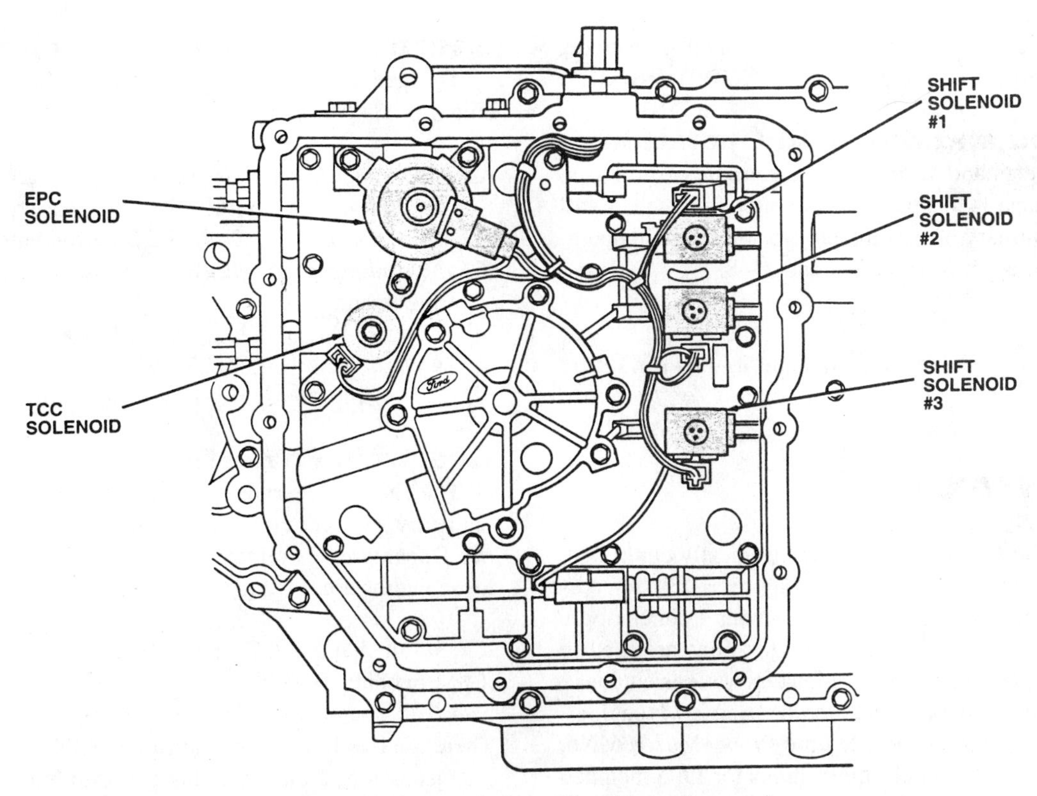

FIGURE 23–67 AX4N solenoid locations. (Reprinted with the permission of Ford Motor Company)

A = APPLIED

X = APPLIED/ INEFFECTIVE

H = HOLDING

OR = OVERRUNNING

GEAR	POSITION	OVERDRIVE BAND	COAST BAND	FORWARD CLUTCH	DIRECT CLUTCH	INTERMEDIATE CLUTCH	REVERSE CLUTCH	LOW-INT CLUTCH	LOW ONE-WAY CLUTCH		DIRECT ONE-WAY CLUTCH		LOW-INT ONE-WAY CLUTCH	
									DRIVE	COAST	DRIVE	COAST	DRIVE	COAST
PARK	P			A					H					
REVERSE	R			A			A		H	OR				
NEUTRAL	N			A					H					
1ST	OD, D			A				A	H	OR			H	OR
2ND	OD, D			X		A		A	OR	OR			H	OR
3RD	OD				A	A		X			H	OR	OR	OR
4TH	OD	A			X	A		X			OR	OR	OR	OR
M-3RD	D			A	A	A		X		H	H		OR	OR
M-2ND	1		A	X		A		A	OR	OR			H	
M-1ST	1		A	A	A			A	H			H	H	

CHART 23–7 AX4N range reference and clutch/band apply summary. (Reprinted with the permission of Ford Motor Company)

- Second-gear operation in OD, D, and 1 positions.
- TCC release in all positions.

The AX4N operates in second gear when in fail-safe mode, rather than in third gear as in the AX4S (AXODE) transaxle.

Clutch/Band and Solenoid Operation

New components and variations in electronic control used in the AX4N cause it to operate differently from the AX4S (AXODE). A summary of the clutch/band and solenoid operation can be reviewed in Charts 23–7 and 23–8.

Pressure Taps

Line pressure and EPC taps are available, as shown in Figure 23–68.

CD4E TRANSAXLE

New for 1994, the CD4E is a fully electronically controlled, four-speed, transverse FWD transaxle (Figure 23–69). It incorporates a three-element torque converter with a clutch apply plate. Different converter housing configurations are used to accommodate starter mounting positions for various engines.

The CD4E is teamed with the Zetec 1.8- and 2.0-L engines. Initial vehicle applications included the Probe, Mazda MX6, and Mazda 626. Added vehicle applications for 1995 included the new model Ford Contour and Mercury Mystique. The Ford CD4E is also used with the all-aluminum, 2.3-L, DOHC V-6 engine.

Geartrain Mechanical Components

The geartrain has the following mechanical components (Figures 23–70 and 23–71).

- *Two integrated simple planetary gearsets:* low/intermediate and reverse/overdrive.
- *One friction band:* 2-4 band (intermediate/overdrive).
- *Five multiple-disc clutches:* reverse, forward, low/reverse, direct, and coast.
- *Two one-way clutches:* forward sprag and low roller.
- *A chain drive* (Figure 23–72).
- *A planetary final drive and differential* (Figure 23–73).

Geartrain Powerflow

The chain drive transfers output torque from the combined planetary gearset to the final drive. It has two widths (3/4 or 1 in). Depending on vehicle application, the sprocket ratios for these widths are:

- 3/4 in chain: 0.982:1 and 0.910:1.
- 1 in chain: 0.982:1.

There are two final drive gear ratios: 3.846:1 and 4.308:1.

Discussion of the following geartrain powerflow summary is based on the operating characteristics of the Ford Con-

TRANSMISSION RANGE SELECTOR LEVER POSITION	POWERTRAIN CONTROL MODULE (PCM) GEAR COMMANDED	AX4N SOLENOIDS			
		ENG BRAKE	SS1	SS2	SS3
P/N (PARK/NEUTRAL)	P/N	NO	OFF[A]	OFF	OFF
R (REVERSE)	R	YES	OFF	OFF	OFF
Ⓓ(OVERDRIVE)	1	NO	OFF	ON	OFF
	2	NO	OFF	OFF	OFF
	3	NO	ON	OFF	ON
	4	YES	ON	ON	ON
D (DRIVE)	1	NO	OFF	ON	OFF
	2	NO	OFF	OFF	OFF
	3	YES	ON	OFF	OFF
L (LOW)	1	YES	OFF	ON	OFF
	2[B]	YES	OFF	OFF	OFF
	3[B]	YES	ON	OFF	OFF

CHART 23–8 AX4N solenoid operation summary. (Reprinted with the permission of Ford Motor Company)

tour/Mercury Mystique six-position range selector (P-R-N-D-2-1). Refer to Figure 23–71 and Chart 23–9 (page 563).

Park and Neutral. In both positions, the turbine shaft input is rotating. With no clutches applied, there is no power transfer to the planetary gearset.

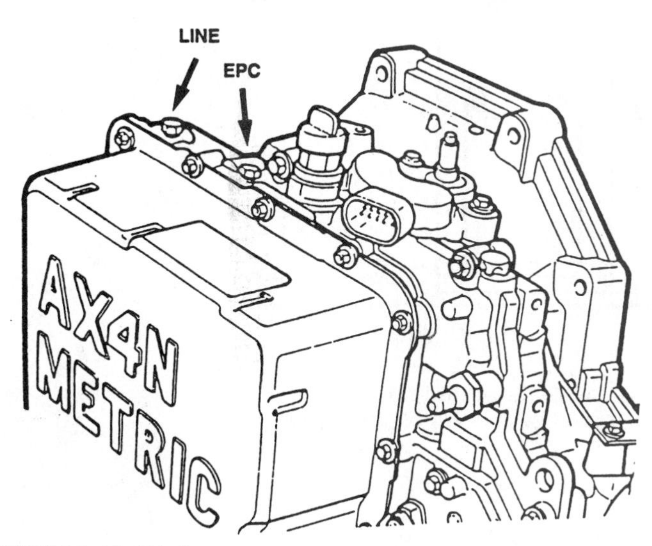

FIGURE 23–68 AX4N pressure taps. (Reprinted with the permission of Ford Motor Company)

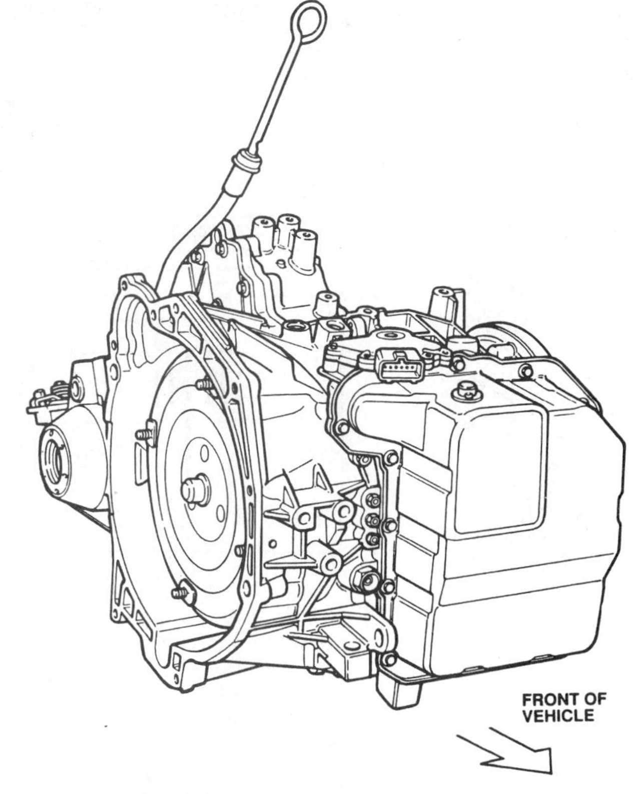

FIGURE 23–69 CD4E transaxle. (Reprinted with the permission of Ford Motor Company)

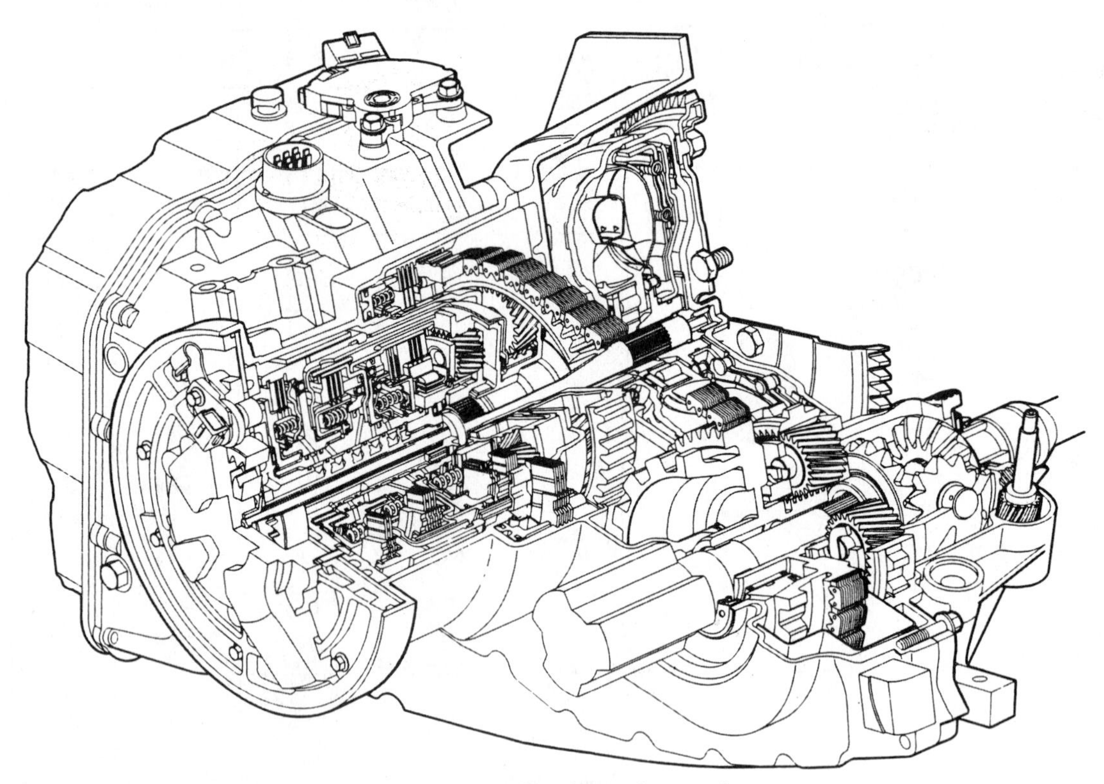

FIGURE 23–70 CD4E sectional view. (Reprinted with the permission of Ford Motor Company)

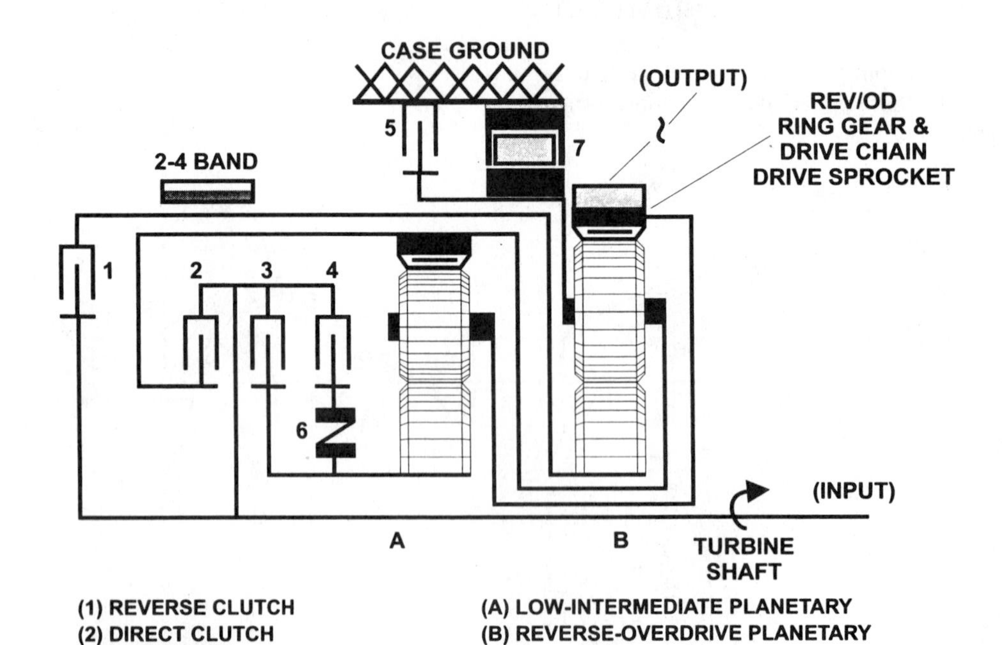

FIGURE 23–71 Ford CD4E powertrain schematic.

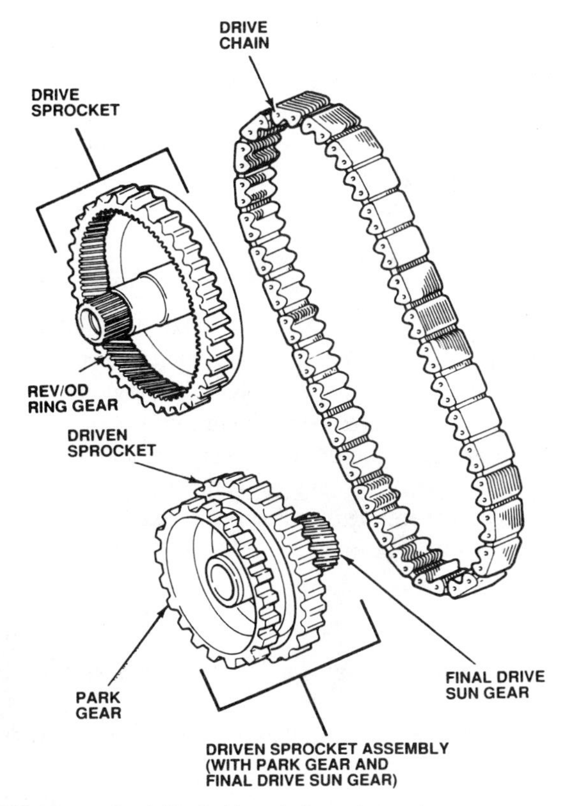

FIGURE 23–72 Ford CD4E drive chain and sprockets.

FIGURE 23–73 CD4E output carrier and differential unit. (Reprinted with the permission of Ford Motor Company)

	INT/OD (2/4) BAND	REVERSE CLUTCH	DIRECT CLUTCH	FORWARD CLUTCH	FORWARD ONE-WAY CLUTCH		COAST CLUTCH	LOW/REV CLUTCH	LOW ONE-WAY CLUTCH	
GEAR					DRIVE	COAST			DRIVE	COAST
REV		X						X		
1ST				X	X	OR			X	OR
2ND	X			X	X	OR			OR	OR
3RD			X	X	X	OR			OR	OR
4TH	X		X	●	OR	OR			OR	OR
MANUAL-3RD			X	X	X		X		OR	OR
MANUAL-2ND	X			X	X		X		OR	OR
MANUAL-1ST				X	X		X	X	X	

X = TRANSMITS TORQUE OR = OVERRUNNING

• = APPLIED BUT DOES NOT TRANSMIT TORQUE

CHART 23–9 CD4E range reference and clutch/band apply chart. (Reprinted with the permission of Ford Motor Company)

First Gear—D Range.

- The forward clutch is applied, and the forward one-way clutch locks the turbine shaft input to the low/intermediate sun gear.
- Interaction between the low/intermediate and reverse/overdrive planetaries allows the low one-way clutch to hold the reverse/overdrive carrier and low/intermediate ring gear.
- First gear ratio is 2.889:1.
- TCS (transmission control switch) ON or OFF: engine braking not available because the coast clutches are not applied, and the one-way clutches overrun.

First Gear—1 Range.

- The low/reverse clutch is applied and allows the case to hold the reverse carrier, bypassing the low one-way clutch during deceleration.
- The coast clutch is applied, bypassing the forward one-way clutch during deceleration. This allows the low/intermediate sun gear to drive the turbine shaft.
- In manual 1, the one-way clutches transmit drive torque only.

Second Gear—D Range.

- The forward clutch remains applied, and the forward one-way clutch locks the turbine shaft input to the low/intermediate sun gear.
- The intermediate/overdrive band applies and holds the reverse/overdrive sun gear stationary.
- Interaction between the low/intermediate and reverse/overdrive planetaries provides second gear. The low one-way clutch overruns.
- Second-gear ratio is 1.571:1.
- TCS OFF: engine braking is not available, and the forward one-way clutch overruns.
- TCS ON: the coast clutch is applied, bypassing the forward one-way clutch during deceleration. The low/intermediate sun gear is allowed to drive the turbine shaft. The one-way forward clutch transmits drive torque only.

Second Gear—2 Range.

- The coast clutch is applied for engine braking.

Third Gear—D Range.

- The forward clutch remains applied, and the forward one-way clutch locks the turbine shaft input to the low/intermediate sun gear.
- The direct clutch is applied and locks the low/intermediate ring gear to the turbine shaft.
- With two planetary member inputs turning at the same speed and in the same direction, the planetaries lock together for a direct-drive rotation of 1:1.
- TCS OFF: engine braking is not available. The forward one-way clutch overruns, and the low/intermediate carrier turns faster than the low/intermediate sun gear.
- TCS ON: the coast clutch is applied to maintain a connection between the low/intermediate sun gear and turbine shaft during deceleration.

Fourth Gear—D Range.

- The forward clutch remains applied but does not transmit drive torque. The forward one-way clutch overruns.
- The direct clutch is applied, driving the low/intermediate ring gear and the reverse/overdrive carrier.
- The intermediate/overdrive (2-4) band is applied, holding the reverse/overdrive sun gear stationary.
- Fourth-gear overdrive is a function of the reverse/overdrive planetary—the gear ratio is 0.698:1.

Reverse.

- The reverse clutch applies and drives the reverse/overdrive sun gear.
- The low/reverse clutch applies and holds the reverse/overdrive carrier and low/intermediate ring gear.
- Reverse is a function of the reverse/overdrive planetary—the gear ratio is 2.310:1.

Electronic Control

The PCM manages the engine and CD4E operation. The PCM responds to inputs and appropriately signals five output solenoids to control shift feel (line pressure control), shift scheduling, modulated TCC apply, 3-2 shift timing, and coast clutch application (Figure 23–74). Solenoid strategy is shown in Chart 23–10 (A and B), page 566.

The solenoids are grouped in a solenoid assembly mounted on the valve body (Figures 23–75 and 23–76). The solenoid valve body is serviced as an assembly and must not be disassembled for service.

The PCM retains its self-diagnostic ability to warn the driver of some detected faults, record DTCs, and display information through diagnostic service test equipment.

Operating Characteristics

Two operating range quadrant patterns are used and offer differences in shift patterns (Figures 23–77 and 23–78). In the "D, 2, 1" arrangement, the TCS cancels overdrive only. In the "D, S, L" version, the TCS is replaced with a manual hold switch (MHS). The switch in the OFF position offers automatic shift scheduling strategy in D, S, and L. When the MHS button is triggered to the ON position, overdrive is canceled. The driver, however, has manual control over the shift scheduling. Chart 23–11 (A and B), page 568 summarizes the comparative operating characteristics of the TCS and MHS controlled CD4E transaxles.

A loss of electronic control from the PCM causes the CD4E to operate in a fail-safe mode. The following functions occur in fail-safe mode:

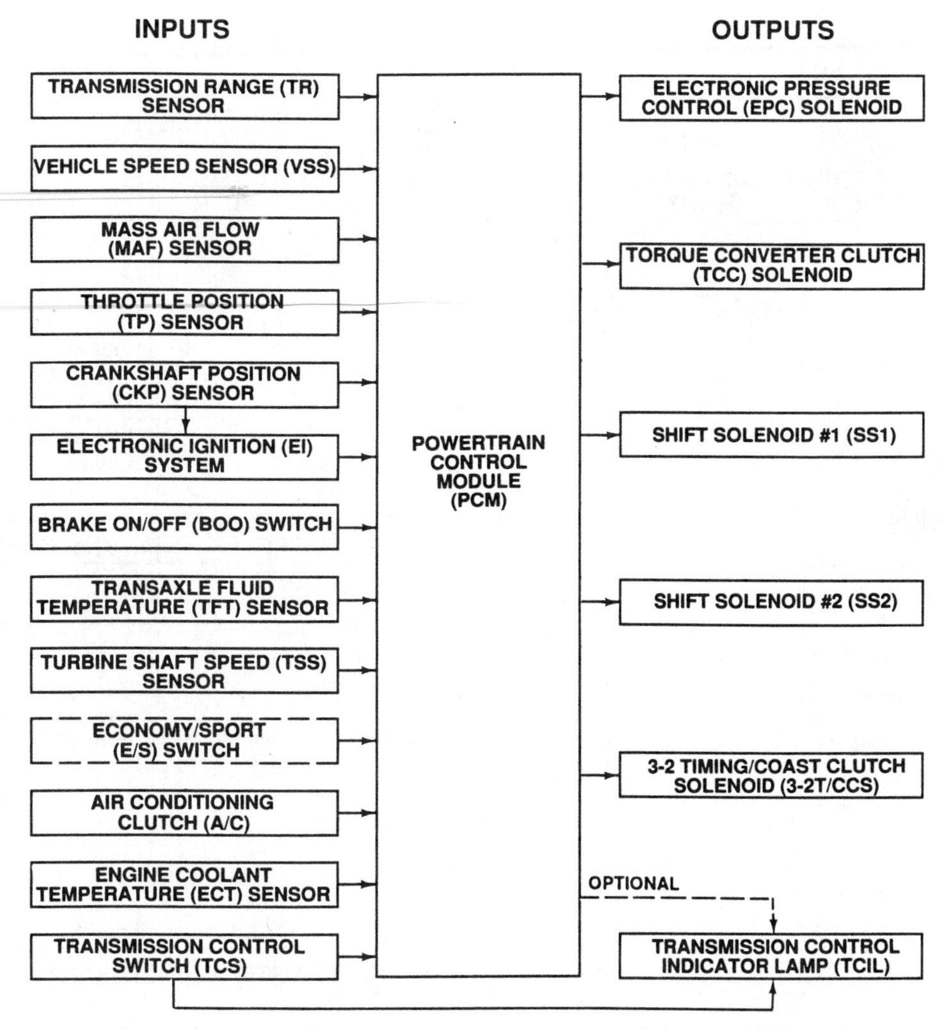

FIGURE 23–74 CD4E input and output devices. (Reprinted with the permission of Ford Motor Company)

- Maximum line pressure in all positions.
- Functional P, R, and N.
- Third-gear operation in D or 2 ranges with coast braking regardless of TCS or MHS status.
- Second-gear operation in manual 1 with coast braking regardless of TCS or MHS status.
- TCC released.

Pressure Taps

The available line pressure tap is shown in Figure 23–79.

SUMMARY

For 1996 North American models, Ford produced three different FWD automatic transaxles and five different RWD automatic transmissions. In addition, Ford used four different FWD units built by Mazda and Nissan. With the exception of the Mazda-designed ATX transaxle (Ford Aspire) and the limited production RWD C-6 transmission (special order E/F series light trucks), all of Ford's automatic transmissions and transaxles are electronically controlled four-speed units. These four-speed transmissions and transaxles come equipped with a conventional torque converter clutch lockup plate.

The 5R55E five-speed rear-wheel drive automatic transmission was introduced in 1997 Ford Aerostar and Ranger vehicles. Electronically and hydraulically controlled clutches and bands operate the planetary gearset members in a pattern similar to that of the 4R44E and 4R55E transmissions.

Ford produces a CVT (continuously variable transmission) belt drive transaxle called the CTX in Bordeaux, France. It is used in Ford's European car line. It is anticipated that this type of transmission will soon find several applications in the North American car line.

Many electronically controlled four-speed automatic transmissions and transaxles used in the mid-1990s will continue to be built for late model Ford passenger cars and light trucks.

A

SOLENOID APPLICATION CHART — VEHICLES WITH MANUAL HOLD SWITCH (MHS)

GEAR SELECTOR POSITION	PCM COMMANDED GEAR	CD4E SOLENOIDS			
		SS1	SS2	3-2 TIMING/ COAST CLUTCH SOLENOID①	TCC
P	—	OFF	ON	ON	OFF#
R	—	OFF	OFF	ON	OFF#
N	—	OFF	ON	ON	OFF#
D (MHS OFF)	1	ON	ON	ON	OFF
	2	OFF	ON	ON	ON or OFF*
	3	OFF	OFF	ON	ON or OFF*
	4	ON	OFF	ON	ON or OFF*
D (MHS ON)	2	OFF	ON	OFF	ON or OFF*
	3	OFF	OFF	OFF	ON or OFF*
S (MHS OFF)	1	ON	ON	ON	OFF
	2	OFF	ON	OFF	ON or OFF*
	3	OFF	OFF	OFF	ON or OFF*
S (MHS ON)	2	OFF	ON	OFF	#
	3**	OFF	OFF	OFF	ON or OFF*
L (MHS OFF)	1	ON	OFF	OFF	#
	2	OFF	OFF	OFF	#
L (MHS ON)	1	ON	OFF	OFF	#

** When a manual pull-in occurs above a calibrated speed the transmission will not downshift from the higher gear until the vehicle speed drops below a calibrated speed.
* Powertrain Control Module (PCM) assembly commanded.
# Not allowed by Hydraulics.
①a. Coast clutch solenoid ON results in coast clutch released (OFF).
b. 3-2 Timing is provided in separate application chart.
☐ Desired steady-state gear.
☐ Fail-safe gear.

B

SOLENOID APPLICATION CHART — VEHICLES WITH TRANSMISSION CONTROL SWITCH (TCS)

GEAR SELECTOR POSITION	PCM COMMANDED GEAR	CD4E SOLENOIDS			
		SS1	SS2	3-2 TIMING/ COAST CLUTCH SOLENOID①	TCC
P	—	OFF	ON	ON	OFF #
R	—	OFF	OFF	ON	OFF #
N	—	OFF	ON	ON	OFF #
D (TCS OFF)	1	ON	ON	ON	OFF
	2	OFF	ON	ON	ON or OFF*
	3	OFF	OFF	ON	ON or OFF*
	4	ON	OFF	ON	ON or OFF*
D (TCS ON)	1	ON	ON	ON	OFF
	2	OFF	ON	OFF	ON or OFF*
	3	OFF	OFF	OFF	ON or OFF*
2	2	OFF	ON	OFF	ON or OFF*
2	3**	OFF	OFF	OFF	ON or OFF*
1	1	ON	OFF	OFF	#
1	2**	OFF	OFF	OFF	#
1	3**	OFF	ON	OFF	#

** When a manual pull-in occurs above a calibrated speed the transmission will not downshift from the higher gear until the vehicle speed drops below a calibrated speed.
* Powertrain Control Module (PCM) assembly commanded.
# Torque converter clutch apply not allowed by Hydraulics.
①a. Coast clutch solenoid ON results in coast clutch released (OFF).
b. 3-2 Timing is provided in separate application chart.
☐ Desired steady-state gear.
☐ Fail-safe gear.

CHART 23–10 CD4E (A) TCS and (B) MHS solenoid applications charts. (Reprinted with the permission of Ford Motor Company).

FIGURE 23–75 Ford CD4E solenoid assembly mounted to the valve body.

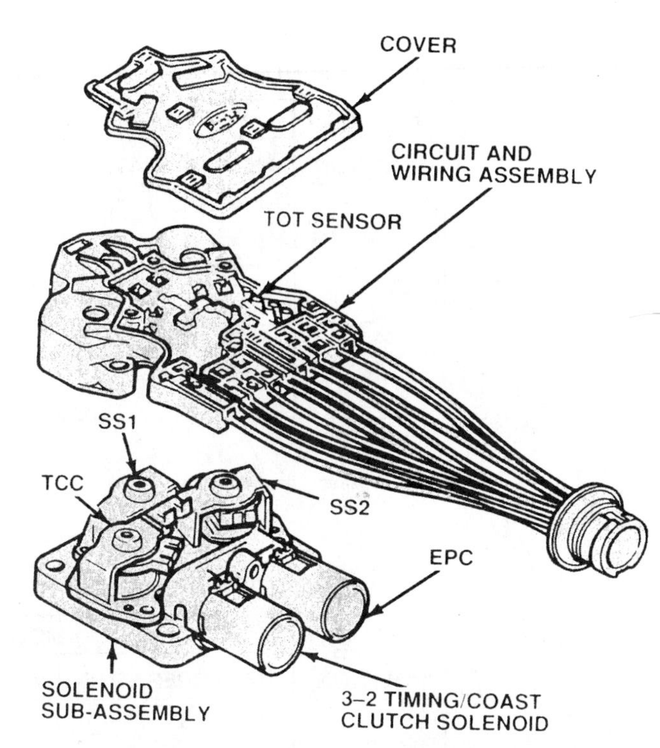

FIGURE 23–76 Separated solenoid body assembly exposing components. (NOTE: Do not disassemble for service.) (Reprinted with the permission of Ford Motor Company)

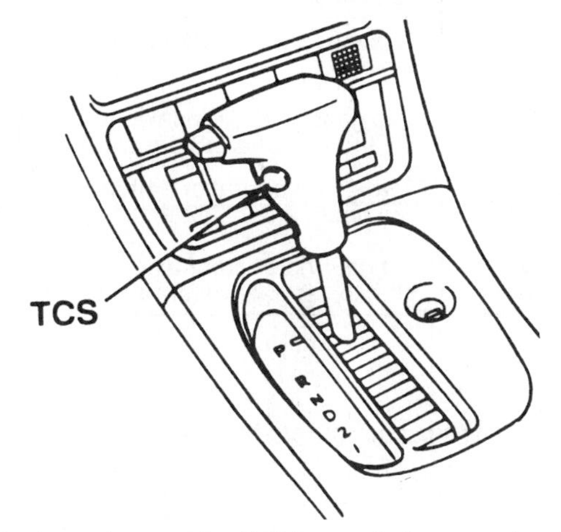

FIGURE 23–77 CD4E selector lever with a TCS (transmission control switch). (Reprinted with the permission of Ford Motor Company)

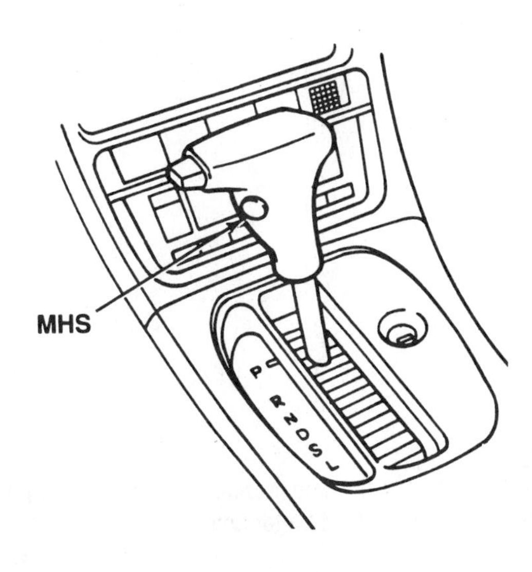

FIGURE 23–78 CD4E selector lever with an MHS (manual hold switch). (Reprinted with the permission of Ford Motor Company)

A

VEHICLE WITH TRANSMISSION CONTROL SWITCH (TCS) – O/D Button

MANUAL LEVER POSITION	MODE SELECT CAPABILITIES		
	AUTO OPERATION	TCS	COMMENT
D	[1⇄2⇄3⇄4]	OFF	Four gears operation.
	[1⇄2⇄3]←4	ON	Three gears operation, 4th gear is inhibited.
2	1→[2]←3←4	OFF or ON	2nd gear hold. Transaxle will shift to 2nd gear automatically.
1	[1]←2←3←4	OFF or ON	1st gear hold. Transaxle will shift to 1st gear automatically.

→ Upshifting
← Downshifting
☐ Desired steady-state gear

B

VEHICLE WITH MANUAL HOLD SWITCH (MHS) – HOLD Button

MANUAL LEVER POSITION	MODE SELECT CAPABILITIES		
	AUTO OPERATION	MHS	COMMENT
D	[1⇄2⇄3⇄4]	OFF	Four gears operation.
	1→[2⇄3]←4	ON	Two gears operation (2nd and 3rd).
S	[1⇄2⇄3]←4	OFF	Three gears operation, 4th is inhibited.
	1→[2]←3←4	ON	2nd gear hold. Transaxle will shift to 2nd gear automatically.
L	[1⇄2]←3←4	OFF	Two gears operation, 4th and 3rd gears are inhibited.
	[1]←2←3←4	ON	1st gear hold. Transaxle will shift to 1st gear automatically.

→ Upshifting
← Downshifting
☐ Desired steady-state gear

CHART 23–11 (A) TCS and (B) MHS operation summaries. (Reprinted with the permission of Ford Motor Company)

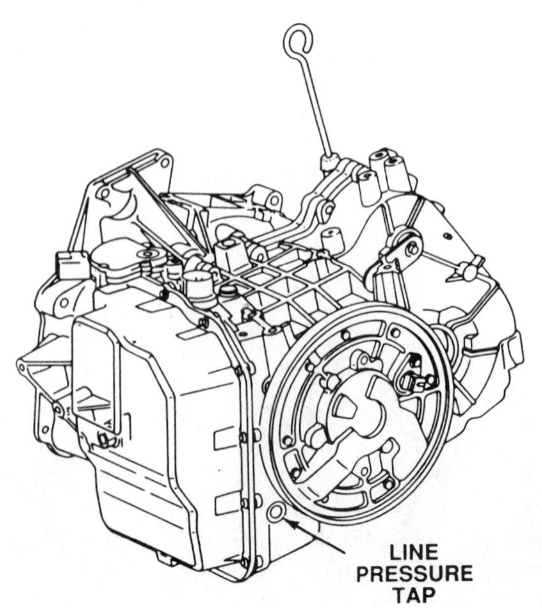

FIGURE 23–79 CD4E line pressure tap location. (Reprinted with the permission of Ford Motor Company)

24 General Motors Automatic Transmission/Transaxle Summaries

INTRODUCTION

In keeping pace with the evolving technology and product designs, General Motors has integrated electronic controls with their new automatic transmissions/transaxles. Electronification now controls shift scheduling, shift quality, line pressure, and modulated converter clutch application.

Popular GM automatic transmissions/transaxles include the 4L60, 4L60-E, 3L80, 4L80-E, 4L80-EHD, 3T40, 4T40-E, 4T60, 4T60-E, 4T65-E, and 4T80-E. The current product designation system is shown in Figure 24–1.

HYDRA-MATIC 4L60-E/4L60 (THM 700-R4)

The 4L60-E was introduced into the Hydra-Matic product line for the 1993 model year (Figure 24–2). It replaced the 4L60 (THM 700-R4), first introduced in 1982. The expanded use of electronics is the major difference between the two transmissions. From a combination of engine and transmission sensor input information, the PCM is used for controlling the 4L60-E.

The 4L60-E and 4L60 (THM 700-R4) are four-speed overdrive, RWD transmissions, equipped with three-element clutching torque converters. Operating pressure is supplied by a vane-style pump. The geartrains have similar planetary gearset designs and clutch/band control (Figures 24–3 and 24–4). Two exceptions are that in the 4L60-E transmission, the low/reverse clutch is applied in the park position, and the selection of the manual 2 position inhibits first-gear operation.

Current 4L60-E powertrain applications range from 2.2-L to 5.7-L gas engines.

Geartrain Mechanical and Clutch/Band Components

Refer to Figures 24–3 and 24–4.

- *Two integrated simple planetary gearsets:* input gearset and reaction gearset.
- *One friction band:* 2-4 band.
- *Five multiple-disc clutches:* reverse input, overrun, forward, 3-4, and low/reverse.
- *Two one-way clutches:* Forward sprag and low roller.

Geartrain Powerflow

The integrated planetary system configuration provides for the traditional three forward speeds, plus an overdrive. In the 4L60-E/4L60 (THM 700-R4) geartrain, the sun gears are independent of one another, and the input internal gear and reaction carrier are coupled together. Planetary principles of operation remain the same and will apply when explaining the powerflow.

The input housing contains the forward, 3-4, and overrun clutches. The drive hub of the reverse input clutch is also a part of the input housing. Whenever the reverse input clutch is applied, the input housing transmits input drive torque through the clutch.

Park/Neutral. The drive clutches are not applied, and there is no input to the planetary unit.

HYDRA-MATIC PRODUCT DESIGNATION SYSTEM

HYDRA-MATIC	3	T	40	— E
	Number of Speeds:	Type:	Series:	Major Features:
	3	T - Transverse	Based on	E - Electronic Controls
	4	L - Longitudinal Plus	Relative	A - All Wheel Drive
	5	M - Manual	Torque	HD - Heavy Duty
	V (CVT)		Capacity	

PREVIOUS DESIGNATIONS	NEW DESIGNATIONS
CURRENT PRODUCTS	
THM 180C	HYDRA-MATIC 3L30 3L30-E
THM R1	HYDRA-MATIC 4L30-E
THM A-1	HYDRA-MATIC 3T40-A
THM 125/125C	HYDRA-MATIC 3T40
THM 700-R4	HYDRA-MATIC 4L60
THM 440-T4	HYDRA-MATIC 4T60 4T60-E
THM 400	HYDRA-MATIC 3L80 4L80-E
THM 475	HYDRA-MATIC 3L80-HD

FIGURE 24–1 Hydra-Matic product designation system. (Courtesy of General Motors Corporation, Service Technology Group)

HYDRA-MATIC 4L60-E TRANSMISSION
RPO M30

Produced at: Toledo, Ohio

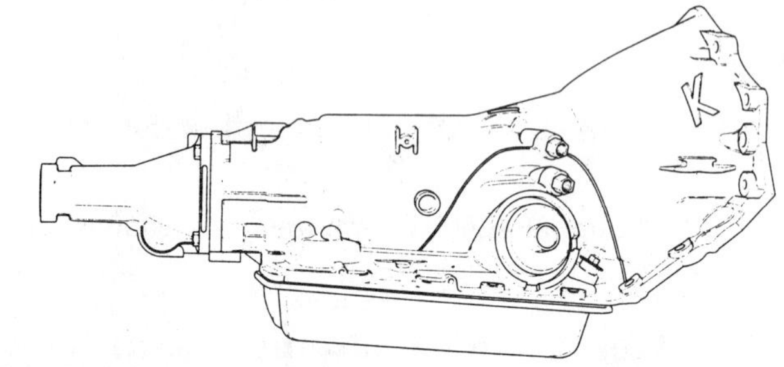

Vehicles used in:

- C/K, S/T-TRUCKS
- G, L, M-VANS
- GM-HOLDEN'S (V-CAR)
- CHEVY CAPRICE, CAMARO, CORVETTE
- CADILLAC FLEETWOOD
- PONTIAC FIREBIRD
- BUICK ROADMASTER

ALSO USED BY: • GM VENEZOLANA

HYDRA-MATIC 4L60-E
(4-SPEED)

FIGURE 24–2 4L60-E applications (1995). (Courtesy of General Motors Corporation, Service Technology Group)

OD Range, First Gear.

- Forward clutch is applied and drives the input sun gear through the forward sprag clutch.
- Forward sprag holding.
- Low roller clutch is effective and holds the input internal gear.

First gear is a function of the input (front) planetary and provides a gear ratio of 3.06:1 (Figure 24–5).

HYDRA-MATIC 4L60-E

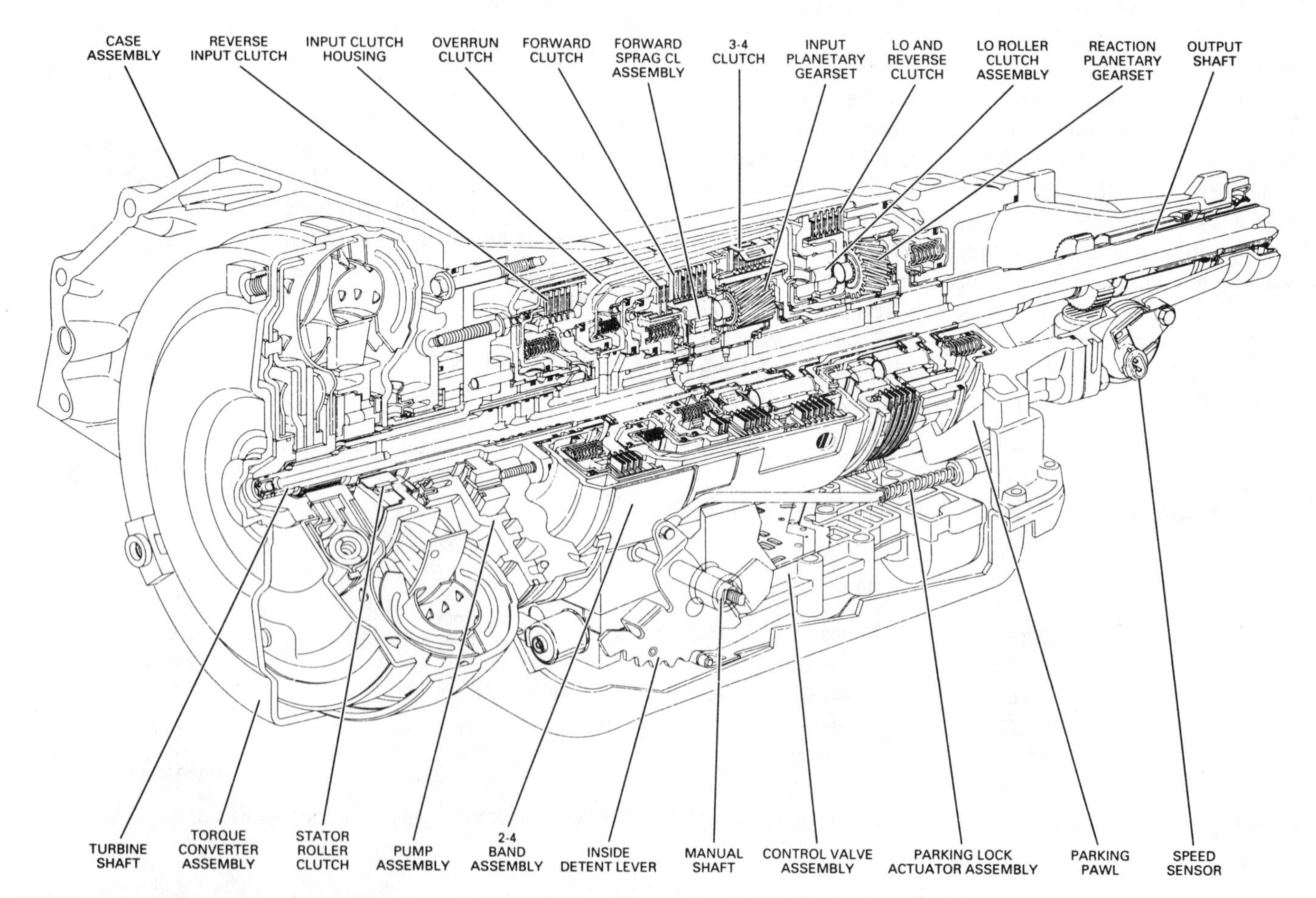

FIGURE 24–3 4L60-E sectional view. (Courtesy of General Motors Corporation, Service Technology Group)

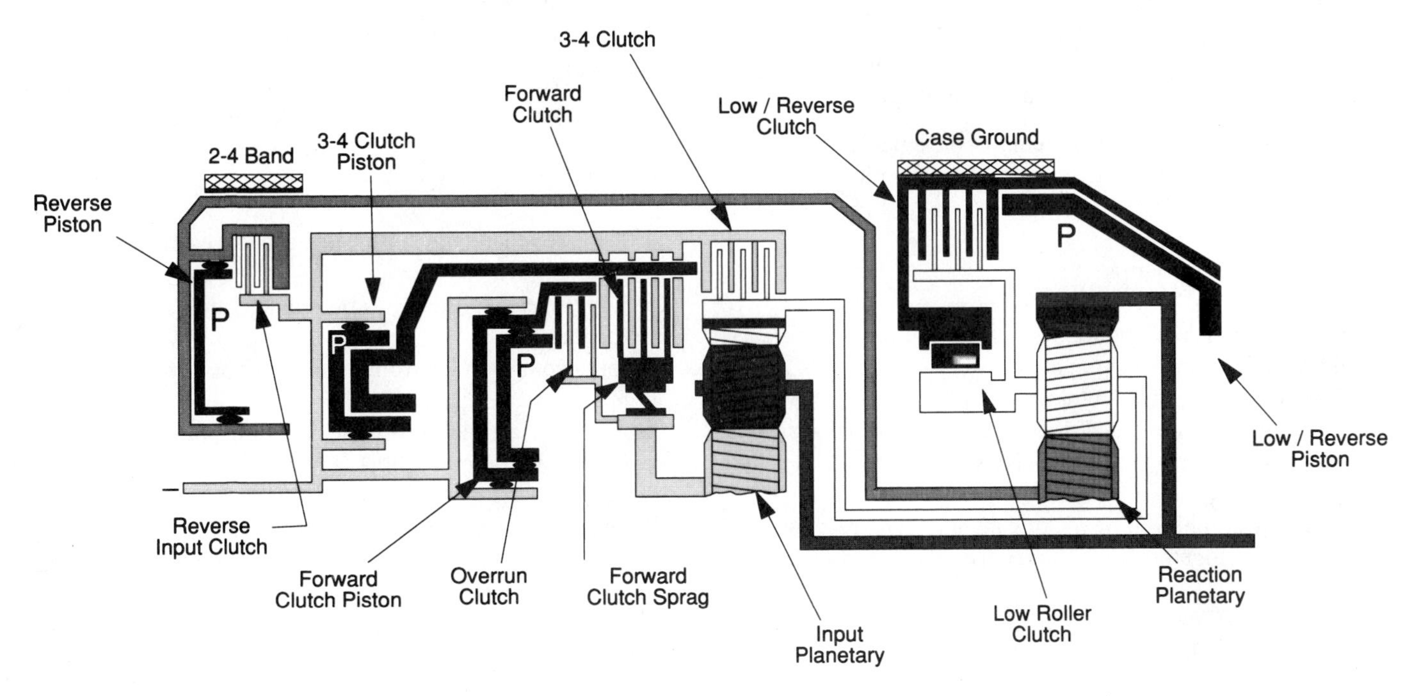

FIGURE 24–4 GM 4L60/4L60-E powertrain schematic.

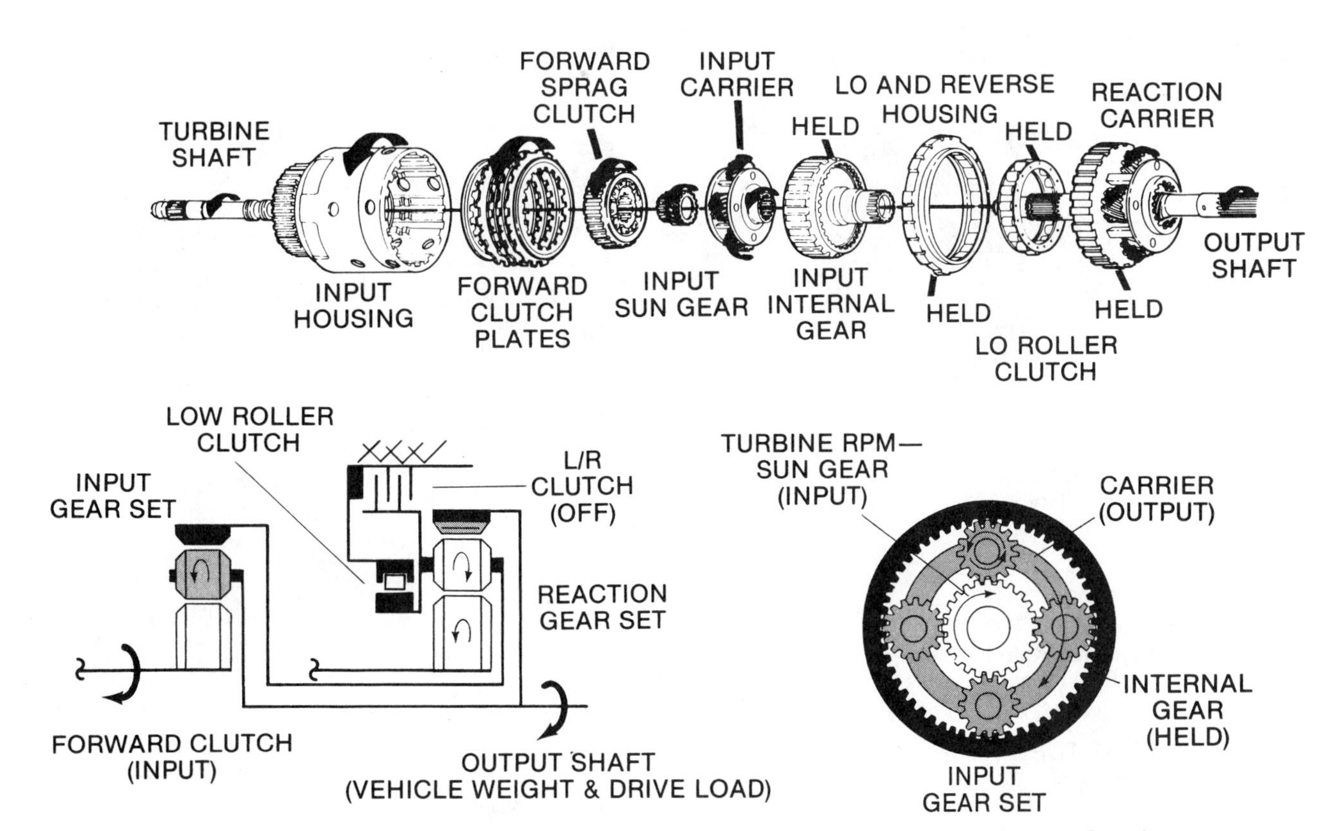

FIGURE 24–5 First-gear powerflow—4L60/4L60-E. (Courtesy of General Motors Corporation, Service Technology Group)

OD Range, Second Gear.

- Forward clutch is applied and drives the input sun gear through the forward sprag clutch.
- Forward sprag is effective.
- 2-4 band is applied and holds the reaction sun gear.
- Low roller clutch freewheels and is ineffective.

Second gear is a combined function of the input and reaction planetary units, providing a compound gear ratio of 1.63:1 (Figure 24–6).

OD Range, Third Gear.

- Forward clutch is applied and drives the input sun gear through the forward sprag clutch.
- Forward sprag clutch holding.
- 3-4 clutch is applied and drives the input internal gear.
- Low roller clutch freewheels and is ineffective.

Third gear is achieved by locking the input and reactionary planetary units together for a direct-drive (1:1) ratio (Figure 24–7).

OD Range, Fourth-gear Overdrive.

- Forward clutch is applied but is not transmitting drive torque.
- Forward sprag clutch freewheels and is ineffective.
- 2-4 band is applied and holds the reaction sun gear.
- 3-4 clutch is applied and drives the front internal gear and reaction carrier.
- Low roller clutch freewheels and is ineffective.

Overdrive is a function of the reaction planetary unit and provides a gear ratio of 0.70:1 (Figure 24–8).

Reverse.

- Reverse input clutch is applied and drives the reaction sun gear.
- Low/reverse clutch is applied and holds the reaction carrier.
- Input sprag and low roller clutch not holding.

Reverse is a function of the reaction planetary unit and provides a gear ratio of 2.30:1 (Figure 24–9).

Manual Ranges 3, 2, and 1. The powerflows in these manual ranges are identical to the powerflows in the OD range. The engagement of the overrun clutch is added to provide engine braking during coast/deceleration. Vehicle deceleration causes the input sun gear to turn faster than the forward clutch drive input, causing the forward sprag to freewheel. The overrun clutch keeps the input sun gear and inner sprag race locked to the input housing and powerflow.

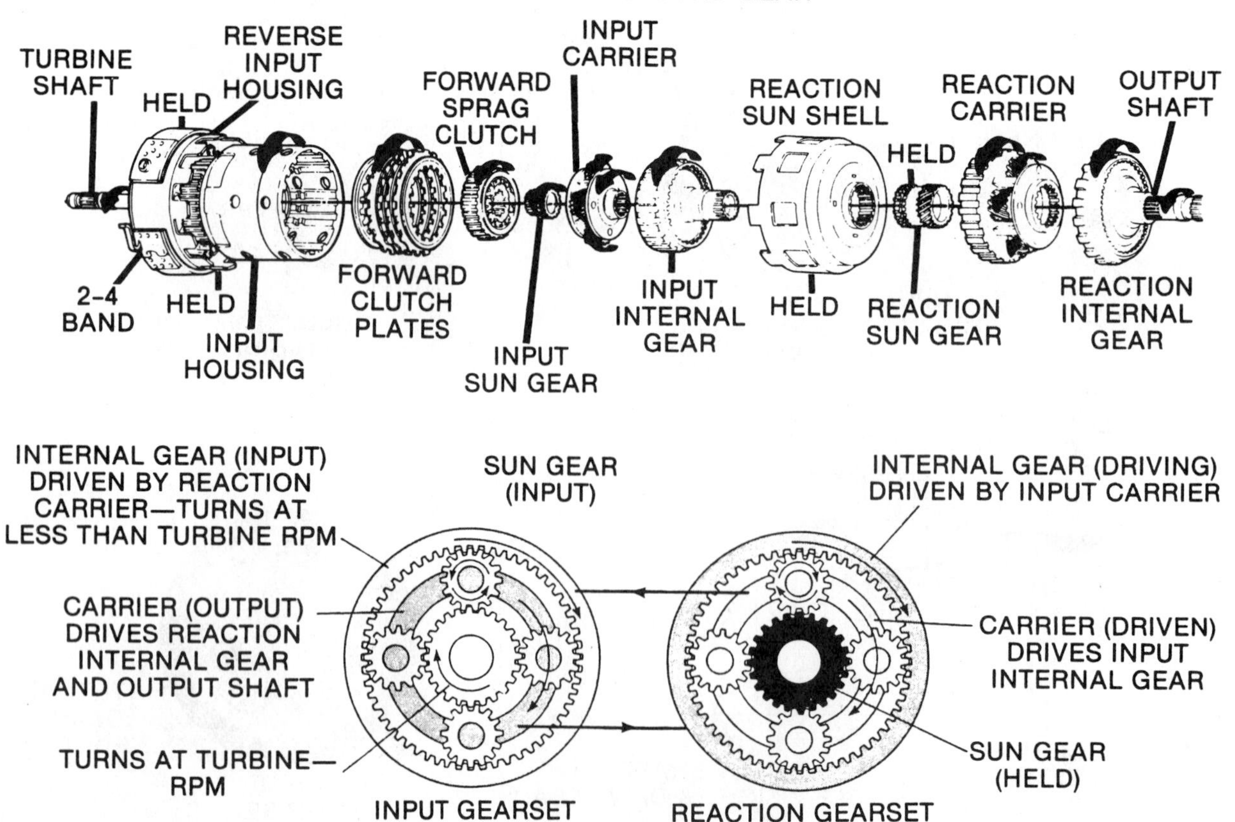

FIGURE 24–6 Second-gear powerflow—4L60/4L60-E. (Courtesy of General Motors Corporation, Service Technology Group)

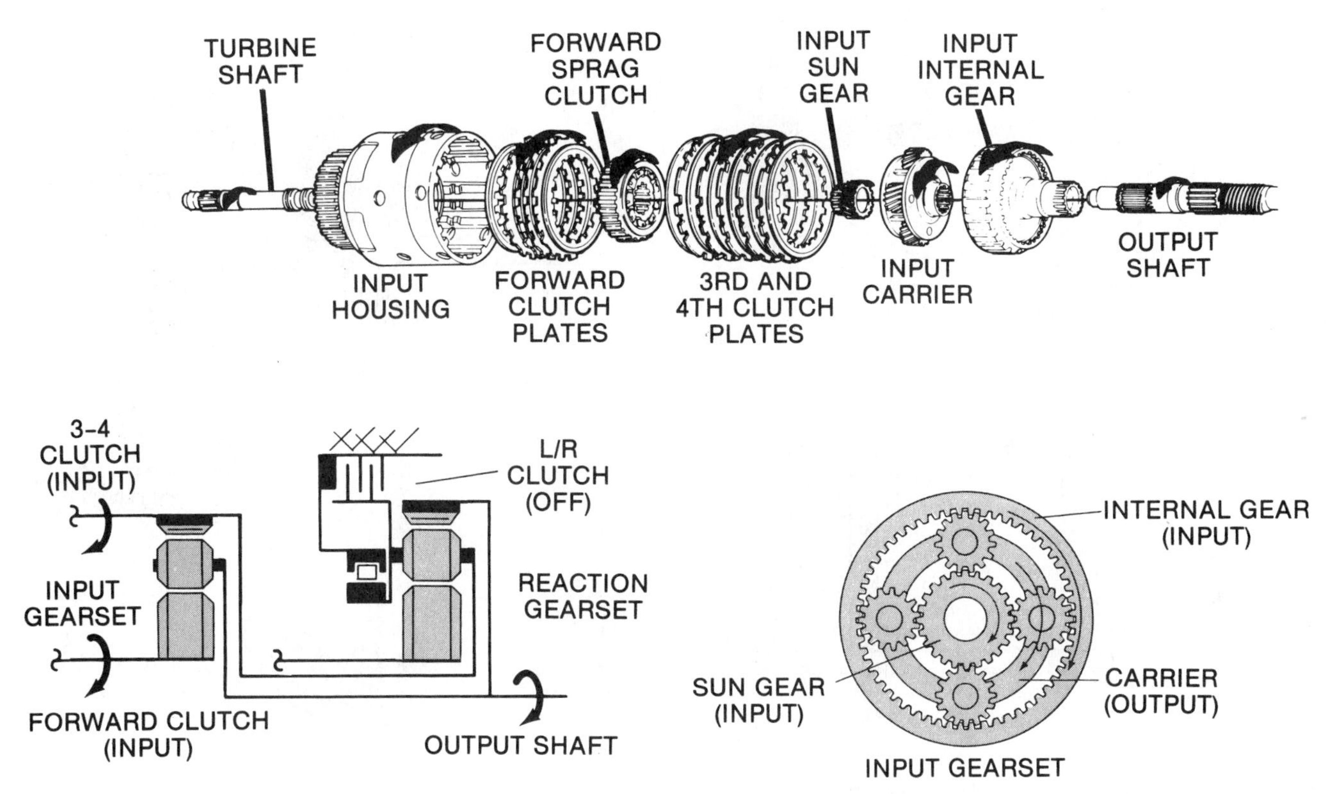

FIGURE 24–7 Third-gear powerflow—4L60/4L60-E. (Courtesy of General Motors Corporation, Service Technology Group)

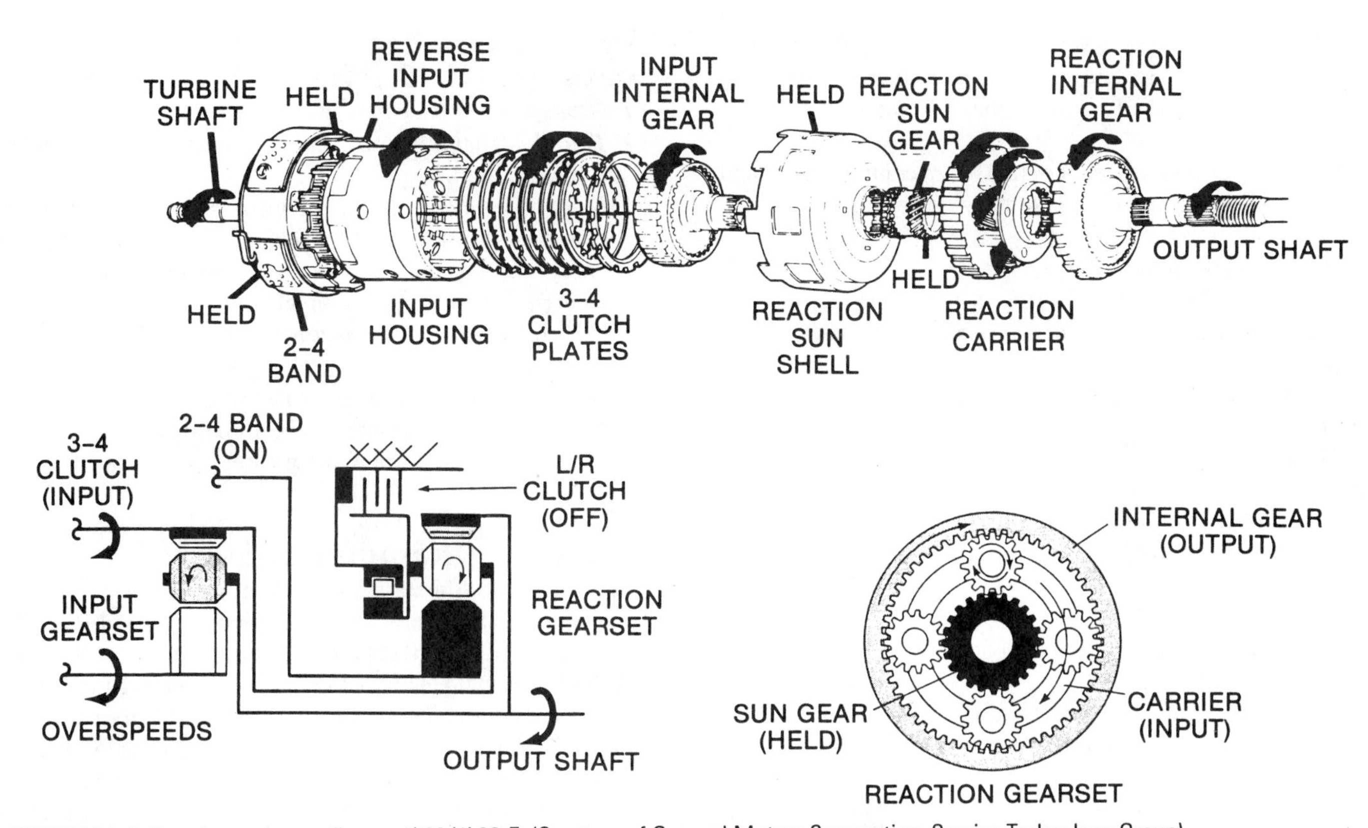

FIGURE 24–8 Fourth-gear powerflow—4L60/4L60-E. (Courtesy of General Motors Corporation, Service Technology Group)

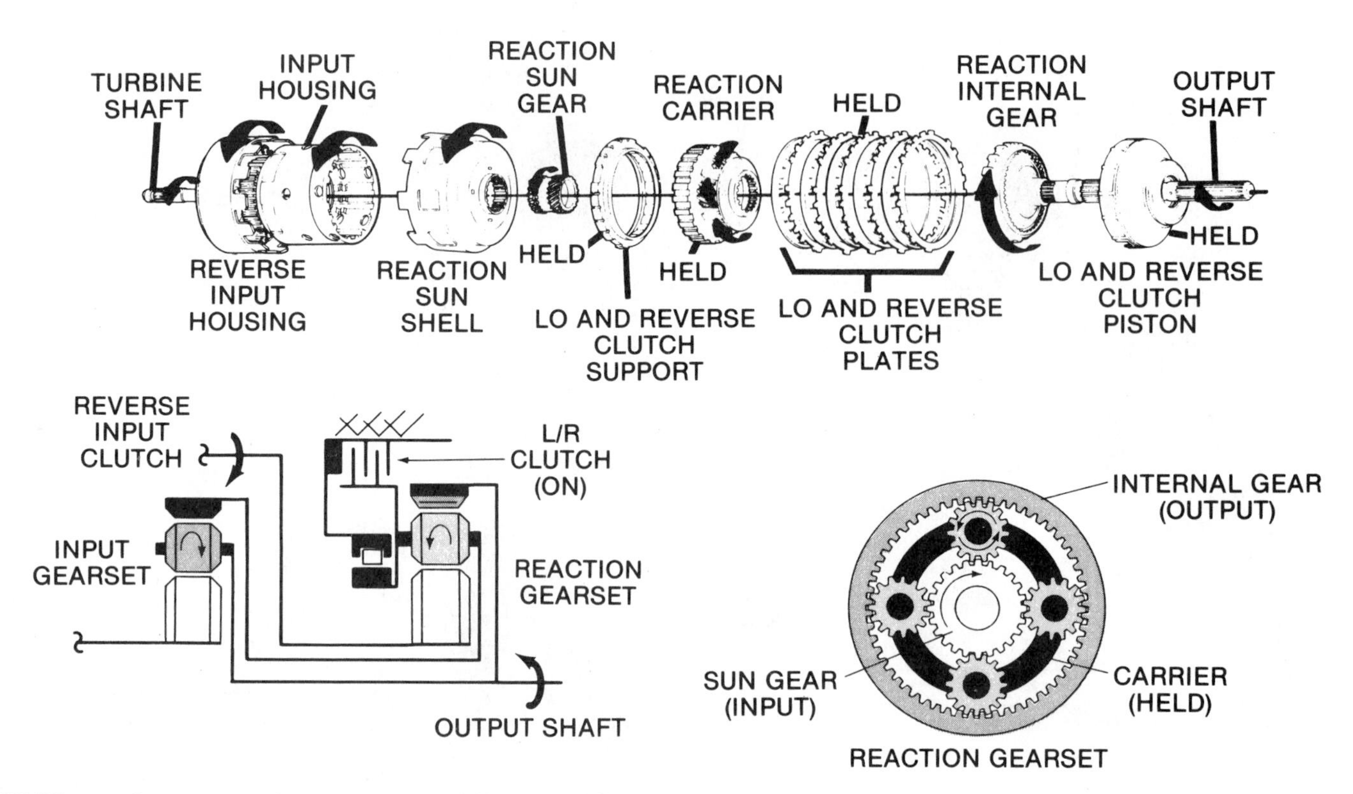

FIGURE 24–9 Reverse powerflow—4L60/4L60-E. (Courtesy of General Motors Corporation, Service Technology Group)

In manual 1, the low/reverse clutch is also applied to keep the reaction carrier locked to the case. Otherwise, the free-wheeling action of the low roller clutch would neutralize the powerflow.

Chart 24–1 summarizes the 4L60-E operating ranges and clutch/band combinations to control the geartrain.

Electronic Control

The PCM provides the control of the engine and 4L60-E operation. It responds to input information and manages five output solenoids for shift scheduling, shift quality (line pressure), TCC, and 3-2 shift timing. The solenoids are shown in Figures 24–10 and 24–11. The PCM has a built-in adaptive learning feature to maintain shift quality over the life of the transmission.

A loss of electronic control from the PCM will cause the transmission to operate in a fail-safe mode with the following features:

- Maximum line pressure in all positions.
- Functional P, R, and N.
- Third-gear operation in OD and D/3 ranges.
- Second-gear operation in manual 2 and 1 ranges.

Refer to Chapter 11 for a detailed discussion of 4L60-E electronic controls.

Transmission External Controls and Features—4L60 (THM 700-R4)

- *Manual range selection:* linkage or cable operated.
- *Throttle system control:* internal throttle lever assembly cable operated.
- *Detent/forced downshift system control:* cable operated in common with the throttle system. TV and detent valve group are in tandem in the same valve body bore.
- *Governor:* case mounted on left side.
- *TCC solenoid case connector:* located on the left side of the case.
- *Line pressure tap:* located on the left front side of the case.
- *Other pressure taps:* second oil, third oil, and fourth oil all located on the right side of the case.

Operating Characteristics

Two slightly different shift quadrant patterns are used, but their operating characteristics are identical (Figure 24–12). The following describes each of the forward ranges.

OD (Overdrive). This is the normal range for most driving conditions and maximum economy. The transmission starts in first and schedules a normal 1-2 and 2-3 shift pattern, with the 3-4 overdrive shift delayed for road cruising. The transmission will not shift into or remain in fourth gear at WOT. The transmission automatically downshifts at closed throttle

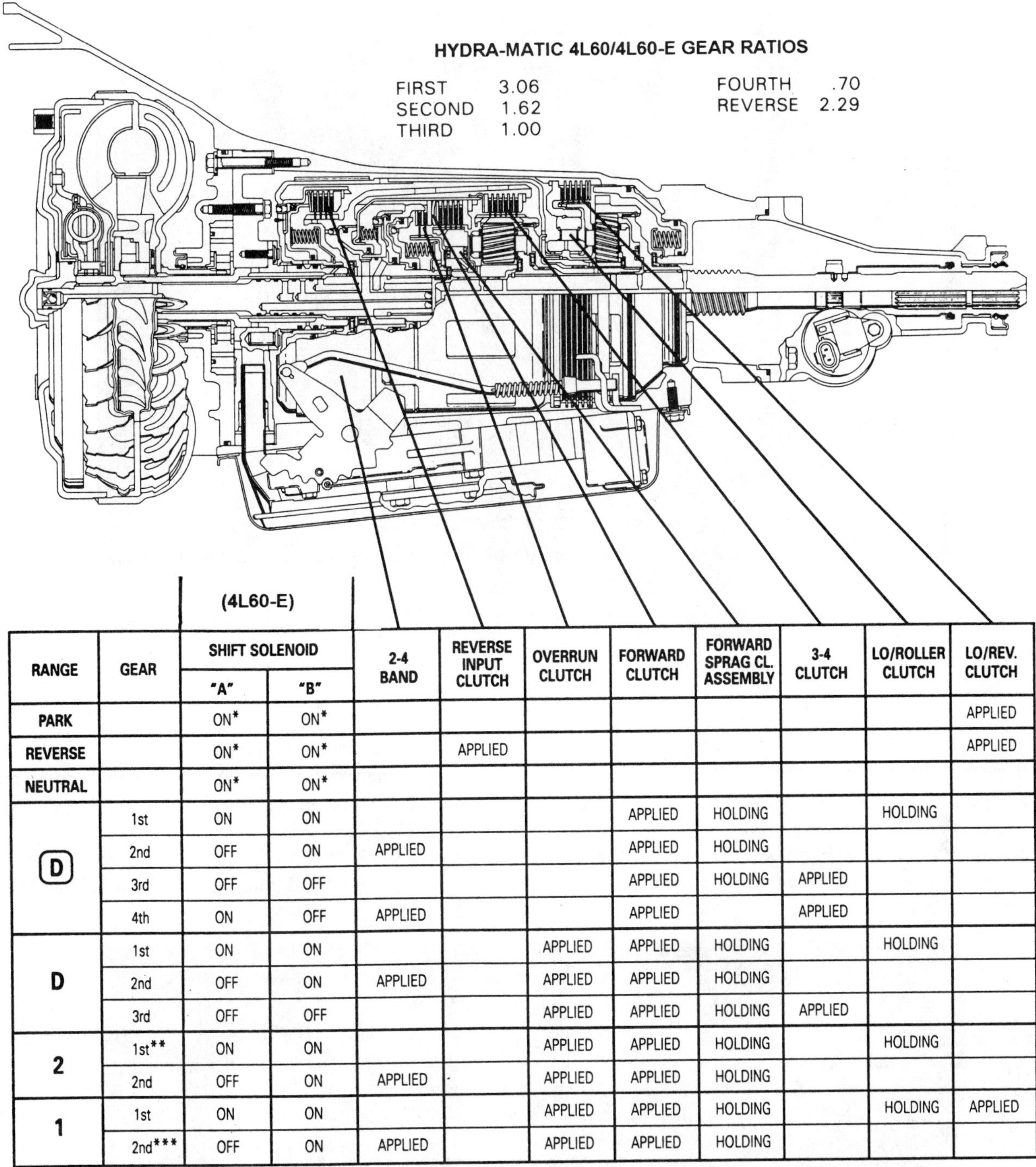

RANGE	GEAR	SHIFT SOLENOID "A"	SHIFT SOLENOID "B"	2-4 BAND	REVERSE INPUT CLUTCH	OVERRUN CLUTCH	FORWARD CLUTCH	FORWARD SPRAG CL. ASSEMBLY	3-4 CLUTCH	LO/ROLLER CLUTCH	LO/REV. CLUTCH
PARK		ON*	ON*								APPLIED
REVERSE		ON*	ON*		APPLIED						APPLIED
NEUTRAL		ON*	ON*								
(D)	1st	ON	ON				APPLIED	HOLDING		HOLDING	
	2nd	OFF	ON	APPLIED			APPLIED	HOLDING			
	3rd	OFF	OFF				APPLIED	HOLDING	APPLIED		
	4th	ON	OFF	APPLIED			APPLIED		APPLIED		
D	1st	ON	ON			APPLIED	APPLIED	HOLDING		HOLDING	
	2nd	OFF	ON	APPLIED		APPLIED	APPLIED	HOLDING			
	3rd	OFF	OFF			APPLIED	APPLIED	HOLDING	APPLIED		
2	1st**	ON	ON			APPLIED	APPLIED	HOLDING		HOLDING	
	2nd	OFF	ON	APPLIED		APPLIED	APPLIED	HOLDING			
1	1st	ON	ON			APPLIED	APPLIED	HOLDING		HOLDING	APPLIED
	2nd***	OFF	ON	APPLIED		APPLIED	APPLIED	HOLDING			

* SHIFT SOLENOID STATE IS A FUNCTION OF VEHICLE SPEED AND MAY CHANGE IF VEHICLE SPEED INCREASES SUFFICIENTLY IN : REVERSE OR NEUTRAL. HOWEVER, THIS DOES NOT AFFECT TRANSMISSION OPERATION.

** MANUAL SECOND – FIRST GEAR IS ELECTRONICALLY PREVENTED UNDER NORMAL OPERATING CONDITIONS.

*** MANUAL FIRST – SECOND GEAR IS ONLY AVAILABLE ABOVE APPROXIMATELY 48 TO 56 KM/H (30 TO 35 MPH).

PH0002-4L60-E

CHART 24–1 4L60-E range reference and clutch/band application chart. (Courtesy of General Motors Corporation, Service Technology Group)

(1) Shift Solenoid A/1-2
(2) Shift Solenoid B/2-3
(3) Pressure Switch Assembly
(4) Pressure Control Solenoid
(5) 1-2 Accumulator
(6) Forward Accumulator

FIGURE 24–10 GM 4L60-E electronic component locations.

(coast), 4-3, 3-2, and 2-1. When coasting in third, second, or low gear, there is no engine braking.

For added performance, forced detent downshift combinations (3-2, 3-1, and 2-1) are available within the proper vehicle speed and engine torque range. For moderate controlled acceleration, the transmission provides a part-throttle 4-3 and 3-2 downshift. The converter clutch engages in third or fourth gear if the engagement criteria have been satisfied. In some models, the converter clutch can also engage in second gear.

Manual D or 3. In the D/3 range, the transmission is fully automatic through third gear—fourth gear is inhibited. Shift

(1) 3-2 Control Solenoid
(2) TCC Solenoid

FIGURE 24–11 GM 4L60-E electronic component locations.

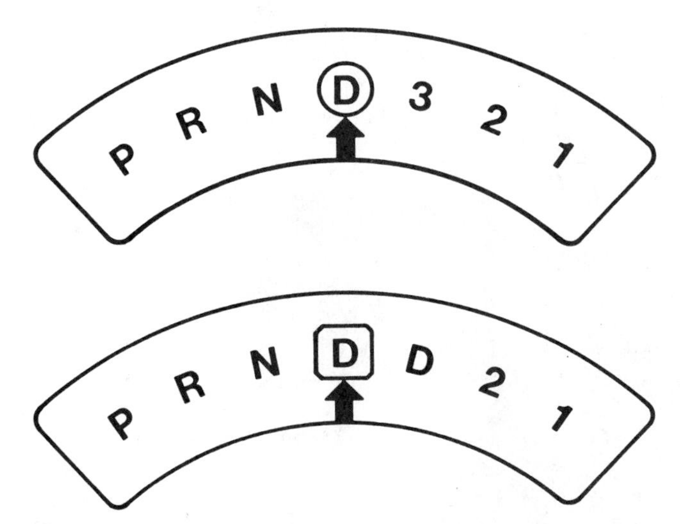

FIGURE 24–12 Shift quadrant patterns. (Courtesy of General Motors Corporation, Service Technology Group)

performance, including the converter clutch, is the same as in OD range for first, second, and third gears. Third and second gears provide for coast engine braking. This range should be used for driving in city traffic or strong headwinds, towing, or operating in hilly/mountainous terrain.

Manual 2. The transmission locks out third and fourth gears for added performance. The 4L60 (700-R4) has a normal first-gear start and a 1-2 shift pattern. The 4L60-E provides a second-gear start and is useful for slippery road conditions. On some 4L60-E applications, a downshift to first gear occurs under heavy-throttle, low-speed conditions. If selected at a high speed (third or fourth gear), the transmission immediately shifts into second gear.

Manual 1. The transmission is locked into first gear for extra pulling power or engine braking. Manual 1 can be selected at any vehicle speed; however, if the vehicle speed is excessive, second gear engages until a safe road speed is attained. The downshift into first gear occurs at approximately 30 mph (48 km/h). Once first gear is locked in, the transmission typically does not upshift. In some vehicles, a 1-2 shift is allowed to avoid damage to the engine and transmission.

HYDRA-MATIC 3L80/3L80-HD (THM 400/475)

The 3L80/3L80-HD (THM 400/475) transmissions are three-speed, high torque capacity units with a long history of use in RWD car/truck and special applications. Built by Hydra-Matic Division, the 3L80 (THM 400) was first produced in 1964 when it was introduced in the Buick and Cadillac. For 1965, it was added to the Chevrolet, Oldsmobile, Pontiac, and truck product lines. A variable-pitch (VP) converter stator was featured in the 1965–1967 Buick, Cadillac, and Oldsmobile car models. The maximum stall ratio of the VP converter was 2.20:1 in the high angle and 1.80:1 in the low-angle stator positions.

In the following years, adaptations of the basic design were engineered for various vehicle and engine applications known as the 375, 425, and 475. The 375, 400, and 475 models essentially are alike in appearance, with some internal changes to adjust the transmission torque handling ability. The 425 model was a longitudinal FWD design used in the 1968 Cadillac Eldorado and Oldsmobile Toronado through production year 1978.

Although eliminated from the domestic passenger car scene by the late 1970s, the 3L80/3L80-HD series remained in production through 1990 for domestic trucks, vans, motor homes, school buses, and some foreign cars (Ferrari, Jaguar, and Rolls Royce). For 1991, it was redesigned as a four-speed transmission with full electronic shift control and converter clutching and was referred to as the 4L80-E/4L80-EHD.

The 3L80/3L80-HD is a fully automatic, three-speed transmission with a three-element, nonlockup converter. Operating pressure is supplied by an IX gear–style pump. The geartrain design uses a compound Simpson planetary gearset. Note that the input planetary is located behind the reaction gearset (Figures 24–13 and 24–14).

External Transmission Controls and Features

- *Manual range selection:* linkage or cable operated.
- *Throttle system control:* engine vacuum; modulator unit located at right side front of the case. On diesel applications, the vacuum source comes from an engine-operated vacuum pump.
- *Detent/forced downshift system control:* 12-V circuit control energizes an internal detent solenoid that activates the detent valve.
- *Electrical connector:* located on the left side of the case.
- *Mainline pressure tap:* located on the left side center of the case.

Geartrain Mechanical and Clutch/Band Components

Refer to Figures 24–13 and 24–14.

- *Compound Simpson planetary gearset:* front (reaction) and rear (input) planetaries.
- *Two friction bands:* front and rear bands.
- *Three multiple-disc clutch units:* forward, direct, and intermediate clutches.
- *Two one-way clutches:* intermediate roller and low roller clutches.

Geartrain Powerflow

Refer to Figure 24–14.

Unlike other three-speed geartrains used in automatic transmissions, the 3L80 series uses a third shaft referred to as the mainshaft. When the forward clutch is applied, the power input is through the rear planetary gearset. The clutch/band functions are as follows:

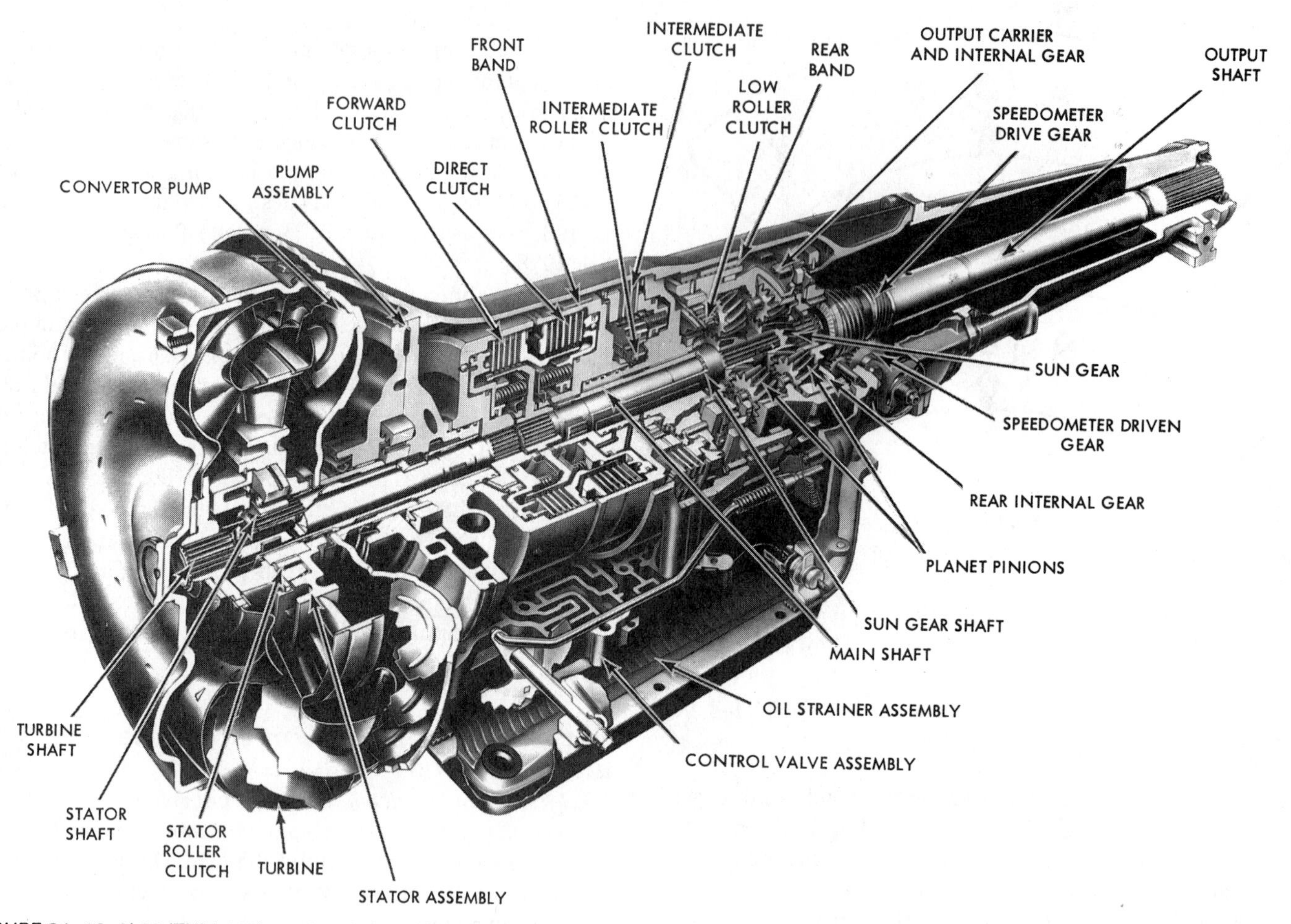

FIGURE 24–13 3L80 (THM 400) sectional view. (Courtesy of General Motors Corporation, Service Technology Group)

- *Forward clutch:* input drive clutch that locks the mainshaft and rear ring gear to the turbine shaft in all forward gears.
- *Direct clutch:* input drive clutch that locks the front and rear sun gears to the turbine shaft in third and reverse gears.
- *Intermediate clutch:* reaction clutch that grounds the outer race of the intermediate one-way roller clutch. This allows the one-way clutch to mechanically lock the sun gears to the transmission case whenever the sun gears want to rotate counterclockwise. The one-way clutch allows for clockwise rotation of the sun gears with the intermediate clutch engaged.

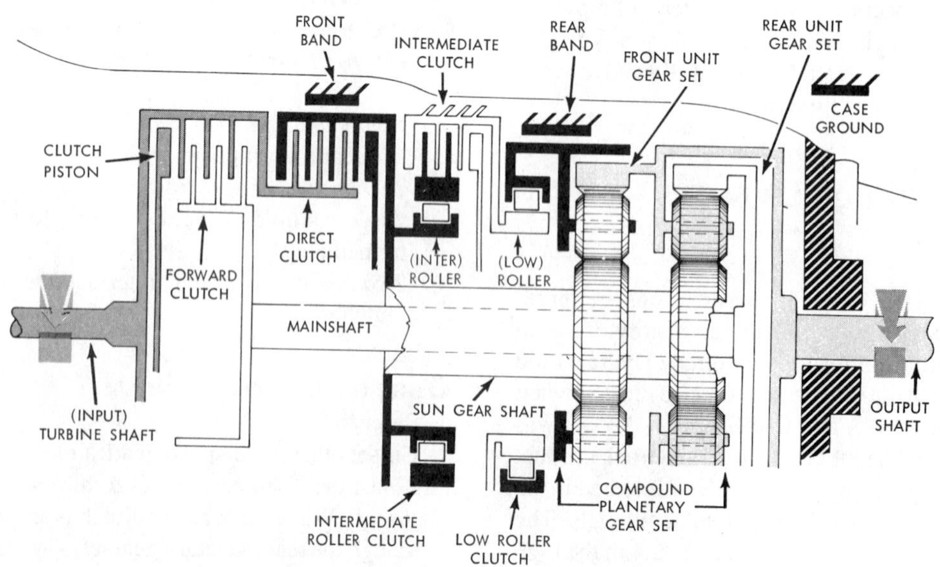

FIGURE 24–14 GM 3L80 (THM 400) powertrain schematic.

- *Front band:* coast band that wraps around the direct clutch drum and holds the sun gears during second-gear deceleration in the intermediate range.
- *Rear band:* wraps around the front carrier drum and holds the carrier reactionary in reverse and during coast/deceleration in manual low.
- *Intermediate one-way roller clutch:* mechanically locks the sun gears to the transmission when the intermediate clutch is engaged and sun gear rotational effort is counterclockwise—second-gear acceleration. Clockwise sun gear rotation releases the one-way clutch.
- *Low roller one-way clutch:* has a permanently grounded inner race to the transmission case through the center support. Mechanically locks the front carrier to the case during first-gear acceleration as it attempts to turn counterclockwise. Clockwise rotation of the front carrier releases the one-way clutch.
- *Neutral:* The powerflow dead-ends at the forward and direct input clutches.
- *First gear:* A combined function of the front and rear planetary units; compound gear ratio of 2.48:1.
- *Second gear:* A function of the rear planetary unit; gear ratio of 1.48:1.
- *Third gear:* Front and rear planetary units locked together for a direct-drive (1:1) ratio.
- *Reverse:* A function of the front planetary unit; gear ratio of 2:1.

Chart 24–2 summarizes the operating ranges and clutch/band combinations to control the geartrain.

Operating Characteristics

The operating range quadrant selector has six positions that are typically designated as: P-R-N-D-2-1.

Park/Neutral. These two quadrant positions provide for engine start. The input drive clutches are OFF, and there is no input to the geartrain. In the P position, the output shaft is mechanically locked by a parking pawl anchored in the case.

Reverse (R). Reverse provides engine braking.

Drive (D). The drive range is used for normal driving and maximum economy. Full automatic shifting and three gear ratios are provided—third gear is a direct drive. The driver can overrule the shift system by depressing the accelerator pedal. Within the proper vehicle speed and engine torque range, three WOT detent downshift combinations are available: 3-2, 3-1, and 2-1. For controlled part-throttle extra acceleration, the transmission downshifts 3-2. The closed-throttle downshift sequence is 3-2 and 2-1.

Intermediate (2). The 2 range provides for added performance in congested traffic or hilly terrain. It starts in first gear and has a normal 1-2 shift pattern. However, the transmission remains in second gear for the desired acceleration and engine braking. Intermediate can be selected at any high-gear vehicle speed, and the transmission will shift to second. A WOT detent downshift (2-1) is available. The close-throttle downshift pattern is 2-1.

Manual Low (1). Low range locks the transmission in first gear for continuous heavy-duty pulling power and engine braking that is demanded for steep grades. Manual low can be selected at any vehicle speed, but the transmission will shift to second gear if the vehicle speed is too high. When vehicle speed is reduced to a safe range below 30 mph (48.5 km/h), manual low first gear engages, and the transmission does not upshift regardless of throttle opening.

RANGE REFERENCE CHART

RANGE	GEAR	FORWARD CLUTCH	DIRECT CLUTCH	FRONT BAND	INTERMEDIATE CLUTCH	INTERMEDIATE SPRAG	LO-ROLLER CLUTCH	REAR BAND
P-N								
D	1st	APPLIED					HOLDING	
	2nd	APPLIED			APPLIED	HOLDING		
	3rd	APPLIED	APPLIED		APPLIED			
2	1st	APPLIED					HOLDING	
	2nd	APPLIED		APPLIED	APPLIED	HOLDING		
1	1st	APPLIED					HOLDING	APPLIED
	2nd	APPLIED		APPLIED	APPLIED	HOLDING		
R	REVERSE		APPLIED					APPLIED

CHART 24–2 3L80/3L80-HD range reference and clutch/band application chart. (Courtesy of General Motors Corporation, Service Technology Group)

HYDRA-MATIC 4L80-E/4L80-EHD* TRANSMISSION

RPO MT1

Produced at: Willow Run
Ypsilanti, MI

HYDRA-MATIC 4L80-E
(4-SPEED)

Vehicles used in:

C/K	PICK-UP, CHASSIS CAB, SUBURBAN, CREW CAB
G	VAN, SPORTVAN, SCHOOL BUS, HI-CUBE / CUTAWAY VAN
P	CHASSIS, VAN, SCHOOL BUS, MOTOR HOME CHASSIS

ALSO USED BY:
- ROLLS-ROYCE / BENTLEY
- JAGUAR / DAIMLER
- ISUZU • GM VENEZOLANA
- SPARTAN MOTORS
- OSHKOSH CHASSIS
- AM GENERAL

FIGURE 24–15 4L80-E/4L80-EHD applications (1995). (Courtesy of General Motors Corporation, Service Technology Group)

HYDRA-MATIC 4L80-E/4L80-EHD

The 4L80-E was introduced into the product line for 1991 models and replaces the 3L80 (THM 400) for heavy-duty applications (Figure 24–15). It is a redesign of the 3L80, using the same compound Simpson planetary gearset and clutch/band control, behind an overdrive planetary gearset. The ratios for first, second, third, and reverse gears remain the same. Fourth-gear overdrive has a 0.75:1 ratio. Refer to Figures 24–13, 24–14, and 24–16. The planetary powerflow is detailed in Chapter 4.

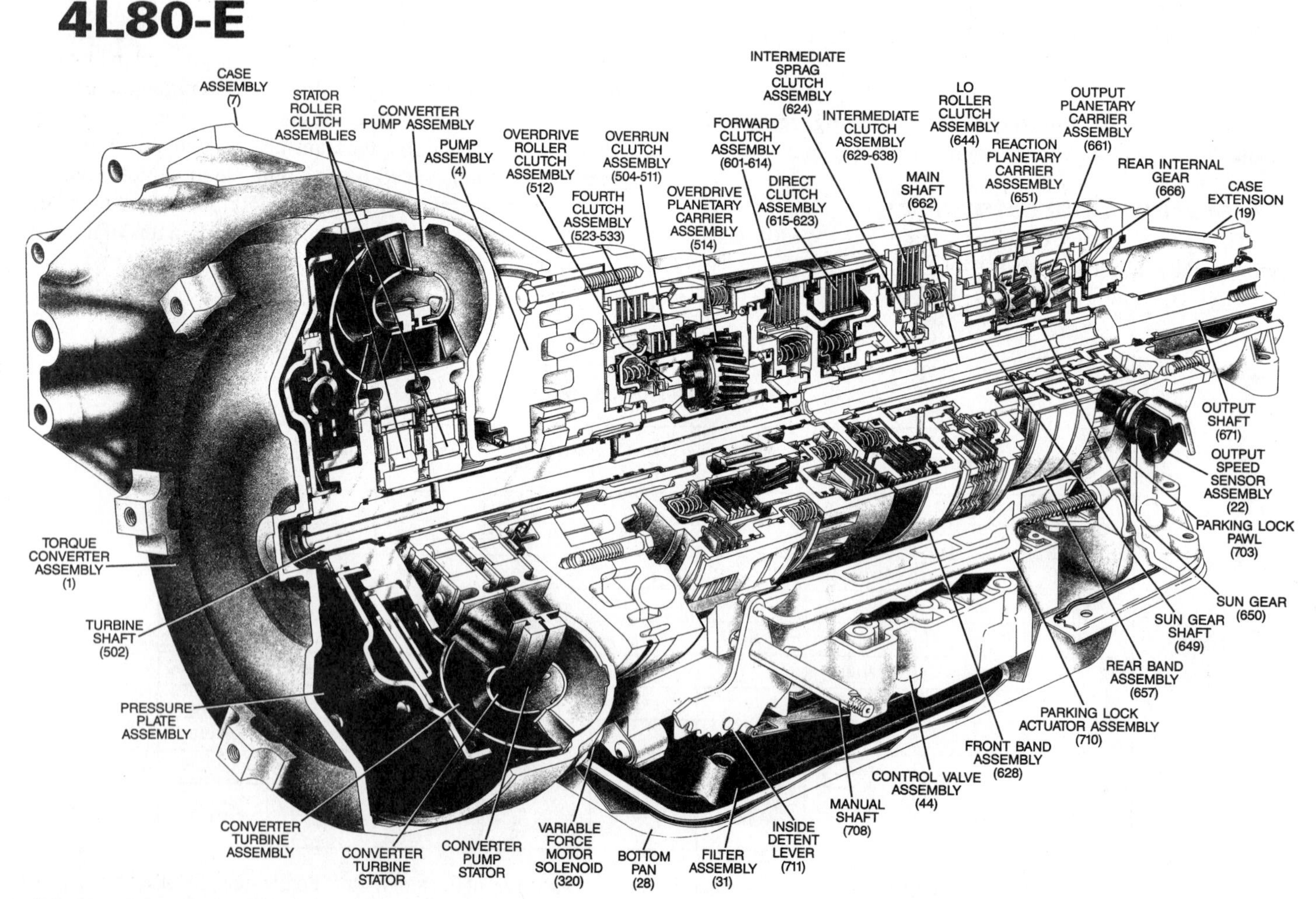

FIGURE 24–16 4L80-E sectional view. (Courtesy of General Motors Corporation, Service Technology Group)

The torque converter is designed with a converter clutch, and certain applications include a dual stator. Depending on engine application, the dual stator torque converter can produce a stall ratio as high as 3.5:1. An IX gear design oil pump supplies the transmission mainline operating pressure. A mainline pressure tap is provided on the left side of the case.

The 4L80-E is electronically controlled by the PCM. The components that make up the system are illustrated in Figure 24–17. The transmission input and output speed sensors are externally mounted and measure the turbine and output shaft speeds (Figure 24–18). The ON/OFF positions of the shift solenoids (A and B) are summarized in Chart 24–3 (page 583).

The operating range quadrant selector has seven positions: P-R-N-OD-D-2-1. The operating characteristics of each range are typical of most other GM four-speed transmissions.

A summary of the operating ranges and clutch/band applications is given in Chart 24–3 (page 583).

HYDRA-MATIC 3T40 (THM 125-C)

The 3T40 (THM 125-C) transaxle is designed for transverse mounting in GM compact and mid-size cars and front-wheel drive APV mini-vans (Figures 24–19 and 24–20). It was introduced into the X body model line in production year 1980. In addition, it was used in mid-engine RWD Pontiac Fieros from 1986 to 1988.

The geartrain is a typical three-speed Simpson planetary system, combined with a planetary final drive differential. The 1980 and 1981 units used a simple three-element torque converter and were designated as THM 125. For 1982, a clutching converter was introduced, and the transaxle was referred to as THM 125-C. The current designation is 3T40. The mainline operating pressure is supplied by a variable-capacity vane pump.

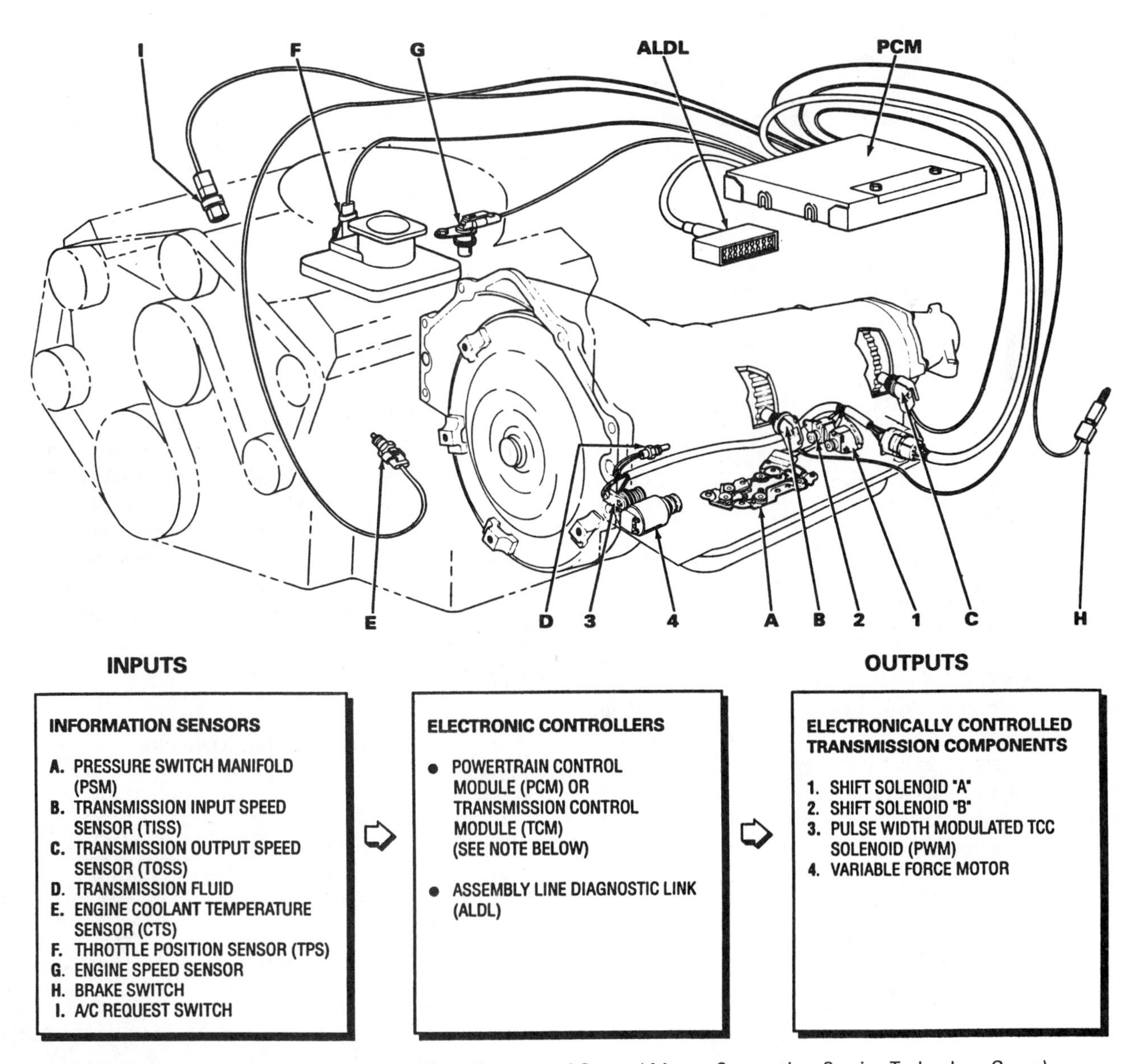

FIGURE 24–17 4L80-E/4L80-EHD input and output devices. (Courtesy of General Motors Corporation, Service Technology Group)

(1) Turbine Speed Sensor
(2) Output Speed Sensor
(3) Mainline Pressure Tap
(4) Electrical Case Connector
(5) Manual Lever Shaft

FIGURE 24–18 GM 4L80-E: left side external features.

The 3T40 (THM 125-C) shift system remains mechanically and hydraulically controlled with a throttle cable and governor. The only internal electrical circuit is for the TCC system.

External Transaxle Controls and Features

- *Manual range selection:* linkage or cable operated.
- *Throttle system control:* internal throttle lever assembly, cable operated.
- *Detent/forced downshift system control:* cable operated in common with throttle system. The TV and detent valve groups are in tandem in the same valve body bore.
- *TCC solenoid case connector:* located in the case cover above the cooler line fittings.
- *Pressure taps:* mainline and governor.
- *Speed sensor housing:* located at the rear of the transaxle case. Used in transaxles equipped with an internal speed sensor that takes the place of the speedometer cable.

Geartrain Powerflow Summary

The torque converter drives the geartrain through a dual sprocket and drive link assembly, with the driven sprocket splined to the forward clutch housing (Figure 24–21). The three-speed planetary gearset is controlled by three multiple-disc clutches, a one-way roller clutch, and a band. Refer to Chapter 4 for a detailed description of the compound Simpson planetary three-speed gearset.

In keeping with SAE standards, the 3T40 (THM 125-C) powerflow input to the forward clutch housing is viewed as being counterclockwise. Chart 24–4 (page 586) summarizes the clutch/band combinations used in controlling the geartrain powerflow in the various operating ranges. The gear ratios are as follows:

First:	2.84:1
Second:	1.60:1
Third:	1.00:1
Reverse:	2.07:1

The final drive assembly is located on the output end of the geartrain (Figure 24–22). The geartrain output drives the final drive sun gear. Reduction is achieved by permanently grounding the internal gear to the case with the carrier as the output member. The planet carrier is mounted to the differential case and turns at a fixed ratio. The planetary final drive is combined with traditional differential components. The chain drive and final drive use a variety of ratio combinations, depending on the engine and vehicle application.

Overall Final Drive Ratio

Final Drive Ratio	Chain Ratio*			
	33/37	35/35	37/33	38/32
2.84	3.18	2.84	2.53	2.39
3.06	3.43	3.06	2.73	2.58
3.33	3.73	3.33	2.97	2.80

*Designates number of teeth on the drive/driven sprockets.

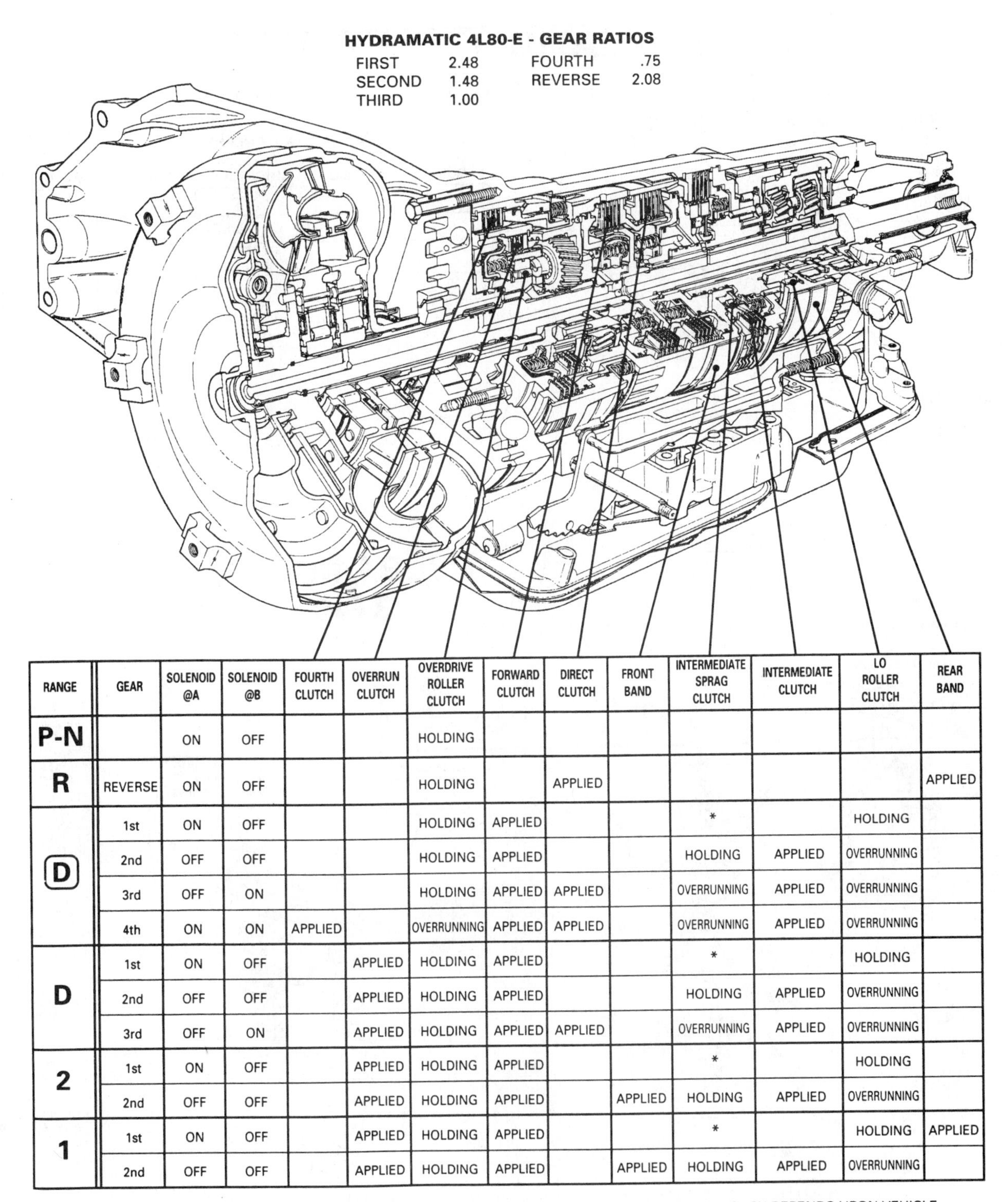

RANGE	GEAR	SOLENOID @A	SOLENOID @B	FOURTH CLUTCH	OVERRUN CLUTCH	OVERDRIVE ROLLER CLUTCH	FORWARD CLUTCH	DIRECT CLUTCH	FRONT BAND	INTERMEDIATE SPRAG CLUTCH	INTERMEDIATE CLUTCH	LO ROLLER CLUTCH	REAR BAND
P-N		ON	OFF			HOLDING							
R	REVERSE	ON	OFF			HOLDING		APPLIED					APPLIED
(D)	1st	ON	OFF			HOLDING	APPLIED			*		HOLDING	
	2nd	OFF	OFF			HOLDING	APPLIED			HOLDING	APPLIED	OVERRUNNING	
	3rd	OFF	ON			HOLDING	APPLIED	APPLIED		OVERRUNNING	APPLIED	OVERRUNNING	
	4th	ON	ON	APPLIED		OVERRUNNING	APPLIED	APPLIED		OVERRUNNING	APPLIED	OVERRUNNING	
D	1st	ON	OFF		APPLIED	HOLDING	APPLIED			*		HOLDING	
	2nd	OFF	OFF		APPLIED	HOLDING	APPLIED			HOLDING	APPLIED	OVERRUNNING	
	3rd	OFF	ON		APPLIED	HOLDING	APPLIED	APPLIED		OVERRUNNING	APPLIED	OVERRUNNING	
2	1st	ON	OFF		APPLIED	HOLDING	APPLIED			*		HOLDING	
	2nd	OFF	OFF		APPLIED	HOLDING	APPLIED		APPLIED	HOLDING	APPLIED	OVERRUNNING	
1	1st	ON	OFF		APPLIED	HOLDING	APPLIED			*		HOLDING	APPLIED
	2nd	OFF	OFF		APPLIED	HOLDING	APPLIED		APPLIED	HOLDING	APPLIED	OVERRUNNING	

*HOLDING BUT NOT EFFECTIVE

@ THE SOLENOID'S STATE FOLLOWS A SHIFT PATTERN WHICH DEPENDS UPON VEHICLE SPEED AND THROTTLE POSTION. IT DOES NOT DEPEND UPON THE SELECTED GEAR.

ON = SOLENOID ENERGIZED
OFF = SOLENOID DE-ENERGIZED

NOTE: DESCRIPTIONS ABOVE EXPLAIN COMPONENT FUNCTION DURING ACCELERATION.

RH0006-4L80-E

CHART 24–3 4L80-E/4L80-EHD range reference and clutch/band application chart. (Courtesy of General Motors Corporation, Service Technology Group)

HYDRA-MATIC 3T40 TRANSAXLE

RPO MD9

Produced at: Willow Run, Ypsilanti, MI

HYDRA-MATIC 3T40
(3-SPEED)

Vehicles used in:

	BUICK	CADILLAC	CHEVROLET	OLDSMOBILE	PONTIAC
A Body	CENTURY			CIERA	
J Body			CAVALIER		SUNFIRE
L Body			CORSICA BERETTA		
N Body	SKYLARK			ACHIEVA	GRAND AM
U Body			LUMINA APV	SILHOUETTE APV	TRANS SPORT APV

ALSO USED BY: • GM de MEXICO • GM do BRAZIL

FIGURE 24–19 3T40 (THM 125-C) applications (1995). (Courtesy of General Motors Corporation, Service Technology Group)

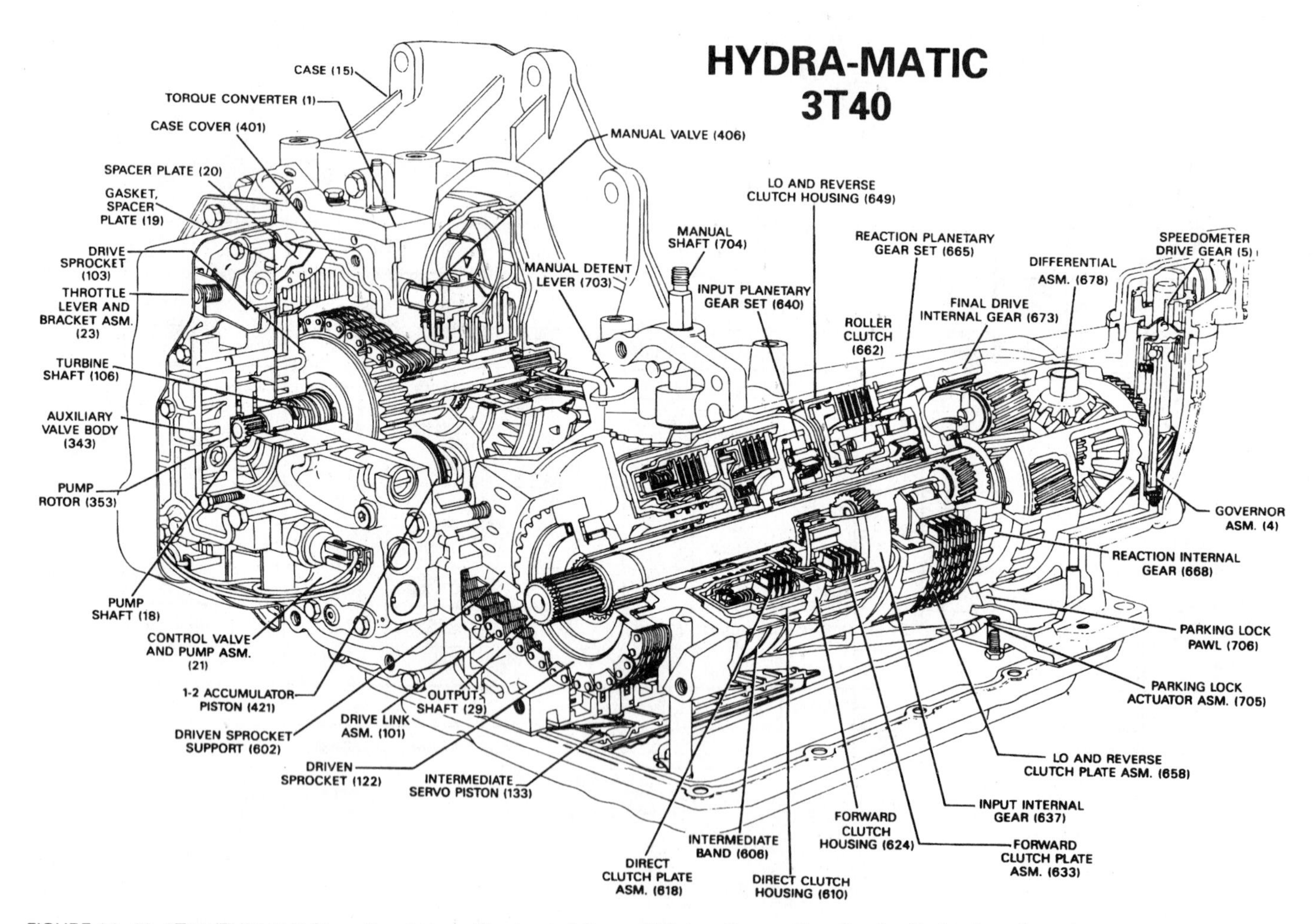

FIGURE 24–20 3T40 (THM 125-C) sectional view. (Courtesy of General Motors Corporation, Service Technology Group)

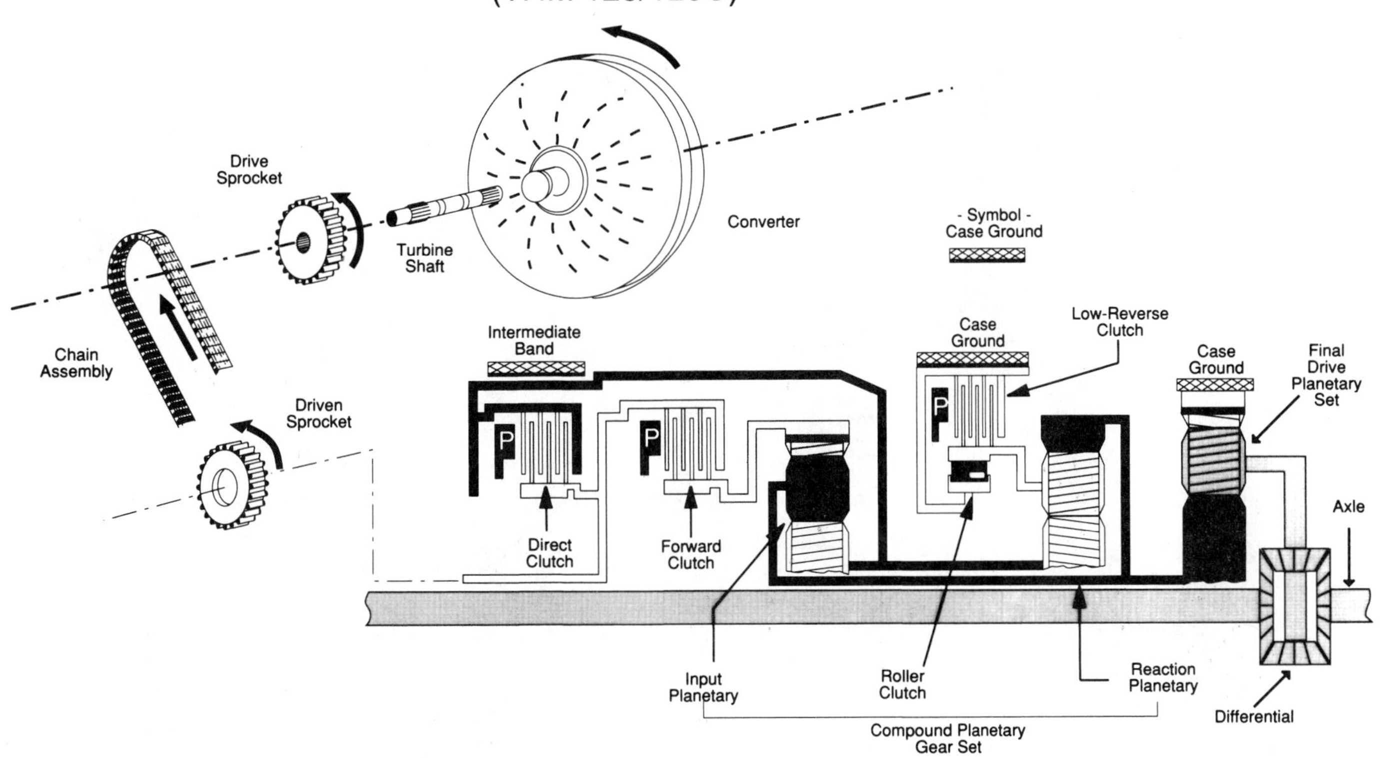

FIGURE 24–21 GM 3T40 (THM 125-C) powertrain schematic.

Operating Characteristics

The operating range selector quadrant has six positions, typically P-R-N-D-2-1 or P-R-N-D-I-L. The operational highlights for D, 2, and 1 are as follows.

Drive. This selector position is used for normal driving and maximum economy with full automatic shifting. The transaxle automatically downshifts at closed throttle (coast), 3-2 and 2-1. When coasting in low gear, there is no braking action. Low-gear coasting occurs below 10 mph (16 km/h). For added performance, forced detent downshift combinations 3-2, 3-1, and 2-1 are available at WOT within the proper vehicle speed and engine torque range. A part-throttle 3-2 downshift feature offers controlled low-speed acceleration.

The converter clutch engages in third gear only, provided that the engagement criterion has been satisfied. Minimum engagement range can vary between 35 and 50 mph (56 to 81 km/h).

Intermediate. The transaxle starts in first gear and has a normal 1-2 shift schedule—third gear is locked out. There is no engine braking when coasting in low gear. A 2-1 forced detent downshift at WOT is available. Manual 2 can be selected at any vehicle speed for immediate second-gear engagement.

Manual Low. The transaxle is locked into low gear for extra pulling power or engine braking. Manual 1 can be selected at any vehicle speed; however, if the vehicle speed is excessive, second gear engages until a safe road speed is attained [below 40 mph (64 km/h)]. Once first gear is locked in, the transaxle does not upshift.

HYDRA-MATIC 4T60 (THM 440-T4)

The 4T60 (THM 440-T4) entered the Hydra-Matic product line in production year 1984. It was designed for FWD mid- and full-size passenger cars (Figures 24–23 and 24–24). In the early 1990s, the 4T60-E began to replace 4T60 applications.

The 4T60 (THM 440-T4) transaxle is a fully automatic transverse four-speed unit combined with a planetary final drive differential assembly. It is equipped with a three-element torque converter utilizing a clutch apply plate. The converter clutch assembly is designed with either a conventional damper plate or a viscous coupling. Combined transaxle hydraulic and PCM electronics control the TCC circuit. The mainline operating pressure is provided by a variable-capacity vane pump.

The 4T60 (THM 440-T4) shift system remains mechanically and hydraulically equipped with a throttle cable control for shift timing and a vacuum modulator to control line pressure and shift quality.

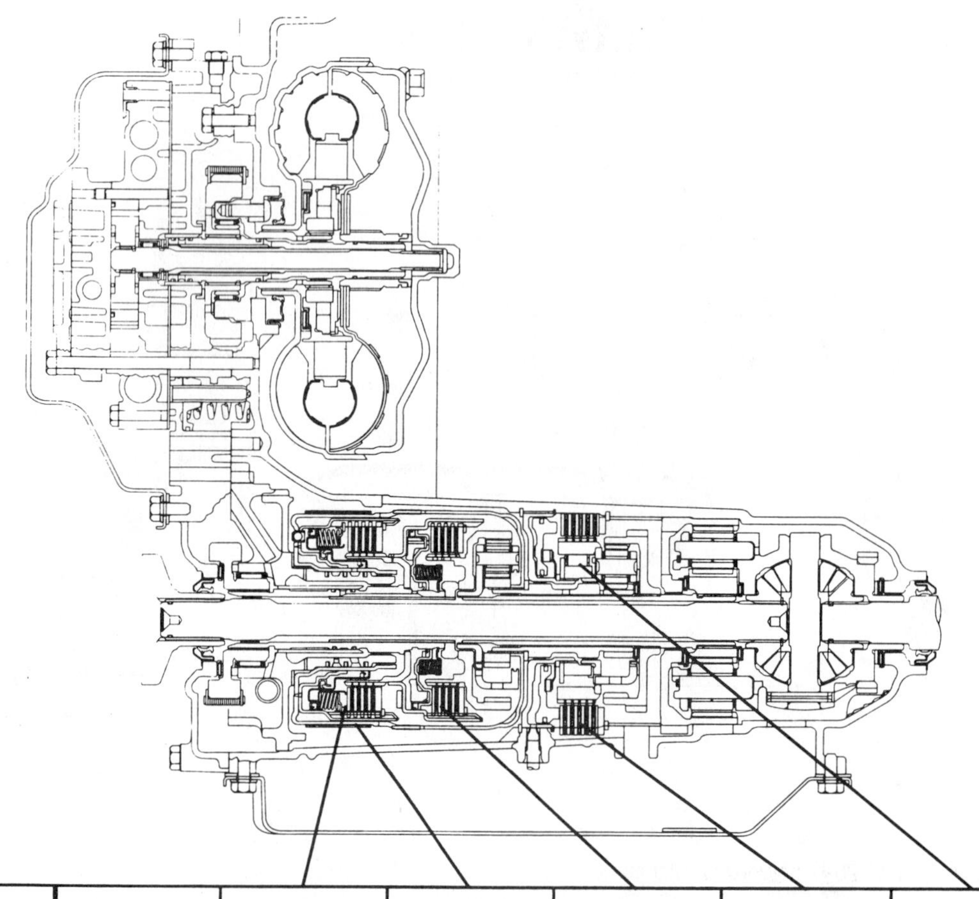

RANGE	GEAR	DIRECT CLUTCH	INTERMEDIATE BAND	FORWARD CLUTCH	LO-REVERSE CLUTCH	ROLLER CLUTCH
P-N						
D	1st			APPLIED		HOLDING
	2nd		APPLIED	APPLIED		
	3rd	APPLIED		APPLIED		
2	1st			APPLIED		*HOLDING
	2nd		APPLIED	APPLIED		
1	1st			APPLIED	APPLIED	HOLDING
R	REVERSE	APPLIED			APPLIED	

* May not be holding in Intermediate (D2), may stay in 2nd gear when vehicle is stopped.

RH0137-3T40

CHART 24–4 3T40 (THM 125-C) range reference and clutch/band application chart. (Courtesy of General Motors Corporation, Service Technology Group)

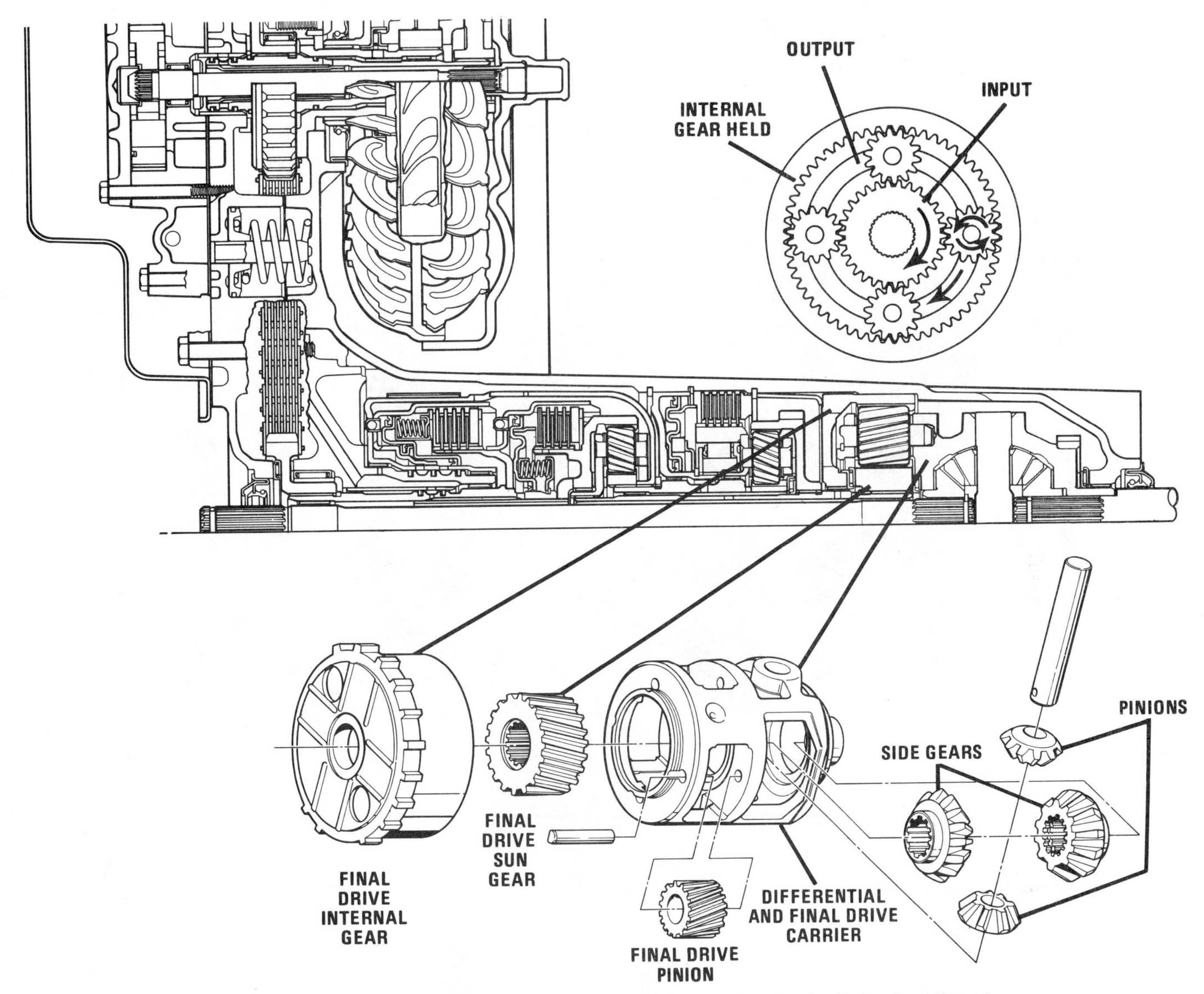

FIGURE 24–22 3T40 (THM 125-C) final drive unit. (Courtesy of General Motors Corporation, Service Technology Group)

FIGURE 24–23 GM 4T60 (THM 440-T4).

4T60

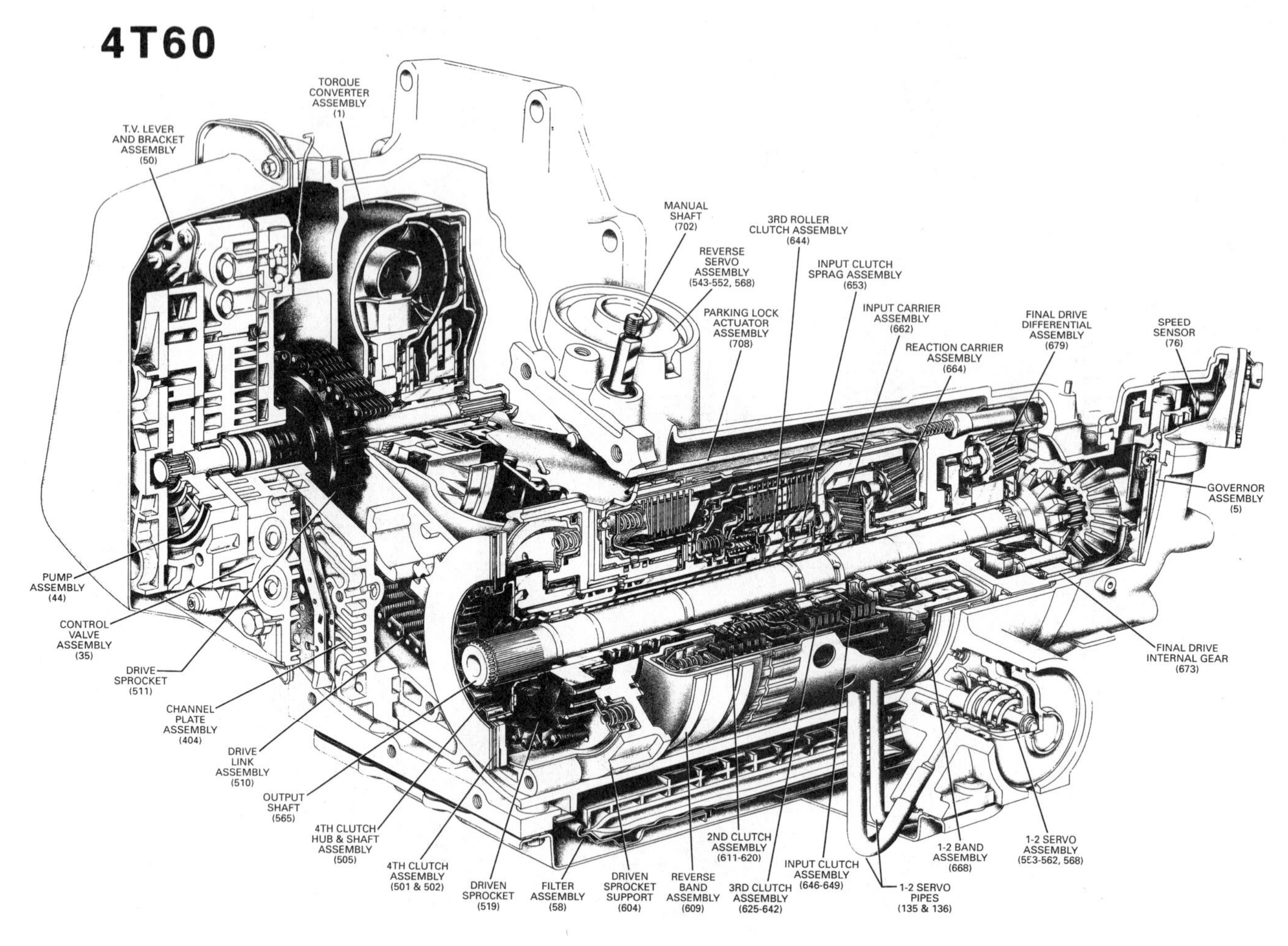

FIGURE 24–24 4T60 sectional view. (Courtesy of General Motors Corporation, Service Technology Group)

External Transaxle Controls and Features

- *Manual range selection:* cable operated.
- *Throttle system control:* internal throttle lever assembly cable operated. The throttle system controls the shift schedule in the 4T60 and does not influence mainline pressure except in detent downshift modes.
- *Detent/forced downshift system control:* cable operated in common with throttle system. The TV and detent valve group are in tandem in the same valve body bore.
- *Vacuum modulator:* either altitude compensating or non–altitude compensating. Modulator oil is used for modifying mainline pressure and accumulator action. The modulator essentially controls the shift firmness or feel.
- *TCC solenoid case connector:* located in channel plate above cooler line fittings.
- *Temperature switch assembly:* for viscous converter clutch applications. Located in channel plate, lower left side.
- *Speed sensor housing:* located at the rear of the transaxle case in units with an internal speed sensor that takes the place of a speedometer cable.

Geartrain Powerflow Summary

The torque converter drives the geartrain through a dual sprocket and drive assembly, with the driven sprocket splined to the input housing (Figures 24–25 and 24–26). The planetary geartrain is controlled by four multiple-disc clutches, one roller clutch, one sprag clutch, and two bands.

Geartrain Clutch/Band Members.

- *Input clutch:* engages the converter drive torque and drives the input sun gear through the input sprag clutch in first, third, and reverse gears.
- *Input sprag clutch:* transmits converter drive torque from the input clutch to the input sun gear. It also allows the input sun gear to overrun and rotate faster than the input clutch.
- *1-2 band:* holds the reaction sun gear stationary for first and second gears.
- *Second clutch:* engages the converter drive torque and drives the input carrier/reaction internal gear assembly in second, third, and fourth gears.

HYDRAMATIC 4T60
(THM 440-T4)

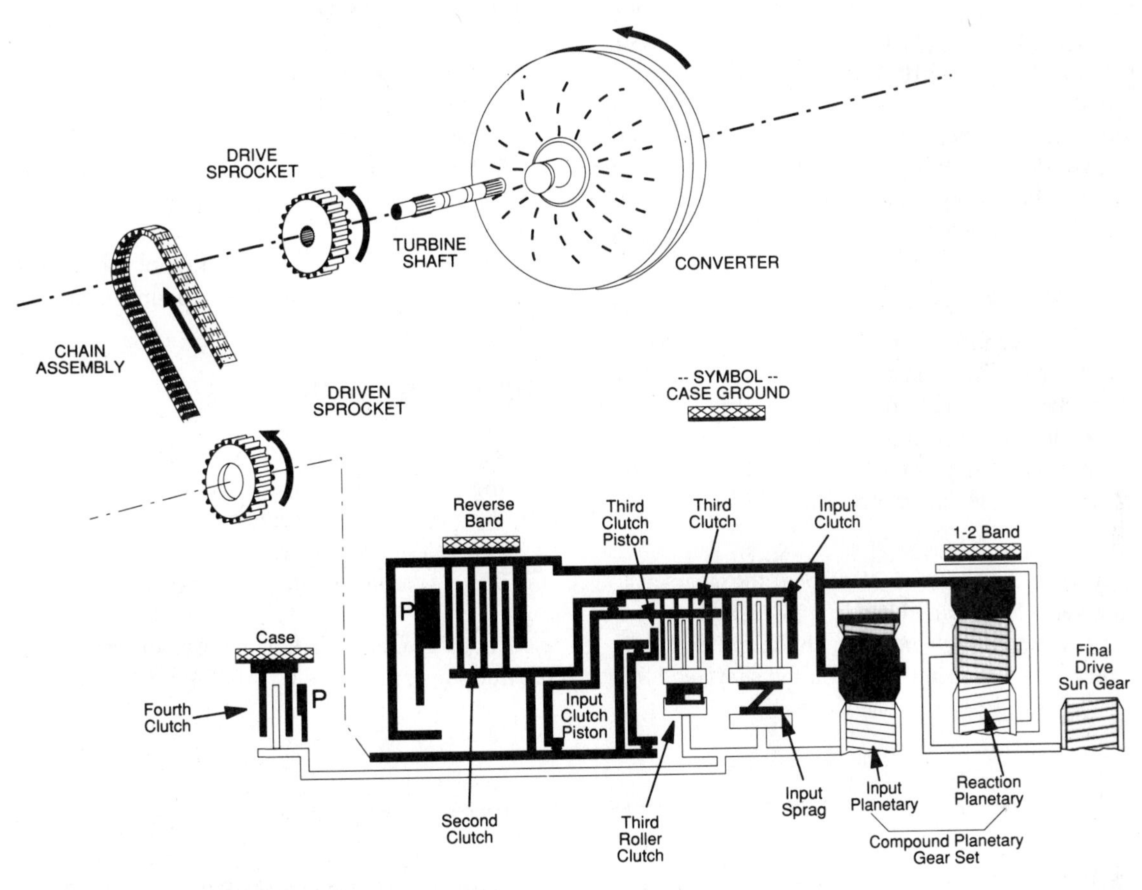

FIGURE 24–25 GM 4T60 (THM 440-T4) powertrain schematic.

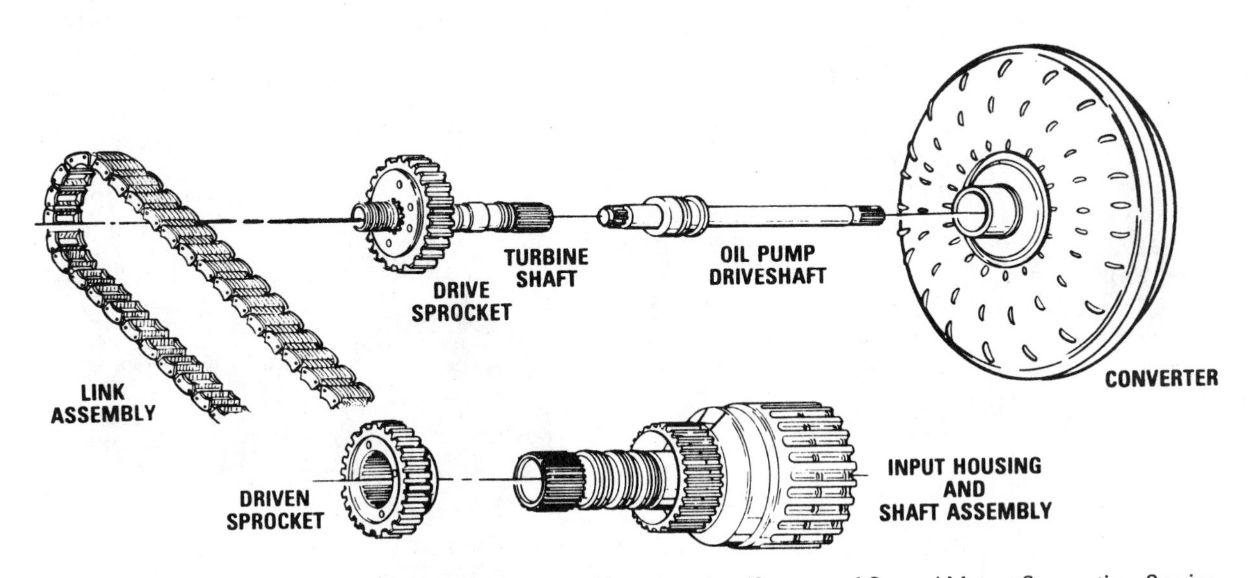

FIGURE 24–26 4T60 (THM 440-T4) converter, drive components, and input housing. (Courtesy of General Motors Corporation, Service Technology Group)

- Third clutch: grounds the third overrun clutch outer race to the input housing. The input sun gear is locked to the input housing through the third roller clutch in third gear.
- *Third roller clutch:* transmits converter drive torque from the third clutch to the input sun gear in third gear. It also allows the third clutch to overrun and rotate faster than the input sun gear.
- *Fourth clutch:* Holds the input sun gear stationary for fourth-gear overdrive.
- *Reverse band:* holds the input carrier/reaction internal gear assembly stationary in reverse.

Geartrain Powerflow

The planetary system configuration eliminates the use of an extra overdrive planetary. Two simple gearsets are integrated and arranged with the sun gears to turn independently of each other. Figure 24–26 shows the input drive to the planetary geartrain. The input housing contains the input and third clutches, plus it drives the second clutch, Figure 24–25.

Park/Neutral. Unlike most P and N modes, the gearset is prepared for first gear. The input clutch is applied, and turbine torque drives the gearset through the sprag clutch and input sun gear. The weight of the car restrains the output shaft, and the planetary gears freewheel but do not transmit power because there is no reactionary member (Figure 24–27).

OD Range: First Gear.

- The input clutch drives the input sun gear through the input sprag.
- Input sprag effective.
- 1-2 band is applied and holds the reaction sun gear.
- Third overrun roller clutch is ineffective.

First gear is a combined function of the input and reaction planetary units, providing a compound gear ratio of 2.92:1 (Figure 24–28).

OD Range: Second Gear.

- Input clutch is applied but ineffective.
- Input sprag clutch overrunning.
- Second clutch is applied and drives the input carrier/reaction internal gear unit.
- Third overrun clutch ineffective.

Second gear is a function of the reaction planetary (rear) and provides a gear ratio of 1.57:1 (Figure 24–29).

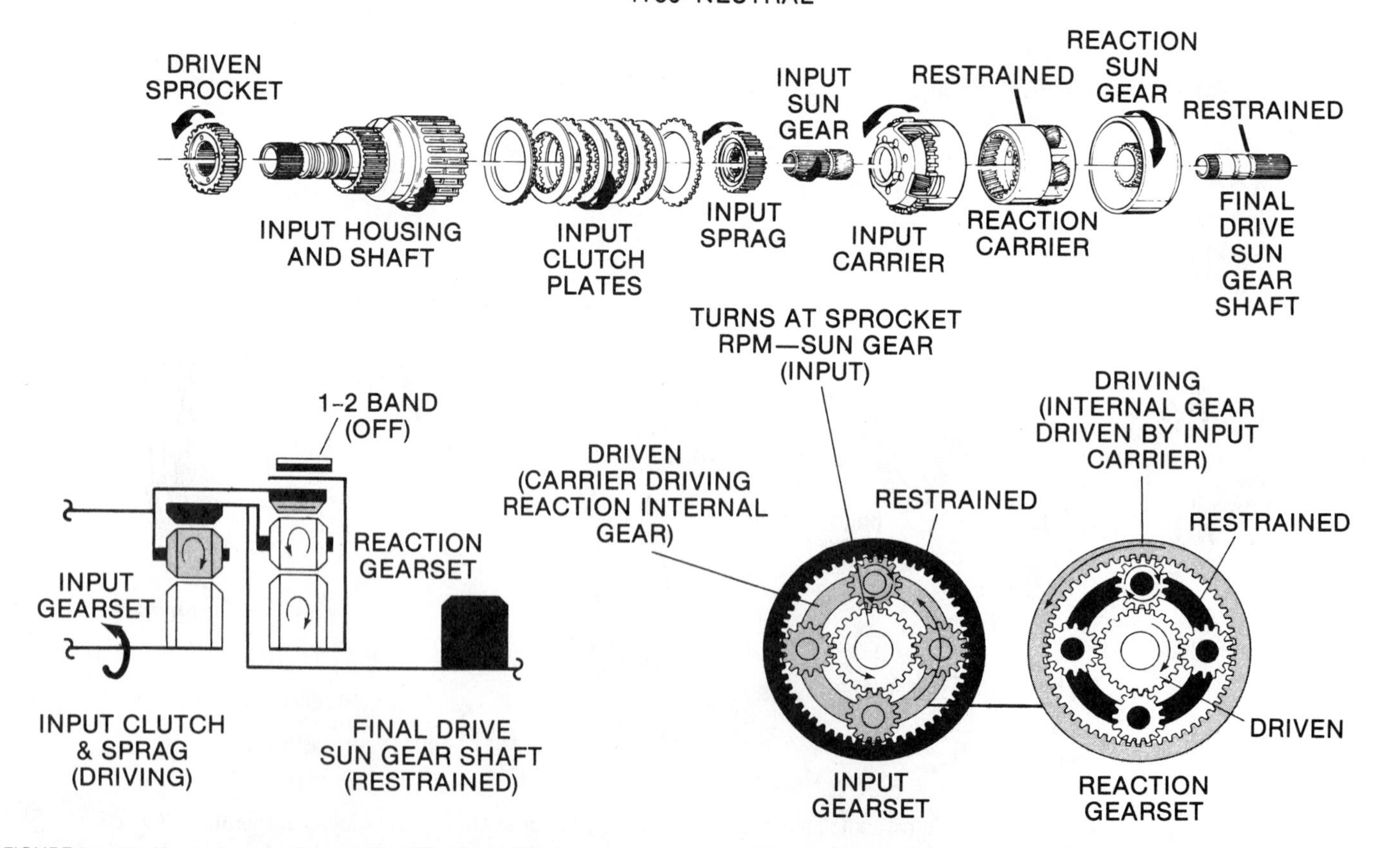

FIGURE 24–27 Neutral powerflow—4T60 (THM 440-T4). (Courtesy of General Motors Corporation, Service Technology Group)

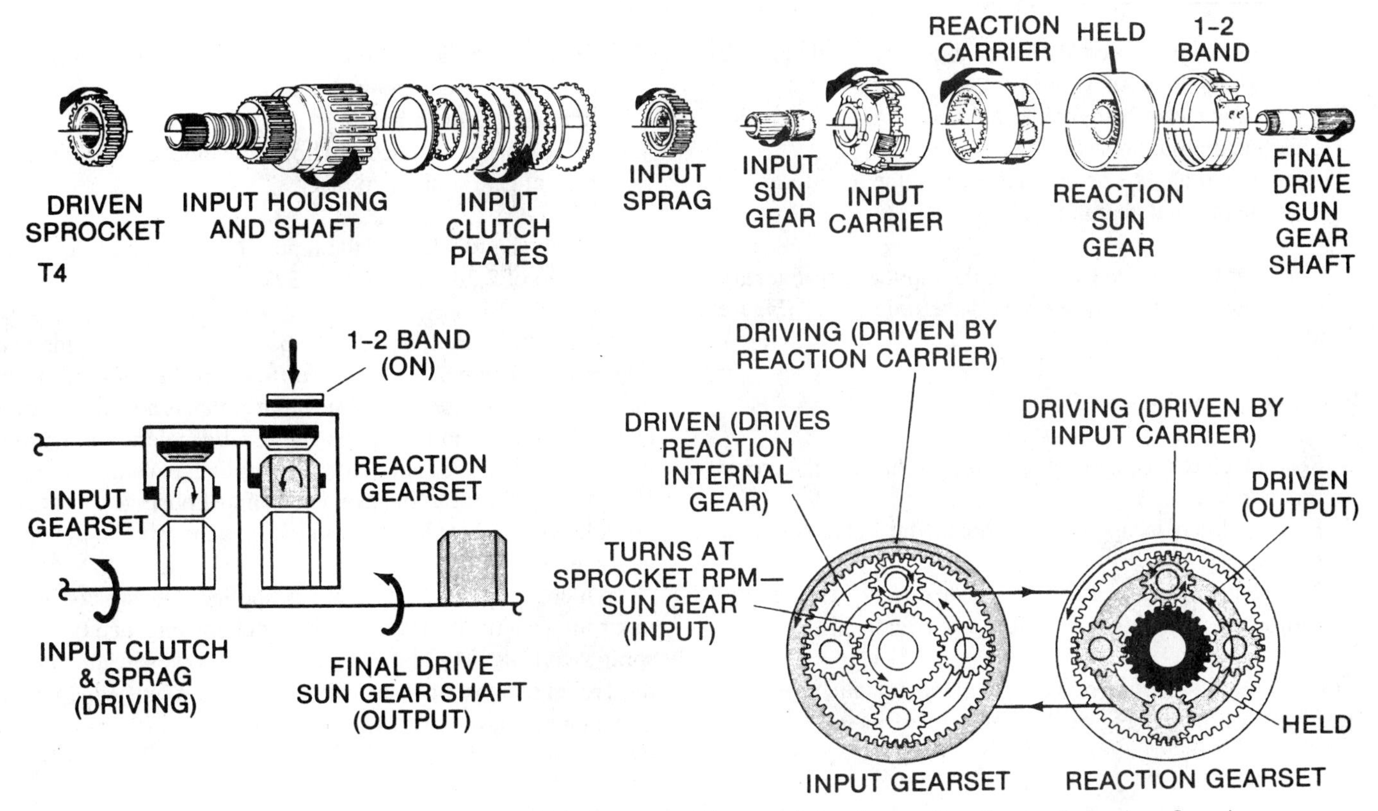

FIGURE 24–28 First-gear powerflow—4T60 (THM 440-T4). (Courtesy of General Motors Corporation, Service Technology Group)

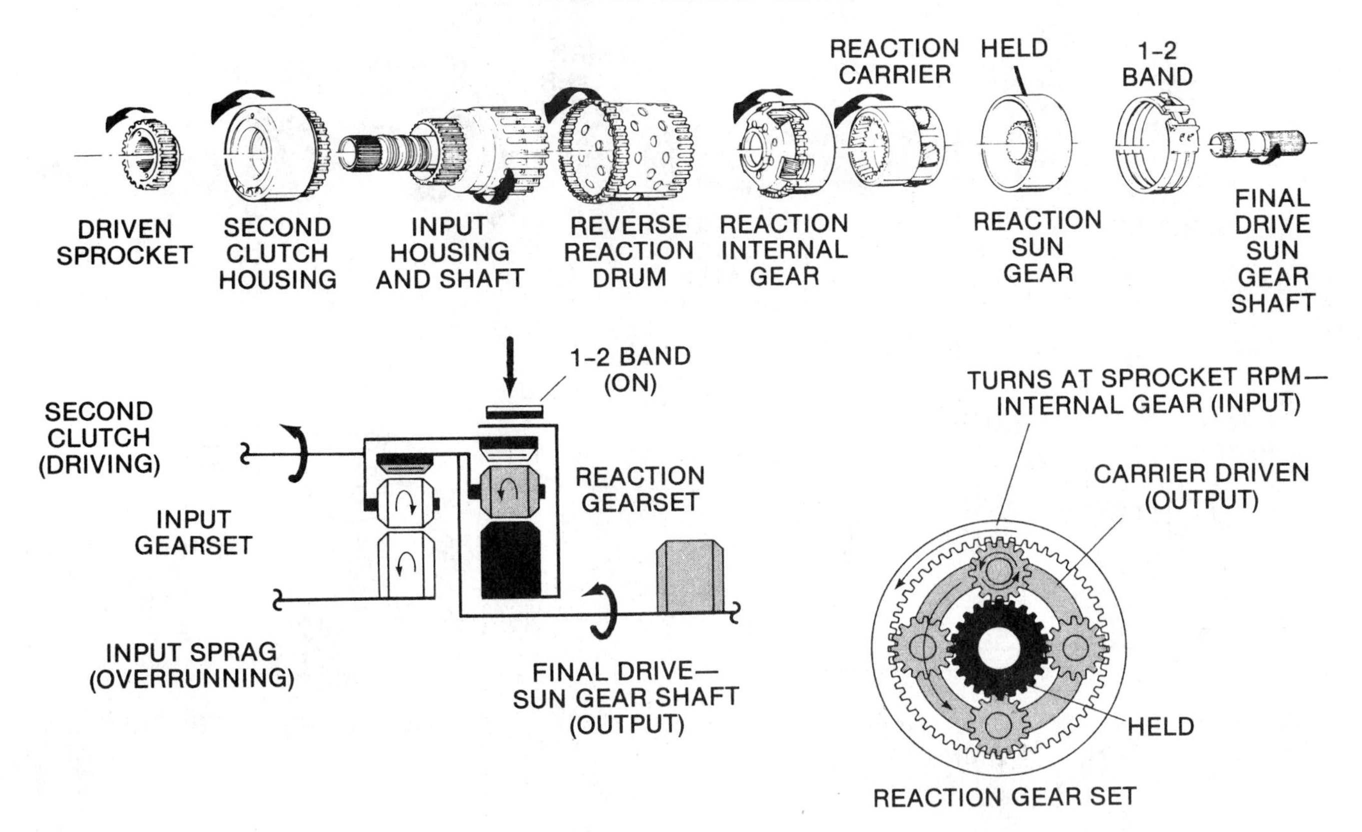

FIGURE 24–29 Second-gear powerflow—4T60 (THM 440-T4). (Courtesy of General Motors Corporation, Service Technology Group)

OD Range: Third Gear.

- ❍ Third clutch is applied and grounds outer race of the third roller clutch to input.
- ❍ Third roller clutch locks the input sun gear to the third clutch and input housing.
- ❍ Second clutch is applied and drives the input carrier/ reaction internal gear unit.

Third gear is achieved by locking the input and reactionary planetary units together for a direct-drive ratio of 1:1 (Figure 24–30).

OD Range: Fourth-Gear Overdrive.

- ❍ Second clutch is applied and drives the input carrier/ reaction internal gear unit.
- ❍ Fourth clutch is applied and holds the input sun gear reactionary.
- ❍ Third clutch is applied but is not transmitting drive torque.
- ❍ Third roller clutch overruns and is ineffective.

Overdrive is a function of the input planetary unit, providing a gear ratio of 0.70:1 (Figure 24–31).

Reverse.

- ❍ Input clutch is applied and drives the input sun gear through the input sprag clutch.
- ❍ Input sprag clutch holding.
- ❍ Reverse band is applied and holds the input carrier/reaction internal gear unit reactionary.

Reverse is a function of the input planetary unit, providing a gear ratio of 2.38:1 (Figure 24–32).

Manual Ranges 3, 2, and 1. The powerflows in these manual ranges are identical to the powerflows in the OD range. In manual 3, the engagement of the input clutch is added to ensure engine braking in third gear during coast/deceleration. During third-gear coast, the input clutch is applied, causing the input sprag to hold the input sun gear to the input housing. This action provides engine braking through the input housing in the manual ranges.

In manual 1, the third clutch is applied. During coast, the input sun gear turns faster than the input housing, and the input sprag overruns. The third clutch apply, however, allows the third roller clutch to lock up and keep the input sun gear coupled to the input housing for engine braking.

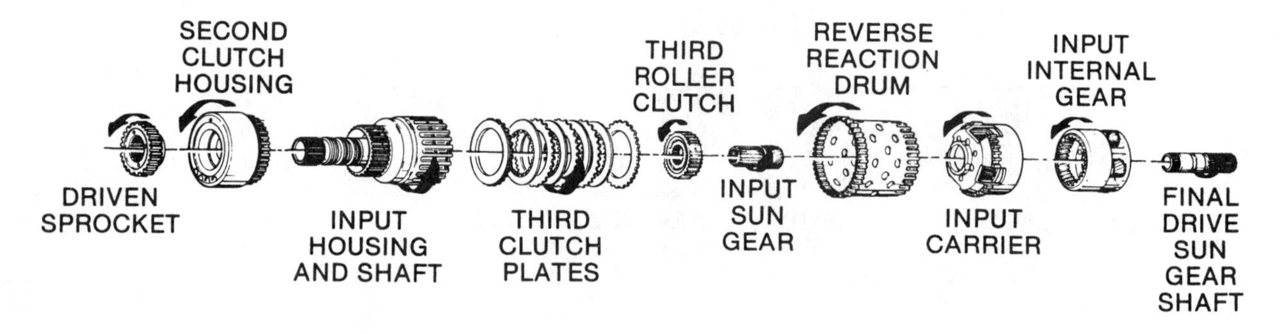

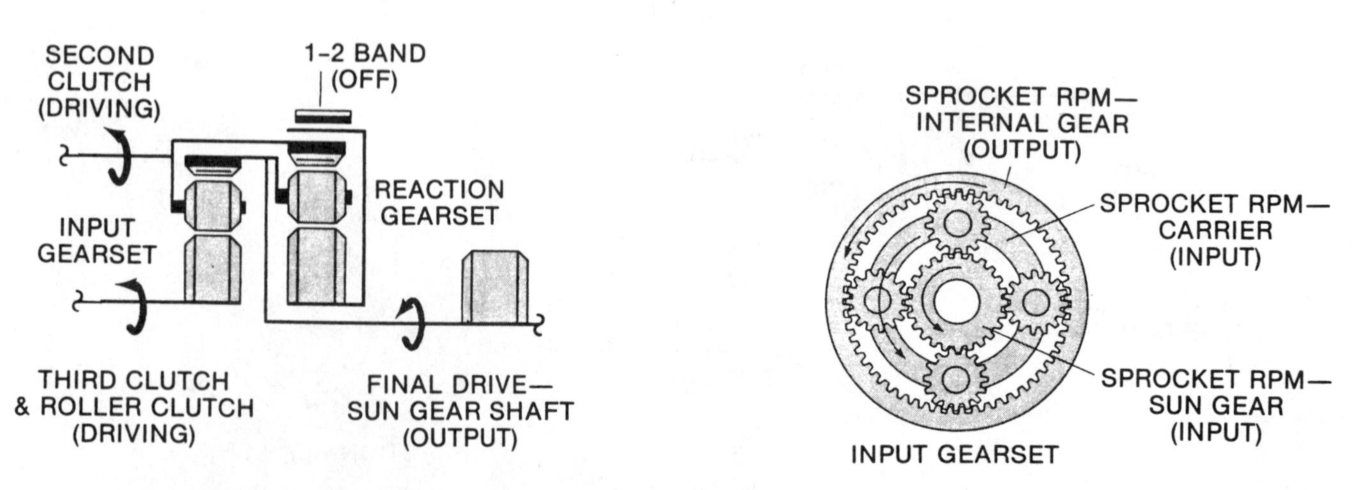

FIGURE 24–30 Third-gear powerflow—4T60 (THM 440-T4). (Courtesy of General Motors Corporation, Service Technology Group)

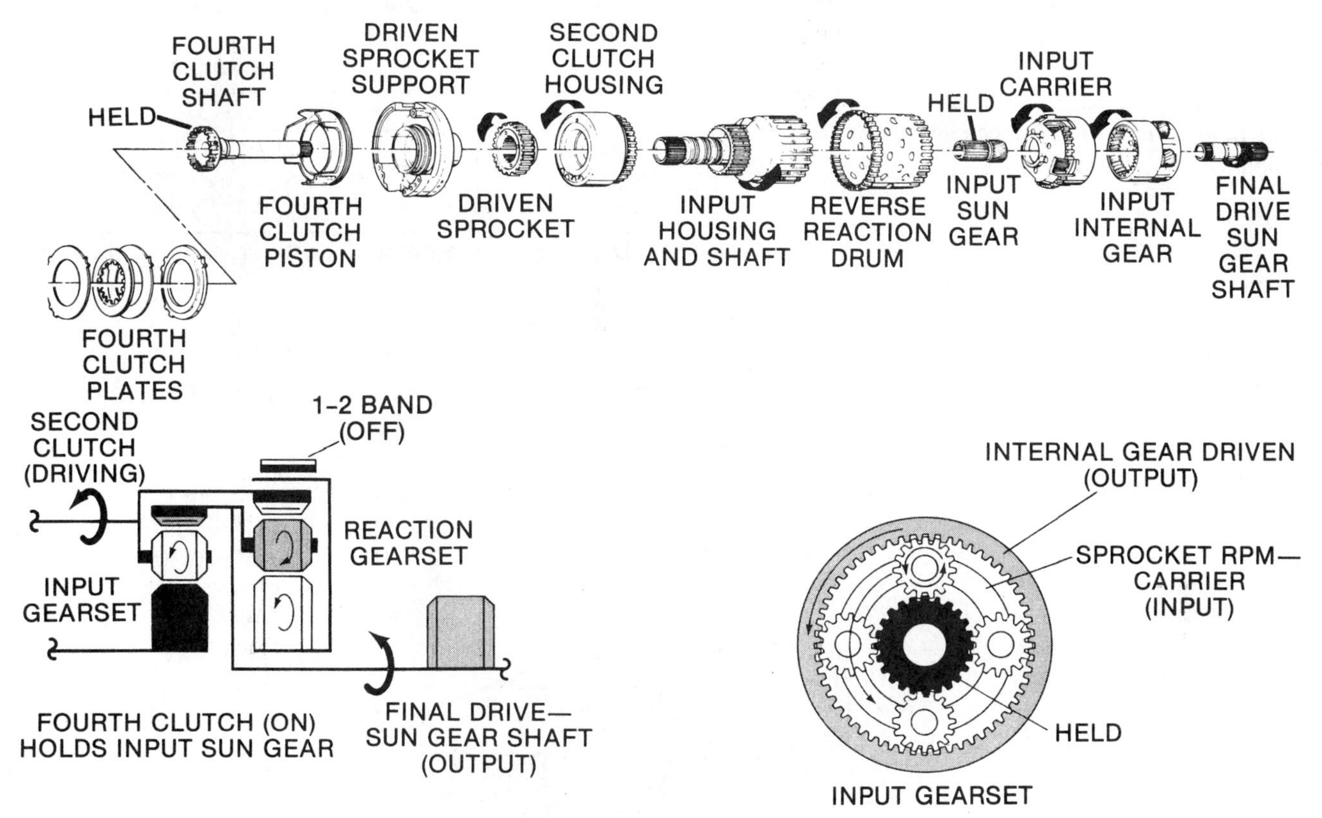

FIGURE 24–31 Fourth-gear powerflow—4T60 (THM 440-T4). (Courtesy of General Motors Corporation, Service Technology Group)

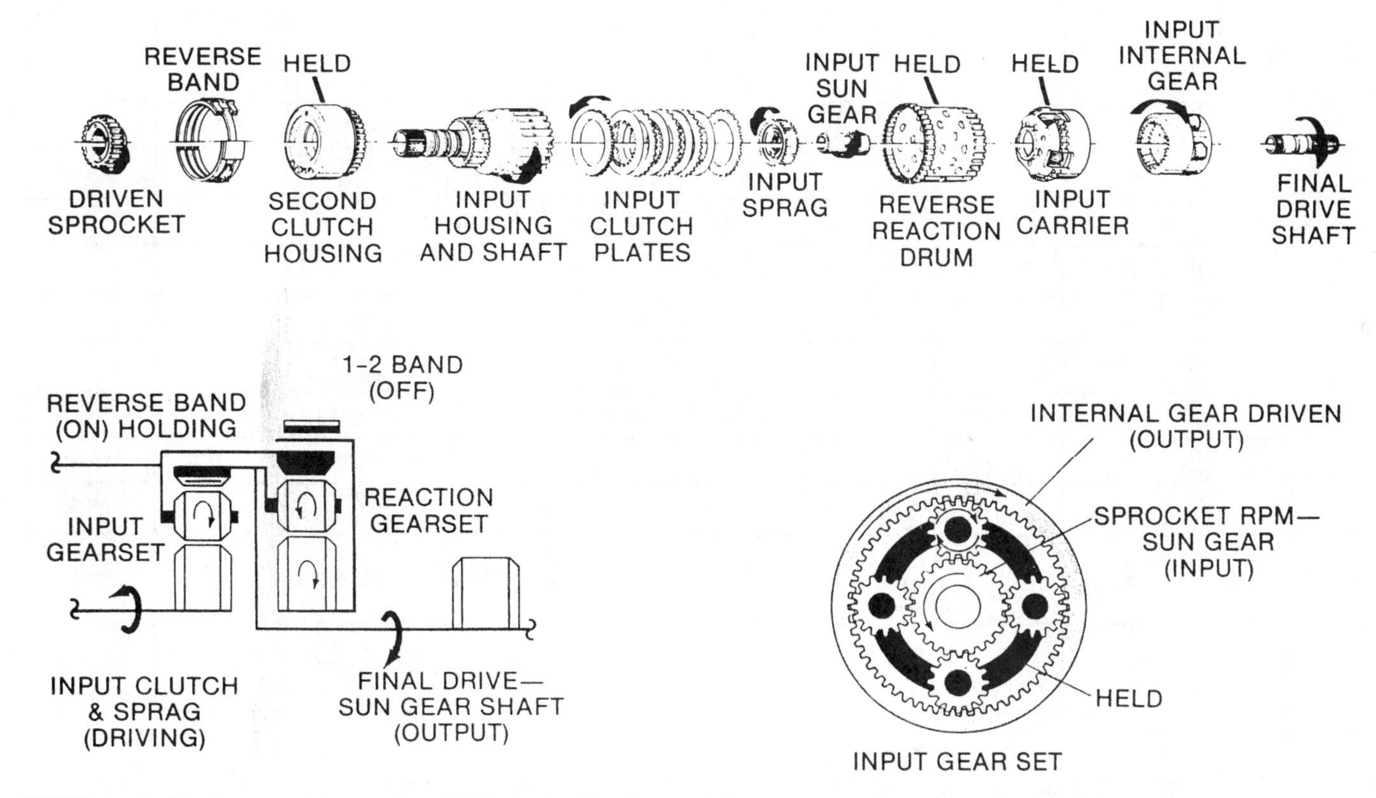

FIGURE 24–32 Reverse powerflow—4T60 (THM 440-T4). (Courtesy of General Motors Corporation, Service Technology Group)

Chart 24–5 features a summary of the operating ranges and clutch/band combinations to control the 4T60 (440-T4) geartrain.

The sprocket chain drive and planetary final drive use a variety of ratio combinations, depending on the engine and vehicle application.

Overall Final Drive Ratio

Final Drive Ratio	Chain Ratio*	
	35/35	37/33
2.84	2.84	2.53
3.06	3.06	2.73
3.33	3.33	2.97

*Designates number of teeth on the drive/driven sprockets.

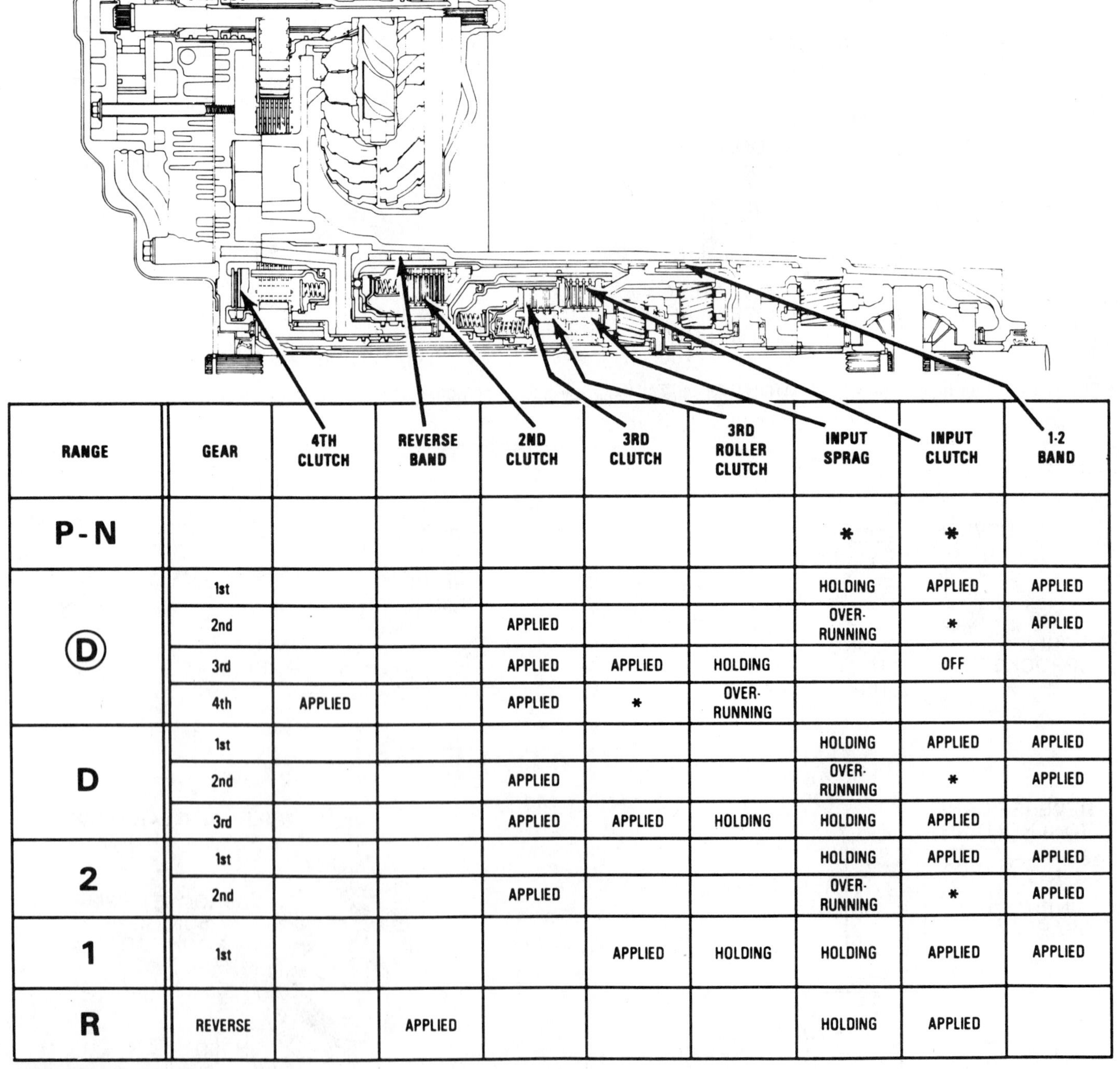

RANGE	GEAR	4TH CLUTCH	REVERSE BAND	2ND CLUTCH	3RD CLUTCH	3RD ROLLER CLUTCH	INPUT SPRAG	INPUT CLUTCH	1-2 BAND
P-N							*	*	
Ⓓ	1st						HOLDING	APPLIED	APPLIED
	2nd			APPLIED			OVER-RUNNING	*	APPLIED
	3rd			APPLIED	APPLIED	HOLDING		OFF	
	4th	APPLIED		APPLIED	*	OVER-RUNNING			
D	1st						HOLDING	APPLIED	APPLIED
	2nd			APPLIED			OVER-RUNNING	*	APPLIED
	3rd			APPLIED	APPLIED	HOLDING	HOLDING	APPLIED	
2	1st						HOLDING	APPLIED	APPLIED
	2nd			APPLIED			OVER-RUNNING	*	APPLIED
1	1st				APPLIED	HOLDING	HOLDING	APPLIED	APPLIED
R	REVERSE		APPLIED				HOLDING	APPLIED	

*APPLIED BUT NOT EFFECTIVE

MH0154-4T60

CHART 24–5 4T60 (THM 440-T4) range reference and clutch/band application chart. (Courtesy of General Motors Corporation, Service Technology Group)

Operating Characteristics

The operating range selector has seven quadrant positions: P-R-N-OD-D-2-1 or P-R-N-D-3-2-1.

Overdrive. Overdrive range is used for most driving conditions and maximum economy. The transmission starts in first and schedules a normal 1-2 and 2-3 shift pattern, with the 3-4 overdrive shift delayed for road cruising. The transmission does not shift into or remain in fourth gear at WOT. At closed-throttle (coast), the downshift pattern is 4-3, 3-2, and 2-1. When coasting in third or low gear, there is no braking action.

For added performance, forced detent downshift combinations (3-2, 3-1, and 2-1) are available within the proper vehicle speed and engine torque range. For moderate controlled acceleration, the transmission provides a part-throttle 4-3 and 3-2 downshift. The converter clutch engages while in third or fourth gear, provided that the engagement criteria have been satisfied.

Manual Third. The transmission is fully automatic through third gear—fourth-gear overdrive is inhibited. Shift performance, including the converter clutch, is the same as in the OD range for first, second, and third gears. Third and second gears provide for coast engine braking.

Manual Second. Manual 2 limits the transaxle to second gear for added performance and engine braking. A normal first-gear start and 1-2 shift pattern is the same as in OD range. The converter clutch has the potential to engage with wide throttle openings. Manual 2 can be selected at any vehicle speed with the transaxle in third or fourth gear, and an immediate shift to second occurs.

Manual Low. The transaxle is locked into low gear for heavy pulling power or to maintain maximum engine braking when descending steep grades. Manual 1 can be selected at any vehicle speed; however, if the vehicle speed is excessive, second gear engages until a safe road speed is attained, usually below 40 mph (64 km/h). Once first gear is locked in, the transaxle does not upshift.

HYDRA-MATIC 4T60-E/4T60-EHD

For the 1991 model year, Hydra-Matic introduced the 4T60-E, an electronically and mechanically enhanced version of the 4T60 (Figures 24–33 and 24–34). The 4T60-E continues to be a popular transaxle within many GM automobiles, such as the 1997 Buick LeSabre.

During the 1996 model year, the 4T60-EHD was introduced. This heavy-duty version of the 4T60-E is coupled to the Series II, super-charged 3.8-L engine. It is found in vehicles such as the Pontiac SSEi and Oldsmobile LSS.

Electronic Control

The electronic system components are shown in Figure 24–35. A PCM manages the shift schedule and torque converter lockup. The mainline pressure remains vacuum modulator controlled. PCM self-diagnostic features for the electronics are incorporated for improved serviceability. Refer to Chapter 11 for a discussion of the shift solenoid operation.

Geartrain Powerflow

The planetary geartrain and gear ratios are identical to the 4T60 (440-T4). The gear ratios are as follows:

First:	2.921:1
Second:	1.568:1
Third:	1.000:1
Fourth:	0.705:1
Reverse:	2.385:1

HYDRA-MATIC 4T60-E TRANSAXLE

RPO M13

Produced at: Warren, MI

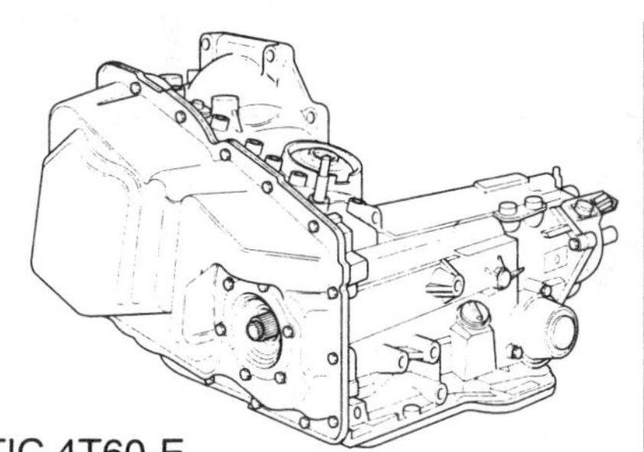

HYDRA-MATIC 4T60-E
(4-SPEED) ALSO USED BY: • GM VENEZOLANA

Vehicles used in:

	BUICK	CADILLAC	CHEVROLET	OLDSMOBILE	PONTIAC
A Body	CENTURY			CIERA	
C Body	PARK AVENUE			98 REGENCY	
G Body	RIVIERA				
H Body	LE SABRE			88 ROYALE	BONNEVILLE
K Body		DE VILLE			
L Body			BERETTA CORSICA		
N Body	SKYLARK			ACHIEVA	GRAND AM
W Body	REGAL		LUMINA	CUTLASS SUPREME	GRAND PRIX
U Van			LUMINA APV	SILHOUETTE	TRANS SPORT

FIGURE 24–33 4T60-E applications (1995). (Courtesy of General Motors Corporation, Service Technology Group)

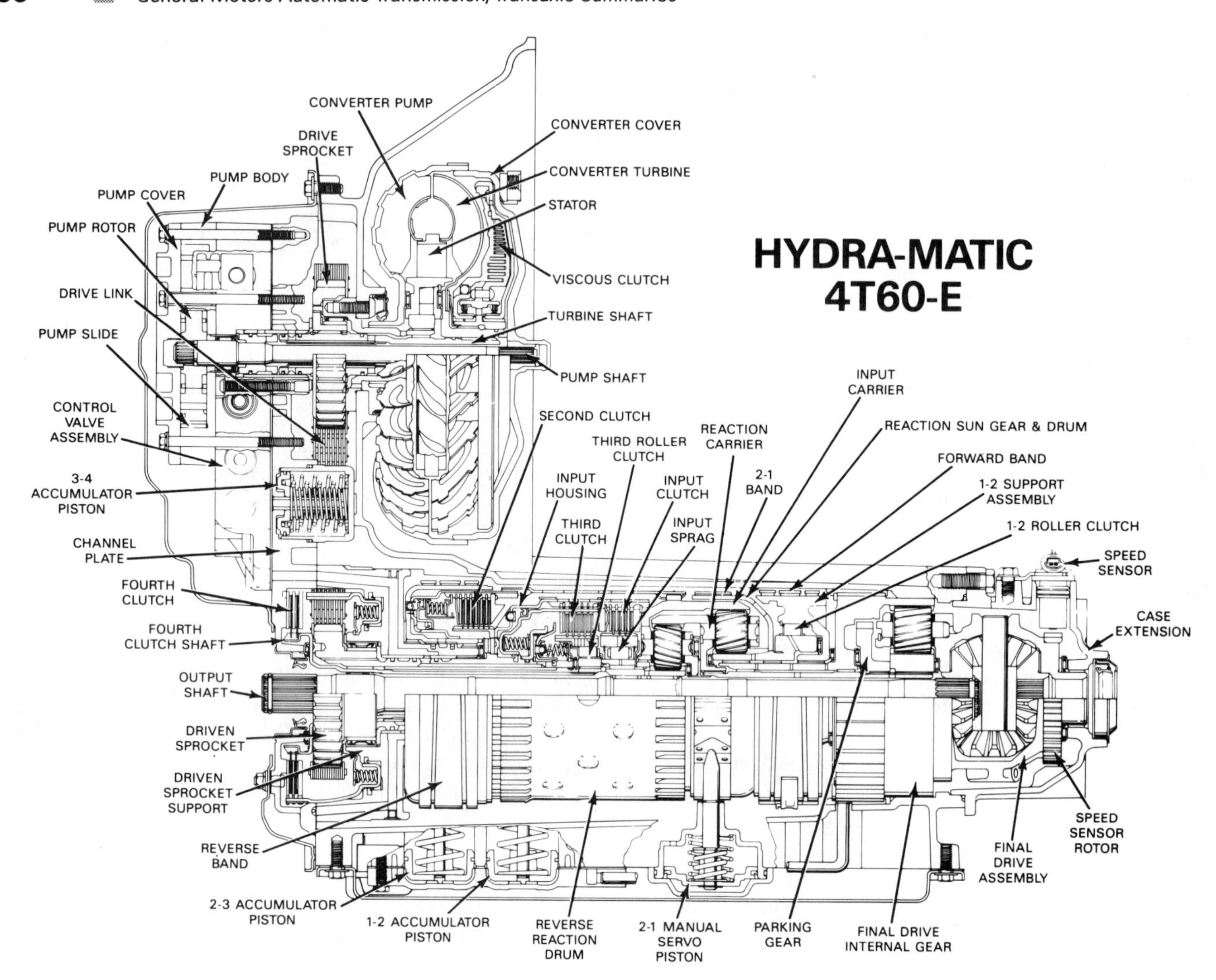

FIGURE 24–34 4T60-E sectional view. (Courtesy of General Motors Corporation, Service Technology Group)

The early final drive ratios were also identical and were teamed with four chain drive ratios for the following overall combinations.

Overall Final Drive Ratio

Final Drive Ratio	Chain Ratio*			
	35/35	37/33	28/27	33/27
2.84	2.84	2.53	2.74	3.18
3.06	3.06	2.73	2.95	3.43
3.33	3.33	2.97	3.21	3.73

*Designates number of teeth on the drive/driven sprockets.

The updated version uses three chain drive ratios with a selection of five final drive ratios.

Overall Final Drive Ratio

Final Drive Ratio	Chain Ratio*		
	35/35	37/33	33/27
2.84	2.84	2.53	3.18
3.05	3.05	2.72	3.42
3.06	3.06	2.73	3.43
3.29	3.29	2.93	3.68
3.33	3.33	2.97	3.73

*Designates number of teeth on the drive/driven sprockets.

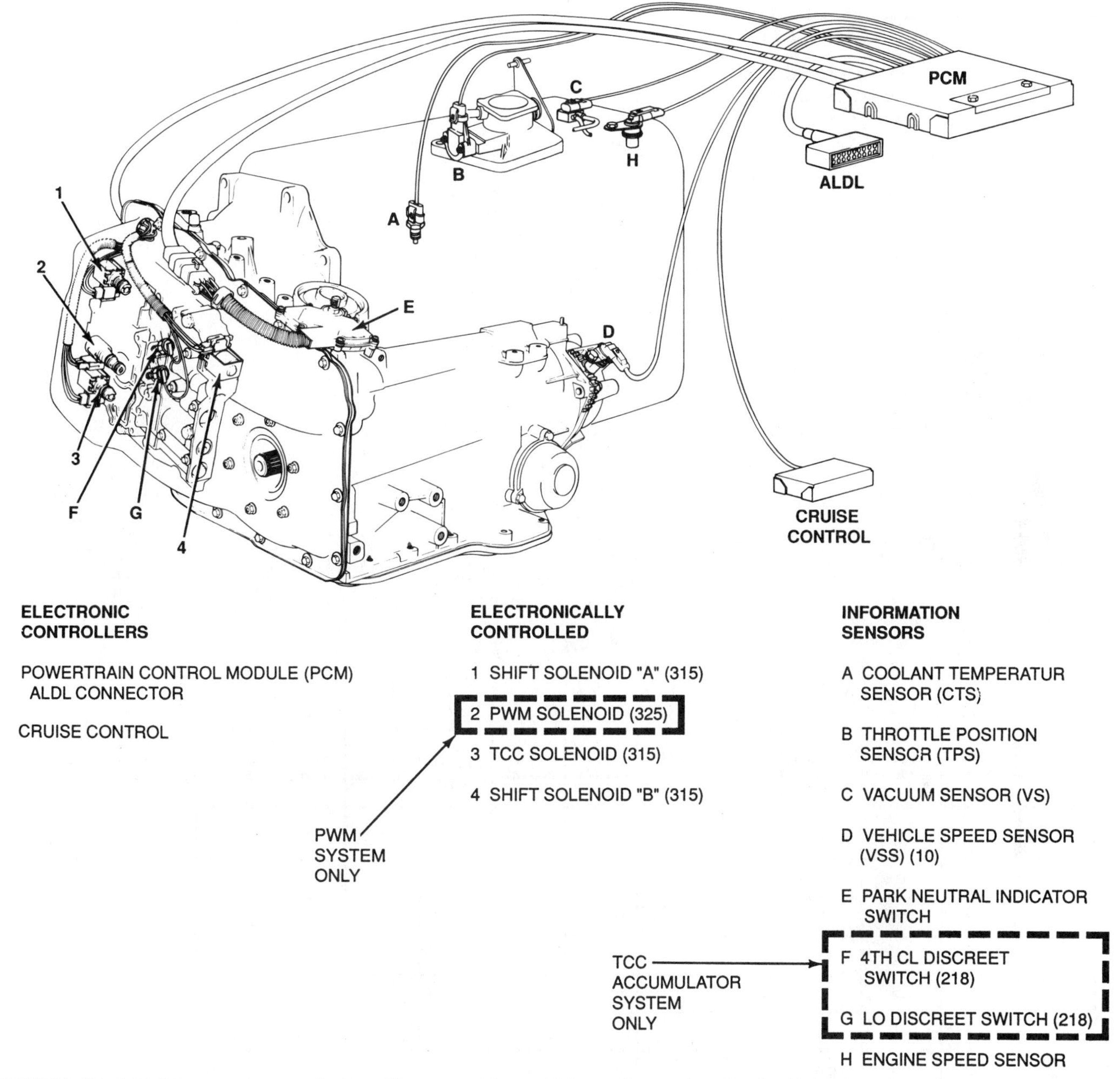

FIGURE 24–35 4T60-E input and output devices. (Courtesy of General Motors Corporation, Service Technology Group)

The major internal design element eliminates the 1-2 band in favor of a 1-2 roller clutch assembly, with an accompanying 2-1 band for coast braking in the manual 2 and manual 1 operating ranges. This eliminates any clutch/band overlap timing concerns that would otherwise be encountered. The 1-2 roller clutch action, therefore, improves the quality of the shift transition for the the 2-3 upshift and the 3-2 downshift.

The case design features a bolt-on case extension that permits the technician to perform repairs to the final drive section without removing the transaxle for a complete disassembly. The operating range and clutch/band application summary is shown in Chart 24–6 (page 598).

Operating Characteristics

The operating range selector uses a typical seven-quadrant arrangement. The following two patterns are used: P-R-N-OD-3-2-1 and P-R-N-D-3-2-1. The characteristics for each operating range are essentially identical to those of the 4T60 (440-T4). Fail-safe operation of the transaxle due to a loss of PCM control allows third-gear operation in OD/D and 3 ranges. Second gear is available in manual 2 during fail-safe mode.

HYDRA-MATIC 4T60-E RANGE REFERENCE CHART

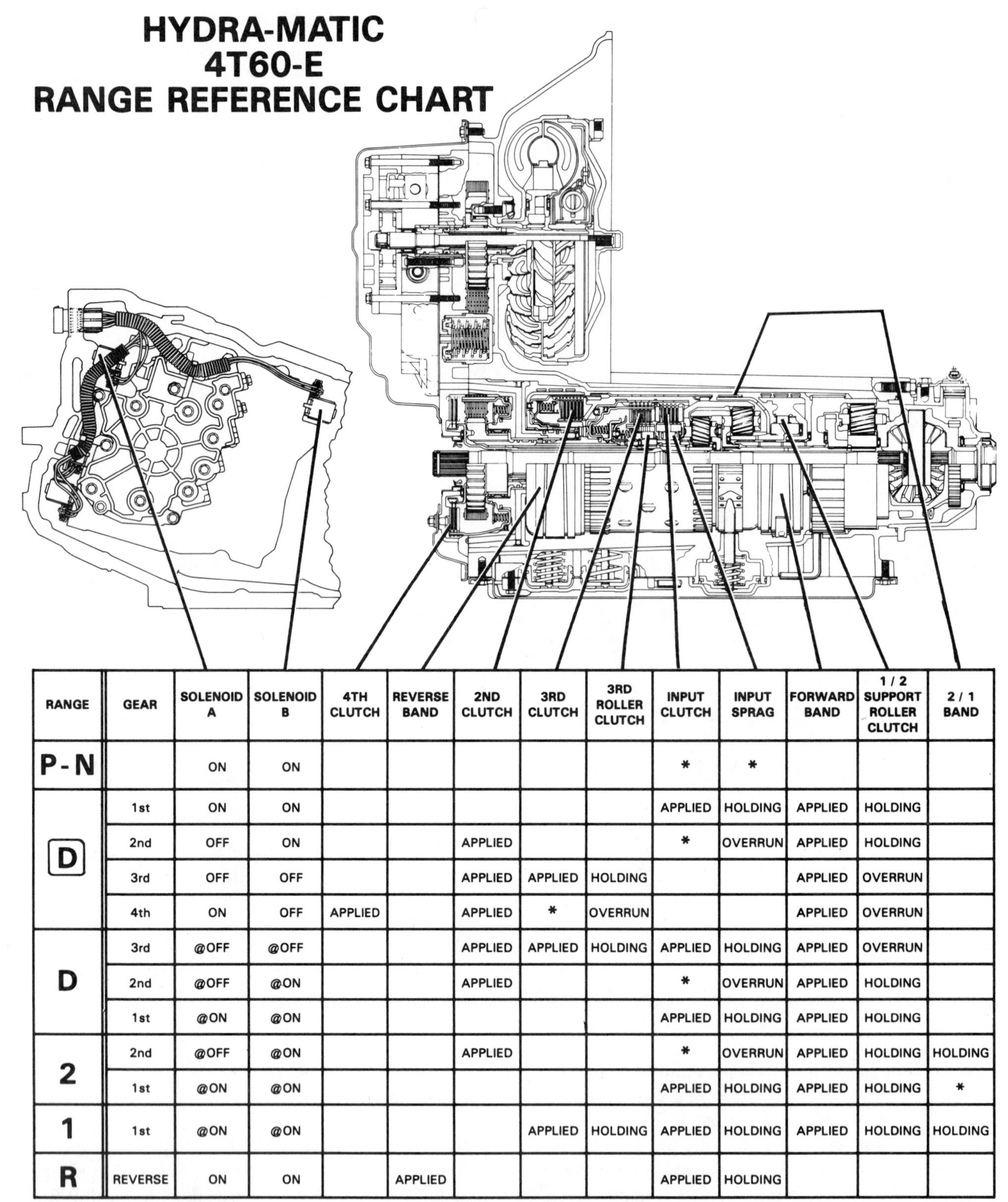

RANGE	GEAR	SOLENOID A	SOLENOID B	4TH CLUTCH	REVERSE BAND	2ND CLUTCH	3RD CLUTCH	3RD ROLLER CLUTCH	INPUT CLUTCH	INPUT SPRAG	FORWARD BAND	1/2 SUPPORT ROLLER CLUTCH	2/1 BAND
P-N		ON	ON						*	*			
[D]	1st	ON	ON						APPLIED	HOLDING	APPLIED	HOLDING	
	2nd	OFF	ON			APPLIED			*	OVERRUN	APPLIED	HOLDING	
	3rd	OFF	OFF			APPLIED	APPLIED	HOLDING			APPLIED	OVERRUN	
	4th	ON	OFF	APPLIED		APPLIED	*	OVERRUN			APPLIED	OVERRUN	
D	3rd	@OFF	@OFF			APPLIED	APPLIED	HOLDING	APPLIED	HOLDING	APPLIED	OVERRUN	
	2nd	@OFF	@ON			APPLIED			*	OVERRUN	APPLIED	HOLDING	
	1st	@ON	@ON						APPLIED	HOLDING	APPLIED	HOLDING	
2	2nd	@OFF	@ON			APPLIED			*	OVERRUN	APPLIED	HOLDING	HOLDING
	1st	@ON	@ON						APPLIED	HOLDING	APPLIED	HOLDING	*
1	1st	@ON	@ON				APPLIED	HOLDING	APPLIED	HOLDING	APPLIED	HOLDING	HOLDING
R	REVERSE	ON	ON		APPLIED				APPLIED	HOLDING			

*APPLIED BUT NOT EFFECTIVE

@ THE SOLENOID'S STATE FOLLOWS A SHIFT PATTERN WHICH DEPENDS UPON VEHICLE SPEED AND THROTTLE POSITION. IT DOES NOT DEPEND UPON THE SELECTED GEAR.

ON = SOLENOID ENERGIZED

OFF = SOLENOID DE-ENERGIZED

MH0163-4T60-E

CHART 24–6 4T60-E range reference and clutch/band application chart. (Courtesy of General Motors Corporation, Service Technology Group)

HYDRA-MATIC 4T65-E

Features and Operating Characteristics

Introduced in certain 1997 models, the 4T65-E powertrain resembles the 4T60-E in basic design and operating characteristics. Many new features and improvements designed into the 4T65-E, however, provide smoother shifts and greater durability.

Expanded PCM input and output parameters enhance shift timing and quality. Rather than using a vacuum modulator to determine line pressure, a PCS (pressure control solenoid) is used. Also, a PSA (pressure switch assembly) provides input signals to the PCM.

Within the newly designed case are stronger components, including the second clutch assembly. In addition, a larger torque converter provides the connection between the engine and the transaxle.

HYDRA-MATIC 4T40-E

The 4T40-E first appeared in the 1995 Chevrolet Cavalier and Pontiac Sunfire with four-cylinder 1.5-L and 2.3-L engines (Figure 24–36). It is a fully automatic and electronically controlled transverse-mounted, four-speed overdrive transaxle. It utilizes an in-line planetary final drive/differential assembly. The torque converter is a three-element unit, with an apply plate for TCC operation, and drives a chain link and sprocket assembly. Figures 24–37 and 24–38 show a sectional view and powertrain schematic of the transaxle.

For mainline operating pressure, the 4T40-E uses a variable-capacity vane pump. Unlike most transmissions/transaxles, this unit does not have a fluid level dipstick. Fluid level is checked at an oil level screw hole below the stub shaft. Fluid is added through the vent cap (Figure 24–39). One pressure tap is available for mainline pressure only.

Electronic Control

The electronic control system is representative of modern design practices. The transaxle electronic system components are illustrated in Figure 24–40. The PCM electronically manages the shift schedule, the TCC operation, and mainline pressure through the pressure control solenoid. Adaptive learning strategies within the PCM are featured for shift timing and line pressure. Self-diagnostics are also incorporated for improved serviceability.

Electronic component functions are similar to those of the Hydra-Matic 4L60-E as detailed in Chapter 11. Fail-safe mode (loss of PCM control) in the 4T40-E allows second-gear operation in the forward-gear shift ranges.

Geartrain Powerflow

Refer to Figures 24–37 and 24–38. The geartrain configuration follows the practice of integrating two simple planetary gearsets, with independent sun gear rotations. To control the action of the geartrain, the 4T40-E incorporates five multiple-disc clutches, two friction bands, and three one-way clutches.

Park/Neutral. Although the turbine torque drives the input gearset through the input sprag, the planetary gears freewheel because of the lack of an effective reactionary member. The weight of the car restrains the output shaft. The low/reverse band is applied to prepare for a smooth reverse garage shift engagement.

Reverse.

- The low/reverse band holds the reaction carrier.
- The reverse input clutch is applied, allowing turbine torque to drive the reaction sun gear.
- Reaction internal gear is the output member.

Reverse is a function of the reaction gearset, providing a gear ratio of 2.310:1.

HYDRA-MATIC 4T40-E TRANSAXLE
RPO MN4

Produced at: Windsor, Ontario Canada

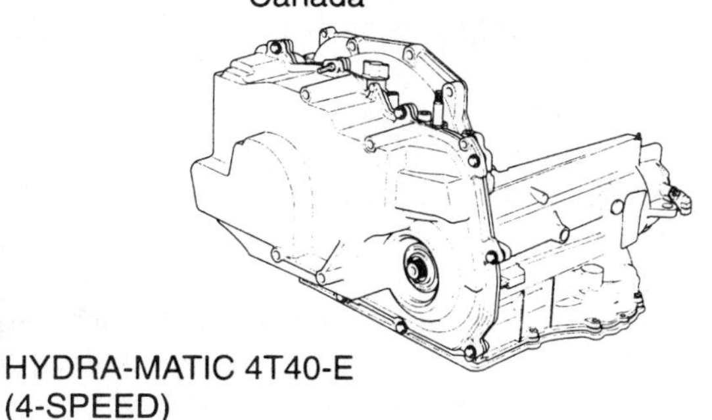

Vehicles used in:

	BUICK	CADILLAC	CHEVROLET	OLDSMOBILE	PONTIAC
J Body			CAVALIER		SUNFIRE

ALSO USED BY: • DAEWOO

HYDRA-MATIC 4T40-E (4-SPEED)

FIGURE 24–36 4T40-E applications (1995). (Courtesy of General Motors Corporation, Service Technology Group)

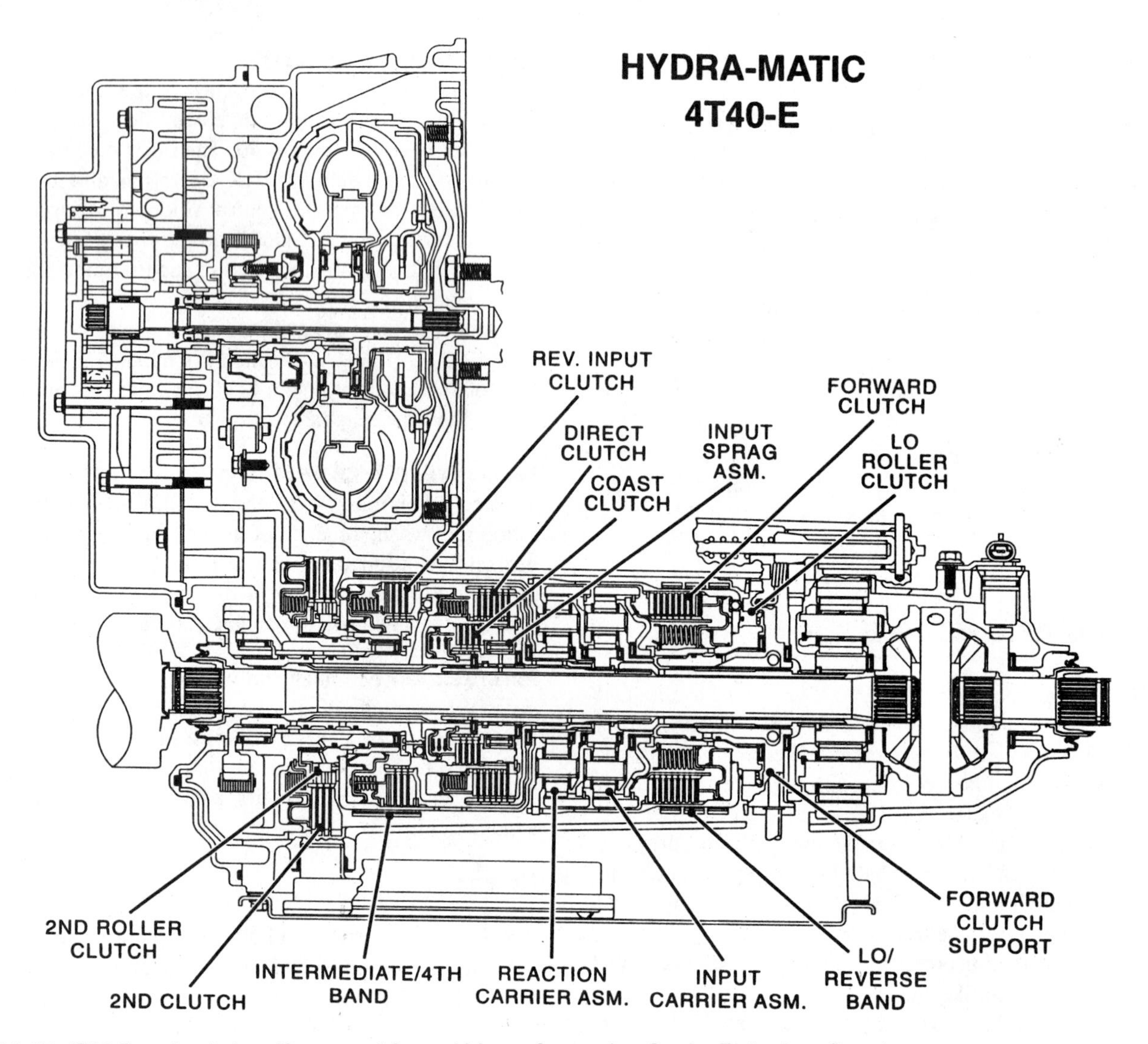

FIGURE 24–37 4T40-E sectional view. (Courtesy of General Motors Corporation, Service Technology Group)

OD Range, First Gear.

- The input sprag is holding, allowing the turbine torque to drive the input sun gear.
- The forward clutch is applied, connecting the input internal gear to the low roller clutch.
- The clockwise effort of the input internal gear locks it to the low roller clutch.
- Input carrier is the output.

First gear is a function of the input gearset, providing a gear ratio of 2.960:1.

OD Range, Second Gear.

- The input sprag is holding, allowing the turbine torque to drive the input sun gear.
- Forward clutch remains applied.
- The second clutch is applied, and the second roller clutch holds. The reaction sun gear clockwise effort locks the roller clutch.
- Low roller clutch overruns.

Second gear is a combined function of the input and reaction planetary set, providing a compound gear ratio of 1.626:1.

OD Range, Third Gear.

- Input sprag holding, allowing the turbine torque to drive the input sun gear.
- Forward clutch remains applied.
- Direct clutch is applied, allowing the turbine torque to drive the reaction planet carrier and input internal gear.
- Second clutch remains applied with no load.
- Second roller clutch and low roller clutch overrun.

Third gear is achieved by locking the input and reaction planetary sets together for a direct-drive ratio of 1:1.

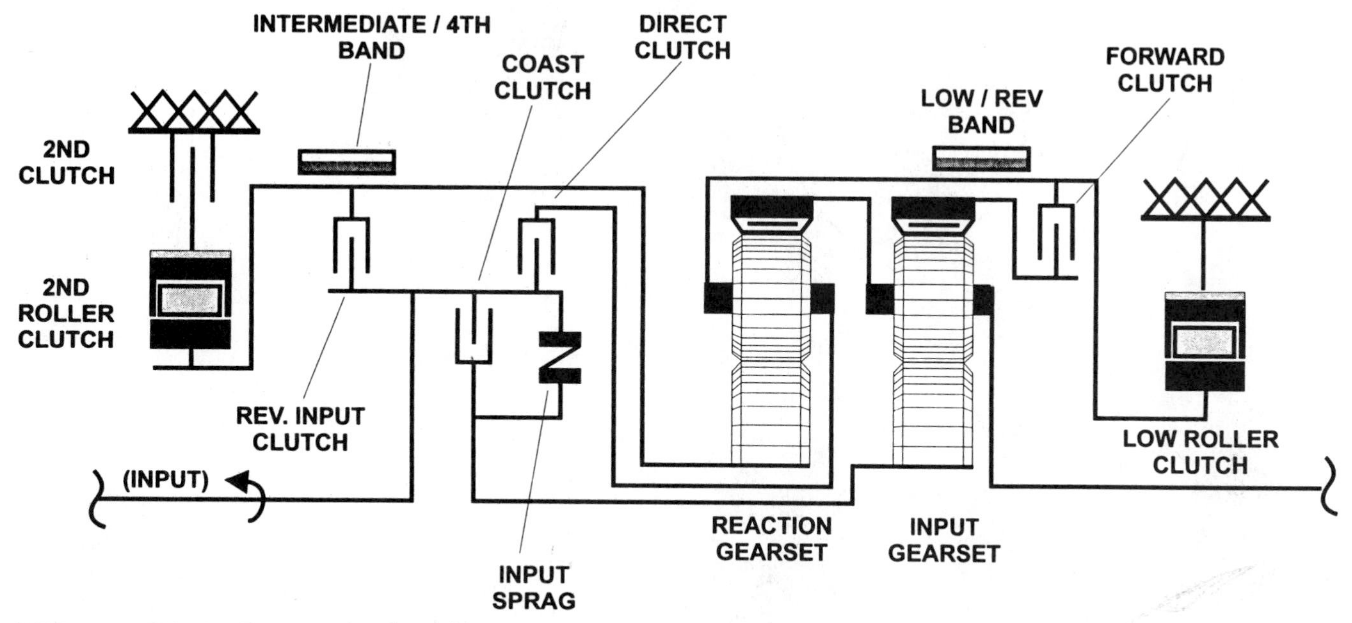

FIGURE 24–38 GM 4T40-E powertrain schematic.

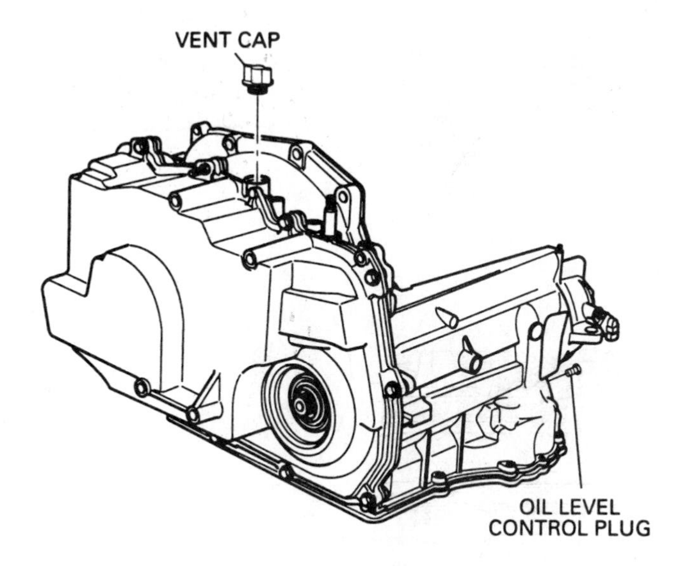

FIGURE 24–39 4T40-E fluid level control plug. (Courtesy of General Motors Corporation, Service Technology Group)

OD Range, Fourth-Gear Overdrive.

- Direct clutch remains applied, allowing the turbine torque to drive the reaction carrier.
- Intermediate/fourth band is applied, holding the reaction sun gear.
- Reaction internal gear is the output overdrive member.
- Input sprag and low roller clutch are overrunning.
- Forward clutch and second clutch remain applied with no load.
- Second roller clutch is in a static, nonoverrunning mode.

Fourth-gear overdrive is a function of the reaction gearset, providing a gear ratio of 0.681:1.

Manual 3, 2, and 1 Ranges. The powerflows in these operating ranges are identical to the powerflows in the OD range. In the manual ranges, the coast clutch is engaged to allow engine braking capabilities for third gear. For manual 2 and 1, the application of the intermediate/fourth band is added for second-

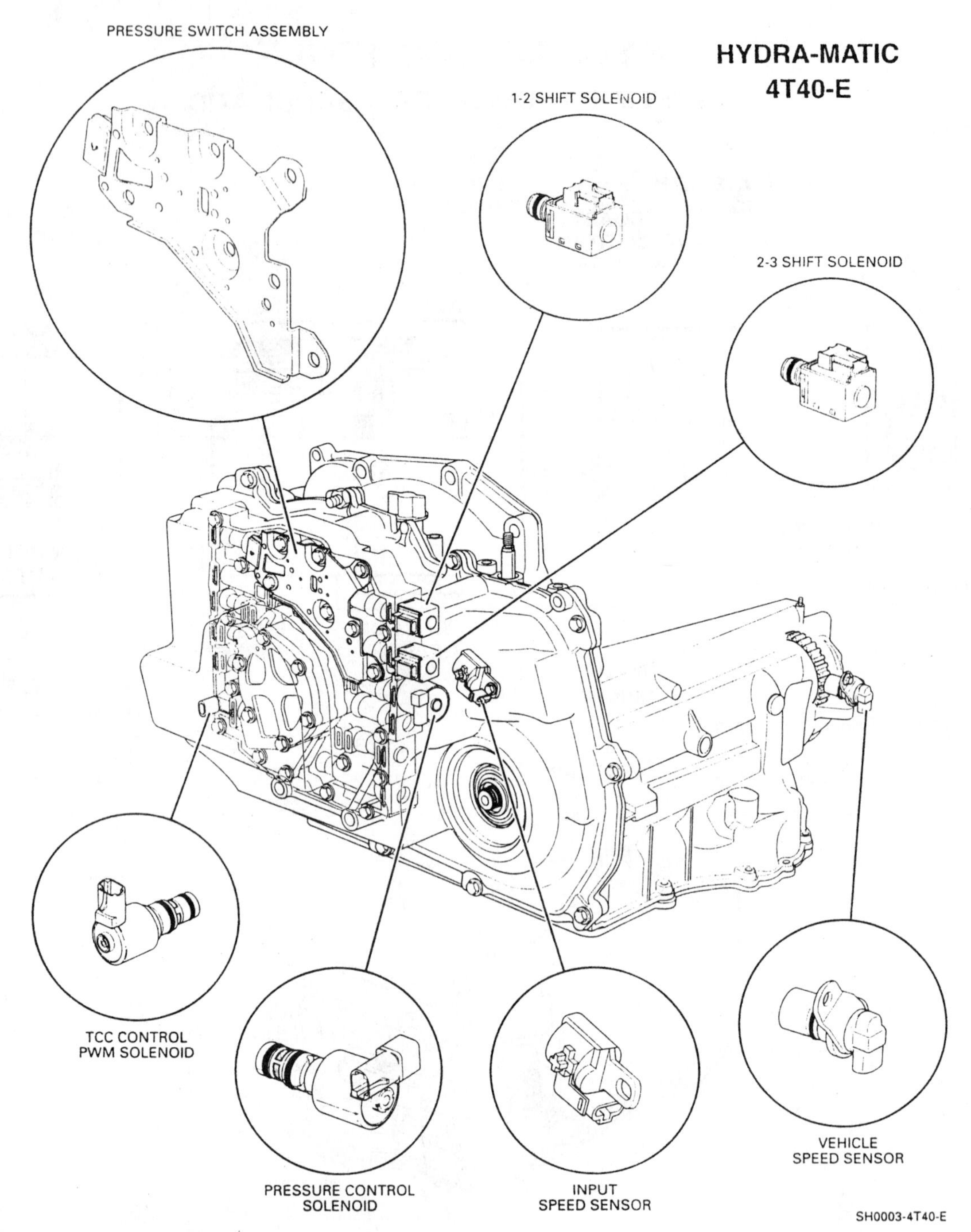

FIGURE 24–40 4T40-E electrical components. (Courtesy of General Motors Corporation, Service Technology Group)

gear coast braking—the second roller clutch and input sprag want to overrun. Manual 1 applies the low/reverse band for first gear – the low roller clutch wants to overrun.

Chart 24–7 features a summary of the operating ranges and clutch/band combinations to control the geartrain.

The sprocket chain drive and planetary final drive offer several ratio combinations, depending on the engine and vehicle application.

Overall Final Drive Ratio

Final Drive Ratio	Chain Ratio*		
	32/38	33/37	35/35
3.05	3.62	3.42	3.05
3.29	3.90	3.59	3.29

*Designates number of teeth on the drive/driven sprockets.

4T40-E **Range Reference Chart**

RANGE	GEAR	1-2 SHIFT SOL	2-3 SHIFT SOL	2ND CLUTCH	2ND ROLLER CLUTCH	INT./4TH BAND	REVERSE CLUTCH	COAST CLUTCH	INPUT SPRAG	DIRECT CLUTCH	FORWARD CLUTCH	LO/REV. BAND	LO ROLLER CLUTCH
PARK	N	ON	OFF									APPLIED	
REV	R	ON	OFF				APPLIED					APPLIED	
NEU	N	ON	OFF									APPLIED	
D	1st	ON	OFF						HOLDING		APPLIED		HOLDING
	2nd	OFF	OFF	APPLIED	HOLDING				HOLDING		APPLIED		OVER-RUNNING
	3rd	OFF	ON	APPLIED*	OVER-RUNNING				HOLDING	APPLIED	APPLIED		OVER-RUNNING
	4th	ON	ON	APPLIED*		APPLIED			OVER-RUNNING	APPLIED	APPLIED*		OVER-RUNNING
3	1st	ON	OFF					APPLIED	HOLDING		APPLIED		HOLDING
	2nd	OFF	OFF	APPLIED	HOLDING			APPLIED	HOLDING		APPLIED		OVER-RUNNING
	3rd	OFF	ON	APPLIED*	OVER-RUNNING			APPLIED	HOLDING	APPLIED	APPLIED		OVER-RUNNING
2	1st	ON	OFF					APPLIED	HOLDING		APPLIED		HOLDING
	2nd	OFF	OFF	APPLIED	HOLDING	APPLIED		APPLIED	HOLDING		APPLIED		OVER-RUNNING
	3rd**	OFF	ON	APPLIED*	OVER-RUNNING			APPLIED	HOLDING	APPLIED	APPLIED		OVER-RUNNING
1	1st	ON	OFF					APPLIED	HOLDING		APPLIED	APPLIED	HOLDING
	2nd***	OFF	OFF	APPLIED	HOLDING	APPLIED		APPLIED	HOLDING		APPLIED		OVER-RUNNING

ON = SOLENOID ENEGERIZED
OFF = SOLENOID DE-ENEGERIZED
* = APPLIED WITH NO LOAD.
** = MANUAL SECOND – THIRD GEAR IS ONLY AVAILABLE ABOVE APPROXIMATELY 100 km/h (62 mph).
*** = MANUAL FIRST – SECOND GEAR IS ONLY AVAILABLE ABOVE APPROXIMATELY 60 km/h (37 mph).
NOTE: MANUAL FIRST – THIRD GEAR IS ALSO POSSIBLE AT HIGH VEHICLE SPEED AS A SAFETY FEATURE.

TH0415-4T40-E

CHART 24–7 4T40-E range reference and clutch/band application chart. (Courtesy of General Motors Corporation, Service Technology Group)

Operating Characteristics

The operating range selector has seven quadrant positions: P-R-N-D-3-2-1. The 4T40-E operating usage and characteristics are similar to other Hydra-Matic four-speed transaxles, but with a couple of variations. Manual 2 and 1 can still be selected when in third or fourth gear. If the vehicle speed is excessive, third gear will be engaged in manual 2 until approximately 62 mph (100 km/h). A high-speed drop into manual 1 causes a similar response, with a downshift into first gear at approximately 37 mph (60 km/h).

HYDRA-MATIC 4T80-E

The 4T80-E entered the Hydra-Matic product line in the 1993 model year and was coupled to the high-torque/high-rpm 4.6-L engine (Figure 24–41). It was first used with Cadillac's Northstar Powertrain system. In the 1994 model year, this transaxle was also applied to the 4.0-L engine in the Oldsmobile Aurora. The 4T80-E is a fully automatic, electronically controlled, transverse four-speed overdrive FWD transaxle. The final drive/differential assembly is in line with the geartrain.

A three-element torque converter incorporates either a spring-dampened or viscous clutch–style apply plate for TCC. The converter drives a chain link and sprocket assembly. Figures 24–42 and 24–43 show a sectional view and a powertrain schematic of the transaxle.

Unique to the 4T80-E is the use of three pump assemblies required to meet transaxle operating demands (Figure 24–44). These are referred to as the scavenge, primary, and secondary pump assemblies. All three pumps are converter driven through a pump drive shaft (Figure 24–45). The oil supply system is designed to operate with a dry sump.

The scavenge pump is an external-to-external positive displacement gear pump design with one drive and two driven gears. It functions to continuously scavenge the fluid from the

HYDRA-MATIC 4T80-E TRANSAXLE

RPO MH1

Produced at: Willow Run, Ypsilanti, MI

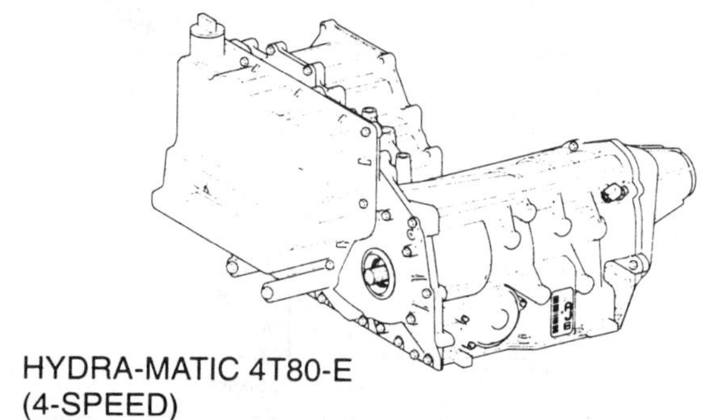

Vehicles used in:

	CADILLAC	OLDSMOBILE
E Body	ELDORADO	
G Body		AURORA
K Body	SEVILLE CONCOURS	
K-Spec	CONCOURS DEVILLE	

FIGURE 24–41 4T80-E applications (1995). (Courtesy of General Motors Corporation, Service Technology Group)

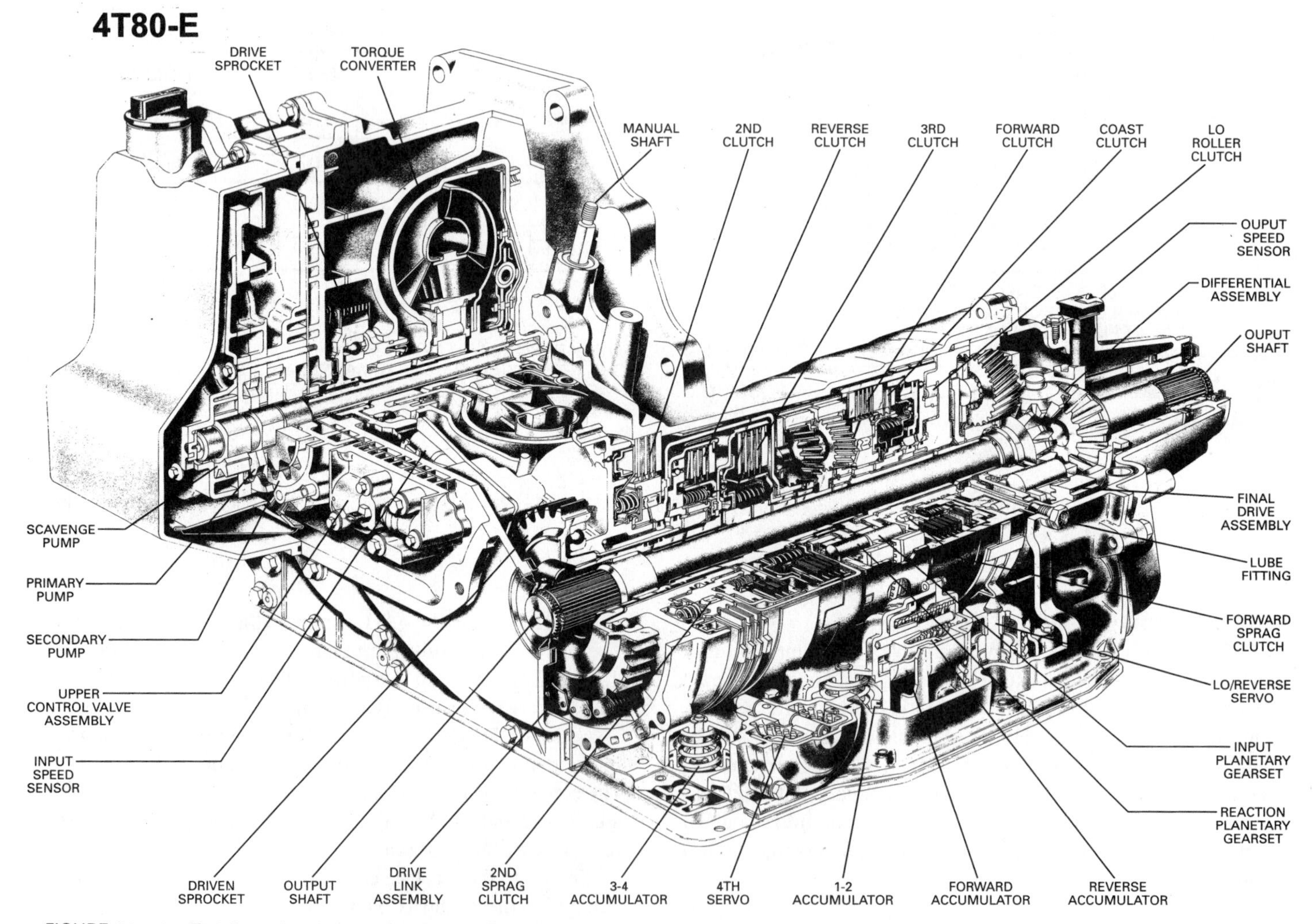

FIGURE 24–42 4T80-E sectional view. (Courtesy of General Motors Corporation, Service Technology Group.)

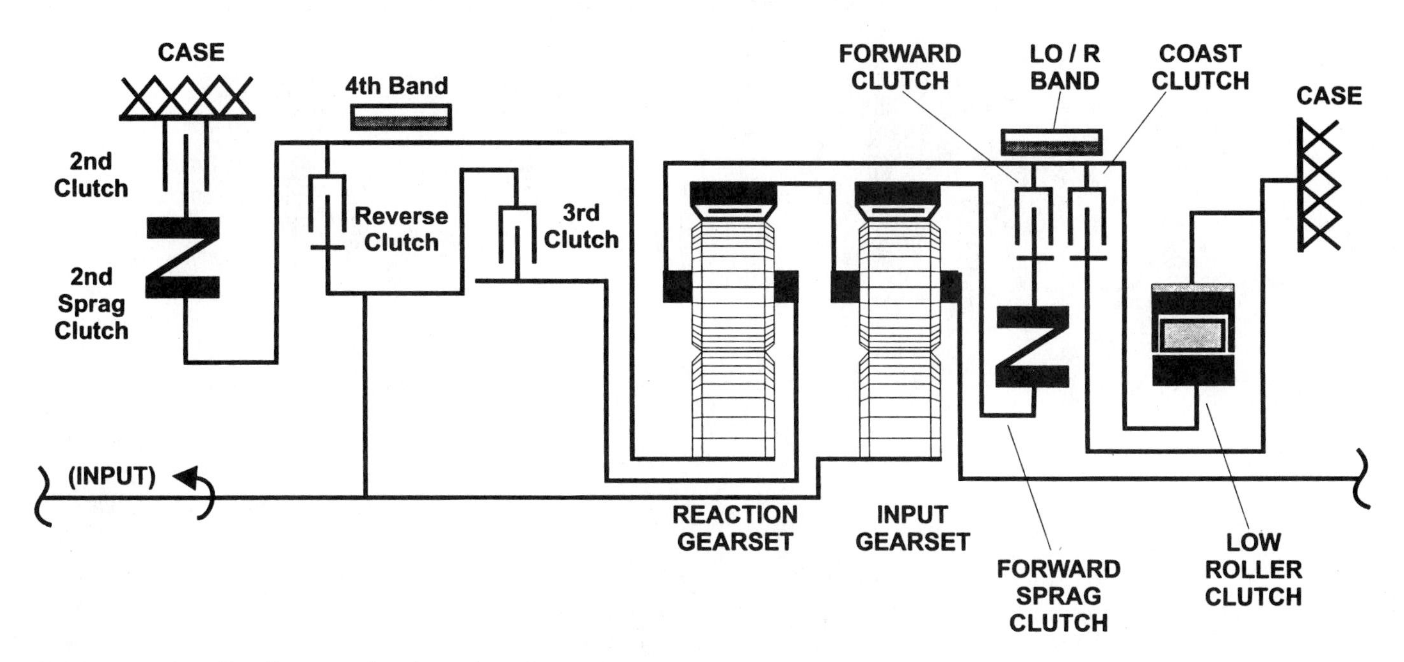

FIGURE 24–43 GM 4T80-E powertrain schematic.

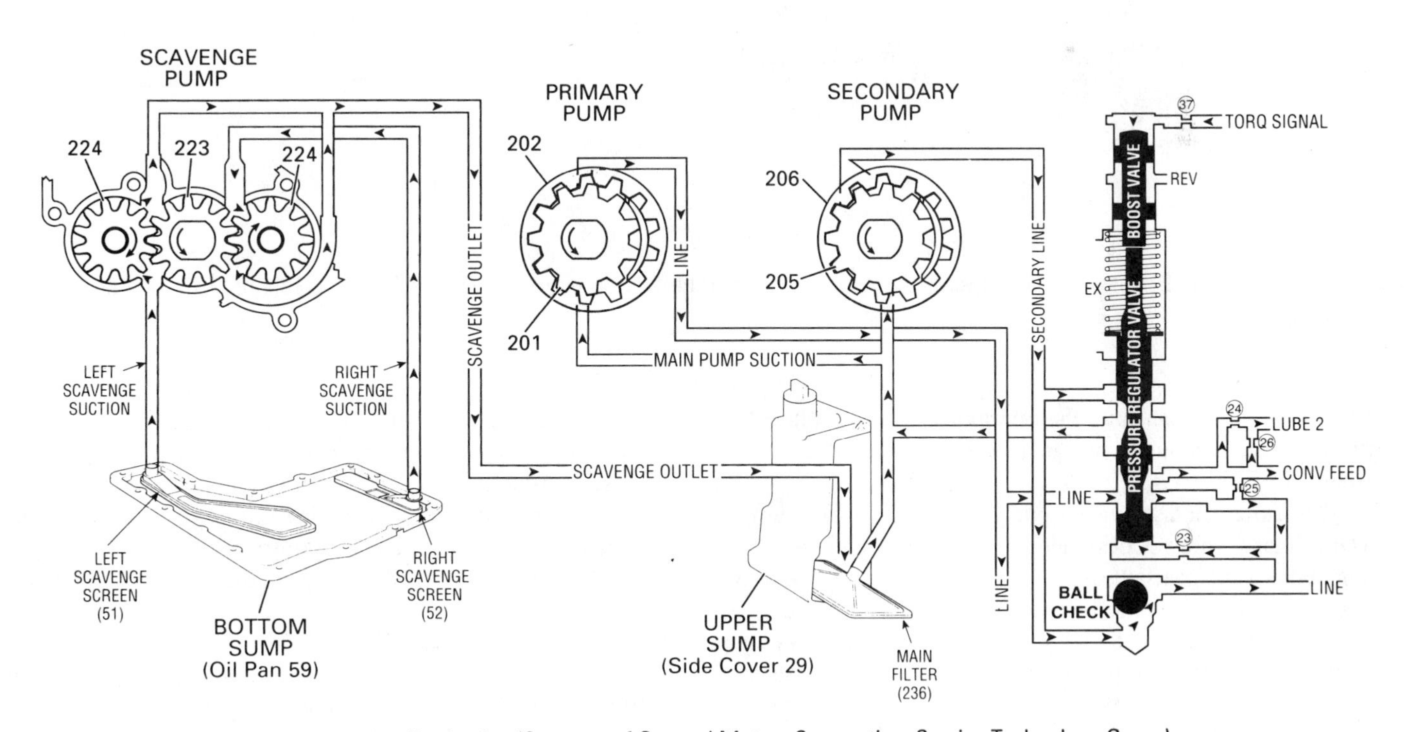

FIGURE 24–44 4T80-E pumps and hydraulic circuits. (Courtesy of General Motors Corporation, Service Technology Group)

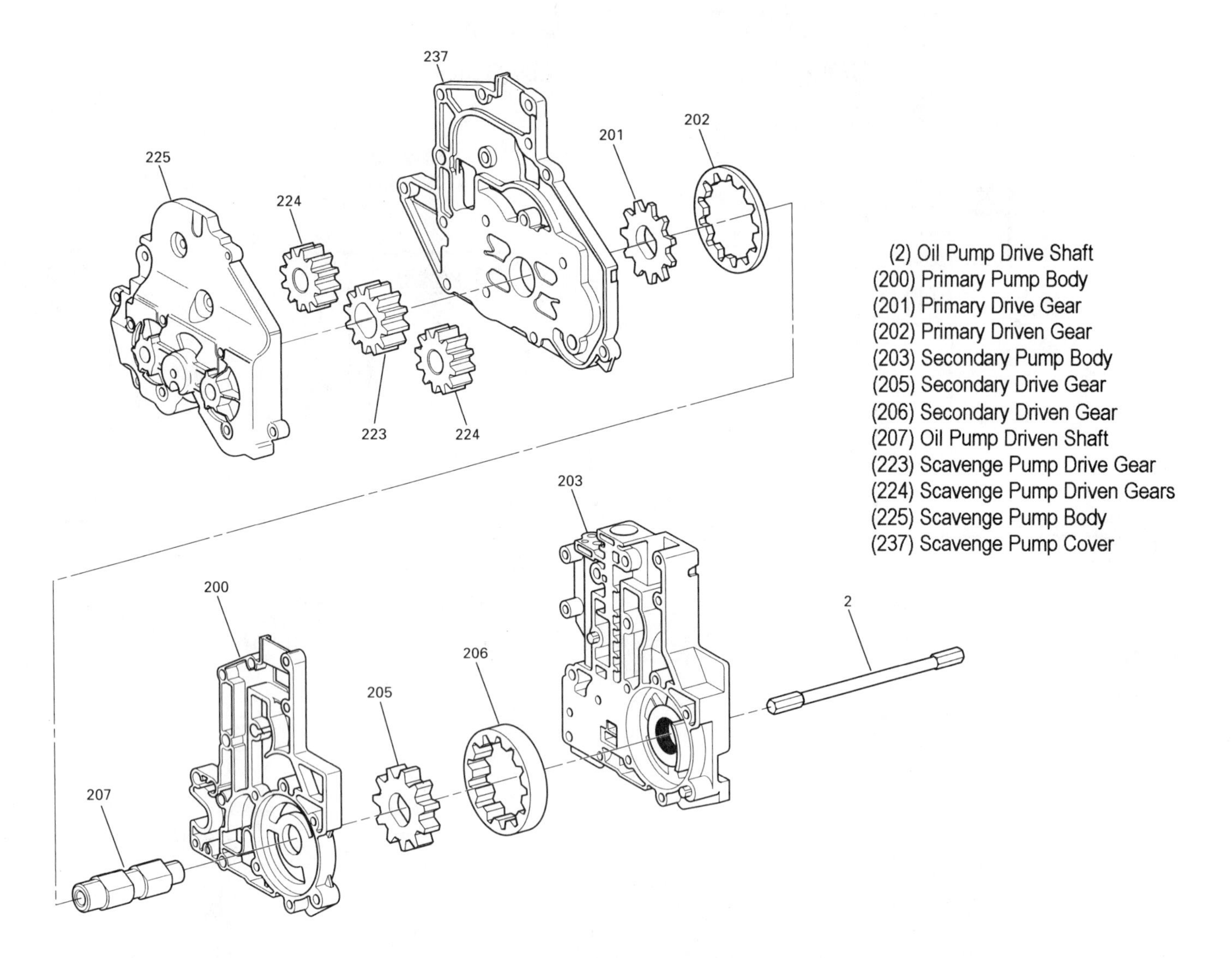

FIGURE 24–45 4T80-E pump components driven by the oil pump drive shaft (#2). (Courtesy of General Motors Corporation, Service Technology Group)

bottom sump (oil pan) into the transaxle side cover where the primary and secondary pumps pick up their common fluid source.

The primary and secondary pumps are IX gear–type pumps. The primary circuit is designed to meet the normal operating demands of the transaxle in all the gear ranges. When extra boost pressure is required during situations such as reverse, wide open throttle, or detent downshifting, the secondary pump assists the primary pump. The secondary pump also assists in situations where extra fluid capacity is demanded during a clutch engagement. Both pumps interact through a common pressure regulator valve and bypass check ball. One pressure tap is available for mainline pressure only.

Electronic Control

The electronic control system is representative of modern Hydra-Matic design practices. The electronic control system components are illustrated in Figure 24–46. The PCM electronically manages the shift schedule, TCC operation, and line pressure. Adaptive learning functions are incorporated into the PCM for shift scheduling and line pressure. Self-diagnostics are also included into the PCM programming.

Electronic component functions are presented in Chapter 11.

Geartrain Powerflow

Refer to Figures 24–42 and 24–43. The planetary configuration follows the practice of integrating two simple sets with independent sun gear rotation. The geartrain is controlled by five multiple-disc clutches, two friction bands, and three one-way clutches (two sprags and one roller clutch).

Park/Neutral. Although the turbine torque drives the input sun gear, the input gearset freewheels because there is not an effective reactionary member. The low/reverse band is

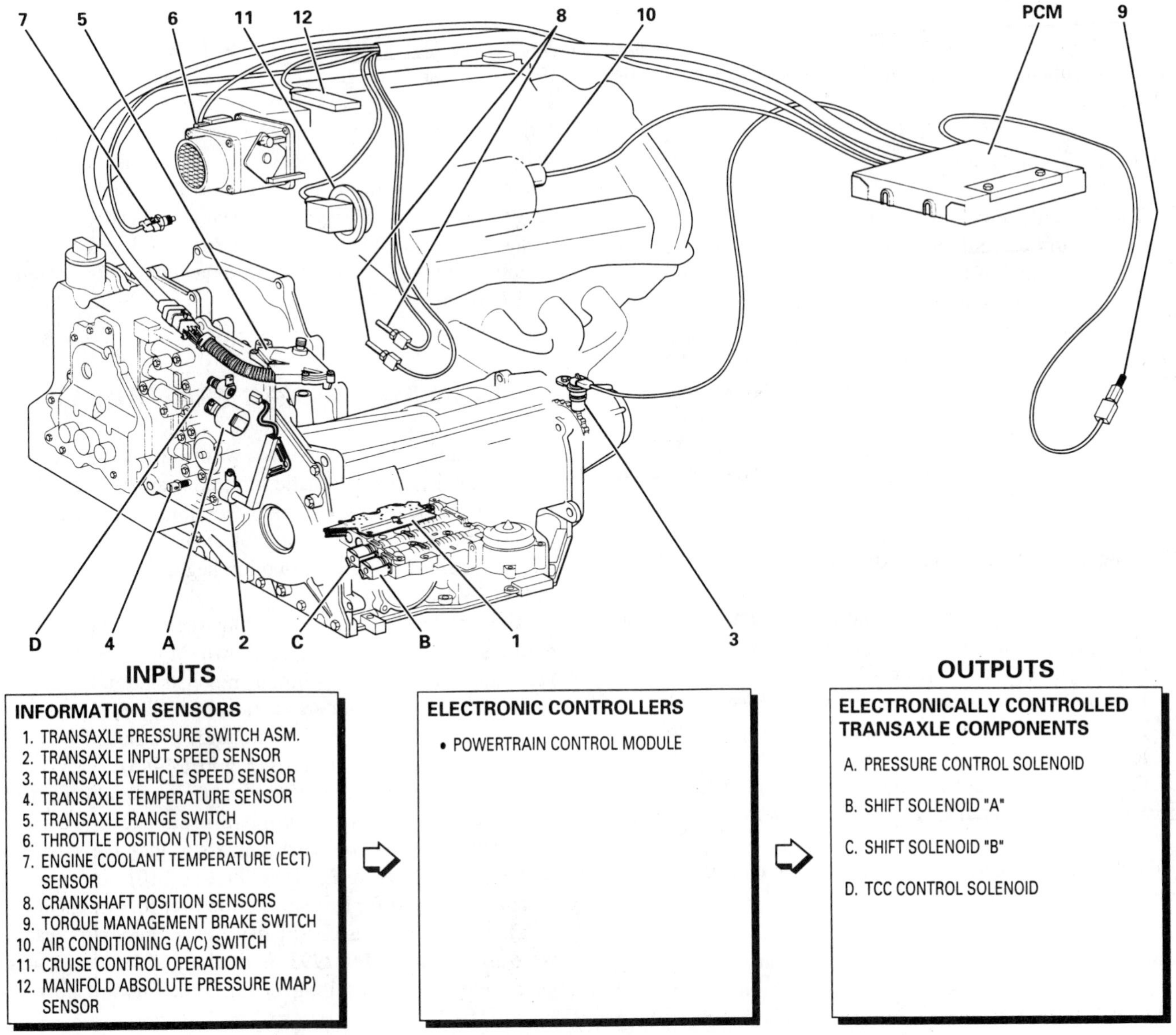

INPUTS

INFORMATION SENSORS

1. TRANSAXLE PRESSURE SWITCH ASM.
2. TRANSAXLE INPUT SPEED SENSOR
3. TRANSAXLE VEHICLE SPEED SENSOR
4. TRANSAXLE TEMPERATURE SENSOR
5. TRANSAXLE RANGE SWITCH
6. THROTTLE POSITION (TP) SENSOR
7. ENGINE COOLANT TEMPERATURE (ECT) SENSOR
8. CRANKSHAFT POSITION SENSORS
9. TORQUE MANAGEMENT BRAKE SWITCH
10. AIR CONDITIONING (A/C) SWITCH
11. CRUISE CONTROL OPERATION
12. MANIFOLD ABSOLUTE PRESSURE (MAP) SENSOR

ELECTRONIC CONTROLLERS

- POWERTRAIN CONTROL MODULE

OUTPUTS

ELECTRONICALLY CONTROLLED TRANSAXLE COMPONENTS

A. PRESSURE CONTROL SOLENOID

B. SHIFT SOLENOID "A"

C. SHIFT SOLENOID "B"

D. TCC CONTROL SOLENOID

FIGURE 24–46 4T80-E input and output devices. (Courtesy of General Motors Corporation, Service Technology Group)

applied to prepare for a smooth reverse or drive garage shift engagement.

Reverse.

- The low/reverse band holds the reaction carrier.
- The reverse clutch is applied, allowing the turbine torque to drive the reaction sun gear.
- Reaction internal gear is the output.

Reverse is a function of the reaction gearset, providing a gear ratio of 2.13:1.

OD Range, First Gear.

- The turbine torque drives the input sun gear.
- The forward clutch is applied, and the forward sprag clutch connects the input internal gear to the low roller clutch.
- The clockwise effort of the input internal gear locks it to the low roller clutch.
- Input carrier is the output.

The applied low/reverse band is not effective during a state of drive torque. The low roller clutch is the prime holder. First gear is a function of the input gearset, providing a gear ratio of 2.96:1.

OD Range, Second Gear.

- The turbine torque drives the input sun gear.
- The forward clutch remains applied, and forward sprag continues to hold.
- The second clutch is applied, and the second roller clutch holds. Clockwise sun gear effort locks the roller clutch.
- Low roller clutch overruns.

Second gear is a combined function of the input and reactionary planetary sets, providing a compound gear ratio of 1.626:1.

OD Range, Third Gear.

- The forward clutch remains applied.
- Turbine torque continues to drive the input sun gear.
- The third clutch is applied, allowing the turbine torque to also drive the input internal gear through the reaction carrier, forward clutch, and forward sprag clutch.
- Second clutch remains applied with no load.
- Second sprag clutch and low roller clutch overrun.

Third gear is achieved by locking the input reaction planetary sets together with a dual path input of equal turbine drive rpm, providing a direct-drive ratio (1:1).

OD Range, Fourth-Gear Overdrive.

- The fourth band is applied, holding the reaction sun gear.
- Direct clutch remains applied, allowing the turbine torque to drive the reaction carrier.
- Reaction internal gear is the output overdrive member.
- Forward clutch and second clutch remain applied with no load.
- Forward sprag and low roller are overrunning.
- Second roller clutch is in a static, nonoverrunning mode.

Fourth-gear overdrive is a function of the reaction gearset, providing a gear ratio of 0.681:1.

Manual 3, 2, and 1 Ranges. The powerflow in these operating ranges is identical to the powerflow in the OD range. In manual 3, 2, and 1, the coast clutch is applied to allow engine braking capabilities. Third-gear engine braking occurs because the coast clutch prevents the input sun gear from freewheeling. In manual 2 and 1, the application of the fourth band is added for second-gear coast braking—the second sprag wants to overrun. Manual 1 applies the low/reverse band for first gear—the low roller clutch wants to overrun.

Chart 24–8 features a summary of the operating ranges and clutch/band combinations to control the 4T80-E geartrain.

The sprocket chain drive and planetary final drive offer several ratio combinations, depending on the engine and vehicle application.

Overall Final Drive Ratio

Final Drive Ratio	Chain Ratio*
	39/39
3.11	3.11
3.71	3.71

*Designates number of teeth on the drive/driven sprockets.

Operating Characteristics

The operating range selector has seven quadrant positions: P-R-N-D-3-2-1. The 4T80-E operating usage and characteristics are similar to other four-speed Hydra-Matic electronically controlled automatic transaxles. When manual 2 is chosen while in third or fourth gear, the unit immediately shifts into second gear. Manual 1 shifts into second gear if selected at an excessive speed. When vehicle speed is reduced to approximately 35 mph (56 km/h), the transaxle downshifts and remains in first gear.

SATURN AUTOMATIC TRANSAXLES

Though Saturn Corporation is a wholly owned subsidiary of General Motors, its automatic transaxles are considerably different in design and operation. Rather than being supplied with units from Hydra-Matic, Saturn has built its own automatic transaxles since vehicle production began for the introduction of the 1991 models.

The electronically controlled four-speed transaxle, utilizing a torque converter clutch, is shown in Figure 24–47. The geartrain arrangement is similar in construction to the design used in Honda automatic transaxles. Instead of using planetary gearsets, Saturn (and Honda) automatic transaxles use helical-cut gears run on parallel shafts (Figure 24–48).

The basic construction of the geartrain resembles a manual transaxle design that uses multiple-disc clutches and a one-way sprag clutch, instead of synchronizers to lock the speed gears to the shafts. Refer to Chart 24–9 (page 610) to examine the clutch apply chart for reverse and the four forward gears.

Three different designs of the Saturn automatic transaxle were used from 1991 to 1993. A variation in the design was the gear teeth counts, resulting in different forward gear ratios (Chart 24–10, page 611).

For additional information on parallel shaft automatic transaxle construction and operation, refer to the section on Honda in Chapter 25.

SUMMARY

The new generation of PCM-managed Hydra-Matic transmissions and transaxles produced by GM's Powertrain Division share similar fundamentals of electronic control and operation. It is anticipated that in the near future, the hydraulically controlled 3T40 (THM 125C) and the vacuum modulator equipped 4T60-E will be phased out and replaced with fullly PCM-managed transaxles. The 4T40-E, introduced in 1995, is currently being utilized in certain vehicle platforms that once used the 3T40, and the 4T65-E will replace 4T60-E applications.

Vehicle fuel efficiency and emission levels continue to be major factors in the design and utilization of new transmissions and transaxles. Alternative fuel vehicles with low and zero emission ratings are currently being developed.

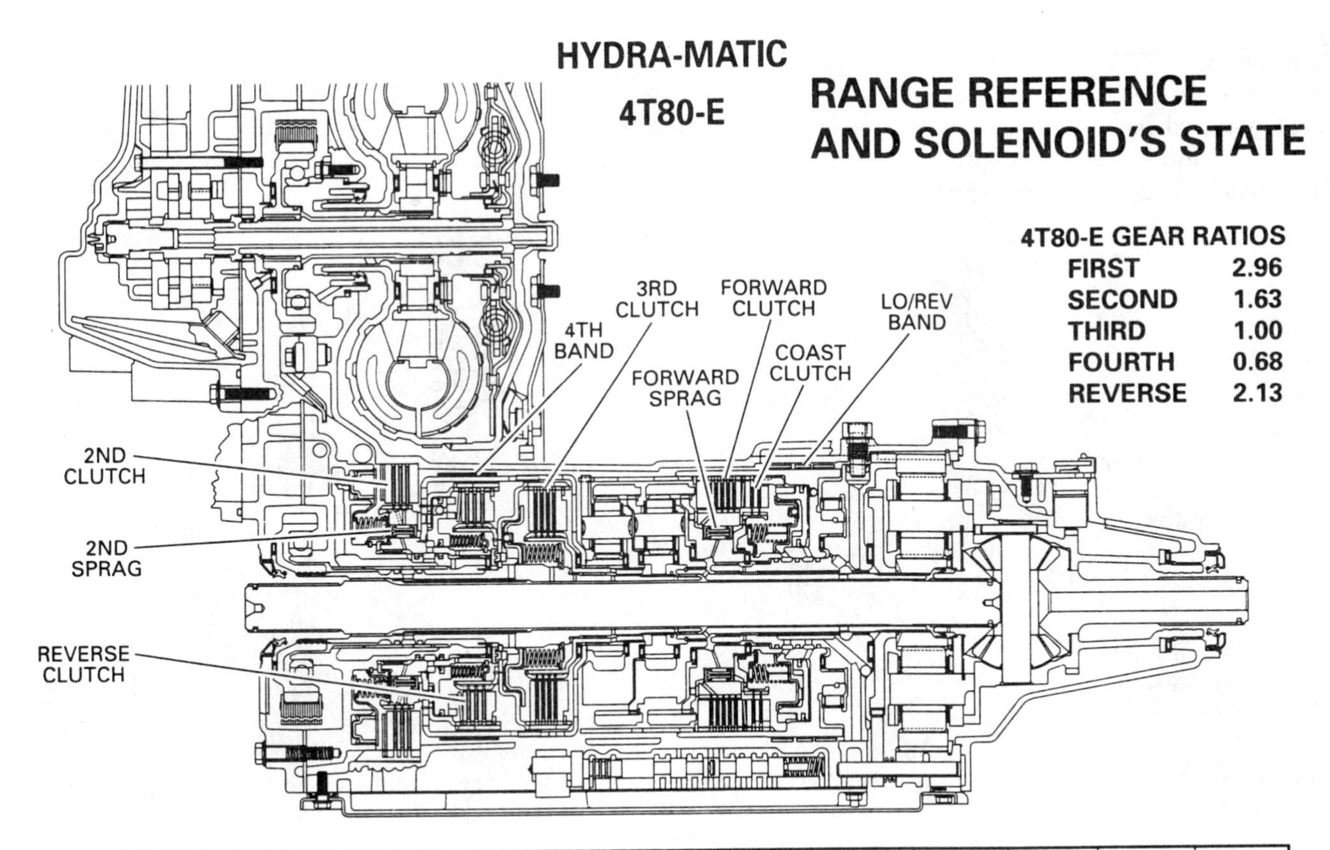

RANGE	GEAR	SHIFT "A" SOL	SHIFT "B" SOL	FORWARD	COAST	3RD CLUTCH	2ND CLUTCH	REVERSE	4TH BAND	LO/REV. BAND	LO ROLLER	2ND SPRAG	FORWARD SPRAG
PARK	N	ON	OFF							APPLIED			
REV	R	ON	OFF					APPLIED		APPLIED			
NEU	N	ON	OFF							APPLIED			
D4	1	ON	OFF	APPLIED						APPLIED	HOLDING		HOLDING
	2	OFF	OFF	APPLIED			APPLIED					HOLDING	HOLDING
	3	OFF	ON	APPLIED		APPLIED	APPLIED						HOLDING
	4	ON	ON	APPLIED		APPLIED	APPLIED		APPLIED				
3	1	ON	OFF	APPLIED						APPLIED	HOLDING		HOLDING
	2	OFF	OFF	APPLIED			APPLIED					HOLDING	HOLDING
	3	OFF	ON	APPLIED	APPLIED	APPLIED	APPLIED						HOLDING
2	1	ON	OFF	APPLIED	APPLIED					APPLIED	HOLDING		HOLDING
	2	OFF	OFF	APPLIED	APPLIED		APPLIED		APPLIED			HOLDING	HOLDING
1	1	ON	OFF	APPLIED	APPLIED					APPLIED	HOLDING		HOLDING
	2	OFF	OFF	APPLIED	APPLIED		APPLIED		APPLIED			HOLDING	HOLDING

ON = SOLENOID ENEGERIZED
OFF = SOLENOID DE-ENEGERIZED
@ THE SOLENOID'S STATE FOLLOWS A SHIFT PATTERN WHICH DEPENDS UPON VEHICLE SPEED AND THROTTLE POSITION. IT DOES NOT DEPEND UPON THE SELECTED GEAR.

SH0002-4T80-E

CHART 24–8 4T80-E range reference and clutch/band application chart. (Courtesy of General Motors Corporation, Service Technology Group)

The EV1, GM's electric-powered two-passenger coupe, is being marketed to the public in California and Arizona through Saturn dealers. Other powertrain designs being developed may, like the EV1, use transmissions that are unique from those found in conventionally powered vehicles.

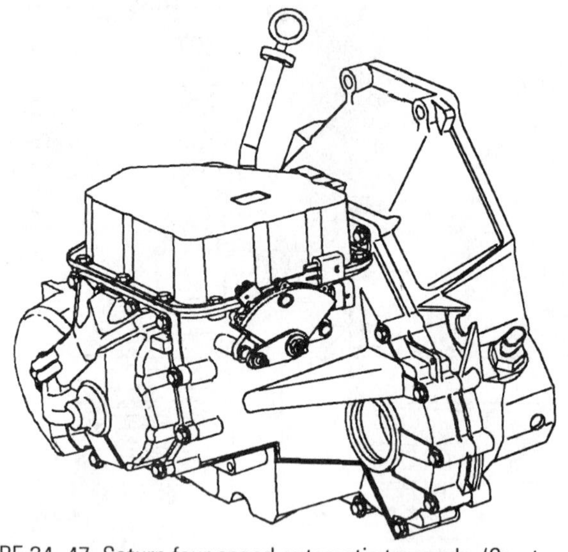

FIGURE 24–47 Saturn four-speed automatic transaxle. (Courtesy of Saturn Corporation)

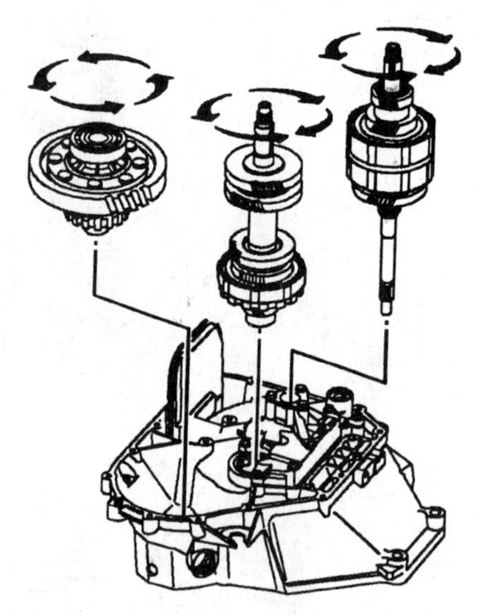

FIGURE 24–48 Saturn automatic transaxle input, output, and differential assembly. (Courtesy of Saturn Corporation)

Clutch	GEAR				
	1ST	2ND	3RD	4TH	REVERSE
1st	ON	ON	ON	ON	
2nd/Reverse		ON			ON
3rd			ON		
4th				ON	
1st Sprag	**ON**				
TCC	ON*	ON	ON	ON	
Fwd/Rev Dog Clutch	To 2nd Driven Gear	To 2nd Driven Gear	To 2nd Driven Gear	To 2nd Driven Gear	To Reverse Driven Gear

* 1993 & 1994 vehicles

CHART 24–9 Saturn automatic transaxle clutch apply chart. (Courtesy of Saturn Corporation)

ALL SOHC USE MP6 TRANSAXLES
ALL DOHC USE MP7 TRANSAXLES

	1991, 1992 1993 1st Design* MP6 (Base)	1993 2nd Design** MP6 (Base)	1991, 1992, 1993 MP7 (Performance)
	Gear Teeth Count	**Gear Teeth Count**	**Gear Teeth Count**
1st Drive:	21	19	19
1st Driven:	47	48	48
2nd Drive:	30	30	27
2nd Driven:	38	38	42
3rd Drive:	37	37	33
3rd Driven:	30	30	34
4th Drive:	42	42	40
4th Driven:	25	25	28
Reverse Drive:	21	21	21
Reverse Driven:	40	40	40
Reverse Idler:	27	27	27
Final Drive:	27	27	27
Output Shaft:	15	15	15
Ring Gear:	62	62	62
	Gear Ratio	**Gear Ratio**	**Gear Ratio**
1st Gear:	2.24	2.53	2.53
2nd Gear:	1.17	1.17	1.56
3rd Gear:	0.81	0.81	1.03
4th Gear:	0.60	0.60	0.70
Reverse Gear:	2.39	2.39	2.39
Final Drive:	4.1333	4.1333	4.1333

* Vehicles built prior to, and including VIN PZ156139

** Vehicles built after, and including VIN PZ156140

CHART 24–10 Saturn gear teeth counts and ratios. (Courtesy of Saturn Corporation)

25 Import and Foreign Automatic Transmission/Transaxle Summaries

INTRODUCTION

Around the world, vehicles equipped with automatic transmissions and transaxles provide drivers and passengers with comfort and convenience. They also contribute to improved fuel economy and reduced vehicle emissions.

Approximately 87% of the North American passenger cars built in 1993 used automatic transmissions. Application of automatic transmissions in domestic vehicles has steadily increased after the oil embargos and energy crisis in the mid- to late-1970s. This trend can be contributed to purchaser demands, as well as the evolution of electronically controlled, four-speed overdrive automatic transmissions utilizing lockup torque converters.

The usage of automatic transmissions in Japanese passenger cars was only 3% in 1970. In 1993, this number was estimated to have increased to 74%. European passenger car buyers, however, continue to prefer manual over automatic transmissions/transaxles. Approximately 9% of 1993 European new car buyers bought passenger vehicles equipped with automatic transmissions.

IMPORT AND FOREIGN TRANSMISSION/TRANSAXLE OVERVIEW

The intent of this chapter is to show that transmissions and transaxles from around the world are more alike than different. Principles of transmission/transaxle construction, operation, diagnosis, and service are common in many aspects. Similarities, as well as differences, are highlighted in this chapter.

To demonstrate typical import and foreign transmission/transaxle design and operation, units of foreign origin are summarized within this chapter. The summaries provide an overview of applications, features, and operation of these units.

TYPICAL CONSTRUCTION AND OPERATION CHARACTERISTICS

Planetary Gearset Design

The basic design and operating characteristics of most transmissions produced and used throughout the world are quite similar (Figure 25–1). There are exceptions, but most transmis-

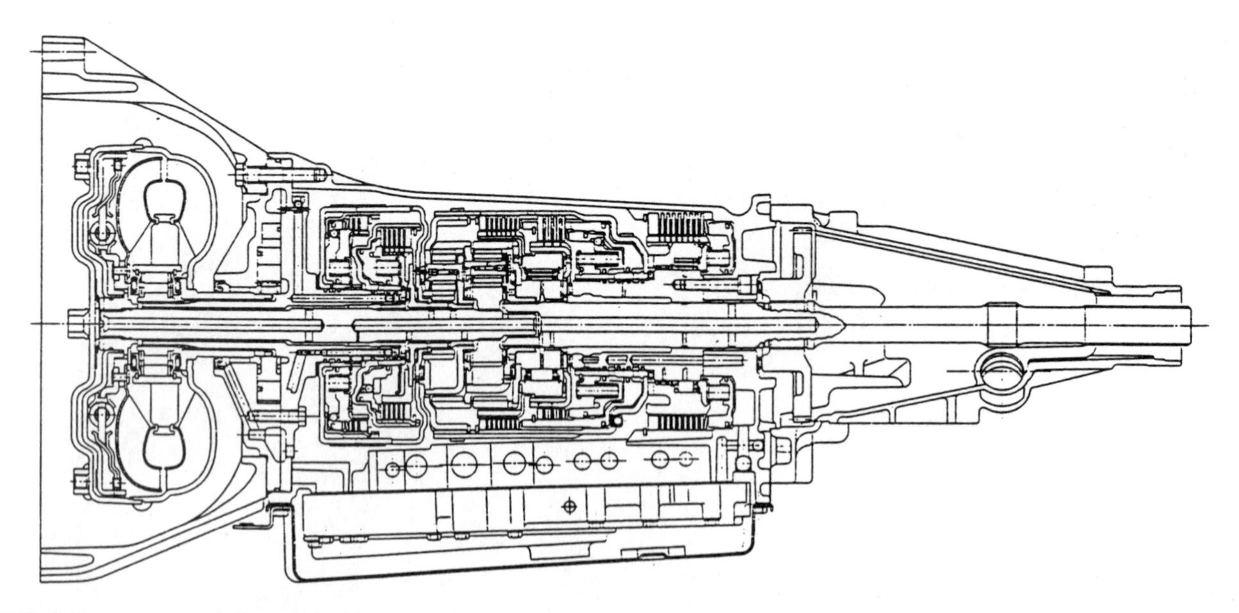

FIGURE 25–1 Cross-sectional view of the Nissan R01A model. (Reprinted with permission from SAE AE-18 © 1994 Society of Automotive Engineers, Inc.)

sions and transaxles utilize planetary gearsets to produce the reverse and forward gears.

Three-speed transmissions use either a compound Simpson or Ravigneaux planetary gearset. Four-speed versions have been developed by adding an overdrive planetary gearset to an existing three-speed design or by designing compound planetary gearsets with independent sun gears. Since 1989, some manufacturers have built five-speed transmissions utilizing a four-speed transmission coupled to a built-in overdrive planetary unit.

Clutches and Bands

Though their names may be different, clutch units and bands serve the same functions as those found in domestically produced transmissions (Figure 25–2). Clutch assemblies, including one-way devices, are used to hold or drive planetary members. Friction bands are also used to hold gearset members reactionary. A device that holds a planetary member stationary is often referred to as a "brake." One-way clutches, both sprag and roller designs, are often called "freewheelers."

Torque Converters and Fluid (ATF)

Torque converter designs are typically the same. In most applications, three-element clutching converters are used (Figure 25–3).

The ATF suggested for most automatic transmissions is Dexron III or an equivalent. Many manufacturers, such as Honda, Ford, Mercedes, and Renault, encourage using their fluid type.

Hydraulic and Electronic Control

Control of automatic upshifts and downshifts is accomplished by a variety of means, as in domestic units. Full hydraulic/mechanical, electronic, and hydraulic/mechanical-electronic control designs are common. The trend is toward full-electronic logic control to operate transmissions for improved vehicle fuel efficiency and reduced emissions. Figure 25–4 shows the relationships between the sensors, control modules, and the powertrain.

While most domestic manufacturers use "hot-side" switching to control transmission shift and torque converter clutch (TCC) solenoids, "ground-side" switching is popular in import and foreign transmission wiring circuits. In addition, TCMs (transmission control modules) rather than PCMs (powertrain control modules) are typically used with electronically controlled import and foreign transmissions (Figure 25–5).

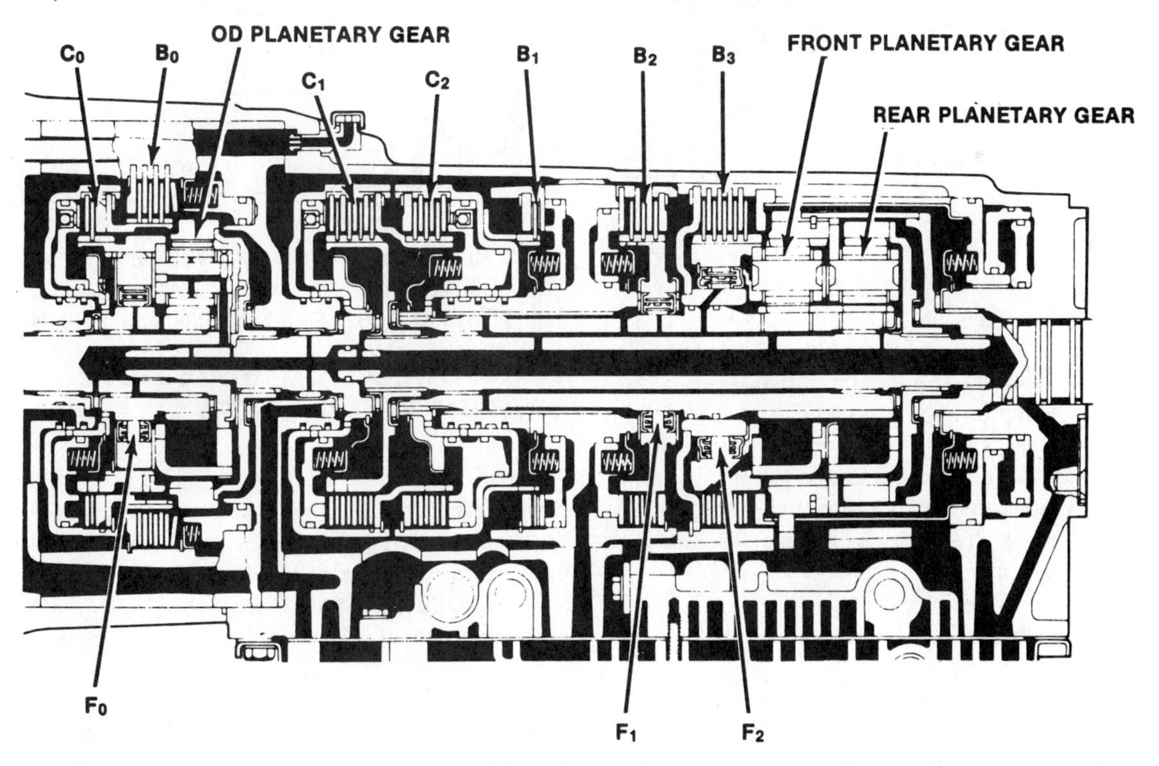

FIGURE 25–2 Mitsubishi 372/KM148 four-speed transmission powertrain and clutch identification. (Courtesy of Chrysler Corp.)

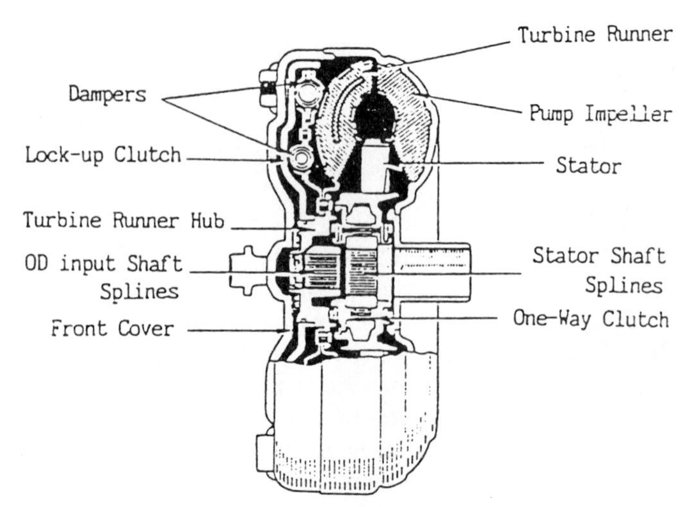

FIGURE 25–3 Three-element lockup torque converter—Toyota A341E. (Reprinted with permission from SAE AE-18 © 1994 Society

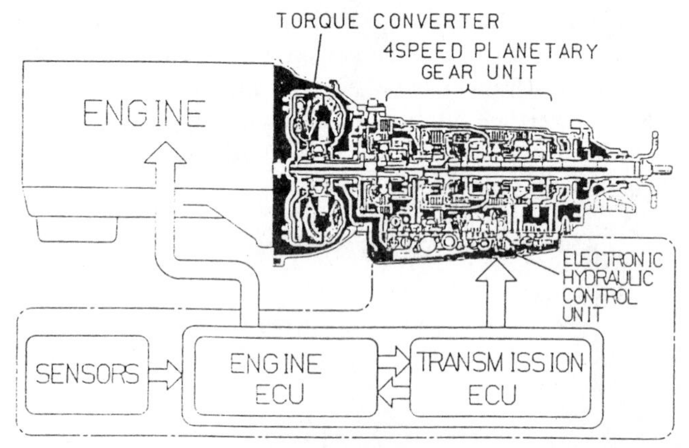

FIGURE 25–4 Basic electronic powertrain control system—Toyota A341E transmission. (Reprinted with permission from SAE AE-18 © 1994 Society of Automotive Engineers, Inc.)

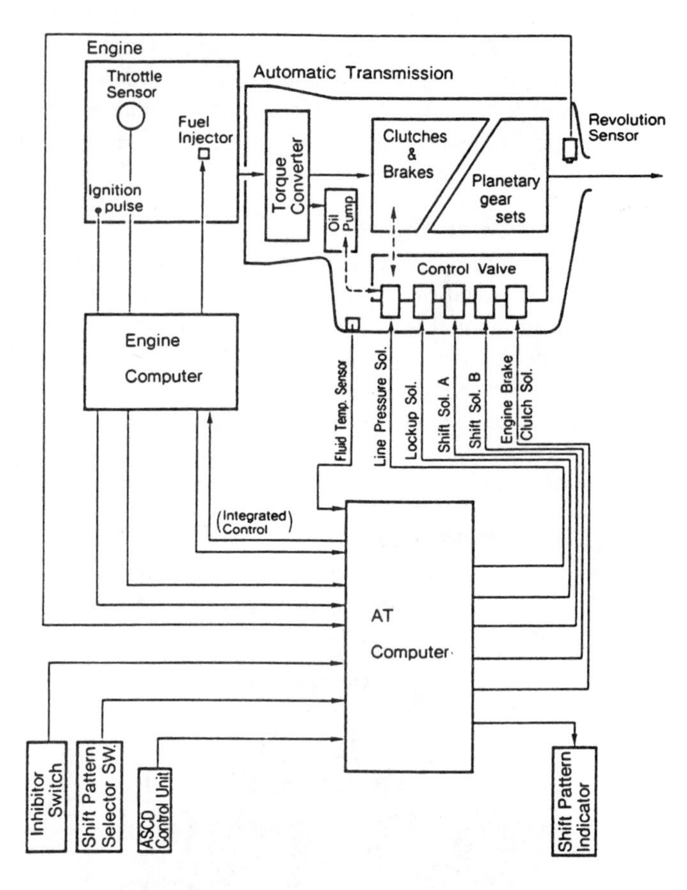

FIGURE 25–5 Simplified electronically controlled circuits of the Nissan R01A automatic transmission. (Reprinted with permission from SAE AE-18 © 1994 Society of Automotive Engineers, Inc.)

Parallel-Shaft/Helical Gear Design

Neither Honda nor Saturn (US) use planetary gearsets in their automatic transaxles. Rather, the powertrain designs use parallel-shaft mounted helical gears to attain the gear ratios (Figure 25–6). [Early Honda automatics used spur-cut (straight tooth) reverse gears]. Honda transaxle designs are discussed in greater detail within this chapter.

CVT (Continuously Variable Transmission)

An additional powertrain design includes automatic transaxles using the CVT concept, as shown in Figure 25–7. The belt drive CVT, using changing pulley dimensions, is used by certain Ford (European), Honda, Subaru, and Suzuki models. ZF Industries, Inc., of Germany, also produces CVTs. Additional coverage of CVTs is highlighted in the final section of this chapter.

THE WORLD TRANSMISSION/ TRANSAXLE MARKET

As mentioned in the introduction, import and foreign transmissions/transaxles share many similar features. Most domestic, European, and Asian transmission manufacturers offer their units to a wide range of vehicle producers. These affiliations are especially common where vehicle manufacturers share ownership (stock) in one another or are partners/co-owners in a joint venture.

Shared Corporate Ownership Affiliations

Numerous vehicle manufacturers share ownership in one another. Joint ownership contributes to shared technology and purchase of components, including transmissions. An example of this concept is the Mazda-produced ATX used in certain Ford

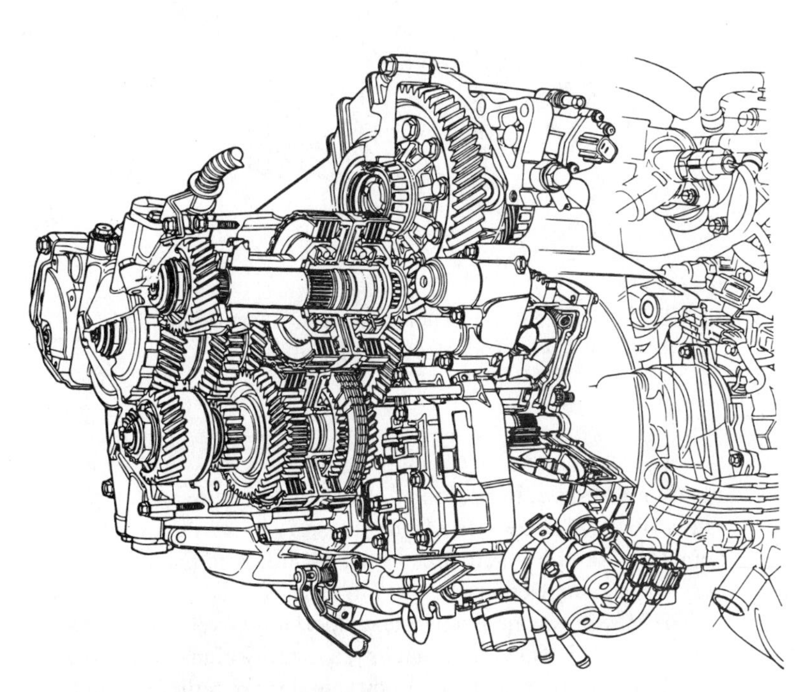

FIGURE 25–6 1995 Honda Odyssey automatic transaxle. (Courtesy of American Honda Motor Co., Inc.)

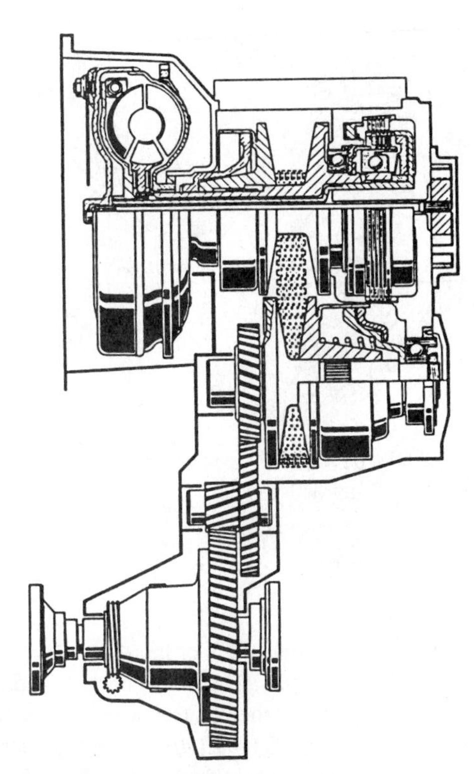

FIGURE 25–7 Basic CVT (continuously variable transmission) design. (Reprinted with permission from SAE AE-18 © 1994 Society of Automotive Engineers, Inc.)

compact passenger cars (Figure 25–8). Following is a partial listing of significant affiliations. (Source: *Automotive Industries,* 1993.)

Chrysler holds 5.9% of Mitsubishi (Japan).
Chrysler owns 15.6% of Maserati (Italy).
Ford owns Jaguar (UK).
Ford owns 10% of Kia (S. Korea); Kia built the Festiva sold by Ford through 1993.
Ford owns 25% of Mazda (Japan).
Fuji Heavy Industries (Subaru of Japan) owns 51% of Subaru-Isuzu America; Isuzu (Japan) is part owner.
GM has owned Adam Opel (Germany) since 1929.
GM holds 5.3% equity in Suzuki (Japan).
GM owns Holden (Australia).
GM owns 38% of Isuzu (Japan).
Honda (UK) and Rover (UK) each hold 20% equity in one another.
Mazda owns 10% of Kia (S. Korea).
Mitsubishi owns 14% of Hyundai (S. Korea).
Nissan owns 5% of Fuji (Japan).
Saab Scania and GM each own 50% of Saab Automobiles AB (Sweden).
Toyota owns Daihatsu (Japan).
Volkswagen owns Audi (Germany).

Corporate Joint Ventures

Many vehicle manufacturers are involved in joint ventures to produce passenger cars and light trucks. Usually an existing powertrain from one of the participating manufacturers is used. Following are descriptions of notable joint ventures.

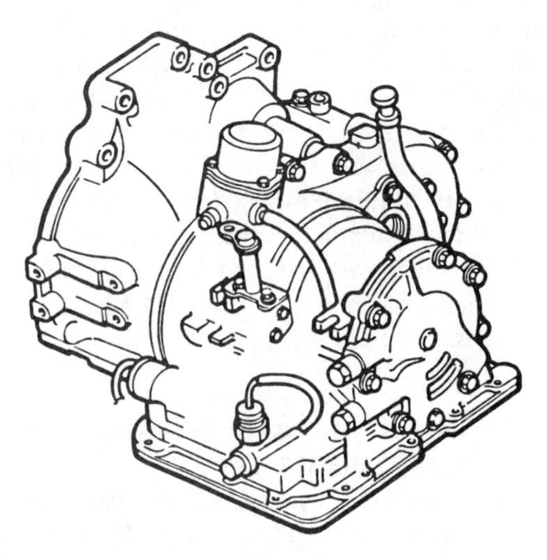

FIGURE 25–8 Mazda-designed ATX. (Reprinted with the permission of Ford Motor Company)

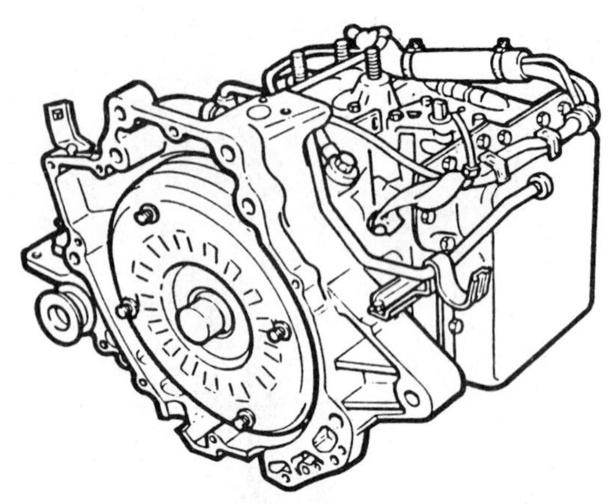

FIGURE 25–9 1996 Ford Probe F4E Type GF automatic transaxle built by Mazda. (Reprinted with the permission of Ford Motor Company)

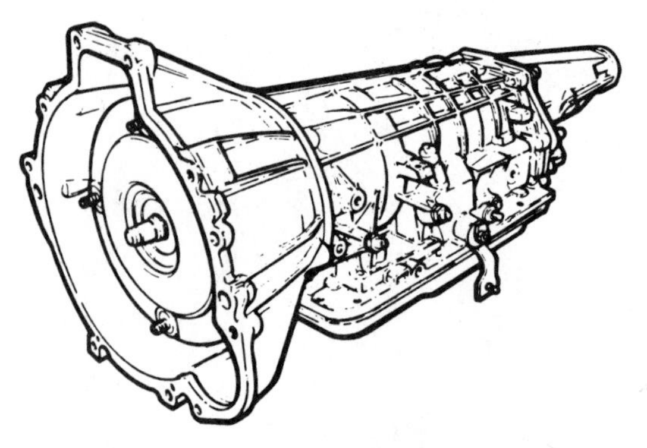

FIGURE 25–10 Ford A4LD, built in Bordeaux, France. (Reprinted with the permission of Ford Motor Company)

Ford and Mazda each own 50% of the AutoAlliance plant located in Flat Rock, Michigan. The Ford Probe, Mazda MX6, and Mazda 626 are assembled there and share similar platforms and powertrains. The Mazda G4A-EL transaxle has been available in these models since production began. Ford refers to the G4A-EL as the 4EAT-F/4FE-type GF transaxle (Figure 25–9). Beginning in model year 1994, certain Probes have been assembled with Ford's CD4E automatic transaxle.

Ford and Nissan supply truck parts to each other and have a joint venture located in Australia to build minivans for the US market. The Mercury Villager uses Nissan RE4F04A/V automatic transaxles.

GM and Suzuki have a 50/50 joint venture named **CAMI (Canada).** CAMI builds two vehicles in Ingersoll, Ontario, that are sold through Chevrolet/Geo dealerships: the Geo Tracker and the Geo Metro. GM Hydra-Matic supplies 3L30 automatic transmissions for the Tracker and the Suzuki Sidekick. Aisin-Seiki provides the Suzuki MX-17 three-speed automatic transaxle for the Metro and Suzuki Swift.

GM and Toyota have two 50/50 ventures. **NUMMI (New United Motors Manufacturing Inc.)** builds the Geo Prizm and the Toyota Corolla in Freemont, California. The subcompacts use a Toyota powertrain, and A131L and A240E/A245E automatic transaxles have been used. **UAAI** is the other joint venture and is located in Australia.

Vehicle Manufacturer—Cooperative Purchase Agreements

Many manufacturers make direct purchases of drivetrain components from other car/truck makers. Some arrangements involve entire vehicles, with slight modifications made along the assembly line to prepare these units to be sold through the buyer's dealer network. This approach is often referred to as "rebadging."

Acura and Isuzu. The new Acura SLX, an upscale sport-utility vehicle, is essentially a rebadged Isuzu Trooper.

Ford and Mazda. In addition to their joint venture and corporate affiliation, Ford uses Mazda transmissions in many of its vehicles. Examples include the Ford Aspire, Ford Escort, Mercury Capri, and Mercury Tracer models.

The Mazda Navajo is a two-door sport utility vehicle assembled at Ford's Louisville, Kentucky, assembly plant, where the Ford Explorer two- and four-door vehicles are built. Through model year 1994, the Ford A4LD automatic transmission was available (Figure 25–10). Beginning in 1995 units, Ford's 4R55E automatic transmission was used. The A4LD and 4R55E are built at Ford's Bordeaux, France, transmission plant. Note: Starting in 1996, some versions of the Explorer use Ford's 4R70W automatic transmission.

Honda and Isuzu. An Isuzu version of the Honda Odyssey minivan, named the Oasis, became available in the 1996 model year. During that same model year, the Isuzu Rodeo sport-utility vehicle was introduced through Honda as the Passport. Refer to the Honda and Isuzu sections in this chapter for additional information.

Rolls Royce and GM Hydra-Matic. Rolls-Royce Motors, Ltd. (UK), does not have major ties with other car builders, but certain models are currently equipped with a GM Hydra-Matic 4L80-E transmission. Prior to the application of the 4L80-E, 3L80 (THM 400) transmissions had been used.

BMW and GM Hydra-Matic/ZF. BMW (Bayerishe Motoren Werke/Bavarian Motor Works) of Germany has no major ties with other car builders. The automobile manufacturer purchases and installs automatic transmissions from different manufacturers, including the GM Hydra-Matic 4L30-E, 4L40-E and the 5L40-E units, built in Strasbourg, France, and transmissions from ZF Industries, Inc.

Jaguar (Ford) and GM Hydra-Matic. Prior to Ford Motor Company acquiring ownership of Jaguar (UK), certain Jaguars used the GM Hydra-Matic 3L80 (THM 400). Beginning in some 1994 models, the Hydra-Matic 4L80-E has been used.

INDEPENDENT TRANSMISSION MANUFACTURERS

A number of independent transmission manufacturers, such as ZF Industries, Inc. (Germany), produce units for worldwide distribution. In Japan, there are three companies that produce only automatic transmissions.

The independent Japanese automatic transmission builders are Aisin-Seiki, Aisin-Warner, and JATCO. Their primary focus is to supply the needs of Asian vehicle manufacturers, but their products are also found in European and domestic cars and trucks.

Aisin-Seiki

Aisin-Seiki is a wholly owned subsidiary of Toyota and produces approximately 5% of the Asian automatic transmissions. Their transmissions have been used by Daihatsu, GM (Geo Prizm), Isuzu, Subaru, Suzuki, and Toyota.

Daihatsu/Aisin-Seiki. Daihatsu depends entirely on Aisin-Seiki units for its vehicle automatic transmission requirements. It procures more transmissions from Aisin-Seiki than any other car/truck builder.

Suzuki, CAMI, and Chevrolet-Geo/Aisin-Seiki. Suzuki is the second most active user of Aisin-Seiki transmissions. Approximately 55% of the automatic transmissions used by Suzuki are produced by Aisin-Seiki. The remainder of Suzuki automatic transmission needs are supplied by JATCO (40%) and GM (5%).

When a Geo Metro is equipped with an automatic transaxle, a Suzuki MX17 (Aisin-Seiki A-210) three-speed has been used (Figure 25–11). This is the same automatic transaxle offered in the Suzuki Swift and Chevrolet Sprint. The MX17/A-210 consists of a three-element (nonlocking) torque converter and three multiple-disc clutches, a one-way clutch, and a band to operate a Simpson compound planetary gearset. The MX-17/A-210 in these applications began to be used in 1985 and is current in 1996 models. Refer to Figure 25–12 for a powerflow schematic and a range reference and clutch/band application summary.

Aisin-Warner

Aisin-Warner is 90% owned by Toyota and provides approximately 21% of the needs of the Asian automatic transmission market. Applications of Aisin-Warner units include Chrysler, Isuzu, Mitsubishi, NUMMI (US), and Toyota vehicles.

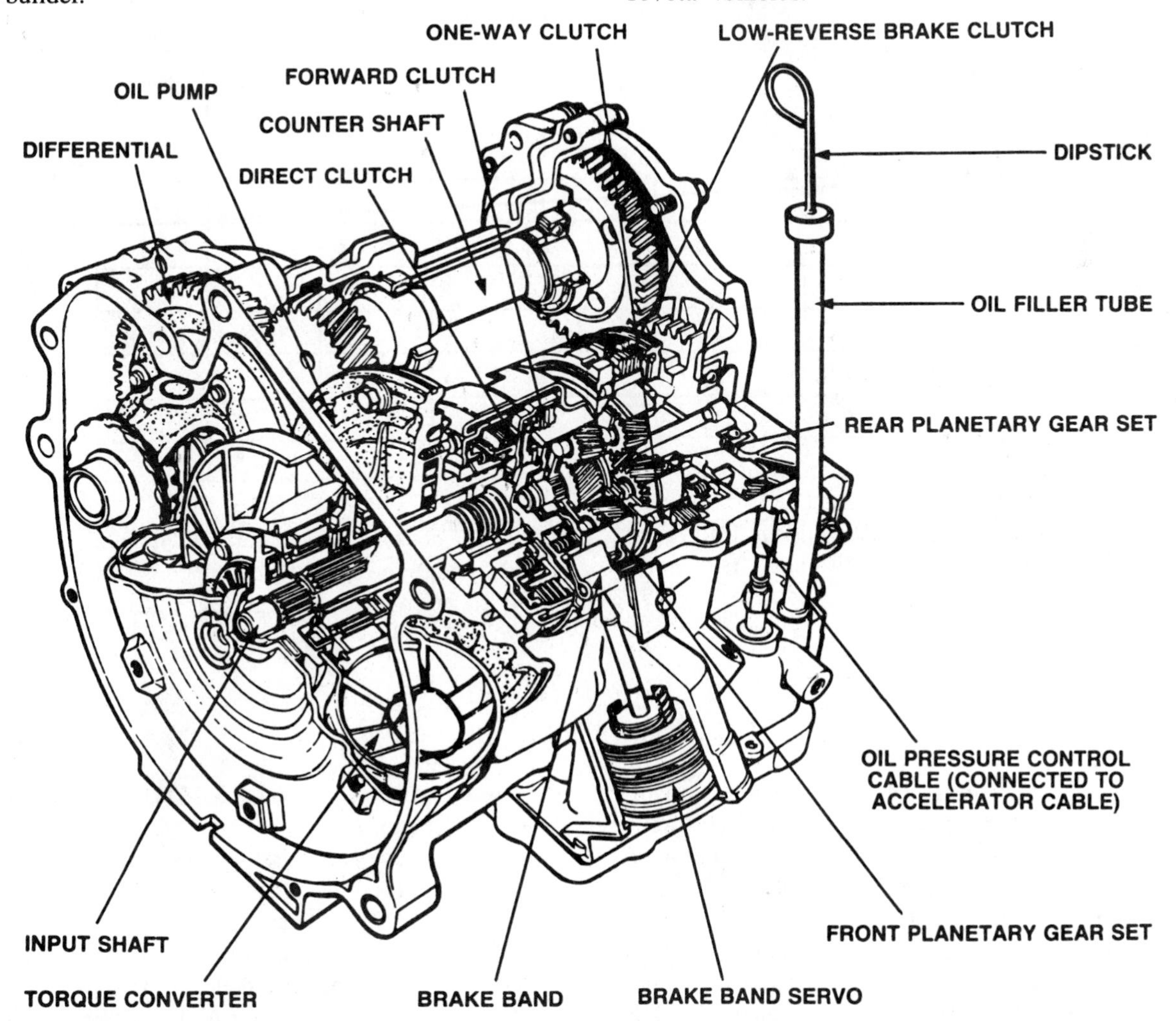

FIGURE 25–11 Three-speed Aisin-Seiki transaxle used in the Geo Metro. (Courtesy of General Motors Corporation, Service Technology Group)

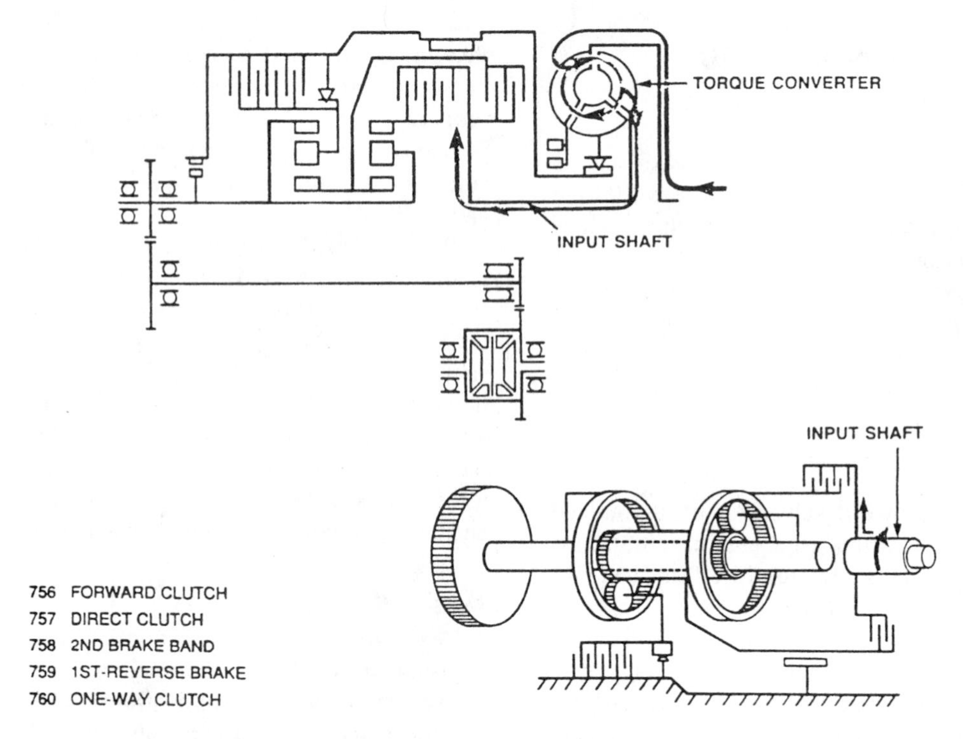

INPUT SHAFT

756 FORWARD CLUTCH
757 DIRECT CLUTCH
758 2ND BRAKE BAND
759 1ST-REVERSE BRAKE
760 ONE-WAY CLUTCH

RANGE	GEAR	FORWARD CLUTCH	DIRECT CLUTCH	2ND BRAKE BAND	1ST AND REVERSE BRAKE	ONE-WAY CLUTCH	PARKING LOCK PAWL
P	PARKING				**O		O
R	REVERSE		O		O		
N	NEUTRAL						
D	1ST	O				O	
	2ND	O		O			
	3RD	O	O				
2	1ST	O				O	
	2ND	O		O			
L	1ST	O			O	O	
	*2ND	O		O			

O APPLIED
* TO PREVENT OVERREVOLUTION OF ENGINE, THIS 2ND GEAR IS OPERATED ONLY WHEN SELECTOR LEVER IS SHIFTED TO "L" RANGE AT THE SPEED OF MORE THAN 55 km/h (34 mph)
** WHEN ENGINE IS RUNNING

MBS0127A1

FIGURE 25–12 Aisin-Seiki/Geo Metro transaxle powertrain schematic; range reference and clutch/band application summary. (Courtesy of General Motors Corporation, Service Technology Group)

Chrysler/Aisin-Warner. Chrysler has used a RWD Aisin-Warner four-speed electronic shift transmission in many Jeep/XJ models (1987–current), including the 1996 Jeep Grand Cherokee. Chrysler refers to this TCM-controlled unit as the AW-4 (Figure 25–13). The AW-4 transmission is very similar to the AW 30-80LE used in certain Isuzu Troopers. Refer to Figure 25–14 for a powerflow schematic and component function chart and to Chart 25–1 (page 620) for a range reference and clutch/band application summary.

Chevrolet Nova-Geo Prizm (NUMMI)/Aisin-Warner. Aisin-Warner FWD transaxles are used in the Geo Prizm (1989–current) and its predecessor, the subcompact Chevrolet Nova (1985–1988). The three-speed A131L (Figure 25–15) and the electronic shift A240E have been used in the Prizm and Nova. The 1996 Geo Prizm automatic transaxle offerings include the A131L and A245E four-speed electronic shift transaxle.

Refer to Figure 25–16 for a A131L powerflow schematic, and a range reference and clutch/band application summary.

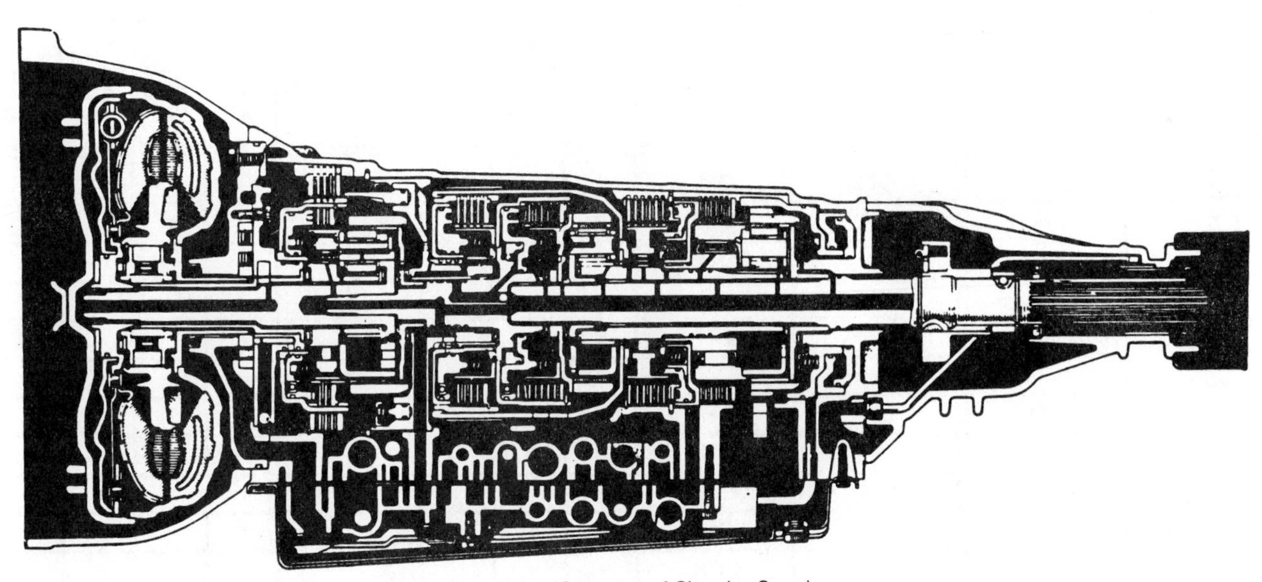

FIGURE 25–13 1995 AW-4 four-speed automatic transmission. (Courtesy of Chrysler Corp.)

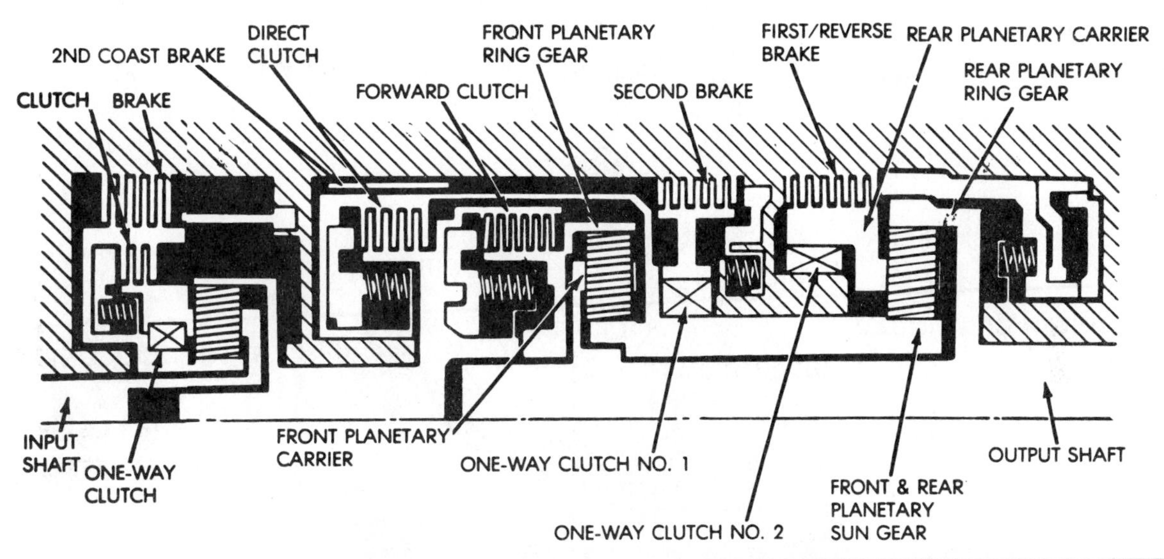

NOMENCLATURE	FUNCTION
Overdrive Direct Clutch	Connects overdrive sun gear and overdrive carrier
Overdrive Brake	Prevents overdrive sun gear from turning either clockwise or counterclockwise
Overdrive One-Way Clutch	When transmission is driven by engine, connects overdrive sun gear and overdrive carrier
Forward Clutch	Connects input shaft and front ring gear
Direct Clutch	Connects input shaft and front and rear sun gear
Second Coast Brake	Prevents front and rear sun gear from turning either clockwise or counterclockwise
Second Brake	Prevents outer race of No. 1 one-way clutch from turning either clockwise or counterclockwise, thus preventing front and rear sun gear from turning counterclockwise
First/Reverse Brake	Prevents rear planetary carrier from turning either clockwise or counterclockwise
One-Way Clutch No. 1	When second brake is operating, prevents front and rear sun gear from turning counterclockwise
One-Way Clutch No. 2	Prevents rear planetary carrier from turning counterclockwise

J8921-404

FIGURE 25–14 AW-4 powertrain schematic and components. (Courtesy of Chrysler Corp.)

Shift Lever Position	Gear	Valve Body Solenoid No. 1	Valve Body Solenoid No. 2	OVERDRIVE CLUTCH	FORWARD CLUTCH	DIRECT CLUTCH	OVERDRIVE BRAKE	SECOND COAST BRAKE	SECOND BRAKE	FIRST/ REVERSE BRAKE	OVERDRIVE ONE-WAY CLUTCH	NO.1 ONE-WAY CLUTCH	NO.2 ONE-WAY CLUTCH
P	Park	ON	OFF	•									
R	Reverse	ON	OFF	•		•				•	•		
N	Neutral	ON	OFF	•									
D	First	ON	OFF	•	•						•		•
	Second	ON	ON	•	•				•		•	•	
	Third	OFF	ON	•	•	•			•		•		
	OD	OFF	OFF		•	•	•		•				
3	First	ON	OFF	•	•						•		•
	Second	ON	ON	•	•			•	•		•	•	
	Third	OFF	ON	•	•	•			•		•		
1-2	First	ON	OFF	•	•					•	•		•
	Second	ON	ON	•	•			•	•		•	•	

• = Applied   J8921-405

CHART 25–1 AW-4 range reference and clutch/band application chart. (Courtesy of Chrysler Corp.)

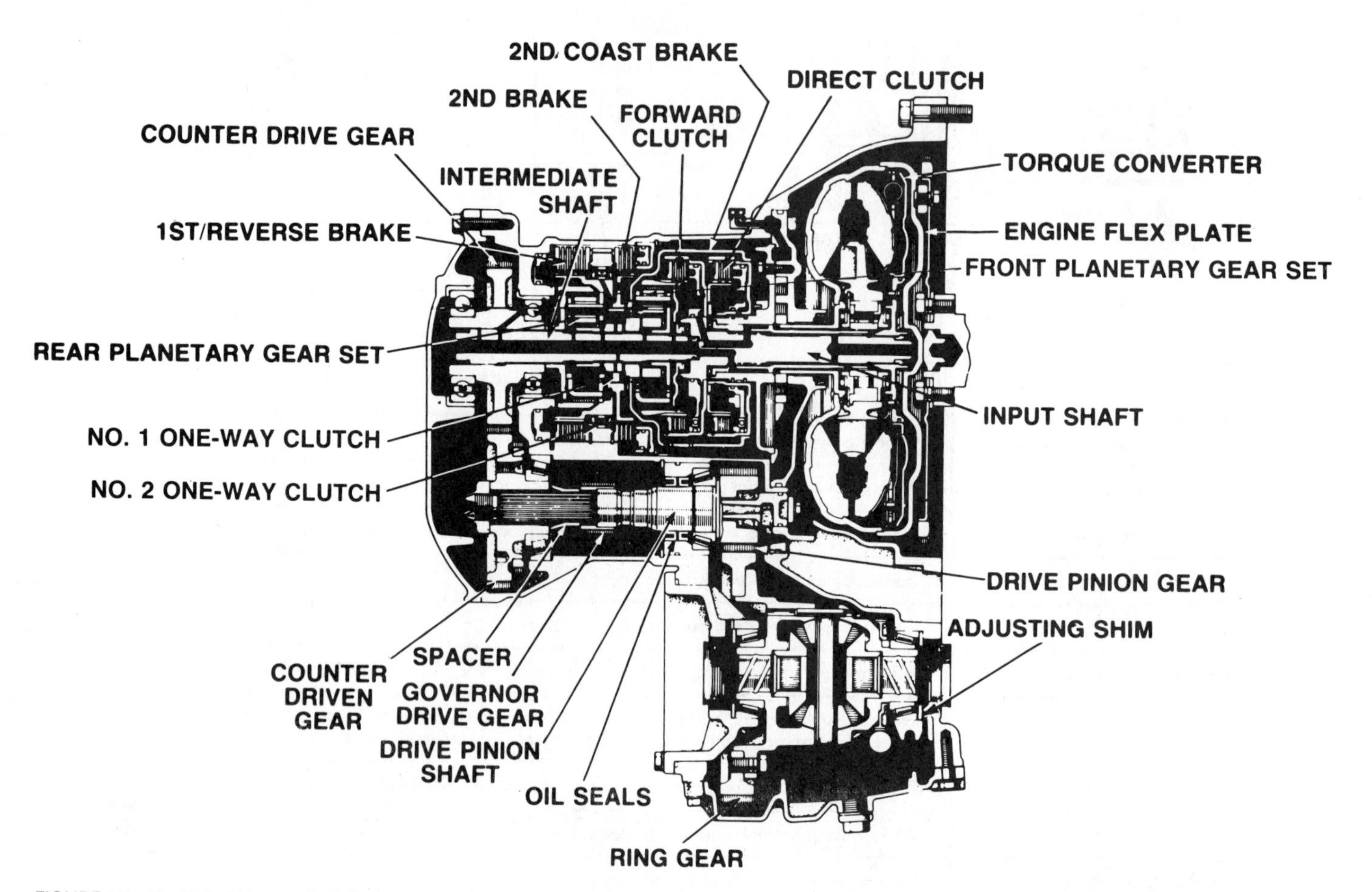

FIGURE 25–15 Aisin-Warner A131L three-speed transaxle used in the Geo Prizm. (Courtesy of General Motors Corporation, Service Technology Group)

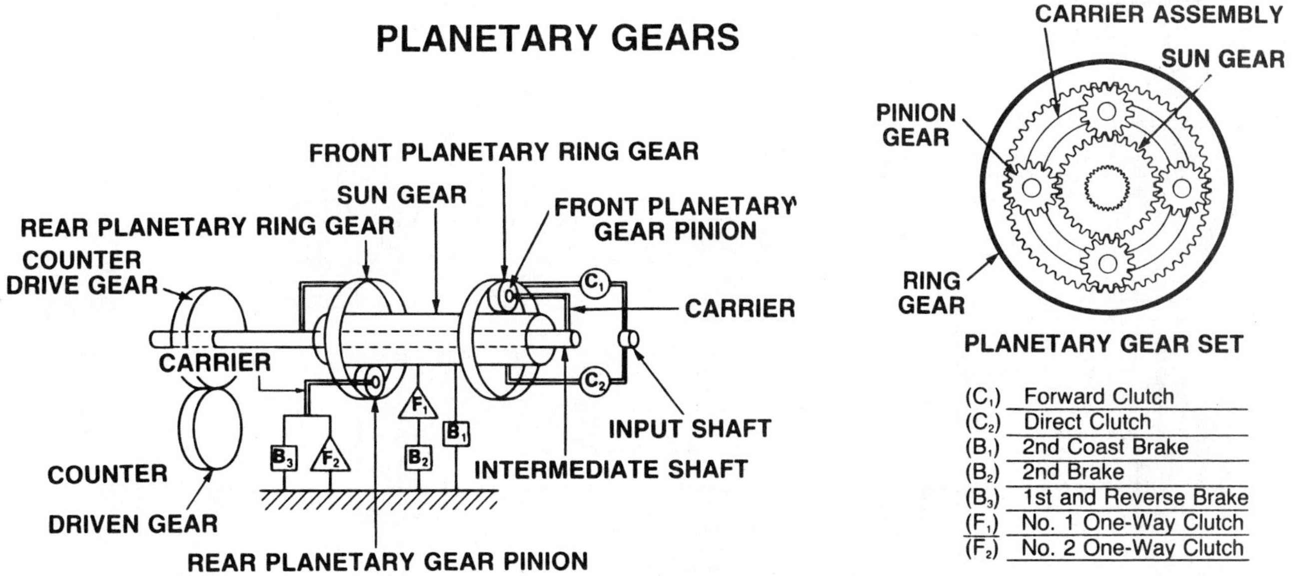

PLANETARY GEARS

FRONT PLANETARY RING GEAR
SUN GEAR
REAR PLANETARY RING GEAR
FRONT PLANETARY GEAR PINION
COUNTER DRIVE GEAR
CARRIER
CARRIER
INPUT SHAFT
INTERMEDIATE SHAFT
COUNTER DRIVEN GEAR
REAR PLANETARY GEAR PINION

CARRIER ASSEMBLY
SUN GEAR
PINION GEAR
RING GEAR
PLANETARY GEAR SET

(C_1) Forward Clutch
(C_2) Direct Clutch
(B_1) 2nd Coast Brake
(B_2) 2nd Brake
(B_3) 1st and Reverse Brake
(F_1) No. 1 One-Way Clutch
(F_2) No. 2 One-Way Clutch

HYDRAULIC RANGE	C_1 Forward Clutch	C_2 Direct Clutch	B_1 2nd Coast Brake (Band)	B_2 2nd Brake (Clutch)	B_3 1st/Rev Brake (Clutch)	F_1 No. 1 One-Way Clutch	F_2 No. 2 One-Way Clutch	GEAR RATIO
Park								—
Reverse		●			●			2.296
Neutral								—
"D" or "2" – 1st Gear	●						●	2.811
"D" – 2nd Gear	●			●		●		1.549
"D" – 3rd Gear	●	●		●				1.000
"2" – 2nd Gear	●		●	●		●		1.549
"L" – 1st Gear	●				●		●	2.811
Pinion Gear								0.945
Final Drive								3.421

FIGURE 25–16 Aisin-Warner A131L powertrain schematic; range reference and clutch/band application summary. (Courtesy of General Motors Corporation, Service Technology Group)

JATCO

JATCO (Japanese Automatic Transmission Company) is currently 40% owned by Nissan and 60% by Mazda. At its inception in 1970, it was a joint venture between Nissan, Toyo Kogyo (now Mazda), and Ford. In 1981, Ford transferred its entire stock to Nissan and Toyo Kogyo.

The company provides approximately 17% of the Asian manufacturers' automatic transmissions, which are used by Isuzu, Mazda, Mitsubishi, Nissan, Suzuki, and Subaru (Fuji Heavy Industries). JATCO has also supplied North American manufacturers and BMW (Germany) with automatic transmissions. As of May 1990, JATCO had produced over ten million units.

Chrysler/JATCO. Chrysler uses the JM600 (JATCO model code: MR600), a four-speed RWD automatic transmission, in its Dodge (Mitsubishi) Conquest. Internally, an overdrive gearset provides power input to a Simpson compound planetary gearset (Figure 25–17). Hydraulic controls operate three multiple-disc clutches, two brake bands, and a one-way clutch. Heat generated from the torque converter (equipped with a lockup plate) is dissipated by an oil-to-air type cooler. Refer to Chart 25–2 (page 622) for a range reference and clutch/band application summary.

Ford/JATCO. Ford used a three-speed JATCO 3N71B transmission from 1977 to 1980 in mid-size RWD passenger cars,

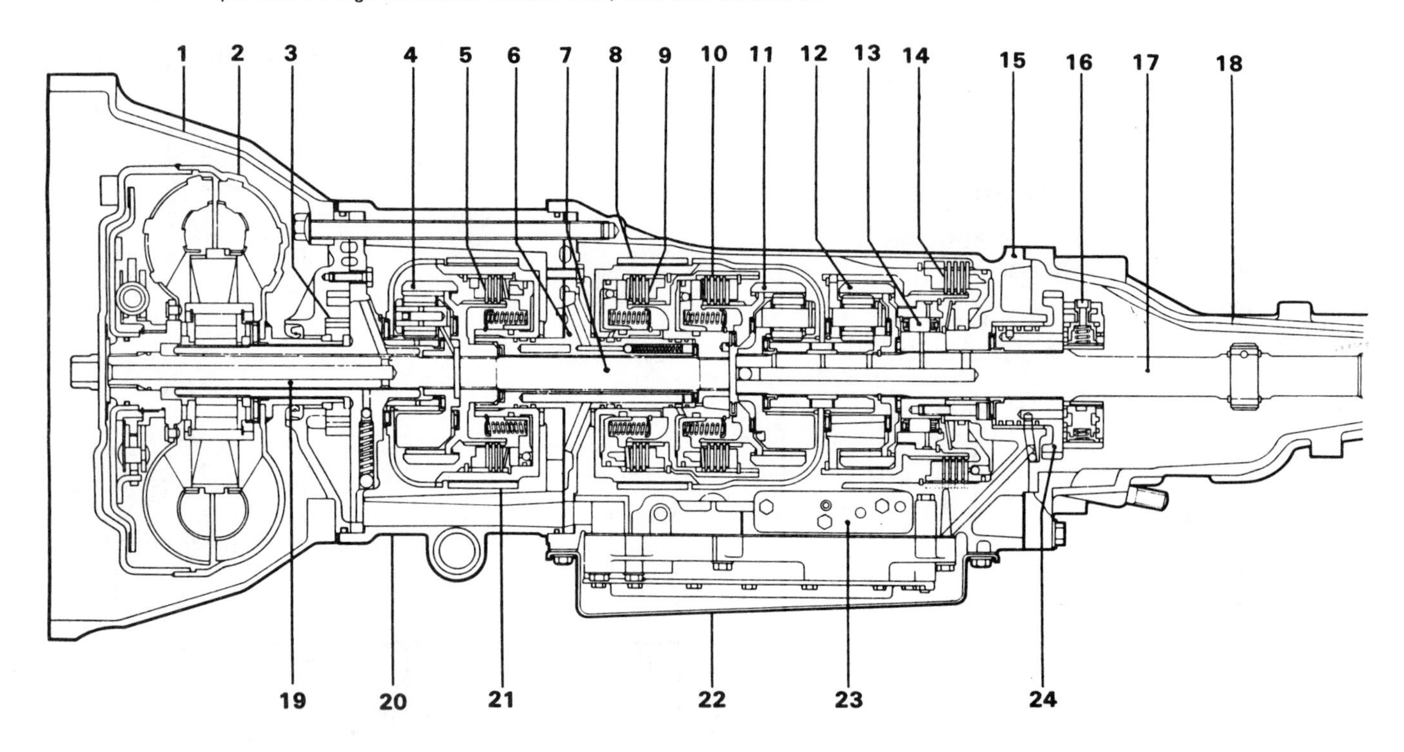

1. Converter housing
2. Torque converter
3. Oil pump
4. O.D. planetary gear
5. Direct clutch
6. Drum support
7. Intermediate shaft
8. Second band brake
9. High – reverse clutch (Front)
10. Forward clutch (Rear)
11. Front planetary gear
12. Rear planetary gear
13. One-way clutch
14. Low-reverse clutch
15. Transmission case
16. Governor valve
17. Output shaft
18. Rear extension
19. Input shaft
20. O.D. case
21. O.D. brake band
22. Oil pan
23. Control valve assembly
24. Oil distributor

FIGURE 25–17 JATCO JM600 automatic transmission sectional view. (Courtesy of Chrysler Corp.)

Range		Direct clutch	O.D. band servo		High-reverse clutch (Front)	Forward clutch (Rear)	Low & reverse brake	2nd band servo		One-way clutch	Parking Pawl
			Apply	Release				Apply	Release		
Park		ON	(ON)	ON			ON				ON
Reverse		ON	(ON)	ON	ON		ON		ON		
Neutral		ON	(ON)	ON							
D	D_1 (Low)	ON	(ON)	ON		ON				ON	
D	D_2 (Second)	ON	(ON)	ON		ON		ON			
D	D_3 (Top)	ON	(ON)	ON	ON	ON		(ON)	ON		
D	D_4 (O.D.)		ON		ON	ON		(ON)	ON		
2	Second	ON	(ON)	ON		ON		ON			
1	1_2 (Second)	ON	(ON)	ON		ON		ON			
1	1_1 (Low)	ON	(ON)	ON		ON	ON			ON	

The low & reverse brake is applied in "1_1" range to prevent free wheeling when coasting and allows engine braking.

CHART 25–2 JM600 range reference and clutch/band application chart. (Courtesy of Chrysler Corp.)

such as the Ford Granada and Mercury Monarch. This transmission has also found application in 1973–1981 Ford Courier pickup trucks. Couriers were assembled by Mazda.

Chevrolet-GEO Storm and Spectrum (Isuzu)/JATCO. Geo Storm GSI, a car line produced by Isuzu, Ltd., of Japan, may come equipped with the optional FWD JATCO JF403E. The JF403E automatic transaxle is a four-speed unit using both electronic and hydraulic shift control.

Isuzu used the FWD JATCO F10/F3A from 1985 to 1989. This unit was also applied to the Chevrolet/Geo Spectrum.

Nissan and Infiniti/JATCO. Nissan and Infiniti models have used numerous JATCO automatic transmissions and transaxles. These include the RWD 3N71B (1971–1987); the RWD RE4R01A (1987–1982); the RWD RL4F03A (1990–1992), used in both Nissan and Infiniti models; the FWD RN3F01A/RL3F01A (1981–1992), used in Sentra and Pulsar models; and the RWD E/L4N71B (1983–1990). The RL4F02A is found in Maxima and Stanza models.

Mazda/JATCO. Mazda used the RWD JATCO 3N71B in many of its vehicles from 1971 to 1987, including the Cosmo, RX-4 and RX-7 models (1974–1984). The 4N71B/N4AHL (RWD) found application in the Miata two-seater sports car (1990–1992), "B" series trucks (1982–1992), and the MPV minivan (1989–1992). The electronic shift R4AEL was also available from 1989 to 1992. The JATCO F10/F3A was available from 1981 to 1987 in the following models: 323, 626, and GLC. It was also available in 1988 and 1989 323 station wagons.

Subaru/JATCO. Subaru used a Jatco three-speed transaxle in its AWD vehicles from 1975 to 1989. A JATCO four-speed unit was used in all Subaru vehicles except the Legacy from 1987 to 1991.

JATCO JR50E Five-speed Transmission. The world's first full-range electronically controlled automatic transmission for passenger cars was produced by JATCO. The JR502E was originally intended for the Japanese market, but it may find future applications in North America.

The RWD transmission uses a seven-range quadrant pattern (P-R-N-D-3-2-1). A variable displacement vane oil pump is used to supply the hydraulic needs of the unit. The transmission is rated to handle a maximum input torque of 217 ft-lb (294 N-m) and is designed for medium-size RWD passenger cars.

Features of the JR502E include three different electronic shift patterns that can be chosen by the driver: economy, power, and snow modes. Standing starts in second gear are available. In snow mode, standing starts in third gear are allowed, which reduces overall torque and helps prevent breaking traction at the drive wheels. A torque converter clutch plate can be applied in fifth gear.

The powertrain arrangement for the five forward speeds and reverse are unique. The main body of the transmission contains a Simpson compound planetary gearset with independent sun gears. Torque leaving the compound gearset is driven through a simple planetary gearset located within the removable extension housing. The rear planetary gearset provides a gear reduction and a direct drive. The gear ratios of the JR502E are as follows:

First gear: 3.857:1
Second gear: 2.410:1
Third gear: 1.384:1
Fourth gear: 1.000:1
Fifth gear: 0.694:1
Reverse: 3.146:1

ZF Industries, Inc.

ZF Industries, Inc., of Germany produces both manual and automatic transmissions for worldwide distribution. Audi (Germany), BMW (Germany), Peugeot/Citroen (France), Porsche (Germany), Rover (UK), Saab (Sweden), and Volvo (Sweden) use ZF transmissions.

The transmission and gear producer was founded in 1915 by Count Frederick Von Zeppelin. Products were first applied to the development of airships. The company was named Zahnradfabrik Friedrichshafen GmgH, loosely translated as "factory for producing gears, Frederick's Harbor." Currently, the company is represented in eighty-six countries around the world.

ZF has produced three-, four-, and five-speed automatic transmissions for the world market (Figure 25–18). In addition, the company has worked on the development of a CVT (continuously variable transmission). CVTs are discussed in the final section of this chapter.

BMW and Peugeot/ZF (Three-speed Transmission). A three-speed transmission, the ZF-3HP-22, was built from 1976 to 1985. It was used by both BMW and Peugeot.

Ford/ZF (Four-speed Transmission). The ZF-4HP-22 (1985–1989) is a four-speed overdrive transmission that was used by Ford Motor Company in the mid-1980s. Ford recommended that dealership service departments exchange these units rather than perform major overhauls. Applications included 1984 and 1985 Lincoln Continental and Mark VII with a 2.4-L turbo diesel engine.

BMW and Porsche/ZF (Four-speed Electronic Shift Transmission). An electronically controlled version of the four-speed ZF-4HP-22 hydraulically controlled transmission is the ZF-4HP22-EH. It is used in a variety of import vehicles, including the BMW 745i series vehicles. (Note: 1996 BMW 745iL models use a five-speed automatic transmission.) TCM operation is dependent upon numerous input and output devices. Valve body mounted output solenoids include two shift solenoids, a converter clutch solenoid, a reverse lockout solenoid, and an electronic pressure regulator solenoid.

A modified version of the ZF-4HP22-EH is used by Porsche in the Carrera 2. The transmission is referred to as the Tiptronic and offers both automatic and manual-controlled shifts. Refer to the Porsche section in this chapter for additional information.

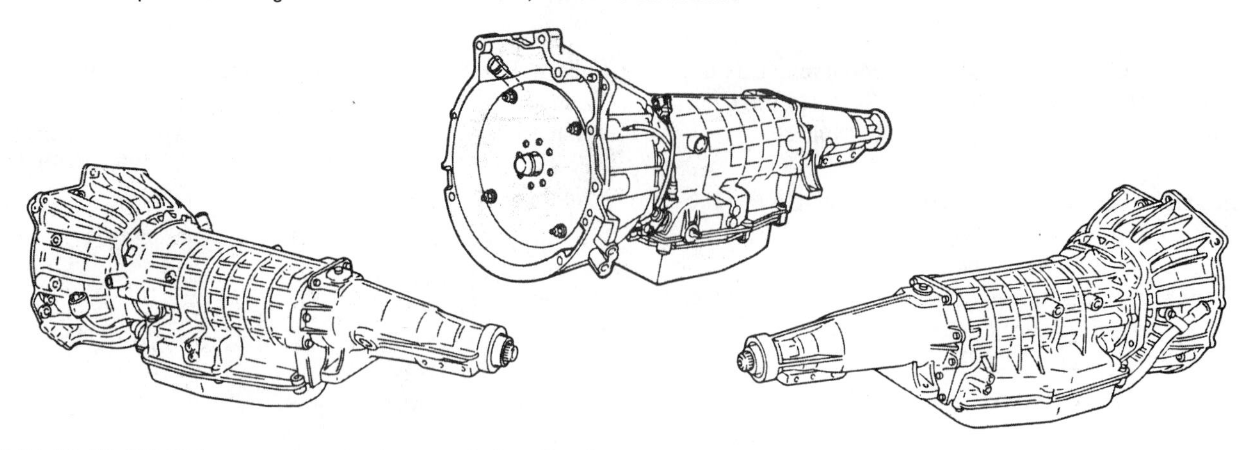

FIGURE 25–18 ZF-4HP-22 four-speed automatic transmission. (Reprinted with the permission of Ford Motor Company)

BMW M3/ZF (Five-speed Electronically Controlled Automatic Transmission). Refer to the BMW section in this chapter for information on this new automatic transmission.

Chrysler/ZF. The ZF-4 (ZF-4HP-18)—a longitudinal-mounted, FWD, four-speed transaxle—incorporates a lockup torque converter and a hypoid ring and pinion final drive unit.The Ravigneaux planetary gearset is operated by five multiple-disc clutches, two one-way clutches, and a band (Figure 25–19). The clutch and band application is displayed in Chart 25–3. Through the 1991 model year, Chrysler used the ZF-4 automatic transaxle in Premier and Monaco vehicles.

CLUTCH, BRAKE, and BAND ELEMENTS

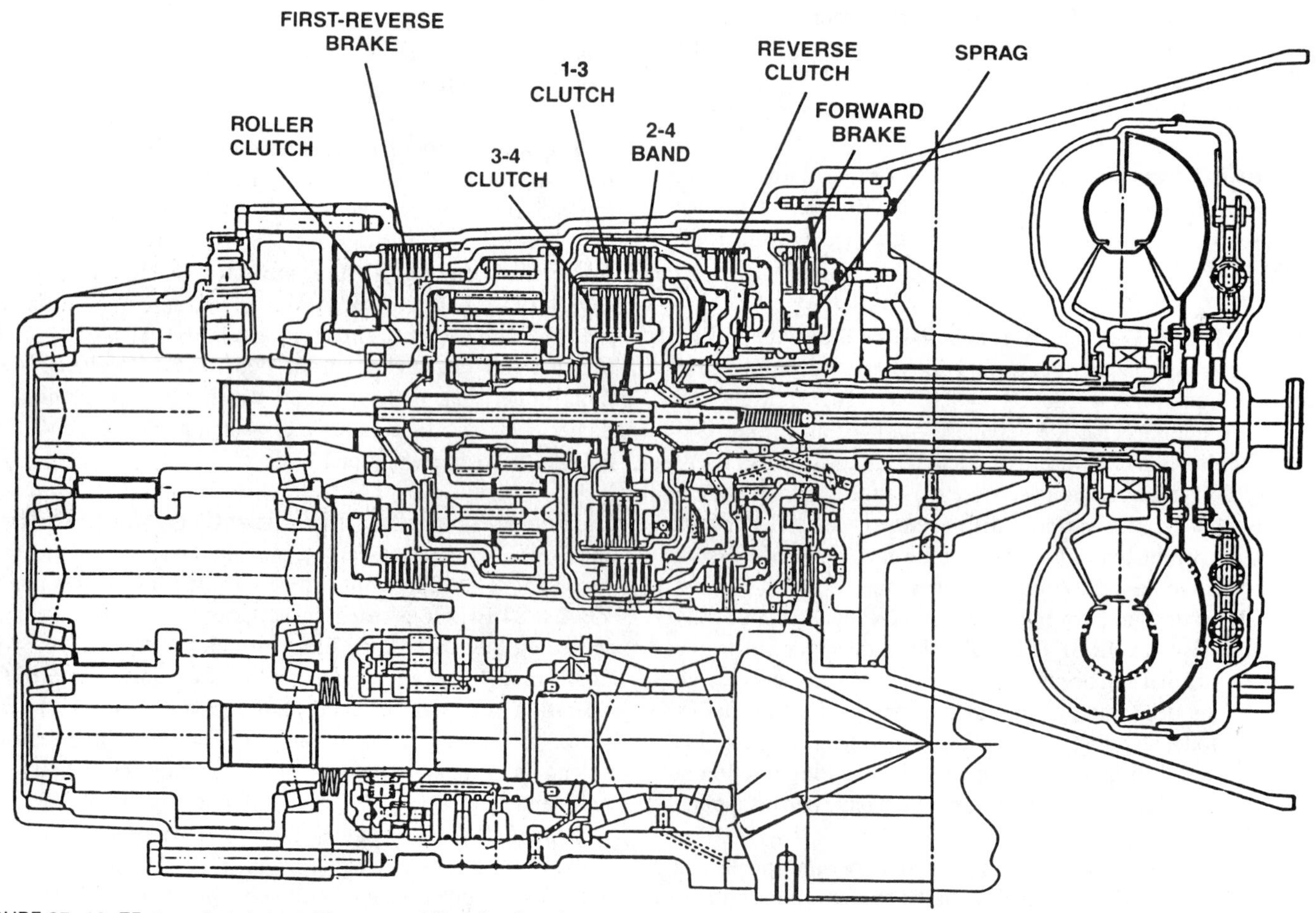

FIGURE 25–19 ZF-4 sectional view. (Courtesy of Chrysler Corp.)

Component	Gear Drive				
	First	Second	Third	Fourth	Reverse
1-3 Clutch	●	●	●		
Reverse Clutch					●
Forward Brake		●			
2-4 Band		●		●	
First-Reverse Brake	●				●
3-4 Clutch			●	●	
Roller Clutch	●				
Sprag Clutch		●			

E8921-763

CHART 25–3 ZF-4 clutch/band application chart. (Courtesy of Chrysler Corp.)

Saab/ZF. Beginning in 1989, Saab adapted the ZF-4HP-18 four-speed unit to their model 9000. The 9000 uses a transverse FWD powertrain.

ZF-5HP-18 Five-speed Automatic Transmission. The ZF-5HP-18 five-speed automatic transmission, as shown in Figure 25–20, is based on the design of the RWD ZF-4HP-18 four-speed system. Its overall length is the same as the four-speed unit. Features include a converter clutch lockup plate and electronic transmission controls. It is adaptable for two- or four-wheel drive applications.

Specifications
Forward gear ratios: 3.65, 2.0, 1.41, 1.0, 0.74
Maximum input torque: 229 ft-lb (310 N-m)
Maximum rpm: 6400
Weight: 163 lb (74 kg)

IMPORT AND FOREIGN VEHICLE TRANSMISSION/TRANSAXLE SUMMARIES

Following are summaries concerning the transmissions and transaxles used by various import and foreign vehicle manufacturers. Included are a few comments concerning their applications, operation, and design features.

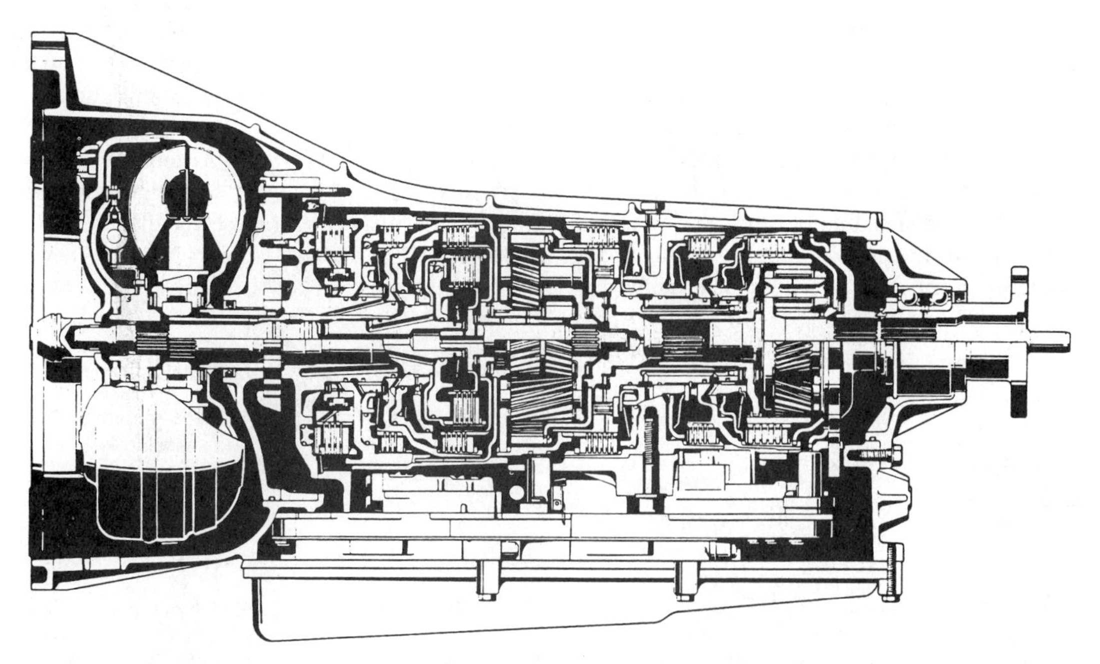

FIGURE 25–20 ZF-5HP-18 five-speed automatic transmission cross-sectional view. (Reprinted with permission from SAE AE-18 © 1994 Society of Automotive Engineers, Inc.)

Acura

Since 1986, Honda four-speed automatic transmissions have been used in Integra and Legend models. Refer to the Honda/Acura section in this chapter for details concerning these transmissions.

Audi

Audi is supplied with transmissions from its parent company, Volkswagen. The VW Type 3 and Type 4 automatic transmissions were used in all models from 1969 to 1976, except for the "100." The FWD VW Type 010 transmissions use a compound Simpson planetary gearset. They have been used in the 1974 to 1977 "100," the 1977 Fox, and other Audi vehicles from 1977 to 1992. Refer to the Volkswagen section in this chapter for additional information.

1996 Audi Models. 1996 Audi models use four- and five-speed automatic transmissions. The A4 Sedan (2WD) and A4 Quattro (4WD) may be equipped with Audi's new five-speed automatic transmission, utilizing a clutching torque converter. All A6 models (2WD wagons and 4WD Quattros) and the Cabriolet convertible came standard with a four-speed automatic transmission.

BMW

ZF supplied BMW (Bayerishe Motoren Werke/Bavarian Motor Works) with ZF-3HP-22 three-speed RWD transmissions from 1976 to 1985. A four-speed version, the ZF-4HP-22, was used from 1984 to 1989. The ZF-4HP-22-EH is an electronically controlled version of the ZF-4HP-22 and is used in the BMW 745i series vehicles.

GM Hydra-Matic 4L30-Es, built in Strasbourg, France, have been used in some 1992 and newer BMW cars.

The 1996 BMW M3 automatic is a five-speed transmission built by ZF, with three electronically controlled shift modes. The driver-controlled shift modes are E (economy), S (sport), and M (manual). The E mode is the default setting and provides the best fuel economy. The S mode produces delayed shift points and lighter throttle forced downshifts. In addition, fifth gear is available in S mode, unlike other "sport-programmed" electronically controlled five-speed automatic transmissions. The M mode allows the driver to pick the gear for pulling away on a slippery surface and manually controlling the upshifts and downshifts. Note: M3 vehicles are speed-governed to 137 mph (220 km/h).

1996 BMW Models. 1996 BMW 328i, 328iC, 328iS, and Z3 Roadster model features included an optional four-speed automatic transmission. The 1996 750iL came standard with a five-speed automatic.

The newly developed "Steptronic" automatic transmission offers a manual up and downshifting gate for driver control of shift timing. This concept is similar to that found in certain Porsche, Audi, and Eagle Vision models. The Steptronic is found in BMW's 840Ci models.

Chrysler

Chrysler-produced RWD A-727 automatic transmissions have been used by Maserati (Italy). The RWD A-904 (three-speed) and A-500 (four-speed) automatics have been applied to some Mitsubishi (Japan) trucks. The Chrysler A-404, a FWD three-speed transaxle, has been used by Fiat (Italy).

Ferrari

Certain Ferrari models, such as the GT 400, used GM Hydra-Matic 3L80 (THM 400) automatic transmissions from 1977 to 1987.

Ford

Ford's transmission assembly plant in Bordeaux, France, produces popular RWD/4WD four-speed overdrive transmissions for "domestic" vehicles. The A4LD, 4R44E, 4R55E, and 5R55E transmissions are used in vehicles such as the Aerostar, Ranger, Explorer, Mountaineer, and F150 series pickup trucks (Figure 25–21). Detailed descriptions of these transmission models are located in Chapters 4 and 23.

Mazda (4EAT and ATX) and Nissan (4F20E) automatic transaxles are built in Japan for Ford's North American market vehicles. Additional coverage of these transaxles is provided in the Mazda and Nissan sections in this chapter.

Ford European Market—Automatic Transmissions. Ford produces numerous vehicles for the European market. Automatic transmissions for these vehicles are produced by Ford in Batavia, Ohio, and in Bordeaux, France. They are also supplied by Volkswagen in Kassel, Germany.

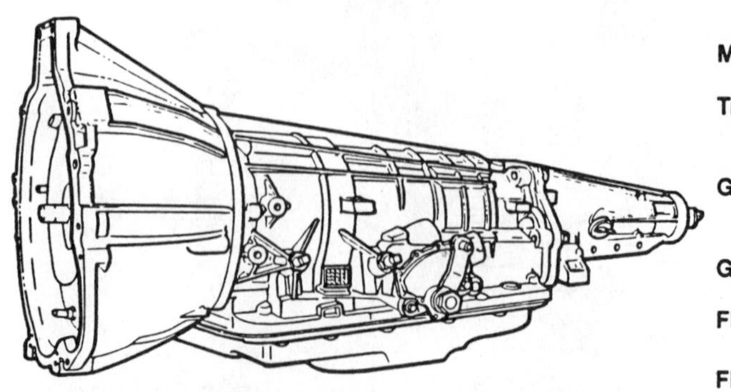

Manufacturer/Assy. Plant	Ford Motor Company/Bordeaux, France
Transmission Type	4-Speed Automatic Overdrive Transmission — Electronic
Gear Ratios	1st 2nd 3rd 4th Rev. 2.47 1.47 1.00 0.75 2.10
Gear Range Selection	P - R - N - OD - 2 - 1
Fluid Refill Capacity	9.0 Liters (9.5 qts.) — 4x2 9.3 Liters (9.8 qts.) — 4x4
Fluid Recommendation	MERCON®

FIGURE 25–21 Ford 4R44E/4R55E automatic transmissions, built in Bordeaux, France. (Reprinted with the permission of Ford Motor Company)

A continuously variable transmission, identified as the CTX, is built at the Bordeaux plant. It is available in 1996 European Escorts, a different vehicle than the domestic Escort. The CTX and other continuously variable transmissions are discussed in the final section of this chapter.

For 1996 Ford European models, the following six automatic transmissions were used:

- CTX: continuously variable transaxle (CVT). Built by Ford in Bordeaux, France.
- CD4E: four-speed automatic overdrive transaxle; electronically controlled. Built by Ford in Batavia, Ohio.
- AG4: four-speed automatic overdrive transmission; electronically controlled. Built by Volkswagen AG in Kassel, Germany.
- A4LD: four-speed automatic overdrive transmission. Built by Ford in Bordeaux, France.
- A4LDe: four-speed automatic overdrive transmission; partial electronically controlled. Built by Ford in Bordeaux, France.
- A4LDE: four-speed automatic overdrive transmission; electronically controlled. Built by Ford in Bordeaux, France (Figure 25–22).

Fiat

From 1969 to 1984, the Trimatic automatic transmission was available. Certain models sold by Fiat used GM Hydra-Matic 3L30 (THM 180), VW Type 010, and Chrysler A-404 automatic transmissions/transaxles.

General Motors Hydra-Matic

GM Hydra-Matic transmissions are distributed to numerous automobile and truck manufacturers around the world. At peak production, it is estimated that Hydra-Matic produces an estimated 22,000 transmissions daily.

1995 Hydra-Matic Applications. The 1995 applications included the following:

Bentley (4L80-E)
BMW (4L30-E/4L40-E/5L40-E)
CAMI (3L30)
Daewoo (4T40-E)
Daimler (4L80-E)
Isuzu (4L30-E/4L80-E)
Jaguar (4L80-E)
Opel (4L30-E)
Saxby (3L30)
Sepma (3L30)
Sisu (3L80)
Suzuki (3L30)
GM Brazil (3T40/3L30/4L30-E/4L60-E)
GM Holden (4L60-E)
GM Venezolana (3T40/4L60-E/4L80-E)

3L30 Three-speed Transmission. The General Motors Hydra-Matic transmission plant in Strasbourg, France, has produced three- and four-speed transmissions for domestic and foreign applications since 1969. The 3L30 (THM 180/180C) uses a Ravigneaux compound planetary gearset. The THM 180C, equipped with a torque converter clutch, was introduced in 1982. The 3L30 was used in the imported Buick Opel (1969–1975), Chevrolet Chevette, and Pontiac T1000 (1977–1986). Recent applications include the 1996 Geo Tracker and GM S/T truck chassis US Postal Service vehicles (Figure 25–23).

3L30 Powertrain Assembly. A sectional view of the GM 3L30 is provided in Figure 25–24. The 3L30 uses a Ravigneaux planetary gearset with the planet carrier as the output member. A simplified schematic of the Ravigneaux arrangement is shown in Figure 25–25.

3L30 Operating Characteristics. The planetary operation provides neutral, three forward gear ratios, and reverse. The manual range selector has six positions (P-R-N-D-2-1), with the operating characteristics typical of General Motors three-speed automatic transmissions with the exception of manual low (1). If manual low is selected with the vehicle in motion, the transmission shifts to first gear immediately. It is important to avoid engaging manual low at vehicle speeds that exceed the manufacturer's recommendation.

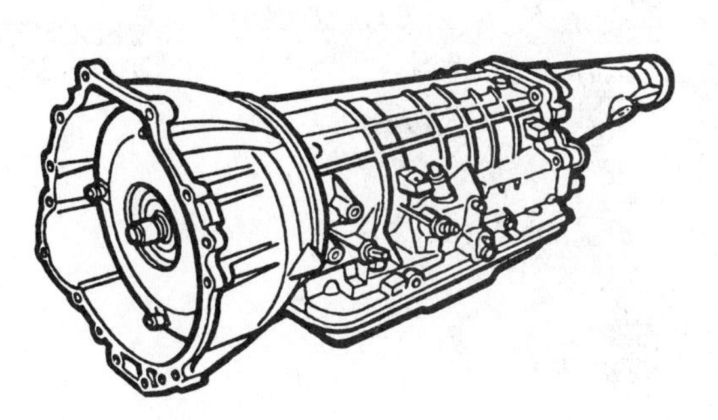

Manufacturer/Assy. Plant	Ford Motor Company Bordeaux, France				
Transmission Type	4-Speed Automatic Transmission, electr. controlled				
Gear Ratios	1st	2nd	3rd	4th	Rev.
2.0 DOHC 16V	2.474	1.474	1.000	0.750	2.111
2.9L V6 24V Cosworth	2.474	1.474	1.000	0.750	2.111
Gear Range Selection	P–R–N–D–2–1				
Fluid Refill Capacity	9.3 Liters				
Fluid Recommendation	ESP-M2C166-H				
Availability	Scorpio				

FIGURE 25–22 Ford A4LDE automatic transmission, built in Bordeaux, France. (Reprinted with the permission of Ford Motor Company)

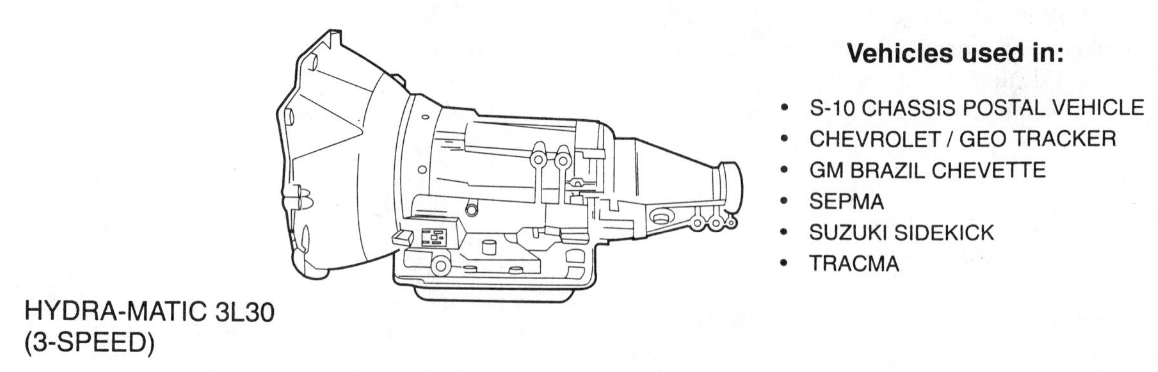

FIGURE 25–23 GM Hydra-Matic 3L30 (THM 180-C) applications. (Courtesy of General Motors Corporation, Service Technology Group)

3L30 Planetary Powerflow. The Ravigneaux planetary operation in the 3L30 (180/180C) is similar to the Ravigneaux planetary operation used in the Ford ATX FWD (1981–1994) presented in Chapter 23. The main difference is in the power feed to planetary. On the 3L30 (180/180C), the power feed drives the planetary from the front end, while on the ATX, the power feed is from the rear of the planetary.

Using the powertrain schematic (Figure 25–25) and the range reference and clutch/band apply chart (Chart 25–4, page 630), the planetary operation can be easily followed. The carrier is the constant output member.

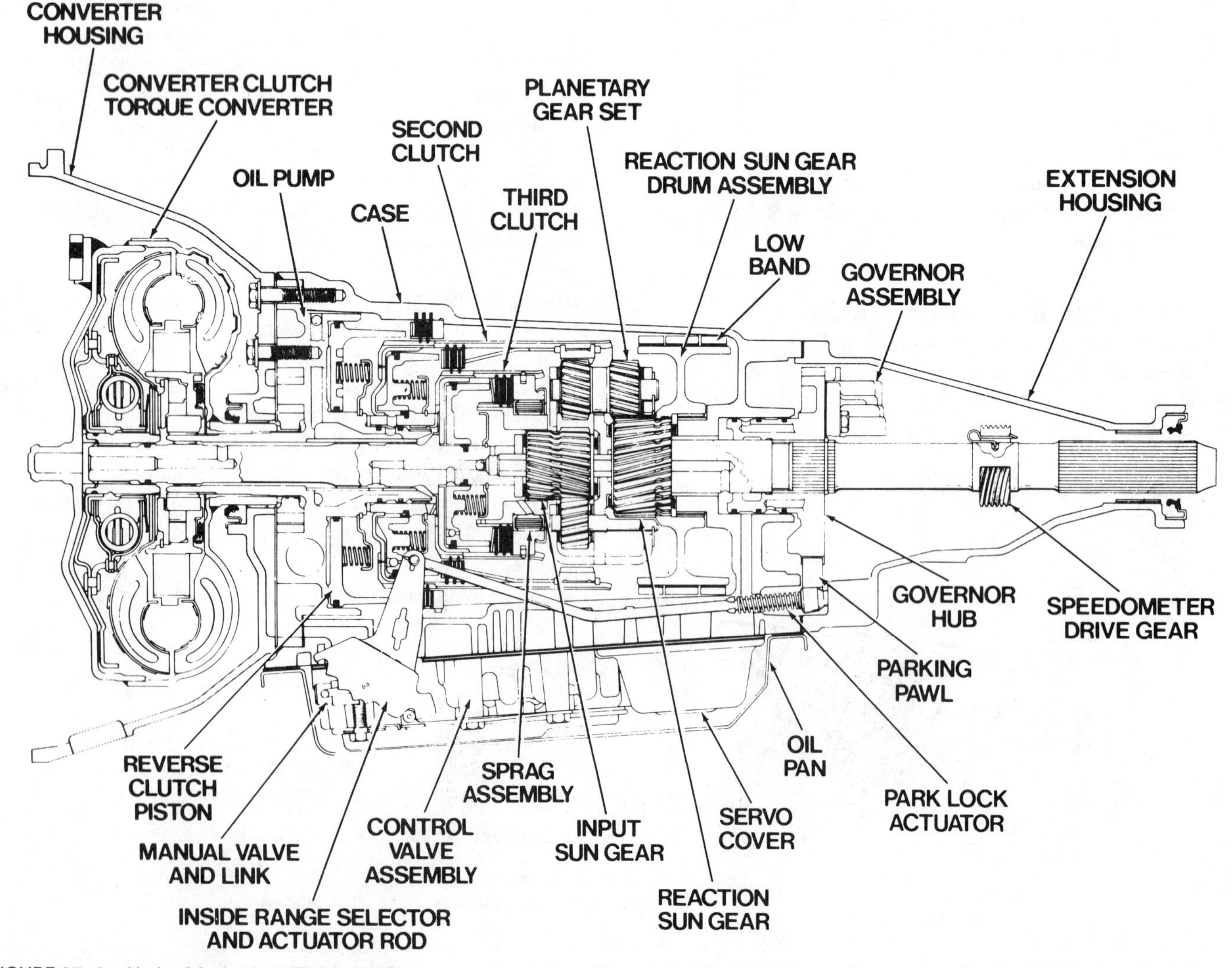

FIGURE 25–24 Hydra-Matic 3L30 (THM 180-C) cross-sectional view. (Courtesy of General Motors Corporation, Service Technology Group)

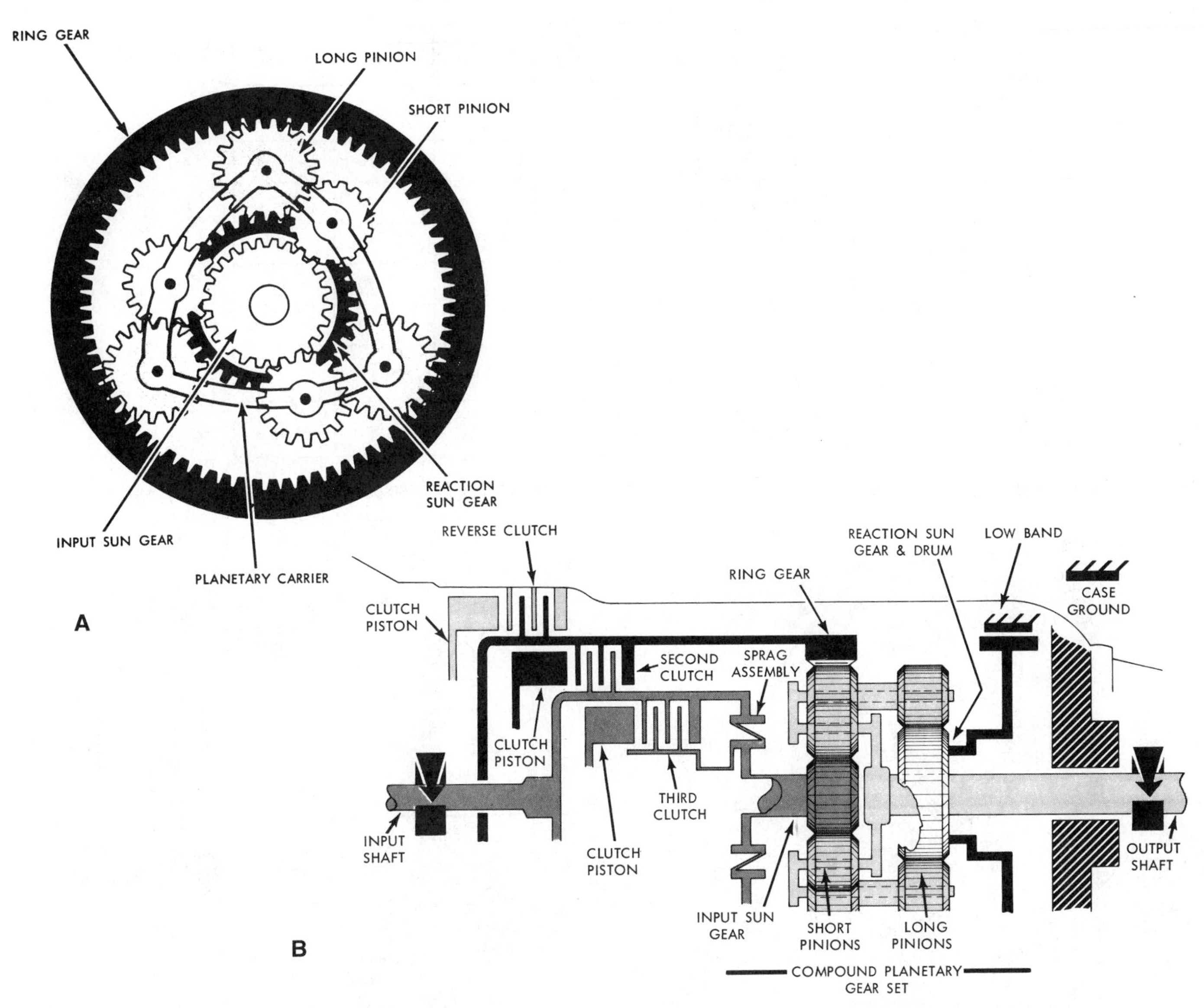

FIGURE 25–25 (a) 3L30 (THM 180-C) Ravigneaux planetary set. (Courtesy of General Motors Corporation, Service Technology Group); (b) 3L30 powertrain schematic.

Neutral. The band and clutches are not applied. The sprag is locked, and there is an input to the planetary set. Because there is no reactionary member, the planetary gears spin freely and cannot provide an output. The planet carrier is held stationary by the weight of the vehicle (Figure 25–26).

First Gear. The reaction sun gear is held stationary. The input sun gear rotates in a clockwise direction, turning the short pinions counterclockwise and the long pinions clockwise. The long pinions walk around the held reaction sun gear, driving the planet carrier and output shaft in a forward reduction. A compound gear ratio of 2.40:1 is provided. The ring gear is in mesh with the long pinions and rotates clockwise but has no function in the powerflow. First gear is illustrated in Figure 25–27.

Second Gear. The reaction sun gear remains stationary. The ring gear is the input member that drives the long pinions in a clockwise direction. The long pinions walk around the reaction sun gear and drive the planet carrier and output shaft at a forward reduction. A simple gear ratio of 1.48:1 is provided. The long pinions are also driving the short pinions counterclockwise. The input sun gear in mesh with the short pinions is driven in a clockwise direction. The effect of the planet carrier rotation causes the short pinions to drive the input sun gear at an overdrive speed, which results in the release of the sprag clutch. Second gear is illustrated in Figure 25–28.

Third Gear. The ring gear and input sun gear are common input members turning at the same speed. The long and short pinions are trapped between the ring gear and input sun gear and cannot rotate on their centers. This results in a lockup of the planetary members and a 1:1 ratio. Third gear is illustrated in Figure 25–29.

Reverse. The ring gear is held and the input sun gear drives the planetary pinion gears. The short pinions turn counterclock-

RANGE	GEAR	INPUT SPRAG	2ND CLUTCH	3RD CLUTCH	LOW BAND	REVERSE CLUTCH
P-N		HOLDING				
D	1st	HOLDING			APPLIED	
	2nd		APPLIED		APPLIED	
	3rd	HOLDING	APPLIED	APPLIED		
2	1st	HOLDING			APPLIED	
	2nd		APPLIED		APPLIED	
1	1st	HOLDING		APPLIED	APPLIED	
R	REVERSE	HOLDING		APPLIED		APPLIED

CHART 25–4 GM 3L30 (THM 180-C) range reference and clutch/band application chart. Typical selector pattern: P-R-N-D-2-1.

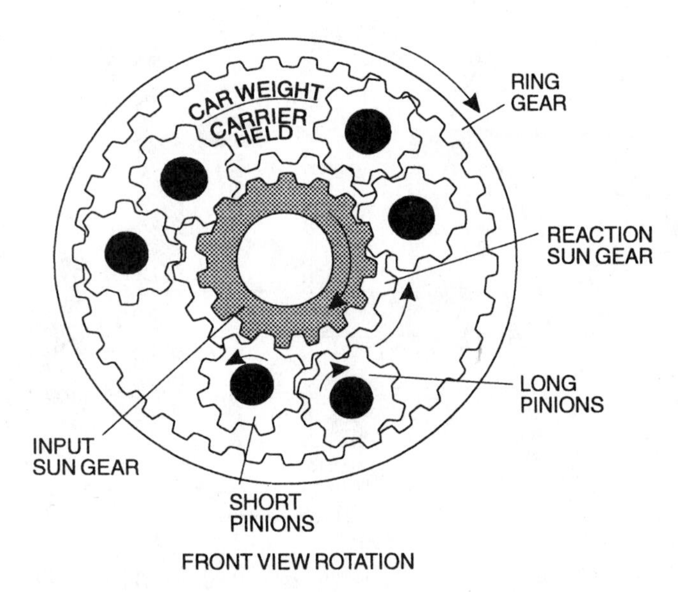

FIGURE 25–26 Neutral planetary gear rotation—GM 3L30 (THM 180-C).

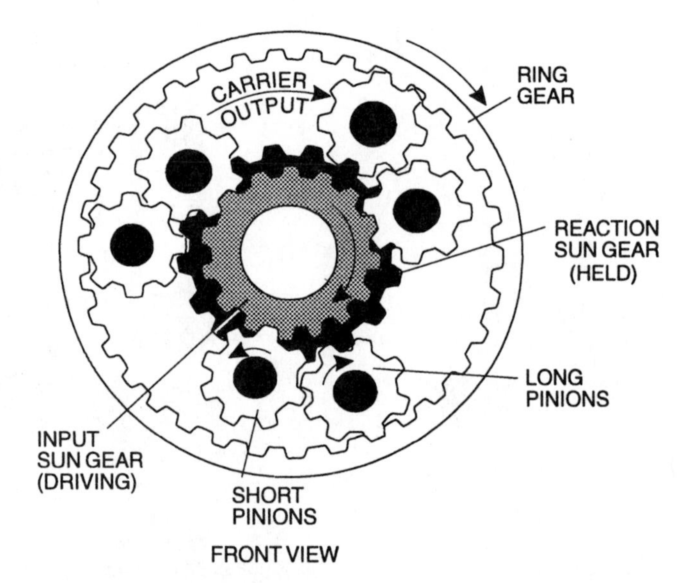

FIGURE 25–27 First-gear planetary gear rotation—GM 3L30 (THM 180-C).

wise, and the long pinions turn clockwise. The long pinions walk around the inside of the stationary ring gear and drive the planet carrier and output shaft in a counterclockwise direction. A simple reverse reduction of 1.91:1 is provided. Reverse is illustrated in Figure 25–30.

4L30-E Four-speed Electronic Shift Transmission. The 4L30-E is produced in the Hydra-Matic Strasbourg, France, transmission plant. This transmission is an adaptation of the 3L30 (THM 180C), with an overdrive planetary that converts the geartrain into a four-speed layout. Full electronic transmission control is used with a self-diagnostic feature. The 4L30-E and its current applications are displayed in Figure 25–31. The new Cadillac Catera, built in Germany and based on the Opel Omega, uses a 4L30-E automatic transmission.

The overdrive planetary and supporting clutch units are positioned up front, as in the 4L80-E and THM 200-4R transmissions. (Refer to Figure 25–32 for a cutaway view of the 4L30-E.) Operation and clutch control of the overdrive planetary are also identical to these transmissions. The 4L30-E range reference and clutch/band application summary is shown in Chart 25–5 (page 632).

Within the three-speed geartrain structure, the 4L30-E dual pinion Ravigneaux planetary gearset has been redesigned for a new set of gear ratios.

	3L30	4L30-E
Reverse	1.92	2.00
First gear	2.40	2.86
Second gear	1.48	1.62
Third gear	1.00	1.00
Fourth gear	N/A	0.72

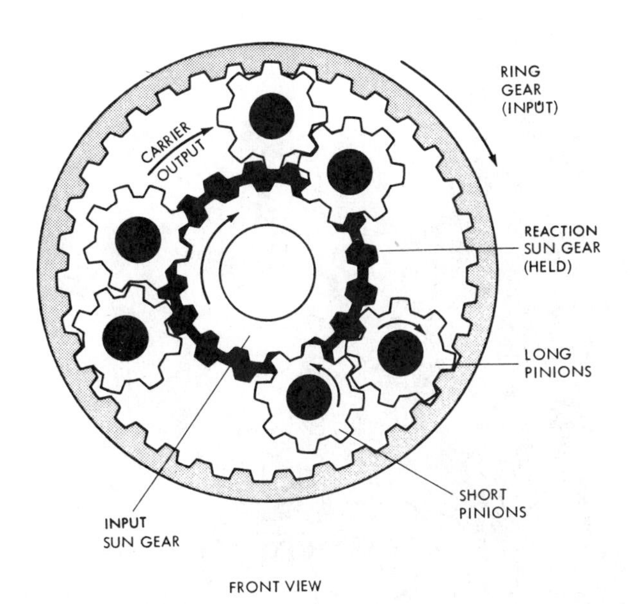

FIGURE 25–28 Second-gear planetary gear rotation—GM 3L30 (THM 180-C).

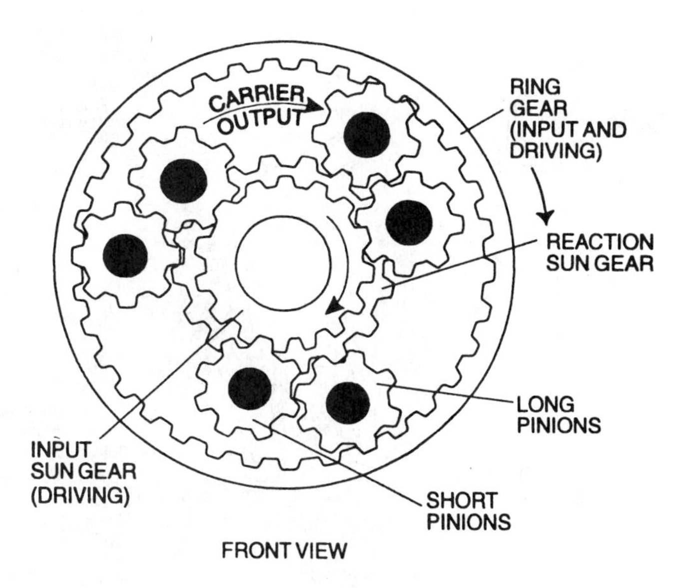

FIGURE 25–29 Third-gear planetary gear rotation—GM 3L30 (THM 180-C).

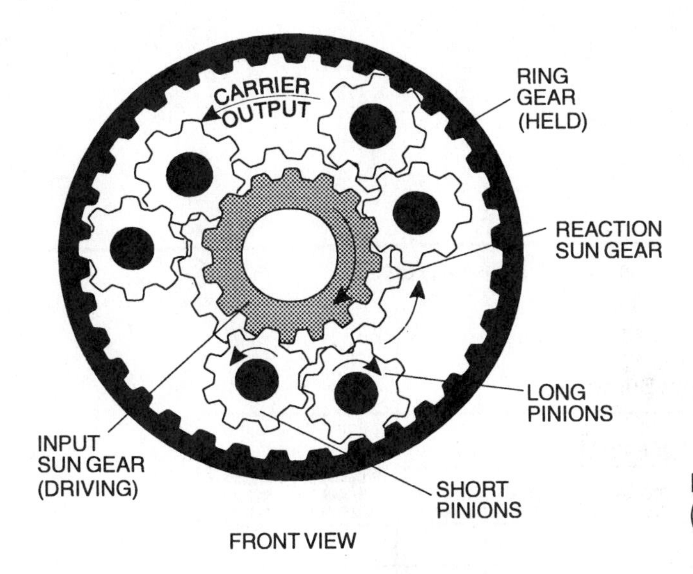

FIGURE 25–30 Reverse planetary gear rotation—GM 3L30 (THM 180-C).

HYDRA-MATIC 4L30-E TRANSMISSION

Produced at: Strasbourg, France

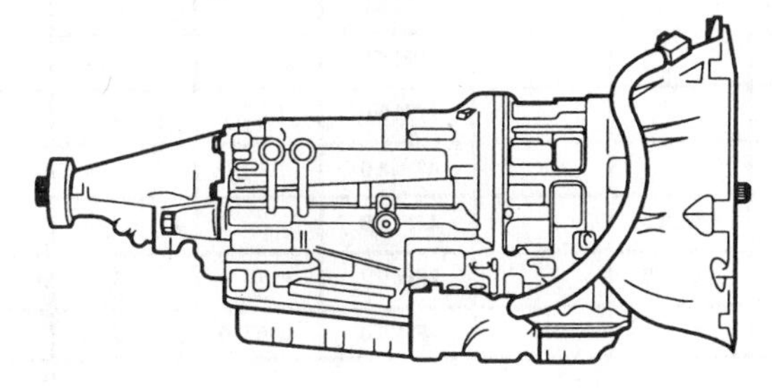

Vehicles used in:

- OPEL OMEGA
- OPEL SENATOR
- ISUZU TROOPER / RODEO
- BMW

HYDRA-MATIC 4L30-E
(4-SPEED)

FIGURE 25–31 GM Hydra-Matic 4L30-E applications. (Courtesy of General Motors Corporation, Service Technology Group)

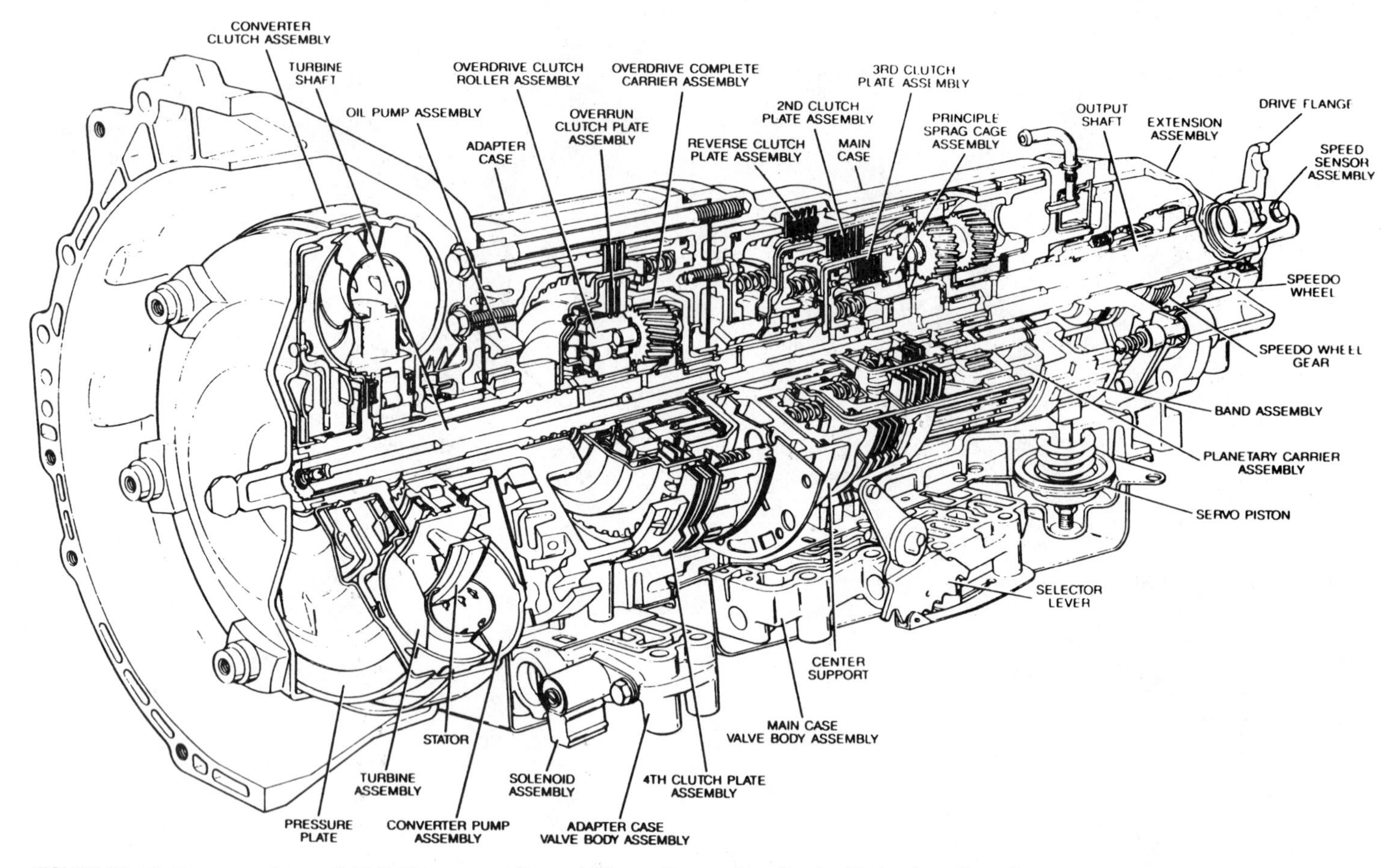

FIGURE 25–32 Cutaway view—4L30-E. (Courtesy of General Motors Corporation, Service Technology Group)

4L30-E Operating Range and Clutch/Band Apply Summary

RANGE REFERENCE CHART

RANGE	GEAR	BAND ASSEMBLY	SECOND CLUTCH	THIRD CLUTCH	FOURTH CLUTCH	OVERRUN CLUTCH	REVERSE CLUTCH	PRINCIPLE SPRAG ASSEMBLY	OVERDRIVE ROLLER CLUTCH
P-N						APPLIED			
D	4th		APPLIED	APPLIED	APPLIED			LC	
	3rd		APPLIED	APPLIED		APPLIED		LC	LD
	2nd	APPLIED	APPLIED			APPLIED			LD
	1st	APPLIED				APPLIED		LD	LD
3	3rd		APPLIED	APPLIED		APPLIED		LC	LD
	2nd	APPLIED	APPLIED			APPLIED			LD
	1st	APPLIED				APPLIED		LD	LD
2	2nd	APPLIED	APPLIED			APPLIED			LD
	1st	APPLIED		APPLIED		APPLIED		LD	LD
1	1st	APPLIED		APPLIED		APPLIED		LD	LD
R	REVERSE					APPLIED	APPLIED	LD	LD

LD = LOCKED IN DRIVE LC = LOCKED IN COAST

CHART 25–5 4L30-E range reference and clutch/band application chart. (Courtesy of General Motors Corporation, Service Technology Group)

The electronic controls to operate the transmission are managed by the TCM (Figure 25–33). A shift mode button allows the driver to request "sport" driving with later shift scheduling. In addition, third-gear standing starts are available for winter road conditions.

Honda/Acura

Honda automatic transaxles, as well as domestically produced Saturn transmissions, do not use planetary gearsets. Rather, their internal construction and layout resemble manual transaxle construction practices.

Powerflow and gear ratios through the transaxle are accomplished by locking constant-mesh helical-cut gears run on parallel shafts. Shaft-mounted hydraulic clutch packs hold and release various speed gear combinations in place of mechanically moved synchronizer units. Early design transaxles used a sliding spur-cut idler for reverse operation.

Honda's first automatic transaxle was a two-speed unit (1974–1980). Three-speed automatics were introduced in 1980, and four-speeds in 1983. Since 1986, Honda has built only four-speed versions of automatic transmissions.

Honda torque converters offer lockup plates that are pulse-width–modulated by the TCM for application in third and fourth gears. Since these transaxles are unique, it is highly recommended that Honda ATF be used when servicing these units.

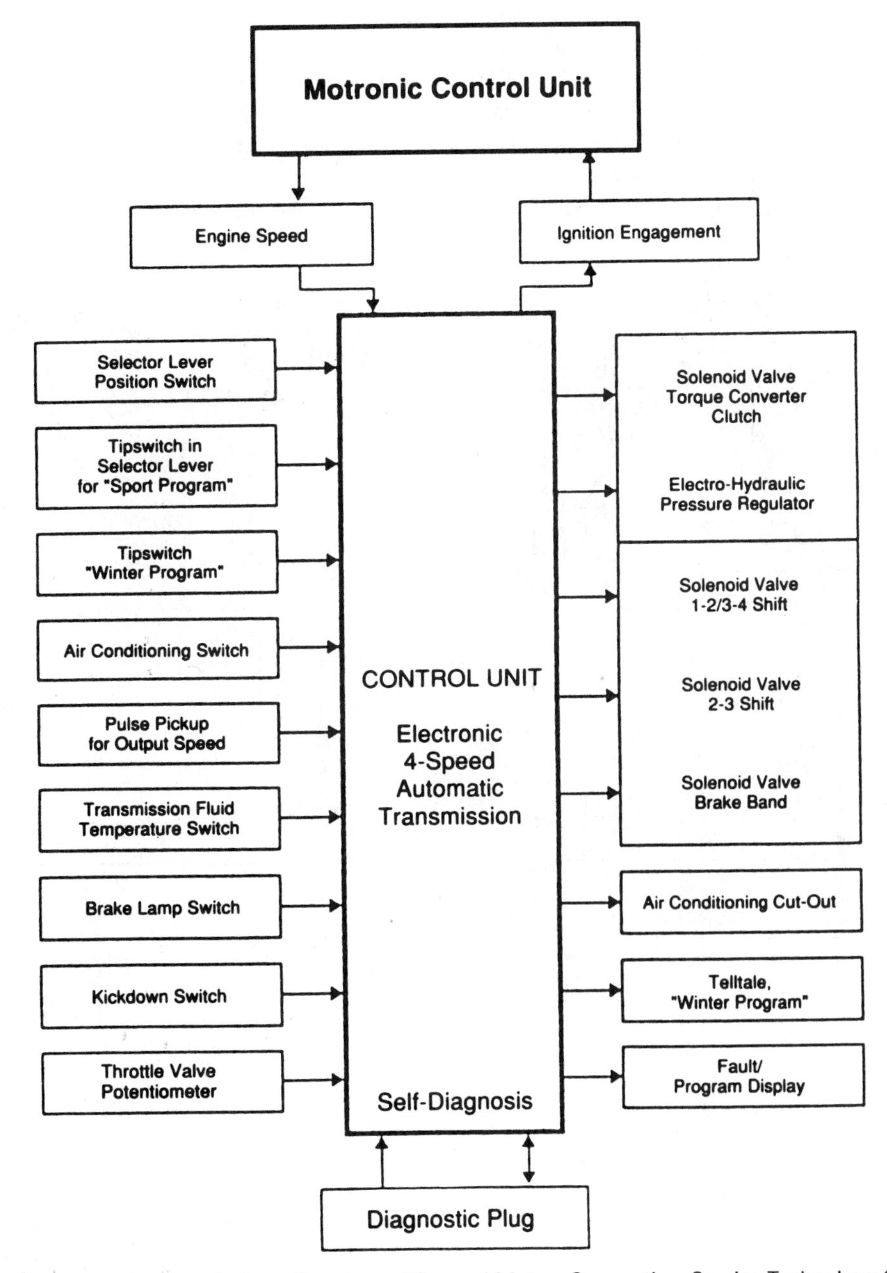

FIGURE 25–33 4L30-E TCM input and output devices. (Courtesy of General Motors Corporation, Service Technology Group)

Honda 4-A/T Single Mode (Two-shaft) Transaxle. The 4-A/T single mode, two-shaft automatic transaxle used in the 1986 to 1988 Accord and Civic and in the 1986 to 1987 Prelude is hydraulically controlled. The shift selector quadrant pattern has six positions: P, R, N, D4 (1-4), D3 (1-3), and 2 (second gear). The geartrain is shown in Figure 25–34. The two-shaft and three-shaft operating range and clutch application summaries are essentially the same and are displayed in Figure 25–35.

The torque converter is equipped with a lockup clutch. The lockup converter can be partially applied in D4 (second, third, and fourth gears) or fully applied in D4 fourth gear. Lockup is not available in D3 or 2 ranges.

Honda 4-A/T Dual Mode (Two-shaft) Transaxle. Beginning with the 1988 Prelude, the 4-A/T dual mode transaxle was available and is based on the single mode 4-A/T. The electronically controlled transaxle uses a unique six-position selector quadrant pattern: P, R, N, D (drive), S (sport), and 2 (second).

In D range, first through fourth gears are available for normal operation. The lockup clutch functions are the same as those of the 4-A/T single mode in D4.

In S (sport S3 and S4) range, rapid acceleration or pulling a heavy load is possible. In S3 mode, 1-2 and 2-3 upshifts occur as much as 1000 rpm later. Fourth gear is eliminated, and lockup can only occur during third-gear deceleration. The S4 mode is engaged by pressing the S4 button. Transmission operation is similar to D range except that the shifts can occur at 1000 rpm higher engine speeds.

The "2" range engages second gear with engine braking. This position is useful for driving at low speeds on slippery roads. Lockup and other forward gears are not available.

Honda 4-A/T Three-shaft Transaxles. The three-shaft, four-speed automatic transaxles were introduced in the 1989 4WD Civic Wagon. The third shaft is constructed with the clutch and gears to provide engine braking for low hold. Overall length remains the same. Also, a seven-position selector quadrant pattern is used. Refer to Figure 25–35 to view a typical three-shaft geartrain schematic and a range reference and clutch application chart.

A similar version was introduced in the 1990 Accord. A large-capacity torque converter with a lockup plate is used. The electronically controlled transaxle uses a unique seven-position selector quadrant pattern: P, R, N, D4 (1-4), D3 (1-3), 2 (sec-

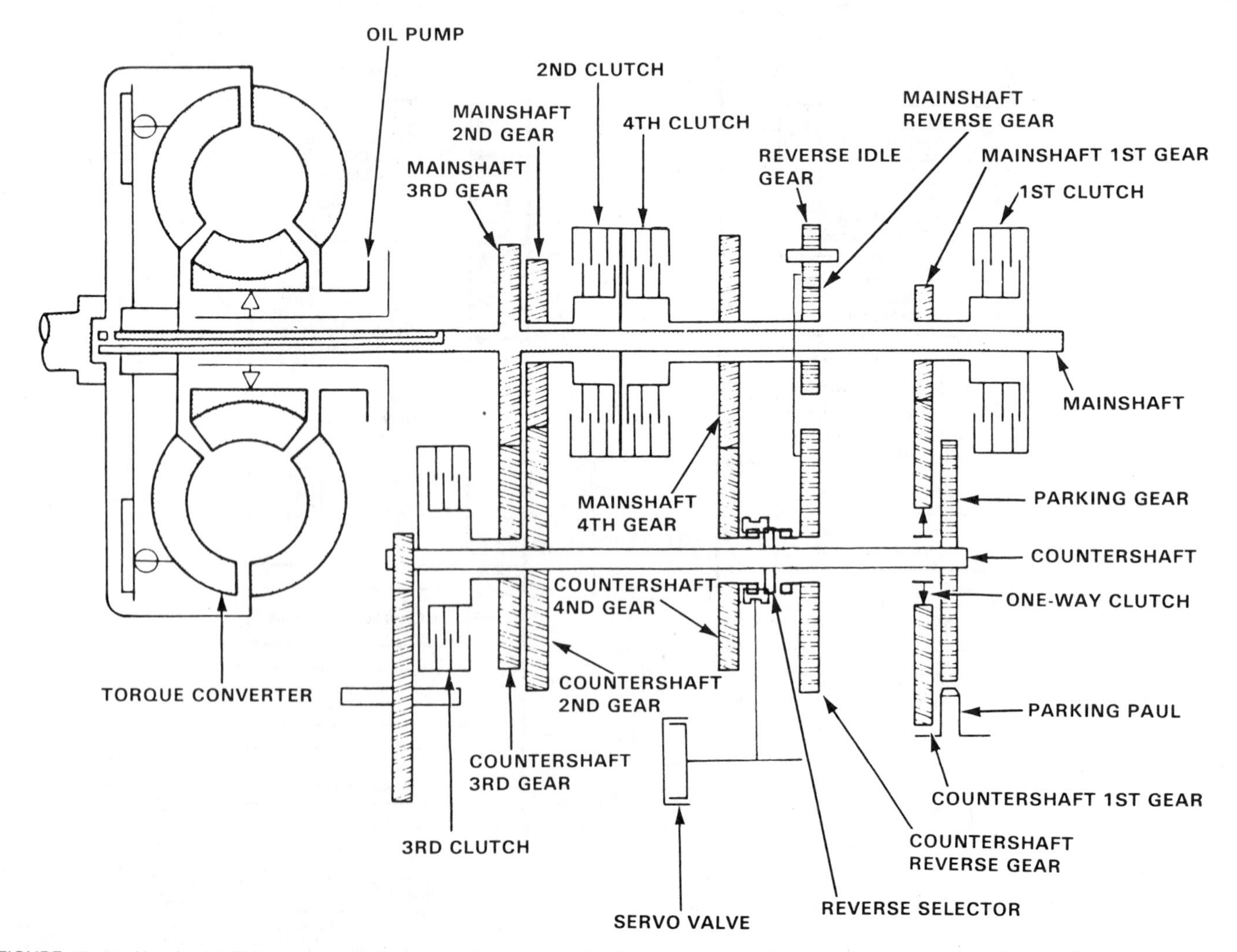

FIGURE 25–34 Honda 4-A/T (two-shaft design) powerflow schematic. (Courtesy of American Honda Motor Co., Inc.)

RANGE \ PART		TORQUE CONVERTER	1ST GEAR 1ST CLUTCH	1ST GEAR ONE-WAY CLUTCH	2ND GEAR 2ND CLUTCH	3RD GEAR 3RD CLUTCH	4TH		REVERSE GEAR	PARKING GEAR
							GEAR	CLUTCH		
P		O	X	X	X	X	X	X	X	O
R		O	X	X	X	X	X	O	O	X
N		O	X	X	X	X	X	X	X	X
S3	1ST	O	O	O	X	X	X	X	X	X
	2ND	O	*O	X	O	X	X	X	X	X
	3RD	O	*O	X	X	O	X	X	X	X
S4 or D	1ST	O	O	O	X	X	X	X	X	X
	2ND	O	*O	X	O	X	X	X	X	X
	3RD	O	*O	X	X	O	X	X	X	X
	4TH	O	*O	X	X	X	O	O	X	X
2	2RD	O	*O	X	O	X	X	X	X	X

O: Operates, X: Doesn't operate, *: Although the 1st clutch engages, driving power is not transmitted as the one-way clutch slips.

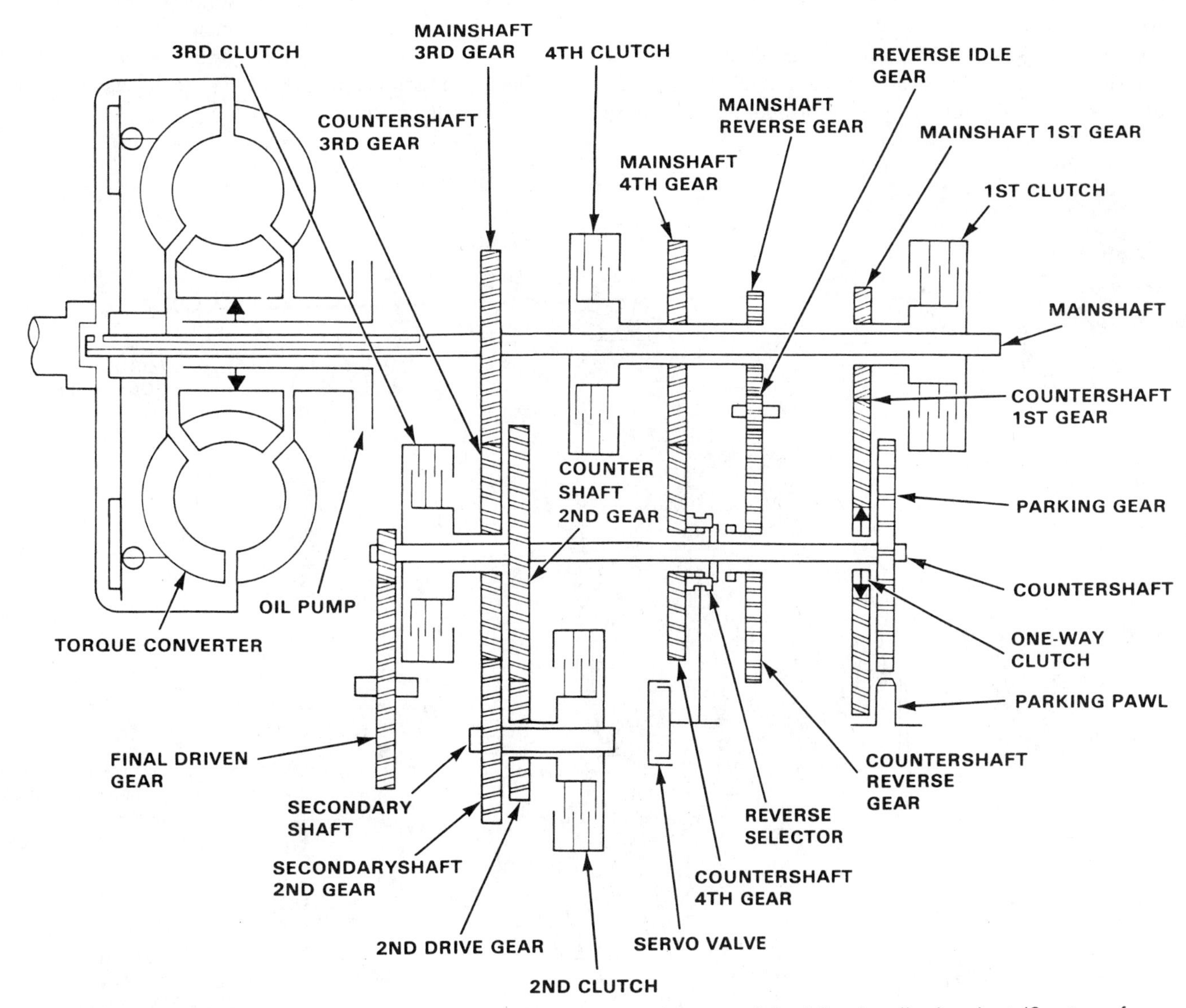

FIGURE 25–35 Honda 4-A/T (three-shaft design) powerflow schematic; range reference and clutch/band application chart. (Courtesy of American Honda Motor Co., Inc.)

ond), and 1 (low hold). A "sport" mode button provides aggressive shift scheduling in D3 and D4 ranges.

Featured in Figure 25–36 is the 1990 Acura Integra electronically controlled four-speed transaxle.

The 1991 Acura Legend introduced a longitudinal FWD version of Honda's four-speed transverse transaxle. The fully electronically controlled unit uses five multiple-disc clutches and a one-way clutch. The final drive assembly is of the ring and pinion style.

Three-shaft automatic transaxles became available in 1992 Prelude and Civic models and are similar to the unit introduced in the 1989 4WD Civic Wagon. Hydraulic and electronic controls are used to operate this style of transaxle.

The 1995 Honda Odyssey minivan three-shaft automatic transaxle is similar to that of the 1990 Accord. Expanded electronic control for improved operation is incorporated. The gear ratios in second gear, third gear, fourth gear, and the final drive unit are lower.

All V-6 equipped 1995 Accords use an all-new automatic transaxle. The design is basically the same as other three-shaft units (Figure 25–37). Internal features include helical reverse gears for quieter reverse operation and design changes to handle the higher torque of the V-6 engines.

Honda powertrains with four-speed automatic transmissions have also been used by other manufacturers. They have been used in Sterling (UK) models since 1987 and are in the new Isuzu Oasis minivan, a rebadged Honda Odyssey.

Honda Passport. Honda's sport-utility vehicle, the Passport, is a rebadged Isuzu Rodeo. Refer to the Isuzu section in this chapter for additional information.

Honda Civic HX CVT. A continuously variable transmission (CVT) was introduced as an option in the 1996 Honda Civic HX coupe. This unique design does not use planetary or helical-cut shaft-mounted gears to cause ratio changes. Rather, a special metallic V-belt drives between an input and output pulley that change widths. The gear selector pattern is traditional in design and function: D (drive), S (sport), and L (low). The transmission ratio range varies from 2.45:1 to 1.45:1 and is followed by a fixed final drive ratio of 5.81:1. This transmission and other CVTs are discussed in the final section of the chapter.

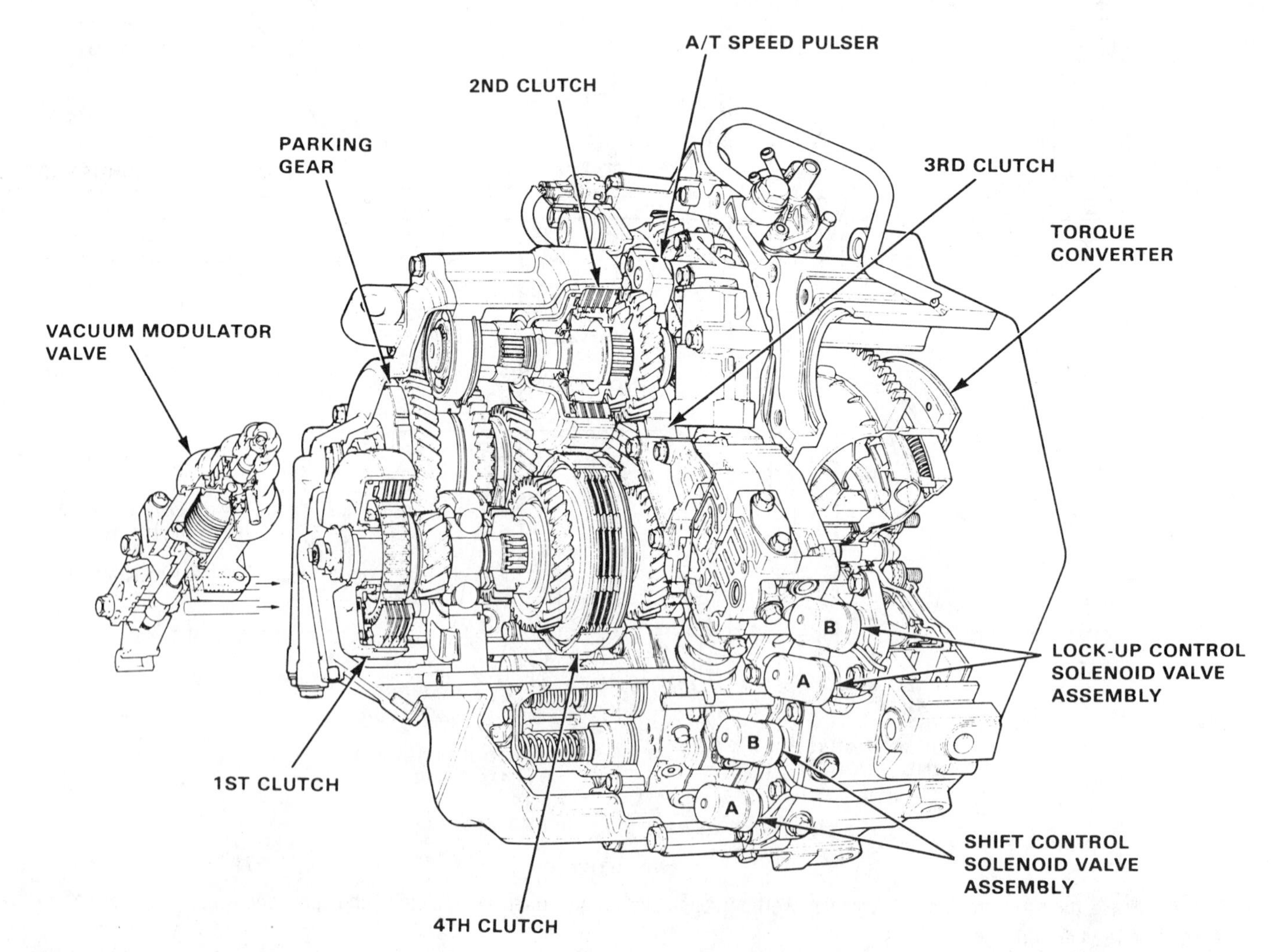

FIGURE 25–36 1990 Acura Integra automatic transaxle. Note the positions of the lockup control and shift solenoids. (Courtesy of American Honda Motor Co., Inc.)

Acura SLX. The 1996 Acura SLX, a 4WD sport-utility vehicle, came equipped with a four-speed automatic transmission as a standard feature. Essentially, it is a rebadged Isuzu Trooper. Refer to the Isuzu section in this chapter for driveline information.

Hyundai

Front-wheel drive automatic transaxles used by Hyundai have been supplied by Mitsubishi. The three-speed KM-170 and KM171 automatic transaxles were used in 1986 to 1989 vehicles. Beginning with 1989 models, four-speed units have been available. These include the F4A21 (KM175), F4A22 (KM176), and F4A23 (KM177). Refer to the Mitsubishi section in this chapter for additional coverage.

For 1996 Hyundai Accent and Sonata models, four-speed automatic transmissions were an optional feature.

Infiniti

Refer to the Nissan/Infiniti section in this chapter for information on Infiniti models.

Isuzu

Automatic transmissions for Isuzu vehicles are supplied by Aisin-Warner, GM Hydra-Matic, and JATCO. The JATCO F10/F3A and KF-100 FWD transaxles were used from 1985 to 1989.

Isuzu/Chevrolet-GEO Storm and Spectrum/JATCO. Certain Geo Storm GSI models, a car line produced by Isuzu, Ltd., of Japan, used an optional FWD JATCO JF403E. The JF403E automatic transaxle is a four-speed unit that uses both electronic and hydraulic shift control.

Isuzu used the FWD JATCO F10/F3A from 1985 to 1989 in Isuzu models and in the Chevrolet/Geo Spectrum.

Isuzu/Hydra-Matic Applications. The Isuzu Trooper and Rodeo use the Hydra-Matic 4L30-E (1990–current), built at the GM Hydra-Matic transmission plant in Strasbourg, France. Isuzu also uses the 4L80-E. Beginning in 1981, certain RWD Isuzu pickup trucks used the THM 200. The Isuzu Hombre pickup truck, introduced in the 1996 model year, is mechanically identical to the Chevrolet S-10 and uses the Hydra-Matic 4L60-E automatic transmission.

Jaguar

Prior to Ford Motor Company acquiring ownership of Jaguar (UK), certain Jaguars used the GM Hydra-Matic 3L80 (THM 400) and the THM 200C. ZF has supplied Jaguar with the ZF-4HP-22 RWD transmission since 1987.

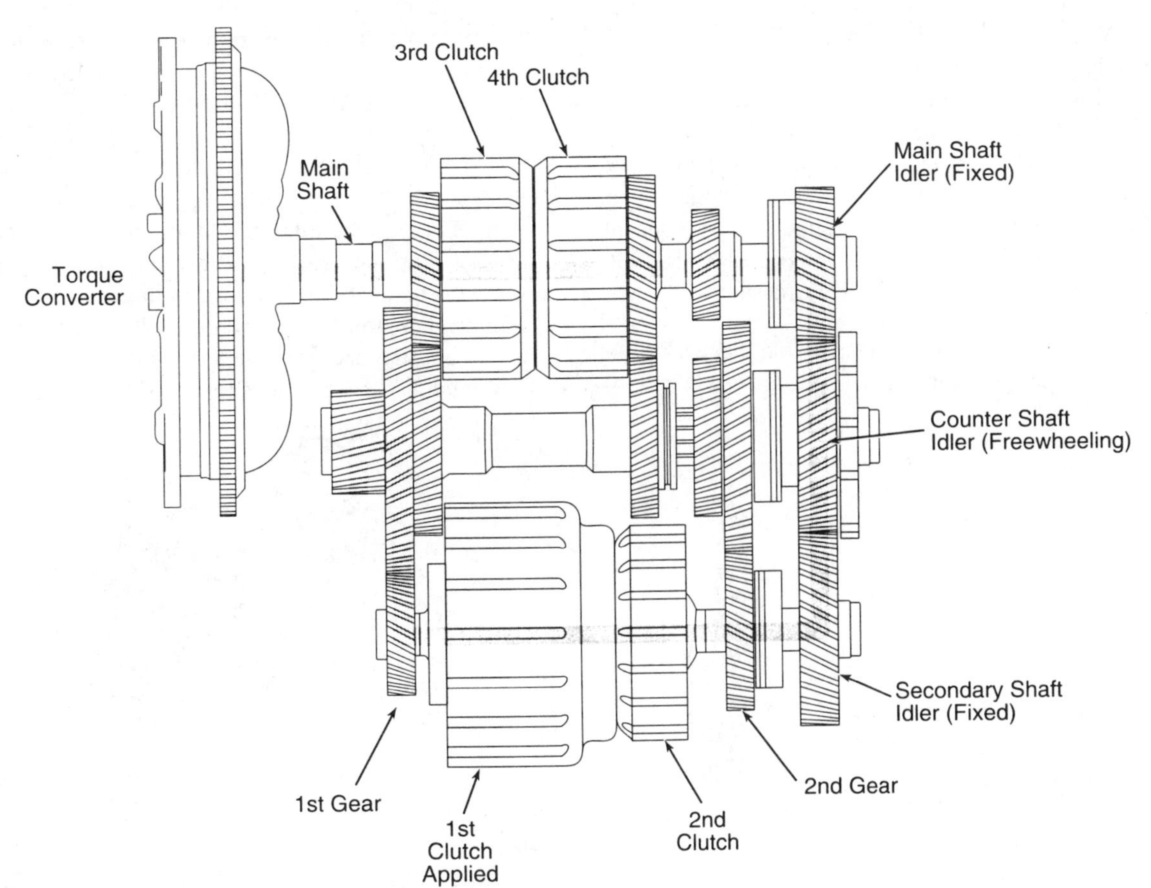

FIGURE 25–37 Powertrain schematic—1995 Honda Accord, Civic, and Prelude. (Note: Reverse idler gear not shown.) (Courtesy of American Honda Motor Co., Inc.)

Beginning in some 1994 models, the Hydra-Matic 4L80-E has been used. All 1996 Jaguar Vanden Plas, XJ6, XJ12, and XJR sedans, as well as XJS convertibles, feature a four-speed automatic transmission as standard equipment. Refer to Chapters 4 and 24 for a detailed description of the 4L80-E.

Lexus

Refer to the Toyota/Lexus section in this chapter for information on Lexus models.

Maserati

Beginning in 1982, some Maserati (Italy) models have been equipped with the RWD Chrysler A-727 (TF-8) three-speed automatic transmission.

Mazda

JATCO supplied Mazda with numerous automatic transmissions and transaxles from 1971 to 1992. Refer to the JATCO section in this chapter for discussion of these units.

Mazda (Ford) ATX Three-speed Transaxle. The Mazda (Japan) designed ATX, shown in Figure 25–38, has many similarities to the Ford (Batavia, Ohio) designed ATX. It has been used in import models sold by Ford, including the Mercury Tracer (1987–1990), Ford Festiva (1987–1990), and the Ford Aspire (1993–current).

Refer to Figure 25–39 for a powerflow schematic of the Mazda-built ATX. Chart 25–6 displays the ATX's range reference and clutch/band apply summary.

The torque converter used in the Aspire utilizes an internally mounted drive plate for clutching action. The plate has weighted, centrifugally actuated clutch lining segments that come into contact with the converter cover's machined inner surface (Figure 25–40). This converter clutch design concept was used in Ford C-5 (1982–1986) transmissions and in 1986 and 1987 Taurus/Sable ATX (Ford design) transaxles.

Mazda RWD/4WD Four-speed Transmission (Ford). Beginning with model year 1991, the Mazda Navajo used the Ford A4LD. The two-door Navajo, a two-or four-wheel drive sport-utility vehicle, is quite similar to the Ford Explorer and is built at Ford's Louisville, Kentucky, assembly plant. The A4LD four-speed overdrive transmission is built at Ford's Bordeaux, France, transmission plant.

In 1995, Explorers began using the Ford 4R55E automatic transmission. Compared to the A4LD, the 4R55E has a higher torque rating and is designed with full electronic controls. Look for these French-built units in the Navajo.

Mazda's B series light-duty trucks are essentially re-badged Ford Rangers. 1994 and earlier models used the A4LD transmission, and beginning in 1995, the 4R44E/4R55E automatic transmissions have been available. Refer to Chapters 4 and 23 for additional coverage of these Ford transmissions.

F4AEL/G4AEL Four-speed Transaxles (Ford 4EAT Series). These Japanese-built, four-speed automatic transaxles are electronically controlled FWD units. There are many similarities between the two units, but the F4AEL is smaller to accommodate the vehicles in which it is used. These transverse-mounted units use a Ravigneaux planetary gearset and a converter lockup clutch (Figure 25–41).

The electronic system controls the transmission shifting and torque converter lockup through four solenoid valves mounted on the valve body. A TCM communicates with the PCM/ECM through a two-way data link. Many engine management sensors and switches are related to transmission shift and lockup operation (Figures 25–42 and 25–43). Self-diagnosis capability is incorporated into the system to simplify troubleshooting procedures.

A six-position selector quadrant pattern is used: P-R-N-OD-D-L.

In OD range, first through fourth gears are available. Converter lockup can engage in third and fourth gears. Engine braking is realized in third and fourth gears.

In D range, fourth gear is eliminated, as is converter lockup. Engine braking occurs in third gear only.

Manual L (low) provides first- and second-gear operation.

Manufacturer/Assy. Plant	Mazda/Japan	
Transmission Type	ATX (Mazda)	
Gear Ratios	1st	2.84
	2nd	1.54
	3rd	1.00
	Rev.	2.40
Gear Range Selection	P - R - N - D - 2 - 1	
Fluid Refill Capacity	5.3 Liters (5.6 qts.)	
Fluid Recommendation	MERCON®	
Availability	Aspire	

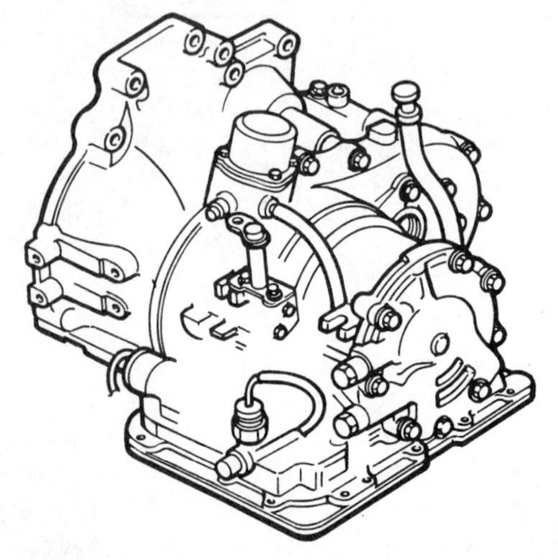

FIGURE 25–38 Mazda-designed ATX. (Reprinted with the permission of Ford Motor Company)

Automatic Transaxle — Mechanical Schematic

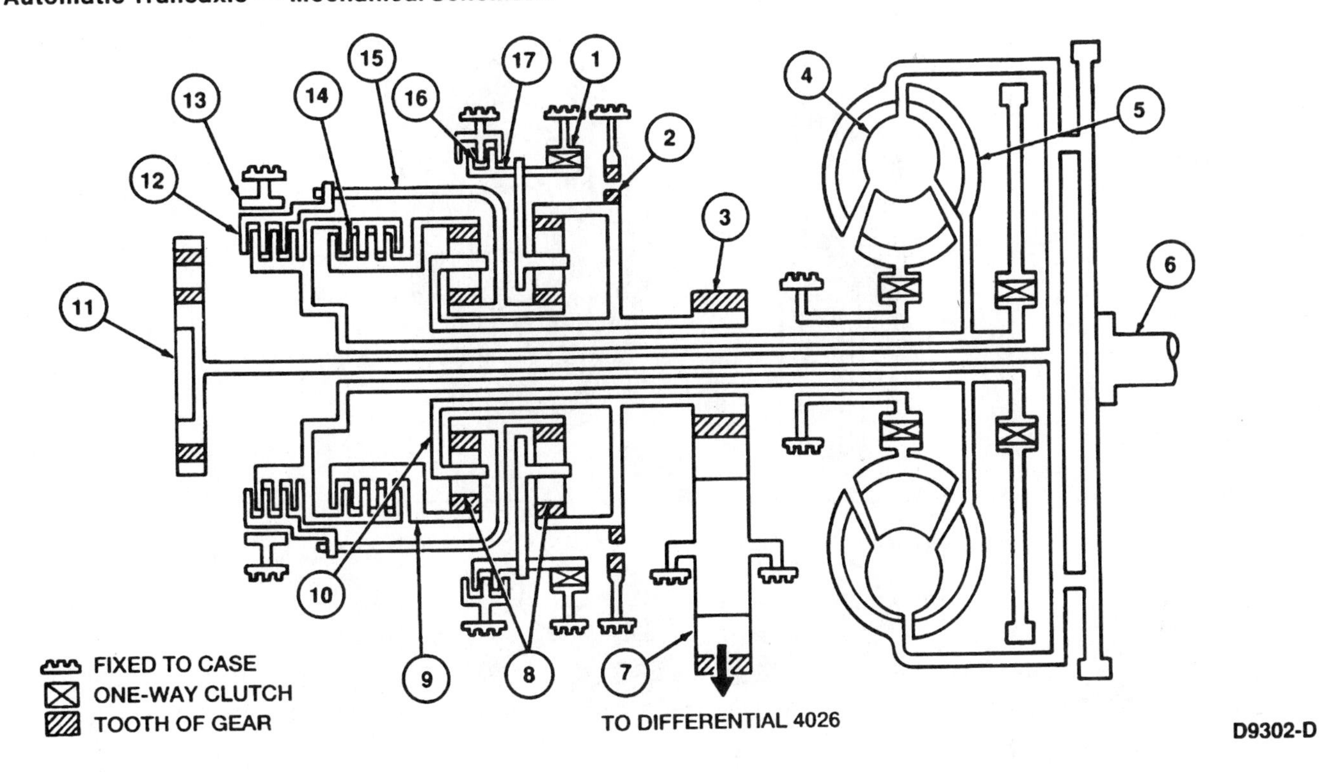

Item	Part Number	Description
1	7A089	One-Way Clutch
2	7A233	Park Gear
3	—	Output Gear
4	—	Impeller (Part of 7902)
5	—	Turbine (Part of 7902)
6	6303	Crankshaft
7	7F475	Idler Gear
8	—	Planetary Carriers

Item	Part Number	Description
9	—	Front Ring Gear
10	7A398	Front Planet Carrier
11	7A103	Oil Pump
12	—	Direct Clutch
13	7D034	Intermediate Band
14	—	Forward Clutch
15	7D064	Sun Shell
16	—	Low/Reverse Clutch
17	—	Rear Planet Carrier

FIGURE 25–39 Mazda ATX powertrain schematic. (Reprinted with the permission of Ford Motor Company)

Shift Position		Clutch		Low & Reverse Clutch	Clutch		One-Way Clutch	Gear Ratio
		Front	Rear		Applied	Released		
P				○				—
R		○		○		⊗		2.400
N								—
D	1		○				○	2.841
	2		○		○			1.541
	3	○	○			⊗		1.000
2			○		○			1.541
1	2		○		○			1.541
	1		○	○				2.841

⊗ Component operates but does not transmit power.

CHART 25–6 Mazda ATX range reference and clutch/band application chart. (Reprinted with the permission of Ford Motor Company)

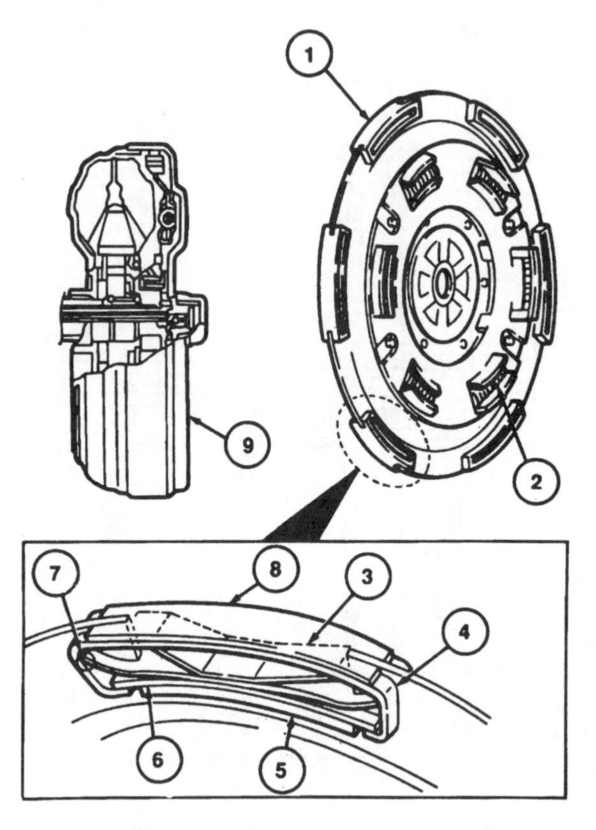

Item	Part Number	Description
1	—	Drive Plate (Part of 7902)
2	—	Torsional Dampening Springs (Part of 7902)
3	—	Main Spring (Part of 7902)
4	—	Shoe Bracket (Part of 7902)
5	—	Weight (Part of 7902)
6	—	Retractor Spring (Part of 7902)
7	—	Pin (Part of 7902)
8	—	Clutch Linings (Part of 7902)
9	7902	Torque Converter

FIGURE 25–40 Centrifugal force applied TCC plate incorporated in the Mazda-designed ATX; used in the 1996 Ford Aspire. (Reprinted with the permission of Ford Motor Company)

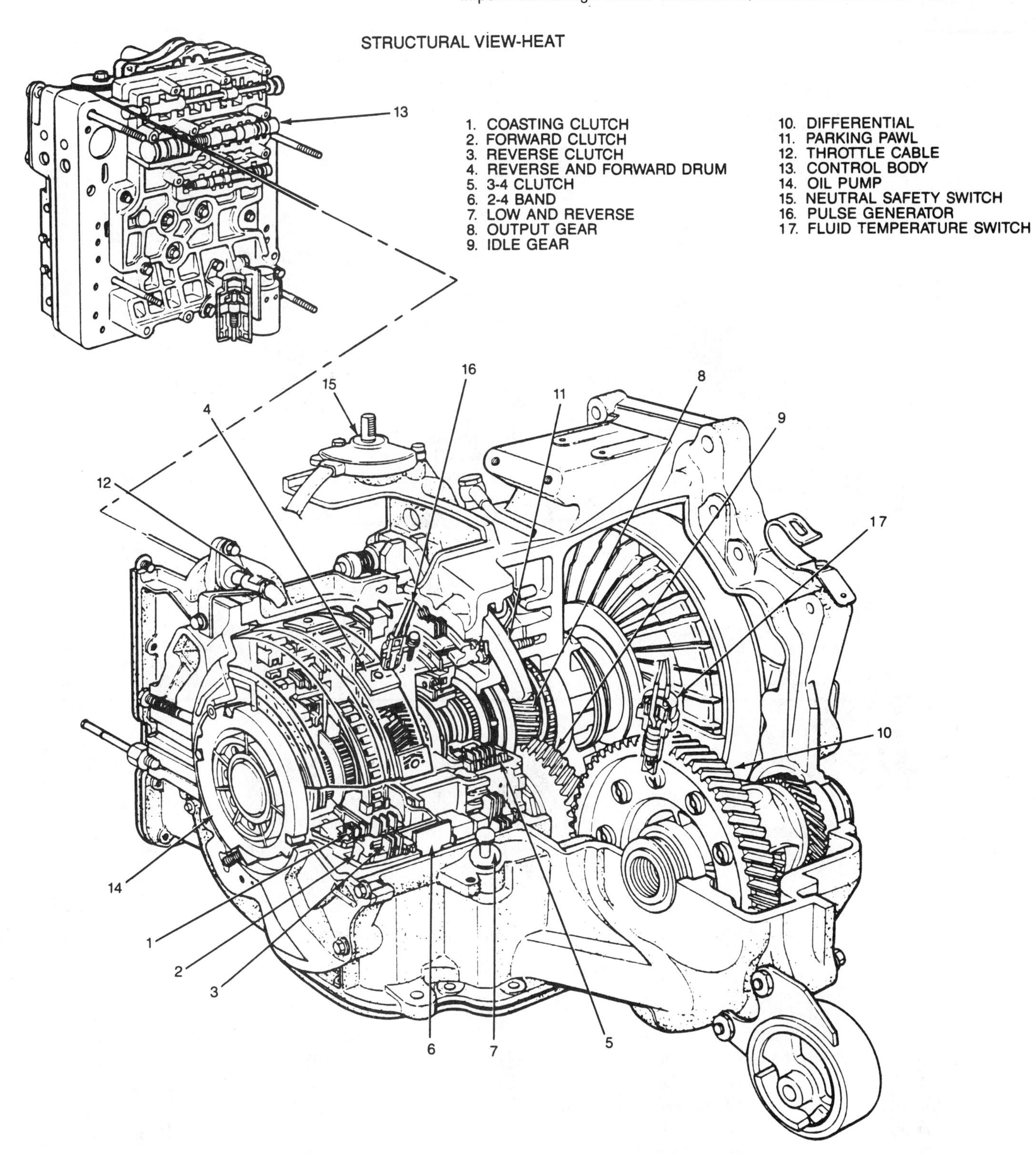

FIGURE 25–41 Structural view—4EAT. (Reprinted with the permission of Ford Motor Company)

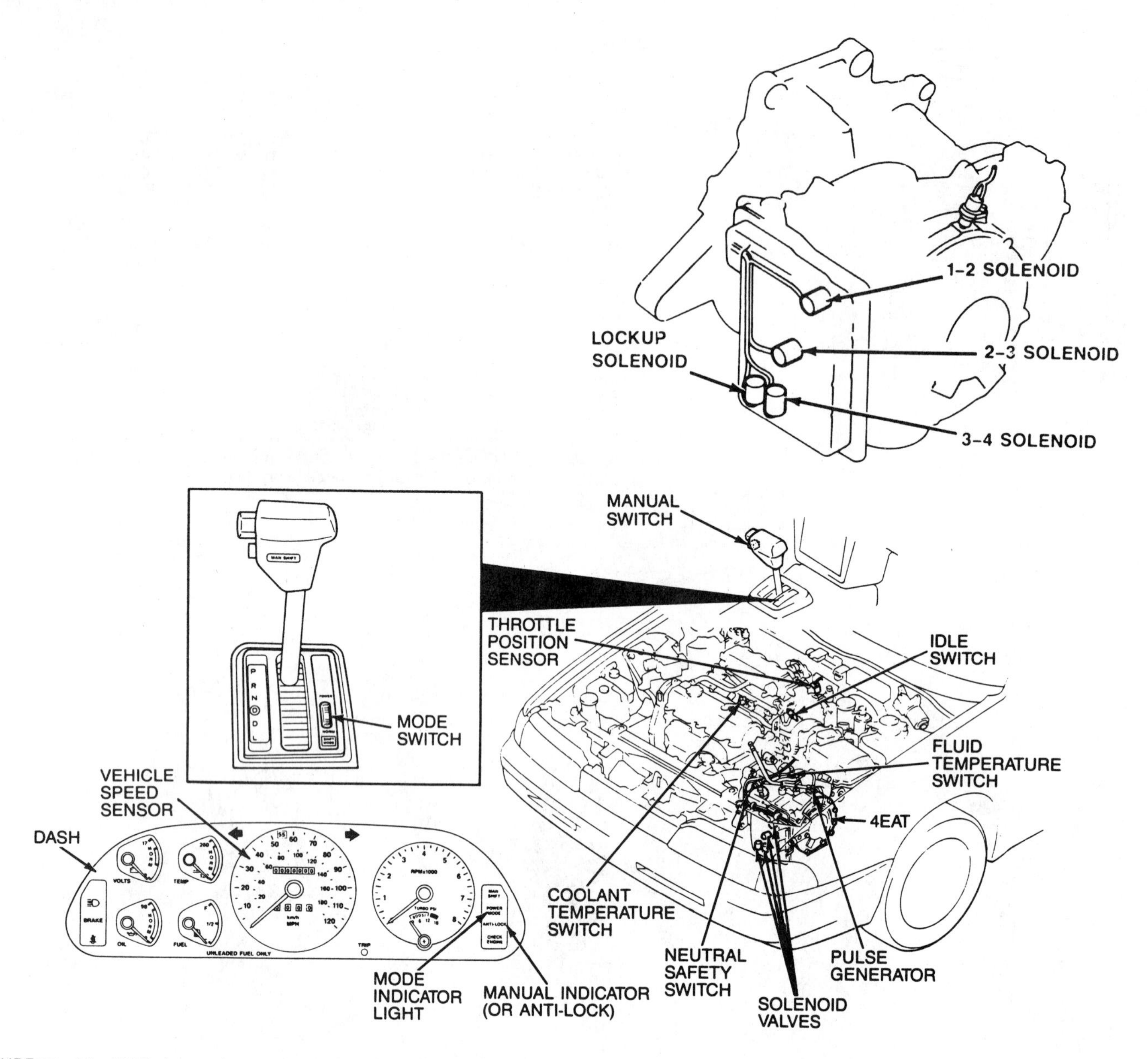

FIGURE 25–42 4EAT electronic component location. (Reprinted with the permission of Ford Motor Company)

The Ford Probe, when equipped with a 4EAT Type G, is designed with automatic and manual mode operation (Figure 25–44). Manual mode locks the transmission in first gear when in L and second gear when in D, as shown in Chart 25–7 (page 644). Note: Abusive driver habits will allow upshifts to occur.

The version used in Ford Probes (4EAT Type G) also comes equipped with a shift mode switch. When activated from normal to power, sportier performance is realized. Shift scheduling is delayed in this mode.

The F4AEL was introduced in the 1990 Mazda 323 and in the Protege 2WD and MX-3 (except for 4WD). Within the Ford product line, the (4EAT Type F) transaxle has been offered in Ford Escorts and Mercury Tracers since 1989 (Figure 25–45).

The G4AEL was introduced in the 1988 MX-6 and 626 (except turbo) models. The Ford Probe used the transaxle, referred to by Ford as the 4EAT Type G, from 1989 to 1994 (Figure 25–46). Note: Ford's CD4E began to be introduced in 1994 Probes. The 4EAT Type G was used in 1994 Ford Probes, and the slightly modified version, Type GF, was used in 1995 and 1996 Probe GT models. The Mazda MX-6 and 626 and the Ford Probe share similar platforms.

The G4AEL/4EAT Type G and F4AEL/4EAT Type F reverse and forward gears operation is presented in Figures 25–47 to 25–53. The range reference and clutch/band application summaries are shown in Charts 25–8 and 25–9 (page 647).

The Mercury Capri, a compact convertible, used the 4EAT Type G transaxle until the vehicle was discontinued after the 1994 model year.

1996 Mazda Models. 1996 Mazda 626, B2300, B3000, B4000, MX-5 Miata, MX-6, and Protege models were available with an optional four-speed automatic transmission. In the Millenia

4EAT ELECTRONIC CONTROL

	PART NAME		FUNCTION
4EAT control unit			Regulates shift points and lockup points according to signals from various sensors; sends ON/OFF signals to solenoid valves
Input	Pulse generator		Detects reverse and forward drum revolution speed
	Vehicle speed sensor		Detects vehicle speed
	Throttle position sensor		Detects amount of throttle valve opening
	Idle switch		Detects throttle valve fully-closed position
	Manual lever position switch		Detects position (range) of gear selector
	Stoplamp switch		Detects use of brakes
	Coolant temperature sensor		Indicates engine coolant temperature
	Fluid temperature sensor		Detects automatic transaxle fluid temperature
Output	Solenoid valves		Switched ON-OFF by electrical signals from 4EAT control unit; regulates shifting and lockup actuation by switching oil passages
		1-2	For 1-2 shift (1st→2nd gear: OFF-ON)
		2-3	For 2-3 shift (2nd gear→3rd gear: ON-OFF)
		3-4	For 3-4 shift (3rd gear→OD: OFF-ON)
		Lockup	For lockup (lockup at ON)

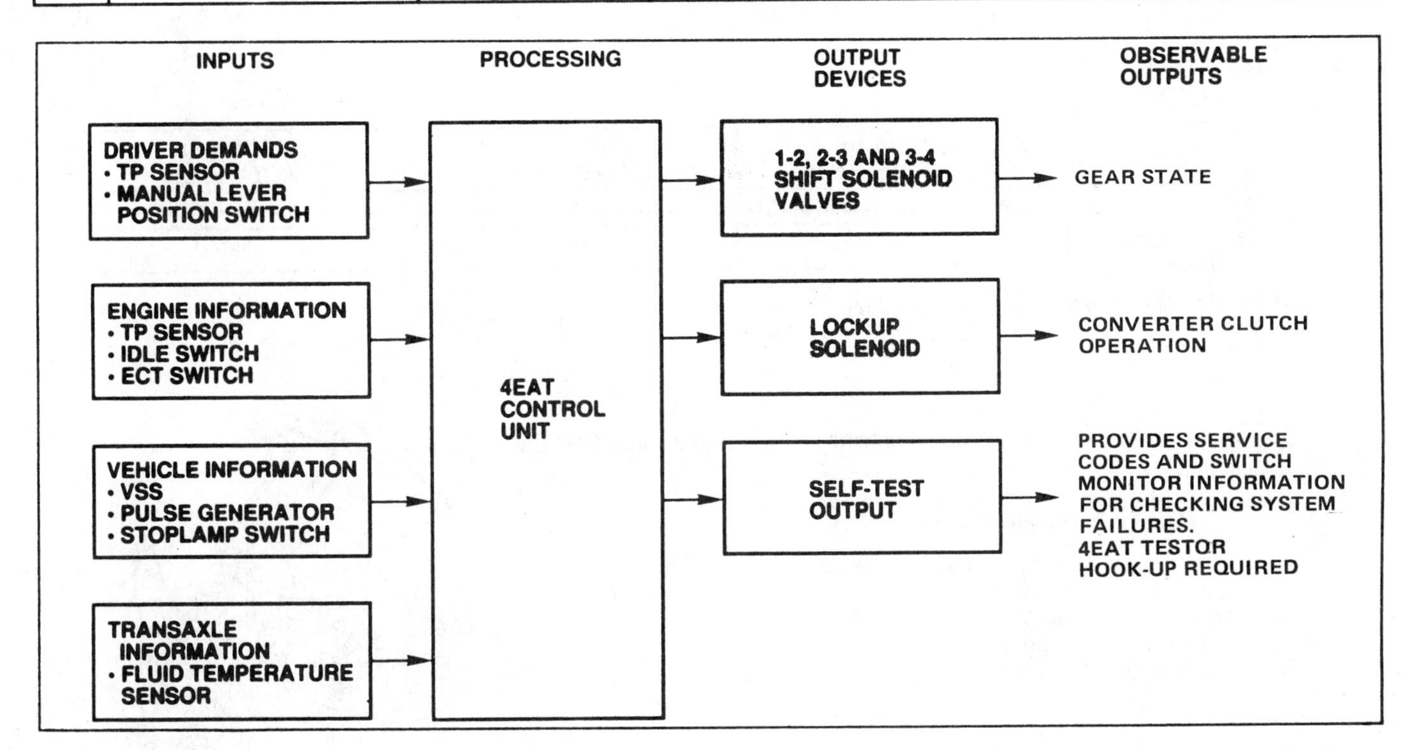

FIGURE 25–43 4EAT component functions and control chart. (Reprinted with the permission of Ford Motor Company)

and MPV models, a four-speed automatic transmission is a standard feature.

Mercedes-Benz

Three- and Four-speed Transmissions. Three- and four-speed RWD transmissions have been built by Mercedes for use in passenger cars for many years. The W3A040 and W4TB025 have been used in the 450 series (1972–1981) and in the 220, 230, 240, 280, and 300 series (1967–1977). The four-speed W4A020 (1984–1991) and W4A040 (1981–1991) transmissions can be easily identified by the shape of the bottom pans. Operation of the W4A040 is controlled by a governor and vacuum modulator.

Mercedes 722.5 Five-speed Transmission. The Mercedes 722.5 first appeared in 1989 and is similar in design and appearance to the four-speed 722.3 (W4A040) and 722.4

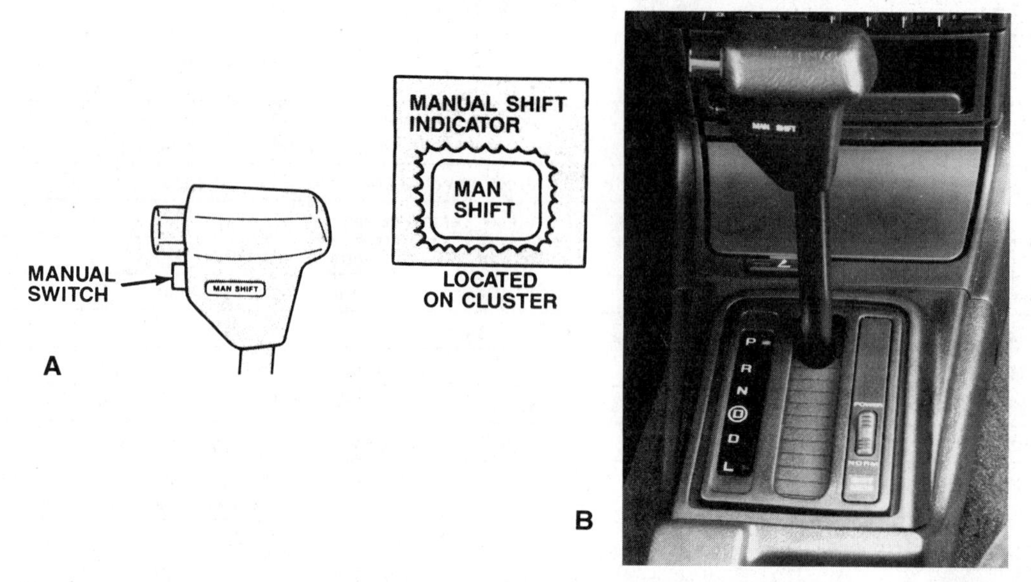

FIGURE 25–44 (a) 4EAT manual switch. (Reprinted with the permission of Ford Motor Company); (b) selector positions and power/normal mode switch.

AUTOMATIC SHIFTING—TYPE F&G
MANUAL SHIFTING—TYPE G

	Shifting In OVERDRIVE Range	Shifting In DRIVE Range	Shifting In LOW Range
"Automatic" Position	1st↔2nd↔3rd↔4th	1st↔2nd↔3rd	1st↔2nd
"Manual" Position	2nd↔3rd	2nd	1st

CHART 25–7 4EAT automatic and manual shift operation.

4-SPEED ELECTRONIC AUTOMATIC OVERDRIVE TRANSAXLE - 4EAT TYPE F

Standard: None

Optional: Ford Escort, Ford Escort LX, Ford Escort GT, Mercury Tracer, Mercury Tracer LTS

Gear ratios: 1st = 2.80, 2nd = 1.54, 3rd = 1.00, 4th = 0.70, Rev = 2.33

Gear ranges: P, R, N, OD, D, L

Made by Mazda in Japan

Fluid refill capacity: 6.1 Liters

Recommended fluid: MERCON®

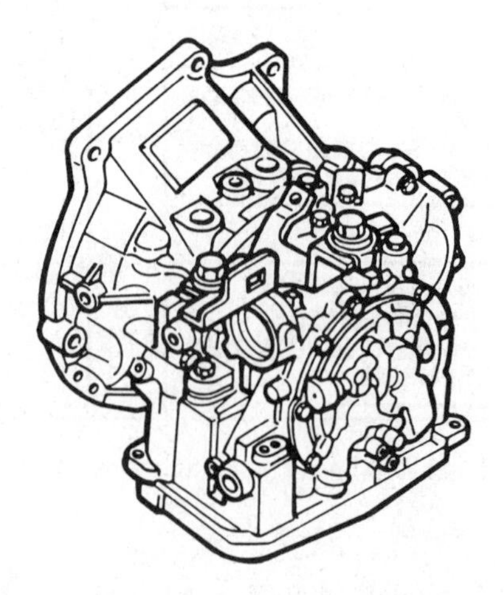

FIGURE 25–45 Mazda-produced 4EAT Type F automatic transaxle. (Reprinted with the permission of Ford Motor Company)

4-SPEED ELECTRONIC AUTOMATIC OVERDRIVE TRANSAXLE - 4EAT TYPE G

Standard: None
Optional: Mercury Capri, Ford Probe GT 2.5L

Gear ratios: 1st = 2.86, 2nd = 1.54, 3rd = 1.00, 4th = 0.70, Rev = 2.33
Gear ranges: P, R, N, OD, D, L
Made by Mazda in Japan
Fluid refill capacity: 6.8 Liters
Recommended fluid: MERCON®

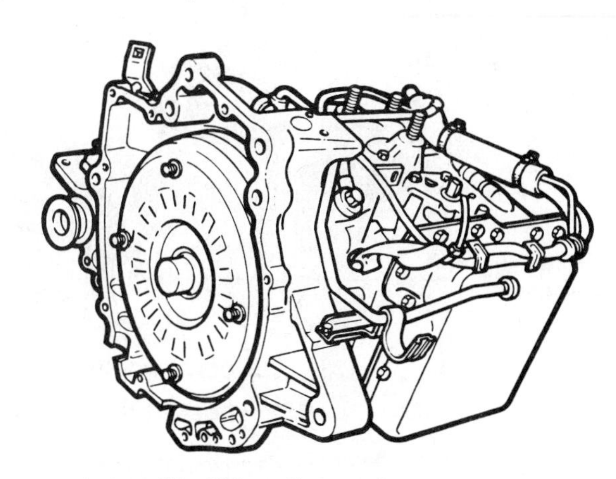

FIGURE 25–46 Mazda-produced 4EAT Type G automatic transaxle. (Reprinted with the permission of Ford Motor Company)

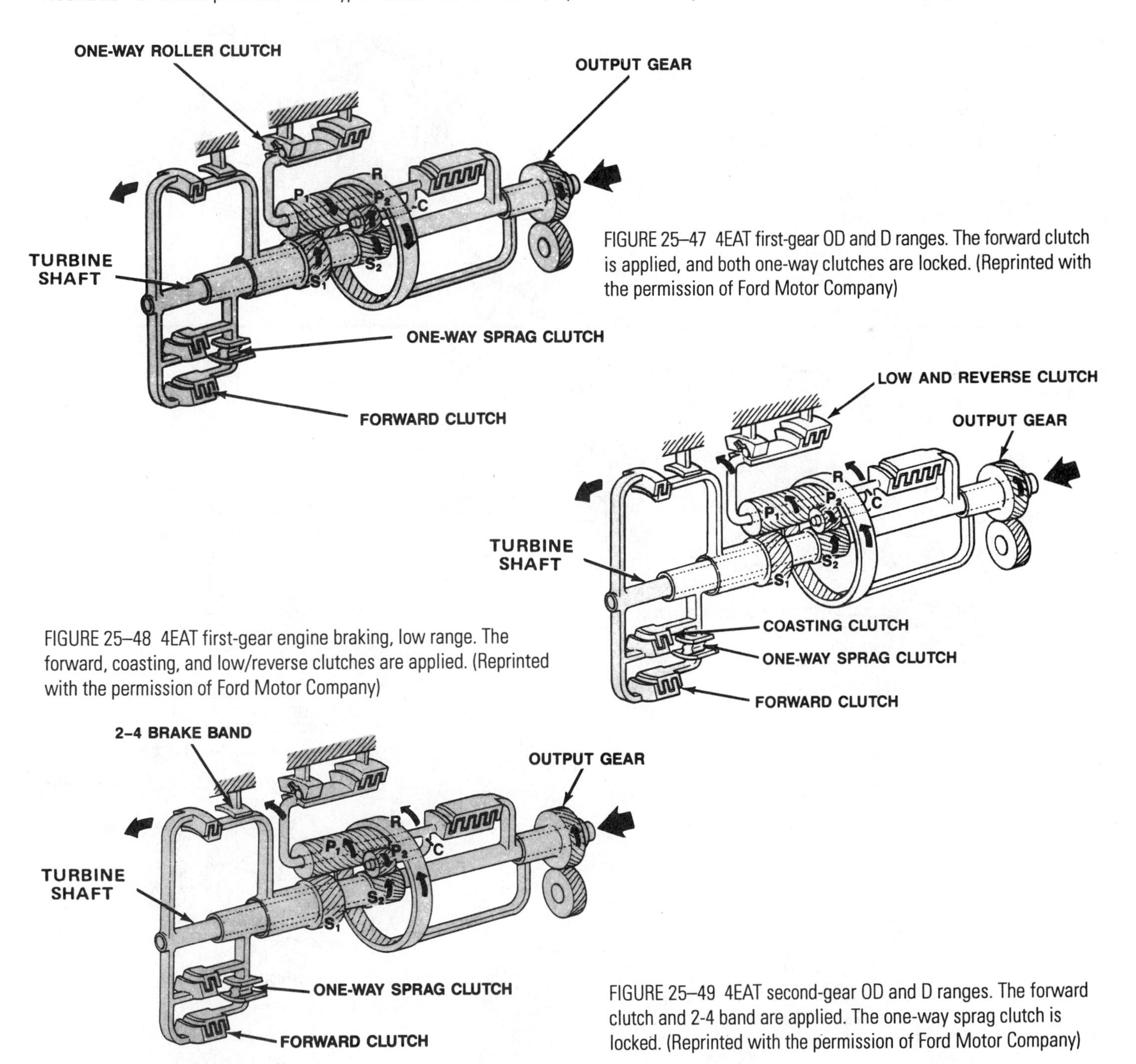

FIGURE 25–47 4EAT first-gear OD and D ranges. The forward clutch is applied, and both one-way clutches are locked. (Reprinted with the permission of Ford Motor Company)

FIGURE 25–48 4EAT first-gear engine braking, low range. The forward, coasting, and low/reverse clutches are applied. (Reprinted with the permission of Ford Motor Company)

FIGURE 25–49 4EAT second-gear OD and D ranges. The forward clutch and 2-4 band are applied. The one-way sprag clutch is locked. (Reprinted with the permission of Ford Motor Company)

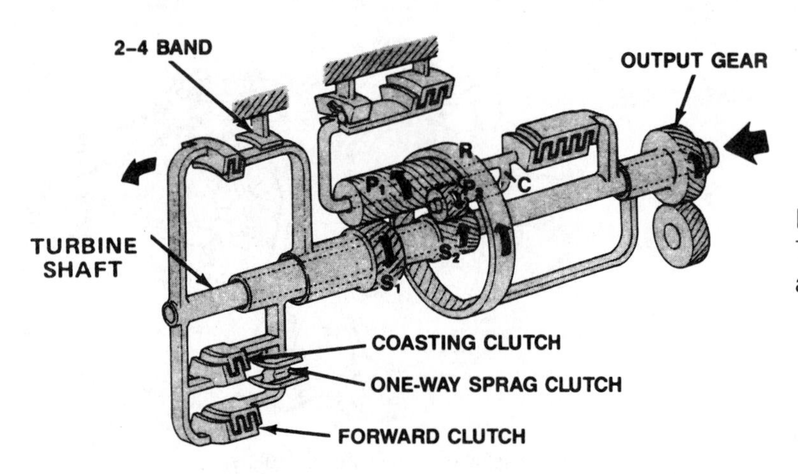

FIGURE 25–50 4EAT second-gear engine braking, low range. The forward and coasting clutches, and the 2-4 band, are applied. (Reprinted with the permission of Ford Motor

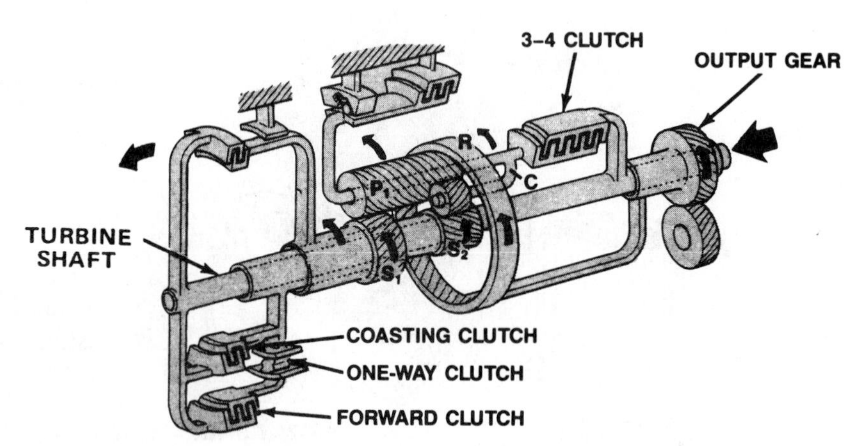

FIGURE 25–51 4EAT third-gear OD and D ranges. The forward, coasting, and 3-4 clutches are applied. The one-way sprag clutch is locked. (Reprinted with the permission of Ford Motor Company)

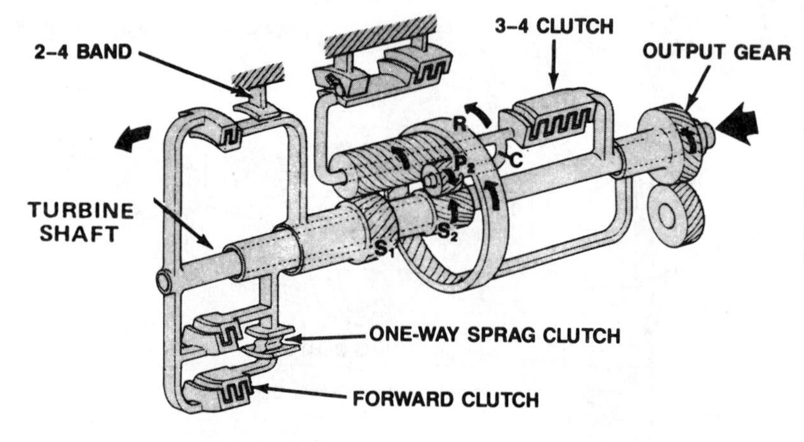

FIGURE 25–52 4EAT fourth-gear OD range. The forward and 3-4 clutches, and the 2-4 band, are applied. (Reprinted with the permission of Ford Motor Company)

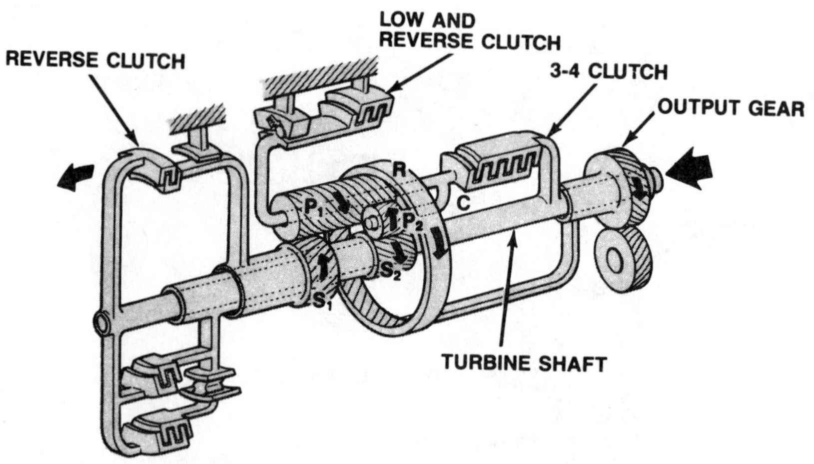

FIGURE 25–53 4EAT reverse gear range. The reverse and the low/reverse clutches are applied. (Reprinted with the permission of Ford Motor Company)

4EAT (Type G) Probe—Operating Range and Clutch/Band Apply Summary

Position	Range	Gear		Engine Braking Effect	Operation Elements								
					Forward Clutch	Coasting Clutch	3-4 Clutch	Reverse Clutch	2-4 Band Applied	2-4 Band Released	Low & Reverse Clutch	One-Way Sprag Clutch	One-Way Roller Clutch
	P	—		—									
	R	Reverse		Yes				O			O		
"Automatic" Position	N	—		—									
	OD	1st		No	O							O	O
		2nd		No	O				O			O	
		3rd	Below approx. 40 km/h	Yes	O	O	O			O		O	
			Above approx. 40 km/h	Yes	O	O	O		ⓧ	O		O	
		4th		Yes	ⓒ		O		O				O
	D	1st		No	O							O	
		2nd		No	O				O			O	
		3rd	Below approx. 40 km/h	Yes	O	O				O		O	
			Above approx. 40 km/h	Yes	O	O	O		ⓧ	O		O	
	L	1st		No	O		O				O	O	O
		2nd		Yes	O	O			O			O	
"Manual" Position	OD	2nd		No	O				O			O	
		3rd	Below approx. 40 km/h	Yes	O	O	O			O		O	
			Above approx. 40 km/h	Yes	O	O	O		ⓧ	O		O	
	D	2nd		Yes	O	O			O			O	
		3rd	Below approx. 40 km/h	Yes	O	O	O			O		O	
			Above approx. 40 km/h	Yes	O	O	O		ⓧ	O			
	L	1st		Yes	O	O					O	O	O
		2nd		Yes	O	O			O			O	

ⓧ Indicates fluid pressure to servo but band not applied due to pressure difference in servo

ⓒ Indicates that it does not function to transmit power.

CHART 25–8 4EAT Type G Probe range reference and clutch/band application chart. (Reprinted with the permission of Ford Motor Company)

4EAT (Type F) Escort/Tracer—Operating Range and Clutch/Band Apply Summary

RANGE	GEAR		ENGINE BRAKING EFFECT	FORWARD CLUTCH	COASTING CLUTCH	3-4 CLUTCH	REVERSE CLUTCH	2-4 BAND APPLIED	2-4 BAND RELEASED	LOW AND REVERSE CLUTCH	ONE-WAY CLUTCH #1	ONE-WAY CLUTCH #2
P	—		—									
R	Reverse		Yes				O			O		
N	—	Below approx. 4 km/h (2.5 mph)	—									
		Above approx. 5 km/h (3.1 mph)	—									
Ⓓ	1st		No	O							O	O
	2nd		No	O				O			O	
	3rd	Below approx. 5 km/h (3 mph) at operating temperature	Yes	O	O	O			O		O	
		Above approx. 5 km/h (3 mph) or cold engine	Yes	O	O	O		ⓧ	O		O	
	OD		Yes	O		O		O			◎	
D	1st		No	O							O	O
	2nd		No	O				O			O	
	3rd	Below approx. 5 km/h (3 mph) at operating temperature	Yes	O	O	O			O		O	
		Above approx. 5 km/h (3 mph) or cold engine	Yes	O	O	O		ⓧ	O		O	
L	1st		Yes	O						O	O	
	2nd	Below approx. 110 km/h (68 mph)	Yes	O	O			O			O	
		Above approx. 110 km/h (68 mph)										

ⓧ **Fluid pressure to servo but band not applied due to pressure difference in servo.**

◎ **Does not transmit power.**

CHART 25–9 4EAT Type F Escort/Tracer range reference and clutch/band application chart. (Reprinted with the permission of Ford Motor Company)

(W4A020) automatics, but with a longer extension housing. At the rear of the 722.5 is an overdrive planetary to provide fifth gear. The four-speed main body is constructed with a powertrain incorporating a Ravigneaux design coupled to a simple (middle) planetary gearset. The torque converter is a conventional three-element unit, without a lockup plate.

The selector pattern is as follows: P-R-N-D-4-3-2. The forward ranges provide the following gears:

D range: 1-2-3-4-5.
4 range: 1-2-3-4.
3 range: 1-2-3.
2 range: 1-2.

First-gear engagement in D, 4, and 3 ranges is only available during full-throttle or kickdown situations.

1996 Mercedes-Benz Models. The new five-speed automatic transmission is used in 1996 S- and SL-class and E320 models, as well as the new E420. The compact, electronically controlled transmission is Mercedes-Benz's first application of a torque converter clutch. To absorb engine vibrations, the clutch plate is programmed to allow between a 20 to 80 rpm slip. Also, the new five-speed automatic transmission consists of 630 parts, compared to 1160 parts in the earlier five-speed version.

Four-speed automatic transmissions are standard in 1996 C-class and E300 (diesel) models.

Mitsubishi

In addition to Mitsubishi's own vehicles, Mitsubishi automatic transmissions have been used in numerous Chrysler import vehicles (Colt, Vista, Eagle, Stealth, and Summit) as well as by Hyundai and other vehicle manufacturers (Figure 25–54). Torque converters have been either lockup or nonlockup designs.

Code Designation. Beginning in 1990, the familiar KM numbering system was dropped, and a designation format similar to that used by GM and Chrysler was adopted. In the original format, the "K" meant Kyoto (manufacturing location), and the "M" meant transaxle/transmission/transfer case. The next two digits described the unit:

13 = RWD
14 = 4WD
16 = FWD manual transaxle
17 = FWD automatic transaxle

The third digit identified variations in design.

Mitsubishi Transmission Codes

Old	New
KM171	F3A21-2
KM172	F3A22-A
KM175	F4A22-2
KM176	F4A21-2
KM177	F4A22-3-2

A new code example is F3A21. These designations are translated as follows:

F = front-wheel drive (W = 4WD)
3 = three-speed
A = automatic
2 = series code
1 = torque handling capacity (1 = light, 2 = standard, 3 = heavy-duty).

Three-speed Automatic Transmissions. Three-speed versions, such as the KM170, KM171 (F3A21-2), and KM172 (F3A22-2), use Ravigneaux planetary gearsets. These transverse-mounted transaxles are unique in the fact that the engine

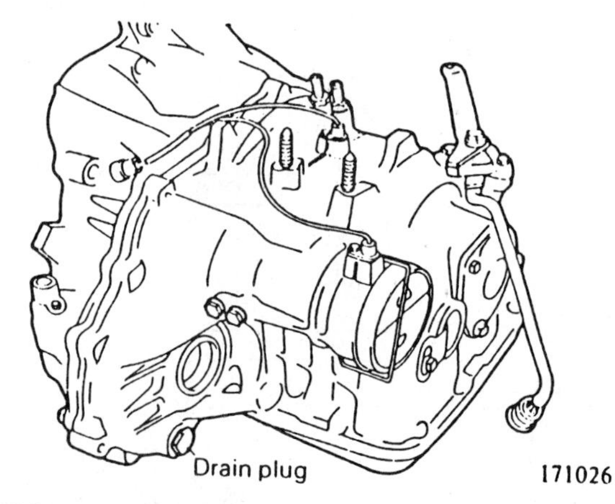

FIGURE 25–54 Mitsubishi-produced three-speed transaxle. (Courtesy of Chrysler Corp.)

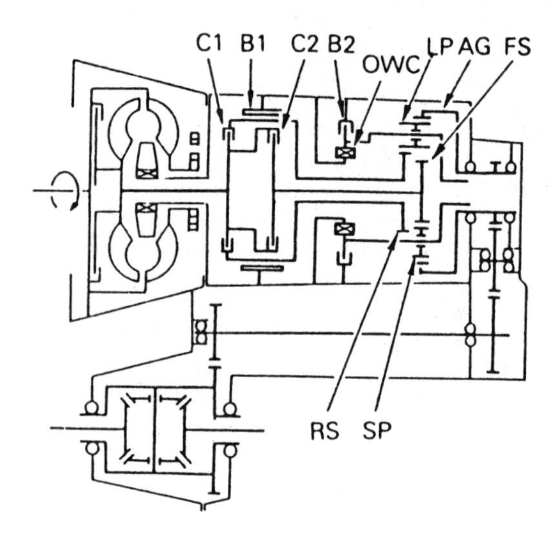

FIGURE 25–55 KM170, F3A21-2 (KM171), and F3A22-2 (KM172) powertrain schematic. (Courtesy of Chrysler Corp.)

is on the driver side and the transaxle is on the passenger side (similar to most Hondas and Acuras). Refer to Figure 25–55 to view a typical three-speed transaxle powerflow schematic. Chart 25–10 displays the range reference and clutch/band application summary.

Four-speed Automatic Transmissions. Mitsubishi transmissions have been used in Chrysler/Dodge Raider sport-utility vehicles and Ram 50 pickups. The KM148 RWD four-speed overdrive transmission is essentially the same as the AW-372 version. The difference is that the KM148 incorporates a transfer case, and the AW-372 does not. Refer to Figure 25–56 to view the four-speed transmission powerflow schematic. Chart 25–11 (page 650) displays the range reference and clutch/band application summary.

Selector lever position	Gear position	Gear ratio	Engine start	Parking mechanism	Clutches			Brakes	
					C1	C2	OWC	B1	B2
P	Neutral	—	O	O					
R	Reverse	2.176			O				O
N	Neutral	—	O						
D	First	2.846				O	O		
	Second	1.581				O		O	
	Third	1.000			O	O			
2	First	2.846				O	O		
	Second	1.581				O		O	
L	First	2.846				O			O

CHART 25–10 Range reference and clutch/band application chart—KM170, F3A21-2 (KM171), and F3A22-2 (KM172). (Courtesy of Chrysler Corp.)

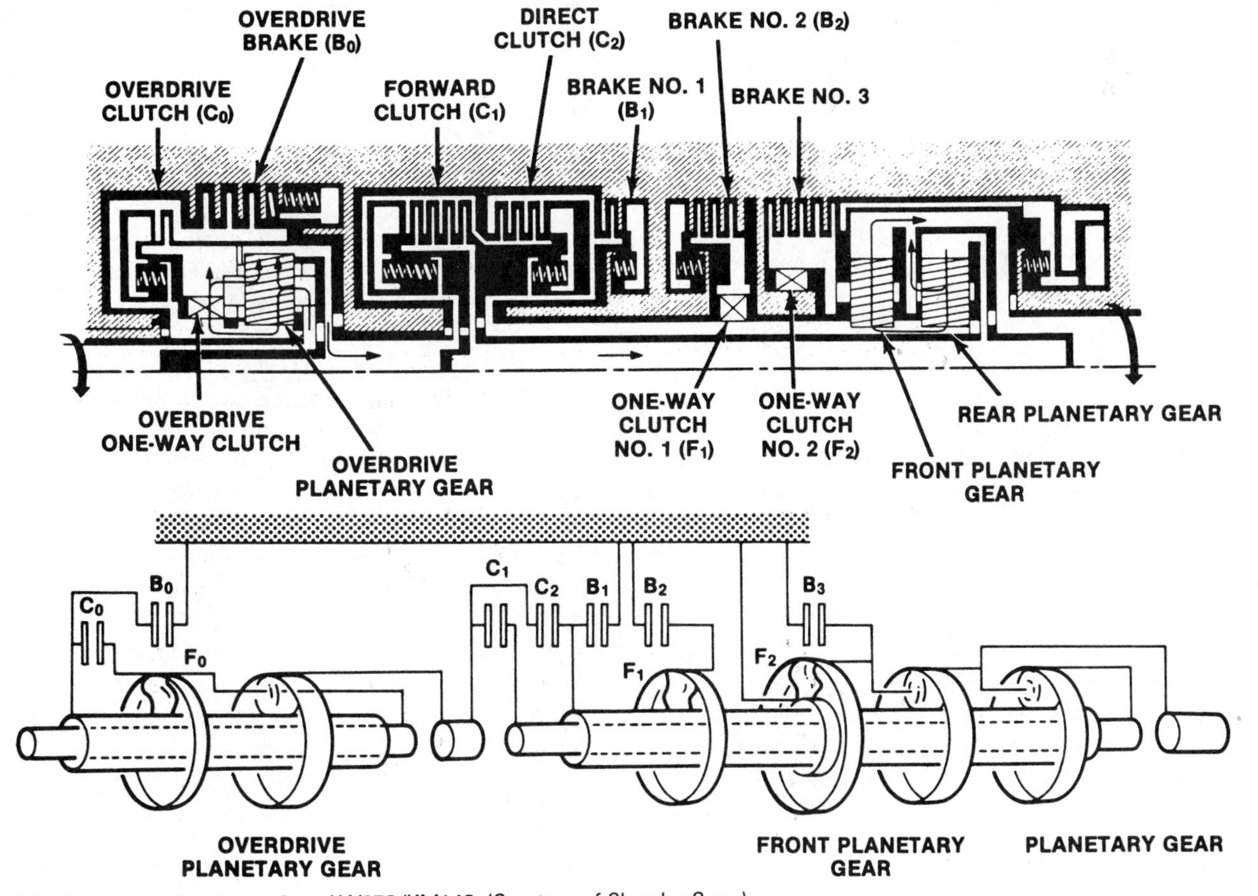

FIGURE 25–56 Powertrain schematic—AW372/KM148. (Courtesy of Chrysler Corp.)

SELECTOR LEVER POSITION	OD-OFF SWITCH	GEAR POSITION	GEAR RATIO	ENGINE START	PARKING MECH-ANISM	CLUTCH				BRAKE					ONE—WAY CLUTCH		
						C_0	C_1	C_2		B_0	B_1	B_2	B_3				
								IP	OP				IP	OP	F_0	F_1	F_2
P	—	NEUTRAL	—	X	X	X								X			
R	—	REVERSE	2.703	—		X		X	X				X	X	X		
N	—	NEUTRAL	—	X		X											
D	ON	FIRST	2.826	—		X	X								X		X
		SECOND	1.493	—		X	X					X			X	X	
		THIRD	1.000	—		X	X		X			X			X		
		FOURTH	0.688	—			X		X	X		X					
D	OFF	FIRST	2.826	—		X	X								X		X
		SECOND	1.493	—		X	X					X			X	X	
		THIRD	1.000	—		X	X					X			X		
2	—	FIRST	2.826	—		X	X								X		X
		SECOND	1.493	—		X	X				X	X			X	X	
L	—	FIRST	2.826	—		X	X						X	X	X		X

NOTE: IP = INNER PISTON OP = OUTER PISTON

CHART 25–11 AW372/KM148 range reference and clutch/band application chart. (Courtesy of Chrysler Corp.)

The F4A33 and W4A33 transverse-mounted transaxles are used in the Diamante, Stealth, and 3000GT. The W4A33 utilizes a viscous coupling unit (VCU) to minimize rotational speed differences between the front and rear axles for improved handling during weight shift (hard acceleration) and during wheel slip. During normal traction conditions, torque split to the front and rear wheels is equally distributed. A bevel gearset is used to transfer torque to the rear differential unit.

Mitsubishi/Chrysler Transmissions. Chrysler TorqueFlite RWD transmissions have been used in certain Mitsubishi trucks. The three-speed A-904, referred to as the "Baby 904," was used from 1975 to 1986. Mitsubishi-built Arrow, Colt, Challenger, and Sapporo models (1980–1986) may be equipped with the "Baby 904." The four-speed overdrive A-500 found applications in 1989 to 1992 trucks. Refer to Chapters 4 and 22 for additional coverage of these transmissions.

1996 Mitsubishi Models. 1996 Mitsubishi 3000GT, Eclipse, and Mighty Max models offered four-speed automatic transmissions as an option. 1996 Galants, Marages, and Montero model versions used four-speed automatic transmissions as either a standard or an optional feature, with the exception of the Mirage S coupe, which used a three-speed automatic as an option to a five-speed manual transmission.

Nissan/Infiniti

Nissan, formerly known as Datsun, produces approximately 55% of its own automatic transmissions, with the remainder being JATCO units. Typically, the geartrains are designed with Simpson planetary design gearsets. Refer to the JATCO section in this chapter for additional coverage of Nissan-JATCO transmissions.

Four-speed RWD transmissions, such as the L4N71B/E4N71B series introduced in 1983, incorporate an overdrive planetary gearset in front of a Simpson compound planetary gearset to achieve the four forward speeds. The "E" model provides converter lockup in third and fourth gears. This series of transmissions has been used by Nissan and Mazda.

The RWD/4WD RE4R01A automatic transmission, introduced in 1987, is used in Nissan and Infiniti vehicles (Figure 25–57). The powerflow schematic of this four-speed electronically controlled transmission is shown in Figure 25–58. The range reference and clutch/band application summary is displayed in Chart 25–12.

Nissan RE404A/Ford 4F20E. The Mercury Villager, a FWD minivan assembled in Avon Lake, Ohio, comes equipped with a Nissan powertrain. The 3.0-L engine is coupled to a 4F20E automatic transaxle (Figure 25–59). The Villager uses the same driveline as the Nissan Quest minivan. Nissan refers to the transaxle as the RE404A.

The Nissan/Ford 4F20E is an electronically controlled, four-speed overdrive transaxle using a locking torque converter. Five multiple-disc clutches, two one-way clutches (a sprag and a roller), and a band operate a Simpson compound planetary gearset. Though the clutch and band names and sequence pattern are similar to the Hydra-Matic 4L60-E/4L60 (THM 700-R4), the internal operation of these components to the geartrain is quite different. The oil pump utilizes a trochoidal design to reduce noise and increase displacement. This newly styled pump closely resembles the IX rotor design.

A 4F20E powertrain schematic is shown in Figure 25–60. The range reference and clutch/band application summary is displayed in Chart 25–13 (page 653).

1996 Nissan Models. All automatic transmissions used in 1996 Nissan models are four-speed units. Automatic transmissions were an option in the 200SX, 240SX, 300ZX, and Sentra models. In Altima and Maxima models, four-speed automatics

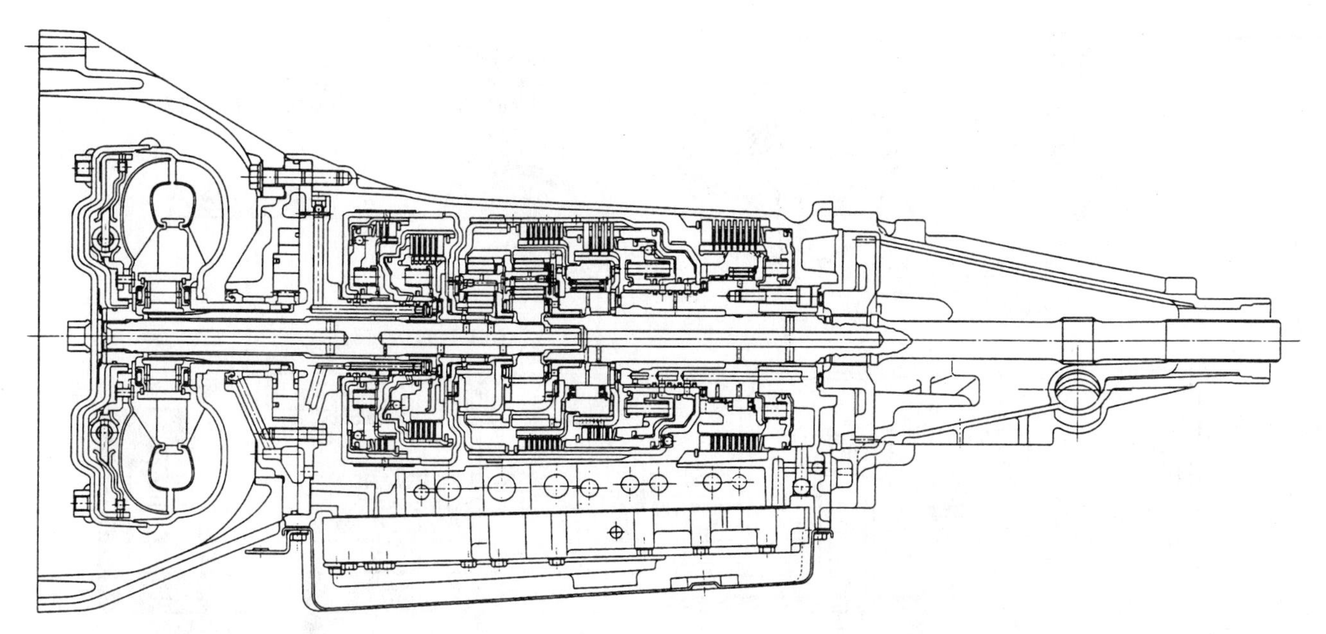

FIGURE 25–57 Cutaway view—Nissan RE4R01A series four-speed automatic transmission. (Reprinted with permission from SAE AE-18 © 1994 Society of Automotive Engineers, Inc.)

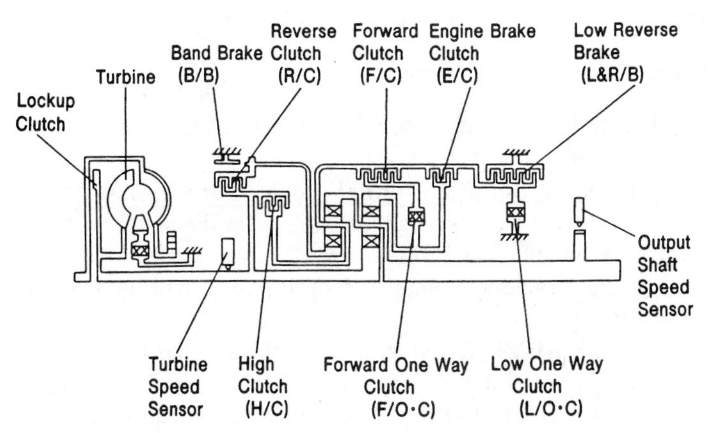

FIGURE 25–58 Nissan RE4R01A powertrain schematic. (Reprinted with permission from SAE AE-18 © 1994 Society of Automotive Engineers, Inc.)

		R/C	H/C	F/C	E/C	B/B 2nd	B/B 4th	F/O·C	L/O·C	L&R/B
P										
R		●								●
N										
D	1st			●				●	●	
	2nd			●		●		●		
	3rd		●	●				●		
	4th		●	●			●			
2	1st			●				●	●	
	2nd			●	●	●		●		
1	1st			●	●			●		●
	2nd			●	●	●		●		

CHART 25–12 Nissan RE4R01A range reference and clutch/band application chart. (Reprinted with permission from SAE AE-18 © 1994 Society of Automotive Engineers, Inc.)

4-Speed Automatic Overdrive Transaxle — Electronic

Manufacturer/Assy. Plant Nissan/Japan

Transmission Type 4F20E

Gear Ratios
1st 2.78
2nd 1.54
3rd 1.00
4th 0.69
Rev. 2.27

Gear Range Selection P - R - N - D - 2 - 1

Fluid Refill Capacity 8.3 Liters (8.8 qts.)
Fluid Recommendation MERCON®

Availability Villager

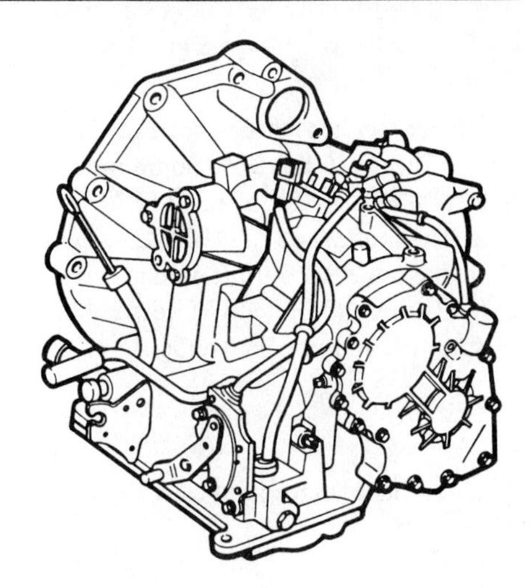

FIGURE 25–59 Nissan-produced 4F20E automatic transaxle. (Reprinted with the permission of Ford Motor Company)

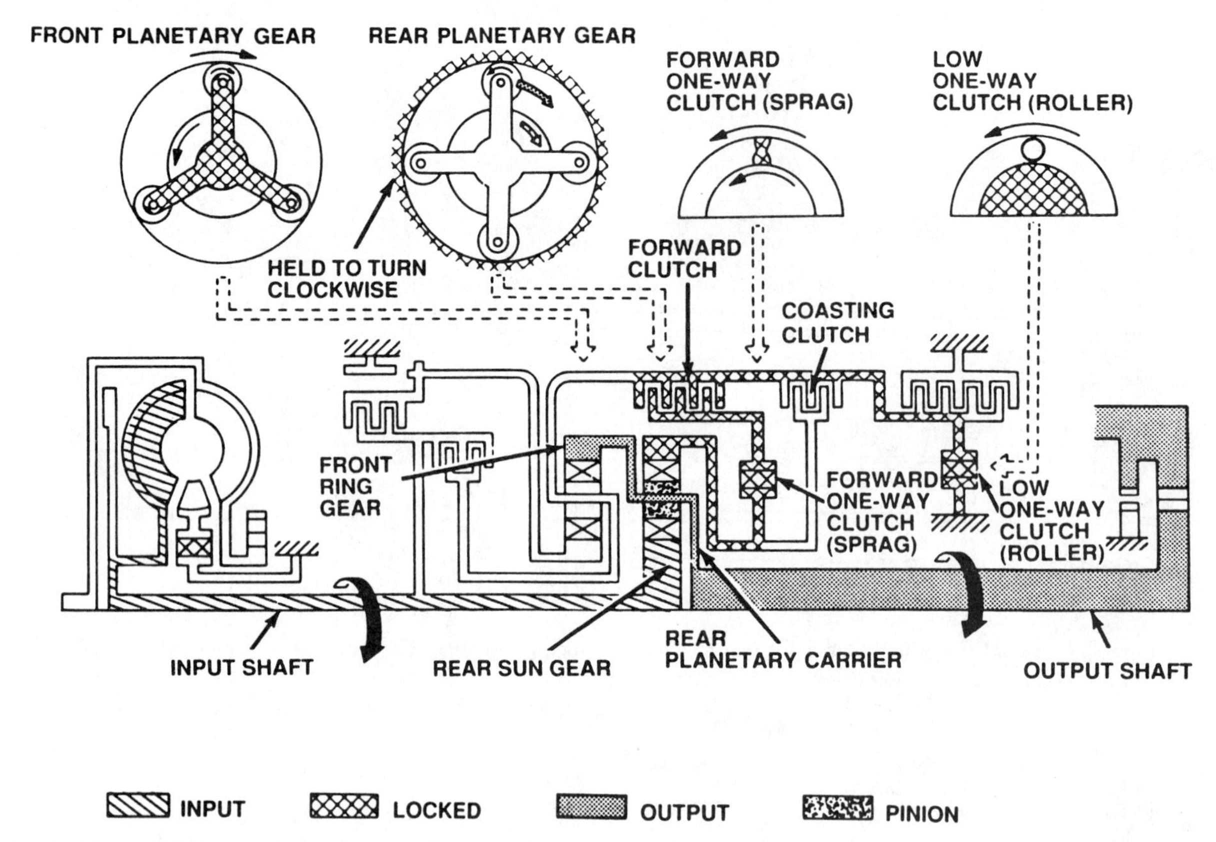

FIGURE 25–60 Nissan 4F20E powerflow schematic (first gear, D range). (Reprinted with the permission of Ford Motor Company)

were either a standard or an optional feature, depending upon the model version. All Quest minivans were equipped with a four-speed automatic as a standard feature.

Infiniti Models. Automatic transmissions were available for all 1996 Infiniti models. The G20 and I30 models came standard with a five-speed manual transmission or an optional four-speed automatic transmission. The J30 and Q45 standard features included a four-speed automatic transmission.

The T30 is essentially an upscale version of the Nissan Pathfinder. Look for similar driveline components, such as an RE4R03A/01A automatic transmission, in these sport-utility vehicles.

Peugeot

ZF supplied Peugeot (France) with the three-speed ZF-3HP-22 automatic transmission from 1976 to 1985. Some models from 1977 to 1981 used the GM Hydra-Matic 3L30 (THM 180) built in Strasbourg, France. The four-speed ZF-4HP-22 became available in 1985. These are the same RWD units used in certain BMWs. Refer to the ZF Industries and General Motors Hydra-Matic sections in this chapter for additional coverage.

Porsche

Certain Porsche 944 models have been equipped with Volkswagen Type 087 transaxles.

Tiptronic Automatic/Manual Shift Transmission. The Tiptronic transmission is an alternative for the manual transmission in the Porsche 911 Carrera 2. The unit provides both an automatic and a manual shift option for driver-controlled performance.

The ZF-4HP22-EH automatic transmission has been modified for the rear-mounted engine, RWD, longitudinal design powertrain. The Tiptronic and the ZF-4HP22-EH are electronically controlled, four-speed planetary gearset transmissions. The Porsche unit is equipped with a lockup converter clutch that can be engaged in second gear. The oil-to-air heat exchanger for the ATF is located in the front of the vehicle and makes use of the engine oil cooler blower.

The Tiptronic transmission offers many modern electronic and operating features. Figure 25–61 shows the powertrain and controls of a 911 Carrera 2 equipped with a Tiptronic automatic transmission. Two parallel-mounted transmission selector/shift levers are located on the floorboard/transmission tunnel. One lever is for automatic shift operation, and the other for manual control.

The automatic shift lever is conventional in design. Seven selector quadrant positions are used (P-R-N-D-3-2-1), and the selector functions similarly to that of most other automatic transmissions.

The manual lever can be used when the automatic shift lever is positioned in the D range. Tipping the manual lever forward causes an upshift, and tipping it to the rear commands a downshift. When the lever is tipped forward or backward, electrical switches are activated as inputs to the TCM (TCU/EGS). The

Range	Gear	Coasting Clutch	Forward Clutch	Forward One-Way Clutch (Sprag) 7A089		Low/Reverse Clutch	Low One-Way Clutch (Roller) 7A089		3-4 Clutch	Reverse Clutch	2-4 Band 7D034
				Drive	Coast		Drive	Coast			
Reverse	Reverse					H				D	
Drive	First		D	D	OR		H	OR			
	Second		D	D	OR						H
	Third		D	D	OR				D		
	Overdrive		*	OR	OR				D		H
Overdrive (O/D) Off	Third	A■	D	D					D		
Second	Second	A■	D	D							H
Manual Low	First	A■	D	D		H■	H				
Planetary Member		Ring Gear 7A153				Front Planet 7A398				Primary Sun Gear 7A399	

■ = For engine braking only
* = Applied but does not transfer power
OR = Overrunning
D = Driving
H = Held
A = Applied

D12885-A

CHART 25–13 Nissan 4F20E range reference and clutch/band application chart. (Reprinted with the permission of Ford Motor Company)

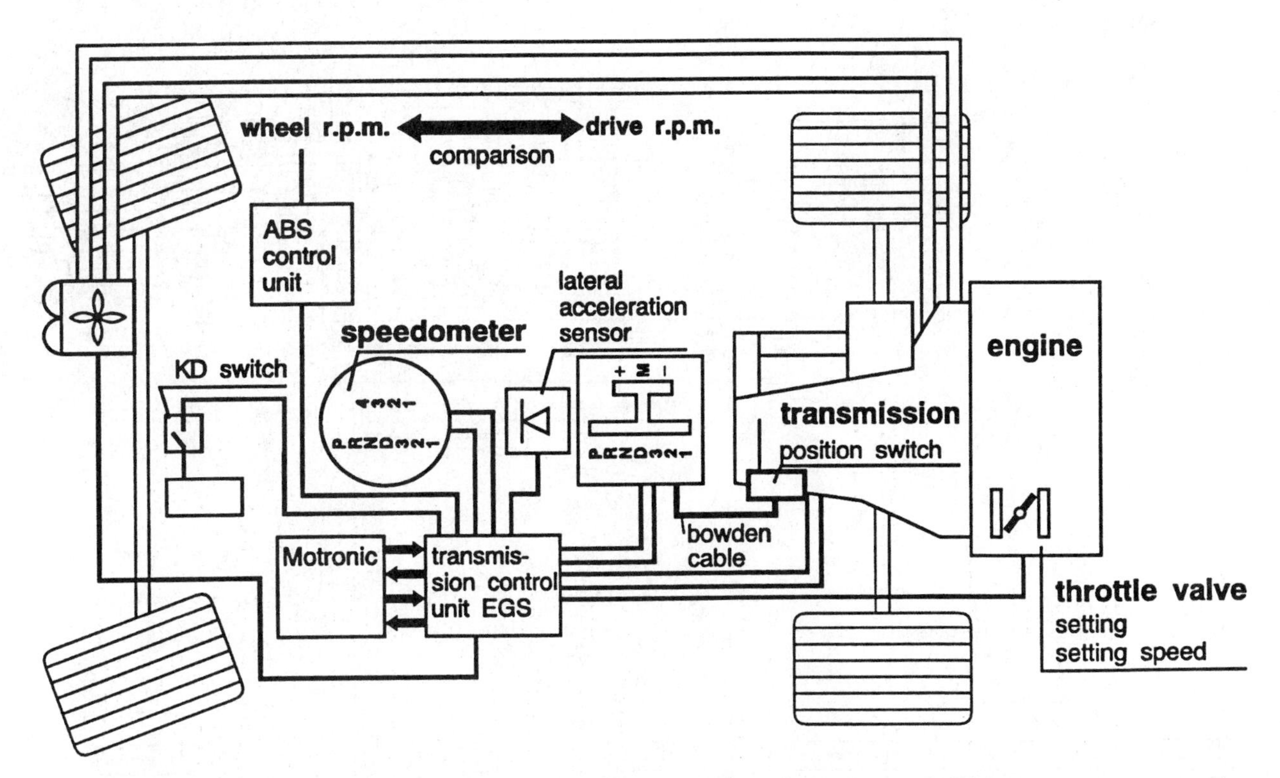

FIGURE 25–61 Porsche Tiptronic powertrain layout and transmission controls. Automatic and manual shift lever gate patterns are shown in the center box. (Reprinted with permission from SAE AE-18 © 1994 Society of Automotive Engineers, Inc.)

TCM commands the appropriate shift solenoid action to cause the transmission to upshift or downshift. To prevent the engine from being over-revved, the TCM will command an upshift if needed.

Renault

The model MB1 transaxle is a transverse, three-speed unit and is found in the AMC/Renault Alliance and Encore (1983–1987). A Simpson compound planetary gearset is controlled by four multiple-disc clutches and a one-way roller. The MB1 features electronic shift control.

The Fuego and Renault 18 and 18i have used the three-speed MJ1/MJ3 transaxle. This is a longitudinal-designed transaxle equipped with a Ravigneaux planetary gearset. The final drive unit is a ring and pinion design and is located between the torque converter and planetary geartrain. Fluid recommendations include ELF Renaultmatic D2, Mobil ATF 220, or Dexron II/III.

Saab

Saab-Scania AB (Sweden) has been supplied with a modified ZF-4HP-18 automatic transmission since 1989. It is used in the Saab 9000, a transverse powertrain–equipped FWD vehicle. Through the 1996 model year, four-speed automatic transmissions were an optional feature on 9000 and 900 model versions.

Sterling (UK)

Sterlings have been built with Honda four-speed automatics since 1987.

Subaru

JATCO provided three-speed automatic transaxles to Subaru from 1975 to 1989. Four-speed JATCO transaxles were provided from 1979 to 1991. Subaru refers to this unit as the 4EAT. Though they use a similar designator, the Subaru and Ford 4EAT transaxles are different. Note: The four-speed JATCO/4EAT was not available in the Legacy model.

Figure 25–62 shows a cross-sectional view of a Subaru AWD automatic transmission. In this longitudinal layout, the turbine shaft transmits torque to the back of the automatic transmission section. In reverse and forward gears, the power flows toward the front, to the reduction drive gear. This gear drives the

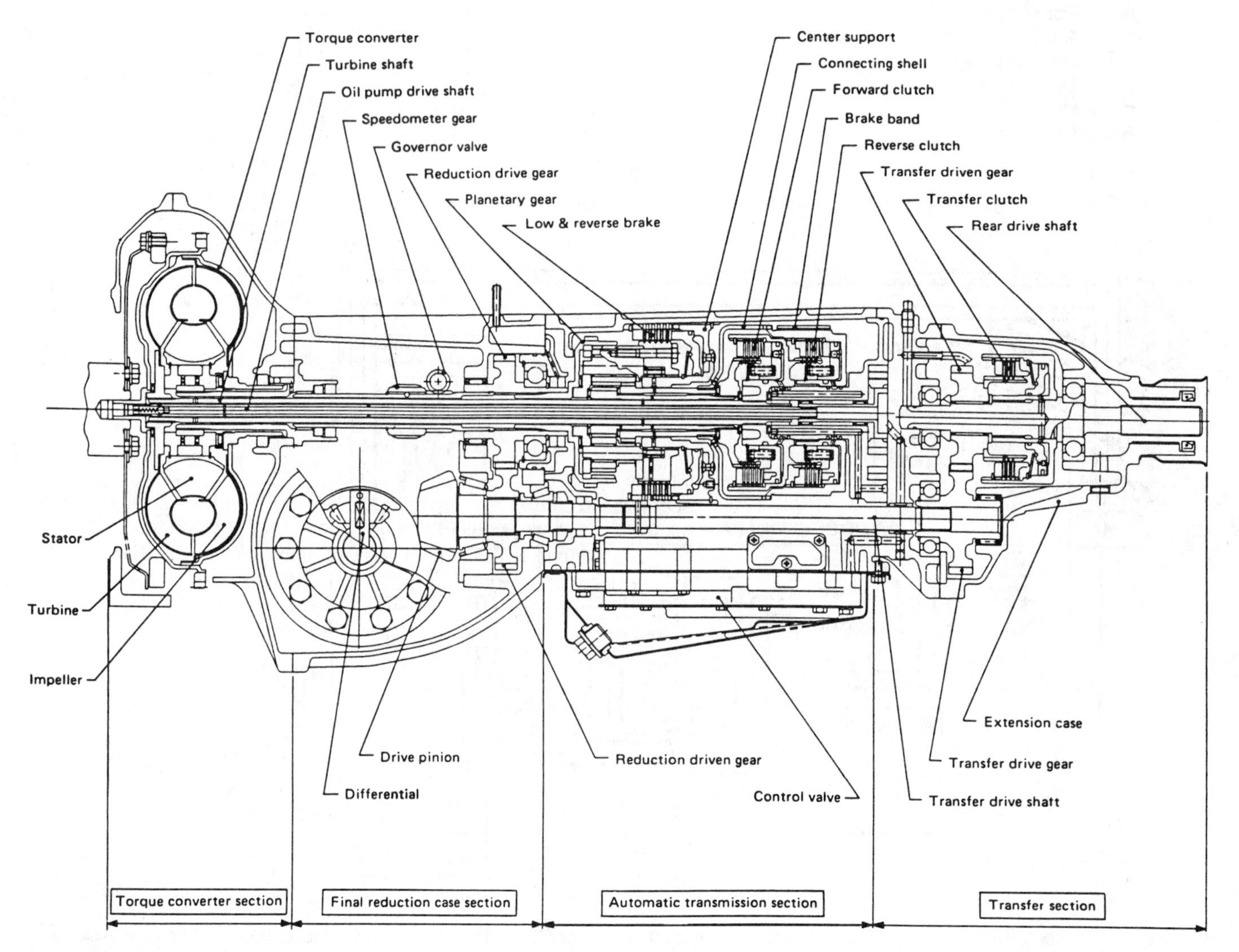

FIGURE 25–62 Subaru longitudinal AWD automatic transmission. (Courtesy of Subaru of America, Inc.)

reduction driven gear, sending power to the hypoid ring and pinion gearset for FWD operation.

To engage AWD, the transfer clutch is pressurized in the rear extension case. Torque from the reduction driven gear drives both the hypoid ring and pinion and the transfer drive shaft. Through the transfer drive and driven gears, torque is sent to the rear wheels when the transfer clutch is applied.

Some Subaru Justy models used an ECVT—an electronically controlled continuously variable transmission (CVT). This transmission and other CVTs are discussed in the final section of this chapter.

1996 Subaru Models. Four-speed automatic transmissions were used in 1996 Subaru models. In Impreza model versions, it was an optional feature. In Legacy model versions, the automatic transmission was either an option or a standard feature. Four-speed automatic transmissions were standard equipment in SVX model versions.

Suzuki

GM Hydra-Matic supplies 3L30 automatic transmissions for the Geo Tracker and the Suzuki Sidekick. Aisin-Seiki provides the Suzuki MX-17 three-speed transaxle for the Geo Metro and Suzuki Swift.

The 1996 X-90, a 4WD sport-utility vehicle, used a four-speed automatic transmission as an option.

Suzuki builds and uses a CVT (continuously variable transmission) in its Cultus, a foreign subcompact. This transmission and other CVTs are discussed in the final section of this chapter.

Toyota/Lexus

Toyota uses a wide variety of automatic transmissions and transaxles. Three- and four-speed designs include a Simpson compound planetary gearset. Four-speed designs are produced with the addition of an overdrive (or underdrive) gearset.

Following is a listing of common rear-, front-, and four-wheel drive units. The designation system uses these letters:

A (automatic)
D (fourth gear/overdrive)
E (electronically controlled)
F (four-wheel drive)
L (lockup torque converter)

The numbers identify relative torque handling capabilities.

A40 Series Transmissions (RWD and 4WD).

3-speed:	A40, A41
4-speed:	A40D, A42D, A43D
4-speed:	A42DL, A43DL, A44DL, A45DL/A45DF
4-speed (with lockup):	A43DE

A100 and A200 Series Transaxles (FWD).

3-speed:	A131, A132
3-speed:	A130L, A131L, A132L
4-speed:	A140L, A240L, A241L
4-speed (with lockup):	A140E, A141E A240E, A241E

Toyota A140E. The A140E is a popular four-speed automatic transaxle and has been available in the Camry (1982–current) and Celica (1985–current). The A140E is shown in Figure 25–63. The electronically controlled powerflow operation and a range reference and clutch/band application summary are shown in Figures 25–64 to 25–66.

1996 Toyota Models. All 1996 Toyota model versions, except for two, were available with a four-speed automatic transmission as either a standard or an optional feature. A three-speed automatic transaxle was optional on Corolla and Tercel base sedans.

Toyota A350E Five-speed Automatic Transmission. The A350E is a compact, high-performance, five-speed RWD transmission used in 1996 Lexus GS300 vehicles (Figure 25–67). It uses the same geartrain as the conventional four-speed A240E.

The geartrain is composed of a front gear unit and a Simpson three-speed compound planetary gearset. The front unit is a simple planetary gearset that provides two ratios. Its basic construction and operational design, as shown in Figure 25–68, are similar to the overdrive planetary gearset used in GM's 4L80-E, 4L30-E, and THM 200-4R. Chart 25–14 (page 660) displays the clutch and band application summary.

Engaging various combinations of the front gear unit (two ratios) and the rear Simpson gear unit (three forward ratios) can allow six forward speeds. Based on performance and gear ratios available, five combinations are used. Refer to Tables 25–1 and 25–2 (page 660).

Three transmission speed sensors, a fluid temperature sensor, and numerous other powertrain sensors and switches are inputs to the TCM. The TCM manages the A350E transmission operation by controlling two shift solenoids, a lockup solenoid, an accumulator control solenoid, and a mainline pressure control solenoid.

Lexus Models. The 1996 Lexus ES300, LS400, SC300, and SC400 models use an electronically controlled, four-speed overdrive transmission. The geartrain and clutch/brake layout is shown in Figure 25–69.

A five-speed Toyota A350E automatic transmission is a standard feature of the 1996 Lexus GS300. On the shift console is a "5ECTi" designation.

The upscale version of Toyota's 4WD sport-utility Land Cruiser is the Lexus LX450, which has an approximate base price of $48,000. The four-speed automatic transmission comes equipped with a clutching torque converter.

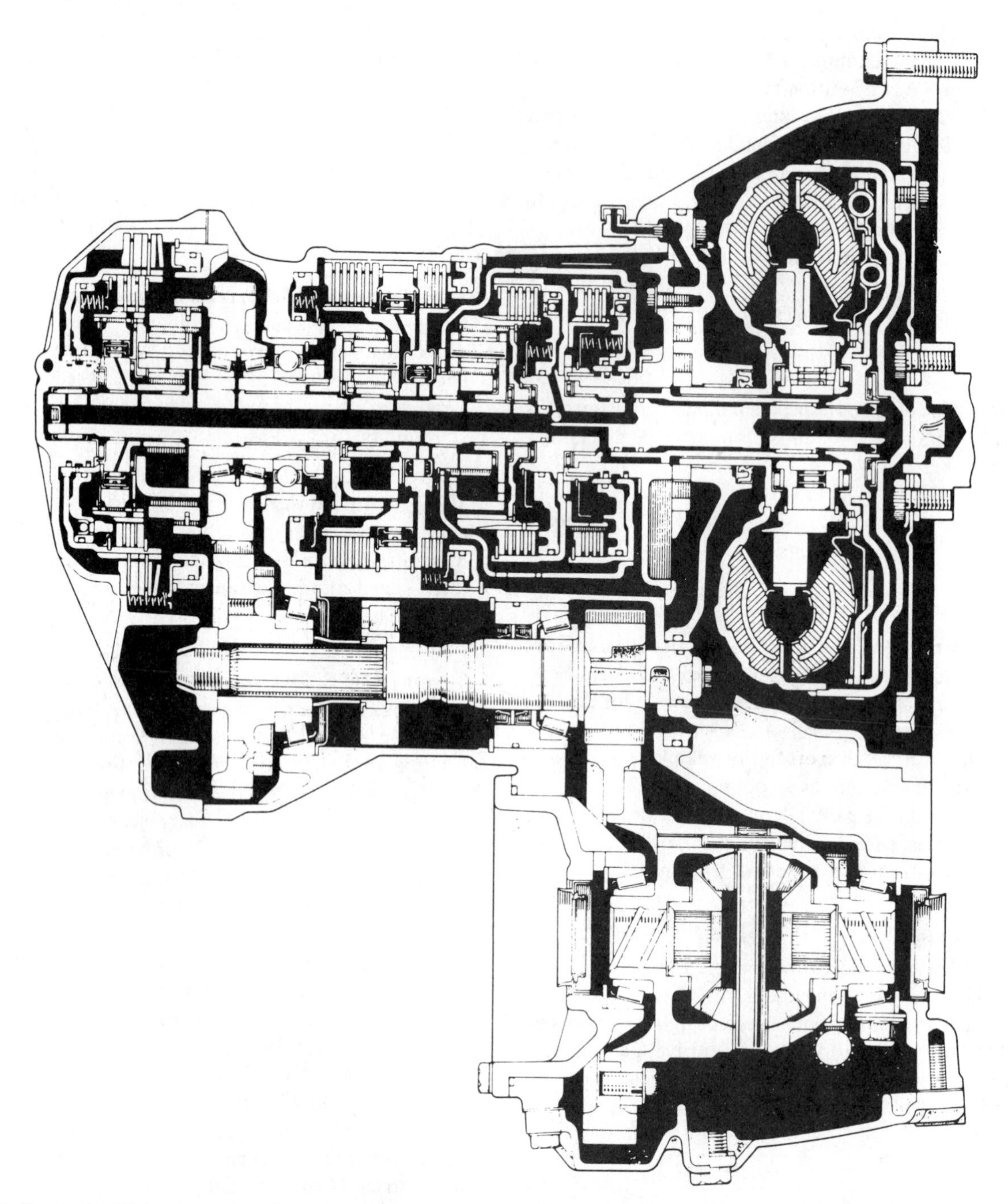

FIGURE 25–63 Toyota A140E four-speed automatic transaxle cutaway view. (Courtesy of General Motors Corporation, Service Technology Group.)

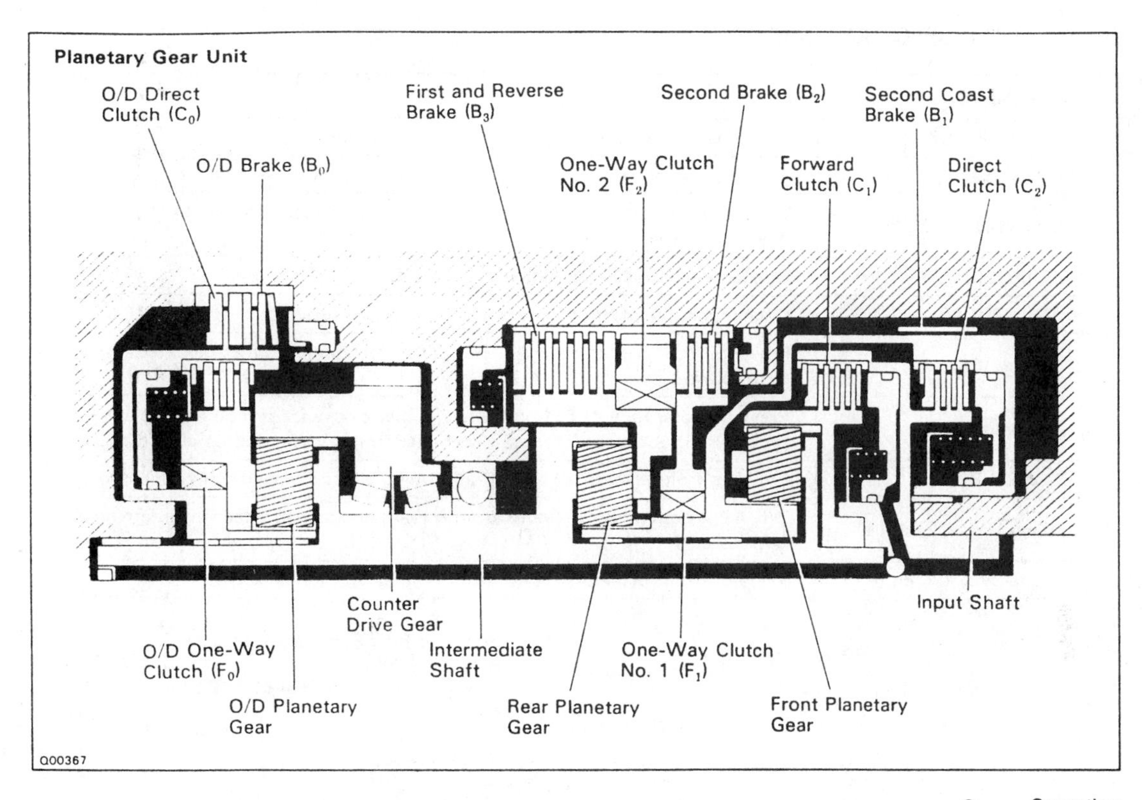

○ ..... Operating

Shift lever position	Gear Position	C_0	C_1	C_2	B_0	B_1	B_2	B_3	F_0	F_1	F_2
P	Parking	○									
R	Reverse	○		○				○			
N	Neutral	○									
D	1st	○	○						○		○
	2nd	○	○				○		○	○	
	3rd	○	○	○			○		○		
	O/D		○	○	○		○				
2	1st	○	○						○		○
	2nd	○	○			○	○		○	○	
	*3rd	○	○	○			○		○		
L	1st	○	○					○	○		○
	*2nd	○	○			○	○		○	○	

*Down-shift only – no up-shift

FIGURE 25–64 Toyota A140E powertrain schematic; range reference and clutch/band application chart. (Courtesy of General Motors Corporation, Service Technology Group.)

FUNCTION OF COMPONENTS

NOMENCLATURE	OPERATION
O/D Direct Clutch (C_0)	Connects overdrive sun gear and overdrive carrier
O/D Brake (B_0)	Prevents overdrive sun gear from turning either clockwise or counterclockwise
O/D One-Way Clutch (F_0)	When transmission is being driven by engine, connects overdrive sun gear and overdrive carrier
Front Clutch (C_1)	Connects input shaft and intermediate shaft
Rear Clutch (C_2)	Connects input shaft and front & rear planetary sun gear
No. 1 Brake (B_1)	Prevents front & rear planetary sun gear from turning either clockwise or counterclockwise
No. 2 Brake (B_2)	Prevents outer race of F_1 from turning either clockwise or counterclockwise, thus preventing front & rear planetary sun gear from turning counterclockwise
No. 3 Brake (B_3)	Prevents front planetary carrier from turning either clockwise or counterclockwise
No. 1 One-Way Clutch (F_1)	When B_2 is operating, prevents front & rear planetary sun gear from turning counterclockwise
No. 2 One-Way Clutch (F_2)	Prevents front planetary carrier from turning counterclockwise

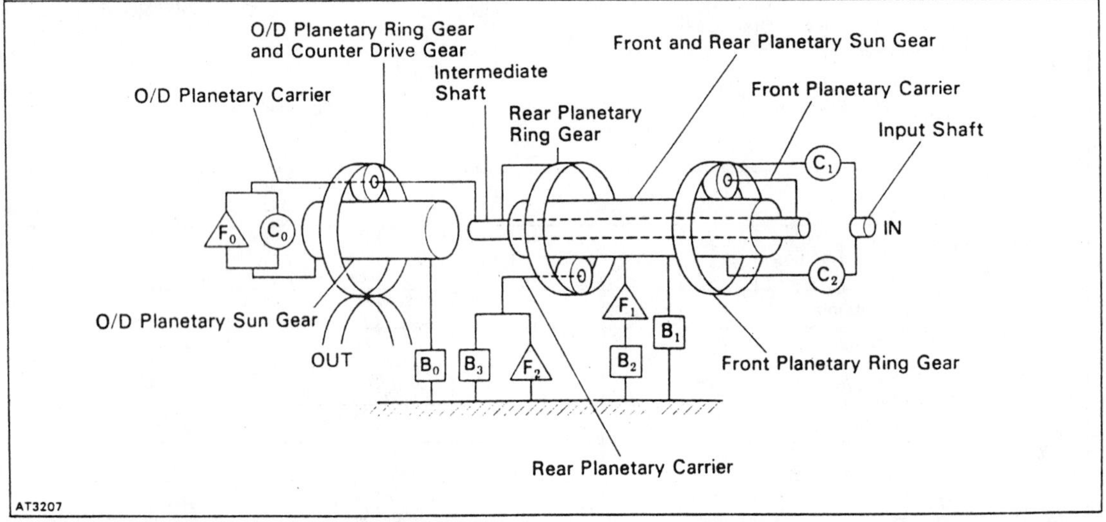

FIGURE 25–65 Toyota A140E component functions. (Courtesy of General Motors Corporation, Service Technology Group.)

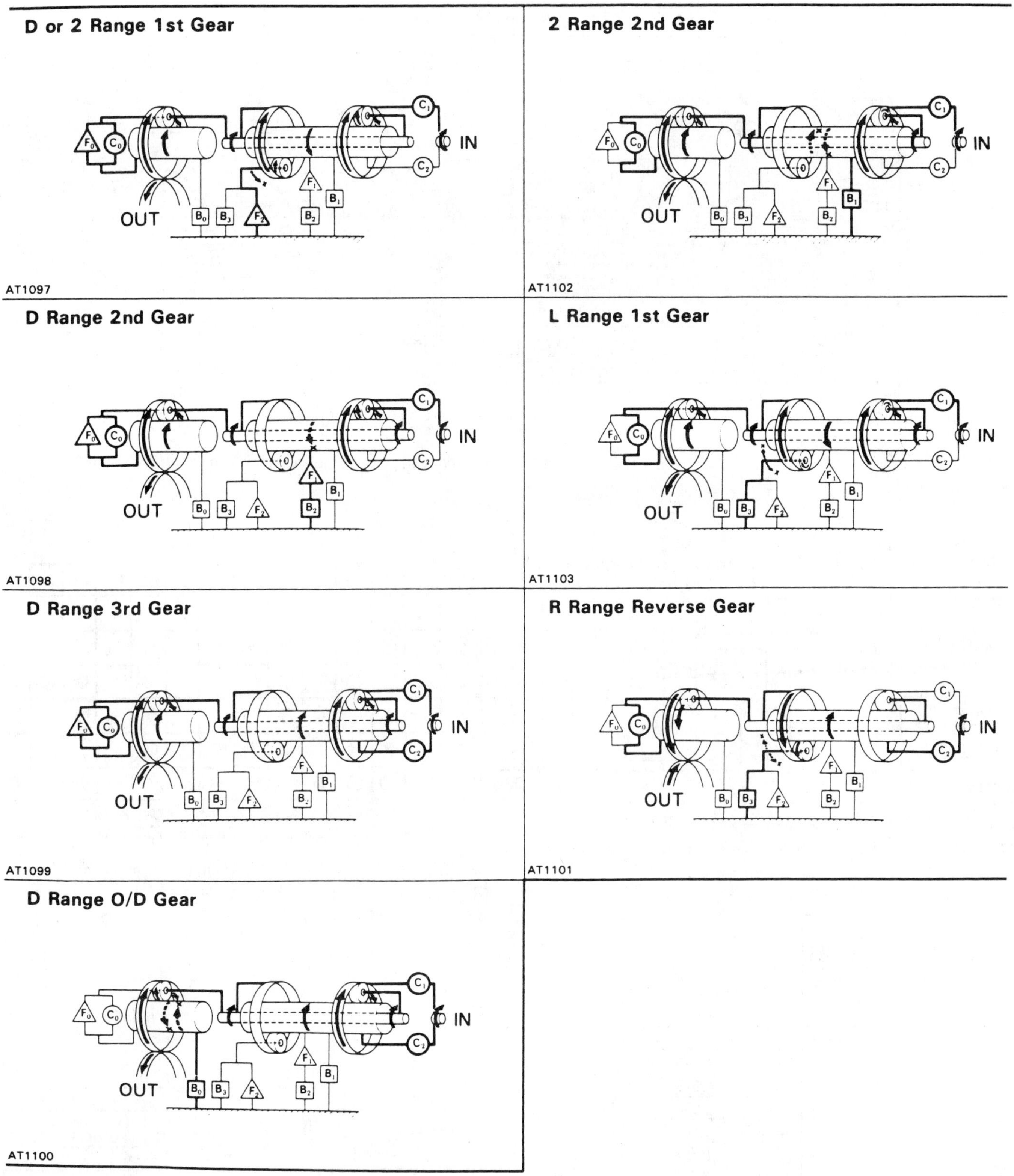
D or 2 Range 1st Gear
AT1097
2 Range 2nd Gear
AT1102
D Range 2nd Gear
AT1098
L Range 1st Gear
AT1103
D Range 3rd Gear
AT1099
R Range Reverse Gear
AT1101
D Range O/D Gear
AT1100
OUT
IN
F_0
C_0
C_1
C_2
B_0
B_1
B_2
B_3
F_1
F_2

FIGURE 25–66 Toyota A140E powerflow schematics. (Courtesy of General Motors Corporation, Service Technology Group.)

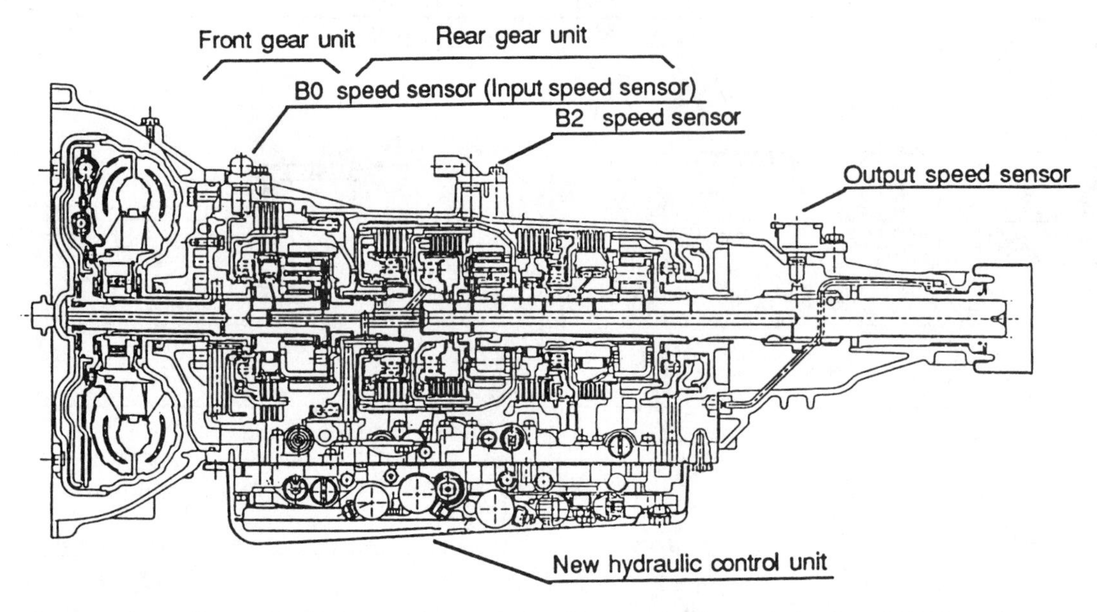

FIGURE 25–67 Cutaway view—Toyota A350E five-speed transmission. (Reprinted with permission from SAE AE-18 © 1994 Society of Automotive Engineers, Inc.)

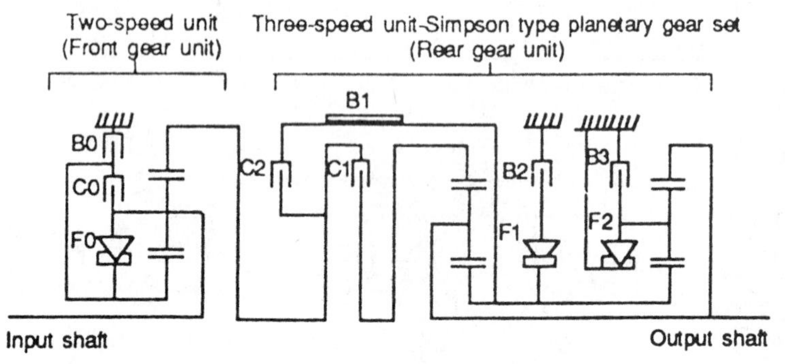

FIGURE 25–68 Powertrain schematic—Toyota A350E. (Reprinted with permission from SAE AE-18 © 1994 Society of Automotive Engineers, Inc.)

Gears	S1	S2	S3	C0	C1	C2	B0	B1	B2	B3	F0	F1	F2
1	X	○	○	●	●					▲	●		●
2	X	○	X		●		●						●
3	○	○	○	●	●			▲	●		●	●	
4	○	X	○	●	●	●			●		●		
5	X	X	X		●	●	●		●				
Rev.	X	○	○	●		●				●	●		●

S1, S2, S3 :Shift solenoid No.1, No.2, No.3, ○: ON, X : OFF
●: Operating, ▲: Operating only in the selector position for engine brake

CHART 25–14 Toyota A350E clutch/band application chart. (Reprinted with permission from SAE AE-18 © 1994 Society of Automotive Engineers, Inc.)

			A350E
Torque Capacity			300 Nm
Torque Converter	Type		3-Element, 2-Phase with Lock-up Clutch
	Size		254 mm
Gear Train	Type		Simpson
	Gear Ratios	1st	2.804
		2nd	1.978
		3rd	1.531
		4th	1.000
		5th	0.705
		Rev.	2.393
Friction Element			3 Multiple Disc Clutches 3 Multiple Disc Brakes 1 Band Brake
Mass(Wet)			77.5 kg

TABLE 25–1 Toyota A350E specifications. (Reprinted with permission from SAE AE-18 © 1994 Society of Automotive Engineers, Inc.)

Gears	1	2	3	4	5
2-speed unit (Front gear unit)	LOW (1.000)	HIGH (0.705)	LOW (1.000)	LOW (1.000)	HIGH (0.705)
3-speed unit (Rear gear unit)	1st (2.804)	1st (2.804)	2nd (1.531)	3rd (1.000)	3rd (1.000)
Total gear ratios	2.804	1.978	1.531	1.000	0.705

↗ UP-SHIFT ↘ DOWN-SHIFT ----→ NOT SHIFT

TABLE 25–2 Toyota A350E front and rear gear unit sequencing. (Reprinted with permission from SAE AE-18 © 1994 Society of Automotive Engineers, Inc.)

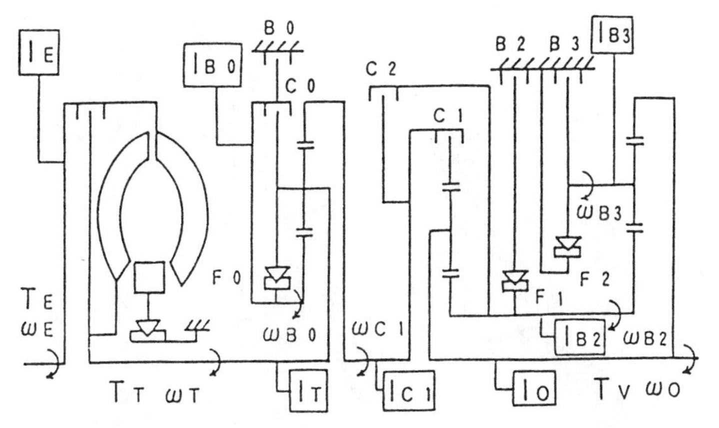

Subscripts

B0	: overdrive brake
B2	: 2nd brake
B3	: 1st reverse brake
C0	: overdrive clutch
C1	: forward clutch
C2	: direct clutch
E	: engine
F2	: one-way clutch
O	: output
P	: pump member of converter
T	: turbine member of converter
V	: vehicle
1C, 1R, 1S	: rear planetary carrier, ring gear, sun gear
2C, 2R, 2S	: front planetary carrier, ring gear sun gear
3C, 3R, 3S	: overdrive planetary carrier, ring gear sun gear

FIGURE 25–69 Lexus LS400 automatic transmission powertrain schematic and component key. (Reprinted with permission from SAE AE-18 © 1994 Society of Automotive Engineers, Inc.)

Volkswagen

Volkswagen produces both transverse- and longitudinal-mounted automatic transaxles. Each uses a Simpson compound planetary gearset.

The VW Type 3 and Type 4 automatic transaxles were used in all 1969 to 1976 models, except for the Rabbit and the Scirocco.

Type 010 is a three-speed FWD transaxle used in the VW Rabbit (1975–1976) and in the Scirocco and other vehicle designs (1975–1992). Audi used this unit from 1977 to 1992.

Types 087, 089, and 090 are three-speed longitudinal versions using a ring and pinion style final drive assembly. Type 090 is used in Vanagons, and Types 087 and 089 are used in certain Audi vehicles and the Volkswagen Quantum. In addition, Porsche 944 models have been equipped with Type 087 transaxles.

Diesel engine powered vehicles using Type 010, 087, or 089 have an "E" selector lever quadrant position. This position allows for freewheeling action during deceleration to improve fuel economy.

Passat and Corrado models (1990–1993) may be equipped with the FWD four-speed VW AG-4 or 095 automatic transaxle.

1996 Volkswagen Models. An optional four-speed automatic transmission is available in all 1996 Volkswagen models: Cabrio convertible, Golf, GTI, Jetta, and Passat.

VW/Ford (Europe) AG4 Automatic Transaxle. The AG4 automatic transaxle, shown in Figure 25–70, is built by Volkswagen in Kassel, Germany. It is used in the 1996 Ford Galaxy sold in Europe. Electronic controls operate the AG4 in this "MPV" styled minivan.

Volvo

Aisin-Warner and ZF have supplied automatic transmissions to Volvo (Sweden). Volvo vehicles have been equipped with

Manufacturer/Assy. Plant	Volkswagen AG Kassel, Germany						
Transmission Type	4-Speed Automatic Transmission, electr. controlled						
Gear and Differential Ratios	1st	2nd	3rd	4th	Rev.	Int.	Diff.
2.0L 8V DOHC	2.714	1.441	1.000	0.743	2.884	1.159	4.235
2.8i CD-V6	2.714	1.441	1.000	0.743	2.884	1.159	3.944
Gear Range Selection	P–R–N–D–3–2–1						
Fluid Refill Capacity	3.5 Liters (Transmission) 0.8 Liters (Differential)						
Fluid Recommendation	N052162 VX00 (Transmission) N052145 VX00 (Differential)						
Availability	Galaxy						

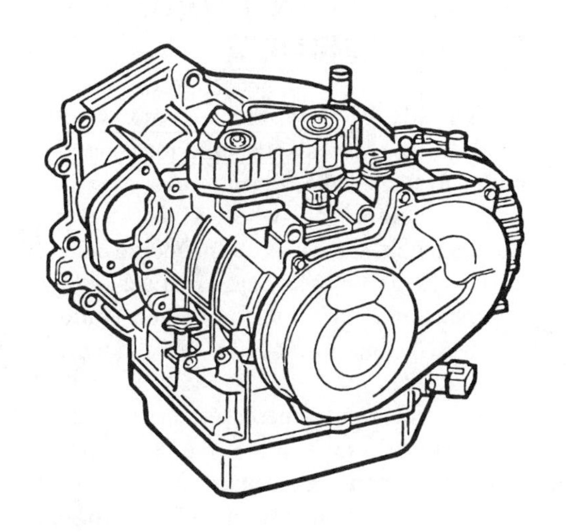

FIGURE 25–70 Volkswagen AG4 four-speed automatic transaxle. (Reprinted with the permission of Ford Motor Company)

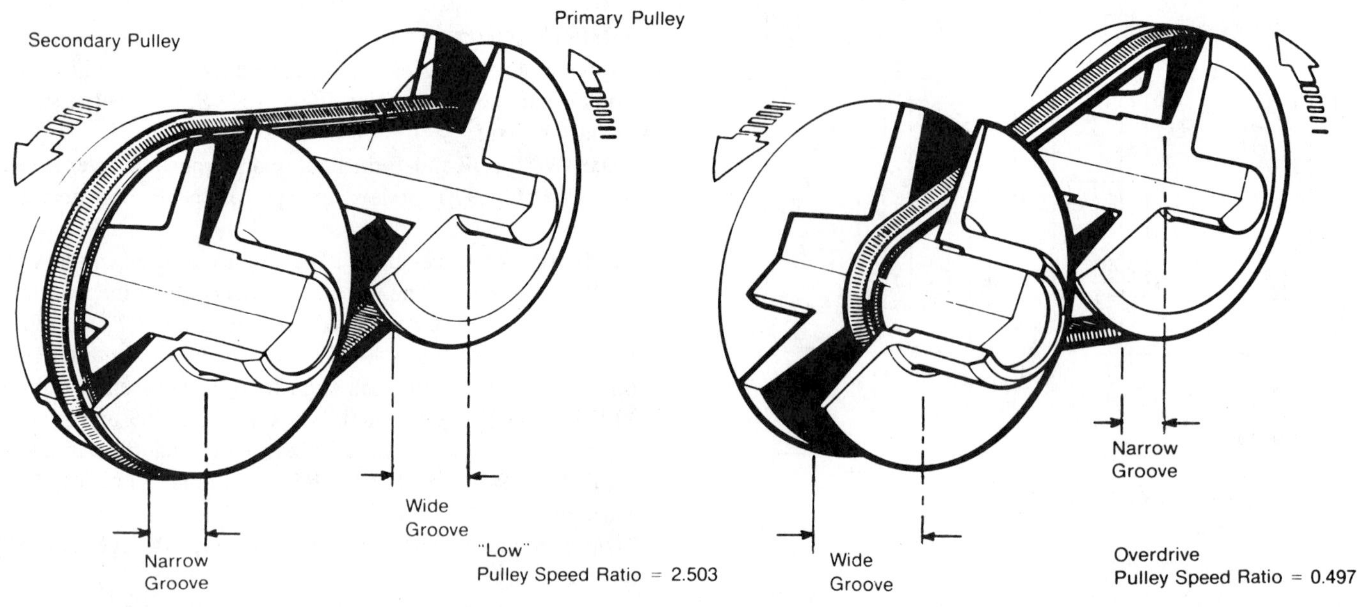

FIGURE 25–71 Changing the width of the drive and driven pulleys provides a continuously variable speed and torque ratio. (Courtesy of Subaru of America, Inc.)

these Aisin-Warner RWD transmissions: the AW-40 (1973–1984) and AW-70/71 (1982–1989). The ZF-4HP-22, four-speed RWD transmission was used from 1985 to 1989 in Volvo vehicles. Refer to the Aisin-Warner and ZF Industries sections in this chapter for additional coverage of these independently produced transmissions.

1996 Volvo Models. Versions of the 850 model use a four-speed automatic transmission as a standard or optional feature. Four-speed automatic transmissions are a standard feature in 1996 960 models.

CVTs—CONTINUOUSLY VARIABLE TRANSMISSIONS

Continuously variable transmissions (CVTs), sometimes referred to as shiftless transmissions, operate on a principle similar to that of the transmissions used in snowmobiles and some recreational four-wheel ATVs. A specially designed flexible metal belt, known as a van Doorne belt, is the link between a drive and driven pulley system. Each pulley is able to change its dimension to affect the drive (gear) ratio (Figure 25–71). A prototype design developed by ZF Industries, Inc., is shown in Figure 25–72.

As vehicle speed increases, the radius of the drive pulley increases, and the radius of the driven pulley decreases. The pulley movements are caused by hydraulic pressures. Due to the two pulleys changing in width, an infinite number of gear ratios are available. Pulley operation can be electronically/hydraulically controlled to match various driver and vehicle demands (Figure 25–73). The action of the pulleys provides seamless ratio changes. A final drive differential unit supplies torque to the drive axles. In certain CVTs, the gear ratios range from 2.53:1 to 14.77:1.

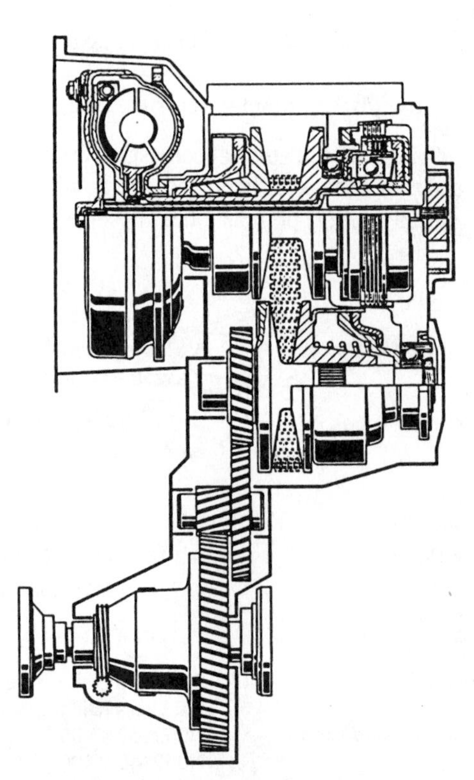

FIGURE 25–72 CVT prototype design developed by ZF Industries, Inc. (Reprinted with permission from SAE AE-18 © 1994 Society of Automotive Engineers, Inc.)

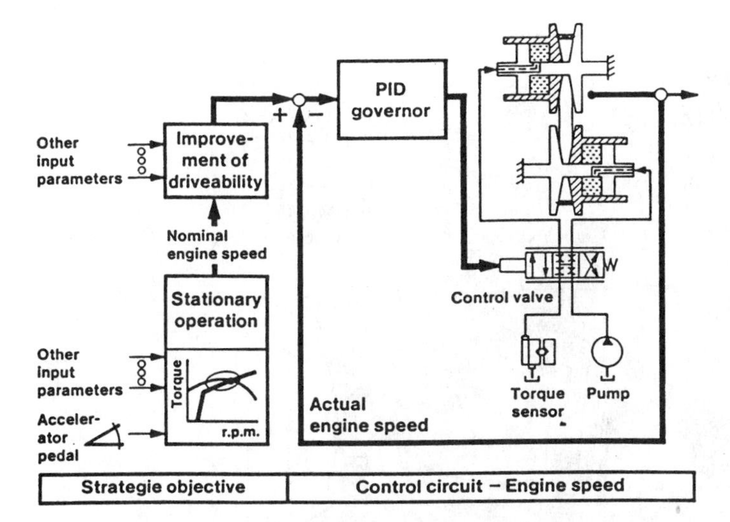

FIGURE 25–73 Typical CVT electronic control system. (Reprinted with permission from SAE AE-18 © 1994 Society of Automotive Engineers, Inc.)

Continuously variable transmissions were first introduced in a compact Dutch vehicle called the DAF in 1958. The CVT used in this vehicle was known as the Variomatic transmission. The transmission drive belt could only handle engines smaller than 1000 cc. As a result, interest in this design declined until fuel economy and reduced vehicle exhaust emissions became concerns in the late 1970s.

Research and development of CVTs by Borg-Warner, Ford, Subaru, Suzuki, and ZF produced units compatible for small cars in the 1980s. Compact passenger car applications of CVTs include the Ford Escort (European version), the Honda Civic HX, the Subaru Justy, and the Suzuki Cultus.

Ford's variable transmission is known as the CTX (Figure 25–74). The 1996 CTX version has a larger fluid capacity than the earlier version: 4.7 L (5.0 qt) versus 4.1 L (4.3 qt).

Subaru's ECVT uses a metal powder/electromagnetic clutch assembly instead of a conventional fluid torque converter with a lockup clutch. A cutaway view of the ECVT is shown in Figure 25–75.

The CVT used in the Suzuki Cultus is the development of a joint effort between Suzuki Motor Corporation and Borg-Warner Automotive. The special drive belt used in this unit is called a Borg-Warner Hy-Vo CVT belt.

Characteristics of CVTs include excellent driveability and acceleration, great fuel economy, and smooth, stepless speed ratio control. As torque holding capabilities and operational functioning is refined, CVT application is expected to increase.

SUMMARY

Though there are many manufacturers and numerous vehicle applications, automatic transmission/transaxle designs and operating characteristics are more alike than different. The basic laws of physics, hydraulics, and electricity/electronics cross international boundaries.

Four-speed electronically controlled automatic transmissions and transaxles, designed with planetary gearsets and lockup torque converter clutches, will continue to be a popular powertrain option. These transmissions will be used by manufacturers to produce passenger cars and light trucks that meet stringent fuel efficiency and vehicle emission levels, and the expectations of modern consumers.

As automobile manufacturers' corporate holdings in one another continue to evolve, and the number of joint ventures increases, cross-application of automatic transmissions/transaxles is expected to increase. The import and foreign, as well as the domestic, automobile and automatic transmission industries can be classified as being elements of today's World Market.

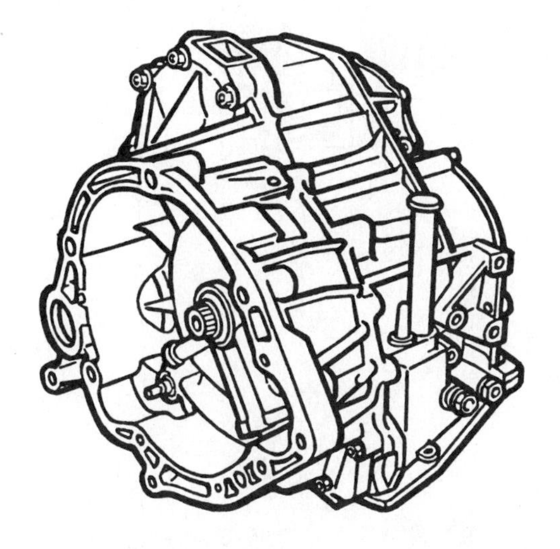

Manufacturer/Assy. Plant	Ford Motor Company Bordeaux, France		
Transmission Type	Continously Variable Transaxle		
Gear and Differential Ratios	Variable	Rev.	Diff.
Fiesta	14.77 – 2.53	4.27	3.84
Escort	13.36 – 2.41	4.27	3.84
Gear Range Selection	P–R–N–D–L		
Fluid Refill Capacity	4.7 Liters		
Fluid Recommendation	ESP-M2C166-H		
Availability	Fiesta '96, Escort '96		

FIGURE 25–74 Ford's CTX automatic transaxle (CVT design), built in Bordeaux, France, for Ford's European car line. (Reprinted with the permission of Ford Motor Company)

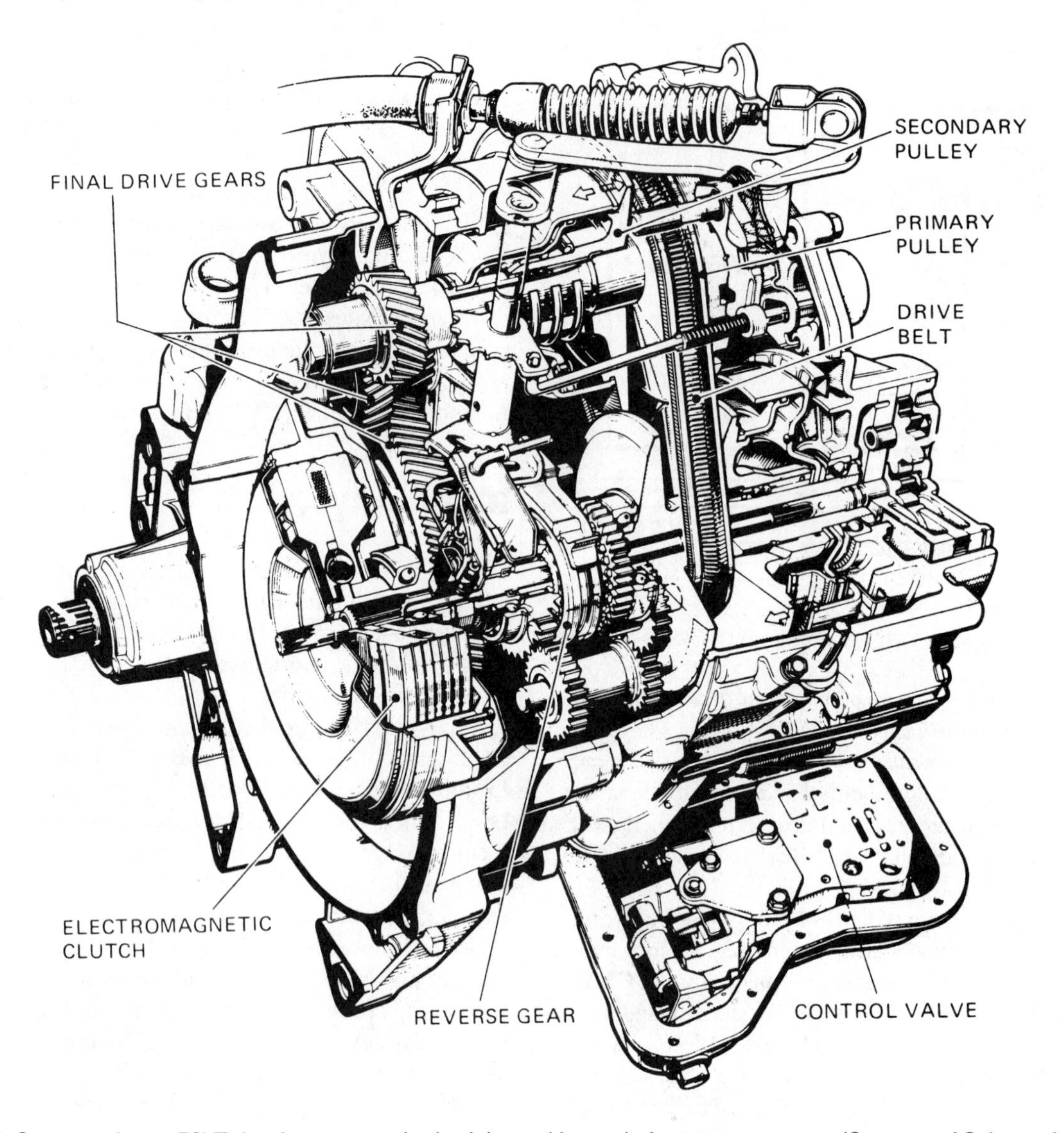

FIGURE 25–75 Cutaway view—ECVT. An electromagnetic clutch is used instead of a torque converter. (Courtesy of Subaru of America, Inc.)

Appendix A English-Metric Conversion Table

Description	Multiply	By	For Metric Equivalent
ACCELERATION	Foot/sec²	0.304 8	metre/sec² (m/s²)
	Inch/sec²	0.025 4	metre/sec²
TORQUE	Pound-inch	0.112 98	newton-metres (N·m)
	Pound-foot	1.355 8	newton-metres
POWER	horsepower	0.746	kilowatts (kw)
PRESSURE or STRESS	inches of water	0.2488	kilopascals (kPa)
	pounds/sq. in.	6.895	kilopascals (kPa)
	pounds/sq. in.	1	bar
ENERGY or WORK	BTU	1 055.	joules (J)
	foot-pound	1.355 8	joules (J)
	kilowatt-hour	3 600 000. or 3.6 × 10⁶	joules (J = one W's)
LIGHT	foot candle	10.76	lumens/metre² (lm/m²)
FUEL PERFORMANCE	miles/gal	0.425 1	kilometres/litre (km/l)
	gal/mile	2.352 7	litres/kilometre (l/km)
VELOCITY	miles/hour	1.609 3	kilometres/hr. (km/h)
LENGTH	inch	25.4	millimetres (mm)
	foot	0.304 8	metres (m)
	yard	0.914 4	metres (m)
	mile	1.609	kilometres (km)
AREA	inch²	645.2	millimetres² (mm²)
		6.45	centimetres² (cm²)
	foot²	0.092 9	metres² (m²)
	yard²	0.836 1	metres²
VOLUME	inch³	16 387.	mm³
	inch³	16.387	cm³
	quart	0.016 4	litres (1)
	quart	0.946 4	litres
	gallon	3.785 4	litres
	yard³	0.764 6	metres³ (m³)
MASS	pound	0.453 6	kilograms (kg)
	ton	907.18	kilograms (kg)
	ton	0.90718	tonne
FORCE	kilogram	9.807	newtons (N)
	ounce	0.278 0	newtons
	pound	4.448	newtons
TEMPERATURE	degree farenheit	0.556 (°F −32)	degree Celsius (°C)

Appendix B Direct Reading Metric-English Conversion Guides

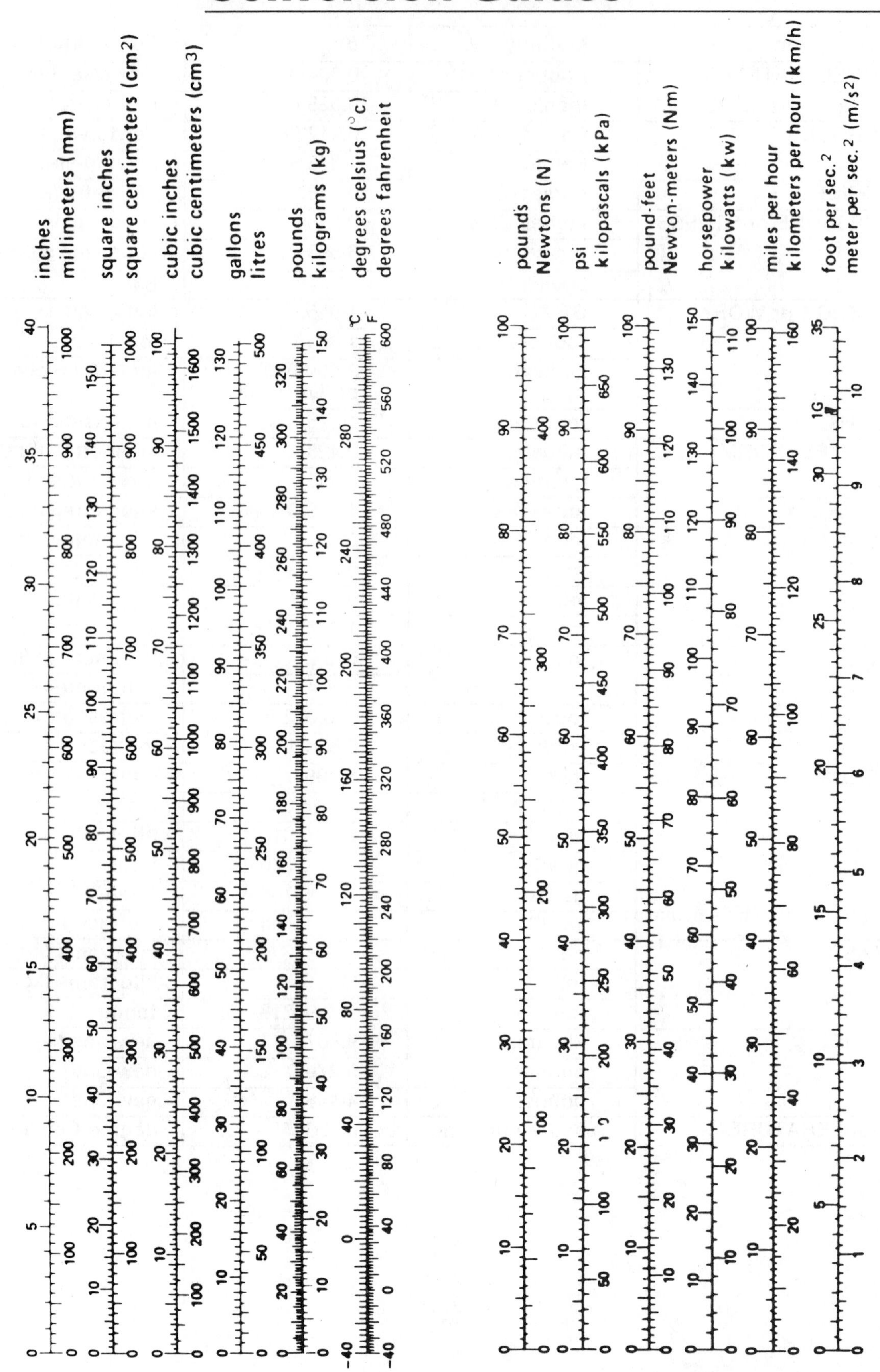

Courtesy of General Motors Corp.

Appendix C NATEF/ASE Task List and Correlation to Chapters

AUTOMATIC TRANSMISSION AND TRANSAXLE

For every task in Automatic Transmission and Transaxle, the following safety concerns must be strictly enforced as a number 1 priority:

Comply with personal and environmental safety practices associated with clothing, eye protection, hand tools, power equipment and handling, storage and disposal of chemicals in accordance with local, state, and federal safety and environmental regulations.

Each task listed below is assigned a priority number: P-1, P-2, or P-3.

A. General Transmission and Transaxle Diagnosis

1. Interpret and verify driver's complaint; verify proper engine operation; determine needed repairs.
   P-2: Chapter 12 (Sections 1, 2, and 3).

2. Diagnose unusual fluid usage, level, and condition problems; determine needed repairs.
   P-2: Chapters 12 (Section 2) and 21.

3. Perform pressure tests; determine needed repairs.
   P-1: Chapter 12 (Section 5).

4. Perform stall tests; determine needed repairs.
   P-2: Chapter 12 (Section 3).

5. Perform lockup converter system tests; determine needed repairs.
   P-2: Chapters 12 (Section 4) and 13.

6. Diagnose electronic, mechanical, and vacuum control systems; determine needed repairs.
   P-2: Chapter 12 (Sections 2 and 5).

7. Diagnose noise and vibration problems; determine needed repairs.
   P-3: Chapter 12 (Section 6).

B. Transmission and Transaxle Maintenance and Adjustment

1. Inspect, adjust, or replace manual shift valve and throttle (TV) linkages or cables (as applicable).
   P-2: Chapter 12 (Section 2).

2. Service transmission; perform visual inspection; replace fluids and filters.
   P-1: Chapters 8, 12 (Section 2), 20, and 21.

C. In-Vehicle Transmission and Transaxle Repair

1. Inspect, adjust, or replace (as applicable) vacuum modulator; inspect and repair or replace lines and hoses.
   P-3: Chapter 12 (Section 5).

2. Inspect, repair, and replace governor assembly.
   P-3: Chapters 19 and 20.

3. Inspect and replace external seals and gaskets.
   P-3: Chapters 20 and 21.

4. Inspect extension housing; replace bushing and seal.
   P-3: Chapters 8 and 19.

5. Inspect, leak test, flush, and replace cooler, lines, and fittings.
   P-3: Chapter 15.

6. Inspect and replace speedometer drive gear, driven gear, vehicle speed sensor (VSS), and retainers.
   P-3: Chapter 19.

7. Inspect, measure, repair, and replace valve body (includes surfaces and bores, springs, valves, sleeves, retainers, brackets, check-balls, screens, spacers, and gaskets); check/adjust valve body bolt torque.
   P-2: Chapters 19 and 20.

8. Inspect servo bore, piston, seals, pin, spring, and retainers; repair or replace as needed.
   P-3: Chapters 17 and 20.

9. Inspect accumulator bore, piston, seals, spring, and retainer; repair or replace as needed.
P-3: Chapter 17.

10. Inspect, test, adjust, repair, or replace transmission-related electrical and electronic components (includes computers, solenoids, sensors, relays, and switches).
P-2: Chapters 13, 14, and 17.

11. Inspect, replace, and align powertrain mounts.
P-3: Chapter 15.

12. Inspect and replace parking pawl, shaft, spring and retainer.
P-3: Chapters 17 and 18.

D. Off-Vehicle Transmission and Transaxle Repair

1. Removal, Disassembly, and Reinstallation

1. Remove and reinstall transmission and torque converter (rear-wheel drive).
P-2: Chapter 15.

2. Remove and reinstall transaxle and torque converter assembly.
P-2: Chapter 15.

3. Disassemble, clean, and inspect transmission/transaxle.
P-1: Chapters 17 and 18.

4. Assemble transmission/transaxle.
P-1: Chapters 18, 19, and 20.

2. Oil Pump and Converter

1. Inspect converter flex plate, attaching parts, pilot and pump drive, and seal areas.
P-2: Chapters 15, 16, and 21.

2. Measure torque converter end play, and check for interference; check stator clutch.
P-2: Chapter 16.

3. Inspect, measure, and replace oil pump housings, shafts, vanes, rotors, gears, valves, seals, and bushings.
P-3: Chapter 19.

4. Check torque converter and transmission cooling system for contamination.
P-3: Chapters 15 and 16.

3. Geartrain, Shafts, Bushings, and Case

1. Check end play or preload; determine needed service.
P-2: Chapters 17 and 20.

2. Inspect, measure, and replace thrust washers and bearings.
P-2: Chapters 12 (Section 4), 17, 18, and 20.

3. Inspect oil delivery seal rings, ring grooves, and sealing surface areas.
P-2: Chapters 18 and 19.

4. Inspect bushings; replace as needed.
P-2: Chapters 16 and 19.

5. Inspect and measure planetary gear assembly (includes sun, ring gear, thrust washers, planetary gears, and carrier assembly); replace as needed.
P-2: Chapters 12 (Section 4), 18, and 19.

6. Inspect cases, bores, passages, bushings, vents, and mating surfaces; replace as needed.
P-2: Chapter 18.

7. Inspect transaxle drive, link chains, sprockets, gears, bearings, and bushings; replace as needed.
P-2: Chapter 17.

8. Inspect, measure, repair, adjust, or replace transaxle final drive components.
P-2: Chapter 20.

9. Inspect and reinstall parking pawl, shaft, spring, and retainer; replace as needed.
P-3: Lab Activity exercise.

4. Friction and Reaction Units

1. Inspect clutch drum, piston, check-balls, springs, retainers, seals, and friction and pressure plates; replace as needed.
P-2: Chapter 19.

2. Measure clutch pack clearance; adjust as needed.
P-2: Chapter 19.

3. Air test operation of clutch and servo assemblies.
P-2: Chapters 19 and 20.

4. Inspect roller and sprag clutch, races, rollers, sprags, springs, cages, and retainers; replace as needed.
P-2: Chapter 18.

5. Inspect bands and drums; replace as needed.
P-3: Chapters 18 and 19.

Task List
Automatic Transmission/Transaxle (Test A2)

Reading this task list will allow you to focus your preparation on those subjects which need your attention. If you feel comfortable with your knowledge about a particular task, you are probably prepared for questions on that subject. If, on the other hand, you have any doubts, this is where you should spend your study time. Any good text book or service manual will give you the information you need. Also, you should consider how many questions will be asked in each content area.

A. General Transmission/Transaxle Diagnosis (19 questions)

1. Listen to driver's complaint and road test vehicle; determine needed repairs.
2. Diagnose noise and vibration problems; determine needed repairs.
3. Diagnose unusual fluid usage, level, and condition problems; determine needed repairs.
4. Perform pressure tests; determine needed repairs.
5. Perform stall tests; determine needed repairs.
6. Perform lock-up converter system tests; determine needed repairs.
7. Diagnose electronic, mechanical, and vacuum control systems; determine needed repairs.

B. Transmission/Transaxle Maintenance and Adjustment (6 questions)

1. Inspect, adjust, and replace manual valve shift linkage.
2. Inspect, adjust, and replace cables or linkages for throttle valve (TV), kickdown, and accelerator pedal.
3. Adjust bands.
4. Replace fluid and filter(s).
5. Inspect, adjust, and replace electronic sensors, wires, and connectors.

C. In-Vehicle Transmission/Transaxle Repair (11 questions)

1. Inspect, adjust, and replace vacuum modulator, valve, lines, and hoses.
2. Inspect, adjust, repair, and replace governor cover, seals, sleeve, valve, weights, springs, retainers, and gear.
3. Inspect and replace external seals and gaskets.
4. Inspect, repair, and replace extension housing.
5. Inspect, test, flush, and replace cooler, lines, and fittings.
6. Inspect and replace speedometer drive gear, driven gear, and retainers.
7. Inspect valve body mating surfaces, bores, valves, springs, sleeves, retainers, brackets, check balls, screens, spacers, and gaskets; replace as necessary.
8. Check/adjust valve body bolt torque.
9. Inspect servo bore, piston, seals, pin, spring, and retainers; repair/replace as necessary.
10. Inspect accumulator bore, piston, seals, spring, and retainers; repair/replace as necessary.
11. Inspect and replace parking pawl, shaft, spring, and retainer.

12. Inspect, test, adjust, or replace electrical/electronic components including computers, solenoids, sensors, relays, switches, and harnesses.
13. Inspect, replace, and align power train mounts.

D. Off-Vehicle Transmission/Transaxle Repair (14 questions)

1. Removal, Disassembly, and Assembly (3 questions)

1. Remove and replace transmission/transaxle.
2. Disassemble, clean, and inspect.
3. Assemble after servicing.

2. Oil Pump and Converter (3 questions)

1. Inspect converter flex (drive) plate, converter attaching bolts, converter pilot, and converter pump drive surfaces.
2. Inspect, flush, measure end play, and test torque converter.
3. Inspect, measure, and replace oil pump housings and parts.

3. Gear Train, Shafts, Bushings, and Case (3 questions)

1. Check end play and/or preload; determine needed service.
2. Inspect, measure, and replace thrust washers and bearings.
3. Inspect and replace shafts.
4. Inspect oil delivery seal rings, ring grooves, and sealing surface areas.
5. Inspect and replace bushings.
6. Inspect and measure planetary gear assembly; replace parts as necessary.
7. Inspect, repair, and replace case(s) bores, passages, bushings, vents, and mating surfaces.
8. Inspect, repair or replace transaxle drive chains, sprockets, gears, bearings, and bushings.
9. Inspect, measure, repair, adjust or replace transaxle final drive components.

4. Friction and Reaction Units (5 questions)

1. Inspect clutch assembly; replace parts as necessary.
2. Measure and adjust clutch pack clearance.
3. Air test the operation of clutch and servo assemblies.
4. Inspect one-way clutch assemblies; replace parts as necessary.
5. Inspect and replace bands and drums. ■

SAMPLE QUESTIONS
AUTOMATIC TRANSMISSION/TRANSAXLE (TEST A2)

1. An automatic transmission does not hold in PARK. Technician A says that misadjusted shift linkage could be the cause. Technician B says that a roller clutch installed backwards could be the cause. Who is right?

   (A) A only
   (B) B only
   (C) Both A and B
   (D) Neither A nor B

Question #1 Explanation:

Technician A's statement is correct. Misadjusted shift linkage could keep the parking pawl from engaging with the park gear.

Technician B's statement is incorrect; the function of the roller clutch is to act as the reaction member when the transmission is in drive.

2. A vehicle with an automatic transaxle and a properly tuned engine accelerates poorly from a stop. Acceleration is normal above 35 mph. Which of these could be the cause?

   (A) A bad torque converter
   (B) A worn front pump
   (C) A bad governor valve
   (D) A low fluid level

3. Which of these act directly on the shift valves in the hydraulic control system?

   (A) Front pump and converter pressures
   (B) Throttle (modulator) and governor pressures
   (C) Servo and clutch pressures
   (D) Main control and compensator pressures

4. When the engine is idling, the transmission line pressure is low in all selector lever positions. Which of these is the most likely cause?

   (A) Vacuum diaphragm unit
   (B) Throttle (modulator) valve
   (C) Connecting vacuum tubes
   (D) Transmission pump

5. With the transmission pan removed, the best way to pin-point an oil pressure leak is to:

   (A) check the filter pick-up tube seal.
   (B) check the line pressure.
   (C) remove and check the valve body.
   (D) perform an air pressure test.

6. A stall test can be used to check:

   (A) planetary gear noise.
   (B) governor pressure.
   (C) convertor operation.
   (D) erratic shifting.

7. The tool in the setup shown above is being used to install a:

   (A) clutch sprag.
   (B) clutch piston return spring and retainer.
   (C) clutch drum selective spacer.
   (D) clutch disc retainer.

8. An automatic transaxle does not work right. What should the technician do first?

   (A) Take a pressure test
   (B) Adjust the bands
   (C) Check the fluid level
   (D) Check the engine vacuum

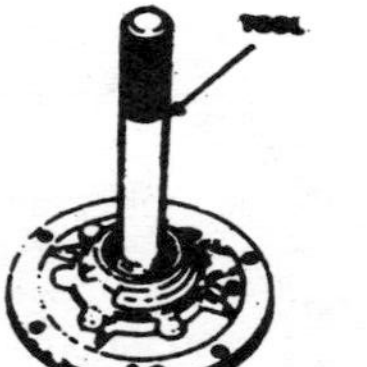

9. The tool in the setup shown above is being used to remove which of these?

   (A) pump bushing
   (B) pump seal
   (C) stator support
   (D) stator seal

Question 10 is not like the ones above.

It has the word **EXCEPT**. For this question, look for the choice that could **NOT** cause the described situation. Read the entire question carefully before choosing your answer.

10. All of these can cause an automatic transmission to slip **EXCEPT**:

    (A) hardened servo seals.
    (B) worn planetary gears.
    (C) plugged sump filter.
    (D) faulty one way clutch.

ANSWER KEY:

1. **A** 2. **A** 3. **B** 4. **D** 5. **D** 6. **C** 7. **B** 8. **C** 9. **A** 10. **B**

TEST SPECIFICATIONS
AUTOMATIC TRANSMISSION/TRANSAXLE (TEST A2)

Content Area	Questions in Test	Percentage of Test
A. General Transmission/Transaxle Diagnosis	19	38%
B. Transmission/Transaxle Maintenance and Adjustment	6	12%
C. In-Vehicle Transmission/Transaxle Repair	11	22%
D. Off-Vehicle Transmission/Transaxle Repair	14	28%
1. Removal, Disassembly, and Assembly (3)		
2. Oil Pump and Converter (3)		
3. Gear Train, Shafts, Bushings, and Case (3)		
4. Friction and Reaction Units (5)		
Total	**50[1]**	**100%**

[1] *Note:* There could be up to ten additional pre-test questions. Your answers to these pre-test questions will not affect your score, but since you do not know which they are, you should answer all questions in the test. The five-year Recertification Test will cover the same content areas as those listed above. However, the number of questions in each content area of the Recertification Test will be reduced by about one-half.

Glossary of Technical Terms

Acronyms and Abbreviations

ACT air charge temperature (sensor)
ALDL assembly line diagnostic link
ATF automatic transmission fluid
AWD all-wheel drive
BHP brake horsepower
BNC British Naval Connector
BP barometric pressure (sensor)
CCC computer command control
CCD/C^{2}D Computer Communication Data link (Chrysler)
CVT continuously variable transmission
DLC data link connector
DRB Digital Readout Box (Chrysler)
DTC diagnostic trouble code
DVOM digital volt-ohmmeter
EATX Electronic Automatic Transaxle (Chrysler)
ECM engine control module
ECT engine coolant temperature (sensor)
EEC-IV electronic engine control, fourth generation (Ford)
EMF electromotive force
EPC electronic pressure control (solenoid)
ESD electrostatic discharge
FWD front-wheel drive
GPM gallons per minute
Hz hertz
IX internal-external
km/h kilometers per hour
kPa kilopascals
LED light-emitting diode
MAF mass air flow (sensor)
MAP manifold absolute pressure (sensor)
MCCC modulated converter clutch control
MIL malfunction indicator lamp
MLPS manual lever position switch
MPH miles per hour
MVLPS manual valve lever position sensor
NGS New Generation STAR (Ford tester)
OBD II on-board diagnostics, second generation
OSS output speed sensor
PCM powertrain control module
PCS pressure control solenoid
PROM programmable read-only memory
PSA pressure switch assembly
psi pounds per square inch
PWM pulse width modulated
R&R remove and replace
RPM revolutions per minute
RTV room temperature vulcanizing (sealant)
RWD rear-wheel drive
SES service engine soon (lamp)
SMEC single-module engine controller
STAR self-test automatic readout (Ford)
TCC torque converter clutch
TCIL transmission control indicator lamp
TCM transmission control module
TCS transmission control switch
TFT transmission fluid temperature (sensor)
TIS transmission input speed (sensor)
TOT transmission oil temperature (sensor)
TP throttle position (sensor)
TRS transmission range selector (switch)
VFS variable force solenoid
VSS vehicle speed sensor
WOT wide open throttle
2WD two-wheel drive
4WD four-wheel drive

Key Terms

absolute pressure: Atmospheric (barometric) pressure plus the pressure gauge reading.

accumulator: A device that controls shift quality by cushioning the shock of hydraulic oil pressure being applied to a clutch or band.

actuating mechanism: The mechanical output devices of a hydraulic system, for example, clutch pistons and band servos.

actuator: The output component of a hydraulic or electronic system.

adaptive memory (adaptive strategy): The learning ability of the TCM or PCM to redefine its decision-making process to provide optimum shift quality.

air charge temperature (ACT) sensor: The temperature of the airflow into the engine is measured by an ACT sensor, usually located in the lower intake manifold or air cleaner.

ALDL (assembly line diagnostic link): Electrical connector for scanning ECM/PCM/TCM input and output devices.

ammeter: An electrical instrument used to measure the rate of current flow in a circuit.

amperage: The total amount of current (amperes) flowing in a circuit.

ampere (amp): A unit of measure for the flow of electrical current in a circuit.

amplifier: A device used in an electrical circuit to increase the voltage of an output signal.

annulus gear: Preferred by Chrysler to describe the internal (ring) gear.

anodized: A special coating applied to the surface of aluminum valves for extended service life.

antifoam agents: Minimize fluid foaming from the whipping action encountered in the converter and planetary action.

antiwear agents: Zinc agents that control wear on the gears, bushings, and thrust washers.

arc: A flow of electricity through the air between two electrodes or contact points that produces a spark.

armature: The rotating part of a motor consisting of a conductor wound around an iron core.

atmospheric pressure: The force exerted on everything around us by the weight of the air. At sea level, atmospheric pressure is 14.7 psi (pounds per square inch), or in the metric system, 1.0355 kg/cm^2 (kilograms per square centimeter).

auxiliary add-on cooler: A supplemental transmission fluid cooling device that is installed in series with the heat exchanger (cooler), located inside the radiator, to provide additional support to cool the hot fluid leaving the torque converter.

auxiliary pressure: An added fluid pressure that is introduced into a regulator or balanced valve system to control valve movement. The auxiliary pressure itself can be either a fixed or a variable value. (See balanced valve; regulator valve.)

axial force: A side or end thrust force acting in or along the same plane as the power flow.

balanced valve: A valve that is positioned by opposing auxiliary hydraulic pressures and/or spring force. Examples include mainline regulator, throttle, and governor valves. (See regulator valve.)

band: A flexible ring of steel with an inner lining of friction material. When tightened around the outside of a drum, a planetary member is held stationary to the transmission case.

BARO (barometric pressure sensor): Measures the change in the intake manifold pressure caused by changes in altitude.

barometric manifold absolute pressure (BMAP) sensor: Operates similarly to a conventional MAP sensor; reads intake manifold pressure and is also responsible for determining altitude and barometric pressure prior to engine operation.

barometric pressure: (See atmospheric pressure.)

boost valve: Used at the base of the regulator valve to increase mainline pressure.

brake horsepower (bhp): The actual horsepower available at the engine flywheel as measured by a dynamometer.

breakaway: Often used by Chrysler to identify first-gear operation in D and 2 ranges. In these ranges, first-gear operation depends on a one-way roller clutch that holds on acceleration and releases (breaks away) on deceleration, resulting in a freewheeling coastdown condition.

brinneling: A wear pattern identified by a series of dents/grooves/indentations at regular intervals. This condition is caused by a lack of lube, overload situations, and/or vibrations.

bump: Sudden and forceful apply of a clutch or band.

bypass circuit: (See circuit, bypass.)

capacity: The quantity of electricity that can be delivered from a unit, as from a battery in ampere-hours, or output, as from a generator.

CCD/C^{2}D bus: Chrysler's Computer Communication Data link, which is an electronic communication link used to transmit information signals to electronic components with only two wires. It operates on multiplexing principles of electronics.

centrifugal force: The outward pull of a revolving object, away from the center of revolution. Centrifugal force increases with the speed of rotation.

check valve: A one-way directional valve in a hydraulic circuit, commonly using a "check ball" and seat arrangement.

chuggle: Bucking or jerking condition that may be engine-related and may be most noticeable when converter clutch is engaged; similar to the feel of towing a trailer.

circuit: The path of electron flow from the source through components and connections and back to the source.

circuit breaker: A device for interrupting a circuit when the current flow becomes unsafe. Most automotive circuit breakers will reset themselves when the overload is removed.

circuit, bypass: Another circuit in parallel with the major circuit through which power is diverted.

circuit, closed: An electrical circuit in which there is no interruption of current flow.

circuit, ground: The noninsulated portion of a complete circuit used as a common potential point. In automotive circuits, the ground is composed of metal parts, such as the engine, body sheet metal, and frame and is usually a negative potential.

circuit, hot: That portion of a circuit not at ground potential. The hot circuit is usually insulated and is connected to the positive side of the battery.

circuit, open: A break or lack of contact in an electrical circuit, either intentional (switch) or unintentional (bad connection or broken wire).

circuit, parallel: A circuit having two or more paths for current flow with common positive and negative tie points. The same voltage is applied to each load device or parallel branch.

circuit, series: An electrical system in which separate parts are connected end to end, using one wire, to form a single path for current to flow.

circuit, short: A circuit that is accidentally completed in an electrical path for which it was not intended.

clamping (isolation) diodes: Diodes positioned in a circuit to prevent self-induction from damaging electronic components.

closed circuit: (See circuit, closed.)

clutch, fluid: The same as a fluid coupling. A fluid clutch or coupling performs the same function as a friction clutch by utilizing fluid friction and inertia as opposed to solid friction used by a friction clutch. (See fluid coupling.)

clutch, friction: A coupling device that provides a means of smooth and positive engagement and disengagement of engine torque to the vehicle powertrain. Transmission of power through the clutch is accomplished by bringing one or more rotating drive members into contact with complementing driven members.

coast: Vehicle deceleration caused by engine braking conditions.

coefficient of friction: The amount of surface tension between two contacting surfaces; identified by a scientifically calculated number.

compatibility: The ATF blend must be compatible with the materials used in the transmission. It cannot react chemically with the metals, seal materials, or friction materials.

compound gear: A gear consisting of two or more simple gears with a common shaft.

compound planetary: A gearset that has more than the three elements found in a simple gearset and is constructed by combining members of two planetary gearsets to create additional gear ratio possibilities.

computer: An electronic control module that correlates input data according to prearranged engineered instructions; used for the management of an actuator system or systems.

computer command control (CCC): A General Motors computer-directed system of engine control. Where applicable, it is interfaced with automatic transmission/transaxle control.

conductor: Any material through which electrons can flow; a path for electrical current flow.

continuity: The completeness of a circuit. A circuit having continuity is complete in that it has no interruptions.

converter: (See torque converter.)

converter lockup: The switching from hydrodynamic to direct mechanical drive, usually through the application of a friction element called the converter clutch.

corrosion inhibitor: An inhibitor in ATF that prevents corrosion of bushings, thrust washers, and oil cooler brazed joints.

coupling phase: Occurs when the torque converter is operating at its greatest hydraulic efficiency. The speed differential between the impeller and the turbine is at its minimum. At this point, the stator freewheels, and there is no torque multiplication.

current: The flow (or rate) of electrons moving through a circuit. Current is measured in amperes (amp).

current flow—conventional: Current flows through a circuit from the positive terminal of the source to the negative terminal (plus to minus).

current flow—electron: Current or electrons flow from the negative terminal of the source, through the circuit, to the positive terminal (minus to plus).

cyclic vibrations: The off-center movement of a rotating object that is affected by its initial balance, speed of rotation, and working angles.

data link connector (DLC): Current acronym/term applied to the federally mandated, diagnostic junction connector that is used to monitor ECM/PCM/TCM inputs, processing strategies, and outputs including diagnostic trouble codes (DTCs).

deceleration bump: When referring to a torque converter clutch in the applied position, a sudden release of the accelerator pedal causes a forceful reversal of power through the drivetrain (engine braking), just prior to the apply plate actually being released.

delayed (late or extended): Condition where shift is expected but does not occur for a period of time, for example, where clutch or band engagement does not occur as quickly as expected during part throttle or wide open throttle apply of accelerator or when manually downshifting to a lower range.

detent: A spring-loaded plunger, pin, ball, or pawl used as a holding device on a ratchet wheel or shaft. In automatic transmissions, a detent mechanism is used for locking the manual valve in place.

detent downshift: (See kickdown.)

Dexron: Transmission fluid developed by General Motors and used by numerous other vehicle manufacturers. This fluid contains a friction modifier.

diagnostic trouble codes (DTCs): A digital display from the ECM's/PCM's memory that identifies the input, processor, or output device circuit that is related to the powertrain emission/driveability malfunction detected. Diagnostic trouble codes can be read by commanding the MIL to flash any codes or by using a handheld scanner.

diaphragm: A flexible fabric or rubber membrane, usually spring-loaded, used in a vacuum modulator.

differential: Located within the final drive unit, the differential provides power to the drive wheels for straight-ahead driving and turning conditions.

differential areas: When opposing faces of a spool valve are acted upon by the same pressure but their areas differ in size, the face with the larger area produces the differential force and valve movement. (See spool valve.)

differential force: (See differential areas.)

digital readout: A display of numbers or a combination of numbers and letters.

diode: A solid-state device that permits current to flow in one direction only. It performs as a one-way check valve.

direct drive: The gear ratio is 1:1, with no change occurring in the torque and speed input/output relationship.

dispersants: Suspend dirt and prevent sludge buildup.

double bump (double feel): Two sudden and forceful applies of a clutch or band.

driveline: The drive connection between the transmission and the drive wheels.

DVOM (digital volt-ohmmeter): A device that uses high impedance internal circuit resistance for testing electronic circuitry.

dynamic: A sealing application in which there is rotating or reciprocating motion between the parts.

early: Condition where shift occurs before vehicle has reached proper speed, which tends to labor engine after upshift.

EATX relay: Chrysler's Electronic Automatic Transaxle Relay that, when energized, provides an electrical power input to the transaxle's TCM (controller). When certain malfunctions are noticed, the relay shuts down and de-energizes solenoid assembly, preventing a hazardous driving situation and/or transaxle damage.

EEC-IV module: The computer processing unit used in Ford's electronic engine control system, fourth generation.

electrolysis: A surface etching or bonding of current-conducting transmission components that may occur when grounding straps are missing or in poor condition.

electromagnet: A coil that produces a magnetic field when current flows through its windings.

electromagnetic induction: A method to create (generate) current flow through the use of magnetism.

electromagnetism: The effects surrounding the relationship between electricity and magnetism.

electromotive force (EMF): The force or pressure (voltage) that causes current movement in an electrical circuit.

electronic engine control (EEC): A computer-directed system of engine control, Ford EEC-IV. Where applicable, it is interfaced with automatic transmission/transaxle control.

electronic pressure control (EPC) solenoid: A specially designed solenoid containing a spool valve and spring assembly to control fluid mainline pressure. A variable current flow, controlled by the ECM/PCM, varies the internal force of the solenoid on the spool valve and resulting mainline pressure. (See variable force solenoid.)

electronics: Miniaturized electrical circuits utilizing semiconductors, solid-state devices, and printed circuits. Electronic circuits utilize small amounts of power.

electronification: The application of electronic circuitry to a mechanical device. Regarding automatic transmissions, electronification is incorporated into converter clutch lockup, shift scheduling, and line pressure control systems.

electrostatic discharge (ESD): An unwanted, high-voltage electrical current released by an individual who has taken on a static charge of electricity. Electronic components can be easily damaged by ESD.

element: A device within a hydrodynamic drive unit designed with a set of blades to direct fluid flow.

end bump (end feel or slip bump): Firmer feel at end of shift when compared with feel at start of shift.

endplay: The clearance/gap between two components that allows for expansion of the parts as they warm up, to prevent binding and to allow space for lubrication.

energy: The ability or capacity to do work.

engine braking: Use of engine to slow vehicle by manually downshifting during zero-throttle coast down.

engine control module (ECM): Manages the engine and incorporates output control over the torque converter clutch solenoid. (Note: Current designation for the ECM in late model vehicles is PCM.)

engine coolant temperature (ECT) sensor: Prevents converter clutch engagement with a cold engine; also used for shift timing and shift quality.

fail-safe (backup) control: A substitute value used by the PCM/TCM to replace a faulty signal from an input sensor. The temporary value allows the vehicle to continue to be operated.

feedback: A circuit malfunction whereby current can find another path to feed load devices.

final drive: An essential part of the axle drive assembly where final gear reduction takes place in the powertrain. In RWD applications and north-south FWD applications, it must also change the power flow direction to the axle shaft by ninety degrees.

firm: A noticeable quick apply of a clutch or band that is considered normal with medium to heavy throttle shift; should not be confused with *harsh* or *rough*.

flare (slipping): A quick increase in engine rpm accompanied by momentary loss of torque; generally occurs during shift.

fluid: A fluid can be either liquid or gas. In hydraulics, a liquid is used for transmitting force or motion.

fluid coupling: The simplest form of hydrodynamic drive, the fluid coupling consists of two look-alike members with straight radial vanes referred to as the impeller (pump) and the turbine. Input torque is always equal to the output torque.

fluid drive: Either a fluid coupling or a fluid torque converter. (See hydrodynamic drive units.)

fluid torque converter: A hydrodynamic drive that has the ability to act both as a torque multiplier and fluid coupling. (See hydrodynamic drive units; torque converter.)

fluid viscosity: The resistance of a liquid to flow. A cold fluid (oil) has greater viscosity and flows more slowly than a hot fluid (oil).

force: A push or pull effort measured in pounds or Newtons.

freewheeling: There is a power input with no transmission of power output. The power flow is ineffective.

friction: The resistance that occurs between contacting surfaces. This relationship is expressed by a ratio called the coefficient of friction (μ).

friction, coefficient of: The amount of surface tension between two contacting surfaces; expressed by a scientifically calculated number.

friction modifier: Changes the coefficient of friction of the fluid between the mating steel and composition clutch/band surfaces during the engagement process and allows for a certain amount of intentional slipping for a good "shift-feel."

full throttle detent downshift: A quick apply of accelerator pedal to its full travel, forcing a downshift.

fuse: A device consisting of a piece of wire with a low melting point, inserted in a circuit. It will melt and open the circuit when the system is overloaded.

fusible link: A device to protect the main chassis wiring harness if a short circuit occurs in the unfused part of the wiring. The link is a short piece of copper wire approximately four inches long inserted in series with the circuit; it acts as a fuse. The link is four gauges smaller in size than the circuit wiring it is protecting.

garage shift: Initial engagement feel of transmission, neutral to reverse or neutral to a forward drive.

garage shift feel: A quick check of the engagement quality and responsiveness of reverse and forward gears. This test is done with the vehicle stationary.

gear: A toothed mechanical device that acts as a rotating lever to transmit power or turning effort from one shaft to another. (See gear ratio.)

gear ratio: The number of revolutions the input gear makes to one revolution of the output gear. In a simple gear combination, two revolutions of the input gear to one of the output gear gives a ratio of 2:1.

gear reduction: Torque is multiplied and speed decreased by the factor of the gear ratio. For example, a 3:1 gear ratio changes an input torque of 180 lb-ft and an input speed of 2700 rpm to 540 lb-ft and 900 rpm, respectively. (No account is taken of frictional losses, which are always present.)

geartrain: A succession of intermeshing gears that form an assembly and provide for one or more torque changes as the power input is transmitted to the power output.

governor: A device that senses vehicle speed and generates a hydraulic oil pressure. As vehicle speed increases, governor oil pressure rises.

ground circuit: (See circuit, ground.)

ground side switching: The electrical/electronic circuit control switch is located after the circuit load.

hard codes: DTCs that are present at the time of testing; also called continuous or current codes.

harsh (rough): An apply of a clutch or band that is more noticeable than a *firm* one; considered undesirable at any throttle position.

heavy throttle: Approximately three-fourths of accelerator pedal travel.

hertz (Hz): The international unit of frequency equal to one cycle per second (10,000 Hertz equals 10,000 cycles per second).

high-impedance DVOM (digital volt-ohmmeter): This styled device provides a built-in resistance value and is capable of limiting circuit current flow to safe milliampere levels.

high resistance: Often refers to a circuit where there is an excessive amount of opposition to normal current flow.

horsepower: The time rate of doing work (one horsepower equals 33,000 ft-lb of work in one minute or 550 ft-lb in one second).

hot circuit: (See circuit, hot; hot lead.)

hot lead: A wire or conductor in the power side of the circuit. (See circuit, hot.)

hot side switching: The electrical/electronic circuit control switch is located before the circuit load.

hunting (busyness): Repeating quick series of upshifts and downshifts that causes noticeable change in engine rpm, for example, as in a 4-3-4 shift pattern.

hydraulics: The use of liquid under pressure to transfer force of motion.

hydrodynamic drive units: Devices that transmit power solely by the action of a kinetic fluid flow in a closed recirculating path. An impeller energizes the fluid and discharges the high-speed jet stream into the turbine for power output.

hypoid gearset: The drive pinion gear may be placed below or above the centerline of the driven gear; often used as a final drive gearset.

impeller: Often called a pump, the impeller is the power input (drive) member of a hydrodynamic drive. As part of the torque converter cover, it acts as a centrifugal pump and puts the fluid in motion.

inductance: The force that produces voltage when a conductor is passed through a magnetic field.

initial feel: A distinct firmer feel at start of shift when compared with feel at finish of shift.

input: In an automatic transmission, the source of power from the engine is absorbed by the torque converter, which provides the power input into the transmission. The turbine drives the input (turbine) shaft.

internal gear: The ring-like outer gear of a planetary gearset with the gear teeth cut on the inside of the ring to provide a mesh with the planet pinions.

isolation (clamping) diodes: Diodes positioned in a circuit to prevent self-induction from damaging electronic components.

IX rotary gear pump: Contains two rotating members, one shaped with internal gear teeth and the other with external gear teeth. As the gears separate, the fluid fills the gaps between gear teeth, is pulled across a crescent-shaped divider, and then is forced to flow through the outlet as the gears mesh.

IX rotary lobe pump: Sometimes referred to as a gerotor-type pump. Two rotating members, one shaped with internal lobes and the other with external lobes, separate and then mesh to cause fluid to flow.

kickdown: Detent downshift system; either linkage, cable, or electrically controlled.

kilo: A prefix used in the metric system to indicate one thousand.

late: Shift that occurs when engine is at higher than normal rpm for given amount of throttle.

light-emitting diode (LED): A semiconductor diode that emits light as electrical current flows through it; used in some electronic display devices to emit a red or other color light.

light throttle: Approximately one-fourth of accelerator pedal travel.

limp-in mode: Electrical shutdown of the transmission/transaxle output solenoids, allowing only forward and reverse gears that are hydraulically energized by the manual valve. This permits the vehicle to be driven to a service facility for repair.

lip seal: Molded synthetic rubber seal designed with an outer sealing edge (lip) that points into the fluid containing area to be sealed. This type of seal is used where rotational and axial forces are present.

load device: A circuit's resistance that converts the electrical energy into light, sound, heat, or mechanical movement.

load torque: The amount of output torque needed from the transmission to overcome the vehicle load.

lockup converter: A torque converter that operates hydraulically and mechanically. When an internal apply plate (lockup plate) clamps to the torque converter cover, hydraulic slippage is eliminated.

lubrite coating: A black-colored coating applied to pump components that allows contact surfaces to wear in smoothly. It is normal for the coating to disappear on the contact surfaces as the components are in service.

magnet: Any body with the property of attracting iron or steel.

magnetic field: The area surrounding the poles of a magnet that is affected by its attraction or repulsion forces.

mainline pressure: Often called control pressure or line pressure, it refers to the pressure of the oil leaving the pump and is controlled by the pressure regulator valve.

malfunction indicator lamp (MIL): Previously known as a *check engine light,* the dash-mounted MIL illuminates and signals the driver that an emission or driveability problem with the powertrain has been detected by the ECM/PCM. When this occurs, at least one diagnostic trouble code (DTC) has been stored into the ECM's/PCM's memory.

manifold absolute pressure (MAP) sensor: Reads the amount of air pressure (vacuum) in the engine's intake manifold system; its signal is used to analyze engine load conditions.

manual lever position switch (MLPS): A mechanical switching unit that is typically mounted externally to the transmission to inform the PCM/ECM which gear range the driver has selected.

manual valve: Located inside the transmission/transaxle, it is directly connected to the driver's shift lever. The position of the manual valve determines which hydraulic circuits will be charged with oil pressure and the operating mode of the transmission.

manual valve lever position sensor (MVLPS): The input from this device tells the TCM what gear range was selected.

mass air flow (MAF) sensor: Measures the airflow into the engine.

medium throttle: Approximately one-half of accelerator pedal travel.

mega: A metric prefix indicating one million.

member: An independent component of a hydrodynamic unit such as an impeller, a stator, or a turbine. It may have one or more elements.

Mercon: A fluid developed by Ford Motor Company in 1988. It contains a friction modifier and closely resembles operating characteristics of Dexron.

metal sealing rings: Made from cast iron or aluminum, their primary application is with dynamic components involving pressure sealing circuits of rotating members. These rings are designed with either butt or hook lock end joints.

meter (analog): A linear-style meter representing data as lengths; a needle-style instrument interfacing with logical numerical increments. This style of electrical meter uses relatively low impedance internal resistance and cannot be used for testing electronic circuitry.

meter (digital): Uses numbers as a direct readout to show values. Most meters of this style use high impedance internal resistance and must be used for testing low current electronic circuitry.

micro: A metric prefix indicating one-millionth (0.000001).

milli: A metric prefix indicating one-thousandth (0.001).

minimum throttle: The least amount of throttle opening required for upshift; normally close to zero throttle.

modulated: In an electronic-hydraulic converter clutch system (or shift valve system), the term *modulated* refers to the pulsing of a solenoid, at a variable rate. This action controls the buildup of oil pressure in the hydraulic circuit to allow a controlled amount of clutch slippage.

modulated converter clutch control (MCCC): A pulse width duty cycle valve that controls the converter lockup apply pressure and maximizes smoother transitions between lock and unlock conditions.

modulator pressure (throttle pressure): A hydraulic signal oil pressure relating to the amount of engine load, based on either the amount of throttle plate opening or engine vacuum.

modulator valve: A regulator valve that is controlled by engine vacuum, providing a hydraulic pressure that varies in relation to engine torque. The hydraulic torque signal functions to delay the shift pattern and provide a line pressure boost. (See throttle valve.)

motor: An electromagnetic device used to convert electrical energy into mechanical energy.

multiple-disc clutch: A grouping of steel and friction lined plates that, when compressed together by hydraulic pressure acting upon a piston, lock or unlock a planetary member.

mushy: Same as *soft*; slow and drawn out clutch apply with very little shift feel.

mutual induction: The generation of current from one wire circuit to another by movement of the magnetic field surrounding a current-carrying circuit as its ampere flow increases or decreases.

nonpositive sealing: A sealing method that allows some minor leakage, which normally assists in lubrication.

O_2 sensor: Located in the engine's exhaust system, it is an input device to the ECM/PCM for managing the fuel delivery and ignition system. A scanner can be used to observe the fluctuating voltage readings produced by an O_2 sensor as the oxygen content of the exhaust is analyzed.

OBD II (on-board diagnostics, second generation): Refers to the federal law mandating tighter control of 1996 and newer vehicle emissions, active monitoring of related devices, and standardization of terminology, data link connectors, and other technician concerns.

O-ring seal: Molded synthetic rubber seal designed with a circular cross-section. This type of seal is used primarily in static applications.

ohm: A unit of measurement of electrical resistance.

ohmmeter: Electrical instrument used to measure circuit resistance to current flow. The ohmmeter is self-powered and must not be connected to a circuit with voltage present.

Ohm's law: A law of electricity that states the relationship between voltage, current, and resistance.

Volts = amperes x ohms

$$E = I \times R$$

one-way clutch: A mechanical clutch of roller or sprag design that resists torque or transmits power in one direction only. It is used to either hold or drive a planetary member.

one-way roller clutch: A mechanical device that transmits or holds torque in one direction only.

open circuit: A break or lack of contact in an electrical circuit, either intentional (switch) or unintentional (bad connection or broken wire).

orifice: Located in hydraulic oil circuits, it acts as a restriction. It slows down fluid flow to either create back pressure or delay pressure buildup downstream.

output speed sensor (OSS): Identifies transmission output shaft speed for shift timing and may be used to calculate TCC slip; often functions as the VSS (vehicle speed sensor).

overdrive: Produces the opposite effect of a gear reduction. Torque is reduced, and speed is increased by the factor of the gear ratio. A 1:3 gear ratio would change an input torque of 180 lb-ft and an input speed of 2700 rpm to 60 lb-ft and 8100 rpm, respectively.

overdrive planetary gearset: A single planetary gearset designed to provide a direct drive and overdrive ratio. When coupled to a three-speed transmission configuration, a four-speed/overdrive unit is present.

overrun clutch: Another name for a one-way mechanical clutch. Applies to both roller and sprag designs.

oxidation stabilizers: Absorb and dissipate heat. Automatic transmission fluid has high resistance to varnish and sludge buildup that occurs from excessive heat that is generated primarily in the torque converter. Local temperatures as high as 600°F (315°C) can occur at the clutch plates during engagement, and this heat must be absorbed and dissipated. If the fluid cannot withstand the heat, it burns or oxidizes, resulting in an almost immediate destruction of friction materials, clogged filter screen and hydraulic passages, and sticky valves.

parallel circuit: (See circuit, parallel.)

pinion gear: The smallest gear in a drive gear assembly.

piston: A disc or cup that fits in a cylinder bore and is free to move. In hydraulics, it provides the means of converting hydraulic pressure into a usable force. Examples of piston applications are found in servo, clutch, and accumulator units.

planet carrier: A basic member of a planetary gear assembly that carries the pinion gears.

planet pinions: Gears housed in a planet carrier that are in constant mesh with the sun gear and internal gear. Because they have their own independent rotating centers, the pinions are capable of rotating around the sun gear or the inside of the internal gear.

planetary gear ratio: The reduction or overdrive ratio developed by a planetary gearset.

planetary gearset: In its simplest form, it is made up of a basic assembly group containing a sun gear, internal gear, and planet carrier. The gears are always in constant mesh and offer a wide range of gear ratio possibilities.

planetary gearset (compound): Two planetary gearsets combined together.

planetary gearset (simple): An assembly of gears in constant mesh consisting of a sun gear, several pinion gears mounted in a carrier, and a ring gear. It provides gear ratio and direction changes, in addition to a direct drive and a neutral.

port: An opening for fluid intake or exhaust.

positive sealing: A sealing method that completely prevents leakage.

potential: Electrical force measured in volts; sometimes used interchangeably with voltage.

power: The ability to do work per unit of time, as expressed in horsepower: one horsepower equals 33,000 ft-lb of work per minute, or 550 ft-lb of work per second.

power flow: The systematic flow or transmission of power through the gears, from the input shaft to the output shaft.

powertrain: All drive components from the flywheel to the drive wheels; defined, at times, to include the engine.

powertrain control module (PCM): Current designation for the engine control module (ECM). In many cases, late model vehicle PCMs manage the engine as well as the transmission. In other settings, the PCM controls the engine and is interfaced with a TCM to control transmission functions.

pressure: The amount of force exerted upon a surface area.

pressure control solenoid (PCS): An output device that provides a boost oil pressure to the mainline regulator valve to control line pressure. Its operation is determined by the amount of current sent from the PCM.

pressure gauge: An instrument used for measuring the fluid pressure in a hydraulic circuit.

pressure regulator valve: In automatic transmissions, its purpose is to regulate the pressure of the pump output and supply the basic fluid pressure necessary to operate the transmission. The regulated fluid pressure may be referred to as mainline pressure, line pressure, or control pressure.

pressure switch assembly (PSA): Mounted inside the transmission, it is a grouping of oil pressure switches that inputs to the PCM when certain hydraulic passages are charged with oil pressure.

PROM (programmable read-only memory): The heart of the computer that compares input data and makes the engineered program or strategy decisions about when to trigger the appropriate output based on stored computer instructions.

pulse generator: A two-wire pickup sensor used to produce a fluctuating electrical signal. This changing signal is read by the controller to determine the speed of the object and can be used to measure transmission input speed, output speed, and vehicle speed.

pulse width duty cycle solenoid (pulse width modulated solenoid): A computer-controlled solenoid that turns on and off at a variable rate producing a modulated oil pressure; often referred to as a pulse width modulated (PWM) solenoid. Employed in many electronic automatic transmissions and transaxles, these solenoids are used to manage shift control and converter clutch hydraulic circuits.

pump: A mechanical device designed to create fluid flow and pressure buildup in a hydraulic system.

range reference and clutch/band apply chart: A guide that shows the application of a transmission's/transaxle's clutches and bands for each gear, within the selector range positions. These charts are extremely useful for understanding how the unit operates and for diagnosing malfunctions.

Ravigneaux gearset: A compound planetary gearset that features matched dual planetary pinions (sets of two) mounted in a single planet carrier. Two sun gears and one ring mesh with the carrier pinions.

reaction member: The stationary planetary member, in a planetary gearset, that is grounded to the transmission case through the use of friction and wedging devices known as bands, disc clutches, and one-way clutches.

reaction pressure: The fluid pressure that moves a spool valve against an opposing force or forces; the area on which the opposing force acts. The opposing force can be a spring or a combination of spring force and auxiliary hydraulic force.

reactor, torque converter: The reaction member of a fluid torque converter, more commonly called a stator. (See stator.)

reduction: (See gear reduction.)

regulator valve: A valve that changes the pressure of the oil in a hydraulic circuit as the oil passes through the valve by bleeding off (or exhausting) some of the volume of oil supplied to the valve.

relay: An electromagnetic switching device using low current to open or close a high-current circuit.

relay valve: A valve that directs flow and pressure. Relay valves simply connect or disconnect interrelated passages without restricting the fluid flow or changing the pressure.

relief valve: A spring-loaded, pressure-operated valve that limits oil pressure buildup in a hydraulic circuit to a predetermined maximum value.

reservoir: The storage area for fluid in a hydraulic system; often called a sump.

residual magnetism: The magnetic strength stored in a material after a magnetizing field has been removed.

resistance: The property of an electrical circuit that tends to prevent or reduce the flow of current. Resistance is measured in ohms.

resistor: A device installed in an electrical circuit to permit a predetermined current to flow with a given applied voltage.

resultant force: The single effective directional thrust of the fluid force on the turbine produced by the vortex and rotary forces acting in different planes.

rheostat: A device for regulating a current by means of a variable resistance.

ring gear: (See internal gear.)

road load: A function of vehicle weight and speed, plus road grade.

roller clutch: A type of one-way clutch design using rollers and springs mounted within an inner and outer cammed race assembly.

rotary flow: The path of the fluid trapped between the blades of the members as they revolve with the rotation of the torque converter cover (rotational inertia).

RTV: A gasket making compound that cures as it is exposed to the atmosphere. It is used between surfaces that are not perfectly machined to one another, leaving a slight gap that the RTV fills and in which it hardens. The letters RTV represent room temperature vulcanizing.

seal swell controllers: Control swelling, hardness, and tensile strength of synthetics and keep the seals pliable.

self-induction: The generation of voltage in a current-carrying wire by changing the amount of current flowing within that wire.

semiconductor: A material (silicon or germanium) that is neither a good conductor nor an insulator; used in diodes and transistors.

series circuit: (See circuit, series.)

servo: In an automatic transmission, it is a piston in a cylinder assembly that converts hydraulic pressure into mechanical force and movement; used for the application of the bands and clutches.

shift busyness: When referring to a torque converter clutch, it is the frequent apply and release of the clutch plate due to uncommon driving conditions.

shift valve: Classified as a relay valve, it triggers the automatic shift in response to a governor and a throttle signal by directing fluid to the appropriate band and clutch apply combination to cause the shift to occur.

short circuit: A circuit that is accidentally completed in an electrical path for which it was not intended.

shudder: Repeated jerking or stick-slip sensation, similar to *chuggle* but more severe and rapid in nature, that may be most noticeable during certain ranges of vehicle speed; also used to define condition after converter clutch engagement.

Simpson gearset: A compound planetary geartrain that integrates two simple planetary gearsets referred to as the front planetary and the rear planetary.

slipping: Noticeable increase in engine rpm without vehicle speed increase; usually occurs during or after initial clutch or band engagement.

SMEC (single-module engine controller): In certain Chrysler applications, it interfaces with the transmission to control selected shifting strategies.

soft: Slow, almost unnoticeable clutch apply with very little shift feel.

soft codes: DTCs that have been set into the PCM memory but are not present at the time of testing; often referred to as history or intermittent codes.

solenoid: An electromagnetic device containing a coil winding and movable core. As current is sent through the winding, the core is moved. When applied to hydraulic oil circuits, pressure in that circuit can be controlled electrically.

solenoid (duty cycle): A computer-controlled solenoid whose cycling on-off action, which controls the pressure in a hydraulic clutch apply circuit, can be varied.

spalling: A wear pattern identified by metal chips flaking off the hardened surface. This condition is caused by foreign particles, overloading situations, and/or normal wear.

split torque drive: In a torque converter, it refers to parallel paths of torque transmission, one of which is mechanical and the other hydraulic.

spool valve: A precision-machined, cylindrically shaped valve made up of lands and grooves. Depending on its position in the valve bore, various interconnecting hydraulic circuit passages are either opened or closed.

sprag clutch: A type of one-way clutch design using cams or contoured-shaped sprags between inner and outer races. (See one-way clutch.)

square-cut seal: Molded synthetic rubber seal designed with a square- or rectangular-shaped cross-section. This type of seal is used for both dynamic and static applications.

stage: The number of turbine sets separated by a stator. A turbine set may be made up of one or more turbine members. A three-element converter is classified as a single stage.

stall: In fluid drive transmission applications, stall refers to engine rpm with the transmission engaged and the vehicle stationary; throttle valve can be in any position between closed and wide open.

stall speed: In fluid drive transmission applications, stall speed refers to the maximum engine rpm with the transmission engaged and vehicle stationary, when the throttle valve is wide open. (See stall; stall test.)

stall test: A procedure recommended by many manufacturers to help determine the integrity of an engine, the torque converter stator, and certain clutch and band combinations. With the shift lever in each of the forward and reverse positions and with the brakes firmly applied, the accelerator pedal is momentarily pressed to the wide open throttle (WOT) position. The engine rpm reading at full throttle can provide clues for diagnosing the condition of the items listed above.

stall torque: The maximum design or engineered torque ratio of a fluid torque converter, produced under stall speed conditions. (See stall speed.)

static: A sealing application in which the parts being sealed do not move in relation to each other.

stator (reactor): The reaction member of a fluid torque converter that changes the direction of the fluid as it leaves the turbine to enter the impeller vanes. During the torque multiplication phase, this action assists the impeller's rotary force and results in an increase in torque.

substitution: Replacing one part suspected of a defect with a like part of known quality.

sump: The storage vessel or reservoir that provides a ready source of fluid to the pump. In an automatic transmission, the sump is the oil pan. All fluid eventually returns back to the sump for recycling into the hydraulic system.

sun gear: In a planetary gearset, it is the center gear that meshes with a cluster of planet pinions.

surge: Repeating engine-related feeling of acceleration and deceleration that is less intense than *chuggle.*

switch: A device used to open, close, or redirect the current in an electrical circuit.

tachometer: An instrument used for measuring engine revolutions per minute (rpm).

Teflon sealing rings: Teflon is a soft, durable, plastic-like material that is resistant to heat and provides excellent sealing. These rings are designed with either scarf-cut joints or as one-piece rings. Teflon sealing rings have replaced many metal ring applications.

terminal: A device attached to the end of a wire or cable to make an electrical connection.

test light, circuit-powered: Uses available circuit voltage to test circuit continuity.

test light, self-powered: Uses its own battery source to test circuit continuity.

thermistor: A special resistor used to measure fluid temperature; it decreases its resistance with increases in temperature.

thermostatic element: A heat-sensitive, spring-type device that controls a drain port from the upper sump area to the lower sump. When the transaxle fluid reaches operating temperature, the port is closed and the upper sump fills, thus reducing the fluid level in the lower sump.

throttle position (TP) sensor: Reads the degree of throttle opening; its signal is used to analyze engine load conditions. The ECM/PCM decides to apply the TCC, or to disengage it for coast or load conditions that need a converter torque boost.

throttle pressure/modulator pressure: A hydraulic signal oil pressure relating to the amount of engine load, based on either the amount of throttle plate opening or engine vacuum.

throttle valve: A regulating or balanced valve that is controlled mechanically by throttle linkage or engine vacuum. It sends a hydraulic signal to the shift valve body to control shift timing and shift quality. (See balanced valve; modulator valve.)

tie-up: Condition where two opposing clutches are attempting to apply at same time, causing engine to labor with noticeable loss of engine rpm.

torque: A twisting effort or turning force applied to a radius.

Torque = force × lever length = lb-ft/N-m.

torque converter: Positioned between the engine and the transmission input shaft, the torque converter is a hydrodynamic drive unit that acts as a fluid coupling and has the ability to multiply torque.

torque converter clutch: The apply plate (lockup plate) assembly used for mechanical power flow through the converter.

torque phase: Sometimes referred to as *slip phase* or *stall phase*, torque multiplication occurs when the turbine is turning at a slower speed than the impeller, and the stator is reactionary (stationary). This sequence generates a boost in output torque.

torque rating (stall torque): The maximum torque multiplication that occurs during stall conditions, with the engine at wide open throttle (WOT) and zero turbine speed.

torque ratio: An expression of the gear ratio factor on torque effect. A 3:1 gear ratio or 3:1 torque ratio increases the torque input by the ratio factor of 3:

Input torque (100 lb-ft) × 3 = output torque (300 lb-ft)

torsional: Twisting or turning effort; often associated with engine crankshaft rotation.

traction: The amount of usable tractive effort before the drive wheels slip on the road contact surface.

tractive effort: The amount of force available to the drive wheels, to move the vehicle.

transaxle: A transmission and final drive unit combined into a single unit; may be designed in a longitudinal or transverse configuration for either a front- or rear-wheel drive vehicle.

transducer: A device that changes energy from one form to another. For example, a transducer in a microphone changes sound energy to electrical energy. In automotive air-conditioning controls used in automatic temperature systems, a transducer changes an electrical signal to a vacuum signal, which operates mechanical doors.

transmission: A powertrain component designed to modify torque and speed developed by the engine; also provides direct drive, reverse, and neutral.

transmission control module (TCM): Manages transmission functions. These vary according to the manufacturer's product design but may include converter clutch operation, electronic shift scheduling, and mainline pressure.

transmission fluid temperature (TFT) sensor: Originally called a transmission oil temperature (TOT) sensor, this input device to the ECM/PCM senses the fluid temperature and provides a resistance value. It operates on the thermistor principle.

transmission input speed (TIS) sensor: Measures turbine shaft (input shaft) rpm's and compares to engine rpm's to determine torque converter slip. When compared to the transmission output speed sensor or VSS, gear ratio and clutch engagement timing can be determined.

transmission oil temperature (TOT) sensor: (See transmission fluid temperature (TFT) sensor.)

transmission range selector (TRS) switch: Tells the module which gear shift position the driver has chosen.

turbine: The output (driven) member of a fluid coupling or fluid torque converter. It is splined to the input (turbine) shaft of the transmission.

turbulence: The interference of molecules of a fluid (or vapor) with each other in a fluid flow.

Type F: Transmission fluid developed and used by Ford Motor Company up to 1982. This fluid type provides a high coefficient of friction.

Type 7176: The preferred choice of ATFs for Chrysler automatic transmissions and transaxles. Developed in 1986, it closely resembles Dexron and Mercon. Type 7176 is the recommended service fill fluid for all Chrysler products utilizing a lockup torque converter dating back to 1978.

upshift: A shift that results in a decrease in torque ratio and an increase in speed.

vacuum: A negative pressure; any pressure less than atmospheric pressure.

vacuum gauge: An instrument used for measuring the existing vacuum in a vacuum circuit or chamber. The unit of measure is inches (of mercury in a barometer).

vacuum modulator: Generates a hydraulic oil pressure in response to the amount of engine vacuum.

valve body assembly: The main hydraulic control assembly of the transmission that contains numerous valves, check balls, and other components to control the distribution of pressurized oil throughout the transmission.

valves: Devices that can open or close fluid passages in a hydraulic system and are used for directing fluid flow and controlling pressure.

variable displacement (variable capacity) vane pump: Slipper-type vanes, mounted in a revolving rotor and contained within the bore of a movable slide, capture and then force fluid to flow. Movement of the slide to various positions changes the size of the vane chambers and the amount of fluid flow. Note: GM refers to this pump design as variable displacement, and Ford terms it variable capacity.

variable force solenoid (VFS): Commonly referred to as the electronic pressure control (EPC) solenoid, it replaces the cable/linkage style of TV system control and is integrated with a spool valve and spring assembly to control pressure. A variable computer-controlled current flow varies the internal force of the solenoid on the spool valve and resulting control pressure.

variable orifice thermal valve: Temperature-sensitive hydraulic oil control device that adjusts the size of a circuit path opening. By altering the size of the opening, the oil flow rate is adapted for cold to hot oil viscosity changes.

vehicle speed sensor (VSS): Provides an electrical signal to the computer module, measuring vehicle speed, and affects the torque converter clutch engagement and release.

Vespel sealing rings: Hard plastic material that produces excellent sealing in dynamic settings. These rings are found in late versions of the 4T60 and in all 4T60-E and 4T80-E transaxles.

viscosity index improvers: Keeps the viscosity nearly constant with changes in temperature. This is especially important at low temperatures, when the oil needs to be thin to aid in shifting and for cold-weather starting. Yet it must not be so thin that at high temperatures it will cause excessive hydraulic leakage so that pumps are unable to maintain the proper pressures.

viscous clutch: A specially designed torque converter clutch apply plate that, through the use of a silicon fluid, clamps smoothly and absorbs torsional vibrations.

volt: A unit of measurement of electrical pressure.

voltage: The electrical pressure that causes current to flow. Voltage is measured in volts (V).

voltage, applied: The actual voltage read at a given point in a circuit. It equals the available voltage of the power supply minus the losses in the circuit up to that point.

voltage drop: The voltage lost or used in a circuit by normal loads such as a motor or lamp or by abnormal loads such as a poor (high-resistance) lead or terminal connection.

voltmeter: An electrical instrument used for reading potential in volts at a particular point in a circuit. It can also be used to test for continuity and resistance.

vortex flow: The crosswise or circulatory flow of oil between the blades of the members caused by the centrifugal pumping action of the impeller.

watt: The unit for measuring electrical power. One watt is the product of one ampere and one volt (watts equals amps times volts). Wattage is the horsepower of electricity (746 watts equal one horsepower).

wide open throttle (WOT): Full travel of accelerator pedal.

work: The force exerted to move a mass or object. Work involves motion; if a force is exerted and no motion takes place, no work is done. Work per unit of time is called power.

Work = force × distance = ft-lbs

33,000 ft-lbs in one minute = 1 horsepower

zero-throttle coast down: A full release of accelerator pedal while vehicle is in motion and in drive range.

Index

D

E

F

G

H

I

J

L

M

N

O

P

R

S

T

U

V

W

Z